Mastering
AutoCAD®
Civil 3D® 2013

Louisa Holland

Kati Mercier, P.E.

WILEY
John Wiley & Sons, Inc.

Senior Acquisitions Editor: Willem Knibbe
Development Editor: Tom Cirtin
Technical Editor: Tommie Richardson
Production Editor: Dassi Zeidel
Copy Editor: Liz Welch
Editorial Manager: Pete Gaughan
Production Manager: Tim Tate
Vice President and Executive Group Publisher: Richard Swadley
Vice President and Publisher: Neil Edde
Book Designer: Maureen Forys, Happenstance Type-O-Rama; Judy Fung
Proofreaders: Louise Watson, James Saturnio, Word One, New York
Indexer: Nancy Guenther
Project Coordinator, Cover: Katherine Crocker
Cover Designer: Ryan Sneed
Cover Image: © mycola / iStockPhoto

ISBN: 978-1-118-28175-8
ISBN: 978-1-118-33356-3 (ebk.)
ISBN: 978-1-118-33071-5 (ebk.)
ISBN: 978-1-118-33468-3 (ebk.)

Library of Congress Control Number: 2012940021

10 9 8 7 6 5 4 3 2 1

Dear Reader,

Thank you for choosing *Mastering AutoCAD Civil 3D 2013.* This book is part of a family of premium-quality Sybex books, all of which are written by outstanding authors who combine practical experience with a gift for teaching.

Sybex was founded in 1976. More than 30 years later, we're still committed to producing consistently exceptional books. With each of our titles, we're working hard to set a new standard for the industry. From the paper we print on, to the authors we work with, our goal is to bring you the best books available.

I hope you see all that reflected in these pages. I'd be very interested to hear your comments and get your feedback on how we're doing. Feel free to let me know what you think about this or any other Sybex book by sending me an email at `nedde@wiley.com`. If you think you've found a technical error in this book, please visit `http://sybex.custhelp.com`. Customer feedback is critical to our efforts at Sybex.

Best regards,

Neil Edde
Vice President and Publisher
Sybex, an Imprint of Wiley

Acknowledgments

A big thank-you to all of the past Mastering Civil 3D authors throughout the years. The book has come a long way and without the long hours put forth every year, it wouldn't be where it is today.

Thanks to the whole team at Wiley: Willem Knibbe for giving us guidance at each step in the process; Pete Gaughan, who promptly answered our technical questions; the editors, Thomas Cirtin, Dassi Zeidel, and Liz Welch, for making sure that the book doesn't just make grammatical sense but has consistency as well.

Thanks to Tommie Richardson, who meticulously combed through the technical jargon making sure the datasets worked, and who offered detailed suggestions to make the book even better (and longer).

Thanks to Dana Probert of Autodesk for her help in pointing us to all the great new features and of course to all of the Civil 3D Autodesk team for making the software more useful each year.

—Louisa Holland and Kati Mercier

Thanking people who have helped along the way is my favorite part of the book-writing process. First and foremost I'd like to thank my husband, Mark, for his encouragement and keeping my office from looking like an episode of *Hoarders: Diet Coke Can Edition*. I'd like to thank my parents for their support and bragging about my book even though they're not 100 percent sure what it's about.

To Russ Nicloy and Ron Butzen: Thanks for your enthusiasm and general awesomeness. Ron, you are the best manager a crazy person can have. To my friends and colleagues across WisDOT: You are like a second family.

Thanks to my Twitter buddies, especially @NickZeeben, @Civil3D_Jedi, @iPinda, @WKFD, and @Sockibear for letting me bounce random metric conversion questions at them.

Lastly, I'd like to thank my dogs, Daisy and Zuzu, for keeping my feet warm while I typed.

—Louisa Holland

A huge thank-you to Lou Holland for inviting me along on this coauthoring journey. We joked from the start that it was exciting that this book was going to be "chick lit" with two female authors. Since the civil engineering field, as well as the technical field, is often led by males, it is refreshing to be an example to girls who dream of one day designing with dirt.

I'd also like to thank my coworker and Civil 3D cohort, Aaron Mortensen, P.E. I could always count on you to be there to bounce ideas off of as we implemented the software, and vent to when needed. And I will definitely not forget my #Civil3D Twitter friends and fellow blog writers whom I could always turn to for some additional insight.

Thank you to all the many hands who have been part of the textbooks and manuals that I have read over the years. I think I took for granted that books just appeared when I was ready to buy them.

But most importantly a thank-you to my wonderful husband, Nick, who is so patient with me and regularly reminds me of all that I am capable.

—Kati Mercier

About the Authors

Louisa "Lou" Holland is a LEED-accredited civil engineer from Milwaukee, WI. She has trained users on Eagle Point Software and AutoCAD® since 2001, and on AutoCAD Civil 3D since 2006. She has worked extensively with the Wisconsin Department of Transportation and various consultants on AutoCAD Civil 3D implementations. Louisa is an Autodesk Approved Instructor (AAI), an AutoCAD Civil 3D Certified Professional, and a regular speaker at Autodesk University, Autodesk User Group International, and other industry events.

Kati Mercier is a professional engineer licensed in the states of Connecticut and New York. Since 2004 she has served as a project engineer at Nathan L. Jacobson & Associates, Inc., a civil and environmental engineering firm that provides engineering, review, and design consultation to local municipalities and other clients. Kati is the firm's Civil 3D protocols and procedures expert, an AutoCAD Civil 3D Certified Professional, speaks at Autodesk University, and can regularly be found discussing Civil 3D and other civil engineering topics on various social networks.

Contents at a Glance

Contents

Introduction

The AutoCAD® Civil 3D® program was introduced in 2004 as a trial product. Over the past few years, AutoCAD Civil 3D series have evolved from the wobbly baby introduced on those first trial discs to a mature platform used worldwide to handle the most complex dynamic engineering designs. With this change, many engineers still struggle with how to make the transition. The civil engineering industry as a whole is an old dog learning new tricks.

We hope this book will help you in this journey. As the user base grows and users get beyond the absolute basics, more materials are needed, offering a multitude of learning opportunities. While this book is starting to move away from the basics and truly become a *Mastering* book, we hope that we are headed in that direction with the general readership. We know we cannot please everyone, but we do listen to your comments — all toward the betterment of this book.

Designed to help you get past the steepest part of the learning curve and teach you some guru-level tricks along the way, *Mastering AutoCAD Civil 3D 2013* is the ideal addition to any AutoCAD Civil 3D user's bookshelf.

Who Should Read This Book

The *Mastering* book series is designed with specific users in mind. In the case of *Mastering AutoCAD Civil 3D 2013*, we expect you'll have some knowledge of AutoCAD in general and some basic engineering knowledge as well. A basic understanding of AutoCAD Civil 3D will be helpful, although there are explanations and examples to please everyone. We expect this book will appeal to a large number of AutoCAD Civil 3D users, but we envision a few primary users:

Beginning Users Looking to Make the Move to Using AutoCAD Civil 3D These people understand AutoCAD and some basics of engineering, but they are looking to learn AutoCAD Civil 3D on their own, broadening their skill set to make themselves more valuable in their firms and in the market.

AutoCAD Civil 3D Users Looking for a Desktop Reference With the digitization of the official help files, many users still long for a book they can flip open and keep beside them as they work. These people should be able to jump to the information they need for the task at hand, such as further information about a confusing dialog or troublesome design issue.

CERT OBJECTIVE

Users Looking to Prepare for the Autodesk Certification Exams This book focuses on the elements you need to pass the Associate and Professional exams with flying colors, and includes margin icons to note topics of interest. Just look for the icon.

Classroom Instructors Looking for Better Materials This book was written with real data from real design firms. We've worked hard to make many of the examples match the real-world problems we have run into as engineers. This book also goes into greater depth than many basic texts, allowing short classes to review the basics and leave the in-depth material for self-discovery, while longer classes can cover the full material presented.

This book can be used front-to-back as a self-teaching or instructor-based instruction manual. Each chapter has a number of exercises and most (but not all) build on the previous exercise. You can also skip to almost any exercise in any chapter and jump right in. We've created a large number of drawing files that you can download from `www.sybex.com/go/masteringcivil3d2013` to make choosing your exercises a simple task.

What You Will Learn

This book isn't a replacement for training. There are too many design options and parameters to make any book a good replacement for training from a professional. This book teaches you to use the tools, explores a large number of the options, and leaves you with an idea of how to use each tool. At the end of the book, you should be able to look at any design task you run across, consider a number of ways to approach it, and have some idea of how to accomplish the task. To use one of our common analogies, reading this book is like walking around your local home-improvement warehouse. You see a lot of tools and use some of them, but that doesn't mean you're ready to build a house.

What You Need

Before you begin learning AutoCAD Civil 3D, you should make sure your hardware is up to snuff. Visit the Autodesk website and review graphic requirements, memory requirements, and so on. One of the most frustrating things that can happen is to be ready to learn, only to be stymied by hardware-related crashes. AutoCAD Civil 3D is a hardware-intensive program, testing the limits of every computer on which it runs.

We also strongly recommend using either a wide format or dual-monitor setup. The number of dialogs, palettes, and so on make AutoCAD Civil 3D a real estate hog. By having the extra space to spread out, you'll be able to see more of your design along with the feedback provided by the program itself.

You need to visit `www.sybex.com/go/masteringcivil3d2013` to download all of the data and sample files. We recommend that you save these files locally on your computer in `C:/Mastering` unless told otherwise. Finally, please be sure to visit the Autodesk website at `www.autodesk.com` to download any service packs that might be available.

The Mastering Series

The *Mastering* series from Sybex provides outstanding instruction for readers with intermediate and advanced skills, in the form of top-notch training and development for those already working in their field and clear, serious education for those aspiring to become pros. Every *Mastering* book includes:

- Real-world scenarios ranging from case studies to interviews that show how the tool, technique, or knowledge presented is applied in actual practice
- Skill-based instruction, with chapters organized around real tasks rather than abstract concepts or subjects
- A self-review section called The Bottom Line, so you can be certain you're equipped to do the job right

What Is Covered in This Book

This book contains 21 chapters and two appendices:

Chapter 1, "The Basics," introduces you to the interface and many of the common dialogs in AutoCAD Civil 3D. This chapter discusses navigating the interface and customizing your drawing's settings. You will also explore various tools for creating linework.

Chapter 2, "Survey," examines the Survey Toolspace and the unique toolset it contains for handling field surveying and fieldbook data handling. You will also look at various surface and surveying relationships.

Chapter 3, "Points," introduces AutoCAD Civil 3D points, and the various methods of creating them. You will also spend some time discussing the control of AutoCAD Civil 3D points with description keys and groups.

Chapter 4, "Surfaces," introduces the various methods of creating surfaces, using free and low-cost data to perform preliminary surface creation. Then you will investigate the various surface edits and analysis methods. The chapter also discusses point clouds and their use.

Chapter 5, "Parcels," examines the best practices for keeping your parcel topology tight and your labeling neat. It examines the various editing methods for achieving the desired results for the most complicated plats.

Chapter 6, "Alignments," introduces the basic horizontal layout element. This chapter also examines using layout tools that maintain the relationships between the tangents, curves, and spiral elements that create alignments.

Chapter 7, "Profiles and Profile Views," examines the sampling and creation methods for the vertical layout element. You will also examine the editing and element-level control. In addition, you will explore how profile views can be customized to meet the required format for your design and plans.

Chapter 8, "Assemblies and Subassemblies," introduces the building blocks of AutoCAD Civil 3D cross-sectional design. You will look at the many subassemblies available in the tool palettes, and look at how to build full design sections for use in any design environment.

Chapter 9, "Custom Subassemblies," introduces you to some of the tools in Autodesk® Subassembly Composer for AutoCAD® Civil 3D®, a new standalone program with an easy-to-use interface to visually create complex subassemblies without the need for advanced programming knowledge.

Chapter 10, "Basic Corridors," introduces the basics of corridors — building full designs from horizontal, vertical, and cross-sectional design elements. You will look at the various components to understand how corridors work before moving to a more complex design set.

Chapter 11, "Advanced Corridors, Intersections, and Roundabouts," further examines using corridors in more complex situations. You will learn about building surfaces, intersections, and other areas of corridors that make them powerful in any design situation.

Chapter 12, "Superelevation," takes a close look at the tools used to add superelevation to roadways and railways. This functionality has changed greatly in the last few years, and you will have a chance to use the new Axis of Rotation subassemblies that can pivot from several design points.

Chapter 13, "Cross Sections and Mass Haul," looks at slicing sections from surfaces, corridors, and pipe networks using alignments and the mysterious sample-line group. Working with the wizards and tools, you will see how to make your sections to order. You will explore Mass Haul to demonstrate the power of AutoCAD Civil 3D for creation of the Mass Haul diagrams.

Chapter 14, "Pipe Networks," gets into the building blocks of the pipe network tools. You will look at modifying an existing part to add new sizes and then building parts lists for various design situations. You will then work with the creation tools for creating pipe networks, and plan and profile views to get your plans looking like they should.

Chapter 15, "Storm and Sanitary Analysis," introduces you to the hydrology and hydraulic design tools first included with AutoCAD Civil 3D 2012. You will look at catchment objects, and the best workflow to export data to this analysis tool.

Chapter 16, "Grading," examines both feature lines and grading objects. You will look at creating feature lines to describe critical areas and then using grading objects to describe mass grading.

Chapter 17, "Plan Production," walks you through the basics of creating view frame groups, sheets, and templates used to automate the plan and profile drawing sheet process. In addition, you will look at creating section views and section sheets.

Chapter 18, "Advanced Workflows," looks at the various ways of sharing and receiving data. We describe the data-shortcut mechanism for sharing data between AutoCAD Civil 3D users. We also consider other methods of importing and exporting, such as XML.

Chapter 19, "Quantity Takeoff," shows you the ins and outs of assigning pay items to corridor codes, blocks, areas, and pipes. You learn how to set up new pay items and generate quantity takeoff reports.

Chapter 20, "Label Styles," is devoted to editing and creating label styles. You learn to navigate the Text Component Editor, and how to master label style conundrums you may come across.

Chapter 21, "Object Styles," examines editing and creating object styles. You will learn how to create styles for surfaces, profile views, and other objects to match your company standards.

Appendix A, "The Bottom Line," gathers together all the Master It problems from the chapters and provides a solution for each.

Appendix B, "AutoCAD Civil 3D Certification," points you to the chapters in this book that will help you master the objectives for the Certified Professional Exam.

How to Contact the Authors

We welcome feedback from you about this book and/or about books you'd like to see from us in the future. You can reach us by emailing `masteringcivil3d@gmail.com`. For more information about our work, please visit our respective websites/blogs http://`civil3detcetera.blogspot.com` (Louisa, owner) and `www.civil4d.com` (Kati, contributor), or contact us on Twitter at @LouisaHollad and @KDinCTPE.

Sybex strives to keep you supplied with the latest tools and information you need for your work. Please check their website at `www.sybex.com/go/masteringcivil3d2013`, where we'll post additional content and updates that supplement this book if the need arises. Enter **Civil 3D** in the Search box (or type the book's ISBN — **9781118281758**) and click Go to get to the book's update page.

Thanks for purchasing *Mastering AutoCAD Civil 3D 2013*. We appreciate it, and look forward to exploring AutoCAD Civil 3D with you!

Chapter 1

The Basics

Before we get into the "Mastering" of the AutoCAD® Civil 3D® program, it is important to understand the basics. There are numerous dialogs, Ribbons, menus, and icons to pore over. They might seem daunting at first glance, but as you use them, you will gain familiarity with their location and use. In this chapter, you will explore the interface and learn terminology that will be used throughout this book.

In addition, we will introduce the Lines and Curves commands, which offer a plethora of options for drawing lines and curves accurately.

In this chapter, you will learn to:

- Find any Civil 3D object with just a few clicks
- Modify the drawing scale and default object layers
- Navigate the Ribbon's contextual tabs
- Create a curve tangent to the end of a line
- Label lines and curves

The Interface

If you have used Civil 3D 2010 or newer, the interface for Civil 3D 2013 is basically the same. If you are new to Civil 3D or are coming from Civil 3D 2009 or prior, this part of the chapter is for you. Civil 3D uses a Ribbon-based interface, which is where you will access many of the tools. The Ribbon consists of tabs and panels that organize tools into logical groups. When working in Civil 3D 2013, you will spend the majority of your time on the Home tab, shown in Figure 1.1.

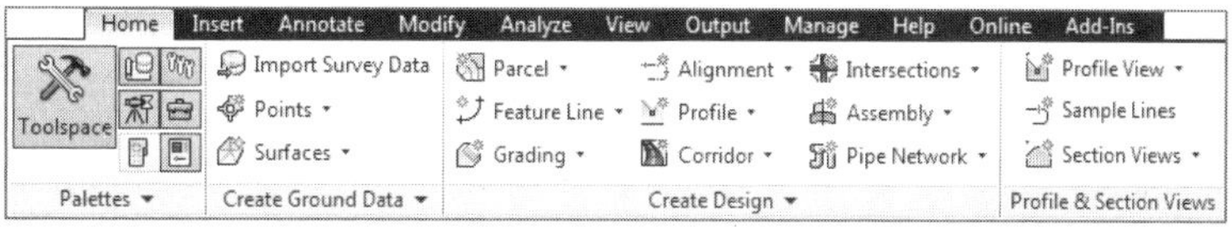

FIGURE 1.1
The Ribbon runs horizontally across the top of your screen and is where you will access many tools.

The interface may change slightly, depending on what task you are performing. When you click on a Civil 3D object, you will see a context-specific *contextual tab* appear in the Ribbon. Figure 1.2 shows the Civil 3D palette sets, along with the AutoCAD tool palettes and Ribbon displayed in a typical environment.

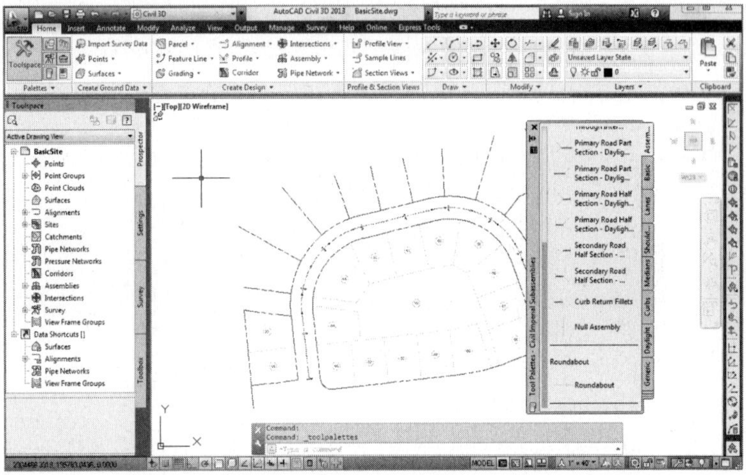

FIGURE 1.2
Overview of the Civil 3D environment. Toolspace is docked to the left, and tool palettes float over the drawing window. The Ribbon is at the top of the workspace.

Panels are subgroups within each tab that further organize your tools. For example, the Palettes panel on the Home tab (shown in Figure 1.3) is where you can toggle on or off the elements you are about to learn.

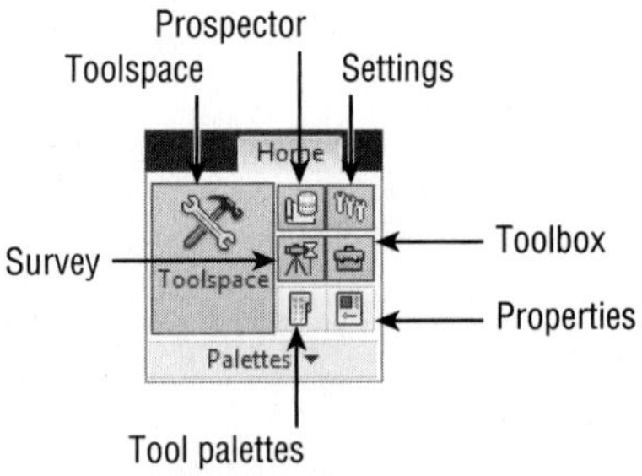

FIGURE 1.3
Palettes panel of the Home tab

Toolspace

Toolspace is a set of palettes that is specific to Civil 3D. You will want to have the palette visible any time you are working in Civil 3D. If you do not see it, click the Toolspace button on the Home tab and select Palettes.

Toolspace can have as many as four tabs to manage user data, as follows:

- Prospector
- Settings
- Survey
- Toolbox

The tabs can be turned on or off by toggling the display on the Palettes panel.

Each tab has a unique role to play in working with Civil 3D. Prospector and Settings will be your most frequently visited tabs. Survey and Toolbox are used for special tasks that you will examine in the following section.

Prospector

Prospector shows your drawing-specific information about what currently exists in your project. Civil 3D objects are listed in workflow order, starting at the top of the listing. Each main grouping under the drawing name is referred to as a *collection*. If you expand a collection by clicking on the plus sign next to the name, you will see the contents of that group. Some of these groups are empty until objects are created.

Because all Civil 3D data is dynamically linked together, you will see object dependencies as well. You can learn details about an individual object by expanding the tree and selecting an object (Figure 1.4).

FIGURE 1.4 A look at the Alignment branch of the Prospector tab. Profiles are linked to alignments; therefore, they appear under alignments.

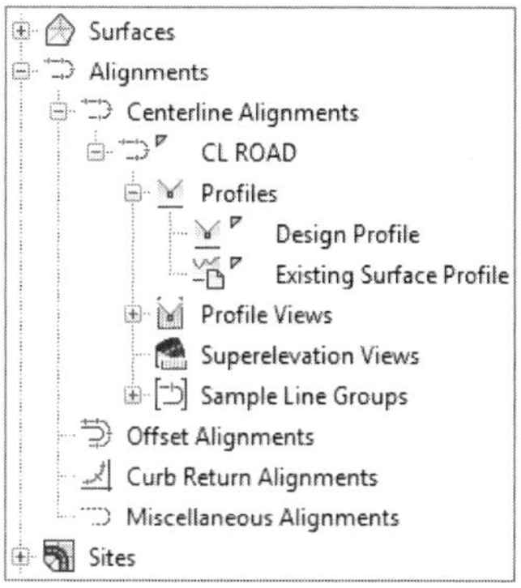

Right-clicking the collection name allows you to select various commands that apply to all the members of that collection. For example, right-clicking the Point Groups collection brings up the menu shown in Figure 1.5.

In addition, right-clicking the individual object in the list view offers many commands unique to Civil 3D: Zoom To Object and Pan To Object are typically included. By using these commands, you can find any parcel, point, cross section, or other Civil 3D object in your drawing almost instantly.

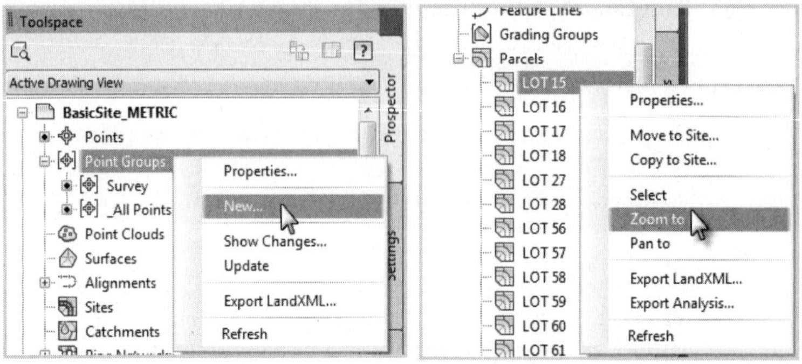

FIGURE 1.5
Context-sensitive menus in Prospector

For example, if you are interested in locating LOT 27 using the Zoom To command, locate the Sites collection on the Prospector tab of Toolspace. Expand the Proposed Site and highlight Parcels. At the bottom of Prospector, you will see all the lots listed. To locate LOT 27 graphically, right-click it and select Zoom To.

Near the top of the Toolspace you will see a pull-down giving you the option of Active Drawing View or Master View.

Active Drawing view will show you:

- The current drawing
- Data shortcuts

Master view will show you:

- Open drawings
- Data shortcuts
- Drawing templates

Master view will list every drawing you have open, as well as its contents and templates. If you use Master view, the name of the drawing you are working with appears at the top of the list in bold. To make a drawing current, right-click its name in Prospector and select Switch To.

Many users prefer to use the *Active Drawing view*. You can have more than one drawing open, but Prospector displays only one set of Civil 3D data at a time. Active Drawing view will change to reflect whichever drawing is current.

In addition to the branches, Prospector has a series of icons across the top that toggle various settings on and off. Some of the Civil 3D icons from previous versions have been removed, and their functionality has been universally enabled for Civil 3D 2013. Let's take a closer look at those icons:

Item Preview Toggle Turns on and off the display of the Toolspace item preview within Prospector. These previews can be helpful when you're navigating drawings in projects (you can select one to check out), or when you're attempting to locate a parcel on the basis of its visual shape. In general, however, you can turn off this toggle — it's purely a user preference.

Preview Area Display Toggle When Toolspace is undocked, this button moves the preview area from the right of the tree view to beneath the tree view area.

Panorama Display Toggle   Turns on and off the display of the Panorama window (which we'll discuss in a bit). To be honest, there doesn't seem to be a point to this button, but it's here nonetheless.

Help   Don't underestimate how helpful Help can be!

> ### HELP USING HELP
>
> The Civil 3D help files are extremely useful. Even for seasoned users, Help provides a comprehensive reference to objects and options. The most difficult part of using Help is knowing what terminology is used to describe the task you are trying to perform. Luckily, you have this book to assist you with that!
>
> At any time during your use of Civil 3D, you can use the F1 key to bring up the help file relevant to the dialog you are working in.

As you navigate the tabs of Toolspace, you will encounter many symbols to help you along the way. Table 1.1 shows you a few that you should familiarize yourself with.

TABLE 1.1:   Common Toolspace symbols and meanings

SYMBOL	MEANING
▽	The object or style is in use.
⊞	Clicking this will expand the branch of Toolspace.
⊟	Clicking this will collapse the branch of Toolspace.
●	Data resides in this branch and more information can be found at the bottom of Toolspace.
⚠	Object needs to be rebuilt or updated. Can also indicate broken data reference.

SETTINGS

The Settings tab of Toolspace controls all things aesthetic and the default behavior of the commands. Text placed by Civil 3D is controlled by *label styles*. *Object styles* control the look of design elements such as surface contours or pipes. The bulk of these settings and styles should be set in your template drawing. Every time you start a project with your company's Civil 3D–specific template, items such as an alignment's color and linetype will already be set. Chapter 20, "Label Styles," and Chapter 21, "Object Styles," are dedicated to building these styles. Later on in this chapter you will learn more about templates.

Drawing Settings

At the top of the Settings tab you will see the name of the drawing. There are some important settings you need to locate before proceeding with a project. Right-click on the name of the drawing and click Edit Drawing Settings to access the Drawing Settings dialog, shown in Figure 1.6.

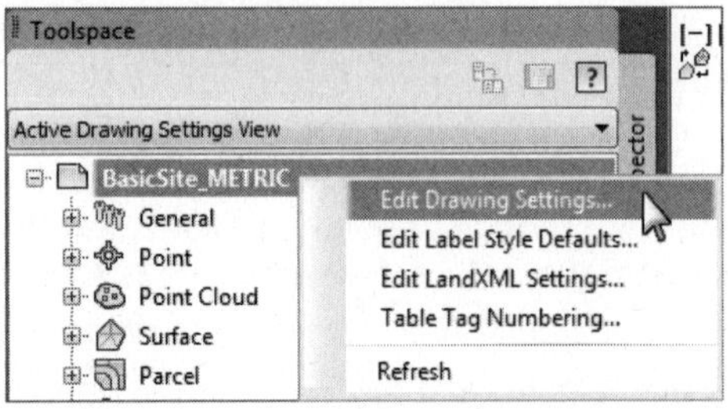

FIGURE 1.6
Accessing the Drawing Settings dialog

Try this quick exercise to make sure you know how to get into the drawing settings:

1. Open the file `BasicSite.dwg` (`BasicSite_METRIC.dwg`) from this book's web page, `www.sybex.com/go/masteringcivil3d2013`.
2. Switch to the Settings tab of Toolspace.
3. Right-click the filename and select Edit Drawing Settings. Switch to the Units And Zone tab to display the options shown in Figure 1.7.

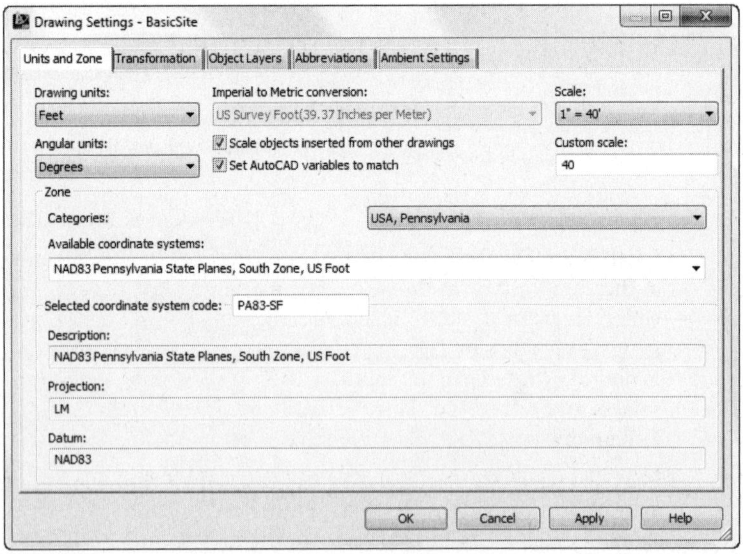

FIGURE 1.7
Before placing any project-specific information in the drawing, set the coordinate system in the Units And Zone tab of the Drawing Settings.

Each tab in this dialog controls a different aspect of the drawing. Most of the time, you'll pick up the Object Layers, Abbreviations, and Ambient Settings from a company-wide template.

However, the drawing scale and coordinate information change for every job, so you'll visit the Units And Zone and Transformation tabs frequently.

Units And Zone Tab

The Units And Zone tab lets you specify metric or Imperial units for your drawing. You can also specify the conversion factor between systems. If you choose a coordinate system with International or US Foot specified, this setting will become unavailable.

This tab also includes the options Scale Objects Inserted From Other Drawings and Set AutoCAD Variables To Match. The Set AutoCAD Variables To Match option sets the base AutoCAD angular units, linear units, block insertion units, hatch pattern, and linetype units to match the values placed in this dialog. As shown in Figure 1.7, you do want these options selected.

The scale that you see on the right side of the Units And Zone tab is the same as your *annotation scale*. You can change it here, but it is much easier to select your annotation scale from the bottom of the drawing window.

If you choose to work in assumed coordinates, you can leave Zone set to No Datum, No Projection. To set the coordinate system for your locale, first set the category from the long list of possibilities. Civil 3D is a worldwide product; therefore, most recognized surveying coordinate systems (including obsolete ones) can be found in the options. Once your coordinate system has been established, you can change it on the Transformation tab if desired.

Try the following quick exercise, working in the drawing from the previous exercise, to practice setting a drawing coordinate system:

1. Select USA, Pennsylvania from the Categories drop-down menu on the Units And Zone tab.

2. Select NAD83 Pennsylvania State Planes, South Zone, US Foot (NAD83 Pennsylvania State Planes, South Zone, Meter) from the Available Coordinate Systems drop-down menu.

   You could have also typed **PA83-SF** (**PA83-S**) in the selected Coordinate System Code box.

Transformation Tab

Earth is neither flat nor a perfect sphere. In reality, Earth is more of a squashed sphere, wider at the equator than at the poles. The mathematical approximation of Earth's physical shape is called an *ellipsoid*. The ellipsoid is constantly being fine-tuned to increase accuracy for precise geographic modeling. To make matters even more complicated, the gravitational pull of Earth is not distributed evenly, creating a slight skew between what a surveyor measures and how it relates to the ellipsoid. The gravitational representation of Earth is referred to as the *geoid*.

Luckily, most engineers and surveyors do not need to worry about all this. Most survey-grade GPS equipment takes care of the transformation to local grid coordinates for you. In the United States, state plane coordinate systems already have regional projections taken into account. In the rare case that surveyors need to transform local observations from geoid to ellipsoid, and ellipsoid to grid manually, the Transformation tab makes the computation quickly.

With a base coordinate system selected, you can do any further refinement you'd like using the Transformation tab, shown in Figure 1.8. The coordinate systems on the Units And Zone tab can be refined to meet local ordinances, tie in with historical data, complete a grid to ground transformation, or account for minor changes in coordinate system methodology. These changes can include the following:

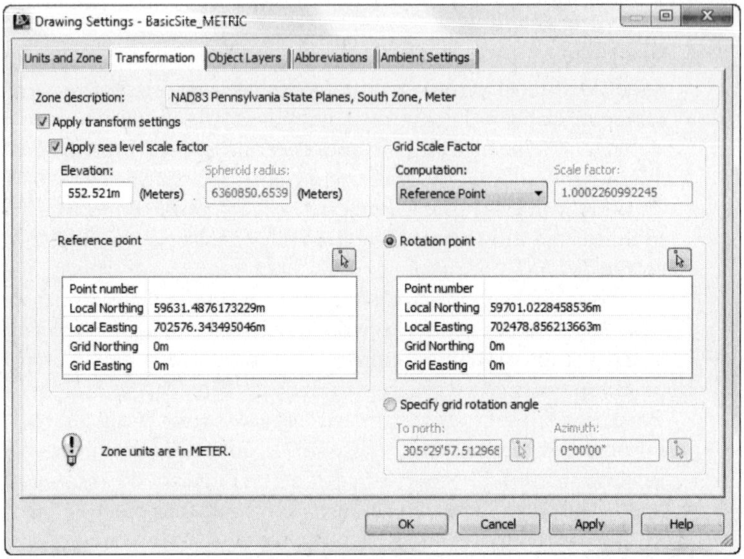

FIGURE 1.8
The Transformation tab

Apply Sea Level Scale Factor This value is known in some circles as *elevation factor* or *orthometric height scale*. Sea Level Factor takes into account the mean elevation of the site and the spheroid radius that is currently being applied as a function of the selected zone ellipsoid.

Grid Scale Factor At any given point on a projected map, there is a distortion between the "flat" measurement and the measurement on the ellipsoid. Grid Scale Factor is based on a 1:1 value, a user-defined uniform scale factor, a reference point scaling, or a prismoidal transformation in which every point in the grid is adjusted by a unique amount.

Reference Point To apply Grid Scale Factor and Sea Level Factor correctly, you need to tell Civil 3D where you are on Earth. Reference Point can be used to set a singular point in the drawing field via pick or point number, local northing and easting, or grid northing and easting values.

Rotation Point Rotation Point can be used to set the reference point for rotation via the same methods as the reference point.

Specify Grid Rotation Angle Some people may know this as the *convergence angle*. This is the angle between Grid North and True North. Enter an amount or set a line to north by picking an angle or deflection in the drawing. You can use this same method to set the azimuth if desired.

Most engineering firms work on either a defined coordinate system or an arbitrary system, so none of these changes are necessary. Given that, this tab will be your only method of achieving the necessary transformation for certain surveying and geographic information system

(GIS)-based and land surveying–based tasks. It should be noted that this is not the place to transform assumed coordinates to a predefined coordinate system. See Chapter 2, "Survey," to learn how to translate a survey.

Object Layers

Civil 3D and AutoCAD layers have a love–hate relationship with each other. Civil 3D is built on top of AutoCAD; therefore, all the objects do reside on layers. However, Civil 3D is not traditional CAD. Your surfaces, corridors, points, profiles, and everything else generated by Civil 3D are dynamic *objects* rather than simple lines, arcs, or circles.

When you create an alignment in Chapter 6, "Alignments," for example, you will not have to think about the current layer. This is because Civil 3D styles "push" objects and labels to the correct layer as part of their intelligence.

Layers are found in several areas of the Civil 3D template. The first location you will examine is the layers found in the Drawing Settings area. The layers listed here represent overall layers where the objects will be created. For those of you who are familiar with AutoCAD blocks, it is useful to think of these layers in the same way as a block's insertion layer.

In the Object Layers tab, every Civil 3D object must have a layer set, as shown in Figure 1.9. Do not leave any object layers set to 0. An optional modifier can be added to the beginning (*prefix*) or end (*suffix*) of the layer name to further separate items of the same type.

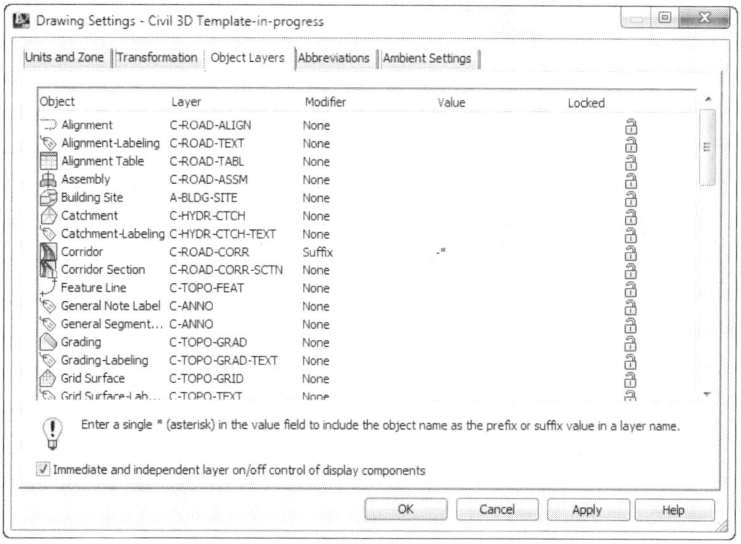

FIGURE 1.9
Every object is on a layer; the corridor layer contains a modifier.

A common practice is to add wildcard suffixes to corridor, surface, pipe, and structure layers to make it easier to manipulate them separately. For example, if the layer for a corridor is specified to be C-ROAD-CORR and a suffix of -* (dash asterisk, as shown in Figure 1.9) is added as the modifier value, a new layer will automatically be created when a new corridor is created. The resulting layer will take on the name of the corridor. If the corridor is called 13th Street, the new layer name will be C-ROAD-CORR-13th Street. This new layer is created once and is not dynamic to the object name. In other words, if you decide to rename "13th Street" to "Stephen Colbert Street," the layer remains C-ROAD-CORR-13th Street.

If the main layer name you are after does not exist in the drawing, you can create it as you work through the Object Layers dialog. Click the New button, as shown in Figure 1.10, and set up the layer as needed including Color, Lineweight, Linetype, and so forth.

FIGURE 1.10
Click New to add a new layer.

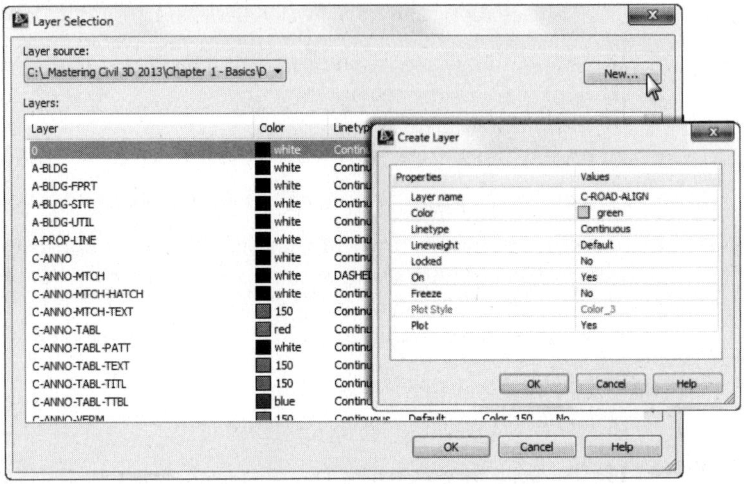

"Immediate And Independent Layer On/Off Control Of Display Components" is a setting you will want to have selected. As mentioned earlier, the layers listed here are like an object's insertion layer. However, you will encounter more layers within the object's display style (these are the object styles you will learn about in Chapter 21). Having this option selected allows you to turn off components within an object's style without turning off the entire object. For example, consider a surface whose Object Layer is set as C-TOPO. When that surface has contours displayed, the major contours might be on C-TOPO-MAJR and the minor contours may be on C-TOPO-MINR. With this option selected, you could turn off the C-TOPO-MINR independent from the overall object.

In the following exercise, you set object layers in a template:

1. Open the drawing `Civil 3D Template-in-progress.dwg` file, which you can download from this book's web page.
2. From the Settings tab of Toolspace, right-click on the name of the drawing and select Edit Drawing Settings.
3. Switch to the Object Layers tab.
4. Click in the Layer field next to Alignment, and New to create a new layer.
5. Create a new layer called **C-ROAD-ALIGN**. Leave other layer settings as default.
6. Set the newly created layer as the layer for the Alignment object.
7. Set the layer for Building Site to A-BLDG-SITE.
8. Set the layer for Catchment-Labeling to C-HYDR-CTCH-TEXT.

9. For the corridor layer, keep the main layer as C-ROAD-CORR.
10. Set the Modifier to Suffix.
11. Set the modifier value to **-***.

    The asterisk acts as a wildcard that will add the corridor name as part of a unique layer for each corridor as previously described.
12. Scroll down to locate the Pipe object listing and then the Structure object listing.
13. Create several new layers and add suffix information:
    - For Pipe, create a layer called **C-NTWK-PIPE** with a Modifier of Suffix and a Value of -*.
    - For Pipe-Labeling, create a new layer called **C-NTWK-PIPE-TEXT**.
    - For Pipe And Structure Table, set the layer to **C-NTWK-PIPE-TABL**.
    - For Pipe Network Section, create a new layer called **C-NTWK-XSEC**.
    - For Pipe Network Profile, create a new layer called **C-NTWK-PROF**.
14. Scroll down a bit further and create a new layer for Structure called **C-NTWK-STRC**.
15. Add a Modifier of Suffix and a Value of -*.
16. For Structure-Labeling, create a new layer called **C-NTWK-STRC-TEXT**.
17. Add a Modifier of Suffix to the TIN Surface object layer and a Value of -*.

    Your layers and suffixes should now resemble Figure 1.11.

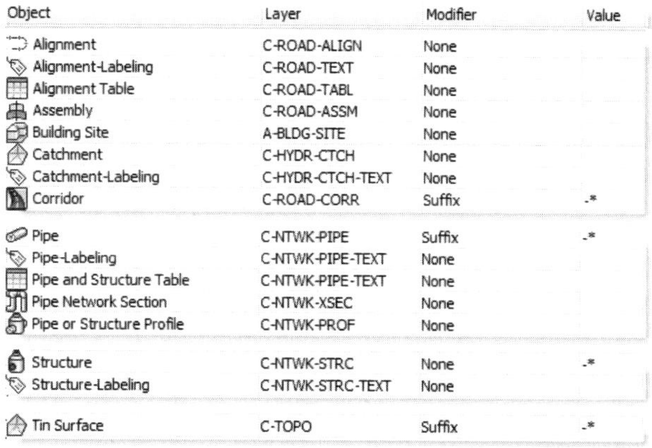

Object	Layer	Modifier	Value
Alignment	C-ROAD-ALIGN	None	
Alignment-Labeling	C-ROAD-TEXT	None	
Alignment Table	C-ROAD-TABL	None	
Assembly	C-ROAD-ASSM	None	
Building Site	A-BLDG-SITE	None	
Catchment	C-HYDR-CTCH	None	
Catchment-Labeling	C-HYDR-CTCH-TEXT	None	
Corridor	C-ROAD-CORR	Suffix	-*
Pipe	C-NTWK-PIPE	Suffix	-*
Pipe-Labeling	C-NTWK-PIPE-TEXT	None	
Pipe and Structure Table	C-NTWK-PIPE-TEXT	None	
Pipe Network Section	C-NTWK-XSEC	None	
Pipe or Structure Profile	C-NTWK-PROF	None	
Structure	C-NTWK-STRC	None	-*
Structure-Labeling	C-NTWK-STRC-TEXT	None	
Tin Surface	C-TOPO	Suffix	-*

FIGURE 1.11 Examples of the completed layer names in the Object Layers tab

18. Place a check mark next to "Immediate And Independent Layer On/Off Control Of Display Components."

    As described previously, this setting will allow you to use the On/Off toggle in Layer Manager to work with Civil 3D objects.

19. Click Apply and then OK.

20. Save the drawing for use in the next exercise.

Abbreviations

When you add labels to certain objects, Civil 3D automatically uses the abbreviations from this area to indicate geometry features. For example, left is *L* and right is *R*.

Civil 3D uses industry-standard abbreviations wherever they are found. If necessary, you can easily change VPI to PVI for Point of Vertical Intersection. In most cases, changing an abbreviation is as simple as clicking in the Value field and typing a new one. Notice that the Alignment Geometry Point Entity Data section has a larger set of values and some special coding attached (as shown toward the bottom of Figure 1.12). These are more representative of other label styles. You will visit the Text Component Editor in Chapter 20.

FIGURE 1.12 Customizable down to the letter, the Abbreviations tab

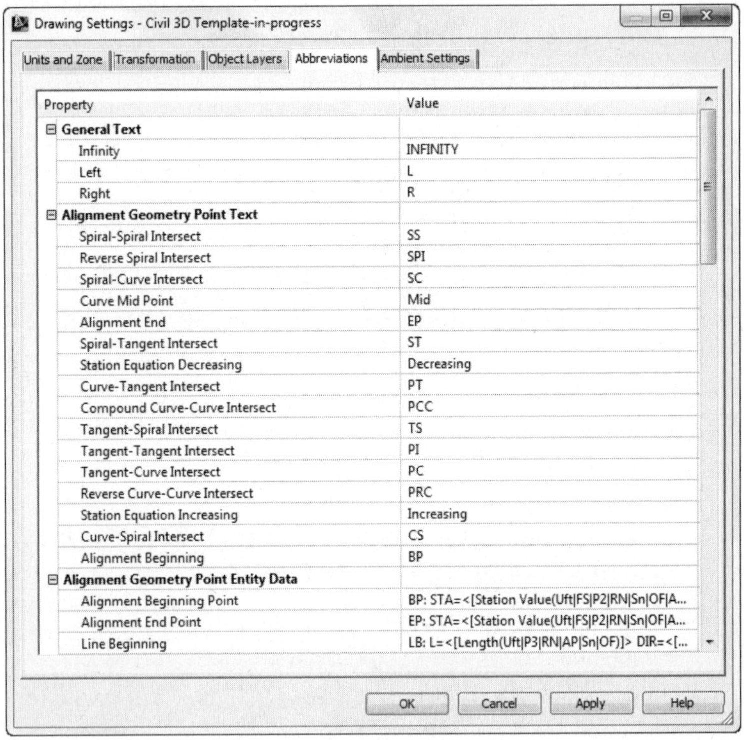

Ambient Settings

Examine the settings in the Ambient Settings tab to see what can be set here. The main options you'll want to adjust are in the General category and the display precision in the subsequent categories.

The precision that you see in this dialog does not change the label precision. The precision you see here is the number of decimal places reported to you in various dialogs.

Being familiar with the way this tab works will help you further down the line, because almost every other settings dialog in the program works like the one shown in Figure 1.13.

FIGURE 1.13
Ambient Settings at the main drawing level

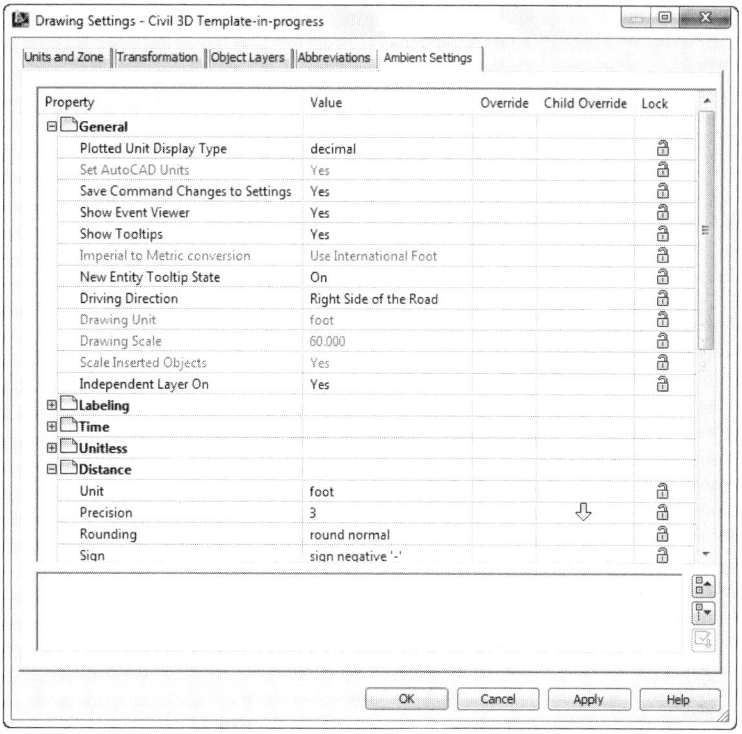

Plotted Unit Display Type Civil 3D knows you want to plot at the end of the day. In this case, it's asking how you would like your plotted units measured. For example, would you like that bit of text to be 0.25" tall or ¼" high? Most engineers are comfortable with the Leroy method of text heights (L80, L100, L140, and so on), so the decimal option is the default.

Set AutoCAD Units This option specifies whether Civil 3D should attempt to match AutoCAD drawing units, as specified on the Units And Zone tab.

Save Command Changes To Settings Set this to Yes. This setting is incredibly powerful but a secret to almost everyone. By setting it to Yes, you ensure that your changes to commands will be remembered from use to use. This means if you make changes to a command during use, the next time you call that Civil 3D command, you won't have to make the same changes. It's frustrating to do work over because you forgot to change one of the five things that needed changing, so this setting is invaluable.

Show Event Viewer Event Viewer is the main Civil 3D feedback mechanism, especially when things go wrong. Event Viewer uses the Panorama interface to display warnings such as when a surface contains crossing breaklines. Event viewer will pop up with informational messages as well. If you have multiple monitors, it is a good idea to leave Panorama on but set aside for review.

Show Tooltips One of the cool features that people remark on when they first use Civil 3D is the small pop-up that displays relevant design information when the cursor is paused on the screen. This includes things such as station-offset information, surface elevation, section information, and so on. Once a drawing contains numerous bits of information, this display can be overwhelming; therefore, Civil 3D offers the option to turn off these tooltips universally with this setting. A better approach is to control the tooltips at the object type by editing the individual feature settings. You can also control the tooltips by pulling up the properties for any individual object and looking at the Information tab.

Imperial To Metric Conversion This setting displays the conversion method specified on the Units And Zone tab. The two options are US Survey Foot and International Foot.

New Entity Tooltip State This setting controls whether the tooltip is turned on at the object level for new Civil 3D objects. If you change this setting to Off partway through a project, the tooltip will not be displayed for any Civil 3D objects created after the change.

Driving Direction Specifies the side of the road that forward-moving vehicles use for travel. This setting is important in terms of curb returns and intersection design.

Drawing Unit, Drawing Scale, and Scale Inserted Objects These settings were specified on the Units And Zone tab but are displayed here for reference, and so that you can lock them if desired.

Independent Layer On This is the same control that was set on the Object Layers tab. Yes is the recommended setting, as described previously.

The Ambient Settings for Direction offer the following choices:

- Unit: Degree, Radian, and Grad.
- Precision: 0 through 8 decimal places.
- Rounding: Round Normal, Round Up, and Truncate.
- Format: Decimal, two types of DDMMSS, and Decimal DMS. In most cases, people want to display DD°MM?SS.SS?. Whether you want spaces between the subdivisions is up to you.
- Direction: Short Name (spaced or unspaced) and Long Name (spaced or unspaced).
- Capitalization: you can display as typed, or force uppercase, lowercase, or title caps.
- Sign: gives you your choice of how negative numbers are displayed. You can use a negative sign to denote negative numbers only, use a parenthesis to denote a negative, or use a sign regardless of value. The latter option will show a plus for positive values and a minus for negative values.
- Measurement Type: Bearings, North Azimuth, and South Azimuth.
- Bearing Quadrant: This should be left at the industry standard. 1 - NE, 2-SE, 3 - SW, 4-NW.

CERT OBJECTIVE

When you're using the Bearing Distance transparent command, for example, these settings control how you input your quadrant, your bearing, and the number of decimal places in your distance.

Explore the other categories, such as Angle, Lat Long, and Coordinate, and customize the settings to how you work.

At the bottom of the Ambient Settings tab is a Transparent Commands category. These settings control how (or if) you're prompted for the following information:

Prompt For 3D Points Controls whether you're asked to provide a *z* elevation after *x* and *y* have been located.

Prompt For Y Before X For transparent commands that require *x* and *y* values, this setting controls whether you're prompted for the y-coordinate before the x-coordinate. Most users prefer this value set to False so they're prompted for an x-coordinate and then a y-coordinate.

Prompt For Easting Then Northing For transparent commands that require Northing and Easting values, this setting controls whether you're prompted for the Easting first and the Northing second. Most users prefer this value set to False, so they're prompted for Northing first and then Easting.

Prompt For Longitude Then Latitude For transparent commands that require longitude and latitude values, this setting controls whether you're prompted for Longitude first and Latitude second. Most users prefer this set to False, so they're prompted for Latitude and then Longitude.

DRAWING PRECISION VS. LABEL PRECISION

The precision and units that you are seeing in the Ambient Settings and commands settings only affect how the values are reported to you in dialogs and command-line display. These settings are not related to the number of decimal places or units displayed in labels.

The same is true for angular display. The vast majority of users prefer to see angular units in degrees, minutes, and seconds. As long as you are not using decimal degrees (or any of the more exotic units such as radians or grads), Civil 3D will expect your input to be in degrees, minutes, and seconds. Be sure to check out the additional information regarding angular entry later in this chapter. You will work in depth with setting precision in labels in Chapter 20.

The settings that are applied here can also be changed at the object levels. For example, you may typically want elevation to be shown to two decimal places, but when looking at surface elevations, you might want just one. The Override and Child Override columns give you feedback about these types of changes. See Figure 1.14.

The Override column shows whether the current setting is overriding something higher up. Because you're at the Drawing Settings level, these are clear. However, the Child Override column displays a down arrow, indicating that one of the objects in the drawing has overridden this setting. After a little investigation of the objects, you'll find the override in the Edit Feature Settings of the Profile view, as shown in Figure 1.15.

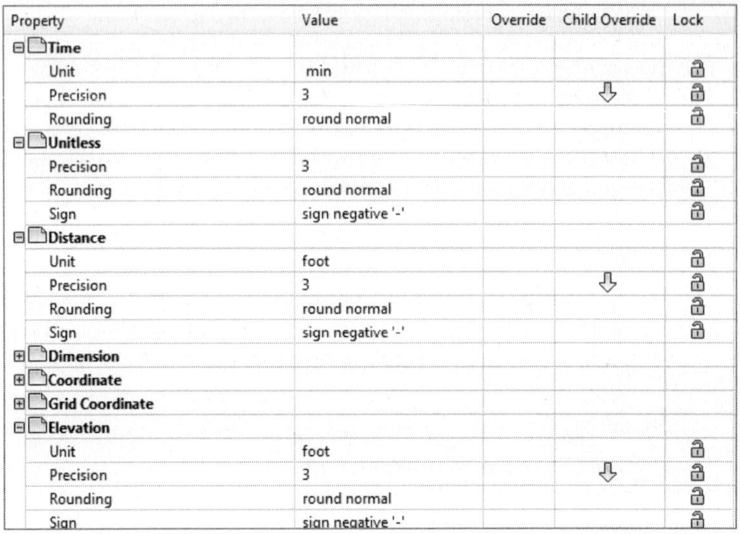

FIGURE 1.14
The Child Override indicator in the Time, Distance, and Elevation values

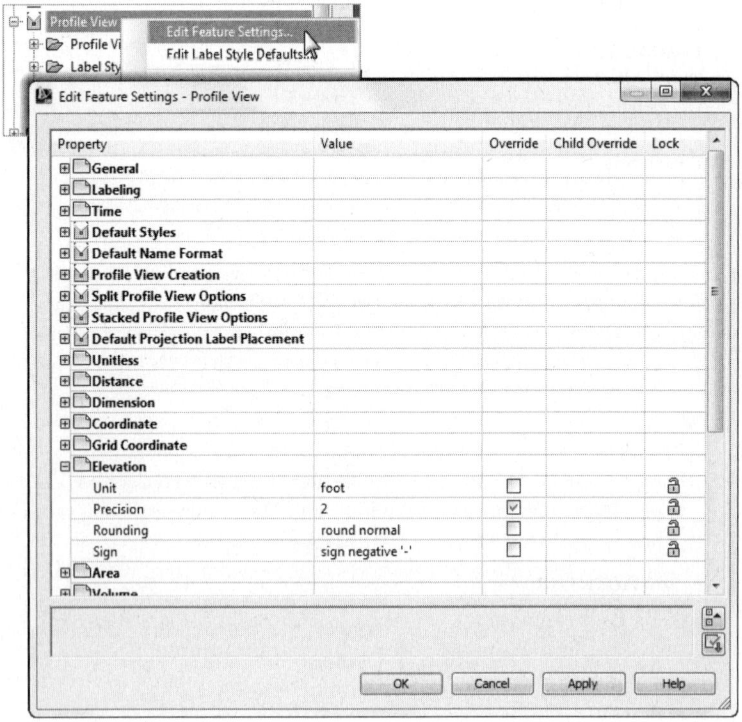

FIGURE 1.15
The Profile Elevation Settings and the Override indicator

Notice that in this dialog, the box is checked in the Override column. This indicates that you're overriding the settings mentioned earlier, and it's a good alert that things have changed from the general Drawing Settings to this Object Level setting.

But what if you don't want to allow those changes? Each Settings dialog includes one more column: Lock. At any level, you can lock a setting, graying it out for lower levels. This can be handy for keeping users from changing settings at the lower level that perhaps should be changed at a drawing level, such as sign or rounding methods.

SURVEY

The Survey palette is displayed optionally and controls the use of the survey, equipment, and figure prefix databases. Survey is an essential part of land-development projects. Because of the complex nature of this tab, all of Chapter 2 is devoted to it.

TOOLBOX

The Toolbox is a launching point for add-ons and reporting functions. To access the Toolbox, from the Home tab in the Ribbon, select Toolspace ➢ Palettes and click the Toolbox icon (as shown previously in Figure 1.3). Out of the box, the Toolbox contains reports created by Autodesk, but you can expand its functionality to include your own macros or reports. The buttons on the top of the Toolbox, shown in Figure 1.16, allow you to customize the report settings and add new content. If you are an Autodesk Subscription customer, new goodies released throughout the year are frequently accessed from this area.

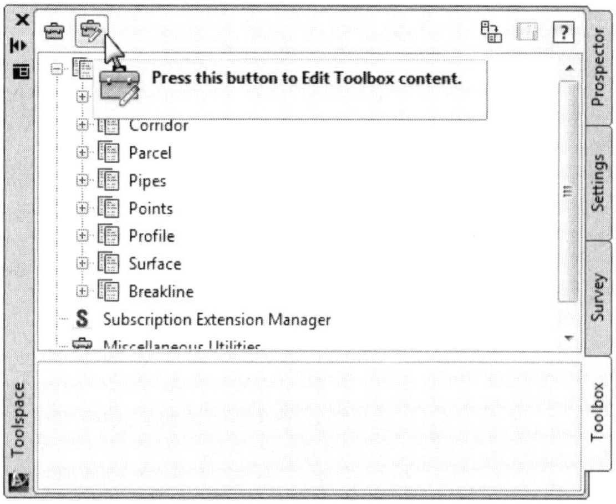

FIGURE 1.16
The Toolbox with the Edit Toolbox Content icon highlighted

Panorama

The Panorama window is the Civil 3D feedback and tabular editing mechanism. It's designed to be a common interface for a number of different Civil 3D–related tasks, and you can use it to provide information about the creation of profile views, to edit pipe or structure information, or to run basic volume analysis between two surfaces. For an example of Panorama in action, open it up by going to the Home tab ➤ Palettes Panel flyout and clicking on the Event Viewer icon. You'll explore and use Panorama more during this book's discussion of specific objects and tasks.

Ribbon

As with AutoCAD, the Ribbon is the primary interface for accessing Civil 3D commands and features. When you select an AutoCAD Civil 3D object, the Ribbon displays commands and features related to that object in a *contextual tab*. If several object types are selected, the Multiple contextual tab is displayed. Use the following procedure to familiarize yourself with the Ribbon:

1. Open the `BasicSite.dwg (BasicSite_METRIC.dwg)`, which you will find at `www.sybex.com/go/masteringcivil3d2013`.
2. Select one of the parcel labels (the labels in the middle of the lot areas).

   Notice that the Labels & Tables, General Tools, Modify, and Launch Pad panels are displayed, as shown in Figure 1.17.

FIGURE 1.17
The contextual tab in the Ribbon

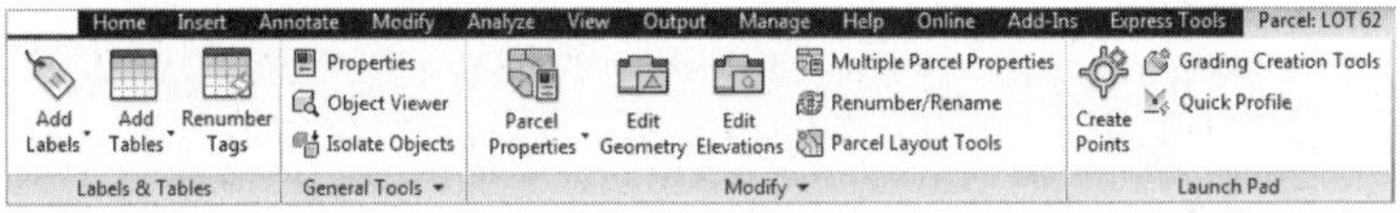

3. Select a parcel line and notice the display of the Multiple contextual tab (Figure 1.18).

FIGURE 1.18
When more than one object is selected, the Multiple tab appears.

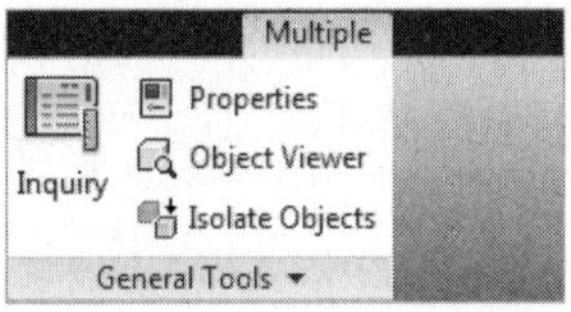

4. Use the Esc key to cancel all selections.
5. Reselect a parcel by clicking one of the numeric labels.
6. Select the down arrow next to the Modify panel name.
7. Click on the pin at the bottom-left corner of the panel to keep it open.
8. Select the Properties command in the General Tools panel to open the AutoCAD Properties palette.

   Notice that the Modify panel remains open and pinned until the current selection set changes.

STYLES AND MORE STYLES

Civil 3D uses *styles* to change the look of objects and labels. Styles control everything from which layer your surface contours will be created on, to the number of decimal places displayed in a label.

Civil 3D has an unbelievable number of options when it comes to how you want your design elements to look. It is easy to get bogged down in the intricacies of object and label style creation. The authors have decided to separate styles into separate chapters so you can focus on learning functionality first. Once you have an understanding of how the tools operate, you can then adjust how your designs are represented graphically.

In this chapter and throughout the book, you will be using styles that have already been created for you. For an in-depth look at styles, refer to Chapter 20 and Chapter 21.

Civil 3D Templates

Styles and settings should come from your template. Ideally, that template will have all the styles you need for the type of project you are working on. If you find you are constantly changing style settings, reexamine your workflow.

When starting a project, or continuing a project from an outside source, it is important to start with a Civil 3D template file. Right after installing the software you will see two usable Civil 3D–specific templates (Figure 1.19).

FIGURE 1.19
Civil 3D template

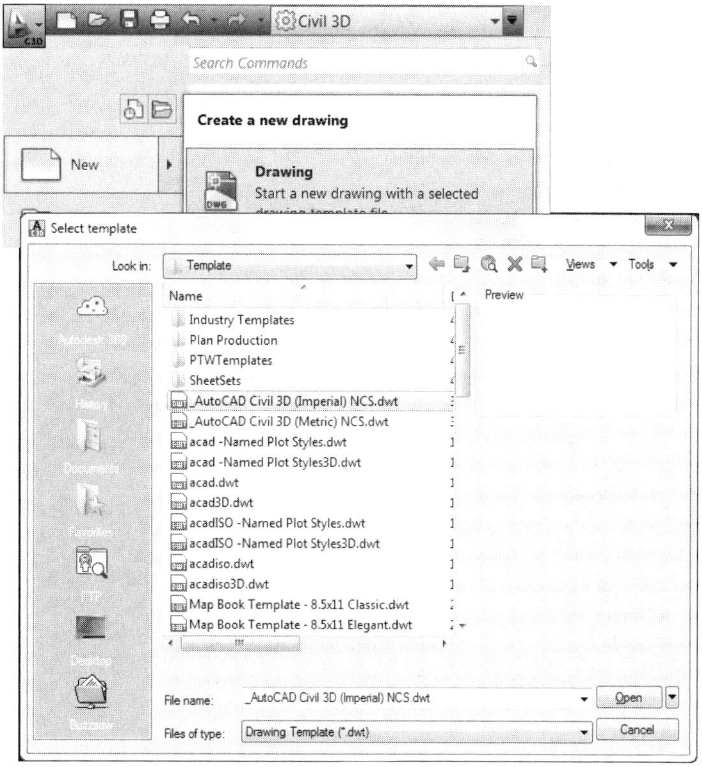

The Civil 3D–specific templates are:

- `_AutoCAD Civil 3D (Imperial) NCS.dwt`
- `_AutoCAD Civil 3D (Metric) NCS.dwt`

Starting New Projects

When you start a project with the correct template (DWT file), the repetitive task of defining the basic framework for your drawing is already completed. As you know, a base AutoCAD DWT file contains the following:

- Unit type (architectural or decimal) and insertion scale (meters or feet)
- Layers and their respective linetypes, colors, and other properties
- Text styles
- Dimension and multileader styles
- Layouts and plot setups
- Block definitions

Civil 3D takes the base AutoCAD template and kicks it up several notches. In addition to the items just listed, a Civil 3D template contains the following:

- More specific unit information (international feet, survey feet, or meters)
- Civil object layers
- Ambient Settings
- Label styles and formulas (expressions)
- Object styles
- Command settings
- Object naming templates
- Report settings
- Description key sets

You must always start new projects with a proper Civil 3D template. If you receive a drawing from a non–Civil 3D user and need to continue it in Civil 3D, you must import the styles and settings. Without suitable styles and settings, all of your object and label styles will show up with the name Standard, as shown in Figure 1.20. You do not want objects and labels to use the Standard style, as it is the Civil 3D equivalent of drawing on layer 0. Items that use the Standard style appear on layer 0, and will contain the most basic display settings.

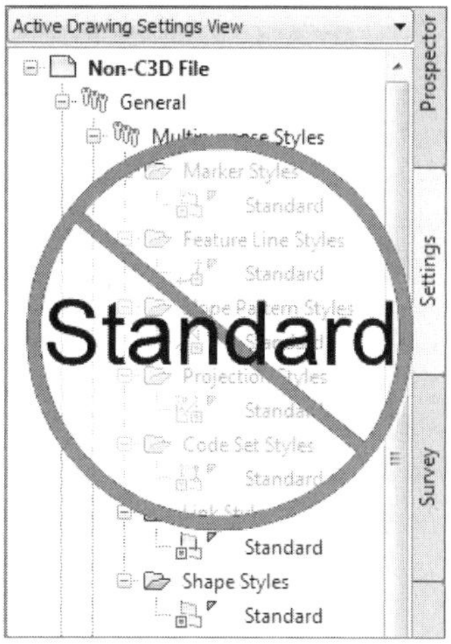

FIGURE 1.20
A non–Civil 3D DWG will list all styles as Standard, which is the Civil 3D equivalent to drawing on layer 0.

TEMPLATES USED IN THIS BOOK

Throughout this book, when you start a file from scratch, you will use one of the templates that come with Civil 3D when you install it. The templates that come with Civil 3D may not be exactly what you want initially, but they are a great starting point when you are customizing your projects.

Real World Scenario

BEST PRACTICES FOR RECEIVING AN OUTSIDE DRAWING

Say someone sends you a drawing that was not done in Civil 3D. Perhaps it was exported from MicroStation, or initially created in Land Desktop, Carlson, or Eagle Point. You now have the task of creating Civil 3D objects, but making this task even more difficult is that there are no Civil 3D styles present. Perhaps the drawing was created in Civil 3D, but your organization's styles look completely different.

The best course of action to take when receiving an outside drawing is to insert it into a blank file that you started with your Civil 3D template. When you insert a drawing, Civil 3D is doing several things to help you:

- The Insert command will detect the units of the incoming drawing and scale it to match your drawing.
- If both drawings have a coordinate system defined, Civil 3D will place the incoming drawing by geographic data.
- All of your styles and settings will stay intact, including a few items that do not get imported using the Import Styles command.

The following exercise walks you through exactly what you need to do in this situation:

1. Start by choosing Application ➤ New ➤ Drawing. Select either `_AutoCAD Civil 3D (Imperial) NCS.dwt` or `_AutoCAD Civil 3D (Metric) NCS.dwt` and click open.
2. Click the Save icon from the Quick Access toolbar.
3. Save the drawing with the rest of your Mastering Civil 3D files as **`My Project - Civil 3D.dwg`**.
4. Go to the Insert tab in the Ribbon. From the Block panel, click Insert.
5. Click Browse and locate the file `Mystery Structure.dwg` that is part of the dataset for this chapter. Click Open.
6. Be sure that the Insertion Point, Scale, and Rotation check boxes are clear.
7. Select the Explode check box. Your Insert dialog should look like this:

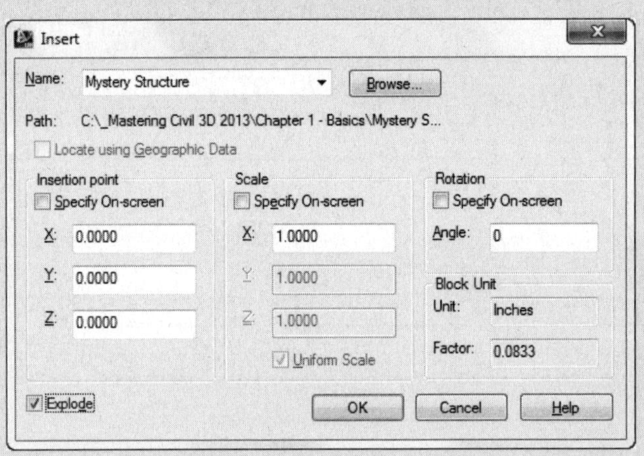

Notice that AutoCAD has picked up the units of the `Mystery Structure.dwg` file and is automatically scaling it as needed. If you used the English units template, you will see the conversion scale factor as 0.0833. If you use the metric template drawing, you will see the scale factor as 0.025.

8. Click OK. Double-click your middle mouse wheel to zoom extents.

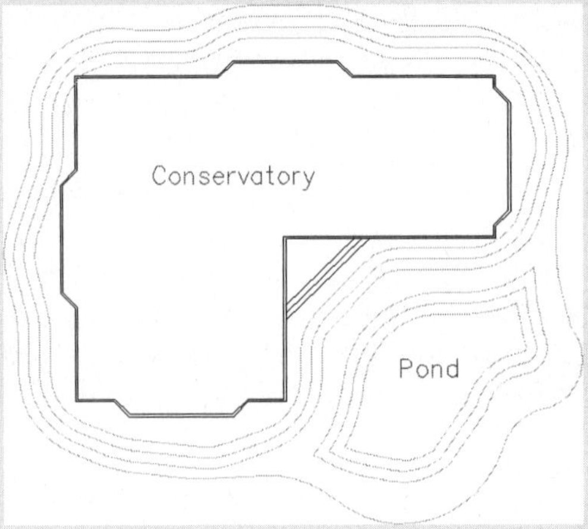

A quick measurement of the northern most wall of the conservatory building should reveal that it is 200' (61 m) in length.

At this point you can now work on the drawing in Civil 3D without re-creating any established standards. Everything from the outside source has come in, such as including blocks, layers, and dimension styles, but they will not override any of yours if they happen to have the same name.

Importing Styles

Luckily, importing styles from an existing drawing or template has gotten much easier. You can find the Import Styles button on the Styles panel of the Manage tab in the Ribbon. When you click the button, you will be prompted to browse for the template (DWT) or drawing (DWG) that contains the styles you are looking for.

You must save the drawing before you can import styles. If you forget, you will be prompted to save the drawing before you can proceed with the import.

Once you select the file whose styles you will import, a dialog similar to Figure 1.21 appears.

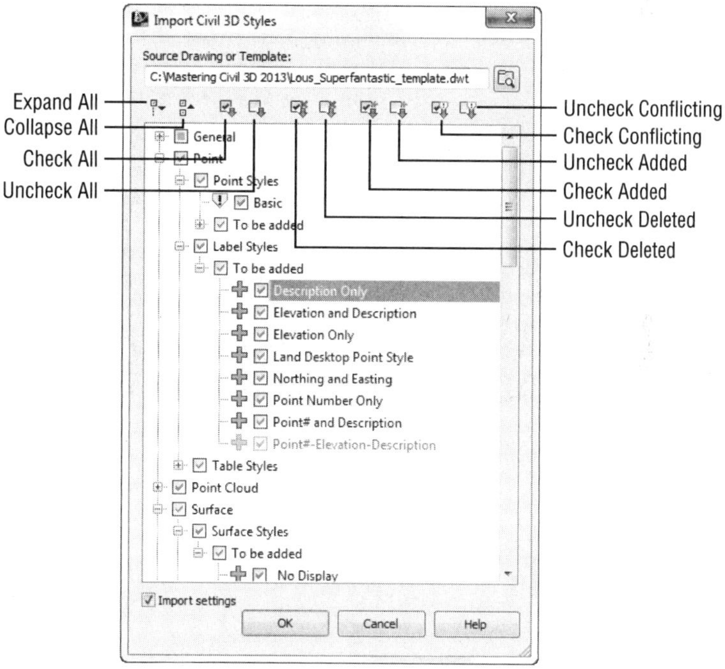

FIGURE 1.21
Import Civil 3D Styles dialog

Import Settings Notice the Import Settings option at the bottom of the dialog. This option is turned on by default. As you look through the list of styles to be imported, you will notice that some items are grayed out and can't be modified.

The grayed-out styles represent items in command settings. Styles referred to in command settings must be imported if the Import Settings option is turned on. You will read about command settings in the upcoming section.

Uncheck Conflicting You will also notice items with a warning symbol (see Basic in Figure 1.21). The warning symbol indicates there is a style in the current drawing with the same name as a style in the batch to be imported. Use the Uncheck Conflicting button if you do not want styles in the destination drawing to be overwritten. If you leave these items selected, the incoming styles "win." If you are not sure if there is a difference between the styles, pause your cursor over the style name and a tooltip will tell you what (if any) difference exists.

Uncheck Added Use the Uncheck Added button if you only want styles with the same name to come in. Wherever possible, Civil 3D will release items in the To Be Added categories. In cases where a style is used by a setting, you will not be able to uncheck it unless you do not import settings.

Uncheck Deleted A style in the current drawing will be deleted if the source drawing does not contain a style with the same name. Use the Uncheck Deleted button to prevent the style from being deleted.

Labeling Lines and Curves

You can draw lines many ways in an AutoCAD-based environment. The tools found on the Draw panel of the Home tab in the Ribbon create lines that are the same as those created by the standard AutoCAD Line command. How the Civil 3D lines differ from those created by the regular Line command isn't in the resulting entity, but in the process of creating them. Figure 1.22 shows the available line commands.

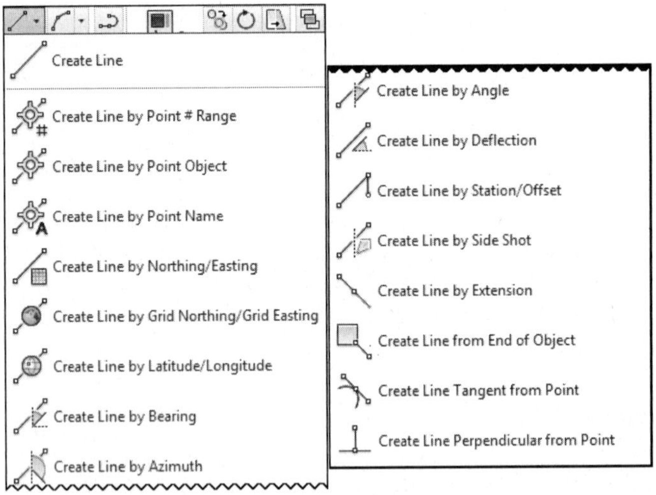

FIGURE 1.22
Line creation tools

Later on in this chapter you will take a look at *transparent commands*. Transparent commands allow you to draw any object using similar "Civil-friendly" techniques.

Coordinate Line Commands

The next few commands discussed in this section help you create a line using Civil 3D points and/or coordinate inputs. Each command requires you to specify a Civil 3D point, a location in space, or a typed coordinate input. These Line tools are useful when your drawing includes Civil 3D points that will serve as a foundation for linework, such as the edge of pavement shots, wetlands lines, or any other points you'd like to connect with a line.

Create Line Command

The Create Line command on the Draw panel of the Home tab in the Ribbon issues the standard AutoCAD Line command. It's equivalent to typing **`line`** on the command line or, clicking the Line tool on the Draw toolbar.

CERT OBJECTIVE

Create Line By Point # Range Command

The Create Line By Point # Range command prompts you for a point number. You can type in an individual point number, press ↵, and then type in another point number. A line is drawn connecting those two points. You can also type in a range of points, such as 601-607. Civil 3D draws a line that connects those lines in numerical order—from 601 to 607, and so on (see Figure 1.23). The line that is created by this method connects point-to-point regardless of the description of type of point.

Figure 1.23
A line created using 601-607 as input

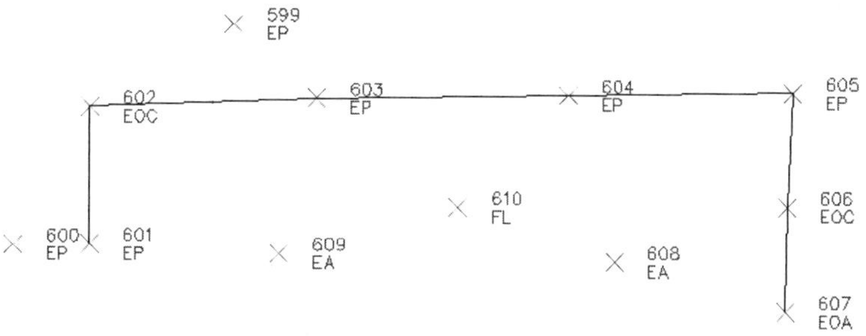

Alternatively, you can enter a list of points such as **601, 603, 610, 605** (Figure 1.24). Civil 3D draws a line that connects the point numbers in the order of input. This approach is useful when your points were taken in a zigzag pattern (as is commonly the case when cross-sectioning pavement), or when your points appear so far apart in the AutoCAD display that they can't be readily identified.

Figure 1.24
A line created using 601, 603, 610, 605 as input

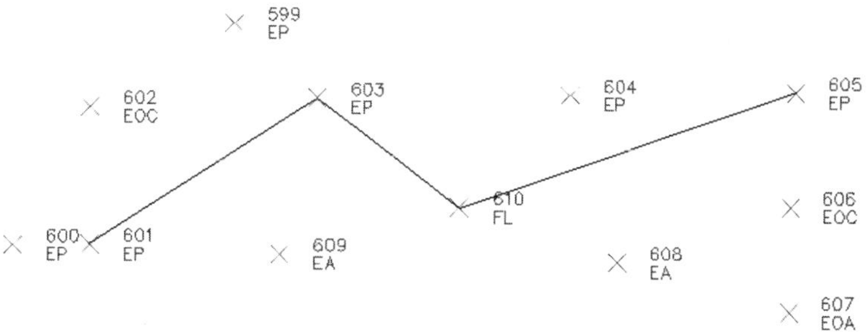

Create Line By Point Object Command

The Create Line By Point Object command prompts you to select a point object. To select a point object, locate the desired start point and click any part of the point. This tool is similar to using the regular Line command and a Node object snap (also known as Osnap); however, it will only work on Civil 3D points.

Create Line By Point Name Command

CERT OBJECTIVE

The Create Line By Point Name command prompts you for a point name. A *point name* is a field in Point properties, not unlike the point number or description. The difference between a point name and a *point description* is that a point name must be unique. It is important to note that some survey instruments name points rather than number points as is the norm.

To use this command, enter the names of the points you want to connect with linework.

Create Line By Northing/Easting and Create Line By Grid Northing/Easting Commands

The Create Line By Northing/Easting and Create Line By Grid Northing/Easting commands let you input northing (y) and easting (x) coordinates as endpoints for your linework. The Create Line By Grid Northing/Easting command requires that the drawing have an assigned coordinate system. This command can be useful when working with known surveyed points or features in a state plane coordinate system (SPCS).

Create Line By Latitude/Longitude Command

The Create Line By Latitude/Longitude command prompts you for geographic coordinates to use as endpoints for your linework. This command also requires that the drawing have an assigned coordinate system. For example, if your drawing has been assigned Delaware State Plane NAD83 US Feet and you execute this command, your Latitude/Longitude inputs are translated into the appropriate location in your state plane drawing. This command can be useful when you are drawing lines between waypoints collected with a standard handheld GPS unit.

Important Notes on Entering Data into Civil 3D

In this chapter, you will get your first taste of keying in data to the command line. Always keep an eye out for what is on your command line (or your tooltip, if dynamic input is on). Sometimes you are asked to confirm an option; other times you are asked for input.

As stated previously, most people prefer to enter angles by degrees, minutes, and seconds, rather than decimal degrees. The Ambient setting format will affect exactly how Civil 3D displays this information but is more flexible when it comes to your data entry. To enter angles into Civil 3D, you can use a *DD.MMSS* format. To input 15°21′35″, you can use **15.2135**. Any numbers beyond four decimal places will be considered decimal seconds. To input 6°5′2″, enter **06.0502**. You could also use **6d5′2″** at the command line, but most people find it faster to use the former method.

When entering station values into Civil 3D, it is not necessary to use station notation. In English units, if you are asked to enter station 3+25, you can simply enter **325**. In metric, a station of 0+110 can be entered as **110**. Similarly, a metric station of 2+450 can be entered as **2450**.

Direction-Based Line Commands

The next few commands help you specify the direction of a line. Each of these commands requires you to choose a start point for your line before you can specify the line direction. You can specify your start point by physically choosing a location, using an Osnap, or using one of the point-related line commands discussed earlier.

Many of these line commands require a line or arc, and will not work with a polyline. Those commands include Create Line By Sideshot, Create Line By Extension, Create Line From End Of Object, Create Line Tangent From Point, and Create Line Perpendicular from Point.

Create Line By Bearing Command

The Create Line By Bearing command will likely be one of your most frequently used line commands.

This command prompts you for a start point, followed by prompts to input the Quadrant, Bearing, and Distance values. You can enter values on the command line for each input, or you can graphically choose inputs by picking them on screen. The glyphs at each stage of input guide you in any graphical selections. After creating one line, you can continue drawing lines by bearing, or you can switch to any other method by clicking one of the other Line By commands on the Draw panel (see Figure 1.25).

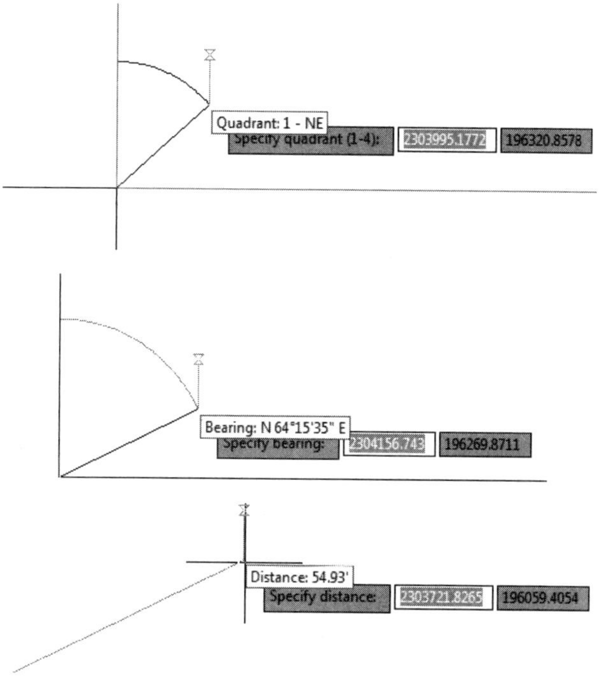

FIGURE 1.25
The tooltips for a quadrant (top), a bearing (middle), and a distance (bottom)

Create Line By Azimuth Command

The Create Line By Azimuth command prompts you for a start point, followed by a north azimuth, and then a distance (Figure 1.26).

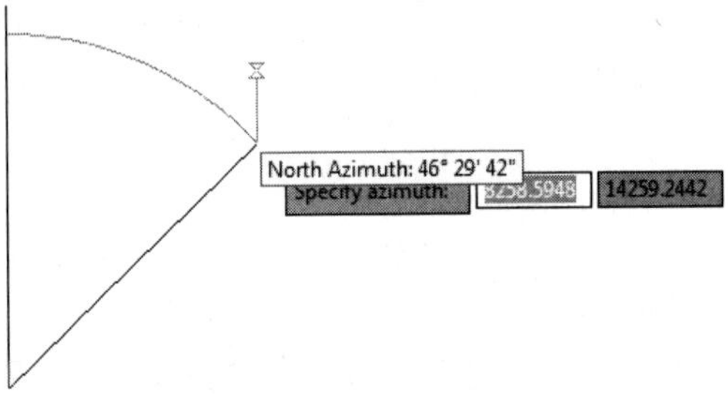

Figure 1.26
The tooltip for the Create Line By Azimuth command

Create Line By Angle Command

The Create Line By Angle command prompts you for a start point, and ending point (think backsight), a turned angle, and then a distance (Figure 1.27). This command is useful when you're creating linework from angles-right (in lieu of angles-left), and distances recorded in a traditional handwritten field book (required by law in many states).

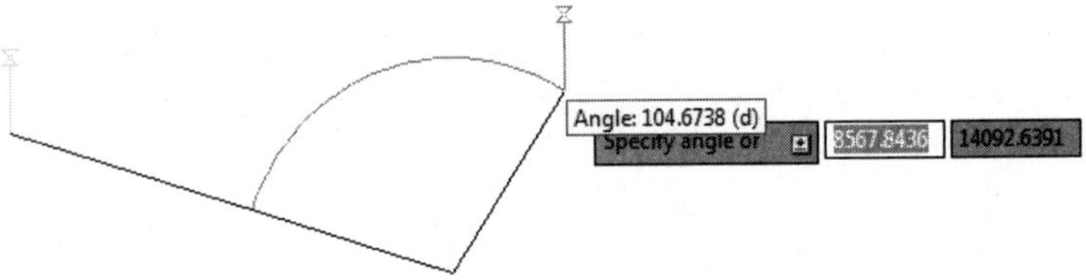

Figure 1.27
The tooltip for the Create Line By Angle command

Create Line By Deflection Command

By definition, a *deflection angle* is the angle turned from the extension of a line from the backsight extending through an instrument. Although this isn't the most frequently used surveying tool in this day of data collectors and GPSs, on some occasions you may need to create this type of line. When you use the Create Line By Deflection command, the command line and tooltips prompt you for a deflection angle followed by a distance (Figure 1.28). In some cases, deflection angles are recorded in the field in lieu of right angles.

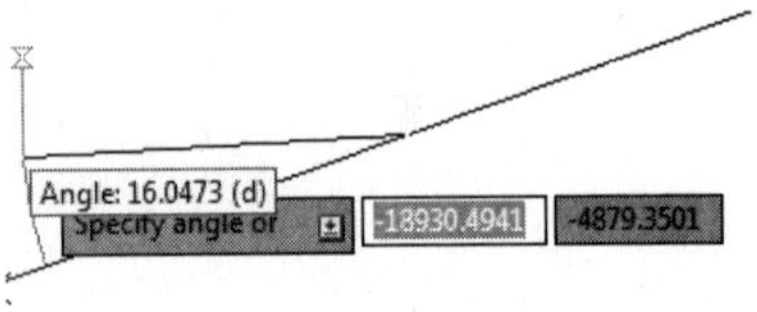

Figure 1.28
The tooltips for the Create Line By Deflection command

Create Line By Station/Offset Command

To use the Create Line By Station/Offset command, you must have a Civil 3D Alignment object in your drawing. The line created from this command allows you to start and/or end a line on the basis of a station and offset from an alignment.

You're prompted to choose the alignment and then input a station and offset value. The line *begins* at the station and offset value. On the basis of the tooltips, you might expect the line to be drawn from the alignment station at offset zero and out to the alignment station at the input offset. This isn't the case.

When prompted for the station, you're given a tooltip that tracks your position along the alignment, as shown in Figure 1.29. You can graphically choose a station location by picking in the drawing (including using your Osnaps to assist you in locking down the station of a specific feature). Alternatively, you can enter a station value on the command line.

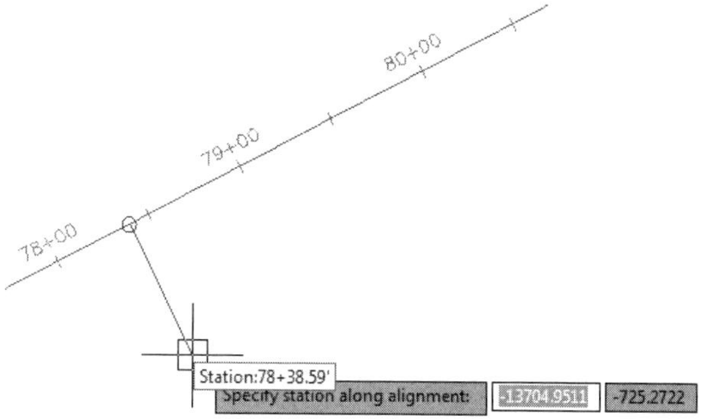

FIGURE 1.29
The Create Line By Station/Offset command provides a tooltip for you to track stationing along the alignment.

Once you've selected the station, you're given a tooltip that is locked on that particular station and tracks your offset from the alignment (see Figure 1.30). You can graphically choose an offset by picking in the drawing, or you can type an offset value on the command line.

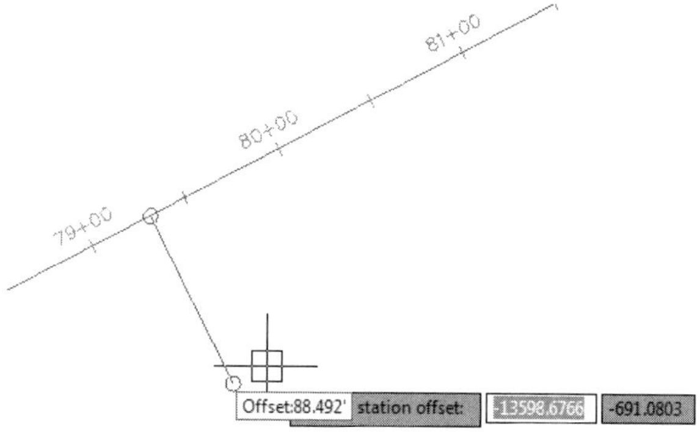

FIGURE 1.30
The Create Line By Station/Offset command provides a tooltip that helps you track the offset from the alignment.

Create Line By Side Shot Command

The Create Line By Side Shot command starts by asking you to select a line (or two points) that will be the basis for creating a new line. After you select the first line, you will see a yellow glyph indicating your occupied point (see Figure 1.31). By default, Civil 3D is looking for an angle-right and distance to establish the first point of the new line. However, you can follow the command prompts to change the angle entry to bearing, direction, or azimuth.

Once you have entered data for the first point, you will be asked a second time for an angle and a distance to place the second point of the new line.

The occupied point represents the setup of your surveying station, whereas the direction to the end point of the first line represents your surveying backsight. This tool may be most useful when you're creating stakeout information or re-creating data from field notes. If you know where your crew set up and you have their side-shot angle measurements, but you don't have electronic information to download, this tool can help.

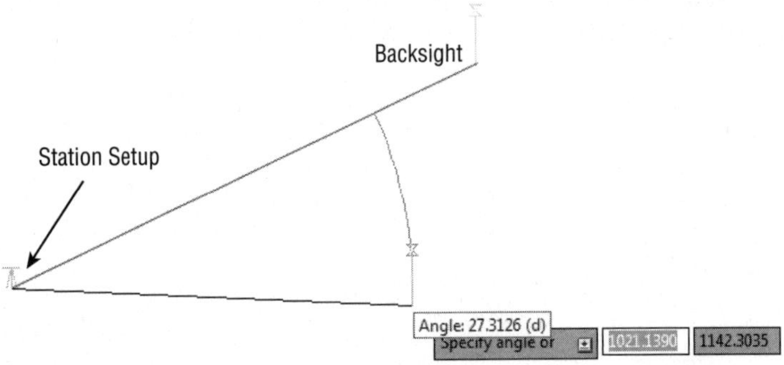

Figure 1.31
The tooltip for the Create Line By Side Shot command tracks the angle, bearing, deflection, or azimuth of the side shot.

Create Line By Extension Command

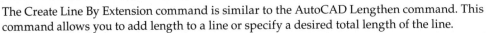

The Create Line By Extension command is similar to the AutoCAD Lengthen command. This command allows you to add length to a line or specify a desired total length of the line.

You are first prompted to choose a line. The command line then displays the prompt `Specify distance to change, or [Total]`. The distance you specify is added to the existing length of the line. The command draws the line appropriately and provides a short summary report on that line. The summary report in Figure 1.32 indicates the beginning and ending coordinates showing that the user input of an additional distance of 50′ (15.5 m) specified with the Create Line By Extension command resulted in a new overall distance of 150′ (46 m). It is important to note that in some cases, it may be more desirable to create a line by a turned angle or deflection of 180 degrees, so as not to disturb linework originally created from existing legally recorded documents.

If instead you specify a total distance on the command line, the length of the line is changed to the distance you specify. The summary report shown in Figure 1.33 shows the same beginning coordinate as in Figure 1.32 but a different end coordinate, resulting in a total length of 100′ (30.5 m).

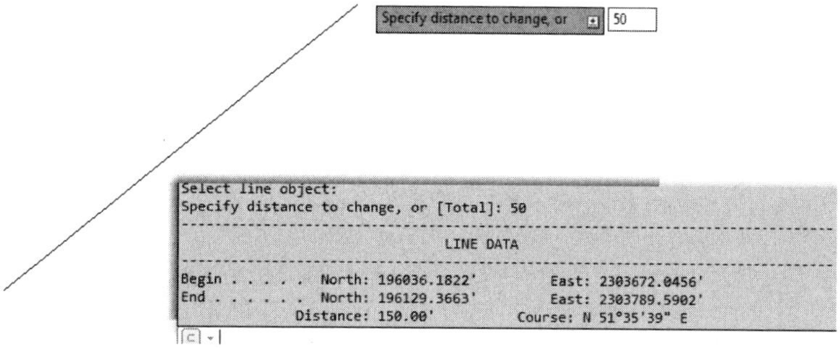

FIGURE 1.32 The Create Line By Extension command provides a summary of the changes to the line.

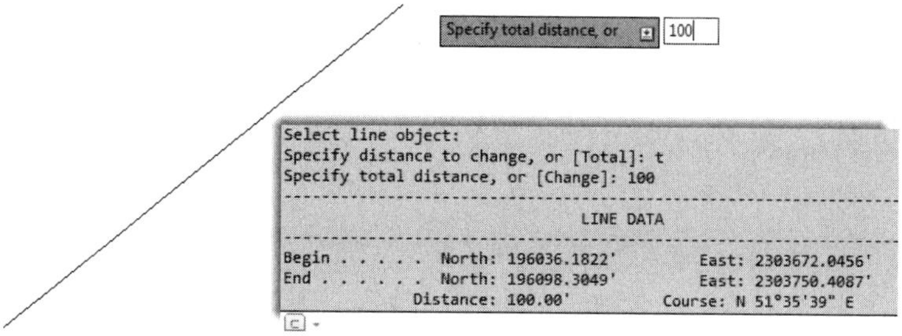

FIGURE 1.33 The summary report on a line where the command specified a total distance

CREATE LINE FROM END OF OBJECT COMMAND

The Create Line From End Of Object command lets you draw a line tangent to the end of a line or arc of your choosing. Most commonly, you'll use this tool when re-creating deeds or other survey work where you have to specify a line that continues a tangent from an arc (see Figure 1.34).

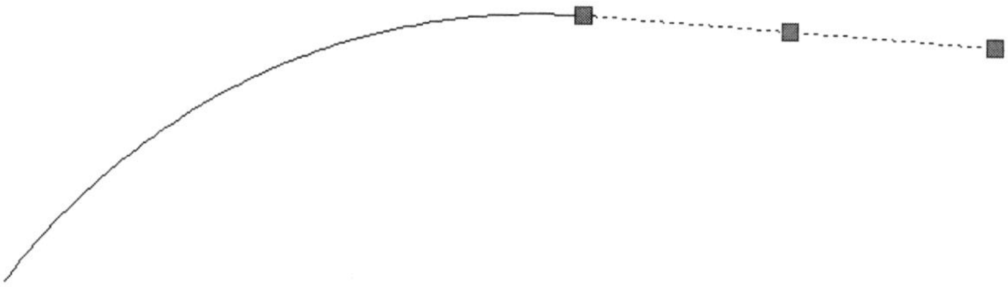

FIGURE 1.34 The Create Line From End Of Object command lets you add a tangent line to the end of an arc.

CERT OBJECTIVE

CREATE LINE TANGENT FROM POINT COMMAND

The Create Line Tangent From Point command is similar to the Create Line From End Of Object command, but Create Line Tangent From Point allows you to choose a point of tangency that isn't the endpoint of the line or arc (see Figure 1.35).

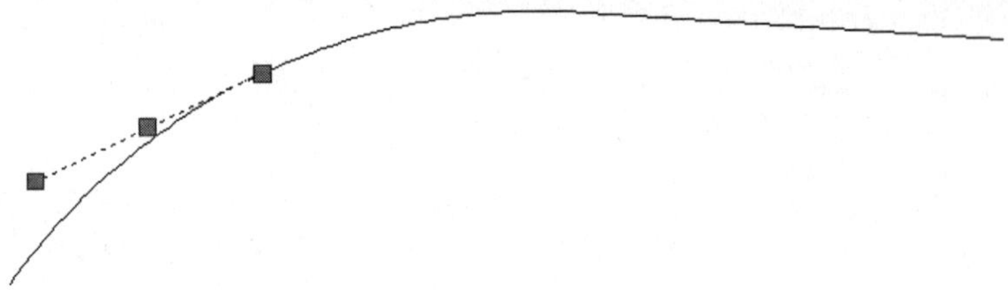

FIGURE 1.35
The Create Line Tangent From Point command can place a line tangent at any point on an arc (or line).

CREATE LINE PERPENDICULAR FROM POINT COMMAND

Using the Create Line Perpendicular From Point command, you can specify that you'd like a line drawn perpendicular to any point of your choosing along a line or arc. In the example shown in Figure 1.36, a line is drawn perpendicular to the endpoint of the arc. This command can be useful when the distance from a known monument, perpendicular to a legally platted line, must be labeled in a drawing.

FIGURE 1.36
A perpendicular line is drawn from the endpoint of an arc, using the Create Line Perpendicular From Point command.

> **WHEN DO WE GET TO THE FUN STUFF?**
>
> You are probably eager to build you first surface or design your first Civil 3D corridor. So far, you have created base AutoCAD lines and arcs. However, much of the interface and terminology will appear again in later chapters. For example, in Chapter 6, you will use tools similar to the line and curve commands to build compound and floating curves. Transparent commands, introduced later in this chapter, can be used in many situations. This chapter may not contain the sexiest information, but you'll be glad you spent the time working through the examples.

Re-creating a Deed Using Line Tools

The upcoming exercise will help you apply some of the tools you've learned so far to reconstruct the overall parcel.

When you open up the file for the exercise, you will see a legal description with the following information (metric will have different lengths, of course):

```
From the POINT OF BEGINNING at a location of N 186156.65',  E 2305474.07'
Thence, S 11° 45' 41.4" E for a distance of 693.77 feet to a point on a line.
Thence, S 73° 10' 54.4" W for a distance of 265.45 feet to a point on a line.
Thence, S 05° 59' 04.4" E for a distance of 185.89 feet to a point on a line.
Thence, S 45° 55' 02.4" W for a distance of 68.73 feet to a point on a line.
Thence, N 06° 04' 37.0" W for a distance of 217.80 feet to a point on a line.
Thence, N 73° 21' 22.5" E for a distance of 4.22 feet to a point on a line.
Thence, N 06° 04' 51.6" W for a distance of 200.14 feet to a point on a line.
Thence, S 87° 32' 10.4" W for a distance of 121.22 feet to a point on a line.
Thence, N 02° 25' 32.2" W for a distance of 168.91 feet to a point on a line.
Thence, N 15° 38' 57.5" E for a distance of 283.16 feet to a point on a line.
Thence, N 06° 19' 22.4" W for a distance of 79.64 feet to a point  on a line.
N 76° 55' 49.8" E a distance of 250.00 feet Returning to the POINT OF BEGINNING;
Containing 5.58 acres (more or less)
```

Follow these steps. (Note: The legal description for metric users is located in the DWG file as text.)

1. Open the `Deed Create Start.dwg` (`Deed Create Start_METRIC.dwg`) file, which you can download from this book's web page at `www.sybex.com/go/masteringcivil3d2013`.
2. Turn off dynamic input by pressing F12, or by toggling the icon off at the status bar.
3. From the Draw panel on the Home tab in the Ribbon, select the Line drop-down and choose the Create Line By Bearing command.
4. At the `Select first point:` prompt, use endpoint object snap to snap to the arrow head of the note indicating POB.
5. At the >>`Specify quadrant (1-4):` prompt, enter **2** to specify the SE quadrant, and then press ↵.
6. At the >>`Specify bearing:` prompt, enter **11.45414**, and press ↵.
7. At the >>`Specify distance:` prompt, enter **693.77′** (**211.4615** m), and press ↵.
8. Repeat steps 4 through 6 for the rest of the courses and distances.
9. Press Esc to exit the Create Line By Bearing command.

   The finished linework should look like Figure 1.37.
10. Save your drawing. You'll need it for the next exercise.

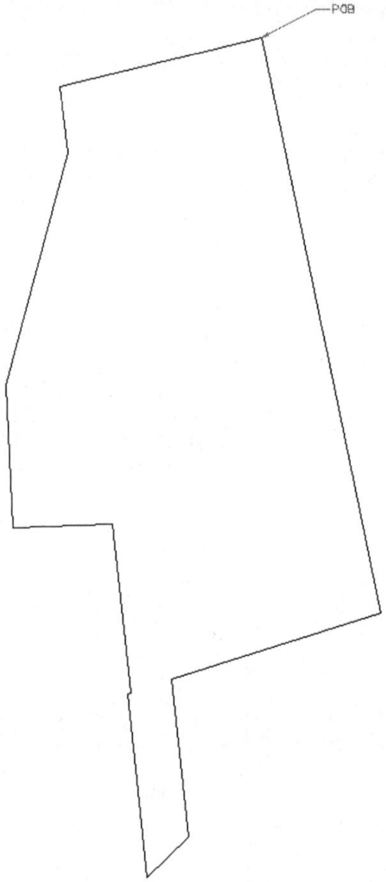

FIGURE 1.37
The finished linework

Creating Curves

Curves are an important part of surveying and engineering geometry. The curves you create in this chapter are no different from AutoCAD arcs. What makes the curve commands unique from the basic AutoCAD commands isn't the resulting arc entity but the inputs used to draw the arc. Civil 3D wants you to provide directions to the arc commands using land surveying terminology rather than with generic Cartesian parameters. Figure 1.38 shows the Create Curves menu options.

Standard Curves

When re-creating legal descriptions for roads, easements, and properties, users such as engineers, surveyors, and mappers often encounter a variety of curves. Although standard AutoCAD arc commands could draw these arcs, the AutoCAD arc inputs are designed to be generic to all industries. The following curve commands have been designed to provide an interface that more closely matches land surveying, mapping, and engineering language.

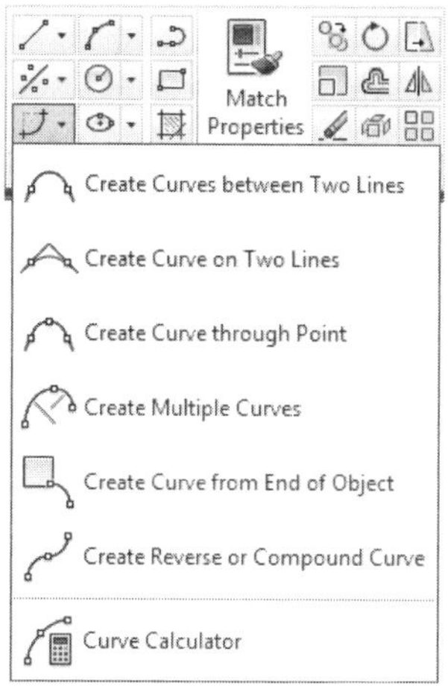

FIGURE 1.38
Create Curves commands

CREATE CURVE BETWEEN TWO LINES COMMAND

The Create Curve Between Two Lines command is much like the standard AutoCAD Fillet command, except that you aren't limited to a radius parameter. The command draws a curve that is tangent to two lines of your choosing. This command also trims or extends the original tangents so their endpoints coincide with the curve endpoints. The lines are trimmed or extended to the resulting PC (point of curve, which is the beginning of a curve) and PT (point of tangency, or the end of a curve). You may find this command most useful when you're creating foundation geometry for road alignments, parcel boundary curves, and similar situations.

The command prompts you to choose the first tangent and then the second tangent. The command line gives the following prompt:

```
Select entry [Tangent/External/Degree/Chord/Length/Mid-Ordinate/
miN-dist/Radius]<Radius>:
```

Pressing ↵ at this prompt lets you input your desired radius. As with standard AutoCAD commands, pressing T changes the input parameter to Tangent, pressing C changes the input parameter to Chord, and so on.

As with the Fillet command, your inputs must be geometrically possible. For example, your two lines must allow for a curve of your specifications to be drawn while remaining tangent to both. Figure 1.39 shows two lines with a 25′ (7.6 m) radius curve drawn between them. Note that the tangents have been trimmed so their endpoints coincide with the endpoints of the curve. If either line had been too short to meet the endpoint of the curve, that line would have been extended.

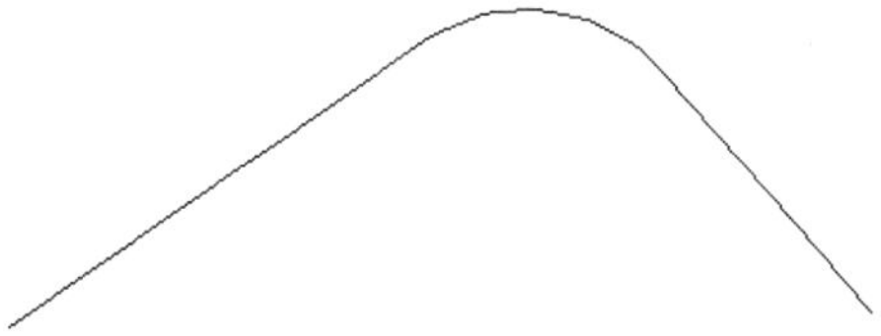

FIGURE 1.39
Two lines using the Create Curve Between Two Lines command

Create Curve On Two Lines Command

The Create Curve On Two Lines command is identical to the Create Curve Between Two Lines command, except that the Create Curve On Two Lines command leaves the chosen tangents intact. The lines aren't trimmed or extended to the resulting PC and PT of the curve.

Figure 1.40, for example, shows two lines with a 25′ (7.6 m) radius curve drawn on them. The tangents haven't been trimmed and instead remain exactly as they were drawn before the Create Curve On Two Lines command was executed.

Figure 1.40
The original lines stay the same after you execute the Create Curve On Two Lines command.

Create Curve Through Point Command

The Create Curve Through Point command lets you choose two tangents for your curve followed by a pass-through point. This tool is most useful when you don't know the radius, length, or other curve parameters but you have two tangents and a target location. It isn't necessary that the pass-through location be a true point object; it can be any location of your choosing.

This command also trims or extends the original tangents so their endpoints coincide with the curve endpoints. The lines are trimmed or extended to the resulting PC and PT of the curve.

Figure 1.41, for example, shows two lines and a desired pass-through point. Using the Create Curve Through Point command allows you to draw a curve that is tangent to both lines and that passes through the desired point. In this case, the tangents have been trimmed to the PC and PT of the curve.

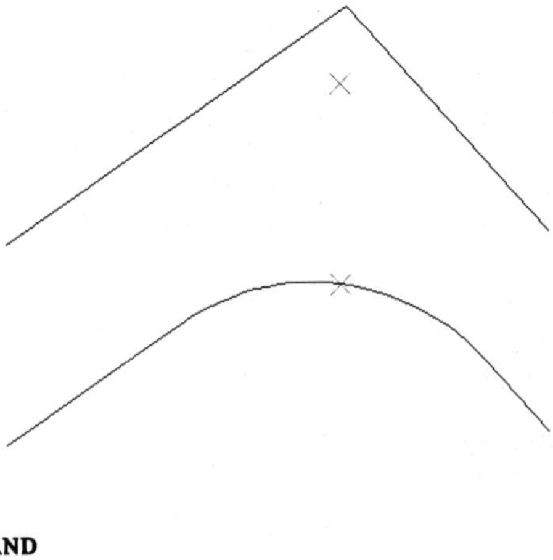

Figure 1.41
The first image shows two lines with a desired pass-through point. In the second image, the Create Curve Through Point command draws a curve that is tangent to both lines and passes through the chosen point.

Create Multiple Curves Command

The Create Multiple Curves command lets you create several curves that are tangentially connected. The resulting curves have an effect similar to an alignment spiral section. This command can be useful when you are re-creating railway track geometry based on field-survey data.

The command prompts you for the two tangents. Then, the command line prompts you as follows:

```
Enter Number of Curves:
```

The command allows for up to 10 curves between tangents.

One of your curves must have a flexible length that's determined on the basis of the lengths, radii, and geometric constraints of the other curves. Curves are counted clockwise, so enter the number of your flexible curve:

```
Enter Floating Curve #:
Enter the length and radii for all your curves:
Enter curve 1 Radius:
Enter curve 1 Length:
```

The floating curve number will prompt you for a radius but not a length.

As with all other curve commands, the specified geometry must be possible. If the command can't find a solution on the basis of your length and radius inputs, it returns no solution (see Figure 1.42).

FIGURE 1.42
Two curves were specified with the #2 curve designated as the floating curve.

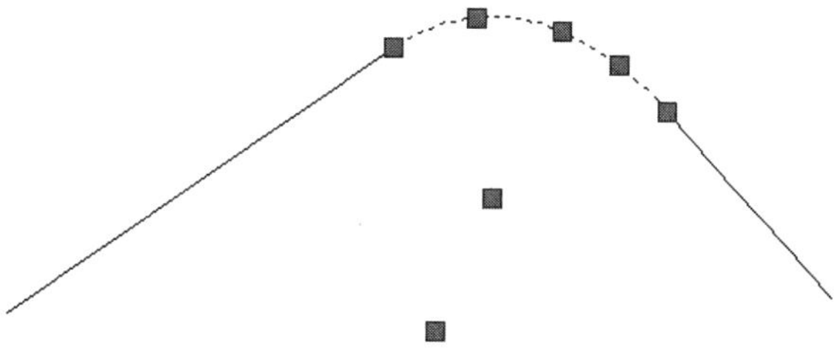

CREATE CURVE FROM END OF OBJECT COMMAND

The Create Curve From End Of Object command enables you to draw a curve tangent to the end of your chosen line or arc.

The command prompts you to choose an object to serve as the beginning of your curve. You can then specify a radius and an additional parameter (such as Delta or Length) for the curve or the endpoint of the resulting curve chord (see Figure 1.43).

FIGURE 1.43
A curve, with a 25′ (7.6 m) radius and a 30′ (9.1 m) length, drawn from the end of a line

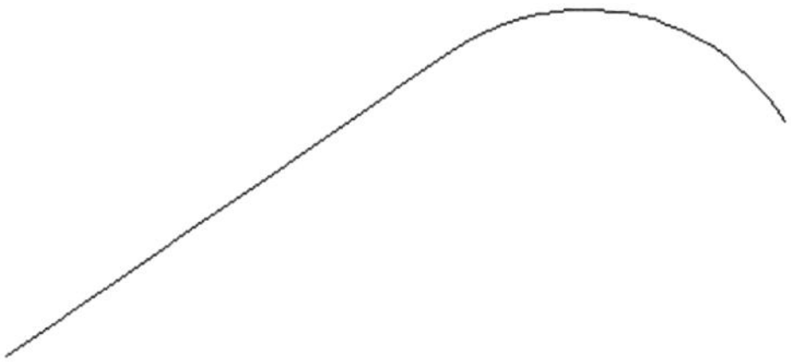

CREATE REVERSE OR COMPOUND CURVES COMMAND

The Create Reverse Or Compound Curves command allows you to add additional curves to the end of an existing curve. Reverse curves are drawn in the opposite direction (i.e., a curve to the right tangent to a curve to the left) from the original curve to form an S shape. In contrast, compound curves are drawn in the same direction as the original curve (see Figure 1.44). This tool can be useful when you are re-creating a legal description of a road alignment that contains reverse and/or compound curves.

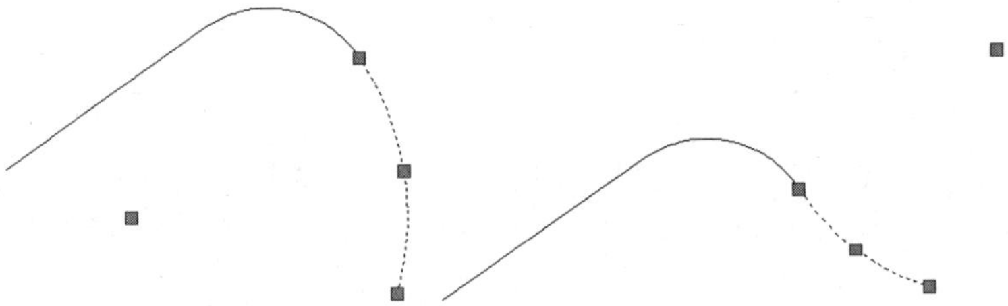

FIGURE 1.44 A tangent and curve before adding a reverse or compound curve (left); a compound curve drawn from the end of the original curve (right).

Best Fit Entities

Although engineers and surveyors do their best to make their work an exact science, sometimes tools like the Best Fit Entities are required.

Roads in many parts of the world have no defined alignment. They may have been old carriage roads or cart paths from hundreds of years ago that evolved into automobile roads. Surveyors and engineers are often called to help establish official alignments, vertical alignments, and right-of-way lines for such roads on the basis of a best fit of surveyed centerline data.

Other examples for using Best Fit Entities include property lines of agreement, road rehabilitation projects, and other cases where existing survey information must be approximated into "real" engineering geometry (see Figure 1.45).

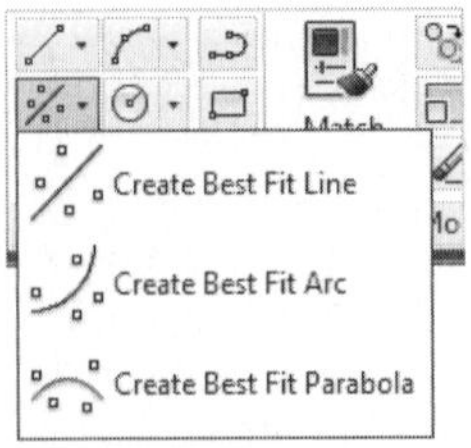

FIGURE 1.45 The Create Best Fit Entities menu options

Create Best Fit Line Command

The Create Best Fit Line command under the Best Fit drop-down on the Draw panel takes a series of Civil 3D points, AutoCAD points, entities, or drawing locations and draws a single best-fit line segment from this information. In Figure 1.46, for example, the Create Best Fit Line command draws a best-fit line through a series of points that aren't quite collinear. Note that the best-fit line will change as more points are picked.

Figure 1.46
A preview line drawn through points that aren't quite collinear

Once you've selected your points, a Panorama window appears with a regression data chart showing information about each point you chose, as shown in Figure 1.47.

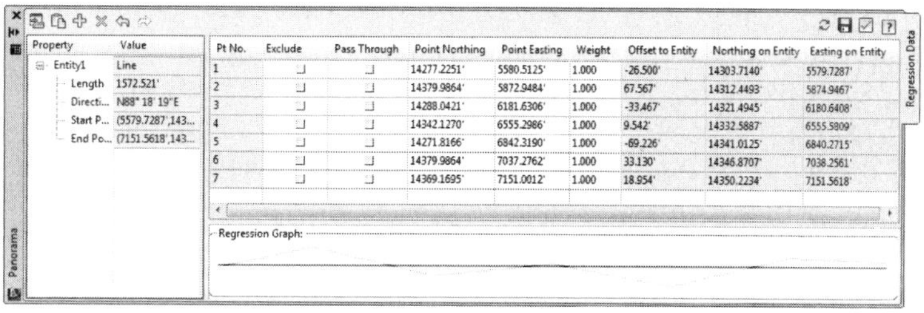

Figure 1.47
The Panorama window lets you optimize your best fit.

This interface allows you to optimize your best fit by adding more points, excluding points, selecting the check box in the Pass Through column to force one of your points on the line, or adjusting the value under the Weight column.

Create Best Fit Arc Command

The Create Best Fit Arc command under the Best Fit drop-down works identically to the Create Best Fit Line command, except that the resulting entity is a single arc segment as opposed to a single line segment (see Figure 1.48)

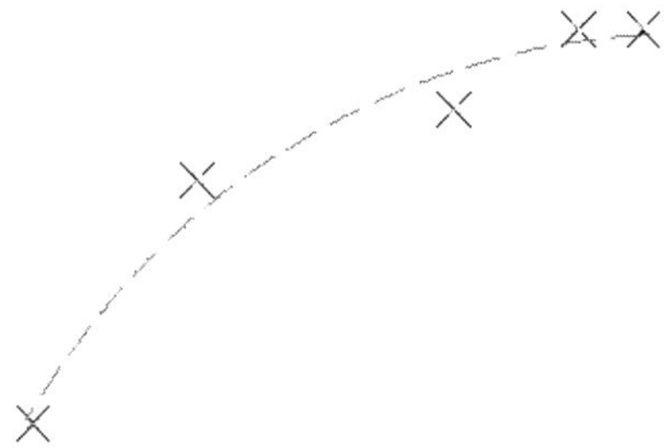

Figure 1.48
A curve created by best fit

CREATE BEST FIT PARABOLA COMMAND

The Create Best Fit Parabola command under the Create Best Fit Entities option works in a similar way to the line and arc commands just described. This command is most useful when you have a triangulated irregular network (also known as TIN) sampled or surveyed road information, and you'd like to replicate true vertical curves for your design information.

After you select this command, the Parabola By Best Fit dialog appears (see Figure 1.49).

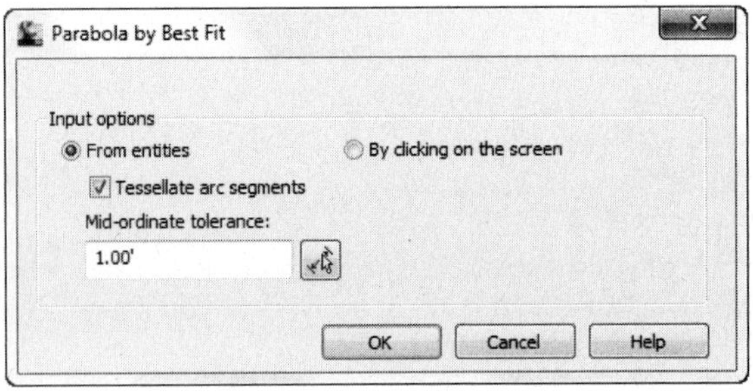

FIGURE 1.49
The Parabola By Best Fit dialog

You can select inputs from entities (such as lines, arcs, polylines, or profile objects) or by clicking on screen. The command then draws a best-fit parabola on the basis of this information. In Figure 1.50, the shots were represented by AutoCAD points; more points were added by selecting the By Clicking On The Screen option and using the Node Osnap to pick each point.

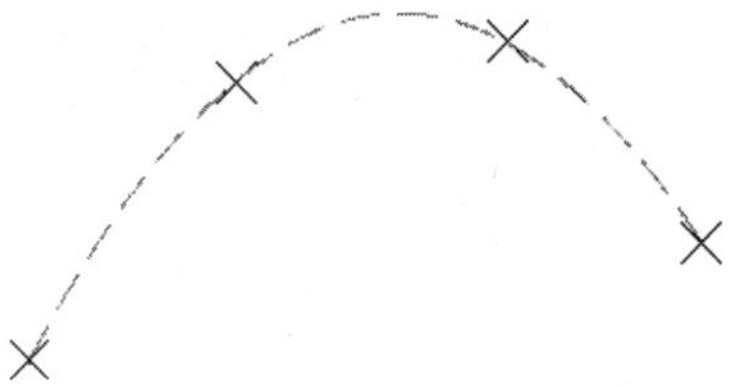

FIGURE 1.50
The best-fit preview line changes as more points are picked.

Once you've selected your points, a Panorama window (shown in Figure 1.51) appears, showing information about each point you chose. Also note the information in the right pane regarding K-value, curve length, grades, and so forth. You can optimize your K-value, length, and other values by adding more points, selecting the check box in the Pass Through column to force one of your points on the line, or adjusting the value under the Weight column.

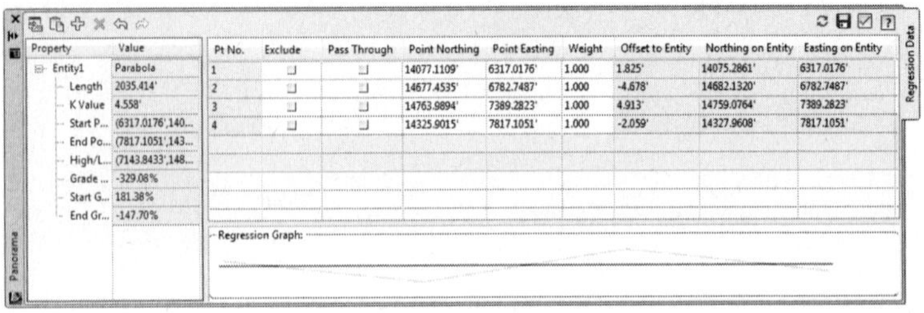

FIGURE 1.51
The Panorama window lets you make adjustments to your best-fit parabola.

Attach Multiple Entities

The Attach Multiple Entities command (found on the Home tab in the Ribbon and extended Draw panel pull-down) is a combination of the Line From End Of Object command and the Curve From End Of Object command. This command is most useful for reconstructing deeds or road alignments from legal descriptions when each entity is tangent to the previous entity. Using this command saves you time because you don't have to constantly switch between the Line From End Of Object command and the Curve From End Of Object command (see Figure 1.52).

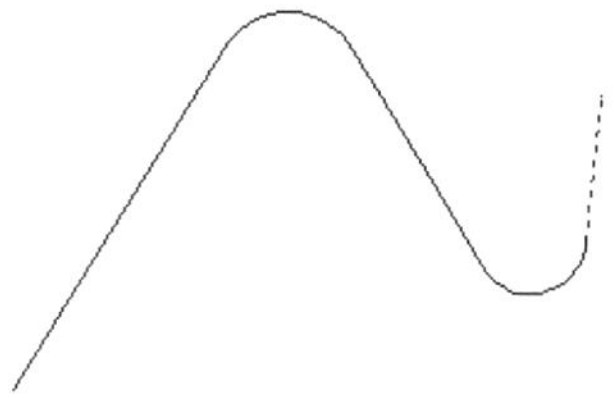

FIGURE 1.52
The Attach Multiple Entities command draws a series of lines and arcs so that each segment is tangent to the previous one.

The Curve Calculator

CERT OBJECTIVE

Sometimes you may not have enough information to draw a curve properly. Although many of the curve-creation tools assist you in calculating the curve parameters, you may find an occasion where the deed you're working with is incomplete.

The Curve Calculator found in the Curves drop-down on the Draw panel helps you calculate a full collection of curve parameters on the basis of your known values and constraints. The units used in the Curve Calculator match the units assigned in your Drawing Settings.

The Curve Calculator can remain open on your screen while you're working through commands. You can send any value in the Calculator to the command line by clicking the button next to that value (see Figure 1.53).

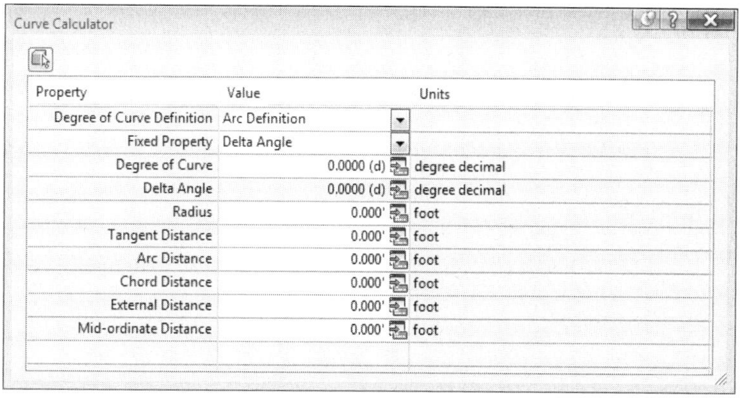

FIGURE 1.53
The Curve Calculator

The button at the upper left of the Curve Calculator inherits the arc properties from an existing arc in the drawing, and the drop-down menu in the Degree Of Curve Definition selection field allows you to choose whether to calculate parameters for an arc or a chord definition.

The drop-down menu in the Fixed Property selection field also gives you the choice of fixing your Radius or Delta Angle when calculating the values for an arc or a chord, respectively (see Figure 1.54). The parameter chosen as the fixed value is held constant as additional parameters are calculated.

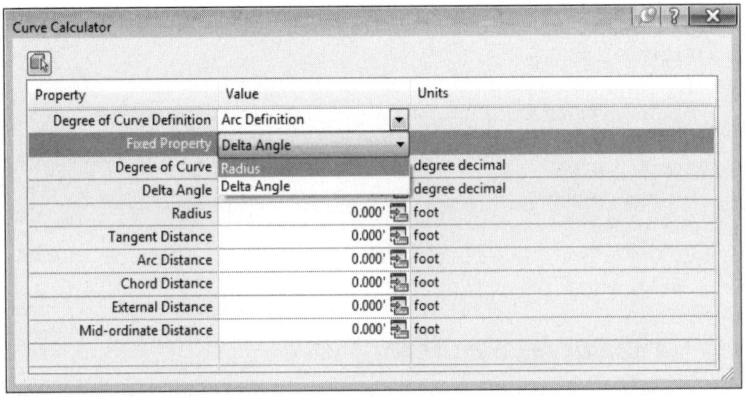

FIGURE 1.54
The Fixed Property drop-down menu gives you the choice of fixing your radius or delta value.

As explained previously, you can send any value in the Curve Calculator to the command line using the button next to that value. This ability is most useful while you're active in a curve command and would like to use a certain parameter value to complete the command.

Test your new line and curve knowledge, follow these steps. (You do not need to have completed the previous exercises to complete this one.)

1. Open the `Deed Create Start.dwg` (`Deed Create Start_METRIC.dwg`) file.
2. Start the Line By Northing/Easting command.
3. For Northing, enter **184898.42** (**56357.038**) and press ↵.
4. For Easting, enter **2305136.46** (**702605.593**) and press ↵.

   You will not see any graphic indication that the line has been started. You will not see the segment until after you complete step 5.
5. You will again be prompted for Northing and Easting. Using the same procedure as you used in step 2, locate the endpoint of the line at Northing = **185059.94** (**56406.270**), Easting = **2305413.52** (**702690.041**).
6. Press Esc once to end the Northing and Easting entry. Press Esc again to exit the line command.
7. Start the Create Curve From End Of Object command.
8. When prompted to select the Line or Arc object, pan to the left, and click on the east end of the line you created in steps 2–4.

You will then be prompted to select the entry.

9. Press ↵ to select the default Radius option.
10. Enter a Radius value of **550′** (**167.6** m), and press ↵.

Next you will see the prompt: Select entry [Tangent/Chord/Delta/Length/External/Mid-ordinate] <Length>:.

11. Enter **D** (for Delta) and press ↵.
12. Enter a delta angle of **40** and press ↵. (Note: The degree symbol is not needed in angular entry.)

    You should now have an arc tangent to the first segment.
13. Return to the Curves menu and select Create Reverse Or Compound Curve.
14. Select the arc you created in the previous steps.
15. When prompted, select [Compound/Reverse] <Compound>:.
16. Enter **R** to specify a reverse curve.
17. Enter a radius of **630′** (**192.0** m) and press ↵.
18. When you see the prompt Select entry [Tangent/Chord/Delta/Length/External/Mid-ordinate] <Length>:, press ↵ to accept the default.
19. Enter a length of curve value of **400′** (**121.9** m), and press ↵.
20. For the last tangent segment, return to the Lines menu and select Create Line From End Of Object.
21. Select the east end of the second arc you created in this exercise.
22. At the Specify Distance: prompt, enter **150′** (**45.7** m) and press ↵.

Your completed lines and arcs will look like Figure 1.55, and they will be located just south of the property you entered in the previous exercise.

FIGURE 1.55
Lines and arcs of the completed exercise

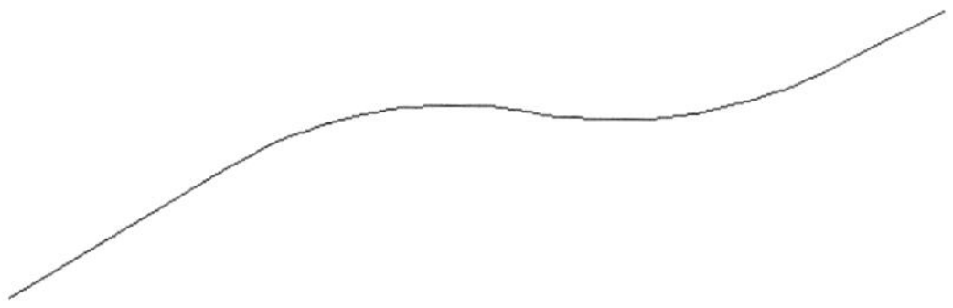

Adding Line and Curve Labels

CERT OBJECTIVE

Although most robust labeling of site geometry is handled using Parcel or Alignment labels, limited line- and curve-annotation tools are available in Civil 3D. The line and curve labels are composed much the same way as other Civil 3D labels, with marked similarities to Parcel and Alignment Segment labels.

Our next exercise leads you through labeling the deed you entered earlier in this chapter:

1. Continue working in the `Deed Create Start.dwg` (`Create Deed Start_METRIC.dwg`) file. If you did not complete the previous exercises, use `Deed Create Finished.dwg` or `Deed Create Finished_METRIC.dwg` from the book's web page.

2. Click the Add Labels button in the Labels & Tables panel on the Annotate tab in the Ribbon.

   The Add Labels dialog appears.

3. Choose Line And Curve from the Feature drop-down menu.

   Your Add Labels dialog will look similar to Figure 1.56.

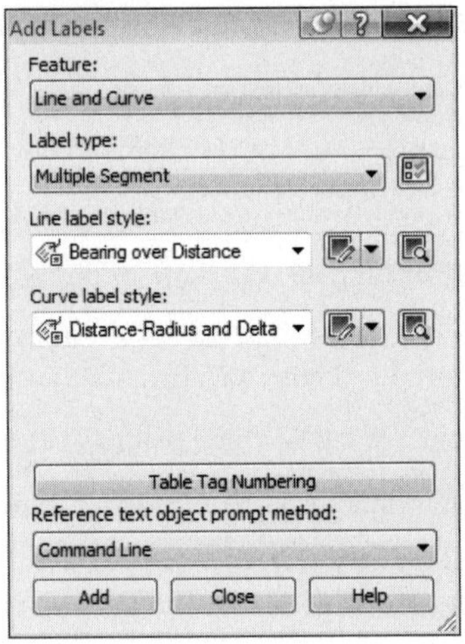

FIGURE 1.56
The Add Labels dialog, with Label Type set to Multiple Segment

4. Choose Multiple Segment from the Label Type drop-down menu.

   The Multiple Segment option places the label at the midpoint of each selected line or arc.

5. Confirm that Line Label Style is set to Bearing Over Distance and that Curve Label Style is set to Distance-Radius And Delta (as shown in Figure 1.56).

6. Click the Add button.

7. At the `Select Entity:` prompt, select each line from the deed and the lines and arcs that you drew in the previous exercises.

   A label appears on each entity at its midpoint, as shown in Figure 1.57.

8. Save the drawing—you'll need it later on.

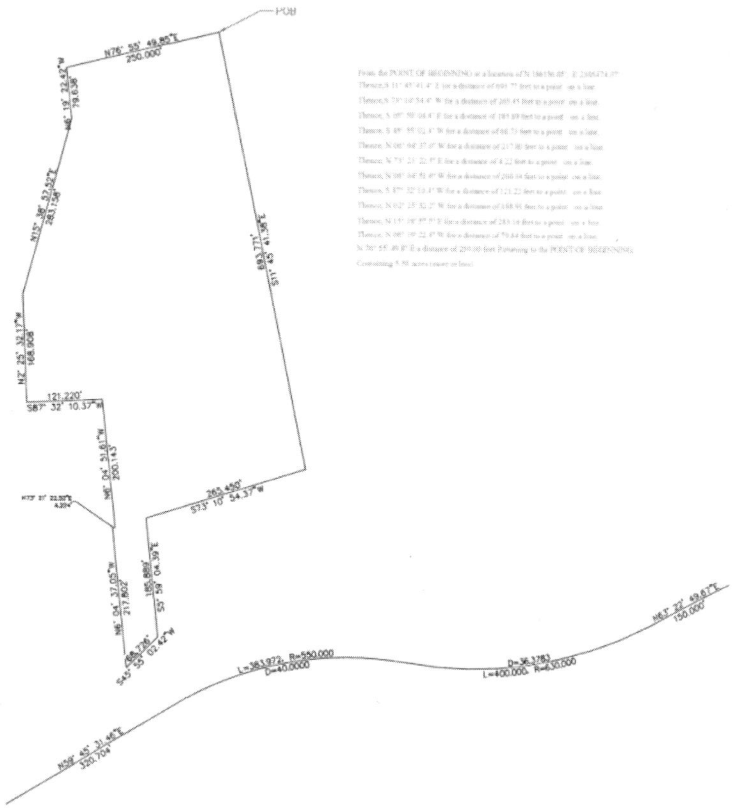

FIGURE 1.57
The labeled linework

Using Transparent Commands

You might be surprised to know that you have already been using a form of the transparent commands. If you used the Line By Point # Range command or the Line By Bearing command, you have already used a transparent command.

Transparent commands are "helper" commands that make base AutoCAD input more surveyor- and civil designer–friendly. Transparent commands can only be used when another command is in progress. They are "commands within commands" that allow you to choose your input based on what information you have.

For example, you may have a plat whose information you are trying to input into AutoCAD. You may need to insert a block at a specific Northing and Easting. While the line command has a built-in option for Northing Easting, the insert command does not, and a transparent command will help you. In this case, you would start the Insert command with the Specify On Screen option checked. Before you click to place the block in the graphic, click the Northing

Easting transparent command. The command line will then walk you through placing the block at your desired location.

While a transparent command is active, you can press Esc once to leave the transparent mode but stay active in your current command. You can then choose another transparent command if you'd like. For example, you can start a line using the Endpoint Osnap, activate the Angle Distance transparent command, draw a line-by-angle distance, and then press Esc, which takes you out of angle-distance mode but keeps you in the line command. You can then draw a few more segments using the Point Object transparent command, press Esc, and finish your line with a Perpendicular Osnap.

You can activate the transparent commands using keyboard shortcuts or using the Transparent Commands toolbar. Be sure you include the Transparent Commands toolbar (shown in Figure 1.58) in all your Civil 3D and survey-oriented workspaces.

FIGURE 1.58
The Transparent Commands toolbar

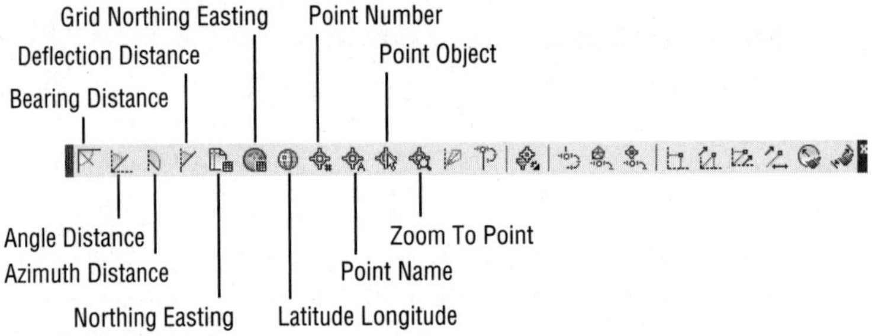

The six profile-related transparent commands will be covered in Chapter 7, "Profiles and Profile Views."

Standard Transparent Commands

CERT OBJECTIVE

The transparent commands shown in Table 1.2 behave identically to their like-named counterparts from the Draw panel (discussed earlier in this chapter). The difference is that you can call up these transparent commands in any appropriate AutoCAD or Civil 3D draw command, such as a line, polyline, alignment, parcel segment, feature line, or pipe-creation command.

TABLE 1.2: The transparent commands used in plan view

TOOL ICON	MENU COMMAND	TOOL ICON	MENU COMMAND
	Angle Distance		Latitude Longitude
	Bearing Distance		Point Number
	Azimuth Distance		Point Name

Tool icon	Menu command	Tool icon	Menu command
	Deflection Distance		Point Object
	Northing Easting		Zoom To Point
	Grid Northing Grid Easting		Side Shot
			Station Offset

Matching Transparent Commands

You may have construction or other geometry in your drawing that you'd like to match with new lines, arcs, circles, alignments, parcel segments, or other entities.

While actively drawing an object that has a radius parameter, such as a circle, an arc, an alignment curve, or a similar object, you can choose the Match Radius transparent command and then select an object in your drawing that has your desired radius. Civil 3D draws the resulting entity with a radius identical to that of the object you chose during the command. You'll save time using this tool because you don't have to first list the radius of the original object and then manually type in that radius when prompted by your circle, arc, or alignment tool.

The Match Length transparent command works identically to the Match Radius transparent command except that it matches the length parameter of your chosen object.

Some people find working with transparent commands awkward at first. It feels strange to move your mouse away from the drawing area and click an icon while another command is in progress. Try a few transparent commands to get the feel for how they operate:

1. Open the drawing `Transparent_Commands.dwg` (`Transparent_Commands_METRIC.dwg`).
2. Start the 3DPOLY command by choosing 3D Polyline from the Draw panel on the Home tab.
3. From the Transparent Commands toolbar, click the Point Number Transparent command.
4. At the Enter Point number prompt, type **100-112** and press ↵.

   The 3D polyline will connect the points in the drawing.
5. Press Esc once.

   Notice you are now back in the 3D polyline command with no transparent commands active.
6. Press Esc a second time.

   This will exit the 3D polyline command.
7. Choose Insert ➢ Block ➢ Insert.

8. From the Name pull-down, locate the block Fire Hydrant 01.

   You do not need to browse because this block is already defined in this example drawing.

9. Check the Specify On-Screen option for Insertion Point.
10. Deselect the Explode option.
11. Verify that your settings match what is shown in Figure 1.59, and click OK.

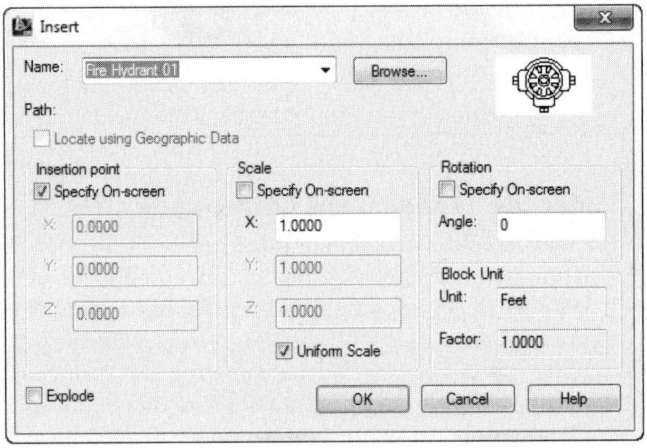

FIGURE 1.59
Using the Insert command to place a block

12. At the `Specify insertion point:` prompt, select the Station Offset Transparent command.

    You are then prompted to select an alignment.

13. Click the alignment to the south of the site. (Hint: Be sure to click the green line, rather than the labels.)
14. At the `Specify station along alignment:` prompt, enter **450′** (**175** m) and press ↵.
15. At the `Specify station offset:` prompt, enter **-30′** (**-10** m) ↵.

You should now see the fire hydrant symbol at the specified station, on the left of the alignment.

The Bottom Line

Find any Civil 3D object with just a few clicks. By using Prospector to view object data collections, you can minimize the panning and zooming that are part of working in a CAD program. When common subdivisions can have hundreds of parcels or a complex corridor can have dozens of alignments, jumping to the desired one nearly instantly shaves time off everyday tasks.

Master It Open `BasicSite.dwg` from `www.sybex.com/go/masteringcivil3d2013`, and find parcel number 18 without using any AutoCAD commands or scrolling around on the drawing screen. (Hint: Take a look at Figure 1.5.)

Modify the drawing scale and default object layers. Civil 3D understands that the end goal of most drawings is to create hard-copy construction documents. By setting a drawing scale and then setting many sizes in terms of plotted inches or millimeters, Civil 3D removes much of the mental gymnastics that other programs require when you're sizing text and symbols. By setting object layers at a drawing scale, Civil 3D makes uniformity of drawing files easier than ever to accomplish.

Master It Change the Annotation scale in the model tab of `BasicSite.dwg` from the 100-scale drawing to a 40-scale drawing. (For metric users: Use `BasicSite_METRIC.dwg` and change the scale from 1:1000 to 1:500.)

Navigate the Ribbon's contextual tabs. As with AutoCAD, the Ribbon is the primary interface for accessing Civil 3D commands and features. When you select an AutoCAD Civil 3D object, the Ribbon displays commands and features related to that object. If several object types are selected, the Multiple contextual tab is displayed.

Master It Continue working in the file `BasicSite.dwg` (`Basic Site_METRIC.dwg`). It is not necessary to have completed the previous exercise to continue. Using the Ribbon interface, access the Alignment properties for Road A.

Create a curve tangent to the end of a line. It's rare that a property stands alone. Often, you must create adjacent properties, easements, or alignments from their legal descriptions.

Master It Open the drawing `MasterIt.dwg` (`MasterIt_METRIC.dwg`). Create a curve tangent to the east end of the line labeled in the drawing. The curve should meet the following specifications:

- Radius: 200.00′ (60 m)
- Arc Length: 66.580′ (20 m)

Label lines and curves. Although converting linework to parcels or alignments offers you the most robust labeling and analysis options, basic line- and curve-labeling tools are available when conversion isn't appropriate.

Master It Add line and curve labels to each entity created in `MasterIt.dwg` or `MasterIt_Metric.dwg`. It is recommended you complete the previous exercise so you will have a curve to work with. Choose a label that specifies the bearing and distance for your lines and length, radius, and delta of your curve.

Chapter 2

Survey

The AutoCAD® Civil 3D® software supports a collaborative workflow in many aspects of the design process, but especially in the survey realm. Accurate data starts outdoors. A survey that has been consistently and correctly coded in the field can save hours of drafting time. Nobody knows the site better than the folks who were out there freezing in their boots, sloshing through the project location. Surveyors can collect line information such as swales, curbs, or even pavement markings and communicate this digitally to the data collectors.

Civil 3D can often eliminate the need for third-party survey software, because it can download and process survey data directly from a data collector. To enter data in a manner that is easily digested by Civil 3D, your survey process should incorporate the information from this chapter.

In this chapter, you will learn to:

- Properly collect field data and import it into Civil 3D
- Set up description key and figure databases
- Translate surveys from assumed coordinates to known coordinates
- Perform traverse analysis

Setting Up the Databases

Before any project-specific data is imported, there is a bit of initial setup that will improve the translation between the field and the office. For this chapter you will need to see your Survey tab in Toolspace. If you do not see this tab, click the Survey button on the Home tab ➢ Palettes panel.

Your survey database defaults, equipment database, linework code set, and the figure prefix database should be in place before you import your first survey. You can find the location of these files by going to the Survey tab in Toolspace, and clicking the Survey User Settings button in the upper-left corner. The dialog shown in Figure 2.1 opens.

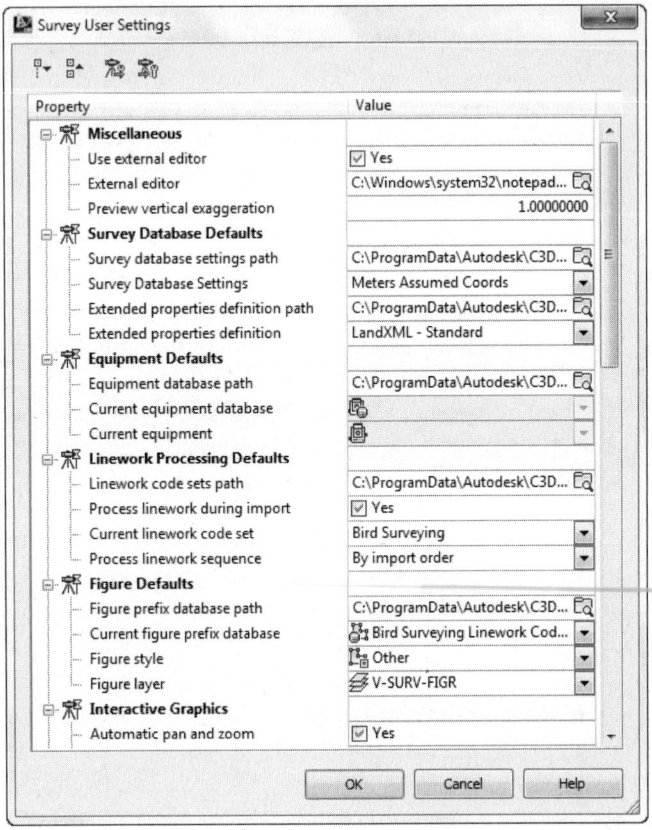

FIGURE 2.1
Survey User Settings dialog

It is common practice to place these files on a network server so your organization can share them. Change the paths in this dialog and they will "stick" regardless of which drawing you have open.

Survey Database Defaults

The Survey User Settings dialog contains these settings:

Units This section is where you set your master coordinate zone for the database. Potentially, your drawing and your incoming survey data may have different coordinate systems. If you insert any information in the database into a drawing with a different coordinate zone, the program will automatically translate that data to the drawing coordinate zone. Your coordinate zone units will lock the distance units in the Units section. You can also set the angle, direction, temperature, and pressure here.

Precision This section is where you define and store the precision information of angles, distance, elevation, coordinates, and latitude and longitude.

Measurement Type Defaults This section lets you define the defaults for measurement types, such as angle type, distance type, vertical type, and target type.

Measurement Corrections This section is used to define the methods (if any) for correcting measurements. Some data collectors and instruments allow you to make measurement corrections as you collect the data, so that needs to be verified — double-correction applications will lead to incorrect data.

Traverse Analysis Defaults This section is where you choose how you perform traverse analyses and define the required precision and tolerances for each. There are four types of 2D traverse analyses: Compass Rule, Transit Rule, Crandall Rule, and Least Squares Analysis.

There are three potential types of 3D traverse analyses: Length Weighted Distribution, Equal Distribution, and Least Squares Vertical. Vertical options for Least Squares will only be available if it is first set as the Horizontal Adjustment Method.

Least Squares Analysis Defaults If you are performing a Least Squares analysis, you must specify 2-Dimensional or 3-Dimensional adjustment type. Use 3-Dimensional if you are performing horizontal and vertical Least Squares adjustment.

Survey Command Window In the rare event that you'll need it, the Survey Command window is the interface for manual survey tasks and for running survey batch files. This section lets you define the default settings for this window.

Error Tolerance Set tolerances for the survey database in this section. If you perform an observation more than one time and the tolerances established here are not met, an error will appear in the Survey Command window and ask you what action you want to take.

Extended Properties This section defines settings for adding extended properties to a survey LandXML file. Extended properties are useful for certain types of surveys, including Federal Aviation Administration (FAA)-certified surveys.

When you first set up your survey settings, it is a good idea to create a test database for setting the survey database defaults. To create the test database:

1. Right-click Survey Databases on the Survey tab.
2. Select New Local Survey Database.
3. Name the new database **Test**, and click OK to continue.
4. Right-click the new Test database and select Edit Survey Database Settings.
5. Set your desired defaults for units, precision, and other options.

6. Click the Export Settings To A File button.
7. Save the settings to the folder specified in the Survey User Settings dialog (see Figure 2.2).

To delete your test database, you will need to close the database from the Survey tab. To do so, locate the database in the Survey tab. Right-click on it and select Close Survey Database. Using Windows Explorer, you can then browse to the working folder containing the database and delete it. There is no way to delete a survey database from within the software.

FIGURE 2.2
Survey Database settings

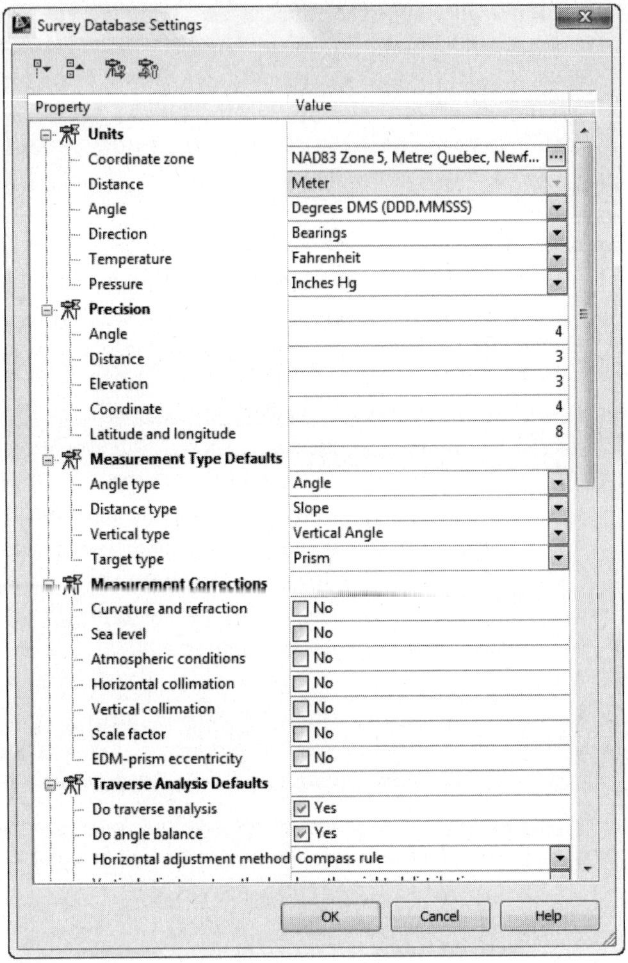

The Equipment Database

The equipment database is where you set up the various types of survey equipment that you are using in the field. Doing so allows you to apply the proper correction factors to your traverse analyses when it is time to balance your traverse. Civil 3D comes with a sample piece of equipment for you to inspect to see what information you will need when it comes time to create your equipment. The Equipment Properties dialog provides all the default settings for the sample equipment in the equipment database. On the Survey tab of Toolspace ➢ Equipment Databases, right-click Sample and select Manage Equipment Database to access this dialog in Toolspace, as shown in Figure 2.3.

You will want to create your own equipment entries and enter the specifications for your particular total station. Add a new piece of equipment to the database by clicking the plus sign at the top of the Equipment Database Manager window. If you are unsure of the settings to enter, refer to the user documentation that you received when you purchased your total station.

FIGURE 2.3
Equipment Database Manager

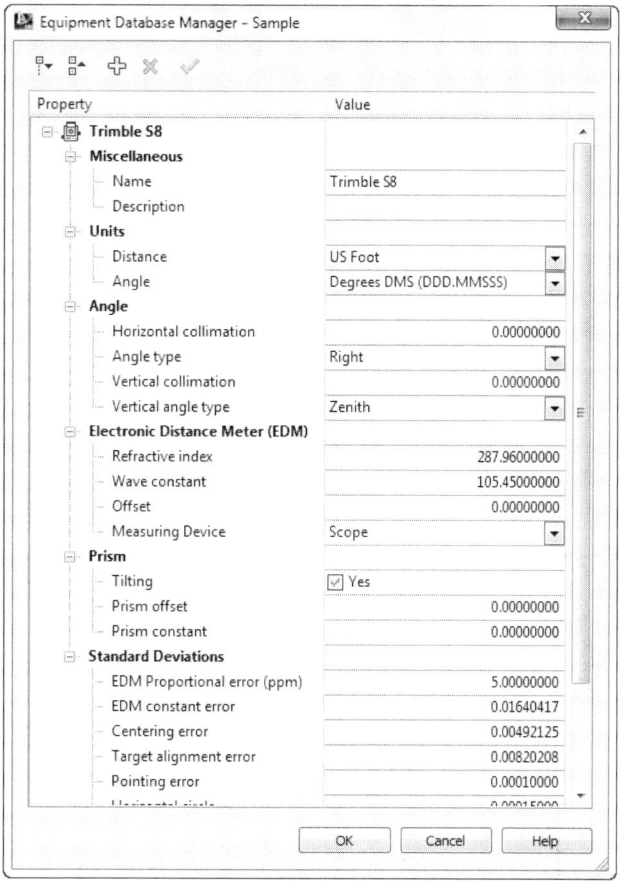

The Figure Prefix Database

The figure prefix database is used to translate descriptions in the field to lines in CAD. These survey-generated lines are called *figures*. If a description matches a listing in the figure prefix database, the figure is assigned the properties and style dictated therein (see Figure 2.4).

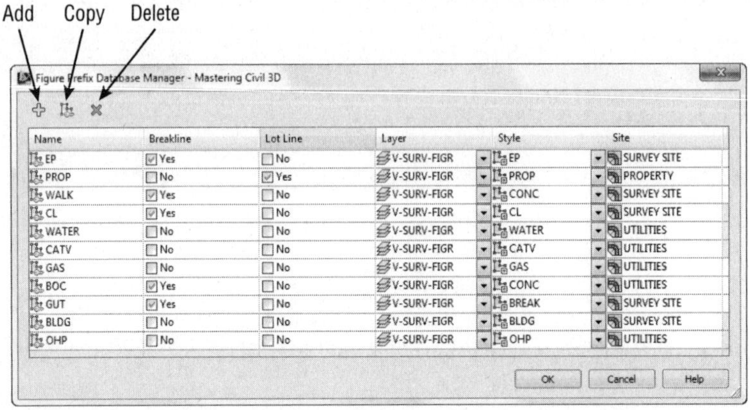

FIGURE 2.4
Figure Prefix Database Manager

The Figure Prefix Database Manager contains these columns:

Breakline The Breakline column will allow you to flag a figure as a breakline. The most powerful use of this setting is in Survey Queries (discussed later in this chapter). If your Survey Query pulls in figures that have this setting on, you can readily add them to the definition of a surface model. Even if you don't flag a figure as a breakline, you can still add it to a surface model manually.

Lot Line This column specifies whether the figure should behave as a parcel segment. If a closed area is formed with figures of this kind, a parcel is formed automatically.

Layer This column specifies the insertion layer of the figure. If the layer already exists in the drawing, the figure will be placed on that layer. If the layer does not exist in the drawing, the layer will be created and the figure placed on the newly created layer.

Style This column specifies the style to be used for each figure. Figure styles will override color and linetype if they are set to something other than ByLayer. See the section on linear object styles in Chapter 21, "Object Styles," for more information on this type of style.

Site This column specifies which site the figures should reside on when inserted into the drawing. As with previous settings, if the site exists in the drawing, the figure will be inserted into that site. If the site does not exist in the drawing, a site will be created with that name and the figure will be inserted into the newly created site.

Remember that figure prefix databases are not drawing specific. The only reason a drawing is needed is to access the styles that the figures will use.

You'll explore these settings in a practical exercise:

1. Open the drawing `figure prefix library.dwg` or `figure prefix library_METRIC.dwg`, which you can download from this book's web page, `www.sybex.com/go/masteringcivil3d2013`.

   This file contains the survey figure styles needed to complete this exercise.

2. In the Survey tab of Toolspace, right-click Figure Prefix Databases and select New.

   The New Figure Prefix Database dialog opens.

3. Enter **Mastering Civil 3D** in the Name text box, and click OK to dismiss the dialog.

   If you expand the Figure Prefix Database listing, you will see Mastering Civil 3D.

4. Right-click the newly created Mastering Civil 3D figure prefix database and select Manage Figure Prefix Database.

   The Figure Prefix Database Manager will appear.

5. Select the white + symbol in the upper-left corner of the Figure Prefix Database Manager to create a new figure prefix.

6. Click the Sample name and rename the figure prefix to **EP** (for Edge of Pavement).

7. Click the check box next to the word no in the Breakline column. This will turn into a Yes, indicating that it will contain the breakline property.

   Leave the box in the Lot Line column unchecked so that the figure will not be treated as a parcel segment.

8. Under the Layer column, select V-SURV-FIGR.

9. Under the Style column, select EP.

10. Under the Site column, leave the name of the site set to Survey Site.

11. Complete the Figure Prefixes table with the values shown in Table 2.1.

    Hint: For figures with similar properties, use the Copy Figure Prefix button.

12. Click OK to dismiss the Figure Prefix Database Manager.

    You have a choice in the linework code set whether lines are automatically formed when they match one of these figure prefixes. When a figure is generated, the site will also be created in the drawing.

TABLE 2.1: FIGURE SETTINGS

NAME	BREAKLINE	LOT LINE	LAYER	STYLE	SITE
PROP	No	Yes	V-SURV-FIGR	PROP	PROPERTY
WALK	Yes	No	V-SURV-FIGR	CONC	SURVEY SITE
CL	Yes	No	V-SURV-FIGR	CL	SURVEY SITE
WATER	No	No	V-SURV-FIGR	WATER	UTILITIES
CATV	No	No	V-SURV-FIGR	CATV	UTILITIES
GAS	No	No	V-SURV-FIGR	GAS	UTILITIES
BOC	Yes	No	V-SURV-FIGR	CONC	SURVEY SITE

TABLE 2.1: FIGURE SETTINGS *(CONTINUED)*

NAME	BREAKLINE	LOT LINE	LAYER	STYLE	SITE
GUT	Yes	No	V-SURV-FIGR	BREAK	SURVEY SITE
BLDG	No	No	V-SURV-FIGR	BLDG	SURVEY SITE
OHP	No	No	V-SURV-FIGR	OHP	UTILITIES

The Linework Code Set Database

The linework code set (Figure 2.5) lists what designators are used to start, stop, continue, or add additional geometry to lines. For example, the B code that is typically used to begin a line can be replaced by a code of your choosing, a decimal (.) can be used for a right-turn value, and a minus sign (–) can be used for a left-turn value. Linework code sets allow a survey crew to customize their data collection techniques based on methods used by various types of software not related to Civil 3D.

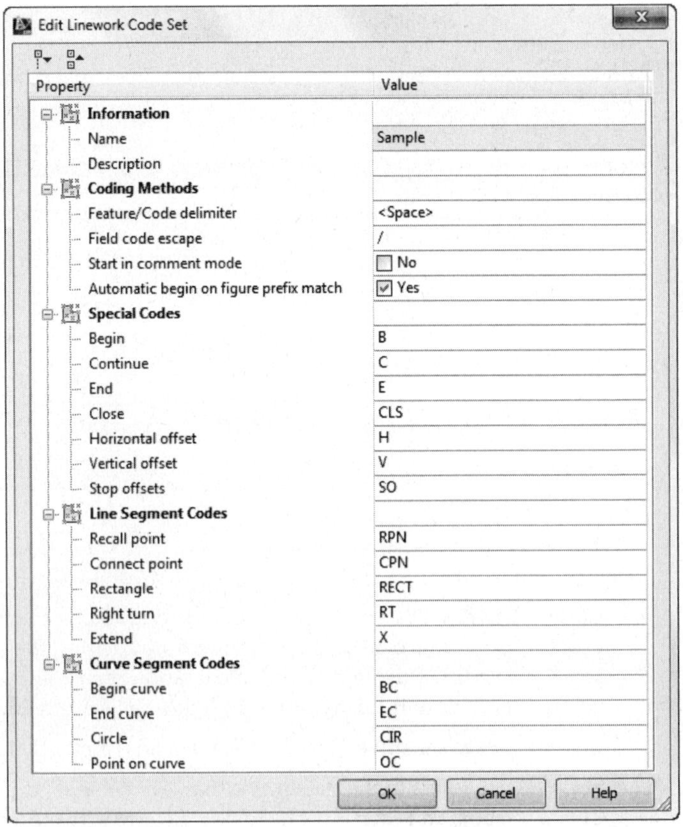

FIGURE 2.5
The Edit Linework Code Set dialog

YOU DO NOT NEED LINEWORK CODES IN YOUR SURVEY TO GENERATE LINES!

There is a setting in the linework code set called Automatic Begin On Figure Prefix Match (as you can see in Figure 2.5). If you select this option, lines will start when a shot description matches one of your figure prefixes (such as EP or CL in Figure 2.4), with no additional coding needed.

Depending on your survey department's method for picking up linework, Automatic Begin On Figure Prefix Match can be a good thing or a bad thing. If the survey crew has fastidiously collected each line with its own name, such as CL1, CL2, EP1, EP2, and so on with no repeats, it can work well. If they use start and stop codes but reuse line names throughout the survey, it can produce figures that connect unintentionally.

Description Keys: Field to Civil 3D

CERT OBJECTIVE

Description keys bridge the gap between the field and the office. Unlike the linework code set and figure prefix database, which are each external files, description keys should be created and saved in your Civil 3D template. *The description key set* is a listing of field descriptions and how they should look and behave once they are imported into Civil 3D.

For example, a surveyor may code in **BM** to indicate a benchmark. When the file is imported into Civil 3D either through the Survey database or by point import from a text file (as you will do in Chapter 3, "Points"), it will be checked against the description key set. If BM exists in the list as it is in Figure 2.6, then the styles, format, layer, and several other parameters are set.

FIGURE 2.6
Description key set

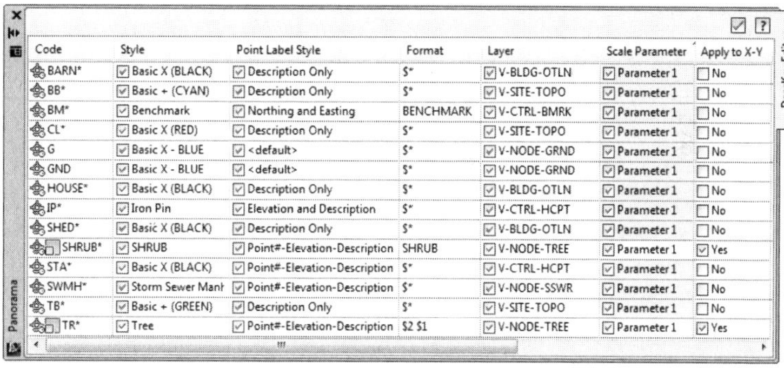

Code	Style	Point Label Style	Format	Layer	Scale Parameter	Apply to X-Y
BARN*	Basic X (BLACK)	Description Only	$*	V-BLDG-OTLN	Parameter 1	No
BB*	Basic + (CYAN)	Description Only	$*	V-SITE-TOPO	Parameter 1	No
BM*	Benchmark	Northing and Easting	BENCHMARK	V-CTRL-BMRK	Parameter 1	No
CL*	Basic X (RED)	Description Only	$*	V-SITE-TOPO	Parameter 1	No
G	Basic X - BLUE	<default>	$*	V-NODE-GRND	Parameter 1	No
GND	Basic X - BLUE	<default>	$*	V-NODE-GRND	Parameter 1	No
HOUSE*	Basic X (BLACK)	Description Only	$*	V-BLDG-OTLN	Parameter 1	No
IP*	Iron Pin	Elevation and Description	$*	V-CTRL-HCPT	Parameter 1	No
SHED*	Basic X (BLACK)	Description Only	$*	V-BLDG-OTLN	Parameter 1	No
SHRUB*	SHRUB	Point#-Elevation-Description	SHRUB	V-NODE-TREE	Parameter 1	Yes
STA*	Basic X (BLACK)	Point#-Elevation-Description	$*	V-CTRL-HCPT	Parameter 1	No
SWMH*	Storm Sewer Manh	Point#-Elevation-Description	$*	V-NODE-SSWR	Parameter 1	No
TB*	Basic + (GREEN)	Description Only	$*	V-SITE-TOPO	Parameter 1	No
TR*	Tree	Point#-Elevation-Description	$2 $1	V-NODE-TREE	Parameter 1	Yes

To access your description key set, you will need to select Toolspace ➤ Settings ➤ Point ➤ Description Key Sets.

Code The raw description or field code entered by the person collecting or creating the points. The code works as an identifier for matching the point with the correct description key. Click inside this field to activate it, and then type your desired code. Wildcards are useful when more information is added to the shot in addition to the field code.

Codes are case sensitive! The code bm is read differently from the description key set as BM.

Style Style refers to the point style that will be applied to points that meet the code criteria. Check the box, and then click inside the field to activate a style selection dialog. By default, styles set here will take precedence over styles set elsewhere. For more information on creating or modifying point styles, see Chapter 21.

Point Label Style The point label style that will be applied to points that meet the code criteria. Check the box, and then click inside the field to activate a style selection dialog. By default, styles set here will take precedence over styles set elsewhere. For more information on creating or modifying point label styles, see Chapter 20, "Label Styles."

Format The Format column can convert a surveyor's shorthand into something that is more drafter-friendly. In Civil 3D terms, the Format column converts the raw description to the full description. The default of $* means the raw description and full description will have the same value. You can also use $+, which means that information after the main description will appear as the full description. In Figure 2.6, Format will convert all codes starting with BM to a full description of BENCHMARK.

If a survey crew is consistent in coding, even fancier formats can be used. The code should always come first, but the crew can use a space to indicate a parameter.

Consider the example raw description: TREE 30 PINE. TREE is the code, or $0. Parameter 1 is 30, or $1 in the Format field. PINE is the second bit of information after the code referred to as parameter 2, or $2. Based on the example description key set in Figure 2.6, this would translate to a full description of PINE 30. You can have up to nine parameters after the code if your survey crew is feeling verbose.

Layer Points that match a description key will be inserted on the layer specified here. Click inside this field to activate a layer selection dialog. The layer set here will take precedence over layer defaults set in the point command settings or the point creation tools.

Scale Parameter The Scale parameter is used to tell Civil 3D which bit of information after the code will be used to scale the symbol. By default it is checked, but it won't do anything unless Apply To X-Y is also selected. Once you enable Apply To X-Y, you can change which parameter contains scale information.

In our example, TREE 30 PINE, 30 is the Scale parameter.

Fixed Scale Factor Fixed Scale Factor is an additional scale multiplier that can be applied to the symbol size. A common use of Fixed Scale Factor is to convert a field measurement of inches to feet. If the 30 in our example represents a dripline measurement and is meant to be feet, no Fixed Scale Factor is needed. However, if the 30 represents inches (i.e., a trunk diameter), you would need to turn on Fixed Scale Factor and set the value to 0.0833.

Use Drawing Scale In most cases, you will leave this option unchecked. By default, marker styles dictate that they will grow or shrink based on the annotative scale of the drawing. Generally, this setting is not needed unless you want to scale your point symbol based on a parameter in addition to the scale factor.

Apply To X-Y If you wish to scale symbols based on information in the field code, you need to turn this option on by placing a check mark in the box. This option works with the marker style and the Scale parameter to increase the size of an item to a scale indicated by the surveyor in the raw description.

Apply To Z In most cases, you will leave this option unchecked. Most marker symbols are 2D blocks, so selecting this option will have no effect on the point. If your marker symbol consists of a 3D block, it will be stretched by the parameter value, which is rarely needed.

Marker Rotate Parameter, Marker Fixed Rotation, Label Rotate Parameter, Label Fixed Rotation, and Rotation Direction These options work similarly to the scale factor parameter except they dictate the rotation of a symbol or label. They are not widely used, however, since it is often more time-effective to have the drafter rotate the points in CAD than to have the surveyor key in a rotation.

Creating a Description Key Set

Description key sets appear on the Settings tab of Toolspace under the Point branch. You can create a new description key set by right-clicking the Description Key Sets collection and choosing New, as shown in Figure 2.7.

FIGURE 2.7 Creating a new description key set on the Settings tab of Toolspace

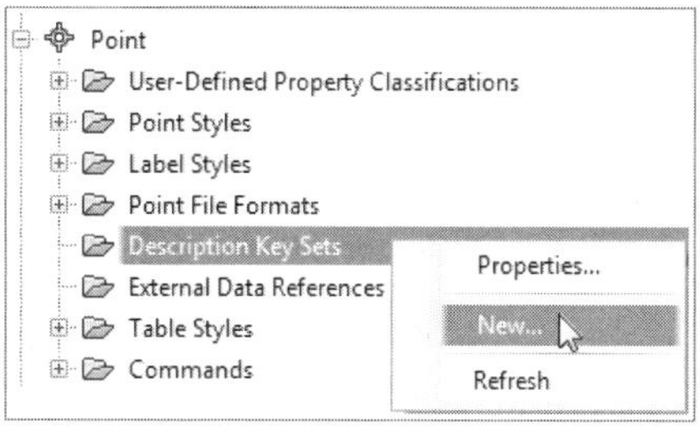

In the resulting Description Key Set dialog, give your description key set a meaningful name, such as your company name, and click OK. You'll create the actual description keys in another dialog.

Creating Description Keys

To enter the individual description key codes and parameters, right-click the description key set, as illustrated in Figure 2.8, and select Edit Keys. The DescKey Editor in Panorama appears.

FIGURE 2.8 Editing a description key set

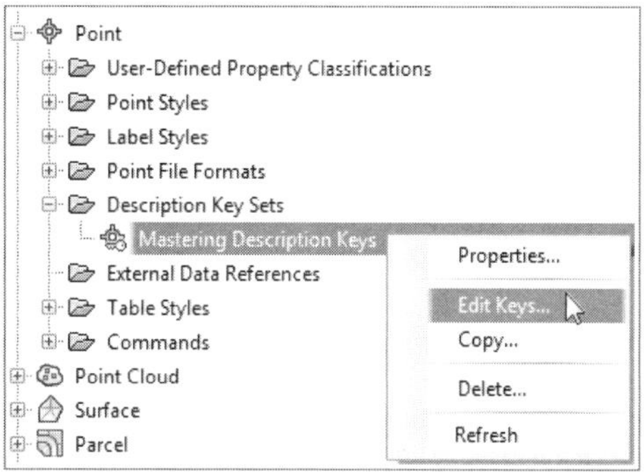

To enter new codes, right-click a row with an existing key in the DescKey Editor, and choose New or Copy from the context menu, as shown in Figure 2.9.

FIGURE 2.9
Creating or copying a description key

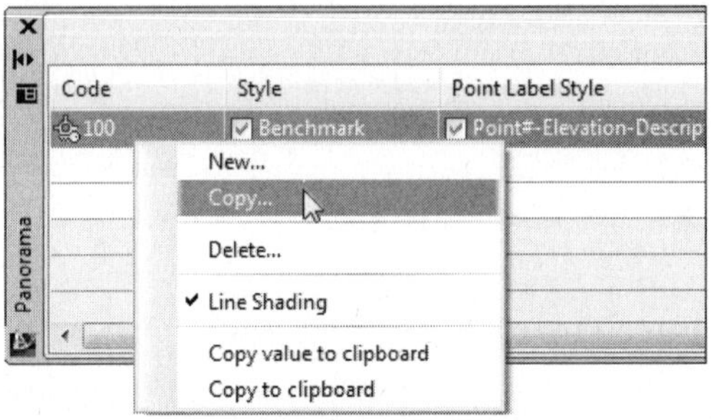

> **USING WILDCARDS**
>
> The asterisk (*) acts as a wildcard in many places in Civil 3D. Two of the most common places to use a wildcard are the DescKey Editor and the Point Group Properties dialog. Whereas a DescKey code of TREE flags any points created with that raw description, a DescKey code of TREE* also picks up raw descriptions of TREE1, TREE2, TREE MAPLE, TREE OAK, and so on. You can use the wildcard the same way in the Point Group Properties dialog when specifying items to include or exclude.

ACTIVATING A DESCRIPTION KEY SET

Once you've created a description key set, you should verify the settings for your commands so that Civil 3D knows to match your newly created points with the appropriate key.

In the Settings tab of Toolspace, choose Point ➢ Commands and right-click CreatePoints. Select Edit Command Settings, as shown in Figure 2.10.

FIGURE 2.10
Right-click CreatePoints and choose Edit Command Settings.

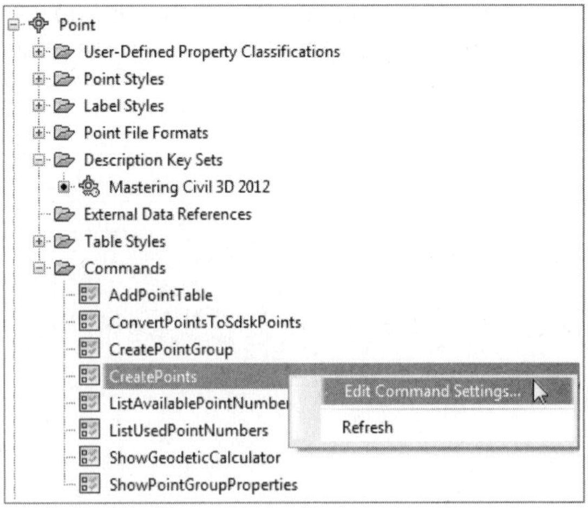

In the Edit Command Settings - CreatePoints dialog, ensure that Match On Description Parameters is set to True and that Disable Description Keys is set to False, as shown in Figure 2.11.

FIGURE 2.11
Verify that Disable Description Keys is set to False.

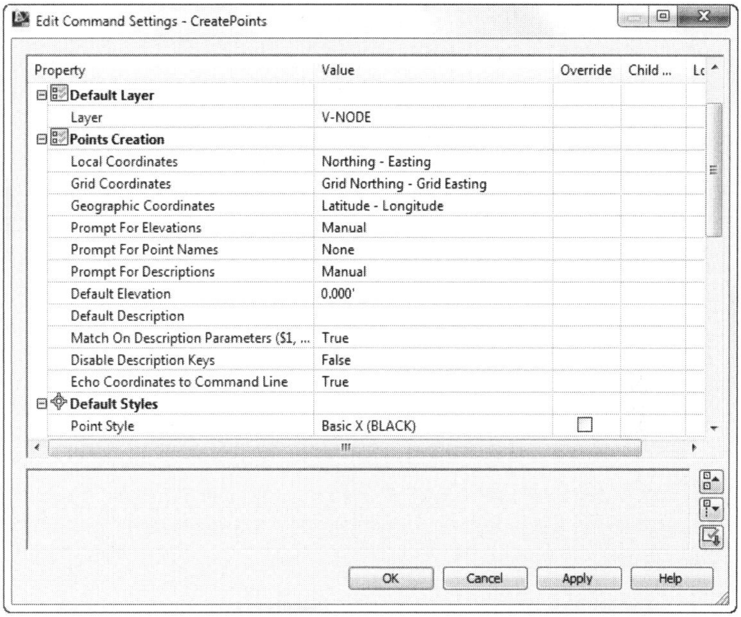

It is not uncommon to have multiple description key sets in your template for multiple clients or external survey firms that you work with. If you have multiple description key sets, they are all active, but if a set has a duplicate key, the first one Civil 3D runs across will take precedence. For example, if one set uses FL for flowline but a second set uses FL for fence line, the second occurrence of the FL key gets ignored.

You can control the search order from the Settings tab of Toolspace by right-clicking on Description Keys Sets and selecting Properties. Figure 2.12 shows the Description Key Sets Search Order dialog box. Use the arrows on the right side of the dialog box to set the order. The set listed first takes first priority, then the second, and so on. Note that the listing in the Settings tab may not reflect the true listing in the properties.

FIGURE 2.12
The Description Key Sets Search Order dialog

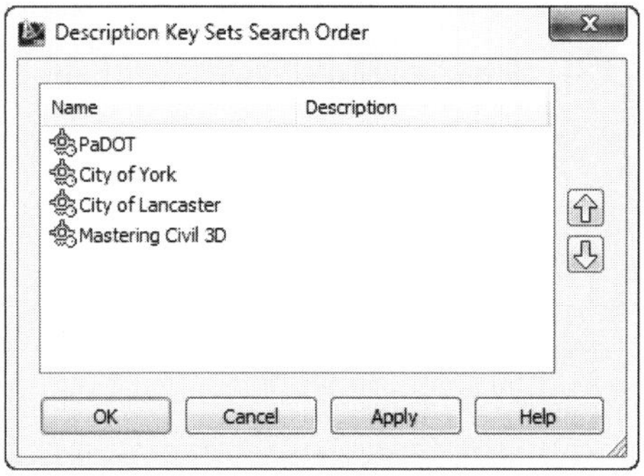

The Main Event: Your Project's Survey Database

Now that you know how to get everything set up, you are probably eager to get some real, live data into your drawing.

First, set your survey working folder to your desired survey storage location. The Civil 3D survey database is a set of external database files that reside in your survey working folder. By default, this folder is located in `C:\Civil 3d Projects\`.

On the Survey tab of Toolspace, right-click on Survey Databases and select Set Working Folder, as shown in Figure 2.13.

FIGURE 2.13
Set the survey working folder.

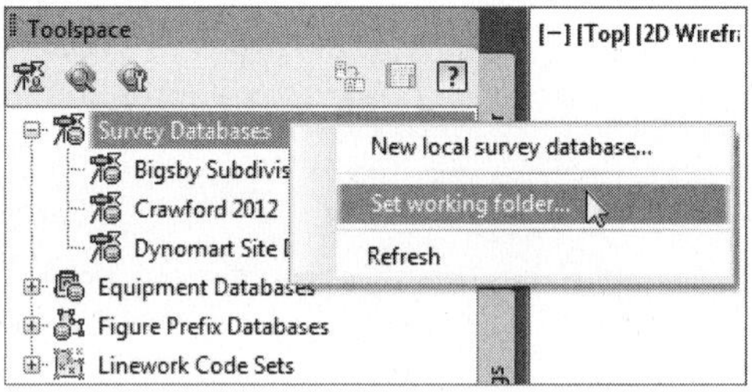

In the current version of Civil 3D, this is a different, independent setting from the working folder for data shortcuts (as discussed in Chapter 18, "Advanced Workflows"). Most folks want this on a network location, tucked neatly into the survey folder for the project.

Keep in mind that the survey database is version specific. You can open a 2012 database in 2013, but once it has been converted, there is no going back.

Import Survey Data

To create a survey database, you can either right-click and select New Local Survey Database as mentioned earlier, or you can select Import Survey Data from the Home tab's Create Ground Data panel.

The contents of a survey database are organized into the following categories:

Import Events Import events provide a framework for viewing and editing specific survey data, and they are created each time you import data into a survey database. The default name for the import event is the same as the imported filename. The Import Event collection contains the networks, figures, and survey points that are referenced from a specific import command, and provides an easy way to remove, re-import, and reprocess survey data in the current drawing.

Survey Queries New to Civil 3D 2013, survey queries allow users to search for specific information within a survey database. You can use a query to locate and group related survey points and figures. For example, you may want all figures representing power lines in a query with survey points representing power poles. If you create a query containing your utilities, you can isolate specific figures and points to insert into another drawing.

The most powerful use of a survey query is when they are used in conjunction with a surface. When a query containing points and figures is added to a surface model, the figures are automatically added as breaklines. Figure 2.14 shows an example query that might be used for generating surface data.

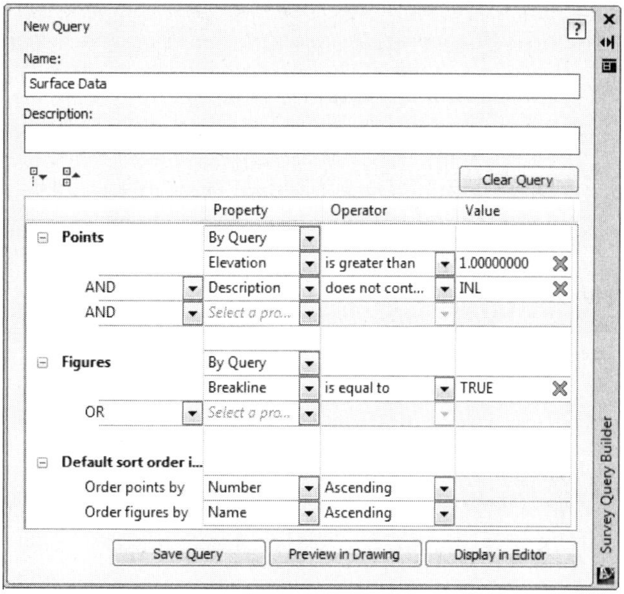

FIGURE 2.14
Survey database query

Networks A survey network is a collection of related data that is collected in the field. The network consists of setups, control points, noncontrol points, known directions, observations, and traverses. A survey database can have multiple networks. For example, you can use different networks for different phases of a project. If working within the same coordinate zone, some users have used one survey database with many networks for every survey that they perform. This is possible because individual networks can be inserted into the drawing simply by dragging and dropping the network from the Survey tab in Toolspace into the drawing. You can hover your cursor over any Survey Network component in the drawing to see information about that component and the survey network. You can also right-click any component of the network and browse to the observation entry for that component.

Network Groups Network groups are collections of various survey networks within a survey database. These groups can be created to facilitate inserting multiple networks into a drawing at once simply by dragging and dropping.

Figures Figures are the linework created by codes and commands entered into the raw data file during data collection. The figure names typically come from the descriptor or description of a point.

Figure Groups Similar to network groups, figure groups are collections of individual figures. These groups can be created to facilitate quick insertion of multiple figures into a drawing.

Survey Points One of the most basic components of a survey database, points form the basis for each and every survey. Survey points look just like regular Civil 3D point objects, and their visibility can be controlled just as easily. However, one major difference is that a survey point cannot be edited within a drawing. Survey points are locked by the survey database, and the only way of editing them is to edit the observation that collected the data for the points. This provides the surveyor with the confidence that points will not be accidentally erased or edited. Like figures, survey points can be inserted into a drawing by either dragging and dropping from the Survey tab of Toolspace or by right-clicking Surveying Points and selecting the Points ➤ Insert Into Drawing option.

Survey Point Groups Just like network groups and figure groups, survey point groups are collections of points that can be easily inserted into a drawing. When these survey point groups are inserted into the drawing, a Civil 3D point group is created with the same name as the survey point group. This point group can be used to control the visibility or display properties of each point in the group.

In the following exercise, you'll create an import event and import an ASCII file with survey data. The survey data includes linework.

1. Create a new drawing from the `_AutoCAD Civil 3D (Imperial) NCS.dwt` or `_AutoCAD Civil 3D (Metric) NCS.dwt` template file. If you did not complete the previous exercise, copy the file `Mastering Civil 3D.fdb_xdef` from the dataset to `C:\ProgramData\Autodesk\C3D 2013\enu\Survey`. Note that this directory is hidden in Windows by default, so you will need to change the folder view options or type this path into the address bar of Windows Explorer.
2. From the Home tab, in the Create Ground Data panel of the Ribbon click Import Survey Data to open the Import Survey Data wizard.
3. Click Create New Survey Database.
4. Enter **Roadway** as the name of the folder in which your new database will be stored. Click OK.

   Roadway is now added to the list of survey databases.
5. With Roadway highlighted, click Edit Survey Database Settings and verify the units are set to US Foot (or meter). Click OK to dismiss the dialog.
6. Click Next.
7. Set the Data Source Type pull-down to Point File.
8. Click the + on the right side of the Selected Files to browse to the `import points.txt` or `import points_METRIC.txt` file, which you can download from this book's web page.
9. Click Open and the name will be listed in the dialog, as shown in Figure 2.15.

10. Under Specify Point File Format, scroll through the list until you find PNEZD (Comma Delimited).
11. Highlight the format.

    The preview will show that the correct data type is selected, as shown at the bottom of Figure 2.15.
12. Click Next to continue.
13. Click Create New Network.

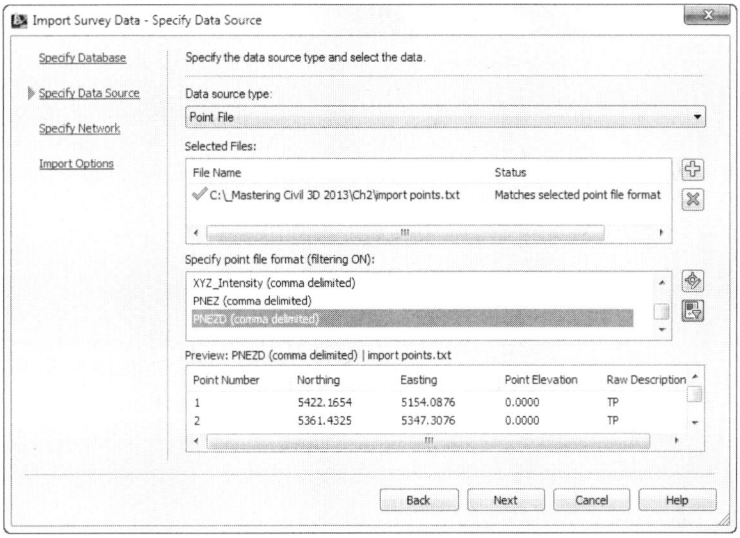

FIGURE 2.15 Select the correct file type, file, and format in the Import Survey Data wizard.

Note that if you forget to create a network to place your points, you will not be able to manipulate this group apart from other points.

14. Enter **Roadway 4-22-2012** as the name of the network. Click OK, and click Next to continue.
15. Set the Current Figure Prefix Database to Mastering Civil 3D. Set the Process Linework During Import option to Yes. Verify that your import options match what is shown in Figure 2.16. Click Finish.
16. Save the file as **`RoadwayFigures.dwg`** to the same location as the rest of your downloaded example files.

    You will need it in the next exercise.

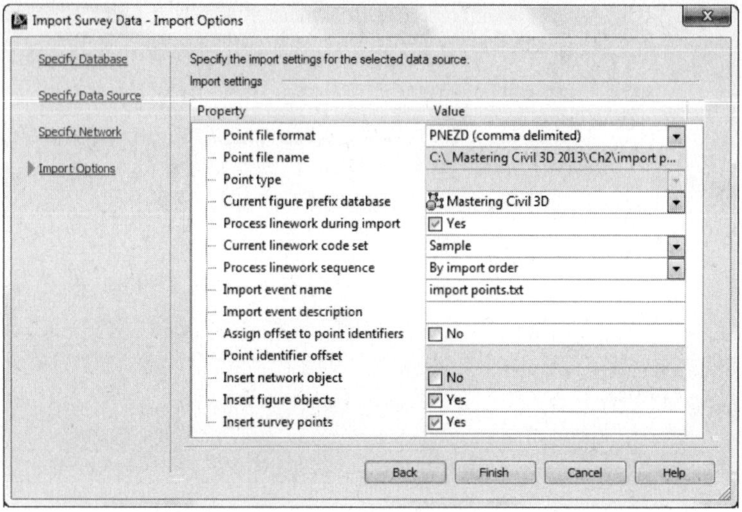

FIGURE 2.16
The Import Options page in the Import Survey Data wizard

The data is imported and the linework is drawn; however, the building is missing the left side. The following steps will resolve this issue:

1. Continue working in the drawing from the previous exercise. In the Survey tab, expand Survey Databases, and then select Roadway ➢ Networks ➢ Roadway ➢ Non-Control Points.

2. Right-click Non-Control Points and select Edit to bring up the Non-Control Points Editor in Panorama.

3. Scroll to the bottom of the point list and notice the last line in the file describing point Number 34. You will be editing the description of this point.

4. Double-click your cursor on the description field. Use the arrow keys on your keyboard to move your cursor to the end of the description. Add a space and type **CLS**↵, as shown in Figure 2.17. This is the default Close Figure command.

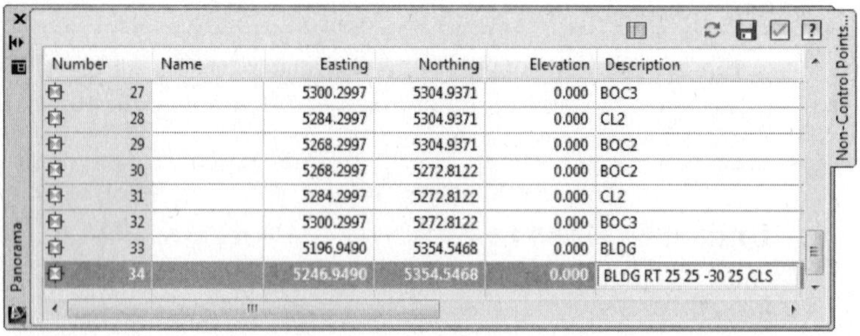

FIGURE 2.17
Editing the import event to add the CLS command and close the building geometry

5. Click the Save button, and then click the green check box in the upper right of the palette to apply changes and save your edits.
6. Click Yes if you are asked to apply your changes.
7. While still working in Survey Databases ➢ Roadway, expand Import Events, and select `import points.txt` (`import points_METRIC.txt`).
8. Right-click the import event, and select Process Linework to bring up the Process Linework dialog.
9. Click OK to reprocess the linework with your updated point description.

The building figure line and your drawing should look something like Figure 2.18.

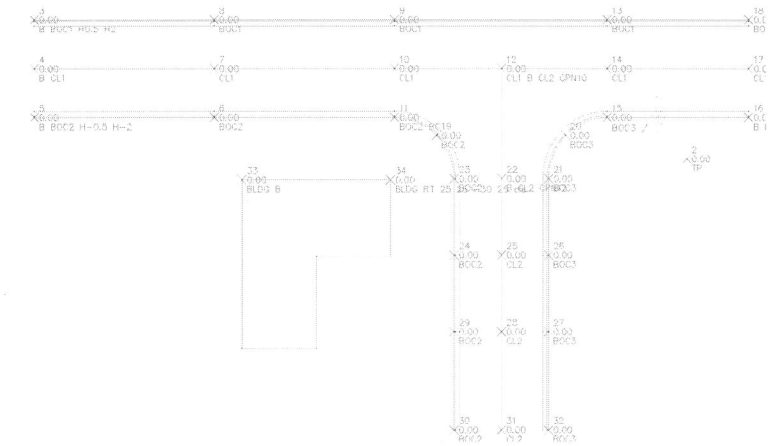

FIGURE 2.18 After editing and reprocessing the linework

EDITING SURVEY DATA

When you make modifications to survey data, only the Civil 3D version is changed. The original file remains untouched. The survey database doesn't use the file unless you re-import the data.

If you edit raw data in the Survey tab, Civil 3D will recalculate all affected information. For example, if you modify an instrument height, all elevations that need to be updated will automatically adjust.

Keep in mind that if you edit the source file and re-import, the Civil 3D survey (and any edits you made) will be overwritten.

Under the Hood in Your Survey Network

Once you import survey data into a network, expand the branch to see how Civil 3D helps you make sense of it. In each network, data is organized by type, as shown in Figure 2.19:

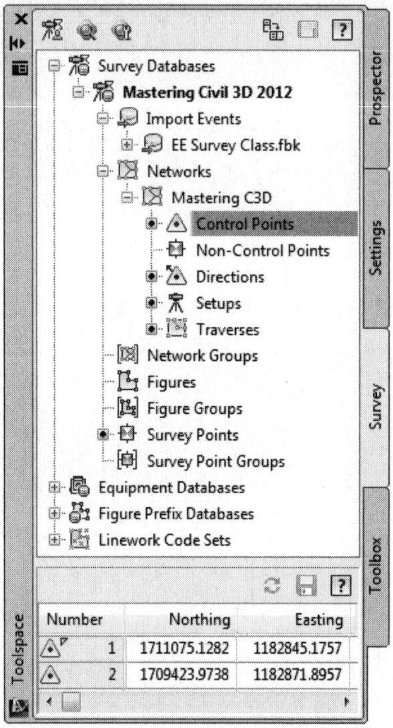

FIGURE 2.19
A typical survey database network with data

Control Points Control points are typically points in your data that have a high confidence factor. These are usually benchmarks or other known points used to lock the survey into its geographical location. In most cases, control points are created when data from an FBK file is imported. Inside the FBK, control points are prefaced by NE, NEZ, or LAT LONG. You can force any point to be listed as a control point by right-clicking this branch and selecting New.

Non-Control Points Keyed-in points, GPS-collected points, and any point brought in through an ASCII file will appear as non-control points. This is the default type if no other information is known about the point.

Directions The direction from one point to another must be manually entered into the data collector for the direction to show up later in the survey network for editing. The direction can be as simple as a compass shot between two initial traverse points that serves as a rough basis of bearings for a survey job.

Setups If you imported data that contains setups and observations, Setups is where the meat of the data is found. Every setup, as well as the points (side shots) located from that setup, can be found listed here. Setups will contain two components: the station (or occupied point) and the backsight. To see or edit the observations located from the setup, right-click Setups and select Edit Setups That Observe. The interface for editing setups is shown in Figure 2.20. Angles and instrument heights can also be changed in this dialog.

FIGURE 2.20
Setups can be changed in the Setups Editor.

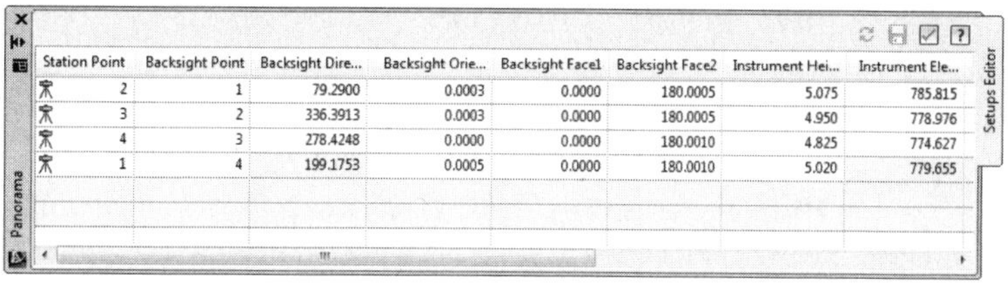

TRAVERSES

The Traverses section is where new traverses are created or existing ones are edited. These traverses can come from your data collector, or they can be manually entered from field notes via the Traverse Editor, as shown in Figure 2.21. You can view or edit each setup in the Traverse Editor, as well as the traverse stations located from that setup.

FIGURE 2.21
The Traverse Editor

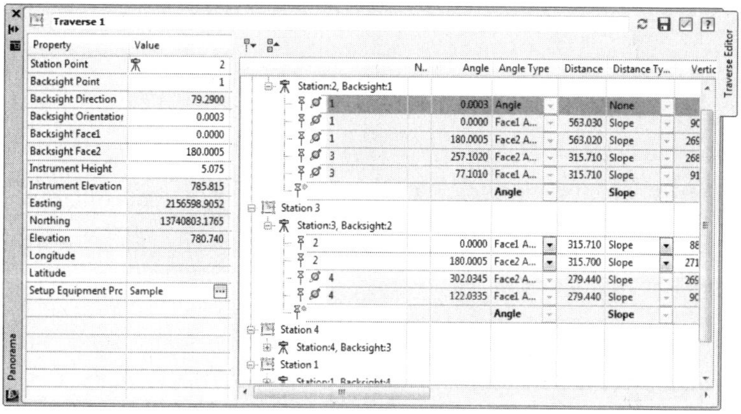

Once you have defined a traverse, you can adjust it by right-clicking its name and selecting Traverse Analysis. You can adjust the traverse either horizontally and/or vertically, using a variety of methods. The traverse analysis can be written to text files to be stored, and the entire network can be adjusted on the basis of the new values of the traverse, as you'll do in the following exercise:

1. Create a new drawing from the `_AutoCAD Civil 3D (Imperial) NCS.dwt` (`_AutoCAD Civil 3D (Metric) NCS.dwt`) template file.
2. Navigate to the Survey tab of Toolspace.
3. Right-click Survey Databases and select New Local Survey Database.

   The New Local Survey Database dialog opens.
4. Enter **Traverse** as the name of the folder in which your new database will be stored.
5. Click OK to dismiss the dialog.

   The Traverse survey database is created as a branch under the Survey Databases branch.

6. Expand the Traverse branch, right-click Networks, and select New.

   The New Network dialog opens.

7. Expand the Network branch in the dialog if needed.

8. Name your new network **Traverse Practice**, and click OK.

   The Traverse Practice network is now listed as a branch under the Networks branch of the Traverse survey database in Prospector.

9. Right-click the Traverse Practice network and select Import ➤ Import Field Book.

10. Select the file `traverse.fbk` (`traverse_METRIC.fbk`), which you can download from this book's web page, and click Open.

    The Import Field Book dialog opens.

11. Make sure you have checked the boxes shown in Figure 2.22.

FIGURE 2.22
The Import Field Book dialog

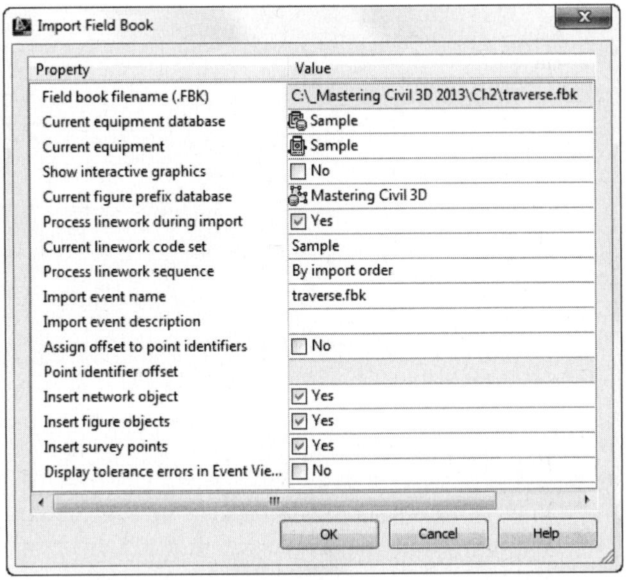

There is no linework in this file, as it is just traverse shots.

12. Click OK.

13. Save the drawing for the next exercise.

Still looking at the Survey tab of Toolspace, expand the network you created earlier and inspect the data. You have one control point in northwest corner that was manually entered into the data collector. There is one direction, and there are four setups. Each setup combines to form a closed polygonal shape that defines the traverse. Notice that there is no traverse definition. In the following exercise, you'll create that traverse definition for analysis.

Continue working in the drawing from the previous exercise.

1. Right-click Traverses under the Traverse Practice network and select New to open the New Traverse dialog.
2. Name the new traverse **Traverse1**.
3. Enter **2** as the value for Initial Station and **1** (if necessary) for Initial Backsight.

   The traverse will now pick up the rest of the stations in the traverse and enter them in the next box.
4. Enter **2** as the value for the Final Foresight if necessary (the closing point for the traverse).

   Your New Traverse dialog box will look like Figure 2.23.
5. Click OK.

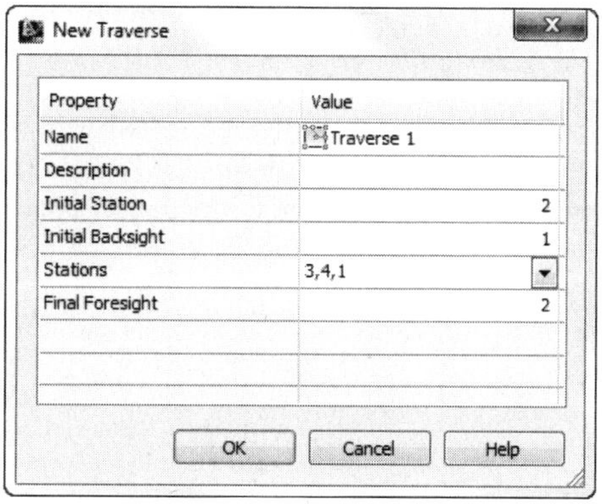

FIGURE 2.23
Defining a new traverse

6. Right-click Traverse1 in the bottom portion of Toolspace. Select Traverse Analysis.
7. In the Traverse Analysis dialog, ensure that Yes is selected for Do Traverse Analysis and Do Angle Balance.
8. Select Least Squares for both the Horizontal Adjustment Method and Vertical Adjustment Method.
9. Set both the Horizontal Closure Limit and Vertical Closure Limit 1:X to **30,000**.
10. Leave Angle Error Per Set at the default.
11. Make sure the option Update Survey Database is set to Yes.

    The Traverse Analysis dialog will look like Figure 2.24.
12. Click OK.

The analysis is performed, and four text files are displayed that show the results of the adjustment. Note that if you look back at your survey network, all points are now control points, because the analysis has upgraded all the points to control point status.

FIGURE 2.24
Specify the adjustment method and closure limits in the Traverse Analysis dialog.

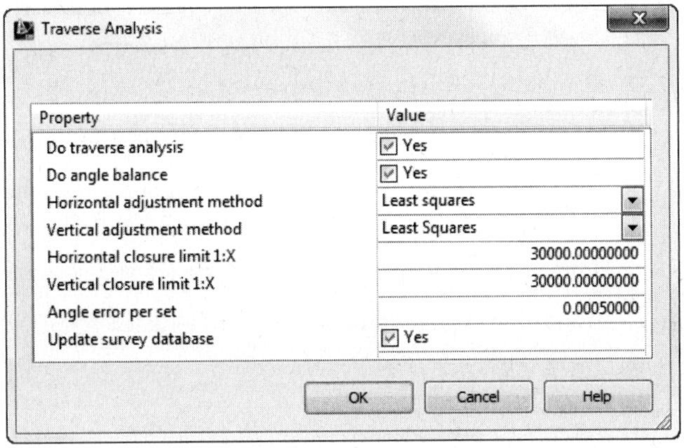

Figure 2.25 shows the `Traverse1 Raw Closure.trv` and `Traverse 1 Vertical Adjustment.trv` files that are generated from the analysis. The raw closure file shows that your new precision is well within the tolerances set in step 8. The vertical adjustment file describes how the elevations have been affected by the procedure.

FIGURE 2.25
Horizontal and vertical traverse analysis results

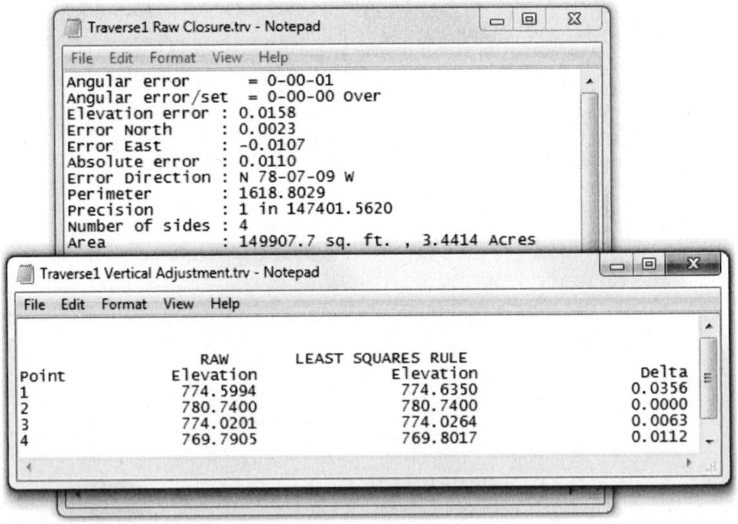

The third text file is shown in three separate portions. The first portion is shown in Figure 2.26. This portion of the file displays the various observations along with their initial measurements, standard deviations, adjusted values, and residuals. You can view other statistical data at the beginning of the file.

FIGURE 2.26
Statistical and observation data text file

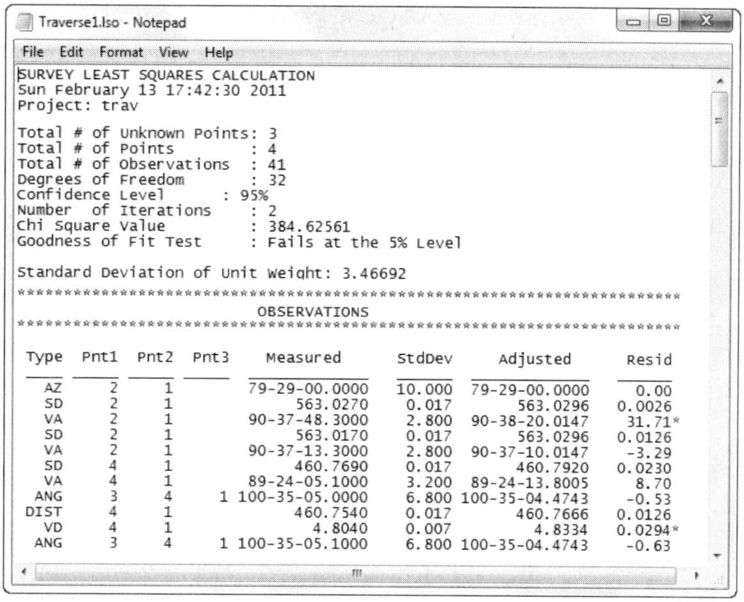

```
Traverse1.lso - Notepad
File  Edit  Format  View  Help
SURVEY LEAST SQUARES CALCULATION
Sun February 13 17:42:30 2011
Project: trav

Total # of Unknown Points: 3
Total # of Points        : 4
Total # of Observations  : 41
Degrees of Freedom       : 32
Confidence Level      : 95%
Number  of Iterations    : 2
Chi Square Value         : 384.62561
Goodness of Fit Test     : Fails at the 5% Level

Standard Deviation of Unit Weight: 3.46692
**************************************************************************
                              OBSERVATIONS
**************************************************************************

 Type  Pnt1  Pnt2  Pnt3     Measured        StdDev     Adjusted       Resid

  AZ     2     1          79-29-00.0000   10.000  79-29-00.0000     0.00
  SD     2     1               563.0270    0.017       563.0296   0.0026
  VA     2     1          90-37-48.3000    2.800  90-38-20.0147    31.71*
  SD     2     1               563.0170    0.017       563.0296   0.0126
  VA     2     1          90-37-13.3000    2.800  90-37-10.0147    -3.29
  SD     4     1               460.7690    0.017       460.7920   0.0230
  VA     4     1          89-24-05.1000    3.200  89-24-13.8005     8.70
 ANG     3     4     1   100-35-05.0000    6.800 100-35-04.4743    -0.53
DIST     4     1               460.7540    0.017       460.7666   0.0126
  VD     4     1                 4.8040    0.007         4.8334   0.0294*
 ANG     3     4     1   100-35-05.1000    6.800 100-35-04.4743    -0.63
```

Figure 2.27 shows the second portion of this text file and displays the adjusted coordinates, the standard deviation of the adjusted coordinates, and information related to error ellipses displayed in the drawing. If the deviations are too high for your acceptable tolerances, first check the instrument settings and tolerances in the Equipment Database. If everything is set correctly, you may need to redo the work or edit the field book.

FIGURE 2.27
Adjusted coordinate information text file

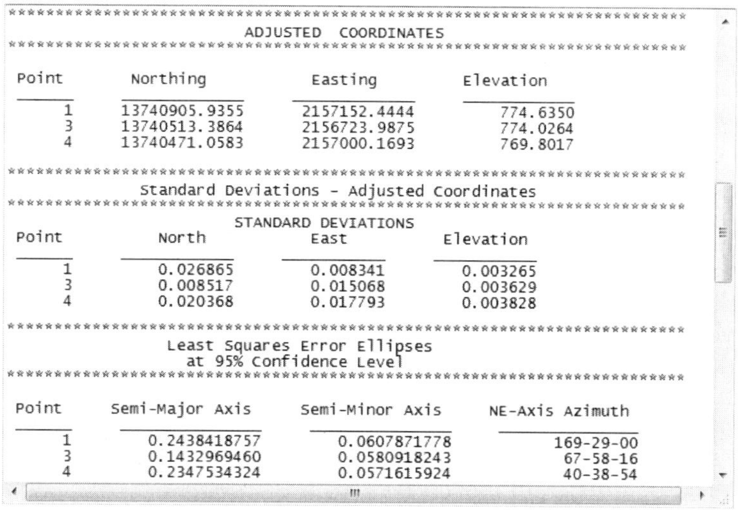

```
**************************************************************************
                         ADJUSTED  COORDINATES
**************************************************************************

Point        Northing              Easting            Elevation

   1      13740905.9355        2157152.4444           774.6350
   3      13740513.3864        2156723.9875           774.0264
   4      13740471.0583        2157000.1693           769.8017

**************************************************************************
             Standard Deviations - Adjusted Coordinates
**************************************************************************
                          STANDARD DEVIATIONS
Point           North               East              Elevation

   1          0.026865            0.008341            0.003265
   3          0.008517            0.015068            0.003629
   4          0.020368            0.017793            0.003828

**************************************************************************
                      Least Squares Error Ellipses
                        at 95% Confidence Level
**************************************************************************

Point      Semi-Major Axis       Semi-Minor Axis       NE-Axis Azimuth

   1        0.2438418757          0.0607871778            169-29-00
   3        0.1432969460          0.0580918243             67-58-16
   4        0.2347534324          0.0571615924             40-38-54
```

Figure 2.28 displays the final portion of this text file — Blunder Detection/Analysis. Civil 3D will look for and analyze data in the network that is obviously wrong and choose to keep it or throw it out of the analysis if it doesn't meet your criteria. If a blunder (or bad shot) is detected, the program will not fix it. You will have to edit the data manually, whether by going out in the field and collecting the correct data or by editing the FBK file.

FIGURE 2.28
Blunder analysis text file

Blunder Detection/Analysis

Type	Pnt1	Pnt2	Pnt3	Adjusted	Resid	Redun	Reliability Tests Estimate	Marg	Ext
AZ	2	1		79-29-00.0000	0.000	0.000	<None>	F	F
SD	2	1		563.030	0.003	0.840	-0.003	P	P
VA	2	1		90-38-20.0147	31.715	0.818	-38.789	F	F
SD	2	1		563.030	0.013	0.840	-0.015	P	P
VA	2	1		90-37-10.0147	-3.285	0.818	4.018	P	P
SD	4	1		460.792	0.023	0.844	-0.027	P	P
VA	4	1		89-24-13.8005	8.701	0.805	-10.802	P	F
ANG	3	4	1	100-35-04.4743	-0.526	0.703	0.747	P	P
DIST	4	1		460.767	0.013	0.844	-0.015	P	P
VD	4	1		4.833	0.029	0.797	-0.037	F	F
ANG	3	4	1	100-35-04.4743	-0.626	0.703	0.890	P	P
SD	3	4		279.439	0.002	0.830	-0.002	P	P
VA	3	4		90-51-35.8216	-11.378	0.804	14.144	P	P
ANG	2	3	4	122-03-35.8498	-4.150	0.754	5.502	P	P
SD	3	4		279.439	0.002	0.830	-0.002	P	P
VA	3	4		90-52-35.8216	18.622	0.804	-23.148	P	F
ANG	2	3	4	122-03-35.8498	0.850	0.754	-1.127	P	P

OTHER METHODS OF MANIPULATING SURVEY DATA

Often, it is necessary to edit the entire survey network at one time. For example, rotating a network to a known bearing or azimuth from an assumed one happens quite frequently. To find this hidden gem of functionality, go to the Survey tab of Toolspace ➢ Survey Databases and right-click the name of the database you wish to modify, as shown in Figure 2.29.

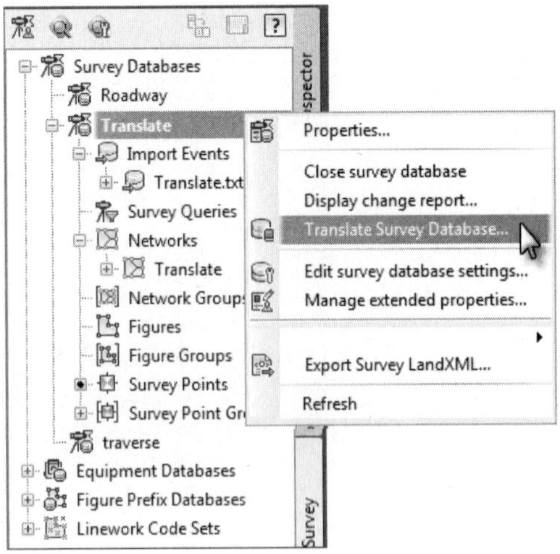

FIGURE 2.29
The elusive yet indispensable Translate Survey Database command

Real World Scenario

MANIPULATING THE NETWORKT

Frequently, surveys performed in assumed coordinate systems need to be adjusted to match a known coordinate system. Along with changing coordinate systems, the survey will probably need to be rotated to the correct orientation. Additionally, networks may need to be adjusted from assumed elevations to a known datum. In this example, the DWG you will be working with contains two known points from the desired coordinate system. The file you will import is initially in an assumed coordinate system. You will use the Translate command to move and rotate the network accordingly.

1. Open the file `Translate.dwg` (`Translate_METRIC.dwg`).
2. In the Survey tab of Toolspace, right-click and select New Local Survey Database.

   The New Local Survey Database dialog opens.
3. Enter **Translate** in the text box.

   This is the name of the folder for the new database.
4. Click OK, and the Translate database will now be listed under the Survey Databases branch.
5. Select Networks under the new Translate branch.
6. Right-click and select New to open the New Network dialog.
7. Enter **Translate** as the name of this new network.
8. Click OK to dismiss the dialog.
9. Right-click the Translate network, and select Import ➢ Import Point File.
10. Navigate to the file `translate.txt` (`translate_METRIC.txt`), which you can download from this book's web page, and click Open.

    The Import Point File dialog opens.
11. Verify that the Point file format is PNEZD (Comma Delimited).
12. Verify that Insert Survey Points option is set to yes.
13. Update the Point Groups collection in Prospector so that the points show up in the drawing if needed.

    Points 1 and 47 from the newly imported network correspond to 10000 and 10001 from the actual coordinate system. This is all the information needed to perform a translation.
14. Make sure you are able to locate these points for the next steps by panning and zooming with your mouse.

    It will also be helpful to turn on your node object snap.
15. Right-click on the name of the database and select Translate Survey Database, as shown earlier in Figure 2.29.

16. For the Base Point, enter **1** and press ↵.

    Civil 3D will pick up the assumed coordinate of this point.

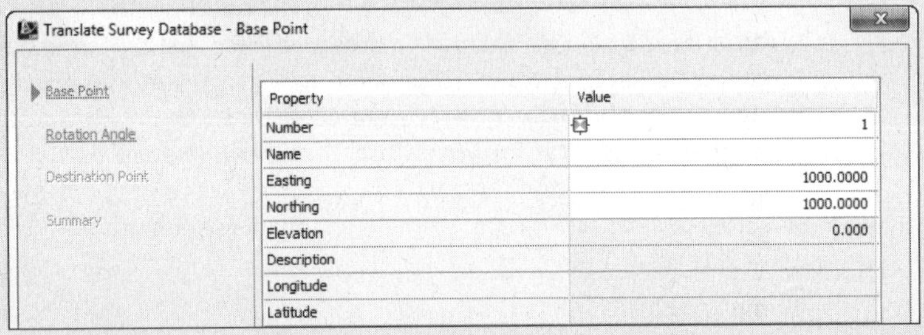

17. Click Next.
18. For the Rotation Angle, click the Pick In Drawing button.

    You will be prompted for an Initial Direction.

19. For Initial Direction, use the node Osnap to pick point 1 and then point 47.

    You will then be prompted to specify a new direction.

20. Use the node Osnap to pick points 10000 and 10001.

    The resulting rotation angle should be 45°.

21. Click Next.
22. In the Destination Point screen, click Pick In Drawing.
23. Use the node Osnap to select point 10000.
24. Click Next.
25. Verify your translation setup on the Summary screen and then click Finish.

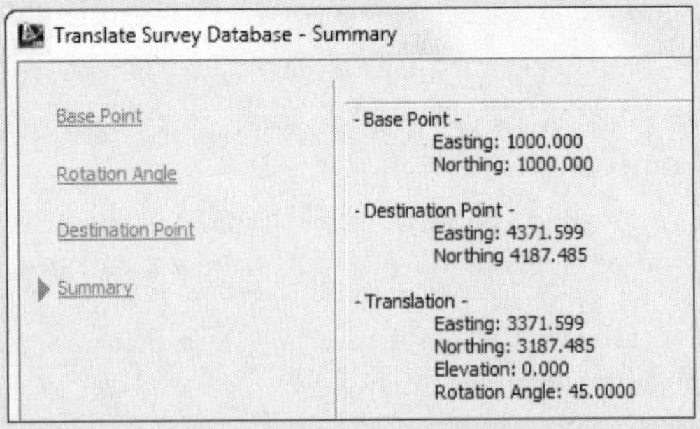

26. Save and close the drawing.

The changes were made directly to the database and the network can be imported into any drawing.

USING A DIRTY WORD, "EXPLODE"

Most of the time in Civil 3D, exploding objects is a huge no-no. Exploding an object removes the underlying Civil 3D intelligence and reduces objects to base AutoCAD components.

In this author's opinion, survey figures are the exception to the no-exploding rule. An exploded survey figure will revert to a 3D polyline, and will become divorced from the survey database that created it. Creating an independent polyline can be useful if your survey figures are going to be the basis for grading feature lines at a later time.

Keep in mind that survey figures have several advantages over 3D polylines. Figures can contain 3D arcs, but 3D polylines cannot. If you explode a figure, the arcs will revert to its corresponding chord. Additionally, figures can be manipulated using the handy Civil 3D elevation and geometry tools.

Other Survey Features

Other components of the survey functionality included with Civil 3D 2013 are the Astronomic Direction Calculator, the Geodetic Calculator, Mapcheck reports, and Coordinate Geometry Editor. All of these features are accessed from Analyze ➤ Ground Data ➤ Survey.

The Astronomic Direction Calculator, shown in Figure 2.30, is used to calculate sun shots or star shots. For the obscure art of these type of observations, Civil 3D has all the ephemeris data built in, making this a very convenient tool.

FIGURE 2.30 The Astronomic Direction Calculator

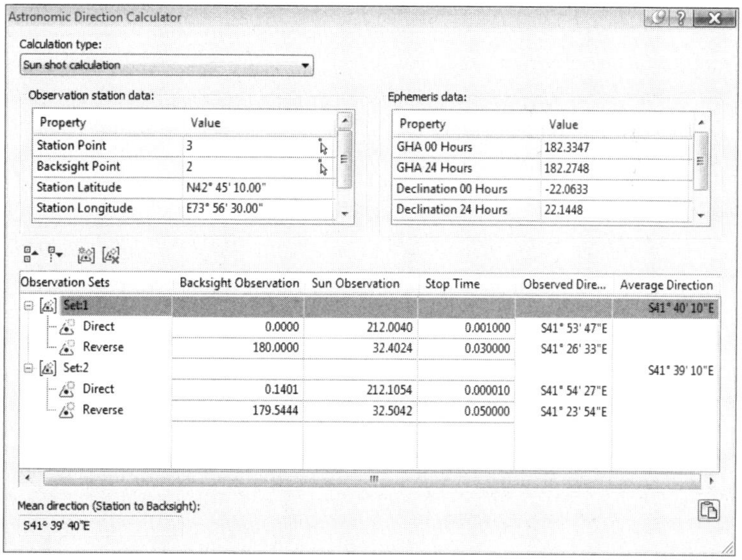

The Geodetic Calculator is used to calculate and display the latitude and longitude of a selected point, as well as its local and grid coordinates. It can also be used to calculate unknown points. If you know the grid coordinates, the local coordinates, or the latitude and longitude of a point, you can enter it in the Geodetic Calculator and create a point at that location. Note that the Geodetic Calculator only works if a coordinate system is assigned to the drawing in the Drawing Settings dialog. In addition, any transformation settings specified in this dialog will be reflected in the Geodetic Calculator, shown in Figure 2.31.

FIGURE 2.31
The Geodetic Calculator

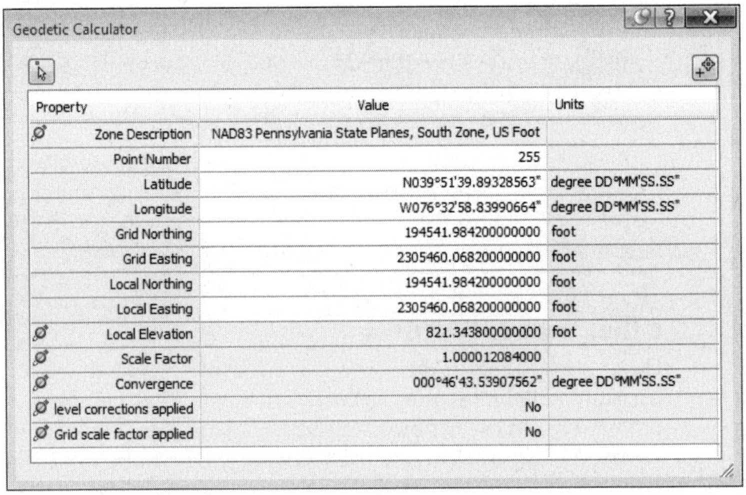

Remember those labels you added to your lines in Chapter 1, "The Basics"? Put them to work in a Mapcheck report. The Mapcheck report computes closure based on line, curve, or parcel segment labels.

1. Open the `Mapcheck-start.dwg` (`Mapcheck-start_METRIC.dwg`) file, which you can download from this book's web page.
2. Change to the Analyze tab, and select Survey ➤ Mapcheck from the Ground Data panel to display the Mapcheck Analysis palette.
3. Click the New Mapcheck button at the top of the menu bar.
4. At the `Enter name of mapcheck:` prompt, type **Record Deed**.
5. At the `Specify point of beginning (POB):` prompt, choose the north endpoint of the line representing the east line of the parcel (the longest line in the file).

A red glyph will appear to represent the POB.

6. Working clockwise, select each parcel label one at a time.

   Be sure not to skip the small segment in the southwest portion of the site.

   In the northwest portion of the site, you will encounter a label whose bearing is flipped. You will use the Mapcheck Reverse command to fix this in your Mapcheck.

7. At the `Select a label or [Clear/Flip/New/Reverse]:` prompt, type **R** and then press ↵ to reverse direction.

The Mapcheck glyph will now appear in the correct location along the segment, as shown in Figure 2.32.

FIGURE 2.32
The Mapcheck glyph verifies your input is correct.

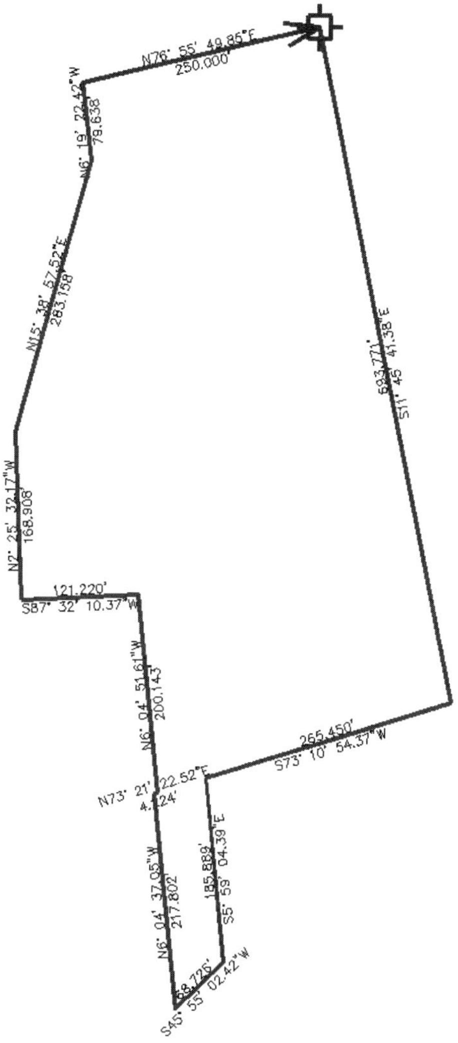

8. Select the last line label along the north of the parcel.
9. Press ↵ to complete the Mapcheck entry.

   The completed parcel should have 12 sides.
10. Select the output view as shown in Figure 2.33 to verify closure.

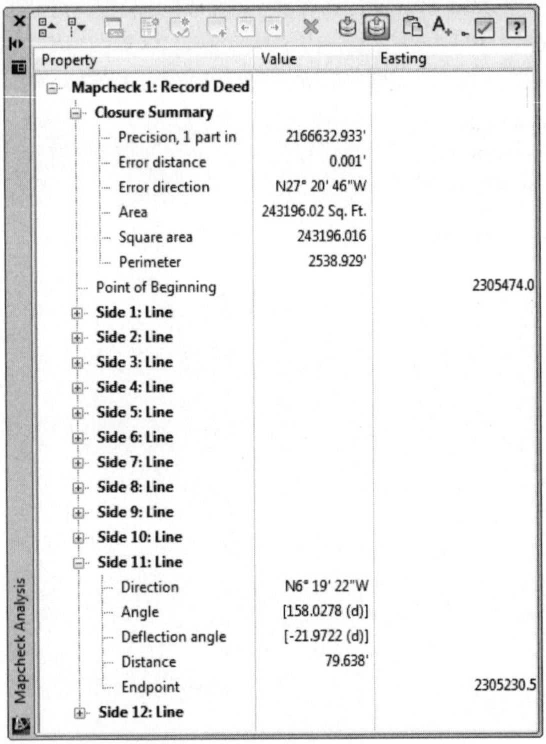

FIGURE 2.33
The completed deed in the Mapcheck Analysis palette

The Coordinate Geometry Editor

The Coordinate Geometry Editor (Figure 2.34) is a powerhouse tool that makes creating and evaluating 2D boundaries easier than before. The functionality introduced with this feature supplants entering parcel data one segment at a time using the Line By Bearing And Distance command. Traverse analysis can be performed on manually entered segments, polylines, or COGO point objects without needing to define them in a survey database.

Boundary data can be entered in the Coordinate Geometry Editor using a mix of methods. As shown in the first line of the traverse seen in Figure 2.34, you can use formulas to enter data. In the example shown, multiple segments with the same bearing have been consolidated into a single entry.

If a value is unknown, such as the distance in line 6 of Figure 2.34, you can enter a **U**. Civil 3D will calculate the unknown value when you generate the traverse report.

To enter data using points, use the Pick COGO Points In Drawing button to select the points in the direction of the traverse side. Note that the direction and distance are entered independently of each other, so you will need to repeat the selection for each column of the table. You can also copy and paste between columns.

If a line has been entered in error, the Coordinate Geometry Editor offers a variety of tools for fixing problems. To remove a line of the table, highlight the row, right-click, and select Delete Row, as shown in Figure 2.35.

FIGURE 2.34 Your new best friend, the Coordinate Geometry Editor

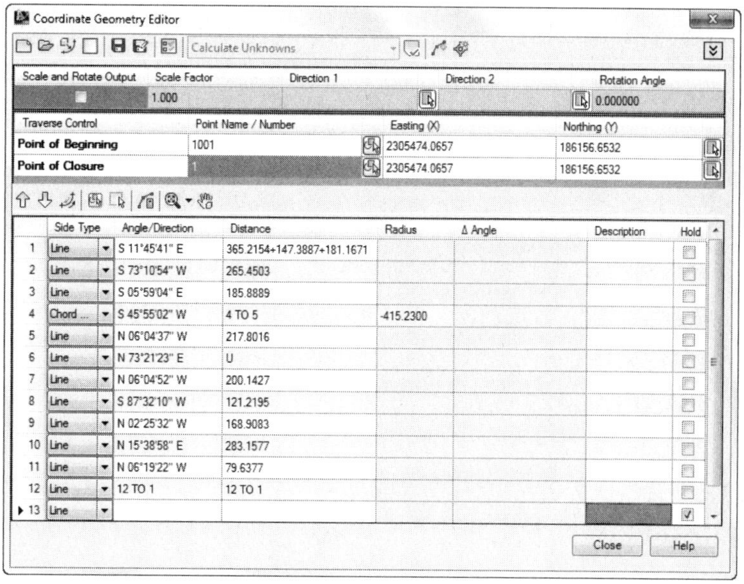

FIGURE 2.35 Removing unwanted traverse data

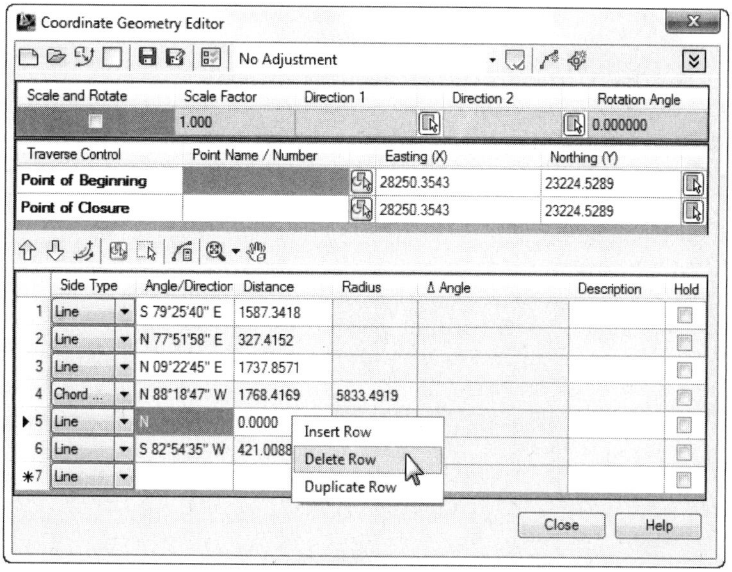

Similar to the glyphs you saw in the Mapcheck command, the Coordinate Geometry glyph will appear in the graphic showing the side directions and point of closure, as seen in Figure 2.36.

FIGURE 2.36
Temporary graphics or "glyphs" to help you identify your boundary

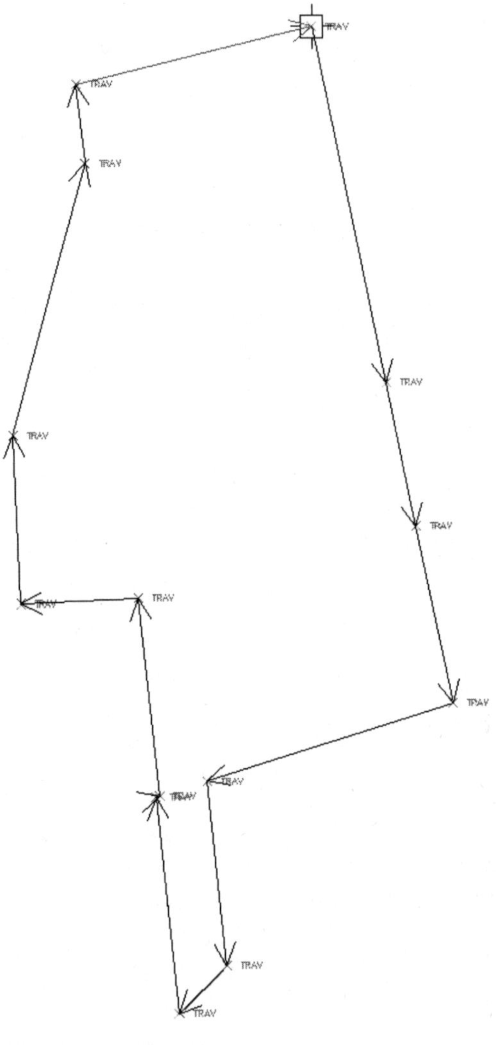

When you want to run a traverse report, set the report type you wish to run from the top of the Coordinate Geometry Editor. If you have unknowns in your traverse, your only option will be to calculate the unknown values. Click the Display Report button to view the results of your entries. Depending on the type of adjustment you chose, your results should resemble Figure 2.37.

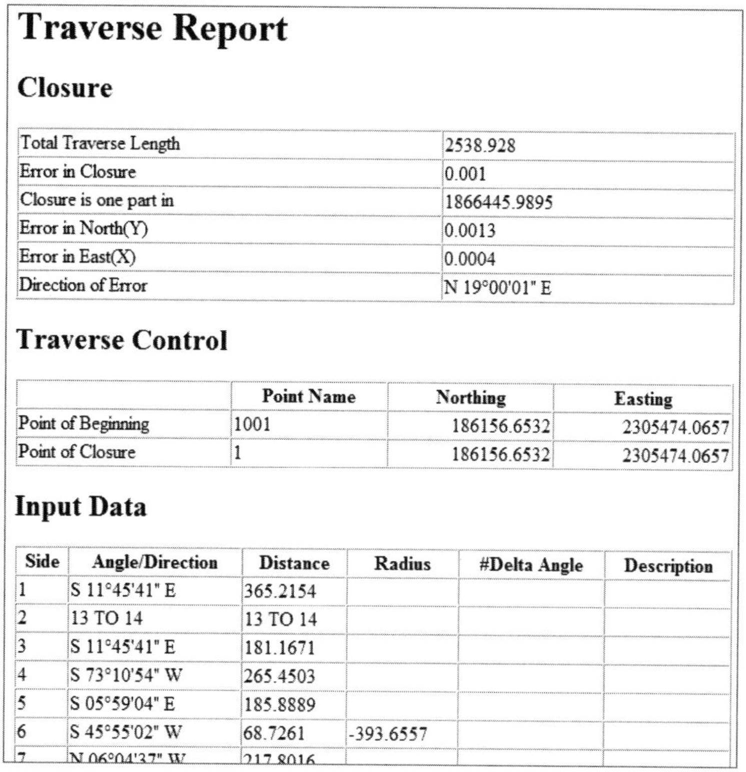

Traverse Report

Closure

Total Traverse Length	2538.928
Error in Closure	0.001
Closure is one part in	1866445.9895
Error in North(Y)	0.0013
Error in East(X)	0.0004
Direction of Error	N 19°00'01" E

Traverse Control

	Point Name	Northing	Easting
Point of Beginning	1001	186156.6532	2305474.0657
Point of Closure	1	186156.6532	2305474.0657

Input Data

Side	Angle/Direction	Distance	Radius	#Delta Angle	Description
1	S 11°45'41" E	365.2154			
2	13 TO 14	13 TO 14			
3	S 11°45'41" E	181.1671			
4	S 73°10'54" W	265.4503			
5	S 05°59'04" E	185.8889			
6	S 45°55'02" W	68.7261	-393.6557		
7	N 06°04'37" W	217.8016			

FIGURE 2.37 Traverse report created by the Coordinate Geometry Editor

Using Inquiry Commands

A large part of a surveyor's work involves querying lines and curves for their length, direction, and other parameters.

The Inquiry commands panel (Figure 2.38) is on the Analyze tab, and it makes a valuable addition to your Civil 3D and survey-related workspaces. Remember, panels can be dragged away from the Ribbon and set in the graphics environment much like a toolbar.

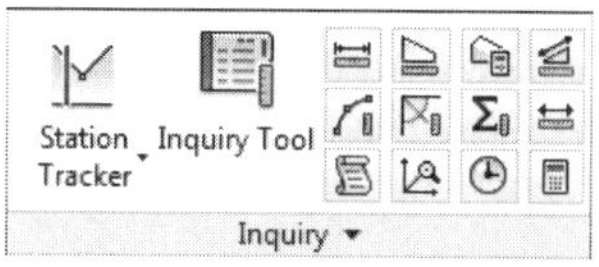

FIGURE 2.38 The Inquiry commands panel

The Inquiry Tool (shown in Figure 2.39) provides a diverse collection of commands that assist you in studying Civil 3D objects. You can access the Inquiry Tool by clicking the Inquiry Tool button on the Inquiry panel.

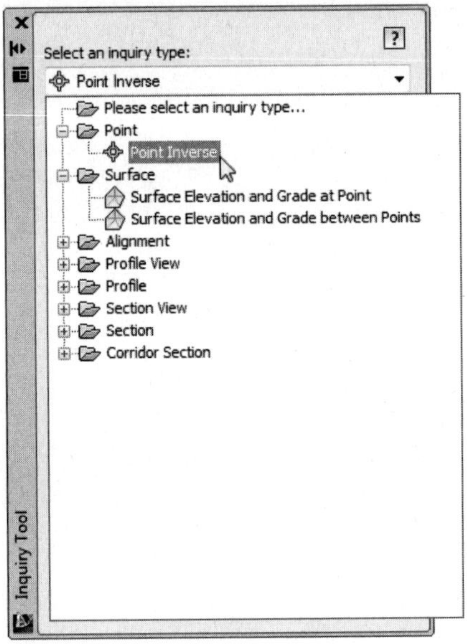

FIGURE 2.39
Choosing an Inquiry type from the Inquiry Tool palette

To use the Point Inverse option in the Inquiry commands, select the Point Inverse option as shown in Figure 2.40.

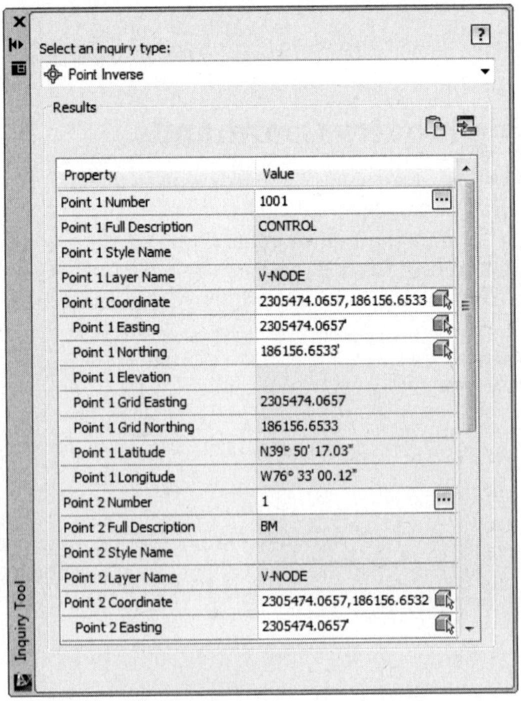

FIGURE 2.40
Point Inverse results

You can enter the point number or use the Pick In CAD icon to select the points you wish to examine.

The other Inquiry commands that are specific to Civil 3D are also handy to the survey process.

The List Slope tool provides a short command-line report that lists the elevations and slope of an entity (or two points) that you choose, such as a line or feature line.

The Line And Arc Information tool provides a short report about the line or arc of your choosing (see Figure 2.41). This tool also works on parcel segments and alignment segments. Alternatively, you can type **P** for points at the command line to get information about the apparent line that would connect two points on screen.

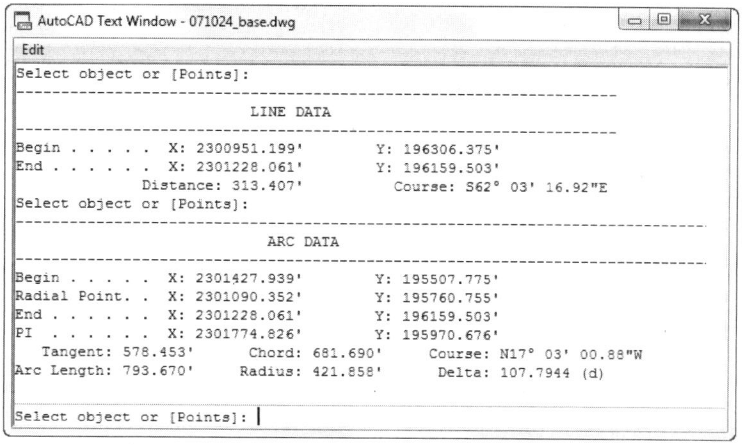

FIGURE 2.41
Command-line results of a line inquiry and arc inquiry

DON'T GET BURNED BY AUTOCAD ANGLES

Do not set base AutoCAD angular units to Surveyors Units. This setting may seem perfectly logical if you have been using base AutoCAD prior to using Civil 3D. However, this angular setting will end up confusing you more than helping you.

When looking for information about a line, use the Line and Arc tool mentioned in this section rather than the base AutoCAD `LIST` command. The Civil 3D Line and Arc tool (`CGLIST`) works on more object types and is not affected by rotated coordinate systems.

When entering angles in base AutoCAD, the full `N50d10'10"E` is needed to denote a bearing. In Civil 3D commands, `N50.1010E` will denote the same thing and is much faster to type. However, as every surveyor knows, there is a huge difference between 50.1010° and 50°10'10".

A setting you will want to change in your Civil 3D template is the angular entry method for general angles, as shown here. This setting mainly affects the Angle-Distance and Bearing Distance Transparent commands. Check back in Chapter 1 for more information on template settings.

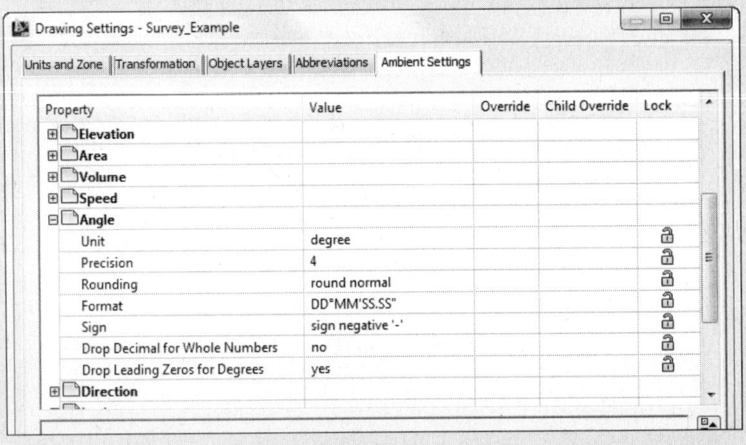

Keep your base AutoCAD angular units set to Decimal Degrees to help you differentiate when you are in base AutoCAD angular entry and the more surveyor-friendly angular entry in Civil 3D.

The Angle Information tool lets you pick two lines (or a series of points on the screen). It provides information about the acute and obtuse angles between those two lines. Again, this also works for alignment segments and parcel segments.

The Continuous Distance tool provides a sum of distances between several points on your screen, or one base point and several points.

The Add Distances tool is similar to the Continuous Distance command, except the points on your screen do not have to be continuous.

The Bottom Line

Properly collect field data and import it into Civil 3D. Once survey data has been collected, you will want to pull it into Civil 3D via the Survey Database. This will enable you to create lines and points that correctly reflect your field measurements.

Master It Create a new drawing based off the template of your choice and a new survey database and import the `MASTER_IT_C2.txt` (or `master_IT_C2_METRIC.txt`) file into the drawing. The format of this file is PNEZD (Comma Delimited).

Set up description key and figure databases. Proper setup is key to working successfully with the Civil 3D survey functionality.

Master It Create a new description key set and the following description keys using the default styles. Make sure all description keys are going to layer V-Node:

- CL*
- EOP*

- TREE*
- BM*

Change the description key search order so that the new description key set takes precedence over the default.

Create a figure prefix database called MasterIt containing the following codes:

- CL
- EOP
- BC

Translate surveys from assumed coordinates to known coordinates. Understanding how to manipulate data once it is brought into Civil 3D is important to making your field measurements match your project's coordinate system.

Master It Create a new drawing and survey database. Start a new drawing based off the template of your choice. Import `traverse.fbk` (or `traverse_METRIC.fbk`). Translate the database based on:

- Base Point 1
- Rotation Angle of 10.3053°

Perform traverse analysis. Traverse analysis is needed for boundary surveys to check for angular accuracy and closure. Civil 3D will generate the reports that you need to capture these results.

Master It Use the survey database and network from the previous Master It. Analyze and adjust the traverse using the following criteria:

- Use an Initial Station of value 2 and an Initial Backsight value of 1.
- Use the Compass Rule for Horizontal Adjustment.
- Use the Length Weighted Distribution Method for Vertical Adjustment.
- Use a Horizontal Closure Limit value of 1:25,000.
- Use a Vertical Closure Limit value of 1:25,000.

Chapter 3

Points

In the previous chapter, "Survey," you looked at a specific method for bringing in points and figures. In this chapter, you will take a closer look at creating and organizing points.

The foundation of any civil engineering project is the simple point, frequently referred to as a shot. Most commonly, points are used to identify the location of existing features, such as trees and property corners; topography, such as ground shots; or stakeout information, such as road geometry points. However, points can be used for much more. This chapter will both focus on traditional point uses and introduce ideas to apply the dynamic power of point editing, labeling, and grouping to other applications.

In this chapter, you will learn to:

- Import points from a text file using description key matching
- Create a point group
- Export points to LandXML and ASCII format
- Create a point table

Anatomy of a Point

AutoCAD® Civil 3D® *points* (see Figure 3.1) are intelligent objects that represent *x*, *y*, and *z* locations in space. Each point has a unique number and, optionally, a unique name that can be used for additional identification and labeling.

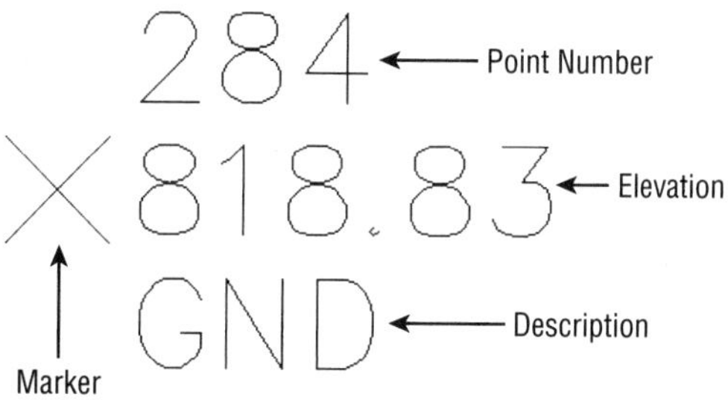

FIGURE 3.1 A typical point object showing a marker, a point number, an elevation, and a description

> **A QUICK WORD ON STYLES**
>
> Separating the point functionality discussed in this chapter from the styles that make them look the way they do is difficult. Chapter 20, "Label Styles," and Chapter 21, "Object Styles," will go into the nitty-gritty of creating and manipulating label styles and point styles. In this chapter, you will work with styles that are already part of a drawing. This is true for points, labels, and tables.

COGO Points vs. Survey Points

In Chapter 2, you imported survey data that contained points. Points brought in through the methods described in Chapter 2 are referred to as *survey points*. In this chapter, you will import points from a delimited text file and place them in CAD using the point creation tools. Points created in this manner are referred to as *COGO points*. Figure 3.2 shows the context tab differences between points brought in as COGO points (top) and points brought in through a database (bottom).

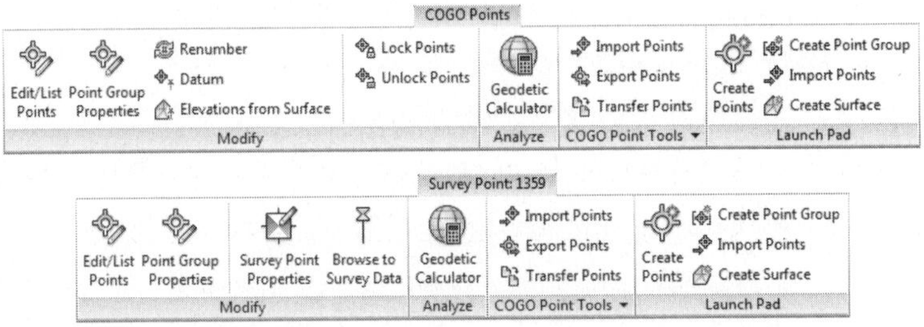

FIGURE 3.2 The context-sensitive Ribbon reflects similarities and differences between COGO points (top) and survey points (bottom).

The differences between COGO points and survey points are subtle but important to note. A COGO point is unlocked by default—meaning it can readily be edited. A survey point, on the other hand, must be unlocked if a user wishes to edit the point. A survey point stays tied to the database from which it came, whereas a COGO point maintains no tie to the originating text file or object it was created against. Regardless of their origin, both COGO points and survey points obey the principles outlined in this chapter.

Creating Basic Points

CERT OBJECTIVE

You can create points many ways, using the Points menu in the Create Ground Data panel on the Home tab. Points can also be imported from text files or external databases or converted from AutoCAD, Land Desktop, or SoftDesk point objects.

Point Settings

Point settings are our first glimpse at what is known as *command settings*. All Civil 3D objects have command settings tucked away in the Settings tab of Toolspace. In the case of points, it is handy to have these settings readily available for on-the-fly modifications. Whether or not the changes you make on the fly are remembered the next time you create points depends on your template settings.

To make sure that the settings you change will hold every time, you create points:

1. On the Settings tab of Toolspace, locate the Point group and click the plus sign to expand the branch.
2. Click the plus sign to expand Commands.
3. As shown in Figure 3.3, right-click the command CreatePoints and select Edit Command Settings.

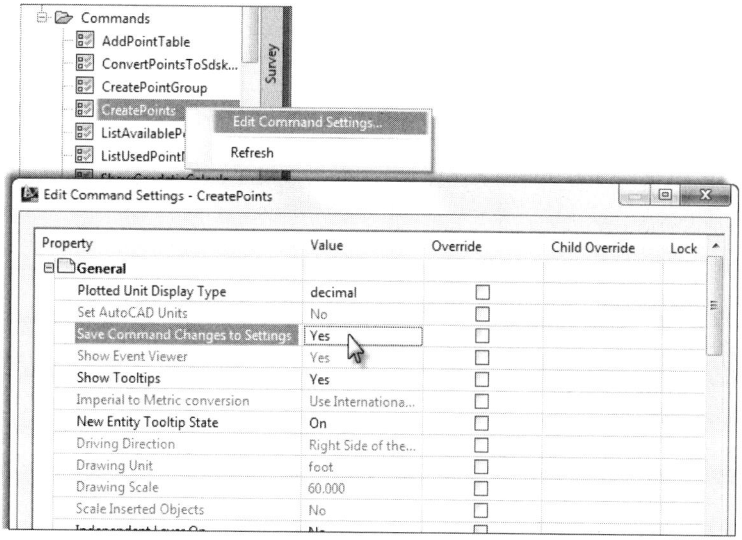

FIGURE 3.3
In your Civil 3D template, make sure Save Command Changes To Settings is set to Yes for Points.

4. Expand the General section and verify that Save Command Changes To Settings is set to Yes.

If you explore the command settings further, you will see the options for Default Layer and Points Creation. By setting this to Yes, you are ensuring that changes you make at the tool level will also be reflected at the command level.

To access the Create Points toolbar:

1. Go to the Home tab.
2. In the Create Ground Data panel, select Points ➢ Point Creation Tools.
3. Expand the toolbar by clicking the chevron button on the far-right side.

DEFAULT LAYER

For most Civil 3D objects, the object layer is established in the drawing settings. In the case of points, the default object layer is set in the command settings for point creation and can be changed in the Create Points dialog (see Figure 3.4).

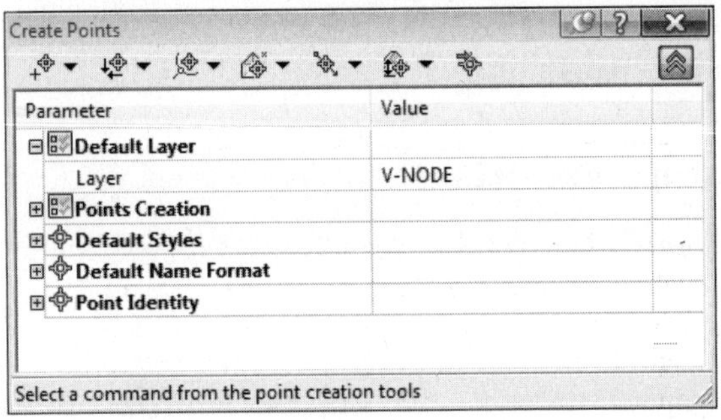

FIGURE 3.4
Verify the point object layer before creating points.

PROMPT FOR ELEVATIONS, NAMES, AND DESCRIPTIONS

When creating points in your drawing, you have the option of being prompted for elevations, names, and descriptions (see Figure 3.5). Initially, the Default description is blank. In many cases, you'll want to leave these options set to Manual. The command line will ask you to assign an elevation and description for every point you create. It is best to avoid using the Point Name option, as described later in this section.

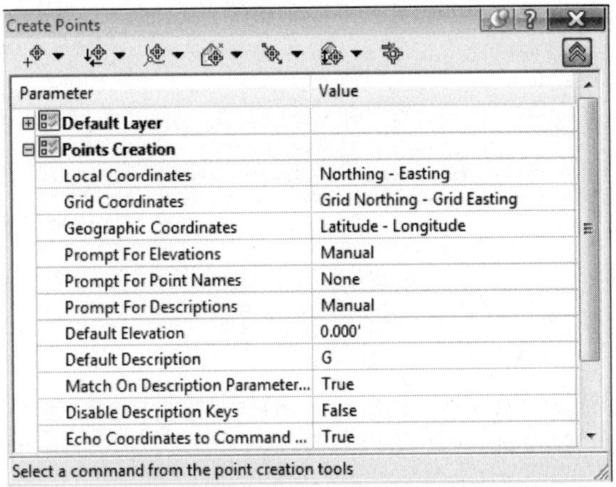

FIGURE 3.5
You can change the elevation, point name, and description settings from Manual to Automatic. You can also use the None option to omit this information.

If you're creating a batch of points that have the same description or elevation, you can change the Prompt toggle from Manual to Automatic and then provide the description and elevation in the default cells. For example, if you're setting a series of trees at an elevation of 10′, you can establish settings as shown in Figure 3.6.

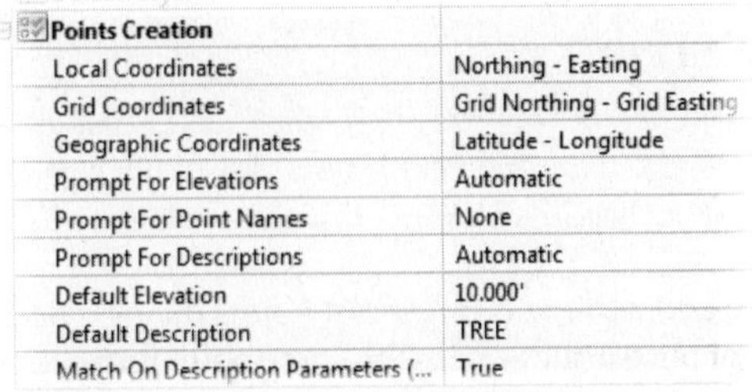

FIGURE 3.6
Default settings for placing tree points at an elevation of 10′

Note that these settings only apply to points created from this toolbar. The settings do not affect the elevation or description of points imported from a file.

WHAT'S IN A NAME?

Users often confuse a point's *name* with its *description*. The name is a unique, alphanumeric sequence that can be used in lieu of a point number. The description refers to the all-important code given to a point out in the field. Most survey data collectors do not use alphanumeric point identification, so this option is usually set to None in Civil 3D. If you do have a point file that uses a name instead of a point number, you will need to create a custom point format, as described later in this chapter.

Importing Points from a Text File

CERT OBJECTIVE

One of the most common means of creating points in your drawing is to import an external text file (see Figure 3.7).

FIGURE 3.7
The Import Points and the Point File Formats dialogs

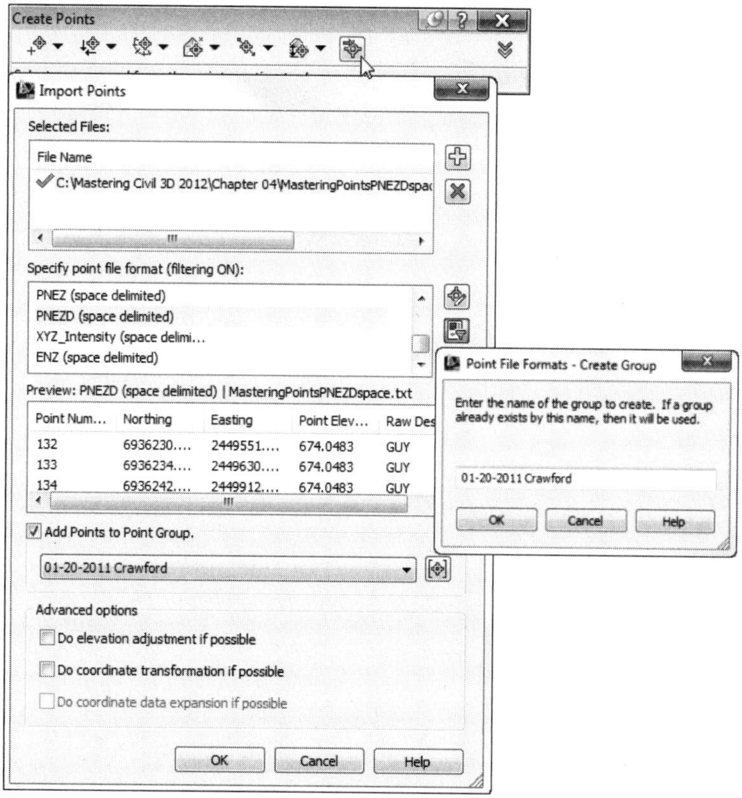

To add a file to your Import Points dialog, click the plus sign to browse. You can add multiple files at once if they are in the same point format—such as PNEZD (Comma Delimited). The import process supports most text formats as well as Microsoft Access database (MDB) files. Later in this chapter you will experience adding your own text format.

When your file is listed in the top of the dialog box, a green check mark will indicate that Civil 3D can parse the information. Be careful, though, because Civil 3D does not know the difference between a Northing and an Easting or a point number and an elevation. You still need to select the correct file format.

The file format filter is there to help you. Civil 3D recognizes how the file is delimited (i.e., tab, comma, space) and only shows you the formats that apply. If you don't want the help, you can turn the filtering off by clicking the Filter icon.

If the file format you need is not available, or you wish to use the adjustment and transformation capabilities, you can do so by clicking the Manage Formats button.

You can make an elevation adjustment if the point file contains additional columns for thickness, Z+, or Z-. You can add these columns as part of a custom format. See the *AutoCAD Civil 3D 2013 User's Guide* section "Using Point File Format Properties to Perform Calculations" for more details.

You can perform a coordinate system transformation if a coordinate system has been assigned both to your drawing (under the drawing settings) and as part of a custom point format. In this case, the program can also do a coordinate data expansion, which calculates the latitude and longitude for each point.

A common use of the formats is to create a point file format for importing a name-based point file. In the following example, you will create a new file format to accommodate names (instead of point numbers):

1. In any drawing, select the Settings tab of Toolspace and choose Point ➢ Point File Formats.

   You can also access this functionality on the fly from the Import Points dialog.

2. Right-click Point File Formats and select New.
3. Select User Point File and click OK.
4. Name the format **Name-NEZD**.
5. Toggle on the Delimited By option and place a comma in the field.
6. Click the first <unused> column heading, select Name from the Column Name pull-down, and click OK.
7. Click the next <unused> column and select Northing from the Column Name pull-down.
8. Leave the Invalid Indicator and Precision fields at their defaults and click OK.
9. Repeat the process for Easting, Point Elevation, and Raw Description.
10. To test the format, click Load, select the file `Test Format.txt` (which you can download from this book's web page at `www.sybex.com/go/masteringcivil3d2013`), and click Open.
11. Click Parse.

    If the format has been created successfully, you will see the file preview, as shown in Figure 3.8.

12. Click OK to complete the format.

The new file format will remain in case you need it. To compare your work with a completed example, see `Format Example_FINISHED.dwg` at the book's web page.

FIGURE 3.8
A completed and tested new point file format

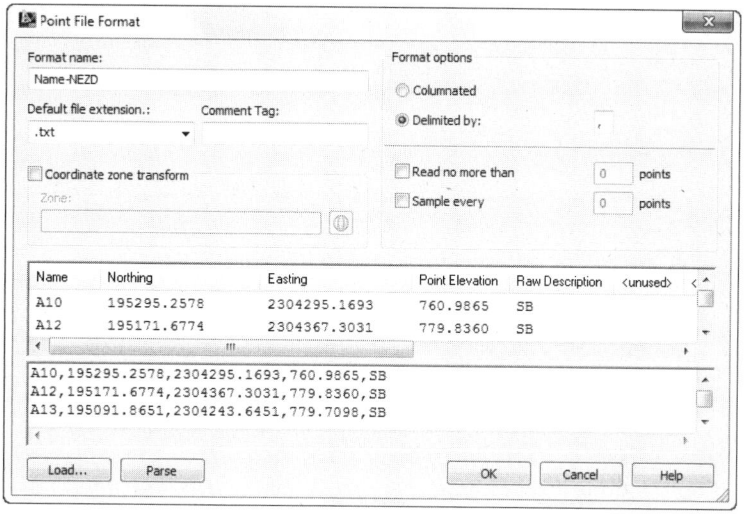

IMPORTING A TEXT FILE OF POINTS

In this exercise, you'll learn how to import a TXT file of points into Civil 3D:

1. Open the `Mastering Points.dwg` (`Mastering Points_METRIC.dwg`) file, which you can download from this book's web page.
2. On the Home tab ➢ Create Ground Data panel ➢ Points, select Point Creation Tools.
3. Click the plus (+) button to the right of the Selected Files field, and navigate out to locate the `Mastering_C3D_Points.txt` (`Mastering_C3D_Points_METRIC.txt`) file.
4. In the Specify Point File Format field, set the format to PNEZD (Comma Delimited).
5. Place a check mark in the box next to Add Points To Point Group and click the Create Point Group icon. Name the point group **Survey 01-20-2012** and click OK.
6. Leave all the other boxes unchecked.
7. Click OK.

You may have to use Zoom Extents to see the imported points. (Hint: Double-click your middle mouse wheel for zooming extents.)

LANDXML AND POINTS: A MATCH MADE IN HEAVEN

LandXML is a file format specifically made to share the type of data Civil 3D and its counterparts create. For points, it works particularly well, since it will also carry point group information. Import LandXML files from the Import panel on the Insert tab. You'll learn much more about LandXML in Chapter 18, "Advanced Workflows."

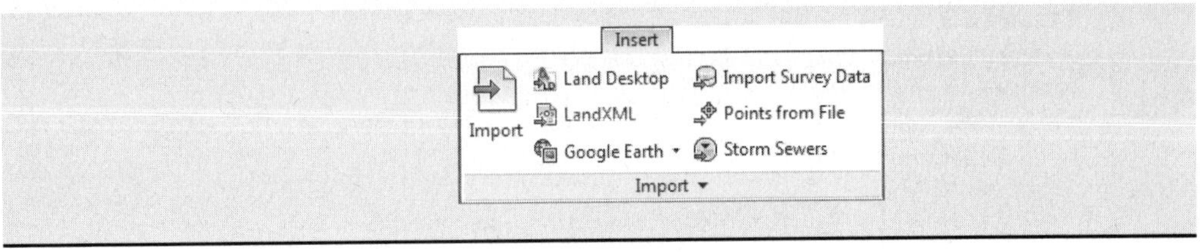

Converting Points from Non–Civil 3D Sources

Civil 3D contains several tools for migrating legacy point objects to the current version. The best results are often obtained from an external point list, such as a text file, LandXML, or an external database. However, if you come across a drawing that contains the original Land Desktop, SoftDesk, AutoCAD, or other types of point objects, tools, and techniques are available to convert those objects into Civil 3D points.

A Land Desktop point database (the `Points.mdb` file found in the `COGO` folder in a Land Desktop project) can be directly imported into Civil 3D in the same interface in which you'd import a text file.

Land Desktop point objects, which appear as AECC_POINTs in the AutoCAD Properties palette, can also be converted to Civil 3D points (see Figure 3.9). Upon conversion, this tool gives you the opportunity to assign styles, create a point group, and more.

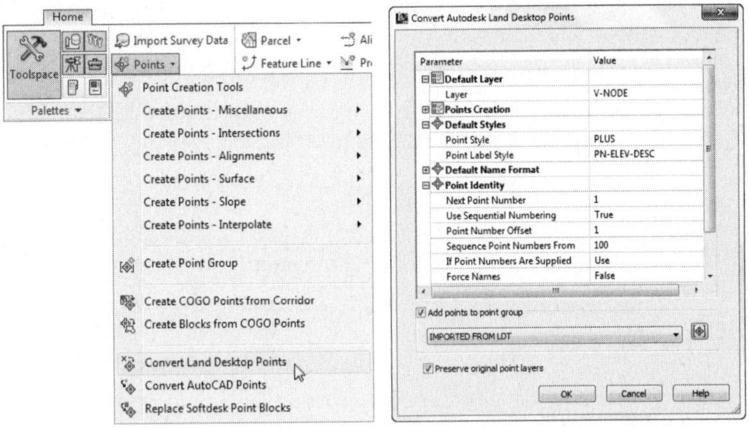

FIGURE 3.9
The Convert Land Desktop Points option (left) opens the Convert Autodesk Land Desktop Points dialog.

Occasionally, you'll receive AutoCAD point objects drawn at elevation from aerial topography information or other sources. It's also not uncommon to receive SoftDesk point blocks from other surveyors. Both of these can be converted to or replaced by Civil 3D points by going to the Home tab ➢ Create Ground Data panel ➢ Points and picking one of the many point conversion options there (see Figure 3.9).

Real World Scenario

USING AUTOCAD ATTRIBUTE EXTRACTION TO CONVERT OUTSIDE PROGRAM POINT BLOCKS

Occasionally, you may receive a drawing that contains point blocks from a third-party program such as Eagle Point or Carlson. These point blocks may look similar to SoftDesk point blocks, but the block attributes may have been rearranged and you can't convert them directly to Civil 3D points using Civil 3D tools.

Whenever possible, your best course of action would be to request the source survey in text format or LandXML. If that is not possible, the exercise that follows will pull the data you need without losing the information from the attributes.

1. Open the file `Mystery Plat.dwg` (`Mystery Plat_METRIC.dwg`).
2. Examine the blocks you are given to determine the names of the blocks with which you are working.

   A glance at AutoCAD properties will give you an idea of your block name and what attributes you are extracting. In this example, you are seeing Carlson survey blocks, which use SRVPNO1 as an anchor for their attributes.

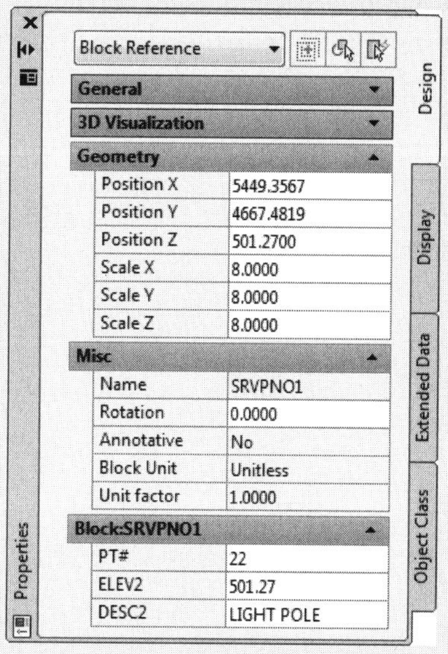

3. Use the Data Extraction tool by typing **EATTEXT↵** in the command line to launch the Data Extraction Wizard.
4. Select the radio button to create a new data extraction and click Next.

   The Save Data Extraction As dialog appears, prompting you to name and save this extraction.
5. Give the extraction a meaningful name, and save it in the appropriate folder. Click Next.
6. Confirm that the drawings to be scanned for attributed blocks are on the list and click Next.

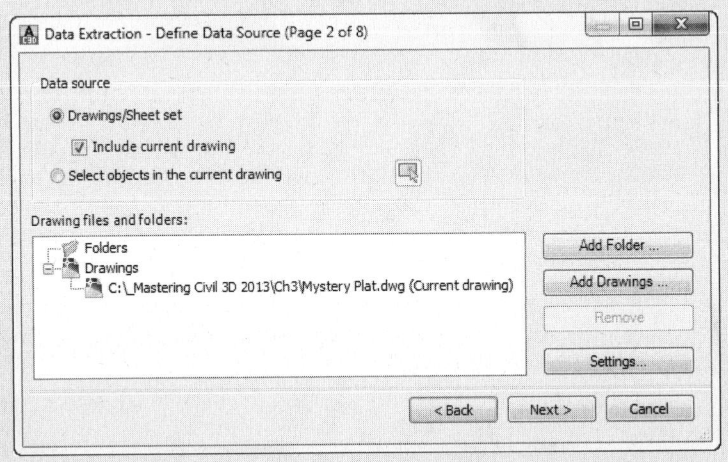

7. In the Select Objects screen of the Data Extraction Wizard, uncheck Display All Object Types. Check Display Blocks With Attributes Only and Display Objects Currently In-Use Only.

   Doing so will filter out most unneeded blocks.

8. Eliminate the other types of attributed blocks by deselecting their boxes until only SRVPNO1 is selected. Click Next.

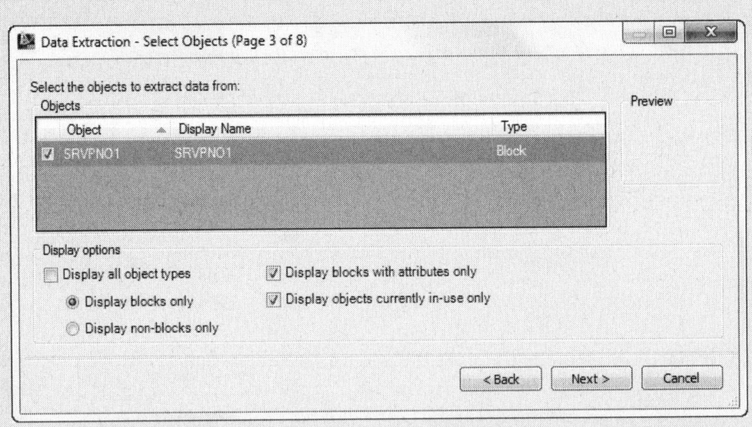

9. On the next screen, uncheck all category filters except Attribute and Geometry and then select these properties:
   - ◆ Point number (PT#)
   - ◆ Elevation (ELEV2)
   - ◆ Description (DESC2)
   - ◆ Position X
   - ◆ Position Y

   In this case, you are relying on the attribute ELEV2 instead of Position Z; you do not need both.

10. Click Next. Click OK when alerted that there were nonuniformly scaled blocks found.

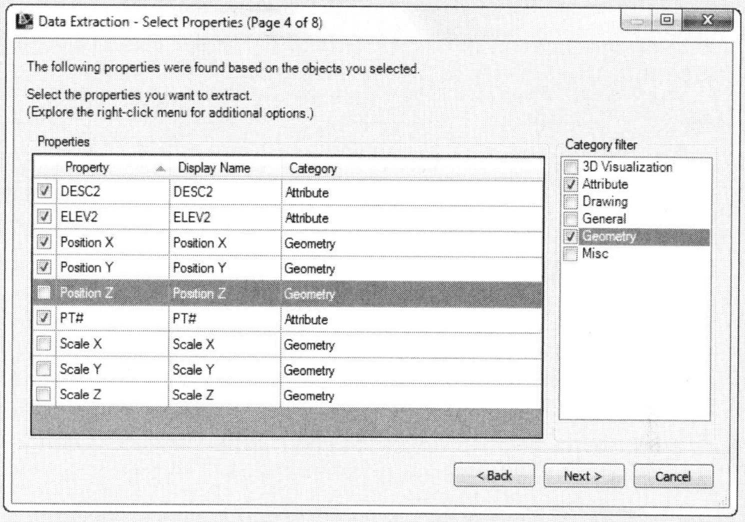

11. On the Refine Data screen, rearrange the columns into a PNEZD format by clicking and dragging the column headers into place. Uncheck Show Count Column and Show Name Column. Click Next.

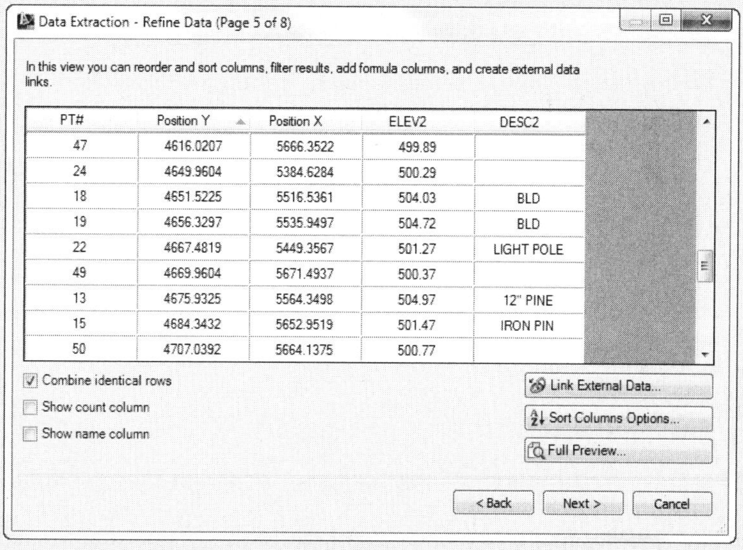

12. On the Choose Output screen, set Output Data to External File, and save your extraction as a CSV file in a logical place.

13. Click Next, and then click Finish.

14. From your Windows Start menu, launch Notepad (or Wordpad).

15. Choose File ➤ Open and locate the new CSV file in Windows.

16. Remove the first line of text (the header information). Save and close the file.

17. In a drawing that contains your Civil 3D styles, use the Import Points tool in the Create Points dialog to import the CSV file.

Converting Points

In this exercise, you'll convert Land Desktop point objects and AutoCAD point entities into Civil 3D points:

1. Open the `Convert LDT Points.dwg` (`Convert LDT Points_METRIC.dwg`) file, which you can download from this book's web page.
2. Use the List command or the AutoCAD Properties palette to confirm that most of the objects in this drawing are AECC_POINTs, which are points from Land Desktop.

   Also, note a cluster of cyan-colored AutoCAD point objects in the western portion of the site.
3. On the Home tab ➤ Create Ground Data panel ➤ Points menu, select Convert Land Desktop Points.

   Note that the Convert Autodesk Land Desktop Points dialog allows you to choose a default layer, point creation settings, and styles.
4. Place a check mark next to Add Points To Point Group.
5. Click the Create A New Point Group button.
6. Name the group **Converted from LDT** and click OK.
7. Clear the Preserve Original Point Layers check box.

   This option will move the resulting points to the layer specified in the description key set if there is a match. If there is no match, the point will go to the default layer.
8. Click OK to complete the conversion process.

   Civil 3D scans the drawing looking for Land Desktop point objects.
9. Once Civil 3D has finished the conversion, zoom in on any of the former Land Desktop points.

   The points should now show as COGO points in the AutoCAD Properties palette, confirming that the conversion has taken place. The Land Desktop points have been replaced with Civil 3D points, and the original Land Desktop points are no longer in the drawing.
10. In Prospector, expand the Point Groups category.

    Notice there is a yellow exclamation shield symbol indicating that the Converted From LDT point group needs to be updated.
11. Right-click on Point Groups and select Update.
12. Zoom in on the cyan AutoCAD point objects.

13. On the Home tab ➤ Create Ground Data panel ➤ Points menu, select Convert AutoCAD Points.

    The command line reads `Select AutoCAD Points`.

14. Use a crossing window to select all the cyan-colored AutoCAD points, and then press ↵.

15. At the command-line prompt, enter a description of **GS** (for Ground Shot) and press ↵ for each point. Hint: Use the up arrow on your keyboard to recall the last typed entry and avoid unnecessary typing.

16. Zoom in on one of the converted points, and confirm that it has been converted to a Civil 3D point.

    Also, note that the original AutoCAD points have been erased from the drawing.

A Closer Look at the Create Points Toolbar

CERT OBJECTIVE

In Civil 3D 2013, you can find point-creation tools directly under the Points drop-down on the Create Ground Data panel of the Home tab, as well as in the Create Points toolbar. The toolbar is *modeless*, which means it stays on your screen even when you switch between tasks. Figure 3.10 shows the toolbar with the point-creation methods labeled.

FIGURE 3.10
The Create Points toolbar

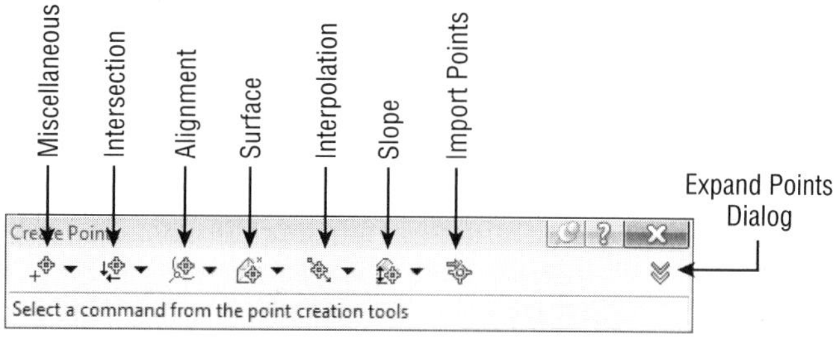

As you place points using these tools, a few general rules apply to all of them. If you place a point on an object with elevation, the point will automatically inherit the elevation of the object. If you use the surface options, the point will automatically inherit the elevation of the surface you choose.

Miscellaneous Point-Creation Options The options in the Miscellaneous category are based on manually selecting a location or on an AutoCAD entity, such as a line, pline, and so on. Some common examples include placing points at intervals along a line or polyline, as well as converting SoftDesk points or AutoCAD points (see Figure 3.11).

Intersection Point-Creation Options The options in the Intersection category allow you to place points at a certain location without having to draw construction linework. For example, if you needed a point at the intersection of two bearings, you could draw two construction lines using the Bearing Distance transparent command, manually place a point where they intersect, and then erase the construction lines. Alternatively, you could use the Direction/Direction tool in the Intersection category (see Figure 3.12).

Alignment Point-Creation Options The options in the Alignment category are designed for creating stakeout points based on a road centerline or other alignments. You can also set Profile Geometry points along the alignment using a tool from this menu. See Figure 3.13.

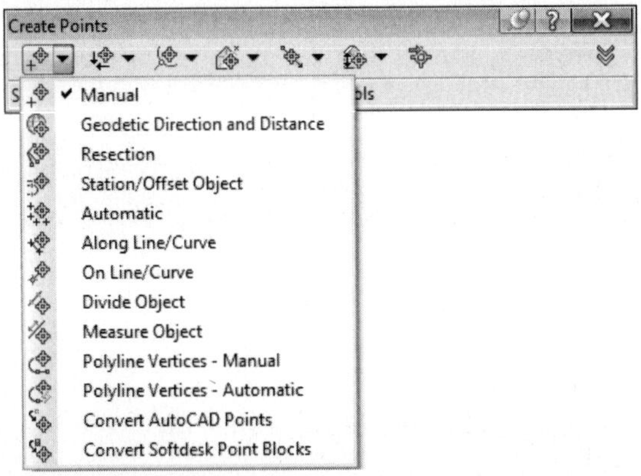

FIGURE 3.11 Miscellaneous point-creation options

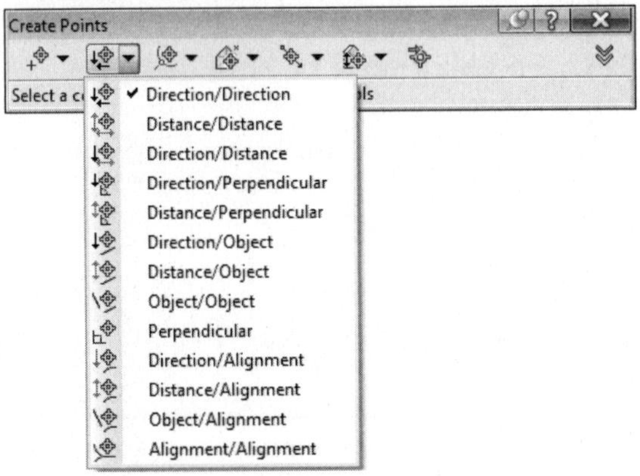

FIGURE 3.12 Intersection point-creation options

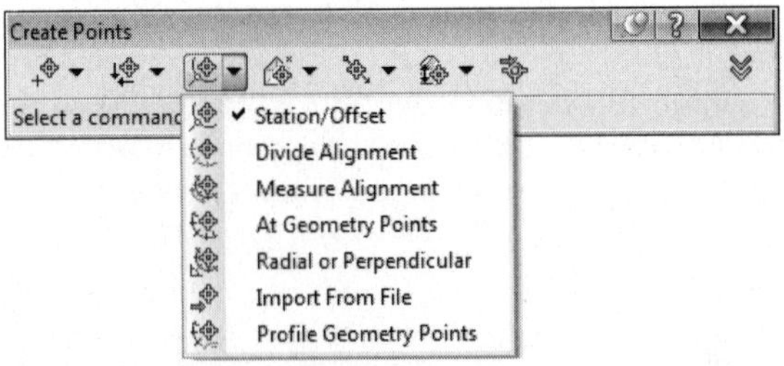

FIGURE 3.13 Alignment point-creation options

AUTOMATIC - OBJECT: UNMASKING THE MYSTERY

The description option in the point settings, Automatic - Object, can only be used when placing points along an alignment. For example, when placing points using the At Geometry Points option, the point will inherit the alignment's name, the station value of the point, and the type of geometry as its description.

For all other point placement options, Automatic - Object will behave exactly the same as Automatic.

Surface Point-Creation Options The options in the Surface category let you set points that harvest their elevation data from a surface. Note that these are points, not labels, and therefore aren't dynamic to the surface. You can set points manually, along a contour or a polyline, or in a grid. See Figure 3.14.

FIGURE 3.14
Surface point-creation options

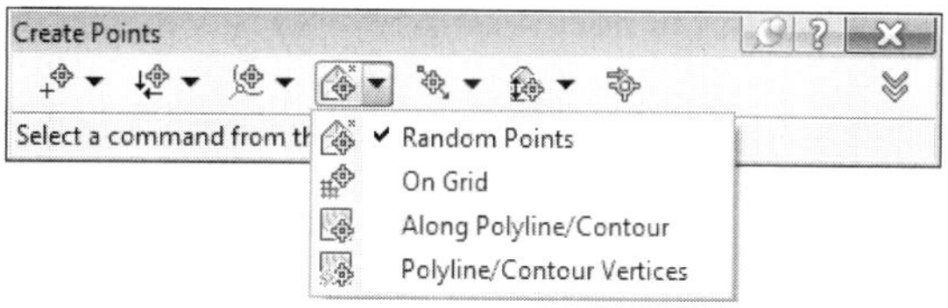

Interpolation Point-Creation Options The Interpolation category lets you fill in missing information from survey data or establish intermediate points for your design tasks. For example, suppose your survey crew picked up centerline road shots every 100′ (30 m), and you'd like to interpolate intermediate points every 25′ (8 m). Instead of doing a manual slope calculation, you could use the Incremental Distance tool to create additional points (see Figure 3.15).

FIGURE 3.15
The Interpolation point-creation options

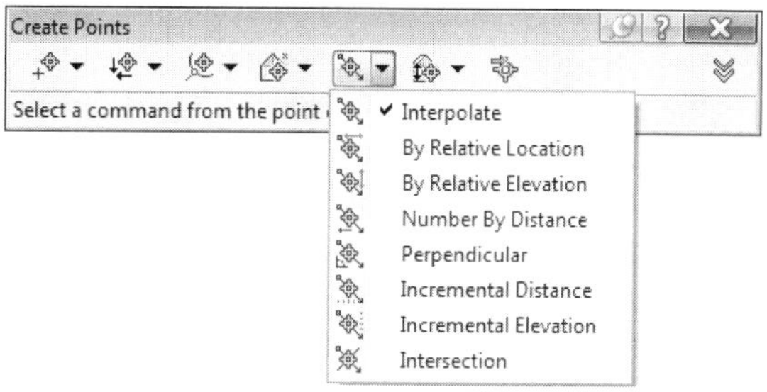

Another use would be to set intermediate points along a pipe stakeout. You could set a point for the starting and ending invert, and then set intermediate points along the pipe to assist the field crew.

Slope Point-Creation Options The Slope category allows you to set points between two known elevations by setting a slope or grade. Similar to the options in the Interpolation and Intersection categories, these tools save you time by eliminating construction geometry and hand calculations (see Figure 3.16).

FIGURE 3.16
Slope point-creation options

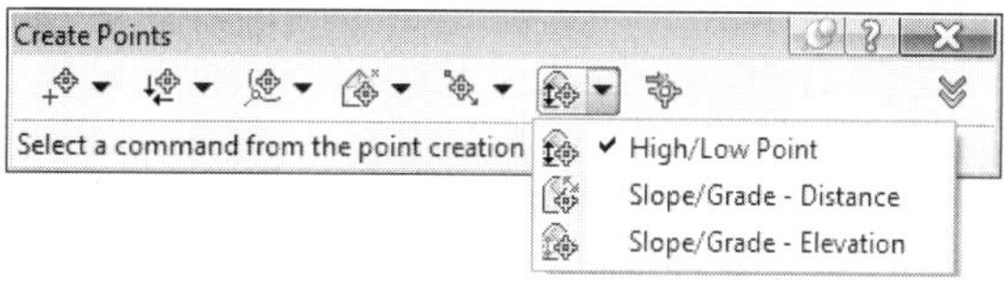

Creating Points

In this exercise, you'll learn how to create points along a parcel segment and along a surface contour:

1. Open `Mastering Point Creation.dwg` (`Mastering Point Creation_METRIC.dwg`), which you can download from this book's web page.

   Note the drawing includes an alignment, a series of parcels, and an existing ground surface.

2. On the Home tab ➢ Create Ground Data panel ➢ Points menu, click Point Creation Tools.
3. Click the chevron icon on the right to expand the dialog.
4. Expand the Points Creation category.
5. Change the Prompt For Elevations value to None and the Prompt For Descriptions value to Automatic by clicking in the respective cell in the value column, clicking the down arrow, and selecting the appropriate option.
6. Enter **LOT** for Default Description (see Figure 3.17).

Figure 3.17
Point-creation settings in the Create Points dialog

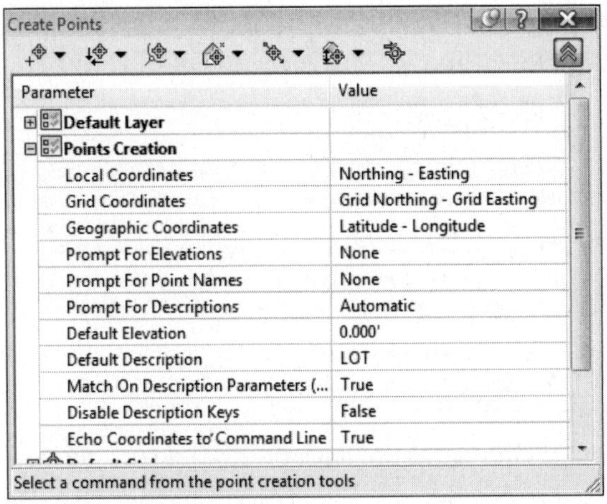

   This will save you from having to enter a description and elevation each time. Because you're setting stakeout points for rear lot corners, you will disregard elevation for now.

7. Select the Automatic tool from the Miscellaneous flyout (the first button flyout on the top left of the Create Points toolbar).
8. Select all five of the blue property lines in the drawing. Press ↵.
9. Press Esc to exit the command.

   A point is placed at each property corner and at the endpoints of each curve.

10. Select the Measure Object tool from the Miscellaneous flyout.
11. Click anywhere on the parcel boundary for Property 10.

After selecting the parcel boundary for Property 10, this tool prompts you for starting and ending stations.

12. Press ↵ twice to accept the default stationing and offset (press ↵ to accept 0).

13. At the Interval prompt, enter **25** if you are working in feet and **10** if you are working in metric.

14. Press Esc to exit the command.

A point is placed at 25′ (10 m) intervals along the property boundary.

Next you'll experiment with the Direction/Direction option from the Intersection Point flyout. Be sure your Endpoint Osnap is on for the next steps.

15. Click the Direction/Direction icon and click the southeast endpoint of the "floating" parcel line.

16. Click the opposite endpoint to establish the direction of the line.

The yellow arrow that appears indicates the direction.

17. Press ↵ to specify a zero offset.

18. Click the southeast corner of Property 6 and then the southwest corner of Property 6.

19. Press ↵ to specify a zero offset.

A point is generated where the two lines would intersect if they were to be extended. Press Esc to exit the command.

20. In the point creation settings, change Prompt For Descriptions to Automatic - Object.

21. From the Alignments flyout, choose At Geometry Points.

22. Select the green centerline alignment.

You are now prompted for a profile.

23. Select Layout (1) from the drop-down and click OK.

24. Press ↵ twice to confirm the starting and ending station values along which you will place points.

25. Press Esc to exit the command.

You should now see points whose names are based on alignment information.

26. Return to the Point Settings category and change Prompt For Elevations to Manual and Default Description to EG.

The next round of points you'll set will be based on the existing ground elevation.

27. Select the Along Polyline/Contour tool in the Surface flyout to create points every 25′ (10 m) along the driveway near HOUSE2.

28. Experiment with the plethora of point placement tools available to you!

Double Troubles

Civil 3D does not allow two points to share the same point number. If a duplicate point is detected, Civil 3D will warn you and ask you how you would like to handle it.

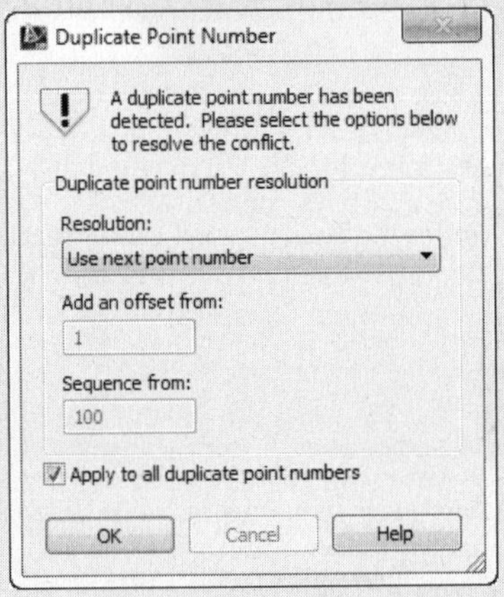

You have the choice of several options:

Add An Offset This option will allow you to add a value to all incoming points. Specifying an offset of 1000 would turn 1, 2, and 5 into 1001, 1002, and 1005.

Merge If the existing point has a description but no elevation, and the incoming point has an elevation and no description, Civil 3D will fill in the gaps with the incoming information. If there is no missing data and the coordinates are identical, the incoming point is ignored. Be careful using merge; it will behave similar to the Overwrite option if the coordinates don't match.

Overwrite This option deletes the existing point and replaces it with the incoming point.

Sequence From This option will restart the numbering at a higher value. Unlike adding an offset, the original point number is ignored. Setting a sequence from 1000 would turn 1, 2, and 5 into 1001, 1002, and 1003.

Use Next Point Number The default option, Use Next Point Number, finds the next available point number and imports the point.

Point numbers are assigned using the Point Identity settings in the Create Points dialog or the point file from which they originated. To list available point numbers, enter **`ListAvailablePointNumbers`** on the command line, or select any point to open the COGO Point contextual tab and choose COGO Point Tools ➢ List Available Point Numbers.

Basic Point Editing

Despite your best efforts, points will often be placed in the wrong location or need additional editing after their initial creation. Points may need to be rotated as a group to match a different horizontal datum. Points may need to be raised or lowered to match a different benchmark.

Physical Point Edits

Points can be moved, copied, rotated, deleted, and more using standard AutoCAD commands and grip edits. When you pause your cursor over a grip, a special grip menu will appear with different options. Figure 3.18 (left) shows the grip menu options for the label. Figure 3.18 (right) shows the options available directly on the point. Using the options shown here, you can move the point and rotate it independently of the text and rotate the text label independently of the marker.

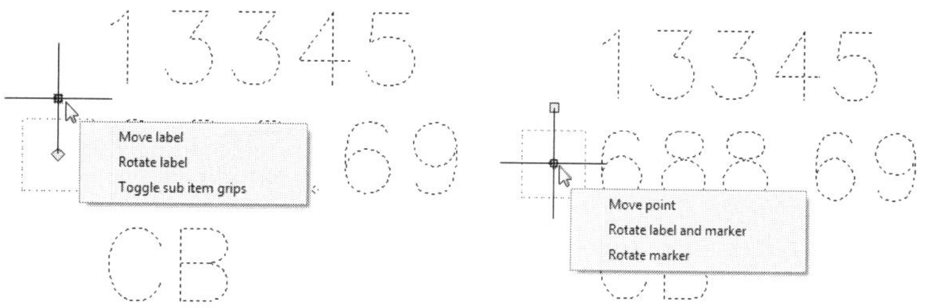

FIGURE 3.18
The top grip allows label modifications (left); the center grip allows marker modifications (right).

Panorama and Prospector Point Edits

You can access many point properties through the Point Editor in Panorama:

1. In any drawing, choose a point (or points).
2. From the COGO Point contextual tab ➢ Modify panel, select Edit/List Points.

   Panorama brings up information for the selected point(s) (see Figure 3.19).

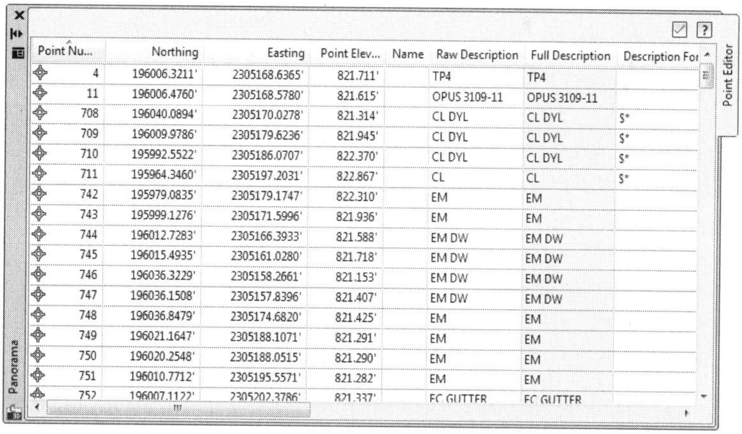

Point Nu...	Northing	Easting	Point Elev...	Name	Raw Description	Full Description	Description For
4	196006.3211'	2305168.6365'	821.711'		TP4	TP4	
11	196006.4760'	2305168.5780'	821.615'		OPUS 3109-11	OPUS 3109-11	
708	196040.0894'	2305170.0278'	821.314'		CL DYL	CL DYL	$*
709	196009.9786'	2305179.6236'	821.945'		CL DYL	CL DYL	$*
710	195992.5522'	2305186.0707'	822.370'		CL DYL	CL DYL	$*
711	195964.3460'	2305197.2031'	822.867'		CL	CL	$*
742	195979.0835'	2305179.1747'	822.310'		EM	EM	
743	195999.1276'	2305171.5996'	821.936'		EM	EM	
744	196012.7283'	2305166.3933'	821.588'		EM DW	EM DW	
745	196015.4935'	2305161.0280'	821.718'		EM DW	EM DW	
746	196036.3229'	2305158.2661'	821.153'		EM DW	EM DW	
747	196036.1508'	2305157.8396'	821.407'		EM DW	EM DW	
748	196036.8479'	2305174.6820'	821.425'		EM	EM	
749	196021.1647'	2305188.1071'	821.291'		EM	EM	
750	196020.2548'	2305188.0515'	821.290'		EM	EM	
751	196010.7712'	2305195.5571'	821.282'		EM	EM	
752	196007.1122'	2305202.3786'	821.337'		FC GUTTER	FC GUTTER	

FIGURE 3.19
Edit points in Panorama

You can access a similar interface in the Prospector tab of Toolspace by following these steps:

1. Highlight the Points collection (see Figure 3.20).
2. In either location, right-click the point or points you wish to examine and select Zoom To.
3. Click the column heading to re-sort the points.

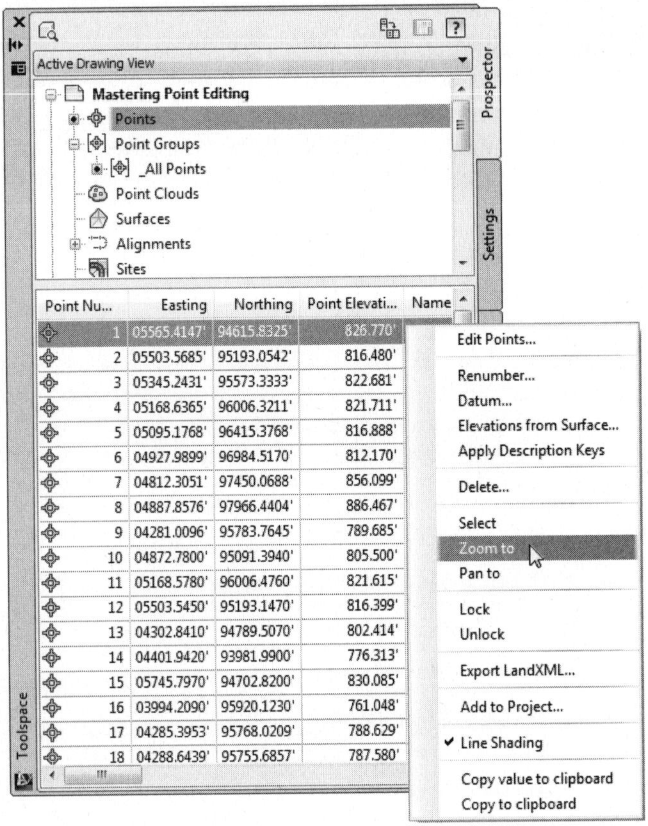

FIGURE 3.20 Prospector lets you view your entire Points collection at once.

For example, if you'd like to list your points alphabetically by description, click the Description column heading.

Point Groups: Don't Skip This Section!

CERT OBJECTIVE

Working with point groups is one of the most powerful techniques you will learn from this chapter. Want to turn all your points off without touching layers? Make a point group! Want to move last week's survey up by the blown instrument height difference? Make a point group! Want to show all your Topo shots as dots rather than Xs? Want to prevent invert shots from throwing off your surface model? Point group! Point group!

A *point group* is a collection of points that has been filtered for a certain criterion. You can use any point property or combination of properties such as description, elevation, and point number, or you can select specific points in the drawing.

Civil 3D creates the _All Points group for you, which contains every point in the drawing. It cannot be renamed or deleted, or have its properties modified to exclude any points. Create point groups for collections of points you might wish to separate from others, as shown in Figure 3.21.

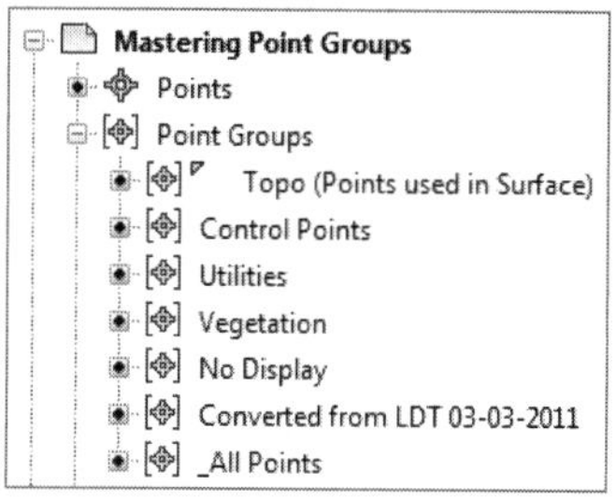

FIGURE 3.21
An example of useful point groups in Prospector

Point groups can (and should!) be created upon import of a text file, as shown previously in Figure 3.7. That way, if a problem comes to light about that group of points (such as incorrect instrument height) they can be isolated and dealt with apart from other points. Create a new point group by right-clicking on the main point group category and selecting New.

In this exercise, you'll learn how to use point groups to separate points into usable categories:

1. Open the drawing `Mastering Point Groups.dwg` (`Mastering Point Groups_METRIC.dwg`), which you can download from this book's web page.
2. In Prospector, right-click Point Groups and select New.
3. On the Information tab, name the point group **Vegetation**.
4. Set Point Style to Tree.
5. Set Point Label Style to Description Only.
6. Switch to the Include tab and place a check mark next to With Raw Descriptions Matching.
7. In the Raw Descriptions Matching field, type **TREE*, SHRUB*, TL***.

   The asterisk acts as a wildcard to include points that may have additional information after the description. You are adding multiple descriptions by separating them with a comma, as shown in Figure 3.22.
8. Switch to the Overrides tab.
9. Place a check mark next to Style and Point Label Style, as shown in Figure 3.23.

   Doing so ensures that the point group will control the styles instead of the description keys.
10. Switch to the Point List tab and examine the points that have been picked up by the group.

    Only points beginning with SHRUB, TREE, and TL should appear in the list.
11. Click OK.
12. Again, right-click on Point Groups and select New.
13. On the Information tab, name the group **NO DISPLAY**, as shown in Figure 3.24.

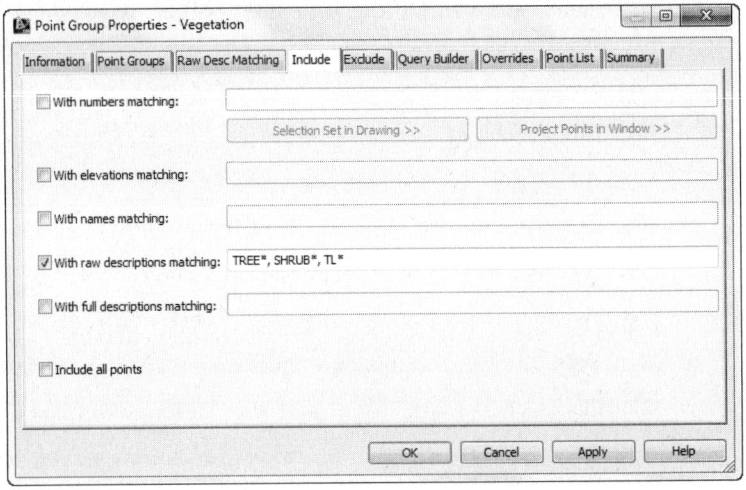

FIGURE 3.22 The Include tab of the Vegetation point group properties

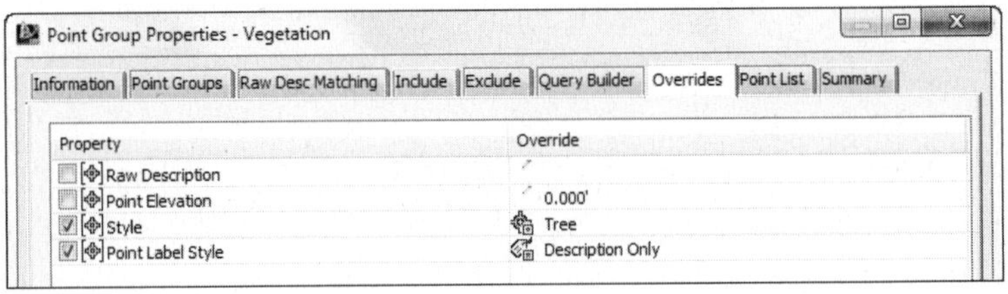

FIGURE 3.23 Overrides forces the styles to conform by point group rather than description key.

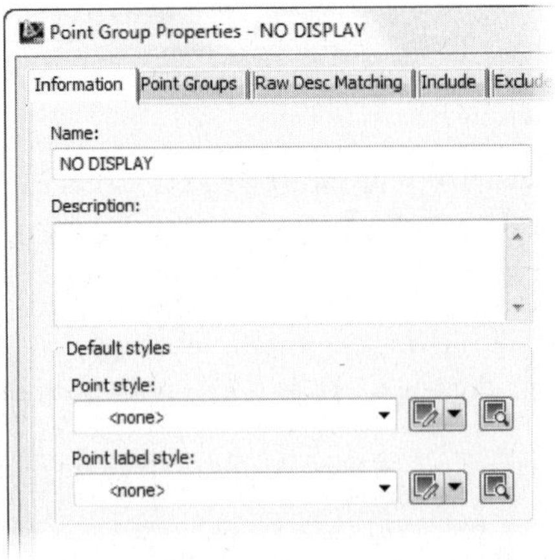

FIGURE 3.24 Most drawings should contain a NO DISPLAY point group with styles set to <none>.

14. Set both Point Style and Point Label Style to <none>.

15. Switch to the Include tab, and at the bottom of the dialog, put a check mark next to Include All Points.
16. Switch to the Overrides tab and place a check mark next to Style and Point Label Style.
17. Click OK.

    All the points are hidden from view as a result of the point group.
18. Create another point group called **Topo**.
19. Set Point Style to Basic X (BLACK) and Point Label Style to Elevation And Description.
20. Switch to the Exclude tab and place a check mark next to With Elevations Matching.
21. Type **<1** in the accompanying field.
22. Place a check mark next to With Raw Descriptions Matching.
23. Type **INV*, HYD*** in the accompanying field, as shown in Figure 3.25, and click OK. You may need to use the AutoCAD `REGEN` command to refresh the graphic and show your points again.

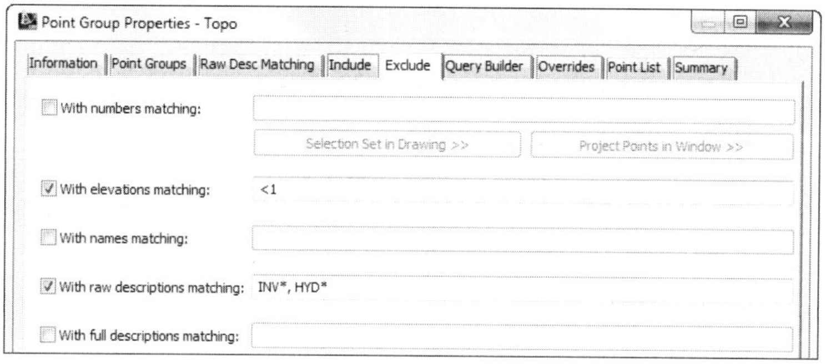

FIGURE 3.25 Use Exclude to create a Topo point group.

BEST PRACTICE: CONTROL POINT DISPLAY USING POINT GROUPS RATHER THAN LAYERS

Civil 3D drawings will have many layers in them. It is much easier to switch the display of the point groups rather than create layer states for each point visibility scenario.

A point can belong to more than one group at once. For instance, a water valve cover with elevation may be in a Topo group, a Utilities group, and the _All Points group. In these cases, the order in which the point group is displayed in Prospector determines which point group a point is "listening to" for its properties.

In this exercise, we will walk you through an example of how point group display order works:

1. Open the drawing `Mastering Point Display.dwg` (`Mastering Point Display_METRIC.dwg`), which you can download from this book's web page.
2. In Prospector, right-click Point Groups and select Properties (see Figure 3.26).

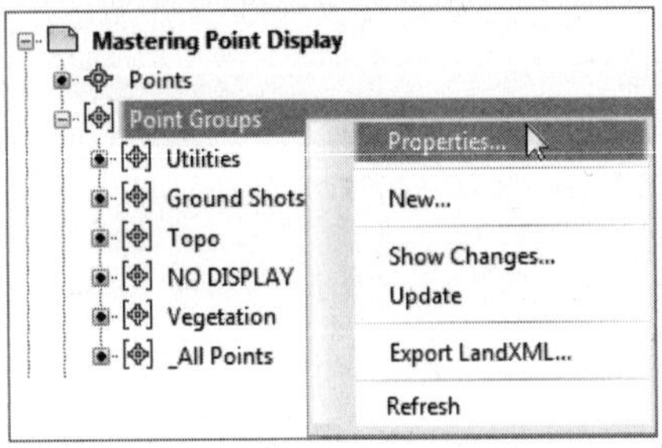

FIGURE 3.26 Select Point Groups ➤ Properties to change point group display precedence.

3. Using the arrows on the far right of the Point Groups listing, move Vegetation to the top of the list.
4. Move NO DISPLAY so that it is listed directly below Vegetation, as shown in Figure 3.27.

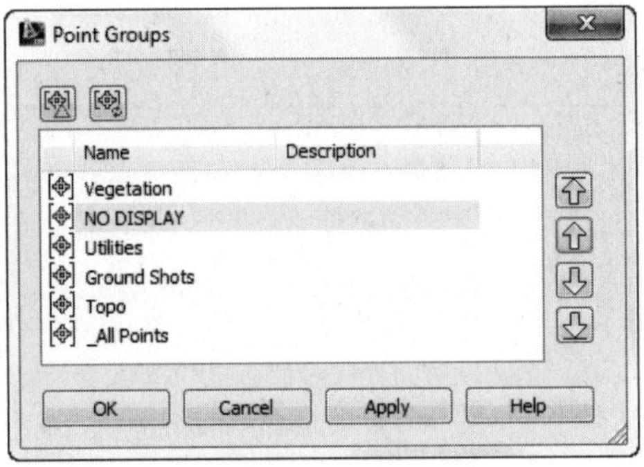

FIGURE 3.27 The order in which the point groups appear in this list controls precedence.

5. Click OK.

   Notice that only the Vegetation group is visible.
6. Experiment with changing the order of the point groups using the properties.

Changing Point Elevations

Points placed on or along an object that has elevation will automatically inherit the object's elevation. Points placed with tools in the Surface flyout will automatically inherit the elevation of a surface model. If you have chosen to place points with manual elevation entry and press ↵ when prompted to specify an elevation, the elevation will be null (no elevation).

You are never stuck with a COGO point's elevation. They can be changed individually or as a group using the Panorama window. Additional tools are available for manipulating points (see Figure 3.28), in the COGO Point contextual tab that opens when you select a point object.

FIGURE 3.28
Point-editing commands in the Ribbon

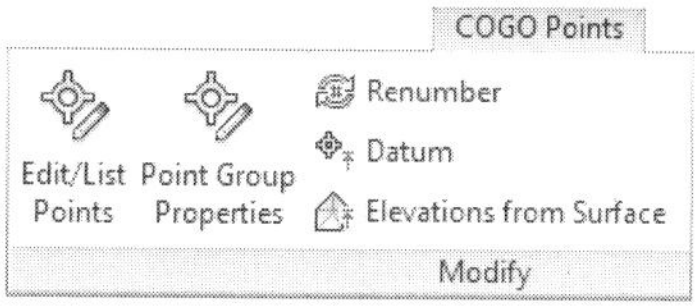

Elevations From Surface is an extremely handy tool for forcing points to a surface elevation (see Figure 3.29).

FIGURE 3.29
Shrub points as placed (a); shrub points moved up to surface elevation (b)

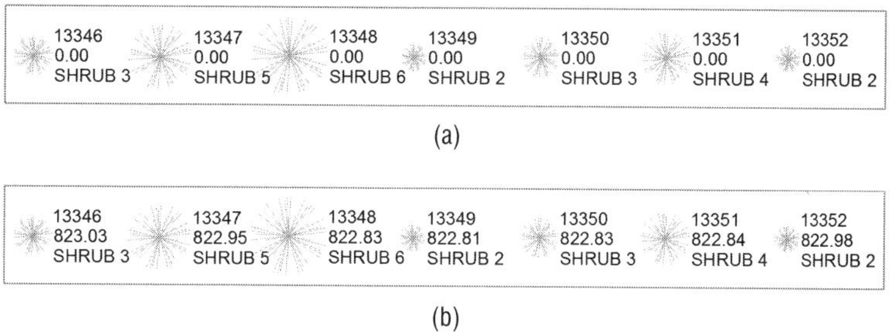

When you change the datum, you are most likely going to move a group of points' elevations. Right-click on the name of the point group in Prospector and select Edit Points. Panorama will appear for your point-editing delight.

Use Windows keyboard tricks to control which points are selected for modification. Pressing Ctrl+A will select all points in the Panorama listing, as shown in Figure 3.30. When you are done selecting points, right-click and choose Datum. The command line will prompt you to specify the change in elevation you require.

FIGURE 3.30
Right-click to access point modification tools from Panorama.

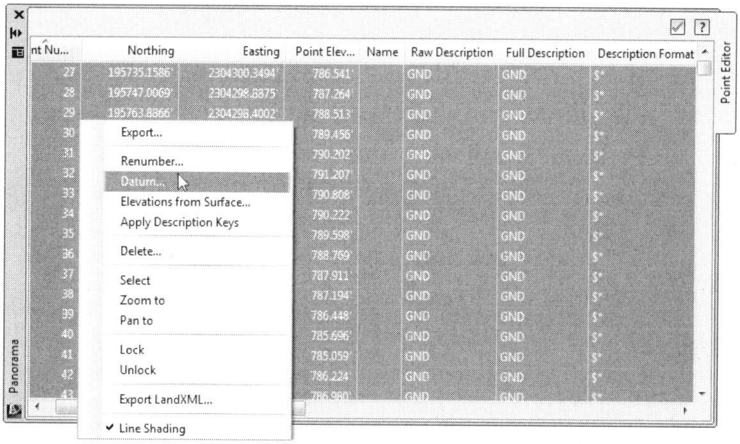

POINT GROUPS OR DESCRIPTION KEYS?

After reading this last section and the section in Chapter 2 on description keys, you're probably wondering which method is better for controlling the look of your points. This question has no absolute answer, but there are some things to take into consideration when making your decision.

Point groups are useful for both visibility control and sorting. They're dynamic and can be used to control the visibility of points that already exist in your drawing.

Both can be standardized and stored in your Civil 3D template.

Your best bet is probably a combination of the two methods. For large batches of imported points or points that require advanced rotation and scaling parameters, description keys are the better tool. For preparing points for surface building, exporting, and changing the visibility of points already in your drawing, point groups will prove most useful.

An interesting fact to note about description keys is that they take over styles and layers set elsewhere. For example, if your point placement options have a layer set but you place a point that matches a description key with a layer set to something different, the description key set "wins." In the point group creation examples, you set the Overrides tab to have Style and Point Label Style selected (Figure 3.22). Those settings wrestle control of the styles away from the description key and into the hands of the point group.

Point Tables

You've seen some of the power of dynamic point editing; now let's look at how those dynamic edits can be used to your advantage in point tables.

Most commonly, you may need to create a point table for survey or stakeout data; it could be as simple as a list of point numbers, northing, easting, and elevation. These types of tables are easy to create using the standard point-table styles and the tools located in the Points menu under the Add Tables option.

1. Open the `Point Table.dwg` (`Point Table_METRIC.dwg`) file, which you can download from this book's web page.

   This file will appear empty, but it isn't. First, you will review reordering point group properties to change the display of points. You don't need to see points to make a point table from a group, but this will help you see that the table reflects the specific group.

2. In Prospector, expand Point Groups.

   You will see that NO DISPLAY is listed on the top, which means the styles set in its properties are taking over the other point groups.

3. Right-click on Point Groups and select Properties.

4. In the listing, move the group Trees To Be Removed to the top by using the arrows on the right. Click OK.

5. Use Zoom Extents to see the cluster of trees you are working with.

6. On the Ribbon, switch to the Annotate tab, and click Add Tables ➢ Add Point Table.

7. Verify that Table Style is set to Tree Removal, and click the Point Group icon.
8. Select Trees To Be Removed, and click OK.
9. Verify that the check box next to Split Table is clear. When all your settings match those in Figure 3.31, click OK.

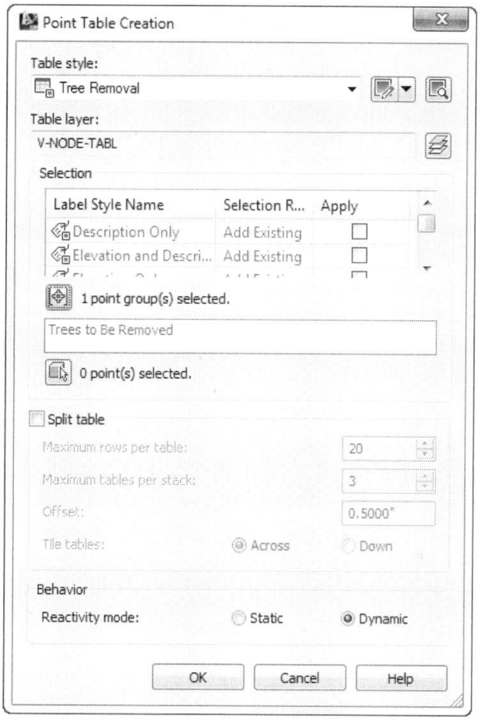

FIGURE 3.31
Point Table Creation options

10. Click anywhere in the graphic to place the table.

User-Defined Properties

Standard point properties include items such as number, easting, northing, elevation, name, description, and the other entries you see when examining points in Prospector or Panorama. But what if you'd like a point to know more about itself?

It's common to receive points from a soil scientist that list additional information, such as groundwater elevation or infiltration rate. Surveyed manhole points often include invert elevations or flow data. Tree points may also contain information about species or caliber measurements. All this additional information can be added as user-defined properties to your point objects. You can then use user-defined properties in point labeling, analysis, point tables, and more.

How Can Civil 3D Work with Soil Boring Data?

In the following example you will add user-defined properties to some soil boring points and leverage point groups to work with the data. The skills you learn in this example can be applied to multiple soil boring values for the purposes of creating subsurface data.

1. Open the file `Soil Borings.dwg` (`Soil Borings_METRIC.dwg`).

   This file contains an existing ground surface along with several point groups, including one containing soil boring points.

2. Take a moment to examine the soil boring points and the current elevation listing.
3. From the Settings tab of Toolspace ➢ Point collection, right-click on User-Defined Property Classifications and select New.
4. Name the new classification **Soil Borings** and click OK.

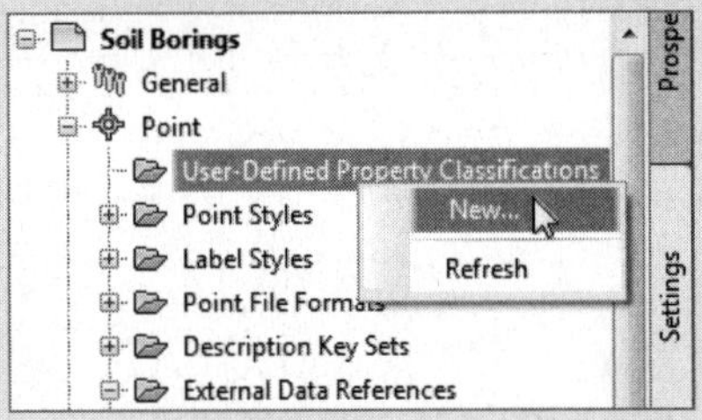

5. Expand the User-Defined Property Classifications category (if it is not already), and right-click on Soil Borings. Select New.
6. Name the new property **Watertable Elevation.**
7. Set Property Field Type to Elevation, deselect Default Value, and then click OK.

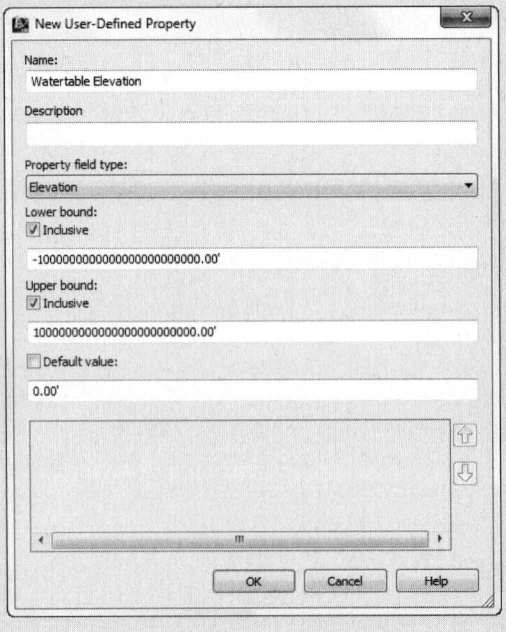

8. Jump back to the Prospector tab, and highlight the main Point Groups listing.

   At the very bottom of Toolspace you will see a listing of all the point groups, as seen here:

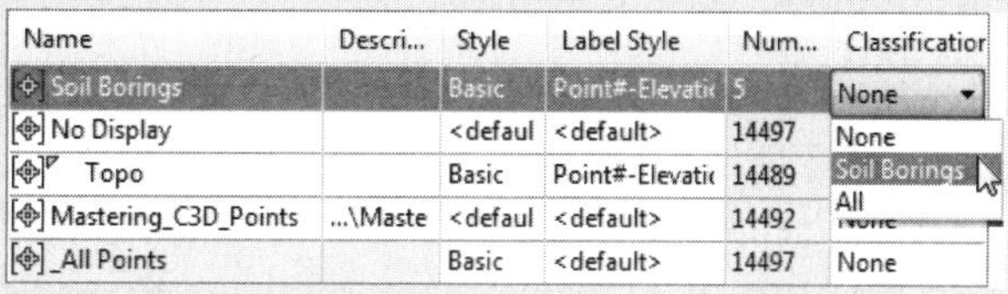

9. Set the classification as shown in the graphic.
10. Right-click the Soil Borings point group and select Edit Points.
11. Scroll over in Panorama until you locate the new classification column.

    This is the information you added in the previous three steps.
12. Add the Watertable Elevation entries in Panorama, as shown here. Dismiss Panorama when you're done.

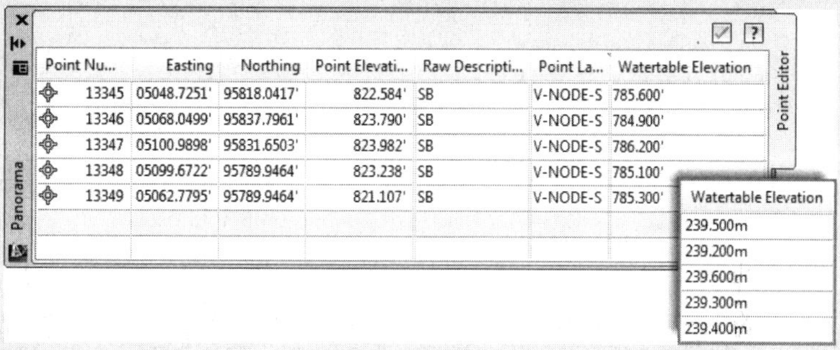

13. Right-click on the Soil Borings group and select Properties. Switch to the Overrides tab, as shown here:

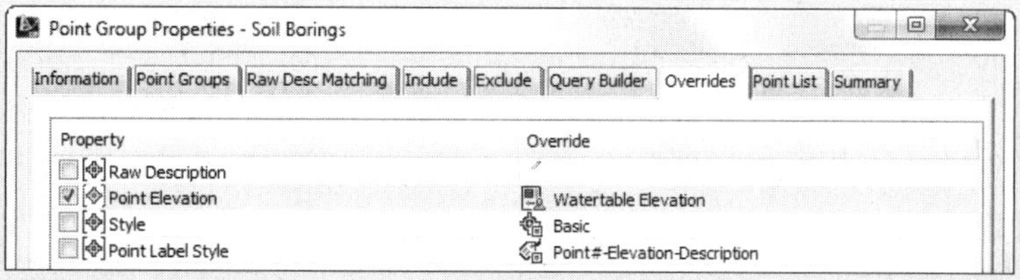

14. Place a check mark next to Point Elevation on the Overrides tab.
15. Click the tiny pencil icon twice, or until it turns into the user-defined property icon. Initially the value will be <none>.
16. Click the field next to the icon to set the value to Watertable Elevation.
17. Click OK to dismiss the Point Group Properties dialog.

    Notice that the elevation labels for the five points are listed as the watertable elevations.

The Bottom Line

Import points from a text file using description key matching. Most engineering offices receive text files containing point data at some time during a project. Description keys provide a way to automatically assign the appropriate styles, layers, and labels to newly imported points.

Master It Create a new drawing from `_AutoCAD Civil 3D (Imperial) NCS.dwt` (or `_AutoCAD Civil 3D (Metric) NCS.dwt`). Revise the Civil 3D description key set to contain only the parameters listed here:

Code	Point Style	Point Label Style	Format	Layer
GS*	Basic	Elevation Only	Ground Shot	V-NODE
GUY*	Guy Pole	Elevation and Description	Guy Pole	V-NODE
HYD*	Hydrant (existing)	Elevation and Description	Existing Hydrant	V-NODE-WATR
TOP*	Basic	Point#-Elevation-Description	Top of Curb	V-NODE
TREE*	Tree	Elevation and Description	Existing Tree	V-NODE-TREE

Import the PNEZD (space delimited) file `Concord.txt` (`Concord_METRIC.txt`). Confirm that the description keys made the appropriate matches by looking at a handful of points of each type. Do the trees look like trees? Do the hydrants look like hydrants?

Save the resulting file.

Create a point group. Building a surface using a point group is a common task. Among other criteria, you may want to filter out any points with zero or negative elevations from your Topo point group.

Master It Create a new point group called Topo that includes all points *except* those with elevations of zero or less. Use the DWG created in the previous Master It or start with `Master_It.dwg` (`Master_It_METRIC.dwg`).

Export points to LandXML and ASCII format. It's often necessary to export a LandXML or ASCII file of points for stakeout or data-sharing purposes. Unless you want to export every point from your drawing, it's best to create a point group that isolates the desired point collection.

Master It Create a new point group that includes all the points with a raw description of TOP. Export this point group via LandXML to a PNEZD comma-delimited text file.

Use the DWG created in the previous Master It or start with `Master_It.dwg` (`Master_It_METRIC.dwg`).

Create a point table. Point tables provide an opportunity to list and study point properties. In addition to basic point tables that list number, elevation, description, and similar options, you can customize point table formats to include user-defined property fields.

Master It Continue working in `Master_It.dwg` (`Master_It_METRIC.dwg`). Create a point table for the Topo point group using the PNEZD format table style.

Chapter 4

Surfaces

One of the most primitive elements in a three-dimensional model of any design is the surface. As you learned in the previous chapter, once survey information is gathered and points are set with elevations, you can proceed to turn some of that information into an intelligent surface. This chapter examines various methods of surface creation and editing. Then it moves into discussing ways to view, analyze, and label surfaces, and explores how they interact with other parts of your project.

In this chapter, you will learn to:

- Create a preliminary surface using freely available data
- Modify and update a TIN surface
- Prepare a slope analysis
- Label surface contours and spot elevations
- Import a point cloud into a drawing and create a surface model

Understanding Surface Basics

A surface in the AutoCAD® Civil 3D® program is generated using the principle of geometric triangulation. At the very simplest, a surface consists of points. In planar geometry, two points can be used to define a line and three points can be used to define a plane. Using this principle, the computer generates a triangular plane using a group of three points (Figure 4.1). Each of these triangular planes shares an edge with another, and a continuous surface is made. This methodology is typically referred to as a *triangulated irregular network (TIN),* as shown in Figure 4.2. On the basis of Delaunay triangulation, this means that for any given (x,y) point, there can be only one unique z value within the surface (since slope is equal to rise over run, when the run is equal to 0 the result is "undefined"). What does this mean to you? It means standard surfaces have two major limitations:

No Thickness Modeled surfaces can be thought of as a sheet draped over a surface; they have no thickness in the vertical direction associated with them.

No Vertical Faces Vertical faces cannot exist in a TIN because two points on the surface cannot have the same (x,y) coordinate pair. At a theoretical level, this limits the ability to handle true vertical surfaces, such as walls or curb structures. You must take this factor into consideration when modeling corridors, as discussed in Chapter 10, "Basic Corridors."

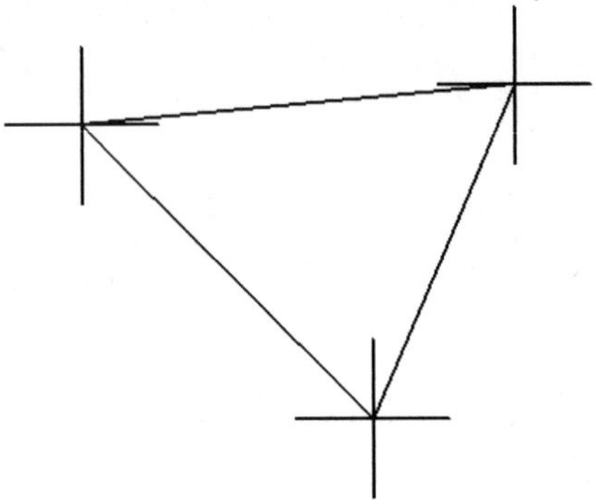

FIGURE 4.1
Three points defining a plane

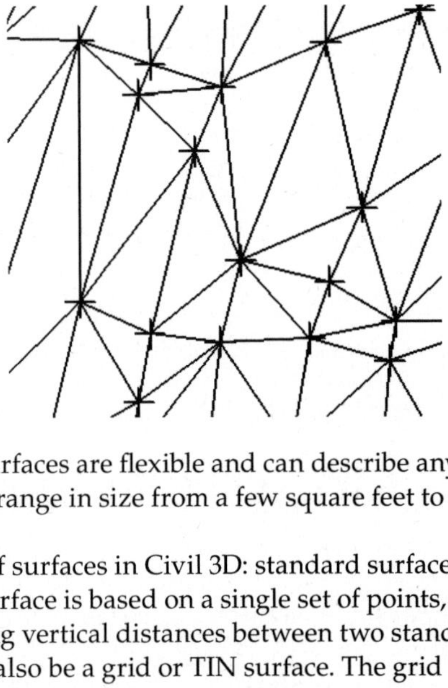

FIGURE 4.2
A triangulated irregular network, or TIN

Beyond these basic limitations, surfaces are flexible and can describe any object's face in astonishing detail. The surfaces can range in size from a few square feet to square miles and generally process quickly.

There are three main categories of surfaces in Civil 3D: standard surfaces, volume surfaces, and corridor surfaces. A standard surface is based on a single set of points, whereas a volume surface builds a surface by measuring vertical distances between two standard surfaces. Each of these two categories of surfaces can also be a grid or TIN surface. The grid version is still a TIN upon calculation of planar faces, but the data points are arranged in a regularly spaced grid of information. The TIN version is made from randomly located points that may or may not follow any pattern to their location. A corridor surface is generated from a corridor and will be discussed further in Chapter 10.

Creating Surfaces

CERT OBJECTIVE

Before you can analyze a surface, you have to make one. To the land development company today, this can mean pulling information from a large number of sources, including Internet sources, old drawings, and fieldwork. Working with each requires some level of knowledge about the reliability of the information and how to handle it in the Civil 3D software. In this section, we'll look at how you can obtain data from a couple of free sources and bring it into your drawing, create new surfaces, and make a volume surface.

Before creating surfaces, you need to know a bit about the components that can be used as part of a surface definition:

LandXML Files These typically come from an outside source or are exported from another project. *LandXML* has become a common means of communicating data in the land development industry. These files include information about points and triangulation, making replication of the original surface as easy as a few mouse clicks.

DEM Files *Digital Elevation Model (DEM)* files are the standard format files from governmental agencies and GIS systems. These files are typically very large in scale but can be great for planning purposes.

TIN Files Typically, a TIN file comes from a land development project on which you or a peer has worked. These files contain the baseline TIN information from the original surface and can be used to replicate it easily.

Boundaries *Boundaries* are closed polylines that determine the visibility of the TIN inside the polyline. The polyline can be a 2D polyline, a 3D polyline, or even a feature line, but only the horizontal information will be used to generate the boundary — any elevation information will not be used. Outer boundaries are often used to eliminate stray triangulation, whereas hide boundaries are used to indicate areas that could perhaps not be surveyed, such as a building pad. Note that only the area within a boundary is utilized in calculations. If the polyline that created the boundary is modified, the surface will become out of date, thus requiring a rebuild unless the surface is set to rebuild automatically.

BOUNDARY VS. BORDER

You may also hear a boundary referred to as a border. While these two terms can often be used interchangeably, a *boundary* is added to a surface whereas a *border* is the limits of a surface. A surface will always have a border; it will not always have a defined boundary.

Breaklines *Breaklines* are used for creating hard-coded triangulation paths, even when those paths violate the Delaunay algorithms for normal TIN creation. They can describe anything from the top of a ridge to the flowline of a curb section. A TIN line may not cross the path of a breakline. A breakline cannot be added to a grid surface. Breaklines can be defined using 2D polylines, 3D polylines, or feature lines. Similar to boundaries, if a breakline is modified the surface will become out of date, thus requiring a rebuild. Breaklines will be discussed in more detail later in this chapter.

Contours Sometimes a specific contour is desired, and it can be inserted into the surface as a 2D polyline at an elevation. Points will be placed along the contour to be used in the triangulation process. This process will be discussed in more detail later in this chapter. Similar to a breakline, a contour cannot be added to a grid surface. Similar to boundaries, if a breakline is modified, the surface will become out of date, thus requiring a rebuild. Adding contour data will be discussed further later in this chapter.

Drawing Objects AutoCAD objects that have an insertion point at an elevation (e.g., text, blocks, lines, polylines, arcs, 3D polylines) can be used to populate a surface with points. It's important to remember that no relationship to the drawing object is maintained; only the point data associated with the drawing object is added to the surface, not the actual drawing object.

Edits Any manipulation after the surface is completed, such as adding or removing triangles or changing the datum, will be part of the edit history. These changes can be viewed in the surface properties, where individual edits can be toggled on and off individually to make reviewing changes simple, or reordered since edits are implemented in the order that they are added.

Point Files Point files work well when you're working with large data sets where the points themselves don't necessarily contain extra information. Examples include laser scanning or aerial surveys. A drawing will stay referenced to a point file. If the point file is moved or deleted, the reference in the drawing will be broken.

Point Groups Civil 3D point groups or survey point groups can be used to build a surface from their respective members and maintain the link between the membership in the point group and being part of the surface. In other words, if a point is removed from a group used in the creation of a surface, it is also removed from the surface.

Working with all these elements, you can model and render almost any surface you'd find in the world. In the next section, you'll start building some surfaces.

THE YELLOW EXCLAMATION POINT FLAG

At some point you are bound to see a yellow exclamation point status icon in Prospector. This is a flag showing you that some elements are out of date and require rebuilding. In the image shown here, the EG surface needs to be rebuilt because the Point Files branch is out of date.

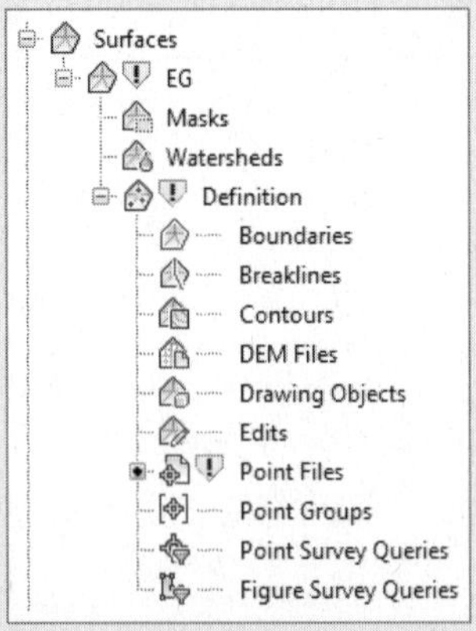

No matter what type of definition in a surface is out of date, to rebuild the surface right-click on the surface's name (in this example that would be EG) and select Rebuild. You could also select Rebuild Automatic, which would result in the surface always rebuilding when required instead of you always having to manually select Rebuild.

Free Surface Information

You can find almost anything on the Internet, including information about your project site. Some of this information may be valuable in generating a surface to use for conceptual design. For most users, free surface information can be gathered from government entities as discussed in the following section.

SURFACES FROM GOVERNMENT DIGITAL ELEVATION MODELS

One of the most common forms of free data is the Digital Elevation Model (DEM). These files have been used by the U.S. Department of the Interior's United States Geological Survey (USGS) for years and are commonly produced by government organizations for their GIS systems. The DEM format can be read directly by Civil 3D, but the USGS typically distributes the data in a complex format called Spatial Data Transfer Standard (SDTS). The files can be converted using a freely available program named `sdts2dem`. This DOS-based program converts the files from the SDTS format to the DEM format you need. Once you are in possession of a DEM file, creating a surface from it is relatively simple, as you'll see in this exercise:

1. Start a new drawing from the `_AutoCAD Civil 3D (Imperial) NCS` template that ships with Civil 3D. For metric users, use the `_AutoCAD Civil 3D (Metric) NCS` template.
2. Switch to the Settings tab of Toolspace, right-click the drawing name, and select Edit Drawing Settings.

The Drawing Settings dialog appears.

3. For Imperial users, set the Zone Category to USA, Pennsylvania and set the Coordinate System to NAD83 Pennsylvania State Planes, South Zone, US Foot (PA83-SF), as shown in Figure 4.3, via the Units And Zone tab of the Drawing Settings dialog. For metric users, set the Zone Category to USA, Pennsylvania and set the Coordinate System to NAD83 Pennsylvania State Planes, South Zone, Meter (PA83-S).

FIGURE 4.3
Imperial coordinate settings for DEM import

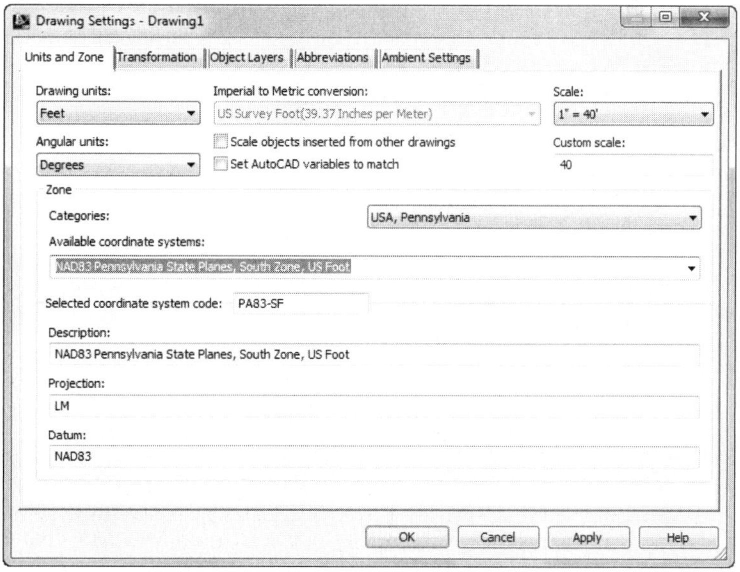

4. Accept all other defaults.
5. For all users, once complete, click OK.

The coordinate system of the DEM file that you will import will be set to adjust to the coordinate system of the drawing.

6. From the Home tab ➢ Create Ground Data panel, choose Surfaces ➢ Create Surface.

The Create Surface dialog appears. You may notice that there is also an option for Create DEM Surface under the Surfaces drop-down. While this may seem like a prudent option since you are using DEM data, the drawback to that method is that no coordinate transformation is possible — and you need it for this example.

7. Accept the options in the dialog, and click OK to create the surface.

This surface is added as Surface1 to the Surfaces collection.

8. In Prospector, expand the Surfaces ➢ Surface1 ➢ Definition branch.
9. Right-click DEM Files and select the Add option (see Figure 4.4).

The Add DEM File dialog appears.

FIGURE 4.4
Adding DEM data to a surface

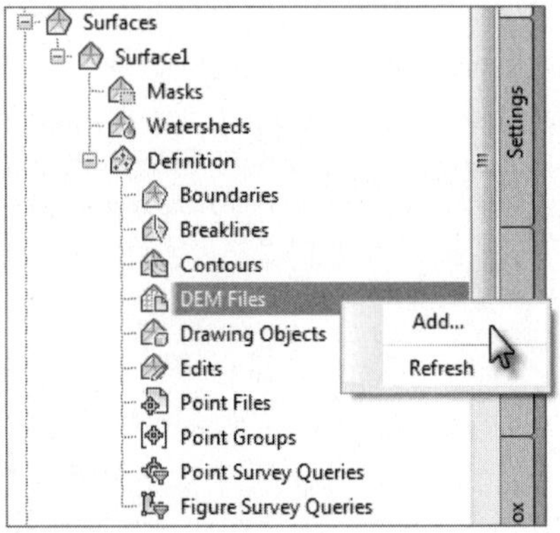

10. Use the button to the right of the DEM File Name area to navigate to the `Stewartstown_PA.DEM` file and click Open.

Remember, all data and drawing files for this book can be downloaded from `www.sybex.com/go/masteringcivil3d2013`. The DEM file information will populate in the Add DEM File dialog showing that the DEM file you are using is UTM Zone 18, NAD27 datum, meters.

11. In the Add DEM File dialog, click in the Value column next to CS Code to display the ellipsis button; click that button to display the Select Coordinate Zone dialog.
12. Set the Coordinate System Code (CS Code) to match the DEM file by selecting UTM with NAD27 datum, Zone 18, Meter; Central Meridian 75d W (UTM27-18) for both Imperial and metric users, as shown in Figure 4.5, and click OK.

FIGURE 4.5
Setting the `Stewartstown_PA.DEM` coordinate zone

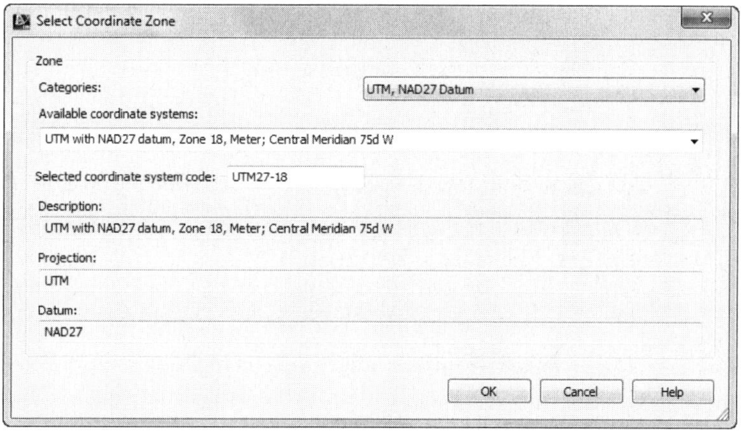

This information is necessary to properly translate the DEM's coordinate system to the drawing's coordinate system.

The Add DEM File dialog should now match the one shown in Figure 4.6.

FIGURE 4.6
Setting the `Stewartstown_PA.DEM` file properties

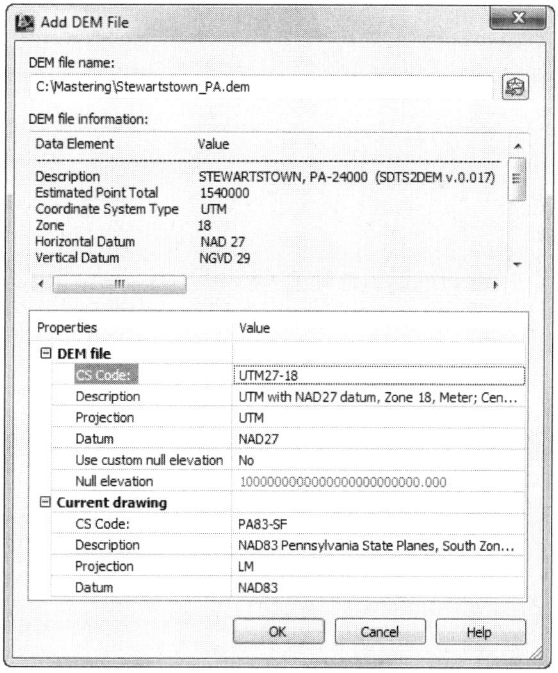

13. Click OK in the Add DEM File dialog. Importing this DEM file might take a few minutes. You can check the status of the import in the lower-left corner of the application window.

The message will be "Reading points from file" and then "Adding points to surface" with a progress gauge.

14. In Prospector, right-click Surface1 and select Zoom To to bring the surface into view.
15. Right-click the surface in your drawing and select Surface Properties.

    The Surface Properties dialog appears. Earlier, in the Create Surface dialog you had allowed the default surface name to be used, which created the name of Surface1. Since the default name does not provide much information, you will now revise the default name to something that offers more information to the user.

16. On the Information tab, change the Name field entry to **Stewartstown PA**.
17. Change the Surface Style drop-down list to Contours and Triangles, and then click OK to accept the settings in the Surface Properties dialog.

Once you have the DEM data imported, you can pause over any portion of the surface and see that feedback showing the surface elevation is provided through a tooltip. This surface can be used for preliminary planning purposes but isn't accurate enough for construction purposes.

The main drawback to DEM data is the sheer bulk of the surface size and point count. The `Stewartstown_PA.DEM` file you just imported contains 1.4 million points and covers more than 55 square miles. This much data can be overwhelming, and it covers an area much larger than the typical site. If you try zooming in and out on the surface, you will notice that the computer will be slow as it tries to regenerate the surface with each change. To ease the processing and activate the Level Of Detail display, do the following:

1. Switch to the View tab.
2. Expand the bottom of the Views panel.
3. Select Level Of Detail.

After these steps are complete you will notice a new icon appears in the upper-left corner of your model space, showing you that Level Of Detail is activated. To turn off Level Of Detail, follow the same steps.

Turning on the Level Of Detail display does not change the data in the surface but simply changes what is viewable at the different zoom levels. Figure 4.7 shows the same area of the surface zoomed out and zoomed in before (left) turning on Level Of Detail and after (right). You'll look at some data reduction methods later in this chapter.

In addition to making a DEM a part of a TIN surface, you can build a surface directly from the DEM:

1. Select the Surfaces branch.
2. Right-click on it.
3. Choose Create Surface From DEM.

The drawback to this approach is that no coordinate transformation is possible. Because one of the real benefits of using georectified data is pulling in information from differing coordinate systems, we're skipping this method to focus on the more flexible method shown in this exercise. When this exercise is complete, you may close the drawing. Due to the large file size, a finished state of this drawing is not available for download on the book's web page.

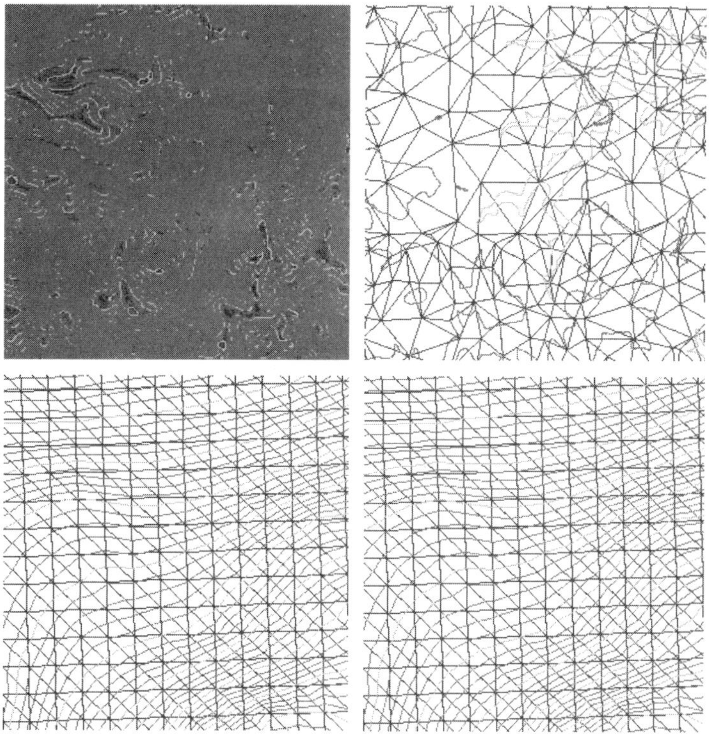

FIGURE 4.7
DEM surface: zoomed out without Level Of Detail (upper left), zoomed out with Level Of Detail (upper right), zoomed in without Level Of Detail (lower left), zoomed in with Level Of Detail (lower right)

Surface from GIS Data

You may run into a situation where an outside firm uses GIS, or perhaps your firm is also using GIS data. Civil 3D understands GIS and can work with the data given. In this section, we'll show you how to import GIS data pertaining to surfaces:

1. Start a new drawing by using the `_AutoCAD Civil 3D (Imperial) NCS` template and set the Coordinate System to NAD83 Georgia State Planes, West Zone, US Foot (GA83-WF). For metric users, use the `_AutoCAD Civil 3D (Metric) NCS` template and set the Coordinate System to NAD83, Georgia State Planes, West Zone, Meter (GA83-W).

2. From the Home tab ➢ Create Ground Data panel, choose Surfaces ➢ Create Surface From GIS Data.

The Create Surface From GIS Data – Object Options page appears.

3. Set Name to **GIS Data**, change Description to **Import from GIS Data**, and set the style to Contours 5′ and 25′ (Background), or Contours 2 m and 10 m (Background) for metric users as shown in Figure 4.8, and click the Next button.

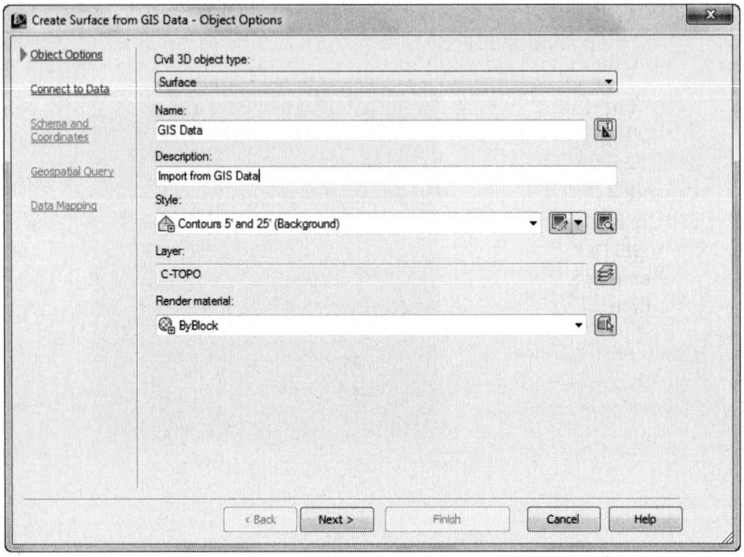

FIGURE 4.8
The Create Surface From GIS Data – Object Options page

The Create Surface From GIS Data – Connect To Data page appears.

4. You are importing a SHP file, so change Data Source Type to SHP.
5. Click the ellipsis next to SHP Path. Locate the `contours2008.shp` file (which you'll find at `www.sybex.com/go/masteringcivil3d2013`).

The path is now populated with the location of the SHP file, as shown in Figure 4.9.

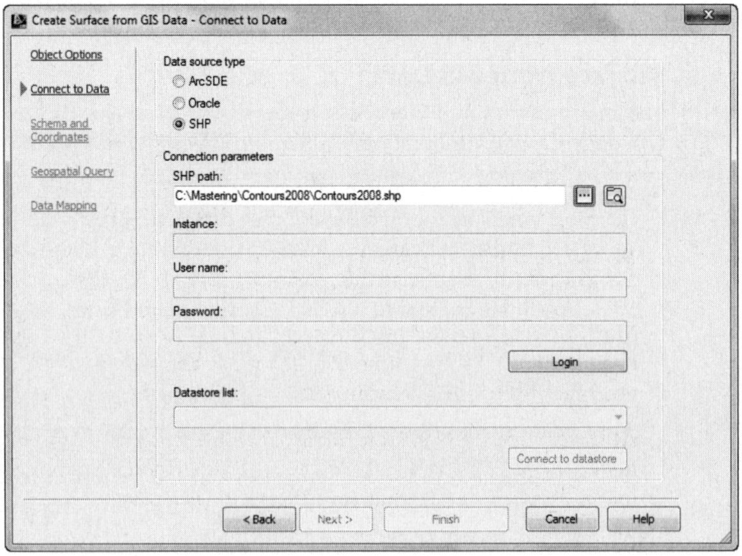

FIGURE 4.9
The Create Surface From GIS Data – Connect To Data page

6. Click the Login button.

Don't worry; you won't actually need a username or password to log in.

The Create Surface From GIS Data – Schema And Coordinates page now appears (Figure 4.10).

FIGURE 4.10
The Create Surface From GIS Data – Schema And Coordinates page

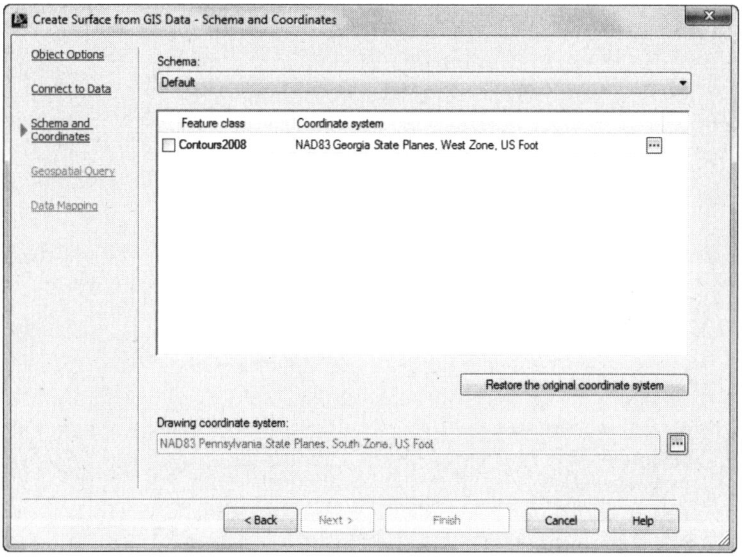

You will notice that the name of the file appears as well as the coordinate system in which the SHP was created. For Imperial users, the NAD83 Georgia State Plane, West Zone, US Foot matches what you set the drawing up with. However, for metric users, the coordinate system of the SHP file is different than the drawing coordinate system as shown in Figure 4.10.

7. Verify that the Contours2008 check box is checked under Feature Class and click Next.

8. On the Create Surface From GIS Data – Geospatial Query page, look at the settings for future reference but do not make any changes (Figure 4.11). Click Next.

9. On the Create Surface From GIS Data – Data Mapping page, click the drop-down list next to GIS Field Elevation and select Elevation, as shown in Figure 4.12.

Many dialogs throughout the software use tables such as those shown on this page for you to input data. If at any time the column is not wide enough for you to view all of the content, you may modify the column width by clicking between the column headings.

At the bottom of the Create Surface From GIS Data – Data Mapping page you will notice a field for File Name as well as Save and Open icons. That means you can save the current data mapping information that you've set for future use.

10. Click the Finish button.

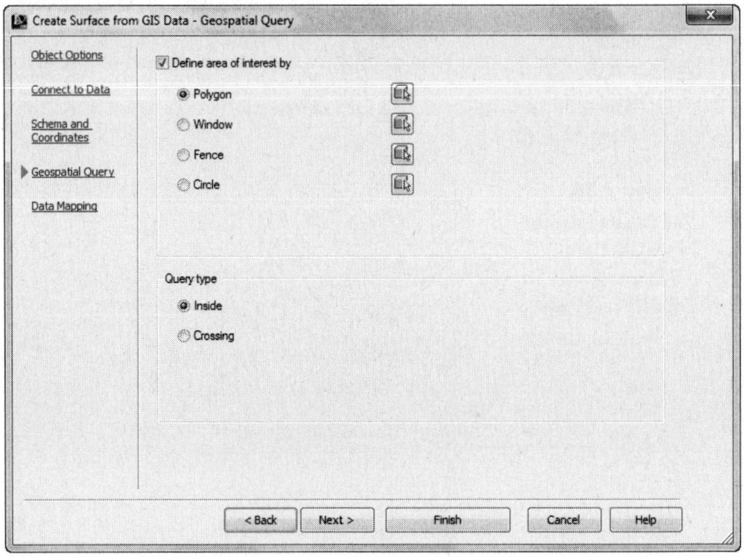

FIGURE 4.11
The Create Surface From GIS Data – Geospatial Query page

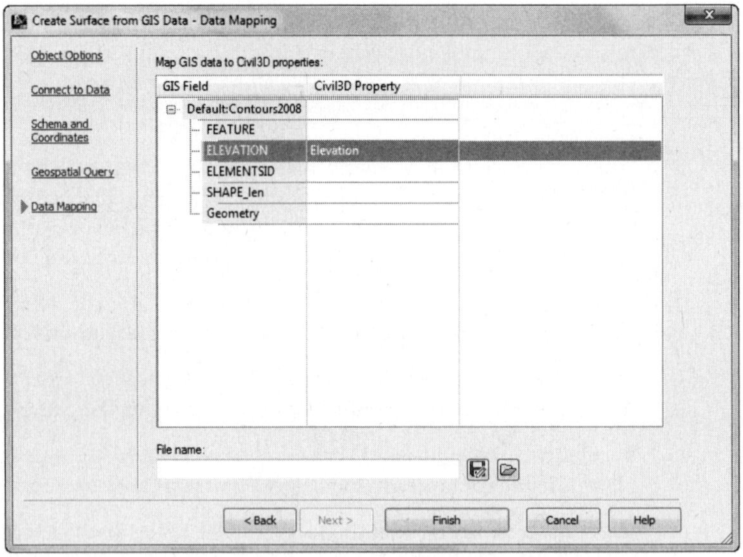

FIGURE 4.12
The Create Surface From GIS Data – Data Mapping page

11. Dismiss the Panorama and zoom extents to see the surface based on the SHP file (Figure 4.13).

When this exercise is complete, you may close the drawing. A saved finished copy of this drawing is available from the book's web page with the filename `SurfaceGIS_FINISHED.dwg` or `SurfaceGIS_METRIC_FINISHED.dwg`.

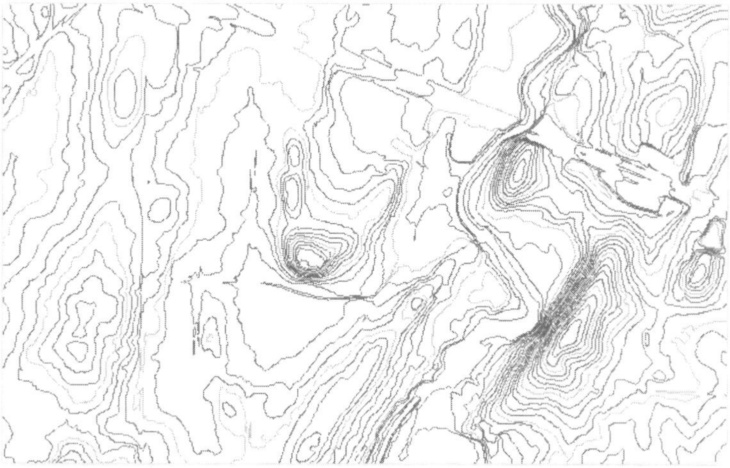

FIGURE 4.13
The finished imported GIS contours

This is just another avenue for getting drawings from other sources into Civil 3D. This topic will be discussed in depth in Chapter 18, "Advanced Workflows."

Surface Approximations

In this section, you'll work with elevated polylines. Later in this chapter, you'll work with a large point cloud delivered as a text file. These polylines are quite common, and historically it can be difficult making an acceptable surface from them.

SURFACES FROM POLYLINE INFORMATION

One common complaint about converting a drawing full of contours at elevation into a working digital surface is that the resulting contours don't accurately reflect the original data. This is because point information is provided along the contour lines but not in between the contour lines, causing the interpolation between the contours to lack accuracy. Civil 3D includes a series of surface algorithms that work very well at matching the resulting surface to the original contour data by providing additional derived data points. You'll look at those surface edits in this series of exercises.

1. Open the `SurfaceFromPolylines.dwg` file (or the `SurfaceFromPolylines_METRIC.dwg` file).

   Note that the contours in this file are composed of polylines with elevation values.

2. In Prospector, right-click the Surfaces branch and select the Create Surface option.

   The Create Surface dialog appears.

3. Leave the Type field set to TIN Surface but change the Name value to **EG-Polylines**.
4. Change Description to **Surface From Polylines**.
5. Click in the Value column next to Style to display the ellipsis button; once it's visible click the ellipsis button to display the Select Surface Style dialog.

6. From the drop-down list, select Contours 1′ and 5′ (Background), or Contours 0.2 m and 1 m (Background) for metric users, and click OK to close the Select Surface Style dialog.
7. Click OK to close the Create Surface dialog.
8. In Prospector, expand the Surfaces ➤ EG-Polylines ➤ Definition branches.
9. Right-click Contours and select the Add option.

The Add Contour Data dialog appears.

10. Set Description to **Polyline**; under Weeding Factors, set Distance to **15** (or **5** for metric users) and Angle to **4** degrees; and under Supplementing Factors, set Distance to **100** (or **30** for metric users) and Mid-Ordinate Distance to **1** (or **0.3** for metric users).
11. Verify that none of the check boxes are checked, as shown in Figure 4.14, and click OK.

FIGURE 4.14 The Add Contour Data dialog

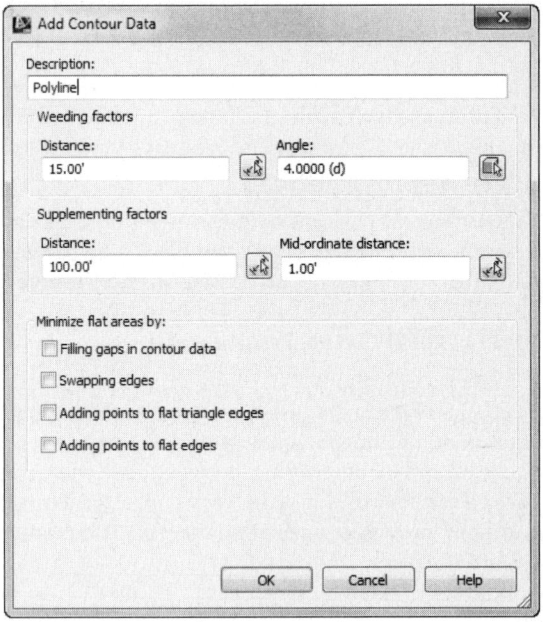

You will return to the Minimize Flat Areas By options in a bit.

12. At the `Select contours:` prompt, enter **ALL**↵ to select all the entities in the drawing and press ↵ again to end the command.

You can dismiss Panorama if it appears and covers your screen. Save and keep the drawing open for the next portion of the exercise.

The contour data has some tight curves and flat spots where the basic contouring algorithms simply fail. Zoom into any portion of the site, and you can see these areas by looking for the blue and cyan original contours not matching the new Civil 3D–generated contour, as shown in Figure 4.15.

FIGURE 4.15
Contour surface without minimizing flat areas

You'll fix that now:

13. In Prospector, expand the Definition branch of the EG-Polylines surface if it's not already open from the previous exercise and right-click Edits.
14. Select the Minimize Flat Areas option to open the Minimize Flat Areas dialog.

Note that the dialog has the same options found in that portion of your original Add Contour Data dialog. The benefit to doing it as two steps instead of one is that you can remove the operation from the Surface Build Operation if it is done separately.

15. Click OK to accept the defaults.

Save and keep the drawing open for the next portion of the exercise.

Now the contours displayed more closely match the original contour information, as shown in Figure 4.16. There might be a few instances where gaps exist between old and new contour lines, but in a cursory analysis, none was off by more than 0.4′ in the horizontal direction — not bad when you're dealing with almost a square mile of contour information.

FIGURE 4.16
Contour surface with minimizing flat areas

You'll see how this was done in this quick exercise:

16. Zoom into an area with a dense contour spacing and select the surface to make the contextual tab associated with the TIN Surface: EG-Polylines appear.

17. From the TIN Surface contextual tab ➤ Modify panel, choose Surface Properties to display the Surface Properties dialog.

18. On the Information tab, set Surface Style to Contours And Points and click OK to see a drawing similar to Figure 4.17.

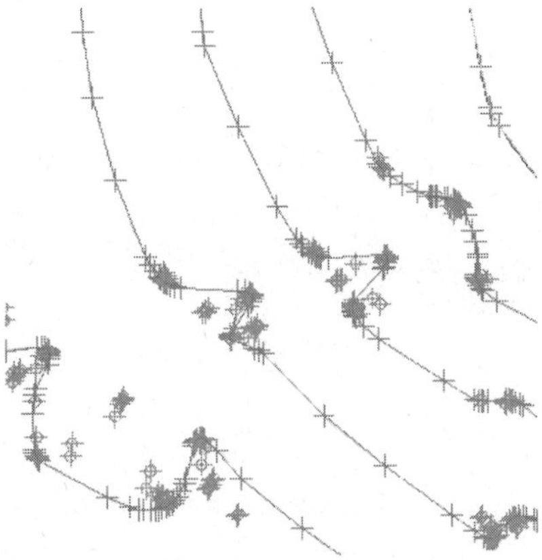

FIGURE 4.17 Surface data points and derived data points

In Figure 4.17, you're seeing the points the TIN is derived from, with some styling applied to help you understand the creation source of the points. Each point shown as a red + symbol is a point picked up from the contour data itself. The magenta points shown with a circle symbol circumscribed over a + symbol are all added data on the basis of the Minimize Flat Areas edits. These points make it possible for the Civil 3D surface to match almost exactly the input contour data.

When this exercise is complete, you may close the drawing. A saved finished copy of this drawing is available from the book's web page with the filename `SurfaceFromPolylines_FINISHED.dwg` or `SurfaceFromPolylines_METRIC_FINISHED.dwg`.

SURFACES FROM POINTS OR TEXT FILES

Besides receiving polylines, it is common for a surveying company to also send a simple text file with points. This isn't an ideal situation because you have no information about breaklines or other surface features, but it is better than nothing or usually more accurate than a surface based on free data. Because you have the same aerial surface described as a series of points, you'll use them to generate a surface in this exercise:

1. Create a folder on your computer called `C:\Mastering`.

2. Place the `Concord Commons.txt` file (or `Concord Commons_METRIC.txt` file for metric users) in your newly created folder.

Doing so ensures that future exercises will function properly.

3. Create a new drawing using the `_AutoCAD Civil 3D (Imperial) NCS` template. For metric users, use the `_AutoCAD Civil 3D (Metric) NCS` template.

4. For Imperial users, change the Coordinate System to NAD83 Pennsylvania, South Zone, US Foot (PA83-SF). For metric users, change the Coordinate System to NAD83 Pennsylvania, South Zone, Meters (PA83-S).
5. From the Home tab ➢ Create Ground Data panel, choose Surfaces ➢ Create Surface.

The Create Surface dialog appears.

6. Change the Name value to **EG**, and click OK to close the dialog.
7. In Prospector, expand the Surfaces ➢ EG ➢ Definition branches.
8. Right-click Point Files and select the Add option.

The Add Point File dialog shown in Figure 4.18 appears.

DEFAULT ADVANCED OPTIONS

Note that some of the default Advanced Options may vary from user to user so your dialog may not have the same options checked as shown in Figure 4.18. We will confirm that the correct options are selected in a later step.

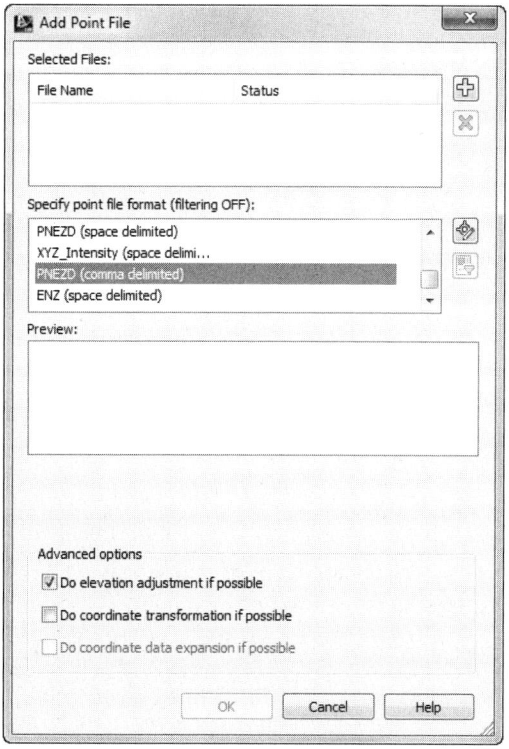

FIGURE 4.18
Adding a point file to the surface definition

9. Set the Specify Point File Format to PNEZD (Comma Delimited).

Make sure it is PNEZD, and not PENZD or your results will be off completely.

10. Click the Browse button.

The Select Source File dialog opens.

11. Navigate to the previously created `C:\Mastering` folder, and select the `Concord Commons.txt` file (or `Concord Commons_METRIC.txt` file). Click OK.

12. In the Advanced Options area of the dialog, verify that "Do elevation adjustment if possible" is checked and "Do coordinate transformation if possible" is not checked.

13. Click OK to accept the settings in the Add Point File dialog and build the surface.

Panorama will appear, but you can dismiss it.

14. Right-click EG Surface in Prospector and select the Zoom To option to view the new surface created.

SURFACE SNAPSHOTS

A surface snapshot captures the surface information in the state when the snapshot was created. If there is a snapshot in a surface, a surface's Rebuild (or surface's Rebuild Automatic) will start at the snapshot since the snapshot summarizes all of the previous build operations.

As mentioned before, a drawing will stay referenced to a point file, and if the point file is moved or deleted, the reference in the drawing will be broken. To prevent this from affecting the surface, you can create a snapshot of the surface while the surface is still working as intended. Then if anything happens to the external data, the surface will not have to go look for it if a snapshot exists.

While referencing an external file is a good reason to create a snapshot, it is not the only reason. Snapshots can be used any time you want to capture the current state of a surface.

If you right-click on a surface's name in Prospector, you will see three options related to snapshots:

Create Snapshot By creating a snapshot, you add a build operation that captures the surface information in the current state. Once a snapshot is created, the icon next to Definition in the Prospector tree will change to a camera icon.

Remove Snapshot This option will remove the snapshot from the build operation.

Rebuild Snapshot Rebuilding the snapshot and rebuilding the surface are in fact two different things. If the operations prior to the snapshot become outdated, you will see a yellow status icon next to its node in the Prospector tree prompting for the snapshot to be rebuilt. You can then choose to rebuild the snapshot if you want the changes to affect the surface or leave the snapshot in place as is if you do not want the changes to affect the surface.

When this exercise is complete, you may save and keep the drawing open to continue on to the next exercise. Or you may use the saved finished copy of this drawing available from the book's web page (`SurfaceFromPoints_FINISHED.dwg` or `SurfaceFromPoints_METRIC_FINISHED.dwg`).

In both the polyline and point file examples, you're making surfaces from the best information available. When you're doing preliminary work or large-scale planning, these types of surfaces are great. For more accurate and design-based surfaces, you typically have to get into field-surveyed information. We'll look at that a little later.

Simply adding surface information to a TIN definition isn't enough. To get beyond the basics, you need to look at the edits and other types of information that can be part of a surface.

Refining and Editing Surfaces

CERT OBJECTIVE

Once a basic surface is built, and, in some cases, even before it is built, you can do some cleanup and modification to the TIN construction that make it much more usable and realistic. Some of these edits include limiting the input data, tweaking the triangulation, adding in breakline information, or hiding areas from view. In this section, you'll explore a number of ways of refining surfaces to end up with the best possible model from which to build.

Surface Properties

The most basic steps you can perform in making a better model are right in the Surface Properties dialog. The surface object contains information about the build and edit operations, along with some values used in surface calculations. These values can be used to tweak your surface to a semi-acceptable state before more manual operations are needed.

In this exercise, you'll go through a couple of the basic surface-building controls that are available. You'll use them one at a time in order to measure their effects on the final surface display.

1. If it's not open from the previous exercise, open the `SurfaceFromPoints_FINISHED.dwg` or the `SurfaceFromPoints_FINISHED_METRIC.dwg` file.

This is the Points From Text drawing that you worked on earlier.

2. In Prospector, expand the Surfaces branch.
3. Right-click EG and select the Surface Properties option.

The Surface Properties dialog appears.

4. Select the Definition tab.

Note the list at the bottom of the dialog.

5. Under the Definition Options at the top of the dialog (Figure 4.19), expand the Build category by clicking on the + symbol.

Save and keep the drawing open for the next portion of the exercise.

The Build options of the Definition tab allow you to tweak the way the triangulation occurs. The basic options are listed here:

Copy Deleted Dependent Objects When you select Yes and an object (such as a surface boundary, breakline, or point group) that is part of the surface definition (such as the polylines you used in your aerial surface, for instance) is deleted, the information derived from that object is copied into the surface definition. Setting this option to Yes in the EG Surface Properties will let you erase the polylines from the drawing file while still maintaining the surface information. If this option is set to No for the EG surface, when the polylines (or any other drawing objects) are deleted they will be removed from the surface definition when the surface is rebuilt.

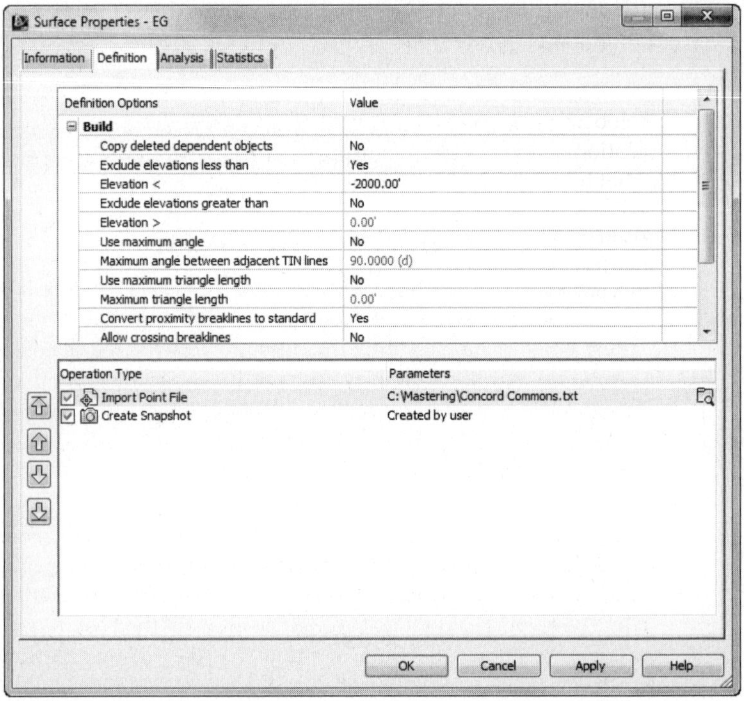

FIGURE 4.19
Surface Properties Definition Options

Exclude Elevations Less Than and Elevation < Setting Exclude Elevations Less Than to Yes puts a floor on the surface. Any point that would be built into the surface but that is lower than the floor is ignored. In the EG surface, there are calculated boundary points with zero elevations, causing real problems that can be solved with this simple click. The floor elevation is controlled by the user by setting the Elevation < value.

Exclude Elevations Greater Than and Elevation > The idea is the same as with the preceding options, but a ceiling value is used.

Use Maximum Angle and Maximum Angle Between Adjacent TIN Lines These settings attempt to limit the number of narrow "sliver" triangles with one large obtuse angle and two acute angles that typically border a site. By not drawing any triangle with an angle greater than the user input value, you can greatly refine the TIN.

Use Maximum Triangle Length and Maximum Triangle Length These settings attempt to limit the number of narrow "sliver" triangles that typically border a site. Similar to using a maximum angle, by not drawing any triangle with a length greater than the user input value, you can greatly refine the TIN.

Convert Proximity Breaklines To Standard Toggling this option to Yes will create breaklines out of the lines and entities used as proximity breaklines. We will look at this more later.

Allow Crossing Breaklines and Elevation To Use These options specify what Civil 3D should do if two breaklines in a surface definition cross each other. An (*x*,*y*) coordinate pair cannot have two *z* values, so some decision must be made about crossing breaklines. If you set Allow Crossing Breaklines to Yes, you can then select whether to use the elevation from the first or the last breakline or to average these elevations.

If you close the Surface Properties dialog and look at the surface, you might not notice the blob area shown in Figure 4.20. It appears that there are a series of blown shots, causing the elevation to dip to zero. In this next portion of the exercise, you'll limit the build options to make that blown surface "disappear."

FIGURE 4.20
EG surface showing blown points

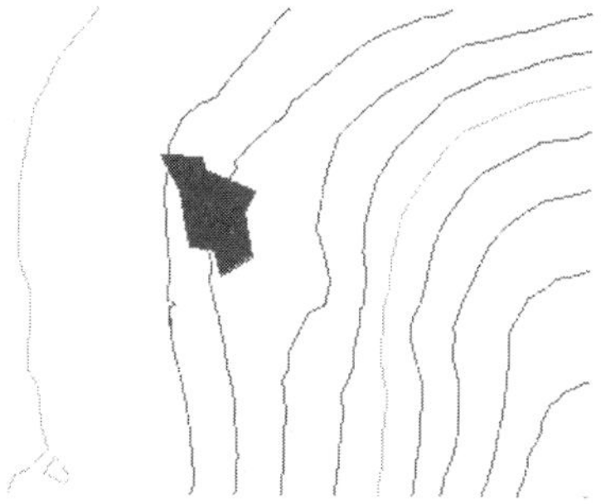

6. If you previously closed the Surface Properties dialog, open it again and on the Definition tab expand the Build category.
7. Verify that the Exclude Elevations Less Than value is Yes.
8. Set the value to **200** (**60** for metric users) and click OK to accept the settings in the dialog.

   Elevations less than 200′ (or 60 m for metric users) will be excluded.

   A warning message will appear. Civil 3D is warning you that your surface definition has changed.
9. Click Rebuild The Surface to rebuild the surface.
10. Zoom extents to view the full surface.

When it's done, it should look similar to Figure 4.21. Save and keep the drawing open for the next portion of the exercise.

FIGURE 4.21
EG surface after ignoring low elevations

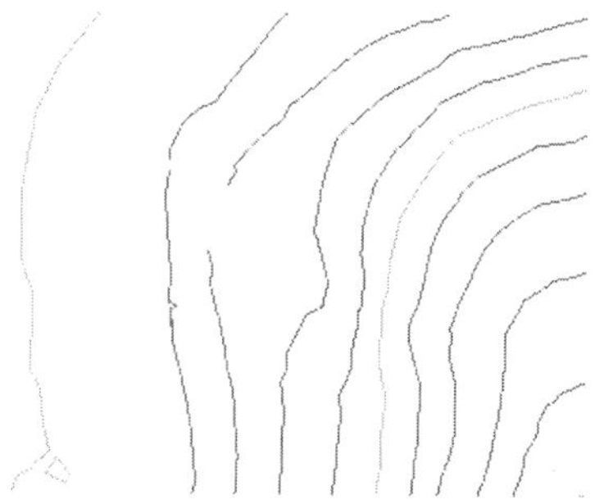

Although this surface is better than the original, there are still large areas being contoured that probably shouldn't be. By changing the style to review the surface, you can see where you still have some issues:

11. Open the Surface Properties dialog again, and switch to the Information tab.
12. Change the Surface Style field to Contours And Triangles.
13. Click Apply. Doing so makes the changes without exiting the dialog.
14. Drag the dialog to the side so you can see the site.

On the outer edges of the site, you can see some long triangles formed in areas where there was no survey taken but the surface decided to connect the triangles anyway (Figure 4.22, left).

FIGURE 4.22
EG surface before Maximum Triangle (left) and after (right)

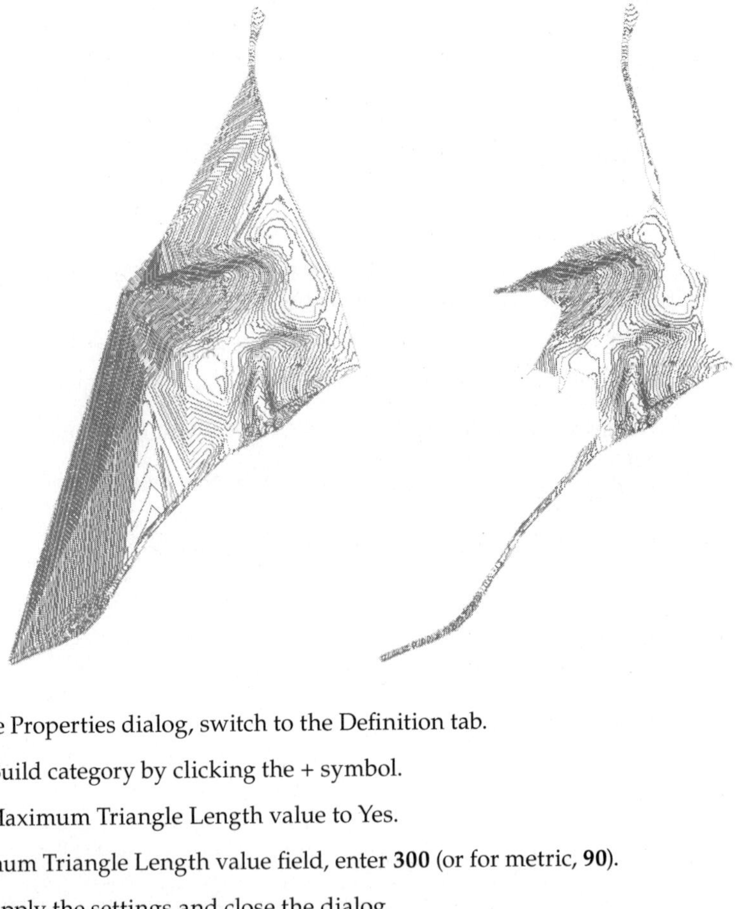

15. In the Surface Properties dialog, switch to the Definition tab.
16. Expand the Build category by clicking the + symbol.
17. Set the Use Maximum Triangle Length value to Yes.
18. In the Maximum Triangle Length value field, enter **300** (or for metric, **90**).
19. Click OK to apply the settings and close the dialog.
20. Click Rebuild The Surface to update and dismiss the warning message to see the revised surface (Figure 4.22, right).

When this exercise is complete, you may close the drawing. A saved finished copy of this drawing is available from the book's web page with the filename `SurfaceProperties_FINISHED.dwg` or `SurfaceProperties_METRIC_FINISHED.dwg`.

The value is a bit high, but it is a good practice to start with a high value and work down to avoid losing any pertinent data. Setting this value to 225′ (or 70 m) will result in a surface that is acceptable because it doesn't lose a lot of important points. Beyond this, you'll need to look at making some edits to the definition itself instead of modifying the build options.

Surface Additions

Beyond the simple changes to the way the surface is built, you can look at editing the pieces that make up the surface. With your drawing so far, you have merely been building from points. Although this is fine for small surfaces, you need to go further with this surface. In this section, you'll add a few breaklines and a border and then perform some manual edits to your site.

Real World Scenario

YOU CAN'T ALWAYS GET WHAT YOU WANT

But sometimes you get what you need. Autodesk has included the ability to reorder the build operations on the Surface Properties Definition tab. If you look at the lower left of the Definition tab shown here, you'll find that there are arrows to the left of the list box showing all the data, edits, and changes you've made to the surface.

As Civil 3D builds a surface, it processes this data and information from top to bottom — in this case, adding points, and then the breaklines, and then a boundary, and so on. If a later operation modifies one of these additions or edits, the later operation takes priority. To change the processing order, select an operation, and then use the arrows at the left to push it up or down within the process. One common example of this is to place a boundary as the last operation to ensure accurate triangulation. You'll look at boundaries in the next section.

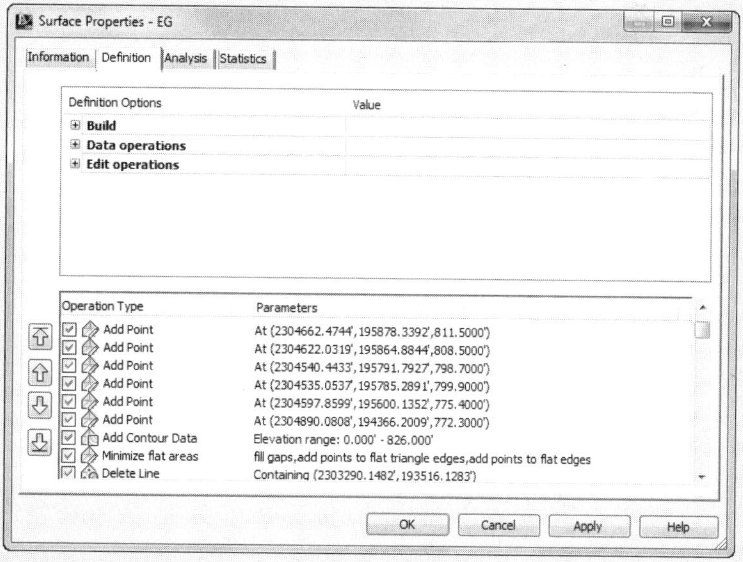

Adding Breakline Information

Breaklines can come from any number of sources. They can be approximated on the basis of aerial photos of the site that help define surface features, or they can be directly input from field book files and the Civil 3D survey functionality. Five types of breaklines are available for use:

Standard Breaklines Built on the basis of 3D lines, feature lines, or polylines, standard breaklines typically connect points already included in the surface definition but can contain their own elevation data. Simple-use cases for connecting the dots include linework from a survey or drawing a building pad to ensure that a flat area is included in the surface. Feature lines and 3D polylines are often used as the mechanism for grading design and include their own vertical information. An example might be the description of a parking lot perimeter or the invert of a V-shaped drainage swale.

Proximity Breaklines These breaklines allow you to force triangulation without picking precise points. They will not add vertical information to the surface. The (x,y,z) coordinates will be based on the surface points in close proximity to the breakline.

Wall Breaklines Wall breaklines define walls in surfaces. Because of the limitation of true vertical surfaces, a wall breakline will let you approximate a wall without having to create two separate breaklines. They are defined on the basis of an elevation at a vertex, and then an elevation difference at each vertex or for the full length of the breakline on a selected side. So a single wall breakline set with elevations will have the same result as creating a breakline with those same elevations and a second breakline that is vertically offset from the first.

From File Breaklines You can select this option if a text file contains breakline information. This file can be the output of another program and can be used to modify the surface without creating additional drawing objects.

Non-destructive Breaklines This type of breakline is designed to maintain the integrity of the original surface while updating triangulation. The (z) coordinate of the breaklines is calculated based on the surface creating new triangle vertexes to be formed.

In most cases, you'll build your surfaces from standard, proximity, and wall breaklines. In this example, you'll add in some breaklines that describe road and surface features:

1. Open the `SurfaceBreaklines.dwg` file or the `SurfaceBreaklines_METRIC.dwg` file.
2. Thaw the _Polylines-Road layer.

The roads are the color red.

3. Select the surface, and then pan and zoom to see that the triangulation lines do not appear along all the breaklines.
4. Unselect the surface by pressing Esc.
5. Select and then right-click on one of the red polylines and choose the Similar option.

All the red polylines are now highlighted.

6. In Prospector, expand the Surfaces ➢ EG ➢ Definition branches.
7. Right-click Breaklines and select the Add option.

The Add Breaklines dialog appears.

8. Enter a description if you wish and the settings as shown in Figure 4.23.
9. Click OK to accept the settings and close the dialog.

You can dismiss Panorama if it appears.

FIGURE 4.23
The Add Breaklines dialog

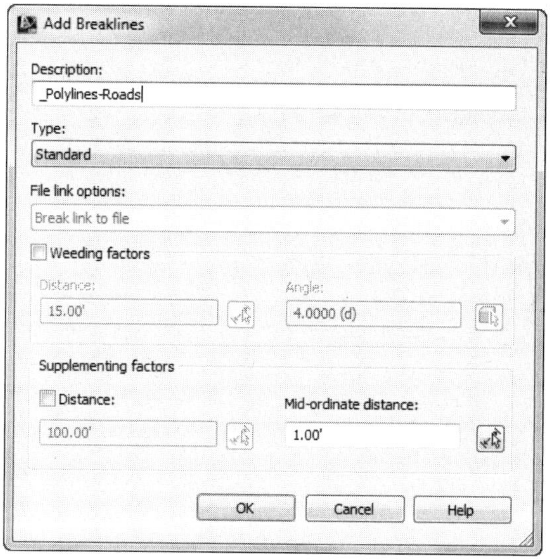

10. Thaw the _Polylines-Surface layer.

The surface polylines are the color green.

11. Repeat steps 5 through 9 but instead of the red polylines select the green lines on the _Polylines-Surface layer. Change the description to **Surface Polylines**.

As shown in Figure 4.24, by adding the breaklines you force the TIN lines to align with the breaklines, thus cleaning up the contours and making them follow the ridgelines of the road centerline, gutterlines, and shoulders as well as the changes in grade around the small detention area.

FIGURE 4.24
EG surface with only contours displayed before breaklines added (left) and after (right)

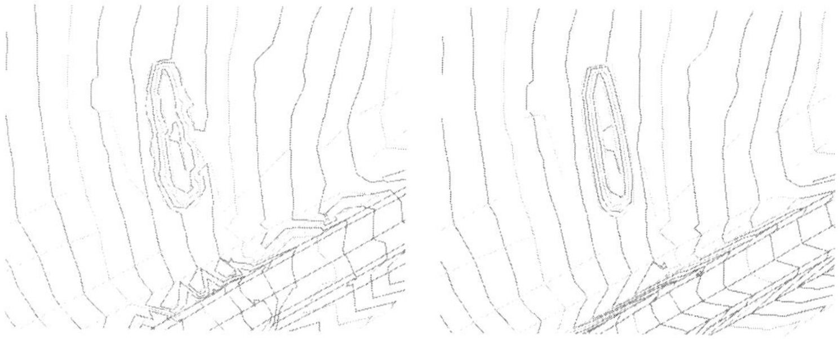

When this exercise is complete, you may close the drawing. A saved finished copy of this drawing is available from the book's web page with the filename `SurfaceBreaklines_FINISHED.dwg` or `SurfaceBreaklines_METRIC_FINISHED.dwg`.

The surface changes reflect the breaklines added. On sites with more extreme grade breaks, such as those that might follow a channel or a site grading, breaklines are invaluable in building the correct surface.

Crossing Breaklines

Invariably, you will see Panorama pop up with a message about crossing breaklines. In general, Civil 3D does not like breaklines that cross themselves. The Resolve Crossing Breaklines tool will let you examine those situations:

1. Click on a surface with breaklines.
2. From the TIN Surface contextual tab ➢ Analyze panel, choose Resolve Crossing Breaklines.
3. At the `Please specify the types of breakline you want to find or [surveyDatabase Figure Surface]:` prompt, enter **S**↵ to select the surface option.

   The Crossing Breaklines tab on Panorama lists the crossing breaklines. You can decide how you want to resolve them using Use Higher Elevation, Use Lower Elevation, Use Average Elevation, or Use Specified Elevation, as shown here:

   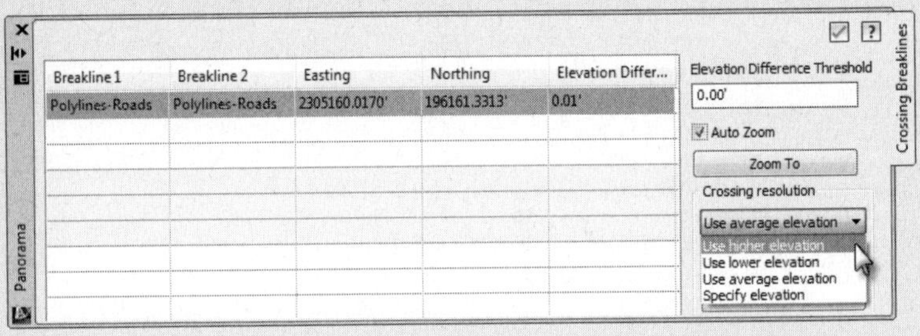

4. Click on each breakline and click the Resolve button, which is located underneath the drop-down list shown here; the conflict disappears from the Crossing Breaklines conflict list.

Adding a Surface Boundary

In the previous exercise, you fixed some breakline issues. However, in the data presented, the bigger issue is still the number of inappropriate triangles that are being drawn along the edge of the site. It is often a good idea to leave these triangles untouched during the initial build of a surface, because they serve as pointers to topographical data (such as monumentation, control, utility information, and so on) that may otherwise go unnoticed. This is a common problem that can be solved by using a surface border. You can sketch in a polyline to approximate a border, but the Extract Objects from Surface utility gives you the ability to use the surface itself as a starting point.

The Extract Objects from Surface utility allows you to re-create any displayed surface element (contours, border, etc.) as an independent AutoCAD entity. It is important to note that only the objects that are currently visible in the surface style are extractable. In this exercise, you'll extract the existing surface boundary as a starting point for creating a more refined boundary that will limit triangulation:

1. Open the `SurfaceBoundary.dwg` or `SurfaceBoundary_METRIC.dwg` file.
2. Select the surface.
3. From the TIN Surface contextual tab ➢ Surface Tools panel, choose Extract Objects to open the Extract Objects From Surface dialog.

Extract Objects

4. Leave the Border object selected and deselect the Major Contour and Minor Contour options, as shown in Figure 4.25.

FIGURE 4.25 Extracting the border from the surface object

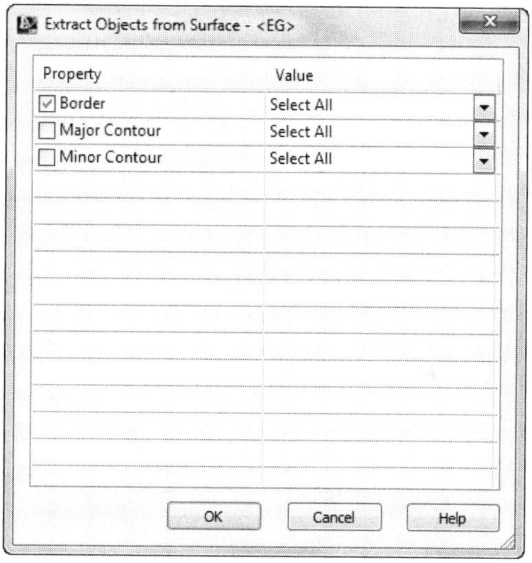

5. Click OK to finish the process. Press Esc to deselect the surface.

A 3D polyline has been created from the surface border.

6. Pick the border 3D polyline.

This polyline will form the basis for your final surface boundary. By extracting the border polyline from the existing surface, you save a lot of time playing connect the dots along the points that are valid. Next, you'll refine this polyline and add it back into the surface as a boundary.

7. In Prospector, right-click Point Groups and select the Properties option. The Point Groups dialog appears.

8. Select _All Points from the list of point groups and then move _All Points to the top of the list using the Move The Selected Item To The Top Of The Order arrow on the right side of the dialog.
9. Click OK to display all your points on the screen.

There are a lot of points, so this might take a minute.

10. Working your way around the site, grip-edit the polyline you created in step 6 to exclude some of the area at the northwest of the site where there are no points.

On a large site, you can see that this is a time-consuming process but worth the effort to clean up the site nicely (Figure 4.26). Thankfully, there are other methods you can use to clean up the surface border; we will discuss these methods in later exercises.

FIGURE 4.26
Using the grips to adjust the border

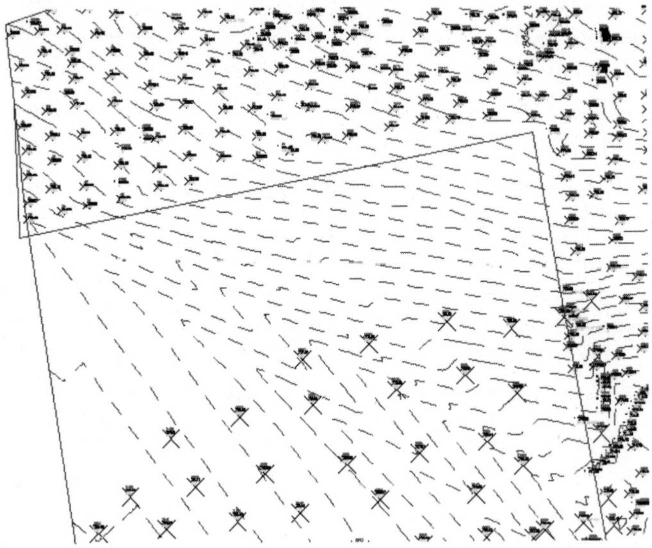

11. In Prospector, right-click Point Groups and select the Properties option.

The Point Groups dialog appears.

12. Select the No Display point group from the list and then move No Display to the bottom of the list using the Move The Selected Item To The Top Of The Order arrow on the right side of the dialog.
13. Click OK to turn off the display of all your points on the screen.

If completed, your polyline might look something like Figure 4.27. A saved finished copy of this drawing with the complete polyline is available from the book's web page with the filename `SurfaceBoundaryPolyline_FINISHED.dwg` or `SurfaceBoundary_METRIC_FINISHED.dwg` for use in the remaining portion of this exercise.

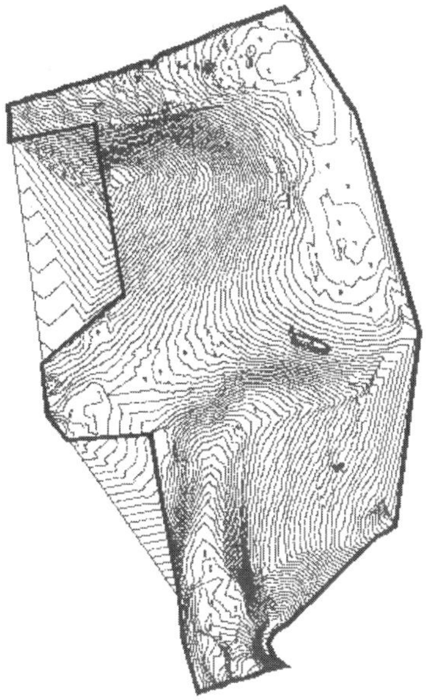

FIGURE 4.27
Revised surface border polyline

Just as with breaklines, there are multiple types of surface boundaries:

Outer Boundaries Use this type to define the outer edge of the shown boundary. When the Non-destructive Breakline option is used, the points outside the boundary are still included in the calculations; then additional points are created along the boundary line where it intersects with the triangles it crosses. This boundary trims the surface for display but does not exclude the points outside the boundary. You'll want to have your outer boundary among the last operations in your surface-building process. Therefore, as future edits are made, you may want to move the Add Boundary build operation back to the bottom of the operations list on the Definition tab in the Surface Properties dialog, as discussed earlier in this chapter.

Show Boundaries Use this type to show the surface inside a hide boundary, essentially creating a reverse donut effect in the surface display.

Hide Boundaries Use this type to punch a hole in the surface display for tasks like building footprints or a wetlands area that are not to be touched by design. Hidden surface areas are *not* deleted but merely not displayed; therefore, the surface inside a hide boundary is still used for calculations such as area or cut/fill.

Data Clip Boundaries Data clip boundaries place limits on data that will be considered part of the surface from that point going *forward*. This type is different from an outer boundary in that the data clip boundary will keep the data from ever being built into the surface as opposed to limiting it after the build. Using data clip boundaries is handy when you are attempting to build Civil 3D surfaces from large data sources such as Light Detection and Ranging (LiDAR) or DEM files. Because they limit data being placed into the surface definition, you'll want to have data clips among the first operations in your surface.

The addition of every boundary is considered a separate part of the building operations. This means that the order in which the boundaries are applied controls their final appearance. For example, a show boundary selected before a hide boundary will be overridden by that hide operation. To finish the exercise, you'll add the outer boundary twice, once as a non-destructive breakline and once with a standard breakline, and observe the difference.

14. Open the `SurfaceBoundaryPolyline_FINISHED.dwg` or `SurfaceBoundaryPolyline_METRIC_FINISHED.dwg` file.
15. In Prospector, expand the Surfaces branch.
16. Right-click EG and select the Surface Properties option.

The Surface Properties dialog appears.

17. On the Information tab, change the Surface Style to 1′ and 5′ TIN Editing (Ex.), or 1 m and 5 m TIN Editing (Ex.) for the metric users, then click OK.
18. In Prospector, expand the Surfaces ➤ EG ➤ Definition branches.
19. Right-click Boundaries and select the Add option.

The Add Boundaries dialog opens (Figure 4.28).

FIGURE 4.28
Add Boundaries dialog

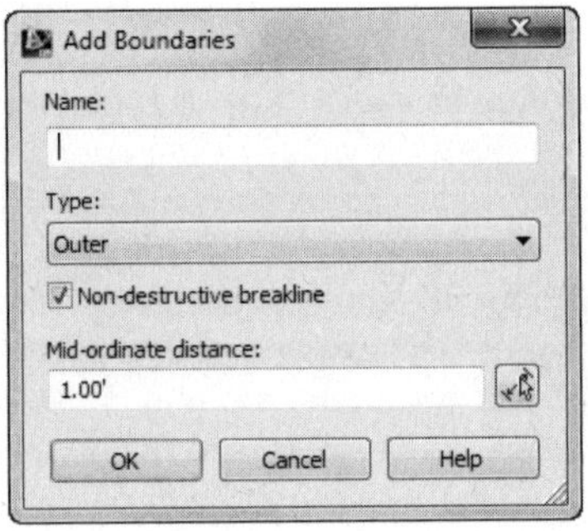

20. Enter a name if you like and verify that there is a check mark next to the Non-destructive Breakline option; then click OK.
21. Pick the polyline border that you previously extracted and edited; then notice the immediate change.
22. Zoom in on the northwest portion of your site, as shown in Figure 4.29.

Save and keep the drawing open for the next portion of the exercise.

Notice how the edge of the triangulation includes points shown as a square with a + symbol; these are the additional points created along the boundary line where it intersects with the triangles it crosses. The points you attempted to exclude from the surface are still being included in the calculation of elevations for this point; they are just excluded from the display and calculations. This isn't the result you were after, so let's fix it now.

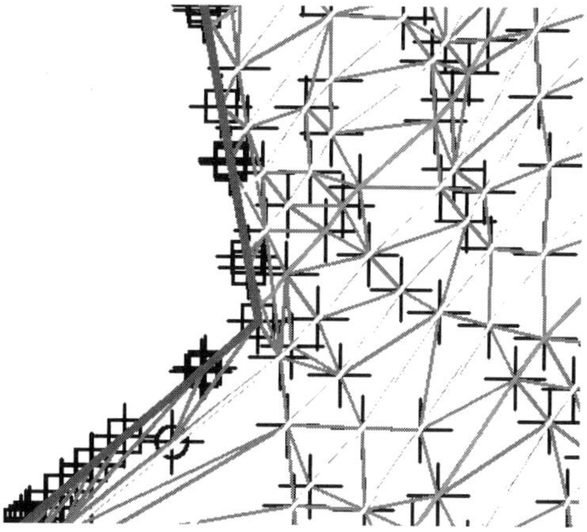

FIGURE 4.29
A non-destructive outer boundary in action

23. In Prospector, expand the Surfaces ➤ EG ➤ Definition branches and select Boundaries.

    A listing of the boundaries appears in the preview area of Toolspace.

24. In the preview area at the bottom of Prospector, right-click the boundary you just created and select the Delete option.

25. Click OK in the warning dialog that tells you the selected definition items will be permanently removed from the surface.

    In Prospector, you will now see the yellow exclamation point status flag next to the EG branch as well as the definition branch. This is because the surface needs to be rebuilt.

26. Right-click the EG branch and select the Rebuild option to rebuild the surface without the boundary definition.

    You can dismiss Panorama if it appears.

27. In Prospector, right-click Boundaries and select the Add option again.

    The Add Boundaries dialog appears.

28. This time, leave the Non-destructive Breakline option unchecked, and click OK.

29. Pick the border 3D polyline again.

    Notice that no triangles intersect your boundary now where it does not connect points, as shown in Figure 4.30.

When this exercise is complete, you may close the drawing. A saved finished copy of this drawing is available from the book's web page with the filename `SurfaceBoundary_FINISHED.dwg` or `SurfaceFromBoundary_METRIC_FINISHED.dwg`.

As you have seen, the non-destructive boundary we looked at first is the peacekeeper. It always seeks to find a "common ground" (interpolated elevation) between the two existing points on either side of a TIN line that it crosses, whereas the destructive boundary is so powerful that it cuts off all communication between points inside and outside the boundary.

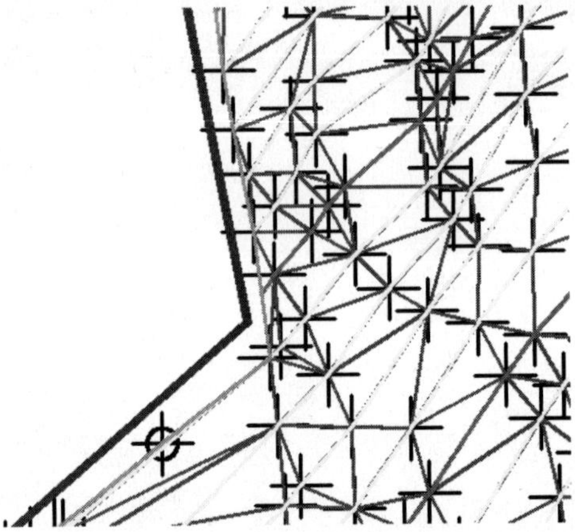

FIGURE 4.30
A destructive outer boundary in action

In spite of adding breaklines and a border, you still have some areas that need further correction or changes.

SURFACE CROPPING

Surface cropping is useful when you are working with a small portion of a much larger surface. A cropped area within a surface becomes a separate surface object (a new surface) to be managed and manipulated on a much smaller scale. In this example, you will create a cropped surface and add it to an existing drawing:

1. Open the `SurfaceCropping.dwg` or `SurfaceCropping_METRIC.dwg` file.
2. Select the surface to activate the contextual tab and expand the bottom of the Surface Tools pane.

Create Cropped Surface

3. From the Surface Tools expanded panel, choose the Create Cropped Surface option.

   You may receive a message box stating that "This drawing must be saved before this command can be used." If so:

   a. Save the drawing.

   b. Reselect the Create Cropped Surface option.

   The Create Cropped Surface dialog is displayed, as shown in Figure 4.31.

4. Click the text that says <Selection Method> in the Value column next to Select Crop Area to display the ellipsis button; once it's visible, click the ellipsis button.
5. At the `Select first corner or [Object Polygon]:` prompt, enter **O**↵.
6. At the `Select objects:` prompt, select the red polyline on the _Cropped layer and press ↵.
7. At the `Select point in the area to crop:` prompt, click inside the polyline you just selected.

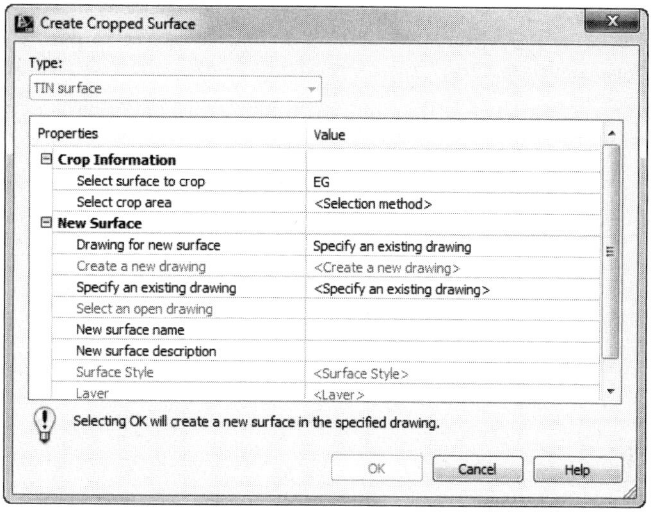

FIGURE 4.31
The Create Cropped Surface dialog

The cropped boundary is now highlighted, and the Create Cropped Surface dialog returns.

8. Left-click inside the Value column next to Drawing For New Surface and select Create A New Drawing from the drop-down list.

9. Click the text that says <Create A New Drawing> in the Value column next to Create A New Drawing to display the ellipsis button and then click that button.

The Select Template dialog opens.

10. Select an appropriate template (either `_AutoCAD Civil 3D (Imperial) NCS.dwt` or `_AutoCAD Civil 3D (Metric) NCS.dwt`) and click Open.

The Value column for Create A New Drawing is populated with the next open drawing name, such as `Drawing1.dwg`. At this point the new drawing has only been created and opened but is not yet saved.

11. In the Value column for New Surface Name, enter **Cropped Area**.

You can also enter a description for the new surface if desired.

12. Set Surface Style to Contours And Triangles.

13. Click OK to complete the Create Cropped Surface command and create a cropped surface in the new drawing.

14. Switch to the drawing just created and zoom extents to see your newly created cropped surface, as shown in Figure 4.32.

When this exercise is complete, you may close the drawing. A saved finished copy of this drawing with the cropped surface is available from the book's web page with the filename `SurfaceCropping_FINISHED.dwg` or `SurfaceCropping_METRIC_FINISHED.dwg`.

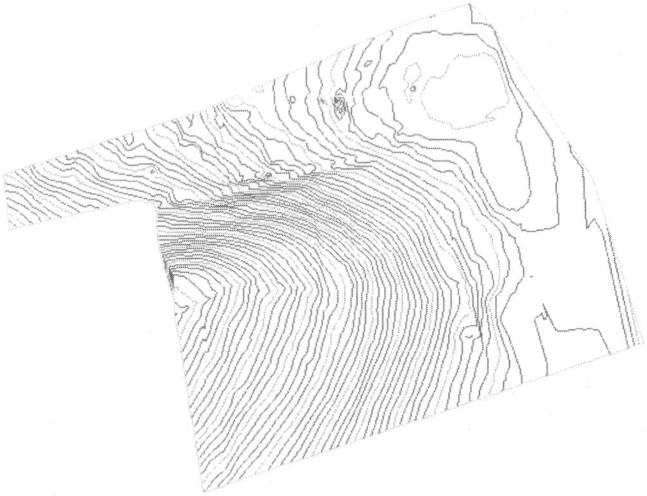

FIGURE 4.32
The completed cropped surface

RENDERING WITH MATERIALS?

A Civil 3D surface must display triangles for rendering materials to be calculated and shown. You will inevitably forget this; it's just one of those frustrating anomalies in the program.

MANUAL SURFACE EDITS

In your surface, you have a few "finger" surface areas where the surveyors went out along narrow paths from the main area of topographic data. The nature of TIN surfaces is to connect dots, and so these fingers often wind up as webbed areas of surface information that's not accurate or pertinent. A number of manual edits can be performed on a surface, including adding a boundary. These edit options are part of the definition of the surface and include the following:

Add Line Connects two points where a triangle did not exist before. This option essentially adds a breakline to the surface, so adding a breakline would generally be a better solution. This option is not available on grid surfaces.

Delete Line Removes the connection between two points. In addition to Outer Boundaries, this option is used frequently to clean up the edge of a surface or to remove internal data where a surface should have no triangulation at all. This can be an area such as a building pad or water surface.

Swap Edge Changes the direction of the triangulation methodology. For any four points, there are two solutions to the internal triangulation, and the Swap Edge option alternates from one solution to the other. The necessity of numerous swap edge operations can be limited by the use of appropriate breaklines. This option is not available on grid surfaces.

Add Point Allows for the manual addition of surface data. This function is often used to add a peak to a digitized set of contours that might have a flat spot at the top of a hill.

Delete Point Allows for the manual removal of a data point from the surface definition. Generally, it's better to fix the source of the bad data, but this option can be a fix if the original data is not editable (in the case of a LandXML file, for example).

Modify Point Modify Point allows for changing the elevation of a surface point. Only the TIN point is modified, not the original data input.

Move Point Move Point is limited to horizontal movement. Like Modify Point, only the TIN point is modified, not the original data input. This option is not available on grid surfaces.

Minimize Flat Areas Performs the edits you saw earlier in this chapter to add supplemental information to the TIN and to create a more accurate surface, forcing triangulation to work in the *z* direction instead of creating flat planes. This option is not available on grid surfaces.

Raise/Lower Surface A simple arithmetic operation that moves the entire surface in the positive or negative *z* direction. This option is useful for testing rough grading schemes for balancing dirt or for adjusting entire surfaces after a new benchmark has been observed.

Smooth Surface Presents a pair of methods for supplementing the surface TIN data (note that this option is not available on grid surfaces). Both smoothing methods work by extrapolating more information from the current TIN data, but they are distinctly different in their methodology:

> **Natural Neighbor Interpolation (NNI)** Adds points to a surface on the basis of the weighted average of nearby points. This data generally works well to refine contouring that is sharply angular because of limited information or long TIN connections. NNI works only within the bounds of a surface; it cannot extend beyond the original data.
>
> **Kriging** Adds points to a surface based on one of five distinct algorithms to predict the elevations at additional surface points. These algorithms create a trending for the surface beyond the known information and can therefore be used to extend a surface beyond even the available data. Kriging is very volatile, and you should understand the full methodology before applying this information to your surface. Kriging is frequently used in subsurface exploration industries such as mining, where surface (or strata) information is difficult to come by and the distance between points can be higher than desired.
>
> **Paste Surface** Pulls in the TIN information from the selected surface and replaces the TIN information in the host surface with this new information while keeping the dynamic relationship to the original surface. This option is helpful in creating composite surfaces that reflect both the original ground and the design intent. This option is not available on grid surfaces. We'll look at pasting in Chapter 16, "Grading."
>
> **Simplify Surface** Allows you to reduce the amount of TIN data being processed while maintaining the accuracy of the surface. This is done using one of two methods: Edge Contraction, wherein Civil 3D tries to collapse two points connected by a line to one point, or Point Removal, which removes selected surface points based on algorithms designed to reduce data points that are similar. This option is not available on grid surfaces.

Manual editing should always be the last step in updating a surface. Fixing the surface is a poor substitution for fixing the underlying data the TIN is built from, but in some cases, it is the quickest and easiest way to make a more accurate surface.

Point and Triangle Editing

In this section, you'll remove triangles manually, and then finish your surface by correcting what appears to be a blown survey shot.

1. Open the `SurfaceEdits.dwg` file (or the `SurfaceEdits_METRIC.dwg` file). Confirm that the EG surface style is set to Contours And Triangles.
2. In Prospector, expand the Surfaces ➤ EG ➤ Definition branches.
3. Right-click Edits and select the Delete Line option.

Note that if you have Level Of Detail turned on, a red circle warning will appear if you are not zoomed in enough to view the true triangulation.

4. At the `Select edges:` prompt, enter **F**↵ to use the Fence selection mode.

In Fence selection mode you will draw a multisegment selection line and any objects that cross the line will be selected. Alternatively you could enter **C**↵ to use the Crossing selection mode, which uses a selection window and selects all objects that have at least a portion within the window.

5. Use a Center Osnap to pick the circle labeled A at the lower right of the pick area, as shown in Figure 4.33; move to the upper-left corner as shown; use a Center Osnap to pick the circle labeled B; and press ↵ twice.

FIGURE 4.33
Using a Crossing Window selection

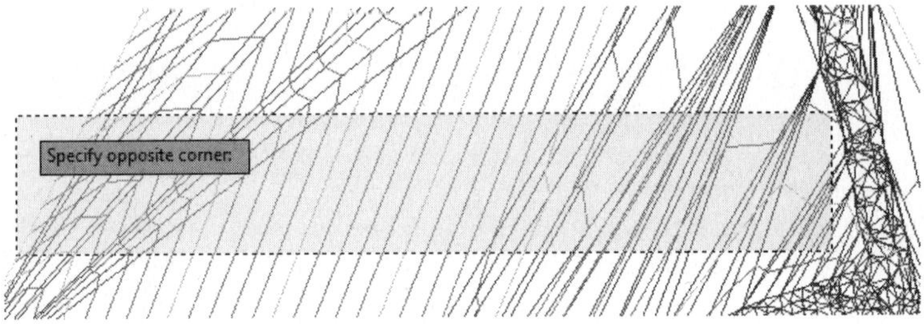

6. Right-click or press ↵ to finish the selection set.
7. Repeat this process, removing triangles until your site resembles the image on the right in Figure 4.34.
8. Zoom to the portion of your site with the red circle labeled C, and you'll notice a collection of contours that seems out of place.
9. Change Surface Style to Contours And Points.
10. In Prospector, right-click Edits again and select the Delete Point option.
11. Zoom in on the area very close.

You will find a series of + point markers in the area with close contours. These are blown shots and the contours are simply obeying the point elevation. In this case, it is 0.

FIGURE 4.34
Surface before removal of extraneous triangles (left) and after (right)

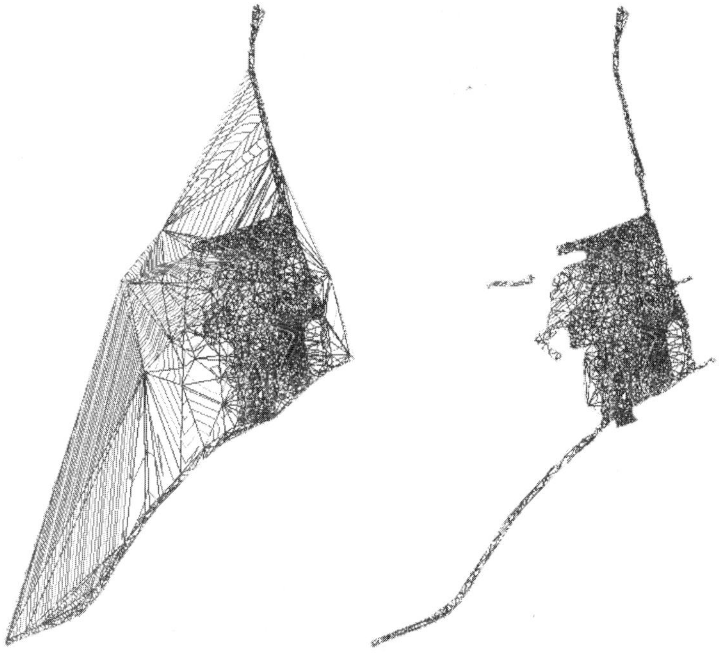

12. Select each of the three markers and delete them; then notice the immediate change in the contouring.

When this exercise is complete, you may save and keep the drawing open to continue on to the next exercise or use the saved finished copy of this drawing available from the book's web page (`SurfaceEdits_FINISHED.dwg` or `SurfaceEdits_METRIC_FINISHED.dwg`).

Surface Smoothing

One common complaint about computer-generated contours is that they're simply too precise. The level of calculations in setting elevations on the basis of linear interpolation along a triangle leg makes it possible for contour lines to be overly exact, ignoring contour line trends in place of small anomalies of point information. Under the eye of a board drafter, these small anomalies were averaged out, and contours were created with smooth flowing lines.

While you can apply object-level smoothing as part of the contouring process, this process smooths the end result but not the underlying data. In this section, you'll use the NNI smoothing algorithm to reduce surface anomalies and create a more visually pleasing contour set:

1. If it's not still open from the previous exercise, open the `SurfaceEdits_FINISHED.dwg` or the `SurfaceEdits_METRIC_FINISHED.dwg` file.
2. In Prospector, expand the Surfaces ➢ EG ➢ Definition branches.
3. Right-click Edits and select the Smooth Surface option.

The Smooth Surface dialog opens.

4. Expand the Smoothing Methods branch, and verify that Natural Neighbor Interpolation is the Select Method value.
5. Expand the Point Interpolation/Extrapolation branch, and click in the Select Output Region value field.
6. Click the ellipsis button.
7. At the `Select region or [rEctangle pOlygon Surface]:` prompt, pick the rectangle located in the center of the surface for smoothing and then press ↵ to return to the Smooth Surface dialog.
8. Enter **5** for the Grid X-Spacing and Grid Y-Spacing values (for metric users, enter **2** for both values), and then press ↵.

Note that Civil 3D will tell you how many points you are adding to the surface immediately below this input area by the value given in the Number Of Output Points field as shown in Figure 4.35. It's grayed out, but it does change on the basis of your input values.

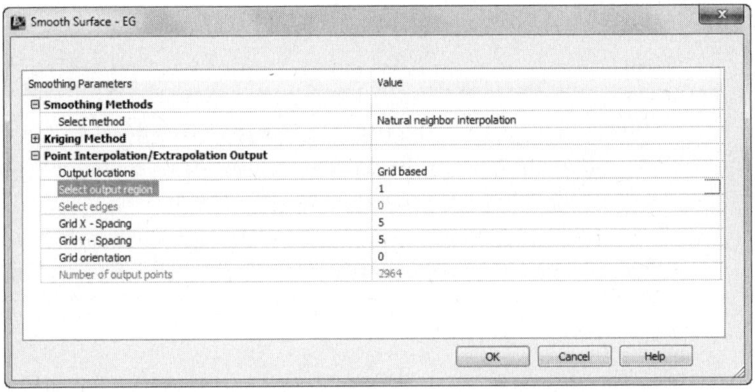

FIGURE 4.35
Smooth Surface dialog

9. Click OK and the surface will be smoothed, similar to what is shown in Figure 4.36.

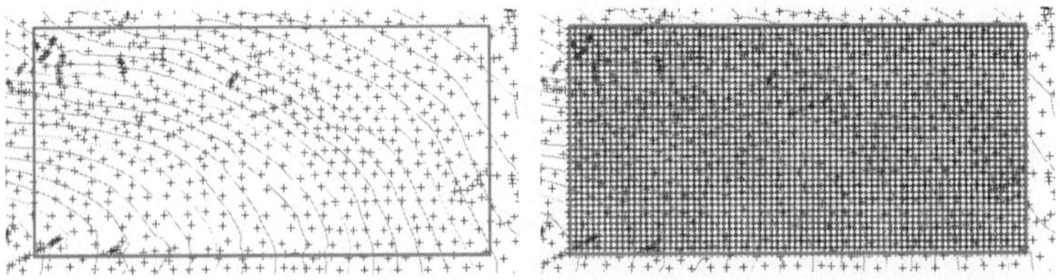

FIGURE 4.36
Surface before NNI smoothing (left) and after (right)

When this exercise is complete, you may close the drawing. A saved finished copy of this drawing is available from the book's web page with the filename `SurfaceSmoothing_FINISHED.dwg` or `SurfaceSmoothing_METRIC_FINISHED.dwg`.

Note that we said the *surface* will be smoothed — not the contours. Contour smoothing will be discussed later. To see the result of the surface smoothing, change Surface Style to Contours And Points to display your image as shown in Figure 4.37.

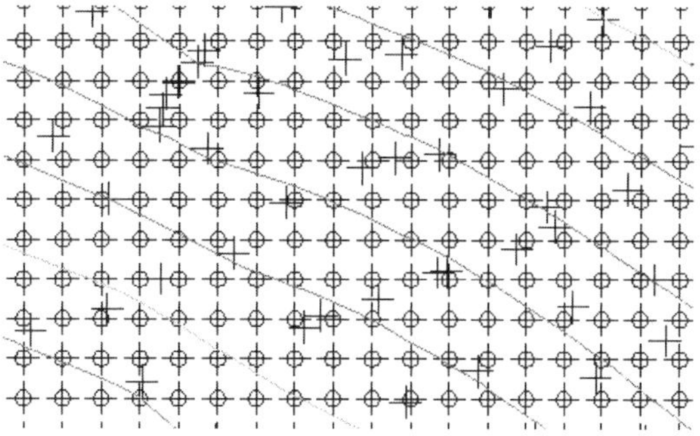

FIGURE 4.37
Points added via NNI surface smoothing

Note all of the points with a circle cross symbol. These points are all new, created by the NNI surface-smoothing operation. The derived points are part of your surface, and the contours reflect the updated surface information.

Surface Simplifying

Because of the increasing use in land development projects of GIS and other data-heavy inputs, it's critical that Civil 3D users know how to simplify the surfaces produced from these sources. In this exercise, you'll simplify the surface created from a drawing earlier in this chapter.

1. Open the `SurfaceSimplifying.dwg` file (or the `SurfaceSimplifying_METRIC.dwg` file). For reference, the surface statistics of the EG-GIS surface are shown in Figure 4.38.

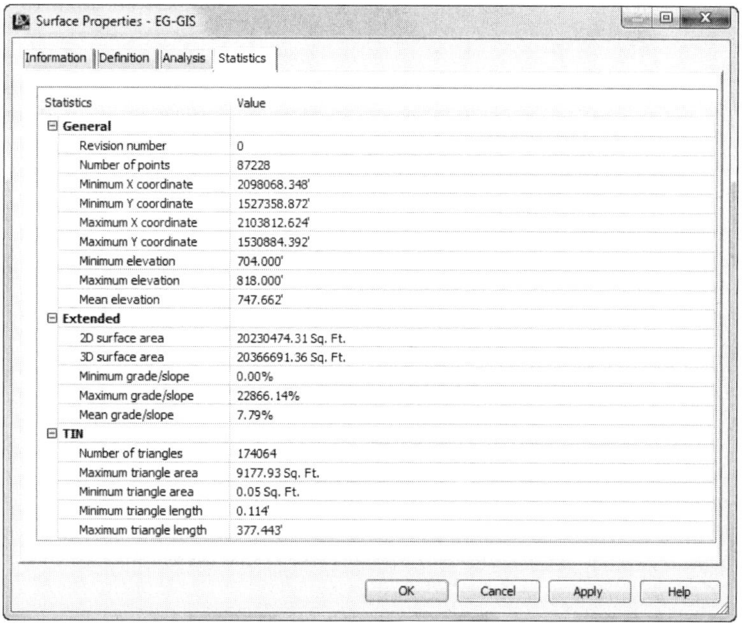

Statistics	Value
General	
Revision number	0
Number of points	87228
Minimum X coordinate	2098068.348'
Minimum Y coordinate	1527358.872'
Maximum X coordinate	2103812.624'
Maximum Y coordinate	1530884.392'
Minimum elevation	704.000'
Maximum elevation	818.000'
Mean elevation	747.662'
Extended	
2D surface area	20230474.31 Sq. Ft.
3D surface area	20366691.36 Sq. Ft.
Minimum grade/slope	0.00%
Maximum grade/slope	22866.14%
Mean grade/slope	7.79%
TIN	
Number of triangles	174064
Maximum triangle area	9177.93 Sq. Ft.
Minimum triangle area	0.05 Sq. Ft.
Minimum triangle length	0.114'
Maximum triangle length	377.443'

FIGURE 4.38
EG-GIS surface statistics before simplification

2. In Prospector, expand Surfaces ➤ EG-GIS ➤ Definition.
3. Right-click Edits and select the Simplify Surface option to launch the Simplify Surface dialog.
4. Select the Point Removal radio button, as shown in Figure 4.39, and click Next to move to the Region Options page.

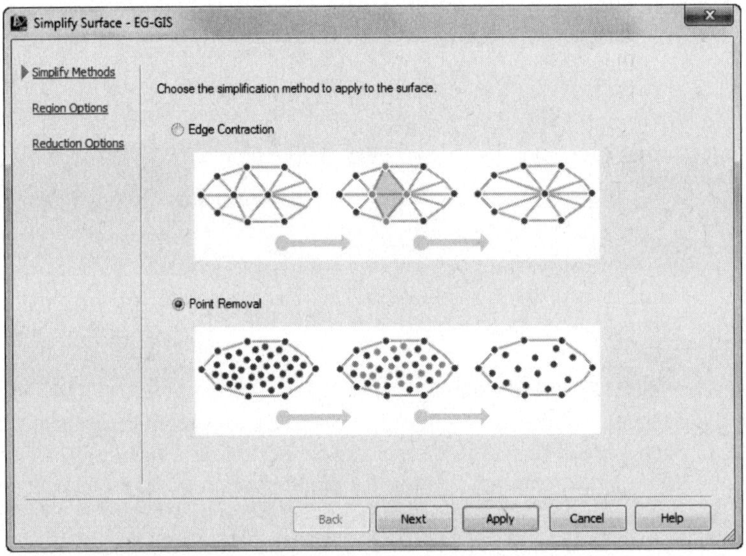

FIGURE 4.39 The Simplify Surface – Simplify Methods page

5. Accept the Region Options defaults as shown in Figure 4.40 and click Next to move to the Reduction Options page.

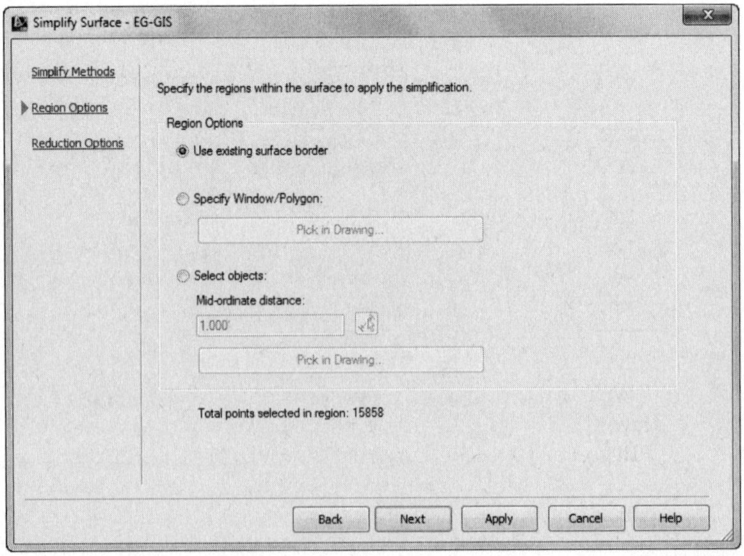

FIGURE 4.40 The Simplify Surface – Region Options page

6. Set Percentage Of Points To Remove to 20 percent and then deselect the Maximum Change In Elevation option.

This value is the maximum change allowed between the surface elevation at any point before or after the simplify process has run.

7. Click Apply.

The program will process this calculation and display a Total Points Removed number, as shown in Figure 4.41. You can adjust the slider or toggle on the Maximum Change In Elevation button to experiment with different values. Note that every time you click Apply or Finished, the number of points decreases by that percentage.

FIGURE 4.41
The Simplify Surface – Reduction Options page

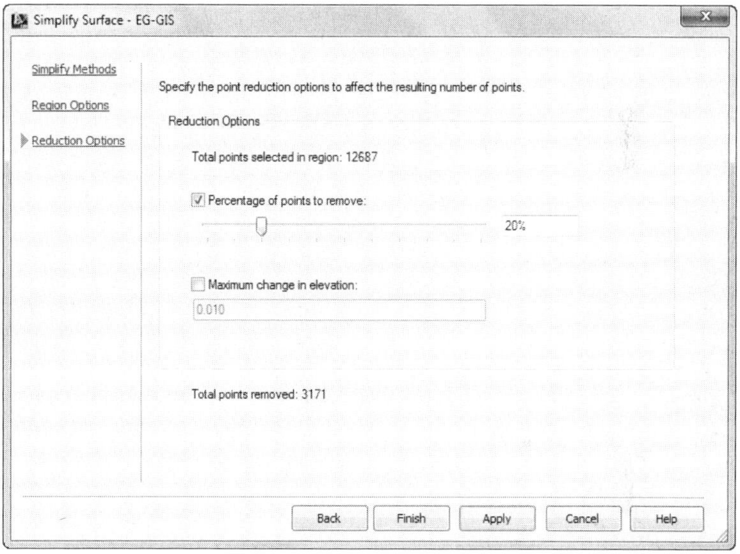

8. Click Finish to close the wizard and fully commit to the Simplify edit.

A Word about the Simplify Build Operation

Notice that the Simplify build operation is listed twice in the preview area of Prospector under Edits. This is because the command was run once when you clicked Apply and a second time when you clicked Finished. If you only want to run the Simplify Surface command once, you should click Apply and then click Cancel or just click Finished without clicking Apply first.

When this exercise is complete, you may close the drawing. A saved finished copy of this drawing is available from the book's web page with the filename `SurfaceSimplifying_FINISHED.dwg` or `SurfaceSimplifying_METRIC_FINISHED.dwg`.

A quick visit to the Surface Properties Statistics tab shows that the number of points has been reduced, as shown in Figure 4.42. On something like an aerial topography or DEM, reducing the point count probably will not reduce the usability of the surface, but this simple point reduction will decrease the file size. Remember, you can always delete the edit or deselect the operation on the Definition tab of the Surface Properties dialog to "un-simplify" the surface.

FIGURE 4.42
EG-GIS surface statistics after simplification

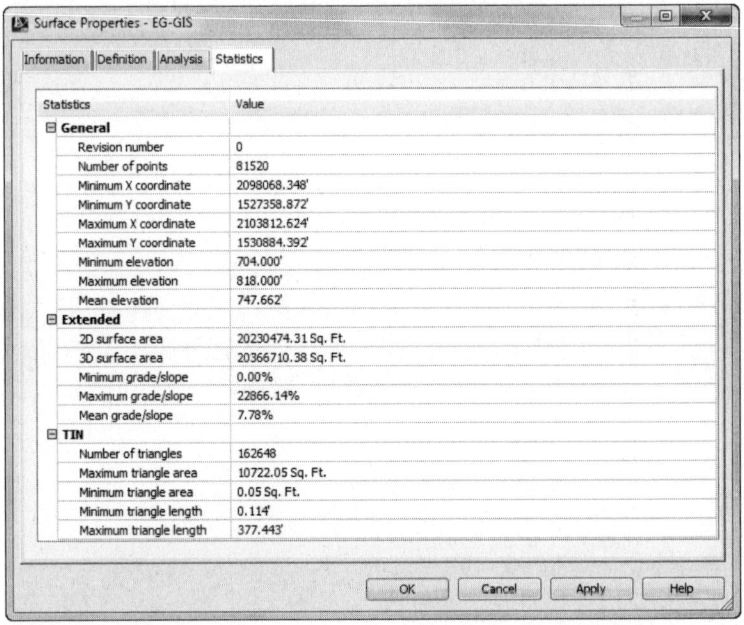

The creation of a surface is merely the starting point. Once you have a TIN to work with, you have a number of ways to view the data using analysis tools and varying styles.

Surface Analysis

CERT OBJECTIVE

Once a surface is created, you can display information in a number of ways. The most common so far has been contours and triangles, but those are the basics. By using varying styles, you can show a large amount of data with one single surface. Surface styles are discussed further in Chapter 21, "Object Styles." While some of the styles are used for generating plans (such as contours), others lend themselves to analyzing the surface during creation.

For a surface object in plan view, Points, Triangles, Border, Major Contour, Minor Contour, User Contours, and Gridded are standard components and are controlled like any other object component. Directions, Elevations, Slopes, and Slope Arrows components are unique to surface styles. Note that the Layer, Color, and Linetype fields are grayed out for these components. Each of these components has its own special coloring schemes, which we'll look at in the next section. In this section, you will explore the elevation and slope analysis styles.

Elevation Banding

Displaying surface information as bands of color is one of the most common display methods for engineers looking to make a high-impact view of the site. Elevations are a critical part of the site design process, and understanding how a site varies in terms of elevation is an important part of making the best design. Elevation analysis typically falls into two categories: showing bands of information on the basis of pure distribution of linear scales or displaying a lesser number of bands to show some critical information about the site. In this first exercise, you'll use a standard style to illustrate elevation distribution along with a prebuilt color scheme that works well for presentations:

1. Open the `SurfaceAnalysis.dwg` file (or the `SurfaceAnalysis_METRIC.dwg` file).
2. Select the surface on your screen to activate the contextual tab.
3. From the TIN Surface contextual tab ➢ Modify panel, choose the Surface Properties option.

The Surface Properties dialog appears.

4. On the Information tab, change the Surface Style field to Elevation Banding (3D).
5. Switch to the Analysis tab for the Elevations analysis type.
6. Verify that Create Ranges By is set to Number Of Ranges and that the value is set to **3**, and then click the Run Analysis arrow in the middle of the dialog to populate the Range Details area.
7. Click OK to close the Surface Properties dialog.

Notice that only the border is currently showing since the surface is being shown in plan view.

[-][Top][2D Wireframe]

You may have noticed three pieces of text in the upper-left corner of your model space. If you left-click on one of these In-Canvas View Controls, you will find that they are drop-down lists that you can use to change what you are looking at.

The first set of bracketed text will either be [-] (denoting that one viewport is being displayed) or [+] (if two viewports are being displayed).

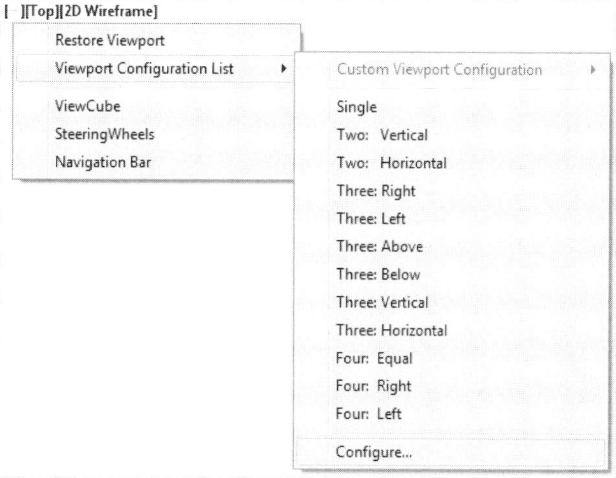

The second set of bracketed text is the View Control and will list the current view within the brackets.

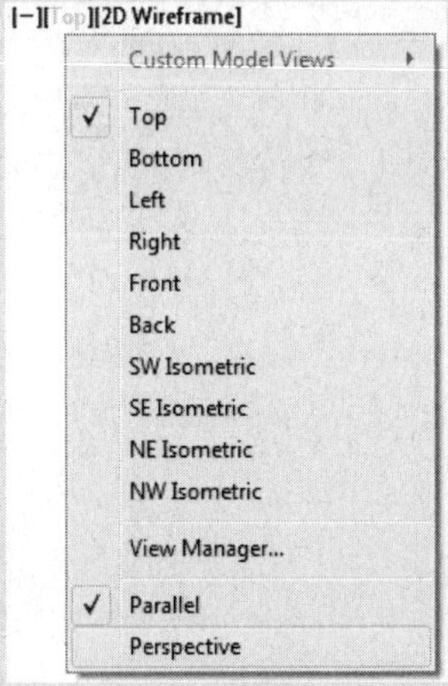

The third set of bracketed text is the Visual Style Control and will list the current visual style within the brackets.

All of these commands can also be changed from the View tab. If the Level Of Detail is currently active, an additional small icon will appear below this line of text.

These three pieces of text provide valuable information for easy reference while you work.

8. From the View Control, select SW Isometric.
9. Zoom in if necessary to get a better view of the surface.
10. From the Visual Style Control, select the Conceptual option to see a semi-rendered view that should look something like Figure 4.43.

Save and keep the drawing open for the next portion of the exercise.

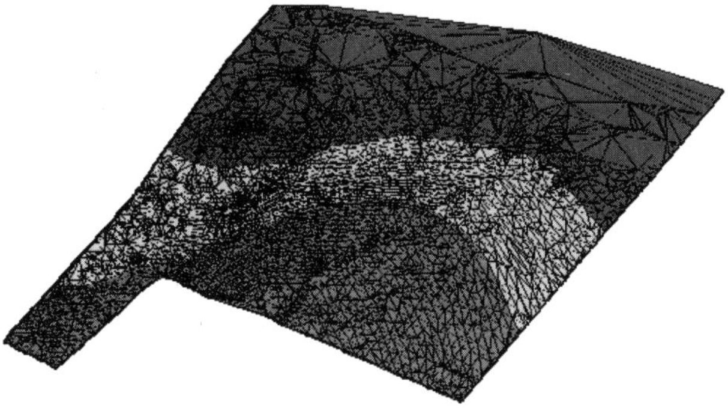

FIGURE 4.43
Conceptual view of the site with the Elevation Banding style

AUTOCAD VISUAL STYLES

The triangles seen are part of the view style and can be modified via the Visual Styles Manager. Turning the Edge mode off will leave you with a nicely gradated view of your site. You can edit the visual style by clicking the Visual Style Control at the upper left of your drawing screen. You can also access the Visual Style Manager on the View tab ➢ Visual Styles panel by clicking the small arrow in the lower-right corner of the panel. The Visual Styles Manager is shown here:

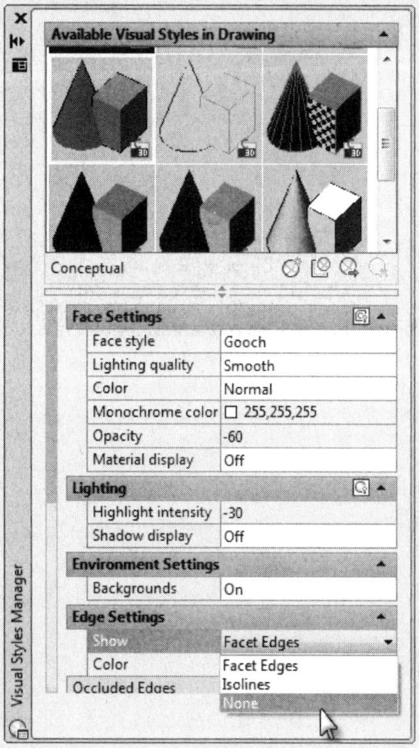

You'll use a 2D elevation to clearly illustrate portions of the site that cannot be developed. Next, you'll manually tweak the colors and elevation ranges on the basis of design constraints from outside the program:

11. Click the View Control and select Top.
12. Click the Visual Style Control and select 2D Wireframe.

This site has a limitation placed in that no development can go below the elevation of 790′ (or 240 m). Your analysis will show you the areas that are below 790′ (or 240 m), a buffer zone to 791′ (or 241 m), and then everything above that.

13. Select the surface by clicking on the border. Right-click and choose the Surface Properties option.

The Surface Properties dialog appears.

14. On the Information tab, change the Surface Style field to Elevation Banding (2D).
15. On the Analysis tab, change Maximum Elevation for ID 1 and Minimum Elevation for ID 2 both to **790** (or **240** for metric users).
16. Change Maximum Elevation for ID 2 and Minimum Elevation for ID 3 both to **791** (or **241** for metric users).
17. Modify your Color Scheme to match Figure 4.44 by double-clicking on each color to open the Select Color dialog, selecting the appropriate color, and clicking OK to close the dialog. (The colors are red, yellow, and green from top to bottom, respectively.)

FIGURE 4.44 The Surface Properties dialog after manual editing

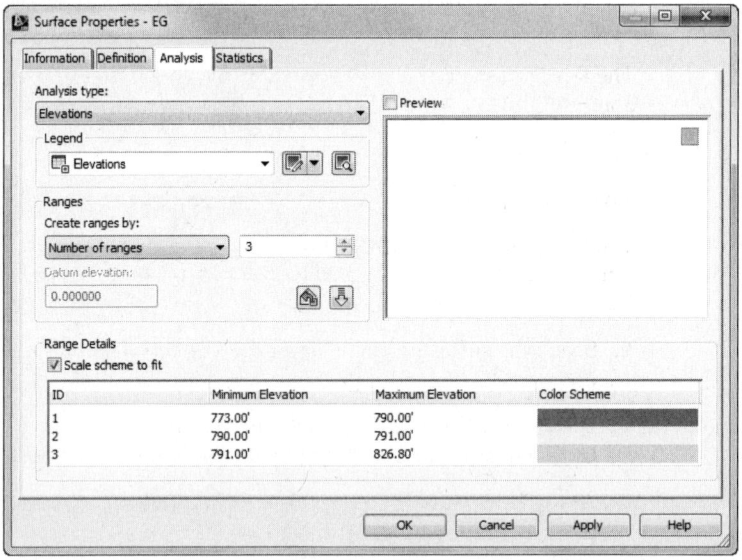

18. Click OK to accept the settings in the Surface Properties dialog.

When this exercise is complete, you may close the drawing. A saved finished copy of this drawing is available from the book's web page with the filename `SurfaceAnalysis_FINISHED.dwg` or `SurfaceAnalysis_METRIC_FINISHED.dwg`.

Understanding surfaces from a vertical direction is helpful, but many times the slopes are just as important. In the next section, you'll take a look at using the slope analysis tools in Civil 3D.

Slopes and Slope Arrows

Beyond the bands of color that show elevation differences in your models, you also have tools that display slope information about your surfaces. This analysis can be useful in checking for drainage concerns, meeting accessibility requirements, or adhering to zoning constraints. Slope is typically shown as areas of color similar to the elevation banding or as colored arrows that indicate the downhill direction and slope. In this exercise, you'll look at a proposed site grading surface and run the two slope analysis tools:

1. Open the `SlopeAnalysis.dwg` file (or the `SlopeAnalysis_METRIC.dwg` file).
2. Select the surface on your screen to activate the contextual tab.
3. From the TIN Surface contextual tab ➤ Modify panel, choose Surface Properties.

   The Surface Properties dialog appears.

4. On the Information tab, change the Surface Style field to Slope Banding (2D).
5. Switch to the Analysis tab of the Surface Properties dialog.
6. Choose Slopes from the Analysis Type drop-down list.
7. Verify that the Number field in the Ranges area is set to **3** and click the Run Analysis arrow in the middle of the dialog to populate the Range Details area.

   The Range Details area will populate. You could change the minimums and maximums as you did in the previous exercise, but this time you'll keep the defaults.

8. Click OK to close the dialog.

   Save and keep the drawing open for the next portion of the exercise.

The colors are nice to look at, but they don't mean much, and slopes don't have any inherent information that can be portrayed by color association. To make more sense of this analysis, you'll add a legend table:

9. Select the surface again, and on the TIN Surface contextual tab ➤ Labels & Tables panel, choose Add Legend.
10. At the `Enter table type [Directions Elevations Slopes slopeArrows Contours Usercontours Watersheds]:` prompt, enter **S**↵ to select Slopes.
11. At the `Behavior [Dynamic Static]: <Dynamic>` prompt, press ↵ again to accept the default value of a Dynamic legend.
12. At the `Select upper left corner:` prompt, pick a point on screen to draw the legend, as shown in Figure 4.45.

FIGURE 4.45
The Slopes legend table

Slopes Table				
Number	Minimum Slope	Maximum Slope	Area	Color
1	0.00%	7.08%	202209.40	
2	7.08%	10.69%	182627.83	
3	10.69%	39000.33%	105351.52	

Save and keep the drawing open for the next portion of the exercise.

By including a legend, you can make sense of the information presented in this view. Because you know what the slopes are, you can also see which way they go.

13. Select the surface, and on the TIN Surface contextual tab ➢ Modify panel, choose Surface Properties.

The Surface Properties dialog appears.

14. On the Information tab, change the Surface Style field to Slope Arrows.
15. Switch to the Analysis tab of the Surface Properties dialog.
16. Choose Slope Arrows from the Analysis Type drop-down list.

17. Verify that the Number field in the Ranges area is set to **3** and click the Run Analysis arrow in the middle of the dialog to populate the Range Details area.
18. Click OK to close the dialog.

The benefit of arrows is in looking for "birdbath" areas that will collect water. These arrows can also verify that inlets are in the right location, as shown in Figure 4.46. Look for arrows pointing to the proposed drainage locations and you'll have a simple design-verification tool.

FIGURE 4.46
Slope arrows pointing to a proposed inlet location

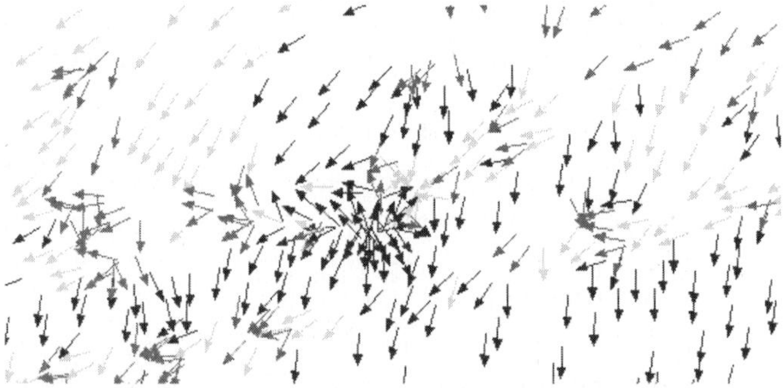

When this exercise is complete, you may close the drawing. A saved finished copy of this drawing is available from the book's web page with the filename `SlopeAnalysis_FINISHED.dwg` or `SlopeAnalysis_METRIC_FINISHED.dwg`.

With these simple analysis tools, you can show a client the areas of their site that meet their constraints. Visually strong and simple to produce, this is the kind of information that a 3D model makes available. Beyond the basic information that can be represented in a single surface, Civil 3D also contains a number of tools for comparing surfaces. You'll compare this existing ground surface to a proposed grading plan in the next section.

Visibility Checker

The Zone of Visual Influence tool allows you to explore what-if scenarios. In this example, a 40′ (12 m) tower has been proposed for the site. A concerned neighbor wants to make sure that it won't obstruct their scenic view. You want to check it from the proposed surface:

1. Open `VisibilityCheck.dwg` or `VisibilityCheck_METRIC.dwg`.
2. In Prospector, expand the Surfaces branch, right-click on the FG surface, and choose the Select option.
3. On the TIN Surface contextual tab ➢ Analyze panel, choose Visibility Check ➢ Zone Of Visual Influence.
4. At the `Specify location of object:` prompt, using the Intersection Osnap select the center of the proposed tower located on the southern portion of the site (denoted by a large white square with an X inside).
5. At the `Specify height of object:` prompt, enter **40**↵ to set the tower height to 40′ (metric users, enter **12**↵).
6. At the `Specify the radius of vision extent:` prompt, pan to and select the endpoint at the upper-right corner of the cyan colored house located at the northeastern corner of the site.

   Save and keep the drawing open for the next portion of the exercise.

The drawing now has bands of color:

- Green near the tower location indicates that the object is completely visible.
- Yellow indicates that the object is partially visible.
- Red indicates that the object is not visible.

So in our example, the homeowner on the upper right will be happy to know that the proposed 40′ tower will not appear in their view.

The next portion of the example will use the Point to Point tool.

7. Using the same drawing, zoom to the intersection of Syrah Way and Cabernet Court, where you will see a car driving on the right-hand side of the road.

   We want to check the sight distance.

8. If it is not already selected, select the FG surface.
9. On the TIN Surface contextual tab ➢ Analyze panel, choose Visibility Check ➢ Point to Point.
10. At the `Specify height of eye:` prompt, enter **3.5**↵ (metric users enter **1**↵).

    This sets the height of the driver's eye while sitting in a typical car.

11. At the `Specify location of eye:` prompt, click where the driver would normally be seated in the vehicle.
12. At the `Specify height of target:` prompt, enter **6**↵ (metric users enter **1.8**↵).

A rubber-banding sightline ray appears.

13. Click along the path where oncoming cars would be seen.

A sightline arrow is drawn on the screen:

- If the arrow is green, it means that the view is unobstructed and the command line will tell you the distance from the eye.
- If any portion of the arrow is red, it indicates that the view is obstructed, and the command line will tell you the distance at which the obstruction occurs.

When this exercise is complete, you may close the drawing. A saved finished copy of this drawing is available from the book's web page with the filename `SurfaceVisibility_FINISHED.dwg` or `SurfaceVisibility_METRIC_FINISHED.dwg`.

Unfortunately, these visual tools are not dynamic; if you change the surface, you will need to rerun the visual tools. Perhaps this will be addressed in a future release.

Comparing Surfaces

Earthwork is a major part of almost every land development project. The money involved with earthmoving is a large part of the budget, and for this reason, minimizing this impact is a critical part of the final design. Civil 3D contains a number of surface analysis tools designed to help in this effort, and you'll look at them in this section. First, a simple comparison provides feedback about the volumetric difference, and then a more detailed approach enables you to perform an analysis on this difference.

For years, civil engineers have performed earthwork using a section methodology. Sections were taken at some interval, and a plot was made of both the original surface and the proposed surface. Comparing adjacent sections and multiplying by the distance between them yields an end-area method of volumes that is generally considered acceptable. The main problem with this methodology is that it ignores the surfaces in the areas between sections. These areas could include areas of major change, introducing some level of error. In spite of this limitation, this method worked well with hand calculations, trading some accuracy for ease and speed.

With the advent of full-surface modeling, more precise methods became available. By analyzing both the existing and proposed surfaces, a volume calculation can be performed that is as good as the two surfaces. At every TIN vertex in both surfaces, a distance is measured vertically to the other surface. These delta amounts can then be used to create a third dynamic surface called a volume surface, which represents the difference between the two original surfaces.

TIN Volume Surface

Using the volume utility for initial design checking is helpful, but quite often contractors and other outside users want to see more information about the grading and earthwork for their own uses. This requirement typically falls into two categories: a cut-fill analysis showing colors or contours or a grid of cut-fill tick marks.

Color cut-fill maps are helpful when reviewing your site for the locations of movement. Some sites have areas of better material or can have areas where the cost of cut is prohibitive (such as rock). In this exercise, you'll use two of the surface analysis methods to look at the areas for cut-fill on your site:

1. Open the `VolumeSurface.dwg` or the `VolumeSurface_METRIC.dwg` file.
2. From the Analyze tab ➢ Volumes And Materials panel, choose the Volumes Dashboard tool to display Panorama open to the Volumes Dashboard tab.
3. Click the Create New Volume Surface button to display the Create Surface dialog.

Notice that Type is already set to TIN Volume Surface. You also have the option to select Grid Volume Surface.

4. Change the name to **VOL-EG-FG** and set the style to Elevation Banding (2D).
5. Click the <Base Surface> field next to Base Surface to display the ellipsis button; once it's visible, click the ellipsis to select the EG surface.
6. Click the <Comparison Surface> next to Comparison Surface to display the ellipsis button; once it's visible, click the ellipsis to select the FG surface.

The Create Surface dialog should now look similar to Figure 4.47.

FIGURE 4.47
Creating a volume surface

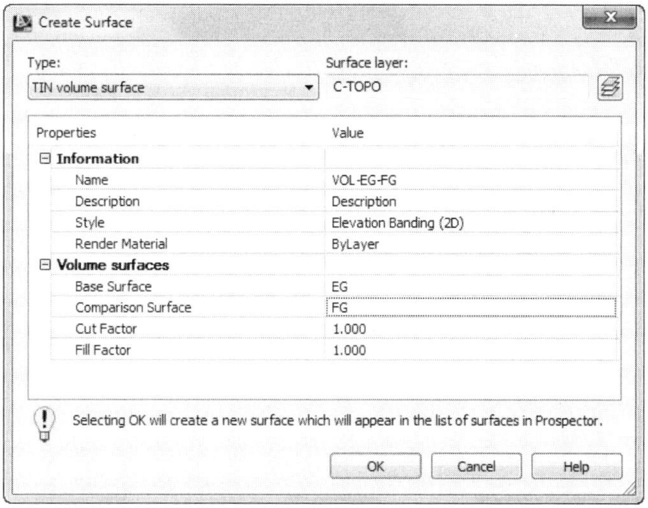

7. Click OK to accept the settings in the Create Surface dialog.

Civil 3D will calculate the volume (Figure 4.48).

FIGURE 4.48
Composite volume calculated

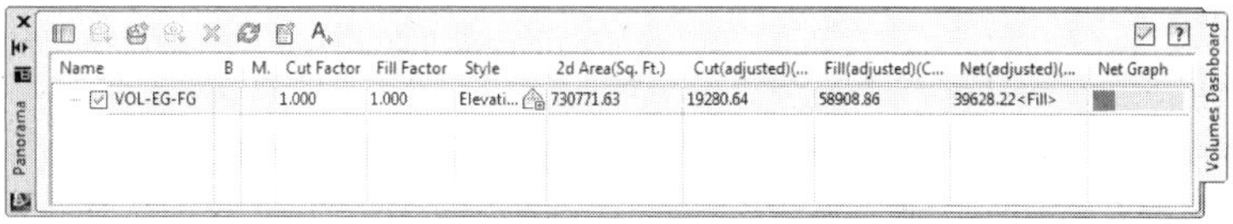

Note that you can scroll right and left in Panorama to display additional information, including the ability to apply a cut or fill factor by typing directly into the cells for these values.

VOL-EG-FG

This new volume surface appears in Prospector's Surfaces collection, but notice that the icon is slightly different, showing two surfaces stacked on each other. The color mapping currently shown is just a default set, though, and does not indicate much.

8. Leave the Panorama open and in Prospector, expand the Surfaces ➢ FG ➢ Definition branches.
9. Right-click Edits and select the Raise/Lower Surface option.
10. Enter **−0.25** at the command line to drop the site 3″ (metric users enter **−0.075**).

Notice that a yellow exclamation point status flag has appeared next to the volume surface in Panorama as well as in Prospector. The Panorama no longer lists the volumes and instead states "Out of date."

11. Right-click the VOL-EG-FG surface in Panorama or Prospector and select the Rebuild option.
12. In Prospector under the Definition branch of the FG surface, click Edits and right-click the Raise/Lower edit in the preview area and select Delete.
13. A dialog will appear warning you that the selected definition will be permanently removed from the surface. Click OK.
14. Return to Panorama and rebuild the volume surface again to return to the original volume calculation.
15. Close Panorama when complete.
16. In Prospector, right-click VOL-EG-FG in the Surfaces branch and select the Surface Properties option.

The Surface Properties dialog appears.

17. In the Surface Properties dialog, switch to the Statistics tab and expand the Volume branch.

The value shown for the Net Volume (Unadjusted) is the same as shown in the Panorama in the first part of this exercise.

18. In the Surface Properties dialog, switch to the Analysis tab for the Elevations analysis type.
19. Verify that the Create Ranges By is set to Number Of Ranges and that the value is set to 3; then click the Run Analysis arrow in the middle of the dialog to populate the Range Of Details area.
20. Change Maximum Elevation for ID 1 and Minimum Elevation for ID 2 to **-0.5** (or **-0.15** for metric users).
21. Change Maximum Elevation for ID 2 and Minimum Elevation for ID 3 to **0.5** (or **0.15** for metric users).

22. Modify your Color Scheme to match Figure 4.49.

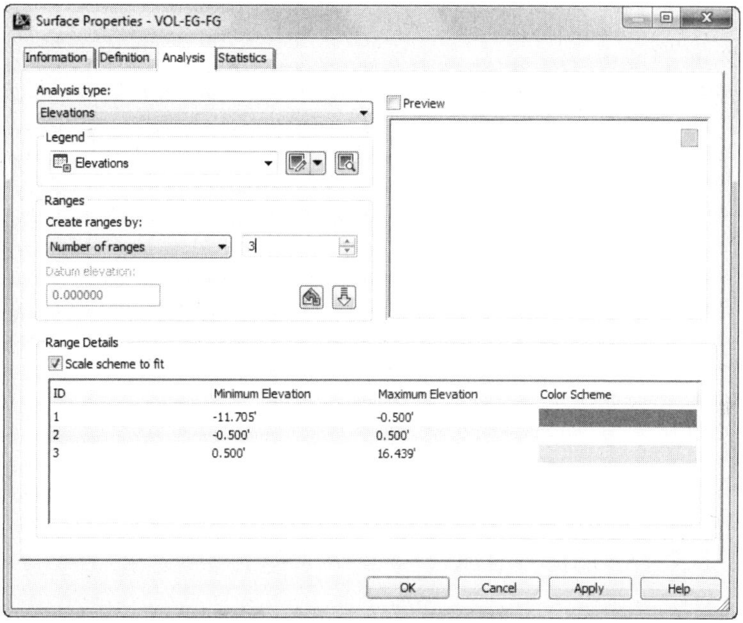

FIGURE 4.49 Elevation analysis settings for earthworks

The recommended colors are red, yellow, and green, where red indicates the worst case cut, green represents the worst case fill, and yellow represents a balance.

23. Click OK to close the Surface Properties dialog.

Save and keep the drawing open for the next portion of the exercise.

The volume surface now indicates areas of cut, fill, and areas near balancing, similar to Figure 4.50. If you leave a small range near the balance line, you can more clearly see the areas that are being left nearly undisturbed.

To show where large amounts of cut or fill could incur additional cost (such as compaction or excavation protection), you would simply modify the analysis range as required.

The Elevation Banding surface is great for onscreen analysis, but the color fills make it hard to plot or use in many applications. In these steps, you use the Contour Analysis tool to prepare cut-fill contours in these same colors:

24. In Prospector, right-click VOL-EG-FG in the Surfaces branch and select the Surface Properties option to display the Surface Properties dialog again.

25. On the Analysis tab, choose Contours from the Analysis Type drop-down list.

26. Verify that the Number field in the Ranges area is set to **3** and click the Run Analysis arrow in the middle of the dialog to populate the Range Details area.

27. Change the ranges to match those you entered in the previous portion of the exercise (as shown in Figure 4.49).

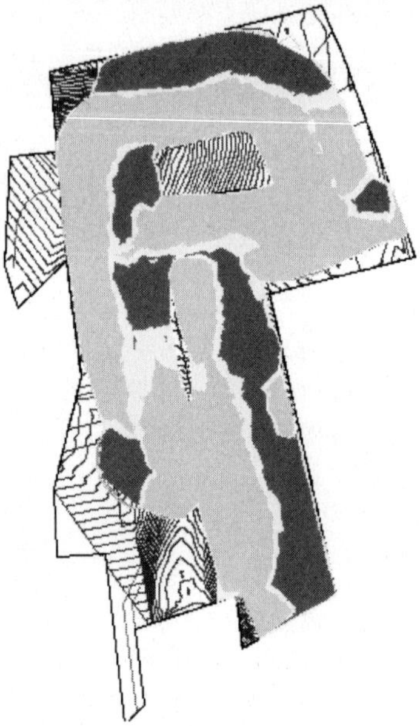

FIGURE 4.50
Completed elevation analysis

28. In the Major Contour column, click the small button to the far right to display the AutoCAD Select Color dialog and set a color for each ID.

Typical contour colors are

- Shades of red for cut
- A yellow for the balance line
- Shades of green for fill

29. Switch to the Information tab on the Surface Properties dialog, and change the Surface Style to Contours 1′ and 5′ (Design) or Contours 0.2 m and 1.0 m (Design).

30. Click the down arrow next to the Style field and select the Copy Current Selection option.

The Surface Style Editor appears.

31. On the Information tab, change the Name field to **Contours 1′ and 5′ (Earthwork)** or **Contours 0.2 m and 1.0 m (Earthwork)**.

32. On the Contours tab, expand the Contour Ranges branch.

33. Change the value of the Use Color Scheme property to True.

It's safe to ignore the values here because you hard-coded the values in your surface properties.

34. Click OK to close the Surface Style Editor and click OK again to close the Surface Properties dialog.

When this exercise is complete, you may close the drawing. A saved finished copy of this drawing is available from the book's web page with the filename `VolumeSurface_FINISHED.dwg` or `VolumeSurface_METRIC_FINISHED.dwg`.

The volume surface can now be analyzed on a lot-by-lot basis or labeled using the surface-labeling functions to show the depths of cut and fill, which you'll look at in the next sections.

Labeling the Surface

CERT OBJECTIVE

Once the three-dimensional surface model has been created, it is time to communicate the model's information in various formats. This includes labeling contours, creating legends for the analysis you've created, adding spot labels, or labeling the slope. These exercises work through these main labeling requirements and building styles for each.

Contour Labeling

The most common requirement is to place labels on surface-generated contours. In Land Desktop, this was one of the last steps because a change to a surface required erasing and replacing all the labels. Once labels have been placed, their styles can be modified.

Contour labels in Civil 3D are created by special lines that understand their relationship with the surface. Everywhere one of these lines crosses a contour line, a label is placed. This label's appearance is based on the style applied and can be a major, minor, or user-defined contour label. Each label can have styles selected independently, so using some AutoCAD selection techniques can be crucial to maintaining uniformity across a surface. In this exercise, you'll add labels to your surface and explore the interaction of contour label lines and the labels themselves.

1. Open the `SurfaceLabeling.dwg` or `Surface Labeling_METRIC.dwg` file.
2. Select the EG surface in the drawing to display the TIN Surface contextual tab.
3. From the TIN Surface contextual tab ➢ Labels & Tables panel, choose Add Labels ➢ Contour – Single.
4. Pick any spot on a major contour to add a label.
5. Press ↵ when complete and press Esc to deselect the surface.
6. From the Annotate tab ➢ Labels & Tables panel, choose Add Labels ➢ Surface ➢ Contour – Multiple.
7. Pick a point to the west of the site and then a second point across the site to the south of the site, crossing a number of contours in the process.
8. Press ↵ to end the picking.
9. From the Annotate tab ➢ Labels & Tables panel, click on the Add Labels button instead of the drop-down to display the Add Labels dialog.
10. Set Feature to Surface and Label Type to Contour – Multiple At Interval.
11. Click the Add button.
12. Pick a point to the west of the site and then a second point across the site to the north of the site.

13. Enter **200** (for metric **60**) at the command line for an interval value.

  Save and keep the drawing open for the next portion of the exercise.

You've now labeled your site in three ways to get contour labels in a number of different locations. You will need additional labels in the northeast and southwest to complete the labeling, because you did not cross these contour objects with your contour label line. You can add more labels by clicking Add, but you can also use the labels created already to fill in these missing areas. By modifying the contour line labels, you can manipulate the label locations and add new labels. Next, you'll fill in the labeling to the northeast:

14. Zoom to the northeast portion of the site, and notice that some of the contours are labeled only along the boundary or not at all, as shown in Figure 4.51.

FIGURE 4.51
Contour labels applied

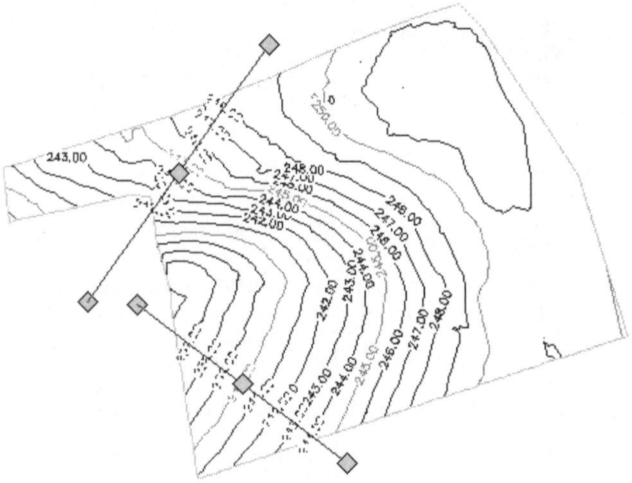

15. Zoom in to any contour label placed using the Contour – Multiple At Interval button, and pick the text.

  Three grips will appear. The original contour label lines are quite apparent, but in reality, every label has a hidden label line beneath it.

16. Grab the northernmost grip and drag across an adjacent contour northeast of the original label, as shown in Figure 4.52.

  New labels will appear everywhere your dragged line now crosses a contour.

FIGURE 4.52
Grip-editing a contour label line

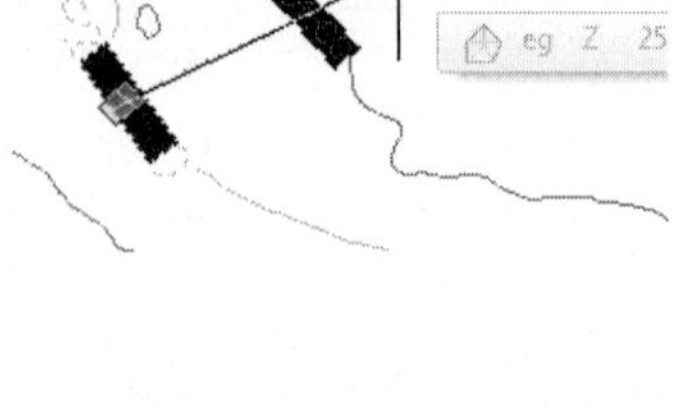

17. Drop the grip somewhere to create labels as desired. Be sure to press Esc a few times when you are done labeling the contours to end the command.

Save and keep the drawing open for the next portion of the exercise.

By using the created label lines instead of adding new ones, you'll find it easier to manage the layout of your labels.

Surface Point Labels

In every site, there are points that fall off the contour line but are critical. In an existing surface, this can be the low point in a swale or a driveway that has to be matched. When you're working with a paper set of plans, the spot grade is the most common review element. One of the most time-consuming issues in land development is the preparation of grading plans with hundreds of individual spot grades. Every time a site grading scheme changes, these were typically updated manually, leaving lots of opportunities for error.

With Civil 3D's surface modeling, spot labels are dynamic and react to changes in the underlying surface. By using surface labels instead of points or text callouts, you can generate a grading plan early on in the design process and begin the process of creating sheets. In this section, you'll label surface slopes in a couple of ways, create a single spot label for critical information, and conclude by creating a grid of labels similar to many estimation software packages.

Labeling Slopes

Beyond the specific grade at any single point, most grading plans use slope labels to indicate some level of trend across a site or drainage area. Civil 3D can generate the following two slope labels:

One-Point Slope Labels One-point slope labels indicate the slope of an underlying surface triangle. These work well when the surface has large triangles, typically in pad or mass grading areas.

Two-Point Slope Labels Two-point slope labels indicate the slope trend on the basis of two points selected and their locations on the surface. A two-point slope label works by dividing the surface elevation distance between the points by the planar distance between the pick points. This works well in existing ground surface models to indicate a general slope direction but can be deceiving in that it does not consider the terrain between the points.

In this next exercise, you'll apply both types of slope labels, and then look at a minor style modification that is commonly requested:

1. Using the drawing from the previous portion of the exercise, select the surface to display the TIN Surface contextual tab.

2. From the TIN Surface contextual tab ➢ Labels & Tables panel, choose Add Labels ➢ Surface ➢ Slope.

3. At the `Create Slope Labels or [One-point Two-point]: <One-point>` prompt, press ↵ to select the default one-point label style.

4. Zoom in on the circle drawn on the western portion of the site and use a Center Osnap to place a label at its center, similar to that shown in Figure 4.53.

FIGURE 4.53
A one-point slope label

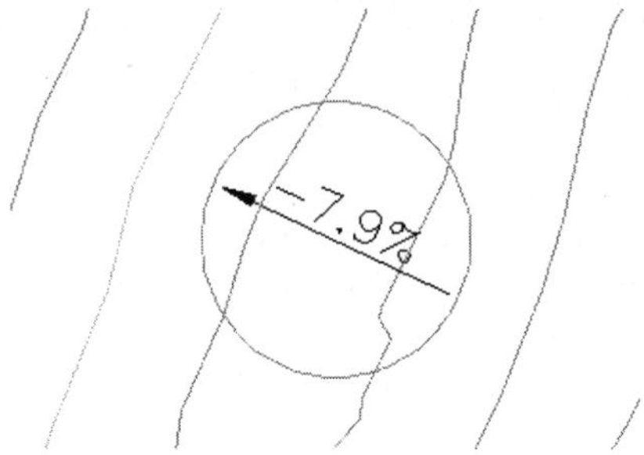

5. Press Esc or ↵ to exit the command.
6. From the TIN Surface contextual tab ➤ Labels & Tables panel, choose Add Labels ➤ Slope.
7. At the `Create Slope Labels or [One-point Two-point]: <One-point>` prompt, enter **T**↵ to switch to a two-point label style.
8. Pan to the central portion of the site, and use an Endpoint Osnap to pick the left end of the line.
9. Use an Endpoint Osnap to select the right end of the line to complete the label.
10. Press Esc or ↵ to exit the command and view the label, as shown in Figure 4.54.

FIGURE 4.54
A two-point slope label

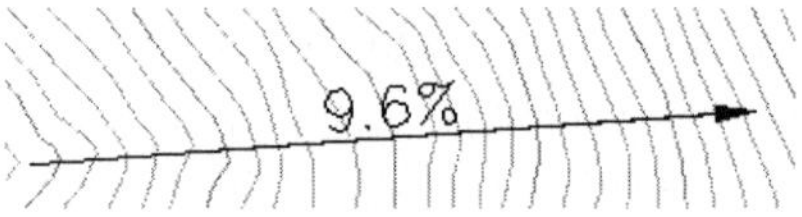

This second label indicates the average slope of the property. By using a two-point label, you get a better understanding of the trend, as opposed to a specific point.

When this exercise is complete, you may close the drawing. A saved finished copy of this drawing is available from the book's web page with the filename `SurfaceLabeling_FINISHED.dwg` or `SurfaceLabeling_METRIC_FINISHED.dwg`.

CRITICAL POINTS

A typical grading plan is a sea of critical points that drive the site topography. In the past, much of this labeling and point work was done by creating *coordinate geometry (COGO)* points and simply displaying their properties. Although this is effective, it has two distinct disadvantages. First, these points are not reflective of the design but part of the design. This makes the sheet creation a part of the grading process, not a parallel process. Second, the addition of COGO points to any drawing and project when they're not truly needed just weighs down the design model. Point management is a mentally intensive task, and anything that can limit extraneous data is worth investigating.

Surface labels react dynamically to the surface and to the point of insertion. Moving any of these labels would update the information to reflect the surface underneath. This relationship makes it possible for one user to place labels on a grading plan while the final surface is still in flux. A change in the proposed surface is reflected in an update from the project, and an updated sheet can be on the plotter in minutes.

Surface Grid Labels

Sometimes, more than a few points are requested. Estimation software typically creates a grid of point labels that can be easily reviewed or passed to a contractor for fieldwork. In this exercise, you'll use the volume surface you generated earlier in this chapter to create a set of surface labels that reflect this requirement:

1. Open the SurfaceVolumeSurface_FINISHED.dwg (or the SurfaceVolumeSurface_METRIC_FINISHED.dwg file).
2. From the Annotate tab ➤ Labels & Tables panel, choose Add Labels ➤ Surface ➤ Spot Elevations On Grid.
3. At the Select a surface <or press enter key to select from list>: prompt, press ↵ to display the Select Surface dialog.
4. Select VOL-EG-FG and click OK.
5. At the Specify a grid basepoint: prompt, pick a point southwest of the surface to set a base point for the grid.
6. At the Grid rotation: prompt, enter **0**↵.
7. At the Grid X spacing: prompt, enter **25**↵ (7↵ for metric users).
8. At the Grid Y spacing: prompt, enter **25**↵ (7↵ for metric users).
9. At the Specify the upper right location for the grid: prompt, pick a point northeast of the surface to set the area for the labels.
10. Verify that the preview window encompasses the VOL-EG-FG surface and press ↵ at the command line to continue.

Wait a few moments as Civil 3D generates all the labels just specified. Your drawing should look similar to Figure 4.55.

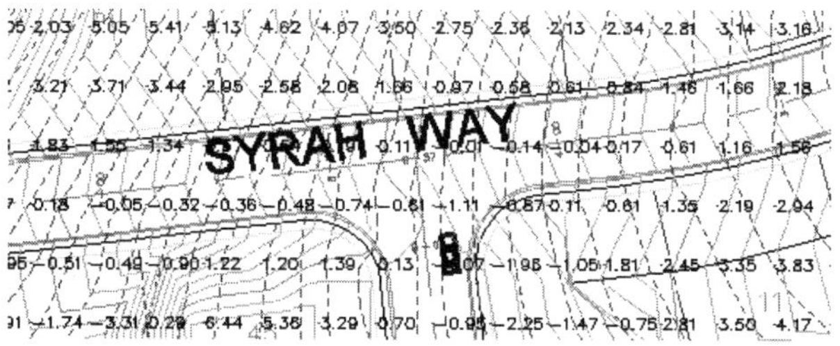

FIGURE 4.55 Volume surface with grid labels

Labeling the grid is imprecise at best. Grid labeling ignores anything that might happen between the grid points, but it presents the surface data in a familiar way for engineers and contractors. By using the tools available and the underlying surface model, you can present information from one source in an almost infinite number of ways.

When this exercise is complete, you may close the drawing. A saved finished copy of this drawing is available from the book's web page with the filename `SurfaceVolumeGridLabels_FINISHED.dwg` or `SurfaceVolumeGridLabels_METRIC_FINISHED.dwg`.

Point Cloud Surfaces

A point cloud is a huge bunch of 3D points, usually collected by laser scanner or *Light Detection and Ranging (LiDAR)*. In a geographic information system (GIS), point clouds are often used as a source for a *Digital Elevation Model (DEM)*. The technology has gotten less expensive and more accurate over the last few years, allowing LiDAR to quickly take over from traditional methods of collecting photogrammetry data.

Point clouds in many formats can be imported to Civil 3D. The most common format is the Log ASCII Standard (LAS) file. This binary format is a public format and at minimum contains *x*, *y*, and *z* data. LAS data can also include the following:

- Coordinate system of scanned area.
- Color — true color on RGB format.
- Intensity; LiDAR depends on lasers bouncing off objects and back to the scanning device and the intensity refers to the strength of the returned information. This value directly relates to the material from which the laser is bouncing. For example, concrete will have a stronger intensity than grass.
- Classification, which will be a number, most frequently between 0 and 9, that categorizes points based on material (i.e., ground, vegetation, or buildings).

For more information on the LAS standard, visit `www.asprs.org`.

Civil 3D can import a point cloud and use it in several ways. For instance, a laser scan of a bridge can be imported and placed for reference when designing a road through an existing abutment. In the example that follows, you will convert LiDAR data into a Civil 3D surface. It is important to note that point clouds often contain millions of points and require a beefy computer (and a little patience on your part) to process.

Importing a Point Cloud

A typical point cloud contains millions of points. These large files are kept external to Civil 3D in a point cloud database. After the LAS has been imported, the data is passed to three files: PRMD, IATI, and ISD. The ISD file contains the points themselves and is the only file needed by CAD if the point cloud would need to be re-created or used in base AutoCAD. By default these files get created in the same directory as the DWG but can be changed when importing the information.

If the point cloud you are working with contains coordinate system information (as all the examples in this book do), the software will automatically convert the point cloud to the units and coordinate system of the drawing. For the exercises in this chapter, it does not matter whether you choose the Metric or the Imperial template.

Civil 3D may take a long time to process these files, and you must ensure you have sufficient disk space to store them, as you will find in the following exercise. It is also a good idea to close the software and then reopen it to help clear some memory before starting memory-intensive tasks such as working with point clouds. The following exercise will show you how to import and work with a point cloud:

1. Start a new file by using the default Civil 3D template of your choice. Save the file before proceeding as **PointCloud.dwg**.
2. For Imperial users, set the Coordinate System to UTM Zone 17, NAD83, US Foot (UTM83-17F). For metric users, set the Coordinate System to UTM Zone 17, NAD83, Meters (UTM83-17).
3. In Prospector, right-click the Point Clouds branch and select the Create Point Cloud option to display the Create Point Cloud wizard, shown in Figure 4.56.

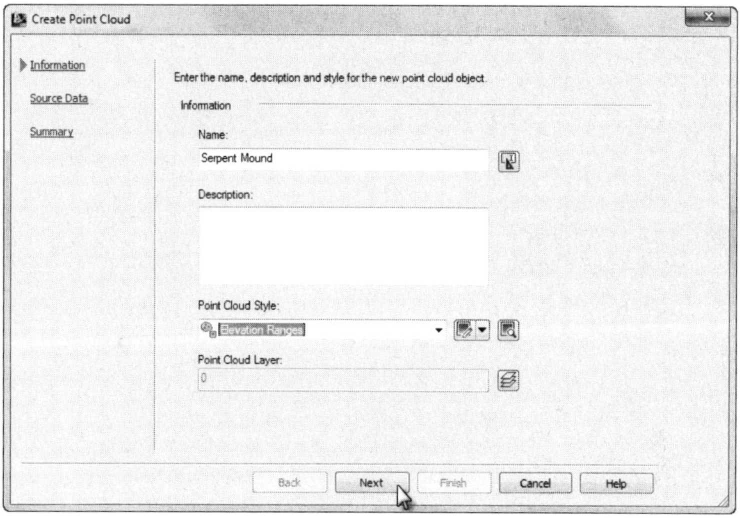

FIGURE 4.56
Create Point Cloud – Information page

4. Set the name of the point cloud to **Serpent Mound**.
5. Set the Point Cloud Style to Elevation Ranges, as shown in Figure 4.56, and click the Next button.

The Source Data page is displayed, as shown in Figure 4.57.

6. Using the white plus sign, browse to the `Serpent Mound.las` file (both Imperial and metric users can use this file).

Remember, all data and drawing files for this book can be downloaded from `www.sybex.com/go/masteringcivil3d2013`.

This is a large (90 MB) file containing roughly 1.4 million points and may require a minute or two to process.

7. Click the Next button to display the Summary page, as shown in Figure 4.58.

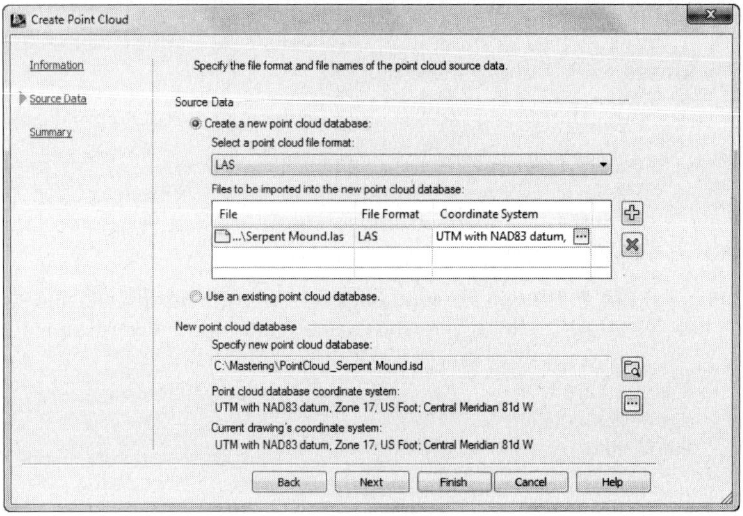

FIGURE 4.57
The Create Point Cloud – Source Data page

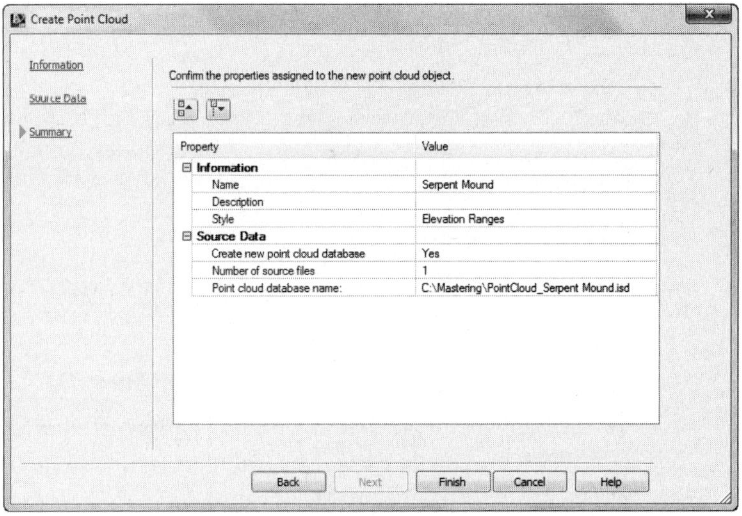

FIGURE 4.58
The Create Point Cloud – Summary page

8. Accept the defaults as shown in Figure 4.58 and click Finish to process the point cloud.
9. If the New Point Cloud Database – Processing In Background dialog appears, click Close to dismiss it.

You may notice a pop-up message in the lower-right corner of your screen indicating that the Point Cloud database is being created in the background.

10. Once the point cloud is created, it will disappear and a new pop-up message will appear stating that the point cloud has been created and a link is provided that says Click Here To Zoom. Click this link.

When complete, a portion of a bounding box outlining a portion of the point cloud is displayed in the center of the screen. Save the drawing but leave it open to complete the next exercise.

Working with Point Clouds

Once the point cloud is visible in your drawing, you'll want to follow a few rules of thumb to prevent performance problems. The key-in POINTCLOUDDENSITY value controls what percentage of the full point cloud displays on the screen at once. You can also access this value using a slider bar in the Point Cloud contextual tab. However, it is easier to hit the percentage you want on the first try if you use the key-in value. The lower this value, the fewer points are visible; hence the easier it will be to navigate your drawing. The POINTCLOUDDENSITY value does not have any effect on the number of points used when generating a surface model (this is similar to the Level Of Detail value used to aid surface processing).

When you are changing view directions on a point cloud, we recommend that you use preset views and named views to flip around the object. The orbit commands should not be used, as they are a surefire way to max out your computer's RAM. If you used the default template, your surface will be located on the V-SITE-SCAN layer. We suggest that you freeze the layer if you do not need to see the point cloud. Use Freeze instead of Off for layer management so the point cloud is not accounted for during pan, zoom, and regen operations (this is true for all AutoCAD objects, but it makes a huge difference when working with point clouds).

Creating a Point Cloud Surface

By specifying either an entire point cloud or a small region of a point cloud, you can create a new TIN surface in your drawing. Any changes to the point cloud object will render the surface definition out of date. In the following exercise, a new TIN surface is created from the point cloud previously imported:

1. Continue using the `PointCloud.dwg` file from the previous exercise.

2. Select the bounding box representing the point cloud to display the Point Cloud: Serpent Mound contextual tab, as shown in Figure 4.59.

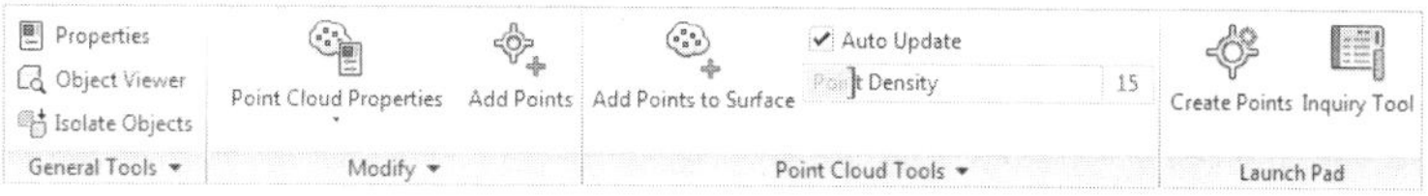

FIGURE 4.59 The Point Cloud contextual tab

3. From the Point Cloud contextual tab ➢ Point Cloud Tools panel, choose Add Points To Surface to display the Add Points To Surface wizard, as shown in Figure 4.60.

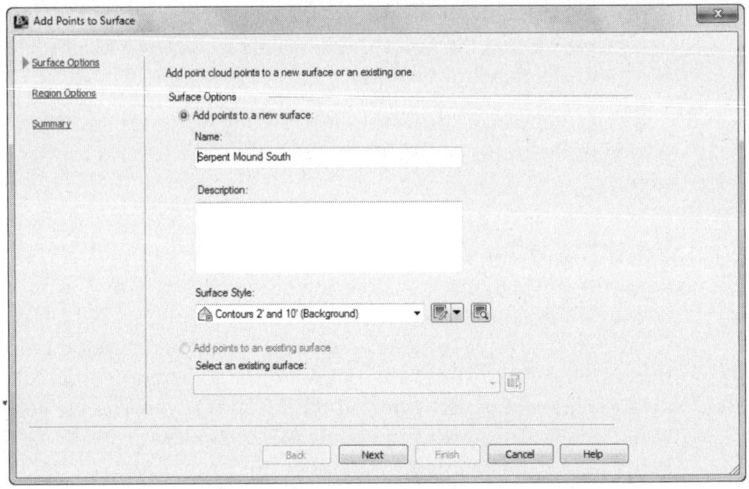

FIGURE 4.60
The Add Points To Surface – Surface Options page

4. Name the surface **Serpent Mound South**. Leave the style set to the default.
5. Click Next and the Region Options page is displayed, as shown in Figure 4.61.

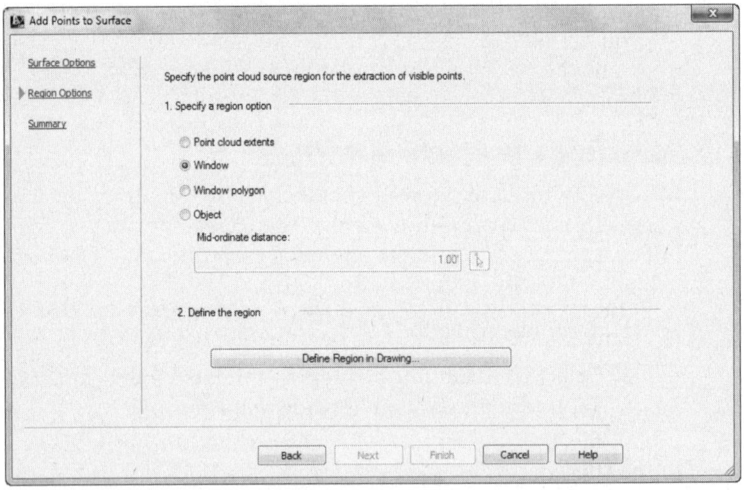

FIGURE 4.61
The Add Points To Surface – Region Options page

6. Choose the Window radio button, and click Define Region In Drawing.
7. Define the region by creating a window around the southern half of the point cloud.
8. Click Next to see the Summary page, as shown in Figure 4.62, and click the Finish button.

When this exercise is complete, you may close the drawing. Due to the large file size, a finished state of this drawing is not available for download.

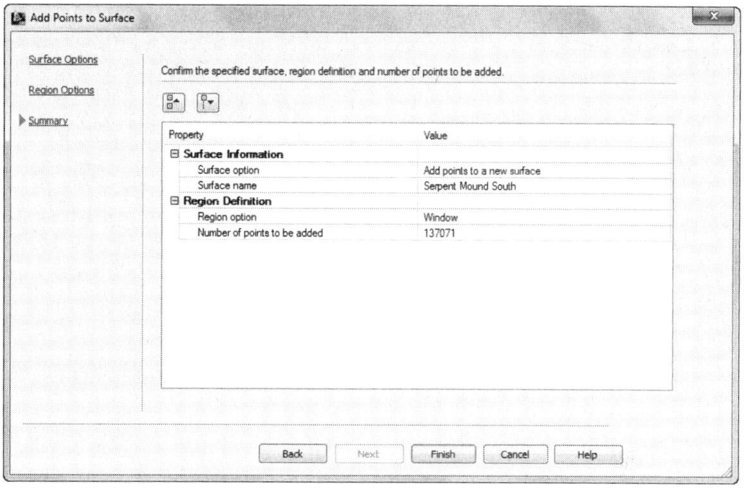

FIGURE 4.62 The Add Points To Surface – Summary page

The Bottom Line

Create a preliminary surface using freely available data. Most land development projects involve a surface at some point. During the planning stages, freely available data can give you a good feel for the lay of the land, allowing design exploration before money is spent on fieldwork or aerial topography. Imprecise at best, this free data should never be used as a replacement for final design topography, but it's a great starting point.

Master It Create a new drawing from the Civil 3D template and set the Coordinate System to NAD83 Connecticut State Plane Zone, US Foot (CT83F) or NAD83 Connecticut State Plane Zone, Meter (CT83). Create a surface named MarlboroughCT_DEM. Add the `Marlborough_CT.DEM` file (UTM Zone 18, NAD27 datum, meters) downloadable from the book's web page.

Modify and update a TIN surface. TIN surface creation is mathematically precise, but sometimes the assumptions behind the equations leave something to be desired. By using the editing tools built into Civil 3D, you can create a more realistic surface model.

Master It Open the `MasteringBoundary.dwg` or the `MasteringBoundary_METRIC.dwg` file. Use the irregular-shaped polyline and apply it to the surface as an outer boundary of the surface.

Prepare a slope analysis. Surface analysis tools allow users to view more than contours and triangles in Civil 3D. Engineers working with nontechnical team members can create strong meaningful analysis displays to convey important site information using the built-in analysis methods in Civil 3D.

Master It Open the `MasteringSlopeAnalysis.dwg` or the `MasteringSlopeAnalysis_METRIC.dwg` file. Create a Slope Banding analysis showing slopes under and over 10 percent and insert a legend to help clarify the image.

Label surface contours and spot elevations. Showing a stack of contours is useless without context. Using the automated labeling tools in Civil 3D, you can create dynamic labels that update and reflect changes to your surface as your design evolves.

Master It Open the `MasteringLabelSurface.dwg` or the `MasteringLabelSurface_METRIC.dwg` file. Label the major contours on the surface at 2′ and 10′ (Background) or 1 m and 5 m (Background).

Import a point cloud into a drawing and create a surface model. As point cloud data becomes more common and replaces other large-scale data-collection methods, the ability to use this data in Civil 3D is critical. Intensity helps postprocessing software determine the ground cover type. While Civil 3D can't do postprocessing, you can see the intensity as part of the point cloud style.

Master It Import an LAS format point cloud `Denver.las` into the Civil 3D template (with a coordinate system) of your choice. As you create the point cloud file, set the style to Elevation Ranges. Use a portion of the file to create a Civil 3D surface model.

Chapter 5

Parcels

Land development projects often involve the subdivision of large pieces of land into smaller lots. Even if your projects don't directly involve subdivisions, you're often required to show the legal boundaries of your site and the adjoining sites.

AutoCAD® Civil 3D® parcels give you a dynamic way to create, edit, manage, and annotate these legal land divisions. If you edit a parcel segment to make a lot larger, all of the affected labels will update — including areas, bearings, distances, curve information, and table information.

In this chapter, you will learn to:

- Create a boundary parcel from objects
- Create a right-of-way parcel using the right-of-way tool
- Create subdivision lots automatically by layout
- Add multiple-parcel segment labels

Introduction to Sites

In Civil 3D, a *site* is a collection of parcels, alignments, grading objects, and feature lines that share a common topology. In other words, Civil 3D objects that are in the same site are related to, as well as interact with, one another. The objects that react to one another are called *site geometry* objects.

The following objects must be placed on a site:

- Feature lines
- Grading groups
- Parcels

Feature lines and grading groups are discussed in depth in Chapter 16, "Grading." Alignments, discussed in the next chapter, do not need to go on a site.

Think Outside of the Lot

You will want to separate by site any objects you don't want interfering with one another. For example, you may have a set of parcels that represent impervious areas for drainage calculations. In the same location, you may have parcels representing property boundaries. By keeping these items on separate sites, you will be able to keep area information separate.

Dynamic area labels are useful for delineating and analyzing soil boundaries; paving, open space, and wetlands areas; and any other region enclosed with a boundary.

Like all Civil 3D objects, parcels utilize styles. Parcel styles can be used to color-code different parcel types.

It's important to understand how site geometry objects react to one another. Figure 5.1 shows a typical parcel that might represent a property boundary.

FIGURE 5.1
A typical property boundary

When an alignment is drawn and placed on the same site as the property boundary, the parcel splits into two parcels, as shown in Figure 5.2.

FIGURE 5.2
An alignment that crosses a parcel divides the parcel in two if the alignment and parcel exist on the same site.

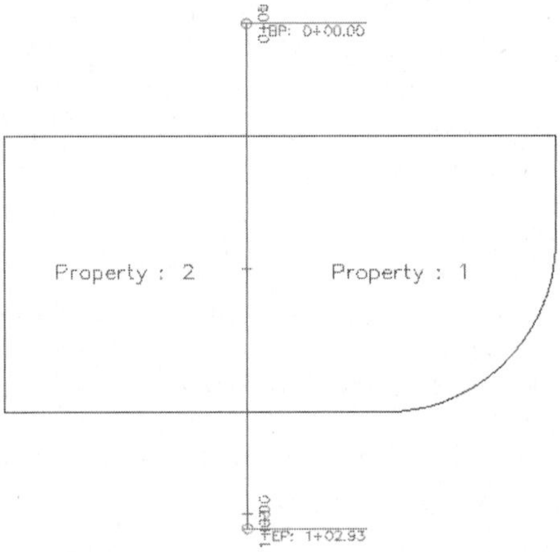

You must plan ahead to create meaningful sites based on interactions between the desired objects. For example, if you want a road centerline, a road right-of-way (ROW) parcel, and the lots in a subdivision to react to one another, they need to be in the same site (see Figure 5.3).

The alignment (or road centerline), ROW parcel, and lots all relate to one another. A change in the centerline of the road should prompt a change in the ROW parcel and the subdivision lots.

If you'd like to avoid the interaction between site geometry objects, place them in different sites. Figure 5.4 shows an alignment that has been placed in a different site from the boundary parcel. Notice that the alignment doesn't split the boundary parcel.

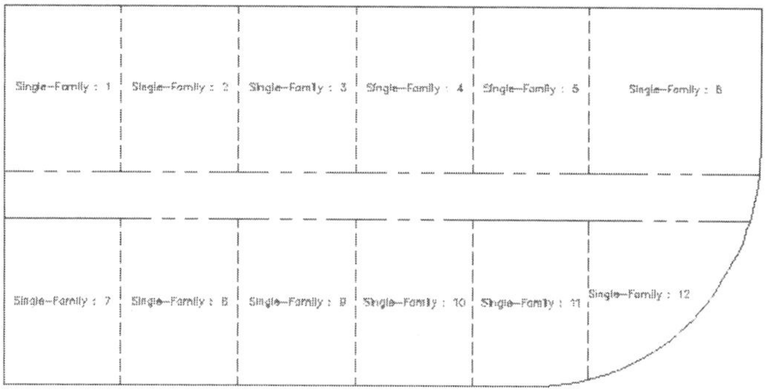

FIGURE 5.3
Alignments, ROW parcels, open space parcels, and subdivision lots react to one another when drawn on the same site.

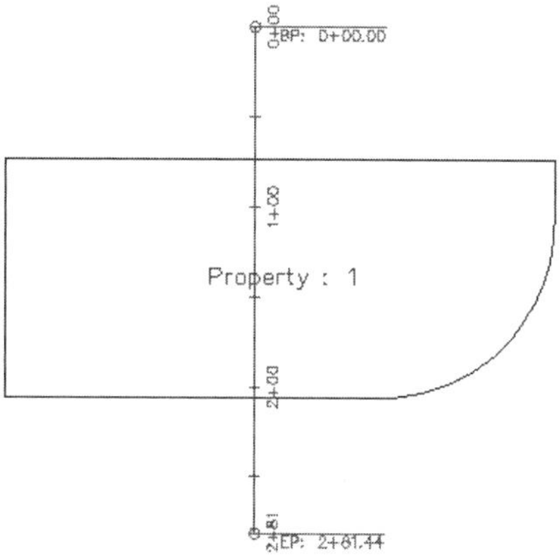

FIGURE 5.4
An alignment that crosses a parcel won't interact with the parcel if they exist on different sites.

It's important that only objects that are intended to react to each other be placed in the same site. For example, in Figure 5.5 you can see parcels representing both subdivision lots and soil boundaries. Because it wouldn't be meaningful for a soil boundary parcel segment to interrupt the area or react to a subdivision lot parcel, the subdivision lot parcels have been placed in a Subdivision Lots site, and the soil boundaries have been placed in a Soil Boundaries site.

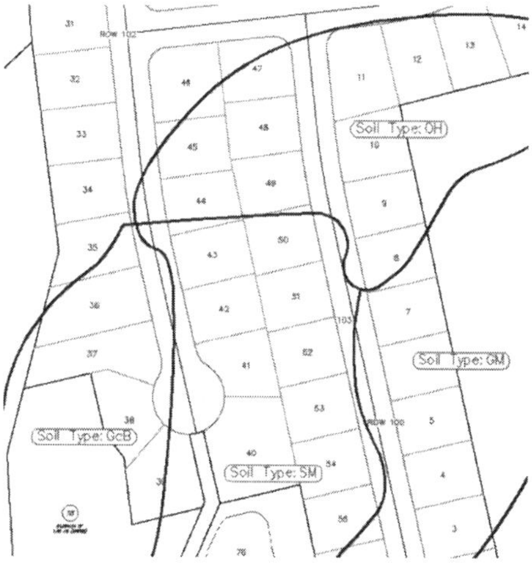

FIGURE 5.5
Parcels can be used for subdivision lots and soil boundaries as long as they're kept in separate sites.

If you didn't realize the importance of site topology, you might create both your subdivision lot parcels and your soil boundary parcels in the same site and find that your drawing looks similar to Figure 5.6. This figure shows the soil boundary segments dividing and interacting with subdivision lot parcel segments, which doesn't make any sense.

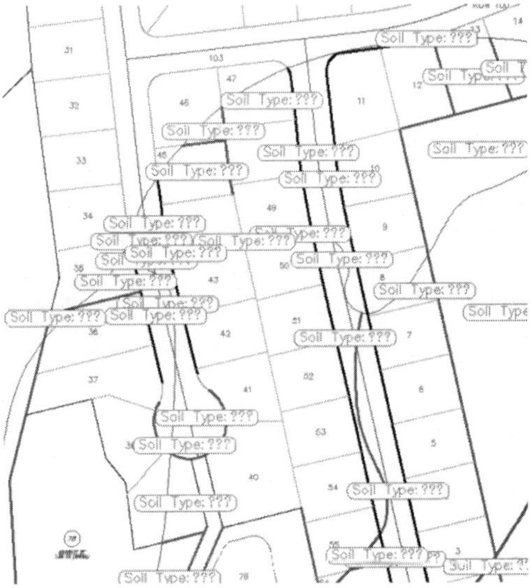

FIGURE 5.6 Subdivision lots and soil boundaries react inappropriately when placed in the same site.

Another way to avoid site geometry problems is to do site-specific tasks in different drawings and use a combination of external references and data references to share information. For example, you could have an existing base drawing that housed the Soil Boundaries site, XRef'd into a subdivision plat drawing that housed the Subdivision Lots site instead of separating the two drawings onto two different sites.

You should consider keeping your legal site plan in its own drawing. Because of the interactive and dynamic nature of Civil 3D parcels, it might be easy to accidentally grab a parcel segment when you meant to grab a manhole, and unintentionally edit a portion of your plat.

You'll see additional examples and drawing divisions later in this chapter, as well as in Chapter 18, "Advanced Workflows."

If you decide to have sites in the same drawing, here are some sites you may want to create. These suggestions are meant to be used as a starting point. Use them to help find a combination of sites that works for your projects:

Roads and Lots This site could contain road centerlines, ROWs, platted subdivision lots, open space, adjoining parcels, utility lots, and other aspects of the final legal site plan.

Grading Feature lines and grading objects are considered part of site geometry. If you're using these tools, you must make at least one site for them. You may even find it useful to have several grading sites.

Easements If you'd like to use parcels to manage, analyze, and annotate your easements, you may consider creating a separate site for easements.

Impervious Areas If you'd like to use parcels to manage, analyze, and annotate paved areas, a separate site will be useful. Drainage areas can be tracked using a separate object called a catchment, so there is no need to create a site for drainage areas. You will learn more about catchments in Chapter 15, "Storm and Sanitary Analysis."

As you learn new ways to take advantage of alignments, parcels, and grading objects, you may find additional sites that you'd like to create at the beginning of a new project.

WHAT ABOUT THE "SITELESS" ALIGNMENT?

As mentioned earlier in this chapter, you have a choice whether or not to place your alignments on a site. There are many situations where having the alignment independent from other objects is desirable, and therefore the <none> site can be used.

However, you can still create alignments in traditional sites, if you desire, and they will react to other site geometry objects. For example, if you wish to use the Create Right Of Way tool, the alignment you are working with must be on the same site as the main parcel.

You'll likely find that best practices for most alignments are to place them in the <none> site. For example, if road centerlines, road transition alignments, swale centerlines, and pipe network alignments are placed in the <none> site, you'll save yourself quite a bit of site geometry management.

If you decide you'd like to move the alignment to a site or to <none>, you can do so at any time. Click the alignment you wish to reassign. On the Alignment contextual tab ➢ Modify panel flyout, click Move To Site. See Chapter 6, "Alignments," for more information about alignments.

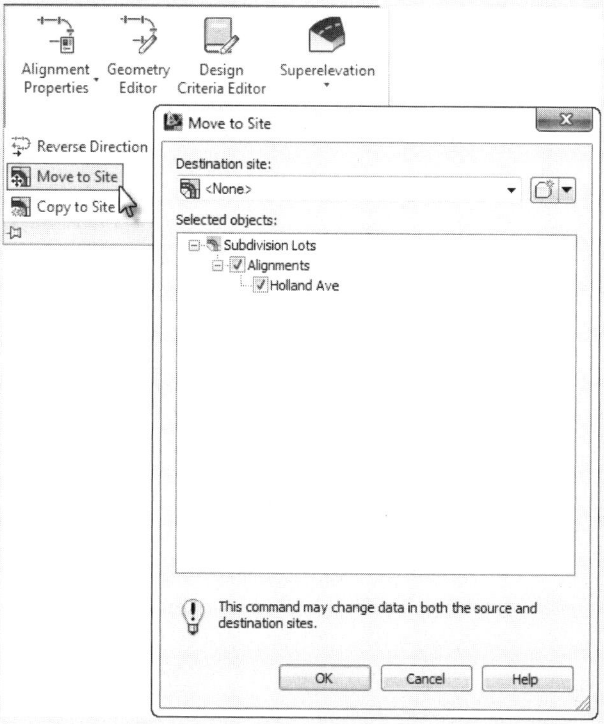

Creating a New Site

You can create a new site in Prospector. You'll find the process easier if you brainstorm potentially needed sites at the beginning of your project and create those sites right away — or, better yet, save them as part of your standard Civil 3D template. You can always add or delete sites later in the project.

The Sites collection is stored in Prospector, along with the other Civil 3D objects in your drawing.

The following exercise will lead you through creating a new site that you can use for creating subdivision lots:

1. Open the `CreateSite.dwg` (`CreateSite_METRIC.dwg`) file, which you can download from this book's web page at `www.sybex.com/go/masteringcivil3d2013`.

   Note that the drawing contains alignments and a boundary parcel, as shown in Figure 5.7.

FIGURE 5.7
The Create Site drawing contains alignments and a boundary parcel.

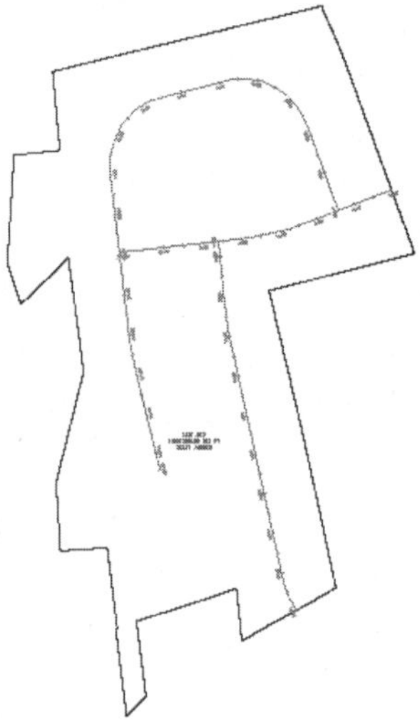

2. On the Prospector tab of Toolspace, go to Sites.
3. Right-click the Sites collection, and select New to open the Site Properties dialog.
4. On the Information tab of the Site Properties dialog, enter **Subdivision Lots** for the name of your site.
5. Confirm that the settings on the 3D Geometry tab match what is shown on Figure 5.8.

   As you create parcels, Civil 3D will automatically number them for you. The values in the Numbering tab are the starting point.

6. Confirm that both values on the Numbering tab are set to **1**. Click OK.
7. Locate the Sites collection on the Prospector tab of Toolspace, and note that your Subdivision Lots site appears on the list, as shown in Figure 5.9. You can repeat the process for all the sites you anticipate needing over the course of the project.

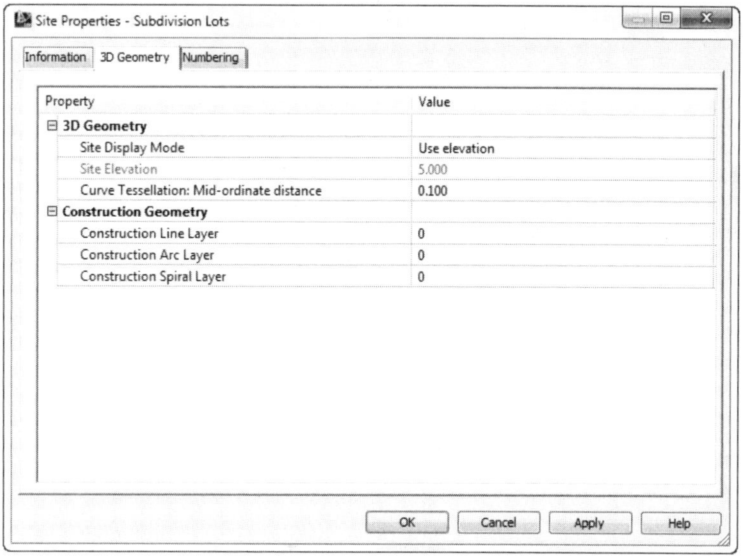

FIGURE 5.8
Confirm the settings on the 3D Geometry tab.

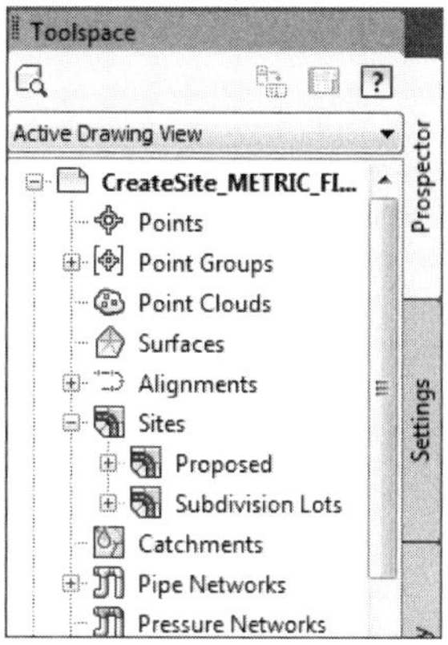

FIGURE 5.9
Your new site is listed in Prospector.

Creating a Boundary Parcel

The Create Parcel From Objects tool allows you to create parcels by choosing AutoCAD entities in your drawing or in an XRef'd drawing. In a typical workflow, it's common to encounter a boundary created by AutoCAD entities, such as polylines, lines, and arcs.

When you're using AutoCAD geometry to create parcels, it's important that the geometry be created carefully and meet certain requirements. The AutoCAD geometry must be lines, arcs, polylines, 3D polylines, or polygons. It can't include blocks, ellipses, circles, or other entities. Civil 3D may allow you to pick objects with an elevation other than zero, but you'll find you get better results if you flatten the objects so all objects have an elevation of zero. Sometimes the geometry appears sound when elevation is applied, but you may notice this isn't the case once the objects are flattened. Flattening all objects before creating parcels can help you prevent frustration when creating parcels.

This exercise will teach you how to create a parcel from Civil 3D objects:

1. Open the `CreateBoundaryParcel.dwg` (`CreateBoundaryParcel_METRIC.dwg`) file, which you can download from this book's web page.

   This drawing has several alignments, which were created on the Proposed site, and some AutoCAD lines representing a boundary. A parcel automatically formed on the site because the alignments form a closed area.

2. On the Home tab ➤ Create Design panel, select Parcel ➤ Choose Create Parcel From Objects.

3. At the `Select lines, arcs, or polylines to convert into parcels or [Xref]:` prompt, pick the red lines that represent the site boundary, and press ↵.

   The Create Parcels – From Objects dialog appears.

4. From the drop-down menus, select Proposed; Lot (Prop); and Name Square Foot & Acres (or Name Square Meter & Ha if you are working in metric units) in the Site, Parcel Style, and Area Label Style selection boxes, respectively.

   Leave everything else set to the defaults, as shown in Figure 5.10.

5. Click OK to dismiss the dialog.

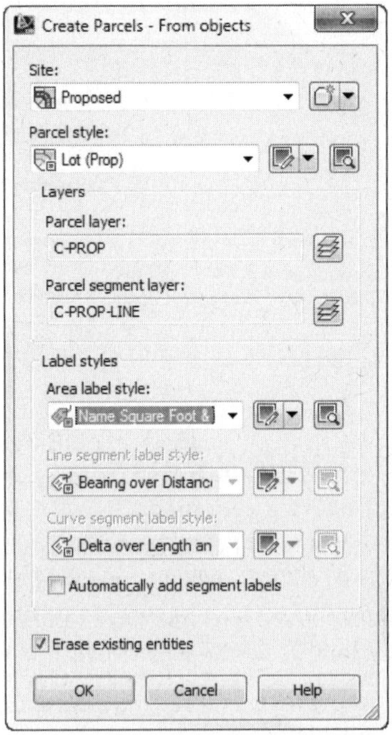

FIGURE 5.10
Site and Style settings for your new boundary parcel

The boundary polyline forms parcel segments that react with the alignments. Area labels are placed at the newly created parcel centroids, as shown in Figure 5.11.

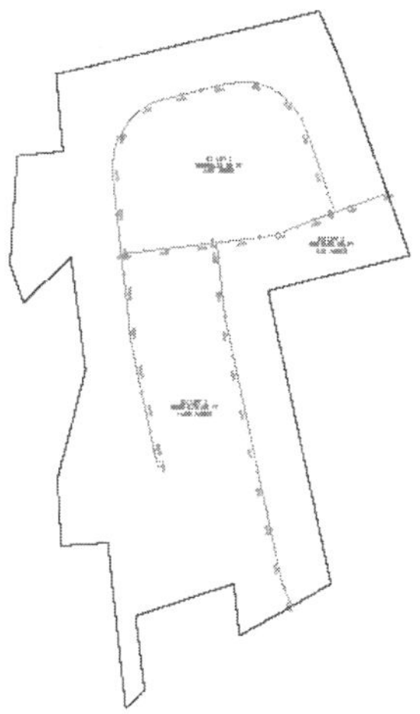

FIGURE 5.11
The boundary parcel segments, alignments, and area labels

Using Parcel Creation Tools

When you don't have existing lines to work with, the best option is to draw parcel segments using the Parcel Layout tools. Figure 5.12 shows the many commands available to you.

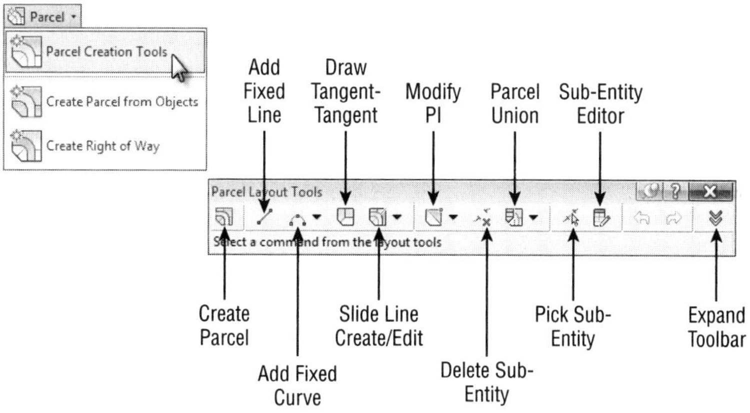

FIGURE 5.12
Selecting Parcel Creation Tools

Although you may never have thought of things like wetland areas or easements as parcels in the past, you can take advantage of the parcel tools to assist in labeling, stylizing, and analyzing these features for your plans.

This exercise will teach you how to create a parcel representing wetlands using the transparent commands and Draw Tangent-Tangent With No Curves tool from the Parcel Layout Tools toolbox:

1. Open the `WetlandsParcel.dwg` (`WetlandsParcel_METRIC.dwg`) file, which you can download from this book's web page.

   Note that this drawing has several alignments, parcels, and a series of points that represent a wetlands delineation.

2. Choose Parcel ➢ Parcel Creation Tools on the Create Design panel.

The Parcel Layout Tools toolbar appears.

3. Click the Draw Tangent-Tangent With No Curves tool on the Parcel Layout Tools toolbar.

   The Create Parcels – Layout dialog appears.

4. From the drop-down menus, select Proposed; Wetland; and Name Square Foot & Acres (or Name Square Meters & HA) in the Site, Parcel Style, and Area Label Style selection boxes, respectively.

   Keep the default settings for all other options.

5. Click OK.

6. At the `Specify start point:` prompt, click the Point Number Transparent Command.
7. At the `Enter Point Number:` prompt, enter **1-6**↵.

   You will see a line form through the wetland boundary points in the northwest corner of the project, and immediately form the parcel.

8. Press Esc once to exit the Transparent Command, and press Esc a second time to complete the parcel.
9. Repeat steps 3–8 for the other wetland points near the south of the project using points **7–18**.
10. Press Esc twice to exit the command.

    Your drawing should look similar to Figure 5.13.

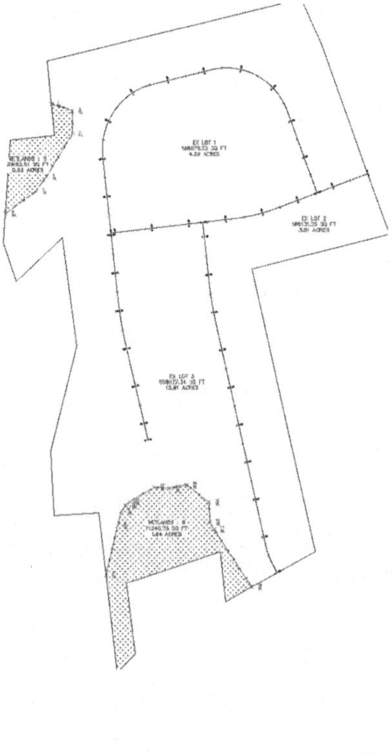

FIGURE 5.13
The wetlands defined on the site

To illustrate how to change the styles that are in use on a parcel, you will next change the parcel style and the area selection label style.

11. Select the parcel's area label. From the Parcel contextual tab ➤ Modify panel, click Parcel Properties.

The Parcel Properties dialog appears.

12. On the Information tab, select Mitigated Wetland from the drop-down menu in the Object Style selection box, and then click Apply to observe the change.

Remain in the Parcel Properties dialog. The parcel hatch pattern will turn blue and only be applied to the perimeter of the parcel.

13. To change the parcel area label style, switch to the Composition tab in Parcel Properties.

14. From the Area Selection Label Style pull-down, select Wetland Area Label, and click OK.

15. Change the Area Selection Label Style setting for the wetland parcel to the south.

16. Select the parcel by clicking on its area label, and click Parcel Properties from the Parcel contextual tab as you did in step 14.

17. In the Composition tab, change the Area Selection Label Style setting to Wetland Area Label and click OK.

Your parcels will look like Figure 5.14.

FIGURE 5.14
The wetlands parcel with the appropriate label styles applied

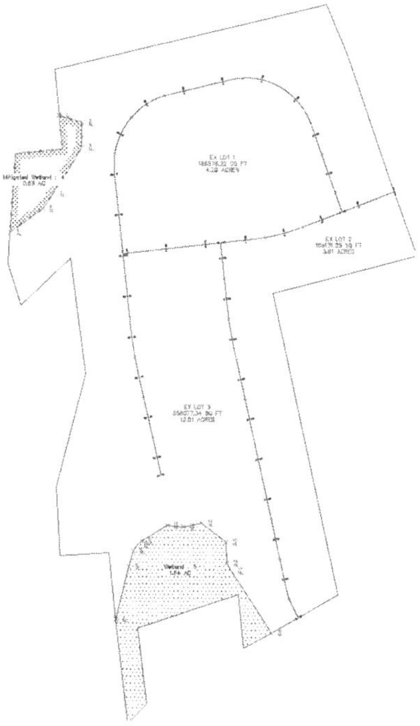

Creating a Right-of-Way Parcel

The Create ROW tool creates ROW parcels on either side of an alignment based on your specifications. The Create ROW tool can be used only when alignments are placed on the same site as the boundary parcel. The resulting ROW parcel will look similar to Figure 5.15.

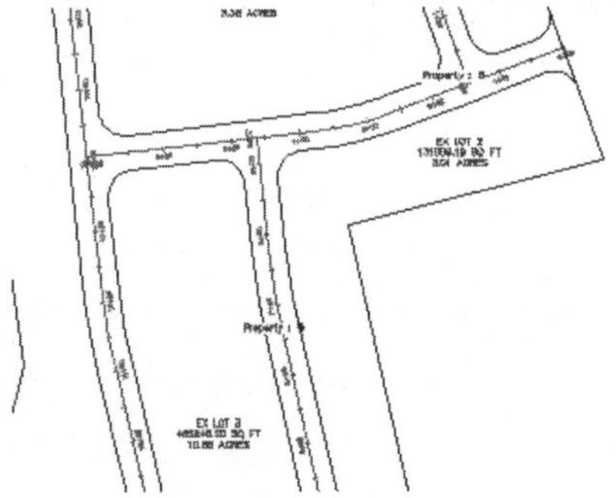

FIGURE 5.15 The resulting parcels after application of the Create ROW tool

Options for the Create ROW tool include:

- Offset distance from alignment
- Fillet or chamfer cleanup at parcel boundaries
- Alignment intersections

Figure 5.16 shows an example of chamfered cleanup at alignment intersections.

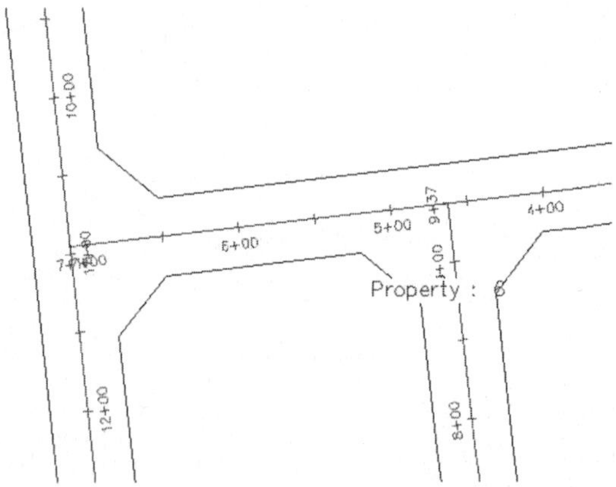

FIGURE 5.16 A ROW with chamfer cleanup at alignment intersections

Once the ROW parcel is created, it's no different from any other parcel. For example, it doesn't maintain a dynamic relationship with the alignment that created it. A change to the alignment will require the ROW parcel to be edited or, more likely, re-created.

This exercise will teach you how to use the Create ROW tool to automatically place a ROW parcel for each alignment on your site:

1. Open the `CreateROWParcel.dwg` (`CreateROWPARCEL_METRIC.dwg`) file, which you can download from this book's web page.

   Note that this drawing has some alignments on the same site as the boundary parcel, resulting in several smaller parcels between the alignments and boundary.

2. On the Home tab ➢ Create Design panel, choose Parcel ➢ Create Right Of Way.
3. At the `Select parcels:` prompt, pick EX LOT 1, EX LOT 2, and EX LOT 3 on the screen.
4. Press ↵ to stop picking parcels.

   The Create Right Of Way dialog appears, as shown in Figure 5.17.

FIGURE 5.17
The Create Right Of Way dialog

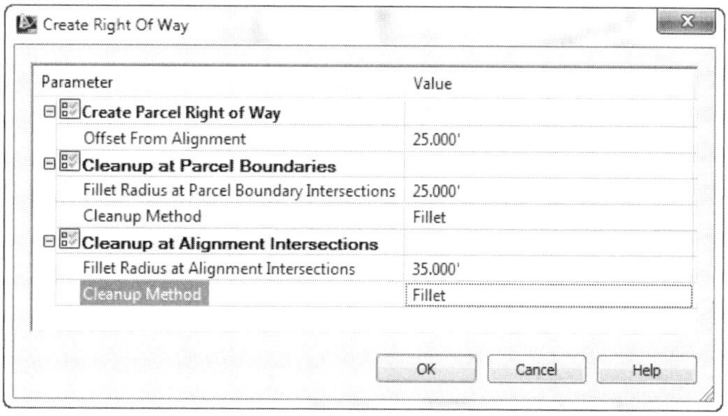

5. Expand the Create Parcel Right of Way parameter, and enter **25′** (**8** m) as the value for Offset From Alignment.
6. Expand the Cleanup At Parcel Boundaries parameter.
7. Enter **25′** (**8** m) as the value for Fillet Radius At Parcel Boundary Intersections.
8. Select Fillet from the drop-down menu in the Cleanup Method selection box.
9. Expand the Cleanup At Alignment Intersections parameter.
10. Enter **35′** (**10** m) as the value for Fillet Radius At Alignment Intersections.
11. Select Fillet from the drop-down menu in the Cleanup Method selection box.
12. Click OK to dismiss the dialog and create the ROW parcels.

    Your drawing should look similar to Figure 5.18.

FIGURE 5.18
The completed ROW parcels

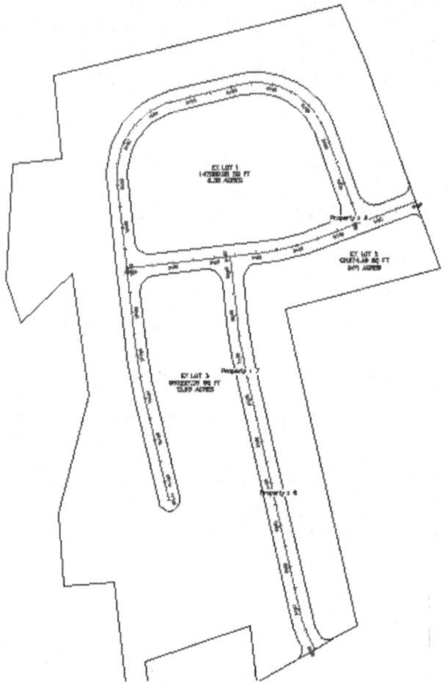

Adding a Cul-de-Sac Parcel

If you look at your drawing, on the lower-left side of your right of way it looks incomplete. You will add a cul-de-sac here. Rather than go through all the mechanics of a cul-de-sac, a block has been created for you and you will insert it and turn it into a parcel. This section also introduces some editing tools.

You will need to have completed the previous exercise before continuing or open the "FINISHED" version of the drawing.

1. Continue working in the drawing from the previous exercise.
2. Insert the `CulDeSacBlock.dwg` (`CulDeSacBlock_METRIC.dwg`) file using the settings shown in Figure 5.19.

FIGURE 5.19
Inserting the cul-de-sac block settings

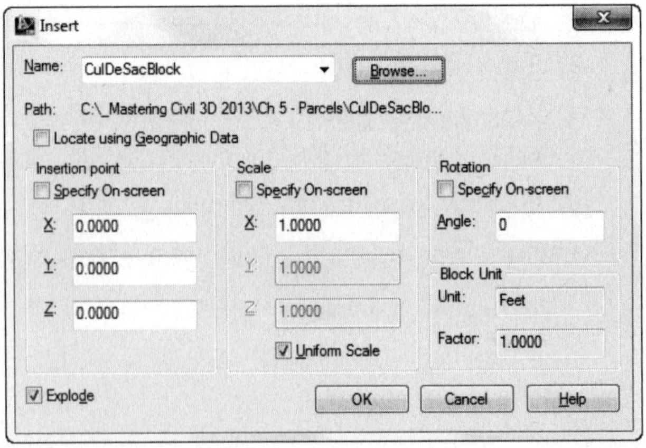

3. From the Home tab ➢ Create Design panel, expand Parcel ➢ Create Parcel From Objects.
4. At the `Select lines, arcs, or polylines to convert into parcels or [Xref]:` prompt, draw a window around all of the cul-de-sac objects, and press ↵.
5. In the Create Parcels – From Objects dialog, make sure that:
   - Site is set to Subdivision Lots
   - Parcel Style is set to Property
   - Area Label Style is set to Name Square Foot & Acres (Name Square Meter & HA)
   - Erase Existing Entities is checked
6. Click OK.

   Your drawing should look like Figure 5.20.

FIGURE 5.20
The cul-de-sac turned into a parcel

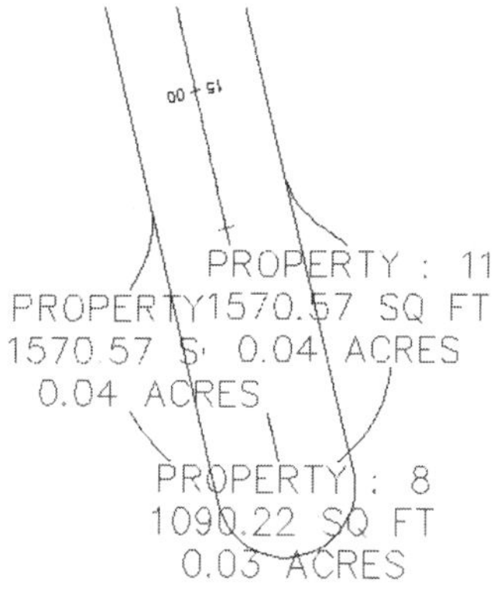

The cul-de-sac is now a parcel, but there are some extra lines that need to be taken care of. Let's see how to clean this up a bit.

7. From the Home tab ➢ Create Design panel ➢ Parcel, select Parcel Creation Tools.

   The Parcel Layout Tools palette opens.

8. Select the Delete Sub-Entity tool.
9. At the `Select subentity to remove:` prompt, select the right of way that interferes with the cul-de-sac, as shown in Figure 5.21.
10. Press Esc twice to exit the command.

Your cul-de-sac is complete, as shown on Figure 5.22.

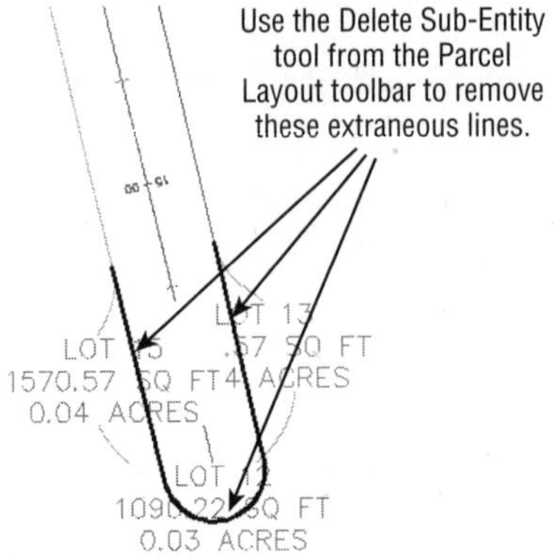

FIGURE 5.21 Delete these portions.

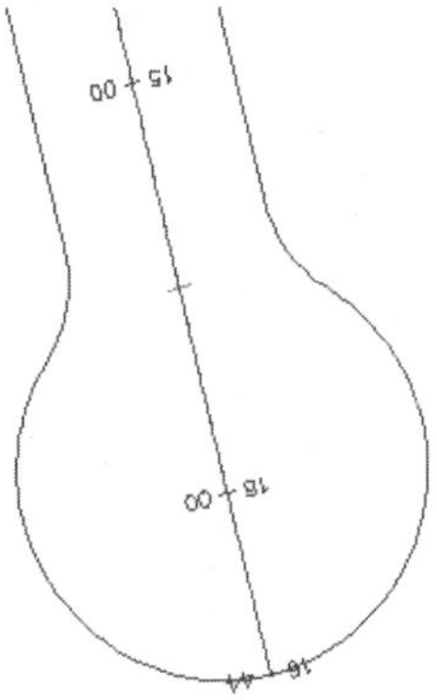

FIGURE 5.22 The finished cul-de-sac

Creating Subdivision Lot Parcels Using Precise Sizing Tools

The precise sizing tools allow you to create parcels to your exact specifications. You'll find these tools most useful when you have your roadways established and understand your lot-depth requirements. These tools provide automatic, semiautomatic, and freeform ways to control frontage, parcel area, and segment direction.

Attached Parcel Segments

Parcel segments created with the precise sizing tools are called *attached segments*. Attached parcel segments have a start point that is attached to a frontage segment and an endpoint that is defined by the next parcel segment they encounter. Attached segments can be identified by their distinctive diamond-shaped grip at their start point and no grip at their endpoint (see Figure 5.23).

In other words, you establish their start point and their direction, but they seek another parcel segment to establish their endpoint. Figure 5.23 shows a series of attached parcel segments. You can tell the difference between their start and endpoints because the start points have the diamond-shaped grips.

FIGURE 5.23
A series of attached parcel segments, with their endpoints at the front lot line

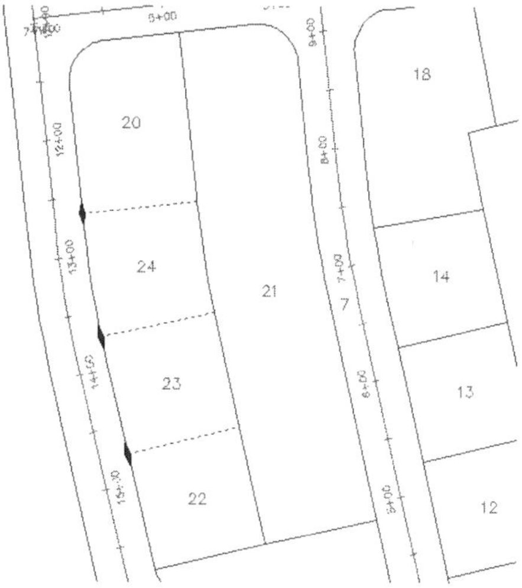

You can drag the diamond-shaped grip along the frontage to a new location, and the parcel segment will maintain its angle from the frontage. If the rear lot line is moved or erased, the attached parcel segments find a new endpoint (see Figure 5.24) at the next available parcel segment.

FIGURE 5.24
The endpoints of attached parcel segments extend to the next available parcel segment if the initial parcel segment is erased.

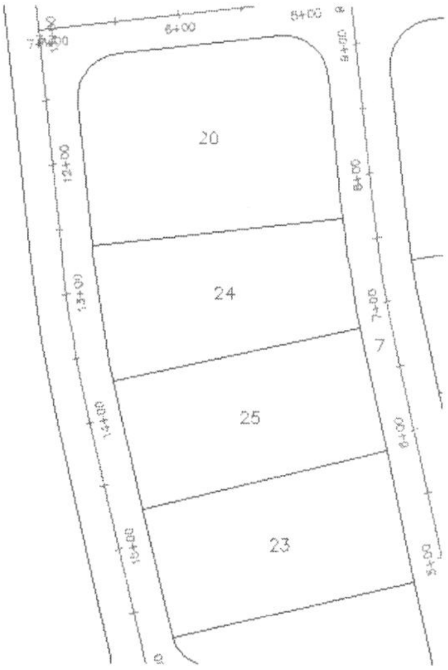

Precise Sizing Settings

The precise sizing tools consist of the Slide Angle, Slide Direction, and Swing Line tools (see Figure 5.25).

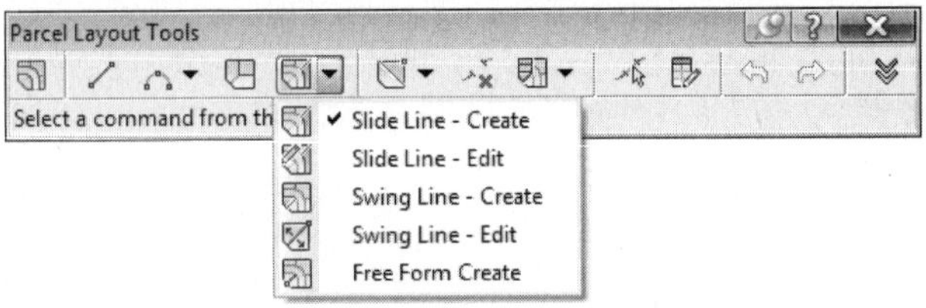

FIGURE 5.25 Helpful parcel segment creation tools on the Parcel Layout Tools toolbar

The Parcel Layout Tools toolbar can be expanded so that you can establish settings for each of the precise sizing tools (see Figure 5.26). Each of these settings is discussed in detail in the following sections.

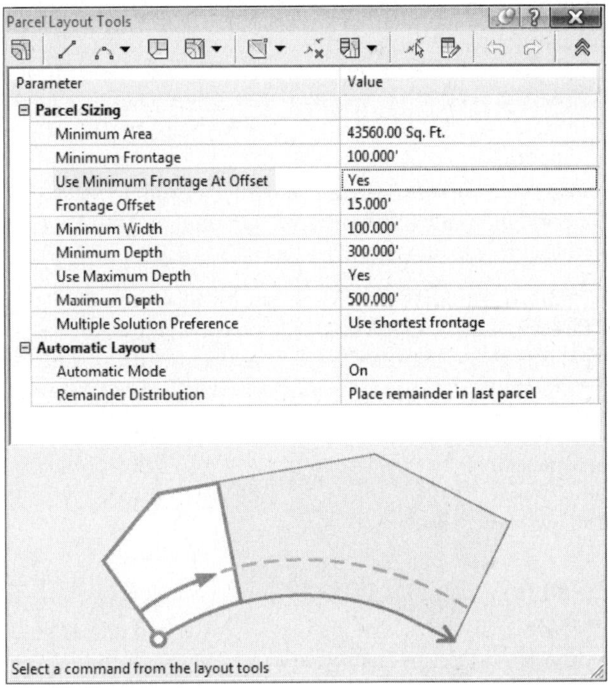

FIGURE 5.26 Automated sizing options on the expanded Parcel Layout Tools toolbar

NEW PARCEL SIZING

CERT OBJECTIVE

When you create new parcels, the tools respect your default area and minimum frontage (measured from either a ROW or a building setback line). The program always uses these numbers as a minimum; it bases the actual lot size on a combination of the geometry constraints (lot depth, frontage curves, and so on) and the additional settings that follow. Keep in mind that the numbers you establish under the New Parcel Sizing option must make geometric sense. For example, if you'd like a series of 7,500 square foot (700 square meter) lots that have 100′ (30 m) of frontage, you must make sure that your rear parcel segment allows for at least 75′ (23 m) of depth; otherwise, you may wind up with much larger frontage values than you desire or a situation where the software can't return a meaningful result.

Automatic Layout

Automatic Layout has two parameters when the list is expanded: Automatic Mode and Remainder Distribution. The Automatic Mode parameter can have the following values:

On Automatically follows your settings and puts in all the parcels, without prompting you to confirm each one.

Off Allows you to confirm each parcel as it's created. In other words, this option provides you with a way to semiautomatically create parcels.

The Remainder Distribution parameter tells Civil 3D how you'd like "extra" land handled. This parameter has the following options:

Create Parcel From Remainder Makes a last parcel with the leftovers once the tool has made as many parcels as it can to your specifications on the basis of the settings in this dialog. This parcel is usually smaller than the other parcels.

Place Remainder In Last Parcel Adds the leftover area to the last parcel once the tool has made as many parcels as it can to your specifications on the basis of the settings in this dialog.

Redistribute Remainder Takes the leftover area and pushes it back through the default-sized parcels once the tool has made as many parcels as it can to your specifications on the basis of the settings in this dialog. The resulting lots aren't always evenly sized because of differences in geometry around curves and other variables, but the leftover area is absorbed.

There aren't any rules per se in a typical subdivision workflow. Typically the goal is to create as many parcels as possible within the limits of available land. To that end, you'll use a combination of AutoCAD tools and Civil 3D tools to divide and conquer the particular tract of land with which you are working.

Slide Line – Create Tool

The Slide Line – Create tool creates an attached parcel segment based on an angle from frontage. You may find this tool most useful when your jurisdiction requires a uniform lot-line angle from the right of way.

This exercise will lead you through using the Slide Line – Create tool to create a series of subdivision lots:

1. Open the `CreateSubdivisionLots.dwg` (`CreateSubdivisionLots_METRIC.dwg`) file, which you can download from this book's web page.

   Note that this drawing has several alignments on the same site as the boundary parcel, resulting in several smaller parcels between the alignments and boundary.

2. Choose Parcel ➢ Parcel Creation Tools on the Create Design panel. The Parcel Layout Tools toolbar appears.

3. Expand the toolbar by clicking the Expand The Toolbar button.
4. In the Parcel Sizing section, change the value of the following parameters by clicking in the Value column and typing in the new values if they aren't already set. Notice how the preview window changes to accommodate your preferences:
   - Minimum Area: **7500.00** sq. ft. (**700** m^2)
   - Minimum Frontage: **75.000′** (**25** m)
   - Use Minimum Frontage At Offset: **Yes**
   - Frontage Offset: **25.000′** (**10** m)
   - Minimum Width: **75.000′** (**25** m)
   - Minimum Depth: **50.000′** (**15** m)
   - Use Maximum Depth: **No**
   - Maximum Depth: (leave default value as this will not be used)
   - Multiple Solution Preference: **Use shortest frontage**
5. In the Automatic Layout section, change the following parameters by clicking in the Value column and selecting the appropriate option from the drop-down menu, if they aren't already set:
   - Automatic Mode: **On**
   - Remainder Distribution: **Redistribute remainder**
6. Click the Slide Line – Create tool. The Create Parcels – Layout dialog appears.
7. From the drop-down menus, select Subdivision Lots; Lot (Prop); and Name Square Foot & Acres (Name Square Meter & HA) in the Site, Parcel Style, and Area Label Style selection boxes, respectively.

   Leave the rest of the options set to their defaults.
8. Click OK to dismiss the dialog.
9. At the `Select parcel to be subdivided or [Pick]:` prompt, click the EX LOT 1 area label.
10. At the `Select start point on frontage:` prompt, use your Endpoint Osnap to pick the point of curvature along the ROW parcel segment for Property 1 (see Figure 5.27).

    The parcel jig appears.
11. Move your mouse slowly along the ROW parcel segment, and notice that the parcel jig follows the parcel segment.
12. At the `Select end point on frontage:` prompt, use your Endpoint Osnap to pick the point of curvature along the ROW parcel segment for Property 1 (see Figure 5.28).

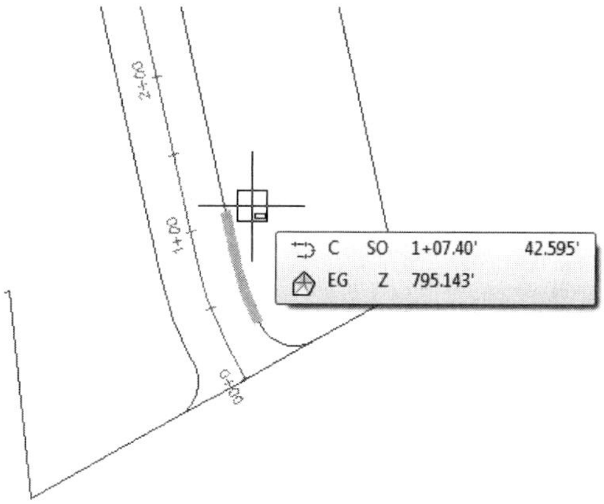

FIGURE 5.27
Pick the point of curvature along the ROW parcel segment.

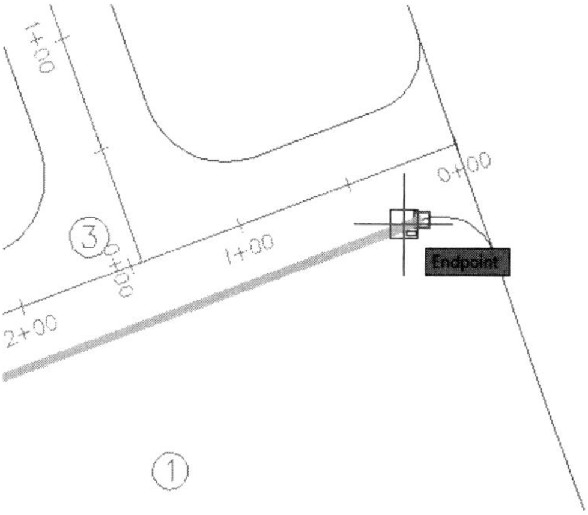

FIGURE 5.28
Allow the parcel-creation jig to follow the parcel segment, and then pick the point of curvature along the ROW parcel segment.

13. At the `Specify angle or [Bearing/aZimuth]:` prompt, enter **90** ↵.

    Notice the preview (see Figure 5.29).

14. At the `Accept result? [Yes/No] <Yes>:` prompt, press ↵ to accept the default Yes.
15. At the `Select parcel to be subdivided or [Pick]:` prompt, press ↵.
16. Click the X to close the Parcel Layout Tools toolbar.

    Your drawing should look similar to Figure 5.30.

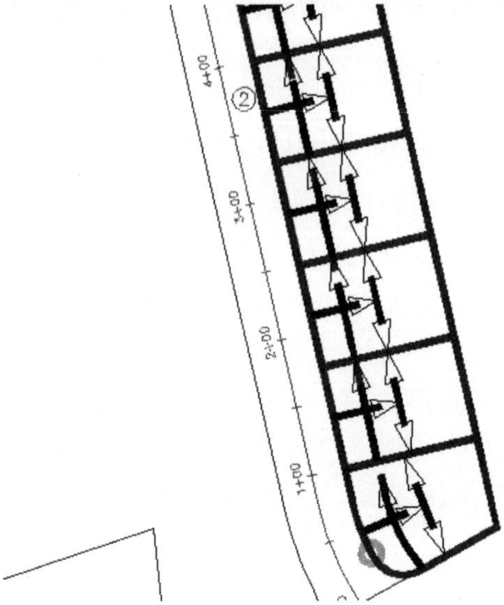

FIGURE 5.29
A preview of the results of the automatic parcel layout

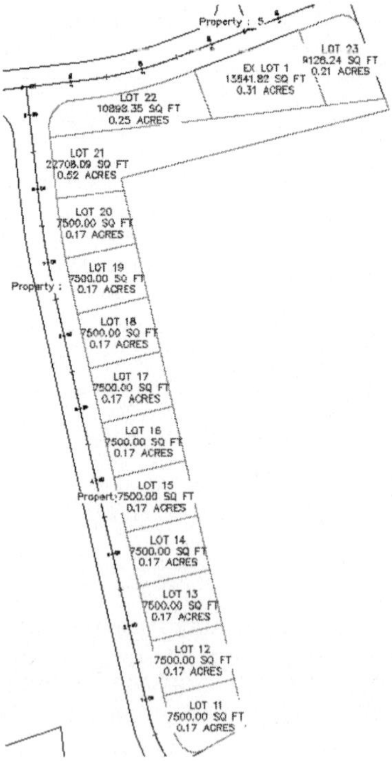

FIGURE 5.30
The automatically created lots

The parcels at the north end just don't look right. We'll address that in the next section of this chapter.

CURVES AND THE FRONTAGE OFFSET

In most cases, the frontage along a building setback is graphically represented as a straight line behind the setback. When you specify a minimum width along a frontage offset (the building setback line) in the Parcel Layout Tools dialog, and when the lot frontage is curved, the distance you enter is measured along the curve, as shown by the dashed line here.

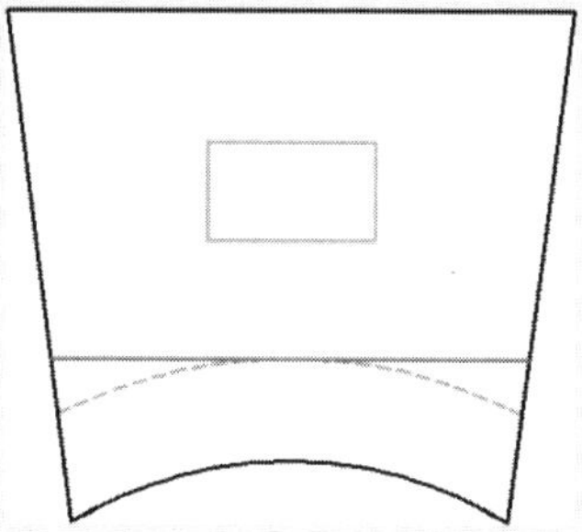

In most cases, this result may be insignificant, but in a large development, the error could be the defining factor in your decision to add or subtract a parcel from the development.

Swing Line – Create Tool

The Swing Line – Create tool creates a "backward" attached parcel segment where the diamond-shaped grip appears not at the frontage but at a different location that you specify. The tool respects your minimum frontage, and it adjusts the frontage so it is larger, if necessary, in order to respect your default area. The Swing Line – Create tool is semiautomatic because it requires your input of the swing point location.

You may find this tool most useful around a cul-de-sac or in odd-shaped corners where you must hold frontage but have a lot of flexibility in the rear of the lot.

Using the Free Form Create Tool

A site plan is more than just single-family lots. Areas are usually dedicated for open space, stormwater-management facilities, parks, and public utility lots. The Free Form Create tool can be useful when you're creating these types of parcels. This tool, like the precise sizing tools, creates an attached parcel segment with the special diamond-shaped grips.

NOTE The lot numbers were designed by the authors for the exercises. Your lot numbers may vary from those shown in the exercises.

In the following exercise, you'll use the Free Form Create tool to create a new parcel:

1. Open the `CreateFreeForm.dwg` (`CreateFreeForm_METRIC.dwg`) file.

   Note that this drawing contains a series of subdivision lots.

2. Pan over to EX LOT 20.

   You can see the lot line that was drawn automatically in the previous exercise that obviously will not work (Figure 5.31).

FIGURE 5.31
Delete the highlighted parcel line.

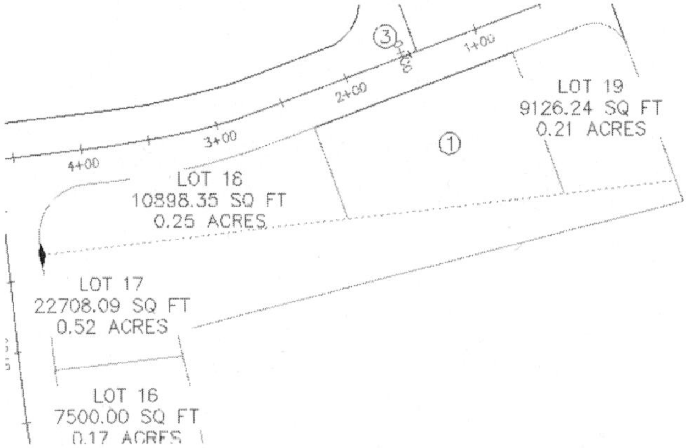

3. Delete the parcel line highlighted in Figure 5.31.

   The parcels readjust but the resulting lot is now much larger than needed. Let's add a line using the Free Form Create tool. You will explore deleting parcel lines in more depth later in this chapter.

4. Select Parcel ➤ Parcel Creation Tools on the Create Design panel, and select the Free Form Create tool.

   The Create Parcels – Layout dialog appears.

5. Select Subdivision Lots; Lot (Prop); and Name Square Foot & Acres (Name Square Meters & HA) from the drop-down menus in the Site, Parcel Style, and Area Label Style selection boxes, respectively.

   Keep the default values for the remaining options.

6. Click OK to dismiss the dialog.

7. Slide the Free Form Create attachment point around the Lot: 20 area (your lot number may differ).

8. At the `Select attachment point:` prompt, use your Endpoint Osnap to pick the endpoint, as shown in Figure 5.32.

9. At the `Specify lot line direction:(ENTER for perpendicular) or [Bearing/aZimuth]:` prompt, press ↵ to specify a perpendicular lot line direction.

FIGURE 5.32 Use the Free Form Create tool to select an attachment point.

A new parcel segment is created perpendicular to the ROW parcel segment, as shown in Figure 5.33 (your resulting lot numbers may differ). Note that a new lot parcel has formed.

FIGURE 5.33 Attach the parcel segment to the marker point provided.

10. Press ↵ to exit the Free Form Create command.
11. Click the X or press Esc to exit the Parcel Layout Tools toolbar.
12. Pick the new parcel segment so that you see its diamond-shaped grip.
13. Grab the grip, and slide the segment along the ROW parcel segment (see Figure 5.34).

FIGURE 5.34 Sliding an attached parcel segment

Notice that when you place the parcel segment at a new location, the segment endpoint snaps back to the rear parcel segment. This is typical behavior for an attached parcel segment.

Editing Parcels by Deleting Parcel Segments

One of the most powerful aspects of Civil 3D parcels is the ability to perform many iterations of a site plan design. Typically, this design process involves creating a series of parcels and then deleting them to make room for iteration with different parameters, or deleting certain segments to make room for easements, public utility lots, and more.

You can delete parcel segments using the AutoCAD Erase tool as shown in the previous exercise, or the Delete Sub-Entity tool on the Parcel Layout Tools toolbar.

It's important to understand the difference between these two methods. The AutoCAD Erase tool behaves as follows:

- If the parcel segment was originally created from a polyline (or similar parcel layout tools, such as the Tangent-Tangent With No Curves tool), the AutoCAD Erase tool erases the entire segment (see Figure 5.35).

FIGURE 5.35 The segments indicated by the blue grips will be erased after using the AutoCAD Erase tool.

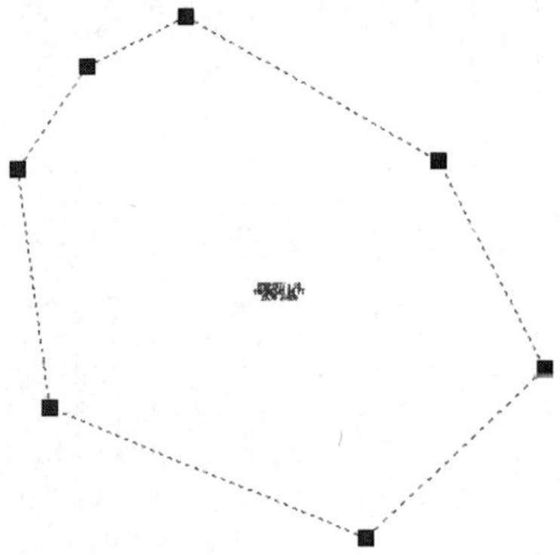

- If the parcel segment was originally created from a line or arc (or similar parcel layout tools, such as the precise sizing tools), then AutoCAD Erase erases the entire length of the original line or arc (see Figure 5.36).

FIGURE 5.36 The AutoCAD Erase tool will erase the entire segment indicated by the blue grips.

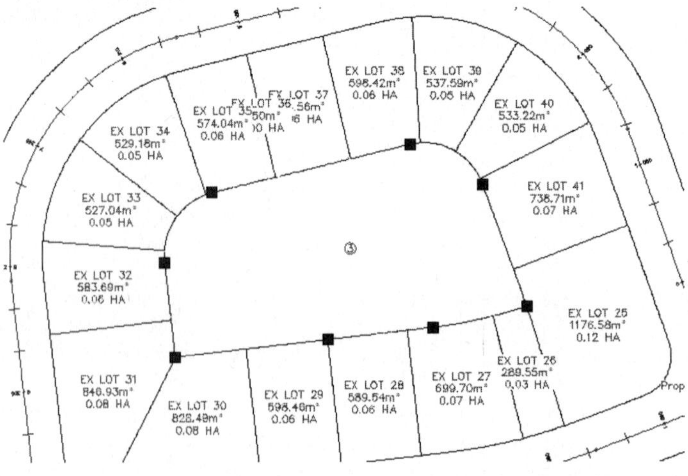

The Delete Sub-Entity tool acts more like the AutoCAD Trim tool. The Delete Sub-Entity tool only erases the parcel segments between parcel vertices. For example, if Lot 26, as shown in Figure 5.37, must be absorbed into Lot 3 to create a public utility easement, you'd want to only erase the segment at the rear of Lot 26 and not the entire segment shown previously in Figure 5.36.

FIGURE 5.37
Use the Delete Sub-Entity tool to erase the rear parcel segment for Lot 26.

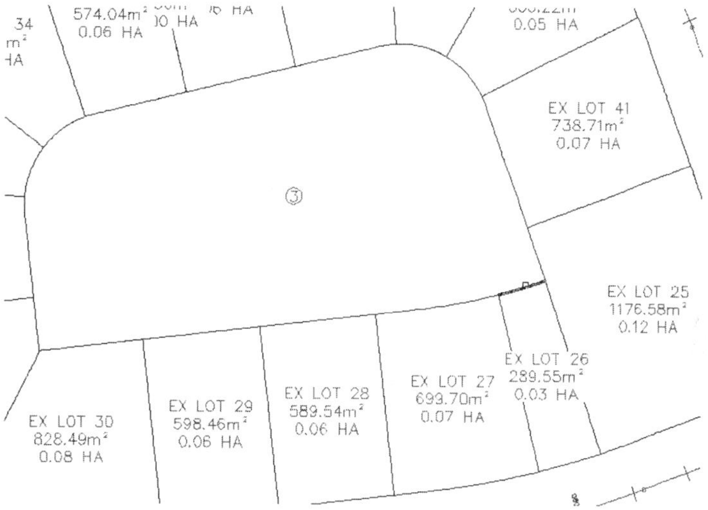

As an alternate to launching from the Home tab ➤ Create Ground Data panel ➤ Parcel ➤ Parcel Creation tools, you can access the Parcel Creation tools by selecting a lot label. From the Parcel contextual tab ➤ Modify panel, select Parcel Layout Tools. Selecting the Delete Sub-Entity tool allows you to pick only the small rear parcel segment for Lot 26. Figure 5.38 shows the result of this deletion.

FIGURE 5.38
The rear lot line for Lot 26 was erased using the Delete Sub-Entity tool, thus creating a larger Lot 3.

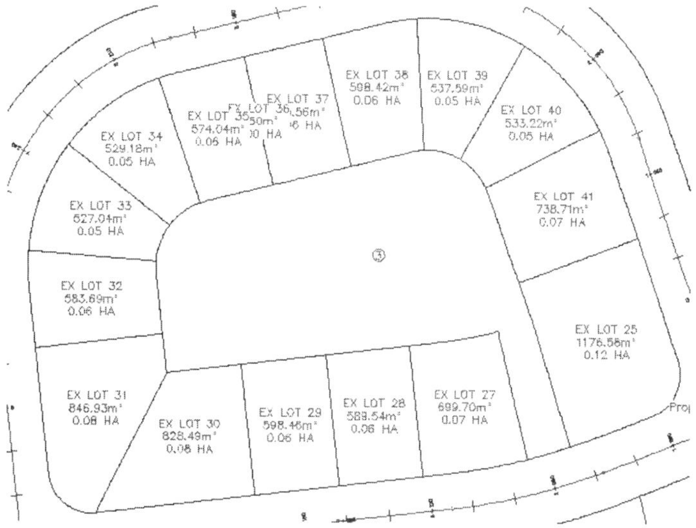

The following exercise will lead you through deleting a series of parcel segments using both the AutoCAD Erase tool and the Delete Sub-Entity tool:

1. Open the `DeleteSegments.dwg` (`DeleteSegments_Metric.dwg`) file.

   Note that this drawing contains a series of subdivision lots, along with a wetlands boundary.

   Let's say you just received word that you are allowed to build on the wetland area.

2. Use the AutoCAD Erase tool to erase the parcel segments that define the wetlands parcel and then freeze the wetlands points.

   Note that the entire parcel disappears as soon as the first segment is removed, causing a "hole" in the parcel.

   Now the developer has decided to enlarge the lots.

3. Erase the lot lines, as shown in Figure 5.39.

FIGURE 5.39
The parcel lines to be deleted

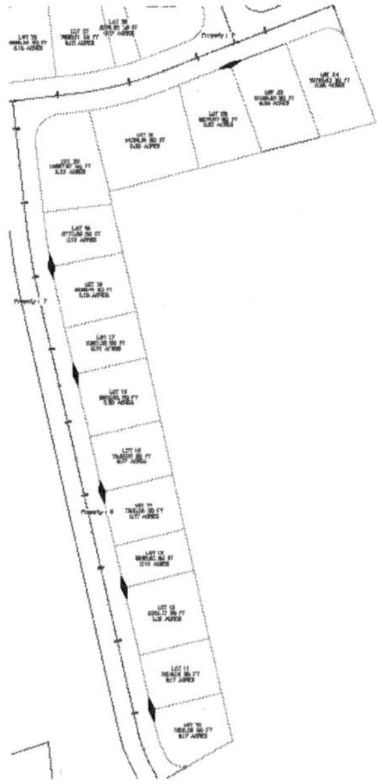

   After deleting parcel segments, your lots should look similar to Figure 5.40.

   Next, you discover that Lot 26 needs to be removed and absorbed into Lot 3.

4. From the Parcel Layout Tools toolbar, click the Delete Sub-Entity tool.

5. At the `Select subentity to remove:` prompt, pick the rear lot line of the newly combined Lot 26.

6. Press ↵ to exit the command and then click the X to exit the Parcel Layout Tools toolbar.

The resulting parcel is shown in Figure 5.41.

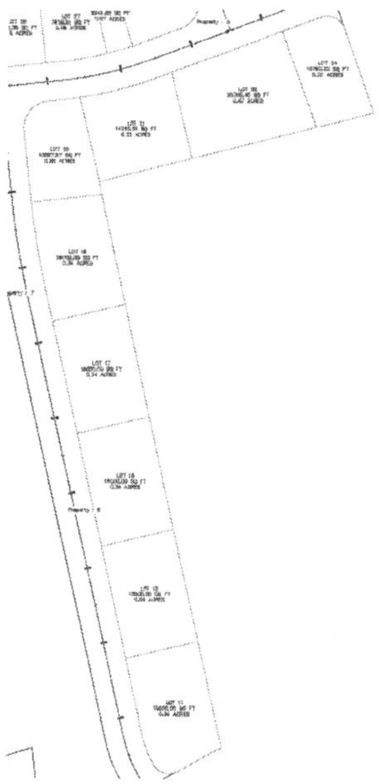

FIGURE 5.40
The re-created lots

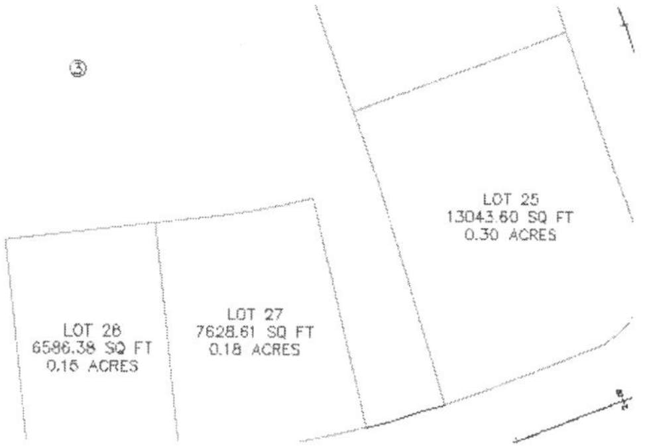

FIGURE 5.41
The parcel after erasing the rear lot line

Best Practices for Parcel Creation

Now that you have an understanding of how objects in a site interact and you've had some practice creating and editing parcels in a variety of ways, we'll take a deeper look at how parcels must be constructed to achieve topology stability, predictable labeling, and desired parcel interaction.

Forming Parcels from Segments

Earlier in this chapter, you saw that parcels are created only when parcel segments form a closed area (see Figure 5.42).

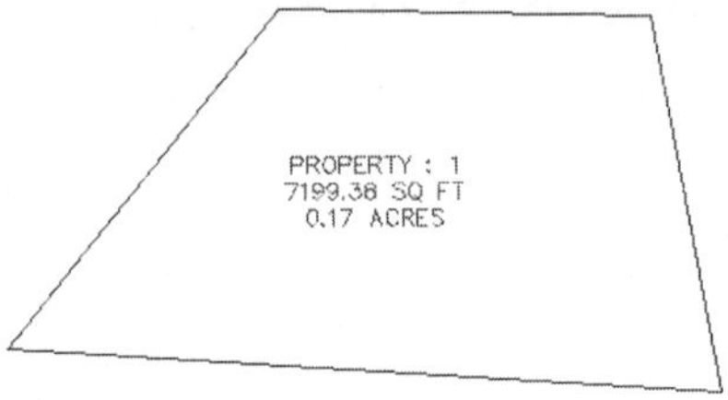

FIGURE 5.42
A parcel is created when parcel segments form a closed area.

Parcels must always close. Whether you draw AutoCAD lines and use the Create Parcel From Objects menu command or use the parcel segment creation tools, a parcel won't form until there is an enclosed polygon. Figure 5.43 shows four parcel segments that don't close; therefore, no parcel has been formed.

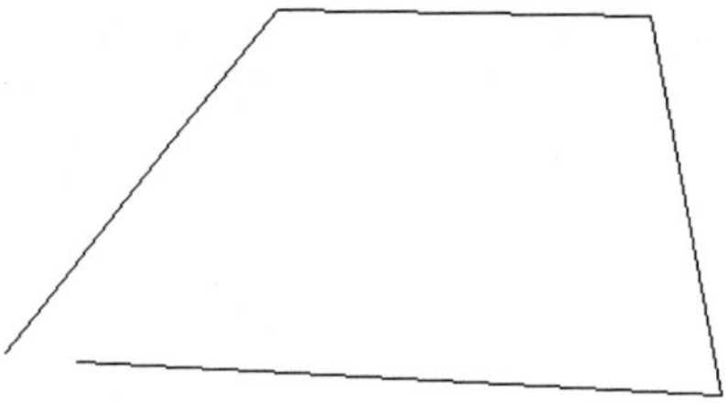

FIGURE 5.43
No parcel will be formed if parcel segments don't completely enclose an area.

There are times in surveying and engineering when parcels of land don't necessarily close when created from legal descriptions. In this case, you must work with your surveyor to perform an adjustment or find some other solution to create a closed polygon.

You also saw that even though parcels can't be erased, if you erase the appropriate parcel segments, the area contained within a parcel is assimilated into neighboring parcels.

Parcels Reacting to Site Objects

Parcels require only one parcel segment to divide them from their neighbor (see Figure 5.44). This behavior eliminates the need for duplicate segments between parcels, and duplicate segments must be avoided.

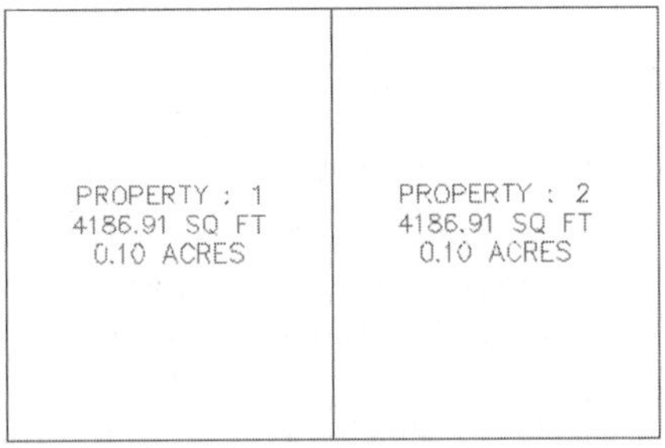

FIGURE 5.44
Two parcels, with one parcel segment between them

As you saw in the section on site interaction, parcels understand their relationships to one another. When you create a single parcel segment between two subdivision lots, you have the ability to move one line and affect two parcels. Figure 5.45 shows the moved parcel segment from Figure 5.44 once the parcel segment between them has been shifted to the left. Note that both areas change in response.

FIGURE 5.45
Moving one parcel segment affects the area of two parcels.

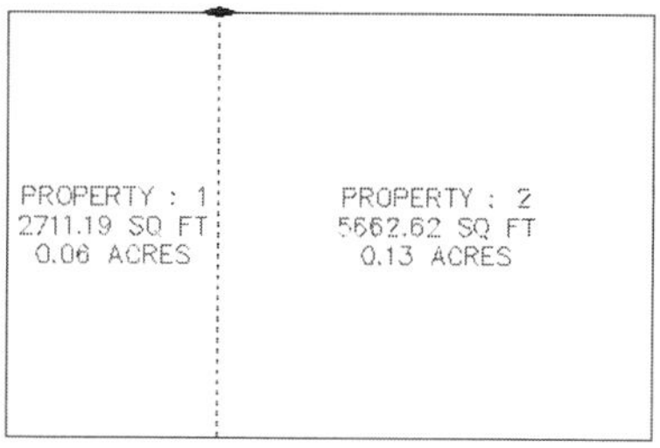

A mistake that many people new to Civil 3D make is to create parcels from closed polylines, which results in a duplicate segment between parcels. Figure 5.46 shows two parcels created from two closed polylines. These two parcels may appear identical to the two seen in the previous example, because they were both created from a closed polyline rectangle; however, the segment between them is actually two segments.

FIGURE 5.46
Adjacent parcels created from closed polylines create overlapping or duplicate segments.

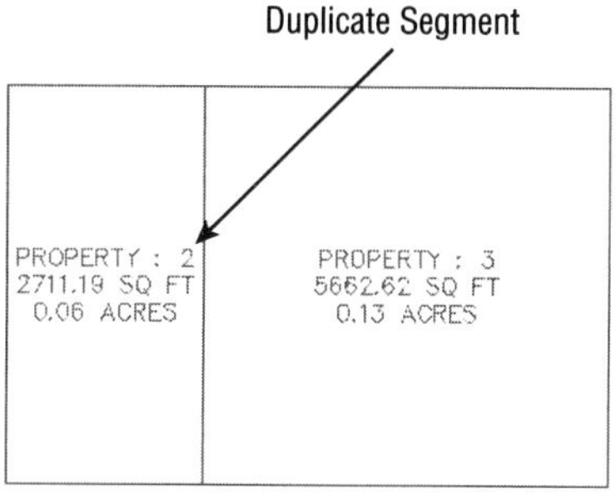

The duplicate segment becomes apparent when you attempt to grip-edit the parcel segments. Moving one vertex from the common lot line, as seen in Figure 5.47, reveals the second segment. Also note that a sliver parcel is formed. Duplicate site geometry objects and sliver parcels make it difficult for Civil 3D to solve the site topology and can cause unexpected parcel behavior.

FIGURE 5.47
Duplicate segments become apparent when they're grip-edited and a sliver parcel is formed.

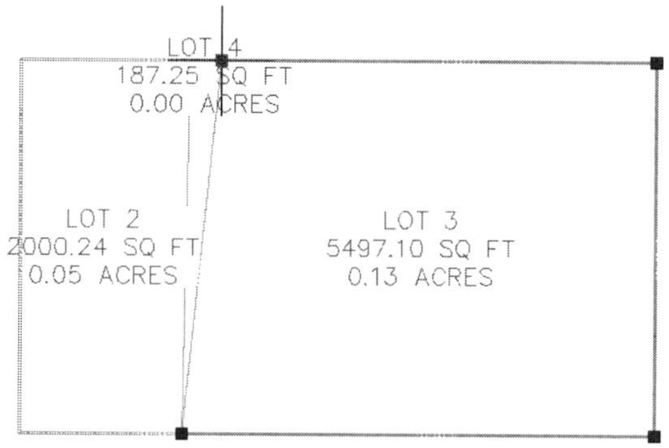

Creating a subdivision plat of parcels this way almost guarantees that your labeling won't perform properly and could lead to inaccurate data.

PARCELS AND LANDXML

The best method for importing lot data from other programs such as LandDesktop is LandXML. Parcels imported from LandXML will automatically clean up duplicate lines to prevent some of the pitfalls mentioned in this section.

Parcels form to fill the space contained by the original outer boundary. You should always begin a parcel-division project with an outer boundary of some sort (see Figure 5.48).

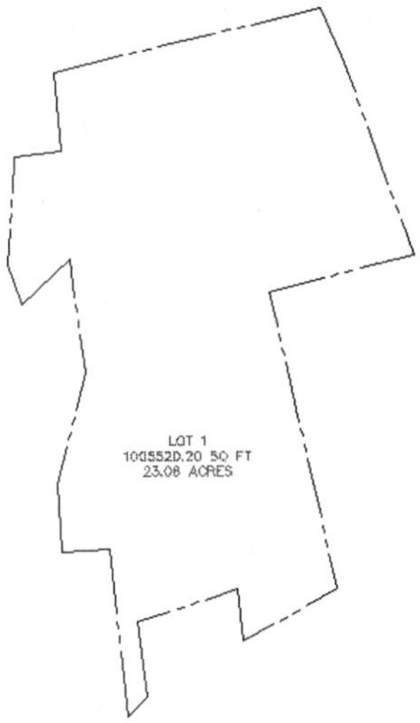

FIGURE 5.48
An outer boundary parcel

You can then add road centerline alignments to the site, which divides the outer boundary, as shown in Figure 5.49.

It's important to note that the boundary parcel no longer exists intact. As you subdivide this site, Parcel 1 is continually reallocated with every division. As road ROW and subdivision lots are formed from parcel segments, more parcels are created. Every bit of space that was contained in the original outer boundary is accounted for in the mesh of newly formed parcels (see Figure 5.50).

From now on, you'll consider ROW, wetlands, parkland, and open space areas as parcels, even if you didn't before. You can make custom label styles to annotate these parcels however you like, including a "no show" or none label.

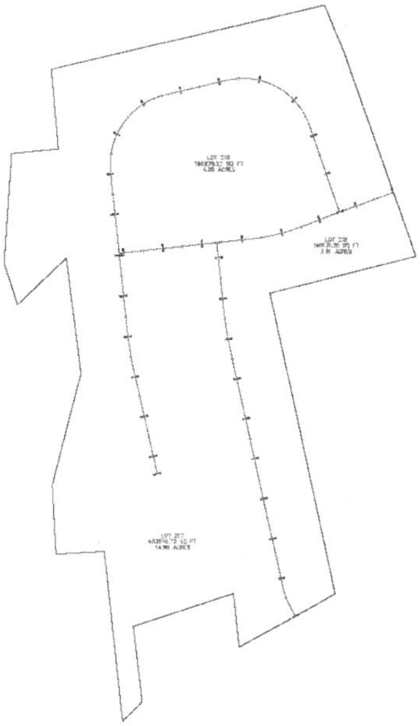

FIGURE 5.49
Alignments added to the same site as the boundary parcel divide the boundary parcel.

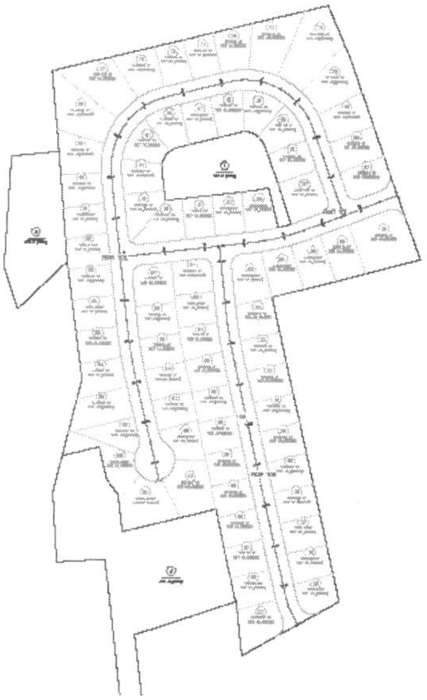

FIGURE 5.50
The total area of parcels contained within the original boundary sums to equal the original boundary area.

Real World Scenario

Overkill Is Just Right for Parcels

Frequently, parcel data comes from GIS sources or from existing plats. The quality of this data can vary, making a direct conversion to Civil 3D parcels difficult.

An excellent tool for ensuring that the conversion of these lines, arcs, and polylines goes smoothly is the Overkill command. Overkill is a base AutoCAD command that will clean up drawings based on the options and tolerance that you choose.

Find the Overkill command on the Home tab ➢ Modify panel flyout.

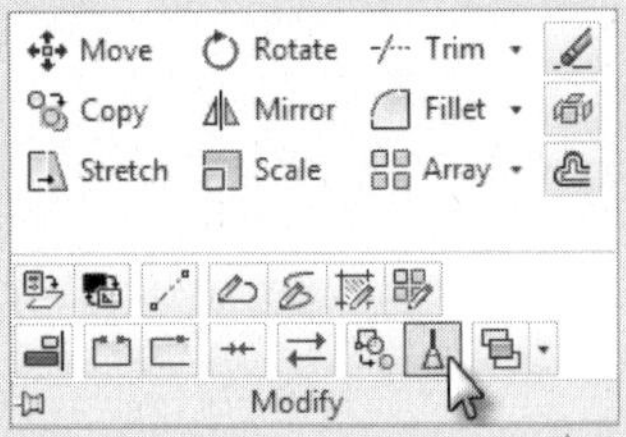

Once you have selected the objects you want the Overkill command to analyze, you are prompted to choose how dramatic of a cleanup you want. The default settings, shown here, are rather conservative.

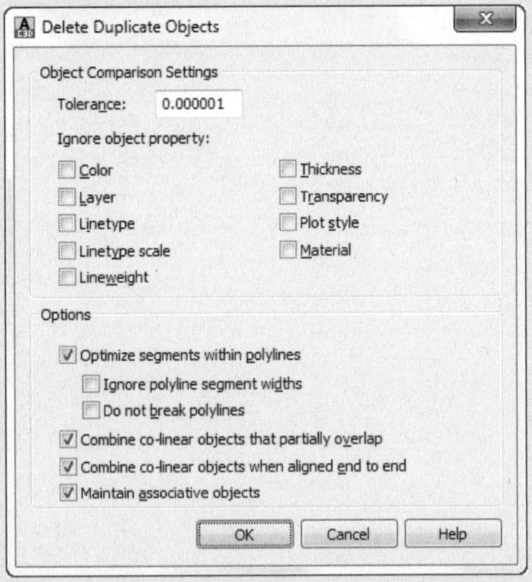

The tolerance determines how close two lines can be before they are considered to be duplicates. The higher this value, the more overlap and duplicates the command will find.

The check boxes indicate properties Overkill will disregard when performing its edits. With all of the check boxes clear, the objects are not considered duplicates unless all of the listed properties match.

Before running the Overkill command, a selected batch of lines slated to become parcels show many extraneous vertices. The plus signs next to the grips indicate where multiple grips overlap.

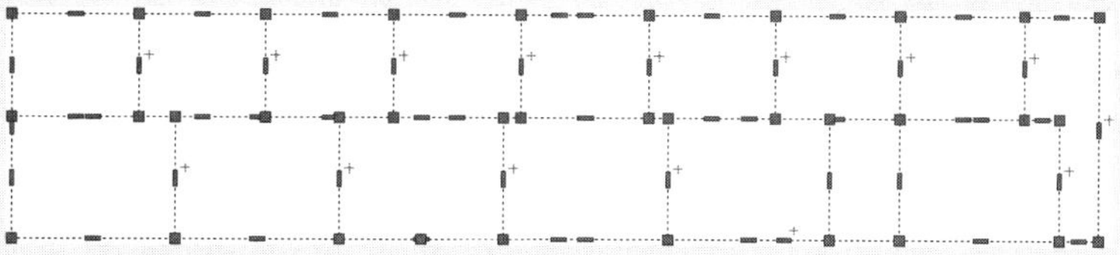

Before Overkill Command

After you run the Overkill command, the selection shows fewer visible grips and no indication that there are overlapping objects.

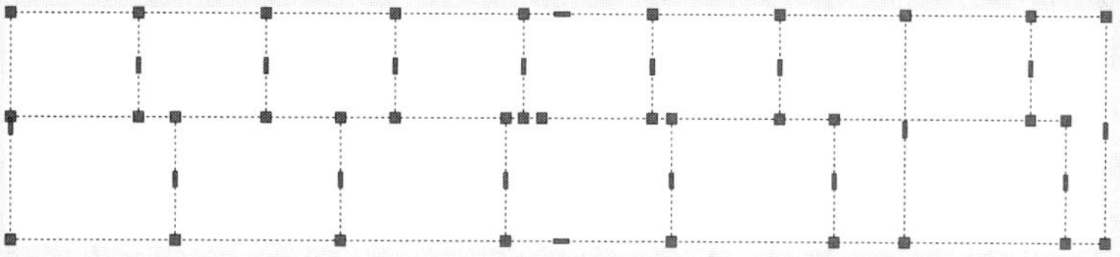

After Overkill Command

As you can see, the Overkill command can save you valuable time cleaning up drawings.

Constructing Parcel Segments with the Appropriate Vertices

Parcel segments should have natural vertices only where necessary and split-created vertices at all other intersections. A natural vertex, or point of intersection (PI), can be identified by picking a line, polyline, or parcel segment and noting the location of the grips (see Figure 5.51).

FIGURE 5.51
Natural vertices on a parcel segment

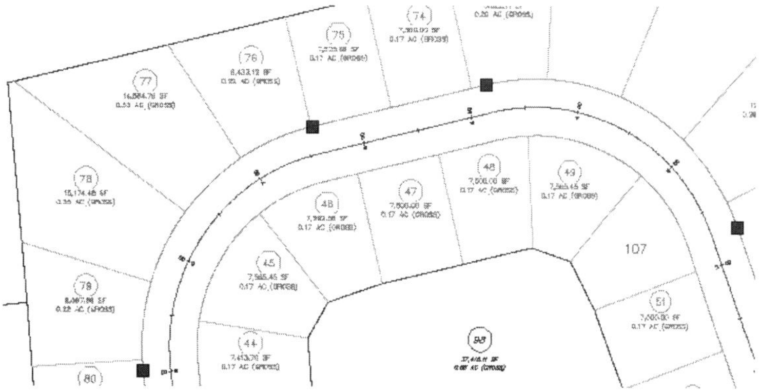

A split-created vertex occurs when two parcel segments touch or cross each other. Note that in Figure 5.52, the parcel segment doesn't show a grip even where each individual lot line touches the ROW parcel.

FIGURE 5.52
Split-created vertices on a parcel segment

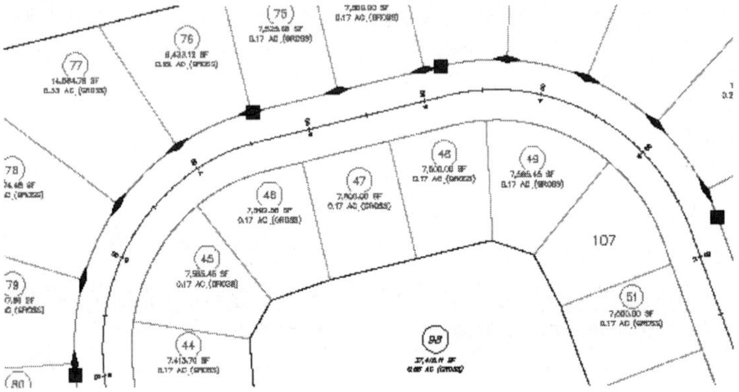

It's desirable to have as few natural vertices as possible. In the example shown in Figure 5.52, the ROW frontage line can be expressed as a single bearing and length from the end of the arc through the beginning of the next arc, as opposed to having several smaller line segments.

If the foundation geometry was drawn with a natural vertex at each lot line intersection, the resulting parcel segment won't label properly and may cause complications with editing and other functions. This subject will be discussed in more detail in the section "Labeling Spanning Segments," later in this chapter.

Parcel segments must not overhang. Spanning labels are designed to overlook the location of intersection-formed (or T-shaped) split-created vertices. However, these labels won't span a crossing-formed (X- or + [plus]-shaped) split-created vertex. Even a very small parcel segment overhang will prevent a spanning label from working and may even affect the area computation for adjacent parcels. The overhanging segment in Figure 5.53 would prevent a label from returning the full spanning length of the ROW segment it crosses.

FIGURE 5.53
Overhanging segment

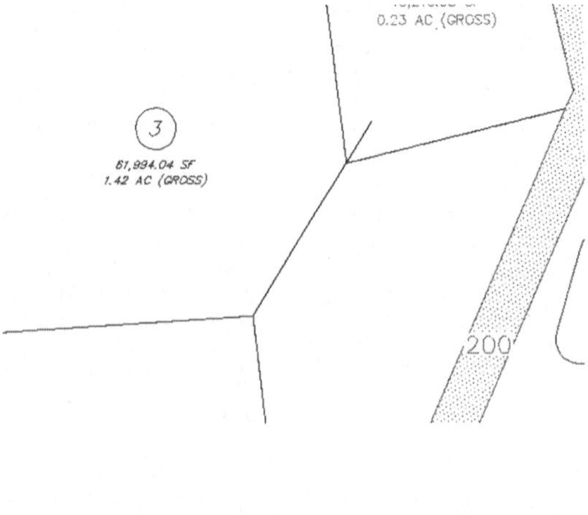

Labeling Parcel Areas

A parcel area label is placed at the parcel centroid by default, and it refers to the parcel in its entirety. When asked to pick a parcel, you pick the area label. An area label doesn't necessarily have to include the actual area of the parcel.

Area labels can be customized to suit your fancy. Figure 5.54 shows a variety of customized area labels.

FIGURE 5.54
Sample area labels

Area labels often include the parcel name or number. After selecting a parcel, on the Parcel contextual tab ➤ Modify panel choose Renumber/Rename.

The following exercise will teach you how to renumber a series of parcels:

1. Open `ChangeAreaLabel.dwg` (`ChangeAreaLabel_METRIC.dwg`). Note that this drawing contains many subdivision lot parcels.
2. Near the southeast corner of the project, select Lot 25. On the Parcel contextual tab ➤ Modify panel, select Renumber/Rename. The Renumber/Rename Parcels dialog appears.
3. In the Renumber/Rename Parcels dialog, make sure Subdivision Lots is selected from the drop-down menu in the Site selection box. Change the value of the Starting Number selection box to **1**. Click OK.
4. At the `Specify start point or [Polylines/Site]:` prompt, pick a point on the screen anywhere inside the Lot 25 parcel, which will become your new Lot 1 parcel at the end of the command.
5. At the `End point or [Undo]:` prompt, pick a point on the screen anywhere inside the Lot 35 parcel, almost as if you were drawing a line; then pick a point inside Lot 39. Press ↵ to complete choosing parcels. Press ↵ again to end the command.

Note that your parcels have been renumbered from 1 through 15. Repeat the exercise with other parcels in the drawing for additional practice if desired.

The next exercise will lead you through one method of changing an area label using the Edit Parcel Properties dialog:

1. Continue working in the `ChangeAreaLabel.dwg` (`ChangeAreaLabel_METRIC.dwg`) file.
2. Select Parcel 1 and select Multiple Parcel Properties from the Modify panel.

3. At the `Specify start point or [Polylines/All/Site]:` prompt, pick a point on the screen anywhere inside Parcel 1.
4. At the `End point or [Undo]:` prompt, pick a point on the screen anywhere inside Parcel 11, then Parcel 15, using the same technique that you used in step 2 of the previous exercise.
5. Press ↵ to complete parcel selection, and press ↵ again to open the Edit Parcel Properties dialog.
6. In the Area Selection Label Styles portion of the Edit Parcel Properties dialog, use the drop-down menu to choose the Parcel Number area label style, as shown in Figure 5.55.

FIGURE 5.55
The Edit Parcel Properties dialog

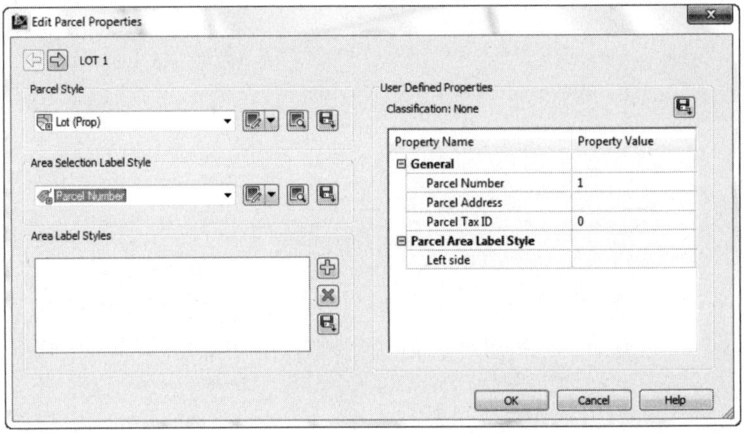

7. Click the Apply To All Parcels button to the right of the Parcel Number listing.
8. Click Yes in the dialog displaying the question "Apply the area selection label style to the 15 selected parcels?"
9. Click OK to exit the Edit Parcel Properties dialog.

The 15 parcels now have parcel area labels that call out numbers only. Note that you could also use this interface to add a second area label to certain parcels if required.

This section's final exercise will show you how to use Prospector to change a group of parcel area labels at the same time:

1. Continue working in the `ChangeAreaLabel.dwg` (`ChangeAreaLabel_METRIC.dwg`) file.
2. In Prospector, expand Sites ➤ Subdivision Lots and select the Parcels collection.
3. Hold down the Ctrl key, and select all of the lots whose names begin with Lot.
4. Release the Ctrl key, and your parcels should remain selected.
5. Slide over to the Area Label Style column, right-click the column header, and select Edit (see Figure 5.56).

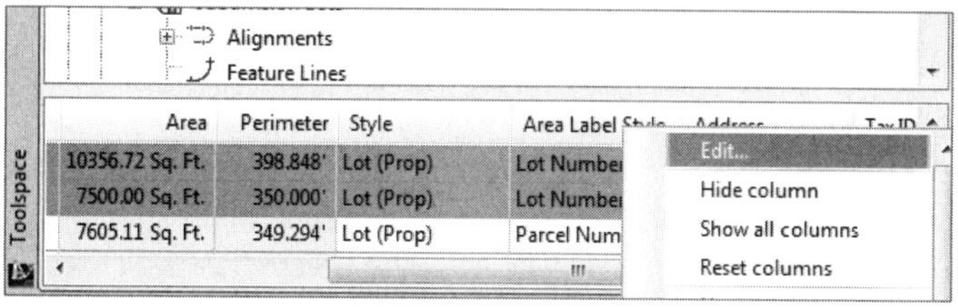

FIGURE 5.56
Right-click the Area Label Style column header and select Edit.

6. In the Select Label Style dialog, select Parcel Number from the drop-down menu in the Label Style selection box.
7. Click OK to dismiss the dialog.

   The drawing will process for a moment.
8. Once the processing is finished, minimize Prospector and inspect your parcels.

   All the Single-Family parcels should now have the Parcel Number area label style.

Real World Scenario

WHAT IF THE AREA LABEL NEEDS TO BE SPLIT ONTO TWO LAYERS?

You may have a few different types of plans that show parcels. Because it would be awkward to have to change the parcel area label style before you plot each sheet, it would be best to find a way to make a second label on a second layer so that you can freeze the area component in sheets or viewports when it isn't needed. Here's an example where the square footage has been placed on a different layer so it can be frozen in certain viewports:

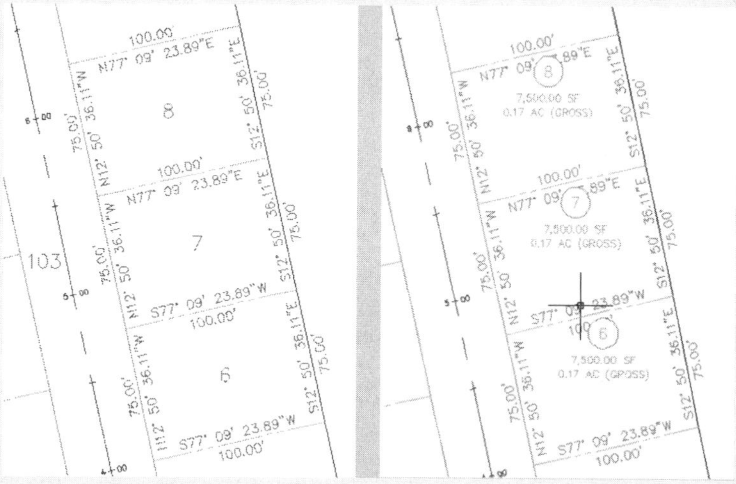

You can accomplish this by creating a second parcel area label that calls out the area only:

1. In any drawing containing parcels, change to the Annotate tab.
2. From the Labels & Tables panel, select Add Labels ➢ Parcel ➢ Add Parcel Labels.
3. Select Area from the drop-down menu in the Label Type selection box, and then select an area style label that will be the second area label.
4. Click Add, and then pick your parcel on screen.

You'll find a second parcel area label to be a little more automatic when you place it (it already knows what parcel to reference).

You can also use the Multiple Parcel Properties dialog, as shown in the "Editing Parcels by Deleting Parcel Segments" section earlier in the chapter, to add a second label.

Labeling Parcel Segments

Although parcels are used for much more than just subdivision lots, most parcels you create will probably be used for concept plans, record plats, and other legal subdivision plans. These plans, such as the one shown in Figure 5.57, almost always require segment labels for bearing, distance, direction, crow's feet, and more.

FIGURE 5.57 A fully labeled site plan

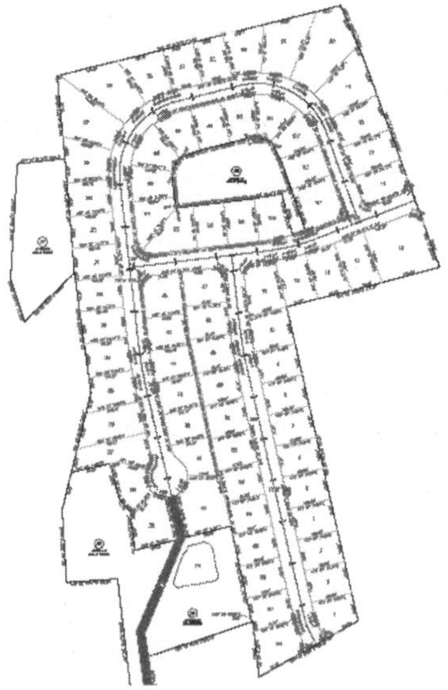

Labeling Multiple-Parcel Segments

The following exercise will teach you how to add labels to multiple-parcel segments:

1. Open the `SegmentLabels.dwg` (`SegmentLabels_METRIC.dwg`) file, which you can download from this book's web page.

Note that this drawing contains many subdivision lot parcels.

2. On the Annotate tab ➤ Labels & Tables panel, click Add Labels.
3. From the drop-down menus in the Add Labels dialog, select Parcel, Multiple Segment, Bearing Over Distance, and Delta Over Length And Radius in the Feature, Label Type, Line Label Style, and Curve Label Style selection boxes, respectively, as shown in Figure 5.58.

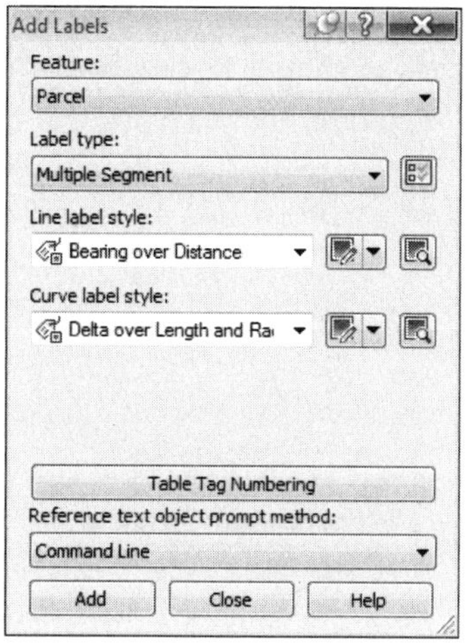

FIGURE 5.58
The Add Labels dialog

4. Click Add.
5. At the `Select parcel to be labeled by clicking on area label:` prompt, pick the area label for Parcel 1.
6. At the `Label direction [CLockwise/COunterclockwise]<CLockwise>:` prompt, press ↵ to accept the default and again to exit the command.

   Each parcel segment for Parcel 1 should now be labeled.
7. Continue picking Parcels 2 through 15 in the same manner.

   Note that segments are never given a duplicate label, even along shared lot lines.
8. If time permits, label all of the parcels in the drawing.
9. Press ↵ to exit the command. Close the Add Labels dialog.

The following exercise will show you how to edit and delete parcel segment labels:

1. Continue working in the `SegmentLabels.dwg` (`SegmentLabels_METRIC.dwg`) file.
2. Zoom in on the label along the frontage of Parcel 8.
3. Select the label.

   You'll know your label has been picked when you see a diamond-shaped grip at the label midpoint (see Figure 5.59).

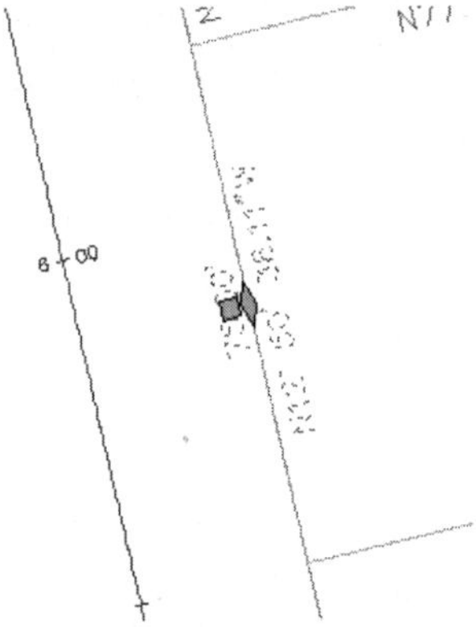

FIGURE 5.59
A diamond-shaped grip appears when the label has been picked.

4. On the Parcel Segment Label contextual tab ➢ Modify panel, click Flip Label, as shown in Figure 5.60.

   The label flips so that the bearing component is on top of the line and the distance component is underneath the line.

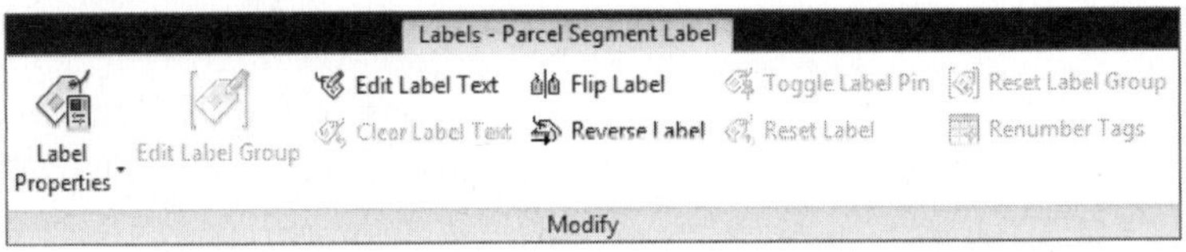

FIGURE 5.60
The Parcel Segment Label contextual tab

5. Select the label again. On the Parcel Segment Label contextual tab ➢ Modify panel, click Reverse.

   The label reverses so that the bearing now reads SE instead of NW.

6. Repeat steps 3 through 5 for several other segment labels, and note their reactions.

7. Select any label, and once the label is picked, execute the AutoCAD Erase tool or press the Delete key.

   Note that the label disappears.

8. Erase all the outer parcel labels in preparation for the next exercise.

Labeling Spanning Segments

Spanning labels are used where you need a label that spans the overall length of an outside segment, such as the example in Figure 5.61.

Spanning labels require that you use the appropriate vertices, as discussed in detail in the earlier section "Constructing Parcel Segments with the Appropriate Vertices." Spanning labels have the following requirements:

- Spanning labels can only span across split-created vertices. Natural vertices will interrupt a spanning length.

- Spanning label styles must be composed to span the outside segment.
- Spanning label styles must be composed to attach the desired spanning components (such as length and direction arrow) on the outside segment (as shown previously in Figure 5.61), with perhaps a small offset.

FIGURE 5.61
A spanning label

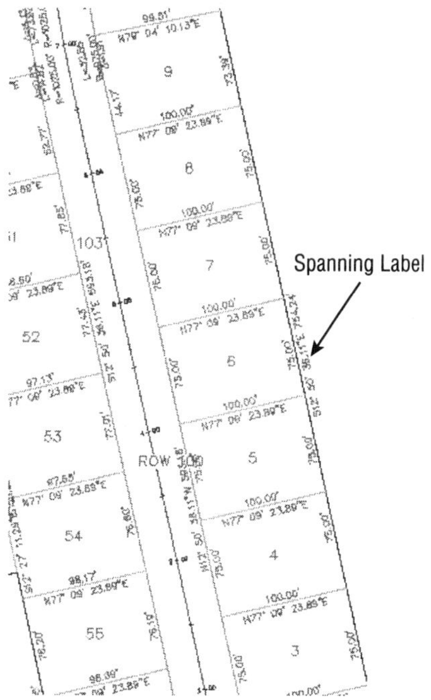

Once you've confirmed that your geometry is sound and your label is properly composed, you're set to span. The following exercise will teach you how to add spanning labels to single-parcel segments:

1. Continue working in the `SegmentLabels.dwg` (`SegmentLabels_METRIC.dwg`) file.
2. Zoom in on the outer parcel segment that runs from Parcel 1 through Parcel 10.
3. Change to the Annotate tab and select Add Labels ➢ Parcel ➢ Add Parcel Labels from the Labels & Tables panel.
4. From the drop-down menus in the Add Labels dialog, select Single Segment, (Span) Bearing And Distance With Crows Feet, and Delta Over Length And Radius in the Label Type, Line Label Style, and Curve Label Style selection boxes, respectively.
5. Click Add.
6. At the `Select label location:` prompt, pick somewhere near the middle of the outer parcel segment that runs from Parcel 1 through Parcel 10.

   A label that spans the full length between natural vertices appears (see Figure 5.61).

> **Flip It, Reverse It**
>
> If your spanning label doesn't seem to work on your first try and you've followed all the spanning label guidelines, try flipping your label to the other side of the parcel segment, reversing the label, or using a combination of both flipping and reversing.

Adding Curve Tags to Prepare for Table Creation

To keep plans tidy, it is common to show labels that reference a table for curves and lines. Civil 3D parcels provide tools for creating dynamic line and curve tables. You can keep lines and curves together or create separate tables for each.

Parcel segments must be labeled before they can be used to create a table. They can be labeled with any type of label, but you'll likely find it to be best practice to create a tag-only style for segments that will be placed in a table.

The following exercise will show you how to replace curve labels with tag-only labels, and then renumber the tags:

1. Continue working in the `SegmentLabels.dwg` (`SegmentLabels_METRIC.dwg`) file.

   Note that the labels along tight curves, such as the cul-de-sac, would be better represented as curve tags.

2. Change to the Annotate tab, and select Add Labels ➤ Parcel ➤ Add Parcel Labels from the Labels & Tables panel.

   The Replace multiple-segment tool only works if the parcel is already labeled. If you did not get a chance to label the parcels in the previous exercise, use the multiple-label segment instead for step 3.

3. From the drop-down menus in the Add Labels dialog, select Replace Multiple Segment, Bearing Over Distance, and Spanning Curve Tag Only in the Label Type, Line Label Style, and Spanning Curve Tag Only selection boxes, respectively.

4. Click Add. At the `Select parcel to be labeled by clicking on area label:` prompt, pick the area label for Parcel 1.

   Note that the line labels for Parcel 1 are reset and the curve labels convert to tags.

5. Repeat step 4 for Parcels 2 through 15.

6. Press ↵ to exit the command.

Now that each curve label has been replaced with a tag, it's desirable to have the tag numbers be sequential. The following exercise will show you how to renumber tags:

1. Continue working in the `SegmentLabels.dwg` (`SegmentLabels_METRIC.dwg`) file.

2. Zoom into the curve on the upper-left side of Parcel 11 (see Figure 5.62).

   Your curve may have a different number from the figure.

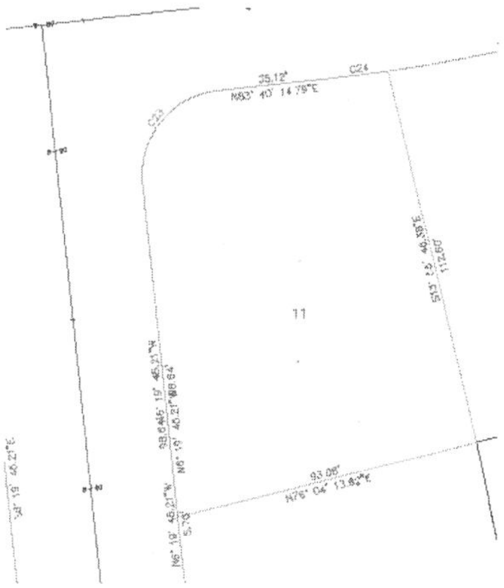

FIGURE 5.62
Curve tags on Parcel 11

3. From the Parcel contextual tab ➤ Labels & Tables panel, select Renumber Tags.
4. At the `Select label to renumber tag or [Settings]:` prompt, type **S**, and then press ↵.

   The Table Tag Numbering dialog appears (see Figure 5.63).

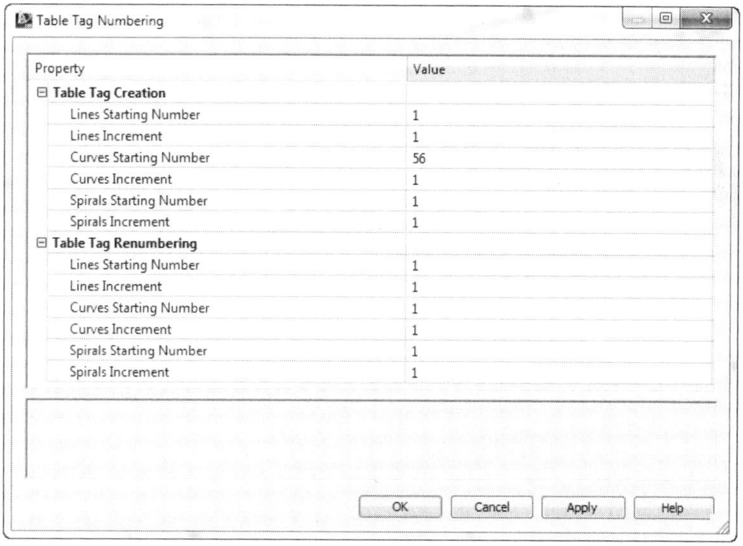

FIGURE 5.63
The Table Tag Numbering dialog

5. Change the value in the Curves Starting Number selection box to **1**, and click OK.
6. Click each curve tag in the drawing at the `Select label to renumber tag or [Settings]:` prompt.

The command line may say `Current tag number is being used, press return to skip to next available or [Create duplicate]`, in which case you should press ↵ to skip the used number.

7. When you're finished, press ↵ to exit the command.

Creating a Table for Parcel Segments

The following exercise demonstrates how to create a table from curve tags:

1. Continue working in the `SegmentLabels.dwg` (`SegmentLabels_METRIC.dwg`) file.

   You should have several curves labeled with the Spanning Curve Tag Only label.

2. Select a parcel. From the Parcel contextual tab ➢ Labels & Tables panel, select Add Curve.

3. In the Table Creation dialog, select Length Radius & Delta from the drop-down menu in the Table Style selection box.

4. In the Select By Label Or Style area of the dialog, click the Apply check box for the Parcel Curve: Spanning Curve Tag Only entry under Label Style Name.

5. Now that the Selection Rule is active for Parcel Curve: Spanning Curve Tag Only, change it to Add Existing And New.

   The Add Existing And New option will ensure that the table updates as more labels fitting this criteria are added to the drawing. Keep the default values for the remaining options. The dialog should look like Figure 5.64.

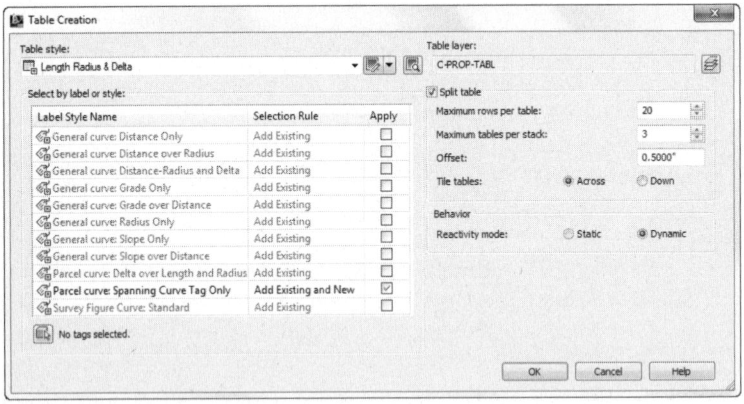

FIGURE 5.64
The Table Creation dialog

6. Click OK.

7. At the `Select upper left corner:` prompt, pick a location in your drawing for the table.

   A curve table appears, as shown in Figure 5.65.

FIGURE 5.65
A curve table

Curve Table					
Curve #	Length	Radius	Delta	Chord Direction	Chord Length
C27	57.21	35.00	93.65	N71° 55' 15"W	51.05
C28	32.07	150.00	12.25	N18° 58' 08"W	32.01
C26	32.55	975.00	1.91	N11° 53' 13"W	32.55
C25	78.30	975.00	4.60	N8° 37' 48"W	78.28
C23	39.27	25.00	90.00	N38° 40' 16"E	35.36
C24	17.25	525.00	1.88	N82° 43' 47"E	17.25
C22	74.38	525.00	8.12	N77° 43' 48"E	74.31
C21	35.72	525.00	3.90	N71° 43' 21"E	35.71
C20	54.98	35.00	90.00	S85° 13' 34"E	49.50
C43	39.27	25.00	90.00	S24° 46' 25"W	35.36
C44	17.70	475.00	2.14	S70° 50' 28"W	17.70
C45	78.26	150.00	29.89	S35° 10' 21"E	77.37
[illegible]	[illegible]	[illegible]	[illegible]	[illegible]	[illegible]

The Bottom Line

Create a boundary parcel from objects. The first step to any parceling project is to create an outer boundary for the site.

Master It Open the `MasteringParcels.dwg` (`MasteringParcels_METRIC.dwg`) file, which you can download from `www.sybex.com/go/masteringcivil3d2013`. Convert the line segments in the drawing to a parcel.

Create a right-of-way parcel using the right-of-way tool. For many projects, the ROW parcel serves as frontage for subdivision parcels. For straightforward sites, the automatic Create ROW tool provides a quick way to create this parcel. A cul-de-sac serves as a terminal point for a cluster of parcels.

Master It Continue working in the `Mastering Parcels.dwg` (`MasteringParcels_METRIC.dwg`) file. Create a ROW parcel that is offset by 25′ (10 m) on either side of the road centerline with 25′ (10 m) fillets at the parcel boundary and alignment ends. Then add the circles representing the cul-de-sac as a parcel.

Create subdivision lots automatically by layout. The biggest challenge when creating a subdivision plan is optimizing the number of lots. The precise sizing parcel tools provide a means to automate this process.

Master It Continue working in the `Mastering Parcels.dwg` (`MasteringParcels_METRIC.dwg`) file. Create a series of lots with a minimum of 8,000 sq. ft. (700 m^2) and 75′ (20 m) frontage. Set the Use Minimum Offset option to No. Leave all other options at their defaults.

Add multiple-parcel segment labels. Every subdivision plat must be appropriately labeled. You can quickly label parcels with their bearings, distances, direction, and more using the segment labeling tools.

Master It Continue working in the `MasteringParcels.dwg` (`MasteringParcels_METRIC.dwg`) file. Place Bearing Over Distance labels on every parcel line segment and Delta Over Length And Radius labels on every parcel curve segment using the Multiple Segment Labeling tool.

Chapter 6

Alignments

Some roads were laid out hundreds of years ago by herds of cows; others were microdesigned by planners with careful consideration for each bend and twist. Either way, when working on a linear project, whether it be the renovation of an old New England cow path or a network of new subdivision roads, horizontal layout information will need to be conveyed to the contractor. This horizontal layout is the alignment and drives much of the design. This chapter shows you how alignments can be created, how they interact with the rest of the design, how to edit and analyze them, and finally, how they work with the overall project.

In this chapter, you will learn to:

- Create an alignment from an object
- Create a reverse curve that never loses tangency
- Replace a component of an alignment with another component type
- Create alignment tables

Alignment Concepts

Before you can efficiently work with alignments, you must understand two major concepts: the interaction of alignments and sites, and the idea of geometry that is fixed, floating, or free.

Alignments and Sites

Prior to the AutoCAD® Civil 3D® 2008 release, alignments were always a part of a site and interacted with the topology contained in that site. This interaction led to the pickle analogy: alignments are like pickles in a Mason jar. You don't put pickles and peppers in the same jar unless you want hot pickles, and you don't put lots and alignments in the same site unless you want subdivided lots.

Civil 3D now has two ways of handling alignments in terms of sites: they can be contained in a site as before, or they can be independent of a site.

Both the alignments contained in a site and those independent of a site can be used to cut profiles or control corridors, but only the alignments contained in a site will react with and create parcels as members of a site topology.

Unless you have good reason for them to interact (as in the case of an intersection), it makes sense to create alignments outside of any site object. They can be moved later if necessary. For the purpose of the exercises in this chapter, you won't place any alignments in a site.

Alignment Entities

Civil 3D recognizes five types of alignments: centerline alignments, offset alignments, curb return alignments, rail alignments, and miscellaneous alignments. Each alignment type can consist of three types of entities or segments: lines, arcs, and spirals. These segments control the horizontal alignment of your design. Their relationship to one another is described by the following terminology:

Fixed Segments Fixed segments are fixed in space (see Figure 6.1). They're defined by connecting points in the coordinate plane and are independent of the segments that occur either before or after them in the alignment. Fixed segments may be created as tangent to other components, but their independence from those objects lets you move them out of tangency during editing operations.

FIGURE 6.1: Alignment fixed segments

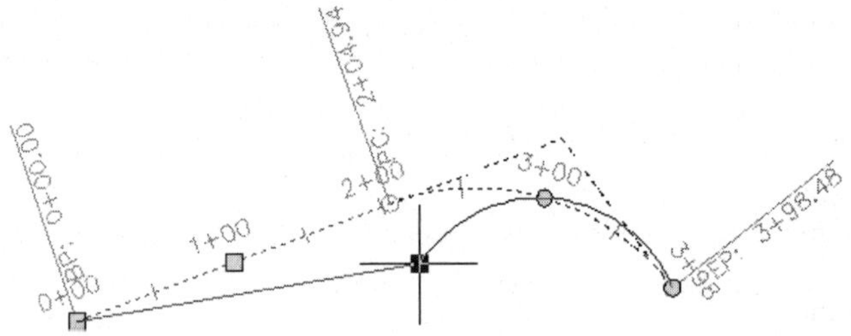

Floating Segments Floating segments float in space but are attached to a point in the plane and to some segment to which they maintain tangency (see Figure 6.2). Floating segments work well in situations where you have a critical point but the other points of the horizontal alignment are flexible.

FIGURE 6.2: Alignment floating segments

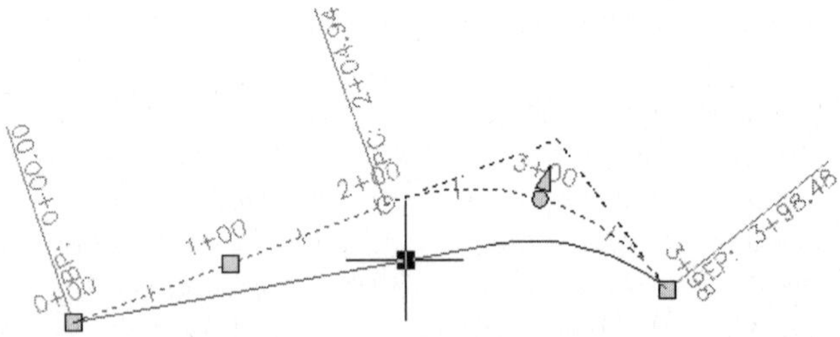

Free Segments Free segments are functions of the entities that come before and after them in the alignment structure (see Figure 6.3). Unlike fixed or floating segments, a free segment must have segments that come before and after it. Free segments maintain tangency to the segments that come before and after them and move as required to make that happen. Although some geometry constraints can be put in place, these constraints can be edited and are user dependent.

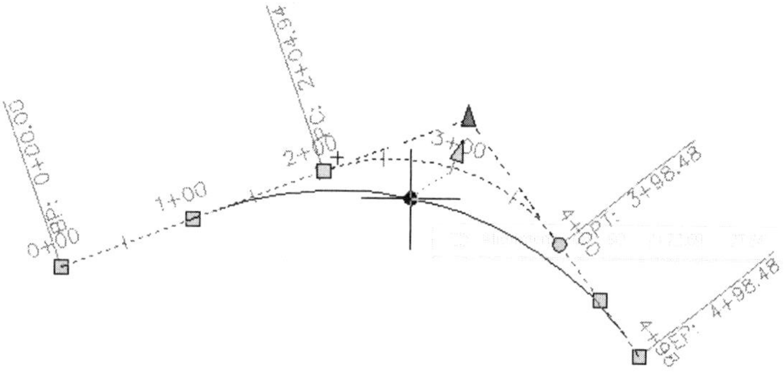

FIGURE 6.3: Alignment free segments

During the exercises in this chapter, you'll use a mix of these entity types to understand them better.

Creating an Alignment

CERT OBJECTIVE

Alignments can be created from AutoCAD objects (lines, arcs, or polylines) or by layout. This section looks at both ways to create an alignment and discusses the advantages and disadvantages of each. The exercise will use the street layout shown in Figure 6.4 as well as the different methods to achieve your designs.

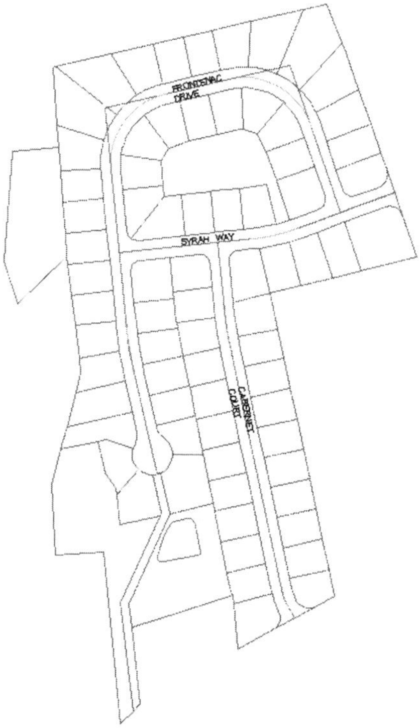

FIGURE 6.4: Proposed street layout

Creating from a Line, Arc, or Polyline

Most designers have used either polylines or lines and arcs to generate the horizontal control of their projects. It's common for surveyors to generate polylines to describe the center of a right-of-way or for an environmental engineer to draw a polyline to show where a new channel should be constructed. These team members may or may not have Civil 3D software, so they use their familiar friends — the line, arc, and polyline — to describe their design intent.

Although these objects are good at showing where something should go, they don't have much data behind them. To make full use of these objects, you should convert them to native Civil 3D alignments that can then be shared and used for myriad purposes. Once an alignment has been created from a polyline, offsets can be created to represent rights of way, building lines, and so on. In this exercise, you'll convert a polyline to an alignment and create offsets:

1. Open the `AlignmentsFromPolylines.dwg` file (or for metric users, the `AlignmentsFromPolylines_METRIC.dwg` file).

   You can download this file from the book's web page at `www.sybex.com/go/masteringcivil3d2013`. You will see the red polylines representing the centers of roads, rights-of-way, and parcels.

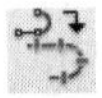

2. From the Home tab ➤ Create Design panel, choose Alignment ➤ Create Alignment From Objects.
3. Pick the two lines and arc labeled Syrah Way, shown previously in Figure 6.4, and press ↵.
4. Verify that the direction is from left to right and press ↵.

   The Create Alignment From Objects dialog appears.
5. Change the Name field to **Syrah Way,** and select the Centerline type, as shown in Figure 6.5.
6. Verify that Alignment Style is set to Proposed and Alignment Label Set is set to Major And Minor Only.
7. Accept the other settings, and click OK.

A Word About Constraints

In previous versions, creating from objects held no constraints (everything was fixed). In current versions, Civil 3D tries to find tangency associates if a curve is between two lines. Lines will always be Not Constrained (Fixed), but if the curve is tangent it will be Constrained On Both Sides (Free).

Keep this drawing open for the next portion of the exercise.

FIGURE 6.5: The settings used to create the Syrah Way alignment

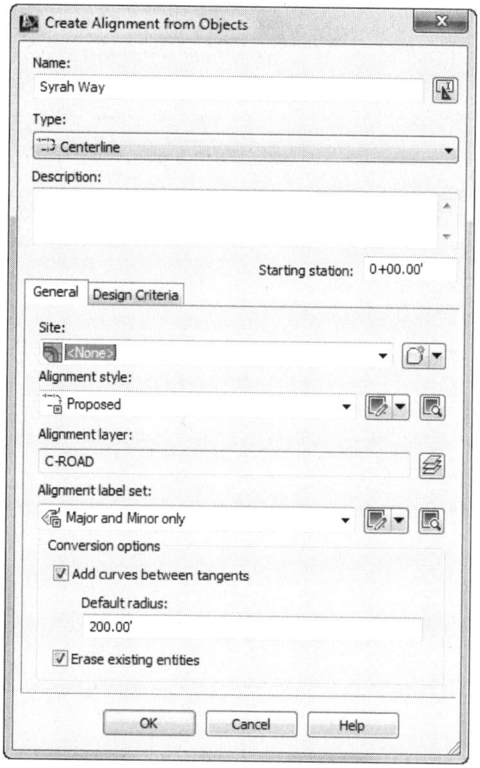

THE CREATE ALIGNMENT FROM OBJECTS DIALOG

In the Create Alignment From Objects dialog (shown in Figure 6.5), there are many settings that you can adjust when creating an alignment:

Name This is the alignment name. No alignment name can be duplicated in a drawing.

Type The alignment types can be thought of as places for objects that are alike. They can react differently depending on which type is selected.

Centerline Used mainly for centers of roads, streams, or swales. Civil 3D places this type of alignment in the Alignments ➤ Centerline Alignments collection.

Offset Used for offset alignments. The difference between this and the centerline alignment is that you have the option in the Alignment Properties dialog to set Offset parameters, such as naming a parent alignment and offset values. Civil 3D places this type of alignment in the Alignments ➤ Offset Alignments collection.

Curb Return Used for curb returns, which are the radii at intersections. The difference between this and the offset alignment is that instead of offset, you have the option in the Alignment Properties dialog to set Curb Return parameters, such as setting two parent alignments and offsets. Civil 3D places this type of alignment in the Alignments ➤ Curb Return Alignments collection.

Rail Used for rails. The difference between this and the other alignments is that the alignment is set using typical rail geometry such as degree of curvature and cant. In addition, a unique tab is added to the Alignment Properties dialog called Rail Parameters, where you can specify the Track Width. Civil 3D places this alignment type in the Alignments ➢ Rail Alignments collection.

Miscellaneous This is a stripped-down alignment type that only contains Information, Stationing, Masking, Point Of Intersection, and Constraint Editing tabs. Civil 3D places this alignment type in the Alignments ➢ Miscellaneous Alignments collection.

Description An optional field where you can be as verbose as you want with information describing your alignment.

Starting Station Setting this with a number, either positive or negative, will be the starting stationing for the alignment. This is handy if you need to start your alignment to coincide with existing stationing, or if you wish to have your 0+00 stationing at an intersection of a road. If you forget to set this value here, you can change it later in the Alignment Properties dialog.

The General tab contains the following options:

Site A place to keep Civil 3D objects that you want to interact with each other. As previously mentioned, all of your alignments in these exercises will be put on the <None> site.

Alignment Style You can set different styles to visually show your alignment. For more on styles, refer to Chapter 21, "Object Styles."

Alignment Layer Overrides the layer that is specified in the Drawing Settings for alignments.

Alignment Label Set As with the Alignment Style, you can choose how your alignment will be labeled. For more on label styles, refer to Chapter 20, "Label Styles."

Conversion Options Depending on your selections, these will add curves or erase the original entities.

The Design Criteria tab contains these options:

Starting Design Speed Specify the design speed of the alignment for the starting station. If no additional design speeds are applied to a different section of the alignment, this speed will be used for the entire alignment.

Use Criteria-Based Design When this check box is selected, the Use Design Criteria File and Use Design Check Set options can be used.

Use Design Criteria File Here you can define the design criteria, such as the American Association of State Highway and Transportation Officials (AASHTO) *A Policy on Geometric Design of Highways and Streets, 2001* design manual.

Use Design Check Set Here you can set rules, or expressions for lines, curves, spirals, and tangent intersections.

Many of these settings will be discussed further throughout this chapter.

You've created your first alignment and attached stationing and geometry point labels. It is common to create offset alignments from a centerline alignment to begin to model rights-of-way.

In the following exercise, you'll create offset alignments and mask them where you don't want them to be seen:

1. Using the drawing from the previous exercise, from the Home tab ➢ Create Design panel, choose Alignment ➢ Create Offset Alignment.
2. Pick the Syrah Way alignment to open the Create Offset Alignments dialog shown in Figure 6.6.

FIGURE 6.6: The Create Offset Alignments dialog

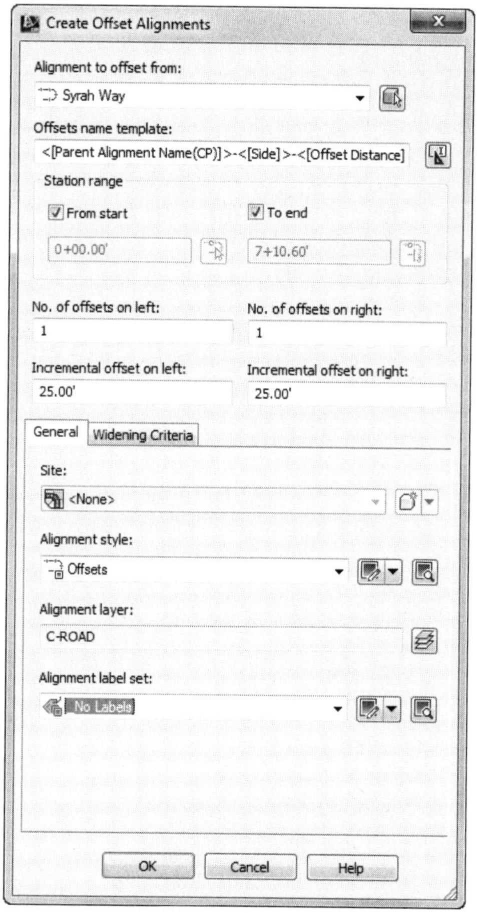

3. Change Incremental Offset On The Left to **25′** (metric users, **7.5**).
4. Change Incremental Offset On The Right to **25′** (metric users, **7.5**).
5. Verify that Alignment Label Set is set to _No Labels, and click OK to accept the rest of the defaults shown in Figure 6.6.
6. Select the offset alignment just created along the northerly right-of-way of Syrah Way to activate the contextual tab.

7. From the contextual tab ➤ Modify panel, choose Alignment Properties to open the Alignment Properties dialog.

8. Change to the Masking tab and click the Add Masking Region button.

9. Enter **0** for the first station and **50** for the second station when prompted (metric users, enter **0** and **15**), and click OK.

   Notice that the alignment is now masked at the intersection of Syrah Way and Frontenac Drive at the east end.

10. Repeat the process for the rest of the intersections, starting at the end of the arc on both right-of-way alignments on Syrah Way.

    While you entered the information using text in the previous step, notice that you could alternatively use the station picker button, which becomes visible when you click into the cell to enter a value.

    Once complete, the Masking tab of the Alignment Properties dialog should look like Figure 6.7.

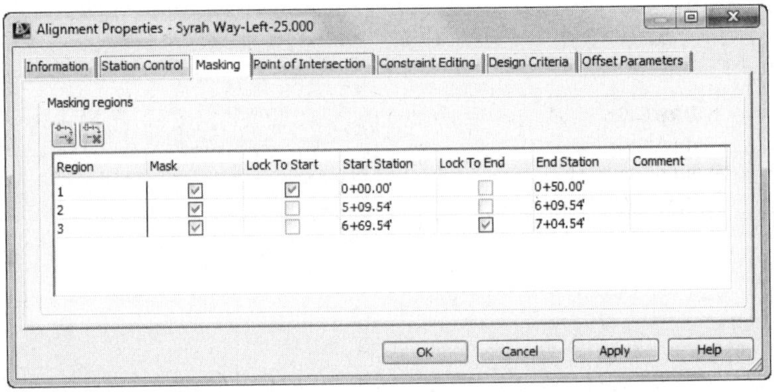

FIGURE 6.7: Creating an alignment mask

When this exercise is complete, you may close the drawing. A saved finished copy of this drawing is available from the book's web page with the filename `AlignmentsFromPolylines_FINISHED.dwg` or `AlignmentsFromPolylines_METRIC_FINISHED.dwg`. Note that when you selected the beginning and end of the offset alignment, the Lock To Start and Lock To End boxes were checked automatically.

Offset alignments are simple to create, and they are dynamically linked to a centerline alignment. To test this, grip the centerline alignment, select the endpoint grip, and stretch the alignment to the west. Notice the change, and then undo this change to return to the original state.

OFFSET GRIPS AND MORE

Offset alignments have two special grips: the arrow and the plus sign. The arrow is used to change the offset value, and the plus sign is used to create a transition, called a *widening*, such as a turning lane. To access the Create Widening command, from the Home tab ➤ Create Design panel, select the Alignment drop-down. Widening criteria can also be found in the Create Offset Alignments dialog.

Even offset alignments with widening remain dynamic to their host alignment. Offset alignment objects can be found in Prospector in the Alignments collection.

You created an alignment from polylines and two offsets. It's ready for use in corridors, in profiling, or for any number of other uses.

Creating by Layout

Now that you've made an alignment from polylines, let's look at another creation option: Create By Layout. You'll use the same street layout (Figure 6.4) that was provided by a planner, but instead of converting from polylines, you'll trace the alignments. Although this seems like duplicate work, it will pay dividends in the relationships created between segments:

1. Open the `AlignmentsByLayout.dwg` or the `AlignmentsByLayout_METRIC.dwg` file.

2. From the Home tab ➢ Create Design panel, choose Alignment ➢ Alignment Creation Tools.

   The Create Alignment–Layout dialog appears.

3. Change the Name field to **Cabernet Court** if it is not already set.
4. Verify that Alignment Style is set to Proposed and Alignment Label Set is set to Major And Minor Only.
5. Click OK to accept the other settings shown in Figure 6.8.

FIGURE 6.8: Create Alignment – Layout dialog

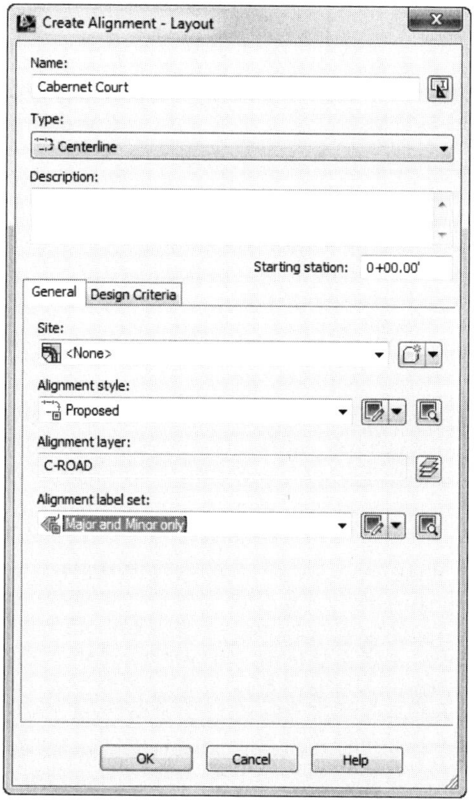

The Alignment Layout Tools toolbar for Cabernet Court appears (Figure 6.9).

FIGURE 6.9: The Alignment Layout Tools toolbar for Cabernet Court

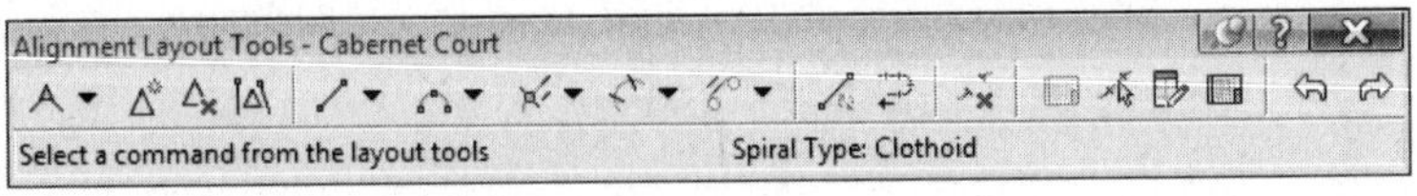

6. Click the down arrow next to the Tangent-Tangent (No Curves) tool at the far left, and select the Tangent-Tangent (With Curves) option (see Figure 6.10).

   The tool places a curve automatically; you'll later adjust the curve, watching the tangents extend as needed.

FIGURE 6.10: The Tangent-Tangent (With Curves) tool

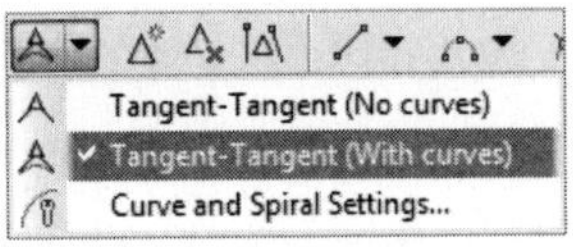

7. Turn on the Endpoint Osnap and snap to the southernmost end of the Cabernet Court centerline.
8. Continue to pick the endpoints of the red lines to finish creating this alignment.
9. Press ↵ to end the command.
10. Click the X button at the upper-right on the Alignment Layout Tools toolbar for Cabernet Court to close it.

    Keep this drawing open for the next portion of the exercise.

    Zoom in on the southern arc. Notice that it follows closely with the desired arc radius. Now zoom in on the northern arc, and notice that it doesn't match the arc the planner put in for you to follow. That's okay — you will fix it in a later exercise.

DESIGN AND THEN REFINE

It bears repeating that in dealing with Civil 3D objects, it is good practice to get something in place and *then* refine. With Land Desktop or other packages, you didn't want to define the object until it was fully designed. In Civil 3D, you design and then refine.

The alignment you just made is one of the most basic. Let's move on to some of the others and use a few of the other tools to complete your initial layout. In this exercise, you build the alignment at the north end of the site, but this time you use a floating curve to make sure the two segments you create maintain their relationship:

11. From the Home tab ➤ Create Design panel, choose Alignment ➤ Alignment Creation Tools.

    The Create Alignment – Layout dialog appears.

12. In the Create Alignment – Layout dialog, do the following:
    A. Change the Name field to **Frontenac Drive**.
    B. Set the Alignment Style field to Layout.
    C. Set the Alignment Label Set field to Major And Minor Only.
13. Click OK to display the Alignment Layout Tools toolbar for Frontenac Drive.
14. Select the Fixed Line (Two Points) tool, as shown in Figure 6.11.

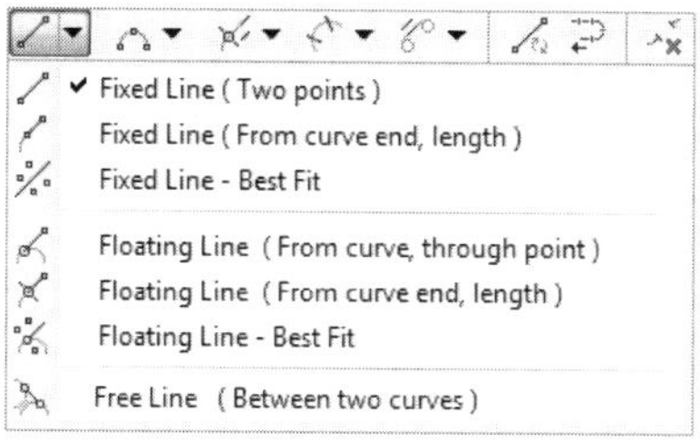

FIGURE 6.11: The Fixed Line (Two Points) tool

15. Using Endpoint Osnaps, select the eastern end of Frontenac Drive and, working south to north, draw the first fixed-line segment to the starting endpoint of the cyan arc.

    When you've finished, the command line will read `Specify start point:` in case you want to draw another line. You can either press ↵ to end the command or draw the next entity without ending the command.
16. Click the down arrow next to the Add Fixed Curve (Three Point) tool on the toolbar, and select More Floating Curves ➤ Floating Curve (From Entity End, Through Point), as shown in Figure 6.12.

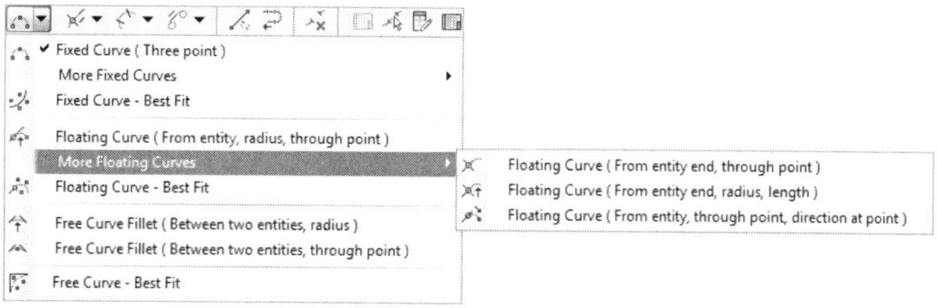

FIGURE 6.12: Selecting the Floating Curve tool

17. Select the fixed-line segment you drew in steps 14 and 15.

    Make sure you select the line segment somewhere north of the segment's midpoint in order to connect to the northern endpoint instead of the southern endpoint.

    Move your cursor around and a blue rubber band should appear, indicating that the alignment of the curve segment is being floated off the endpoint of the fixed segment.
18. Pick the endpoint of the arc of the Frontenac Drive polyline arc segment.

Notice that you are generating the segments from low station to high station. If you perform these steps backward, you can reverse a segment using the Reverse Sub-Entity Direction button on the Alignment Layout Tools toolbar.

19. Right-click or press ↵ to end the command.
20. Close the Alignment Layout Tools toolbar.
21. Pick the Frontenac Drive alignment and then pick the grip on the upper end of the line and pull it away from its location.

    Notice that the line and the arc move in sync and tangency is maintained (see Figure 6.13).

FIGURE 6.13: Floating curves maintain their tangency.

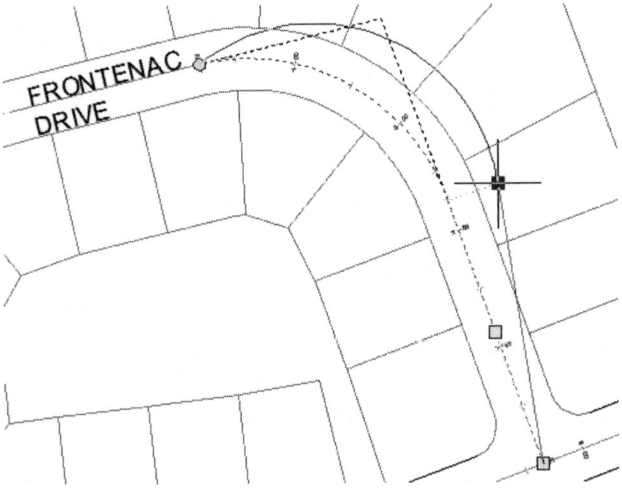

22. Press Esc to cancel the grip edit.

When this exercise is complete, you may close the drawing. A saved finished copy of this drawing is available from the book's web page with the filename `AlignmentsByLayout_FINISHED.dwg` or `AlignmentsByLayout_METRIC_FINISHED.dwg`.

All Aboard! Rail Alignments

New to Civil 3D 2013 are rail alignments. Since railways often have very large sweeping curves, the alignments are often defined slightly differently than typical roadway alignments. Instead of radii, railway plans often refer to the degree of curvature, and rather than defining the curve by length, Civil 3D 2013 defines the curve by chords. This terminology, which is common in the rail world, is included in the alignment components. Let's look at a quick, simple example:

1. Start a new blank drawing from the `_AutoCAD Civil 3D (Imperial) NCS` template that ships with Civil 3D. Metric users should use the `_AutoCAD Civil 3D (Metric) NCS` template.
2. From the Home tab ➤ Create Design panel, choose Alignment ➤ Alignment Creation.

   The Create Alignment – Layout dialog appears.

3. Change the Name to **Mastering Railway** and set Type to Rail.
4. On the General tab, make these changes:
   - Set Alignment Style to Proposed.
   - Set Alignment Label Set to All Labels.
5. Switch to the Design Criteria tab.
6. Change the starting design speed to **30 mi/hr** (or **50 km/hr** for metric users).
7. Select the Use Criteria-Based Design check box.

   If the Design Criteria File doesn't default to the Railway Design Standards, you may need to click the ellipsis next to the design criteria file to browse to `_Autodesk Civil 3D Imperial Rail Cant Design Standards.xml` or `_Autodesk Civil 3D Metric Rail Cant Design Standards.xml`.
8. Verify that Use Design Criteria File is also checked.
9. Click OK.

   The Alignment Layout Tools toolbar appears.
10. Select the Fixed Line (Two Points) tool.
11. Click anywhere on the screen to define the start point and define the second point by entering **@330,0** on the command line (or **@100,0** for metric users).
12. Press ↵ to end the command.
13. Select Floating Curve (From Entity End, Radius, Length).
14. At the `Select entity to attach to:` prompt, select the line segment that you just drew.
15. At the `Specify curve direction [Clockwise cOunterclockwise] <Clockwise>:` prompt, press ↵ to accept the default, Clockwise.
16. At the `Specify radius or [Degree of curvature]:` prompt, enter **D** ↵ to switch to entering the radius by degree of curvature.
17. At the `Specify degree of curvature or [Radius]:` prompt, enter **3** ↵.
18. At the `Specify curve length or [deltaAngle Tanlen Chordlen midOrd External]:` prompt, enter **C** ↵ to define by chord length (which is common in rail alignments).
19. At the `Specify chord length or [curveLen deltaAngle Tanlen midOrd External]:` prompt, enter **100** ↵.
20. Press ↵ to end the command.

    Note that this will result in a different curve for Imperial and metric users.

When this exercise is complete, you may close the drawing. A saved finished copy of this drawing is available from the book's web page with the filename `Rail_FINISHED.dwg` or `Rail_METRIC_FINISHED.dwg`.

Best Fit Alignments

Often designers have to re-create an alignment for an existing road that does not have true horizontal geometry. Civil 3D has multiple tools to re-create the alignment using a best fit algorithm. You can either re-create a full best fit alignment or use best fit lines or best fit curves in an alignment. We will look at both methodologies in this section.

The Create Best Fit Alignment command can use AutoCAD blocks, entities, points, COGO points, or feature lines. You can also simply click on the screen. The Line and Curve drop-down menus on the Alignment Layout toolbar include options for Floating and Fixed Lines By Best Fit, as well as Best Fit curves in all three flavors: Fixed, Float, and Free. It is similar to what we covered in Chapter 1, "The Basics," in the "Best Fit Entities" section. Let's see how it works with alignments:

1. Open the `AlignmentsBestFit.dwg` or `AlignmentsBestFit_METRIC.dwg` file, which you can download from this book's web page.
2. From the Home tab ➢ Create Design panel, choose Alignment ➢ Alignment Creation Tools.

   The Create Alignment – Layout dialog appears.
3. Enter **Best Fit Lines** in the Name field.
4. Leave the rest at their defaults and click OK.

   The Alignment Layout Tools – Best Fit Lines palette opens.
5. Click the down arrow next to the Fixed Line (Two Points) tool on the toolbar, and select the Fixed Line – Best Fit option.

   The Tangent By Best Fit dialog opens (Figure 6.14).

FIGURE 6.14: The Tangent By Best Fit dialog

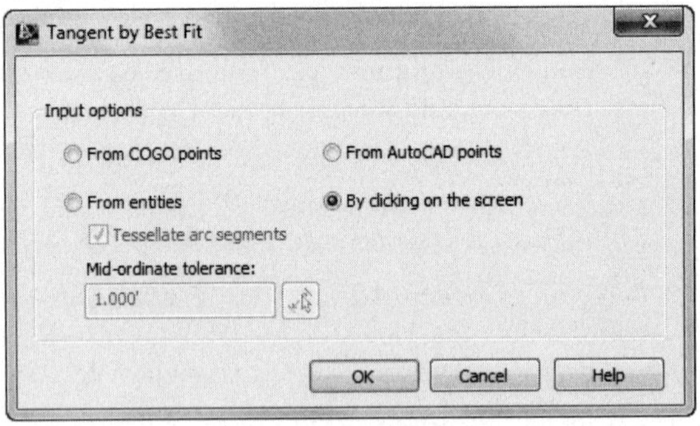

Here, you can choose various methods to create a best fit line alignment:

- From COGO Points
- From Entities
- From AutoCAD Points
- By Clicking On The Screen

6. Pick the By Clicking On The Screen radio button and click OK.
7. With the running Endpoint Osnap, click on all the endpoints of the polyline.

   As you progress, you see a red dashed line being formed. In your selections, this line looks at all the endpoints selected in order to create the best fit line alignment (Figure 6.15).

FIGURE 6.15: The best fit line being formed

8. When you get to the last endpoint, press ↵ to open the Regression Data vista.
9. In the Regression Data vista, you can choose to exclude endpoints or force them to be a pass-through endpoint by checking the appropriate boxes (Figure 6.16).

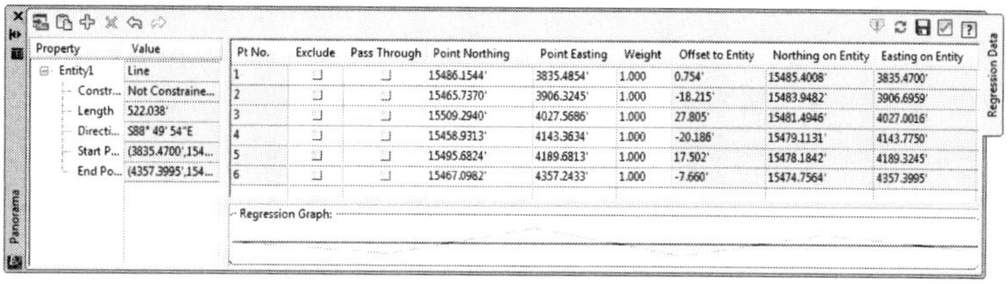

FIGURE 6.16: Regression Data vista

   As you do, notice the changes that occur on your best fit line alignment.

10. Click the green check mark on the upper-right side of the Regression Data vista to accept and dismiss the Panorama window.

    The best fit line alignment is complete. You can also create a similar command for a best fit curve sub-entity that can be either fixed or floating. The procedure is similar to that of the best fit line and can be performed on the polyline that vaguely resembles a curve in the exercise drawing.

    Keep this drawing open for the next portion of the exercise.

    One of the benefits to doing individual best fit segments (lines or curves) is that it is easy to exclude points. However, sometimes you will want a single "rough and dirty" alignment full of curves and lines without having to define them individually. Next you will create a full best fit alignment.

11. Zoom and pan to the area with the 13 points.

12. From the Home tab ➤ Create Design panel, choose Alignment ➤ Create Best Fit Alignment. The Create Best Fit Alignment dialog appears.
13. Set the input type to AutoCAD Points and click the Selection button next to Path 1 Points.
14. In modelspace, use a crossing window to select the 13 points and press ↵.
15. Deselect the Create Spirals check box.

16. Change the name to **Best Fit Alignment**.

17. Verify that the Alignment style is set to Layout and the Alignment label set is set to _No Labels, as shown in Figure 6.17, and click OK to accept all the other defaults.

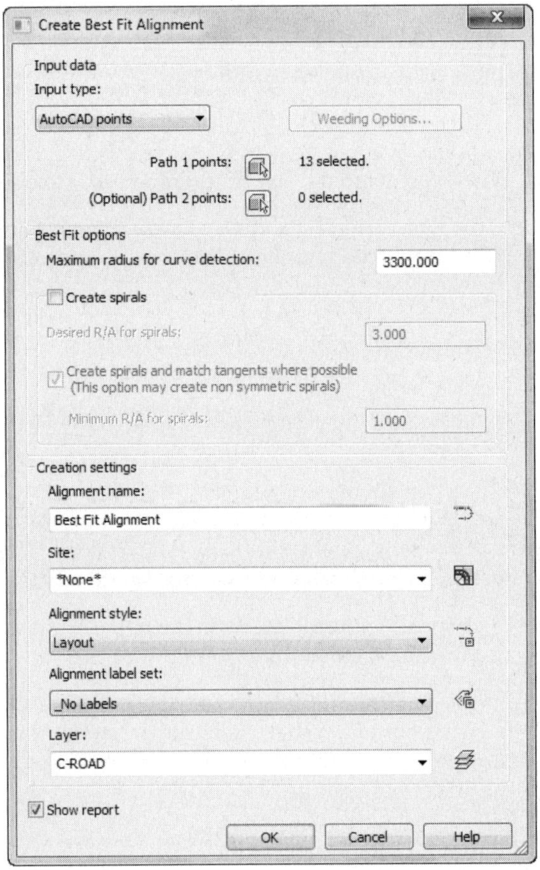

FIGURE 6.17: The Create Best Fit Alignment dialog

The Best Fit Report dialog opens, as shown in Figure 6.18.

Review the results in the Best Fit Report. Notice that if you select a row it will zoom to the location on the regression graph.

18. Click the close X button to dismiss the dialog.

Unlike the Regression Data vista in the Panorama window, the Best Fit Report is purely informational and does not allow you to select to exclude or pass through a specific point. While the Create Best Fit Alignment procedure is fast, it may not be precise. Therefore, this command may be useful for providing a draft alignment and then manually creating your final alignment using similar results to those shown in the Best Fit Report that meet your specific design criteria.

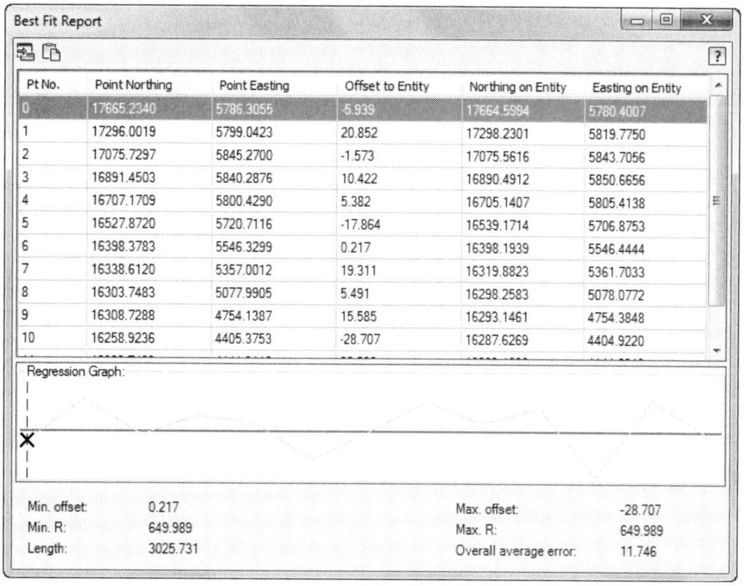

Pt No.	Point Northing	Point Easting	Offset to Entity	Northing on Entity	Easting on Entity
0	17665.2340	5786.3055	-5.939	17664.5994	5780.4007
1	17296.0019	5799.0423	20.852	17298.2301	5819.7750
2	17075.7297	5845.2700	-1.573	17075.5616	5843.7056
3	16891.4503	5840.2876	10.422	16890.4912	5850.6656
4	16707.1709	5800.4290	5.382	16705.1407	5805.4138
5	16527.8720	5720.7116	-17.864	16539.1714	5706.8753
6	16398.3783	5546.3299	0.217	16398.1939	5546.4444
7	16338.6120	5357.0012	19.311	16319.8823	5361.7033
8	16303.7483	5077.9905	5.491	16298.2583	5078.0772
9	16308.7288	4754.1387	15.585	16293.1461	4754.3848
10	16258.9236	4405.3753	-28.707	16287.6269	4404.9220

FIGURE 6.18: The Best Fit Report dialog

The completed best fit alignment is shown in Figure 6.19. Although a best fit alignment may not give you exactly what you are looking for (especially if you like nice, whole-number radii), it generates a good starting point that you can then edit to fit your design needs.

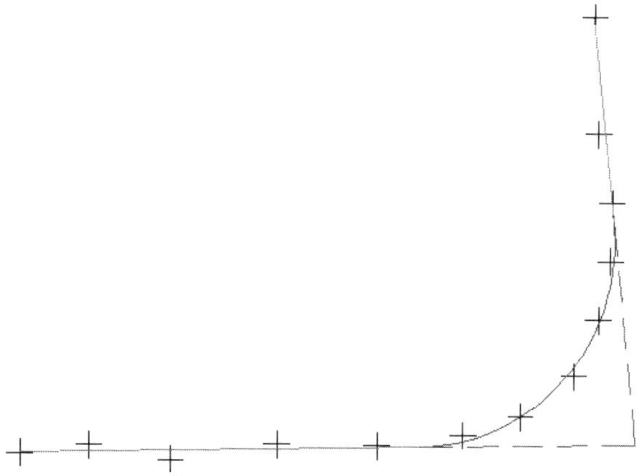

FIGURE 6.19: The best fit alignment

When this exercise is complete, you may close the drawing. A saved finished copy of this drawing is available from the book's web page with the filename `AlignmentsBestFit_FINISHED.dwg` or `AlignmentsBestFit_METRIC_FINISHED.dwg`.

Reverse Curve Creation

Next, let's look at a more complicated alignment construction — building a reverse curve connecting two curves:

1. Open the `AlignmentReverse.dwg` or `AlignmentReverse_METRIC.dwg` file.
2. For this exercise, confirm that your Endpoint Osnap is activated and the other Osnaps can be deactivated.
3. From the Home tab ➤ Create Design panel, choose Alignment ➤ Alignment Creation Tools.

   The Create Alignment – Layout dialog appears.
4. In the Create Alignment – Layout dialog, do the following:
   A. Change the Name field to **Reverse**.
   B. Set the Alignment Style field to Layout.
   C. Set the Alignment Label Set field to Major Minor And Geometry Points.
5. Click OK to display the Alignment Layout Tools toolbar for the reverse alignment.
6. Start by drawing a fixed line from the north end of the western portion to its endpoint using the same Fixed Line (Two Points) tool as used in an earlier exercise.
7. Use the Floating Curve (From Entity, Radius, Through Point) tool to connect a curve from the southern end of this segment.
8. At the `Specify radius:` prompt, enter **500** ↵ (for metric, **150** ↵).
9. At the `Is curve solution angle [Greaterthan180 Lessthan180] <Lessthan180>:` prompt, press ↵ to accept the default.
10. At the `Specify end point:` prompt, click on the other end of the arc.
11. The command repeats; at the `Select entity to attach to:` prompt, select the curve that you just created.
12. At the `Specify radius or [Degree of curvature]:` prompt, enter **400** ↵ (for metric, **120** ↵).
13. At the `Is curve solution angle [Greaterthan180 Lessthan180] <Lessthan180>:` prompt, press ↵ to accept the default.
14. The program detects that you are attaching a curve to a curve; at the `Is curve compound or reverse to curve before? [Compound Reverse] <Compound>:` prompt, enter **R** ↵ to specify that it is a reverse curve.
15. At the `Specify end point:` prompt, click the endpoint of the southern arc.
16. Press ↵ to end the command.
17. Use the Floating Line (From Curve, Through Point) tool to connect a line from the second curve and to finish the alignment by selecting the two-point line (Figure 6.20).

FIGURE 6.20:
Segment layout for the reverse curve alignment

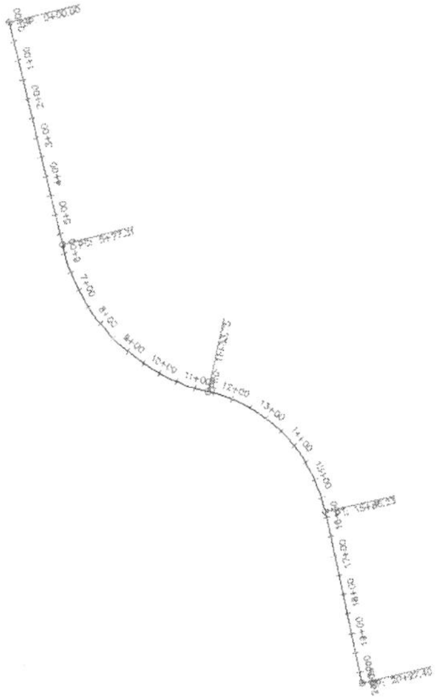

The alignment now contains a perfect reverse curve. Move any of the pieces, and you'll see the other segments react to maintain the relationships shown in Figure 6.21. The flexibility of the Civil 3D tools allows you to explore an alternative solution (the reverse curve) as opposed to the basic solution. Flexibility is one of the strengths of Civil 3D.

FIGURE 6.21:
Curve relationships during a grip edit

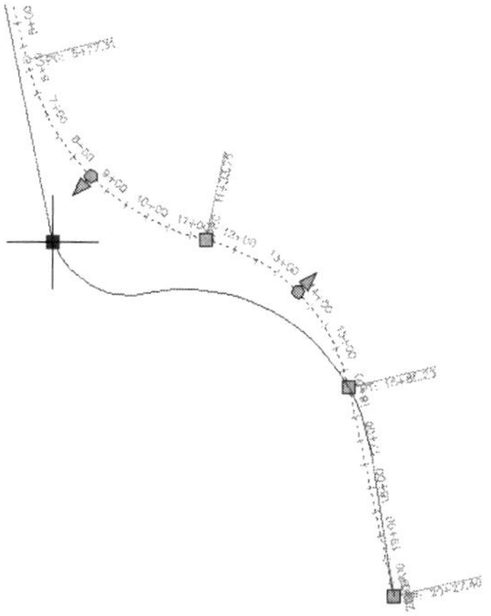

You've completed your initial reverse curve layout. Unfortunately the reverse curve may not be acceptable to the designer, but you'll look at those changes later in the section "Component-Level Editing."

When this exercise is complete, you may close the drawing. A saved finished copy of this drawing is available from the book's web page with the filename `AlignmentReverse_FINISHED.dwg` or `AlignmentReverse_METRIC_FINISHED.dwg`.

Creating with Design Constraints and Check Sets

Starting in Civil 3D 2009, users have the ability to create and use design constraints and design check sets during the process of aligning and creating design profiles. Typically, these constraints check for things like curve radius, length of tangents, and so on. Design constraints use information from AASHTO or other design manuals to set curve requirements. Check sets allow users to create their own criteria to match local requirements, such as subdivision or county road design. First, you'll make one quick set of design checks:

1. Open the `CreatingChecks.dwg` or `CreatingChecks _METRIC.dwg` file.
2. On the Settings tab in Toolspace, expand the Alignment ➤ Design Checks branch.
3. Right-click the Line folder, and select New to display the New Design Check dialog.
4. Change the name to **Subdivision Tangent**.
5. Click the Insert Property drop-down menu, and select Length.

6. Click the greater-than/equals symbol (>=) button, and then enter **100** (for metric, **30**) in the Expression field as shown.

   When complete, your dialog should look like Figure 6.22. Click OK to accept the settings in the dialog.

FIGURE 6.22: The completed Subdivision Tangent design check

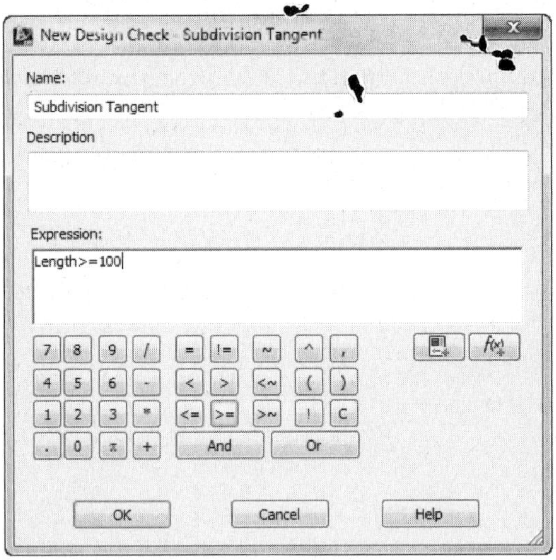

7. Right-click the Curve folder, and select New to display the New Design Check dialog.
8. Change the name to **Subdivision Radius**.
9. Click the Insert Property drop-down menu, and select Radius.
10. Click the greater-than/equals symbol (>=) button, and then enter **200** (for metric, **60**) in the Expression field, as shown in Figure 6.23.

FIGURE 6.23: The completed Subdivision Radius design check

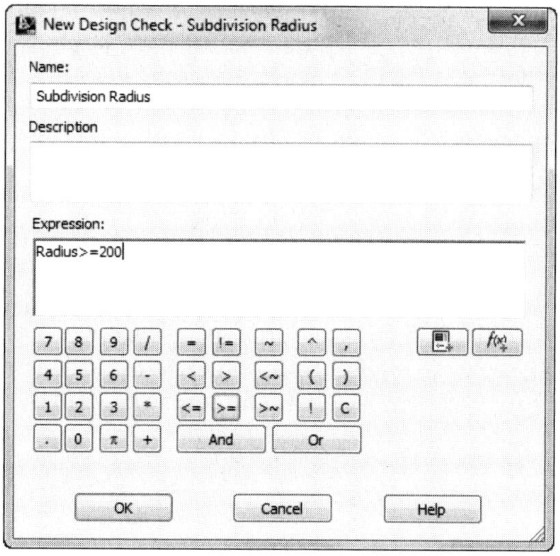

11. Click OK to accept the settings in the dialog.
12. Right-click the Design Check Sets folder, and select New to display the Alignment Design Check Set dialog.
13. On the Information tab, change the name to **Mastering Subdivision**, and then switch to the Design Checks tab.
14. Choose Line from the Type drop-down list and select the Subdivision Tangent line check that you just created.
15. Click the Add button to add the Subdivision Tangent check to the set.
16. Choose Curve from the Type drop-down list and select the Subdivision Radius curve check that you just created.
17. Click the Add button again to complete the set, as shown in Figure 6.24.

FIGURE 6.24: The completed Mastering Subdivision design check set

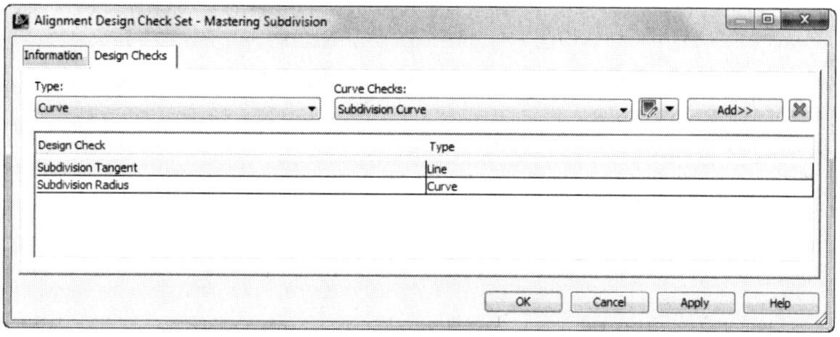

18. Click OK to accept the settings in the dialog.

You may keep this drawing open to continue to the next exercise or use the saved finished copy of this drawing available from the book's web page (`CreatingChecks_FINISHED.dwg` or `CreatingChecks_METRIC_FINISHED.dwg`).

Once you've created a number of design checks and design check sets, you can apply them as needed during the design and layout stage of your projects.

DESIGN CHECKS VS. DESIGN CRITERIA

In typical fashion, the language used for this feature isn't clear. What's the difference? A design check uses basic properties such as radius, length, grade, and so on to check a particular portion of an alignment or profile. These constraints are generally dictated by a governing agency based on the type of road involved. Design criteria use speed and related values from design manuals such as AASHTO to establish these geometry constraints. Think of design criteria as a suite of check sets with different sets for each city, type of street, design speed, and so on.

In the next exercise, you'll see the results of your Mastering Subdivision check set in action:

1. If not still open from the previous exercise, open the `CreatingChecks_FINISHED.dwg` or `CreatingChecks_METRIC_FINISHED.dwg` file.
2. Change to the Prospector tab in Toolspace, and expand the Alignments ➢ Centerline Alignments branch.
3. Right-click on Frontenac Drive and select Properties.
4. Change to the Design Criteria tab, and set the start station design speed to **30 mi/h** (or **50 km/h**), as shown in Figure 6.25.

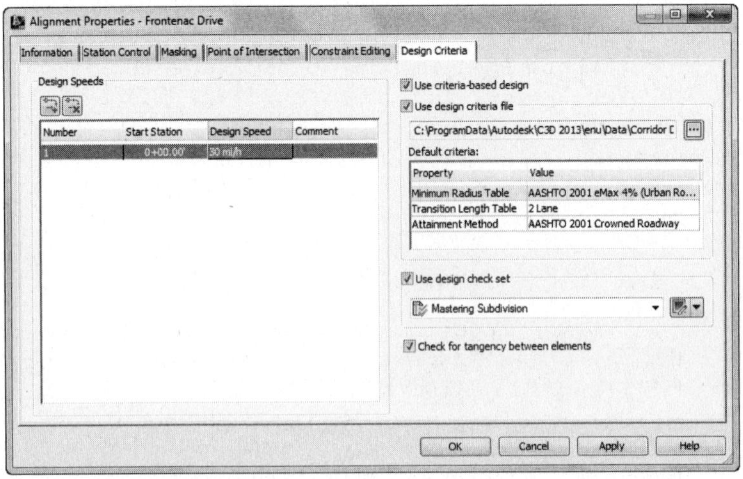

FIGURE 6.25: Setting up design checks from Alignment Properties

5. Verify that all of the check boxes are selected and that Design Check Set is set to Mastering Subdivision.

   Note that the Use Criteria-Based Design check box must be selected to activate the other two.

6. Click OK to accept the settings in the dialog.

   You could have alternatively set the design criteria when you were originally creating the alignment.

   The curve radius on the northernmost part of Frontenac Drive is less than the Subdivision Radius design check, as illustrated by the design check failure indicators.

7. In Prospector, right-click on the alignment named Frontenac Drive and select the Select option to activate the Alignment contextual tab.

8. From the contextual tab ➢ Modify panel, choose Geometry Editor to display the Alignment Layout Tools toolbar.

9. Select the Tangent-Tangent With Curves option on the left side of the Alignment Layout Tools toolbar.

10. Starting at the end of the northeastern Frontenac Drive curve, connect the endpoints of the tangent lines to finish creating the Frontenac Drive alignment.

    Notice that the two curves generated do not match the ones that the planner had originally laid out; you will fix them in the next exercise.

    If you hover over the exclamation-point symbol, as shown in Figure 6.26, it will indicate which design criteria and design checks have been violated.

FIGURE 6.26: Completed alignment layout with design criteria and design checks failure indicator

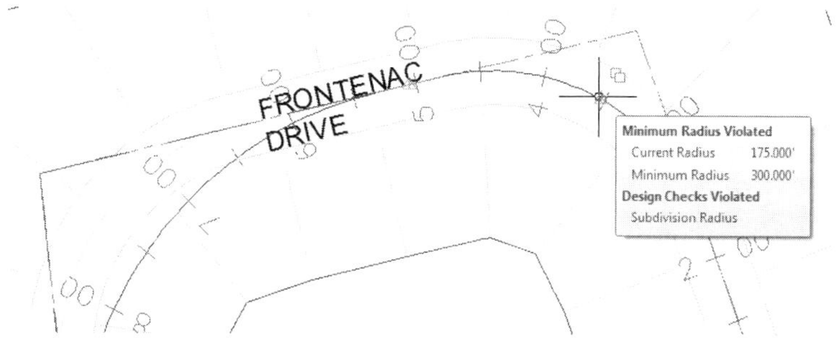

11. Close the Alignment Layout Tools toolbar.

When this exercise is complete, you may close the drawing. A saved finished copy of this drawing is available from the book's web page with the filename `AlignmentsChecked_FINISHED.dwg` or `AlignmentsChecked_METRIC_FINISHED.dwg`.

Now that you know how to create an alignment that doesn't pass the design checks, let's look at different ways of modifying alignment geometry. As you correct and fix alignments that violate the assigned design checks, those symbols will disappear.

Editing Alignment Geometry

The general power of Civil 3D lies in its flexibility. The documentation process is tied directly to the objects involved, so making edits to those objects doesn't create hours of work in updating the documentation. With alignments, there are three major ways to edit the object's horizontal geometry without modifying the underlying construction:

Graphical Grip Editing Select the object, and use the various grips to move critical points. This method works well for realignment, but precise editing for things like a radius or direction can be difficult without construction elements.

Tabular Design Use Panorama to view all the alignment segments and their properties; type in values to make changes. This approach works well for modifying lengths or radius values, but setting a tangent perpendicular to a screen element or placing a control point in a specific location is better done graphically.

Component-Level Editing Use the Alignment Layout Parameters dialog to view the properties of an individual piece of the alignment. This method makes it easy to modify one piece of an alignment that is complicated and that consists of numerous segments, whereas picking the correct field in a Panorama view can be difficult.

In addition to these methods, you can use the Alignment Layout Tools toolbar to make edits that involve removing components or adding to the underlying component count. The following exercises look at the three simple edits and then explain how to remove components from and add them to an alignment without redefining it.

Grip Editing

You already used graphical editing techniques when you created alignments from polylines, but those techniques can also be used with considerably more precision than shown previously. The alignment object has a number of grips that reveal important information about the elements' creation (see Figure 6.27).

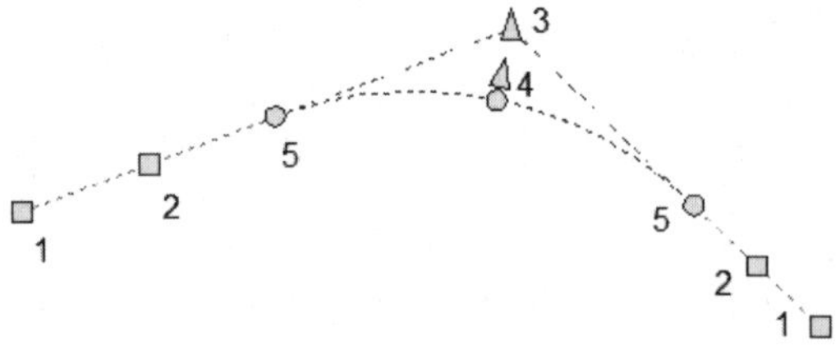

FIGURE 6.27: Alignment grips

You can use the grips in Figure 6.27 to do the following actions:

- The square pass-through grip at the beginning of the alignment, Grip 1, indicates a segment point that can be moved at will to change the length and angle of the line. This grip doesn't attach to any other components.
- The square pass-through grip in the middle of the tangents, Grip 2, allows the element to be translated while maintaining the length and angle. Other components attempt to hold their respective relationships, but moving the grip to a location that would break the alignment isn't allowed.

- The triangular grip at the intersection of tangents, Grip 3, indicates a Point of Intersection (PI) relationship. The curve shown is a function of these two tangents and is free to move on the basis of incoming and outgoing tangents, while still holding a radius.
- The triangular and circular radius grips near the middle of the curve, Grip 4, allow the user to modify the radius directly. The tangents must be maintained, so any selection that would break the alignment geometry isn't allowed.
- The circular pass-through grip on the end of the curve, Grip 5, allows the radius of the curve to be indirectly changed by changing the point of the Point of Curvature (PC) of the alignment. You make this change by changing the curve length, which in effect changes the radius.

In the following exercise, you'll use grip edits to make one of your alignments match the planner's intent more closely:

1. Open the `AlignmentsChecked_FINISHED.dwg` or `AlignmentsChecked_METRIC_FINISHED.dwg` file.
2. In Prospector, expand the Alignments ➢ Centerline Alignments branch.
3. Right-click on Cabernet Court and select Zoom To.
4. Zoom in on the northern curve of the Cabernet Court alignment.

   This curve was inserted in a previous exercise using the default settings and doesn't match the guiding polyline well.
5. Select the Cabernet Court alignment to activate the grips.
6. Select the circular grip shown in Figure 6.28, and use an Endpoint Osnap to place it on the original cyan arc. Doing so changes the radius without changing the PI.

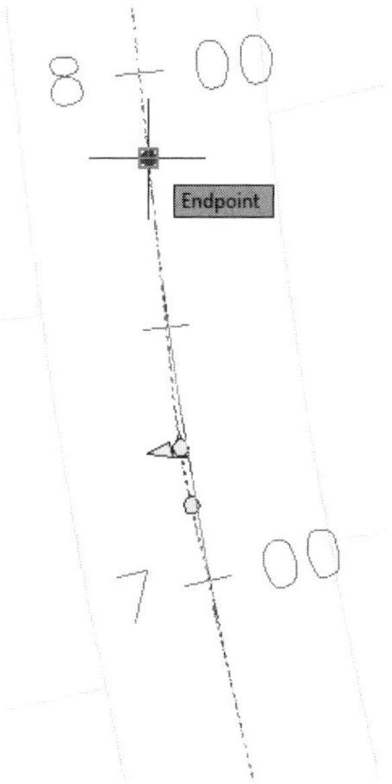

FIGURE 6.28: Grip-editing the Cabernet Court curve

Keep this drawing open for the next portion of the exercise.

Your alignment of Cabernet Court now follows the planned layout. With no knowledge of the curve properties or other driving information, you've quickly reproduced the design's intent.

Tabular Design

When you're designing on the basis of governing requirements, one of the most important elements is meeting curve radius requirements. It's easy to work along an alignment in a tabular view, verifying that the design meets the criteria. In this exercise, you'll verify that your curves are suitable for the design:

1. Using the drawing from the previous exercise, zoom to the Frontenac Drive alignment, and select it to activate the contextual tab.
2. From the Alignment contextual tab ➤ Modify panel, choose Geometry Editor.

   The Alignment Layout Tools toolbar opens.

3. Select the Alignment Grid View tool.

   Panorama appears as shown in Figure 6.29, with all the elements of the alignment listed along the left. You can use the scroll bar along the bottom to review the properties of the alignment if necessary. Note that the columns can be resized as well as toggled off by right-clicking the column headers. The segment selected in the alignment grid view is also highlighted in the model, which can also make identifying the segment easier. Note that the design check failure indicators that are shown on the plan also appear in the table.

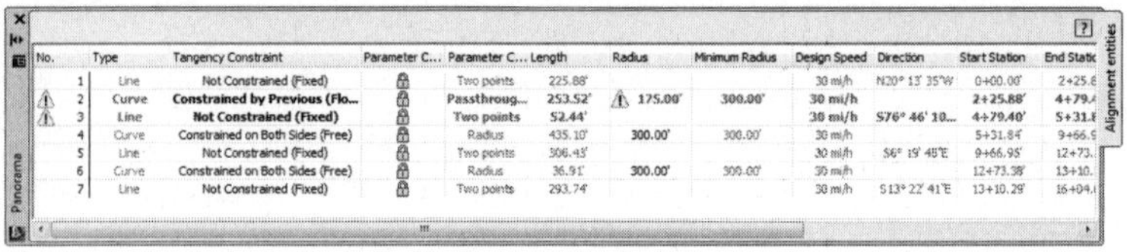

FIGURE 6.29: Alignment Entities vista for Frontenac Drive

> **CREATING AND SAVING CUSTOM PANORAMA VIEWS**
>
> If you right-click a column heading in Panorama and select the Customize Columns option at the bottom of the menu, you're presented with a Customize Columns dialog. This dialog allows you to set up any number of column views, such as Road Design or Stakeout, to show a set of different columns. These views can be saved, allowing you to switch between views easily.

Notice that you can click many of the fields to edit them.

4. Change the radius value of segment No. 6 to **1000** (or **300** for metric users).
5. Click the close X button (notice that in the Panorama window, there is no green check mark to accept the changes) to dismiss Panorama, and then close the toolbar.

   Keep this drawing open for the next portion of the exercise.

Panorama allows for quick and easy review of designs and for precise data entry, if required. Grip editing is commonly used to place the line and curve of an alignment in an approximate working location, but then you use the tabular view in Panorama to make the values more reasonable — for example, to change a radius of 292.56 to 300.00.

Component-Level Editing

Once an alignment gets more complicated, the tabular view in Panorama can be hard to navigate, and deciphering which element is which can be difficult. In this case, reviewing individual elements by picking them on screen can be easier:

1. Using the drawing from the previous exercise, zoom to Frontenac Drive, and select it to activate the contextual tab if not still active from the previous exercise.
2. If the Alignment Layout Tools toolbar for Frontenac Drive is not still open from the previous exercise, from the contextual tab ➢ Modify panel, choose Geometry Editor.

3. Select the Sub-Entity Editor tool to open the Alignment Layout Parameters dialog for Frontenac Drive.

4. Select the Pick Sub-Entity tool on the Alignment Layout Tools toolbar.
5. Pick the first line of Frontenac Drive on the east side of the site to display its properties in the Alignment Layout Parameters dialog (see Figure 6.30).

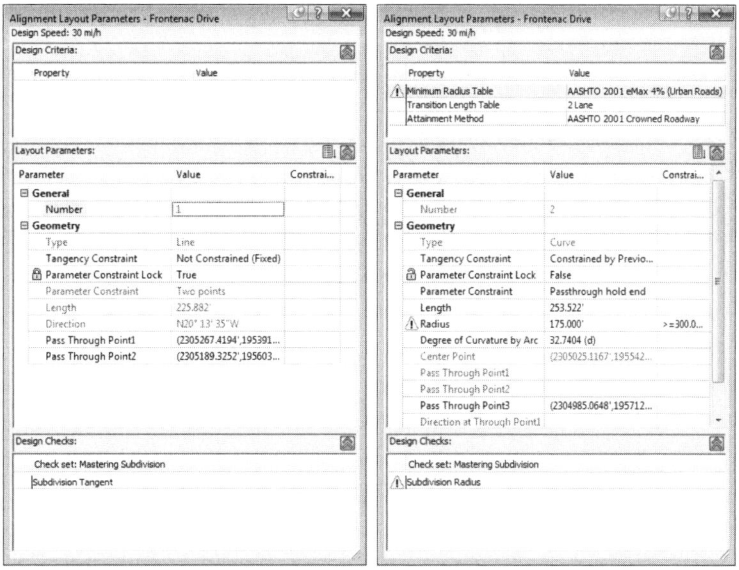

FIGURE 6.30: The Alignment Layout Parameters dialogs for the first line (on the left) and the first curve (on the right) on Frontenac Drive

Since this alignment was created from objects, all line segments are not constrained — that is, they do not hold any relationship between the other segments that make up the alignment.

6. Zoom in, and pick the first curve.

Notice that the Radius Tangency now reports that it is Constrained By Previous (Floating).

7. Change the Parameter Constraint Lock to False.

   Notice that this will enable the ability to set many of the values that previously were grayed out and could not be set.

8. Change the value in the Radius field to **300** (for metric, **100**), and watch the screen update.

   This value is too far from the original design intent to be a valid alternative and the tangency with the next line was not maintained. We will reinvestigate the tangency constraints in the next exercise.

9. Change the value in the Radius field back to **175** (for metric, **55**), and again watch the update.

   This value is closer to the design and is acceptable despite being less than the minimum radius for the 30 mi/hr design speed.

10. Change the radius of the second curve to match the first curve by using the Pick Sub-Entity tool and changing the value in the Alignment Layout Parameters dialog.

11. Close the Alignment Layout Parameters dialog and the Alignment Layout Tools toolbar.

You may keep this drawing open to continue on to the next exercise or use the saved finished copy of this drawing available from the book's web page (`EditingAlignments_FINISHED.dwg` or `EditingAlignments_METRIC_FINISHED.dwg`).

By using the Alignment Layout Parameters dialog, you can concisely review all the individual parameters of a component. In each of the editing methods discussed so far, you've modified the elements that were already in place. We will look at how to change the makeup of the alignment itself, not just the values driving it. But first, let's look at some of the constraints and understand how they work.

Understanding Alignment Constraints

In the previous exercise you were exposed to constraints. The various constraints will help keep geometry together to maintain tangency or to maintain the radius.

When an alignment is created from objects, the lines are not constrained (Fixed). This is the same if you select the Fixed Lines (Two Points) from the Alignment Layout Tools toolbar. The curves are all constrained on both sides (Free). When you grip-edit a line or arc, the lines maintain tangency to the arc, but the arc loses its original radius, as shown in the figures at the beginning of this chapter. You may have noticed in Panorama the Tangency Constraint field that we looked at briefly in the previous exercise. You can click on any segment and change constraints (Figure 6.31). You can also change the constraints in the Sub-Entity Editor.

FIGURE 6.31: The tangency constraints in Panorama

No.		Type	Tangency Constraint
	1	Line	Not Constrained (Fixed)
⚠	2	**Curve**	**Constrained by Previous (Floating)**
	3	Line	Not Constrained (Fixed)
⚠	4	**Curve**	**Constrained on Both Sides (Free)**
	5	Line	Not Constrained (Fixed)
	6	Curve	Constrained on Both Sides (Free)
	7	Line	Not Constrained (Fixed)

1. If not still open from the previous exercise, open the `EditingAlignments_FINISHED.dwg` or `EditingAlignments_METRIC_FINISHED.dwg` file.
2. Select the Frontenac Drive alignment to activate the contextual tab.
3. From the Alignment contextual tab ➤ Modify panel, choose Geometry Editor.
4. In the Alignment Layout Tools toolbar, select the Alignment Grid View.
5. Click in the Tangency Constraint field for the first curve and change it to Not Constrained (Fixed).

   Notice that when you change the first curve from Constrained By Previous (Floating) to Not Constrained (Fixed), the first line changes from Not Constrained (Fixed) to Constrained By Next (Floating).
6. Grip-edit the curve and notice how it does not maintain tangency or radius with the second line but does maintain tangency with the first line because of the first line's tangency constraint (Figure 6.32). Also notice that the angle of the first line (set to Floating) does change.

   You may need to move Panorama out of the way to do this, but don't close it yet.

FIGURE 6.32: Gripping on an alignment with the first line set to Constrained By Next (Floating) and the first curve set to Not Constrained (Fixed)

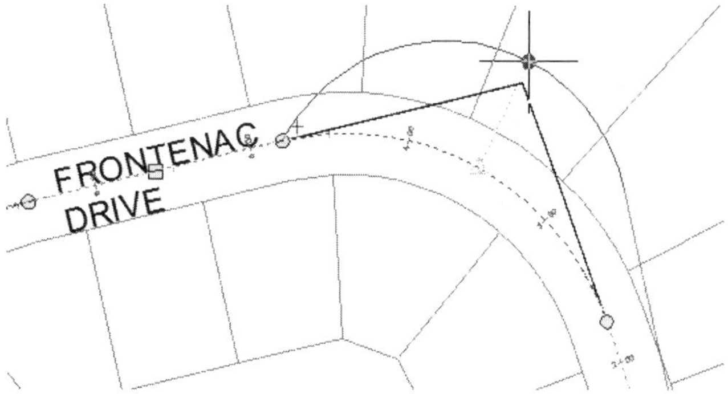

7. Press Esc on your keyboard to cancel the grip edit. If you inadvertently moved the curve, you can use the Undo command.
8. In Panorama, click in the Tangency Constraint field for the first curve and change it back to Constrained By Previous (Floating). The first line will change back to Not Constrained (Fixed).
9. Grip-edit the curve again and notice that the curve still maintains its tangency with the previous line but does not for the following line, just like it did in the previous steps (Figure 6.33). But unlike the previous steps, the first line (set to Fixed) does not change its angle.
10. Press Esc on your keyboard to cancel the grip edit. If you inadvertently moved the curve, you can use the Undo command.

FIGURE 6.33: Gripping on an alignment with the first line set to Not Constrained (Fixed) and the first curve using Constrained By Previous (Floating)

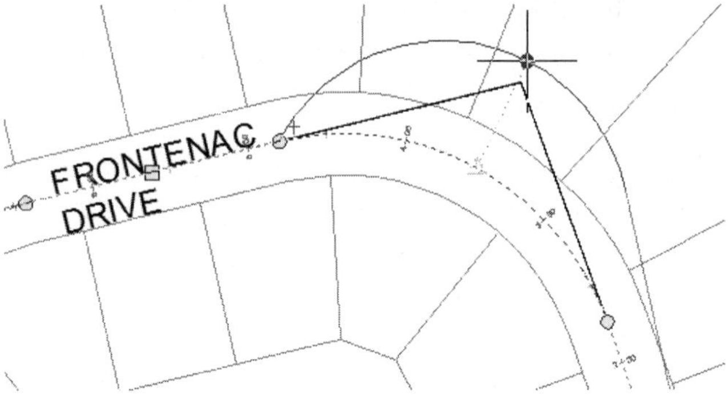

11. Click in the Tangency Constraint field for the first curve and change it to Constrained By Next (Floating).

    The lines before it and after it will be set to Not Constrained (Fixed).

12. Grip-edit the curve and notice that the curve now maintains its tangency with the following line but not the previous line (Figure 6.34).

FIGURE 6.34: Gripping on an alignment with the first curve set to Constrained By Next (Floating) and the following line set to Not Constrained (Fixed)

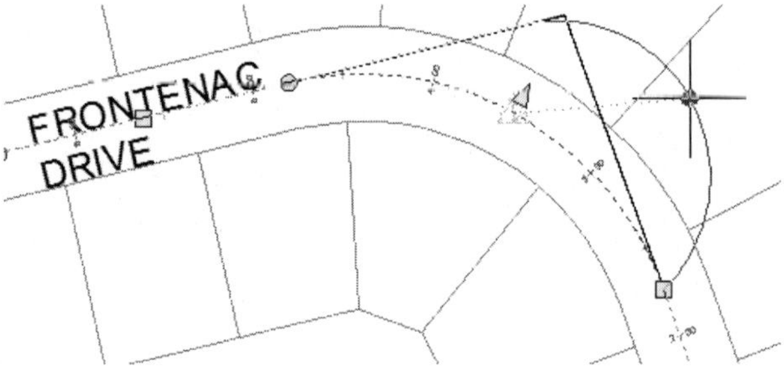

13. Press Esc on your keyboard to cancel the grip edit action. If you inadvertently moved the curve, you can use the Undo command.
14. Click in the Tangency Constraint field for the first curve and change it to Constrained On Both Sides (Free).
15. Grip-edit the curve. See that the curve now maintains its tangency with both the previous and the following lines (Figure 6.35).
16. Press Esc on your keyboard to cancel the grip edit. If you inadvertently moved the curve, you can use the Undo command.
17. Close the Alignment Layout Tools toolbar.

    When you click on an alignment that contains curves and select Alignment Properties, some additional options become available that were not available when the alignment was originally created. In the Point Of Intersection tab, you can select whether you want to visually show points of intersection by a change in alignment direction or by individual curves and curve group. The default is to not display any implied points of intersection (Figure 6.36).

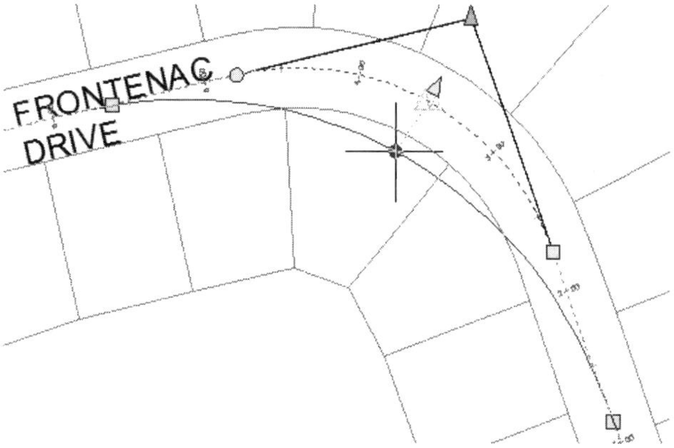

FIGURE 6.35: Gripping on an alignment with the first curve set to Constrained On Both Sides (Free) with both adjoining lines set to Not Constrained (Fixed)

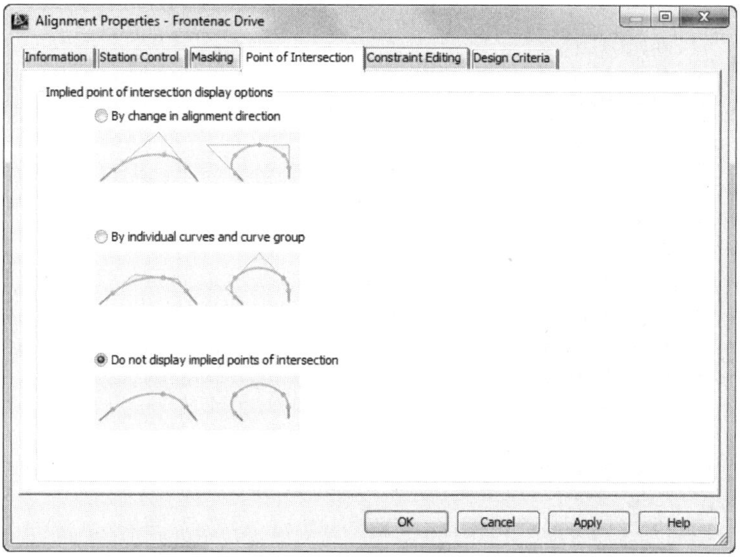

FIGURE 6.36: The Point Of Intersection tab

In the Constraint Editing tab, you can select if you wish to always perform any implied tangency constraint swapping and whether to lock all parameter constraints (Figure 6.37). The Always Perform Implied Tangency Constraint Swapping check box is what caused the swapping seen in this exercise.

When this exercise is complete, you may close the drawing. A saved finished copy of this drawing is available from the book's web page with the filename `AlignmentConstraints_FINISHED.dwg` or `AlignmentConstraints_METRIC_FINISHED.dwg`.

FIGURE 6.37: The Constraint Editing tab

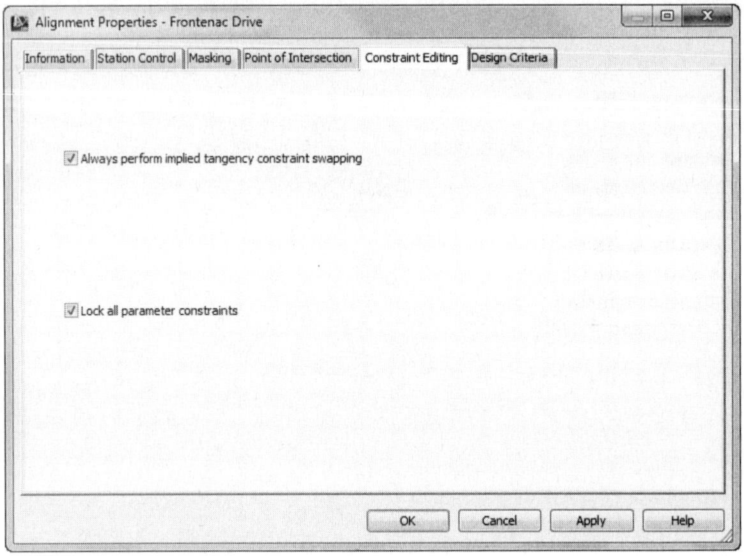

Changing Alignment Components

One of the most common changes is adding a curve where there was none before or changing the makeup of the curves and tangents already in place in an alignment. Other design changes can include swapping out curves for tangents or adding a second curve to smooth a transition area.

It turns out that your perfect reverse curve isn't allowed by the current ordinances for subdivision design! In this example, you'll go back to the design the planner gave you and place a minimum length tangent between the curves:

1. Open the `AlignmentReverse_FINISHED.dwg` or `AlignmentReverse_METRIC_FINISHED.dwg` file from a previous exercise in this chapter.
2. Select the Reverse alignment to activate the contextual tab.
3. From the contextual tab ➢ Modify panel, choose Geometry Editor.

   The Alignment Layout Tools toolbar appears.

4. Select the Delete Sub-Entity tool.
5. Pick the northern curve to remove it and press ↵ to end the command.

   Note that the last curve and tangent is still part of the alignment — it just isn't connected.
6. Select the Floating Line (From Curve End, Length) option.
7. Click the northern end of the southern curve, and then specify a length of **100** ↵ (for metric, **30** ↵).

   If you inadvertently select the polyline instead of the alignment, you may want to turn on selection cycling by entering **Selectioncycling** on the command line and entering **2** ↵.
8. Select the Free Curve Fillet (Between Two Entities, Through Point) option.

9. Select the first line segment to the north and then the line just drawn.
10. For the through point, use an Endpoint Osnap to select the free end of the tangent line just drawn and press ↵ to end the command.
11. Close the Alignment Layout Tools toolbar.

    Your reverse curve with tangent section is complete, as shown in Figure 6.38.

FIGURE 6.38: Reverse curve with tangent segment

When this exercise is complete, you may close the drawing. A saved finished copy of this drawing is available from the book's web page with the filename `AlignmentReverseEdit_FINISHED.dwg` or `AlignmentReverseEdit_METRIC_FINISHED.dwg`.

So far in this chapter, you've created and modified the horizontal alignments, adjusted them on screen to look like what your planner delivered, and tweaked the design using a number of different methods. Now let's look beyond the lines and arcs and get into the design properties of the alignment.

Alignments As Objects

Beyond the simple nature of lines and arcs, alignments represent other things such as highways, streams, sidewalks, or even flight patterns. All these items have properties that help define them, and many of these properties can also be part of your alignments. In addition to obvious properties like names and descriptions, you can include functionality such as superelevation, station equations, reference points, and station control. This section will look at other properties that can be associated with an alignment and how to edit them.

Alignment Properties

While the properties of the alignment were originally assigned during creation, later in design there are often changes that are required. In this exercise, you'll learn an easy way to change the object style and how to add a description.

Most of an alignment's basic properties can be modified in Prospector. In this exercise, you'll change the name in a couple of ways:

1. Open the `AlignmentConstraints_FINISHED.dwg` or `AlignmentConstraints_METRIC_FINISHED.dwg` file, and make sure Prospector is open.
2. In Prospector, expand the Alignments ➤ Centerline Alignments branch, and select the Centerline Alignments text.

   Notice that Cabernet Court, Frontenac Drive, and Syrah Way are listed as members.
3. Click the Centerline Alignments branch, and the individual alignments appear in the Toolspace preview area (see Figure 6.39).

FIGURE 6.39: The Alignments collection listed in the preview area of Prospector

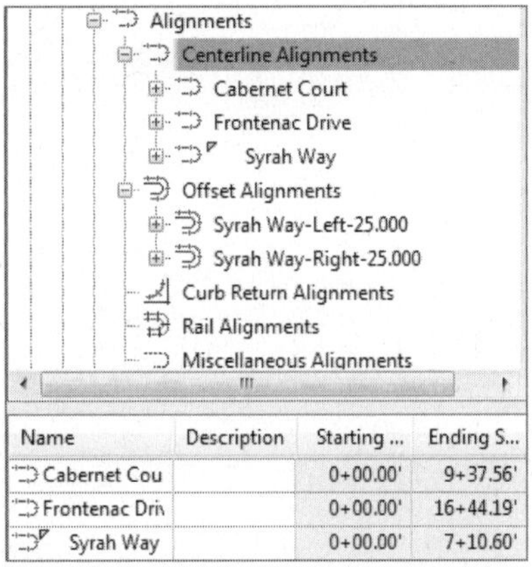

4. Down in the Prospector preview area, click in the Name field for Cabernet Court, and pause briefly before clicking again.

   The text highlights for editing.
5. Click in the Description field, and enter a description (such as **North South**). Press ↵.
6. Click the Style field (you may have to scroll left/right to find the Style field), and the Select Label Style dialog appears.
7. Select Basic from the drop-down menu, and click OK .

   The screen updates. Keep this drawing open for the next portion of the exercise.

   That's one method to change alignment properties. The next is to use the Properties palette:
8. Open the Properties palette by using the Ctrl+1 keyboard shortcut or some other method.
9. Select Syrah Way alignment in the drawing.

   The Properties palette looks like Figure 6.40.

FIGURE 6.40: Syrah Way in the Properties palette

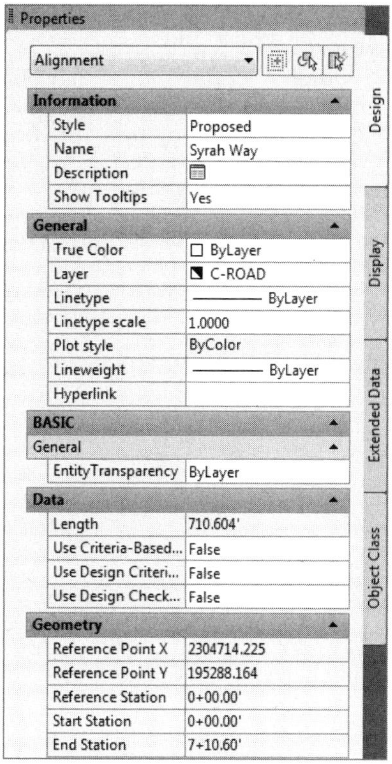

10. Click in the Description field and the Description dialog will open. Enter a description (such as **Central East West**), and click OK.

11. Press Esc on your keyboard to deselect all objects, and close the Properties palette if you'd like.

    Keep this drawing open for the next portion of the exercise.

    The final method involves getting into the Alignment Properties dialog, your access point to information beyond the basics:

12. In the Prospector, right-click Frontenac Drive and select Properties. This is the alignment at the north end of the site.

    The Alignment Properties dialog for Frontenac Drive opens.

13. Change to the Information tab if it isn't selected.

14. Enter a description in the Description field (such as **Loop Cul-de-sac**).

15. Set Object Style to Existing.

16. Click the Apply button.

    Notice that the display style in the drawing updates immediately.

17. Click OK to accept the settings in the dialog.

    Keep this drawing open for the next portion of the exercise.

    Now that you've updated your alignments, let's make them all the same style for ease of viewing. The best way to do this is in the Prospector preview area:

18. In Prospector, expand the Alignments ➤ Centerline Alignments branch, and select the Centerline Alignments text.
19. Select one of the alignments in the preview area.
20. Press Ctrl+A to select them all, or pick the top and then Shift+click the bottom item.

    The idea is to pick *all* of the alignments.

21. Right-click the Style column header and select Edit (see Figure 6.41).

FIGURE 6.41: Editing alignment styles en masse via Prospector

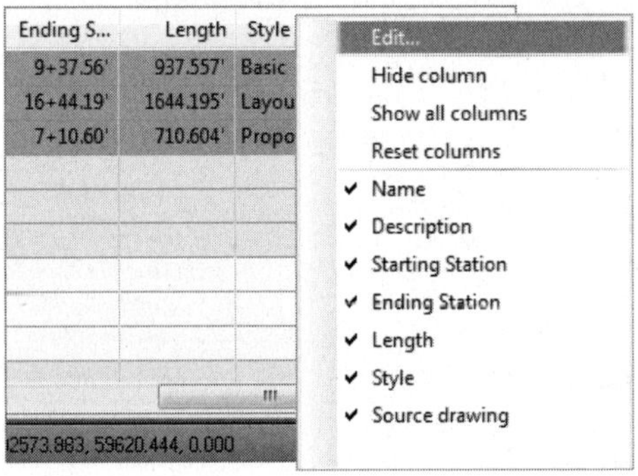

22. Select Layout from the drop-down list in the Select Label Style dialog that appears, and click OK.
23. Press Esc to deselect all of the alignments.

    Notice that all alignments pick up this style. Although the dialog is named Edit Label Style, you are editing the object style.

You may keep this drawing open to continue on to the next exercise or use the saved finished copy of this drawing available from the book's web page (`AlignmentProperties_FINISHED.dwg` or `AlignmentProperties_METRIC_FINISHED.dwg`).

DON'T FORGET THIS TECHNIQUE

This technique of selecting all, right-clicking the column heading in the preview area, and editing the value works on every object that displays in the List Style preview: parcels, pipes, corridors, assemblies, and so on. It can be painfully tedious to change a large number of objects from one style to another using any other method. If you left-click the column header instead of right-click, the alignments will sort alphabetically based on the values in that column.

The alignments now look the same, and they all have a name and description. Let's look beyond these basics at the other properties you can modify and update.

The Right Station

At the end of the process, every alignment has stationing applied to help locate design information. This stationing often starts at zero, but it can also tie to an existing object and may start at some arbitrary value. Stationing can also be fixed in both directions, requiring station equations that help translate between two disparate points that are the basis for the stationing in the drawing.

One common problem is an alignment that was drawn in the wrong direction. Thankfully, Civil 3D has a quick edit command to fix that:

1. Open the `AlignmentStations.dwg` file or the `AlignmentStations_ METRIC.dwg` file, and make sure Prospector is open.

2. Pick the Syrah Way alignment to activate the contextual tab.

3. From the Alignment contextual tab ➢ Modify drop-down panel, choose Reverse Direction.

   A warning dialog appears, reminding you of the consequences of such a change.

4. Click OK to dismiss the warning dialog.

   The stationing reverses, with 0+00 (or 0+000 for metric users) now at the east end of the street.

   Notice that the masking is no longer applied. This is one of the reasons that warning message was given. You can reapply the masking if you would like by going to the Station Control tab of the Alignment Properties dialog.

5. Press Esc to deselect the alignment.

   Keep this drawing open for the next portion of the exercise.

   This technique allows you to reverse an alignment almost instantly. The warning that appears is critical, though! When an alignment is reversed, the information that was derived from its original direction may not translate correctly, if at all. One prime example of this is design profiles: they don't reverse themselves when the alignment is reversed, and this can lead to serious design issues if you aren't paying attention.

   Beyond reversing, it's common for alignments to not start with zero. For example, the Cabernet Court alignment may be a continuation of an existing street, and it makes sense to make the starting station for this alignment the end station from the existing street. In this next portion of the exercise, you'll set the beginning station:

6. Select the Cabernet Court alignment to activate the contextual tab.

7. From the contextual tab ➢ Modify panel, choose Alignment Properties.

8. Switch to the Station Control tab of the Alignment Properties dialog.

   This tab controls the base stationing and lets you create station equations.

9. Enter **315.62** (for metric, **96.20**) in the Station field in the Reference Point section (see Figure 6.42).

FIGURE 6.42: Setting a new starting station on the Cabernet Court alignment

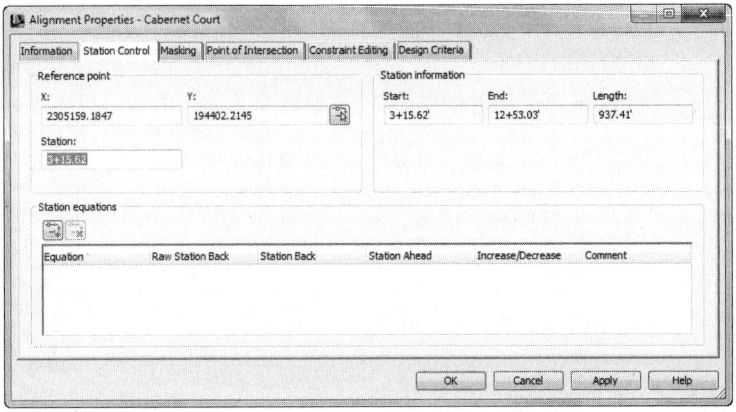

10. Dismiss the warning message that appears, and click the Apply button.

11. Press Esc to deselect the alignment.

    The Station Information section in the top right will update. These options can't be edited but provide a convenient way for you to review the alignment's length and station values.

    Keep this drawing open for the next portion of the exercise.

    In addition to changing the value for the start of the alignment, you could use the Pick Reference Point button, as shown in Figure 6.42, to select another point as the stationing reference point.

    Station equations can occur multiple times along an alignment. They typically come into play when plans must match existing conditions or when the stationing has to match other plans, but the lengths in the new alignment would make that impossible without some translation. In this last portion of the exercise, you'll add a station equation on Frontenac Drive at the intersection with Syrah Way.

12. Select the Frontenac Drive alignment to activate the contextual tab.

13. From the contextual tab ➢ Modify panel, choose Alignment Properties.

14. On the Station Control tab of the Alignment Properties dialog, click the Add Station Equations button.

15. Use an Endpoint Osnap to select the intersection at the west end of Syrah Way.

16. Change the Station Ahead value to **5000** (for metric, **1500**).

17. Click the Apply button, and notice the change in the Station Information section as shown in Figure 6.43.

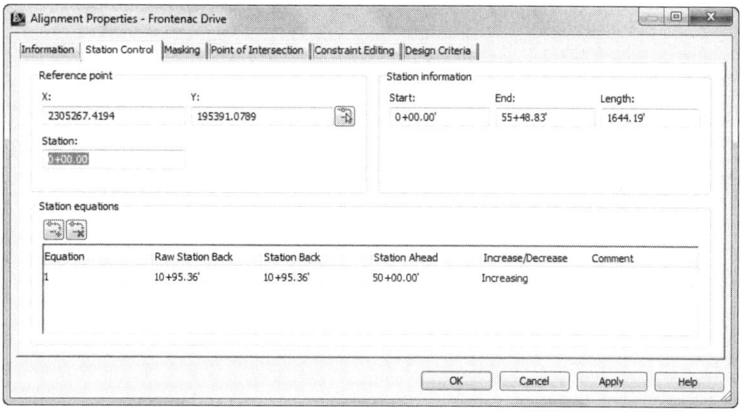

FIGURE 6.43: Frontenac Drive station equation in place

18. Click OK to accept the settings in the dialog, and review the stationing that has been applied to the alignment.

You may keep this drawing open to continue on to the next exercise or use the saved finished copy of this drawing available from the book's web page (`AlignmentStations_FINISHED.dwg` or `AlignmentStations_METRIC_FINISHED.dwg`).

Stationing constantly changes as alignments are modified during the initial stages of a development or as late design changes are pushed back into the plans. With the flexibility shown here, you can reduce the time you spend dealing with minor changes that seem to ripple across an entire plan set.

Assigning Design Speeds

One driving part of transportation design is the design speed. Civil 3D considers the design speed a property of the alignment, which can be used in labels or calculations as needed. In this exercise, you'll add a series of design speeds to Frontenac Drive. Later in the chapter, you'll label these sections of the road using a label set:

1. If not still open from the previous exercise, open the `AlignmentStations_FINISHED.dwg` file or the `AlignmentStations_METRIC_FINISHED.dwg` file.
2. Bring up the Alignment Properties dialog for Frontenac Drive to the north using any of the methods discussed.
3. Switch to the Design Criteria tab.
4. Verify that the Design Speed field for Number 1 is 30 mi/hr (or 50 km/hr for metric users).

   This speed is typical for a subdivision street.

5. Click the Add Design Speed button.
6. Click in the Start Station field for Number 2.

A small Pick On Screen button appears to the right of the Start Station value, as shown in Figure 6.44.

FIGURE 6.44: Setting the design speed for a Start Station field

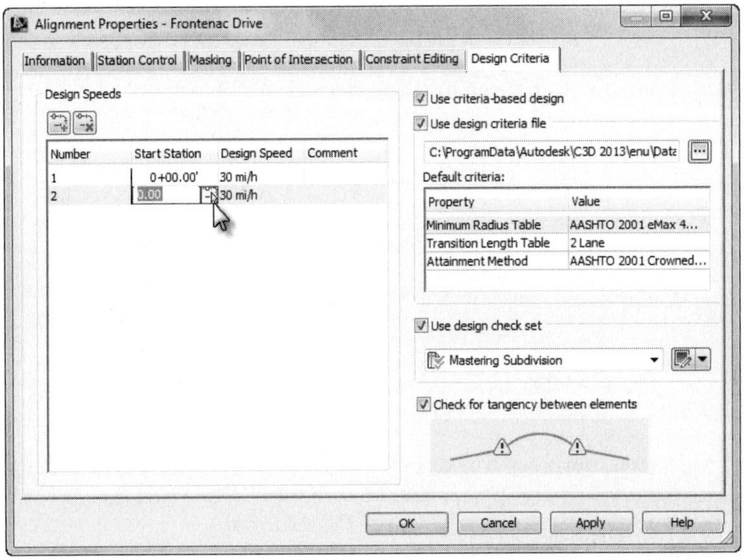

7. Click the Pick On Screen button, and then use an Osnap to pick the PC on the east portion of the site, near station 2+25.88 (or station 0+067.38 for metric users).
8. Enter a value of **20** (for metric, **40**) in the Design Speed field for the line you just added.
9. Click the Add Design Speed button again.

   Notice that the additional row is added to the top of the list.
10. Click in the Start Station field for the new row, click the Pick On Screen button again to add one more design speed portion, and enter **900** ↵ (for metric, **275** ↵).
11. Enter a value of **30** (for metric, **50**) for this design speed.

    When complete, the tab should look like Figure 6.45.
12. Click OK to accept the settings in the Alignment Properties dialog.

When this exercise is complete, you may close the drawing. A saved finished copy of this drawing is available from the book's web page with the filename `DesignSpeed_FINISHED.dwg` or `DesignSpeed_METRIC_FINISHED.dwg`.

In a subdivision, these values can be inserted for labeling purposes. In a highway design, they can be used to drive the superelevation calculations that are critical to a working design. Chapter 11, "Superelevations," looks at this subject.

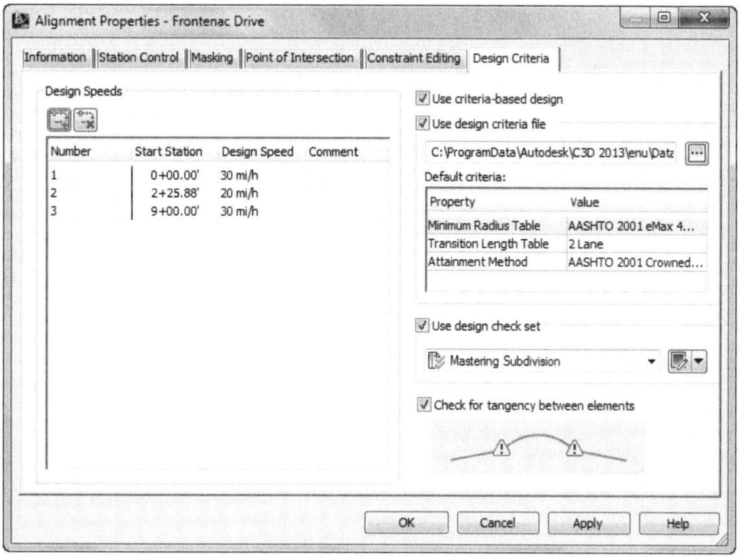

FIGURE 6.45: The design speeds assigned to Frontenac Drive

Labeling Alignments

Labeling in Civil 3D is one of the program's strengths, but it's also an easy place to get lost. There are myriad options for every type of labeling situation under the sun, and keeping them straight can be difficult.

THE POWER OF LABEL SETS

When you think about it, any number of items can be labeled on an alignment. These include major and minor stations, geometry points, design speeds, and profile information. Each of these objects can have its own style. Keeping track of all these individual labeling styles and options would be burdensome and uniformity would be difficult, so Civil 3D features the concept of *label sets*.

A label set lets you build up the labeling options for an alignment, picking styles for the labels of interest, or even multiple labels on a point of interest, and then save them as a set. These sets are available during the creation and labeling process, making the application of individual labels less tedious. Out of the box, a number of sets are available, primarily designed for combinations of major and minor station styles along with geometry information.

You'll learn how to create individual label styles in Chapter 20. We will use the out-of-the-box styles over the next couple of exercises and then pull them together with a label set. At the end of this section, you'll apply your new label set to the alignments.

Major Station

Major station labels typically include a tick mark and a station callout.

Geometry Points

Geometry points reflect the PC, PT, and other points along the alignment that define the geometric properties.

In this exercise, you'll apply a label set to all of your alignments and then see how an individual label can be changed from the set:

1. Open the `AlignmentLabels.dwg` file or the `AlignmentLabels_METRIC.dwg` file.
2. Select the Cabernet Court alignment on the screen.
3. Right-click, and select Edit Alignment Labels to display the Alignment Labels dialog shown in Figure 6.46.

FIGURE 6.46: The Alignment Labels dialog for Cabernet Court

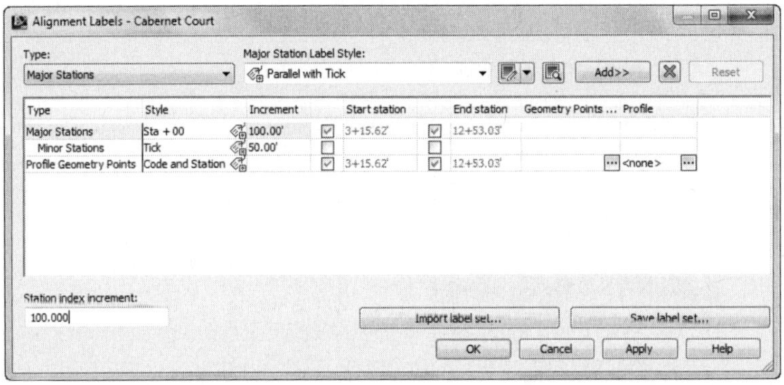

4. Click the Import Label Set button near the bottom of this dialog to display the Select Label Set dialog.
5. In the Select Label Set drop-down list, select the All Labels label set and click OK.

   The Alignment Labels dialog now populates with the additional labels imported from the label set, as shown in Figure 6.47.

FIGURE 6.47: The All Labels label set

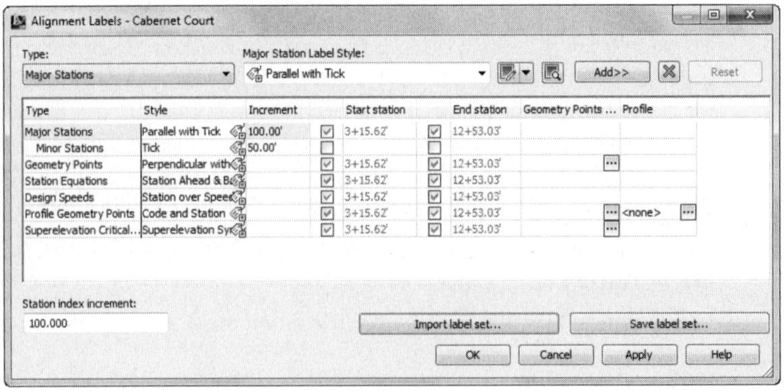

6. Click OK to accept the settings in the Alignment Labels dialog.
7. Repeat this process to import All Labels for the rest of the centerline alignments.
8. When you've finished, zoom in on any of the major station labels (for example, on Frontenac Drive, the 1+00 label or the 0+020 label for metric users).
9. Select the label and notice how all of the label set group labels are selected.
10. Press Esc to deselect.
11. Hold down the Ctrl key, and select the label.

    Notice that a single label is selected, not the label set group, and the Label contextual tab is activated.

CTRL+CLICK? WHAT'S THAT ABOUT?

Prior to AutoCAD Civil 3D 2008, clicking an individual label picked the label and the alignment. Because labels are part of a label set object now, Ctrl+click is the *only* way to access the Flip Label and Reverse Label functions!

12. With the individual major station selected, right-click and select Label Properties.

    The Properties palette for the label appears.
13. Use the Major Station Label Style drop-down list to change from <default> to Perpendicular With Link.
14. Change the Flip Label value to True to move the label from the left side of the alignment to the right side.

    This command can be helpful if your plans become busy and you need to move a label to the other side for better visibility.
15. Press Esc to deselect the label item and dismiss this dialog.

Keep this drawing open for the next portion of the exercise.

By using alignment label sets, you'll find it easy to standardize the appearance of labeling and stationing across alignments. Building label sets can take some time, but it's one of the easiest, most effective ways to enforce standards. Building label sets will be discussed further in Chapter 20.

STATION OFFSET LABELING

Beyond labeling an alignment's basic stationing and geometry points, you may want to label points of interest in reference to the alignment. Station offset labeling is designed to do just that. In addition to labeling the alignment's properties, you can include references to other object types in your station offset labels. The objects available for referencing are as follows:

- Alignments
- COGO points

- Parcels
- Profiles
- Surfaces

In this last portion of the exercise, you'll use an alignment reference to create a label suitable for labeling the station information for the center point of the cul-de-sac at the end of Frontenac Drive.

1. Using the drawing from the previous exercise, from the Annotate tab ➤ Add Labels drop-down, choose Alignment ➤ Add Alignment Labels. The Add Labels dialog appears with Feature set to Alignment.
2. In the Label Type drop-down list, select Station Offset (which is different from Station Offset – Fixed Point).
3. In the Station Offset Label Style drop-down, select Station And Offset.
4. Leave the Marker Style field at the default setting, but remember that you could use any of these styles to mark the selected point.
5. Click the Add button.
6. At the `Select Alignment:` prompt, select the Frontenac Drive alignment.
7. Use the Center Osnap to snap to the center of the right-of-way arc at the cul-de-sac of Frontenac Drive.
8. At the `Specify station offset:` prompt, enter **0** ↵ for the offset amount, and press ↵ to end the command.
9. Close the Add Labels dialog.
10. Select the label and use the square grip to drag the label to a convenient location (Figure 6.48). Label grips will be discussed further in Chapter 20.

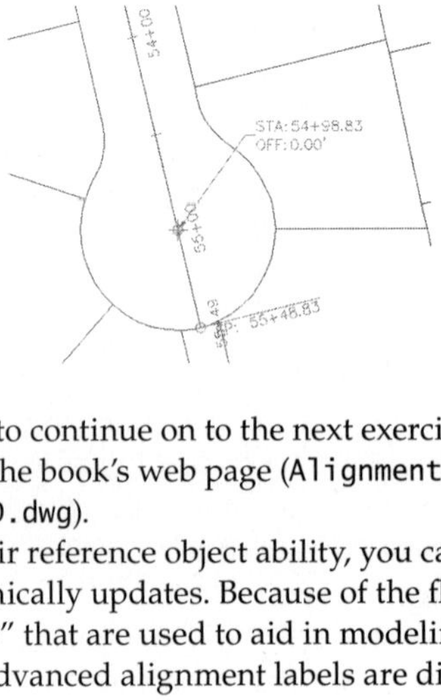

FIGURE 6.48: The alignment station offset label in use

You may keep this drawing open to continue on to the next exercise or use the saved finished copy of this drawing available from the book's web page (`AlignmentLabels_FINISHED.dwg` or `AlignmentLabels_METRIC_FINISHED.dwg`).

Using station offset labels and their reference object ability, you can label most site plans quickly with information that dynamically updates. Because of the flexibility of labels in terms of style, you can create "design labels" that are used to aid in modeling yet never plot and aren't seen in the final deliverables. More advanced alignment labels are discussed in Chapter 20.

Alignment Tables

There isn't always room to label alignment objects directly on top of them. Sometimes doing so doesn't make sense, or a reviewing agency wants to see a table showing the radius of every curve in the design. Documentation requirements are endlessly amazing in their disparity and seeming randomness. Beyond labels that can be applied directly to alignment objects, you can also create tables to meet your requirements.

You can create four types of tables:

- Line
- Curve
- Spiral
- Segment

Each of these is self-explanatory except perhaps the segment table. That table generates a mix of all the lines, curves, and spirals that make up an alignment, essentially re-creating the alignment in a tabular format. In this section, you'll generate a new line table and draw the segment table that ships with the product.

All the tables work in a similar fashion. To add a table:

1. From the Annotate tab ➢ Labels & Tables panel, choose the drop-down from the Add Tables button.
2. From the Add Tables drop-down, choose Alignment.
3. Pick a table type (such as Add Segment) that is relevant to your work.

   The Alignment Table Creation dialog appears (see Figure 6.49).

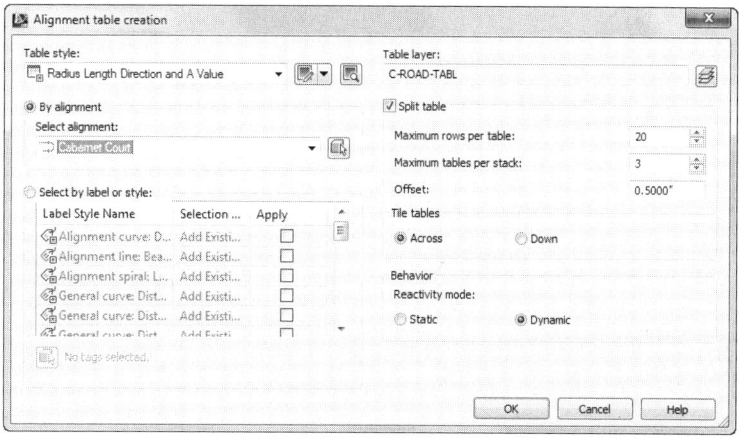

FIGURE 6.49: The Alignment Table Creation dialog for the Add Segment table type

4. Select a table style from the drop-down list or create a new one.

The Select By Label Or Style area determines how the table is populated. All the label style names for the selected type of component are presented, with a check box to the right of each one. Applying one of these styles enables the Selection Rule, which has the following two options:

Add Existing Any label using this style that currently exists in the drawing is converted to a tag format, substituting a key number such as L1 or C27, and added to the table. Any labels using this style created in the future will *not* be added to the table.

Add Existing And New Any label using this style that currently exists in the drawing is converted to a tag format and added to the table. In addition, any labels using this style created in the future will be added to the table.

To the right of the Select section is the Split Table section, which determines how the table is stacked up in modelspace once it's populated. You can modify these values after a table is generated, so it's often easier to leave them alone during the creation process.

Finally, the Behavior section provides two options for the Reactivity Mode: Static and Dynamic. This section determines how the table reacts to changes in the labeled objects. In some cases in surveying, this disconnect is used as a safeguard to the platted data, but in general, the point of a 3D model is to have live labels that dynamically react to changes in the object. Be sure to cancel out of the Alignment Table Creation dialog before proceeding to the next exercise.

Before you draw any tables, you need to apply labels so the tables will have data to populate. In this exercise, you'll place some labels on your alignments, and then you'll move on to drawing tables:

1. If not still open from the previous exercise, open the `AlignmentLabels_FINISHED.dwg` or the `AlignmentLabels_METRIC_FINISHED.dwg` file.
2. From the Annotate tab ➢ Labels & Tables panel choose Add Labels to open the Add Labels dialog.
3. In the Feature field, select Alignment.
4. In the Label Type field, select Multiple Segment from the drop-down list.

   With these options, you'll click each alignment one time, and every subcomponent will be labeled with the style selected here.
5. Verify that the Line Label Style field is set to Line Label Style: Bearing Over Distance (not General Line Label Style: Bearing Over Distance).

   You won't be left with these labels — you just want them for selecting elements later.
6. Verify that the Curve Label Style field is set to Delta Over Length & Radius.

   Since you have no spirals, no Spiral label style needs to be specified.
7. Click the Add button, and select each of the three alignments; then press ↵ to end the command.
8. Click the Close button to dismiss the Add Labels dialog.

Keep this drawing open for the next portion of the exercise.

Now that you've got labels to play with, let's build some tables.

CREATING A LINE TABLE

Most line tables are simple: a line tag, a bearing, and a distance. You'll also see how Civil 3D can translate units without having to change anything at the drawing level:

1. Using the drawing from the previous exercise, from the Annotate tab ➤ Labels & Tables panel click the Add Tables drop-down arrow, and choose Alignment ➤ Add Line to open the Table Creation dialog.
2. Verify that a check mark appears in the Apply column for the Alignment Line: Bearing Over Distance label, as shown in Figure 6.50.

FIGURE 6.50: Creating an alignment line table

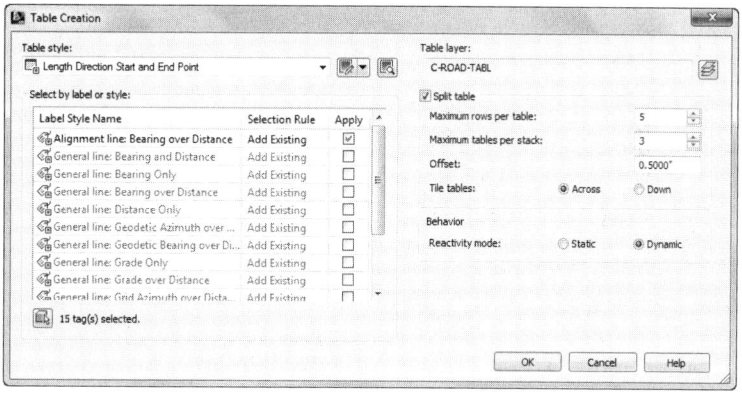

3. Click the Pick On-Screen button and make a crossing window across the entire project.
4. Press ↵ to complete the selections set.
5. When presented with the Create Table – Convert Child Styles warning dialog, select Convert All Selected Label Styles To Tag Mode.

   Notice that there are now 15 tags selected.
6. Click OK to accept the settings in the Table Creation dialog.
7. At the `Select upper left corner:` prompt, select an insertion point on screen, and the table will generate.

Keep this drawing open for the next portion of the exercise.

Pan back to your drawing, and you'll notice that the line labels have turned into tags on the line segments. After you've made one table, the rest are similar. Be patient as you create tables — a lot of values must be tweaked to make them look just right. By drawing one on screen and then editing the style, you can quickly achieve the results you're after.

AN ALIGNMENT SEGMENT TABLE

An individual segment table allows a reviewer to see all the components of an alignment. In this exercise, you'll draw the segment table for Frontenac Drive:

1. Using the drawing from the previous exercise, from the Annotate tab ➤ Labels & Tables panel choose the Add Tables drop-down arrow, and choose Alignment ➤ Add Segment to open the Table Creation dialog.
2. In the Select Alignment field, choose the Frontenac Drive alignment from the drop-down list (Figure 6.51), and click OK.

FIGURE 6.51: Creating an alignment segment table

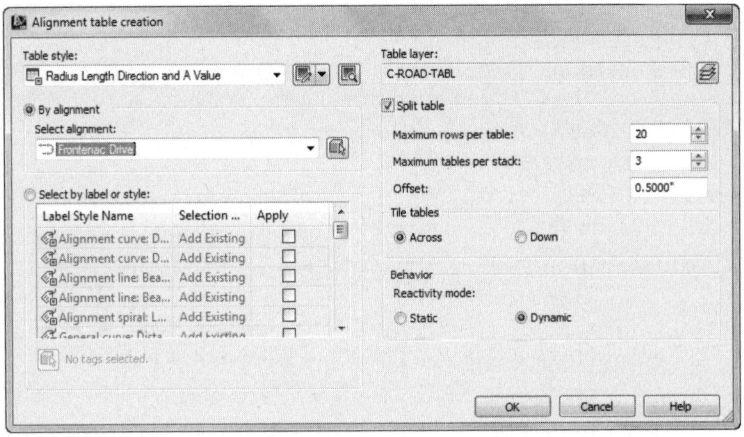

3. At the `Select upper left corner:` prompt, select an insertion point on the screen, and the table will generate.

When this exercise is complete, you may close the drawing. A saved finished copy of this drawing is available from the book's web page with the filename `AlignmentTables_FINISHED.dwg` or `AlignmentsTables_METRIC_FINISHED.dwg`.

The Bottom Line

Create an alignment from an object. Creating alignments based on polylines is a traditional method of building engineering models. With built-in tools for conversion, correction, and alignment reversal, it's easy to use the linework prepared by others to start your design model. These alignments lack the intelligence of crafted alignments, however, and you should use them sparingly.

Master It Open the `MasteringAlignments-Objects.dwg` or `MasteringAlignments-Objects_METRIC.dwg` file, and create alignments from the linework found there with the All Labels label set.

Create a reverse curve that never loses tangency. Using the alignment layout tools, you can build intelligence into the objects you design. One of the most common errors introduced to engineering designs is curves and lines that aren't tangent, requiring expensive revisions and resubmittals. The free, floating, and fixed components can make smart alignments in a large number of combinations available to solve almost any design problem.

Master It Open the `MasteringAlignments-Reverse.dwg` or the `MasteringAlignments-Reverse_METRIC.dwg` file, and create an alignment from the linework on the right. Create a reverse curve with both radii equal to 200 (or 60 for metric users) and with a pass-through point at the intersection of the two arcs.

Replace a component of an alignment with another component type. One of the goals in using a dynamic modeling solution is to find better solutions, not just the first solution. In the layout of alignments, this can mean changing components out along the design path, or changing the way they're defined. The ability of Civil 3D to modify alignments' geometric construction without destroying the object or forcing a new definition lets you experiment without destroying the data already based on an alignment.

Master It Convert the reverse curve indicated in the `MasteringAlignments-Rcurve.dwg` or the `MasteringAlignments-Rcurve_METRIC.dwg` file to a floating arc that is constrained by the next segment. Then change the radius of the curves to 150 (or 45 for metric users).

Create alignment tables. Sometimes there is just too much information that is displayed on a drawing, and to make it clearer, tables are used to show bearings and distances for lines, curves, and segments. With their dynamic nature, these tables are kept up to date with any changes.

Master It Open the `MasteringAlignments-Table.dwg` or `MasteringAlignments-Table_METRIC.dwg` file, and generate a line table, a curve table, and a segment table. Use whichever style you want to accomplish this.

Chapter 7

Profiles and Profile Views

Profile information is the backbone of vertical design. The AutoCAD® Civil 3D® software takes advantage of sampled data, design data, and external input files to create profiles for a number of uses. Profiles will be an integral part of corridors, as we'll discuss in Chapter 10, "Basic Corridors." In this chapter we'll look at using profile creation tools, editing profiles, and generating and editing profile views, and you'll learn ways to get your labels just so.

In this chapter, you will learn to:

- Sample a surface profile with offset samples
- Lay out a design profile on the basis of a table of data
- Add and modify individual entities in a design profile
- Apply a standard band set

The Elevation Element

The whole point of a three-dimensional model is to include the elevation element that's been missing for years on two-dimensional plans. But to get there, designers and engineers still depend on a flat 2D representation of the vertical dimension as shown in a profile view (see Figure 7.1).

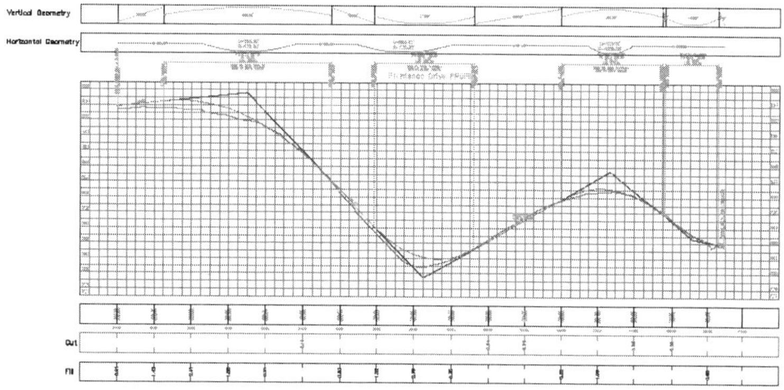

FIGURE 7.1
A typical profile view of the surface elevation along an alignment

A profile is nothing more than a series of data pairs in a station, elevation format. There are basic curve and tangent components, but these are purely the mathematical basis for the paired data sets. In AutoCAD Civil 3D, you can generate profile information in one of the following four ways:

Sampling from a Surface Sampling from a surface involves taking vertical information from a surface object every time the sampled alignment crosses a TIN line of the surface. This is perfect for generating a profile for the existing ground.

Using a Layout to Create a Profile Using a layout to create a profile allows you to input design information, setting critical station and elevation points, calculating curves to connect linear segments, and typically working within design requirements laid out by a reviewing agency.

Creating a Profile from a File Creating from a file lets you reference a specially formatted text file to pull in the station and elevation pairs. Doing so can be helpful in dealing with other analysis packages or spreadsheet tabular data.

Creating a Best Fit Profile Similar to the ability to generate a best fit alignment that we discussed in Chapter 6, "Alignments," you can also create a best fit profile. You may find yourself using this method when you are trying to generate defined geometry for an existing road.

This section looks at all four methods of creating profiles.

Surface Sampling

CERT OBJECTIVE

Working with surface information is the most elemental method of creating a profile. This information can represent any of the surfaces already in your drawing, such as an existing surface or any number of other surface-derived data sets. Surfaces can also be sampled at offsets, as you'll see in the next series of exercises. Follow these steps:

1. Open the `ProfileSampling.dwg` file (or the `ProfileSampling_METRIC.dwg` file for metric users) shown in Figure 7.2. Remember, all data files can be downloaded from this book's web page at `www.sybex.com/go/masteringcivil3d2013`.

FIGURE 7.2
The drawing you'll use for this exercise

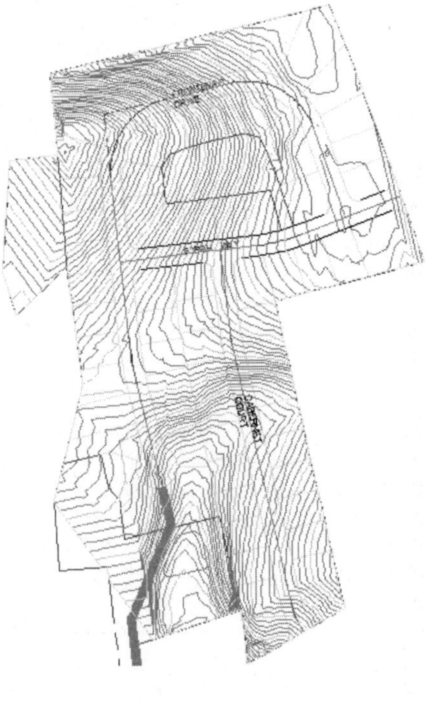

2. From the Create Design panel on the Home tab, choose Profile ➤ Create Surface Profile to display the Create Profile From Surface dialog (Figure 7.3).

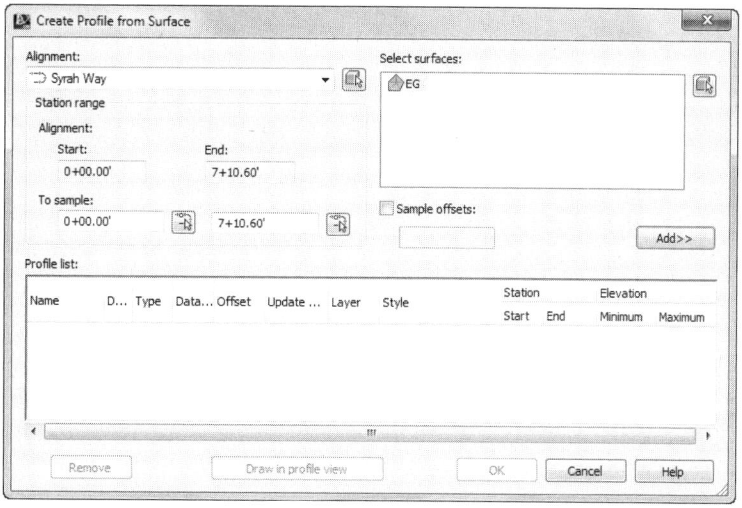

FIGURE 7.3
The Create Profile From Surface dialog

A QUICK TOUR AROUND THE CREATE PROFILE FROM SURFACE DIALOG

This dialog has a number of important features, so take a moment to see how it breaks down:

- The upper-left quadrant is dedicated to information about the alignment. You can select the alignment from a drop-down list, or you can click the Pick On Screen button. The Station Range area shows the starting and ending stations of the alignment and sets the To Sample range automatically to run from the beginning to the end of the alignment. You can control it manually by entering the station ranges in the To Sample text boxes or by using the Pick A Station buttons.
- The upper-right quadrant controls the selection of the surface that will be sampled as well as the offsets. You can select a surface from the list, or you can click the Pick On Screen button. Beneath the Select Surfaces box is a Sample Offsets check box. The offsets aren't applied in the left and right direction uniformly. You must enter a negative value to sample to the left of the alignment or a positive value to sample to the right. You can add multiple offsets in a comma-delimited list here, for example: -50, 25, -10, -5, 0, 5, 10, 25, 50. In all cases, whether or not you are sampling offsets, the profile isn't generated until you click the Add button.
- The Profile List box displays all profiles associated with the alignment currently selected in the Alignment drop-down menu. This area is generally static (it won't change), but you can modify the Update Mode, Layer, and Style columns by clicking the appropriate cells in this table. You can stretch and rearrange the columns to customize the view. The columns can only be modified in this location as they are created. Upon returning to the Create Profile From Surface dialog, the profiles previously created will have static values and can be changed in the Profile Properties dialog instead.

---

3. Select the Syrah Way option from the Alignment drop-down menu if it isn't already selected.
4. In the Select Surfaces box, select EG.
5. Click Add to add the centerline profile to the Profile list.
6. Select the Sample Offsets check box to make the entry box active.
7. Enter **-25, 25** (or **-7.5, 7.5** for metric users) to sample at the left and right right-of-way lines and click Add again.
8. In the Profile List, select the cell in the Style column that corresponds to the negative (left offset) value (see Figure 7.4) to activate the Pick Profile Style dialog. If you need to widen the columns, you can do so by clicking the line between the column headings.

FIGURE 7.4
The Create Profile From Surface dialog with styles assigned on the basis of the Offset value

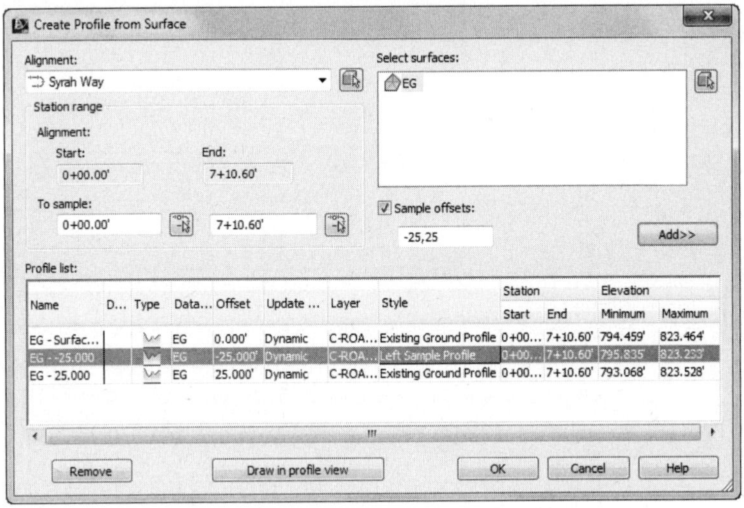

9. Select the Left Sample Profile option from the drop-down list, and click OK to dismiss the Pick Profile Style dialog.

The style changes from Existing Ground Profile to Left Sample Profile in the table.

10. Select the cell in the Style column that corresponds to the positive (right offset) value to activate the Pick Profile Style dialog.
11. Select the Right Sample Profile option from the drop-down list, and click OK to dismiss the Pick Profile Style dialog.
12. Click Draw In Profile View to dismiss this dialog and open the Create Profile View wizard, as shown in Figure 7.5.
13. Verify that the Select Alignment drop-down list shows Syrah Way and Profile View is selected in the Profile View Style drop-down list, and click Next.

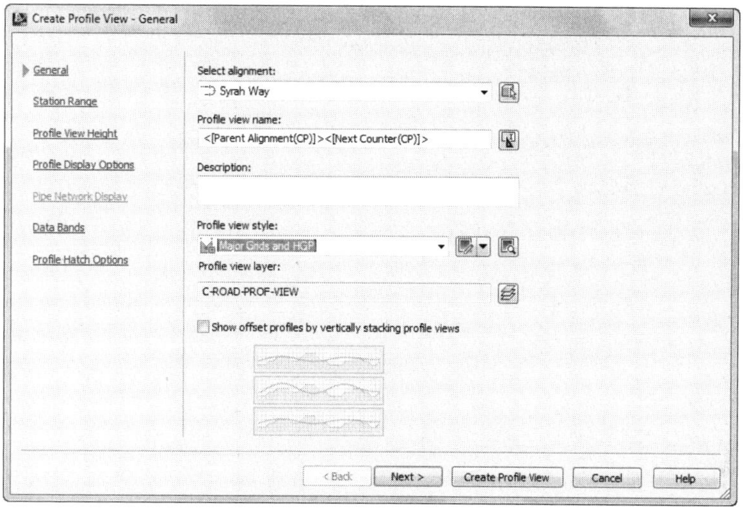

FIGURE 7.5
The Create Profile View – General wizard page

14. On the Create Profile View – Station Range wizard page, verify that the Automatic option has been selected (Figure 7.6). Click Next.

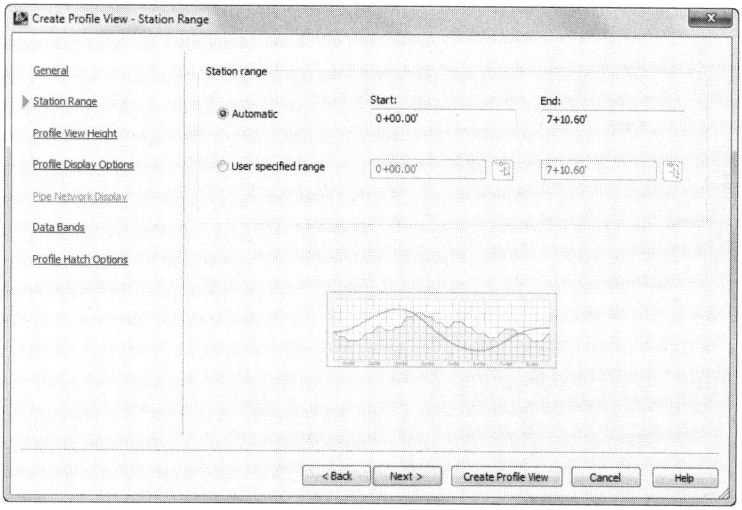

FIGURE 7.6
The Create Profile View – Station Range wizard page

15. On the Create Profile View – Profile View Height wizard page, verify that the Automatic option has been selected (Figure 7.7). Click Next.

We will examine split profile views in a later exercise.

16. On the Create Profile View – Profile Display Options wizard page, look at the settings but do not make any changes (Figure 7.8). Click Next.

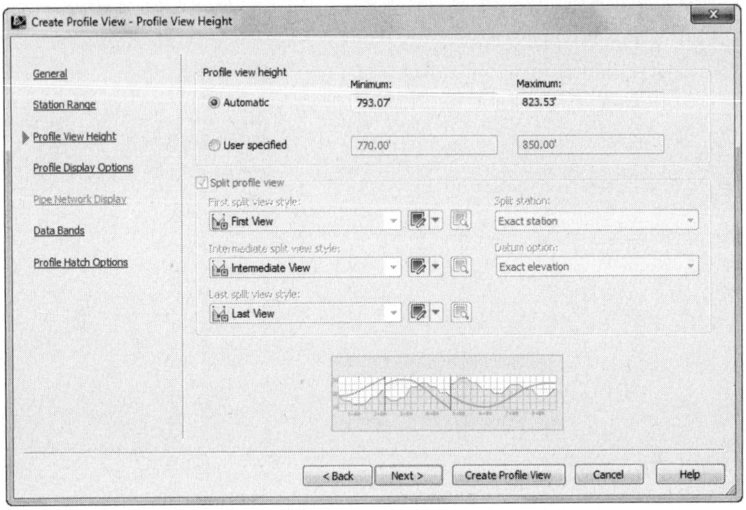

FIGURE 7.7
The Create Profile View – Profile View Height wizard page

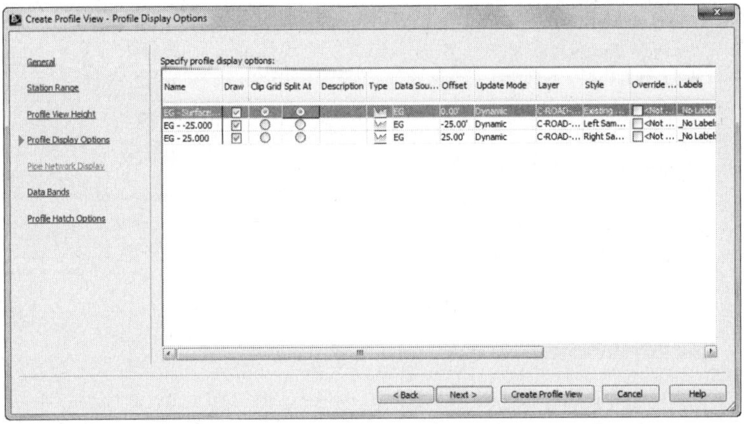

FIGURE 7.8
The Create Profile View – Profile Display Options wizard page

17. On the Create Profile View – Data Bands Options wizard page, verify that the band set is set to EG-FG Elevations And Stations (Figure 7.9). Notice that in the Set Band Properties area both profiles are set to the same profile by default. We will look at data bands in greater detail a bit later.

Notice that you could continue to click Next to step through the remainder of the wizard; however, you have no need to adjust further options at this time.

18. Click the Create Profile View button to dismiss the dialog.

19. Pick a point on the screen somewhere to the right of the site to draw the profile view, as shown in Figure 7.10.

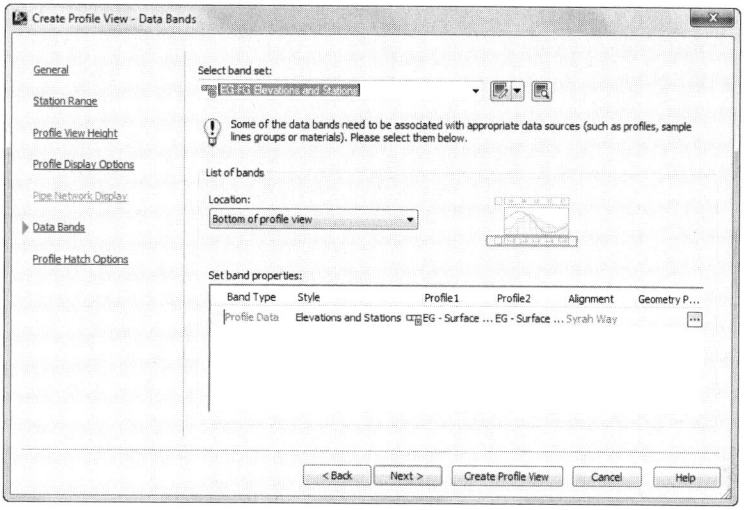

FIGURE 7.9
The Create Profile View – Data Bands wizard page

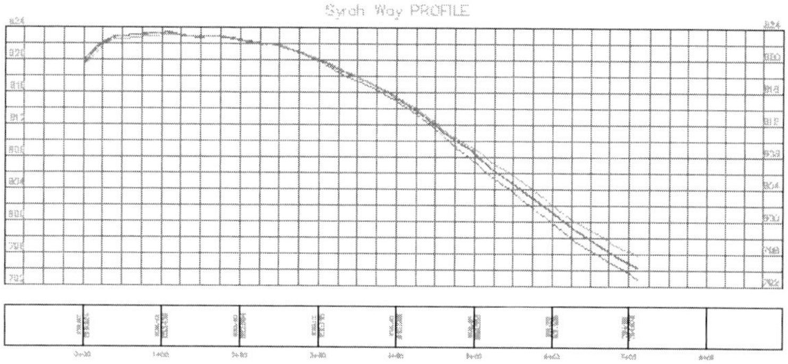

FIGURE 7.10
The complete profile view for Syrah Way

If the Events tab in Panorama appears, telling you that you've sampled data or if an error in the sampling needs to be fixed, then click the green check mark or the X to dismiss Panorama.

Keep the drawing open for the next portion of the exercise.

Profiles are dependent on the alignment they're derived from, so they're stored as profile branches under their parent alignment on the Prospector tab, as shown in Figure 7.11.

By maintaining the profiles under the alignments, you make it simpler to review what has been sampled and modified for each alignment. Note that the profiles from surface that you just created are dynamic and continuously update, as you'll see in this next portion of the exercise:

20. From the View tab ➢ Model Viewports panel, choose the drop-down list on the Viewport Configuration button and select Two: Horizontal.

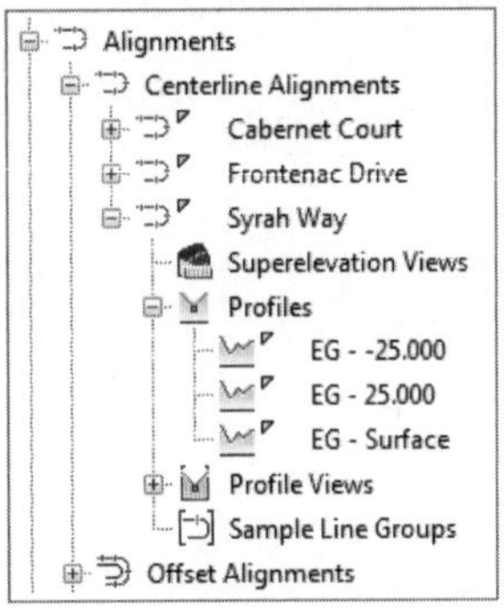

FIGURE 7.11 Alignment profiles on the Prospector tab

21. Click in the top viewport to activate it.
22. On the Prospector tab, expand the Alignments branch to view the alignment types, and expand the Centerline Alignments branch.
23. Right-click Syrah Way, and select the Zoom To option.
24. Click in the bottom viewport to activate it.
25. Expand the Alignments ➤ Centerline Alignments ➤ Syrah Way ➤ Profile Views branches.
26. Right-click the profile view named Syrah Way1, and select Zoom To.

Your screen should now look similar to Figure 7.12.

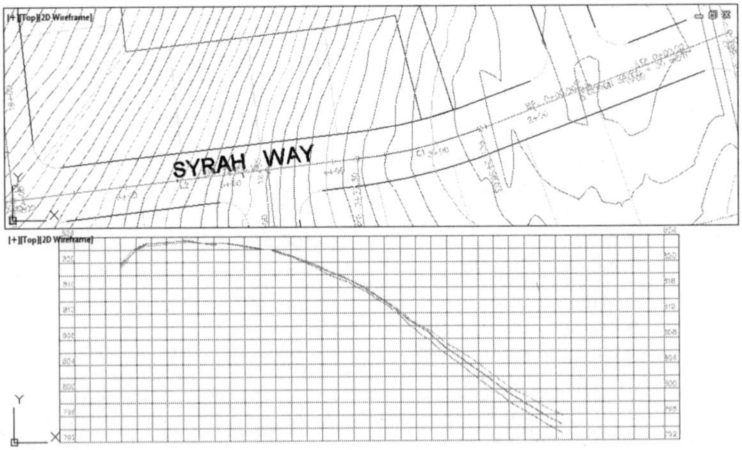

FIGURE 7.12 Splitting the screen for plan and profile editing

27. Click in the top viewport.
28. Zoom out so you can see more of the plan view.
29. Pick the alignment to activate the grips, and stretch the western end grip to lengthen and/or move the alignment, as shown in Figure 7.13.

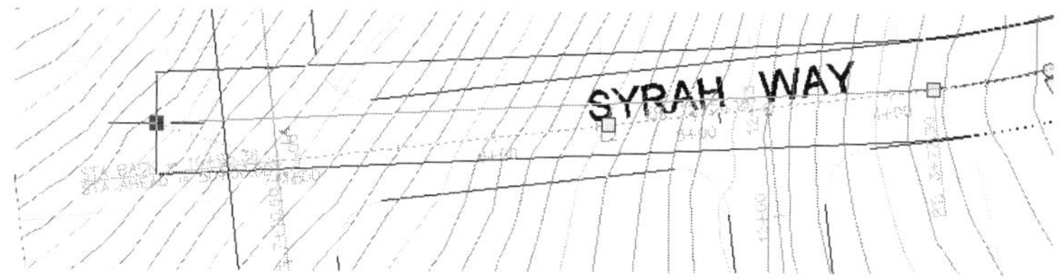

FIGURE 7.13
Grip-editing the alignment

30. Click to complete the edit.

The alignment profile automatically adjusts to reflect the change in the starting point of the alignment. Note that the offset profiles move dynamically as well.

31. Press Ctrl+Z enough times to undo the movement of your alignment and return it to its original location.

32. Select the top viewport and then switch back to a single viewport by once again clicking the Viewport Controls in the upper-left corner of one of the modelspace viewports and selecting Viewport Configuration List ➤ Single.

Keep the drawing open for the next brief exercise.

By maintaining the relationships between the alignment, the surface, the sampled information, and the offsets, the software creates a much more dynamic feedback system for designers. This system can be useful when you're analyzing a situation with a number of possible solutions, where the surface information will be a deciding factor in the final location of the alignment. Once you've selected a location, you can use this profile view to create a vertical design, as you'll see in the next section.

LEFT TO RIGHT AND RIGHT TO LEFT?

You may have noticed that the alignment for Syrah Way is drawn right to left but the profile shows it left to right. It is often desirable to have the plan and profile go in the same direction.

One option is to rotate the plan view 180 degrees. If you have your labeling all set to be plan-readable, it will follow along nicely.

Another option is to generate a profile view object style that is set to read right to left instead of the default left to right.

In the following short exercise you will generate a profile view which displays right to left:

1. Using the file from the previous exercise, select and then right-click the profile view (grid); choose Profile View Properties.

2. On the Information tab, click the drop-down edit button to the right of the current object style (Profile View) and select Copy Current Selection.

3. On the Information tab, change the Name to **Profile View: Right to Left**.

You may revise the description if desired.

4. In the Profile View Style dialog, on the Graph tab, change the profile view direction to Right To Left.
5. Click OK to dismiss the Profile View Style dialog.
6. Click OK to dismiss the Profile View Properties dialog.

You may need to move the profile view since the insert point of the profile view will now be in the lower-right corner instead of the lower-left corner as it was previously.

When this exercise is complete, you may close the drawing. A finished copy of this drawing is available from the book's website with the filename `ProfileSampling_FINISHED.dwg` or `ProfileSampling_METRIC_FINISHED.dwg`.

Changing a profile view style is straightforward, but because of the large number of settings in play with a profile view style, the changes can be dramatic. A profile view style includes information such as labeling on the axis, vertical scale factors, grid clipping, and component coloring.

Using various styles lets you make changes to the view to meet requirements without changing any of the design information associated with the profile. To learn more about editing and creating profile styles, refer to Chapter 21, "Object Styles."

Layout Profiles

CERT OBJECTIVE

Working with sampled surface information is dynamic, and the improvement over previous generations of Autodesk® civil design software is profound. Moving into the design stage, you'll see how these improvements continue as you look at the nature of creating design profiles. By working with layout profiles as a collection of components that understand their relationships with each other as opposed to independent finite elements, you will realize the power of the AutoCAD Civil 3D software as a design tool in addition to being a drafting tool.

You can create layout profiles in two basic ways:

PVI-Based Layout PVI-based layouts are the most common, using tangents between points of vertical intersection (PVIs) and then applying curve parameters to connect them. PVI-based editing allows editing in a more conventional tabular format.

Entity-Based Layout Entity-based layouts operate like horizontal alignments in the use of free, floating, and fixed entities. The PVI points are derived from pass-through points and other parameters that are used to create the entities. Entity-based editing allows for the selection of individual entities and editing in an individual component dialog.

You'll work with both methods in the next series of exercises to illustrate a variety of creation and editing techniques. First, you'll focus on the initial layout, and then you'll edit the various layouts.

Layout by PVI

PVI layout is the most common methodology in transportation design. Using long tangents that connect PVIs by derived parabolic curves is a method most engineers are familiar with, and it's the method you'll use in the first exercise:

1. Open the `LayoutByPVI.dwg` file or the `LayoutByPVI_METRIC.dwg` file.
2. From the Home tab ➢ Create Design panel, choose Profile ➢ Profile Creation Tools.

3. At the `Select profile view to create profile:` prompt, pick the Syrah Way profile view by clicking one of the grid lines to display the Create Profile – Draw New dialog.
4. Set Name to **Syrah Way FG**.
5. On the General tab, set Profile Style to Design Profile and Profile Label Set to Complete Label Set, as shown in Figure 7.14.

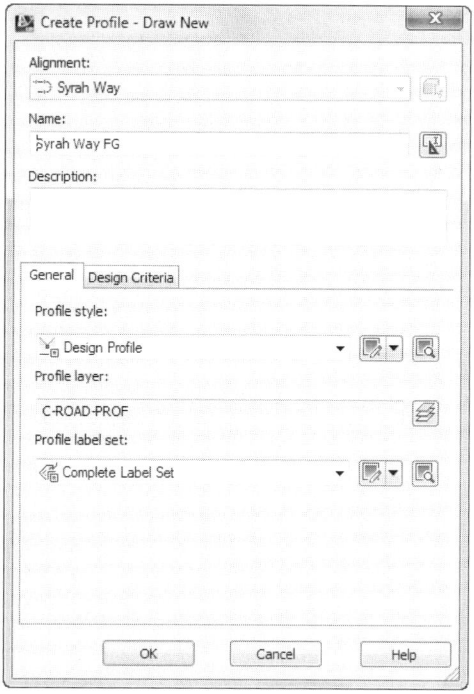

FIGURE 7.14
The Create Profile – Draw New dialog, General tab

6. Switch to the Design Criteria tab to examine the options provided.

Criteria-based design operates in profiles similar to Chapter 6 in that the software compares the design speed to a selected design table (typically AASHTO 2001 in the North America releases) and sets minimum values for curve K values. This can be helpful when you're laying out long highway design projects, but most site and subdivision designers have other criteria to design against. We won't be using the option in this exercise so you can leave everything unchecked.

7. Click OK to display the Profile Layout Tools toolbar shown in Figure 7.15.

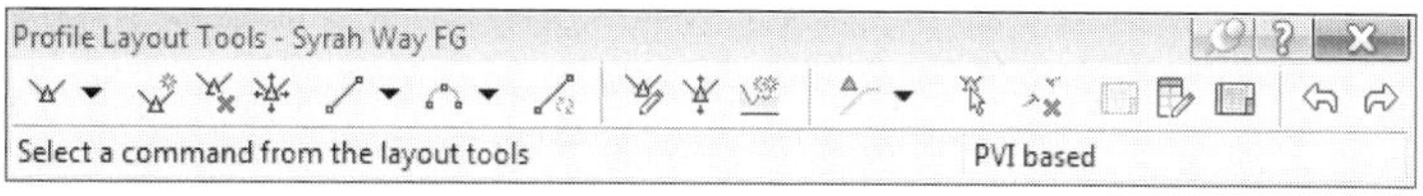

FIGURE 7.15
Profile Layout Tools toolbar

Notice that the toolbar is *modeless*, meaning it stays open even if you do other AutoCAD operations such as Pan or Zoom.

8. On the Profile Layout Tools toolbar, click the drop-down arrow next to the Draw Tangents button on the far left.

9. Select the Curve Settings option.

The Vertical Curve Settings dialog opens, as shown in Figure 7.16.

FIGURE 7.16
The Vertical Curve Settings dialog

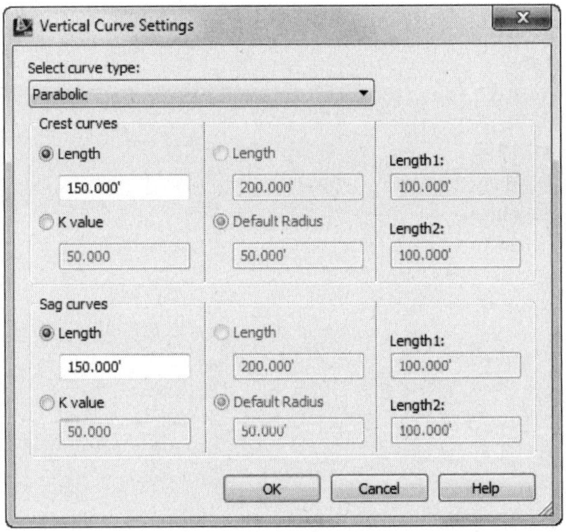

The Select Curve Type drop-down menu should be set to Parabolic, and the Length values in both the Crest Curves and Sag Curves areas should be **150.000′** (or **45.000** m for metric users), as shown in Figure 7.15. Selecting a Circular or Asymmetric curve type activates the other options in this dialog.

TO K OR NOT TO K

You don't have to choose. Realizing that users need to be able to design using both, the software lets you modify your design based on what's important. You can enter a K value to see the required length, and then enter a length with a nice round value that satisfies the K. The choice is up to you.

10. Click OK to dismiss the Vertical Curve Settings dialog.

11. On the Profile Layout Tools toolbar, click the drop-down arrow next to the Draw Tangents button on the far left again. This time, select the Draw Tangents With Curves option.
12. Use a Center Osnap to pick the center of the circle at the far right in the profile view.

A jig line, which will be your layout profile, appears. Remember, the profile is reversed so that station 0+00 is lined up with the plan. Therefore, station 0+00 is on the right of the profile. You need to draw your profile from low station to high station, which in this case is left to right.

13. Continue working your way across the profile view, picking the center of each circle right to left with a Center Osnap.
14. Right-click or press ↵ after you select the center of the last circle.
15. The profile labels will default to a location; however, you can click on any of the profile labels and use the grips to move them to a more legible location.

Your drawing should look similar to Figure 7.17.

FIGURE 7.17
A completed layout profile with labels

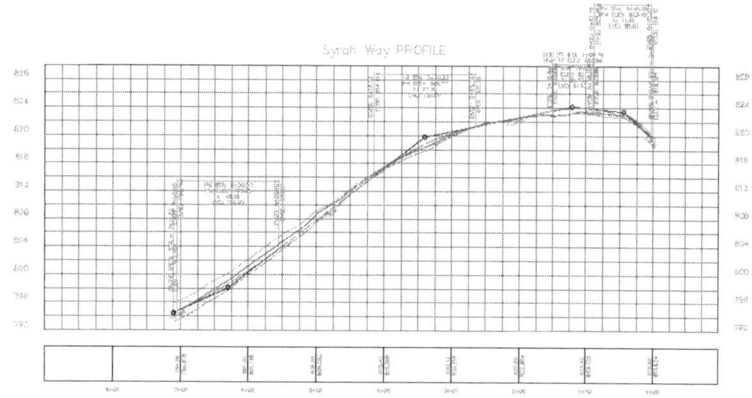

16. Close the Profile Layout Tools toolbar.

When this exercise is complete, you may close the drawing. A finished copy of this drawing is available from the book's web page with the filename `LayoutByPVI_FINISHED.dwg` or `LayoutByPVI_METRIC_FINISHED.dwg`.

WHAT IS A JIG?

A *jig* is a temporary line shown on screen to help you locate your pick point. Jigs work in a similar way to osnaps in that they give feedback during command use but then disappear when the selection is complete. The jigs will help you locate information on the screen for alignments, profiles, profile views, and a number of other places where a little feedback goes a long way.

The layout profile is labeled with the Complete Label Set you selected in the Create Profile dialog. As you'd expect, this labeling and the layout profile are dynamic. If you select the profile and then zoom in on this profile line, not the labels or the profile view, you'll see something like Figure 7.18.

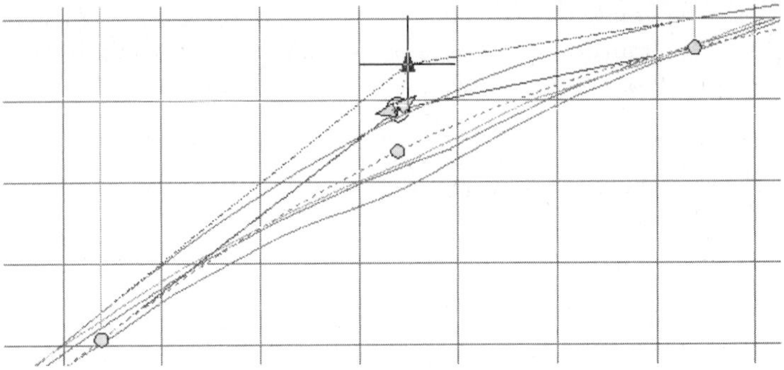

FIGURE 7.18
The types of grips on a layout profile

The PVI-based layout profiles include the following unique grips:

Vertical Triangular Grip The vertical triangular grip at the PVI point is the PVI grip. Moving this alters the inbound and outbound tangents, but the curve remains in place with the same design parameters of length and type.

Angled Triangular Grips The angled triangular grips on either side of the PVI are sliding PVI grips. Selecting and moving either moves the PVI, but movement is limited to along the tangent of the selected grip. The curve length isn't affected by moving these grips, but the PVI station and elevation will be, as well as the grade of the other tangent.

Circular Pass-Through Grips The circular pass-through grips near the PVI and at each end of the curve are curve grips. Moving any of these grips makes the curve longer or shorter without adjusting the inbound or outbound tangents or the PVI point.

Although this simple pick-and-go methodology works for preliminary layout, it lacks a certain amount of control typically required for final design. For that, you'll use another method of creating PVIs:

1. Open the `LayoutByPVITransparent.dwg` or `LayoutByPVITransparent_METRIC.dwg` file.
2. Verify that the Transparent Commands toolbar (Figure 7.19) is displayed somewhere on your screen. If it is not shown, from the View tab ➢ User Interface panel, choose Toolbars ➢ CIVIL ➢ Transparent Commands.

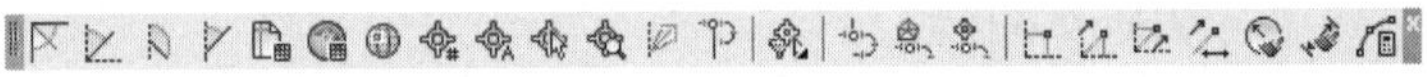

FIGURE 7.19
The Transparent Commands toolbar

3. From the Home tab ➢ Create Design panel, choose Profile ➢ Profile Creation Tools.
4. Pick the Cabernet Court profile view (located at the lower right inside a rectangle) by clicking on one of the grid lines to display the Create Profile – Draw New dialog.
5. Set the name to **Cabernet Court FG**.
6. On the General tab, set Profile Style to Design Profile and Profile Label Set to Complete Label Set, and then click OK to display the Profile Layout Tools toolbar.

7. On the Profile Layout Tools toolbar, click the drop-down arrow next to the Draw Tangents button on the far left, and select the Draw Tangents With Curves option, as in the previous exercise.

8. Use a Center Osnap to snap to the center of the circle near the left edge of the profile view.

Unlike in the previous exercise, this profile view is set up left to right.

9. On the Transparent Commands toolbar, select the Profile Station Elevation transparent command.

For those who prefer using the command line, the key-in command for this transparent command is **'PSE**.

10. Pick a grid line on the Cabernet Court profile view to select it.

If you move your cursor within the profile grid area, a vertical red line, or *jig*, appears; it moves up and down and from side to side. Notice the tooltips currently show the station values of the jig location.

11. Enter **365↵** (or **112.8↵** for metric users) at the command line for the station value.

If you move your cursor within the profile grid area, a horizontal and vertical jig appears (see Figure 7.20), but it can only move vertically along the station just specified.

FIGURE 7.20
A jig appears when you use the Profile Station Elevation transparent command

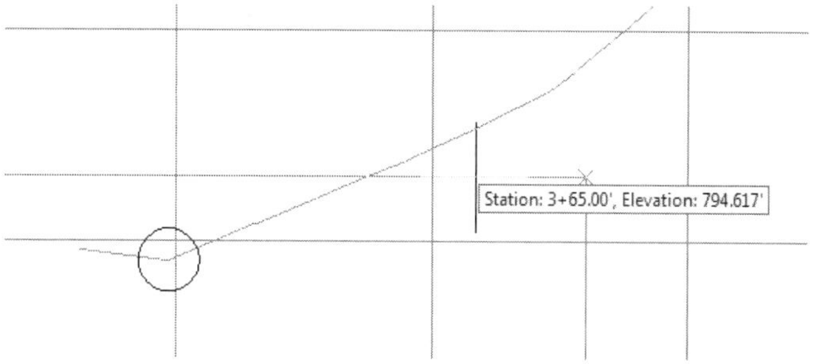

12. Enter **795↵** (or **242.5↵** for metric users) at the command line to set the elevation for the second PVI.

13. Press Esc only once.

The Profile Station Elevation transparent command is no longer active, but the Draw Tangents With Curves button that you previously selected on the Profile Layout Tools toolbar continues to be active.

14. On the Transparent Commands toolbar, select the Profile Grade Station transparent command.

For those who prefer using the command line, the key-in command for this transparent command is **'PGS**.

Notice that you did not need to select a profile view this time; that's because you are still in the same command (Draw Tangents With Curves in this case). The transparent command will default to the same profile view that was previously selected.

15. Enter **10**↵ at the command line for the profile grade.

16. Enter **430**↵ (or **131**↵ for metric users) for the station value at the command line.

17. Press Esc only once to deactivate the Profile Grade Station transparent command.

18. On the Transparent Commands toolbar, select the Profile Grade Length transparent command.

For those who prefer using the command line, the key-in command for this transparent command is **'PGL**.

19. Enter **-1.5**↵ at the command line for the profile grade.

20. Enter **120**↵ (or **36.5**↵ for metric users) for the profile grade length.

21. Press Esc only once to deactivate the Profile Grade Length transparent command.

22. Continue defining the profile using the Profile Grade Station transparent command using the following input:

    a. **-3.5**↵, **850**↵ (**-3.5**↵, **259**↵ for metric users)

    b. **10**↵, **1035**↵ (**10**↵, **315.5**↵ for metric users)

    c. **3.5**↵, **1165**↵ (**3.5**↵, **355**↵ for metric users)

23. Press Esc only once to deactivate the Profile Grade Station transparent command and to continue using the Draw Tangent With Curves command.

24. Use a Center Osnap to select the center of the circle along the far-right side of the profile view.

25. Press ↵ to complete the profile.

Your profile should look like Figure 7.21.

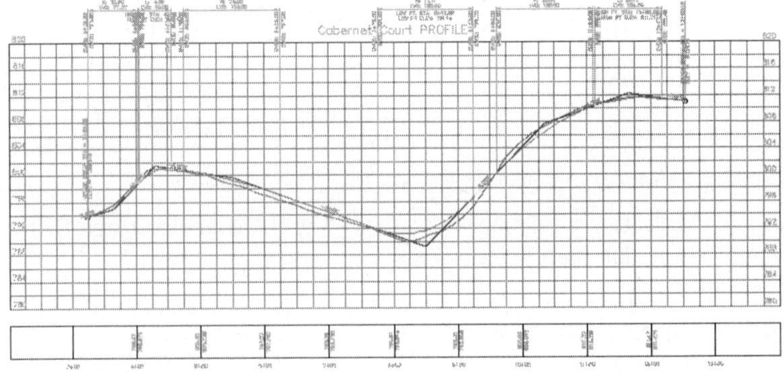

FIGURE 7.21 Using the Transparent Commands toolbar to create a layout profile

26. Close the Profile Layout Tools toolbar.

When this exercise is complete, you may close the drawing. A finished copy of this drawing is available from the book's web page with the filename `LayoutByPVITransparent_FINISHED.dwg` or `LayoutByPVITransparent_METRIC_FINISHED.dwg`.

Using PVIs to define tangents and fitting curves between them is the most common approach to create a layout profile, but you'll look at an entity-based design in the next section.

Layout by Entity

Working with the concepts of fixed, floating, and free entities as you did with alignments in Chapter 6, you'll lay out a design profile in this exercise:

1. Open the `LayoutByEntity.dwg` or the `LayoutByEntity_METRIC.dwg` file.
2. From the Home tab ➢ Create Design panel, choose Profile ➢ Profile Creation Tools.
3. Pick the Frontenac Drive profile view (located at the lower right inside a rectangle) by clicking on one of the grid lines to display the Create Profile – Draw New dialog.
4. Set the name to **Frontenac Drive FG**.
5. On the General tab, set Profile Style to Design Profile and Profile Label Set to Complete Label Set; then click OK to display the Profile Layout Tools toolbar.

Oops, You Closed the Profile Layout Tools Toolbar!

If at any point you inadvertently close the Profile Layout Tools toolbar, have no fear. You can reopen it by selecting the profile that you were editing and selecting Geometry Editor from the Profile contextual tab. If you created the profile but there aren't any entities to select, you can select it in Prospector by expanding the Alignments ➢ Centerline Alignments ➢ *Alignment Name* ➢ Profiles branch. Right-click the profile and choose Delete to start again from scratch. Unfortunately there isn't a Select option like you have with alignments, so deleting the profile is the only option.

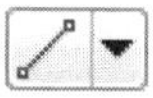

6. On the Profile Layout Tools toolbar, click the drop-down arrow next to the Tangent Creation button, and select the Fixed Tangent (Two Points) option.
7. Using a Center Osnap, pick the circle at the left edge of the profile view labeled A.

   A rubber-banding line appears.

8. Using a Center Osnap, pick the circle labeled B.

   A tangent is drawn between these two circles.

9. Using a Center Osnap, pick the circle labeled B again as the start point and the circle labeled C as the endpoint.

   A tangent is drawn between these two circles. Notice that the tangent does not automatically continue from the previous two-point fixed tangent; therefore, you have to select the B circle again.

10. Using a Center Osnap, pick the circle labeled D as the start point and the circle labeled E as the endpoint.

A tangent is drawn between these two circles.

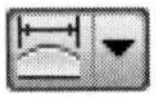

11. On the Profile Layout Tools toolbar, click the drop-down arrow next to the Vertical Curve Creation button and select the More Free Vertical Curves ➤ Free Vertical Curve (PVI Based) option.

Notice that the image shown to the left of the drop-down arrow for the Tangent Creation button and Vertical Curve Creation button will match the last type of entity you selected from the drop-down menu.

12. At the `Pick point near PVI or curve to add curve:` prompt, pick the circle labeled B as the PVI.

13. At the `Specify curve length or [Passthrough K]:` prompt, enter **450**↵ (or **137**↵ for metric users) as the curve length.

14. Press ↵ to end the command.

Your drawing should now look similar to Figure 7.22. Notice that although you have added the tangent between D and E, it is not yet labeled since it is not connected with the main portion of the profile created up until this point.

FIGURE 7.22
Some tangent and vertical curve entities placed on Frontenac Drive

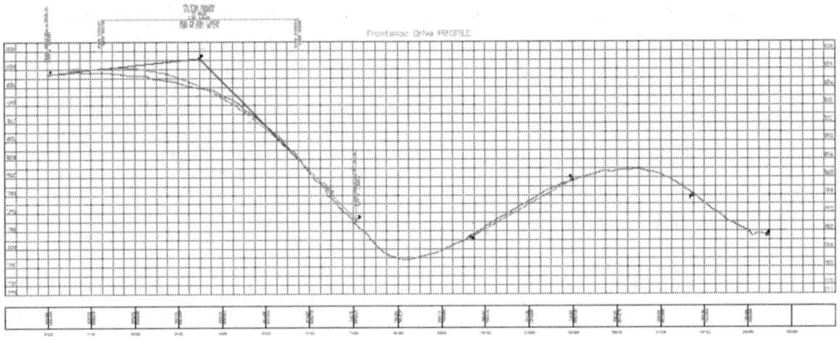

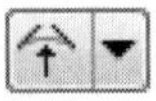

15. On the Profile Layout Tools toolbar, click the drop-down arrow next to the Vertical Curve Creation button and select the Free Vertical Curve (Parameter) option.

16. Select the tangent between B and C as the first entity and the tangent between D and E as the next entity.

Remember to pick the tangent line and not an end circle.

17. At the `Specify curve length or [Radius K]:` prompt, enter **300**↵ (or **90**↵ for metric users) as the curve length and press ↵ again to end the command.

Notice with this command the tangents do not have to meet at a PVI, unlike the previous Free Vertical Curve (PVI Based) curve.

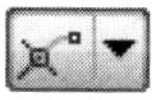

18. On the Profile Layout Tools toolbar, click the drop-down arrow next to the Vertical Curve Creation button and select the More Fixed Vertical Curves ➢ Fixed Vertical Curve (Entity End, Through Point) option.

19. At the `Select entity to attach to:` prompt, select the tangent between D and E to attach the fixed vertical curve.

Remember to pick the tangent line and not the end circle. Also, you have to select the tangent between the midpoint and the endpoint of the tangent at the circle labeled E. Selecting too close to the endpoint at the circle labeled D will give a result of `End of selected entity already has an attachment`. A rubber-banding curve appears. If you move the cursor on the wrong side of the tangent endpoint, it will become a large red circle with an X across it indicating that you cannot select that point.

20. At the `Specify end point:` prompt, using a Center Osnap, pick the circle labeled F and press ↵ to end the command.

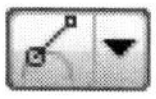

21. On the Profile Layout Tools Toolbar, click the drop-down arrow next to the Tangent Creation button and select the Float Tangent (Through Point) option.

22. At the `Select entity to attach to:` prompt, select the curve between E and F to attach the floating tangent.

A rubber-banding curve appears.

23. At the `Select through point:` prompt, using a Center Osnap, select the circle labeled G.

24. Press ↵ or right-click to end the Fixed Tangent (Through Point) command; then close the Profile Layout Tools toolbar.

Your drawing should look like Figure 7.23.

FIGURE 7.23
Completed profile built using entities

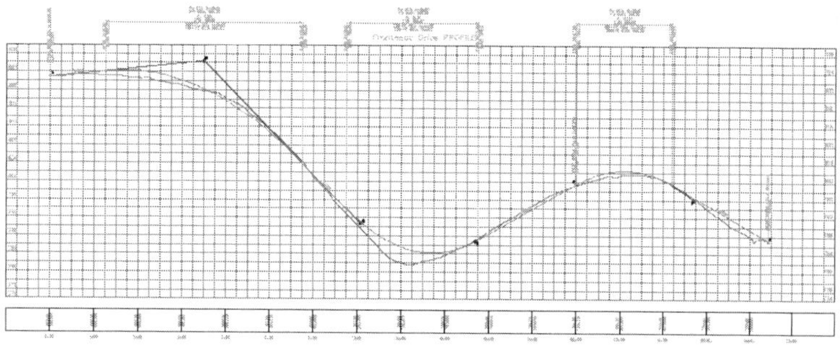

When this exercise is complete, you may close the drawing. A finished copy of this drawing is available from the book's web page with the filename `LayoutByEntity_FINISHED.dwg` or `LayoutByEntity_METRIC_FINISHED.dwg`.

With the entity-creation method, grip editing works in a similar way to other layout methods based on the fixed, floating, and free constraints.

Profile Layout Tools

CERT OBJECTIVE

Although we have touched on many of the available tools in the Profile Layout Tools toolbar, as shown in Figure 7.24, there are still many that we have not.

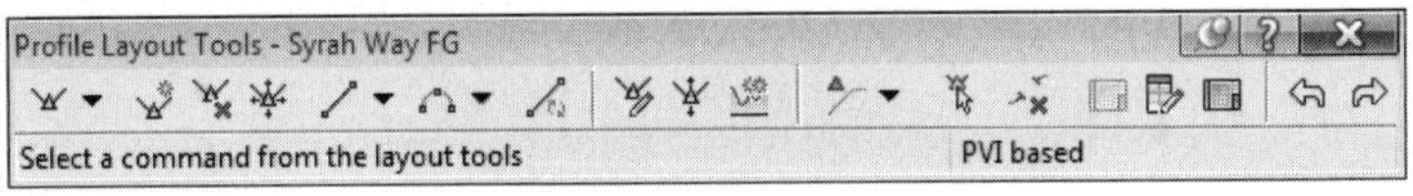

Figure 7.24 Profile Layout Tools toolbar

Draw Tangent Drop-Down The Draw Tangent drop-down button contains three options: Draw Tangents, Draw Tangents With Curves, and Curve Settings. Draw Tangents lays out a profile point to point with no curves. Draw Tangent With Curves lays out a profile from point to point with the curve type and length determined from the Curve Settings. Curve Settings sets the type of curve (parabolic, circular, asymmetric), default crest and sag K values, and default crest and sag lengths.

Insert PVI The Insert PVI button adds a new PVI at the specified location, consequently breaking an existing tangent and generating two new tangents connected to the new PVI.

Delete PVI The Delete PVI button removes an existing PVI at the specified location, consequently taking two tangents and replacing them with a single tangent.

Move PVI The Move PVI button allows you to select an existing PVI and relocate it to a specified location while keeping the two existing tangents. You can get the same result by grip-editing the PVI with the vertical triangular grip, as described earlier.

Tangent Creation Drop-Down The Tangent Creation Drop-down button contains six tools:

- Fixed Tangent (Two Points)
- Fixed Tangent – Best Fit
- Float Tangent (Through Point)
- Float Tangent – Best Fit
- Free Tangent
- Solve Tangent Intersection

The Fixed, Floating, and Free terminology is consistent with those discussed in Chapter 6 when we were generating alignments, and therefore many of these options should be self-explanatory. The Solve Tangent Intersection option extends two tangents that do not currently connect to form a PVI.

Vertical Curve Creation Drop-Down The Vertical Curve Creation Drop-down button contains 14 options:

- Fixed Vertical Curve (Three Points)
- Fixed Vertical Curve (Two Points, Parameter)
- Fixed Vertical Curve (Entity End, Through Point)
- Fixed Vertical Curve (Two Points, Grade At Start Point)

- Fixed Vertical Curve (Two Points, Grade At End Point)
- Fixed Vertical Curve – Best Fit
- Floating Vertical Curve (Parameter, Through Point)
- Floating Vertical Curve (Through Point, Grade)
- Floating Vertical Curve – Best Fit
- Free Vertical Curve (Parameter)
- Free Vertical Parabola (PVI Based)
- Free Asymmetrical Parabola (PVI Based)
- Free Circular Curve (PVI Based)
- Free Vertical Curve – Best Fit

Once again the fixed, floating, and free terminology should be familiar from Chapter 6. By trying these various options, you will become comfortable with their capabilities and you will find the ones that best fit your design needs.

Convert AutoCAD Line And Spline The Convert AutoCAD Line And Spline button takes a singular line/spline and converts it into a profile object, either a tangent or a three-point vertical curve as applicable.

Insert PVIs – Tabular The Insert PVIs – Tabular button allows you to enter PVI station and elevation information in a table-like dialog, which is helpful if you want to create multiple PVIs at once using station and elevation information. This table allows you to insert PVIs and curves anywhere geometrically possible in the profile. The order in which you insert the PVIs isn't important either when using this method of entry.

Raise/Lower PVIs The Raise/Lower PVIs button allows you to raise or lower a profile at a specified distance for either the entire profile or for the PVIs within a specified station range. This button will be discussed in a later exercise on Other Profile Edits.

Copy Profile The Copy Profile button allows you to copy either the entire profile or within a specified station range. This button will be discussed in a later exercise in the "Other Profile Edits" section.

PVI or Entity Based Drop-Down The PVI or Entity Based drop-down button allows you to choose to select and display profile layout parameters based on either PVI or Entity.

Select PVI or Select Entity The Select PVI or Select Entity button opens the Profile Layout Parameters dialog for the selected PVI or Entity.

Delete Entity The Delete Entity button removes a selected curve or tangent.

Edit Best Fit Data For All Entities The Edit Best Fit Data For All Entities button turns on the display of a table of the regression data for a profile that was created by best fit. A discussion of Best Fit Profile is provided in the next section.

Profile Layout Parameters The Profile Layout Parameters button opens the Profile Layout Parameters dialog, which shows numeric data for editing the selected entity or PVI.

Profile Grid View The Profile Grid View button opens the Profile Entities tab in Panorama, showing information about all the entities and PVIs in the profile. This is where you have access to make edits on all the entities of the profile.

Undo/Redo The Undo and Redo buttons reverse the last command or the last undo operation.

The Best Fit Profile

You've surveyed along a centerline, and you need to closely approximate the tangents and vertical curves as they were originally designed and constructed.

The Create Best Fit Profile option is found in the Home tab ➢ Create Design panel, on the Profile drop-down. Once you select a profile view, the Create Best Fit Profile dialog appears, as shown in Figure 7.25.

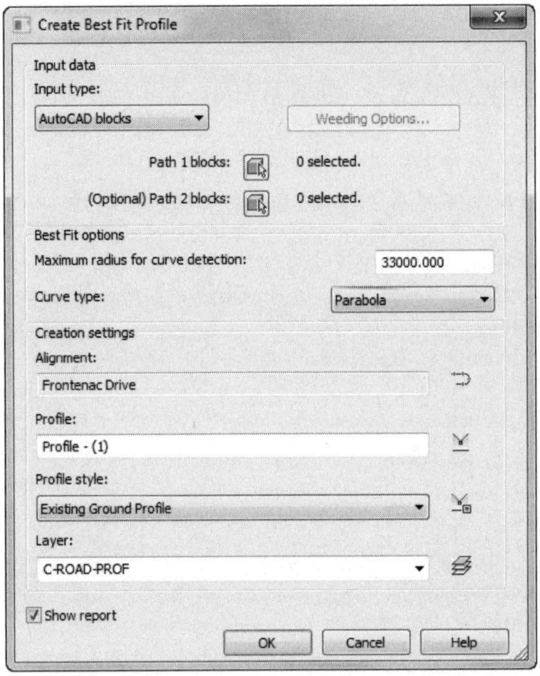

FIGURE 7.25
Create Best Fit Profile dialog

Similar to the best fit alignment we discussed in Chapter 6, a best fit profile can be based on an Input Type setting of AutoCAD Blocks, AutoCAD 3D Polylines, AutoCAD Points, COGO Points, Surface Profile, or Feature Lines. The most common option is Surface Profile.

The command attempts to run a complex algorithm to determine the best fit profile, including both tangents and vertical curves. However, the only best fit option available for determining vertical curves is the maximum curve radius. The maximum curve radius is a formula rarely used in the design of parabolic curves, the most common curves found in roadway design. Once the analysis is run, a Best Fit Report is provided; however, unlike the Best Fit command for lines and curves as seen in Chapter 1, "The Basics," this command has no options for selecting or deselecting points, making the command somewhat cumbersome.

In some cases, the commands you choose to use in a given scenario, as promising as the name may sound, may actually prove to be more of a hindrance than a help. The Best Fit Profile command may be one of these cases. The command attempts to perform an analysis on all the sampled existing ground profile points in a profile view, the options are limited, and in the end, you may find a good pair of eyes and graphical analysis using other profile creation methods are less time consuming and produce an equally valid result.

It is important to understand all the commands at your disposal, but you are not required to use each one in a production environment. If results don't meet your needs and you've spent considerable time with a single method, find another method, or methods, that accomplish the same task. Remember, at the end of the day, you still have sheets to cut and plans to get out the door. You're limited only by your selection of commands and your creativity.

Creating a Profile from a File

Working with profile information in the AutoCAD Civil 3D environment is nice, but it isn't the only place where you can create or manipulate this sort of information. Many programs or analysis packages generate profile information. One common case is the plotting of a hydraulic grade line against a stormwater network profile of the pipes. When information comes from outside the program, it is often output in myriad formats. If you convert this format to the format required by AutoCAD Civil 3D, the profile information can be input directly.

There is a specific format that is required for creating a profile from a text file. Each line is a PVI definition (station and elevation) listed in ascending order. The station should not include the plus character (use 100, not 1+00 or 0+100). Curve information is an optional third bit of data on any line except for the first and last lines in the file. The vertical curve that is created will be a parabolic curve, which is the most popular type of vertical curve. Note that each line is space delimited. Here's one example of a profile text file:

```
0 550.76
127.5 552.24
200.8 554 100
256.8 557.78 50
310.75 561
```

In this example, the third and fourth lines include the curve length as the optional third piece of information. The only inconvenience of using this input method is that the information in Civil 3D doesn't directly reference the text file. Once the profile data is imported, no dynamic relationship exists with the text file, but other methods can be used to edit the profile once imported.

In this exercise, you'll import a small text file to see how the function works:

1. Open the `ProfilefromFile.dwg` file (or the `ProfilefromFile_METRIC.dwg` file).

2. From the Home tab ➢ Create Design panel, choose Profile ➢ Create Profile From File.

The Import Profile From File – Select File dialog appears.

3. Browse to the `ProfileFromFile.txt` file (or `ProfileFromFile_METRIC.txt` file for metric users), and click Open to display the Create Profile – Draw New dialog.

4. For Alignment choose Frontenac Drive, and set Name to **Frontenac Drive FG**.

5. On the General tab, set Profile Style to Design Profile and Profile Label Set to Complete Label Set; then click OK.

Your drawing should look like Figure 7.26. The Frontenac profile view is located at the lower right inside a rectangle.

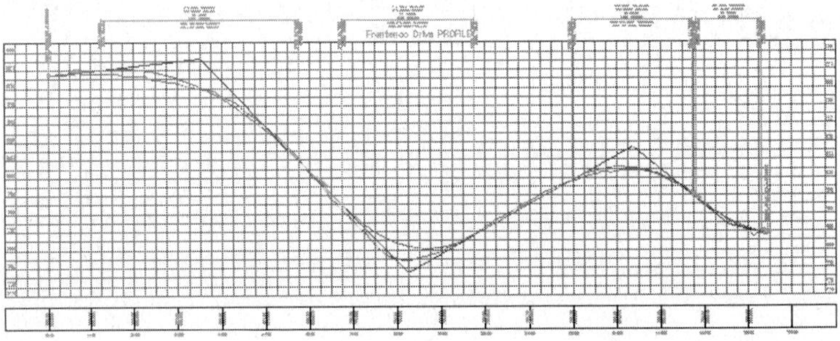

FIGURE 7.26
Completed profile created from a file

When this exercise is complete, you may close the drawing. A finished copy of this drawing is available from the book's web page with the filename `ProfileFromFile_FINISHED.dwg` or `ProfileFromFile_METRIC_FINISHED.dwg`.

Now that you've tried the three main ways of creating profile information, you'll edit a profile in the next section.

Editing Profiles

The methods just reviewed let you quickly create profiles. You saw how sampled profiles reflect changes in the parent alignment and how to lay out a design profile using a few different methodologies. You also imported a text file with profile information. In all these cases, you just left the profile as originally designed with no analysis or editing.

This section will begin to look at some of the profile editing methods available. The most basic is a more precise grip-editing methodology, which you'll learn about first. Then you'll see how to modify the PVI-based layout profile, how to change out the components that make up a layout profile, and how to use some other miscellaneous editing functions.

Grip Profile Editing

Once a profile layout is in place, sometimes a simple grip edit will suffice. But for precision editing, you can use a combination of the grips and the buttons on the Transparent Commands toolbar, as in this short exercise:

1. Open the `GripEditingProfiles.dwg` file (or the `GripEditingProfiles_METRIC.dwg` file).
2. Zoom to the Cabernet Court profile view (located inside a rectangle) and pick the Cabernet Court FG profile (the blue line) to activate its grips.
3. Locate the PVI around Sta. 8+50 (or 0+260 for metric users) and pick the vertical triangular grip on the vertical sag curve to begin a grip stretch of the PVI, as shown in Figure 7.27.

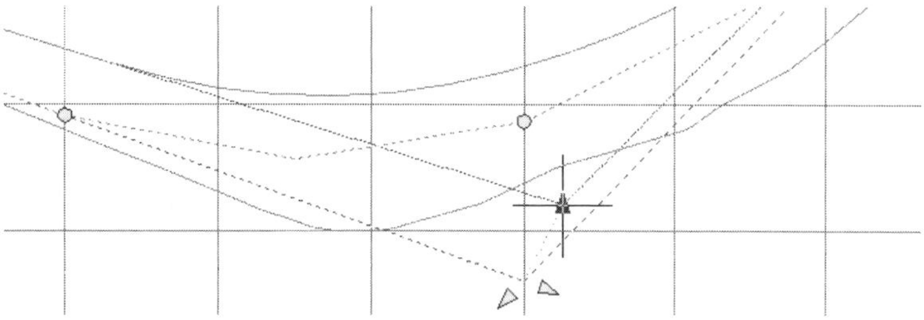

FIGURE 7.27
Grip-editing a PVI

The command line states `Specify stretch point or [Base point Copy Undo eXit]:`.

4. On the Transparent Commands toolbar, select the Profile Station Elevation command. That's **'PSE** for the command-line users.
5. At the `Select a profile view:` prompt, pick a grid line on the Cabernet Court profile view to select.
6. At the `Specify station:` prompt, enter **835**↵ (or **255**↵ for metric users).
7. At the `Specify elevation:` prompt, enter **788**↵ (or **240.5**↵ for metric users).

When this exercise is complete, you may close the drawing. A finished copy of this drawing is available from the book's web page with the filename `GripEditingProfiles_FINISHED.dwg` or `GripEditingProfiles_METRIC_FINISHED.dwg`.

The grips can go from quick-and-dirty editing tools to precise editing tools when you use them in conjunction with the transparent commands in the profile view. They lack the ability to precisely control a curve length, though, so you'll look at editing a curve next.

Parameter and Panorama Profile Editing

Beyond the simple grip edits, but before changing out the components of a typical profile, you can modify the values that generate an individual component. In this exercise, you'll use the Profile Layout Parameter dialog and the Panorama to modify the curve properties on your design profile:

1. Open the `ParameterEditingProfiles.dwg` file (or the `ParameterEditingProfiles_METRIC.dwg` file).
2. Zoom to the Cabernet Court profile view (located inside a rectangle) and pick the Cabernet Court FG profile (the blue line) to activate the Profile contextual tab.
3. From the Profile contextual tab ➤ Modify Profile panel, choose the Geometry Editor.

4. On the Profile Layout Tools toolbar, click the Profile Layout Parameters button to open the Profile Layout Parameters dialog and place the dialog somewhere on your screen so that you can still see the profile view.

5. On the Profile Layout Tools toolbar, click the Select PVI button.

If the Select Entity button is showing on the toolbar instead, from the PVI or Entity Based drop-down button select the PVI Based option.

6. Zoom in to click near the PVI at station 11+65 (or 0+355 for metric users) to populate the Profile Layout Parameters dialog (Figure 7.28).

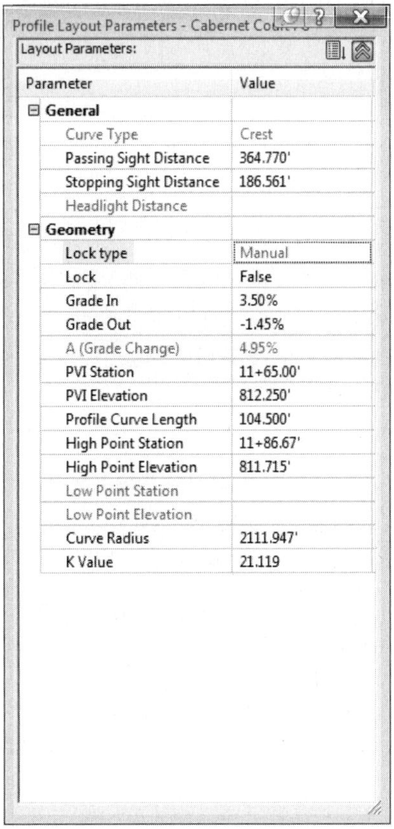

FIGURE 7.28
The Profile Layout Parameters dialog

Values that can be edited are in black; the rest, shown grayed out, are mathematically derived and can be of some design value but can't be directly modified. The two buttons at the top of the dialog adjust how much information is displayed. The one on the left is the Show More/Show Less button and the one on the right is the Collapse All Categories/Expand All Categories button.

7. In the Profile Layout Parameters dialog, change the K Value to **19**↵ (or **7**↵ for metric users). Notice that the curve changes but the label does not update. This is because you are still in the command. Once you end the command, all appropriate labels will update.

8. Click near the PVI at station 10+35 (or 0+315.5 for metric users) to repopulate the Profile Layout Parameters dialog with a different curve.

9. In the Profile Layout Parameters dialog, change Profile Curve Length to **130**↵ (or **42**↵ for metric users) and press ↵ or right-click to end the command and apply the changes to the profile.

10. Close the Profile Layout Parameters dialog by clicking the X in the upper-right corner.

11. On the Profile Layout Tools toolbar, click the Profile Grid View tool to activate the Profile Entities tab in Panorama.

Panorama allows you to view all the profile components at once, in a compact form.

12. Scroll right in Panorama until you see the Profile Curve Length column.

You can show and hide columns by right-clicking on one of the column headings. You can also resize the columns by dragging on the breaks between the columns or by double-clicking on the break between columns to auto-size to the column contents.

13. Double-click in the cell for the Entity 5 value in the Profile Curve Length column (see Figure 7.29), and change the value from 150′ to **250′** (or from 45 to **78** for metric users).

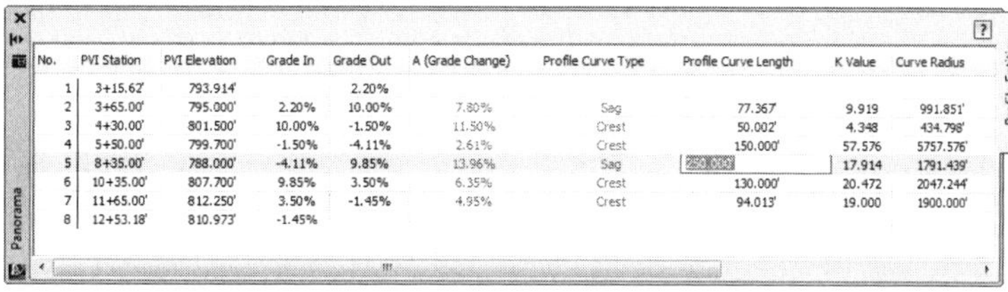

No.	PVI Station	PVI Elevation	Grade In	Grade Out	A (Grade Change)	Profile Curve Type	Profile Curve Length	K Value	Curve Radius
1	3+15.62'	793.914'		2.20%					
2	3+65.00'	795.000'	2.20%	10.00%	7.80%	Sag	77.367'	9.919	991.851'
3	4+30.00'	801.500'	10.00%	-1.50%	11.50%	Crest	50.002'	4.348	434.798'
4	5+50.00'	799.700'	-1.50%	-4.11%	2.61%	Crest	150.000'	57.576	5757.576'
5	8+35.00'	788.000'	-4.11%	9.85%	13.96%	Sag	250.000	17.914	1791.439'
6	10+35.00'	807.700'	9.85%	3.50%	6.35%	Crest	130.000'	20.472	2047.244'
7	11+65.00'	812.250'	3.50%	-1.45%	4.95%	Crest	94.013'	19.000	1900.000'
8	12+53.18'	810.973'	-1.45%						

FIGURE 7.29
Direct editing of the curve length in Panorama

14. Close Panorama and the Profile Layout Tools toolbar, and zoom out to review your edits.

Your complete profile should now look like Figure 7.30.

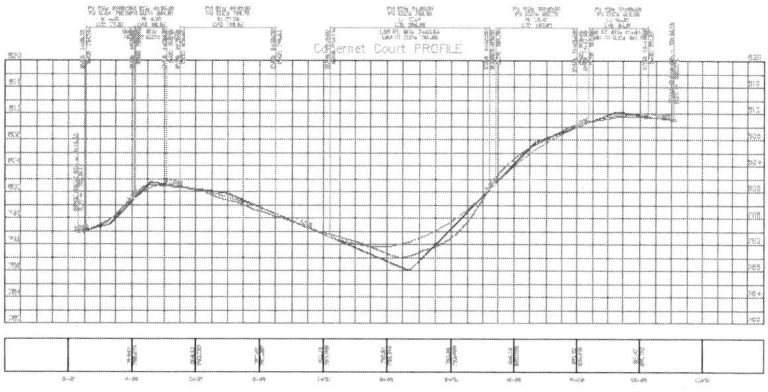

FIGURE 7.30
The completed editing of the curve length in the layout profile

When this exercise is complete, you may close the drawing. A finished copy of this drawing is available from the book's web page with the filename `ParameterEditingProfiles_FINISHED.dwg` or `ParameterEditingProfiles_METRIC_FINISHED.dwg`.

You can use these tools to modify the PVI points or tangent parameters, but they won't let you add or remove an entire component. You'll do that in the next section.

PVIs in Lockdown

You may have noticed in Figure 7.29 that the last column is named Lock. You can also lock a PVI at a specific station and elevation in the Profile Layout Parameters dialog. PVIs that are locked cannot be moved with edits to adjacent entities. However, it's important to note that a PVI can be unlocked with editing grips by clicking on the PVI in the profile view.

Component-Level Editing

In addition to editing basic parameters and locations, sometimes you have to add or remove entire components. In this exercise, you'll delete a PVI, remove a curve from an area in order to revise a nearby PVI location, and insert a new curve into the layout profile:

1. Open the `ComponentEditingProfiles.dwg` file (or the `ComponentEditingProfiles_METRIC.dwg` file).
2. Zoom to the Cabernet Court profile view (located inside a rectangle) and pick the Cabernet Court FG profile (the blue line) to activate the Profile contextual tab.
3. From the Profile contextual tab ➤ Modify Profile panel, choose the Geometry Editor to display the Profile Layout Tools toolbar.

4. On the Profile Layout Tools toolbar, click the Delete PVI button.
5. Pick a point near the 4+30 station (or the 0+131 station for metric users) and right-click or press ↵ to end the command.

The profile is adjusted accordingly. The vertical curves that were in place are also modified to accommodate this new geometry.

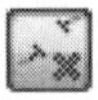

6. On the Profile Layout Tools toolbar, click the Delete Entity button.
7. Zoom in, and pick the curve entity near station 3+65 (or 0+112.8 for metric users) to delete it and right-click or press ↵ to end the command.

Notice that the incoming and outgoing tangents remain.

8. Using one of the methods discussed previously, change the PVI currently at Station 5+50 (or 0+167.5 for metric users) to PVI Station **4+70** and PVI Elevation **802.30** (or PVI Station **0+143** and PVI Elevation **244.650** for metric users).

9. On the Profile Layout Tools toolbar, click the drop-down arrow next to the Vertical Curve Creation button.
10. Select the More Free Vertical Curves ➤ Free Vertical Parabola (PVI Based) option.
11. Pick near the PVI where you removed the vertical curve in step 7 to re-add a curve.
12. At the `Specify curve length or [Passthrough K]:` prompt, enter **50**↵ (or **25**↵ for metric users) at the command line to set the curve length.

13. Right-click or press ↵ to end the command and update the profile display.

14. Close the Profile Layout Tools toolbar.

Your drawing should look similar to Figure 7.31.

FIGURE 7.31
The completed editing of the curve using component-level editing

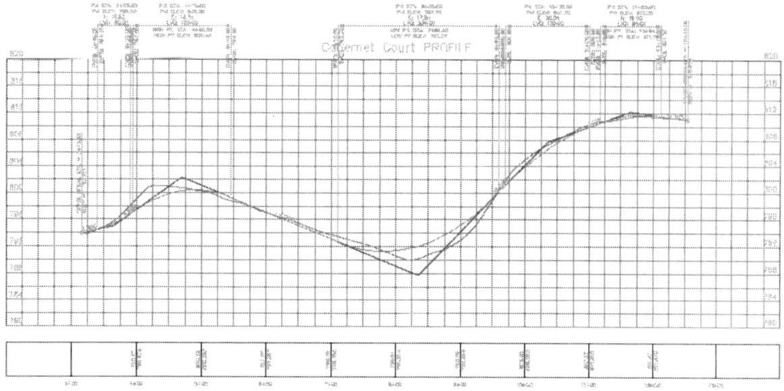

When this exercise is complete, you may close the drawing. A finished copy of this drawing is available from the book's web page with the filename `ComponentEditingProfiles_FINISHED.dwg` or `ComponentEditingProfiles_METRIC_FINISHED.dwg`.

Editing profiles using any of these methods gives you precise control over the creation and layout of your vertical design. In addition to these tools, some of the tools on the Profile Layout Tools toolbar are worth investigating and somewhat defy these categories. You'll look at them next.

Other Profile Edits

Some handy tools exist on the Profile Layout Tools toolbar for performing specific actions. These tools aren't normally used during the preliminary design stage, but they come into play as you're working to create a final design for grading or corridor design. They include raising or lowering a whole layout in one shot, as well as copying profiles. Try this exercise:

1. Open the `OtherProfileEdits.dwg` file (or the `OtherProfileEdits_METRIC.dwg` file).

2. Zoom to the Cabernet Court profile view (located inside a rectangle) and pick the Cabernet Court FG profile (the blue line) to activate the Profile contextual tab.

3. From the Profile contextual tab ➤ Modify Profile panel, choose the Geometry Editor to display the Profile Layout Tools toolbar.

4. On the Profile Layout Tools toolbar, click the Copy Profile button to display the Copy Profile Data dialog, shown in Figure 7.32.

5. Click OK to create a new layout profile directly on top of Cabernet Court FG.

FIGURE 7.32
The Copy Profile Data dialog

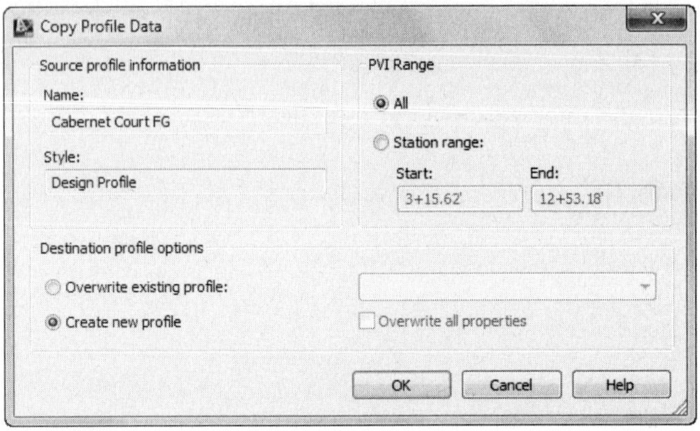

6. In Prospector, expand the Alignments ➤ Centerline Alignments ➤ Cabernet Court ➤ Profiles branch to see that a profile named Cabernet Court FG [Copy] has been added.

7. Select the Cabernet Court FG [Copy] profile in the profile view on the screen to have the Profile Layout Tools toolbar reference the new profile.

8. On the Profile Layout Tools toolbar, click the Raise/Lower PVIs button to display the Raise/Lower PVI Elevation dialog, shown in Figure 7.33.

FIGURE 7.33
The Raise/Lower PVI Elevation dialog

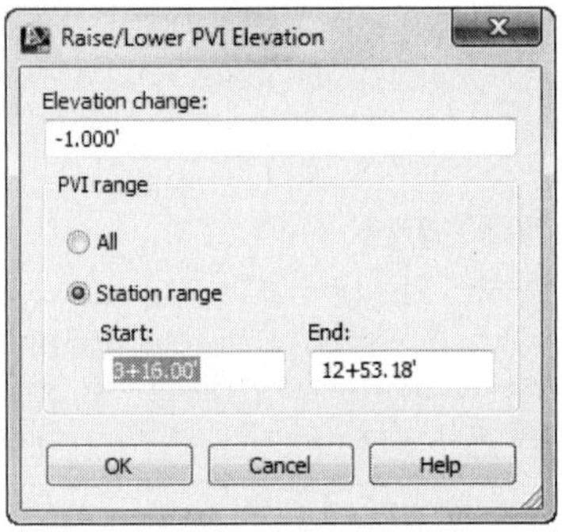

9. Set Elevation Change to **-1** (or **-0.3** for metric users).

10. Click the Station Range radio button, and set the Start value to **3+16** (or **0+097** for metric users) and leave the default End value to modify all the PVIs after the starting PVI.

11. Click OK to dismiss the Raise/Lower PVI Elevation dialog.

12. Close the Profile Layout Tools toolbar.

When this exercise is complete, you may close the drawing. A finished copy of this drawing is available from the book's web page with the filename `OtherProfileEdits_FINISHED.dwg` or `OtherProfileEdits_METRIC_FINISHED.dwg`.

Generating a copy is useful if you want to remember a conceptual profile layout but would like to experiment with a different layout. The copies do not stay dynamically related to one another.

Using the layout and editing tools discussed in these sections, you should be able to design and draw any combination of profile information presented to you.

Matching Profile Elevations at Intersections

Up to now, you have learned how to use some of the available tools for modifying profiles, but you might be wondering about the intersecting roads and how the profiles will interact with one another. Have no fear; we are going to discuss those later in Chapter 10 when we discuss corridor intersections.

Profile Views

CERT OBJECTIVE

Working with vertical data is an integral part of building the model. Once profile information has been created in any number of ways, displaying it to make sense is another task. It can't be stated enough that profiles and profile views are not the same thing. The profile view displays the profile data. A single profile can be shown in an infinite number of views, with different grids, exaggeration factors, labels, or linetypes. In this section, you'll look at the various methods available for creating profile views.

Creating Profile Views during Sampling

The easiest way to create a profile view is to draw it as an extended part of the surface sampling procedure as shown in the first exercise. By combining the profile sampling step with the creation of the profile view, you have avoided one more trip to the menus. This is the most common method of creating a profile view, but we'll look at a manual creation in the next section.

Creating Profile Views Manually

Once an alignment has profile information associated with it, any number of profile views might be needed to display the proper information in the right format. To create a second, third, or even tenth profile view once the sampling is done, you must use a manual creation method. In this exercise, you'll create a profile view manually for an alignment that already has a surface-sampled profile associated with it:

1. Open the `ProfileViews.dwg` file (or the `ProfileViews_METRIC.dwg` file).

2. From the Home tab ➤ Profile & Section Views panel, choose Profile View ➤ Create Profile View to display the Create Profile View wizard.

This is the same wizard that was discussed in the surface sampling example.

3. In the Select Alignment text box, select Cabernet Court from the drop-down list.
4. Set the Profile View name as **Cabernet Court Full Grid**.
5. In the Profile View Style drop-down list, select the Full Grid style.
6. Click the Create Profile View button and pick a point on screen to draw the profile view, as shown in Figure 7.34.

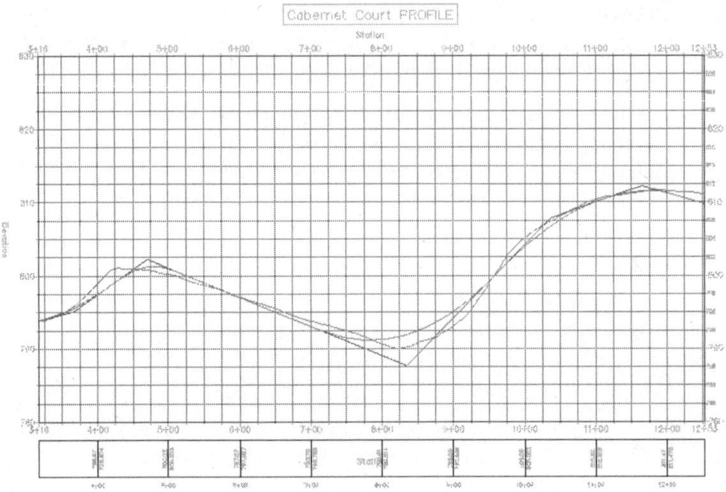

FIGURE 7.34
The completed profile view of Cabernet Court using the Full Grid profile view style

When this exercise is complete, you may close the drawing. A finished copy of this drawing is available from the book's web page with the filename `ProfileViews_FINISHED.dwg` or `ProfileViews_METRIC_FINISHED.dwg`.

Using these two creation methods, you've made short, simple profile views, but in the next exercise we will look at a longer alignment, as well as some more of the options available in the Create Profile View wizard.

SO YOU WANT TO DELETE A PROFILE VIEW?

Getting rid of a profile view is easy, but be careful not to inadvertently delete your profiles at the same time. If you use a crossing window on your profile view and delete, then you will also be deleting your profiles that hold the data shown in your profile view.

The easiest way to get rid of a profile view is to click on one of the grid lines to activate the profile view object and press the Delete key on your keyboard.

Alternatively, you can delete the profile view by expanding the branches in Prospector until you see the profile view you want to get rid of and then right-clicking on it and selecting the Delete option.

Splitting Views

Dividing up the data shown in a profile view can be time consuming. The Profile View wizard is used for simple profile view creation, but the wizard can also be used to create manually limited profile views, staggered (or stepped) profile views, multiple profile views with gaps between the views, and stacked profiles (aka three-line profiles). You'll look at these variations on profile view creation in this section.

Creating Manually Limited Profile Views

Continuous profile views like you made in the exercises prior to this point work well for design purposes, but they are often unusable for plotting or exhibiting purposes. In this exercise, you'll use the Profile View wizard to create a manually limited profile view. This variation will allow you to control how long and how high each profile view will be, thereby making the views easier to plot.

1. Open the `ProfileViewsSplit.dwg` file (or the `ProfileViewsSplit_METRIC.dwg` file).
2. From the Home tab ➢ Profile & Section Views panel, choose Profile View ➢ Create Profile View to display the Create Profile View wizard.
3. Verify that the Select Alignment drop-down list shows Frontenac Drive, Profile Name is set to **Frontenac Drive Limited Full Grid**, and Full Grid is selected in the Profile View Style drop-down list; then click Next.
4. On the Station Range wizard page, select the User Specified Range radio button.
5. Enter **0** for the Start station and **800** for the End station (or **0** and **245** for metric users), as shown in Figure 7.35. It isn't necessary to include the + when entering station data.

Notice the preview picture now shows a clipped portion of the total profile.

6. Click Next.

FIGURE 7.35
The start and end stations for the user-specified profile view

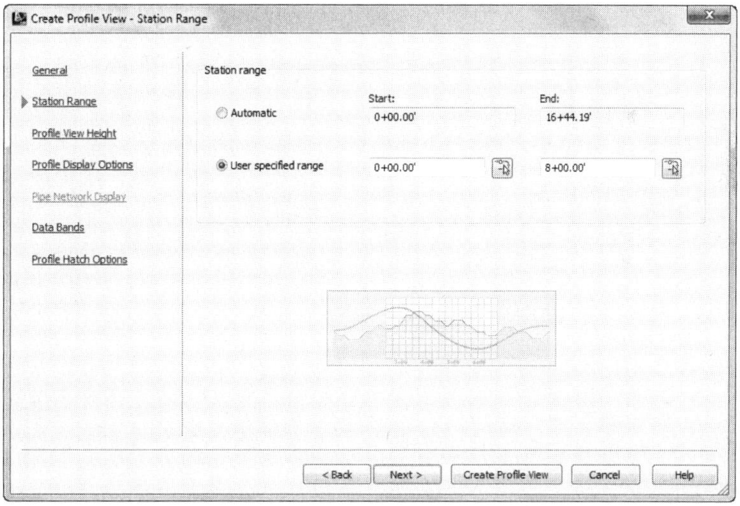

7. On the Profile View Height wizard page, select the User Specified radio button.
8. Set the Minimum height to **780** and the Maximum height to **820** (or **238** and **250** for metric users). It isn't necessary to include the foot mark (') or m for meters when entering elevations.

9. Click the Create Profile View button and pick a point on screen to draw the profile view.

Your screen should look similar to Figure 7.36. When this exercise is complete, you may save and keep the drawing open to continue on to the next exercise. If you would like to view the result of this exercise, it is included in the finished drawing (along with the finished portions of the next two exercises) available from the book's web page (`ProfileViewsSplit_FINISHED.dwg` or `ProfileViewsSplit_METRIC_FINISHED.dwg`).

FIGURE 7.36
Applying user-specified station and height values to a profile view

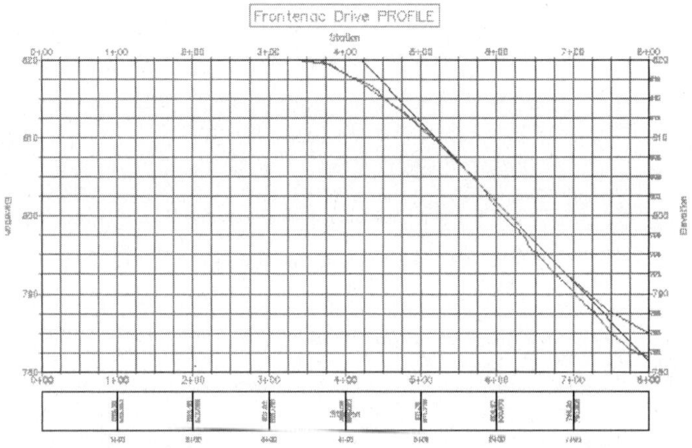

CREATING STAGGERED PROFILE VIEWS

When large variations occur in profile height, the profile view must often be split just to keep from wasting much of the page with empty grid lines. In this exercise, you use the Profile View wizard to create a staggered, or stepped, view:

1. You may continue using the file from the previous exercise or start this exercise with the `ProfileViewsSplit.dwg` file (or the `ProfileViewsSplit_METRIC.dwg` file).
2. From the Home tab ➤ Profile & Section Views panel, choose Profile View ➤ Create Profile View to display the Create Profile View wizard.
3. Verify that the Select Alignment drop-down list shows Syrah Way, Profile Name is set to **Syrah Way Staggered Full Grid**, and Full Grid is selected in the Profile View Style drop-down list; then click Next.
4. Verify that Station Range is set to Automatic to allow the view to show the full length, and click Next.
5. In the Profile View Height field, select the User Specified option and set the values to **790.00′** and **805.00′** (or **240** and **246** for metric users), as shown in Figure 7.37. These heights are only important in that they set the height of the profile view (15′ or 6 m in this case).
6. Check the Split Profile View option and set the view styles to First View, Intermediate View, and Last View, as shown in Figure 7.37.

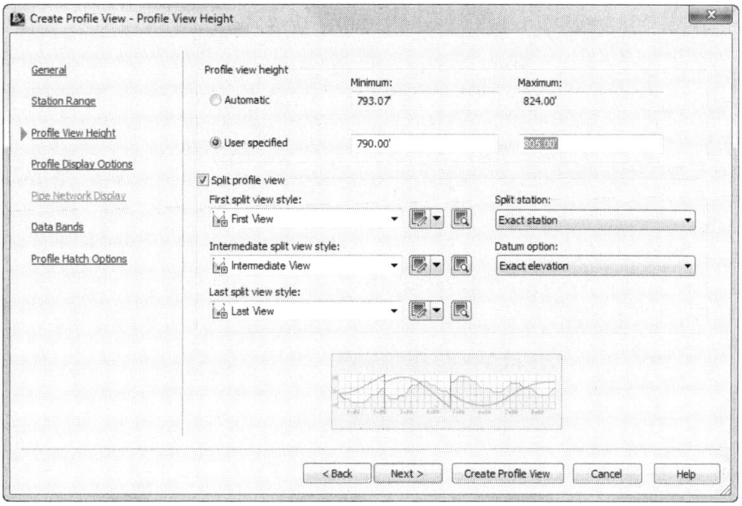

FIGURE 7.37
Split Profile View settings

7. Click the Create Profile View button and pick a point on screen to draw the staggered display, as shown in Figure 7.38.

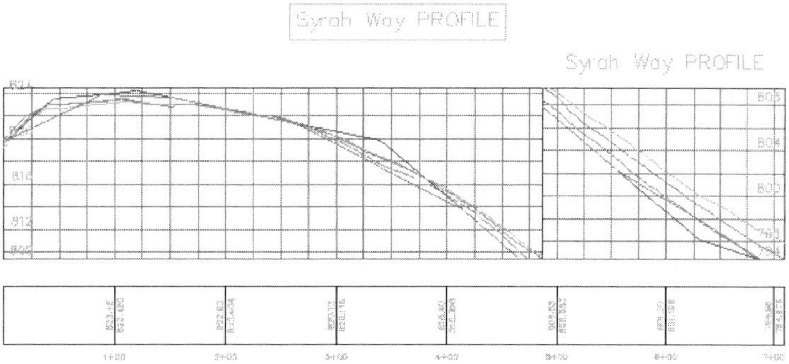

FIGURE 7.38
A staggered (stepped) split profile view created via the wizard

The profile view is split into views according to the settings that were selected in the Create Profile View wizard in step 6. The first portion of the profile view shows the profile from 0 to the station where the elevation change of the profile exceeds the limit for height. The next portion of the profile view displays the same, and so on for the rest of the profile. Each of these portions is part of the same profile view and can be adjusted via the Profile View Properties dialog.

When this exercise is complete, you may save and keep the drawing open to continue on to the next exercise. If you would like to view the result of this exercise, it is included in the finished drawing (along with the finished portions of the previous exercise and next exercise) available from the book's web page (`ProfileViewsSplit_FINISHED.dwg` or `ProfileViewsSplit_METRIC_FINISHED.dwg`).

Creating Gapped Profile Views

Profile views must often be limited in length and height to fit a given sheet size. Gapped views are a way to show the entire length and height of the profile, by breaking the profile into different sections with "gaps" or spaces between each view.

When you are using the Plan and Production tools (covered in Chapter 17, "Plan Production"), the gapped profile views are automatically created.

In this exercise, you will use a variation of the Create Profile View wizard called the Create Multiple Profile Views wizard to create gapped views automatically:

1. You may continue using the file from the previous exercise or start this exercise with the `ProfileViewsSplit.dwg` file (or the `ProfileViewsSplit_METRIC.dwg` file).

2. From the Home tab ➢ Profile & Section Views panel, choose Profile View ➢ Create Multiple Profile Views to display the Create Multiple Profile Views wizard.
3. Verify that the Select Alignment drop-down list shows Frontenac Drive, Profile Name is set to **Frontenac Drive Gapped Full Grid**, and Full Grid is selected in the Profile View Style drop-down list, as shown in Figure 7.39; then click Next.

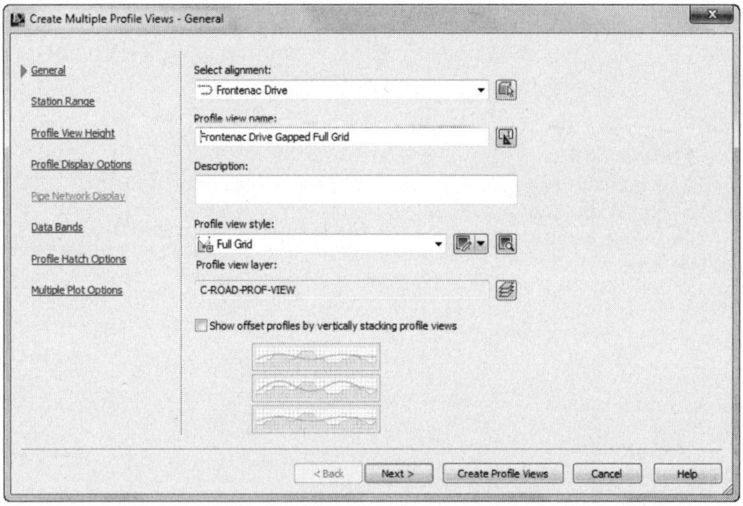

Figure 7.39
The Create Multiple Profile Views – General wizard page

4. On the Station Range wizard page, verify that the Automatic option is selected.
5. Set the length of each view to **650** (or **200** for metric users), and click Next.
6. On the Profile View Height wizard page, verify that the Automatic option is selected.

Note that you could use the Split Profile View options from the previous exercise here as well if you use the User Specified profile view height.

7. Click Next.
8. On the Profile Display Options wizard page, scroll across until you get to the Labels column and verify Style is set to _No Labels on both profiles; then click Next.

9. On the Data Bands wizard page, verify that the band set is EG-FG Elevations And Stations, and click the Multiple Plot Options link in the left sidebar of the wizard to jump ahead to that wizard page. We will look at the data bands in further depth a little later in this chapter.

The Multiple Plot Options wizard page shown in Figure 7.40 is unique to the Create Multiple Profile Views wizard. This wizard page step controls whether the gapped profile views will be arranged in a column, a row, or a grid. The Frontenac Drive alignment is fairly short, so the gapped views will be aligned in a row. However, it could be prudent with longer alignments to stack the profile views in a column or a compact grid, thereby saving screen space.

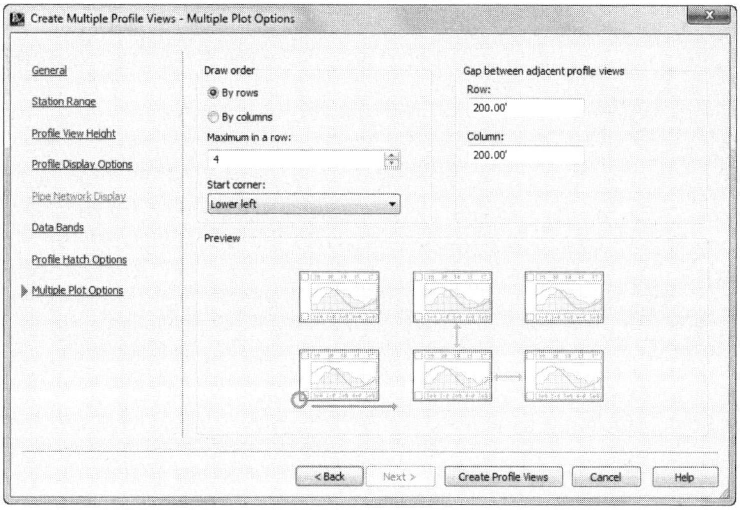

FIGURE 7.40 The Create Multiple Profile Views – Multiple Plot Options wizard page

10. Click the Create Profile Views button and pick a point on screen to create a view similar to Figure 7.41.

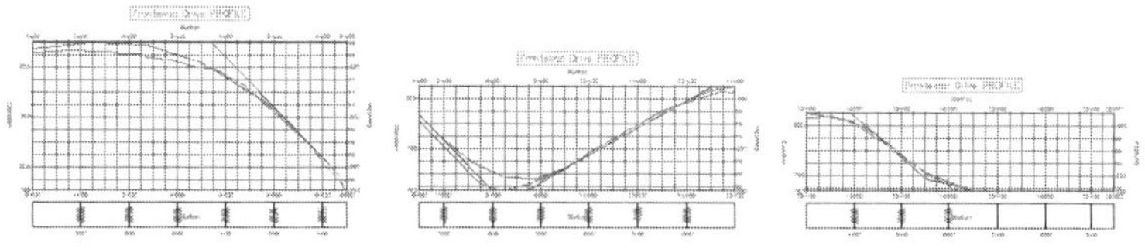

FIGURE 7.41 The staggered and gapped profile views of the Frontenac alignment

The gapped profile views are the three profile views on the bottom of the screen and, just like the staggered profile view, show the entire alignment from start to finish. Unlike the staggered view, however, the gapped view is separated by a "gap" into three profile views. In addition, the gapped profile views are independent of each other so they can be modified to have their own styles, properties, and labeling associated with them, making them useful when you don't want a view to show information that is not needed on a particular section. This is also the primary way to create divided profile views for sheet production.

When this exercise is complete, you may close the drawing. A finished copy of this drawing showing the results of the previous three exercises is available from the book's web page with the filename `ProfileViewsSplit_FINISHED.dwg` or `ProfileViewsSplit_METRIC_FINISHED.dwg`.

Creating Stacked Profile Views

In some parts of the United States, a three-line profile view is a common requirement. In this situation, the centerline is displayed in a central profile view, with left and right offsets shown in profile views above and below the centerline profile view. These are then typically used to show top-of-curb design profiles in addition to the centerline design. In this exercise, you look at how the Create Profile View wizard makes generating these stacked views a simple process:

1. Open the `StackedProfiles.dwg` file (or the `StackedProfiles_METRIC.dwg` file). This drawing has sampled profiles for the Syrah Way alignment at center as well as left and right offsets.
2. From the Home tab ➢ Profile & Section Views panel, choose Profile View ➢ Create Profile View to display the Create Profile View wizard.
3. Verify that the Select Alignment drop-down list shows Syrah Way, Profile Name is set to **Syrah Way Stacked Full Grid**, and Full Grid is selected in the Profile View Style drop-down list.
4. Check the Show Offset Profiles By Vertically Stacking Profile Views option on the General wizard page.

   Notice that by checking this box an additional link named Stacked Profile is added to the left sidebar of the wizard.
5. Click Next.
6. On the Create Profile View – Station Range wizard page, verify that the Station Range is set to Automatic, and click Next.
7. On the Create Profile View – Profile View Height wizard page, verify that Profile View Height is set to Automatic, and click Next.
8. On the Create Profile View – Stacked Profile wizard page, set the gap between views to **10** (or **4** for metric users).
9. Set the view styles to Top Stacked View, Middle Stacked View, and Bottom Stacked View, as shown in Figure 7.42.
10. Set the Gap Between Views value to **15** (or **5** for metric users) and click Next.
11. On the Create Profile View – Profile Display Options wizard page, select Top View in the Select Stacked View To Specify Options For list box.
12. Toggle the Draw option for the left offset profile (EG - Syrah Way -25.000 or EG - Syrah Way -7.500), as shown in Figure 7.43. If you need to widen the columns, you can do so by clicking the line between the column headings.

    Remember that the negative offset denotes a left profile whereas a positive offset denotes a right profile.

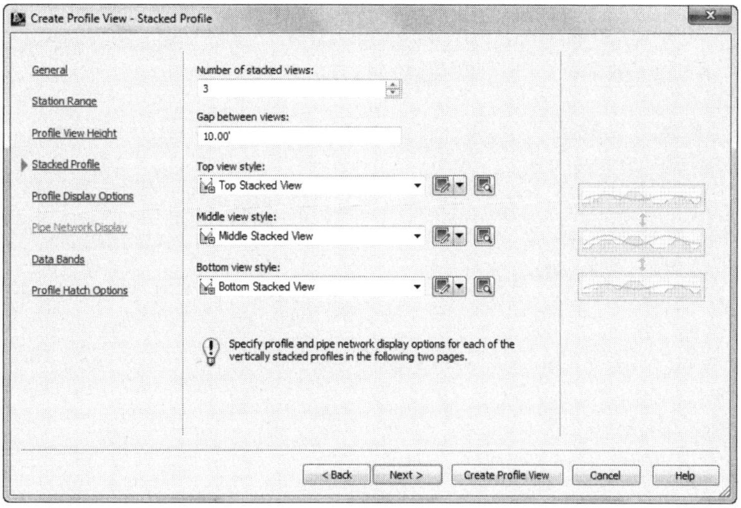

FIGURE 7.42
The Create Profile View – Stacked Profile wizard page

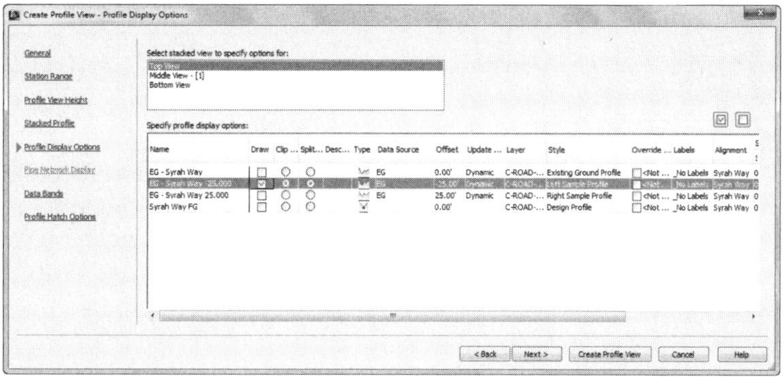

FIGURE 7.43
Setting the stacked view options for each view

13. Select Middle View - [1] in the Select Stacked View To Specify Options For list box.
14. Toggle the Draw option for the sampled centerline profile (EG - Syrah Way) as well as the layout centerline profile (Syrah Way FG).
15. Select Bottom View in the Select Stacked View To Specify Options For list box.
16. Toggle the Draw option for the right offset profile (EG - Syrah Way 25.000 or EG - Syrah Way -7.500) profile, and click Next.
17. On the Data Bands wizard page, verify that the band set is set to EG-FG Elevations And Stations.
18. Click the Create Profile View button and pick a point on the screen to draw the stacked profiles, as shown in Figure 7.44.

When this exercise is complete, you may close the drawing. A finished copy of this drawing is available from the book's web page with the filename StackedProfiles_FINISHED.dwg or StackedProfiles_METRIC_FINISHED.dwg.

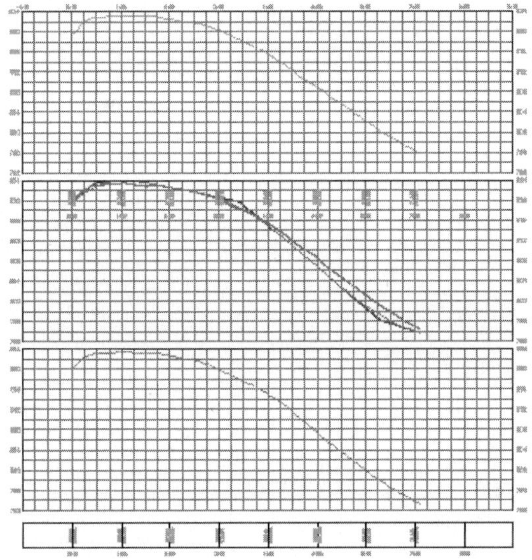

FIGURE 7.44
Completed stacked profiles

Like the gapped profile views that you generated in a previous exercise, the profile views are independent of one another so they can be modified to have their own styles, properties, and labeling associated with them. The stacking here simply automates a process that many users previously found tedious. At this point you do not have finished grade information at the offsets, but you can add it to these views later by editing the Profile View Properties for those profile views.

When you create a profile, the profile will automatically be visible in all profile views based on that same alignment by default. In the Profile View Properties dialog, you can always turn the Draw option off for any profile that you don't want visible in a profile view.

STYLES: WHERE TO LOOK

The exercises in this chapter have many different styles created to show variety. You'll learn more about styles in Chapter 20, "Label Styles," and Chapter 21. It's okay to take a peek ahead once in a while.

Editing Profile Views

Once profile views have been created, things get interesting. Any number of modifications to the view itself can be applied, even before editing the styles, which makes profile views one of the most flexible objects of the AutoCAD Civil 3D package. In this series of exercises, you'll look at a number of changes that can be applied to any profile view in place.

Profile View Properties

Picking a profile view from the Profile View contextual tab ➤ Modify View panel and choosing Profile View Properties yields the dialog shown in Figure 7.45. The properties of a profile include the style applied, station and elevation limits, the number of profiles displayed, and the bands associated with the profile view. If a pipe network is displayed, a tab labeled Pipe Networks will appear. These tabs should look very similar to the links in the sidebar of the Create Profile View wizard.

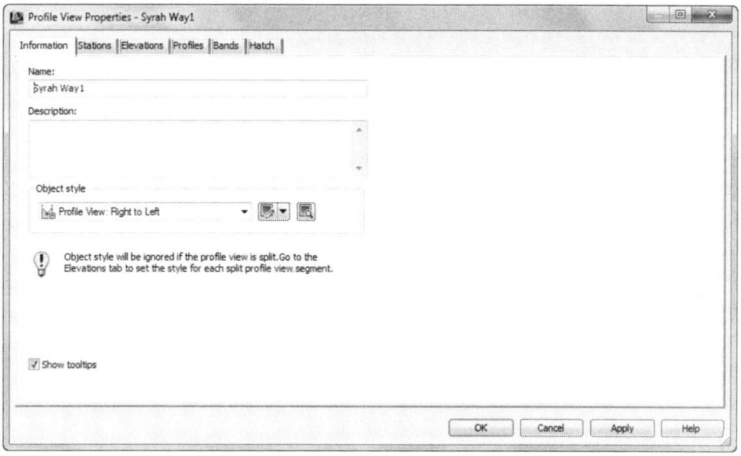

FIGURE 7.45
Typical Profile View Properties dialog

ADJUSTING THE PROFILE VIEW STATION LIMITS

In spite of the wizard, there are often times when a profile view needs to be manually adjusted. For example, the most common change is to limit the length of the profile view that is being shown so it fits on a specific size of paper or viewport. You can make some of these changes during the initial creation of a profile view (as shown in a previous exercise), but you can also make changes after the profile view has been created.

One way to do this is to use the Profile View Properties dialog to make changes to the profile view. The profile view is an AutoCAD Civil 3D object, so it has properties and styles that can be adjusted in this dialog to make the profile view look like you need it to.

1. Open the `ProfileViewProperties.dwg` file (or the `ProfileViewsProperties_METRIC.dwg` file).
2. Zoom to the Cabernet Court Full Grid profile view (located inside a rectangle labeled "Cabernet Court Adjust Stations").
3. Pick a grid line, and from the Profile View contextual tab ➤ Modify View panel, choose Profile View Properties to display the Profile View Properties dialog.
4. On the Stations tab, click the User Specified Range radio button, and set the value of the End station to **700** (or **200** for metric users), as shown in Figure 7.46. Note that you do not need to type the + symbol.

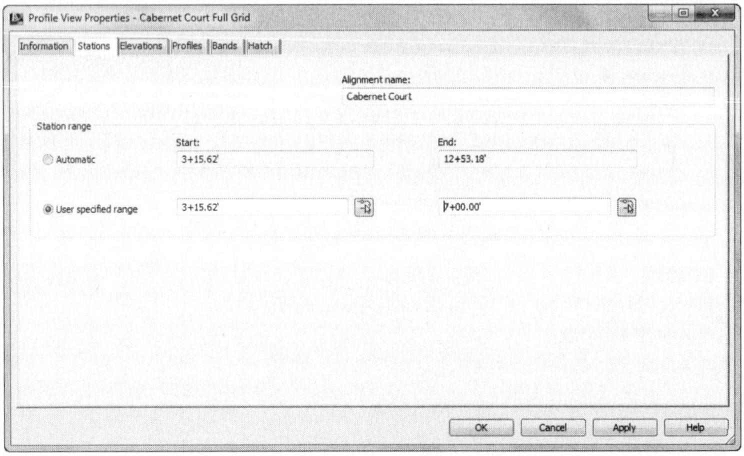

FIGURE 7.46
Adjusting the end station values for Cabernet Court

5. Click OK to dismiss the dialog.

The profile view will now reflect the updated end station value.

One of the nice things about Civil 3D is that copies of a profile view retain the properties of that view, making a gapped view easy to create manually if it was not created with the wizard.

6. Press F8 on your keyboard to enable Ortho mode.
7. Enter **Copy**↵ on the command line. Pick the Cabernet Court profile view you just modified. Make sure you are selecting the grid representing the profile view and not the linework that represents the profile.
8. Pick a base point and move the crosshairs to the right.
9. When the crosshairs reach a point where the two profile views do not overlap, pick that as your second point, and press ↵ to end the Copy command.
10. Pick the copy just created and from the Profile View contextual tab ➤ Modify View panel, choose Profile View Properties to display the Profile View Properties dialog.
11. On the Stations tab, change the stations again.
12. This time, set the Start field to **700** (or **200** for metric users) and the End field to **1253.18** (or **381.94** for metric users).
13. Click OK to dismiss the dialog.

The total length of the alignment will now be displayed on the two profile views, with a gap between the two views at station 7+00 (or 0+200 for metric users).

You may want to move the copied profile view since it held the station location, thus shifting it further right than originally.
Once this exercise is complete, your drawing will look like Figure 7.47.

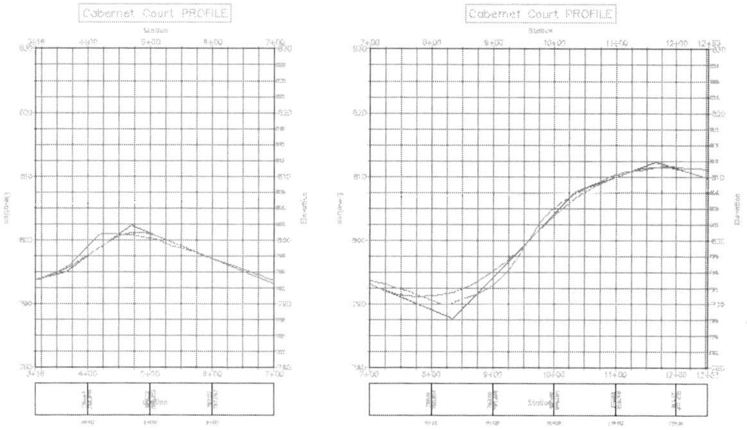

FIGURE 7.47
A manually created gap between profile views

When this exercise is complete, you may save and keep the drawing open to continue on to the next exercise. If you would like to view the result of this exercise, it is included in the finished drawing (along with the finished portions of the next two exercises) available from the book's web page (`ProfileViewProperties_FINISHED.dwg` or `ProfileViewProperties_METRIC_FINISHED.dwg`).

In addition to creating gapped profile views by changing the profile properties, you could show phase limits by applying a different style to the profile in the second view.

ADJUSTING THE PROFILE VIEW ELEVATIONS

Another common issue is the need to control the height of the profile view. Civil 3D automatically sets the datum and the top elevation of profile views on the basis of the data to be displayed. In most cases this is adequate, but in others, this simply creates a view too large for the space allocated on the sheet or does not provide the adequate room for layout PVIs to be laid out.

1. You may continue using the file from the previous exercise or start this exercise with the `ProfileViewProperties.dwg` file (or the `ProfileViewProperties_METRIC.dwg` file).
2. Zoom to the Syrah Way Full Grid profile view (located inside a rectangle labeled "Cabernet Court Adjust Elevation Step 2").
3. Pick a grid line, and from the Profile View contextual tab ➢ Modify View panel, choose Profile View Properties to display the Profile View Properties dialog.
4. On the Elevations tab, in the Elevation Range section, check the User Specified Height radio button and set the maximum height to **830** (or **255** for metric users), as shown in Figure 7.48.
5. Click OK to dismiss the dialog.

The profile view of Syrah Way should reflect the updated elevations, as shown in Figure 7.49.

The Elevations tab can also be used to split the profile view and create the staggered view that you previously created with the wizard.

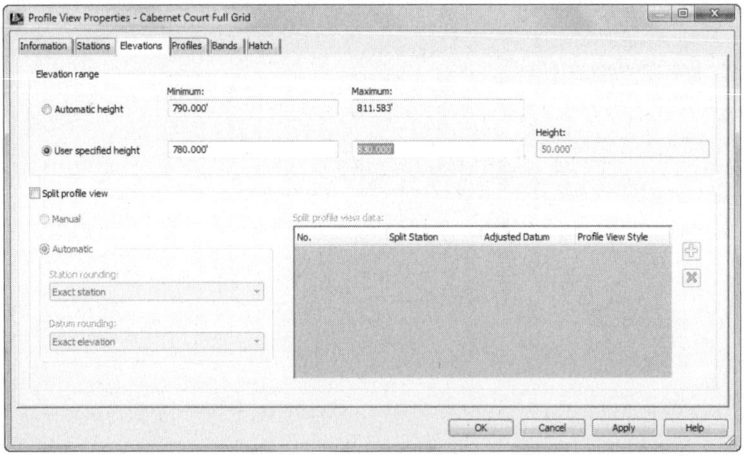

FIGURE 7.48
Modifying the height of the profile view

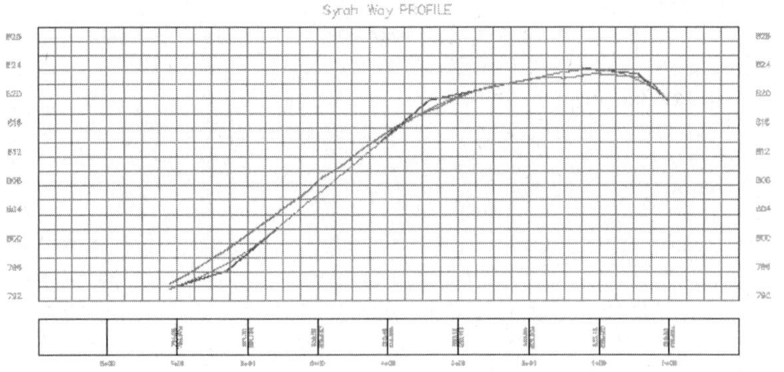

FIGURE 7.49
The updated profile view with the heights manually adjusted

6. Pick the Syrah Way Full Grid (1) profile view (located inside a rectangle labeled "Cabernet Court Adjust Elevations Step 6").
7. Right-click a grid line, and select the Profile View Properties option to open the Profile View Properties dialog.
8. Switch to the Elevations tab.
9. In the Elevations Range area, click the User Specified Height radio button.
10. Check the Split Profile View option, and verify that the Automatic radio button is selected.

Notice that the Height field is now active.

11. Set the Height to **16** (or **7** for metric users), as shown in Figure 7.50.
12. Click OK to exit the dialog.

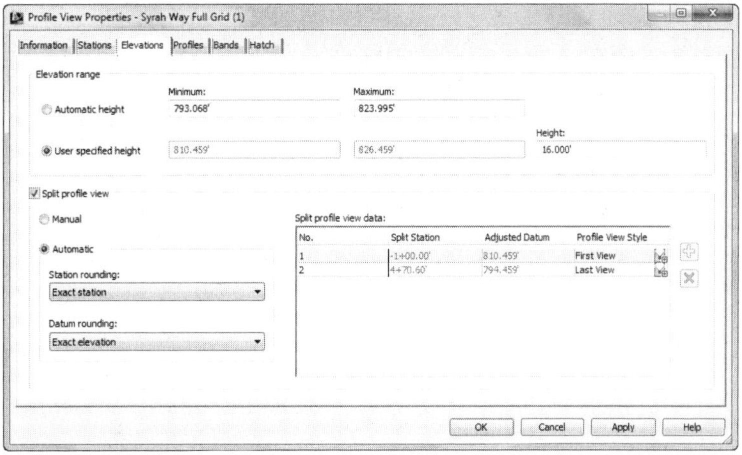

FIGURE 7.50
Defining a Split Profile View on the Elevations tab

13. Enter **REGEN** on the command line.

The profile view should look like Figure 7.51.

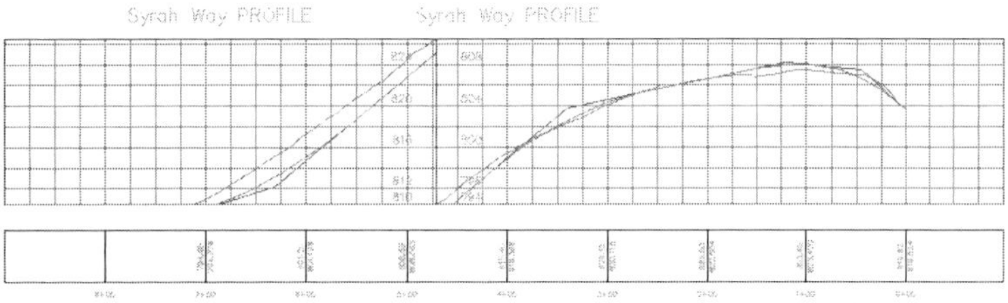

FIGURE 7.51
A split profile view for the Syrah Way alignment

When this exercise is complete, you may save and keep the drawing open to continue on to the next exercise. If you would like to view the result of this exercise, it is included in the finished drawing (along with the finished portions of the previous exercise and next exercise) available from the book's web page (`ProfileViewProperties_FINISHED.dwg` or `ProfileViewProperties_METRIC_FINISHED.dwg`).

Automatically creating split views is a good starting point, but you'll often have to tweak them as you've done here. The selection of the proper profile view styles is an important part of the Split Profile View process. We'll look at object styles in Chapter 21.

PROFILE DISPLAY OPTIONS

AutoCAD Civil 3D allows the creation of literally hundreds of profiles for any given alignment. Doing so makes it easy to evaluate multiple design solutions, but it can also mean that profile views get very crowded. In this exercise, you'll look at some profile display options that allow the toggling of various profiles within a profile view:

1. You may continue using the file from the previous exercise or start this exercise with the `ProfileViewProperties.dwg` file (or the `ProfileViewProperties_METRIC.dwg` file).

2. Pick the left Cabernet Court Full Grid profile view from the adjusting stations exercise (located inside a rectangle labeled "Cabernet Court Adjust Stations") to activate the Profile View contextual tab.
3. From the Profile View contextual tab ➤ Modify View panel, choose Profile View Properties to display the Profile View Properties dialog.
4. Switch to the Profiles tab.
5. Uncheck the Draw option in the EG Surface row and click OK.

Your profile view should look similar to Figure 7.52.

FIGURE 7.52
The Cabernet Court profile view with the Draw option toggled off for the EG profile

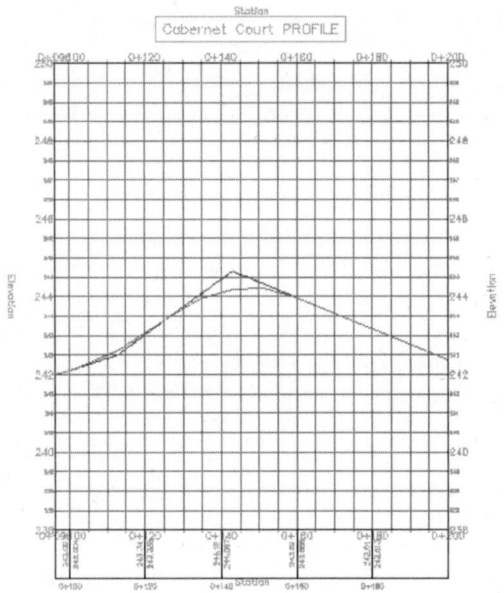

Toggling off the Draw option for the EG surface has created a profile view style in which a profile of the existing ground surface will not be drawn on the profile view.

The sampled profile from the EG surface still exists under the Cabernet Court alignment; it simply isn't shown in the current profile view.

When this exercise is complete, you may close the drawing. A finished copy of this drawing showing the results of the previous three exercises is available from the book's web page with the filename `ProfileViewProperties_FINISHED.dwg` or `ProfileViewProperties_METRIC_FINISHED.dwg`.

Now that you've modified a number of styles, let's look at another option that is available on the Profile View Properties dialog: bands.

PROFILE VIEW BANDS

Data bands are horizontal elements that display additional graphical and numerical information about the profile or alignment that is referenced in a profile view. Bands can be applied to both the top and bottom of a profile view, and there are six different band types:

Profile Data Bands Display information about the selected profile. This information can include simple elements such as elevation, or more complicated information such as the cut-fill between two profiles at the given station.

Vertical Geometry Bands Create an iconic view of the elements making up a profile. Typically used in reference to a design profile, vertical data bands make it easy for a designer to see where vertical curves are information about the tangents and vertical curves located along the alignment.

Horizontal Geometry Bands Create a simplified view of the horizontal alignment elements, giving the designer or reviewer information about line, curve, and spiral segments and their relative location to the profile data being displayed.

Superelevation Bands Display the various options for Superelevation values at the critical points along the alignment.

Sectional Data Bands Can display information about the sample line locations, the distance between them, and other sectional-related information.

Pipe Data Bands Can display specific information such as part, offset, elevation, or direction about each pipe or structure being shown in the profile view.

In this exercise, you'll add bands to give feedback on the EG and layout profiles, as well as horizontal and vertical geometry:

1. Open the `ProfileViewBands.dwg` file (or the `ProfileViewBands_METRIC.dwg` file).
2. Zoom to the Frontenac Drive profile view.
3. Pick a grid line, and from the Profile View contextual tab ➤ Modify View panel, choose Profile View Properties to display the Profile View Properties dialog.
4. On the Bands tab (Figure 7.53), in the List Of Bands area, verify that the Location drop-down is set to Bottom Of Profile View and notice that an Elevations And Stations band has already been set during the creation of this profile view.

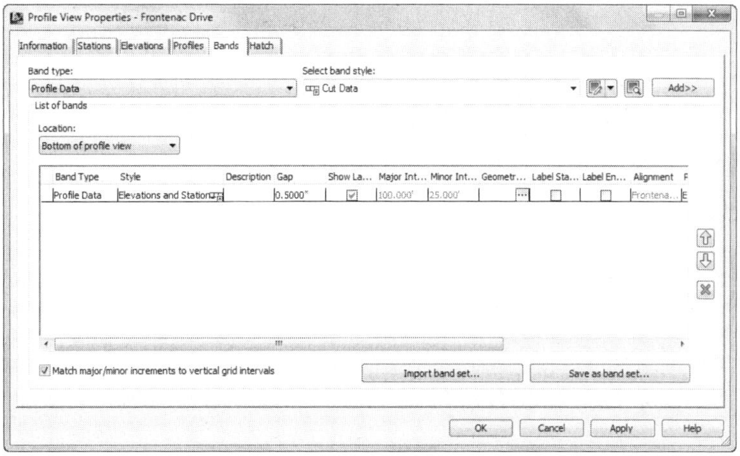

FIGURE 7.53
The Bands tab of the Profile View Properties dialog

Selecting the type of band from the Band Type drop-down menu changes the Select Band Style drop-down menu to include styles that are available for that band type. Next to the Select Band Style drop-down menu are the usual Style Edit/Copy button and a preview button. Once you've selected a style from the Select Band Style drop-down, clicking the Add button places it on the profile. The middle portion of this dialog allows you to switch between the bands shown at the Bottom of Profile View or the Top of Profile View; you'll look at that in a moment.

5. Change the Band Type drop-down to the Profile Data option and choose the Cut Data option from the Select Band Style drop-down.
6. Click the Add button to add the Profile Data band to display the Geometry Points To Label In Band dialog shown in Figure 7.54.

FIGURE 7.54
The Geometry Points To Label In Band dialog showing the Alignment Points tab (left) and the Profile Points tab (right)

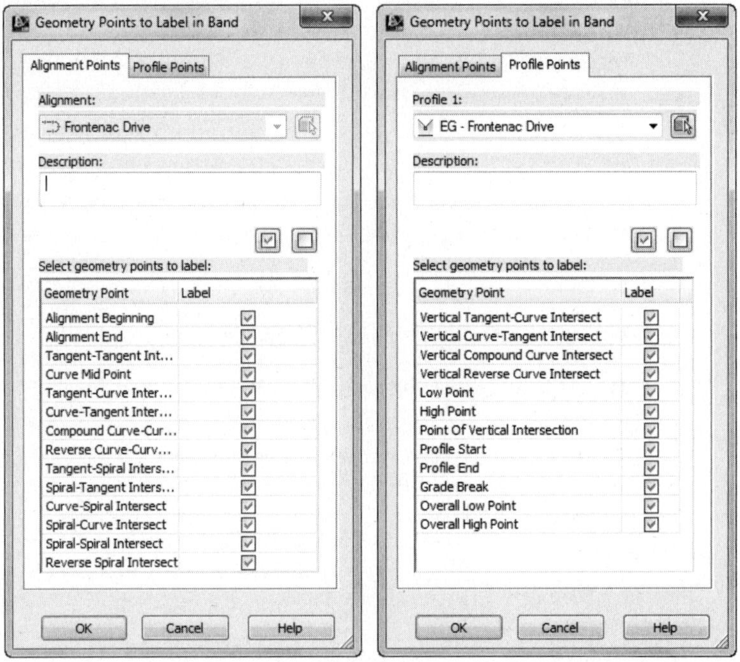

7. Click OK to accept the defaults in the Geometry Points To Label In Band dialog.
8. Leave Band Type set to Profile Data, and choose the Fill Data option from the Select Band Style drop-down.
9. Click the Add button to add the Profile Data band to display the Geometry Points To Label In Band dialog.
10. Click OK to accept the defaults in the Geometry Points To Label In Band dialog.
11. Change the Location drop-down to Top Of Profile View.
12. Change the Band Type drop-down to the Horizontal Geometry option and choose the Geometry option from the Select Band Style drop-down.

13. Click the Add button to add the Horizontal Geometry band to the table in the List Of Bands area.
14. Change the Band Type drop-down to the Vertical Geometry option.

    Do not change the Select Band Style field from its current selection (Geometry).

15. Click the Add button to also add the Vertical Geometry band to the table in the List Of Bands area.
16. Click OK to exit the dialog.

    Your profile view should look like Figure 7.55.

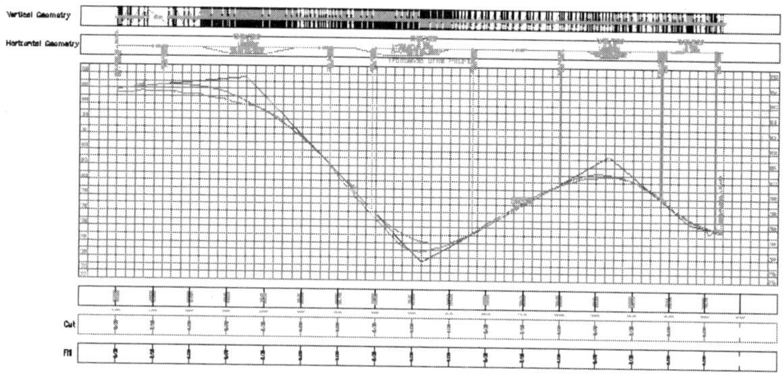

FIGURE 7.55
Applying bands to a profile view

There are obviously problems with the bands. The Vertical Geometry band is a mess and is located above the title of the profile view, whereas the Horizontal Geometry band overwrites the title. In addition, the elevation information has only the existing ground profile being referenced. Next, you'll fix those issues:

17. Pick the Frontenac Drive profile view.
18. From the Profile View contextual tab ➢ Modify View panel, choose Profile View Properties to display the Profile View Properties dialog.
19. On the Bands tab, verify that the Location drop-down in the List Of Bands area is set to Bottom Of Profile View.
20. Verify that the "Match major/minor increments to vertical grid intervals" option at the bottom of the page is selected.

    Checking this option ensures that the major/minor intervals of the profile data band match the major/minor profile view style's major/minor grid spacing.

    Three Profile Data bands are listed in the table in the List Of Bands area (Elevations And Stations, Cut Data, and Fill Data). If you need to widen the columns, you can do so by clicking the line between the column headings.

21. Scroll right in the Profile Data row and notice the two columns labeled Profile1 and Profile2.
22. For all three rows change the value of Profile2 to Frontenac Drive FG, as shown in Figure 7.56.

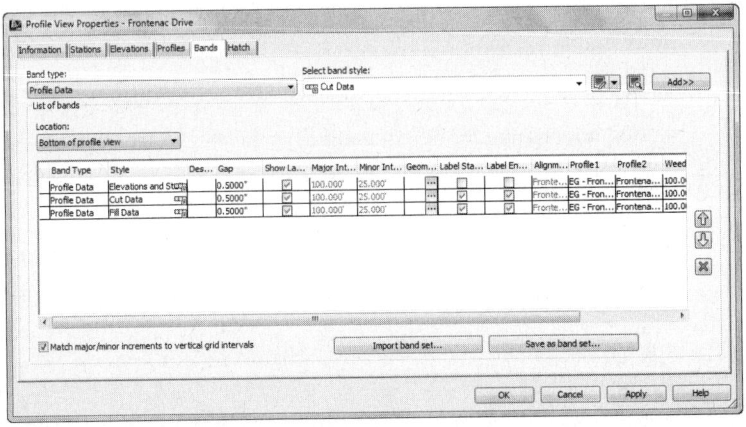

FIGURE 7.56 Setting the profile view bands to reference the Frontenac Drive FG profile

23. Change the Location drop-down to Top Of Profile View.

The Horizontal Geometry and the Vertical Geometry bands are now listed in the table as well.

24. Scroll to the right again, and set the value of Profile1 in the Vertical Geometry band to Frontenac Drive FG.

Notice that some of the Profile1 and Profile2 boxes are not available for editing, such as those in the horizontal geometry band in the Top Of Profile View; this is because profile information isn't needed for this band.

25. Scroll back to the left and set the Gap value for the Horizontal Geometry band to **1.5"** (or **35 mm** for metric users).

This value controls the distance from one band to the next or to the edge of the profile view itself and will move the bands to above the profile view title.

26. Click OK to dismiss the dialog.

Your profile view should now look like Figure 7.57.

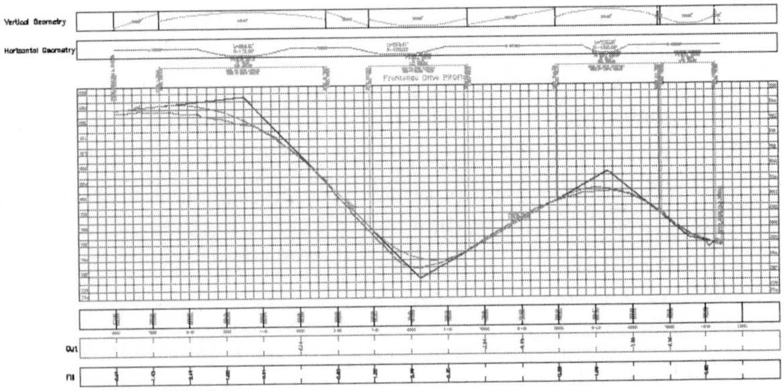

FIGURE 7.57 Completed profile view with the Bands set appropriately

When this exercise is complete, you may close the drawing. A finished copy of this drawing is available from the book's web page with the filename `ProfileViewBands_FINISHED.dwg` or `ProfileViewBands_METRIC_FINISHED.dwg`.

Bands use the Profile1 and Profile2 designation as part of their style construction. By changing the profile referenced as Profile1 or Profile2, you change the values that are calculated and displayed (e.g., existing versus proposed elevations). These bands are just more items that are driven by object styles, which you will learn more about in Chapter 21.

We will look at using Band Sets a little later in this chapter.

Profile View Hatch

Sometimes it is necessary to shade cut/fill areas in a profile view. The settings on the Hatch tab of the Profile View Properties dialog are used to specify upper and lower cut/fill boundary limits for associated profiles (see Figure 7.58). Shape styles from the General Multipurpose Styles collection found on the Settings tab of Toolspace can also be selected here. These settings include the following:

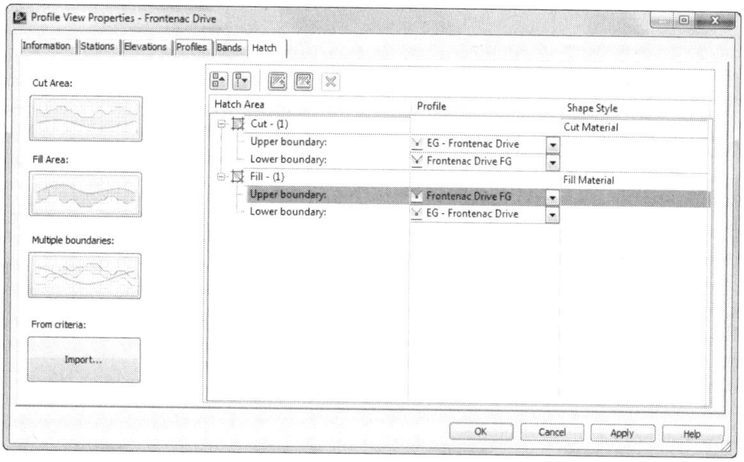

FIGURE 7.58 Shape style selection on the Hatch tab of the Profile View Properties dialog

Cut Area Click this button to add hatching to a profile view in areas of the cut (the layout profile is at a lower elevation than the sampled surface profile).

Fill Area Click this button to add hatching to a profile view in areas of the fill (the layout profile is at a higher elevation than the sampled surface profile).

Multiple Boundaries Click this button to add hatching to a profile view in areas of a cut/fill where the area must be averaged between two existing profiles (for example, finished ground at the centerline vs. the left and right top of a curb).

From Criteria Click this button to add hatch in areas where quantity takeoff criteria are used to define a hatch region.

Figure 7.59 shows a cut and fill hatched profile based on the criteria shown previously in Figure 7.58.

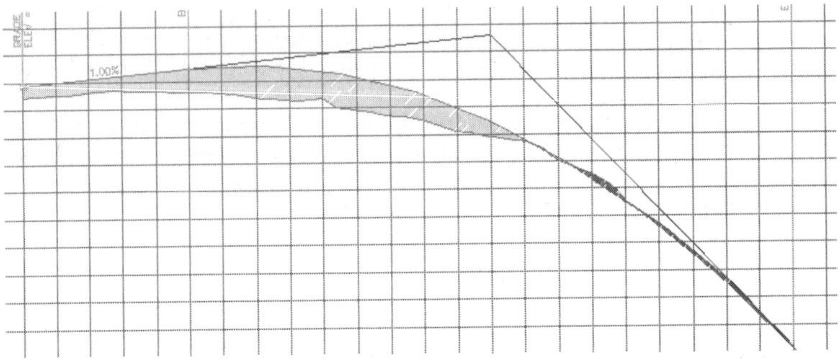

FIGURE 7.59
A portion of the Frontenac Drive Profile shown with cut and fill shading

MASTERING PROFILES AND PROFILE VIEWS

One of the most difficult concepts to master in AutoCAD Civil 3D is the notion of which settings control which display property. Although the following two rules may sound overly simplistic, they are easily forgotten in times of frustration:

- Every object has a label and an object style.
- Every label has a label style.

Furthermore, if you can remember that there is a distinct difference between a profile object and the profile view object you place it in, you'll be well on your way to mastering profiles and profile views. When in doubt, select an object, right-click, and pay attention to the Civil 3D commands available. Label styles and object styles will be discussed further in Chapters 20 and 21, respectively.

Profile View Labeling Styles

Now that the profile view is created, the profile view grid spacing is set, and the titles all look good, it's time to add some specific callouts and detail information. Civil 3D uses profile view labels and bands for annotating. The specific label styles will be discussed further in Chapter 20, but for now we will just discuss how to apply the labels.

VIEW ANNOTATION

Profile view annotations label individual points in a profile view, but they are not tied to a specific profile object. Profile view labels can be station elevation, depth labels, or projection labels. Station elevation labels can be used to label a single point or the depth between two points in a profile while recognizing the vertical exaggeration of the profile view and applying the scaling factor to label the correct depth. In this exercise, you'll use both the station elevation label and the depth label:

1. Open the `ProfileViewLabels.dwg` file (or the `ProfileViewLabels_METRIC.dwg` file).
2. Zoom to the Cabernet Court profile view located in the rectangle.
3. From the Annotate tab ➤ Labels & Tables panel, choose Add Labels (not the drop-down list) from the Labels & Tables panel to open the Add Labels dialog.

4. In the Feature drop-down, select Profile View.
5. In the Label Type drop-down, verify that Station Elevation is selected.
6. In the Station Elevation Label Style drop-down, verify that Station And Elevation is selected.
7. Verify that the Marker style is set to Basic Circle With Cross.
8. Click the Add button.
9. When prompted to select a profile view, click a grid line in the Cabernet Court profile view and a jig will appear.
10. Zoom in around station 7+85 (or 0+235 for metric users) so that you can see the point where the EG and layout profiles cross.
11. Pick this profile crossover point visually; then pick the same point to set the elevation and press ↵.

Your label should look like Figure 7.60.

FIGURE 7.60
An elevation label for a profile station

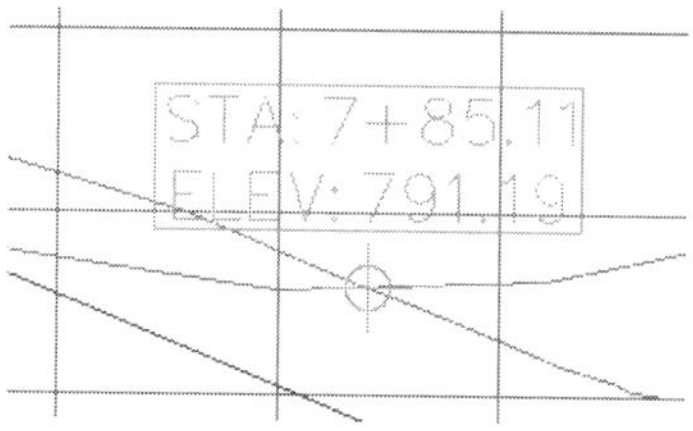

SNAPPING TO PROFILES AND PROFILE VIEWS

For a number of releases now, users have been asking for the ability to simply snap to the intersection of two profiles. We mention this because you'll try to snap and wonder if you've lost your mind. You haven't—it just doesn't work. If you are after a solution (it isn't elegant), you can draw lines on top of the profiles.

12. In the Add Labels dialog, change both Label Type and Depth Label Style to the Depth option.
13. Click the Add button.
14. Pick the Cabernet Court profile view by clicking one of the grid lines.
15. Pick a point along the layout profile; then pick a point along the EG profile and press ↵.

The depth between the two profiles will be measured, as shown in Figure 7.61.

16. Close the Add Labels dialog.

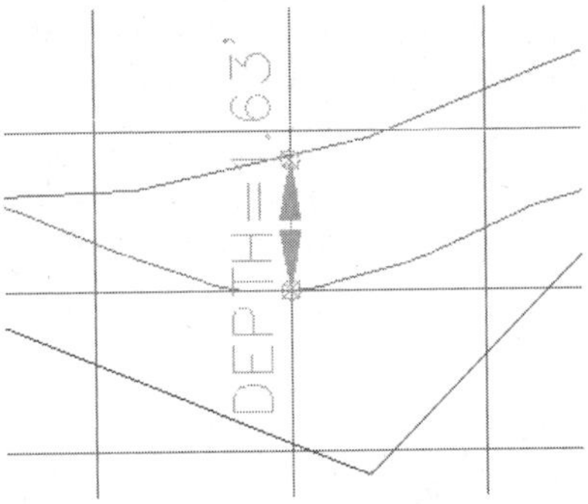

FIGURE 7.61
A depth label applied to the Cabernet Court profile view

When this exercise is complete, you may close the drawing. A finished copy of this drawing is available from the book's web page with the filename `ProfileViewLabels_FINISHED.dwg` or `ProfileViewLabels_METRIC_FINISHED.dwg`.

Depth labels can be handy in earthworks situations where cut and fill become critical, and individual spot labels are important to understanding points of interest, but most design documentation is accomplished with labels placed along the profile view axes in the form of data bands. The next section describes these band sets.

BAND SETS

Band sets are simply collections of bands, much like the profile label sets or alignment label sets. In this exercise, you'll save a band set and then apply it to a second profile view:

1. Open the `ProfileViewBandSets.dwg` file (or the `ProfileViewBandSets_FINISHED.dwg` file).
2. Pick the Frontenac Drive profile view (at the lower right located in a rectangle), and from the Profile View contextual tab ➤ Modify View panel, choose Profile View Properties to display the Profile View Properties dialog.
3. On the Bands tab, click the Save As Band Set button to display the Band Set – New Profile View Band Set dialog.
4. On the Information tab, in the Name field, enter **Cut Fill Elev Station and Horiz Vert Geometry**, as shown in Figure 7.62.

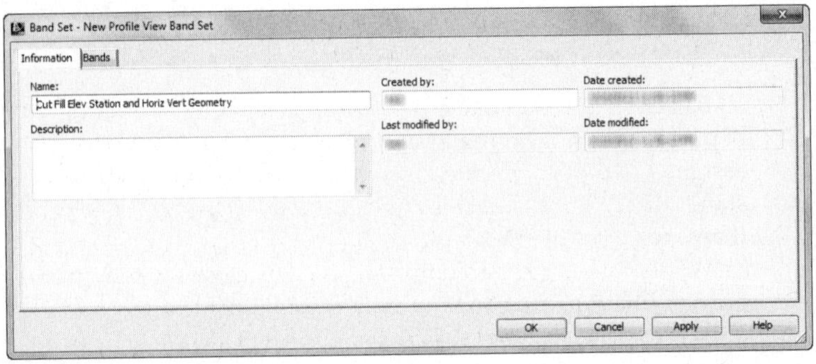

FIGURE 7.62
The Information tab for the Band Set – New Profile View Band Set dialog

5. Click OK to dismiss the Band Set – New Profile View Band Set dialog.
6. Click OK to dismiss the Profile View Properties dialog.
7. Pick the Syrah Way profile view (at the upper right located in a rectangle), and from the Profile View contextual tab ➤ Modify View panel, choose Profile View Properties to display the Profile View Properties dialog.
8. On the Bands tab, click the Import Band Set button, and the Band Set dialog opens.
9. Select the Cut Fill Elev Station and Horiz Vert Geometry option from the drop-down list and click OK.
10. Select Top Of Profile View from the Location drop-down list.
11. Scroll over on the Vertical Geometry and set Profile1 to Syrah Way FG.
12. Select the Bottom Of Profile View from the Location drop-down list.
13. Scroll over and change the Profile2 to Syrah Way FG for all three rows.
14. Click OK to exit the Profile View Properties dialog.

Your Syrah Way profile view (Figure 7.63) now looks like the Frontenac Drive profile view.

FIGURE 7.63
Completed profile view after importing the band set and matching properties

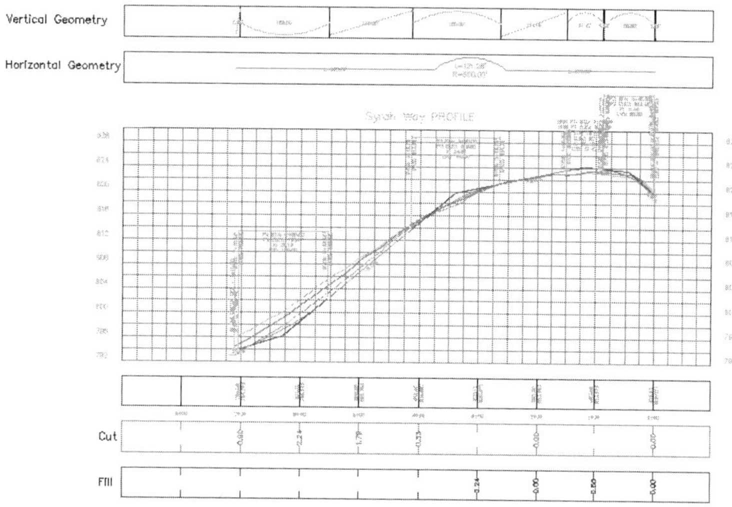

Band sets allow you to create uniform labeling and callout information across a variety of profile views. By using a band set, you can apply myriad settings and styles that you've assigned to a single profile view to a number of profile views. The simplicity of enforcing standard profile view labels and styles makes using profiles and profile views simpler than ever.

When this exercise is complete, you may close the drawing. A finished copy of this drawing is available from the book's web page with the filename `ProfileViewBandSets_FINISHED.dwg` or `ProfileViewBandSets_METRIC_FINISHED.dwg`.

BRING OUT THE BAND

Once design speeds have been assigned to an alignment and superelevation has been calculated, you'll find the Create Superelevation View command on the Alignment contextual tab ➤ Modify panel. Although not the focus of this chapter, superelevation views behave much like profile views, and you can access their properties via the right-click menu after selecting a view.

Profile Labels

It's important to remember that the profile and the profile view aren't the same thing. The labels discussed in this section are those that relate directly to the profile but are visible for a specific profile view. This usually means station-based labels, individual tangent and curve labels, or grade breaks.

APPLYING LABELS

As with alignments, you can apply labels as a group of objects separate from the profile. In this portion of the exercise, you'll learn how to add labels along a profile object:

1. Open the `ApplyingProfileLabels.dwg` file (or the `ApplyingProfileLabels_METRIC.dwg` file).
2. Zoom to the Cabernet Court profile view in the rectangle and pick the Cabernet Court FG profile (the blue line) to activate the Profile contextual tab.
3. From the Profile contextual tab ➤ Labels panel, choose Edit Profile Labels to display the Profile Labels dialog (see Figure 7.64).

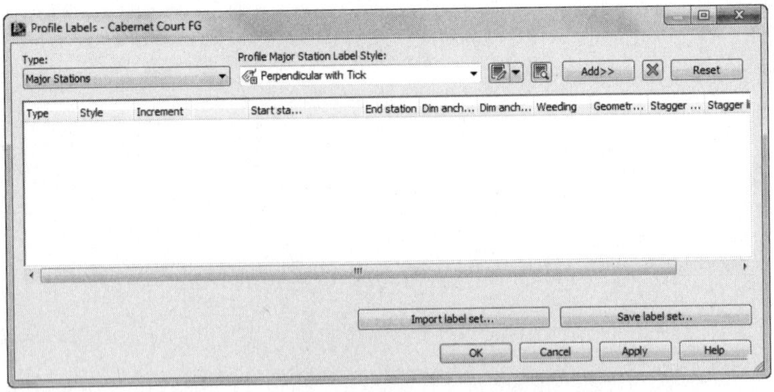

FIGURE 7.64
An empty Profile Labels dialog

Selecting the type of label from the Type drop-down menu changes the Style drop-down menu to include styles that are available for that label type. Next to the Style drop-down menu are the usual Style Edit/Copy button and a preview button. Once you've selected a style from the Style drop-down menu, clicking the Add button places it on the profile. The middle portion of this dialog displays information about the labels that are being applied to the profile selected; you'll look at that in a moment.

4. Choose the Major Stations option from the Type drop-down menu.

The name of the second drop-down menu changes to Profile Major Station Label Style to reflect this option.

5. Verify that Perpendicular With Tick is selected in this menu.
6. Click the Add button to apply this label to the profile.
7. Choose Horizontal Geometry Points from the Type drop-down menu.

The name of the Style drop-down menu changes to Profile Horizontal Geometry Point.

8. Select the Horizontal Geometry Station option, and click the Add button again to display the Geometry Points dialog.

This dialog lets you apply different label styles to different geometry points if necessary.

9. Deselect the Alignment Beginning and Alignment End rows, as shown in Figure 7.65, and click OK to dismiss the dialog.

FIGURE 7.65
The Geometry Points dialog appears when you apply labels to horizontal geometry points.

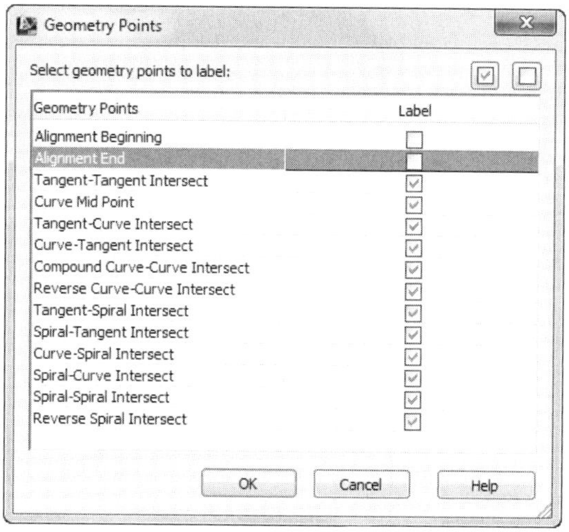

10. On the Profile Labels dialog, click the Apply button.
11. Drag the dialog out of the way to view the changes to the profile, as shown in Figure 7.66.
12. In the middle of the Profile Labels dialog, change the Increment value in the Major Stations row to **50** (or **10** for metric users), as shown in Figure 7.67.

This modifies the labeling increment only, not the grid or other values.

13. Click OK to dismiss the Profile Labels dialog.

When this exercise is complete, you may close the drawing. A finished copy of this drawing is available from the book's web page with the filename `ApplyingProfileLabels_FINISHED.dwg` or `ApplyingProfileLabels_METRIC_FINISHED.dwg`.

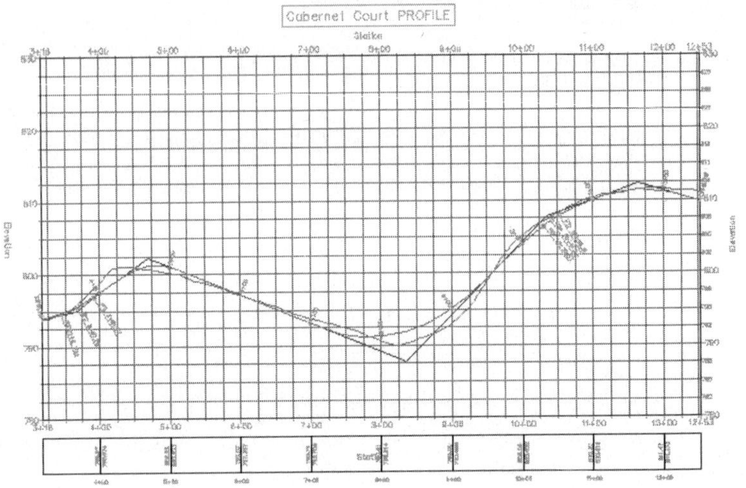

FIGURE 7.66
Labels applied to major stations and alignment geometry points

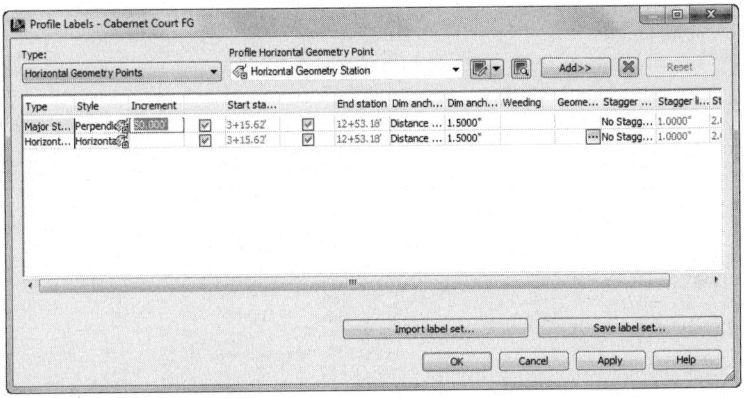

FIGURE 7.67
Modifying the major station labeling increment

As you can see, applying labels one at a time could turn into a tedious task. After you learn about the types of labels available, you may want to revisit this dialog and use the Import Label Set and Save Label Set buttons, which work similarly to the Import Band Set and Save Band Set buttons that we discussed previously. A further discussion, including more of the profile labels that are available, is provided in Chapters 20 and 21.

PROFILE LABEL SETS

Applying labels to both crest and sag curves, tangents, grade breaks, and geometry with the label style selection and various options can be monotonous. Thankfully, Civil 3D gives you the ability to use label sets, as in alignments, to make the process quick and easy. Just as you saved and imported band sets in a previous example, you can save and import label sets as well.

In all of the previous examples you used either the Complete Label Set or the _No Labels label set, which you specified on the Profile Display Options wizard page.

SOMETIMES YOU DON'T WANT TO SET EVERYTHING

Resist the urge to modify the beginning or ending station values in a label set. If you save a specific value, that value will be applied when the label set is imported. For example, if you set a station label to end at 15+00 because the alignment is 15+15 long, that label will always stop at 15+00, even if the target profile is 5,000′ long!

Label sets are the best way to apply profile labeling uniformly. When you're working with a well-developed set of styles and label sets, it's quick and easy to go from sketched profile layout to plan-ready output. We will discuss profile label sets in further detail and go over an example in Chapter 20.

Profile Utilities

One common requirement is to compare profile data for objects that are aligned similarly but not parallel. Another is the ability to project objects from a plan view into a profile view. The abilities to superimpose profiles and project objects are both discussed in this section.

Superimposing Profiles

In a profile view, a profile is sometimes superimposed to show one profile adjacent to another (e.g., a ditch adjacent to a road centerline). In this brief exercise, you'll superimpose one of your street designs onto the other to see how they compare over a certain portion of their length:

1. Open the `SuperimposeProfiles.dwg` file (or the `SuperimposeProfiles_METRIC.dwg` file).
2. From the Home tab ➢ Create Design panel, choose Profile ➢ Create Superimposed Profile.
3. At the `Select source profile:` prompt, zoom to the Frontenac Drive profile view at the bottom right and pick the Frontenac Drive FG profile (the blue line).
4. At the `Select destination profile view:` prompt, pick the Cabernet Court profile view at the middle right by clicking one of the grid lines to display the Superimpose Profile Options dialog shown in Figure 7.68.

FIGURE 7.68 The Superimpose Profile Options dialog

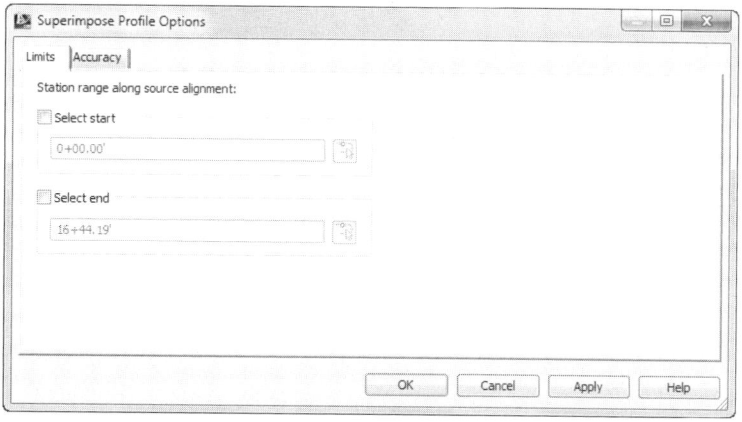

5. Click OK to dismiss the dialog, accepting the default settings.
6. Zoom in on the Cabernet Court profile view to see the superimposed data, as shown in Figure 7.69.

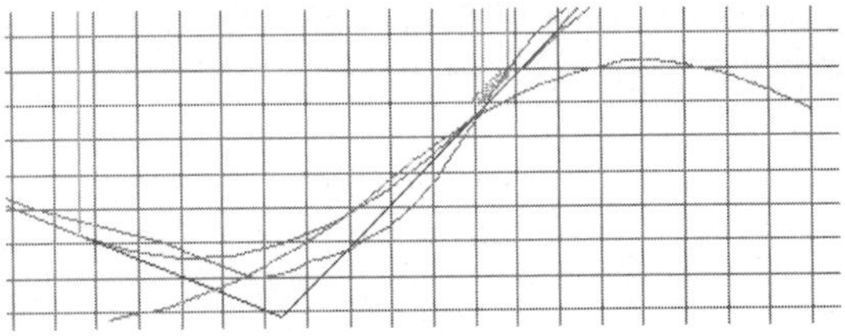

FIGURE 7.69
The Frontenac Drive layout profile superimposed on the Cabernet Court profile view

When this exercise is complete, you may close the drawing. A finished copy of this drawing is available from the book's web page with the filename `SuperimposeProfiles_FINISHED.dwg` or `SuperimposeProfiles_METRIC_FINISHED.dwg`.

Note that the vertical curve in the Frontenac Drive layout profile has been approximated on the Cabernet Court profile view, using a series of PVIs. Superimposing works by projecting a line from the target alignment (Cabernet Court) to a perpendicular intersection with the other source alignment (Frontenac Drive).

The target alignment is queried for an elevation at the intersecting station and a PVI is added to the superimposed profile. Note that this superimposed profile is still dynamic! A change in the Frontenac Drive layout profile will be reflected on the Cabernet Court profile view.

Object Projection

Some AutoCAD and some AutoCAD Civil 3D objects can be projected from a plan view into a profile view. The list of available AutoCAD objects includes points, blocks, 3D solids, and 3D polylines. The list of available AutoCAD Civil 3D objects includes COGO points, feature lines, and survey figures. These objects can be projected to the objects elevation, a manually selected elevation, a surface, or a profile. In the following exercise, you'll project a 3D object into a profile view:

1. Open the `ObjectProjection.dwg` file (or the `ObjectProjection_METRIC.dwg` file).

2. From the Home tab ➤ Profile & Section Views panel, choose Profile View ➤ Project Objects To Profile View.
3. At the `Select objects to add to profile view:` prompt, select the Fire Hydrant object located in the center of the circle near the eastern intersection of Syrah Way and Frontenac Drive and press ↵.
4. At the `Select a profile view:` prompt, select the Syrah Way profile view (located at the upper right inside a rectangle) by clicking one of the grid lines to display the Project Objects To Profile View dialog.
5. Verify that Style is set to Fire Hydrant, Elevation Options is set to Surface ➤ EG, and Label Style is set to Projection Dimension Below, as shown in Figure 7.70.
6. Click OK to dismiss the dialog, and review your results as shown in Figure 7.71.

FIGURE 7.70
A completed Project Objects To Profile View dialog

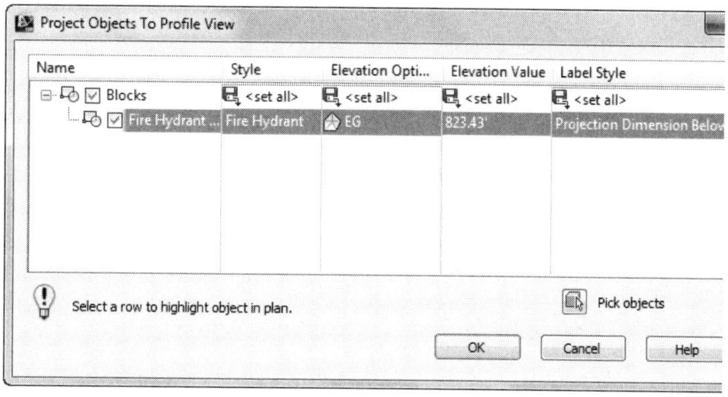

FIGURE 7.71
The COGO point object projected into a profile view

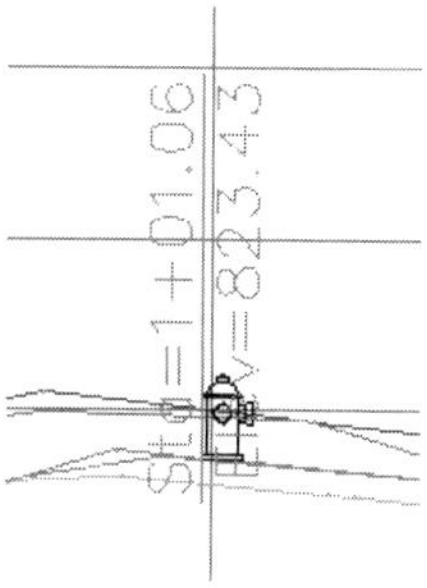

We actually wanted the Fire Hydrant to show on the Proposed surface. No problem. Follow these steps:

7. Click on the Fire Hydrant in profile view to activate the Projected Object contextual tab.

8. From the Projected Object contextual tab ➤ Modify Projected Object panel, choose Projection Properties to display the Projection View Properties dialog.

9. In the Projection View Properties dialog, click the <set all> option in the Elevation Options column and select Syrah Way FG, as shown in Figure 7.72. Click OK.

FIGURE 7.72
Select the Syrah Way FG elevation.

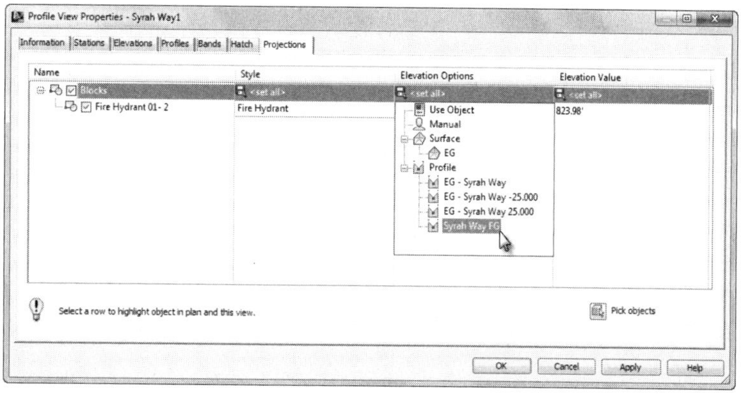

The Fire Hydrant is now reprojected to match the elevation of the Syrah Way Finished Grade (FG) alignment, as shown in Figure 7.73.

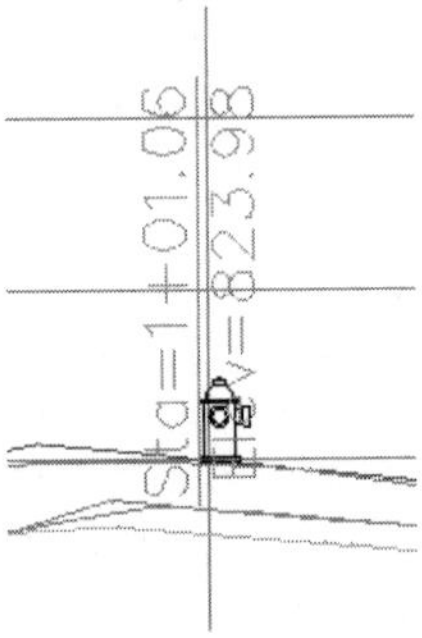

FIGURE 7.73
The COGO point object projected onto the Syrah Way FG profile

When this exercise is complete, you may close the drawing. A finished copy of this drawing is available from the book's web page with the filename `ObjectProjection_FINISHED.dwg` or `ObjectProjection_METRIC_FINISHED.dwg`.

Once an object has been projected into a profile view, the Profile View Properties dialog will display a new Projections tab. Projected objects will remain dynamically linked with respect to their plan placement. If you move them manually after placing them dynamically, a warning will appear to confirm that you want to break the dynamic setting. Because profile views and section views are similar in nature, objects can be projected into section views in the same fashion. However, projecting a feature line onto a profile view will give you a different result than projecting it onto a section view. On a profile view, it looks more like a superimposed profile. On a section view, it's more like a pipe crossing since it only appears where it intersects the section line.

Quick Profile

There are going to be times when all you want is to quickly look at a profile and not keep it for later use. When this is the case, instead of creating an alignment, a profile, and a profile view, you can instead create a quick profile. A quick profile is a temporary object that will not be saved with the drawing. You can create a quick profile for 2D or 3D lines or polylines, lot lines, feature lines, survey figures, or even a series of points.

From the Home tab ➤ Create Design panel, choose Profile ➤ Quick Profile. The command line will state `Select object or [by Points]:`. Once you select your object (or points), the Create Quick Profiles dialog is displayed, as shown in Figure 7.74.

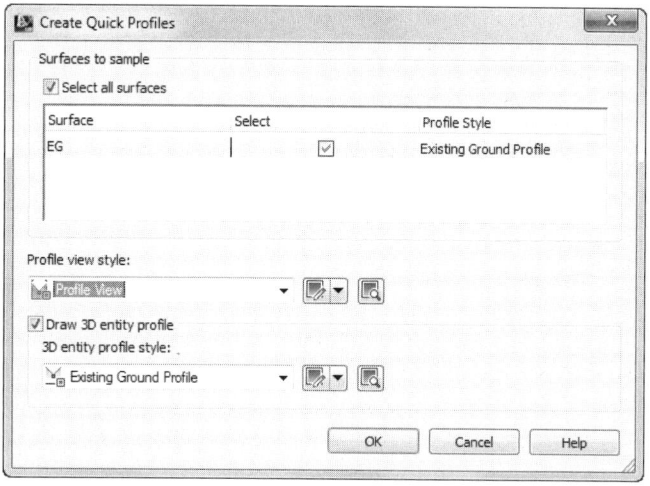

FIGURE 7.74
The Create Quick Profiles dialog

You can select which surface you want to sample as well as what profile view style and 3D entity profile style to use.

The Bottom Line

Sample a surface profile with offset samples. Using surface data to create dynamic sampled profiles is an important advantage of working with a three-dimensional model. Quick viewing of various surface centerlines and grip-editing alignments makes for an effective preliminary planning tool. Combined with offset data to meet review agency requirements, profiles are robust design tools in Civil 3D.

Master It Open the `MasteringProfiles.dwg` file (or `MasteringProfiles_METRIC.dwg` file) and sample the ground surface along Alignment A, along with offset values at 15′ left and 15′ right (or 4.5 m left and 4.5 m right) of the alignment. Generate a profile view showing this information using the Major Grids profile view style with no data band sets.

Lay out a design profile on the basis of a table of data. Many programs and designers work by creating pairs of station and elevation data. The tools built into Civil 3D let you input this data precisely and quickly.

Master It In the `MasteringProfiles.dwg` file (or the `MasteringProfiles_METRIC.dwg` file), create a layout profile on Alignment A using the Layout profile style and a Complete Label Set with the following information for Imperial users:

Station	PVI Elevation	Curve Length
0+00	822.00	
1+80	825.60	300′
6+50	800.80	

Or the following information for metric users:

Station	PVI Elevation	Curve Length
0+000	250.400	
0+062	251.640	100 m
0+250	244.840	

Add and modify individual entities in a design profile. The ability to delete, modify, and edit the individual components of a design profile while maintaining the relationships is an important concept in the 3D modeling world. Tweaking the design allows you to pursue a better solution, not just a working solution.

Master It In the `MasteringProfiles.dwg` file (or the `MasteringProfiles_METRIC.dwg` file) used in the previous exercise, on profile A modify the original curve so that it is 200′ (or 60 m for metric users). Then insert a PVI at Station 4+90, Elevation 794.60 (or at Station 0+150, Elevation 242.840 for metric users) and add a 300′ (or 96 m for metric users) parabolic vertical curve at the newly created PVI.

Apply a standard band set. Standardization of appearance is one of the major benefits of using styles in labeling. By applying band sets, you can quickly create plot-ready profile views that have the required information for review.

Master It In the `MasteringProfiles.dwg` (or the `MasteringProfiles_METRIC.dwg`) file, apply the Cut and Fill band set to the layout profile created in the previous exercise with the appropriate profiles referenced in each of the bands.

Chapter 8

Assemblies and Subassemblies

Roads, ditches, trenches, and berms usually follow a predictable pattern known as a *typical section*. Assemblies are how you tell the AutoCAD® Civil 3D® software what these typical sections look like. Assemblies are made up of smaller components called *subassemblies*. For example, a typical road section assembly contains subassemblies such as lanes, sidewalks, and curbs.

In this chapter, the focus will be on understanding where these assemblies come from and how to build and manage them.

In this chapter, you will learn to:

- Create a typical road assembly with lanes, curbs, gutters, and sidewalks
- Edit an assembly
- Add daylighting to a typical road assembly

Subassemblies

A *subassembly* is a building block of a typical section, known as an *assembly*. Examples of subassemblies include lanes, curbs, sidewalks, channels, trenches, daylighting, and any other component required to complete a typical corridor section.

An extensive catalog of subassemblies has been created for use in Civil 3D. More than a hundred subassemblies are available in the standard catalogs, and each subassembly has a list of adjustable *parameters*. There are also about a dozen generic links you can use to further refine your most complex assembly needs. From ponds and berms to swales and roads, the design possibilities are almost infinite.

The Tool Palettes

You will add subassemblies to a design by clicking on them from the subassembly tool palette, as you'll see later in this chapter. By default, Civil 3D has several tool palettes created for corridor modeling.

You can access these tool palettes from the Home tab and clicking the Tool Palettes button on the Palettes panel or by pressing Ctrl+3.

When Civil 3D is installed, you have an initial set of the most commonly used assemblies and subassemblies ready to go. The Tool Palettes window consists of multiple customizable tabs that run down the right side. These tabs categorize the assemblies and subassemblies so that they are easy to find and organize.

The top default tab in the Tool Palettes window is the Assemblies tab. On this tab you will find a selection of predefined, completed assemblies (Figure 8.1). These are a great starting point for beginners who are looking for examples of how subassemblies are put together into an assembly. There are examples of simple roadway sections as well as more advanced items, such as intersection and roundabout examples. To use one, click on the desired assembly, and then click to place it in your drawing and press ↵ to end the command.

FIGURE 8.1
Tool Palettes predefined assemblies

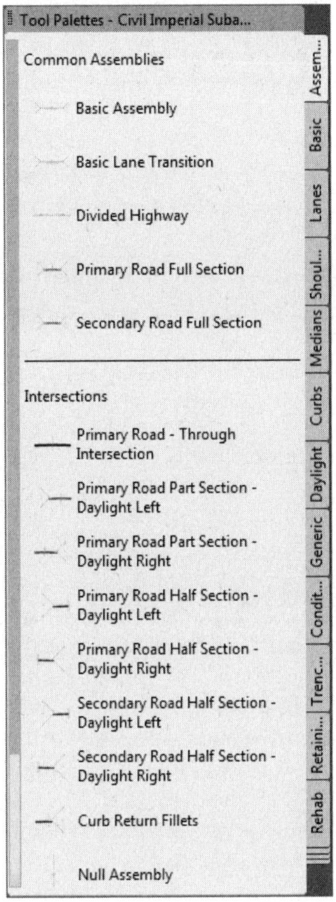

GETTING TO THE TOOL PALETTES

The exercises in this chapter depend heavily on the use of the Tool Palettes window of AutoCAD® when pulling together assemblies from subassemblies. To avoid some redundancy, we will omit the initial step of opening the Tool Palettes window and positioning it so that the baseline is viewable in every single exercise. If it's open, leave it open; if it's closed, open it. In case you need a reminder, the easiest way to open the Tool Palettes window is either from the Home tab ➤ Palettes panel or by pressing Ctrl+3 on your keyboard.

The Corridor Modeling Catalog

If the default set of subassemblies in the Tool Palettes window are not adequate for your design situation, check the Corridor Modeling Catalog for one that will work.

The Corridor Modeling Catalog is installed by default on your local hard drive. On the Home tab, expand the Palettes panel and click the Content Browser button to open a content browser interface.

Choose either the Metric or Imperial catalog to explore the entire collection of subassemblies available in each category (see Figure 8.2).

FIGURE 8.2
The front page of the Corridor Modeling Catalog

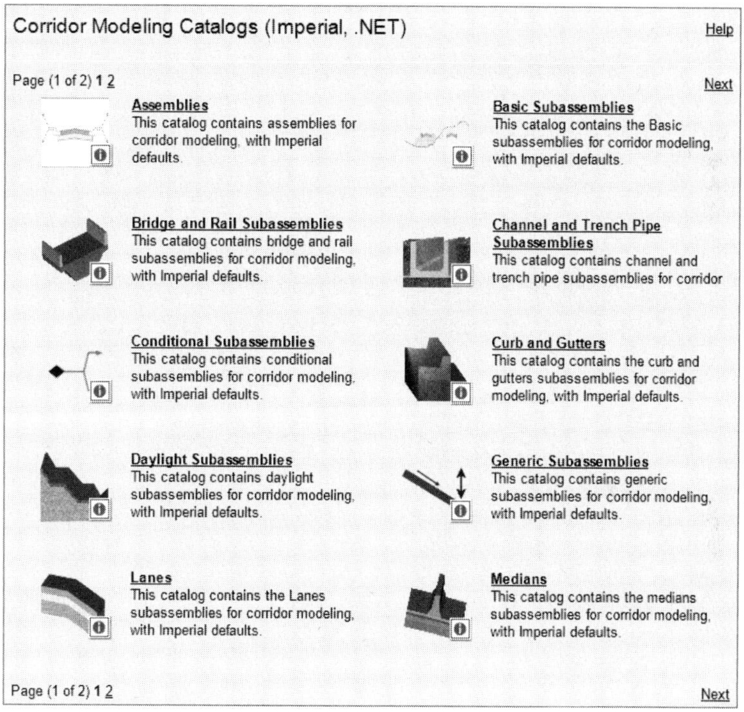

Adding Subassemblies to a Tool Palette

If you'd like to add additional subassemblies to your Tool Palettes window, you can use the i-drop to grab subassemblies from the catalog and drop them onto the Tool Palettes window. To use the i-drop:

1. Click the small blue *i* next to any subassembly, and continue to hold down your left mouse button until you're over the desired tool palette.
2. Release the button, and your subassembly should appear on the tool palette (see Figure 8.3).

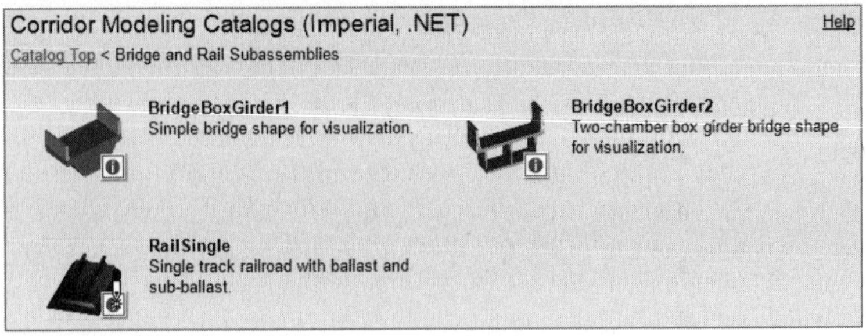

FIGURE 8.3
Using the i-drop to add the RailSingle subassembly to a tool palette

> **ACCESSING SUBASSEMBLY HELP**
>
> Later, this chapter will point out other shortcuts to access the extensive subassembly documentation. You can get quick access to information by right-clicking any subassembly entry on the Tool Palettes window or the Corridor Modeling Catalog page and selecting the Help option.
>
> The Subassembly Reference page in the help file provides a detailed breakdown of each subassembly, examples for its use, its parameters, a coding diagram, and more. While you're searching the catalog for the right parts to use, you'll find the Subassembly Reference page infinitely useful.

Building Assemblies

You build an assembly from the Home tab ➤ Create Design panel by choosing Assembly ➤ Create Assembly. The result is the main assembly baseline marker. This is the point on the assembly that gets attached to your design alignment and profile. A typical assembly baseline is shown in Figure 8.4.

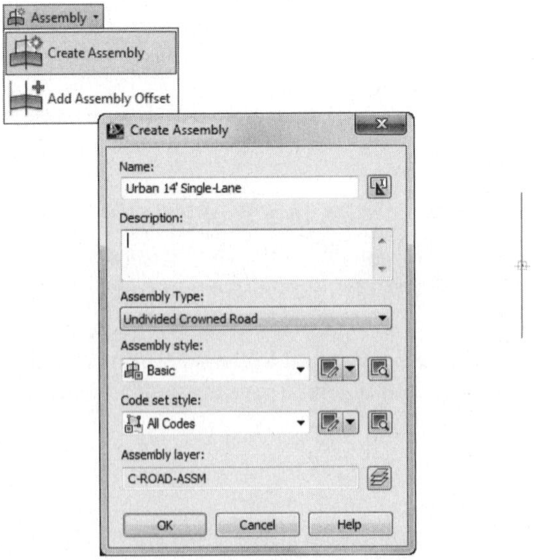

FIGURE 8.4
Creating an assembly (left); an assembly baseline marker (right)

When an assembly is created, you have the option of telling Civil 3D what type of assembly this will be:

- Undivided Crowned Road
- Undivided Planar Road
- Divided Crowned Road
- Divided Planar Road
- Railway
- Other

These categories will help the software determine the axis of rotation options in superelevation, if needed.

Once an assembly is created and assigned a type, you start piecing it together using various subassemblies to meet your design intent. In the next section we will look at how you can create the most common assembly type, an undivided crowned road.

Creating a Typical Road Assembly

The process for building an assembly requires the use of the Tool Palettes window (accessible using Ctrl+3) and the AutoCAD Properties palette (accessible using Ctrl+1), both of which can be docked. You'll quickly learn how to best orient these palettes with your limited screen real estate. If you run dual monitors, you may find it useful to place both of these palettes on your second monitor.

The exercise that follows builds a typical assembly, as shown in Figure 8.5, using LaneSuperelevationAOR, UrbanCurbGutterGeneral, UrbanSidewalk, and DaylightMaxOffset subassemblies.

FIGURE 8.5
A typical road assembly

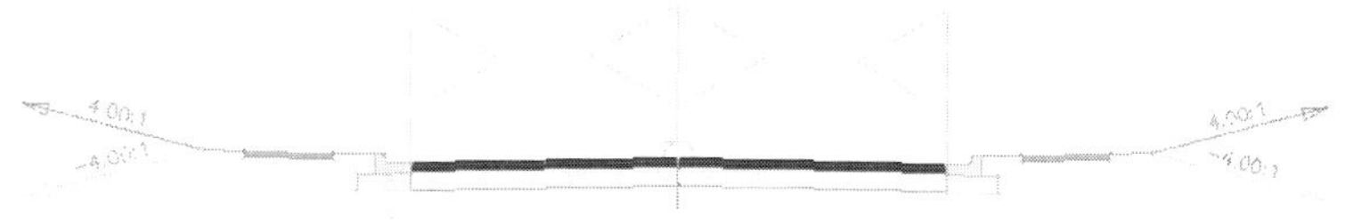

Let's have a more detailed look at each component you'll use in the following exercise. A quick peek into the subassembly help file will give you a breakdown of attachment options; input parameters; target parameters; output parameters; behavior; layout mode operation; and the point, link, and shape codes.

The LaneSuperelevationAOR Subassembly The LaneSuperelevationAOR subassembly is the best all-purpose subassembly for lanes. It can superelevate for an inside or outside lane if needed, and allows for up to four layers of materials. The input parameters available are Side, Width, Default Slope, Pave1 Depth, Pave2 Depth, Base Depth, Subbase Depth, Use Superelevation, Slope Direction, Potential Pivot, Inside Point Code, and Outside Point Code. The default width of 12′ (3.6 m) can be adjusted in the parameters or can be used with an offset alignment to control its width. Figure 8.6 shows the image provided in the subassembly help file for this subassembly.

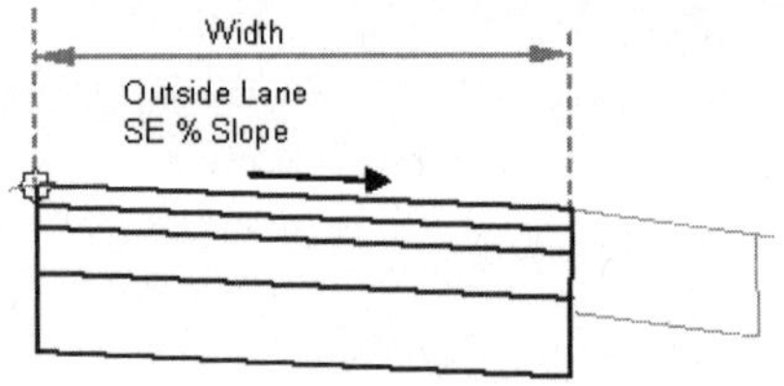

FIGURE 8.6
The Lane SuperelevationAOR subassembly help diagram

The UrbanCurbGutterGeneral Subassembly The UrbanCurbGutterGeneral subassembly is another standard component that creates an attached curb and gutter. Looking into the subassembly help file, you'll see a diagram of UrbanCurbGutterGeneral with input parameters for Side, Insertion Point, Gutter Slope Method, Gutter Slope, Gutter Slope Direction, Subbase Depth, Subbase Extension, Subbase Slope Method, Subbase %Slope, and the subassembly's seven dimensions. You can adjust these parameters to match many standard curb-and-gutter configurations. Figure 8.7 shows the image provided in the subassembly help file for this subassembly.

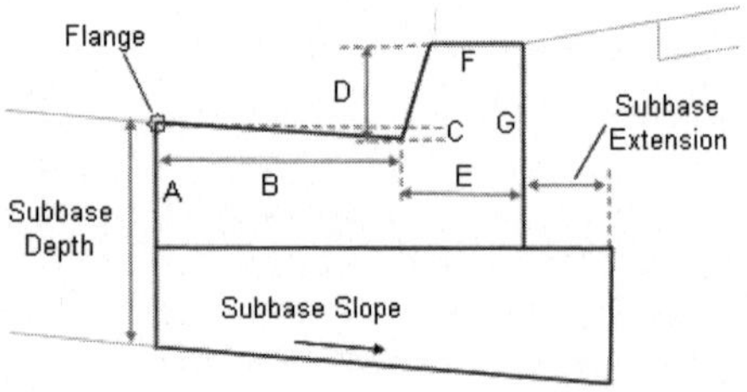

FIGURE 8.7
The UrbanCurb GutterGeneral subassembly help diagram

The UrbanSidewalk Subassembly The UrbanSidewalk subassembly creates a sidewalk and terrace buffer strips. The help file lists the following six input parameters for the UrbanSidewalk subassembly: Side, Inside Boulevard Width, Sidewalk Width, Outside Boulevard Width, %Slope, and Depth. These input parameters let you adjust the sidewalk width, material depth, and buffer widths to match your design specification. Figure 8.8 shows the image provided in the subassembly help file for this subassembly.

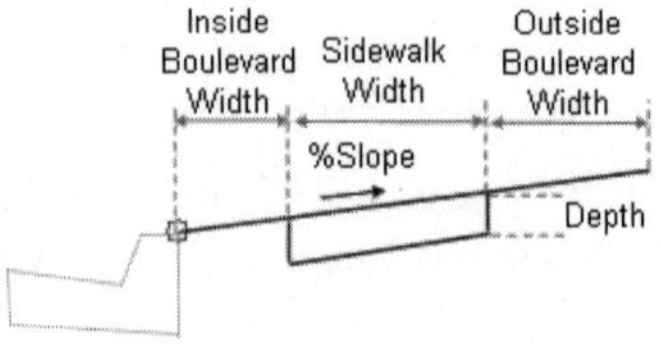

FIGURE 8.8
The UrbanSidewalk subassembly help diagram

The UrbanSidewalk subassembly can return quantities of concrete (or other sidewalk construction material), but not gravel bedding or other advanced material layers.

The DaylightMaxOffset Subassembly The DaylightMaxOffset subassembly is a nice "starter" for creating simple, single-slope daylight instructions for your corridor. In Civil 3D, an offset dimension is measured from the baseline, and a width is measured from the attachment point. Therefore, the maximum offset in our example is measured from the centerline of the road, which is the baseline. The slope will attempt a default of 4:1, but it will adjust if it needs to in order to keep inside your specified maximum offset (such as a right-of-way line). Options are also available for rounding. Figure 8.9 shows the image provided in the subassembly help file for this subassembly.

FIGURE 8.9
The DaylightMax Offset subassembly help diagram for the cut scenario

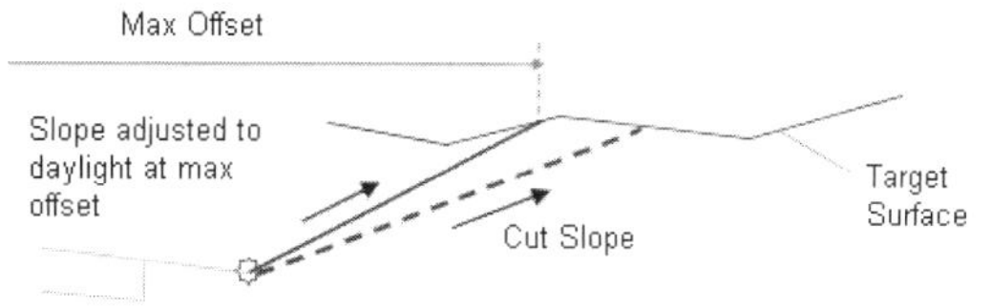

WHAT'S WITH THE FUNNY NAMES?

You'll notice that all subassemblies have names with no spaces. This is because of the underlying .NET coding that makes up a subassembly. When you place one of these in your project, it will retain the name from the tool palette.

Prior to Civil 3D 2013, each subassembly was required to have a unique name; therefore, it was suffixed with a number. This is no longer the case. Nonetheless, later in this chapter you'll see how to rename them to something more user-friendly if you so desire.

In the following exercise, you'll build a typical road assembly using these subassemblies. Follow these steps:

1. Start a new blank drawing from the `_AutoCAD Civil 3D (Imperial) NCS` template that ships with Civil 3D. For metric users, use the `_AutoCAD Civil 3D (Metric) NCS` template.

2. Confirm that your Tool Palettes window is showing the subassembly set appropriate for your drawing units (or you may end up with monster 12 meter lanes!).

   If you need to change your active Tool Palettes window from metric to Imperial or vice versa, right-click the Tool Palettes control bar located at the top of the window, as shown in Figure 8.10.

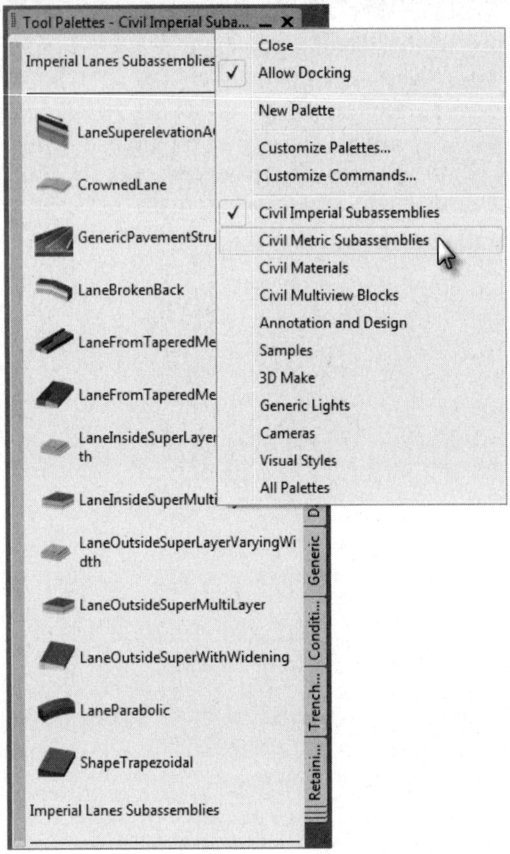

FIGURE 8.10
Right-click the Tool Palettes control bar to change assembly sets if needed.

3. Verify that your drawing scale is set to 1″=10′ (1:50 for metric users).
4. From the Home tab ➤ Create Design panel choose Assembly ➤ Create Assembly.

   The Create Assembly dialog opens.
5. Enter **Urban 14′ Single-Lane** (or **Urban 4.5m Single-Lane**) in the Name text box.
6. Set Assembly Type to Undivided Crowned Road.
7. Confirm that Assembly Style is set to Basic and Code Set Style is set to All Codes, and click OK.
8. Pick a location in your drawing for the assembly; somewhere in the center of your screen is fine to place your red assembly baseline marker.
9. Locate the Lanes tab on the Tool Palettes window, and position the palette on your screen so that you can clearly see the assembly baseline.
10. Click the LaneSuperelevationAOR button on the Tool Palettes window.

The AutoCAD Properties palette appears.

11. Locate the Advanced Parameters section on the Design tab of the AutoCAD Properties palette (Figure 8.11).

FIGURE 8.11
Advanced Parameters on the Properties palette

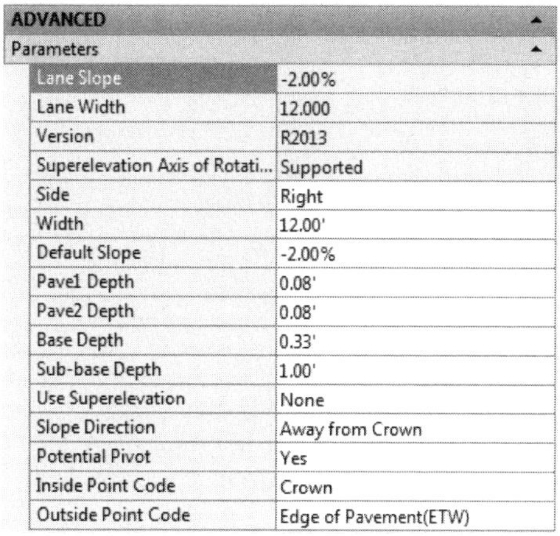

This section lists the LaneSuperelevationAOR parameters.

12. Change the Width parameter to **14'** (**4.5** m).

    Your Properties palette should resemble Figure 8.11.

13. At the `Select marker point within assembly or [Insert Replace Detached]:` prompt, select anywhere on the red assembly baseline marker to place the first lane.

    Note that it is placed on the right side as you had specified in the Advanced Parameters section.

14. Before ending the command, click the red assembly baseline marker again to place the left lane.

    Civil 3D 2013 now has the intelligence to autodetect the side and places a left lane even though you did not change the side specified in the Advanced Parameters.

15. Press ↵ to end the command.

16. Switch to the Curbs tab in the Tool Palettes window.

17. Click the UrbanCurbGutterGeneral button on the Tool Palettes window.

    The AutoCAD Properties palette appears.

18. You will accept the parameter defaults, so no changes are needed. Remember that the Side parameter will automatically be detected so there is no need to change it.

19. At the `Select marker point within assembly or [Insert Replace Detached]:` prompt, select the circular marker located at the top right of the right LaneSuperelevationAOR subassembly.

This marker represents the top-right edge of pavement (see Figure 8.12).

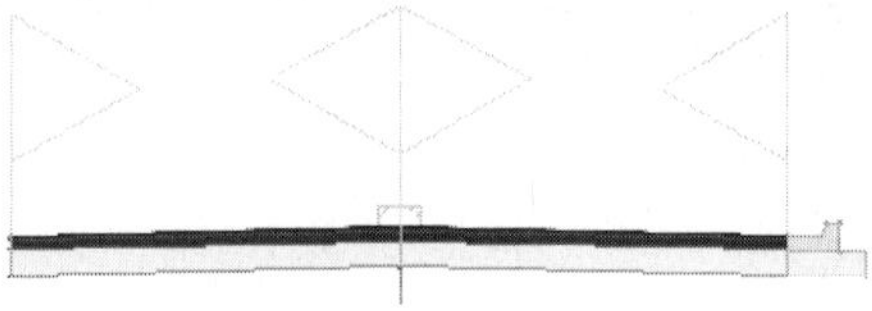

FIGURE 8.12
The UrbanCurb-GutterGeneral subassembly placed on the Lane SuperelevationAOR subassembly

20. Press ↵ to end the command.

You will add the left curb later. Keep this drawing open for the next portion of the exercise.

IF YOU GOOF...

Often, the first instinct when a subassembly is misplaced is to Undo or erase the wayward piece. However, if you have spent a lot of time diligently tweaking parameters, there is a way to fix things without redoing the subassembly.

MOVING A SUBASSEMBLY

Select the errant subassembly component and use the Move option from the contextual tab ➤ Modify Subassembly panel. Use this instead of the base AutoCAD Move tool to get the best results. Using regular AutoCAD Move may cause unexpected results in the corridor.

INSERTING A SUBASSEMBLY

Sometimes you forget to place a subassembly component, or your design changes and you want to include a subassembly that wasn't there before. In previous versions of Civil 3D, you had to delete the subassembly components from the outside in until you got to where you wanted to insert your missing subassembly and then had to re-create the deleted subassembly pieces.

In the latest version of Civil 3D, you may have noticed that every time you place an assembly the command line states `Select marker point within assembly or [Insert Replace Detached]:`. If you enter **I**, the command line will state `Select the subassembly to insert after or [Before]:`. This new ability to insert a subassembly will come in useful when the planner decides to add a sidewalk at the shoulder of your road.

REPLACING A SUBASSEMBLY

Similar to the insert operation, you can also replace a subassembly component with another component. Again, you will find this infinitely helpful when the planner decides to make changes to your design.

DELETING A SUBASSEMBLY

To delete a subassembly component, you can simply select the subassembly component and press the Delete key. The assembly will connect the subassemblies on either side at the connection points previously used with the deleted component.

CHANGING SUBASSEMBLY PARAMETERS

If you placed everything correctly but forgot to change a parameter or two, there's an easy fix for that, too. Cancel out of any active subassembly placement and select the subassembly you wish to change. Most subassembly parameters can be changed from the AutoCAD Properties palette. For more heavy-duty modifications (such as specifying the side), you will want to get into the Subassembly Properties discussed later in this chapter.

---

21. In the Curbs tab, click the UrbanSidewalk button on the Tool Palettes window.

22. In the Advanced section of the Design tab on the AutoCAD Properties palette, change the following parameters, leaving all other parameters at their default values:

    Sidewalk Width: **5′** (**1.5** m)

    Inside Boulevard Width: **2′** (**0.7** m)

    Outside Boulevard Width: **2′** (**0.7** m)

    It may be hard to ignore the Side parameter, but setting it is not required with Civil 3D side autodetection.

23. At the `Select marker point within assembly or [Insert Replace Detached]:` prompt, select the circular marker on the UrbanCurbGutterGeneral subassembly that represents the top rear of the curb to attach the UrbanSidewalk subassembly (see Figure 8.13).

FIGURE 8.13
The BasicSidewalk subassembly placed on the UrbanCurb-GutterGeneral subassembly

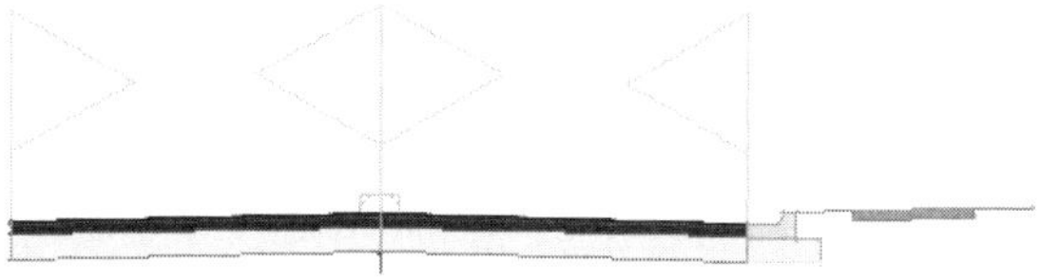

24. Switch to the Daylight tab on the subassemblies Tool Palettes window, and select the DaylightMaxOffset subassembly.

25. In the Advanced Parameters area, change Max Offset From Baseline to **50′** (**17** m), leaving all other parameters at their default values.

26. At the `Select marker point within assembly or [Insert Replace Detached]:` prompt, select the circular marker on the outermost point of the sidewalk subassembly.

    Your drawing should now resemble Figure 8.14.

FIGURE 8.14
The complete right side of the assembly with DaylightMaxOffset

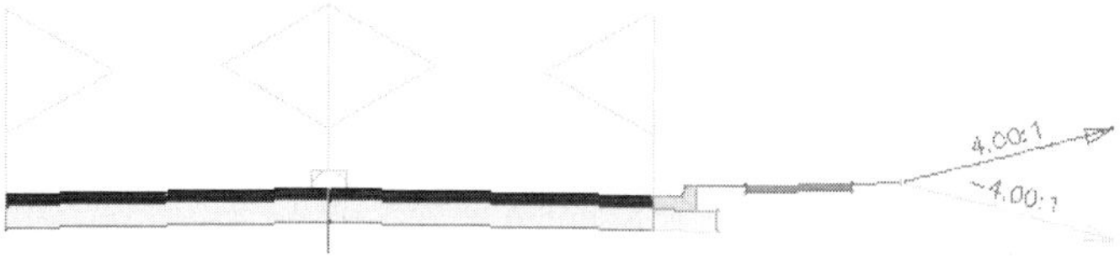

To complete the left side, you will use the Mirror Subassemblies command.

27. Select the curb, sidewalk, and daylight subassemblies on the right side of the baseline.

    The Subassembly contextual tab will show a variety of tools, including Mirror (Figure 8.15).

ASSEMBLY LABELS

You may notice the 4.00:1 label shown in Figure 8.14, which may or may not show up in your drawing as you work through this exercise. These labels are governed by the code-set style, which in this exercise is set to All Codes. In the Imperial template, no labels are assigned to the All Codes code set style, but in the metric template there is a label style assigned to the Daylight link. If you would like to add this label or other labels, follow these simple steps:

1. Switch to the Settings tab in Toolspace.
2. Expand the General ➤ Multipurpose Styles ➤ Code Set Styles branch.
3. Right-click on All Codes and select Edit to display the Code Set Style – All Codes dialog.
4. On the Codes tab, expand the Link branch.
5. In the Label Style column of the Daylight link row, click the Style button to display the Pick Style dialog.

   You may need to widen the column headings in order to view the full names.
6. Use the drop-down list to select Steep Grades.
7. Click OK to dismiss the Pick Style dialog.
8. Click OK to dismiss the Code Set Style – All Codes dialog.

Now all Daylight links in any assembly that uses the All Codes code-set style will be labeled with the grade. You may find it helpful to provide other labels on your subassemblies to be able to easily differentiate visually between the assemblies that are similar. For example, now you can tell the difference between the assembly that uses the 4:1 daylight and the one that uses the 5:1 daylight.

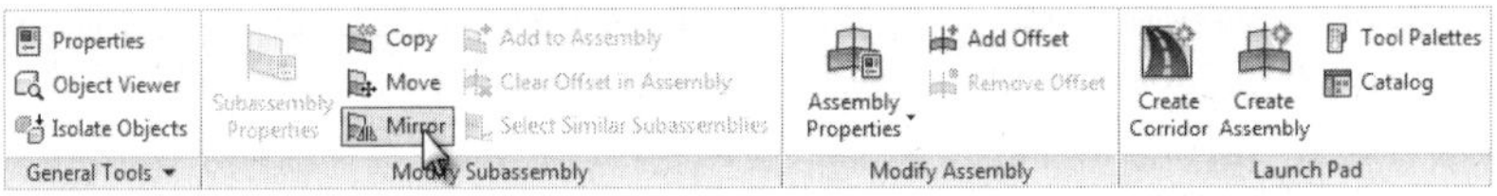

FIGURE 8.15 The Subassembly contextual tab with subassembly modification tools

28. From the Subassembly contextual tab ➤ Modify Subassembly panel choose Mirror, and then click the circular point marker located at the top left of the left LaneSuperelevationAOR subassembly.
29. Your assembly should now resemble Figure 8.5 from earlier in the chapter.

    You have now completed a typical road assembly.

You may keep this drawing open to continue on to the next exercise, or use the saved copy of this drawing available from the book's web page (`TypicalRoadAssembly_FINISHED.dwg` or `TypicalRoadAssembly_METRIC_FINISHED.dwg`).

Subassembly Components

A subassembly is made up of three basic parts: *links, marker points,* and *shapes*, as shown in Figure 8.16. Each piece plays a role in your design and is used for different purposes at each stage of the design process.

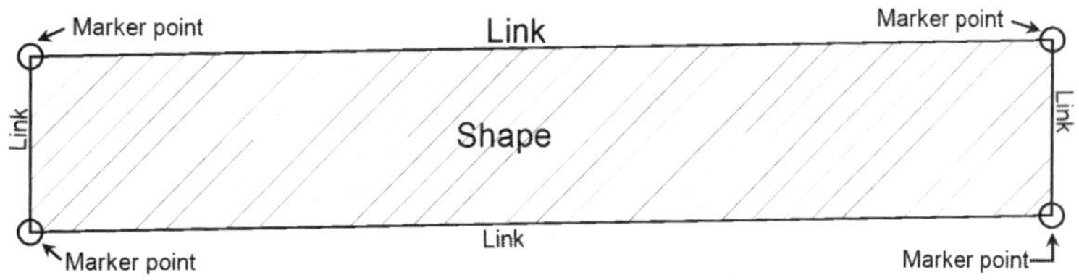

FIGURE 8.16 Schematic showing parts of a subassembly

LINKS

Links are the linear components to your assembly. A link usually represents the top or bottom of a material but can also be used as a spacer between subassemblies.

Links can have codes assigned to them that Civil 3D uses to build the design. Think of these codes as nicknames. In the example assembly you created in the previous exercise, each of the subassembly components contained numerous coded links. As shown in Figure 8.17, on the sidewalk the topmost link has the codes Top and Sidewalk and on the lane subassembly the topmost codes are Top and Pave.

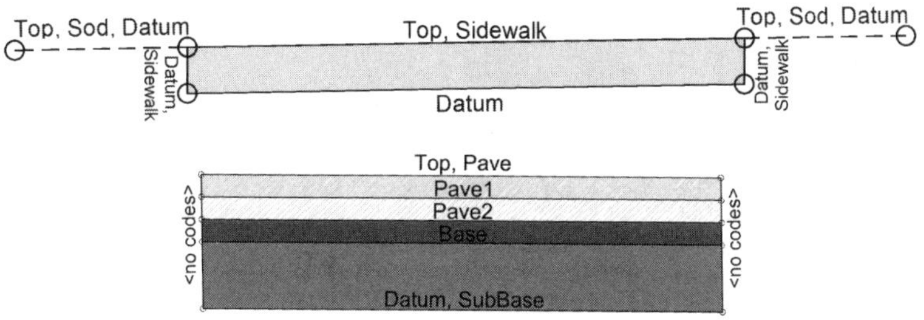

FIGURE 8.17 Link codes on the UrbanSidewalk subassembly (top) and link codes on the Lane-SuperelevationAOR subassembly (bottom)

Coded links will be your primary source of data when creating proposed surfaces from your corridors.

MARKER POINTS

Marker points are located at the endpoints of every link and usually are represented by the circles you see on the subassemblies, as shown in Figure 8.18. As you experienced in the previous exercise, the markers are used in assembly creation to "click" subassemblies together and will also "hook" to attach to alignments and/or profiles, known as *targets*.

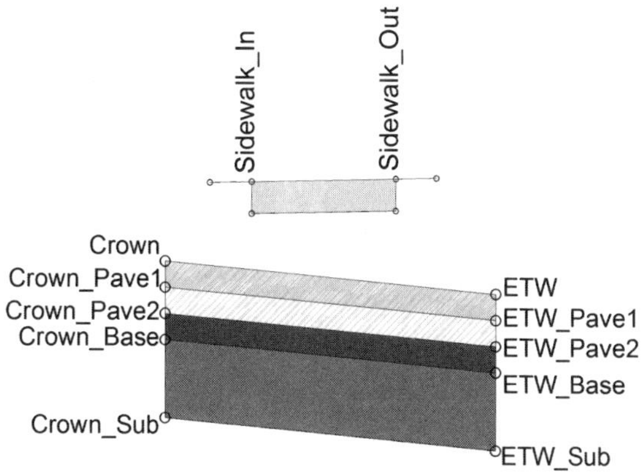

FIGURE 8.18 Point codes on the UrbanSidewalk subassembly (top) and point codes on the Lane-SuperelevationAOR subassembly (bottom)

Coded markers are the starting point for *feature lines* generated by the corridor, which are used for a variety of purposes that we will discuss in the upcoming chapters.

Shapes

Shapes are the areas inside a closed formation of links. For example, Figure 8.19 shows different subassemblies with shape codes labeled. Shapes are used in end-area material quantity calculations. At the time an assembly is created, you do not need to consider what material these shapes represent. After your corridor is complete, you will specify what materials the codes represent upon computing materials.

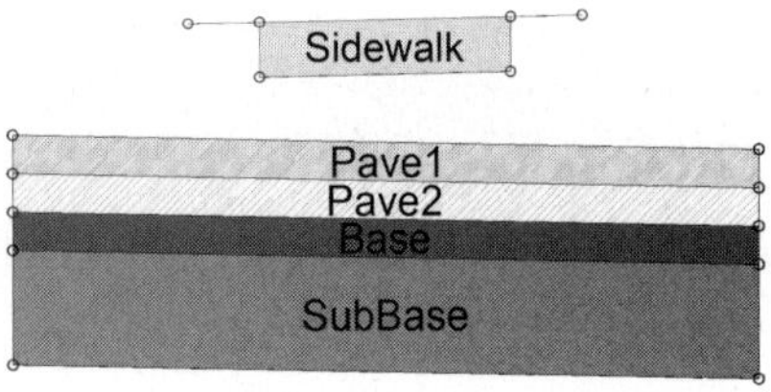

FIGURE 8.19 Shape codes on the UrbanSidewalk subassembly (top) and shape codes on the Lane-SuperelevationAOR subassembly (bottom)

Jumping into Help

Each subassembly is capable of accomplishing different tasks in your design. There is no way to tell just by looking at the icon all the acrobatics that an assembly can do. For a detailed rundown of each parameter, and what can be done with a subassembly, you will need to pop into the help files.

Subassembly Help is extremely — well, helpful! There are many doors into the help files, including from the Corridor Modeling Catalog as you saw earlier. Another way to access the help files is to right-click on any subassembly in the tool palette and select Help, as shown in Figure 8.20.

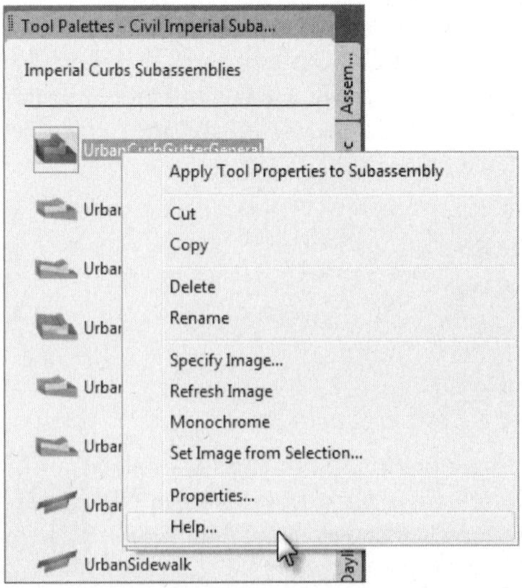

FIGURE 8.20 Getting to the subassembly help file for UrbanCurbGutterGeneral

ATTACHMENT AND INPUT PARAMETERS

When you access Subassembly Help, it will take you to the help file specific to the subassembly you are working with. At the top, you will see a diagram showing the location of the numeric parameters that can be edited in the Properties palette, as shown in Figure 8.21.

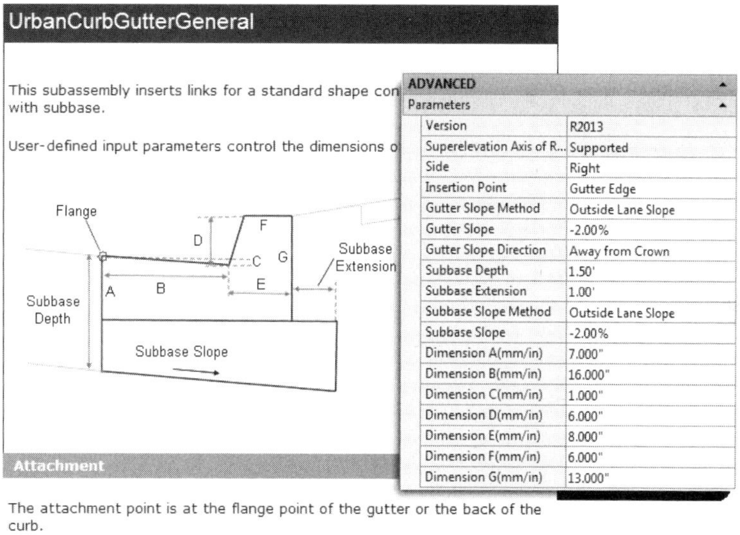

FIGURE 8.21
The top portion of Subassembly Help shown with subassembly parameters

For most subassemblies, the default attachment point will be the topmost-inside marker point. The help file will tell you if this differs for the subassembly you are looking at. Scroll further down to see detailed explanation of each input parameter.

TARGET PARAMETERS

Target parameters are a listing of what attachments can be set for a subassembly. There are three types of targets: a target surface, a target elevation, and a target offset. The help file will also tell you whether the target is optional or required. We will look at target parameters and setting targets in Chapter 10, "Basic Corridors."

OUTPUT PARAMETERS

Output parameters are values calculated on corridor build, such as the cross-slope of a lane. In several subassemblies, there is an advanced option called Parameter Reference that can use an output parameter from a previous subassembly in the assembly instead of using the value entered in the subassembly properties. We will discuss this concept further in Chapter 9, "Custom Subassemblies."

READING A CODING DIAGRAM

The coding diagram gives you a list of all the codes used on the subassembly you are working with. Every coded point, link, and shape is listed here. Not all subassembly components have explicit names, such as L9 shown in Figure 8.22. If the point, shape, or link is not included in the table, it is considered uncoded.

FIGURE 8.22
Coding diagram and name table for UrbanCurb-GutterGeneral

Point / Link	Code	Description
P1	Flange	Flange point of the gutter
P2	Flowline_Gutter	Gutter flowline point
P3	TopCurb	Top-of-curb
P4	BackCurb	Back-of-curb
L1 - L3	Top, Curb	Finish grade on the curb and gutter
L7	Subbase Datum	
S1	Curb	Curb-and-gutter concrete area
S2	Subbase	

Coding Diagram

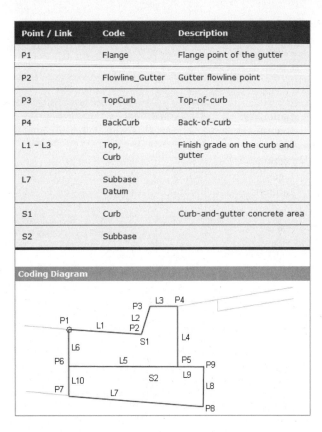

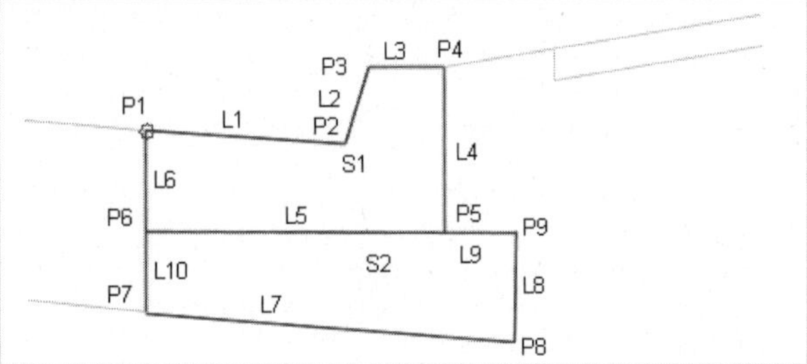

Commonly Used Subassemblies

Once you gain some skills in building assemblies, you can explore the Corridor Modeling Catalog to find subassemblies that have more advanced parameters so that you can get more out of your corridor model. For example, if you must produce detailed schedules of road materials such as asphalt, coarse gravel, fine gravel, subgrade material, and so on, the catalog includes lane subassemblies that allow you to specify those thicknesses for automatic volume reports.

The following section includes some examples of different components you can use in a typical road assembly. Many more alternatives are available in the Corridor Modeling Catalog. The help file provides a complete breakdown of each subassembly in the catalog; you'll find this useful as you search for your perfect subassembly.

Each of these subassemblies can be added to an assembly using the same process specified in the first exercise in this chapter. Choose your alternative subassembly instead of the basic parts specified in the exercise, and adjust the parameters accordingly.

Common Lane Subassemblies

The LaneSuperelevationAOR subassembly is suitable for many roads, including undivided roads as shown in the previous example, and divided roads as shown in Figure 8.23. However, you may need different road lanes for your locality or design situation:

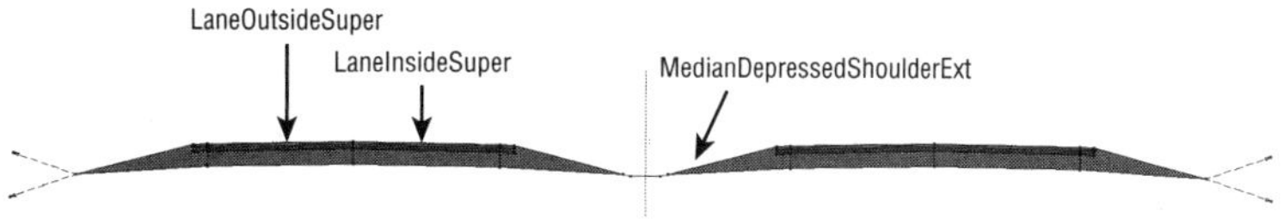

FIGURE 8.23
Use of Lane-SuperelevationAOR in a divided highway

LaneParabolic The LaneParabolic subassembly (Figure 8.24) is used for road sections that require a parabolic lane in contrast to the linear grade of LaneSuperelevationAOR. The LaneParabolic subassembly also adds options for four material depths. This is useful in jurisdictions that require two lifts of asphalt, base material, and sub-base material; taking advantage of these additional parameters gives you an opportunity to build corridor models that can return more detailed quantity takeoffs and volume calculations.

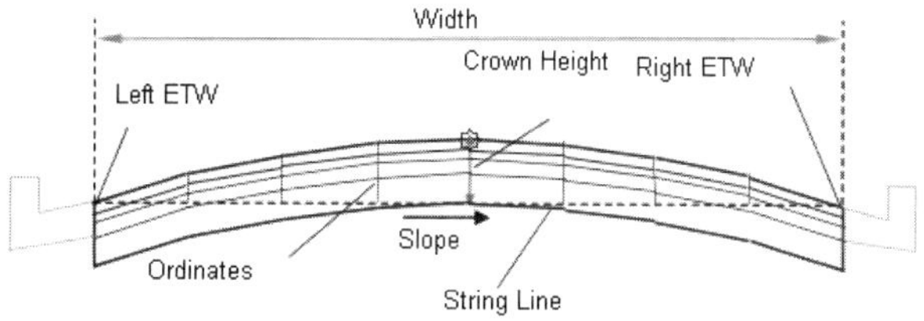

FIGURE 8.24
The LaneParabolic subassembly help diagram

Note that the LaneParabolic subassembly doesn't have a Side parameter. The parabolic nature of the component results in a single attachment point that would typically be the assembly centerline marker.

LaneBrokenBack For designs that call for two lanes, and those lanes must each have a unique slope, the LaneBrokenBack subassembly (Figure 8.25) can be used. This subassembly provides parameters to change the road-crown location and specify the width and slope for each lane. Like LaneParabolic, the LaneBrokenBack subassembly provides parameters for additional material thicknesses.

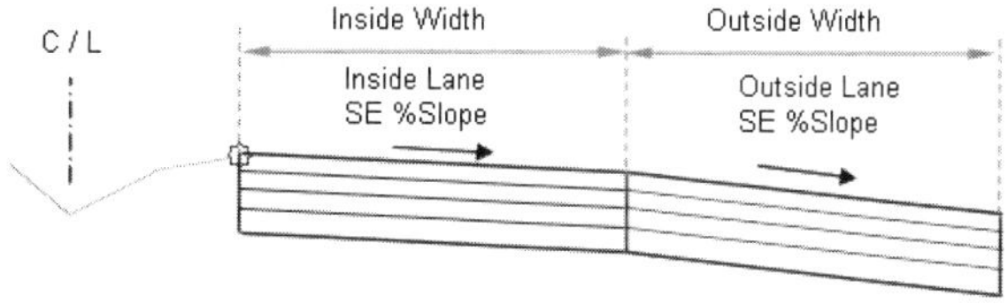

FIGURE 8.25
The LaneBrokenBack subassembly and parameters

The LaneBrokenBack subassembly, like LaneSuperelevationAOR, allows for the use of target alignments and profiles to guide the subassembly horizontally and/or vertically for both of the lanes.

COMMON SHOULDER AND CURB SUBASSEMBLIES

There are many types of curbs, and the UrbanCurbGutterGeneral subassembly can't model them all. Sometimes you may need a mountable curb, or perhaps you need a shoulder instead. In those cases, the Corridor Modeling Catalog provides many alternatives:

UrbanCurbGutterValley (1, 2, or 3) The UrbanCurbGutterValley subassemblies are great if you need mountable curbs. UrbanCurbGutterValley 1, 2, and 3, shown in Figure 8.26, Figure 8.27, and Figure 8.28 respectively, vary slightly in how they handle the sub-base slope. UrbanCurbGutterValley 1 also differs because it comes to a point instead of offering a width at the top of curb.

FIGURE 8.26
The UrbanCurb GutterValley1 subassembly help diagram

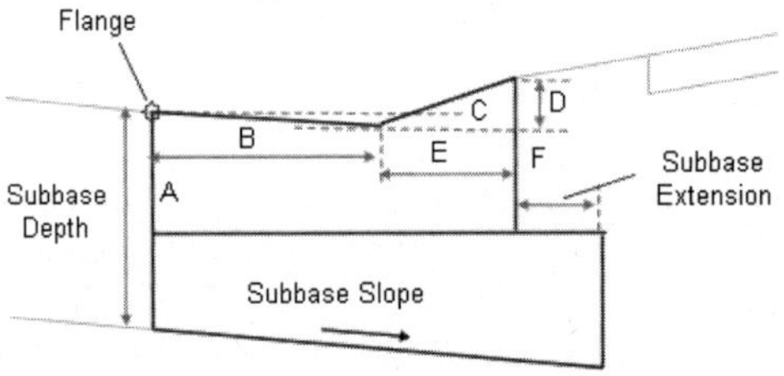

FIGURE 8.27
The UrbanCurb GutterValley2 subassembly help diagram

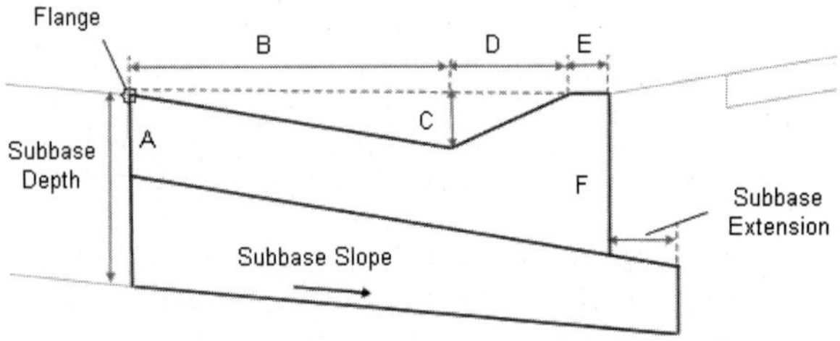

FIGURE 8.28
The UrbanCurb GutterValley3 subassembly help diagram

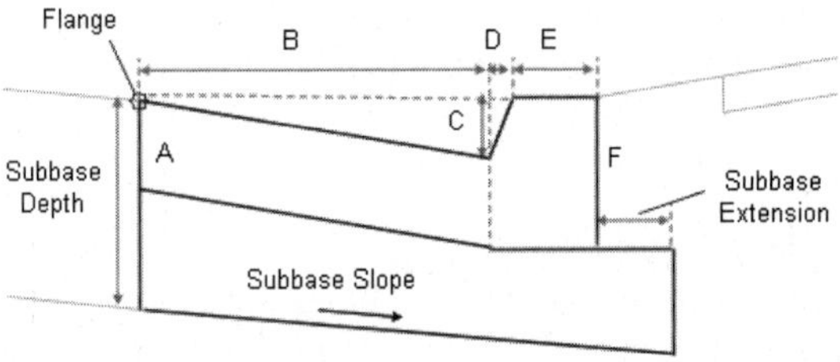

BasicShoulder BasicShoulder (see Figure 8.29) is another simple yet effective subassembly for use with road sections that require a shoulder. The predefined shape for this subassembly is Pave1, which is good if you are planning to treat this as a paved shoulder and quantify the material with the Pave1 from a lane.

FIGURE 8.29
The BasicShoulder subassembly help diagram

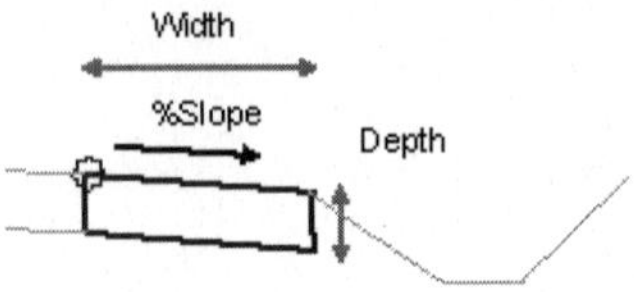

ShoulderExtendSubbase and ShoulderExtendAll Shoulders that can work with your lanes in a superelevation situation, as these two do, are extremely helpful. These two subassemblies, shown in Figure 8.30, will "play nice" with your breakover-removal settings, as you will see in Chapter 12, "Superelevation."

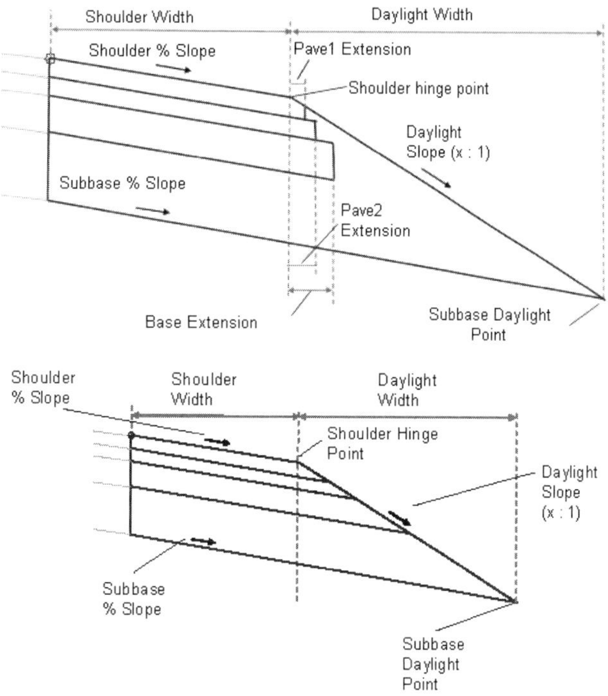

FIGURE 8.30
ShoulderExtend-Subbase subassembly help diagram (top) and ShoulderExtendAll subassembly help diagram (bottom)

Editing an Assembly

As you saw earlier in this chapter, the AutoCAD Properties palette is an option for changing subassembly parameters for one or more subassemblies of the same type. However, there are a handful of settings that can only be controlled in the Civil 3D Subassembly Properties. For example, the side (left or right) is a parameter must be changed in the Subassembly Properties.

EDITING A SINGLE SUBASSEMBLY'S PARAMETERS

Once your assembly is created, you can edit individual subassembly components as follows:

1. Pick the subassembly component you'd like to edit.

   This will bring up the Subassembly contextual tab.

Subassembly Properties

2. From the contextual tab, ➤ Modify Subassembly panel choose the Subassembly Properties option.

The Subassembly Properties dialog appears.

3. Switch to the Parameters tab, as shown in Figure 8.31, to access the same parameters you saw in the AutoCAD Properties palette when you first placed the subassembly.

FIGURE 8.31
Subassembly Properties – Parameters tab

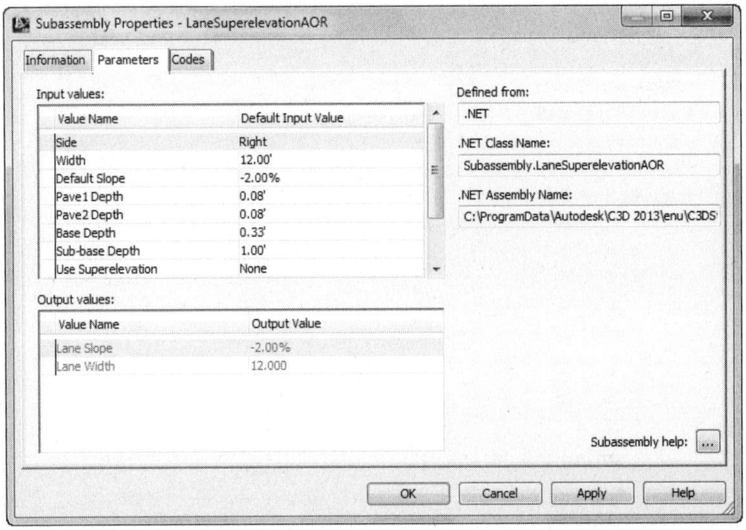

4. Click the Subassembly Help button at the bottom right of the dialog if you want to access the help page that gives detailed information about the use of this particular subassembly.

   Do not confuse the Subassembly Help button with the plain Help button, which will just give you help on the Subassembly Properties dialog.

5. Close the help file when you've finished viewing it.

6. On the Parameters tab of the Subassembly Properties, click inside any field in the Default Input Value column to make changes.

EDITING THE ENTIRE ASSEMBLY

Sometimes it's more efficient to edit all the subassemblies in an assembly at once. To do so, pick the assembly baseline marker, or any subassembly that is connected to the assembly you'd like to edit. This time, select the Assembly Properties option from the Modify Assembly panel of either the Subassembly or Assembly contextual tab to display the Assembly Properties dialog, as shown in Figure 8.32.

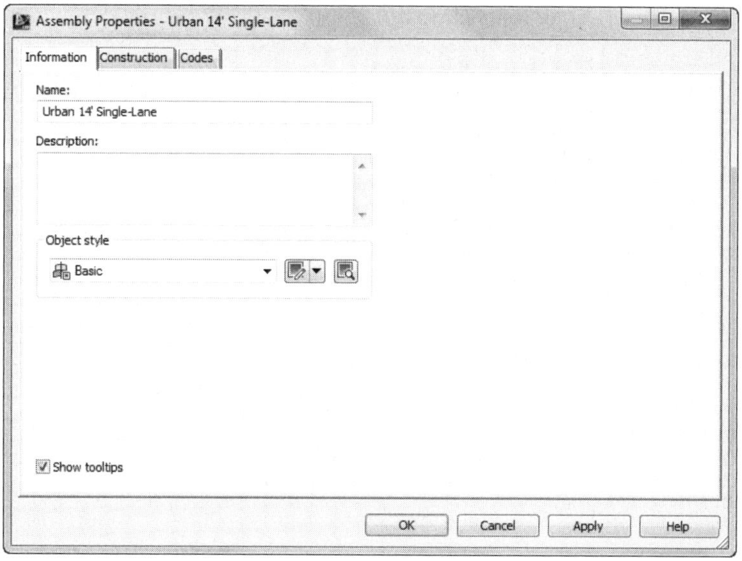

FIGURE 8.32 Assembly Properties – Information tab

Renaming the Assembly

The Information tab on the Assembly Properties dialog shown in Figure 8.32 gives you an opportunity to rename your assembly and provide an optional description. It is good practice to be consistent and detailed in your assembly names (for example, **Divided 4-Lane 12′ w Paved Shoulder**). With informative assembly names, you will eliminate much of the guesswork when it comes to building corridors later on.

Changing Parameters

The Construction tab in the Assembly Properties dialog, as shown in Figure 8.33, houses each subassembly and its parameters. At the top of the dialog you can change the Assembly Type setting using the drop-down list. In addition, you can change the parameters for individual subassemblies by selecting the subassembly in the Item pane on the left side of the Construction tab, and changing the desired parameter in the Input Values pane on the right side.

Renaming Groups and Subassemblies

Note that the left side of the Construction tab displays a list of groups. Under each group is a list of the subassemblies in use in your assembly. A new group is formed every time a subassembly is connected directly to the assembly baseline marker.

With the new side autodetection implemented in Civil 3D 2013, you will notice that the groups have already been named Right and Left with the appropriate symbol next to the group name, as shown in Figure 8.33. The subassemblies in each group appear in the same order in which they were originally placed, usually from the inside out. The first subassembly under the Right group is LaneSuperelevationAOR. If you dig into its parameters on the right side of the dialog, you'll learn that this lane is attached to the right side of the assembly marker, the UrbanCurbGutterGeneral is attached to right side of the LaneSuperelevationAOR, and the UrbanSidewalk is attached to the right side of the UrbanCurbGutterGeneral. In this example, the next group, Left, is identical but attached to the left side of the assembly baseline marker.

FIGURE 8.33 Assembly Properties – Construction tab

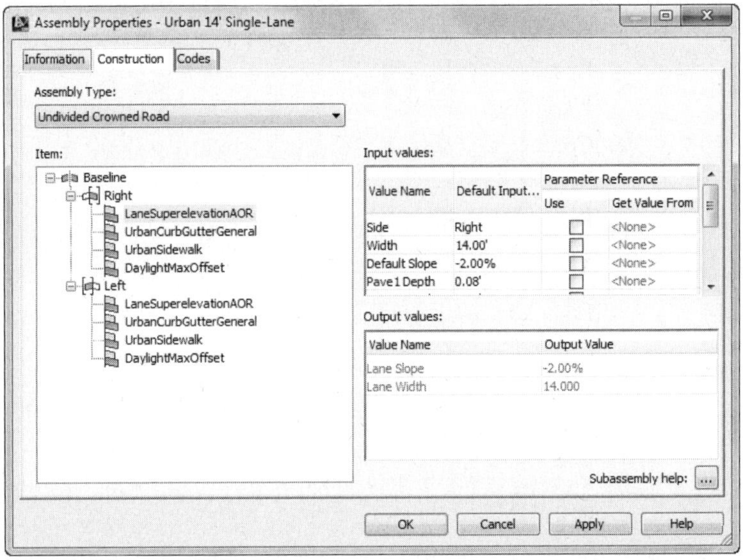

RENAMING TO INCLUDE LEFT OR RIGHT?

The automatic naming conventions are somewhat simple but usually provide enough information. In previous editions of Civil 3D, many users would change the subassembly names in order to reference what side they were on; this approach was convenient so users did not have to dig into the subassembly parameters to determine which side of the assembly a certain group was on when it came time to attach targets to a corridor. However, with AutoCAD Civil 3D 2013 both the subassembly and the assembly group are listed in the Corridor Target Mapping dialog, as you will see in Chapter 10, so there is no longer any need to add the left or right information to the subassembly.

If you want, you can rename any of the groups or subassemblies on the Construction tab of the Assembly Properties dialog by right-clicking on the group or subassembly you wish to rename and choosing Rename. From this same right-click menu you can also delete the group or subassembly.

There is no official best practice for renaming your groups and subassemblies, but you may find it useful to keep the designation of what type of subassembly it is or other distinguishing features. For example, if a lane is to be designated as a transition lane or a generic link used as a ditch foreslope, it would be useful to name them descriptively.

Creating Assemblies for Nonroad Uses

There are many uses for assemblies and their resulting corridor models aside from road sections. The Corridor Modeling Catalog also includes components for retaining walls, rail sections, bridges, channels, pipe trenches, and much more. In Chapter 10, you'll use a channel assembly and a pipe-trench assembly to build corridor models. Let's investigate how those assemblies are put together by building a channel assembly for a stream section:

1. Start a new blank drawing from the `_AutoCAD Civil 3D (Imperial) NCS` template that ships with Civil 3D. For metric users, use the `_AutoCAD Civil 3D (Metric) NCS` template, or continue working in your drawing from the first exercise in this chapter.
2. Confirm that your Tool Palettes window is showing the subassembly set (Imperial or metric) appropriate for your drawing units.
3. From the Home tab, ➤ Create Design panel choose Assembly ➤ Create Assembly.

   The Create Assembly dialog opens.
4. Enter **Channel** in the Name text box.
5. Set Assembly Type to Other.
6. Confirm that Assembly Style is set to Basic and that Code Set Style is set to All Codes, and click OK.
7. Pick a location in your drawing for the assembly — somewhere in the center of your screen where you have room to work is fine — to place your red assembly baseline marker.
8. Locate the Trench Pipes tab on the Tool Palettes window.
9. Click the Channel button on the Tool Palettes window.

   The AutoCAD Properties palette appears.
10. Locate the Advanced Parameters section of the Design tab on the AutoCAD Properties palette.

    You'll place the channel with its default parameters and make adjustments through the Assembly Properties dialog, so don't change anything for now. Note that there is no Side parameter. This subassembly will be centered on the assembly baseline marker.
11. At the `Select marker point within assembly or [Insert Replace Detached]:` prompt, select the red assembly baseline marker, and a channel is placed on the assembly (see Figure 8.34).

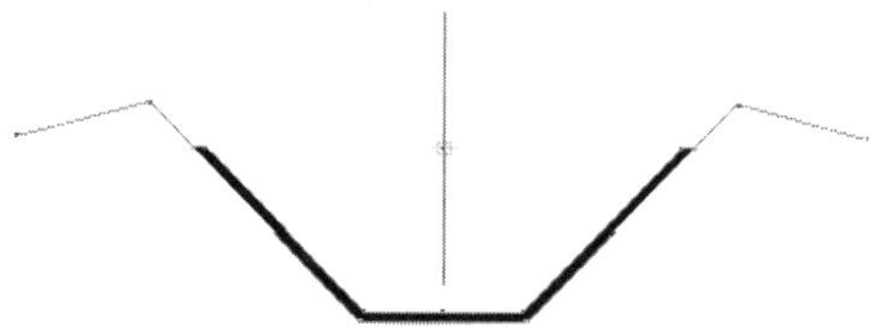

FIGURE 8.34
The Channel subassembly with default parameters

12. Press Esc to leave the assembly creation command and dismiss the palette.
13. Right-click the assembly baseline marker and select Assembly Properties.

    The Assembly Properties dialog appears.
14. Switch to the Construction tab.

    Notice that while the typical road assembly in the previous exercise generated a Left group and a Right group, the Channel subassembly generated a Centered group.
15. Select the Channel entry on the left side of the dialog (under the Centered group).
16. Click the Subassembly Help button located at the bottom right on the dialog's Construction tab.

    The Subassembly Reference page of the AutoCAD Civil 3D 2013 help file appears.
17. Familiarize yourself with the diagram, as shown in Figure 8.35, and the input parameters for the Channel subassembly.

    Especially note the Attachment Point, Bottom Width, Depth, and Sideslope parameters. The attachment point indicates where your baseline alignment and profile will be applied.

FIGURE 8.35 The Channel subassembly help diagram

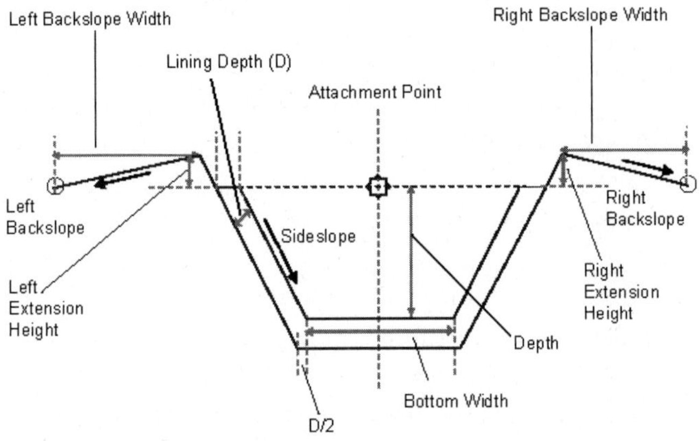

18. Minimize or close the help file.

    To match the engineer's specified design, you need a stream section 6′ (2 m) deep with a 3′ (1 m)-wide bottom, 1:1 sideslopes, and no backslopes.
19. Change the following parameters in the Assembly Properties dialog, leaving all other parameters at their default values:

    Depth: **6′** (**2** m)

    Bottom Width: **3′** (**1** m)

    Sideslope: **1** (This value will automatically change to be displayed as 1:1.)

    Left and Right Backslope Width: **0′** (**0** m)

ZERO SUBASSEMBLY VALUES

There are some subassemblies that do not like zero values, so you may be taught to enter 0.001 or some other value that is so small that it is as if you enter 0. However, there are also some subassemblies that like zero values. If you look in the help file for the Channel subassembly, you will notice that in the Behavior section it explains that if a zero value is specified for left or right extensions and backslope widths, those links are omitted or are not drawn. So in this case a zero is what you want, but be sure to check the help files before using a zero in your subassemblies!

20. Click OK, and confirm that your completed assembly looks like Figure 8.36.

FIGURE 8.36 The channel assembly with customized parameters

You may keep this drawing open to continue on to the next exercise, or use the finished copy of this drawing available from the book's web page (`ChannelAssembly_FINISHED.dwg` or `ChannelAssembly_METRIC_FINISHED.dwg`).

Real World Scenario

A PIPE TRENCH ASSEMBLY

Projects that include piping, such as sanitary sewers, storm drainage, gas pipelines, or similar structures, almost always include trenching. The trench must be carefully prepared to ensure the safety of the workers placing the pipe, as well as provide structural stability for the pipe in the form of bedding and compacted fill.

The corridor is an ideal tool for modeling pipe trenching. With the appropriate assembly combined with a pipe-run alignment and profile, you can not only design a pipe trench but also use cross-section tools to generate section views, materials tables, and quantity takeoffs. The resulting corridor model can also be used to create a surface for additional analysis.

The following exercise will lead you through building a pipe trench corridor based on an alignment and profile that follow a pipe run, and a typical trench assembly:

1. Start a new blank drawing from the `_AutoCAD Civil 3D (Imperial) NCS` template that ships with Civil 3D. For metric users, use the `_AutoCAD Civil 3D (Metric) NCS` template, or continue working in your drawing from the previous exercise.
2. Confirm that your Tool Palettes window is showing the subassembly set (Imperial or metric) appropriate for your drawing units.
3. From the Home tab, ➤ Create Design panel choose Assembly ➤ Create Assembly.

   The Create Assembly dialog opens.

4. Enter **Pipe Trench** in the Name text box to change the assembly's name.
5. Set Assembly Type to Other.
6. Confirm that Assembly Style is set to Basic and Code Set Style is set to All Codes, and click OK.
7. Pick a location in your drawing for the assembly; somewhere in the center of your screen where you have room to work is fine.
8. Locate the Trench Pipes tab on the Tool Palettes window.
9. Click the TrenchPipe1 button on the Tool Palettes window.

   The AutoCAD Properties palette appears.
10. Locate the Advanced section of the Design tab on the AutoCAD Properties palette.

    This section lists the TrenchPipe1 parameters. You'll place TrenchPipe1 with its default parameters and make adjustments through the Assembly Properties dialog, so don't change anything for now. Note that similar to the Channel subassembly, there is no Side parameter. This subassembly will be placed centered on the assembly baseline marker.
11. At the `Select marker point within assembly or [Insert Replace Detached]:` prompt, select the assembly baseline marker. A TrenchPipe1 subassembly is placed on the assembly as shown here:

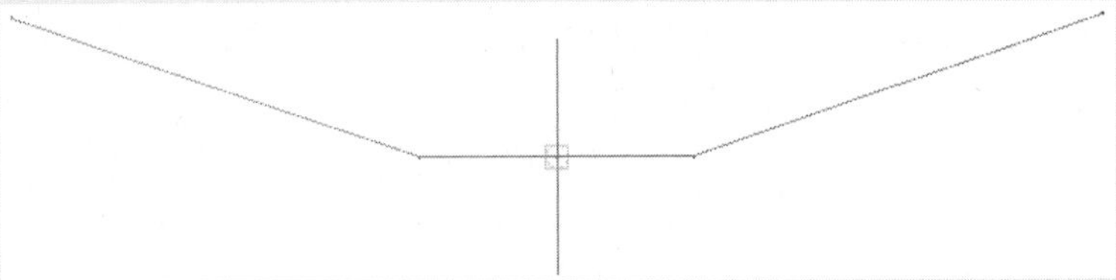

12. Press Esc to leave the assembly creation command and dismiss the AutoCAD Properties palette.
13. Select the assembly baseline marker to activate the Assembly contextual ab.
14. From the Assembly contextual tab ➢ Modify Assembly panel choose Assembly Properties.

    The Assembly Properties dialog appears.
15. On the Construction tab, select the TrenchPipe1 assembly entry on the left side of the dialog.
16. Click the Subassembly Help button located at the bottom right.

    The Subassembly Reference page of the AutoCAD Civil 3D 2013 Help file appears.
17. Familiarize yourself with the diagram, shown in the following graphic, and with the input parameters for the TrenchPipe1 subassembly.

    In this case, the profile grade line will attach to a profile drawn to represent the pipe invert. Because the trench will be excavated deeper than the pipe invert to accommodate gravel bedding, you'll want to provide information for the bedding depth parameter. Also note under the Target Parameters that this subassembly requires a surface target to determine where the sideslopes terminate.

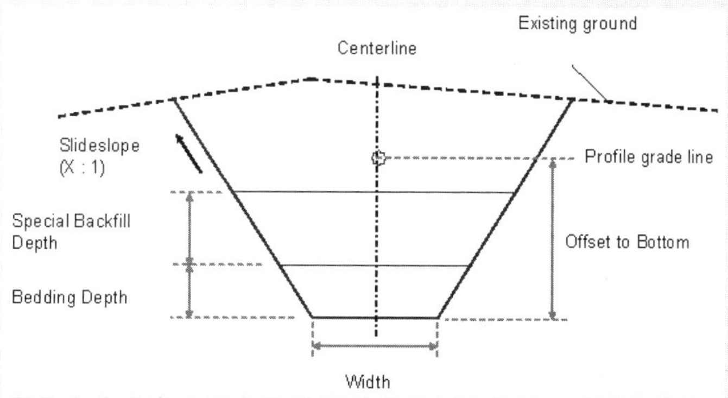

18. Minimize or close the help file.

To match the engineer's specified design, the pipe trench should be 3′ (1 m) deep and 4′ (1.3 m) wide with 2:1 sideslopes and 1′ (0.3 m) of gravel bedding.

19. In the Assembly Properties dialog, change the following parameters, leaving all other parameters at their default values:

Width: **4′** (1.3 m)

Sideslope: **2** (This value will automatically change to be displayed as 2:1.)

Bedding Depth: **1′** (**0.3** m)

Offset To Bottom: **3′** (**1** m)

20. Click OK.

21. Confirm that your completed assembly looks like the graphic shown here.

You may keep this drawing open to continue on to the next exercise, or use the finished copy of this drawing available from the book's web page (`PipeTrenchAssembly_FINISHED.dwg` or `PipeTrenchAssembly_METRIC_FINISHED.dwg`).

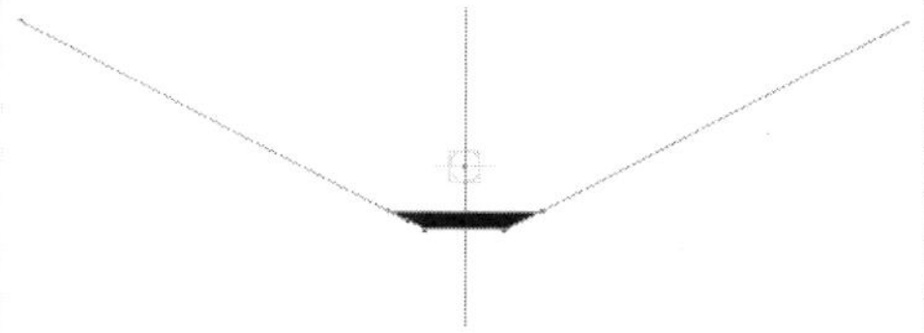

This assembly will be used to build a pipe-trench corridor in Chapter 10.

Specialized Subassemblies

Despite the more than 100 subassemblies available in the Corridor Modeling Catalog, sometimes you may not find the perfect component. Perhaps none of the channel assemblies exactly meet your design specifications, and you'd like to make a more customized assembly, or neither of the sidewalk subassemblies allows for the proper boulevard slopes. Maybe you'd like to try to do some preliminary lot grading using your corridor, or mark a certain point on your subassembly so that you can extract important features easily.

You can handle most of these situations by using subassemblies from the Generic Subassembly Catalog (see Figure 8.37). These simple yet flexible components can be used to build almost anything, although they lack the coded intelligence of some of the more intricate assemblies (such as knowing if they're paved, grass, or similar, and understanding things like sub-base depth, and so on).

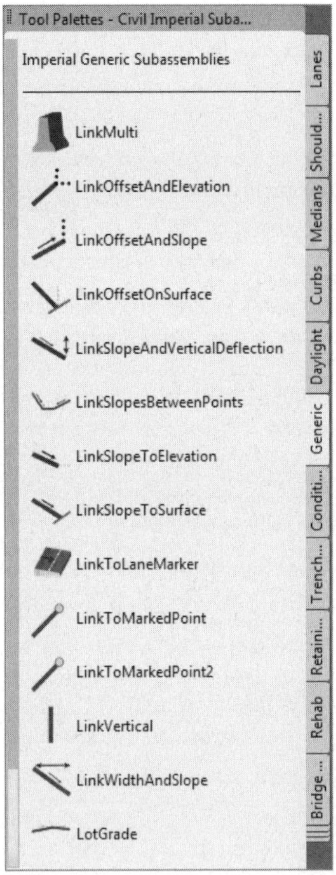

FIGURE 8.37
The Generic Subassembly tool palette

Using Generic Links

Let's look at two examples where you might take advantage of generic links.

The first example involves the typical road section you built in the first exercise in this chapter. You saw that UrbanSidewalk doesn't allow for differing cross-slopes for the inside boulevard (terrace), sidewalk, and outside boulevard (buffer strip). If you need a 3′ (1 m)-wide terrace with a 3 percent slope, and then a 5′ (1.5 m) sidewalk with a 2 percent slope, followed by another buffer strip that is 6′ (2 m) wide with a slope of 5 percent, you can use generic links to assist in the construction of the proper assembly.

In this exercise we will be creating a new assembly based on the typical road assembly made in the first exercise. Therefore, any of the previously saved files (which you can download from this book's web page) can be used, if you do not have one open from a previous exercise.

1. In Prospector, locate and expand the Assemblies group.
2. Right-click on Urban 14′ Single-Lane or Urban 4.5 m Single-Lane and select Zoom To.
3. Select the lane and curb subassemblies as well as the assembly baseline marker; then right-click and select Clipboard ➢ Copy.
4. Right-click to select Clipboard ➢ Paste and pick a location directly under the Urban Single-Lane assembly to paste the copied assembly.

   While you could place the subassembly anywhere, you will find that as you gather more and more assemblies in a drawing, having them organized in a logical manner with similar assemblies in a common area makes them easier to manage.

LABELING ASSEMBLIES

When you start getting multiple similar assemblies in your drawing, you may find it helpful to add an Mtext next to the assembly with the assembly's name so that you know which assembly is which. By using a Field in an Mtext, these labels will remain dynamic to their associated object (i.e., if you change the name of the assembly, the Mtext will change as well). You can do this using the following simple steps:

1. Enter MTEXT on the command line.
2. Specify the location of your Mtext box.
3. From the Text Editor contextual tab ➢ Insert panel, choose the Field tool to display the Field dialog.
4. Verify that Field Category is set to Objects and Field Names is set to Object.
5. Click the Select button next to Object Type.
6. At the `Select object:` prompt, select the assembly baseline marker.
7. Set Property to Name and click OK to dismiss the Field dialog.

You now have a dynamic field that will maintain the name of the associated assembly; however, you may need to run a REGEN in order for the field to update.

5. Select and right-click the assembly baseline marker and select Assembly Properties.

   The Assembly Properties dialog appears.

6. On the Information tab, change Name to **Urban 14′ Single-Lane with Terraced Sidewalk** (or **Urban 4.5 m Single-Lane with Terraced Sidewalk**) and click OK.

   If you added the Field coded Mtext, you may want to run a REGEN to update the label.

7. Locate the Generic tab on the Tool Palettes window.
8. Click the LinkWidthandSlope subassembly, and the AutoCAD Properties palette appears.

9. Scroll down to the Advanced Parameters section of Properties and change the parameters as follows to create the first buffer strip, leaving all other parameters at their default values:

   Width: **3′** (**1** m)

   Slope: **3%**

10. At the `Select marker point within assembly or [Insert Replace Detached]:` prompt, select the circular marker on the right UrbanCurbGutterGeneral subassembly, as well as the circular marker on the left UrbanCurbGutterGeneral subassembly, both of which represent the top back of the curb.

11. Switch to the Curbs tab of the Tool Palettes window, and click the UrbanSidewalk button.

12. In the Advanced Parameters area of Properties, change the parameters as follows to create the sidewalk, leaving all other parameters at their default values:

    Sidewalk Width: **5′** (**1.5** m)

    Slope: **2%**

    Inside Boulevard Width: **0′** (**0** m)

    Outside Boulevard Width: **0′** (**0** m)

13. At the `Select marker point within assembly or [Insert Replace Detached]:` prompt, select the circular marker on the right LinkWidthandSlope subassembly.

14. Switch to the Generic tab of the Tool Palettes window, and click the LinkWidthandSlope button.

    The AutoCAD Properties palette appears.

15. In the Advanced Parameters area of Properties, change the parameters as follows to create the second buffer strip, leaving all other parameters at their default values:

    Width: **6′** (**2** m)

    Slope: **5%**

16. At the `Select marker point within assembly or [Insert Replace Detached]:` prompt, the upper right circular marker on the right UrbanSidewalk subassembly, as well as the upper left circular marker on the left UrbanSidewalk subassembly, both of which represent the outside edge of the sidewalk. When complete, press Esc to end the command.

17. Select and right-click the right daylight subassembly from the Urban Single-Lane assembly and select Copy To.

18. Select the outermost marker on the right side of the new assembly that you are working on. Do the same for the left daylight.

    The completed assembly should look like Figure 8.38 (shown with the typical road assembly from the first exercise for comparison).

You may keep this drawing open to continue on to the next exercise, or use the finished copy of this drawing available from the book's web page (`GenericLinks_FINISHED.dwg` or `GenericLinks_METRIC_FINISHED.dwg`).

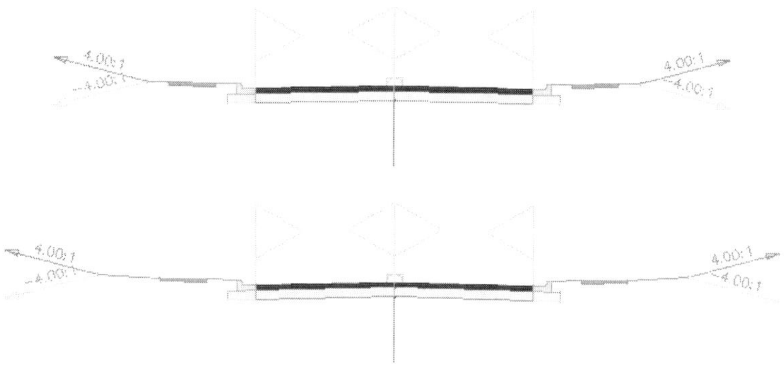

FIGURE 8.38
The completed Urban Single-Lane assembly from the first exercise (top) and the Urban Single-Lane with Terraced Sidewalks assembly (bottom)

You've now created a custom sidewalk terrace for a typical road.

Daylighting with Generic Links

The next example involves the channel section you built earlier in this chapter. This exercise will lead you through using the LinkSlopetoSurface generic subassembly, which will provide a surface target to the channel assembly that will seek the target assembly at a 25 percent slope. For more information about surface targets, see Chapter 10.

In this exercise, you will be creating another new assembly based on the channel assembly made in the second exercise; therefore, any of the previously saved files (which you can download from this book's web page) can be used, if you do not have one open from a previous exercise. You do not need to have the other previous exercises completed to continue.

1. In Prospector, locate and expand the Assemblies group.
2. Right-click on Channel and select Zoom To.
3. Locate the Generic tab on the Tool Palettes window.
4. Click the LinkSlopetoSurface button.
5. In the Advanced Parameters area of Properties, change the Slope parameter to **25%**, leaving all other parameters at their default values.
6. At the `Select marker point within assembly or [Insert Replace Detached]:` prompt, select the circular marker at the upper right on the channel subassembly, as well as the circular marker on the upper left on the channel subassembly.

   A surface target link appears. Press Esc to end the command.

   The completed assembly should look like Figure 8.39.

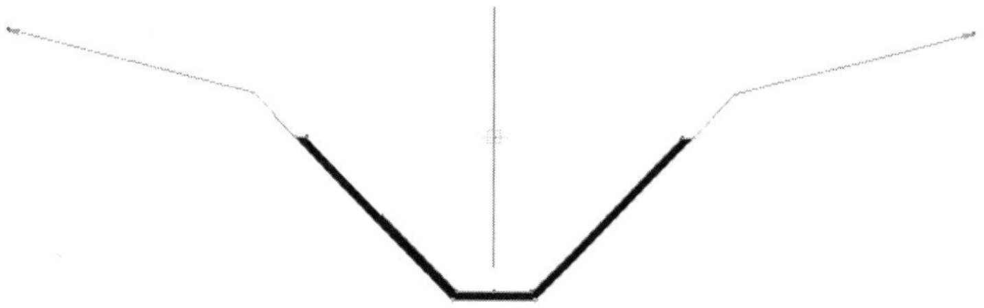

FIGURE 8.39
The completed channel assembly

You may keep this drawing open to continue on to the next exercise, or use the finished copy of this drawing available from the book's web page (`ChannelLinkDaylight_FINISHED.dwg` or `ChannelLinkDaylight_METRIC_FINISHED.dwg`).

Adding a surface link to the channel assembly provides a surface target for the assembly. Now that you've added the LinkSlopetoSurface, you will be able to specify your existing ground as the surface target for a corridor, and the subassembly will grade between the top of the bank and the surface for you. You can achieve additional flexibility for connecting to existing ground with the more complex daylight subassemblies, as discussed in the next section.

Working with Daylight Subassemblies

In previous examples, we worked with a generic daylight subassembly, but now let's take a closer look at what they can do for you.

A daylight subassembly tells Civil 3D how to extend a link to a target surface. The instructions might include a ditch or berm before looking for existing ground. Others provide a straight shot but with contingencies for certain design conditions. Figure 8.40 shows the many options you have for adding a daylight subassembly to an assembly.

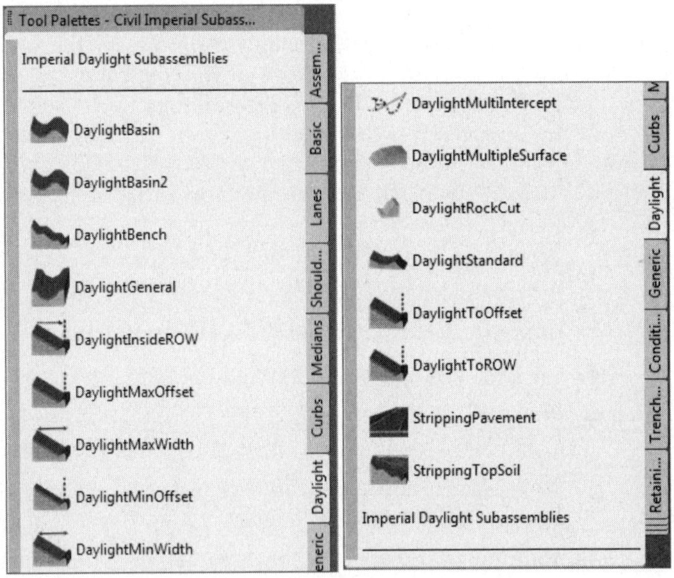

FIGURE 8.40
Daylight subassemblies in the Tool Palettes window

In the following exercise, you'll use the DaylightInsideROW subassembly. This subassembly contains parameters for specifying the maximum distance from the centerline or offset alignments. If the 4:1 slope hits the surface inside the right-of-way (ROW), no adjustment is made to the slope. If 4:1 causes the daylight to hit outside of the ROW, the slope adjusts to stay inside the specified location.

In this exercise you will be creating a new assembly based on the typical road assembly you made in the first exercise; therefore, any of the previously saved files (which you can download from this book's web page) can be used, if you do not have one open from a previous exercise.

1. In Prospector, locate and expand the Assemblies group.
2. Right-click on Urban 14′ Single-Lane or Urban 4.5 m Single-Lane, and select Zoom To.
3. Select the lane, curb, and sidewalk subassemblies as well as the assembly baseline marker, and right-click to select Clipboard ➢ Copy.
4. Pick a location directly under the Urban Single-Lane assemblies to paste the copied assembly.
5. Select and right-click the assembly baseline marker and select Assembly Properties.

   The Assembly Properties dialog appears.
6. On the Information tab, change the name to **Urban 14′ Single-Lane Daylight ROW** (or **Urban 4.5 m Single-Lane Daylight ROW**), and click OK.
7. Locate the Daylight tab on the Tool Palettes window.
8. Right-click the DaylightInsideROW button on the Tool Palettes panel and select Help.

   The Subassembly Reference page opens in a new window.
9. Familiarize yourself with the options for the DaylightInsideROW subassembly, especially noting the optional parameters for a lined material, a mandatory daylight surface target, and an optional ROW offset target that can be used to override the ROW offset specified in the parameters.
10. Click the DaylightInsideROW button on the Tool Palettes window.
11. In the Advanced Parameters area of Properties, change the parameter ROW Offset From Baseline to **33′** (**10** m), leaving all other parameters at their default values.
12. At the `Select marker point within assembly or [Insert Replace Detached]:` prompt, select the circular marker on the farthest-right link.
13. Click the DaylightInsideROW button on the Tool Palettes window again, but this time in the Advanced Parameters area of Properties, change the parameter ROW Offset From Baseline to **-33′** (**-10** m) before placing the left side.

    Notice there is no Left or Right parameter. The negative value in the ROW Offset From Baseline parameter is what tells Civil 3D the daylight is to the left.
14. You can now dismiss the Properties palette.

    The completed assembly should look like Figure 8.41.

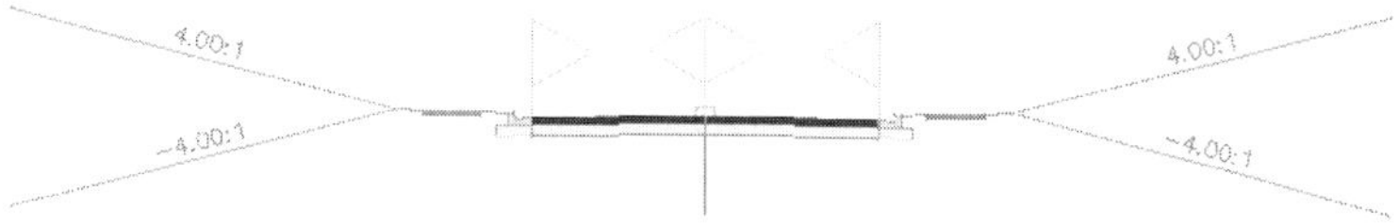

FIGURE 8.41 An assembly with the DaylightInsideROW subassembly attached to each side

You may keep this drawing open to continue on to the next exercise, or use the finished copy of this drawing available from the book's web page (`DaylightROWAssembly_FINISHED.dwg` or `DaylightROWAssembly_METRIC_FINISHED.dwg`).

When to Ignore Daylight Input Parameters

The first time you attempt to use many daylight subassemblies, you may become overwhelmed by the sheer number of parameters.

The good news is that many of these parameters are unnecessary for most uses. For example, many daylight subassemblies, such as DaylightGeneral (shown here), include multiple cut-and-fill widths for complicated cases where the design may call for test scenarios. If your design doesn't require this level of detail, leave those parameters set to zero.

ADVANCED

Parameters

Version	R2013
Side	Left
Daylight Link	Include Daylight link
Cut Test Point Link	3
Cut 1 Width	0.00'
Cut 1 Slope	Horizontal
Cut 2 Width	0.00'
Cut 2 Slope	Horizontal
Cut 3 Width	0.00'
Cut 3 Slope	Horizontal
Cut 4 Width	0.00'
Cut 4 Slope	Horizontal
Cut 5 Width	0.00'
Cut 5 Slope	Horizontal
Cut 6 Width	0.00'
Cut 6 Slope	Horizontal
Cut 7 Width	0.00'
Cut 7 Slope	Horizontal
Cut 8 Width	0.00'
Cut 8 Slope	Horizontal
Flat Cut Slope	6.00:1
Flat Cut Max Height	5.00'
Medium Cut Slope	4.00:1
Medium Cut Max Height	10.00'
Steep Cut Slope	2.00:1
Fill 1 Width	0.00'
Fill 1 Slope	Horizontal
Fill 2 Width	0.00'
Fill 2 Slope	Horizontal
Fill 3 Width	0.00'
Fill 3 Slope	Horizontal
Flat Fill Slope	6.00:1
Flat Fill Max Height	5.00'
Medium Fill Slope	4.00:1
Medium Fill Max Height	10.00'
Steep Fill Slope	2.00:1
Guardrail Width	2.00'
Guardrail Slope	-2.00%
Include Guardrail	Omit Guardrail
Width to Post	1.000
Rounding Option	None
Rounding By	Length
Rounding Parameter	1.50'
Rounding Tessellation	6
Place Lined Material	None
Slope Limit 1	1.00:1
Material 1 Thickness	1.00'
Material 1 Name	Rip Rap
Slope Limit 2	2.00:1
Material 2 Thickness	0.50'
Material 2 Name	Rip Rap
Slope Limit 3	4.00:1
Material 3 Thickness	0.33'
Material 3 Name	Seeded Grass

Some daylight subassemblies include guardrail options. If your situation doesn't require a guardrail, leave the default parameter set to Omit Guardrail and ignore it from then on. Another common, confusing parameter is Place Lined Material, which can be used for riprap or erosion-control matting. If your design doesn't require this much detail, ensure that this parameter is set to None, and ignore the thickness, name, and slope parameters that follow.

If you're ever in doubt about which parameters can be omitted, investigate the help file for that subassembly.

Alternative Daylight Subassemblies

Over a dozen daylight subassemblies are available, varying from a simple cut-fill parameter to a more complicated benching or basin design. Your engineering requirements may dictate something more challenging than the exercise in this section. Here are some alternative daylight subassemblies and the situations where you might use them. For more information on any of these subassemblies and the many other daylighting choices, see the AutoCAD Civil 3D 2013 Subassembly Reference page in the help file.

DaylightToROW and DaylightInsideROW The DaylightToROW subassembly differs slightly from the DaylightInsideROW, as shown in Figure 8.42. DaylightToROW constantly adjusts the slope to stay a certain distance away from your ROW, as specified by the Offset Adjustment input parameter. For example, you can have a ROW alignment specified, but use this subassembly to tell Civil 3D to always stay 3′ inside the ROW line. The DaylightInsideROW uses the typical slope but adjusts up to a maximum slope in order to stay inside of the ROW. In both subassemblies, you must specify an offset value or an offset target to use as the ROW.

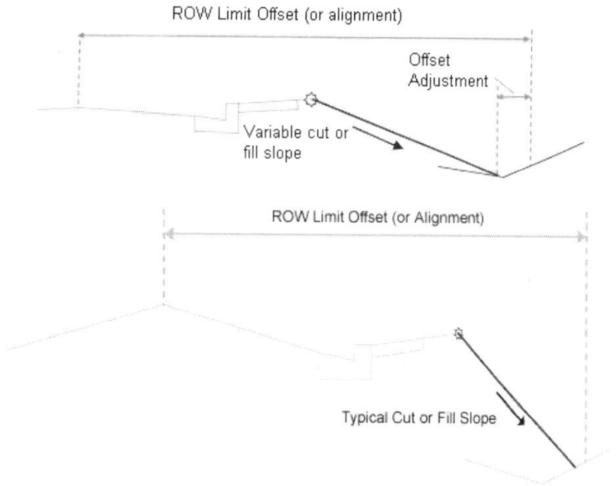

Figure 8.42 DaylightToROW subassembly help diagram (top) and DaylightInsideROW subassembly help diagram (bottom)

BasicSideSlopeCutDitch In addition to including cut-and-fill parameters, the BasicSideSlopeCutDitch subassembly (see Figure 8.43) creates a ditch in a cut condition. This is most useful for road sections that require a roadside ditch through cut sections, but omit it when passing through areas of fill. If your corridor model is revised in a way that changes the location of cut-and-fill boundaries, the ditch will automatically adjust. Note that this subassembly is located on the Basic tab whereas the other subassemblies in this section are located on the Daylight tab.

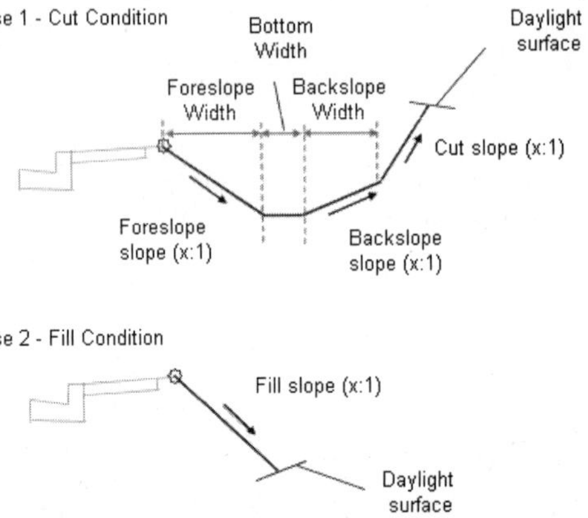

FIGURE 8.43
The BasicSideSlopeCutDitch subassembly help diagram

When you insert this subassembly you will notice that it does not look anything like the help diagram and instead will display the "LayoutMode" text on the design assembly, as shown in Figure 8.44. This will not display on the completed corridor. There are several subassemblies where this will occur — which is another good reason to always check the help file for an accurate representation of what the final product will look like.

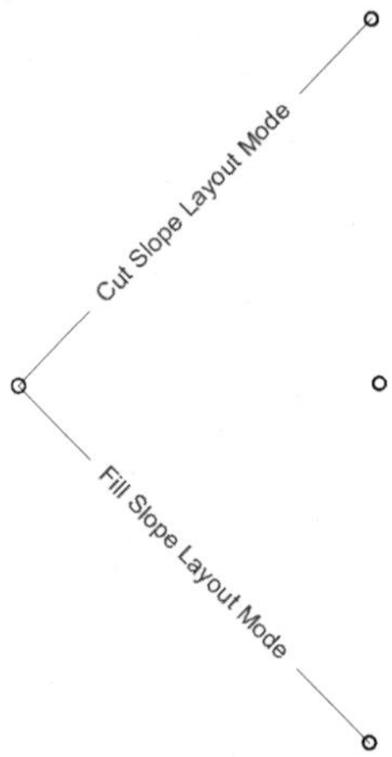

FIGURE 8.44
The BasicSideSlopeCutDitch in layout mode

DaylightBasin Many engineers must design berms to contain roadside swales when the road design is in the fill condition. The process for determining where these berms are required is often tedious. The DaylightBasin subassembly (see Figure 8.45) provides a tool for automatically creating these "false berms." The subassembly contains parameters for the specification of a basin (which can be easily adapted to most roadside ditch cross sections as well) and parameters for containment berms that appear only when the subassembly runs into areas of roadside cut.

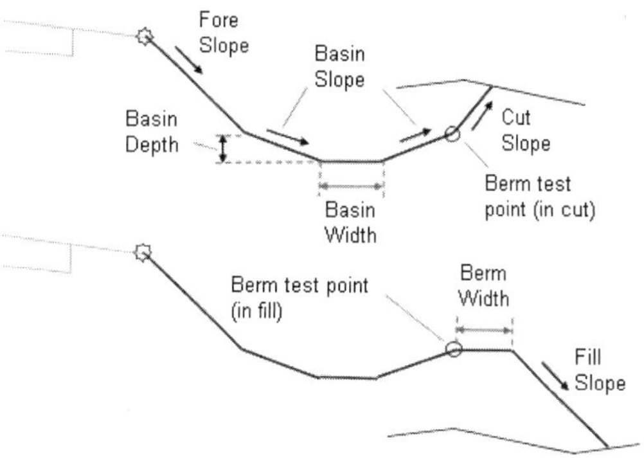

FIGURE 8.45 The DaylightBasin subassembly help diagram

Advanced Assemblies

As you get to know Civil 3D better, you will want it to do more for you. With the tools you are given and your own creativity and problem-solving skills, Civil 3D can create some complex designs. Offset assemblies and marked point assemblies are powerful tools you have at your fingertips.

Offset Assemblies

Offset assemblies are an advanced option when you want to model a coordinating component of the design whose cross section is related to the main assembly. An example of where an offset assembly would be helpful is a main road adjacent to a meandering bike path. The bike path generally follows the main road, but its alignment is not always parallel and the profile might be altogether different. Figure 8.46 shows what the assembly for a bike path to the left of a road would look like.

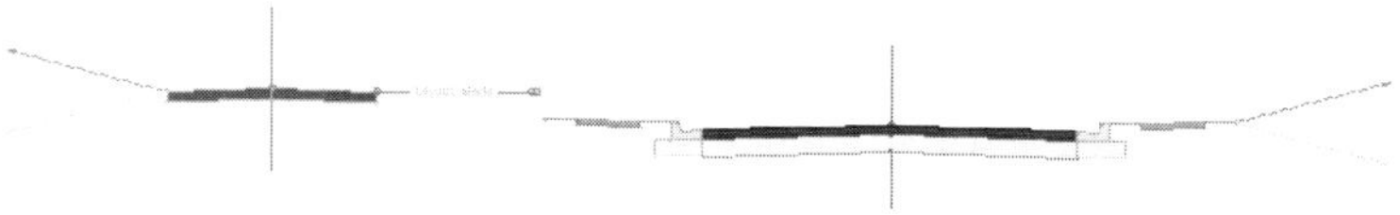

FIGURE 8.46 An example of an assembly with an offset to the left representing a bike path

To use an offset assembly, from the Home tab ➢ Create Design panel choose Assembly ➢ Add Assembly Offset. You will be prompted to select the main assembly and place the offset in the graphic. The location of the offset assembly in relation to the main assembly will have no effect on the final design.

Once the offset assembly is placed, the construction of the offset assembly is identical to any other assembly. We will use an example of an assembly with an offset in Chapter 11, "Advanced Corridors, Intersections, and Roundabouts."

Marked Points and Friends

The marked point assembly is a small but powerful subassembly found in the Generic palette. It consists of a single marker, and you can place it on an assembly to flag a location. You can use the marked point by itself to generate a feature line where no coded marker currently exists, say in the midpoint of a lane link. Where marked points really shine are when used with one of the subassemblies designed to look for a marked point.

When using a marked point, name it right away, and make note of that name for using it with its "friends" (Figure 8.47).

FIGURE 8.47 Name the marked point in the Advanced Parameters

LINKING TO A MARKED POINT

In the example shown in Figure 8.48, a LinkToMarkedPoint2 subassembly is placed on the right side of the bike path pavement. The LinkToMarkedPoint2 subassembly has been created to look for the marked point on the left side of the sidewalk buffer.

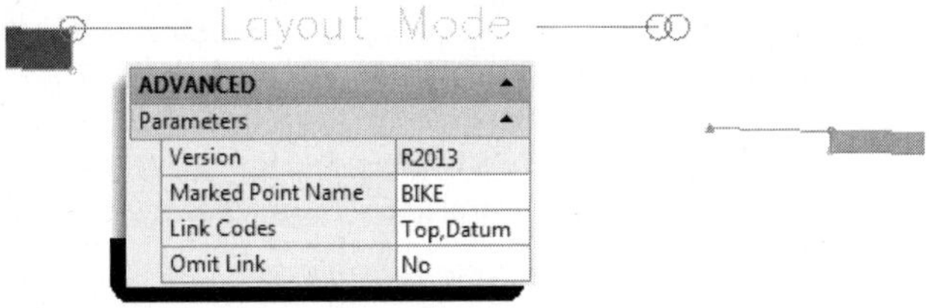

FIGURE 8.48 Add the name of the marked point before you place it on the assembly

Before placing a marked point subassembly, change the name of the marked point in the Advanced Parameters. Before you create a subassembly that references a marked point, make certain that you have properly defined the marked point that it references. Be sure to reference this marked point by name with the subsequent subassemblies that reference it before you place it on your assembly.

At this stage, the geometry for a subassembly using a marked point is not known. The final geometry will be determined when you plug it into a corridor. All subassemblies that use the marked point will appear with the "Layout Mode" placeholder. Subassemblies designed to look for a marked point include the following:

- Channel
- ChannelParabolicBottom
- LinkToMarkedPoint
- LinkToMarkedPoint2
- LinkSlopesBetweenPoints

- MedianDepressed
- MedianRaisedConstantSlope
- MedianRaisedWithCrown
- OverlayBrokenBackBetweenEdges
- OverlayBrokenBackOverGutters
- OverlayParabolic
- UrbanReplaceCurbGutter (1 and 2)
- UrbanReplaceSidewalk

MAKING SURE YOUR MARKED POINT PROCESSES

Always place the marked point before the links that use it to avoid having to reorder subassemblies in the Construction tab of Assembly Properties. If the marked point is listed below the subassembly that needs it in the Construction tab, Civil 3D will not process it.

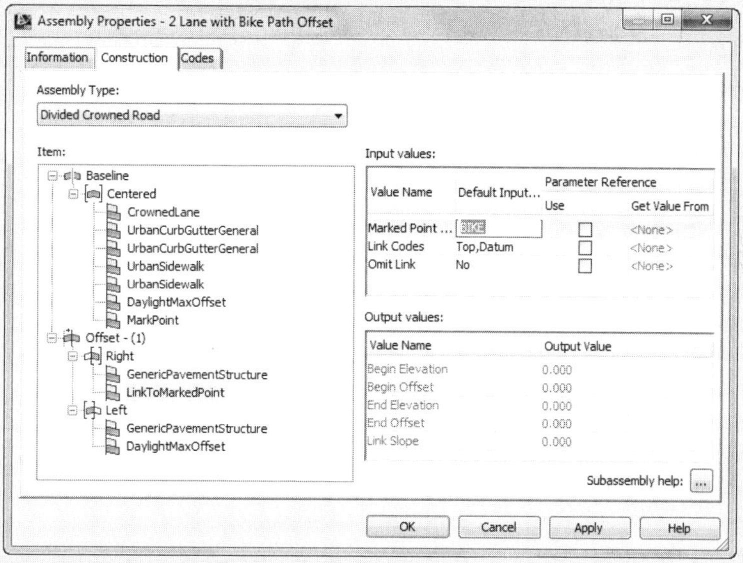

To reorder subassemblies in this dialog, right-click the subassembly and select Move Up or Move Down as needed.

Organizing Your Assemblies

The more geometry changes that occur throughout your corridor, the more assemblies you will have. Civil 3D offers several tools to keep your assemblies organized and available for future use.

Storing a Customized Subassembly on a Tool Palette

Customizing subassemblies and creating assemblies are both simple tasks. However, you'll save time in future projects if you store these assemblies for later use.

A typical jurisdiction usually has a finite number of allowable lane widths, curb types, and other components. It would be extremely beneficial to have the right subassemblies with the parameters already available on your Tool Palettes window.

The following exercise will lead you through storing a customized subassembly on a tool palette.

In this exercise you will be storing some of the subassemblies you made in earlier exercises; therefore, any of the previously saved files (which you can download from this book's web page) can be used, if you do not have one open from a previous exercise.

You can only add a tool from a saved drawing, so make sure you save the drawing you are working in before following these steps:

1. Be sure your Tool Palettes window is displayed.
2. Right-click the Tool Palettes control bar located at the top of the window, and select New Palette to create a new tool palette.
3. Enter **My Road Parts** in the Name text box.
4. Select the sidewalk sub-base from the Urban Single-Lane assembly.

   You'll know it's selected when you can see it highlighted and the grip appears.
5. Click on the dashed portion of the subassembly (i.e., a subassembly marker, *not* the grip point) and drag the assembly into the Tool Palettes window.

   It may take you several tries to get the click-and-drag timing correct, but it will work. You'll know it is working when the cursor appears with a plus sign in the tool palette.

   When you release the mouse button, an entry appears on your tool palette with the name of the subassembly component that you are adding as well as a graphic of the subassembly.
6. Right-click this entry, and select the Properties option.

   The Tool Properties dialog appears (see Figure 8.49).

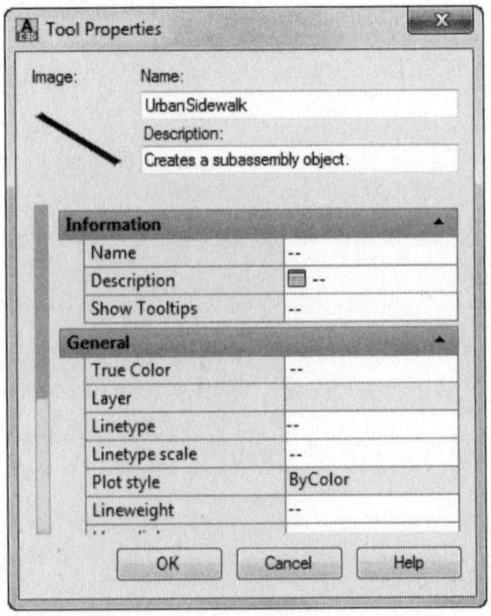

FIGURE 8.49
The Tool Properties dialog

7. If desired, change the image, description, and other parameters in the Tool Properties dialog, and click OK.
8. Try this process for several lanes and curbs in the drawing.

   The resulting tool palette looks similar to Figure 8.50.

FIGURE 8.50
A tool palette with three customized subassemblies

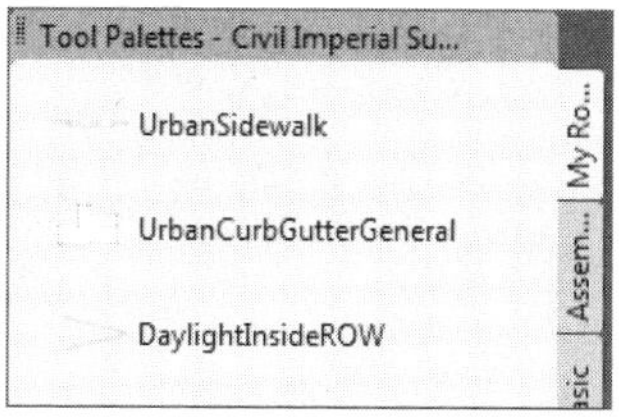

Note the tool palette entries for each subassembly point to the location of the `Subassembly.NET` directory, and not to this drawing. If you share this tool palette, make sure the subassembly directory is either identical or accessible to the person with whom you're sharing.

Storing a Completed Assembly on a Tool Palette

In addition to storing individual subassemblies on a tool palette, it's often useful to warehouse entire completed assemblies. Many jurisdictions have several standard road cross sections; once each standard assembly has been built, you can save time on future similar projects by pulling in a prebuilt assembly.

The process for storing an assembly on a tool palette is nearly identical to the process of storing a subassembly. Simply select the assembly baseline, hover your cursor over the assembly baseline, left-click, and drag to a palette of your choosing.

It's usually a good idea to create a library drawing in a shared network location for common completed assemblies, and to create all assemblies in that drawing before dragging them onto the tool palette. By using this approach, you'll be able to test your assemblies for validity before they are rolled into production. Alternatively, you can right-click on the new palette name and choose Import Subassemblies to display the Import Subassemblies dialog. Here you can choose a source file and then specify whether you want the subassemblies from that source file to import into the palette (optional) and/or the Catalog Library/My Imported Tools.

ORGANIZING ASSEMBLIES WITHIN PROSPECTOR

There are multiple features that help you keep a drawing with many assemblies organized. In Prospector, you will see your listing of assemblies and an Unassigned Subassemblies entry.

Unassigned subassemblies are orphaned parts that are not attached to any main assembly. They may be left over from some assembly customization or they may just be a mistake. In either case, you will want to clean them out. Right-click on the Assemblies collection and select Erase All Unreferenced Assemblies.

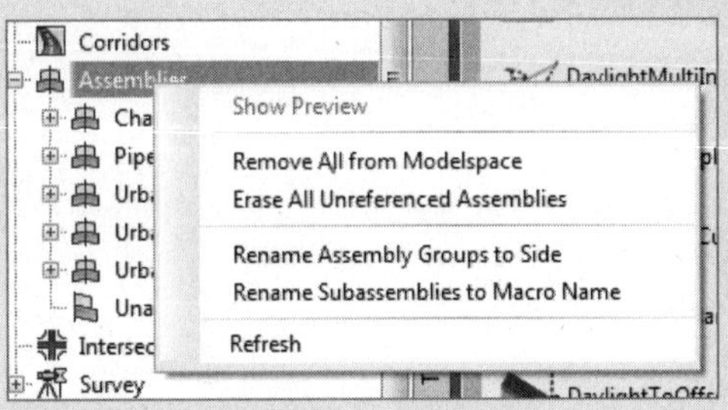

In this same right-click menu, you can also choose to remove the display of the assemblies from modelspace. This hides the display of the assembly but retains its definition in the drawing.

You can still use a hidden assembly in a corridor. If you need it visible again for editing purposes, right-click on the assembly and select Insert To Modelspace.

The niftiest part of this new method of organizing assemblies is that they can now be part of your Civil 3D template without having them visible.

The Bottom Line

Create a typical road assembly with lanes, curbs, gutters, and sidewalks Most corridors are built to model roads. The most common assembly used in these road corridors is some variation of a typical road section consisting of lanes, curb, gutter, and sidewalk.

> **Master It** Create a new drawing from the DWT of your choice. Build a symmetric assembly using LaneSuperelevationAOR, UrbanCurbGutterValley2, and LinkWidthAndSlope for terrace and buffer strips adjacent to the UrbanSidewalk. Use widths and slopes of your choosing.

Edit an assembly Once an assembly has been created, it can be easily edited to reflect a design change. Often, at the beginning of a project you won't know the final lane width. You can build your assembly and corridor model with one lane width, and then change the width and rebuild the model immediately.

> **Master It** Working in the same drawing, edit the width of each LaneSuperelevationAOR to 14′ (4.3 m), and change the cross slope of each LaneSuperelevationAOR to -3.00%.

Add daylighting to a typical road assembly Often, the most difficult part of a designer's job is figuring out how to grade the area between the last engineered structure point in the cross section (such as the back of a sidewalk) and existing ground. An extensive catalog of daylighting subassemblies can assist you with this task.

> **Master It** Working in the same drawing, add the DaylightMinWidth subassembly to both sides of your typical road assembly. Establish a minimum width between the outermost subassembly and the daylight offset of 10′ (3 m).

Chapter 9

Custom Subassemblies

The stock subassemblies available with the AutoCAD® Civil 3D® software are capable of providing many solutions to your corridor needs, as discussed in Chapter 8, "Assemblies and Subassemblies." However, there are going to be instances when what they offer isn't exactly what you need. Sometimes it may be too robust or sometimes not robust enough. When you install Civil 3D 2013, you are given the option to install an additional program called Autodesk® Subassembly Composer for AutoCAD® Civil 3D® (hereafter referred to as Subassembly Composer). This standalone program allows you to create custom subassemblies with an easy-to-use interface without having to know how to write programming code. These custom subassemblies can then be imported into Civil 3D for use in your assembly and corridor creation right alongside the stock subassemblies.

In this chapter, you will learn to:

- Define input and output parameters with default values
- Define target offsets, target elevations, and/or target surfaces
- Generate a flowchart of subassembly logic using the elements in the Tool Box
- Import a custom subassembly made with Subassembly Composer into Civil 3D

The User Interface

The Subassembly Composer is a separate program that is optionally installed when you install the Civil 3D software. Once you've installed it, select Start ➢ All Programs ➢ Autodesk ➢ Subassembly Composer 2013 to launch the Autodesk Subassembly Composer 2013. The Subassembly Composer consists of five individual window panels: Tool Box, Flowchart, Properties, Preview, and Settings and Parameters (Figure 9.1).

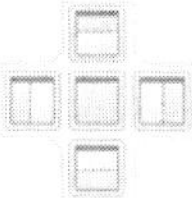

Each of these window panels may be moved around independently by clicking on the panel's title bar and dragging it while using the docking control icons that appear to serve your needs. At any time you can return to the default position by choosing View ➢ Restore Default Layout.

FIGURE 9.1
Subassembly Composer user interface

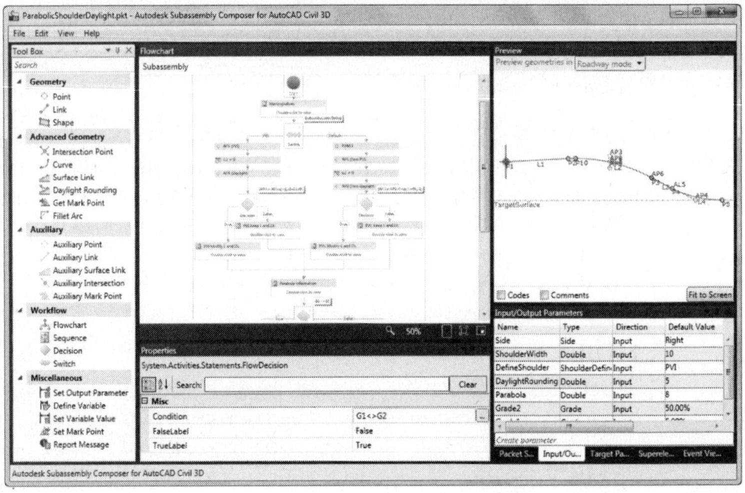

Tool Box

The Tool Box panel, which by default is located along the left side of the user interface, is the storage location for elements available for constructing the subassembly. This is similar to how the Tool Palettes panel in Civil 3D serves as the storage location for subassemblies available for constructing assemblies.

This panel will provide all the elements used to build your flowchart. There are five branches that can each be expanded and contracted to show the elements within that category: Geometry, Advanced Geometry, Auxiliary, Workflow, and Miscellaneous. To use any of these elements you click on the desired element and drag and drop it over into the Flowchart panel. You will look more closely at all the elements available later in this chapter.

Flowchart

The Flowchart panel, which by default is located at the top center of the user interface, is the workspace used to build and organize the subassembly logic and elements. A flowchart could be a simple straight line of logic, or it can be a complex tree of branching decisions. Either way, the subassembly definition always begins at the Start element. If there is a problem with your subassembly's logic, a small red circle with an exclamation point will be displayed in the upper-right corner of this panel.

Properties

The Properties panel, which by default is located at the bottom center of the user interface, is the input location of the parameters that define each geometry element. This is where you will spend most of your time defining the subassembly's geometry.

Preview

The Preview panel, which by default is located at the upper right of the user interface, allows you to view your subassembly as currently defined by the Flowchart panel. There are two preview modes:

Roadway Mode Shows the subassembly built using any target surfaces, target elevations, and/or target offsets.

Layout Mode Shows the subassembly built using only the input parameters (no targets).

At the bottom of this panel are two check boxes: Codes and Comments. If any codes or comments were entered in the properties for the points, links, or shapes and you select these check boxes, this information will be listed next to the applicable geometry in the Preview pane. Codes will be shown in brackets [ ] and comments will be shown in parentheses ( ). You will learn how to define code and comments a little later. Use the rolling wheel on your mouse to zoom in and out or click the Fit To Screen button to zoom extents.

Settings and Parameters

The Settings and Parameters panel, which by default is located at the bottom right of the user interface, consists of five tabs that define the subassembly: Packet Settings, Input/Output Parameters, Target Parameters, Superelevation, and Event Viewer. You will take a closer look at each of these in this chapter.

Creating a Subassembly

Every subassembly that you create using the Subassembly Composer will be built using the same five basic steps:

1. Define the packet settings.
2. Define the input and output parameters.
3. (Optional) Define the target parameters.
4. Build the flowchart that defines the subassembly.
5. Save the packet (PKT) file in the Subassembly Composer and import it into Civil 3D.

The names in each of these steps should sound familiar, as many of them are built into specific sections of the user interface. Generally you will start in the Settings and Parameters panel in the lower-right corner of the user interface and work across the tabs from left to right to define the packet settings, input and output parameters, and then the target parameters. Then you will move over to the Tool Box, Flowchart, and Properties panels to define your subassembly. There are two other tabs in the Settings and Parameters panel: Superelevation and Event Viewer.

Defining the Subassembly

As you saw in Chapter 8, there are some instances when the stock subassemblies just aren't going to do what you need them to do to meet your design. One example of such a scenario (which you learned about in Chapter 8) was the inability to define different cross-slopes in the stock subassembly UrbanSidewalk for the inside boulevard (terrace), sidewalk, and the outside boulevard (buffer strip). To achieve the result you wanted, you used three separate subassemblies: a generic link, a sidewalk, and another generic link, which all worked independently of one another. Instead of always using this combination of subassemblies, you could generate a custom subassembly using Subassembly Composer in order to meet your specific design needs.

As you can see from the five basic steps discussed earlier, the first three are all about defining various components of your subassembly. These first three steps are the first three tabs of the Settings and Parameters panel. In the following example you will go through these three steps and define your subassembly.

Packet Settings

On the Packet Settings tab shown in Figure 9.2, you can define the subassembly name, provide a description, link to a help file, and link to an image. Of these four pieces of information, only the subassembly name is required—the others are optional.

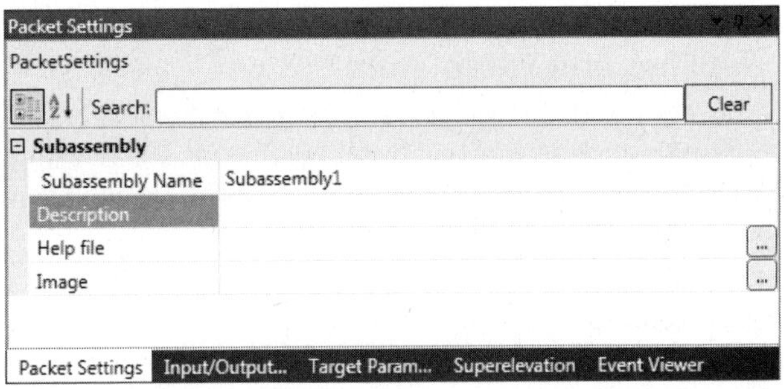

Figure 9.2
The default Packet Settings tab

The subassembly name (which can be different from the name of the PKT file) will be the name that is displayed on the tool palettes once it's imported into Civil 3D. Like the stock subassemblies, the subassembly name that you define is not allowed to have any spaces.

If you include a description, this text will appear on the tool palette as hover text. Similarly, the image, if provided, will display in the tool palette next to the subassembly's name. We recommend that the image you provide be 224×224 pixels (although an image as small as 64×64 is allowed). If you create a help file, supply it in HTM or HTML format. To add an image or a help file, click the ellipsis button in that row and navigate to the file location.

Let's use this information to define the packet settings for this first example in a new subassembly PKT file. On the Packet Settings tab, do the following:

1. Set Subassembly Name to **UrbanSidewalkSlopes** or **UrbanSidewalkSlopesMetric**.

   Remember, just like the stock subassemblies, your subassemblies can't have spaces in their names.

> **Metric or Imperial**
>
> Subassembly Composer is unitless. The same PKT file generated by the Subassembly Composer can be used in either the metric or the Imperial Civil 3D. The only difference is in default values: the Imperial subassembly will have all of these values based on feet and the metric subassembly will have all of these values based on meters. For this reason you are going to name your subassemblies differently so that you know which is which when they are later imported into the tool palettes in Civil 3D.

2. For Description, type **Urban sidewalk with varying cross slopes**.

   You will notice that there are also Help File and Image options, which you will leave blank at this time since they are both optional.

The Packet Settings tab should now look like Figure 9.3.

FIGURE 9.3
Packet Settings for UrbanSidewalkSlopes

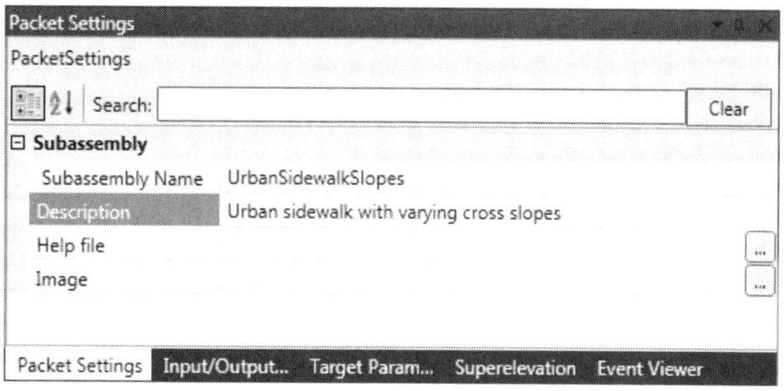

Keep this file open to define the input and output parameters next.

INPUT AND OUTPUT PARAMETERS

On the Input/Output Parameters tab, shown in Figure 9.4, you can define numerous parameters along with their default values. These are the same as the input and output parameters that are available in the stock subassemblies (which you learned about in Chapter 8). The information on this tab is presented in a table. To add a parameter, simply click on the Create Parameter text. To remove a parameter, highlight that row in the table and press Delete on your keyboard.

FIGURE 9.4
The default Input/Output Parameters tab

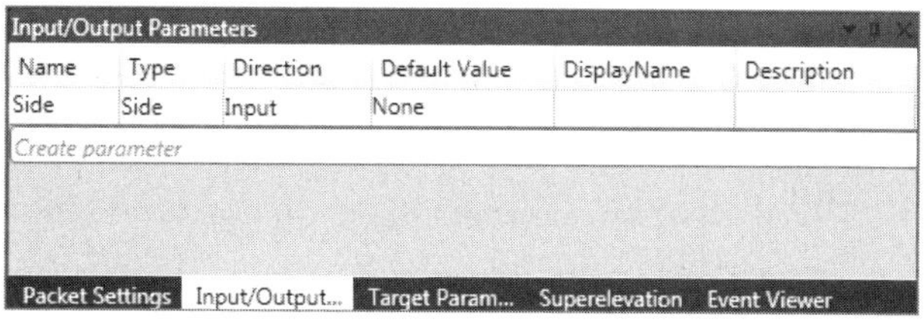

Because this subassembly example is based on the UrbanSidewalk stock subassembly, it is good practice to keep terminology consistent with the stock subassembly for ease of use by the end user in Civil 3D.

STOCK SUBASSEMBLIES IN SUBASSEMBLY COMPOSER

One of your first questions when running into a limitation with one of the stock subassemblies and wanting to make a modified subassembly based on the stock subassembly will likely be "Where can I find the PKT file for the stock subassemblies to use as a starting point?"

Unfortunately, at this time the stock subassemblies are written in a different type of code that does not allow them to be opened in Subassembly Composer. Let's hope that some of the stock subassemblies are made available in PKT format soon. In the meantime, creating a subassembly in Subassembly Composer based on one of the stock subassemblies is a great learning experience.

To stay consistent with the stock subassembly, let's look at the UrbanSidewalk help file for the input and output parameters that it uses. The help file indicates that there are six input parameters:

Side, Inside Boulevard Width, Sidewalk Width, Outside Boulevard Width, %Slope, and Depth. You will use these six input parameters; however, the whole reason you are using Subassembly Composer for this example is to allow the end user to specify different slopes for each of the components. Therefore, in addition to %Slope (which you will instead name SidewalkSlope), you will define an Inside Boulevard Slope and an Outside Boulevard Slope. You will also give the end user the ability to specify a point code at the outside point numbered P4 in the help file.

The help file also indicates that there are no output parameters, so you will also not be defining any output parameters in this example.

As you look at the help file for the UrbanSidewalk subassembly, you may notice that the different input parameters have a type assigned to them. These same types are available in Subassembly Composer.

INPUT/OUTPUT PARAMETER TYPES

Eight input and output parameter types are provided with the program and accessible through the drop-down list in the Type column.

The eight variable types are:

- Integer, which is a whole number
- Double, which is a number that allows decimal precision
- String, which is text
- Grade, which is a percentage slope
- Slope, which is a ratio of horizontal distance to vertical distance
- Yes/No, which is used for Boolean variables
- Side, which can be set to Left, Right, or None
- Superelevation, which can be set to LeftInsideLane, LeftInsideShoulder, LeftOutsideLane, LeftOutsideShoulder, RightInsideLane, RightInsideShoulder, RightOutsideLane, RightOutsideShoulder, or None. The values of these cross slopes can be set on the Superelevation tab in the Settings/Parameters panel.

You can also create custom types, called enumerations, which will be discussed later in this chapter.

Let's use the help file information to define the input parameters for the example started in the previous exercise. On the Input/Output Parameters tab, do the following:

1. Change the default value for the Side input parameter to Right.

   If you leave Default Value as None, this parameter will not be listed in the Advanced Parameters for the subassembly in Civil 3D. In a later symmetric example, you will leave this value set to None.

2. Click Create Parameter to add a new parameter.
3. Change Name to **InsideWidth**.

4. Verify that Type is set to Double by using the drop-down list.
5. Verify that Direction is set to Input by using the drop-down list.
6. Set Default Value to **0**.
7. Set DisplayName to **Inside Boulevard Width**.
8. Click Create Parameter seven times to add seven more parameters. Alternatively you can click Create Parameter before defining each parameter.

   Don't forget that if you add too many parameters, highlight that row in the table that you want to remove and press Delete on your keyboard.

   Since you want all of these to be input parameters, all the Direction values will remain as Input.

9. Define the remaining seven input parameters with the following settings (note that the metric values are shown in parentheses if different from the Imperial values):

NAME	TYPE	DEFAULT VALUE	DISPLAYNAME
SidewalkWidth	Double	5 (1.5)	Sidewalk Width
OutsideWidth	Double	0	Outside Boulevard Width
InsideSlope	Grade	5	Inside Boulevard Slope
SidewalkSlope	Grade	2	Sidewalk Slope
OutsideSlope	Grade	5	Outside Boulevard Slope
Depth	Double	0.333 (0.1)	
OutsidePointCode	String	Hinge	Point Code at Outside Point, P4

   The program will change the parameters with the type of Grade to have a percent sign in the Default Value column. The DisplayName for Depth is intentionally left blank. The Description column is optional and any that are left blank will use Name as the DisplayName.

   The Input/Output Parameters tab should now look like Figure 9.5.

FIGURE 9.5
Input/Output Parameters tab for UrbanSidewalkSlopes

Name	Type	Direction	Default Value	DisplayName	Description
Side	Side	Input	Right		
InsideWidth	Double	Input	0	Inside Boulevard Width	
SidewalkWidth	Double	Input	5	Sidewalk Width	
OutsideWidth	Double	Input	0	Outside Boulevard Width	
InsideSlope	Grade	Input	5.00%	Inside Boulevard Slope	
SidewalkSlope	Grade	Input	2.00%	Sidewalk Slope	
OutsideSlope	Grade	Input	5.00%	Outside Boulevard Slope	
Depth	Double	Input	0.333		

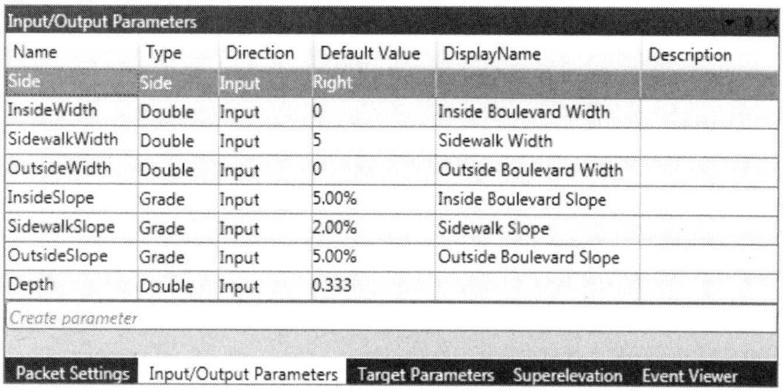

TARGET PARAMETERS

On the Target Parameters tab, shown in Figure 9.6, you can define the targets to be used by the subassembly. You can use three types of targets:

- Offset targets
- Elevation targets
- Surface targets

These are the same types of targets that you learned about in Chapter 8.

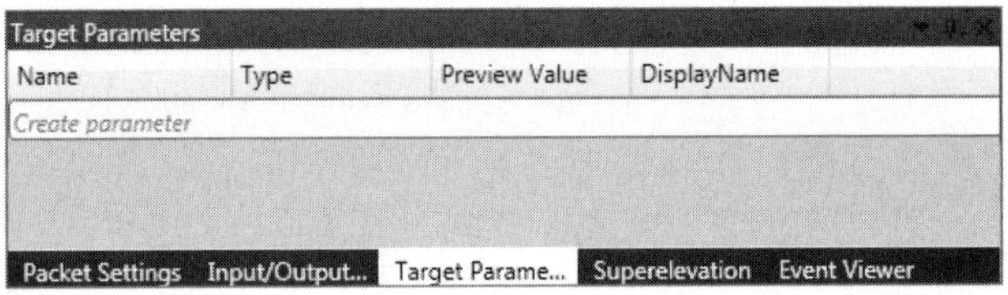

FIGURE 9.6
The default Target Parameters tab

As with the Input/Output Parameters tab, to add a parameter you click on the Create Parameter text, and to remove a parameter you highlight that row in the table and press Delete on your keyboard.

Just as you looked in the UrbanSidewalk help file for the input parameters that it uses, you will look in the help file for the target parameters. The help file indicates that there are three target parameters: Inside Boulevard Width, Sidewalk Width, and Outside Boulevard Width. The help file also indicates that these are optional. While the UrbanSidewalk file called all of these widths, they are actually offsets measured from the subassembly origin. Therefore, you are going to suffix the names of these targets with the word Offset.

Let's use this information to define the target parameters for this example. On the Target Parameters tab, do the following:

1. Click Create Parameter to add a new parameter, and then do the following:
   a. Change Name to **TargetInsideOffset**.
   b. Verify that Type is set to Offset.
   c. Set Preview Value to **1** (or **0.25** for metric).

      The preview value is different from the default value you defined for the input parameters. A preview value is used for graphical purposes and the actual value is generated based on the assigned targets.
2. Click Create Parameter to add a new parameter, and then do the following:
   a. Change Name to **TargetSidewalkOffset**.
   b. Verify that Type is set to Offset.
   c. Set Preview Value to **5** (or **1.5** for metric).
3. Click Create Parameter to add a new parameter, and then do the following:

a. Change Name to **TargetOutsideOffset**.

b. Verify that Type is set to Offset.

c. Set Preview Value to **6** (or **1.75** for metric).

The Target Parameters tab should now look like Figure 9.7.

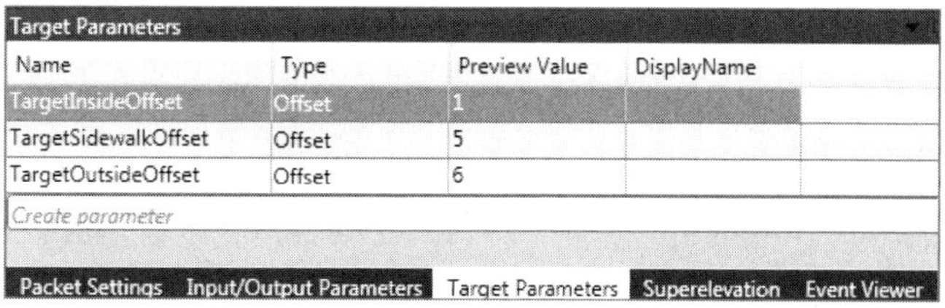

FIGURE 9.7
Target parameters for UrbanSidewalkSlopes

As you add the target parameters, you should notice that they will be added to the Preview panel, as shown in Figure 9.8, if you have the Preview Geometry option set to Roadway Mode. Remember, Roadway Mode will show your subassembly based on the target parameter, and Layout Mode will show your subassembly without the targets considered.

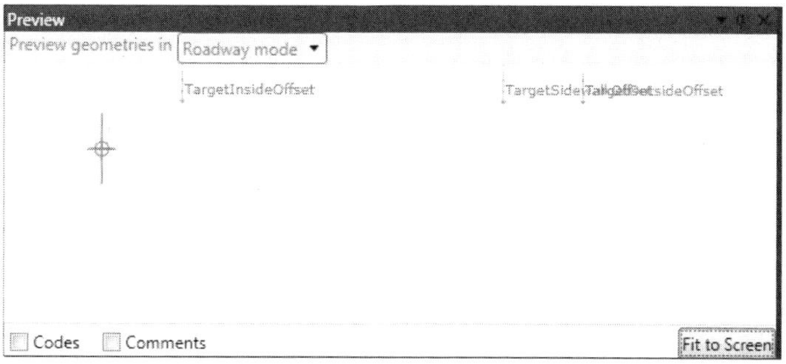

FIGURE 9.8
Target parameters shown in the Preview panel

You can zoom in and out in your Preview panel by scrolling the mouse scroll wheel and pan by clicking the mouse scroll wheel and dragging. Once there are elements shown in the Preview panel, you can also zoom to the extents of your subassembly using the Fit To Screen button at the bottom-right corner of the panel, but the Fit To Screen button does not consider the target parameters.

Now that you have defined the subassembly parameters, you will start building the subassembly. You may keep this file open to continue on to the next exercise or use the finished copy of this file available from the book's web page, `www.sybex.com/go/masteringcivil3d2013`. The file is named `UrbanSidewalkSlopes_Defined.pkt` or `UrbanSidewalkSlopesMetric_Defined.pkt`.

Building the Subassembly Flowchart

Every flowchart will start from the appropriately named Start element. From there the flowchart is built with elements from the Tool Box, which are connected together with arrows. As you will see, every element that you add from the Tool Box has at least two nodes: one node for an incoming connection arrow and one node for an outgoing connection arrow.

Building a flowchart is as simple as dragging and dropping elements from the Tool Box into the Flowchart panel, as shown in this series of steps:

1. If it is still open from the previous exercise, continue working in the PKT file from the previous example or open the `UrbanSidewalkSlopes_Defined.pkt` or `UrbanSidewalkSlopesMetric_Defined.pkt` file.
2. From the Tool Box ➤ Geometry branch, drag a Point element to below the Start element.

   The Start element will automatically connect to the new Point element, which has automatically been numbered P1, as shown in Figure 9.9.

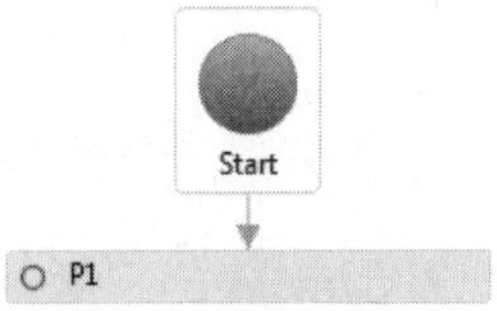

FIGURE 9.9 The Start element connected to the Point element

If you look at the Properties panel while point P1 is selected, you will see that it has been placed on the origin. The origin is the geometry point that your subassembly will attach to when generating an assembly in Civil 3D. While you may have instances where you don't want a point at the origin, it is generally the case that P1 and the origin will be coincident. As such, you do not need to change any of the defaults for this point.

P, AP, L, AL, AND S NUMBER PREFIXES

While there are all sorts of geometric elements available in your Tool Box, they all boil down to three types: points, links, and shapes. These should all be familiar from the discussions in Chapter 8. Of the points and links there are regular points and links as well as auxiliary points and links. Auxiliary elements will be used for the purposes of calculations in Subassembly Composer but will never be displayed in Civil 3D.

In the Subassembly Composer, all points are prefixed with the letter P, all auxiliary points are prefixed with the letters AP, all links are prefixed with the letter L, all auxiliary links are prefixed with the letters AL, and all shapes are prefixed with the letter S. If you try to change this prefix for any of the elements you will get an error dialog.

Every element has to have a unique number. Subassembly Composer therefore will automatically number your elements with the next available appropriate number; if you want, you can change this number to one that is available but not to one that is already in use.

If you want to name your elements so that you can keep track of what they represent, instead of trying to put this information in the Number field, put it in the Comment field located at the bottom of the Properties panel.

3. From the Tool Box ➤ Geometry branch, drag another Point element to below P1.

   This point will be automatically numbered P2, and will represent the end of the Inside Boulevard and the start of the Sidewalk.

4. Define point P2 as follows:

   a. Set Point Codes to **"Sidewalk_In"**.

      Note that the quote marks are required in order to define this text as a string. If you are defining the point codes using a variable that is already defined as a string, the quotes are not needed.

   b. Verify that Type is set to Slope And Delta X, as shown in Figure 9.10.

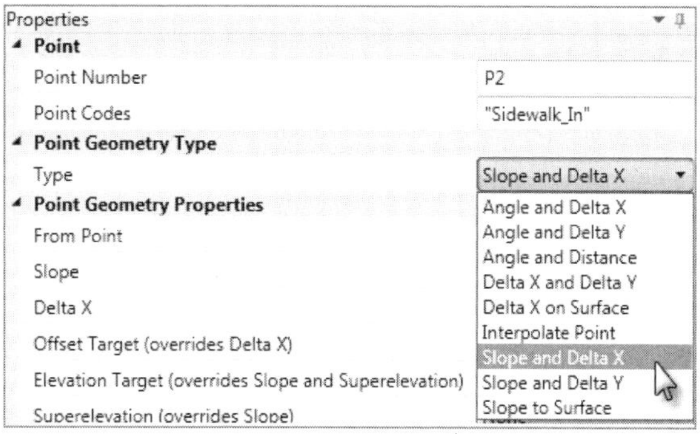

FIGURE 9.10
Point geometry types

   c. Verify that From Point is set to P1.

   d. Set Slope to **InsideSlope**.

   e. Set Delta X to **InsideWidth**.

   f. Set Offset Target (Overrides Delta X) to **TargetInsideOffset**.

      Figure 9.11 shows the current Preview panel with the two point elements being generated. Notice that the element currently being displayed in the Properties panel (in this case point P2) is highlighted in yellow in the Preview panel. If your P2 appears to be in the same location as P1 in the Preview panel, zoom into P1 or click the Fit To Screen button.

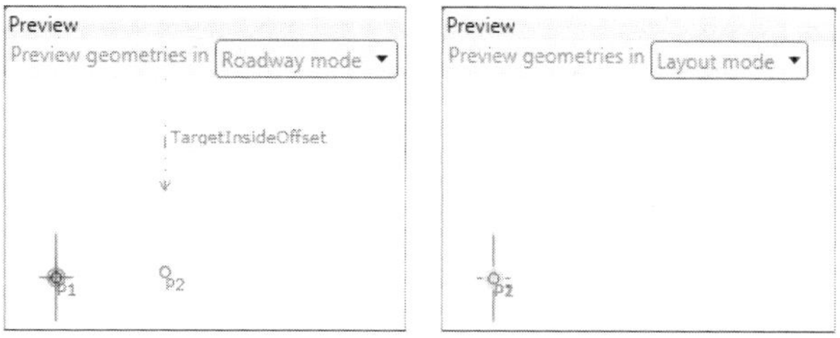

FIGURE 9.11
Point P2 is shown at the TargetInsideOffset when Roadway Mode is on (left) and coincident with P1 when Layout Mode is on, since the InsideWidth default value is currently set to 0 (right).

   g. Verify that the Add Link To From Point check box is selected. This link will automatically be numbered L1.

While you could add a link element from the Tool Box as a separate element after this point, this check box allows you to combine the two elements into one. This option is not available if the From Point is set to Origin or if the From Point is an auxiliary point when you are creating a point, or vice versa.

The P2 element in the Flowchart panel will change to now read P2&L1, as shown in Figure 9.12. In addition, both P2 and L1 will be highlighted in the Preview panel.

FIGURE 9.12
Combined point and link element in the flowchart

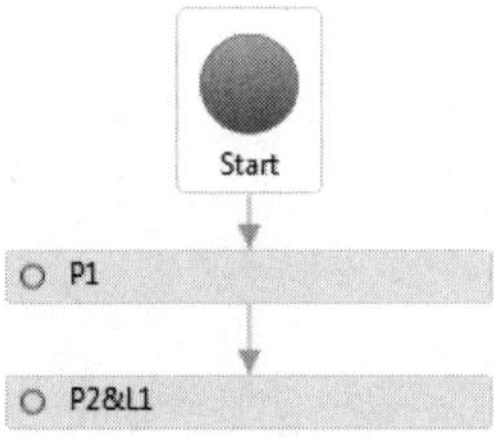

h. Set the Link ➤ Codes to **"Top", "Datum"**.

While you only defined one code for the point, you want two codes for this link to be consistent with the UrbanSidewalk subassembly. You can add multiple codes by generating a comma-delimited list.

The properties for point P2 should now look like Figure 9.13.

FIGURE 9.13
Properties panel for point P2

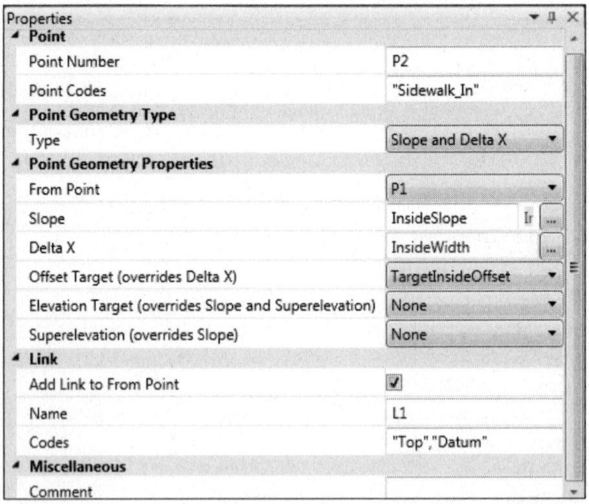

5. From the Tool Box ➤ Geometry branch, drag another Point element to below P2&L1.

This point will be automatically numbered P3 (or P3&L2), and will represent the end of the Sidewalk and the start of the Outside Boulevard.

6. Define point P3 as follows:

   a. Set Point Codes to **"Sidewalk_Out"**.

   b. Verify that Type is set to Slope And Delta X.

   c. Verify that From Point is set to P2.

   d. Set Slope to **SidewalkSlope**.

   e. Set Delta X to **SidewalkWidth**.

   f. Set Offset Target (Overrides Delta X) to **TargetSidewalkOffset**.

   g. Verify that the Add Link To From Point check box is selected.

      This link will automatically be numbered L2.

   h. Set Link ➢ Codes to **"Top", "Sidewalk"**.

      Point P3 and link L2 are shown highlighted in Figure 9.14.

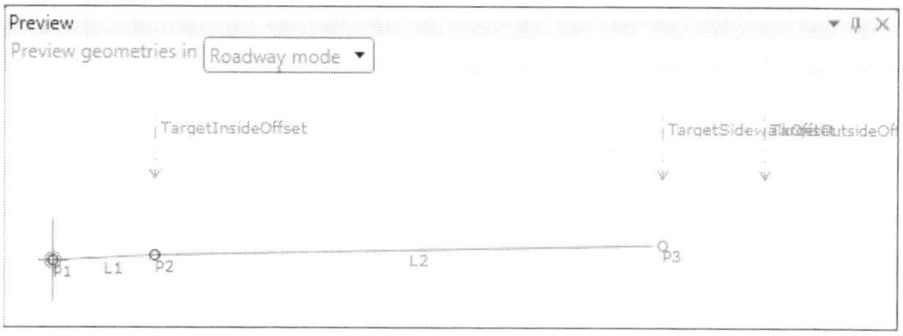

FIGURE 9.14 Preview panel highlighting point P3 and link L2

7. From the Tool Box ➢ Geometry branch, drag another Point element to below P3&L2.

   This point will be automatically numbered P4, and will represent the end of the Sidewalk and the start of the Outside Boulevard.

8. Define point P4 as follows:

   a. Set Point Codes to **OutsidePointCode**.

      The UrbanSidewalk subassembly doesn't provide a code at point P4, but the ability to add codes wherever you want is another one of the benefits of using Subassembly Composer. Notice that unlike the previous point codes, you did not need to specify the quote marks on this one because this is a string variable that you created in the input/output parameters.

   b. Verify that Type is set to Slope And Delta X.

   c. Verify that From Point is set to P3.

   d. Set Slope to **OutsideSlope**.

   e. Set Delta X to **OutsideWidth**.

   f. Set Offset Target (Overrides Delta X) to **TargetOutsideOffset**.

g. Verify that the Add Link To From Point check box is selected.

This link will automatically be numbered L3.

h. Set the Link ➢ Codes to **"Top", "Datum"**.

Point P4 and link L3 are shown highlighted in Figure 9.15.

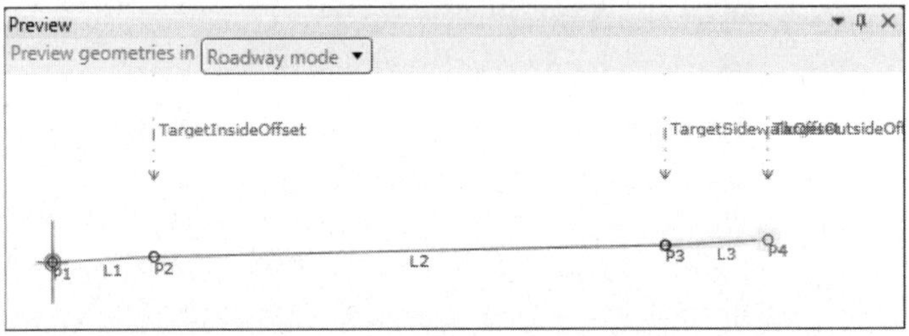

FIGURE 9.15
Preview panel highlighting point P4 and link L3

9. From the Tool Box ➢ Geometry branch, drag another Point element to below point P4 and link L3.

   This point will be automatically numbered P5, and will represent the bottom of the sidewalk shape directly below point P2.

10. Define point P5 as follows:

    a. Verify that Type is set to Delta X And Delta Y.

    b. Verify that From Point is set to P2.

    Because you are basing point P5 on point P2, wherever point P2 is located P5 will always remain in relation to it. So point P5 will follow the target offset already defined for point P2.

    c. Set Delta X to **0**.

    d. Set Delta Y to **–Depth**.

    e. Verify that the Add Link To From Point check box is selected.

    This link will automatically be numbered L4.

    f. Set the Link ➢ Codes to **"Sidewalk", "Datum"**.

    Point P5 and link L4 are shown highlighted in Figure 9.16.

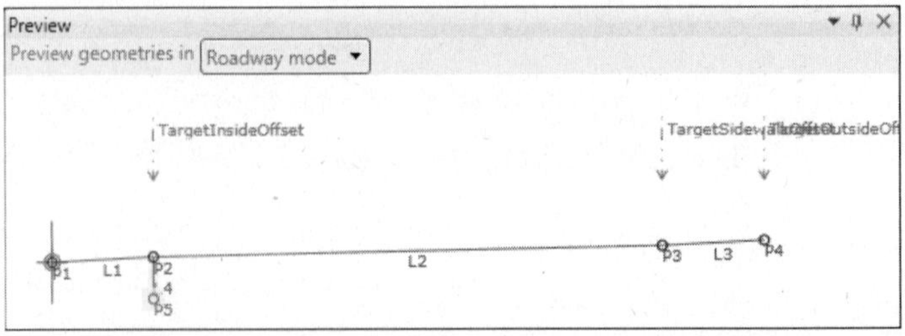

FIGURE 9.16
Preview panel highlighting point P5 and link L4

The last point to define is point P6 at the inside bottom of the sidewalk shape directly under point P3. You could define point P6 in the same way you just defined point P5, as a –Depth from point P3, but in order to show a variety of element types, you will instead define it using an intersection point. An intersection point is classified the same as a regular point, so it will have the same P prefix as all the other points you already defined.

INTERSECTION POINT GEOMETRY TYPES

There are three intersection point geometry types. These types are applicable to both intersection points and auxiliary intersection points. These types are as follows:

Intersection: LinkPointSlope To define an intersection by LinkPointSlope, you will need to have a link and a point with a slope from the point. The new point will be placed at the intersection of the link and the projected slope.

In addition there are three Geometry Properties check boxes: Extend Link, Extend Slope, and Reverse Slope. The Extend Link and Extend Slope check boxes convert the link or slope from a segment (link) or ray (point with slope) into a line that extends beyond its original extents in order to find the desired intersection if no direct intersection is available. The Reverse Slope check box reverses the slope (i.e., a slope of 25% is treated as –25%).

Intersection: TwoLinks To define an intersection by TwoLinks, you will need to have two links. The new point will be placed at the intersection of these two links.

In addition there are two Geometry Properties check boxes: Extend Link 1 and Extend Link 2. These have the same type of behavior as the extending check boxes previously discussed.

Intersection: TwoPointsSlope To define an intersection by TwoPointsSlope, you will need to have two points and a slope from each of these points. The new point will be placed at the intersection of these two projected slopes.

In addition there are four Geometry Properties check boxes: Extend Slope 1, Extend Slope 2, Reverse Slope 1, and Reverse Slope 2. These have the same type of behavior as the extending and reversing check boxes previously discussed.

With all of the intersection types, if an intersection is not found, no point is generated.

11. From the Tool Box ➤ Advanced Geometry branch, drag an Intersection Point element to below point P5 and link L4.

This point will be automatically numbered P6, and will represent the outside bottom of the sidewalk shape directly below point P3.

12. Define intersection point P6 as follows:

a. Verify that Type is set to Intersection: TwoPointsSlope.

b. Verify that Point 1 is set to P5.

c. Set Slope 1 to **SidewalkSlope**.

This assumes that the slope of the sidewalk does not change. If this subassembly had an elevation target that could override the cross-slope of the Top link (L1), you would instead want to set this variable to `L1.slope`. This annotation is called an application programming interface (API) function. You will learn about many other API functions later in this chapter.

d. Verify that Point 2 is set to P3.

e. Set Slope 2 to **-Infinity%**.

You could alternatively set Slope 2 to Infinity% and select the Extend Slope 2 check box.

Point P6 is shown highlighted in Figure 9.17.

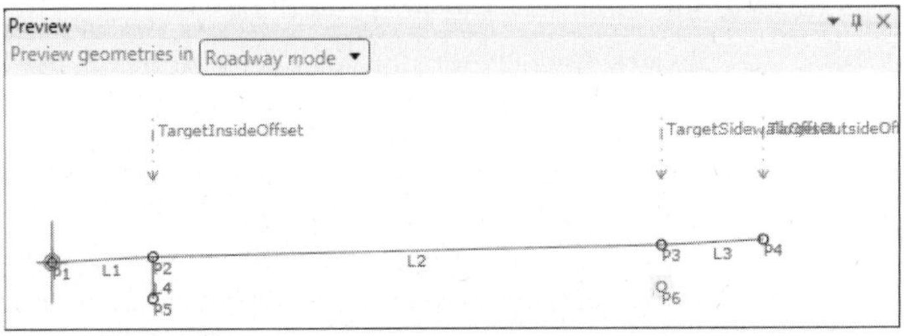

FIGURE 9.17 Preview panel highlighting point P6

13. From the Tool Box ➢ Geometry branch, drag a Link element to below P6.

This link will be automatically numbered L5, and will represent the datum of the sidewalk shape.

14. Define link L5 as follows:

a. Set Link Codes to **"Sidewalk", "Datum"**.

b. Verify that Start Point is set to P5.

c. Verify that End Point is set to P6.

15. From the Tool Box ➢ Geometry branch, drag another Link element to below L5.

This link will be automatically numbered L6, and will define the vertical outside edge of the sidewalk shape, thus closing the shape. When adding a shape element, you must have a closed set of links.

16. Define link L6 as follows:

a. Set Link Codes to **"Sidewalk", "Datum"**.

b. Verify that Start Point is set to P6.

c. Verify that End Point is set to P3.

Link L5 and link L6 are shown highlighted in Figure 9.18.

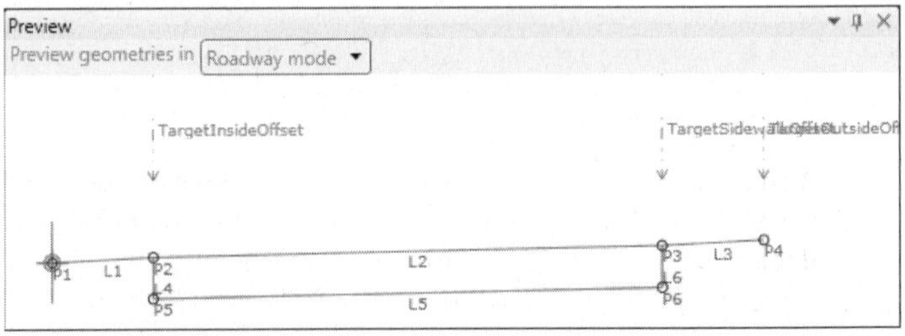

FIGURE 9.18 Preview panel highlighting link L5 and link L6

17. From the Tool Box ➢ Geometry branch, drag a Shape element to below L6.

    This shape will be automatically numbered S1 and will define the sidewalk shape.

18. Define link S1 as follows:

    a. Set Shape Codes to **"Sidewalk"**.

    b. Click the Select Shape In Preview button.

       To pick the shape, do not take time to figure out the name of each link and try to use Add Link and type in the links. Instead, by clicking the green Select Shape In Preview button to the right of the entry field, you can then click inside the shape in the Preview panel, just like you would when defining a hatch in AutoCAD. If you are working with a subassembly that has multiple closed shapes and want to change to a different shape, click the green selection icon again and reselect the new shape—there's no need to delete the previously listed links; the list of links will update for you.

    c. Select inside the closed shape representing the sidewalk, as shown in Figure 9.19.

FIGURE 9.19
Preview panel highlighting shape S1

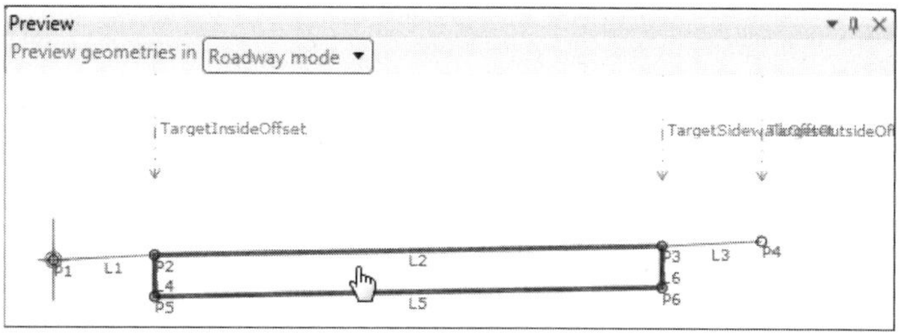

Once you generate all of your points, links, and shapes, it is good practice to select the Codes check box in the bottom-left corner of the Preview panel to confirm that all the desired codes have been assigned, as shown in Figure 9.20. You want to check to make sure that the links have the desired link codes, that the points have the desired point codes, and that the shapes have the desired shape codes.

FIGURE 9.20
UrbanSidewalkSlopes subassembly with codes assigned

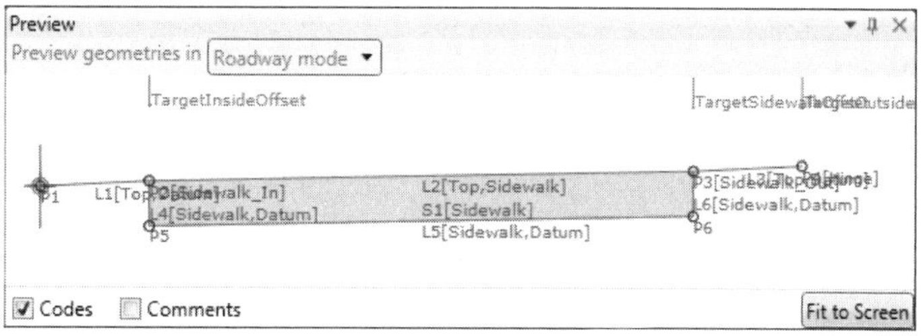

You can keep this file open to continue on to the next exercise or use the finished copy of this file available from the book's web page. The file is called `UrbanSidewalkSlopes_Flowchart.pkt` or `UrbanSidewalkSlopesMetric_Flowchart.pkt`.

Keeping the Flowchart Organized

As you built the subassembly in the previous exercise, you should have noticed all the connection arrows that were automatically created between the elements. If you click and drag any of the elements around, the connection arrows will bend and twist to maintain the connection. Therefore, if you want to reorder your elements you will need to delete the connection arrows, reposition the elements, and redraw the connection arrows.

If you highlight one of the connection arrows and press Delete on your keyboard, the connection will break. As you are building and troubleshooting your subassemblies, you may find it useful to occasionally delete a connection arrow so that you can confirm that everything above an element is in working order. To redo a broken connection, click on the starting element to display the connection nodes and click and drag from one of the connection nodes to the desired node on the next element.

You may notice that with even a simple subassembly such as the previous example, the Flowchart panel starts to fill up quickly. You can enlarge the Flowchart work area by clicking and dragging on the lower-right corner. In addition to the default Flowchart work area, you can nest workflow inside your flowchart, like you nest folders on your computer to organize information.

In general there are two types of flowchart workflows:

Sequence Workflow

This is a straight sequential series of elements.

Flowchart Workflow

This can be a straight series of elements or a branching complex series of decisions and elements.

Both of these workflows can be found in the Workflow branch of the Tool Box. In the previous example, you placed each element one after another into the flowchart, as shown in Figure 9.21.

FIGURE 9.21
Flowchart showing sequential elements

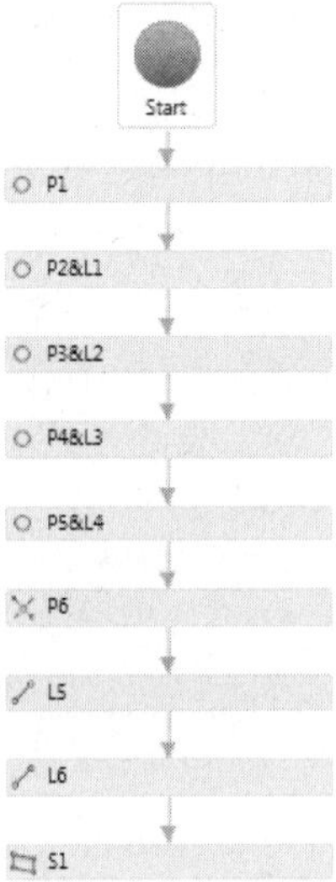

When you have a sequence of elements like this, it is often cleaner to put them into a Sequence element. In our next exercise, you will move the subassembly elements into a sequence:

1. If it is still open from the previous exercise, continue working in the PKT file from the previous example, or open the `UrbanSidewalkSlopes_Flowchart.pkt` or `UrbanSidewalkSlopesMetric_Flowchart.pkt` file.
2. Delete the first connection arrow from the Start to point P1.
3. 
From the Tool Box ➢ Workflow branch, drag a Sequence element to an open area near the Start element.

   Notice that because it is off to the side, the connection arrow connected to the side of the Sequence element instead of the top like in the previous exercise. There are actually four or more nodes on each of the elements that connection arrows can connect to. You can see these connection nodes when you hover over an element.
4. To change the connection node of the connection arrow, click on the connection arrow to display the grips shown in Figure 9.22.

FIGURE 9.22 Connection nodes on the Sequence element (left) and grips on the connection arrow (right)

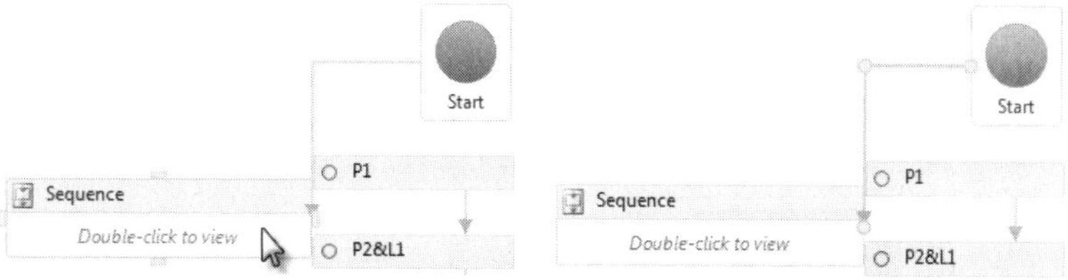

5. Click on the grip at the arrow end of the connection arrow and drag it to the desired connection node at the top of the Sequence element.

   You can do the same thing with the opposite end of the connection arrow to connect out of the bottom of the Start element if desired.
6. Click on one of the elements in your flowchart.
7. Press and hold the Ctrl key on your keyboard and continue to click on each of the elements in your flowchart.
8. Press Ctrl+X to cut the elements from the flowchart.

   If you want to duplicate some of the elements in your flowchart instead of move them, you can use Ctrl+C and Ctrl+V. The duplicated points, links, and shapes will be renumbered with the next available numbers.
9. Double-click on the Sequence element to enter the sequence.

   The sequence will be empty, as shown at the left in Figure 9.23.
10. Press Ctrl+V to paste the elements into the sequence, as shown at the right in Figure 9.23.

FIGURE 9.23
Empty Sequence element (left) and sequence with elements (right)

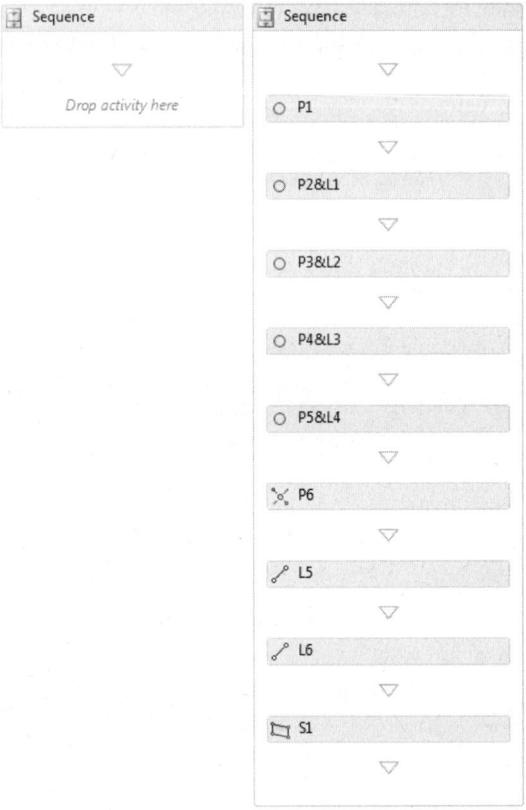

Now you can easily shuffle the elements around if desired.

11. To return to the Flowchart area, click on the word *Flowchart* in the upper-left corner of the Flowchart panel, as shown in Figure 9.24.

FIGURE 9.24
Navigating nested flowchart elements

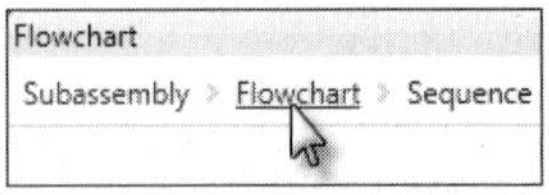

Compare Figure 9.25 to Figure 9.21 shown earlier, and you will see how much room a Sequence element can save.

In addition, you can click on the word *Sequence* on the element to rename the element. This can be helpful if you want to combine and organize groups of common elements, such as a sequence for each layer of pavement in a multiple-layer subassembly.

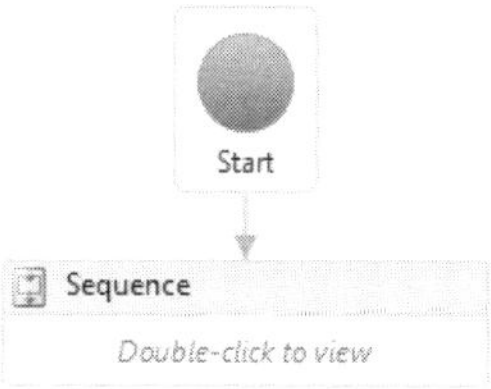

FIGURE 9.25
Navigating nested flowchart elements

When this exercise is complete, you can close the file. A finished copy of this file is available from the book's web page with the filename `UrbanSidewalkSlopes_Sequence.pkt` or `UrbanSidewalkSlopesMetric_Sequence.pkt`.

Importing the Subassembly into Civil 3D

To import your subassembly into the Civil 3D tool palettes, use the following steps:

1. If you haven't already done so, in Subassembly Composer save your PKT file from the previous exercise and close the program.
2. Open Civil 3D and create a new drawing using the `_AutoCAD Civil 3D (Imperial) NCS` or the `_AutoCAD Civil 3D (Metric) NCS` template.
3. From the Insert tab ➢ expanded Import panel, choose Import Subassemblies, as shown in Figure 9.26.

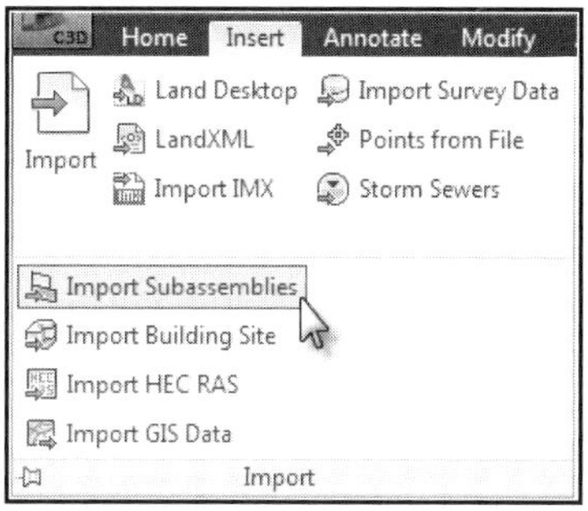

FIGURE 9.26
Choosing Import Subassemblies in the expanded Import panel

The Tool Palettes window will open, if it's not already open, and the Import Subassemblies dialog will be displayed.

4. In the Import Subassemblies dialog, click the folder button to display the Open dialog.
5. Navigate to your saved `UrbanSidewalkSlopes.pkt` or `UrbanSidewalkSlopesMetric.pkt` file and click Open.
6. Verify that the Import To: Tool Palette check box is selected.
7. Using the drop-down list, select Create New Palette from the bottom of the list to display the New Tool Palette dialog.

CUSTOM TOOL PALETTE FOR CUSTOM SUBASSEMBLIES

By creating a new tool palette you can keep all of your custom subassemblies in one easy-to-find location. As you continue to make more subassemblies and test them, you may also find it helpful to have a tool palette named **Testing** in addition to your tool palette of confirmed working subassemblies.

8. For Name enter **Mastering** and click OK to return to the Import Subassemblies dialog.

   The Import Subassemblies dialog should now look like Figure 9.27.

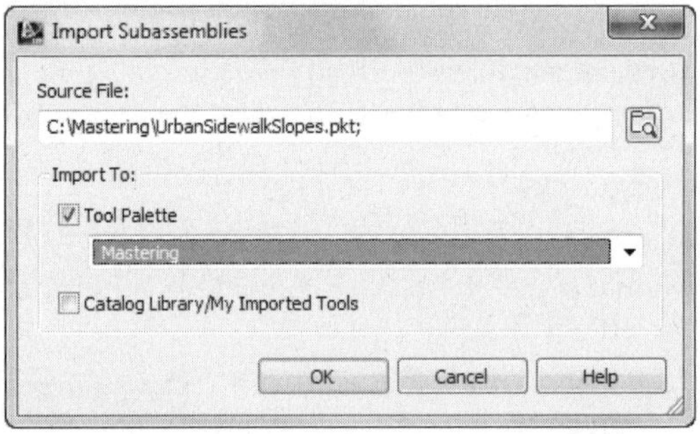

FIGURE 9.27
The UrbanSidewalkSlopes subassembly shown on the Mastering tool palette with the description shown in the tooltip

9. Click OK to accept the settings in the Import Subassemblies dialog.

   A new palette named Mastering will be added and it will contain your UrbanSidewalkSlopes subassembly. Since you did not specify an image, an empty white square is shown. But if you hover over the subassembly name, a tooltip will be displayed showing the description that you provided in Subassembly Composer, as shown in Figure 9.28.

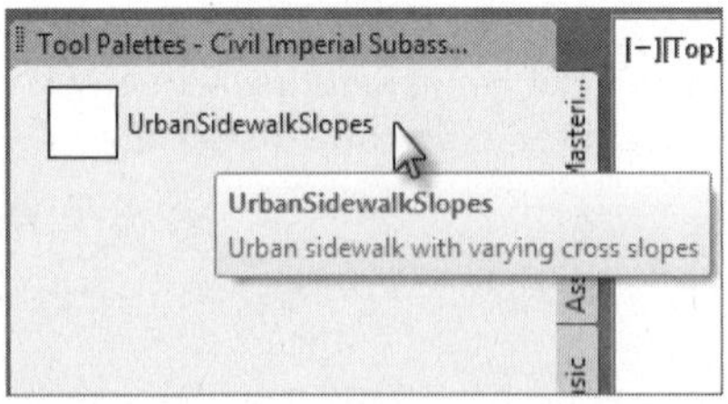

FIGURE 9.28
Custom subassembly in a custom tool palette

You can see how helpful the description can be if you provide your PKT file to a different user. You can now use this subassembly just as you would any other subassembly using the procedures discussed in Chapter 8, "Assemblies and Subassemblies."

Using Expressions

In the UrbanSidewalkSlopes example, all of the dimensions and slopes were provided through the input parameters. There are going to be some instances when you want to perform more detailed calculations for the expressions in your various elements. For these instances you will turn to API functions to generate your expressions. The link API function `L1.slope` was briefly mentioned earlier but there are many more API functions that you have at your fingertips. These API functions are defined by classes. Each API function has at least two parts: one before the period and one after the period. Some API functions also call for additional information provided in parentheses. While you will find you use some API functions more than others, each has an important use. Let's summarize the API functions available in Subassembly Composer.

Point and Auxiliary Point Class API Functions

There are multiple API functions that you can call to provide information about points and auxiliary points. To use these API functions, you type the point you want information about and then a period, followed by the API function. For example, in order to get the horizontal distance from the origin to point P1, you would enter **`P1.X`**. The API functions in the `Point` class are as follows (shown for point P1; however, this can apply to any of the points or auxiliary points previously defined in the flowchart):

`X` `P1.X` returns the horizontal distance from point P1 to the origin. A positive value means that point P1 is to the right of the origin, and a negative value means that point P1 is to the left of the origin.

`Y` `P1.Y` returns the vertical distance from point P1 to the origin. A positive value means that point P1 is above the origin, and a negative value means that point P1 is below the origin.

`Offset` `P1.Offset` returns the horizontal distance from the assembly baseline to point P1. A positive value means that point P1 is to the right of the assembly baseline, and a negative value means that point P1 is to the left of the assembly baseline.

`Elevation` `P1.Elevation` returns the elevation of point P1 relative to elevation 0.

`DistanceTo` `P1.DistanceTo("P2")` returns the distance from point P1 to point P2. This value is always positive. The quote marks are required unless you are defining the second point using another API function, such as the `L1.StartPoint` API function discussed later.

`SlopeTo` `P1.SlopeTo("P2")` returns the slope from point P1 to point P2. If the value is positive it is an upward slope, or if it is negative it is a downward slope. The quote marks are required unless you are defining the second point using another API function such as the `L1.StartPoint` API function discussed later.

`IsValid` `P1.IsValid` returns a value of `True` or `False` if the point P1 is assigned and valid.

`DistanceToSurface` `P1.DistanceToSurface(SurfaceTarget)` returns the vertical distance from the point P1 to the target surface. A positive value means that point P1 is to the above the target surface, and a negative value means that point P1 is below the target surface.

Link and Auxiliary Link Class API Functions

Like points and auxiliary points, there are multiple API functions that you can call to provide information about links and auxiliary links. To use these API functions, you type the link you want information about and then a period, followed by the API function. For example, in order to get the slope of link L1, you would enter **`L1.Slope`**. The API functions in the `Link` class are as follows (shown for link L1; however, this can apply to any of the links or auxiliary links previously defined in the flowchart):

`Slope` `L1.Slope` returns the slope of link L1.

`Length` `L1.Length` returns the length of link L1. This value is always positive.

`XLength` `L1.XLength` returns the horizontal distance between the start and end of link L1. This value is always positive.

`YLength` `L1.YLength` returns the vertical distance between the start and end of link L1. This value is always positive.

`StartPoint` `L1.StartPoint` returns a point located at the start of link L1. This can then be used in the API functions for the `Point` class.

`EndPoint` `L1.EndPoint` returns a point located at the end of link L1. This can then be used in the API functions for the `Point` class.

`MaxY` `L1.MaxY` returns the maximum Y elevation from a link's points.

`MinY` `L1.MinY` returns the minimum Y elevation from a link's points.

`MaxInterceptY` `L1.MaxInterceptY(slope)` returns the highest intercept of a given link's points to the start of another link.

`MinInterceptY` `L1.MinInterceptY(slope)` returns the lowest intercept of a given link's points to the start of another link.

`LinearRegressionSlope` `L1.LinearRegressionSlope` returns the slope calculated as a linear regression on the point in a link to find the best fit slope between all of them.

`LinearRegressionInterceptY` `L1.LinearRegressionInterceptY` returns the Y intercept of the linear regression link.

`IsValid` `L1.IsValid` returns a value of `True` or `False` if the link is assigned and valid to use.

`HasIntersection` `L1.HasIntersection("L2",true,true)` returns a value of `True` or `False` if links L1 and L2 have an intersection. The second and third expressions are optional. If not defined, they both default to false. The second expression defines whether to extend link L1, and the third expression defines whether to extend link L2.

Elevation Target Class API Functions

There are multiple API functions that you can call to provide information about elevation targets. The API functions in the `Elevation Target` class are as follows (shown for an elevation target name `ElevationTarget`; however, this can apply to any of the elevation targets):

`IsValid` `ElevationTarget.IsValid` returns a value of `True` or `False` if the elevation target is assigned and valid to use.

`Elevation` `ElevationTarget.Elevation` returns the elevation of the elevation target relative to elevation 0.

Offset Target Class API Functions

There are multiple API functions that you can call to provide information about offset targets. The API functions in the `Offset Target` class are as follows (shown for an offset target name `OffsetTarget`; however, this can apply to any of the offset targets):

IsValid `OffsetTarget.IsValid` returns a value of `True` or `False` if the offset target is assigned and valid to use.

Offset `OffsetTarget.Offset` returns the horizontal distance from the assembly baseline to the offset target. A positive value means that the offset target is to the right of the assembly baseline, and a negative value means that the offset target is to the left of the assembly baseline.

Surface Target Class API Functions

There is one API function that you can call to provide information about surface targets. The API function in the `Surface Target` class is `IsValid`. It is shown here for a surface target name `SurfaceTarget`; however, this can apply to any of the surface targets:

`SurfaceTarget``.IsValid`

This returns a value of `True` or `False` if the surface target is assigned and valid to use.

Baseline Class API Functions

There are multiple API functions that you can call to provide information about the assembly baseline. Unlike some of the other API functions that use the name of a specific element before the period, with the `Baseline` class API functions, all of them begin with the word `Baseline`. The API functions in the `Baseline` class are as follows:

Station `Baseline.Station` returns the station of the assembly baseline at the current frequency line.

Elevation `Baseline.Elevation` returns the elevation of the assembly alignment at the current frequency line.

RegionStart `Baseline.RegionStart` returns the station at the start of the current corridor region.

RegionEnd `Baseline.RegionEnd` returns the station at the end of the current corridor region.

Grade `Baseline.Grade` returns the instantaneous grade of the assembly baseline at the current frequency line.

TurnDirection `Baseline.TurnDirection` returns the turn direction of the current station on the baseline alignment.

Enumeration Type Class API Functions

There is one API function that you can call to provide information about a user-defined enumeration. The API function in the `Enumeration Type` class is `Value`. It is shown here for the enumeration name `EnumerationType`; however, this can apply to any of the enumerations:

`EnumerationType``.Value`

This returns the string value of the current enumeration item.

Real World Scenario

ENUMERATION?

When you learned about the eight different types for input and output parameters earlier, the term *enumeration* was mentioned. Enumeration allows you to generate a unique type. The enumeration will generate a drop-down list of options in the Advanced Parameters properties category for the subassembly in Civil 3D, similar to the Yes/No drop-down list.

For example, if you had three different types of curbs, you could place a Switch element below the Start element and then build a sequence for each of the types of curbs and connect them to the Switch element as shown here:

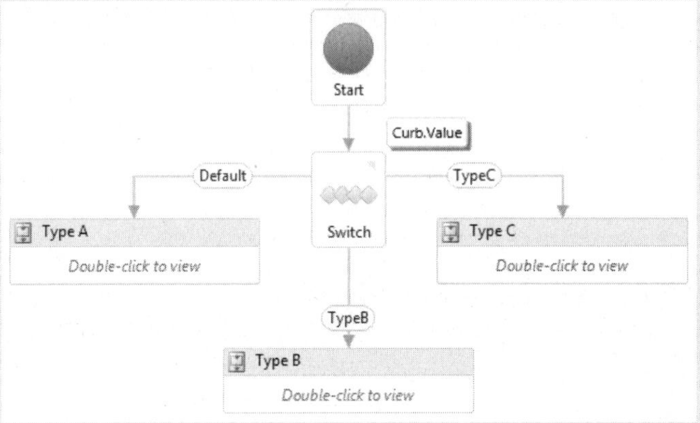

If you create an enumeration group listing the three curb types (enumeration items), the end user can specify which curb they want from the subassembly's properties without having to change the dimensions to match the desired curb shape.

To create a custom enumeration, follow these steps:

1. From the View menu, select Define Enumeration to display the Define Enumeration dialog shown here:

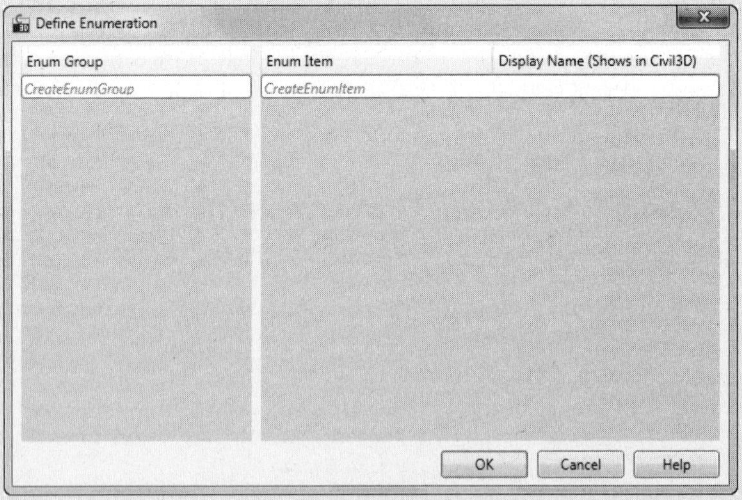

2. Click CreateEnumGroup to add an enumeration group.
3. Change the enumeration group name from Enum0 to **CurbType**.
4. Click CreateEnumItem three times to add three items within the group.
5. Change the enumeration items to match the following image:

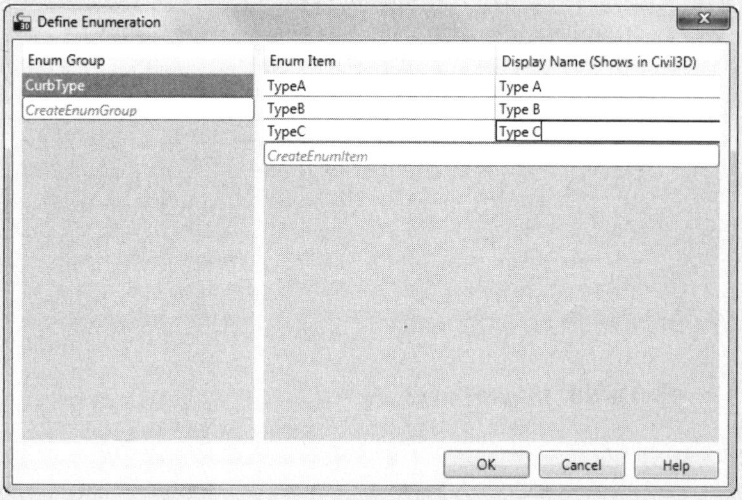

Now you can generate an input parameter with the type as CurbType and specify the default value from the enumeration items:

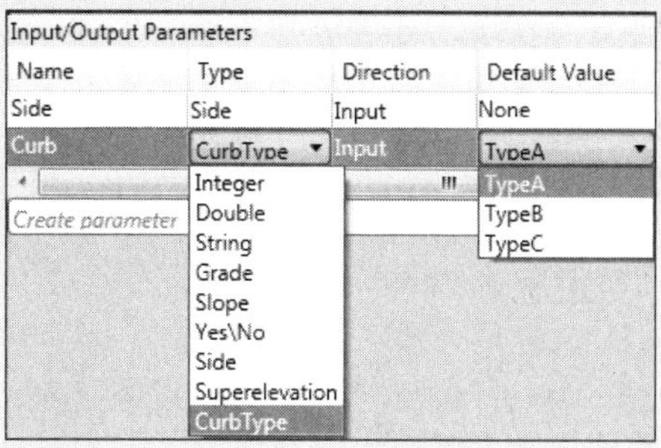

If you look back at the flowchart of the Switch element, you will notice a few things.

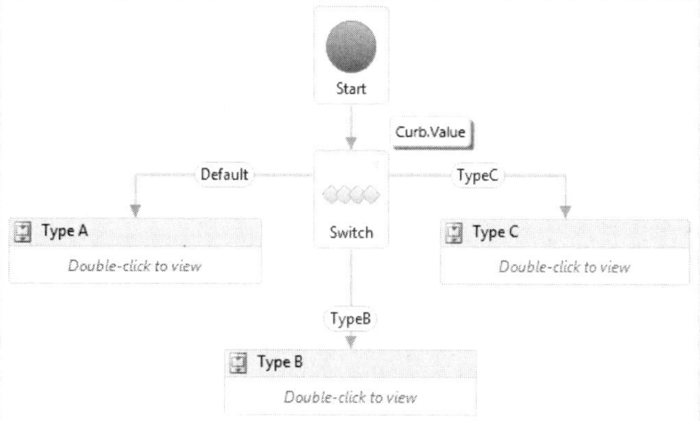

First, you will notice the `Curb.Value` enumeration class API function floating to the upper right of the Switch element. This flyout is available on both the Switch elements as well as the Decision elements; it shows the expression that is being used by the element. In this case, the expression calls for the value of the Curb input parameter.

Second, you will notice the three connection arrows coming out of the Switch element. Decision elements are used for deciding between two options (true or false), whereas Switch elements can actually handle 11 conditions. The first condition is always Default. The next conditions will default to Case1, Case2, and so on. You will want to change these to correspond with the other cases. You can change the case names by clicking on the connection arrow and changing the Case setting in the Properties panel.

Math API Functions

In addition to all the Subassembly Composer API functions, numerous math API functions are available. A few of the common ones you may find a need for are as follows:

`Round` `Math.Round(12.1009,2)` returns the first value rounded to the nearest specified decimal places. If the second value is -2, it will round to the nearest hundreds, -1 will round to the nearest tens, 0 to the nearest integer, 1 to the nearest tenths, 2 to the nearest hundredths, and so on.

`Floor or Ceiling` `Math.Floor(12.1009)` or `Math.Ceiling(12.1009)` returns the next smallest integer (floor) or next largest integer (ceiling). In this example, the floor would be 12 and the ceiling would be 13.

`Max or Min` `Math.Max(12.1009,4.038)` or `Math.Min(12.1009,4.038)` returns the maximum value or minimum value of the two specified values. In this example, the max would be 12.1009 and the min would be 4.038.

`Pi` `Math.pi` returns the value of the constant pi.

`Sin, Cos, Tan, Asin, Acos, or Atan` `Math.Sin(math.pi)` returns sine of the specified value (in this case the constant pi). This similar format could be used for any of the other trigonometric functions. The angles specified or calculated should be provided in radians.

Using an API Function in an Expression

With all of the stock subassemblies, the parameters are constants throughout the region unless a target is used to taper the offset or elevation. By using the API functions in the baseline class, you can now generate a subassembly that knows the start and end of the region that it is in as well as the current station. By using this information, you can taper from a start value to an end value. In the following example, you will use the Subassembly Composer to calculate the current value (a double variable) based on the starting value and end value, which have a linear taper between them. You will then import this simple calculation subassembly into Civil 3D to use in tapering a shoulder using a parameter reference.

1. Start a new subassembly file and save it as **`TaperDouble.pkt`**.
2. On the Packet Settings tab, enter **TaperDouble** for the subassembly name.

3. On the Input/Output Parameters tab, click Create Parameter to add a new parameter for each of the input and output parameters with the following settings:

NAME	TYPE	DIRECTION	DEFAULT VALUE	DISPLAYNAME
Side	Side	Input	None	
StartValue	Double	Input	0	Start Value
EndValue	Double	Input	0	End Value
CalcValue	Double	Output		Calculated Value

4. From the Tool Box ➢ Geometry branch, drag a Point element to below the Start element.

   Leave this point as generated at the origin and do not change any of the default values. A minimum of one point must be in the subassembly in order for the calculation that you will define in the next step to work once imported into Civil 3D.

5. From the Tool Box ➢ Miscellaneous branch, drag a Set Output Parameter element to below P1.
6. Define the Set Output Parameter element as follows:
   a. Verify that Output Parameter is set to CalcValue.
   b. Set the value to **`StartValue+(baseline.station-baseline.regionstart)/(baseline.regionend-baseline.regionstart)*(EndValue-StartValue)`**.

   If you ever want to have a larger text field for entering text, many of the fields throughout the Subassembly Composer have an ellipsis button that you can click to display the Expression Editor dialog, as shown in Figure 9.29.

FIGURE 9.29
Expression Editor can be accessed by clicking the ellipsis button on many of the expression fields.

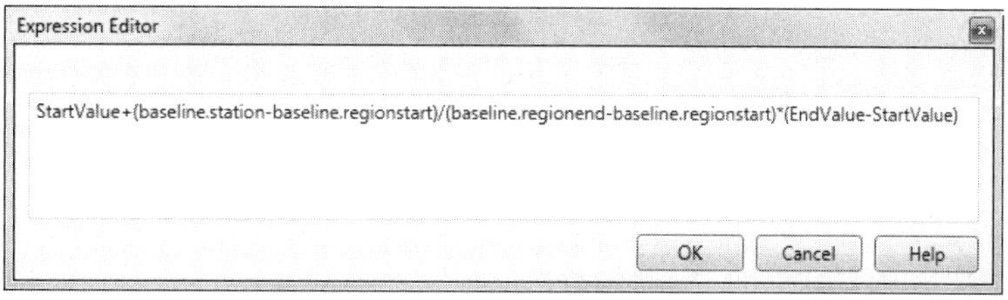

7. In Subassembly Composer, save your PKT file, which should now look like Figure 9.30.

FIGURE 9.30
Simple subassembly with just a Set Output Parameter element

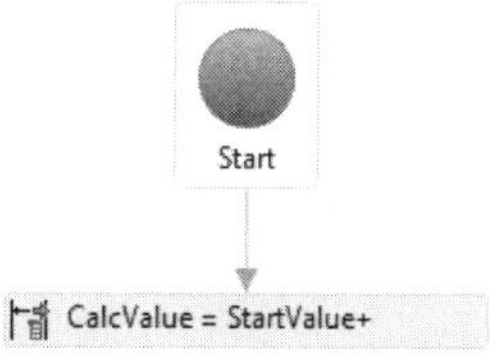

When this portion of the exercise is complete, you can close the file. A finished copy of this file is available from the book's web page with the filename `TaperDouble.pkt`. You will now import this subassembly into Civil 3D and use a parameter reference to pass the output parameter into another subassembly as an input parameter.

8. Open the `TaperShoulder.dwg` or `TaperShoulder_METRIC.dwg` file.

   This file has an assembly of a simple road cross section with lanes, curbs, shoulders, and daylight links, as shown in Figure 9.31.

FIGURE 9.31
Road subassembly

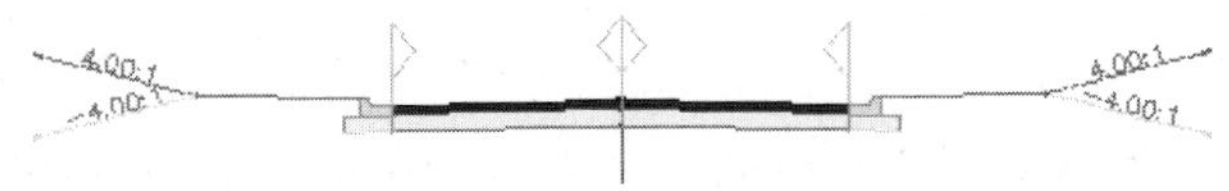

9. From the Insert tab ➢ expanded Import panel, choose Import Subassemblies.

   The Tool Palettes window will open, if it's not already open, and the Import Subassemblies dialog will be displayed.

10. In the Import Subassemblies dialog, click the folder button to display the Open dialog.
11. Navigate to your saved `TaperDouble.pkt` file and click Open.
12. Verify that the Import To: Tool Palette check box is selected and select the Mastering tool palette you created in an earlier exercise.
13. Click OK to accept the settings in the Import Subassemblies dialog and add the subassembly to the Mastering tool palette.

    Now you will add the TaperDouble calculation before the left and right shoulders. The TaperDouble subassembly component will calculate the shoulder width and pass it to the shoulder subassembly using a parameter reference.

14. Click the TaperDouble subassembly on the Tool Palettes window.
15. Locate the Advanced-Parameters section on the Design tab of the AutoCAD Properties palette.
16. Change the Start Value to **3′** (or **1** m) and the End Value to **10′** (or **3** m).
17. At the `Select marker point within assembly or [Insert Replace Detached]:` prompt, press **I** ↵.
18. At the `Select the subassembly to insert after or [Before]:` prompt, press **B** ↵.
19. At the `Select the subassembly to insert ahead of:` prompt, select the left shoulder, which is a LinkWidthAndSlope subassembly.
20. At the `Select a different marker point on the new subassembly, or enter to continue:` prompt, press ↵.

    Now you will do the same thing to add the TaperDouble calculation before the right shoulder.

21. At the `Select the subassembly to insert after or [Before]:` prompt, press **B** ↵.
22. At the `Select the subassembly to insert ahead of:` prompt, select the right shoulder.
23. At the `Select a different marker point on the new subassembly, or enter to continue:` prompt, press ↵.
24. Press ↵ again to end the command.
25. Right-click on the red assembly baseline marker and select Assembly Properties.
26. On the Construction tab, highlight the LinkWidthAndSlope subassembly in the Right group.
27. In the Input Values area, select the check box in the Parameter Reference Use column for the Width value.
28. In the Get Value From column next to the check box, select TaperDouble.Calculated Value, as shown in Figure 9.32.

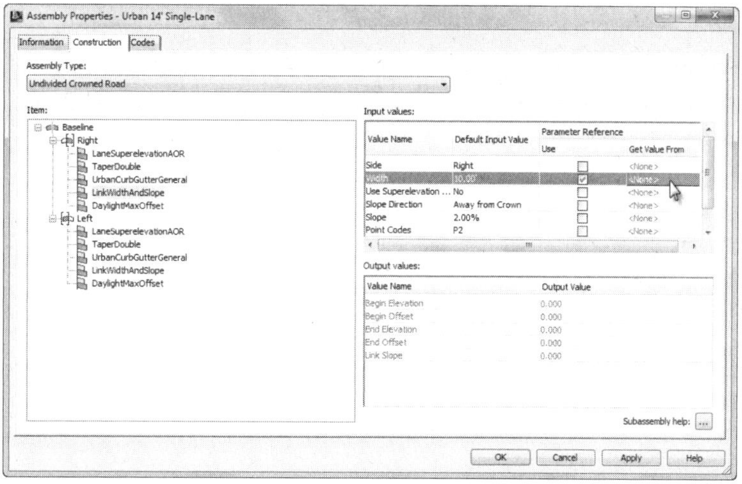

FIGURE 9.32
Setting the shoulder width using a parameter reference of the value calculated by the TaperDouble subassembly

29. Repeat steps 26 through 28 for the Left group.
30. Click OK when complete.

WHERE ARE THE SHOULDERS AND DAYLIGHT LINKS?

You will notice that the shoulders and daylight links appear to have disappeared from the assembly. This is because the shoulder links are trying to calculate based on their location in a region of a corridor and don't yet have anything to calculate. Don't worry—the subassembly elements are still there, as shown in the Assembly Properties that you just looked at. Just because the assembly does not look right doesn't mean that it is broken!

When this exercise is complete, you can close the drawing. A finished copy of this drawing is available from the book's web page with the filename `TaperShoulder.dwg` or `TaperShoulder_Metric.dwg`.

By passing the value to a stock subassembly, you avoid having to re-create the entire stock subassembly just to incorporate an API function into the calculation of one parameter.

Employing Conditional Logic

Decisions are the most common way you will define conditions that will have either a true or a false result. There are two ways you can make decisions: one is with a Decision or Switch element, and the other is by using conditional logic operators in your expression. If the decision is going to affect only one expression, then you will want to use the conditional logic operators. However, if the decision is going to have an effect on the flowchart logic, then you will use a Decision or Switch element.

Conditional Logic Operators

There are numerous conditional logic operators that you will want to be familiar with as you define decision conditions. Examples of these are as follows:

> or < `P1.Y>P2.Y` (or `P1.Y<P2.Y`) returns true if `P1.Y` is greater than (or less than) `P2.Y`.

>= or <= `P1.Y>=P2.Y` (or `P1.Y<=P2.Y`) returns true if `P1.Y` is greater than or equal to (or less than or equal to) `P2.Y`.

= or <> `P1.Y=P2.Y` (or `P1.Y<>P2.Y`) returns true if `P1.Y` is equal to (or not equal to) `P2.Y`.

AND `P1.Y>P2.Y)AND(P2.X>P3.X)` returns true if both the first condition located in a pair of parentheses and the second condition located in a pair of parentheses are true.

OR `(P1.Y>P2.Y)OR(P2.X>P3.X)` returns true as long as at least one of the conditions located in either pair of parentheses is true.

XOR `(P1.Y>P2.Y)XOR(P2.X>P3.X)` returns true if one of the conditions located in a pair of parentheses is true and the other condition located in a pair of parentheses is false.

Any of these operators could also be used in an `If` statement within any expression in order to test a condition and provide one of two different values.

In the following example expression for a slope value, the distance from auxiliary point AP1 to the target surface is calculated. If it is greater than 0, meaning that AP1 is located above the surface (in fill), then the slope value will be –0.25. If the value is not greater than 0, meaning that AP1 is located below the surface (in cut), then the slope value will be 0.33.

```
IF(AP1.DistanceToSurface(TargetSurface)>0,-0.25,0.33)
```

With a combination of the right conditional logic operators, you can make all kinds of decisions.

Decision and Switch Elements

You briefly saw an example of a Switch element in action earlier in this chapter when you learned about enumeration. The Decision elements have a similar behavior but provide only two conditions: True or False. The Decision element is located in the Tool Box ➤ Workflow branch along with the Switch element, Flowchart element, and Sequence element. When you insert a Decision element into your flowchart and hover over it, you will see that the True condition will be out of the left side and the False condition will be out of the right side, as shown in Figure 9.33.

FIGURE 9.33
A Decision element with True on the left and False on the right

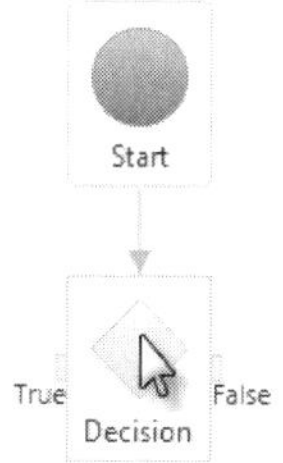

An easy way to remember which side is which is "Right side means the condition is wRong."

By highlighting the element you can define the condition in the Properties panel. For example, you can specify `AP1.DistanceToSurface(TargetSurface)>0` similar to the example expression discussed in the previous section. However, notice that you did not need to specify in the expression what will happen if the answer is true or false; this is because you will specify the true and false through the use of the connection arrows out of the left and right sides of the Decision element.

One of the benefits of using a Decision element in your flowchart instead of burying a conditional expression into an element is that finding them is easy, since the flowchart is a visual representation of the logic. There are a few things you can do to make your decisions easy to follow in the flowchart, as shown in Figure 9.34.

FIGURE 9.34
A defined Decision element with customized True and False connection arrow labels and the condition expression flyout

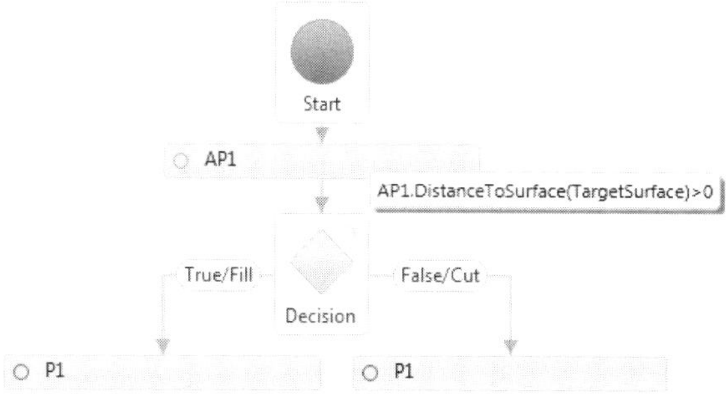

The first thing you can do is click on the triangular flag in the upper-right corner of the element. Doing this will cause the condition expression to be displayed in a flyout.

The second thing you can do is to further annotate the True and False connection arrow labels. You can change these labels through the Decision element's Properties panel.

You will also notice in Figure 9.34 that there are two elements named P1. This may confuse you, as you have learned up until this point that every point, link, and shape must have a unique number, which is true—however, there is a caveat. The elements must have a unique number along any given connection arrow path. Because these elements are not along the same path, they may be duplicated.

Making Sense of Someone Else's Thoughts

There is going to come a time that you will open a PKT file that someone else made or you made long ago. Most likely you are opening up this confusing and complicated PKT file because you have found a mistake in how it behaves in Civil 3D or you want to edit it for different Civil 3D capabilities. So instead of building the subassembly from scratch you will have to understand how the subassembly was built.

In this last example, you will open a complex subassembly that was made to behave similarly to the Channel stock subassembly. You have decided that instead of the baseline attachment point being at a point located above the midpoint of the bottom width, at a height equal to the Depth parameter, you want it at the midpoint of the bottom link.

1. Open the `Channel.pkt` or `Channel_METRIC.pkt` file.

   The channel subassembly in the Preview panel, as shown in Figure 9.35, should look similar to the subassembly shown in the Channel stock subassembly help file.

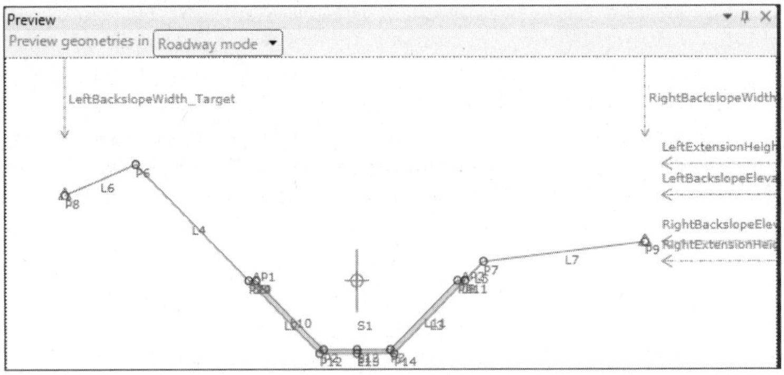

FIGURE 9.35 The Channel subassembly in the Preview panel as shown in Roadway Mode

The flowchart looks like Figure 9.36.

FIGURE 9.36 Complex flowchart for the Channel subassembly

As you can see, the flowchart is full of decisions and sequences and elements. Thankfully, if you take a close look you will see that the sequences are clearly labeled, as are the True and False labels, as shown in Figure 9.37.

FIGURE 9.37
Labeled elements in the complex flowchart

Because the attachment point is typically one of the first elements defined, it should be pretty easy to find.

2. Double-click on the Main Channel Points sequence element.

   Be certain to double-click on the actual element and not the text; otherwise you will just select the name of the sequence for editing.

3. Click on the element for point P1.

   You will notice that it also highlights in the Preview panel.

4. In the Properties panel, change the Delta Y from -math.abs(Depth), as shown in Figure 9.38, to **0**.

FIGURE 9.38
Changing the Delta Y value from the origin for point P1 to 0

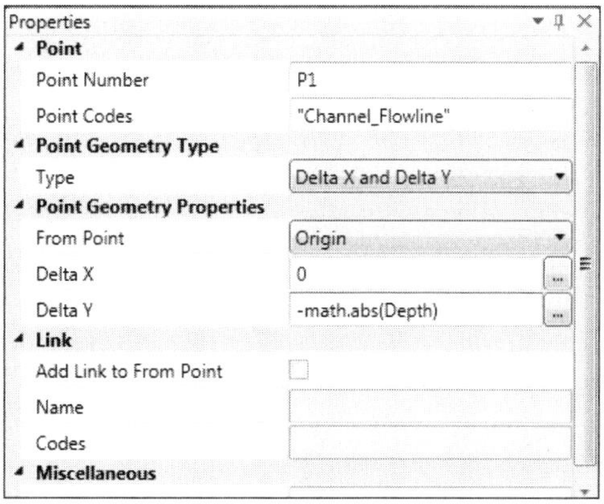

When you do so, you will notice that the whole subassembly jumps up to the origin. You will also notice that the links at right have disappeared in Roadway Mode. This is because Roadway Mode uses the targets that are set at locations which are no longer valid. If you look at Event Viewer in the Settings/Parameters panel you will notice three errors about points that can be placed. These errors will disappear if you choose to change the preview values of the right target parameters. If you switch to Layout Mode, your subassembly will use only the input parameter that matches Figure 9.39.

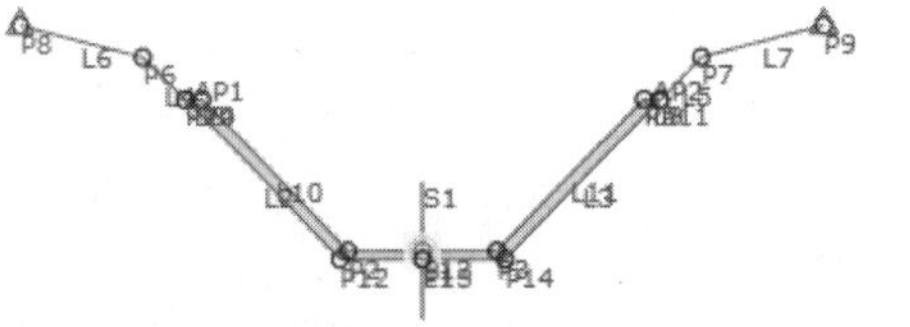

FIGURE 9.39
The Channel subassembly with the bottom of the channel now at the baseline assembly marker shown in Layout Mode

> ***math.abs(Depth)?***
>
> **You may have noticed the original expression that was used to define the Delta Y value of this point and wondered what it was doing. `math.abs()` is an API function that finds the absolute value of the variable in the parentheses. By first finding the absolute value of the input parameter, you avoid having an error occur if one user in Civil 3D thinks the Depth parameter should be a negative value and one user thinks it should be a positive value.**

Although this was a simple change to a complex flowchart, it all starts with finding the right location to change the variable. Now that the attachment point is at the bottom of the channel, let's make one more change to this subassembly. You should add a link representing the water surface elevation and then define a shape for the water, as shown in the next steps.

5. In the Preview panel, zoom in to the subassembly and you will notice that the depth points located at the top of the lining of the channel are at point P4 on the left and point P5 on the right.

6. On the Input/Output Parameters tab, change the value for LiningDepth to **0**.

   You will notice that the lining disappears but the depth points remain as P4 and P5. This is good because it means that you can define one link as the water surface that will be applicable to the subassembly with and without the lining.

7. From the Tool Box ➤ Geometry branch, drag a Link element to below point P5 in the Main Channel Points sequence.

   This link will automatically be numbered L13.

8. Define link L13 as follows:

   a. Set Link Codes to **"WaterSurface"**.

Note that the quote marks are required in order to define this text as a string.

b. Verify that Start Point is P4 and End Point is P5.

9. Click on the word Flowchart in the upper-left corner of the Flowchart panel to return to the main flowchart.

You must now find an appropriate place to define the shape delineating the water.

10. On the No Lining side of the first decision, double-click the Main Channel Links sequence element.

11. From the Tool Box ➢ Geometry branch, drag a Shape element to below link L3. This shape will automatically be numbered S2.

12. Define shape S2 as follows:

a. Set Shape Codes to **"Water"**.

Note that the quote marks are required in order to define this text as a string.

b. Click the Select Shape In Preview button.

c. Select inside of the closed shape representing the water, as shown in Figure 9.40.

FIGURE 9.40
Defining the water shape for the channel subassembly with no lining

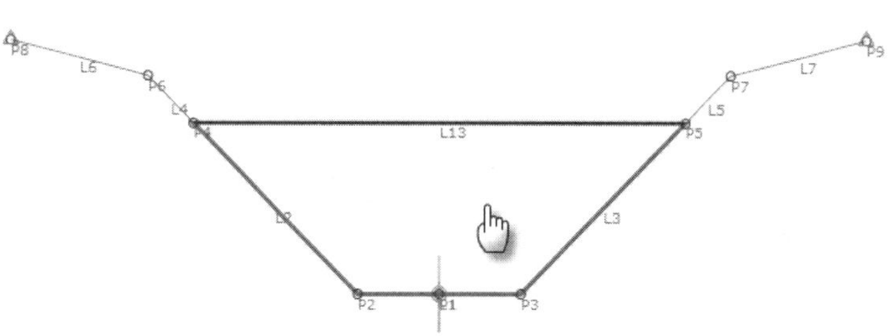

13. On the Input/Output Parameters tab, change LiningDepth back to **0.333** (or **0.1** for metric).

Notice that the water link remains but the shape is no longer present. This is because you defined the shape in a branch of the flowchart that isn't being used when the lining is in place.

14. Click on the word Flowchart in the upper-left corner of the Flowchart panel to return to the main flowchart.

You must now find an appropriate place to define the shape delineating the water for the subassembly when a lining is being used.

15. On the Add Lining side of the first decision, double-click the Lining Links And Shape sequence element.

16. From the Tool Box ➢ Geometry branch, drag a Shape element to below shape S1.

This shape will automatically be numbered S3. Note that if you wanted you could renumber this shape to S2 since it is located in a different branch from the S2 previously defined.

17. Define shape S3 as follows:
    a. Set Shape Codes to **"Water"**.

       Note that the quote marks are required in order to define this text as a string.
    b. Click the Select Shape In Preview button.
    c. Select inside of the closed shape representing the water, as shown in Figure 9.41.

FIGURE 9.41 Defining the water shape for the channel subassembly with a lining

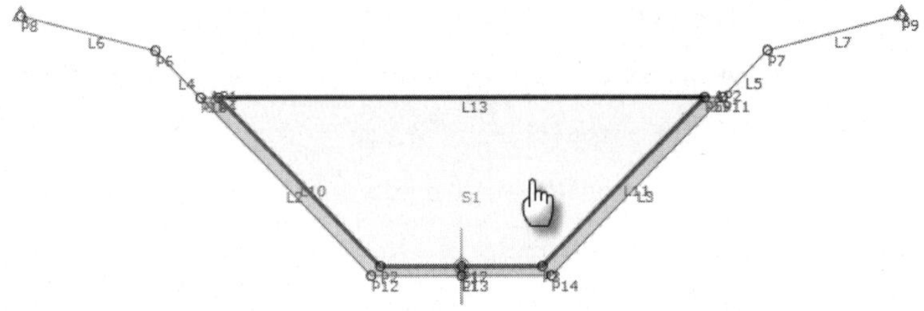

You have now changed the attachment point from the depth location to the bottom of the channel, added a link at the water elevation, and defined a water shape.

When this exercise is complete, you can close the file. A finished copy of this file is available from the book's web page with the filename `Channel_modified.pkt` or `Channel_Metric_modified.pkt`.

SHARING CUSTOM SUBASSEMBLIES

PKT files are not part of the drawing file. So anytime you use a PKT file in a drawing you must make the PKT file available for anyone who has to rebuild the corridor. If a user does not first import the subassembly into the drawing, they will receive an error message that says "One or more subassembly macros (or .NET classes) could not be found. Check Event Viewer for more information. Continue? Yes/No."

In conjunction with the stock subassemblies already available in the Civil 3D and the custom subassemblies you create with the Tool Box in the Subassembly Composer, your corridors will have endless possibilities.

The Bottom Line

Define input and output parameters with default values. By providing detailed input parameters, you let the Civil 3D user edit your custom subassembly. In addition, by providing detailed output parameters you let them use the subassembly's characteristics to edit the subsequent subassemblies in the assembly.

Master It Generate a new subassembly PKT file for a subassembly named MasteringLane (or MasteringLaneMetric). For input parameters, define the following:

- Side with a default value of Right
- LaneWidth provided as a decimal precise number with a default value of 12 feet or 3.6 meters
- LaneSlope provided as a percentage with a default value of -3%
- Depth provided as a decimal precise number with a default value of 0.67 feet or 0.2 meters

For output parameters, define the following:

- CalcLaneWidth provided as a decimal precise number
- CalcLaneSlope provided as a percentage

Define target offsets, target elevations, and/or target surfaces. Targets allow the subassembly to reference unique items within the drawing. They allow for varying widths, tapered elevations, and daylighting to a surface.

Master It For the same subassembly from the previous Master It, define an elevation target named TargetLaneElevation with a preview value of -0.33 feet (or -0.1 meters) and an offset target named TargetLaneOffset with a preview value of 14 feet (or 4 meters).

Generate a flowchart of subassembly logic using the elements in the Tool Box. The Tool Box in Subassembly Composer is full of all sorts of tools. But like a tool box in your garage, the tools are only useful if you know their capabilities and practice using them. The simplest tools are points, links, and shapes, but there are also tools for more complex flow charts.

Master It Define the subassembly using the width, slope, and depth input parameters defined earlier. Make certain that if the slope of the lane changes based on the targets, then the top and the datum links have matching slopes.

The subassembly should have four points: P1 ("Crown") located at the origin, P2 ("ETW") located at the upper-right corner of the lane cross section (with both the TargetLaneElevation and TargetLaneOffset), P3 ("Crown_Subbase") at the lower left of the lane cross section, and P4 ("ETW_Subbase") at the lower right of the lane cross section.

The subassembly should have four links: L1 ("Top", "Pave") connecting P1 and P2, L2 connecting P1 and P3, L3 ("Datum", "Subbase") connecting P3 and P4, and L4 connecting P2 and P4.

The subassembly should have a shape: S1 ("Pave1").

Import a custom subassembly made with Subassembly Composer into Civil 3D. A custom subassembly is no good if it is just made and not used. By knowing how to import your PKT file into Civil 3D, you open up the world of building and sharing tool palettes full of subassemblies to help your office's workflow.

Master It Import your MasteringLane or MasteringLaneMetric subassembly into Civil 3D.

Chapter 10

Basic Corridors

The corridor object is a three-dimensional model that combines the horizontal geometry of an alignment, the vertical geometry of a profile, and the cross-sectional geometry of an assembly.

Corridors range from extremely simple roads to complicated highways and interchanges, but aren't limited to just road travel ways. Corridors can be used to model many linear designs. This chapter focuses on building several simple corridors that can be used to model and design roads, channels, and trenches.

In this chapter, you will learn to:

- Build a single baseline corridor from an alignment, profile, and assembly
- Use targets to add lane widening
- Create a corridor surface
- Add an automatic boundary to a corridor surface

Understanding Corridors

In its simplest form, a corridor combines an alignment, a profile, and an assembly (see Figure 10.1).

FIGURE 10.1
A corridor shown in 3D view

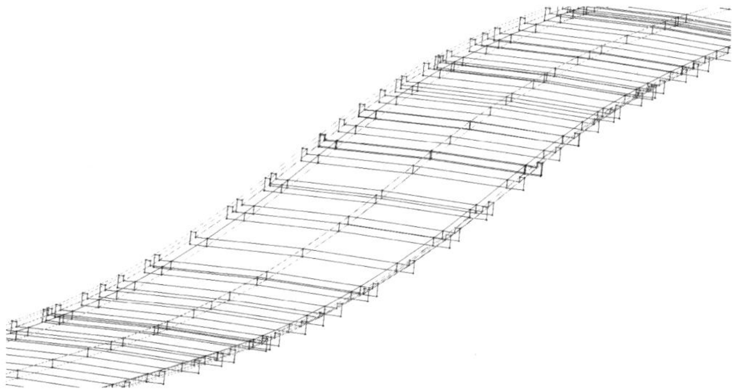

You can also build corridors with complex combinations of alignments, profiles, and assemblies to make complicated intersections, interchanges, or branching streams (see Figure 10.2).

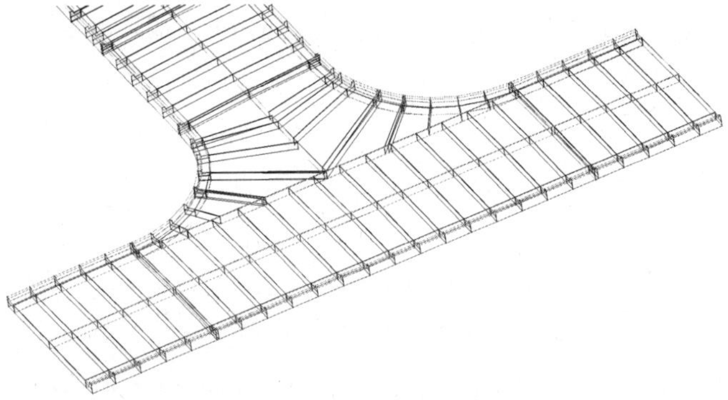

FIGURE 10.2
An intersection modeled with a corridor

The horizontal properties of the alignment, the vertical properties of the profile, and the cross-sectional properties of the assembly are merged together to form a dynamic model that can be used to build surfaces, sample cross sections, generate quantities, and much more.

Most commonly, corridors are thought of as being used to model roads, but they can also be adapted to model berms, streams (see Figure 10.3), trails, and even parking lots.

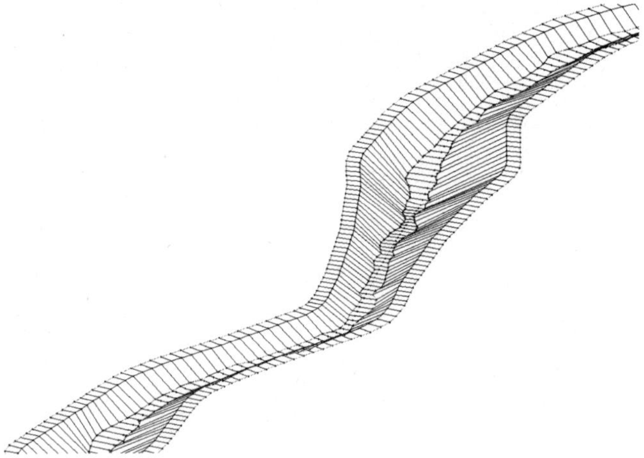

FIGURE 10.3
A stream modeled with a corridor

Corridor Components

CERT OBJECTIVE

First, let's look at some important corridor components you will want to become familiar with before proceeding. *Baseline, regions, assemblies, frequency,* and *targets* are all parts of a corridor that you will encounter even on your first design.

Baseline

The first component for any corridor is a *baseline*. The baseline is actually composed of two AutoCAD® Civil 3D® objects, an alignment providing the horizontal layout and a profile providing the vertical layout. The baseline generates the backbone skeleton on which the assembly can hang.

In most examples in this chapter, the baseline will correspond to the centerline alignment with a profile representing the elevation at the crown of a proposed road. However, this is not always the case, as we will explore in more depth in Chapter 11, "Advanced Corridors, Intersections, and Roundabouts." As your designs become more detailed, you may have corridors with multiple baselines.

Regions

When the geometry along a baseline changes enough to warrant a new assembly, a new *region* is needed. Regions specify the station range where a specific assembly is applied to the design. There may be many regions along a baseline to accommodate design geometry but the regions may not overlap. Therefore, each region has a start station and an end station with the end station of one region often matching the start station of the next region.

Assemblies

Marker points and *links* are coded into the subassemblies that comprise the assembly, as you saw in Chapter 8, "Assemblies and Subassemblies." Assemblies are the third Civil 3D object that is required to generate the corridor by providing cross-sectional information to be applied along some or all of the length of the baseline. Figure 10.4 shows a symmetric typical roadway assembly which consists of lanes, curbs, sidewalks, and daylight links.

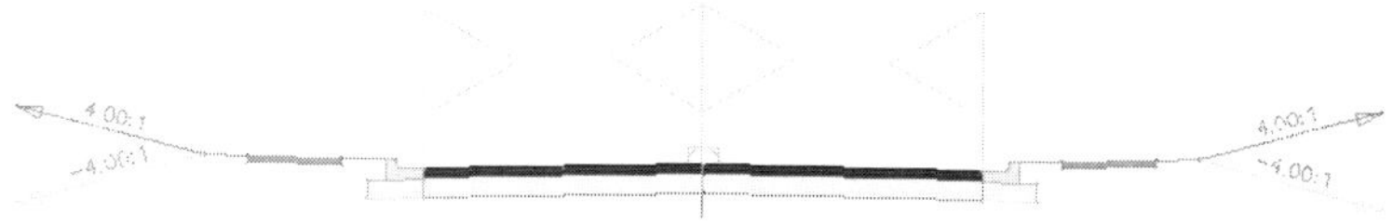

FIGURE 10.4
Typical roadway assembly

Frequency

Frequency refers to how often the assembly is applied to the corridor design. You can set the frequency distance for the corridor as a whole, but in most cases, you should apply it at the region level.

The frequency value will vary depending on the situation. The default frequencies of the stock Civil 3D templates are 25′ in Imperial units and 20 m for metric units. But like many of the settings in Civil 3D, these defaults can be changed in your drawing or template settings by using the following simple steps:

1. On the Settings tab of Toolspace, expand the Corridor ➢ Commands branch.
2. Right-click the CreateCorridor command to display the Edit Command Settings – CreateCorridor dialog.

3. Expand the Assembly Insertion Defaults.
4. Change the Frequency Along Tangents, Frequency Along Curves, Frequency Along Spirals, Frequency Along Profile Curves, or any other of the multitude of default settings associated with creating a corridor.

In addition to the frequencies based on the alignment or profile entity, Civil 3D can place frequency lines at special stations such as horizontal geometry stations, superelevation critical stations, profile geometry stations, profile high/low stations, or offset geometry stations. You can also create additional frequency stations for things like driveways or culvert crossings.

Targets

As you learned in Chapter 8, there are three types of targets that an assembly can reference: a target surface, a target elevation, and a target offset. Some targets are optional whereas others are required. You will want to reference the help files, for the various subassemblies that are used in the assembly, to determine further information on what types of targets you may need or want.

Corridor Feature Lines

After a corridor is created, you will see a series of lines running parallel (or mostly parallel) to the alignment; these lines are called *corridor feature lines*. Corridor feature lines, sometimes referred to simply as feature lines, are the result of Civil 3D playing connect-the-dots with marker points between the frequency lines.

Everywhere the assembly contains a named marker point, Civil 3D creates a feature line with the same name.

These special feature lines are the third dimension that takes a corridor from being simply a collection of cross sections to being a model with meaningful flow (see Figure 10.5).

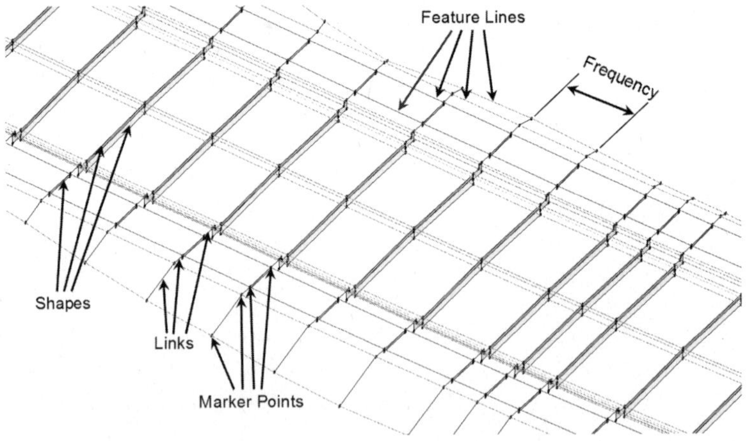

FIGURE 10.5 The anatomy of a corridor

Later on in this chapter, we will take a closer look at corridor feature lines.

This exercise gives you hands-on experience in building a corridor model from an alignment, a profile, and an assembly:

1. Open the `RoadCorridor.dwg` file or `RoadCorridor_METRIC.dwg` file). You can download files from `www.sybex.com/go/masteringcivil3d2013`.

Note that the drawing has an alignment for Cabernet Court, a profile view with two profiles for Cabernet Court, an assembly, and an existing ground surface.

2. From the Home tab ➤ Create Design panel, choose Corridor to display the Create Corridor dialog.
3. In the Name text box, name the corridor **Cabernet Court Corridor**.

   Keep the default values for Corridor Style and Corridor Layer.
4. Verify that Alignment is set to Cabernet Court and Profile is set to Cabernet Court FG.

   Notice that the small green Selection button could also be used to select the object on the screen instead of using the drop-down list if you prefer.
5. Verify that Assembly is set to Urban 14′ Single-Lane (or Urban 4.5 m Single-Lane for metric users).
6. Verify that Target Surface is set to the EG surface.
7. Verify that the Set Baseline And Region Parameters check box is selected.

   The Create Corridor dialog should now look similar to Figure 10.6.

FIGURE 10.6
The Create Corridor dialog

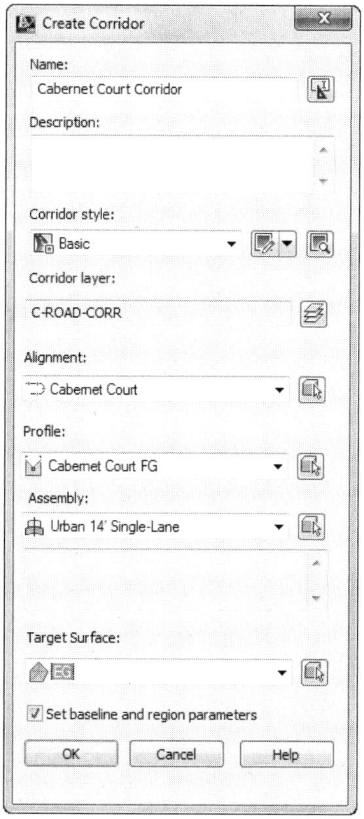

8. Click OK to accept the settings and to display the Baseline And Region Parameters dialog, as shown in Figure 10.7.

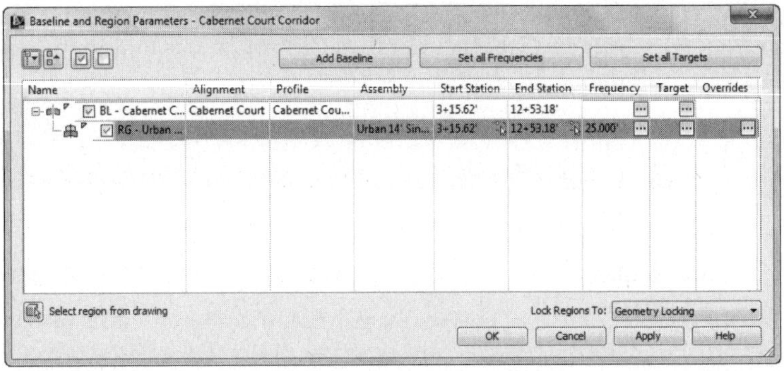

FIGURE 10.7
The Baseline And Region Parameters dialog

Notice that there is currently one baseline with one baseline region. The Start Station and End Station of the region match that of the baseline.

UNDERSTANDING THE LOCKING REGIONS

At the bottom of the Baseline And Region Parameters dialog is a drop-down list for locking regions that is new to the Autodesk® Civil 3D® 2013 software. In previous versions, all regions were locked to the station. However, now there are two options for locking regions: Geometry Locking and Station Locking.

By default on new drawing files this drop-down will be set to Geometry Locking. If you open a drawing created in a previous version, the drop-down will be set to Station Locking, but you can change it to Geometry Locking if you desire. You can update this default in your existing templates as needed by using the following steps:

1. On the Settings tab of Toolspace, expand the Corridor ➢ Commands branch.
2. Right-click the CreateCorridor command to display the Edit Command Settings – CreateCorridor dialog.
3. Expand the Assembly Insertion Defaults.
4. Change the Lock Region To value to the desired setting.

You can experiment with either option to see how it will affect your workflow as you change and modify the baseline and the corridor it is associated with.

Most of the other settings have already been set from the information you provided in the Create Corridor dialog, but let's take a few minutes to look at some of the settings in the Baseline And Region Parameters dialog.

9. Click the ellipsis button in the Frequency column in the first row associated with the baseline (BL) to display the Frequency To Apply Assemblies dialog.

By clicking on the ellipsis in the first row that is associated with the baseline, you will be setting the frequency for all of the regions within that baseline. In this instance you have only one region, but setting them all at once is a good habit to get into when applicable. You could also click the Set All Frequencies button, which would set the frequency for all of the baselines and all of the baseline regions.

10. Examine the settings in the Frequency To Apply Assemblies dialog, as shown in Figure 10.8.

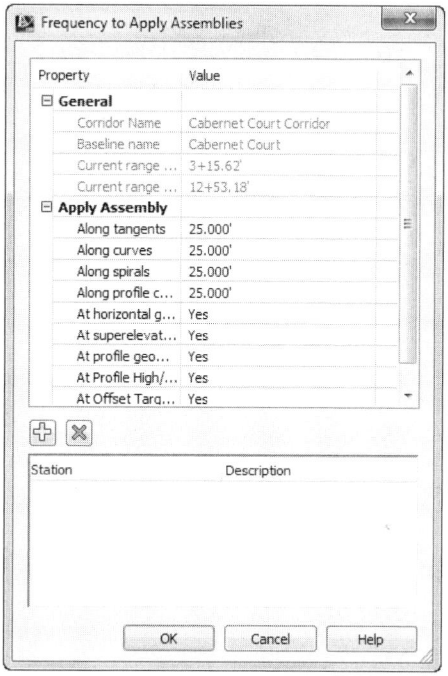

FIGURE 10.8
The Frequency To Apply Assemblies dialog

You can vary the default frequency for the portions of the region along tangents, curves, spirals, and profile curves. In addition, you can add frequency lines at various geometry points. At the bottom of the dialog you can add a user-defined station.

11. Click OK to accept the settings in the Frequency To Apply Assemblies dialog.

12. Click the ellipsis button in the Targets column in the baseline row to display the Target Mapping dialog.

    The ellipsis buttons in this column behave the same way as the ellipsis buttons in the Frequency column. Notice that there is also a Set All Targets button, which can be used to set the targets for all of the baselines and all of the baseline regions.

13. Examine the settings in the Target Mapping dialog, as shown in Figure 10.9.

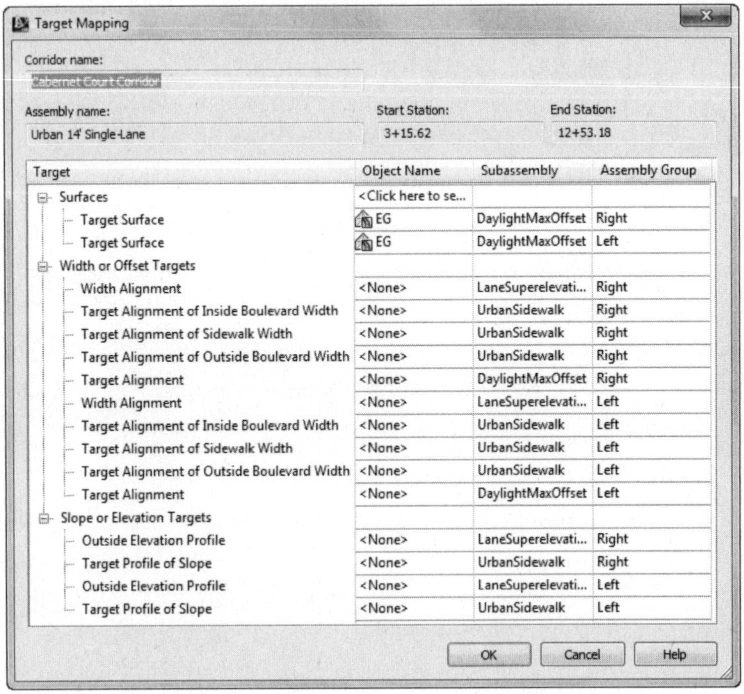

FIGURE 10.9
The Target Mapping dialog

The information in this dialog will vary depending on the composition of the assembly that was used for this corridor baseline but will always be broken into three categories: Surfaces, Width Or Offset Targets, and Slope Or Elevation Targets. Notice that because you selected EG as the target surface when creating the corridor, the EG surface is already set at the target surface for all subassemblies that target a surface.

14. Click OK to accept the settings in the Target Mapping dialog.

15. Click OK to accept the settings in the Baseline And Region Parameters dialog.

If at any point you want to return to the information shown on this dialog, you can do so on the Parameters tab of the Corridor Properties dialog, which we will look at a little later in this chapter.

16. You will see a dialog warning you that the corridor definition has been modified and giving you two options: Rebuild The Corridor or Mark The Corridor As Out-Of-Date. Select the Rebuild The Corridor option.

If you click the check box at the bottom of this warning dialog that says "Always perform my current choice," you will not see this warning dialog again.

UNDERSTANDING THE LOCKING REGIONS

If you select the check box (or any others) to hide a message and later want to receive the warning dialog again, you can reactivate it in the future by entering **`Options`** on the command line. On the System tab, click the Hidden Message Settings button.

You will receive several error messages in Panorama, as shown in Figure 10.10, that read `Intersection with target could not be computed`, `Intersection Point doesn't exist`, or something similar. You will rectify this issue in the following steps.

If Panorama did not automatically display, from the Home tab ➤ Palettes expanded panel, choose Event Viewer.

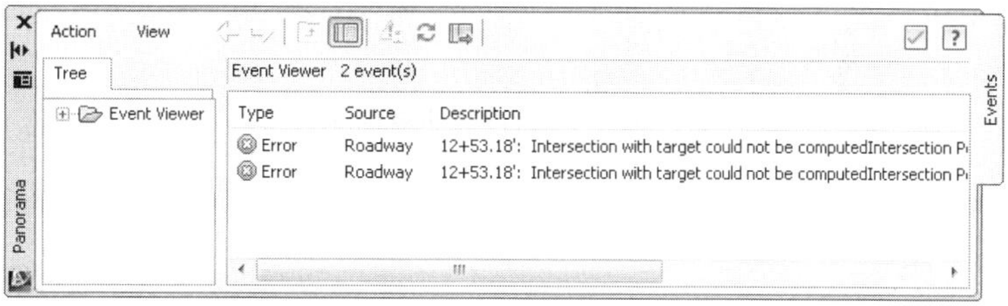

FIGURE 10.10
Corridor errors in the Panorama

17. Dismiss the Panorama.

Your corridor should look similar to Figure 10.11.

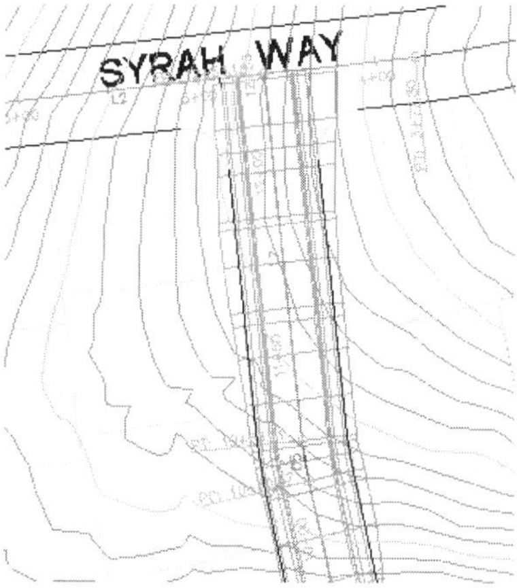

FIGURE 10.11
A portion of the nearly completed corridor

Now, let's try to figure out why you got those messages back in Panorama. In plan view, everything looks fine. However, a look at the corridor in the Object Viewer tells a different story (see Figure 10.12).

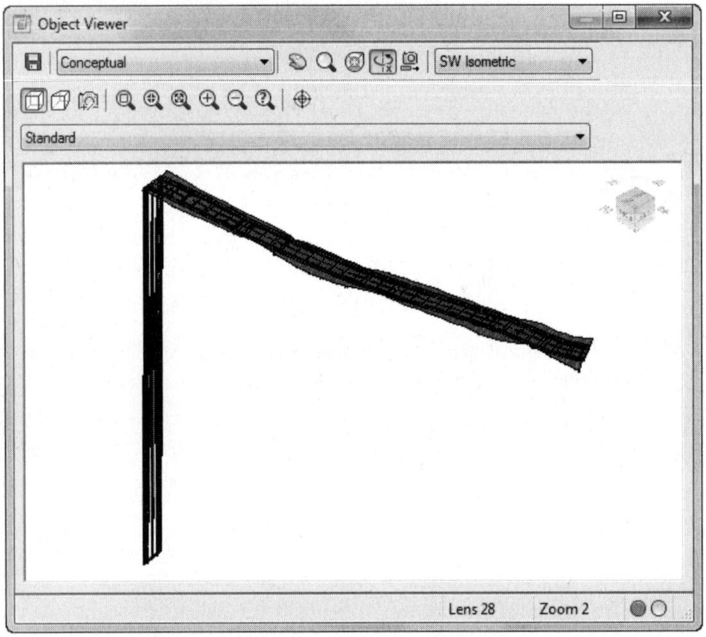

FIGURE 10.12
A "waterfall" at the end of the alignment viewed in the Object Viewer

VIEWING THE CORRIDOR IN THE OBJECT VIEWER

If you'd like to view the corridor in the Object Viewer, do the following:

1. Select one of the corridor lines to activate the contextual tab.
2. From the Corridor contextual tab ➢ General Tools panel, choose Object Viewer.

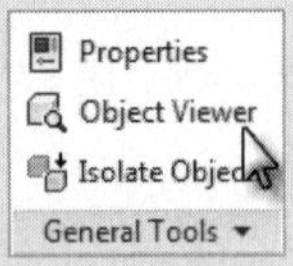

3. Use the View Control drop-down from the top of the Object Viewer to view the corridor from various isometric views, or press and hold the left button on your mouse and move the object around to view it from various custom angles.
4. After you have examined the corridor, click the X in the upper-right corner to dismiss the Object Viewer.

18. Select the corridor to activate the contextual tab.

19. From the Corridor contextual tab ➢ Modify Corridor panel, choose Corridor Properties to display the Corridor Properties dialog.

20. In the Corridor Properties dialog, switch to the Parameters tab.

    Notice in Figure 10.13 that the end station for the region is 12+53.18 (or 0+381.94 for metric users), which is based on the full length of the alignment. However, the design profile ends at 12+50 (or 0+381 for metric users). The difference in these values is causing the waterfall. To resolve this, you need to tell the corridor to stop processing the region at the end of the design profile. Note that even though the display value shows only two (or three) decimal places, the corridor examines up to eight decimal places of precision.

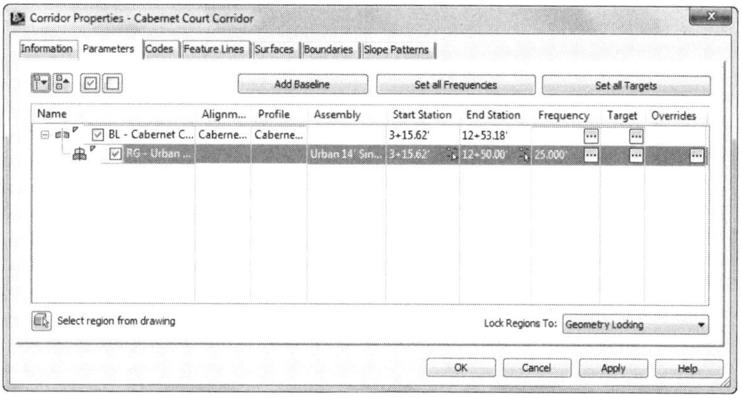

FIGURE 10.13
Corridor Properties dialog, Parameters tab

21. Click into the End Station field for the region and enter **1250** (or **381** for metric users).

    Entering the value with station notation (the plus symbol) is not needed.

    Notice that you could alternatively click the station picker button in this cell to select the station on the plan. This may be preferable if you do not know the exact station at which your profile starts or ends, or if the precision of the value may be an issue.

22. Click OK to accept the settings in the Corridor Properties dialog.

23. If you didn't click the check box at the last warning dialog, you will receive the same warning again. If so, select the Rebuild The Corridor option to allow the corridor to rebuild.

    No new errors will appear in Panorama.

A quick look in the Object Viewer should reveal that the waterfall is gone. When this exercise is complete, you may close the drawing. A finished copy of this drawing is available from the book's website with the filename `RoadCorridor_FINISHED.dwg` or `RoadCorridor_METRIC_FINISHED.dwg`.

Rebuilding Your Corridor

A corridor is a *dynamic* model — which means that if you modify any of the objects used to create the corridor, the corridor must be updated to reflect those changes. For example, if you make a change to the Finished Ground profile, the corridor needs to be rebuilt to reflect the new design. The same principle applies to changes to alignments, assemblies, target surfaces, and any other corridor ingredients or parameters.

You can access the Rebuild command by right-clicking on the corridor name from Prospector, as shown in Figure 10.14.

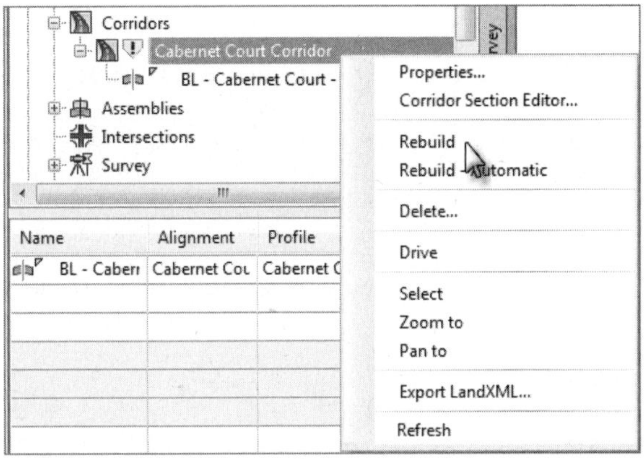

FIGURE 10.14
Right-click the corridor name in the Corridor collection in Prospector to rebuild it.

You can also rebuild the corridor or rebuild all corridors by selecting the corridor object and choosing Rebuild Corridor or Rebuild All Corridors from the contextual tab ➤ Modify Corridor panel, as shown in Figure 10.15.

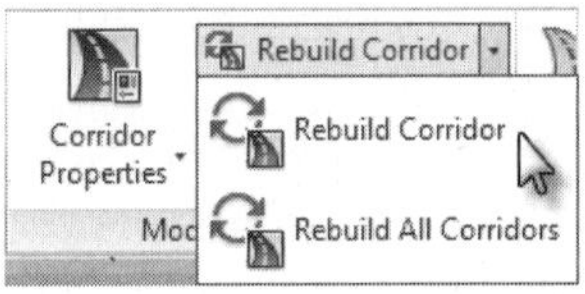

FIGURE 10.15
Rebuilding the corridor from the contextual tab

AVOIDING LENGTHY REBUILDS

For large, complex proposed surfaces, you may want to consider leaving the option for Rebuild Automatic unchecked. Every time a change is made that affects a region in your corridor, those regions that have been modified will go through the rebuilding process. This is an improvement from previous versions of Civil 3D, which rebuilt the full corridor even if a modification affected only one region. Nonetheless, the larger or more complex the corridor and the more regions you are modifying, the longer the rebuild process will take.

Corridor Tips and Tweaks

Your corridor may not be perfect on your first iteration of the design. Get comfortable getting into and working with the Parameters tab of the Corridor Properties dialog. This tab will be your first stop to examine what might be amiss in your corridor model.

Whether you are building your first corridor or your five hundredth, odds are good that you will run into one of the following common issues:

Problem Your corridor seems to fall off a cliff, meaning the beginning or ending station of your corridor drops down to elevation zero.

Typical Cause The range of your corridor is longer than the design profile.

Fix This is exactly what you ran into in the first exercise. The corridor takes the initial station range from your alignment. However, most designers don't tie into existing ground at the exact alignment start and end stations, so you need to adjust the corridor stations accordingly.

If you need to check the exact start and end stations of the design profile, the best place to do so is the Profile Data tab in the Profile Properties dialog (as shown in Figure 10.16). Edit your corridor region to begin and end at the design profile station.

FIGURE 10.16
Check your Profile Properties dialog to verify the station range of the design profile.

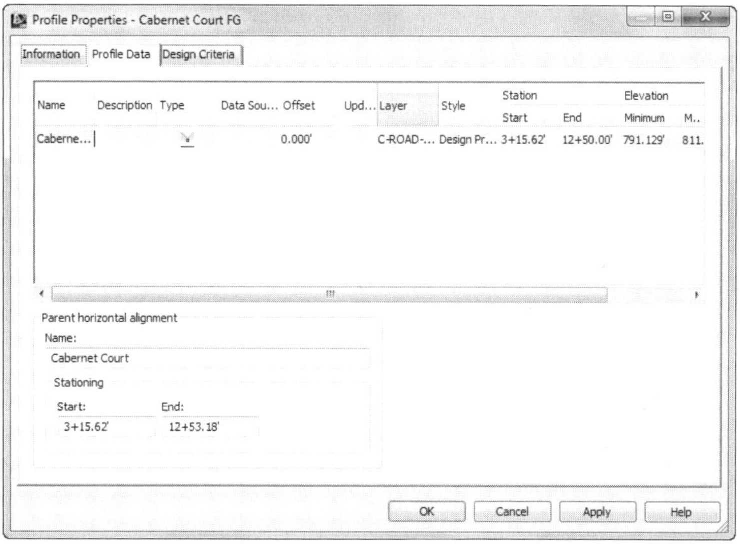

Alternately, you can use the station picker button to select the first and last corridor region stations by snapping to the frequency lines that correspond to your profile geometry.

Problem Your corridor seems to take longer to build and has irregular frequency stations. Also, your daylighting may not extend out to where you expect it (see Figure 10.17).

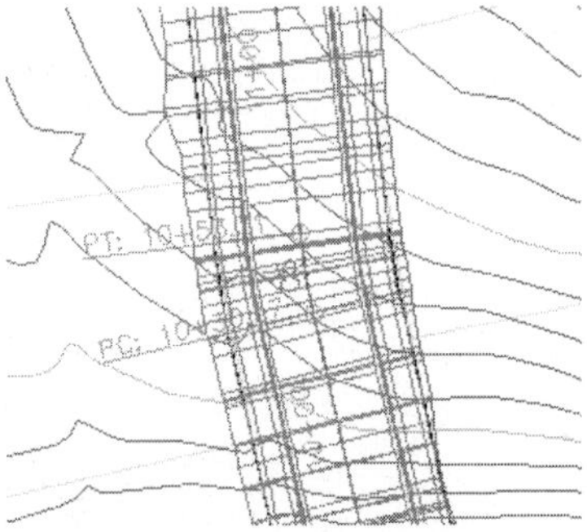

FIGURE 10.17 An example of unexpected corridor frequency

Typical Cause You accidentally chose the Existing Ground profile instead of the Finished Ground profile for your baseline profile. Most corridors are set up to place a frequency line at every vertical geometry point, and a profile, such as the Existing Ground profile, has many more vertical geometry points than a layout profile (in this case, the Finished Ground profile). These additional points on the Existing Ground profile are the cause of the unexpected sample lines and flags that something is wrong.

Fix Always use care to choose the correct profile. Either physically select the profile on screen or make sure your naming conventions clearly define your finished grade as finished grade. If your corridor is already built, select the corridor, right-click, and choose Corridor Properties, as shown in Figure 10.18. On the Parameters tab of the Corridor Properties dialog, change the profile from the existing ground profile to the finished ground profile.

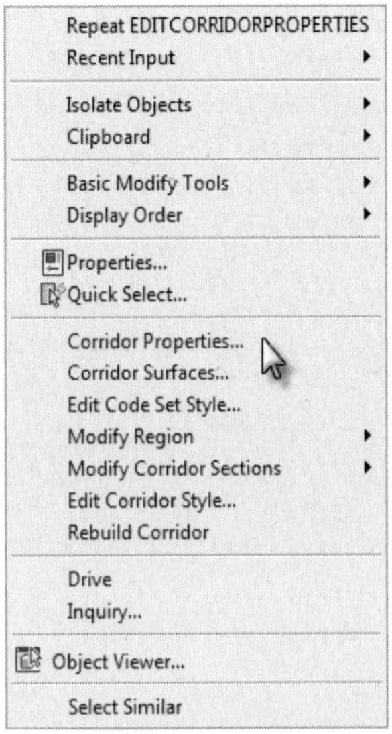

FIGURE 10.18 The right-click menu available on the corridor object

Adding a surface target throws another variable into the mix. Here is a list of some of the most typical problems new users face and how to solve them:

Problem Your corridor doesn't show daylighting even though you have a daylight subassembly on your assembly. You may also get an error message in Event Viewer.

Typical Cause You forgot to set the surface target when you created your corridor.

Fix If your corridor is already built, select the corridor, right-click, and choose Corridor Properties. On the Parameters tab of the Corridor Properties dialog, click the Set All Targets button. The Target Mapping dialog opens, and its first category is Surfaces. Click the <Click here to set all> text in the Object Name column field to display the Pick A Surface dialog. In this dialog, you can choose a surface for the daylight subassembly to target.

Problem Your corridor seems to be missing patches of daylighting. You may also get an error message in Event Viewer.

Typical Cause The Daylight link cannot find the target surface within the grade, width, or other parameters you've set in the subassembly properties. This could also mean that the target surface doesn't fully extend the full length of your corridor, or your target surface is too narrow at certain locations.

Fix Add more data to your target surface so that it is large enough to accommodate daylighting for the full length of the corridor.

You can also revisit your daylight subassembly settings to give the program a narrower offset or steeper grade. Alternatively you can adjust your alignment and/or profile to require less cut or fill and therefore daylight sooner.

If this is not possible, omit daylighting through those specific stations, and once your corridor is built, do hand grading using feature lines or grading objects. You can also investigate other subassemblies such as LinkOffsetAndElevation that will meet your design intention without requiring a surface target.

Corridor Feature Lines

Corridor feature lines are first drawn connecting the same point codes. For example, a feature line will work its way down the corridor and connect all the Back_Curb points defined by an assembly. If there are Back_Curb points on the entire length of your corridor, then the feature line does not have any decisions to make. If your corridor changes from having a curb to having a grassed buffer or ditch, the feature line needs to figure out where to go next.

The Feature Lines tab of the Corridor Properties dialog has a drop-down menu called Branching (see Figure 10.19), with two options — Inward and Outward. Inward branching forces the feature line to connect to the next point it finds toward the baseline. Outward branching forces the feature line to connect to the next point it finds away from the baseline. There is also a Connect Extra Points check box. When checked, this option will force all of the points with the same point code to connect.

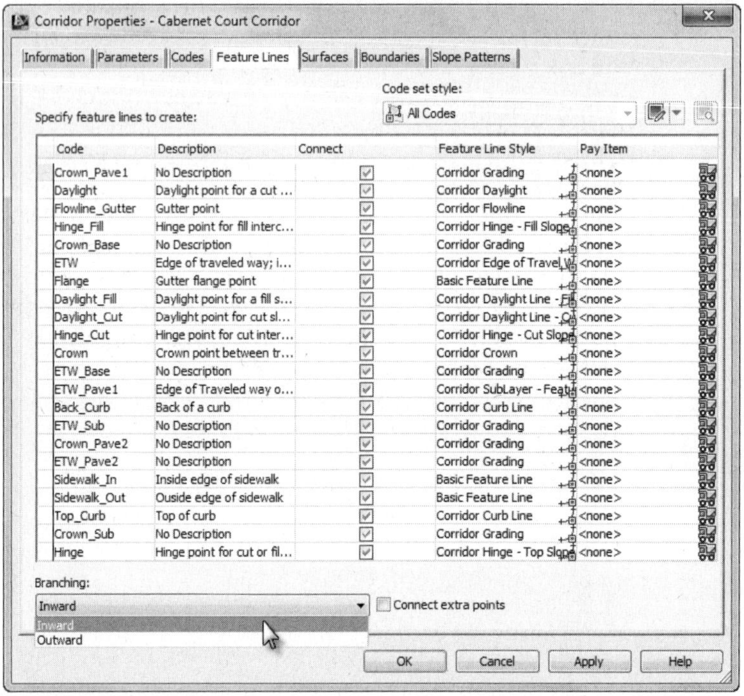

FIGURE 10.19
The Feature Lines tab of the Corridor Properties dialog

As mentioned earlier, a feature line will only connect the same point codes by default. However, the Feature Lines tab of the Corridor Properties dialog allows you to eliminate certain feature lines on the basis of the point code. For example, if for some reason you did not want your Back_Curb points connected with a feature line, you could toggle that feature line off in the Connect column.

When a corridor is built, the feature line that is created is a similar type of object to the feature line you will use in Chapter 16, "Grading," for grading purposes. The difference is that corridor feature lines are locked in the corridor object. There is not much you can do with them until they are extracted.

If you select the corridor to activate the contextual tab, you will notice that the Launch Pad panel, shown in Figure 10.20, offers multiple commands that relate to corridor feature lines.

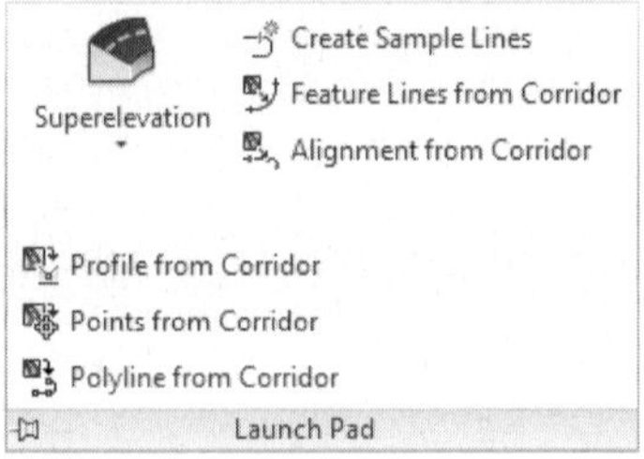

FIGURE 10.20
Launch Pad panel of the Corridor contextual tab

Corridor feature lines can be extracted from a corridor to produce grading feature lines, alignments, profiles, and 3D polylines. A step-by-step explanation of how to extract a corridor feature line is given in the next exercise.

In the case of extracting a feature line, you have the choice of keeping the line dynamically linked to the corridor geometry or making it an entirely separate entity. If the feature line is linked, it can be used for grading and cannot be manipulated in any way other than modifying the corridor (i.e., it cannot be grip-edited). If the feature line is not linked, it can be used for grading or as a target in a corridor. The nonlinked feature line can be grip-edited and is completely divorced from the originating corridor. Both types of feature lines can be used as breaklines in surface models.

Once an alignment, profile, or 3D polyline has been extracted, its geometry can be adjusted independently. In this case, the profile, alignment, or polyline no longer retains a link to the corridor from which it originated. Once a profile, alignment, or 3D polyline has been extracted, you can use it as a target back in the corridor that formed it. In the following exercise, a corridor feature line is extracted to produce an alignment and profile:

1. Open the `CorridorFeatureLine.dwg` or `CorridorFeatureLine_METRIC.dwg` file.

2. From the Home tab ➤ Create Design panel, choose Alignment ➤ Create Alignment From Corridor.

   Alternatively you can select the corridor by selecting the corridor feature line that you want to create an alignment from to activate the contextual tab and from the Launch Pad choose Alignment From Corridor. *If you do this, then skip the next step.*

3. At the `Select a corridor feature line:` prompt, select the feature line labeled in the drawing as East ETW Feature Line (ETW stands for Edge of Traveled Way).

   This opens the Select A Feature Line dialog similar to the one shown in Figure 10.21.

FIGURE 10.21
Selecting the ETW feature line to be extracted as an alignment

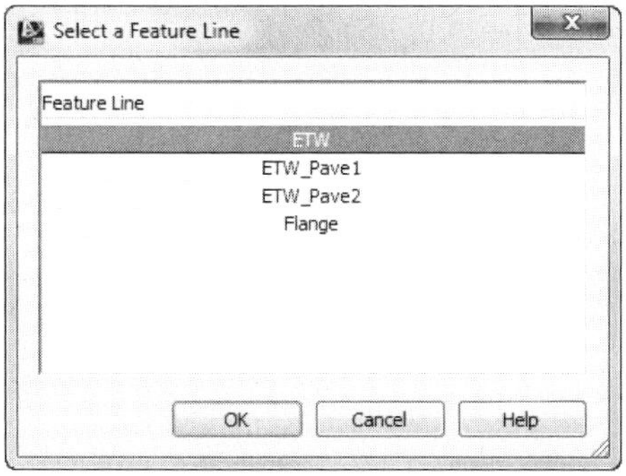

4. Select the ETW feature line from the Select A Feature Line dialog.

5. Click OK to accept the settings in the Select A Feature Line dialog and display the Create Alignment From Objects dialog.

6. In the Name text box of the Create Alignment From Objects dialog, name the alignment **Cabernet Court East ETW**.
7. Verify that Type is set to Offset.
8. On the General tab of the Create Alignment From Objects dialog, verify that:
   - Alignment Style is set to Basic
   - Alignment Label Set is set to _No Labels
   - The Create Profile check box is selected

   The dialog should match Figure 10.22.

FIGURE 10.22
The completed Create Alignments From Objects dialog

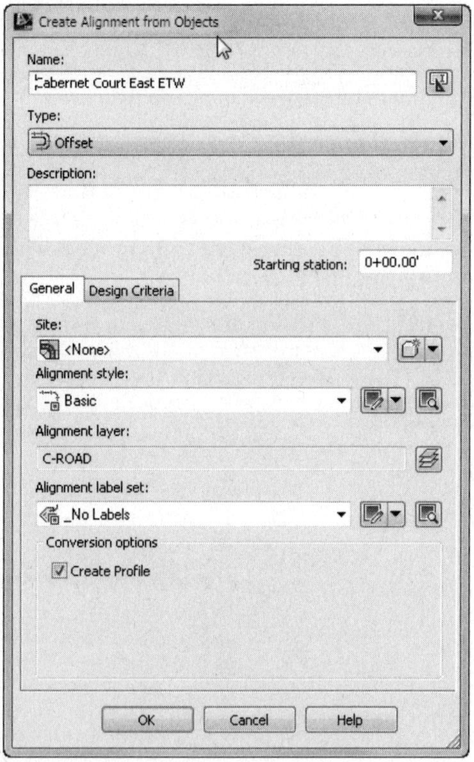

9. Click OK to accept the settings in the Create Alignment From Objects dialog and display the Create Profile – Draw New dialog.
10. In the Name text box of the Create Profile – Draw New dialog, name the profile **Cabernet Court East ETW Profile**.

11. On the General tab of the Create Profile – Draw New dialog, verify that the profile style is set to Design Profile and that the profile label set is set to _No Labels.

    The dialog should match Figure 10.23.

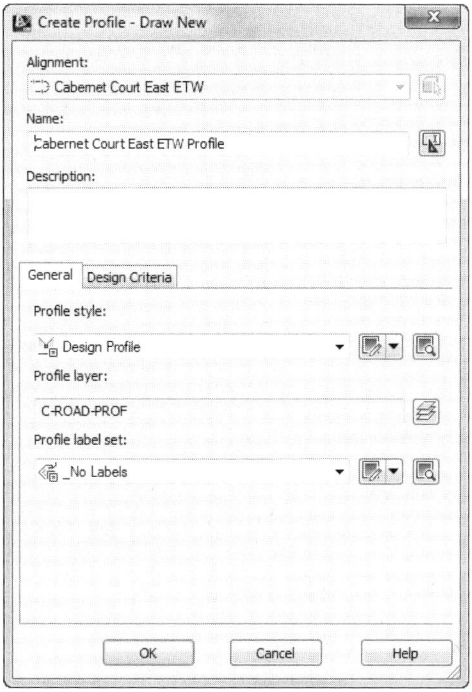

FIGURE 10.23 The Create Profile – Draw New dialog

12. Click OK to accept the settings in the dialog and then press ↵ to exit the command.

13. In Prospector, expand Alignments ➤ Offset Alignments ➤ Cabernet Court East ETW ➤ Profiles to review the completed alignment and profile (see Figure 10.24).

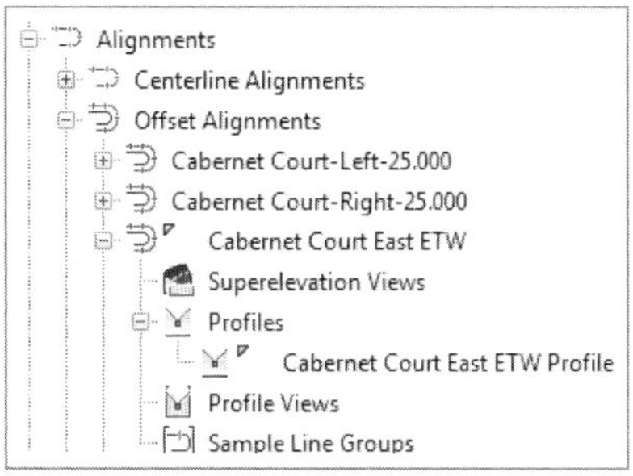

FIGURE 10.24 A completed alignment and profile shown on the Prospector tab of Toolspace

Next, you will extract a corridor feature line and keep it dynamically linked to the corridor.

14. From the Home tab ➤ Create Design panel, choose Feature Line From Corridor.

    Alternatively you can select the corridor by selecting the corridor feature line that you want to create a feature line from to activate the contextual tab, and from the Launch Pad select Feature Line From Corridor. *If you do this, then skip the next step.*

15. At the `Select a corridor feature line:` prompt, select the feature line labeled in the drawing as East Hinge Feature Line.

    This opens the Select A Feature Line dialog.

16. Select the Hinge feature line from the Select A Feature Line dialog.
17. Click OK to close the Select A Feature Line dialog and display the Create Feature Line From Corridor dialog.
18. Select the Name check box and name the feature line **Cabernet Court Hinge East**.
19. Verify that Style is set to Corridor Grading.
20. Verify that the Create Dynamic Link To The Corridor check box is selected.

    All other settings can be left at their defaults. The dialog should match Figure 10.25.

FIGURE 10.25
The Create Feature Line From Corridor dialog with the option to create a dynamic link turned on

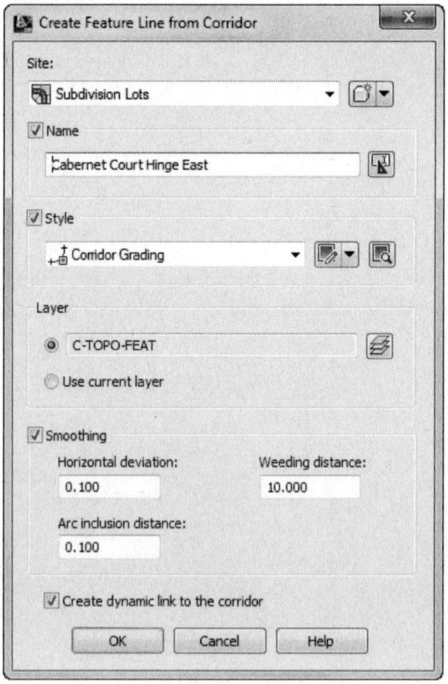

21. Click OK to accept the settings in the dialog and then press ↵ to exit the command.

22. Select the newly created Auto Corridor Feature Line.

    Note that it can be selected but does not have grips.

23. From the Feature Line contextual tab ➤ Modify panel, click the Edit Geometry and Edit Elevations buttons to display the hidden panels. Note which options are grayed out and which are not, as shown in Figure 10.26.

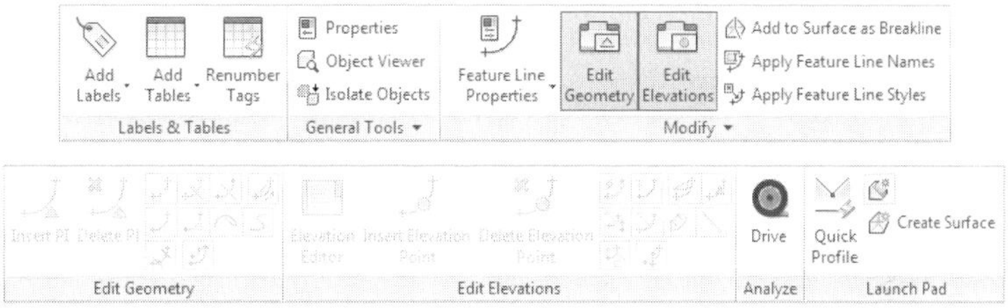

FIGURE 10.26
Auto Corridor Feature Line contextual tab

These alignments and feature lines can now be used for further design. Remember that feature lines created from the corridor are not connected to the corridor and will not update as you change the corridor.

When this exercise is complete, you may close the drawing. A finished copy of this drawing is available from the book's web page with the filename `CorridorFeatureLine_FINISHED.dwg` or `CorridorFeatureLine_METRIC_FINISHED.dwg`.

Understanding Targets

Every subassembly is programmed to have certain capabilities, but it needs some guidance from the user to figure out exactly how to do what is wanted. For example, a daylight subassembly contains instructions for how it is to extend to a surface model, but it is up to you to tell it which surface model to work with. Many of the lane subassemblies can be used to hook into an offset alignment and/or profile but need your help in determining which alignment and/or profile is desired.

These instructions to the subassembly are called *targets*. When you built the corridor in the first exercise, Civil 3D prompted you to add targets. In that case, you specified a surface target and continued. However, surface targets are just the tip of the iceberg.

Using Target Alignments and Profiles

So far, all of the corridor examples you looked at have a constant cross section. In the next section, we'll take a look at what happens when a portion of your corridor needs to transition to a wider section and then transition back to normal.

Many subassemblies have been programmed to allow for not only a baseline attachment point but also additional attachment points on target alignments and/or profiles. Be sure to check the subassembly help file to make sure the subassembly you are using will accept targets if you need them. In Chapter 8, you learned that you can right-click on any subassembly in the Tool Palettes window to enter the help file. For instance, the BasicLane subassembly will show None in the Target Parameters area of the help file, but LaneSuperElevationAOR lists Lane Width and Outside Elevation.

Think of a subassembly as a rubber band that is attached both to the baseline of the corridor (such as the road centerline) and the target alignment. As the target alignment, such as a lane widening, gets further from the baseline, the rubber band is stretched wider. As that target alignment transitions back toward the baseline, the rubber band changes to reflect a narrower cross section.

Figure 10.27 shows what happens to a cross section when various targets are set for the left edge of a traveled way for a lane subassembly. Figure 10.27a shows the assembly as it was originally placed in the drawing. The width from the original assembly is 14′ with a cross slope of 2 percent to the edge of pavement. Figure 10.27b shows how the geometry changes if an alignment target is set for the edge of pavement. Notice that because there is no profile specified to change the elevation, the 2 percent cross slope is held and the lane width is the only geometry that changes. Figure 10.27c shows that if both an alignment and profile are specified for an edge of pavement alignment, the design cross slope and width both change. Lastly, you see the assembly with just a profile target assigned to the edge of pavement in Figure 10.27d. In this case, the width stays at 14′ but the elevation of the edge of pavement is dictated by the profile.

FIGURE 10.27
How geometry changes with a target on the left: Original assembly geometry (a), assembly with width alignment target only (b), assembly with both alignment and profile target (c), and assembly with only profile target set (d).

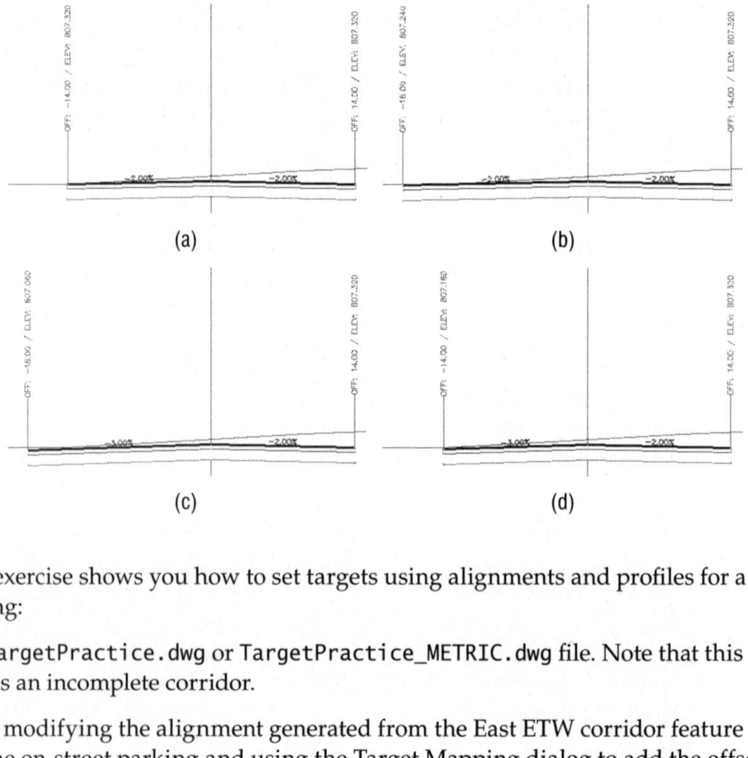

The following exercise shows you how to set targets using alignments and profiles for a corridor lane widening:

1. Open the `TargetPractice.dwg` or `TargetPractice_METRIC.dwg` file. Note that this drawing has an incomplete corridor.

   You will be modifying the alignment generated from the East ETW corridor feature line for some on-street parking and using the Target Mapping dialog to add the offset alignments.

2. From the Home tab ➤ Create Design panel, choose Alignment ➤ Create Widening.
3. At the `Select an alignment:` prompt, select the East ETW Alignment.
4. At the `Select start station:` prompt, enter **690** ↵ (or **210**↵ for metric users).
5. At the `Select end station:` prompt, enter **900** ↵ (or **275**↵ for metric users).
6. At the `Enter widening offset:` prompt, enter **12** ↵ (or **3.75** ↵ for metric users).
7. At the `Specify side: [Left Right]:` prompt, enter **R** ↵.

   The Offset Alignment Parameters palette appears as shown in Figure 10.28. Though you won't be making any changes to the defaults at this time, observe that you can change the transition parameters at the entry and exit if desired.

FIGURE 10.28
Offset Alignment Parameters palette

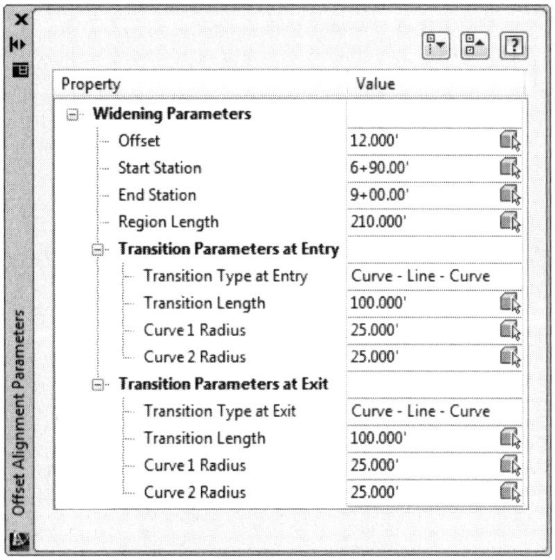

8. Dismiss the Offset Alignment Parameters palette by clicking the X in the upper corner.

   Notice that a new alignment was created named Cabernet Court East ETW-Left-0.000. This is because Add Widening is actually an offset alignment so it uses the Offset Alignment Name Template. Since the widening is created on top of the original alignment, it is offset "Left-0.000" regardless of whether you add the widening to the left or to the right. You could rename this alignment, but for the purposes of this exercise leave it as is.

9. Select the corridor to activate the Corridor contextual tab.
10. From the contextual tab ➤ Modify Corridor panel, choose Corridor Properties.
11. On the Parameters tab, in the Target column for region RG – Urban Single-Lane – (1) row, click the ellipsis button to display the Target Mapping dialog. You may need to make the Name column wider to see the row name.

12. In the Width Or Offset Targets ➤ Width Alignment branch for the Lane SuperelevationAOR in the Right Subassembly Group, click on <None> to display the Set Width Or Offset Target dialog box.

13. Use the Select From The Drawing button to select the alignment that you just added the widening to in the drawing or select Cabernet Court East ETW-Left-0.000 from the list.

14. Click the Add button.

    The alignment should appear in the listing of Selected entities to target at the bottom of the dialog, as shown in Figure 10.29.

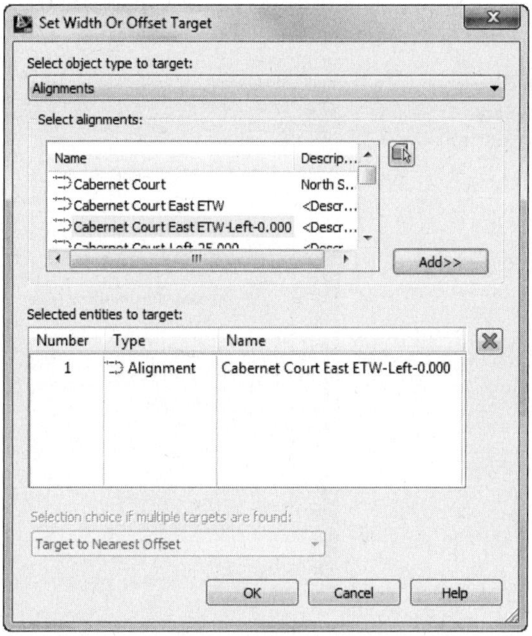

FIGURE 10.29
The Set Width Or Offset Target dialog

15. Click OK to accept the settings in the Set Width Or Offset Target dialog.

    The Target Mapping dialog should now look like Figure 10.30.

16. Click OK to accept the settings in the dialog.

17. Click OK to accept the settings in the Corridor Properties dialog and allow the corridor to rebuild.

    The corridor should now be following the ETW alignment and resemble Figure 10.31.

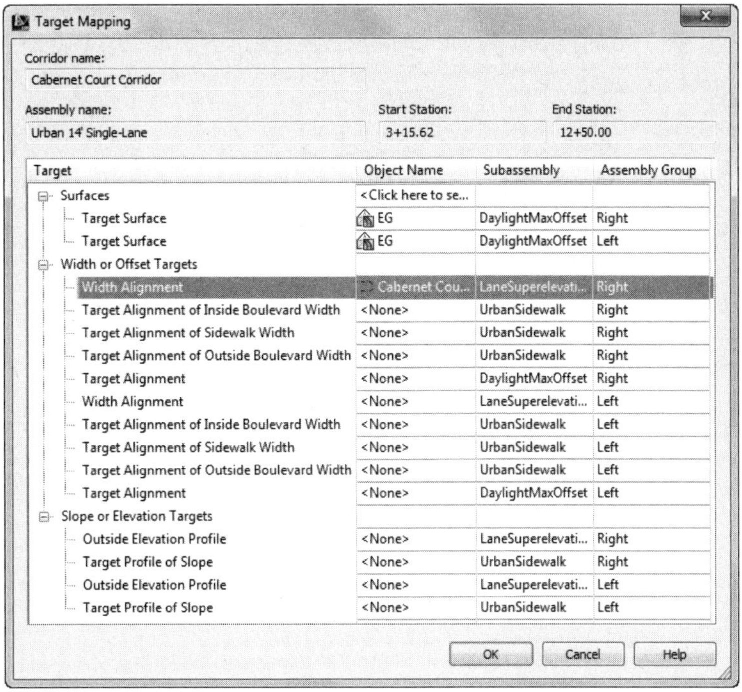

FIGURE 10.30
Targets set for surface and width for the right side

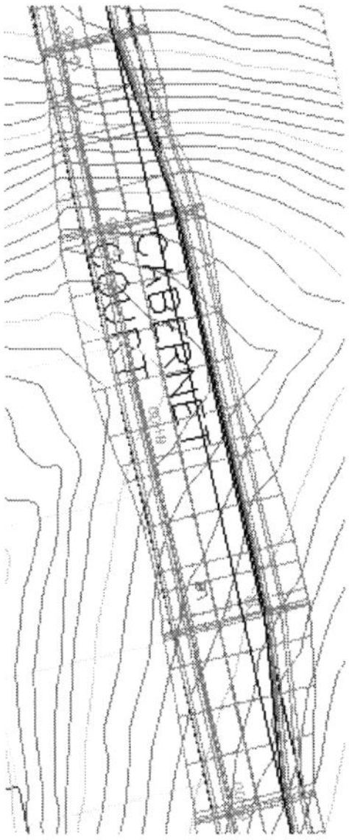

FIGURE 10.31
The completed exercise in plan view

You may keep this drawing open to continue on to the next exercise or use the finished copy of this drawing available from the book's web page (`TargetPractice_FINISHED.dwg` or `TargetPractice_METRIC_FINISHED.dwg`).

Real World Scenario

Corridor Properties Modifications Made Easy

It can be tedious to modify alignments, profiles, starting stations, ending stations, frequencies, and targets along a corridor. Each of these items can be modified in the item view at the bottom of Prospector as shown here:

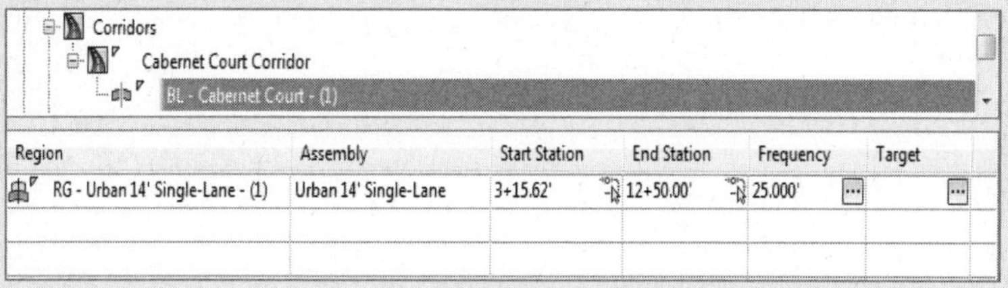

In the following quick exercise, you will use Toolspace to access and modify Frequency values for the Cabernet Court region:

1. If not still open from the previous exercise, open the `TargetPractice_FINISHED.dwg` or `TargetPractice_FINISHED_METRIC.dwg` file.
2. In Prospector, expand and highlight the Cabernet Court Corridor.
3. Click the Frequency ellipsis button for the BL-Cabernet Court Court -(1) to display the Frequency To Apply Assemblies dialog.
4. Set the Frequency values for Along Curves and Along Profile Curves to **5′** (or **1.5** m for metric users).
5. Click OK.

   The corridor will rebuild and the additional frequency lines should be visible in the plan as shown here:

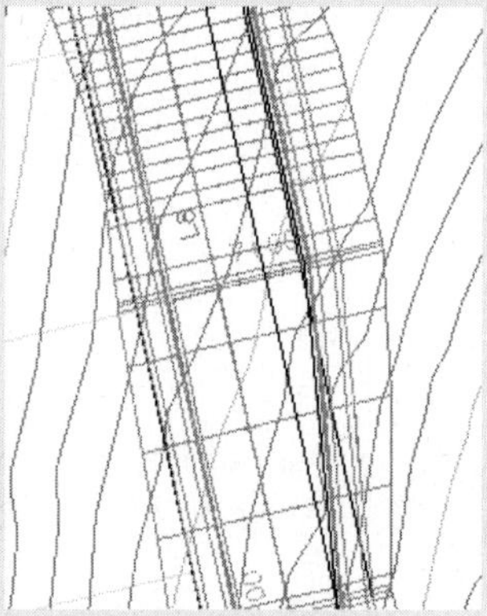

   When this exercise is complete, you may close the drawing. A finished copy of this drawing is available from the book's web page with the filename `CorridorFrequency_FINISHED.dwg` or `CorridorFrequency_METRIC_FINISHED.dwg`.

Editing Sections

Once your corridor is built, chances are you will want to examine it in section view, tweak a station here or there, and check for problems. For a station-by-station look at a corridor, select the corridor and from the Corridor contextual tab ➤ Modify Corridor Sections panel, choose Section Editor (see Figure 10.32).

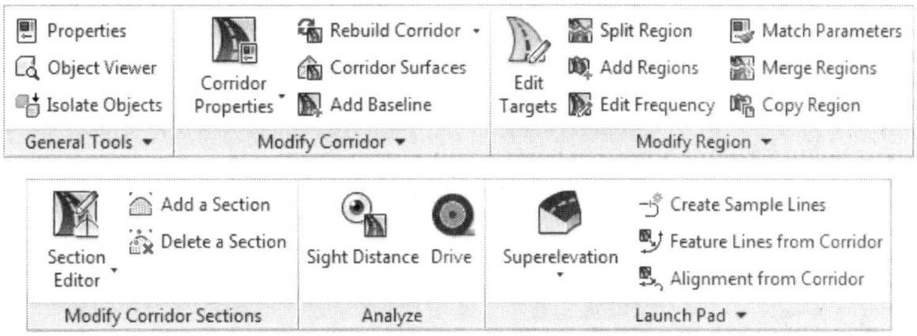

FIGURE 10.32 Some of the many tools, including the Section Editor, available on the Corridor contextual tab

Once you are in the Section Editor, you are in a purely data-driven view. That means that this is a live, editable section of the corridor and is not for plotting purposes. We will discuss plotting cross sections in Chapter 13, "Cross Sections and Mass Haul."

The Section Editor allows for multiple viewport configurations so that you can see the plan, profile, and section all at the same time, as seen in the following exercise:

1. From the Section Editor contextual tab ➤ View Tools panel, choose Viewport Configuration to display the Corridor Section Editor: Viewport Configuration dialog, as shown in Figure 10.33.

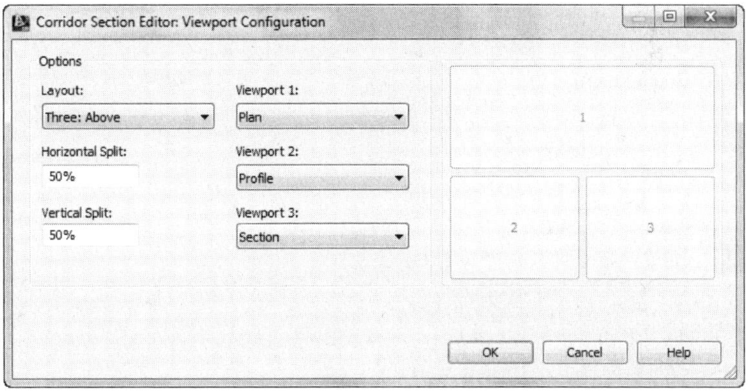

FIGURE 10.33 The Corridor Section Editor: Viewport Configuration dialog

2. Set Layout to Three: Above and Viewport 1 to Plan, Viewport 2 to Profile, and Viewport 3 to Section with a 50% Horizontal Split and a 50% Vertical Split.

3. Click OK to accept the settings in the Viewport Configuration dialog. Your screen should now look similar to Figure 10.34.

4. You may receive a warning dialog asking if you would like to turn on Viewport Configuration; if so, select Yes.

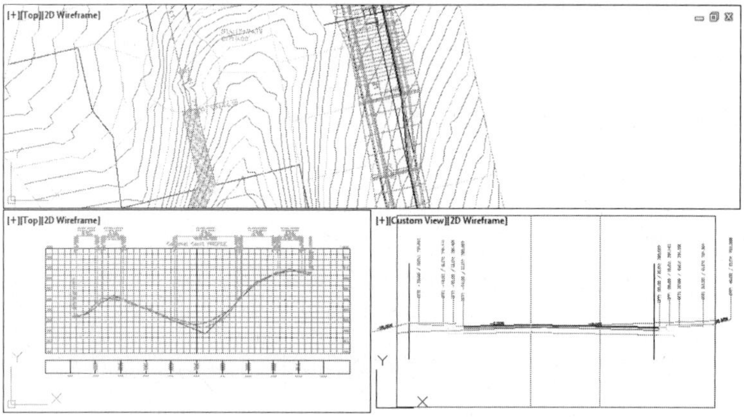

FIGURE 10.34 The Corridor Section Editor with Viewport Configuration on

The Section Editor contextual tab offers many commands other than just Viewport Configuration, as shown in Figure 10.35.

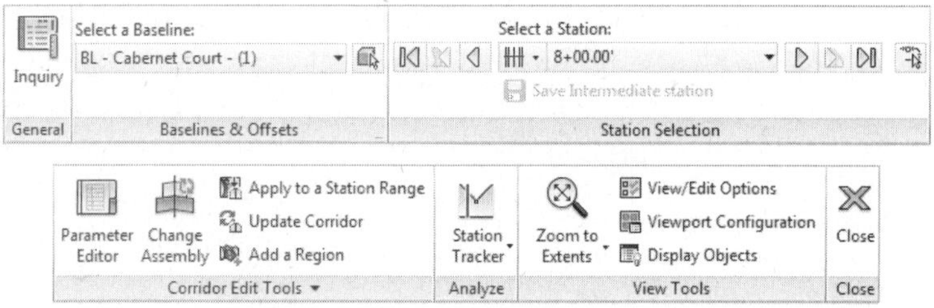

FIGURE 10.35 The Corridor Section Editor contextual tab

The Station Selection panel on the Section Editor contextual tab allows you to move forward and backward through your corridor to see what each section looks like.

If you wish to edit a section, you may do so geometrically in the viewport showing the section in the Section Editor or through the Parameter Editor palette, but not both. To graphically edit a link, hold down the Ctrl key on the keyboard while selecting the item. This will activate grips that you can use to relocate or stretch the link (Figure 10.36).

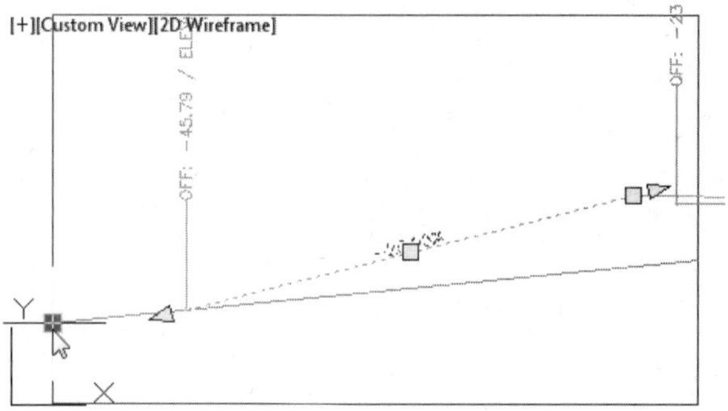

FIGURE 10.36 A Daylight link ready for grip editing in the Section Editor

If you would rather use the Parameter Editor to make changes that are more precise to your section, do the following:

1. From the Section Editor contextual tab ➢ Corridor Edit Tools panel, choose Parameter Editor.
2. Look through the listing of subassemblies and their current values.
3. When you find a value you wish to edit, click on the value field to override the design value, as shown in Figure 10.37.

FIGURE 10.37
The Corridor Parameter Editor

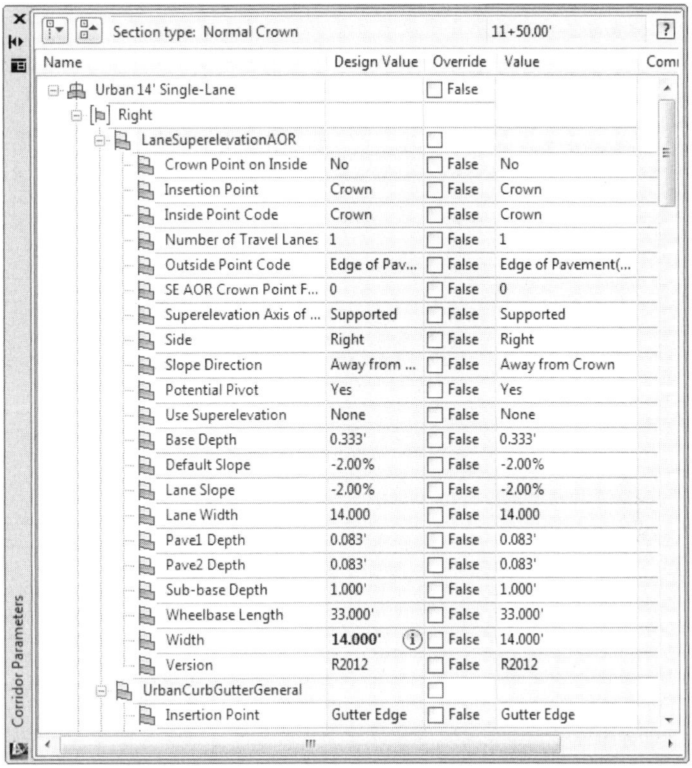

When you change the value, a check mark will appear in the Override column. You will find that some of the values are uneditable.

The changes you make can be applied to just the section you are viewing or to a range of stations in the region you are working in (use the Apply To A Station Range button in the Corridor Edit Tools panel to do so).

To exit the Section Editor, click the Close button on the Close panel.

ACK! STUCK IN BIZZARRO COORDINATE SYSTEM

If you accidentally exit your drawing without exiting the Section Editor, you will return to a drawing in a rotated coordinate system as shown here. It may be difficult to see your design, but don't panic.

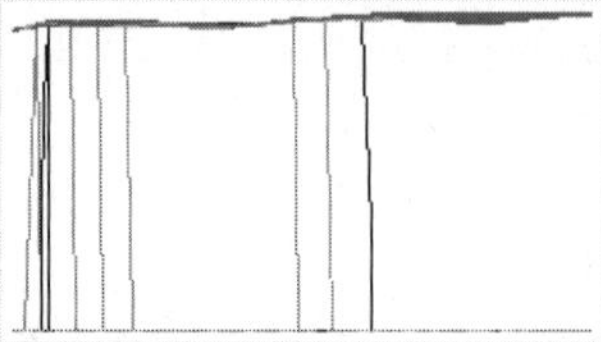

At the command line, enter **Plan** ↵ **W** ↵. You will see your project, but there is still another step.

From the View tab ➢ Coordinates panel, set the current UCS to World, as shown in the following graphic. The UCS icon will return to normal and you can continue working.

If your viewport configuration was set to three views, you can return to a single view on the View tab ➢ Model Viewports panel. Click the drop-down from the Viewport Configuration button and select Single.

Creating a Corridor Surface

CERT OBJECTIVE

A corridor provides the raw components for surface creation. Just as you would use points and breaklines to make a surface, a corridor surface uses corridor points as point data and uses feature lines and links like breaklines.

The Corridor Surface

Civil 3D does not automatically build a corridor surface when you build a corridor. From examining subassemblies, assemblies, and the corridors you built in the previous exercises, you have probably noticed that there are many "layers" of points, links, and feature lines. Some represent the very top of the finished ground of your road design, some represent subsurface gravel or concrete thicknesses, and some represent subgrade, among other possibilities. You can choose to build a surface from any one of these layers or from all of them. Figure 10.38 shows an example of a TIN surface built from the links that are all coded Top, which would represent final finished ground.

When you first create a surface from a corridor, the surface is dependent on the corridor object. This means that if you change something that affects your corridor and then rebuild the corridor, the surface will also update.

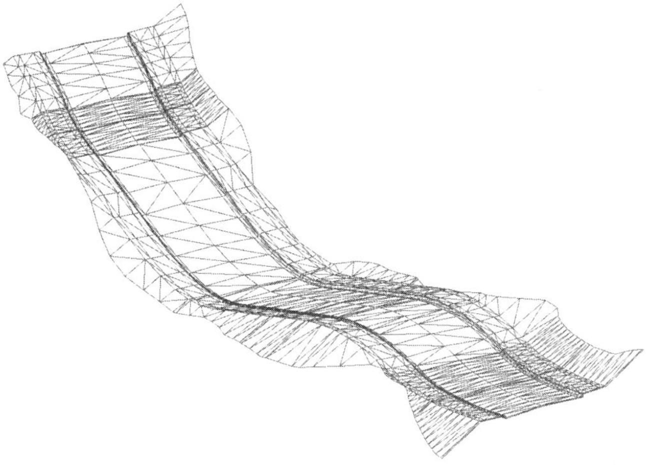

FIGURE 10.38
A surface built from Top code links

A corridor surface shows up as a surface under the Surfaces branch in Prospector with a slightly modified icon that denotes it as related to the corridor.

After you create the initial corridor surface, you can create a static export of the surface by changing to the Home tab ➢ Create Ground Data panel and choosing Surfaces ➢ Create Surface From Corridor. A detached surface will not react to corridor changes and can be used to archive a version of your surface. To remind you that this surface is not related to a corridor, the icon in the Surfaces branch in Prospector will not show the corridor icon.

Corridor Surface Creation Fundamentals

You create corridor surfaces on the Surfaces tab in the Corridor Properties palette using the following two steps (which are examined in detail later in this section):

1. Click the Create A Corridor Surface button to add a surface item.

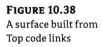

2. Choose the data to add, and click the Add Surface Item button.

You can choose to create your corridor surface on the basis of links, feature lines, or a combination of both.

CREATING A SURFACE FROM LINK DATA

Most of the time, you will build your corridor surface from links. As discussed earlier, each link in a subassembly is coded with a name such as Top, Pave, Datum, and so on. Choosing to build a surface from Top links will create a surface that triangulates between the points at the link vertices that represent the final finished grade.

The most commonly built link-based surfaces are Top and Datum; however, you can build a surface from any link code in your corridor. Figure 10.39 shows a schematic of how links are used to form the most common surfaces — in this case a top surface, as shown in the top image of Figure 10.39, and a datum surface, as shown in the bottom image of Figure 10.39.

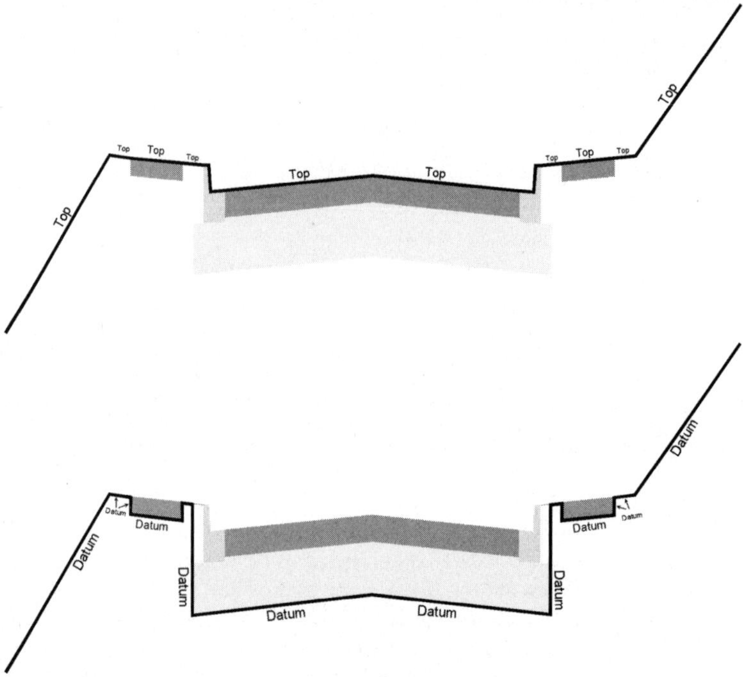

FIGURE 10.39
Schematic of Top links connecting to form a surface (top) and schematic of Datum links connecting to form a surface (bottom)

OVERHANG CORRECTION FOR CONFUSED DATUM SURFACES

A common situation with assemblies is a material that juts out past another material, such as curb sub-base. Triangulated surfaces cannot contain caves or perfectly vertical faces, since both of those scenarios result in two elevation points for a given x, y coordinate, so Civil 3D needs to go around the material in a logical manner. What the software sees as logical and what you actually want from the surface may not always be the same thing, as shown here:

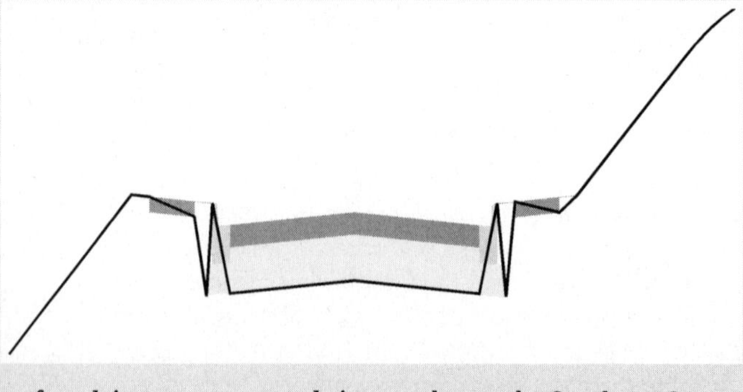

If you have a datum surface doing an unexpected zig-zag, change the Overhang Correction setting to Bottom Links. There is also an option to set the Overhang Correction to Top Links. This overhang-correction setting is in the Corridor Properties dialog on the Surfaces tab. After your corridor rebuilds, the result will resemble the bottom image of Figure 10.39.

When building a surface from links, you have the option of checking a box in the Add As Breakline column. Checking this box will add the actual link lines themselves as additional breaklines to the surface. In most cases, especially in intersection design, checking this box forces better triangulation.

Creating a Surface from Feature Lines

There might be cases where you would like to build a simple surface from your corridor — for example, by using just the crown and edge-of-travel way. If you build a surface from feature lines only or a combination of links and feature lines, you have more control over what Civil 3D uses as breaklines for the surface.

If you added all the topmost corridor feature lines to your surface item and built a surface, you would get a very similar result as if you had added the Top link codes.

Creating a Surface from Both Link Data and Feature Lines

A link-based surface can be improved by the addition of feature lines. A link-based surface does not automatically include the corridor feature lines, but instead uses the link vertex points to create triangulation. Therefore, the addition of feature lines ensures that triangulation occurs where desired along ridges and valleys. This is especially important for intersection design, curves, and other corridor surfaces where triangulation around tight corners is critical. Figure 10.40 shows the Surfaces tab of the Corridor Properties dialog where a Top link surface will be improved by the addition of Back_Curb, ETW, and Top_Curb feature lines.

Figure 10.40
The Surfaces tab indicates that the surface will be built from Top links as well as from several feature lines.

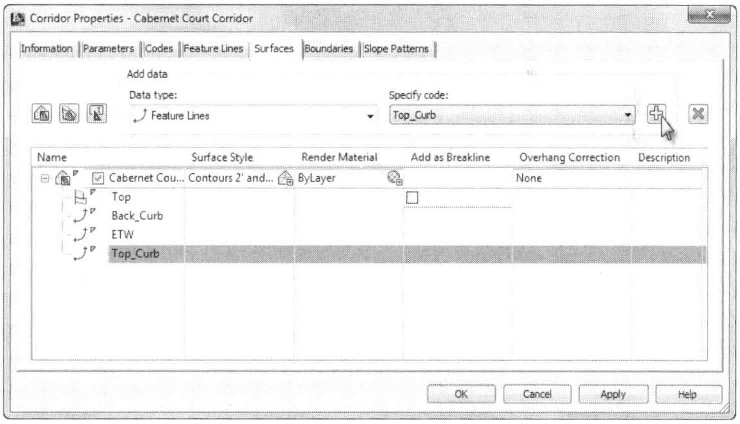

If you are having trouble with triangulation or contours not behaving as expected, experiment with adding a few feature lines to your corridor surface definition.

Create a Corridor Surface for Each Link

To the right of the Create Corridor Surface button is the Create A Corridor Surface For Each Link button. Clicking this button populates the Surface List area with a multitude of corridor surfaces; as the button name would suggest, you will now have a corridor surface for each of the link codes.

This may not be desirable for many people as you rarely need a surface for every link, but once they are created you can always remove the unwanted corridor surfaces using the Delete Surface Item button.

Corridor Surface Name Template

The third button at the upper left of the Surfaces tab of the Corridor Properties dialog is the Surface Name Template button. When you click this button, the Name Template dialog shown in Figure 10.41 appears.

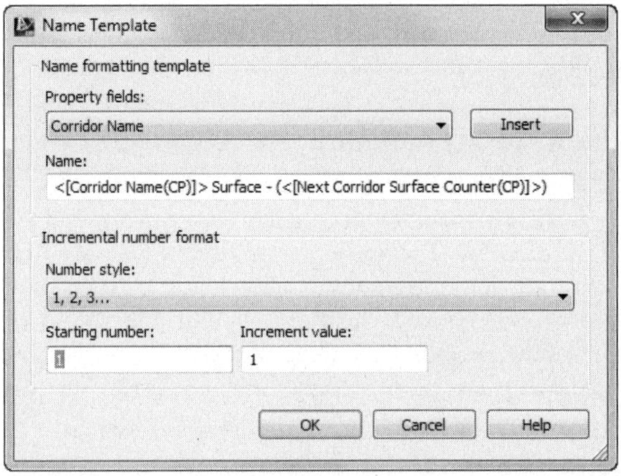

Figure 10.41
The Name Template dialog for Corridor Surfaces

Here you can set the formatting for the corridor name as well as the number style, starting number, and increment value. The Property fields are Corridor Name and Next Corridor Surface Counter. Based on the name shown in Figure 10.41, the next surface created for the Cabernet Court corridor will be named Cabernet Court Surface – (2). The Number style can be set to 1, 2, 3 … or 01, 02, 03 … or a multitude of other styles based on the number of leading zeros.

Other Surface Tasks

You can do several other tasks on the Surfaces tab. For each corridor surface, you can set a surface style, revise the default name assigned by the Name Template, and provide a description for your corridor surface. Alternatively, you can do all those things once the corridor surface appears in the drawing through Prospector.

Adding a Surface Boundary

Surface boundaries are critical to any surface, but especially so for corridor surfaces. Tools that automatically and interactively add surface boundaries, using the corridor intelligence, are available. Figure 10.42 shows a corridor surface before and after the addition of a boundary. Notice how the extraneous contours have been eliminated along the line of intersection between the existing ground and the proposed ground (the daylight line), thereby creating a much more accurate surface.

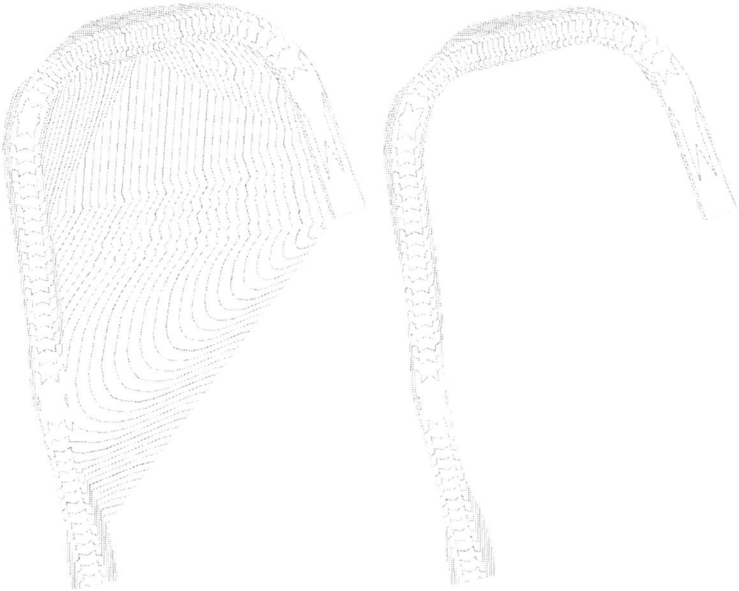

FIGURE 10.42
A corridor surface before the addition of a boundary (left) and after the addition of a boundary (right)

You can create corridor surface boundaries using the Boundaries tab of the Corridor Properties dialog. Each corridor surface will be listed. To add the boundary, right-click on the corridor surface and select the desired boundary type.

BOUNDARY TYPES

There are several tools to assist you in corridor surface boundary creation. They can be automatic, semiautomatic, or manual in nature depending on the complexity of the corridor.

You access these options on the Boundaries tab of the Corridor Properties dialog by right-clicking the name of your surface item, as shown in Figure 10.43.

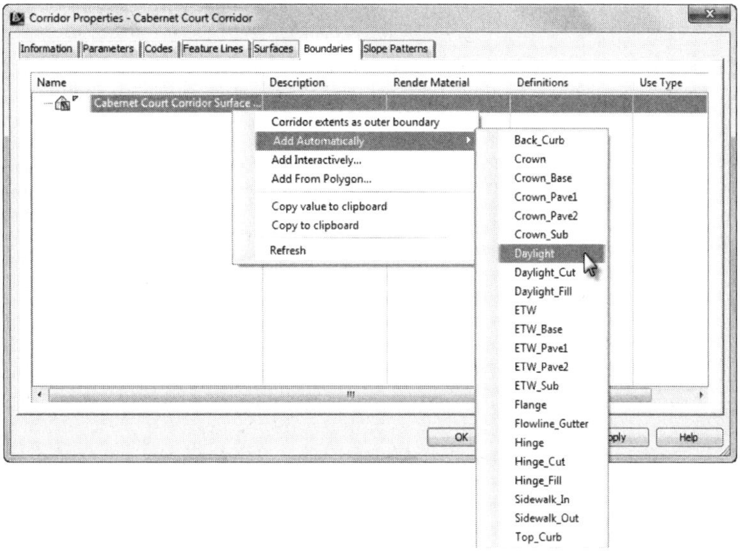

FIGURE 10.43
Corridor Surface Boundary options for a corridor containing a single baseline

The following corridor boundary methods are listed in order of desirability. Corridor Extents As Outer Boundary is the most user-friendly, whereas Add From Polygon is fast but needs constant updating as it is not dynamically linked to the corridor.

Corridor Extents As Outer Boundary This option is available only when you have a corridor with multiple baselines. With this selection, Civil 3D will shrink-wrap the corridor, taking into account intersections and various daylight options on different alignments. Corridor Extents As Outer Boundary will probably be your most-used boundary option.

Add Automatically The Add Automatically boundary tool allows you to pick a point code and use the associated feature lines as your corridor boundary. This tool is available only for single-baseline corridors. Because this tool is the most automatic and easiest to apply, and will remain dynamically linked to the corridor, you will use it almost every time you build a single-baseline corridor.

Add Interactively The Add Interactively boundary tool allows you to work your way around a corridor and choose which corridor feature lines you would like to use as part of the boundary definition.

Choosing this option is better than using Add From Polygon if Add Automatically and Corridor Extents are not available. It takes a bit of patience to trace the corridor, but the result is a dynamically linked boundary that changes if the corridor changes. Using this method, once you select a feature line, a thick line will trace around the corridor following your mouse; in order to switch to a different feature line, simply click the new feature line at the transition location. When complete, you can close the boundary just as you would a polyline.

Add From Polygon The Add From Polygon tool allows you to choose a closed polyline (including 2D or 3D polylines) or polygon in your drawing that you would like to add as a boundary for your corridor surface. This method is quick, but unlike the other methods, the resulting boundary is not dynamic to your design.

The next exercise leads you through creating a corridor surface with an automatic boundary:

1. Open the `CorridorBoundary.dwg` or `CorridorBoundary_METRIC.dwg` file.
2. Select the Frontenac Drive corridor to activate the contextual tab.
3. From the Corridor contextual tab ➢ Modify Corridor panel, choose Corridor Properties.

4. On the Surfaces tab of the Corridor Properties dialog, click the Create A Corridor Surface button in the upper-left corner of the dialog.

   You should now have a surface item in the bottom half of the dialog.
5. Click the surface item under the Name column and change the default name of your surface to **Frontenac-Top Surface**.
6. Verify that Links has been selected from the drop-down list in the Data Type selection box.
7. Verify that Top has been selected from the drop-down list in the Specify Code selection box.

8. Click the Add Surface Item button to add Top Links to the Surface Definition.
9. Click OK to accept the settings in this dialog. Choose Rebuild The Corridor when prompted and examine your surface.

   The road surface should look fine; however, because you have not yet added a boundary to this surface, undesirable triangulation is occurring outside your corridor area.
10. Expand the Surfaces branch in Prospector.

    Note that you now have a corridor surface listed in addition to the EG surface that was already in the drawing.
11. Select the Frontenac Drive corridor to activate the contextual tab.
12. From the Corridor contextual tab ➤ Modify Corridor panel, choose Corridor Properties.

    If you do not see the Corridor Properties button on the Modify Corridor panel, you may have inadvertently chosen the corridor surface.
13. On the Boundaries tab of the Corridor Properties dialog, right-click Frontenac – Top Surface in the listing.
14. Hover over the Add Automatically menu flyout, and select Daylight as the feature line that will define the outer boundary of the surface.
15. Verify that the Use Type column says Outside Boundary to ensure that the boundary definition will be used to define the desired extreme outer limits of the surface.
16. Click OK to accept the settings in the Corridor Properties dialog and choose Rebuild The Corridor when prompted.
17. Examine your surface, and note that the triangulation terminates at the Daylight point all along the corridor model.
18. (Optional) Experiment with making changes to your finished grade profile, assembly, or alignment geometry and rebuilding both your corridor and finished ground surface to see the boundary in action.

REBUILD: LEAVE IT ON OR OFF?

Once you rebuild your corridor, your corridor surface will need to be updated. Typically, the best practice is to leave Rebuild – Automatic off for corridors and keep Rebuild – Automatic on for corridor surfaces. The corridor surface will only want to rebuild when the corridor is rebuilt. For very large corridors, this may become a bit of a memory lag, so try it both ways to see what you like best.

When this exercise is complete, you may close the drawing. A finished copy of this drawing is available from the book's web page with the filename `CorridorBoundary_FINISHED.dwg` or `CorridorBoundary_METRIC_FINISHED.dwg`.

Common Surface Creation Problems

Here are some common problems you may encounter when creating surfaces:

Problem Your corridor surface does not appear or seems to be empty.

Typical Cause You might have created the surface item but not added any data.

Fix Open the Corridor Properties dialog and switch to the Surfaces tab. Select a data type from the drop-down menus in the Data Type and Specify Code selection boxes, and click the Add Surface Item button. Make sure your dialog shows both a surface item and a data type, as shown in Figure 10.44.

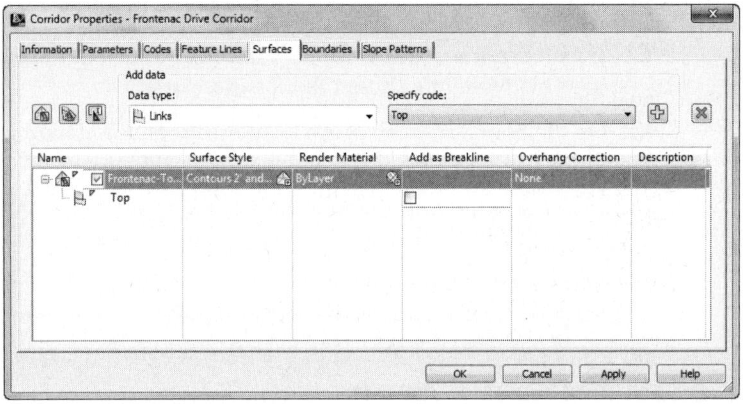

Figure 10.44
A surface cannot be created without both a surface item and a data type.

Problem Your corridor surface does not seem to respect its boundary after a change to the assembly or surface-building data type (in other words, you switched from link data to feature lines).

Typical Cause Automatic and interactive boundary definitions are dependent on the codes used in your corridor. If you remove or change the codes used in your corridor, the boundary needs to be redefined.

Fix Open the Corridor Properties dialog and switch to the Boundaries tab. Remove any boundary definitions that are no longer valid (if any) by right-clicking on the boundary and selecting Remove Boundary. Once the outdated boundary has been removed, you can redefine the corridor surface boundaries using any of the applicable boundary types.

Problem Your corridor surface seems to have gaps at points of curvature (PCs) and points of tangency (PTs) near curb returns.

Typical Cause You may have encountered an error in rounding at these locations, and you may have inadvertently created gaps in your corridor. This is commonly the result of building a corridor using two-dimensional linework as a guide, but some segments of that linework do not touch.

Fix Be sure your corridor region definitions produce no gaps. You might consider using the `PEDIT` command to join lines and curves representing corridor elements that will need to be modeled later. You might also consider setting a COGO point at these locations (PCs, PTs, and so on) and using the Node object snap instead of the Endpoint object snap to select the same location each time you are required to do so.

Performing a Volume Calculation

One of the most powerful aspects of Civil 3D is having instant feedback on your design iterations. Once you create a preliminary road corridor, you can immediately compare a corridor surface to existing ground and get a good understanding of the earthwork magnitude. When you make an adjustment to the finished grade profile and then rebuild your corridor, you can see the effect that this change has on your earthwork within minutes, if not sooner.

Even though volumes were covered in detail in Chapter 4, "Surfaces," it is worth revisiting the subject here in the context of corridors.

This exercise uses a TIN-to-TIN composite volume calculation to compare the existing ground surface and the datum corridor surface; average end area and other section-based volume calculations are covered in Chapter 13.

1. Open the `CorridorVolume.dwg` or `CorridorVolume_METRIC.dwg` file.

   Note that this drawing has a completed Frontenac Drive corridor, as well as a top corridor surface and a datum corridor surface for Frontenac Drive.

2. From the Analyze tab ➤ Volumes And Material panel, choose Volumes Dashboard to display Panorama open to the Volumes Dashboard tab.
3. Click the Create New Volume Surface button to display the Create Surface dialog.
4. Change the name to **VOL-Frontenac-EG-Datum** and set Style to Elevation Banding (2D).
5. Click the <Base Surface> field to display the ellipsis button; once it's visible click the ellipsis button to select EG.
6. Click the <Comparison Surface> field to display the ellipsis button; once it's visible click the ellipsis button to select Frontenac – Datum Surface.
7. Click OK to accept the settings in the Create Surface dialog.

   A Cut/Fill breakdown should appear in the Volume tab of Panorama, as shown in Figure 10.45.

8. Make a note of these numbers.

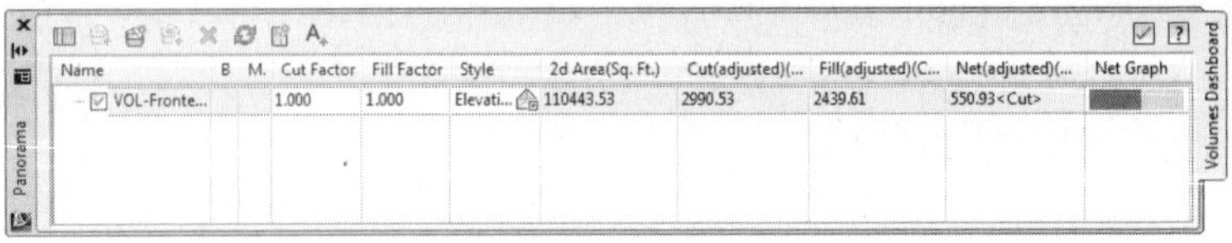

FIGURE 10.45 Panorama showing an example of a volume surface and the cut/fill results

9. Leave Panorama open on your screen (make it smaller, if desired), and pan over to the proposed profile for Frontenac Drive.
10. Select the Finished Ground profile and move the vertical triangular grip on the first PVI to the center of the circle shown in the profile view. Notice that the volume calculations have changed to Out Of Date because the corridor isn't set to rebuild automatically.
11. In Prospector, expand Corridors and select the Frontenac Drive corridor, right-click, and choose Rebuild. If any errors appear in the Panorama they can be ignored for this exercise.

    Notice that the corridor changes, and therefore the corridor surfaces (which are both set to Rebuild Automatic) change as well. However, the volume surface did not automatically rebuild.
12. On the Volumes Dashboard tab of the Panorama, right-click the volume surface and select Rebuild.

    Notice the new values for cut and fill.
13. Close the Panorama using the X in the upper corner.

When this exercise is complete, you may close the drawing. A finished copy of this drawing is available from the book's web page with the filename `CorridorVolume_FINISHED.dwg` or `CorridorVolume_METRIC_FINISHED.dwg`.

Non-Road Corridors

As discussed in the beginning of this chapter, corridors are not just for roads. Once you have the basics about corridors down, your ingenuity can take hold.

Corridors can be used for far more than just road designs. You will explore some more advanced corridor models in Chapter 11, but there are plenty of simple, single-baseline applications for alternative corridors such as channels, berms, retaining walls, and more. You can take advantage of several specialized subassemblies or build your own custom assembly using the Subassembly Composer. Figure 10.46 shows an example of a channel corridor.

One of the subassemblies discussed in Chapter 8 is the channel subassembly. The following exercise shows you how to apply this subassembly to design a simple drainage channel:

1. Open the `CorridorChannel.dwg` or `CorridorChannel_METRIC.dwg` file.

    Note that there is an alignment that represents a drainage channel centerline, a profile that represents the drainage channel normal water line, and an assembly created using the Channel and LinkSlopetoSurface subassemblies.

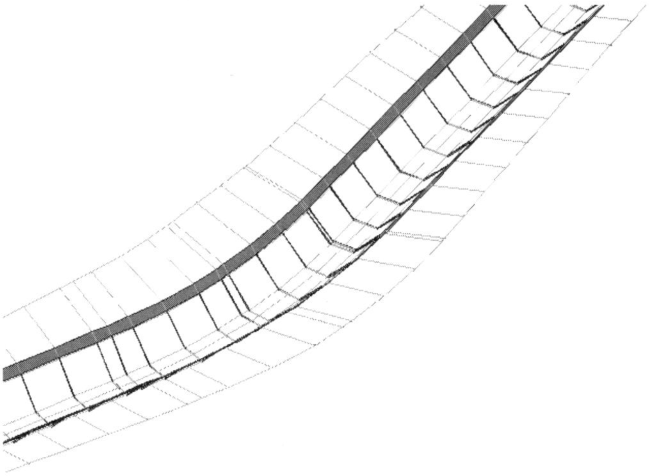

FIGURE 10.46
A simple channel corridor viewed in 3D built from the channel subassembly and a generic link subassembly

2. From the Home tab ➤ Create Design panel, choose Corridor to display the Create Corridor dialog.
3. In the Name text box, name your corridor **Drainage Channel**.

   Keep the default values for Corridor Style and Corridor Layer.
4. Verify that Alignment is set to Channel CL and Profile is set to Channel NWL.

   NWL stands for normal water level.
5. Verify that Assembly is set to Project Channel.
6. Verify that Target Surface is set to Existing Ground Surface.
7. Verify that the Set Baseline And Region Parameters check box is checked.
8. Click OK to accept the settings in the Create Corridor dialog and to display the Baseline And Region Parameters dialog.
9. Click the Set All Frequencies button.
10. Change the values for Along Tangents and Along Curves to **10′** (**3** m for metric users).
11. Click OK to accept the settings in the Baseline And Region Parameters dialog.
12. You may receive a dialog warning that the corridor definition has been modified. If you do, select the Rebuild The Corridor option.
13. Select the corridor to activate the contextual tab.
14. From the Corridor contextual tab ➤ Modify Corridor Sections panel, choose Section Editor.

15. From the Section Editor contextual tab ➢ Station Selection panel, you can navigate through the drainage channel cross sections by clicking the forward or backward arrows.

    The cross section should look similar to Figure 10.47.

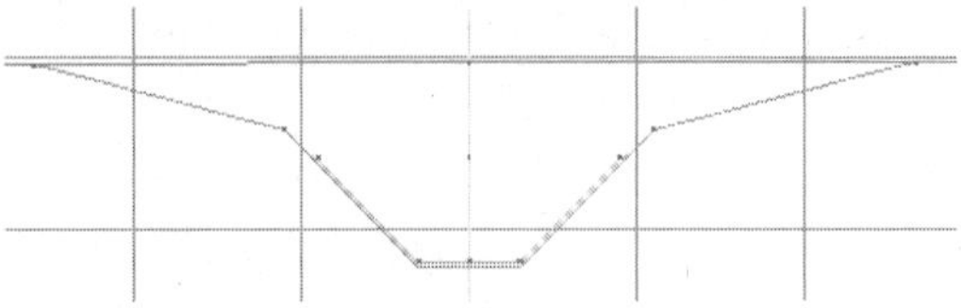

FIGURE 10.47
The completed Drainage Channel corridor viewed in the Section Editor

This corridor can be used to build a surface for a TIN-to-TIN volume calculation or can be used to create sections and generate material quantities, cross-sectional views, and anything else that can be done with a more traditional road corridor.

When you are finished viewing the sections, dismiss the dialog by clicking the X on the Close panel. When this exercise is complete, you may close the drawing. A finished copy of this drawing is available from the book's web page with the filename `CorridorChannel_FINISHED.dwg` or `CorridorChannel_METRIC_FINISHED.dwg`.

Real World Scenario

CREATING A PIPE TRENCH CORRIDOR

Another use for a corridor is a pipe trench. A pipe trench corridor is useful for determining quantities of excavated material, limits of disturbance, trench-safety specifications, and more. This graphic shows a completed pipe trench corridor:

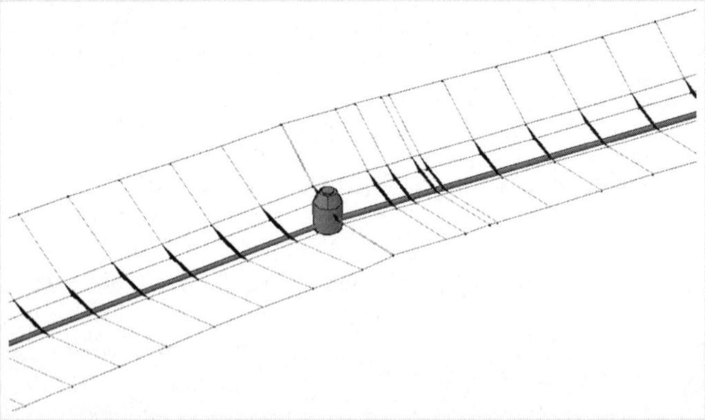

One of the subassemblies discussed in Chapter 8 is the TrenchPipe1 subassembly. The following exercise leads you through applying this subassembly to a pipe trench corridor:

1. Open the `CorridorPipeTrench.dwg` or `CorridorPipeTrench_METRIC.dwg` file.

Note that there is a pipe network, with a corresponding alignment, profile view, and pipe trench assembly. Also note that there is a profile drawn that corresponds with the inverts of the pipe network.

2. From the Home tab ➢ Create Design panel, choose Corridor.
3. In the Name text box, name the corridor **Pipe Trench**.

   Keep the default values for Corridor Style and Corridor Layer.
4. Verify that Alignment is set to Pipe Centerline and Profile is set to Bottom Of Pipe Profile.
5. Verify that Assembly is set to Pipe Trench.
6. Verify that Target Surface is set to Existing Ground.
7. Verify that the Set Baseline And Region Parameters check box is selected.
8. Click OK to accept the settings in the Create Corridor dialog and display the Baseline And Region Parameters dialog.
9. Click the Set All Frequencies button.
10. Change the values for Along Tangents and Along Curves to **10′** (**3** m for metric users).
11. Click OK to accept the settings in the Baseline And Region Parameters dialog.
12. You may receive a dialog warning that the corridor definition has been modified. If you do, select the Rebuild The Corridor option.

    The corridor will build.
13. Select the corridor to activate the contextual tab.
14. From the contextual tab ➢ Modify Corridor Sections panel, choose Section Editor. Browse the cross sections through the trench.
15. When you are finished viewing the sections, dismiss the Section Editor by clicking the X on the Close panel.

You may notice at the sharp bends in the pipe alignment the corridor frequency lines cross one another. This is called a bow-tie and will be discussed further in the next chapter.

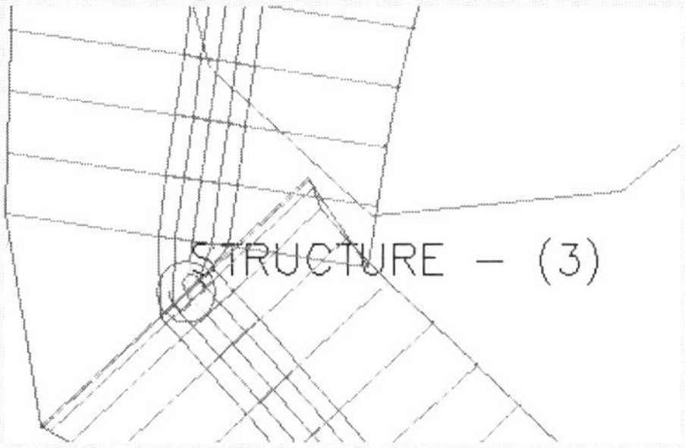

When this exercise is complete, you may close the drawing. A finished copy of this drawing is available from the book's web page with the filename `CorridorPipeTrench_FINISHED.dwg` or `CorridorPipeTrench_METRIC_FINISHED.dwg`.

The Bottom Line

Build a single baseline corridor from an alignment, profile, and assembly. Corridors are created from the combination of alignments, profiles, and assemblies. Although corridors can be used to model many things, most corridors are used for road design.

Master It Open the `MasteringCorridors.dwg` or `MasteringCorridors_METRIC.dwg` file. Build a corridor named Corridor A on the basis of the Alignment A alignment, the Project Road Finished Ground profile, and the Basic Assembly. Set all frequencies to **10′** (or **3** m for metric users).

Use targets to add lane widening. Targets are an essential design tool used to manipulate the geometry of the road.

Master It Open the `MasteringCorridorTargets.dwg` or `MasteringCorridorTargets_METRIC.dwg` file. Set Right Lane to target the Alignment A-Left alignment.

Create a corridor surface. The corridor model can be used to build a surface. This corridor surface can then be analyzed and annotated to produce finished road plans.

Master It Open the `MasteringCorridorSurface.dwg` or `MasteringCorridorSurface_METRIC.dwg` file. Create a corridor surface for the Alignment A corridor from Top links. Name the surface **Corridor A-Top**.

Add an automatic boundary to a corridor surface. Surfaces can be improved with the addition of a boundary. Single-baseline corridors can take advantage of automatic boundary creation.

Master It Open the `MasteringCorridorBoundary.dwg` or `MasteringCorridorBoundary_METRIC.dwg` file. Use the Automatic Boundary Creation tool to add a boundary using the Daylight code.

Chapter 11

Advanced Corridors, Intersections, and Roundabouts

In Chapter 10, "Basic Corridors," you built several simple corridors — including an example where the road lane used an alignment and profile target to add a variable width and elevation to the edge of traveled way — and began to see the dynamic power of the corridor model. The focus of that chapter was to get things started, but it's unrealistic to think that a project would have only one road in the middle of nowhere with no intersections, no adjustments, and no complications. You may be having trouble visualizing how you'll build a corridor to tackle your more complex design projects, such as the one pictured in Figure 11.1.

This chapter focuses on taking your corridor-modeling skills to a new level by introducing more tools to your corridor-building toolbox, such as intersecting roads, cul-de-sacs, advanced techniques, and troubleshooting. It deals with a roadside swale that follows a variable horizontal alignment, as well as a vertical profile that doesn't follow the centerline of the road. This happens frequently when existing culvert crossings must be met, or when you have different slope requirements for the roadside swale.

This chapter assumes that you've worked through the examples in the alignments, profiles, profile view, assemblies, and basic corridor chapters. Without a strong knowledge of the foundation skills, many of the tasks in this chapter will be difficult.

In this chapter, you will learn to:

- Create corridors with noncenterline baselines
- Add alignment and profile targets to a region for a cul-de-sac
- Create a surface from a corridor and add a boundary

Corridor Utilities

To create an alignment and profile for the swale, you'll take advantage of some of the corridor utilities found in the Launch Pad panel (Figure 11.2) by selecting the corridor.

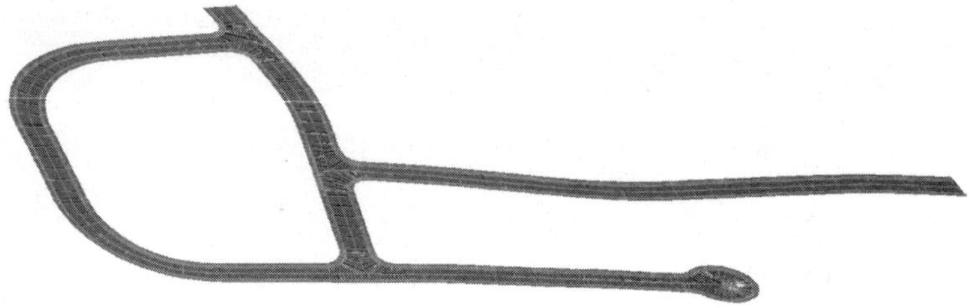

FIGURE 11.1
A corridor model for the example subdivision

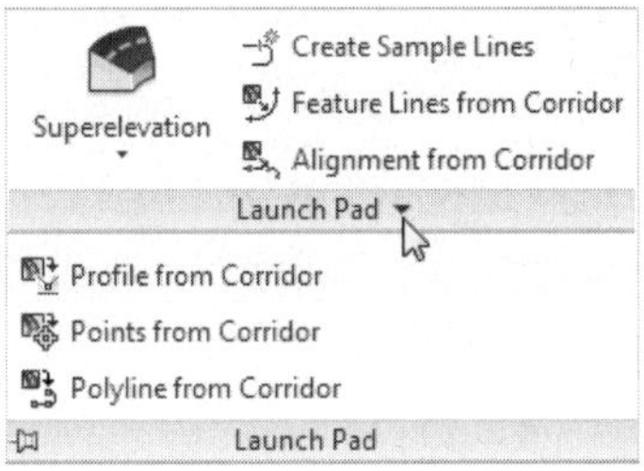

FIGURE 11.2
A bounty of corridor utilities on the Launch Pad panel

The utilities on this panel are as follows:

Superelevation This button will jump you to the alignment superelevation parameters. See Chapter 12, "Superelevation," for a detailed look at how Civil 3D creates banked curves for your design speed.

Create Sample Lines Corridors and sample lines are both linked to alignments. Civil 3D gives you a shortcut to the sample line creation tool. Chapter 13, "Cross Sections and Mass Haul," explores the creation and uses of sample lines.

Feature Lines From Corridor This utility extracts a grading feature line from a corridor feature line. This grading feature line can remain dynamic to the corridor, or it can be a static extraction. Typically, this extracted feature line will be used as a foundation for some feature-line grading or projection grading. If you choose to extract a dynamic feature line, it can't be used as a corridor target due to possible circular references.

Alignments From Corridor This utility creates an alignment that follows the horizontal path of a corridor feature line. You can use this alignment to create target alignments, profile views, special labeling, or anything else for which a traditional alignment could be used. Extracted alignments are not dynamic to the corridor.

Profile From Corridor This utility creates a profile that follows the vertical path of a corridor feature line. This profile appears in Prospector under the baseline alignment and is drawn on any profile view that is associated with that baseline alignment. This profile is typically used to extract edge of pavement (EOP) or swale profiles for a finished profile view sheet or as a target profile for additional corridor design, as you'll see in this section's exercise. Extracted profiles aren't dynamic to the corridor.

Points From Corridor This utility creates Civil 3D points that are based on corridor point codes. You select which point codes to use, as well as a range of corridor stations. A Civil 3D point is placed at every point-code location in that range. These points are a static extraction and don't update if the corridor is edited. For example, if you extract COGO points from your corridor and then revise your baseline profile and rebuild your corridor, your COGO points won't update to match the new corridor elevations.

Polyline From Corridor This utility extracts a 3D polyline from a corridor feature line. The extracted 3D polyline isn't dynamic to the corridor. You can use this polyline as is or flatten it to create road linework.

GETTING CREATIVE WITH CORRIDOR MODELS

No two civil projects are exactly the same. Furthermore, no two corridor designs will be exactly the same. You will want a handful of corridor techniques in your back pocket to handle the myriad design situations you will encounter.

In this book are many corridor examples that cover a good amount of design scenarios. However, the ways that corridors are used in this book are just the beginning. Getting creative with corridor tools, you can create parking lots, berms, trenches, and even water slides.

The best part about Civil 3D is that design iterations are easy to accomplish. If you are not sure if your corridor will work, give it a try. The feedback from Panorama will tell you immediately if something isn't right. If your assembly and alignment combination doesn't work, you can keep trying new geometry and rebuilding until you have exactly what you want.

Using Alignment and Profile Targets to Model a Roadside Swale

The following exercise takes you through revising a model from a symmetrical corridor with roadside swales to a corridor with a transitioning roadside swale centerline. You'll also take advantage of some of the static extraction corridor utilities discussed in the previous section:

1. Open the `Corridor Swale.dwg` (`Corridor Swale_METRIC.dwg`) file, which you can download from this book's web page at `www.sybex.com/go/masteringcivil3d2013`.

   Note that the drawing contains a symmetrical corridor (see Figure 11.3), which was built using an assembly that includes two roadside swales. This drawing has been split into two modelspace views so you can see what is going on in plan and in profile at the same time.

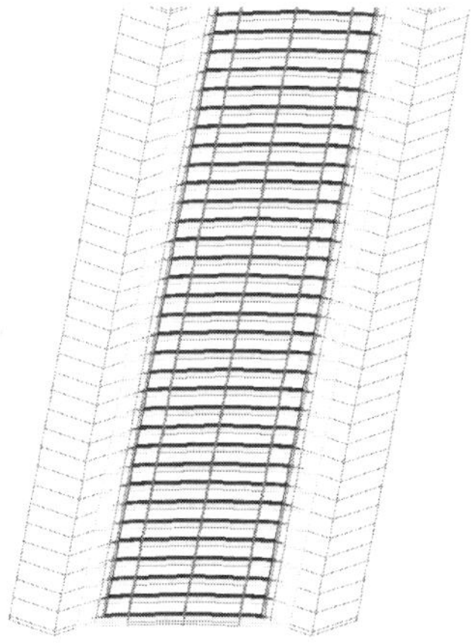

FIGURE 11.3
The initial corridor with symmetrical roadside swales

2. Select the corridor, and choose the Alignment From Corridor tool from the Launch Pad panel.

3. When prompted, pick the corridor feature line that represents the swale on the left side of the road centerline, as noted in the drawing, to create an alignment.

   Note that if you select one of the corridor feature lines to activate the Corridors tab, that feature line will be created by default. You can pick one of the frequency lines to avoid this potential pitfall!

4. In the Create Alignment From Objects dialog (Figure 11.4),

   A. Name the alignment **Swale Left**.

   B. Keep the default value for Site And Style.

   C. Change the Alignment Label Set to All Labels, as shown in Figure 11.4.

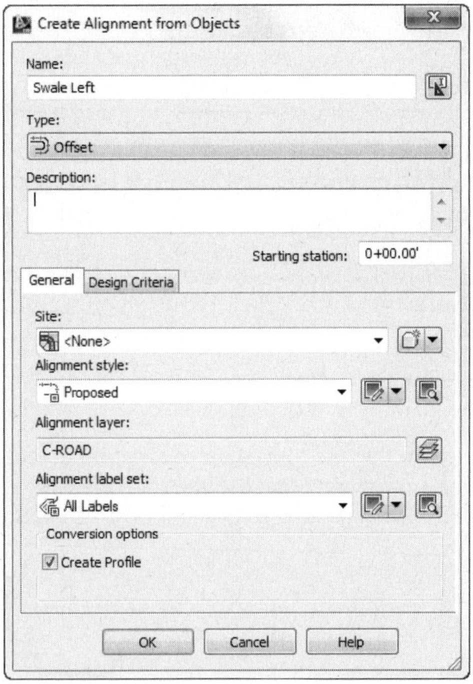

FIGURE 11.4 The Create Alignment From Objects dialog appears when you're extracting a feature line to an alignment.

5. Be sure to uncheck Create Profile, and then click OK to finish.

6. Press Esc to exit the command.

   You should now have a new alignment with labels in the plan view, as shown in Figure 11.5.

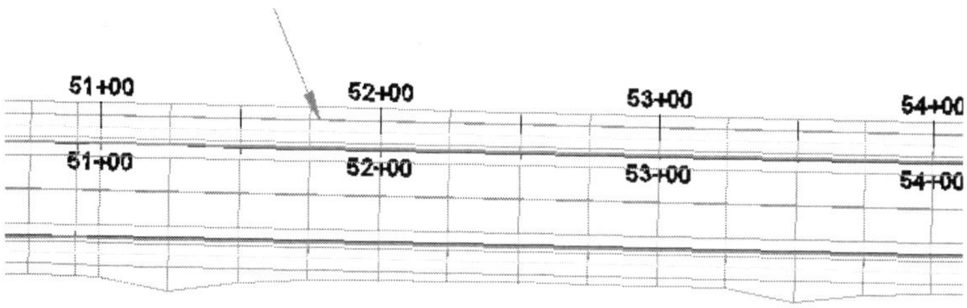

FIGURE 11.5
The extracted alignment

Next, you will extract the profile in a separate step. By performing these actions in two steps, you ensure that the profile will be associated with the main alignment, rather than with the offset alignment you created in the previous step. Keeping the extracted profile with the main alignment will allow you to work on the profile in the same profile view as the proposed centerline.

7. Select the corridor by clicking a frequency line or feature line and select Profile From Corridor from the drop-down list on the Launch Pad panel of the contextual Ribbon tab.
8. Select the same Swale Left feature line as you did in the previous step.

   You may need to use the DRAWORDER command or press Ctrl+W on your keyboard to access the feature line hiding behind the new alignment. When the feature line is selected, you will see the Create Profile dialog, as shown in Figure 11.6.

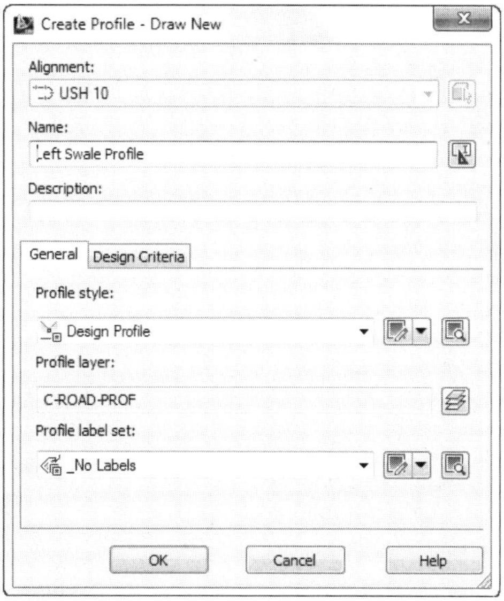

FIGURE 11.6
Create Profile dialog

9. Name the profile **Swale Left Profile** and click OK.

You should now see the profile you just created in the bottom viewport. The profile represents the vertical path of the feature line representing the swale centerline, as shown in Figure 11.7.

FIGURE 11.7
A portion of the resulting extracted profile

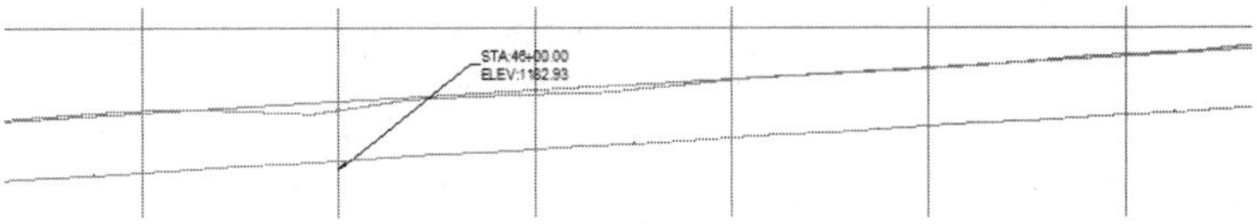

10. Select your newly extracted Left Swale Profile. From the Profile contextual tab ➢ Modify Profile panel, select Geometry Editor.

11. From the Profile Layout Tools Toolbar, click Move PVI.
12. Zoom in to station 46+00 (1+400 for metric users) if necessary, and select the PVI at 1182.93′ (360.72 m) in elevation to elevation 1170′ (356.50 m), as shown in Figure 11.8. Press Esc to exit the command.

FIGURE 11.8
Move a PVI to provide an exaggerated low spot.

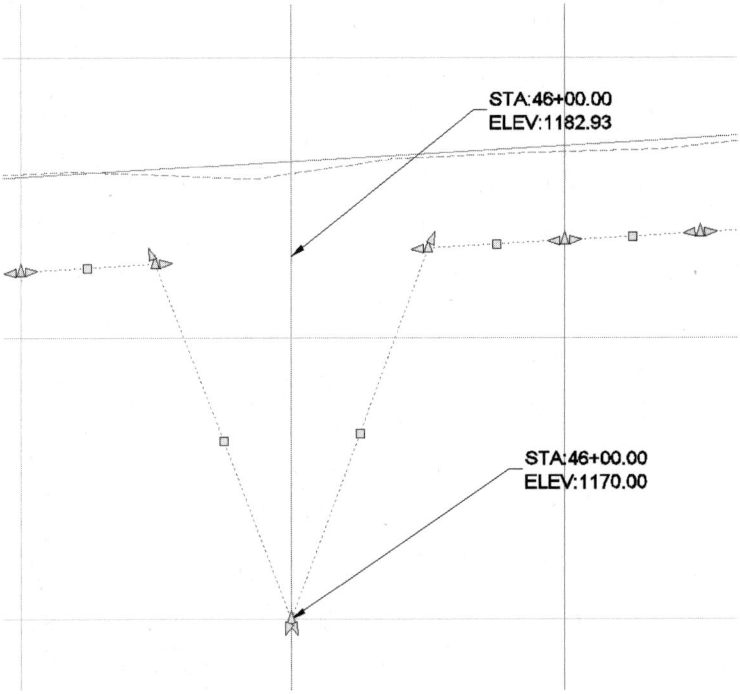

13. Select the corridor and from the Corridor Contextual tab ➢ Modify Panel, select Corridor Properties. Switch to the Parameters tab.

14. There is just one baseline and region in this project. Scroll over to the ellipsis button to open the Target Mapping dialog on the same line as the region.

When setting a surface target for a corridor, going to Set All Targets will work great. As your corridors get more complex, the number of options for elevation and offset targets become more numerous. By getting in the habit of targeting objects at the region level, you'll save yourself a lot of looking.

15. In the Width or Offset Target area (Figure 11.9), set the Foreslope - L subassembly to target the Swale Left Target Alignment.

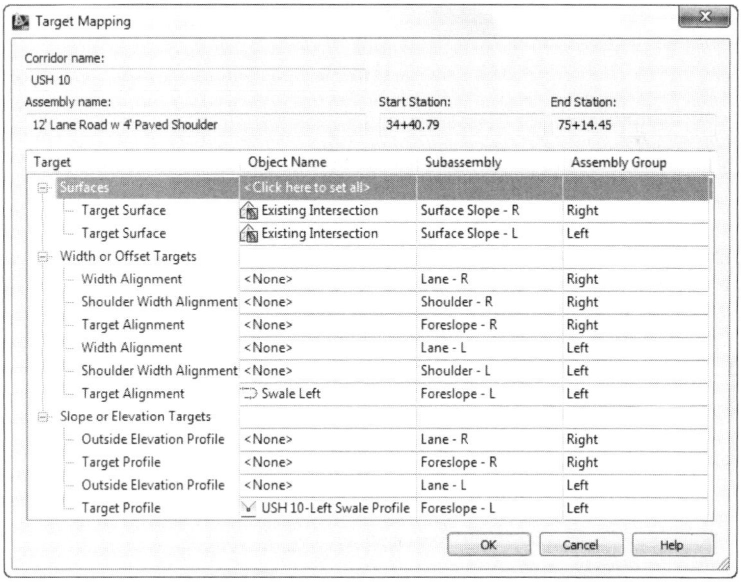

FIGURE 11.9
Set the Foreslope - L subassembly to follow the Swale Left alignment and profile.

16. In the Slope or Elevation Targets area, set the Foreslope - L subassembly to target the Swale Left Profile.

Figure 11.9 shows the Target Mapping dialog with the alignments and profiles appropriately mapped.

17. Click OK to dismiss the Target Mapping dialog, and click OK again to dismiss the Corridor Properties dialog and rebuild the corridor.

The subtle change you made in the corridor can be observed at station 46+00 (or 1+400 for metric users), as shown in Figure 11.10.

Note that the corridor has been adjusted to reflect the new target alignment and profile. Also note that you may want to increase the sampling frequency. You can view the sections using the View/Edit Corridor Section tools. You can also view the corridor in 3D by picking it, right-clicking, and choosing Object Viewer. Use the 3D Orbit tools to change your view of the corridor.

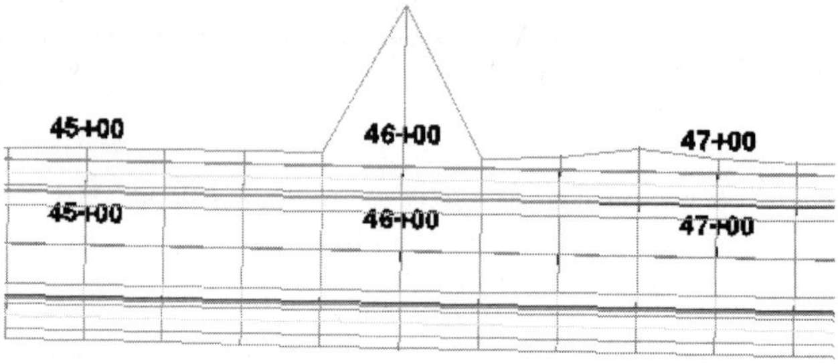

FIGURE 11.10
The adjusted corridor

Multiregion Baselines

A question many people ask when working with corridors is, "At what point do I need another region?" The answer is simple: If you need a different assembly, you need a different region.

In the following example, you will step through adding an additional region to an existing baseline:

1. Open the drawing `Multi-RegionCorridor.dwg` (`Multi-RegionCorridor_METRIC.dwg`).

   This drawing has been split into two modelspace viewports so that you can observe the results of your efforts in 3D.

2. Select the corridor and, from the Corridor contextual tab ➤ Modify Corridor panel, select Corridor Properties.

   On the Parameters tab, notice there is a baseline containing a single region.

3. Right-click on the region and select Split Region, as shown in Figure 11.11.

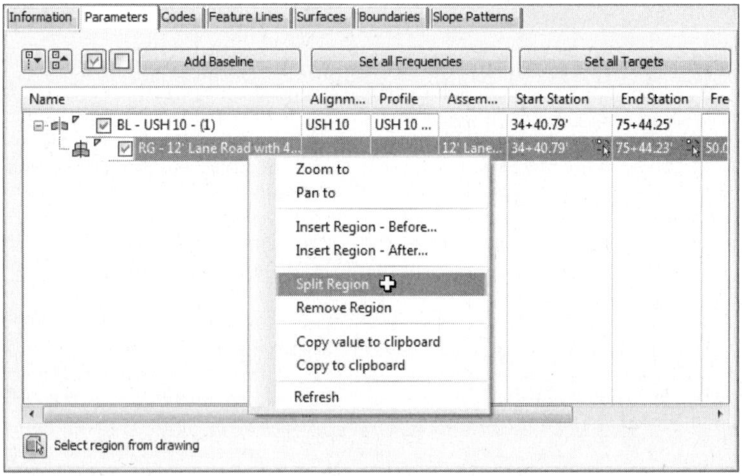

FIGURE 11.11
Right-click to access region creation tools.

4. At the command line, enter **4000** ↵ (**1220** ↵ for metric users) to create a split at 40+00 (1+220 for metric users), and create a second split at 55+00 (1+680 for metric users) by entering **5500** ↵ (**1680** ↵).
5. Press ↵ again to return to the Corridor Parameters dialog.

   If you see a warning dialog referring to 0+00 being outside of the station limits, click OK and continue.

   You should have three regions at this step, as shown in Figure 11.12.

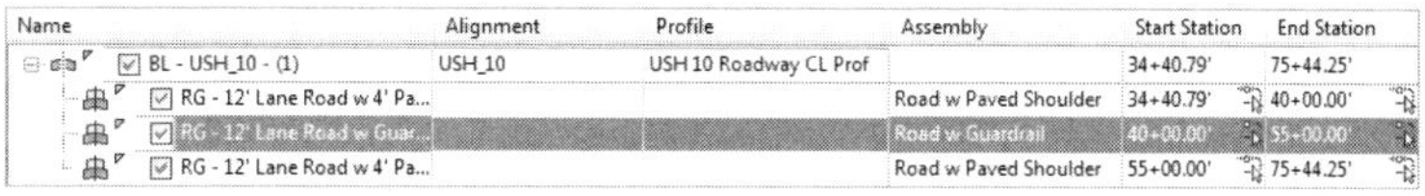

FIGURE 11.12 New regions with different assemblies applied

6. Change the assembly for the middle region to use the Road w Guardrail assembly, and then click OK.
7. Scroll over and click the Target ellipsis button.
8. Verify that the targets have been maintained for this region. If necessary, set the targets as shown in Figure 11.13.

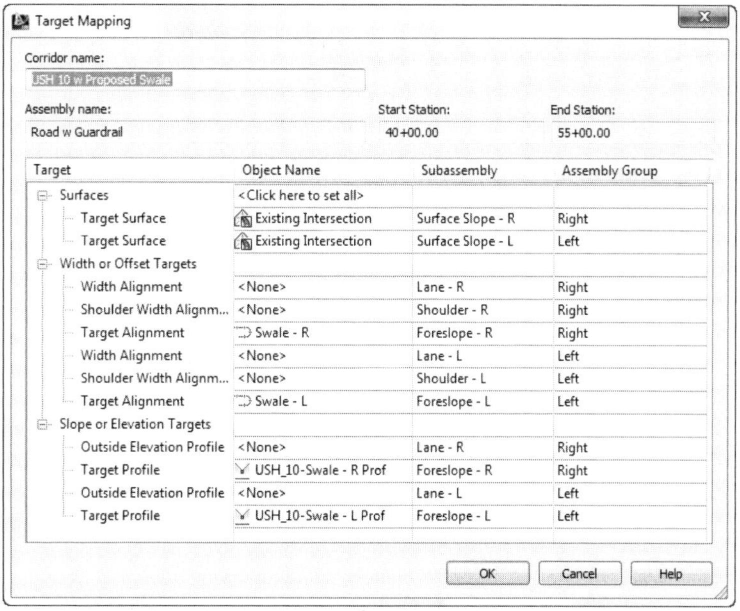

FIGURE 11.13 Targets for the middle region

9. Click OK when Target Mapping is complete, and then click OK.
10. Select Rebuild when prompted.

You should now see additional feature lines in the plan view representing the top of the guardrail. In the right viewport, it will resemble Figure 11.14.

FIGURE 11.14
The completed corridor with guardrails

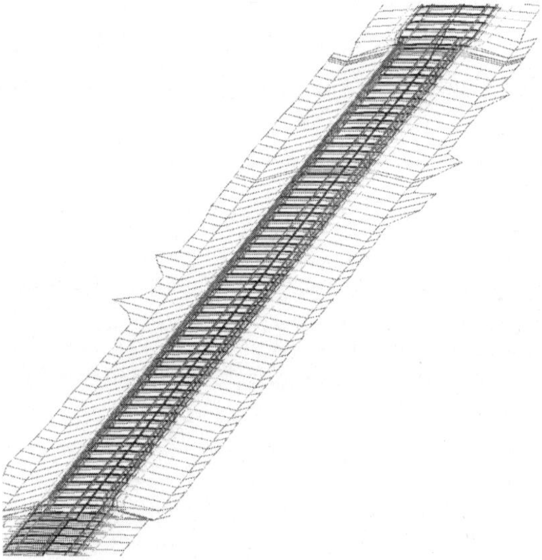

Modeling a Cul-de-Sac

Even if you never plan to design one in real life, understanding what is going on in a cul-de-sac corridor model will set you on the right path for building more complex models. If you truly understand the principles explained in the section that follows, then expanding your repertoire to include intersections and roundabouts will become much easier.

Using Multiple Baselines

Up to this point, every corridor we've examined has had a single baseline. In our examples, the centerline alignment has been the main driving force behind the corridor design. In this section, the training wheels are coming off! You've seen the last of single baseline corridors in this book.

A cul-de-sac by itself can be modeled in two baselines, as shown in Figure 11.15. The procedures that follow will work for most cul-de-sacs, symmetrical, asymmetrical, and hammerhead styles. You will need the centerline alignment and design profile, as well as an edge of pavement alignment and design profile.

FIGURE 11.15
Example cul-de-sac alignment setup

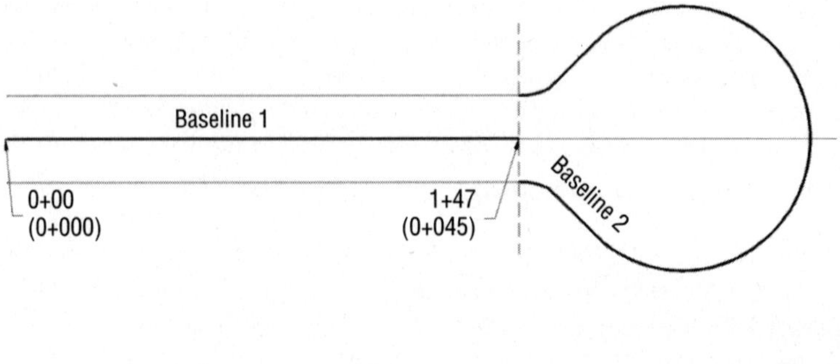

In the section leading to the cul-de-sac bulb, the corridor can be modeled using the tools you learned in Chapter 10. In the example shown in Figure 11.15, the centerline of the road is the baseline, with an assembly using two lanes and with the crown as the baseline marker from beginning station to 1+47 (for metric users 0+045). However, once the curvature of the bulb starts at 1+47.00 (for metric users 0+045), the centerline is no longer an acceptable baseline.

Assemblies are always applied to a baseline and the geometry will be perpendicular to the baseline. To get correct pavement grades in the bulb, you'll need to hop over to the edge of pavement alignment for a baseline. This second baseline will use a partial assembly that is built with the main assembly marker at the outside edge of pavement (Figure 11.16). The crown of the road will stretch to meet the centerline alignment and profile as its targets.

FIGURE 11.16
Assembly used for designing off the edge of pavement

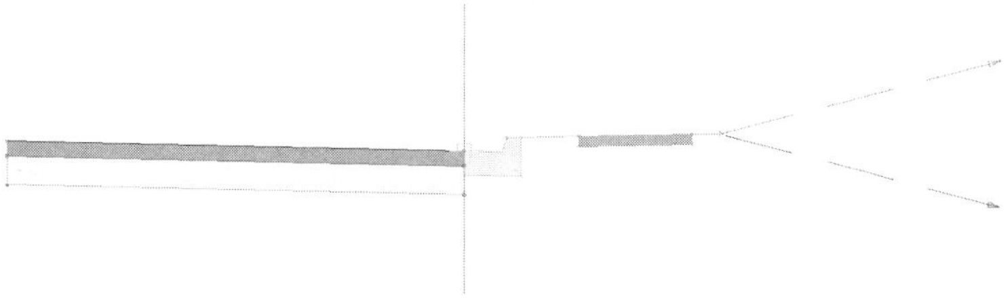

It helps to think of the assembly as radiating away from the baseline, from the assembly base outward, toward a target (Figure 11.17). Because the assemblies are applied to the baseline in a perpendicular manner, using the edge of pavement for a baseline in curved areas (such as cul-de-sac bulbs or curb returns) will result in a smooth, properly graded pavement surface.

FIGURE 11.17
It helps to think of the assemblies radiating away from the baseline toward the targets.

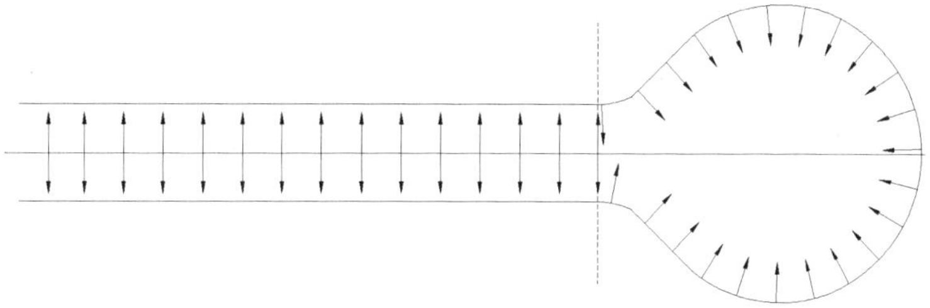

Establishing EOP Design Profiles

One of the most challenging parts of any "fancy" corridor (i.e., cul-de-sac, intersection, or roundabout) is establishing design profiles for noncenterline alignments. You must have design profiles for every alignment used as a target — there's no way around it — but it does not have to be a painful process to obtain them.

Using a simple, preliminary corridor and the profile creation tools you learned in Chapter 7, "Profiles and Profile Views," you'll find that establishing an edge of pavement (EOP) profile can go quickly.

In the exercise that follows, you will work through the steps of creating an EOP profile:

1. Open the `Cul-de-sac_Profile.dwg` (`Cul-de-Sac_Profile_METRIC.dwg`)file, which you can download from this book's web page.

   This drawing contains several alignments and an existing surface whose style is set to show the border only.

2. From the Home tab ➤ Create Design Panel, click Corridor, and do the following:
   - **A.** Name the corridor **PRELIM**.
   - **B.** Set Alignment to Frontenac Drive.
   - **C.** Set Profile to Frontenac Drive - FG.
   - **D.** Set Assembly to PRELIM
   - **E.** Clear the checkbox for Set Baseline And Region Parameters.
3. No targets are needed in this preliminary corridor, so click OK to complete the corridor.
4. Select the new corridor and open the Corridor Properties.
5. On the Surfaces tab, click the leftmost button to start a corridor surface.

6. With Data Type set to Links and Code set to Top, click the plus sign to add data to the surface.
7. On the Boundaries tab, right-click the PRELIM Surface, and select Corridor Extents as Outer Boundary.
8. Click OK and rebuild the corridor.

   You should now see contours in your drawing representing the 2 percent crossfall from the centerline.
9. Select the red EOP alignment for the cul-de-sac, and in the contextual Ribbon tab, click Surface Profile.
10. In the Create Profile From Surface dialog, highlight PRELIM Surface - (1) and click Add.
11. Click the Draw In Profile View button.
12. Leave all the defaults in the Create Profile View dialog and click Create Profile View.
13. Click in the graphic to the right of the surface to place the profile view.

    The profile you are seeing will have gaps in the middle that need grading information. However, the bulb portion of the profile only needs to exist between 16+87.00 (0+514.2 for metric users) and 19+42.89 (0+592.2 for metric users). They are the PC and PT stations for the EOP alignment in plan. In this example, you will create a design profile that starts slightly before, and ends slightly after, these critical stations.
14. Select the profile view and click Profile Creation Tools on the contextual Ribbon tab.
15. Click OK to accept the defaults in the Create Profile dialog.
16. In the Profile Layout toolbar, select Draw Tangents, and snap to the intersection of the preliminary profile and the grid line at 16+50 (0+500 for metric users). The elevation is also determined by this location.
17. Next, snap to the endpoint where the preliminary surface trails off at station 17+02.38 (0+517.13 for metric users) elevation 792.88′ (241.737 m).
18. The next VPI will be the peak of the preliminary profile in the center of the view; station 18+14.94 (0+553.20 for metric users), elevation 790.35′ (240.898 m).

19. Snap to the endpoint where the preliminary profile picks up again; station 19+27.51 (0+589.26 for metric users), elevation 792.88′ (241.737 m).

20. Finally, snap to the grade break at station 19+78.30 (0+605.99 for metric users), elevation 794.68′ (242.351 m).

21. Press ↵ to complete the command and close the Profile Layout tools.

    You now have a proposed profile that is acceptable to use in the cul-de-sac corridor.

Putting the Pieces Together

You have all the pieces in place to perform the first iterations of this cul-de-sac design.

The following exercise will walk you through the steps to put the cul-de-sac together. You will complete several steps and let the corridor build to observe what is happening at each stage. This exercise will also encourage you to get comfortable using Corridor Properties to make design modifications.

1. Open the `Cul-de-sac_Design.dwg` (`Cul-de-Sac_Design_METRIC.dwg`) file, which you can download from this book's web page.

   This drawing contains the cul-de-sac centerline alignment and profile, the EOP alignment and profile, and the assemblies needed to complete the process. The PRELIM corridor layer is frozen. The view of this drawing is twisted 90° to better fit most monitors.

2. From the Home tab, on the Create Design panel select Corridor, and do the following:

   A. Name the corridor **Cul-de-Sac**.

   B. Set Alignment to Frontenac Drive.

   C. Set Profile to Frontenac Drive FG.

   D. Set Assembly to Urban.

   E. Set Target Surface to EG.

   F. Keep the checkmark next to Set Baseline And Region Parameters.

   G. Click OK.

3. In the Baseline And Region Parameters dialog, click the Pick Station button for the start station of the first region.

4. Snap to the start of the EOP alignment on the west side, as shown in Figure 11.18 (left).

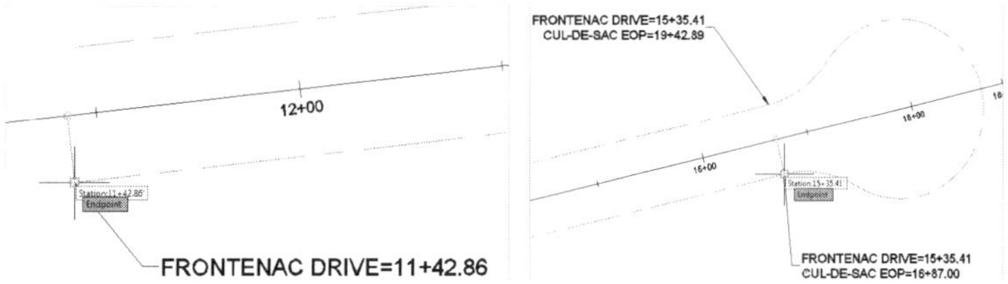

FIGURE 11.18 Setting the start station (left) and end station (right) for the centerline region

This will result in a start station of 11+42.86 (0+348.34 for metric users) in the Create Corridor dialog.

5. Click the Pick Station icon for the end station of the first region.
6. Snap to the PT station on the east side of the cul-de-sac bulb, as shown in Figure 11.18 (right).

   This will result in the region end station of 15+35.41 (0+467.99 for metric users).
7. Click OK to dismiss the Baseline And Region Parameters dialog, and click OK to complete the initial piece of the corridor.

   Examine the corridor you just created. It should start south of the intersection with the Syrah Way alignment, and end before the curvy part of the cul-de-sac bulb.

Corridor Properties

8. Select the corridor and click Corridor Properties.

   If necessary, correct any station problems with the first region.
9. In the Parameters tab of Corridor Properties, click Add Baseline.
10. Select the Cul-de-Sac EOP as the alignment, and click OK.
11. In the Profile Column, click <Click here…>, and select the Cul-de-Sac EOP - FG profile.

    This is the profile that you learned how to develop in the previous exercise.
12. Right-click on the newly added baseline, and select Add Region, as shown in Figure 11.19.

FIGURE 11.19
Right-click the baseline to add a region.

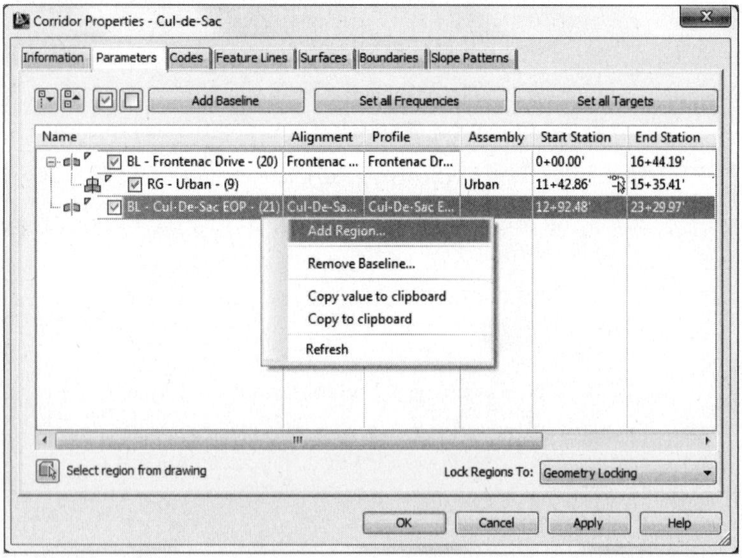

13. Select the Curb Right assembly, and click OK.
14. Expand the Cul-de-Sac baseline and then click the Pick Station button for the Curb Right region start station.

15. Select the PC station of the corridor bulb.

This will result in a start station value of 16+87.00 (0+514.20 for metric users).

16. Click the Pick Station icon for the end station of the region.

17. Select the PT station of the corridor bulb.

This will result in an end station of 19+42.89 (0+592.19 for metric users).

18. Click OK and let the corridor build once again.

Have a look at the corridor in its current state (see Figure 11.20). Even if you are not exactly sure what you are looking at, you should at least see a few things are amiss with the corridor so far.

FIGURE 11.20
The cul-de-sac corridor several steps away from completion

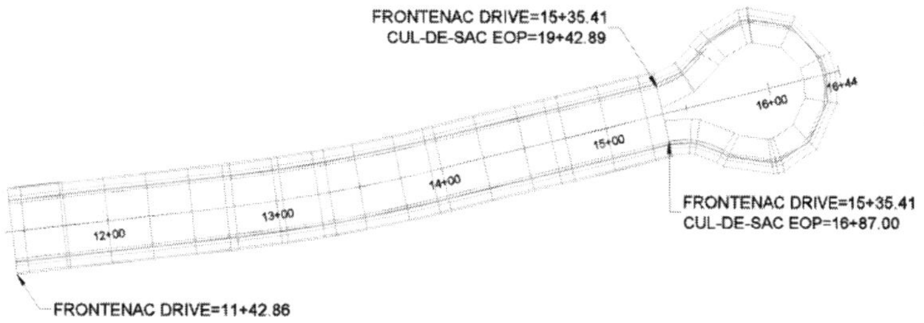

There are three things that need to be modified before the cul-de-sac is complete. The lane needs to extend to the center of the cul-de-sac bulb and the daylight surface needs to be set, both of which can be corrected in the Target Mapping area. Finally, you will want to increase the frequency around the curvy area to get a smoother, more precise design.

In the next steps, you will correct these issues and complete the cul-de-sac.

19. Select the corridor and return to Corridor Properties (last time in this exercise, we promise).

20. Scroll over and click the ellipsis to enter the Target Mapping dialog for the region named RG - Curb Right - (36).

The number following the region name may vary.

A. Set Target Surface to EG, and then click OK.

B. Set Width Alignment for Lane - L to Frontenac Drive. Click Add and then click OK.

C. Set Outside Elevation Profile to Frontenac Drive-FG for Lane - L. Click Add and then click OK.

D. Click OK again to dismiss the Target Mapping dialog.

21. Click the frequency ellipsis next to the current value of 25′ (20 m), and set the frequency for both tangents and curves to **5′** (**1** m). Set the At Offset Target Geometry Points to Yes.

22. Click OK to dismiss the Frequency dialog.

23. Click OK and let the corridor rebuild one last time.

The completed corridor will look like Figure 11.21.

FIGURE 11.21
The completed cul-de-sac corridor. Gorgeous!

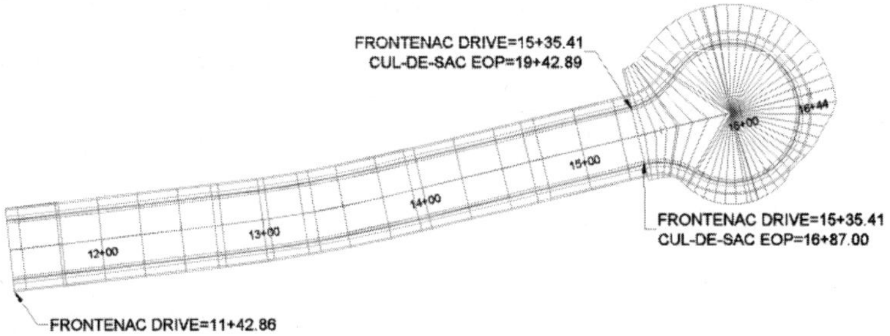

Troubleshooting Your Cul-de-Sac

People make several common mistakes when modeling their first few cul-de-sacs:

Your cul-de-sac appears with a large gap in the center. If your curb line seems to be modeling correctly but your lanes are leaving a large empty area in the middle (see Figure 11.22), chances are pretty good that you forgot to assign targets or perhaps assigned the incorrect targets.

FIGURE 11.22
A cul-de-sac without targets

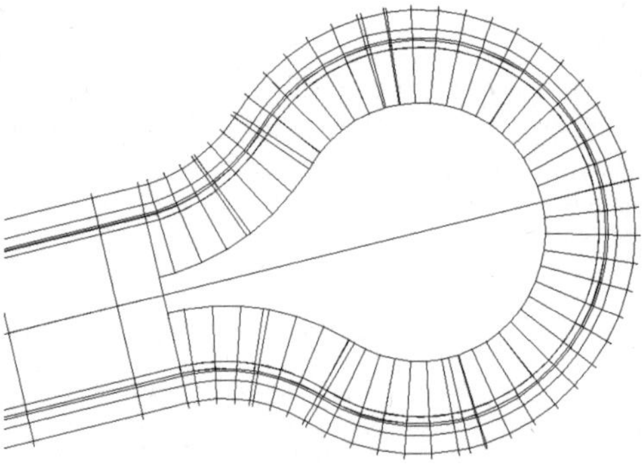

Fix this problem by opening the Target Mapping dialog for your region and checking to make sure you assigned the road centerline alignment and FG profile for your transition lane. If you have a more-advanced lane subassembly, you may have accidentally set the targets for another subassembly somewhere in your corridor instead of the lane for the cul-de-sac transition, especially if you have poor subassembly-naming conventions. Poor naming conventions become especially confusing if you used the Map All Targets button.

Your cul-de-sac appears to be backward. Occasionally, you may find that your lanes wind up on the wrong side of the EOP alignment, as shown in Figure 11.23. The direction of your alignment will dictate whether the lane will be on the left or right side. In the example from the previous exercise, the alignment was running counterclockwise around the cul-de-sac bulb; therefore, the lane was on the left side of the assembly.

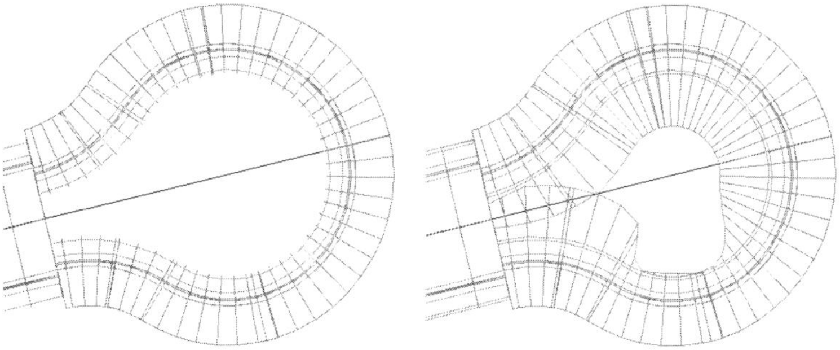

FIGURE 11.23
A cul-de-sac with the lanes modeled on the wrong side without targets (left) and with targets (right)

You can fix this problem by changing the assembly to the correct side.

Your cul-de-sac drops down to 0. A common problem when you first begin modeling cul-de-sacs, intersections, and other corridor components is that one end of your baseline drops down to 0. You probably won't notice the problem in plan view, but once you build your surface (see Figure 11.24, left) or rotate your corridor in 3D (see Figure 11.24, right), you'll see it. This problem will always occur if your region station range extends beyond the proposed profile.

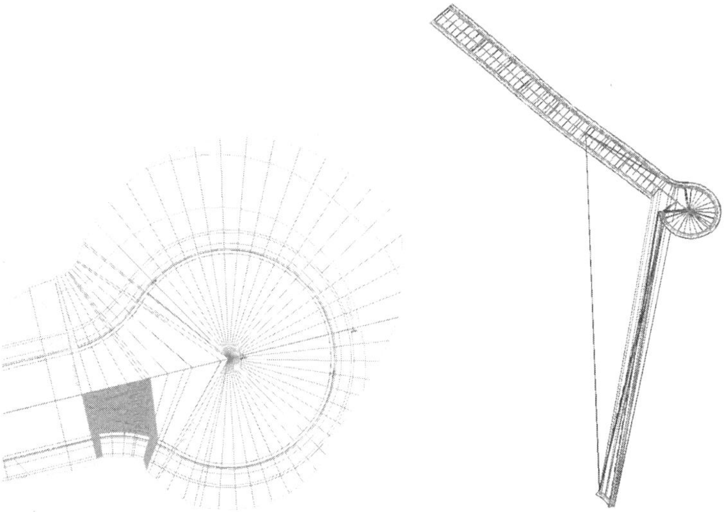

FIGURE 11.24
Contours indicating that the corridor surface drops down to 0 (right), and a corridor viewed in 3D showing a drop down to 0 (left)

The fix for this is the same as what you saw in the first exercise in Chapter 10 — that is, make sure your region station range corresponds with the design profile length. You may need to extend the design profile in some cases, but usually the station range will do the trick.

Your cul-de-sac seems flat. When you're first learning the concept of targets, it's easy to mix up baseline alignments and target alignments. In the beginning, you may accidentally choose your EOP alignment as a target instead of the road centerline. If this happens, your cul-de-sac will look similar to Figure 11.25.

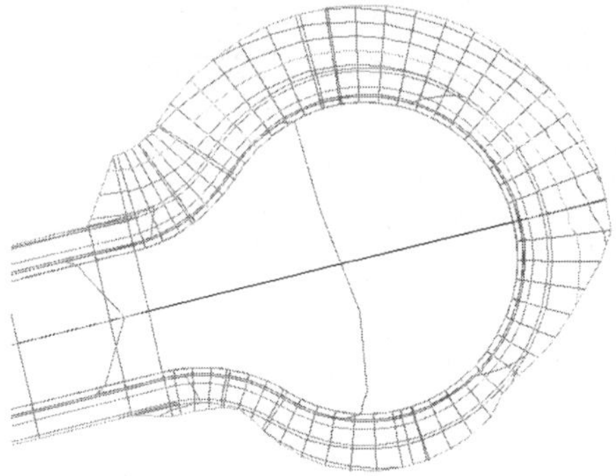

FIGURE 11.25
A flat cul-de-sac with the wrong lane target set

You can fix this problem by opening the Target Mapping dialog for this region and making sure the target alignment is set to the road centerline and the target profile is set to the road centerline FG profile.

Intersections: The Next Step Up

Ask yourself the following question: "Did I understand why we did what we did to build the cul-de-sac in the previous section?" If the answer is "Not really," you may want to review a few topics before proceeding. If the answer is, "Yeah, mostly," then you are ready to move to the next level of corridor complexity: intersections.

The steps that follow apply to all intersections, regardless of whether it is a T-shaped intersection, a four-way intersection, perfectly perpendicular, or skewed at an angle.

CERT OBJECTIVE

Corridor modeling is an iterative process. The more advanced your model, the more iterations it may take to get to the correct design. You will often not know the final design parameters until you see how the model relates to existing conditions or ties into other pieces of the design. Get comfortable jumping in and out of corridor properties and identifying regions within your corridor.

Plan what alignments, profiles, and assemblies you'll need to create the right combination of baselines, regions, and targets to model an intersection that will interact the way you want. It helps to create a simple sketch, as shown in Figure 11.26.

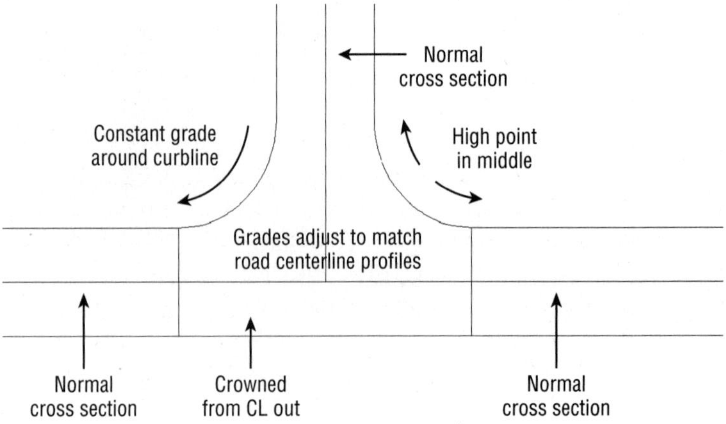

FIGURE 11.26
Plan your intersection model in sketch form.

Figure 11.27 shows a sketch of required baselines. As you saw in the previous example, *baselines* are the horizontal and vertical foundation of a corridor. Each baseline consists of an alignment, and its corresponding finished ground (FG) profile. You may never have thought of edge of pavement (EOP) in terms of profiles, but after you build a few intersections, thinking that way will become second nature. The Intersection tool on the Create Design panel of the Home tab will create EOP baselines as curb return alignments for you, but it will rely on your input for curb return radii.

FIGURE 11.27
Required baselines for modeling a typical intersection

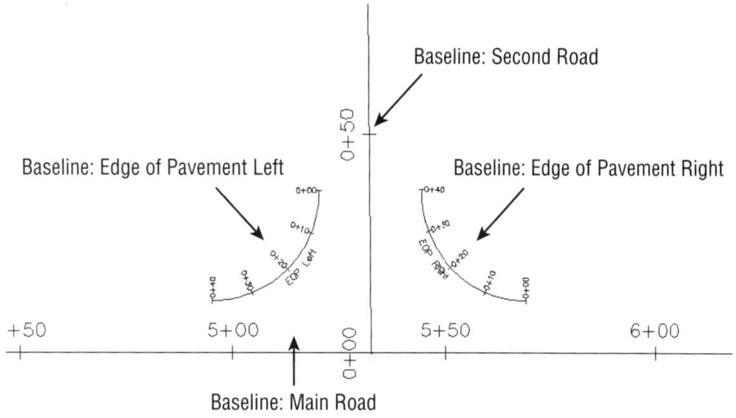

Figure 11.28 breaks each baseline into regions where a different assembly or different target will be applied. Once the intersection has been created, target mapping as well as other particulars can be modified as needed.

FIGURE 11.28
Required regions for modeling an intersection created by the Intersection tool

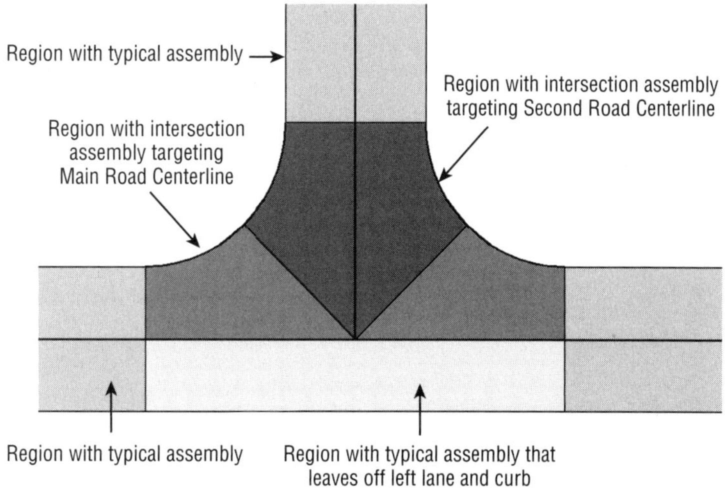

CERT OBJECTIVE

Using the Intersection Wizard

All the work of setting baselines, creating regions, setting targets, and applying the correct frequencies can be done manually for an intersection. However, Civil 3D contains an automated Intersection tool that can handle many types of intersections.

On the basis of the schematic you drew of your intersection, your main road will need several assemblies to reflect the different road cross sections. Figure 11.29 shows the full range of potential assemblies you may need in an intersection and the design situations in which they may arise.

FIGURE 11.29 Various assembly schematics and applications

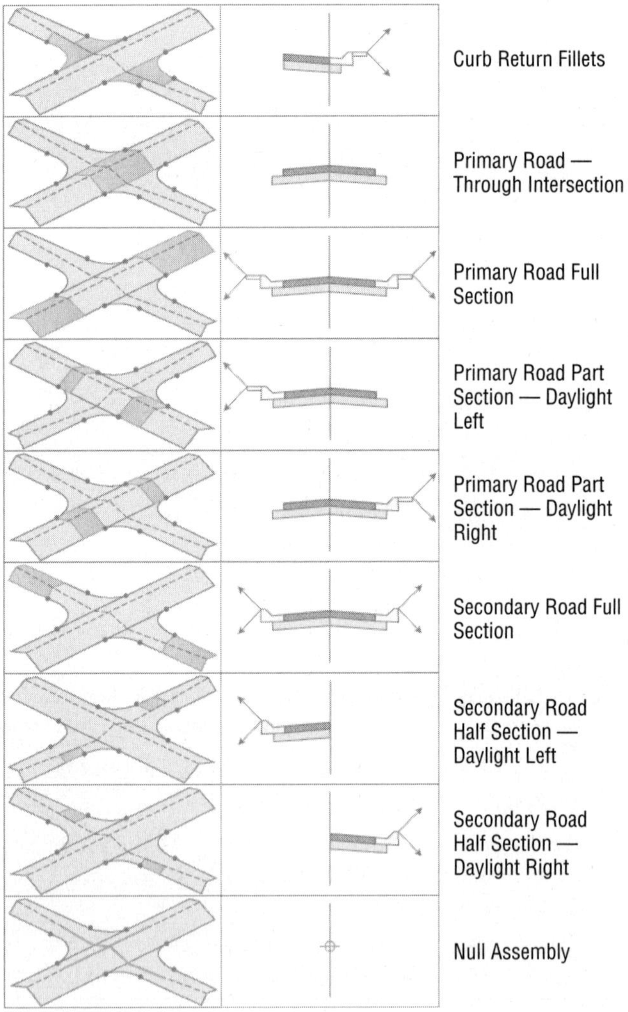

ASSEMBLY SETS

When you are ready to create an intersection, you do not need to have all the special assemblies created ahead of time. On the Corridor Regions page of the Create Intersection wizard, you will see a list of the assemblies Civil 3D plans to use. If the assemblies are not already part of the drawing, they will get pulled in automatically when you click Create Intersection.

The default intersection assemblies are general, and may not work for your design situation. You will want to create and save an assembly set of your own.

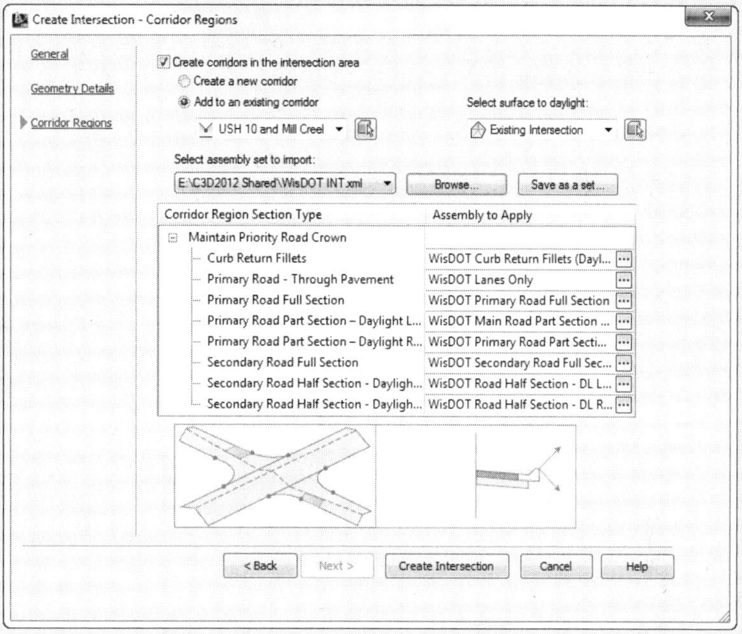

1. In a file that contains all of your desired assemblies, work through the Create Intersection wizard to get to the Corridor Regions page.
2. Click the ellipsis to select the appropriate assembly for each Corridor Region Section Type.
3. Once the listing is complete, click the Save As A Set button.

Civil 3D creates an XML file that stores the listing of the assemblies. It also creates a copy of each assembly as a separate DWG file. Save the set in a network shared location so your office colleagues can use the set as well.

The next time an intersection is created, you can use the assembly set by clicking Browse and selecting the XML file. Civil 3D will pull in your assemblies, saving lots of time!

This exercise will take you through building a typical peer-road intersection using the Create Intersection wizard:

1. Open the `Intersection.dwg` (`Intersection_METRIC.dwg`) file, which you can download from this book's web page.

   The drawing contains two centerline alignments (USH 10 and Mill Creek Drive) that are part of the same corridor.

2. From the Home tab ➤ Create Design panel, choose the Intersections ➤ Create Intersection tool.
3. Using the Intersection object snap, choose the intersection of the two existing alignments.
4. When prompted to select the main road alignment, click the USH 10 alignment that runs vertically in the project.

   The Create Intersection – General dialog will appear (Figure 11.30).

FIGURE 11.30
The Create Intersection – General dialog

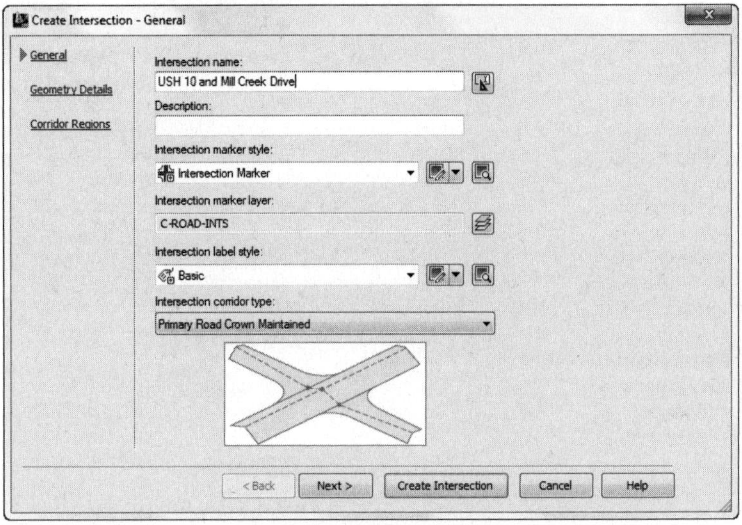

5. Name the intersection **USH 10 and Mill Creek Drive**. Set the Intersection corridor type to Primary Road Crown Maintained, as shown in Figure 11.30, and click Next.
6. In the Geometry Details page (Figure 11.31), verify that USH 10 is the primary road by looking at the Priority listing. If USH 10 is not at the top of the list, use the arrow buttons on the right side to reorder the roads.
7. Click the Offset Parameters button, and do the following:

   A. Set the offset values for both the left and right sides to **24′** (**8** m) for USH 10.

   B. Set the left and right offset values for Mill Creek Drive to **18′** (**5** m).

   C. Select the check box Create New Offsets From Start To End Of Centerlines.

      At this step, the screen should resemble Figure 11.32.

   D. Click OK to close the Intersection Offset Parameters dialog.

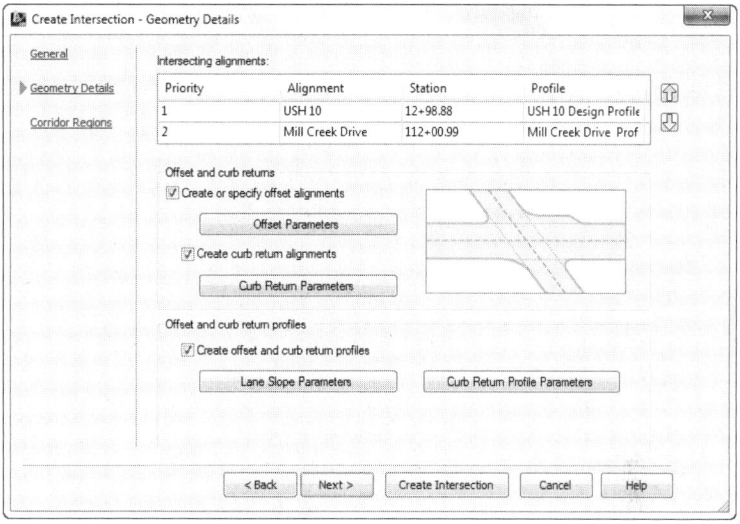

FIGURE 11.31 The Geometry Details page of the Create Intersection wizard

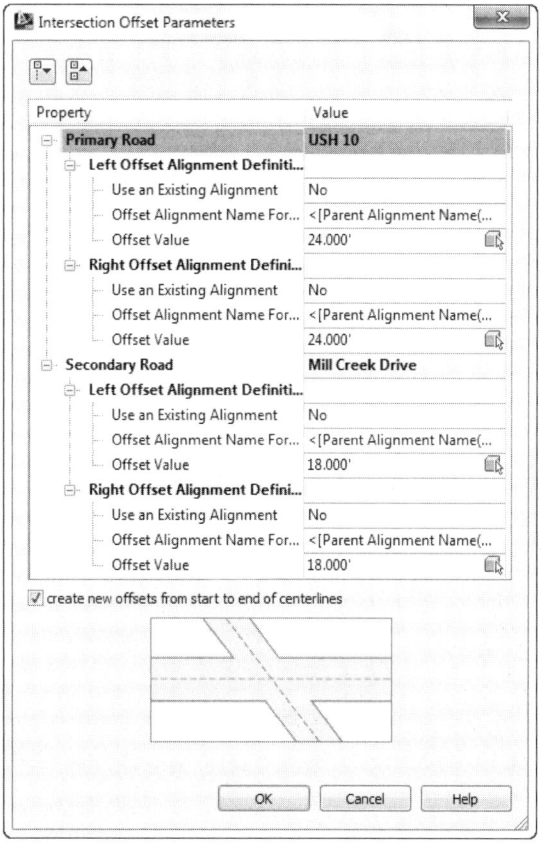

FIGURE 11.32 Intersection Offset Parameters dialog

8. Click the Curb Return Parameters to enter the Curb Return Parameters dialog.

   A. For all four quadrants of the intersection, place a check mark next to Widen Turn Lane For Incoming Road and Widen Turn Lane For Outgoing Road.

   B. Click the Next button at the top of the dialog to move from quadrant to quadrant.

   As shown in Figure 11.33, a temporary glyph will help you determine which quadrant you are currently modifying.

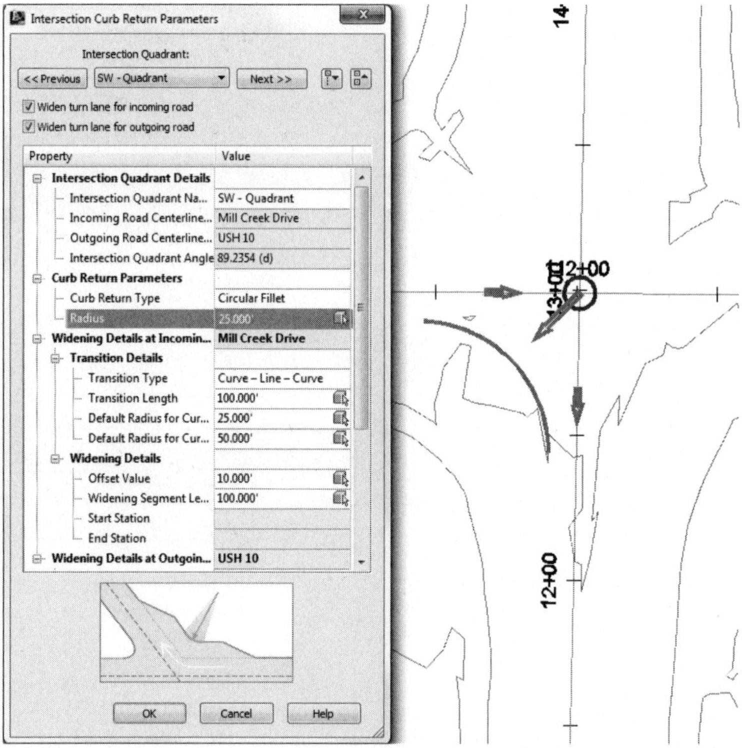

FIGURE 11.33 Adding lane widening to the SW – Quadrant of the intersection

   When you reach the last quadrant, you will see that the Next button is grayed out. This means that you have successfully worked through all four curb returns.

   D. Click OK to return to wizard's Geometry Details page.

Some locales require that lane slopes flatten out to a 1% cross-slope in an intersection. If this is the case for you, you can change the lane slope parameters in the Intersection Lane Slope Parameters dialog (Figure 11.34). In this exercise you will leave this as 2%.

Civil 3D is performing the task of generating the curb return profile. The profile will be at least as long as the rounded curb plus the turn lanes that are added in this exercise. If you wish to have Civil 3D generate even more than the length needed, you can specify that in the Intersection Curb Return Profile Parameters area (Figure 11.35).

FIGURE 11.34
Lane slope parameters control the cross-slope in the intersection.

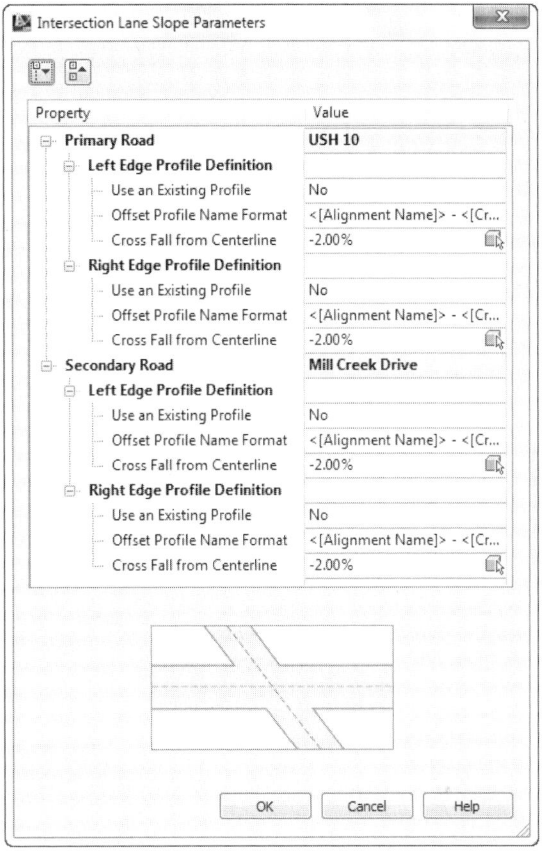

You will be keeping all default settings in both the Lane Slope Parameters area (Figure 11.34) and the Curb Return Parameters area (Figure 11.35).

9. Click Next to continue to the Corridor Regions page (Figure 11.36).

The Corridor Regions page is where you control which assemblies are used for the different design locations around the intersection. Clicking each entry in the Corridor Region Section Type list will give you a clear picture of which assemblies you should use and where (as shown at the bottom of Figure 11.36). If your assemblies have the same names as the default assemblies, as is the case in this example, they will be pulled from the current drawing. Alternately, you can click the ellipsis to select any assembly from the drawing.

If you don't have all the necessary assemblies at this point, you can still create your intersection. Civil 3D will pull in the default set of assemblies. You can always modify these assemblies after they are brought in.

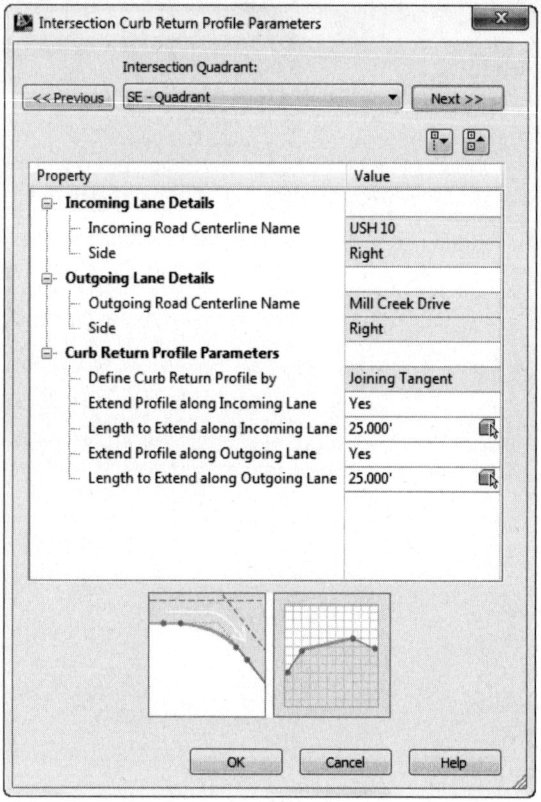

FIGURE 11.35 Intersection Curb Return Profile Parameters options extend the Civil 3D-generated profile beyond the curb returns by the value specified.

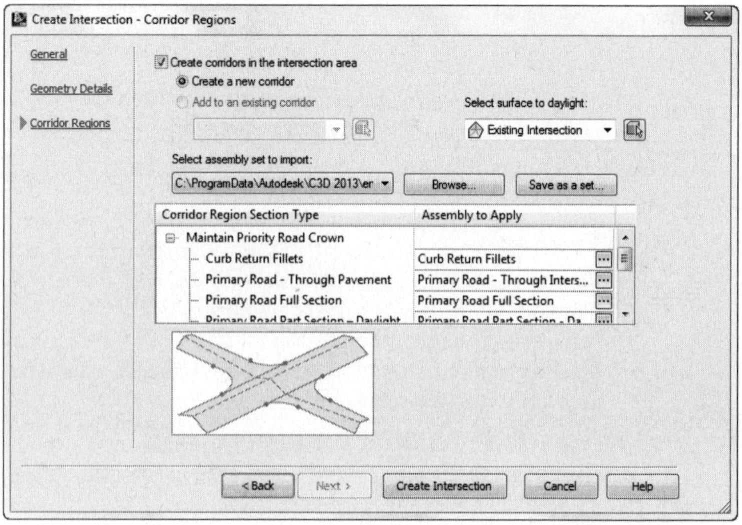

FIGURE 11.36 The Corridor Regions page drives the assemblies used in the intersection.

10. Click through the Corridor Region Section Type list to view the schematic preview for each.

    Do not make any changes to the assembly listing.

11. Click Create Intersection.

    After a few moments of processing, you will see a corridor appear at the intersection of the roads. Use the REGEN command if you do not see the frequency lines. Your corridor should now resemble Figure 11.37.

FIGURE 11.37
The nearly completed intersection

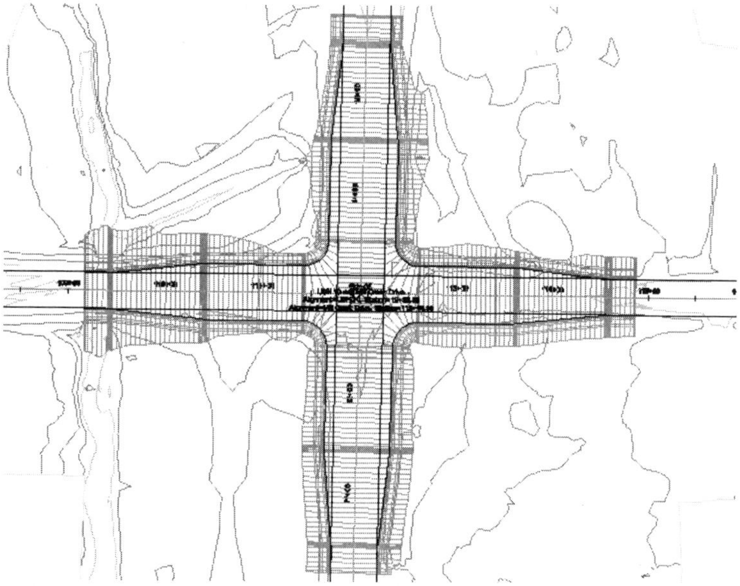

12. Select the corridor and from the Corridor contextual tab ➢ Modify Corridor panel, select Corridor Properties.

13. Switch to the Parameters tab of the Corridor Properties dialog and highlight one of the regions in the listing, as shown in Figure 11.38.

    Notice how the region highlighted in the Parameters tab is outlined in the graphic. This will help you determine which region to edit, even in the largest of corridors.

    To move from region to region without searching through the somewhat daunting list, use the Select Region From Drawing button.

    In the next steps you will leverage the work Civil 3D has done for you and extend the intersection in all four directions.

14. Change the region start station on the west side of Mill Creek Road to 101+00 (3+200 for metric users).

15. Change the east side end station to 121+00 (3+700 for metric users).

16. Change the start station of the south end of USH 10 to 10+00 (0+300 for metric users).
17. Change the north side end station to 16+50 (0+500 for metric users).
18. Click OK and rebuild the corridor.

When the corridor is complete, it will look like Figure 11.39.

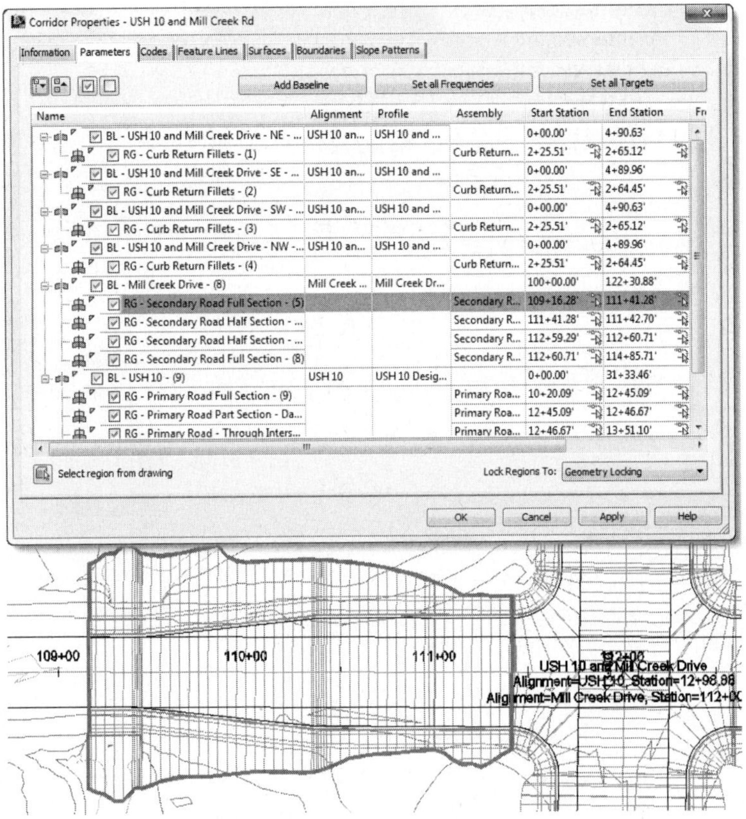

FIGURE 11.38 The best way to modify regions is using the corridor properties. The selected region will highlight graphically.

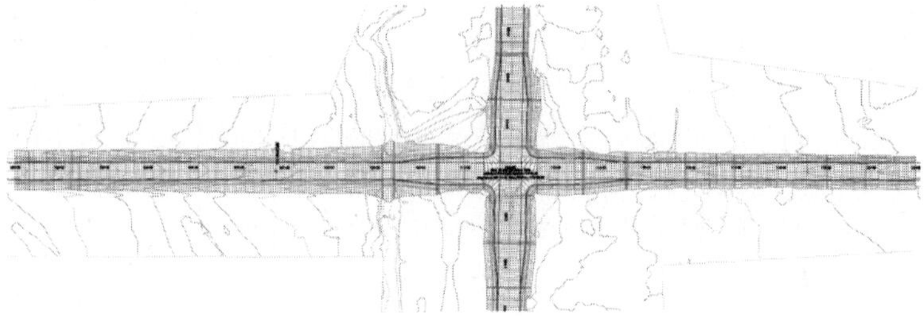

FIGURE 11.39 The completed intersection

Manually Modeling an Intersection

In some cases, you may find it necessary to model an intersection manually. Five-way intersections and intersections containing superelevated curves can't be created with the automated tools alone. You need to understand what the intersection tool is doing behind the scenes before you can be a true corridor guru. The next example will take you through an overview of the manual steps.

At the point where the example begins, a few pieces are already in place: the centerline alignments and profiles, EOP alignments and profiles, and the full road assemblies. The corridor for the intersection is started for you, but it contains only one baseline. For simplicity's sake, daylight subassemblies have been omitted from this example.

In this example, you'll add a baseline for an intersecting road to your corridor:

1. Open the `ManualIntersection.dwg` (`ManualIntersection_METRIC.dwg`) file, which you can download from this book's web page.
2. Select the corridor and from the Corridor contextual tab ➤ Modify Corridor panel, select Corridor Properties.
3. Switch to the Parameters tab and click Add Baseline.

   The Create Corridor Baseline dialog opens.
4. Pick the Syrah Way alignment, and then click OK to dismiss the dialog.
5. Click in the Profile field on the Parameters tab of the Corridor Properties dialog.

   The Select A Profile dialog opens.
6. Pick the Syrah Way FG profile, and then click OK to dismiss the dialog.

   You now have a new baseline, but the region still needs to be added.
7. In the Corridors Properties dialog, right-click BL – Syrah Way and select Add Region.
8. In the Create Corridor Region dialog, select Full Road Section, and then click OK to dismiss the dialog.
9. Expand BL – Syrah Way by clicking the small + sign, and see the new region you just created.
10. Click OK to dismiss the Corridor Properties dialog and select the option to rebuild the corridor.

    Once the corridor finishes rebuilding, it will look like Figure 11.40.

    Notice that the region for the second baseline extends all the way through the intersection. You must now split the Syrah region to accommodate the curb returns.
11. Open the Corridor Properties dialog and switch to the Parameters tab. Select the RG-Full Road Section under the BL – Syrah Way – (10) baseline (number may vary), and then right-click and select Split Region.
12. Enter **377.3′** (**115** m) and press ↵ to create the first region.

13. Enter **537.21′** (**163.74** m) and press ↵ to form the second split.

14. Press ↵ again to return to the Corridor Parameters dialog.

15. Switch the middle subassembly to Daylight Right and then click OK.

    Your corridor parameters will now match Figure 11.41.

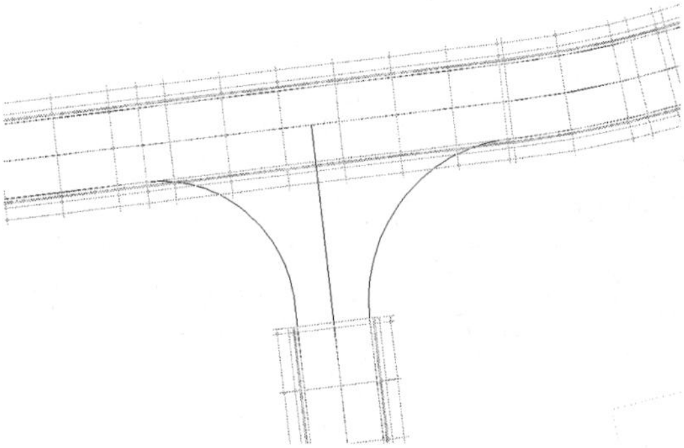

FIGURE 11.40
The Syrah Way baseline added

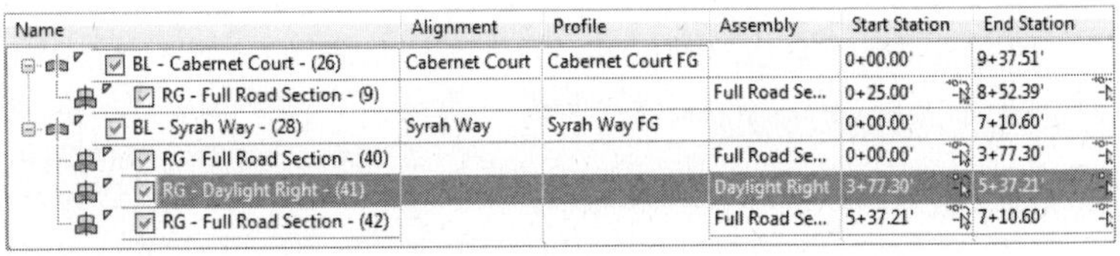

Name	Alignment	Profile	Assembly	Start Station	End Station
BL - Cabernet Court - (26)	Cabernet Court	Cabernet Court FG		0+00.00'	9+37.51'
RG - Full Road Section - (9)			Full Road Se...	0+25.00'	8+52.39'
BL - Syrah Way - (28)	Syrah Way	Syrah Way FG		0+00.00'	7+10.60'
RG - Full Road Section - (40)			Full Road Se...	0+00.00'	3+77.30'
RG - Daylight Right - (41)			Daylight Right	3+77.30'	5+37.21'
RG - Full Road Section - (42)			Full Road Se...	5+37.21'	7+10.60'

FIGURE 11.41
Split the Syrah Way baseline into three regions to accommodate curb returns.

16. Click OK and, as always, choose to rebuild the corridor.

    You will see that the way has been cleared for the curb assemblies to be applied, as shown in Figure 11.42.

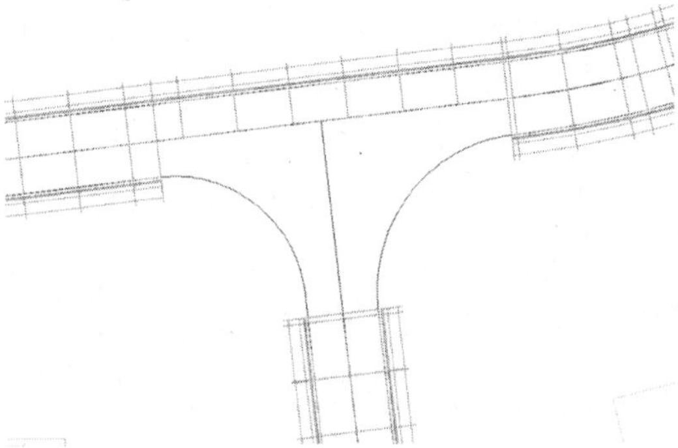

FIGURE 11.42
The intersection corridor...so far

17. Save the drawing for use in the next exercise.

Creating an Assembly for the Intersection

You built several assemblies in Chapter 9, "Subassembly Composer," but most of them were based on the paradigm of using the assembly marker along a centerline. This next exercise leads you through building an assembly that attaches at the EOP.

This exercise uses the LaneSuperelevationAOR subassembly (see Figure 11.43), but you're by no means limited to this subassembly in practice. You can use any lane subassembly that allows for a width and elevation target such as BasicLaneTransition, GenericPavementStructure, or LaneOutsideSuperWithWidening. Be sure the previous exercise is complete before proceeding.

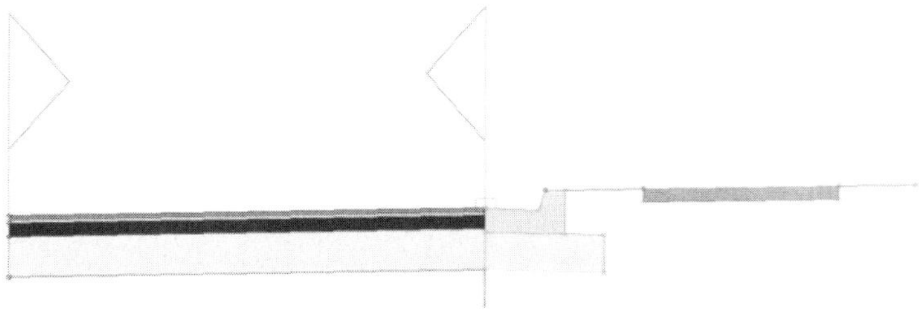

FIGURE 11.43
An assembly for an intersection curb return

To create an assembly suitable for use on the filleted alignments of the intersection, follow these steps:

1. Continue working in your drawing from the previous exercise (`ManualIntersection.dwg` or `ManualIntersection_METRIC.dwg`).
2. From Prospector, expand the Assemblies group and locate Curb Return Fillets.
3. Right-click on this assembly and select Zoom To.

   The assembly base has been placed in the drawing for you, but does not contain any subassemblies.
4. Open your tool palette, and switch to the Lanes palette.
5. Click the LaneSuperelevationAOR subassembly, and on the Properties palette, change the side to Left.
6. Click on the main baseline assembly to add the LaneSuperelevationAOR subassembly to the left side of the assembly.
7. Switch to the Curbs tab of the tool palette, and select UrbanCurbGutterGeneral.
8. Change the side to Right and leave all other values at their defaults.
9. Click anywhere on the baseline to add the curb and gutter to the right side of the assembly.
10. Also on the Curbs tab of the tool palette, add an UrbanSidewalk assembly.
11. Set the boulevard widths to **2′** (**0.5** m) for both inside and outside, and click to place the assembly at the back of the curb assembly.

You could combine steps 7–10 by using the Copy To Assembly option after selecting the curb and gutter along with the sidewalk from the Daylight Right Assembly.

Your assembly should now look like Figure 11.43.

Take the Pebble from My Hand, Grasshopper

If you are a little worried about the downward slope of the lane in the subassembly you just created, you needn't be. Remember that the slope and length of the lane will be controlled by the alignment and profile target in the corridor. The geometry that you see in the assembly is purely preliminary. If this concept makes sense to you, you are ready for bigger and better corridors!

And of course, if you'd rather see that slope go +2% in the subassembly, you can change it in the good old AutoCAD properties, but keep in mind it won't make one bit of difference to the corridor.

Adding Baselines, Regions, and Targets for the Intersections

The Intersection assembly attaches to alignments created along the EOP. Because a baseline requires both horizontal and vertical information, you also need to make sure that every EOP alignment has a corresponding finished ground (FG) profile.

It may seem awkward at first to create alignment and profiles for things like the EOP, but after some practice you'll start to see things differently. If you've been designing intersections using 3D polylines or feature lines, think of the profile as a vertical representation of a feature line and the profile grid view as the feature-line elevation editor. If you've designed intersections by setting points, think of alignment PIs and profile PVIs as points. If you need a low point midway through the EOP, as indicated in the sketch at the beginning of this section, you'll add a PVI with the appropriate elevation to the EOP FG profile.

To avoid getting confused by multiple targets, baselines, and regions, here are a few tips:

Stay Organized On screen, it is helpful to group similar objects. For example, many people find it helpful to group assemblies in the same area of the drawing. As you see in many of the examples in this book, the names of the assemblies can be added with a base AutoCAD MTEXT label.

Group profiles in a way that helps you identify with which alignment they are associated. For example, place your predominant alignment in the drawing with cross-street alignments placed below it in station order.

Naming Conventions If you let Civil 3D defaults have their way, you'll end up with objects that have generic names whose only distinguishing feature is the number indicating how many tries you've done before. Be sure to give your alignments, profiles, and assemblies names that can be easily identified. Rename the subassemblies to have more user-friendly names, as you learned in Chapter 10.

Working at the Region Level There are several buttons you can click to get into the Target Mapping dialog. You can set targets for the entire corridor, at the baseline level or the region level. The simplest place to go is the Targets button for the region you are working with. Use Set All Targets for setting the target surface, but you'll get lost in a long list of subassemblies if you use it for much else.

Using Multiple Targets

Civil 3D allows you to select more than one item in Width Or Offset Target and Slope Or Elevation Target areas. You can use any number of targets. You can even mix and match the types of targets you use, such as alignments with polylines, survey figures, and feature lines.

In the example that follows, you need two alignment targets and two profile targets in each curb region. To help Civil 3D figure out what you want, you need to verify that it will use the target that it finds *first* as it radiates out from the baseline. In most cases, you can click the Target To Nearest Offset option found in the target selection dialogs.

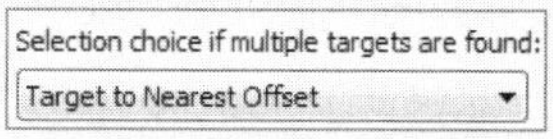

In the following graphic, the arrows represent the assemblies seeking out a target. From station 1+73.40 (0+052.85 for metric users) to 2+19.56 (0+066.92 for metric users), the Syrah Way alignment is closer to the west curb baseline; therefore, Syrah Way's geometry is used as the target. From station 2+19.56 (0+066.92 for metric users) to 2+76.07 (0+084.15 for metric users), Cabernet Court's alignment is closer and is used as the target.

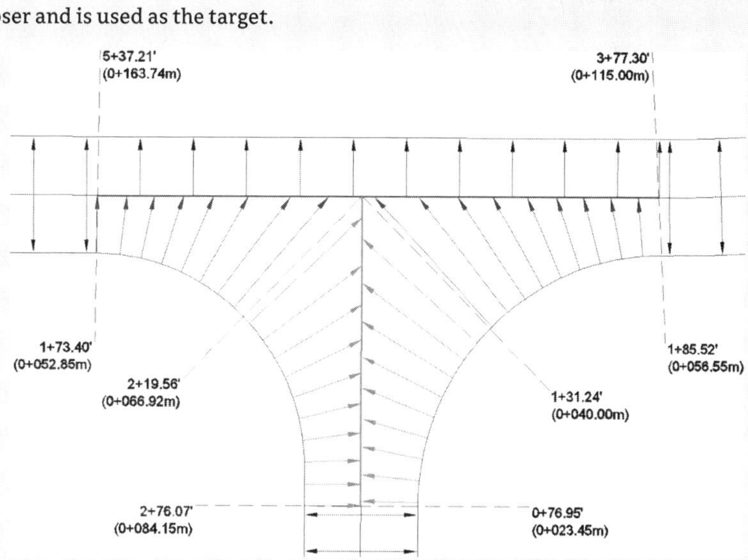

You'll finalize the corridor intersection in the next exercise. Along the way, you'll examine the corridor in various stages of completion so that you'll understand what each step accomplishes. In practice, you'll likely continue working until you build the entire model. Make sure you've completed the last two exercises, and then follow these steps:

1. Continue working in your drawing from the previous exercise (`ManualIntersection.dwg` or `ManualIntersection_METRIC.dwg`).
2. (Optional) Zoom to the area where the profile views are located. Notice that existing and proposed ground profiles are created for both EOP alignments (see Figure 11.44). The step of developing the design profiles has been completed for you.

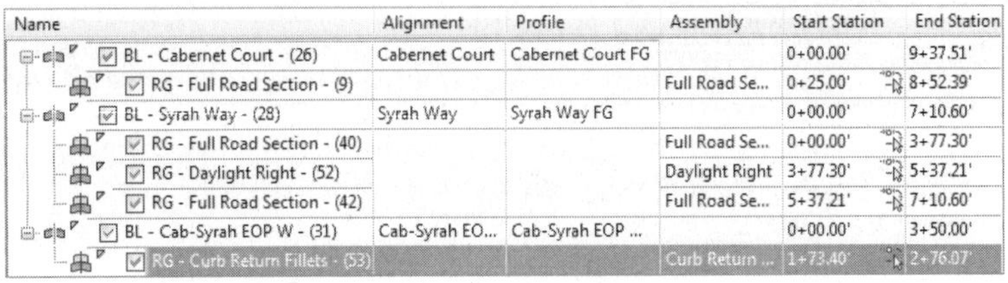

Name	Alignment	Profile	Assembly	Start Station	End Station
BL - Cabernet Court - (26)	Cabernet Court	Cabernet Court FG		0+00.00'	9+37.51'
RG - Full Road Section - (9)			Full Road Se...	0+25.00'	8+52.39'
BL - Syrah Way - (28)	Syrah Way	Syrah Way FG		0+00.00'	7+10.60'
RG - Full Road Section - (40)			Full Road Se...	0+00.00'	3+77.30'
RG - Daylight Right - (52)			Daylight Right	3+77.30'	5+37.21'
RG - Full Road Section - (42)			Full Road Se...	5+37.21'	7+10.60'
BL - Cab-Syrah EOP W - (31)	Cab-Syrah EO...	Cab-Syrah EOP ...		0+00.00'	3+50.00'
RG - Curb Return Fillets - (53)			Curb Return ...	1+73.40'	2+76.07'

FIGURE 11.44
The Parameters tab with a new baseline and region

3. Select the corridor, and then open the Corridor Properties dialog and switch to the Parameters tab.
4. Click Add Baseline.

   The Create Corridor Baseline dialog opens.
5. Select the Cab-Syrah EOP W alignment, and click OK.
6. Click in the Profile field on the Parameters tab of the Corridor Properties dialog where it says <Click here…>.
7. In the Select A Profile dialog, select Cab-Syrah EOP W - FG, and click OK.
8. Right-click baseline BL – Cab-Syrah EOP W and select Add Region. (Your numbering may differ from that shown in Figure 11.44.)
9. In the Create Corridor Region dialog, select Curb Return Fillets, and then click OK to dismiss the dialog.
10. Click the small plus sign to expand the baseline, and see the new region you just created.

    The Parameters tab of the Corridor Properties dialog will look like Figure 10.44.
11. Change the region start station to 1+73.40 (0+052.85 for metric users), and the region end station to 2+76.07 (0+084.15 for metric users).

    These stations represent the PC and PT stations of the curb alignment.
12. Click OK to dismiss the Corridor Properties dialog and rebuild your corridor.

    Your corridor should now look similar to Figure 11.45.

    What is missing from the corridor? You need to set targets, and add more frequency lines along the curves. Of course, you will repeat the process on the east side of the road as well.
13. Select the corridor and from the Corridor Contextual tab ➢ Modify Corridor, click Corridor Properties.
14. On the Parameters tab, set the frequency by clicking the ellipsis button in the Frequency column of the RG – Curb Return Fillets region row. Set both tangents and curves to 5′ (1 m). Click OK.

15. Click the ellipsis for targets of this region, and do the following:

  A. Click the Width Alignment target for LaneSuperelevationAOR. Click the Select From Drawing button. Click both Syrah Way and Cabernet Court centerline alignments.

  B. Press ↵ when complete.

  C. In the Set Width Or Offset Target dialog, click Add. Both alignments will appear in the listing, as shown in Figure 11.46a.

  D. Click OK when complete.

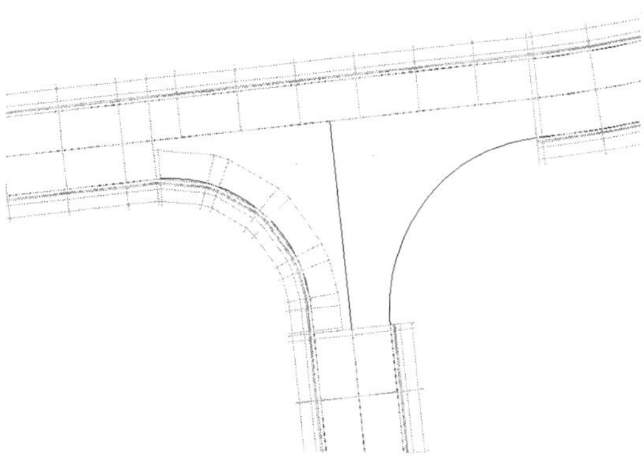

FIGURE 11.45
Your corridor after applying the Curb Return Fillets assembly

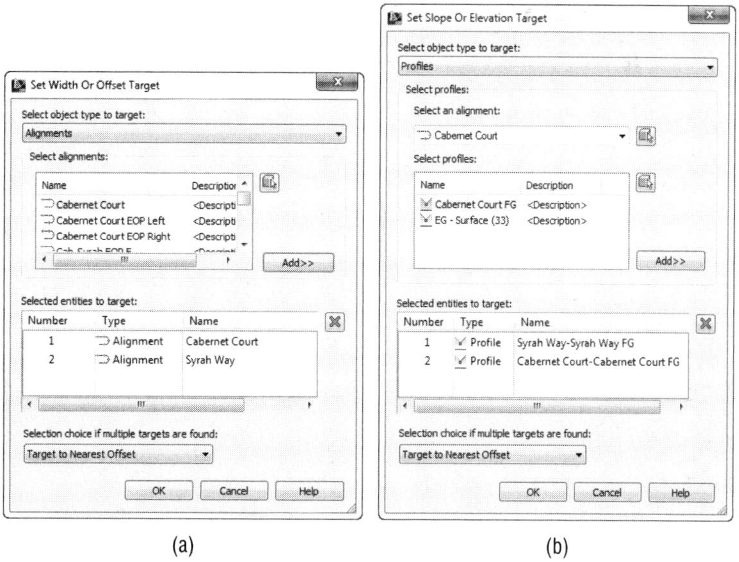

FIGURE 11.46
In the Set Width Or Offset Target dialog, use the Select From Drawing button to pick both Syrah Way and Cabernet Court centerlines (left). In the Set Slope Or Elevation Target dialog, pick alignment first, and then add the FG profile (right)

16. Click the Outside Elevation Profile target for LaneSuperelevationAOR, and do the following:

   A. From the Select an Alignment drop-down, select Syrah Way from the list.

   B. Highlight the Syrah Way FG profile and click Add.

   C. Change the selected alignment to Cabernet Court.

   D. Highlight the Cabernet Court FG profile and again click Add.

   Both profiles will appear in the listing, as shown in Figure 11.46b.

   E. Click OK.

17. Click OK to dismiss target mapping, and click OK to let the corridor rebuild.

   Your intersection should now look like Figure 11.47.

FIGURE 11.47
Three baselines completed — one to go

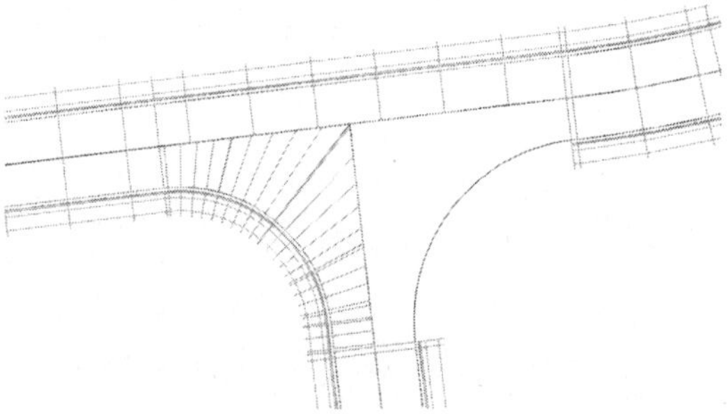

18. Repeat steps 4–17 for the Cab-Syrah EOP E.

   Here are some hints to get you going: The region should go from station 0+76.95 (0+023.45 for metric users) to 1+85.52 (0+056.54 for metric users). The frequency should be 5′ (1 m) for both tangents and curves. The target mapping will be identical to the region you created in steps 15 and 16.

19. Let the corridor rebuild one last time and admire your work.

The completed intersection should look like Figure 11.48.

FIGURE 11.48
The completed intersection

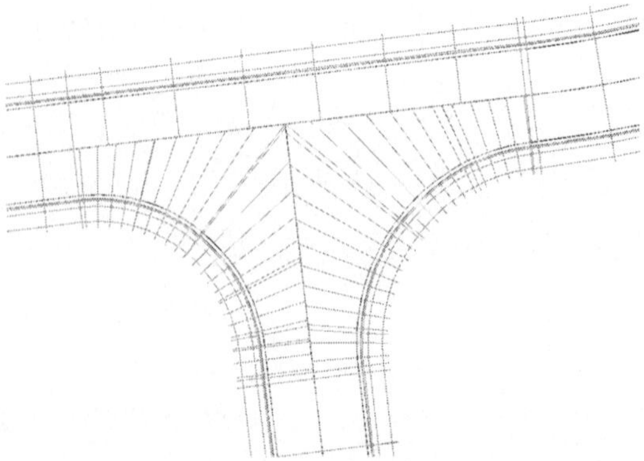

Troubleshooting Your Intersection

The best way to learn how to build advanced corridor components is to go ahead and build them, make mistakes, and try again. This section provides some guidelines on how to "read" your intersection to identify what steps you may have missed.

Your lanes appear to be backward. Occasionally, you may find that your lanes wind up on the wrong side of the EOP alignment, as in Figure 11.49. The most common cause is that your assembly is backwards from what is needed based on your alignment direction.

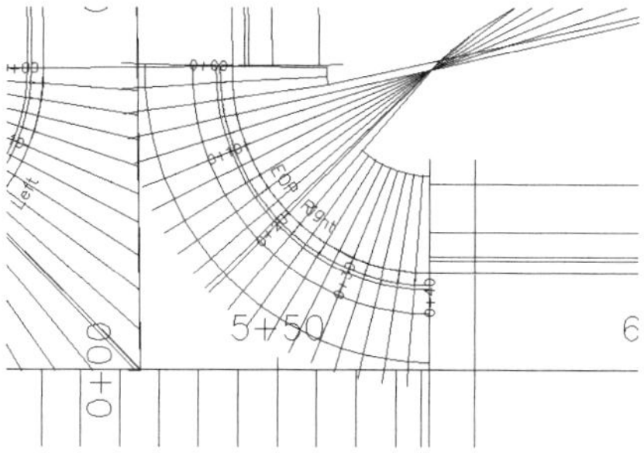

FIGURE 11.49
An intersection with the lanes modeled on the wrong side

Fix this problem by editing your subassembly to swap the lane to the other side of the assembly. If the assembly is used in another region that is correct, just make a new assembly that is the mirror reverse of the other assembly and apply the new one to the alignment.

Since so many design elements rely on the alignment as their base, it is better to add a new assembly rather than reversing the direction of the alignment.

Your intersection drops down to zero. A common problem when modeling corridors is the cliff effect, where a portion of your corridor drops down to zero. You probably won't notice in plan view, but if you rotate your corridor in 3D using the Object Viewer (see Figure 11.50), you'll see the problem. The most common cause for this phenomenon is incorrect region stationing.

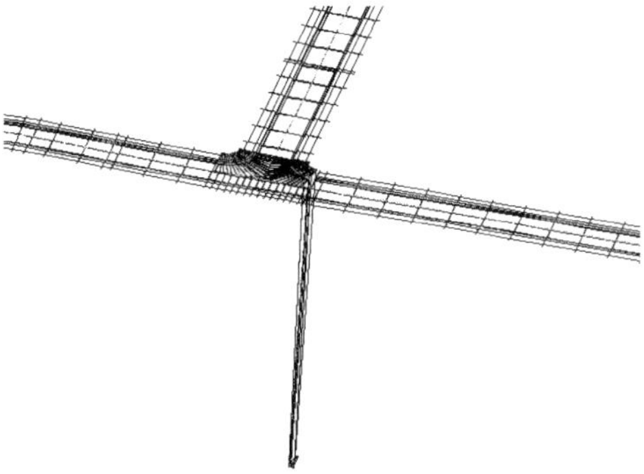

FIGURE 11.50
A corridor viewed in 3D, showing a drop down to zero

Fix this problem by making sure your baseline profile exists where you need it, and make note of the station range. Set your station range in the corridor to be within the correct range.

Your lanes extend too far in some directions. There are several variations on this problem, but they all appear similar to Figure 11.51. All or some of your lanes extend too far down a target alignment, or they may cross one another, and so on.

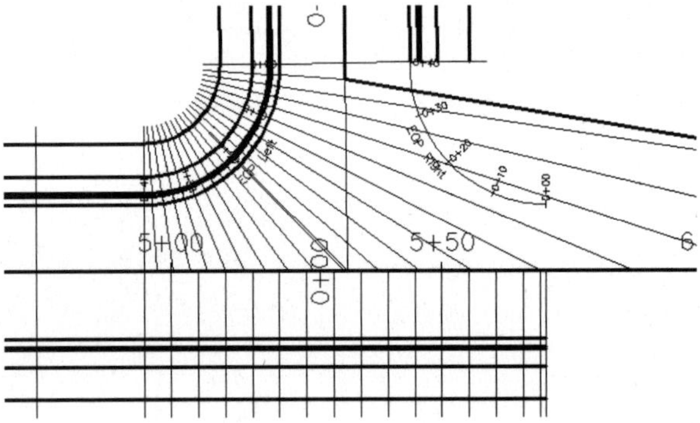

FIGURE 11.51 The intersection lanes extend too far down the main road alignment.

This occurs when a target alignment and profile have been omitted for one or more regions. In the case of Figure 11.51, the EOP Left baseline region was only set to one alignment. In an intersection, you need two targets in a corner region to model the road correctly.

Your lanes don't extend far enough. If your intersection or portions of your intersection look like Figure 11.52, you neglected to set the correct target alignment and profile.

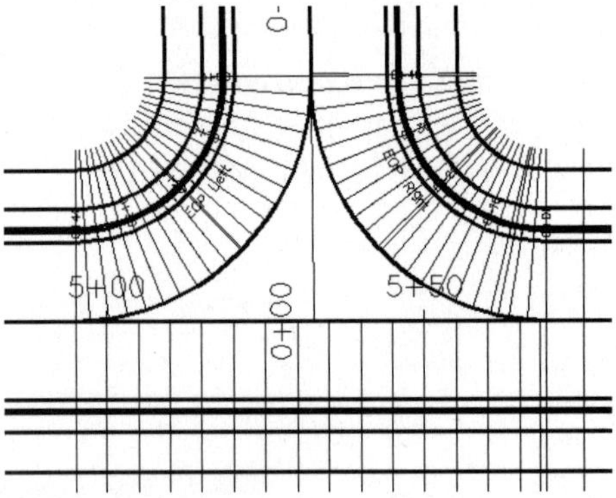

FIGURE 11.52 Intersection lanes don't extend out far enough.

You can fix this problem by opening the Target Mapping dialog for the appropriate regions and double-checking that you assigned targets to the right subassembly. It's also common to accidentally set the target for the wrong subassembly if you use Map All Targets, or if you have poor naming conventions for your subassemblies.

Checking and Fine-Tuning the Corridor Model

CERT OBJECTIVE

Recall that the EOP profiles are developed by creating a preliminary corridor surface model. To create the EOP elevations, you filled in the gaps by creating the EOP design profile where the preliminary surface stops and then picks back up again on the adjacent road (or in the case of the cul-de-sac, the opposite side of the street).

You will want to check the elevations of the corridor model against the profiles you created. The preliminary surface model was a best guess for elevations; now you need to reconcile your corridor geometry with those best-guess profiles. Figure 11.53 shows a common elevation problem that arises in the first iteration of intersection design.

FIGURE 11.53
A potential elevation problem in the intersection

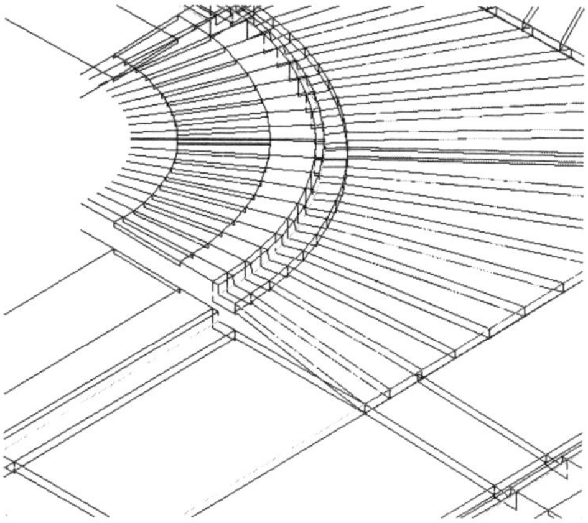

The first step in correcting this is creating a corridor surface model out of Top links. The surface will show you exactly how Civil 3D is interpreting your design, elevation-wise.

Before beginning this exercise, review the basic corridor surface building sections in Chapter 10. When you think you are ready, follow these steps:

1. Open the `ManualFineTune.dwg` (`ManualFineTune_METRIC.dwg`) file, which you can download from this book's web page.
2. Select the corridor in plan view, and from the Corridor Contextual tab ➢ Modify Corridor, select Corridor Properties.
3.  Click the Create A Corridor Surface button.

   A Corridor Surface entry appears.
4. Rename the surface to **Corridor Top** by clicking on the surface name.
5.  Ensure that Links is selected as Data Type and that Top appears under Specify Code, and then click the + button.

   An entry for Top appears under the Corridor Surface entry.
6. Switch to the Boundaries tab.
7. Right-click on Corridor Surface and select Corridor Extents as the outer boundary.
8. Click OK, and select rebuild.
9. Save the drawing for use in the next exercise.

   You should now see a surface model in addition to the corridor.

The next exercise will lead you through placing some design labels to assist in perfecting your model, and then show you how to easily edit your EOP FG profiles to match the design intent on the basis of the draft corridor surface built in the previous section.

Don't forget a few basics that will help you navigate this exercise easier. If you wish to see the drawing without the corridor, then freeze the corridor layer, which is C-ROAD-CORR. Don't forget to thaw this layer when you need it later! You can change the view of your surface around by changing the surface style. Also, remember that display order may be an issue. You will have corridor feature lines overlapping alignments, so it may be best to use the Send To Back option on the corridor.

1. Continue working in `ManualFineTune.dwg` (`ManualFineTune_METRIC.dwg`) from the previous exercise.
2. For the next few steps, either freeze the corridor layer or send your corridor display order to the back.
3. From the Annotate tab on the ribbon ➤ Labels & Tables panel, click the Add Labels button.

   The Add Labels dialog will appear.
4. Set Feature to Alignmentand Label Type to Station Offset.
5. Set Station Offset Label Style to Intersection Centerline Label, as shown in Figure 11.54.

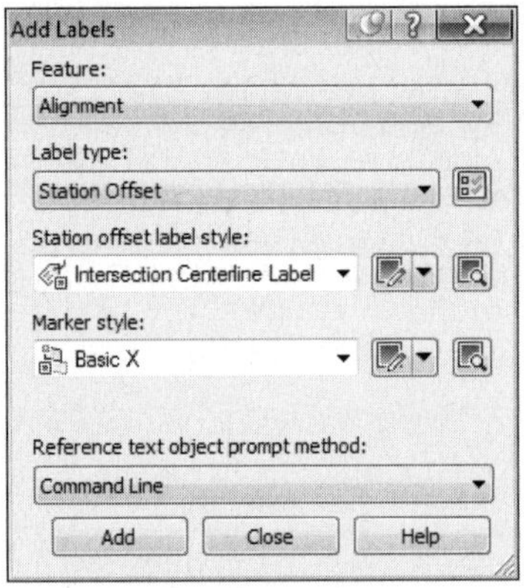

FIGURE 11.54
Add alignment labels to identify potential problem areas.

6. Set Marker Style to Basic X, and click Add.

   This label reference two alignments and two profiles.
7. The command line will read `Select Alignment:`. Click Syrah Way.
8. The command line will read `Specify station along alignment:`. Use object snaps to specify the intersection of Syrah Way and Cabernet Court.
9. At the `Specify station offset:` prompt, type **0** (zero) and press ↵.
10. At the `Select profile for label style component Profile 1:` prompt, right-click to bring up a list of profiles and choose Syrah Way FG.

11. Click OK to dismiss the dialog.
12. At the `Select alignment for label style component Alignment 2:` prompt, pick the Cabernet Court alignment.
13. At the `Select profile for label style component Profile 2:` prompt, right-click to open a list of profiles, and choose Cabernet Court FG.
14. Click OK to dismiss the profile listing, and press Esc to exit the labeling command.
15. Pick the label, and use the square-shaped grip to drag the label somewhere out of the way.

    Keep the Add Labels dialog open for the next part of the exercise.

    This label shows you that the crown elevations of the Cabernet Court FG and Syrah Way FG are both equal to 811.204′ (247.255 m), which is great! If they differed at all, you'd have some adjustments to make on the FG profiles where the roads meet.

    The next part of the exercise guides you through the process of adding a label to help determine what elevations should be assigned to the start and end stations of the EOP Right and EOP Left alignments.

16. In the Add Labels dialog, change the feature to Surface and Label Type to Spot Elevation, and click Add.

    This label compares proposed surface elevation and the corresponding profile elevation.

17. At the `Select a Surface <or press enter to select from list>:` prompt, pick any contour from the corridor surface.
18. At the `Select a Point:` prompt, use your Endpoint osnap to pick the PC station of the right curb alignment.
19. At the `Select surface for label style component Surface2:` prompt, press ↵.
20. Highlight Prelim, and click OK.
21. At the `Select alignment for label style component Alignment:` prompt, press ↵.
22. Highlight Cab-Syrah EOP W and click OK. Remain in the command.

    The question marks that initially came in with the label should now show the elevation differences between proposed and Prelim elevation at the station of interest.

23. Place another label at the PT station.

    Because you have already specified the surface and profile, the label will pop right in.

24. Press Esc and click Add in the Add Labels dialog.

    You need to stop and start the command to switch the alignment used in the label.

25. Using the same techniques you used on the west side of the road, add labels at the PC and PT stations of Cab-Syrah EOP E.
26. Select the labels. Click and drag the square grip and add leaders to make them easier to read.

Your labels should look like Figure 11.55.

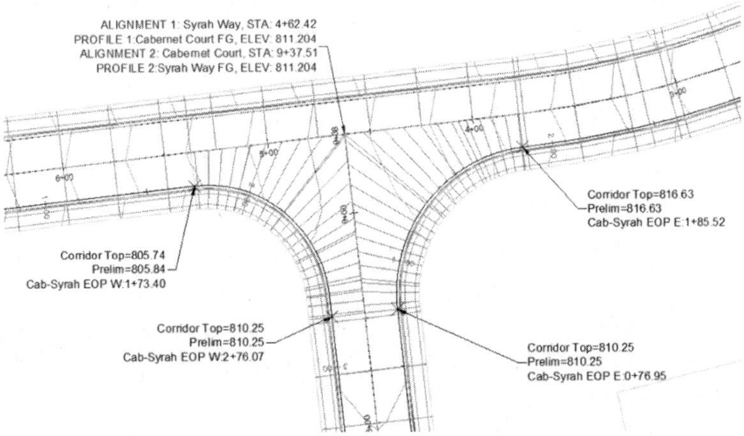

FIGURE 11.55
The corridor with all labels placed

27. Save the drawing for use in the next exercise.

As you can observe from the labels, the profiles and the surface model are perfect except for station 1+73.40 (0+052.85 for metric users) along Cab-Syrah EOP W. In fact, if the elevation difference doesn't bother you, you can skip the next steps. If you want your design to be as perfect as it can be, carry on:

1. Continue working in `ManualFineTune.dwg` (`ManualFineTune_METRIC.dwg`) from the previous exercise.

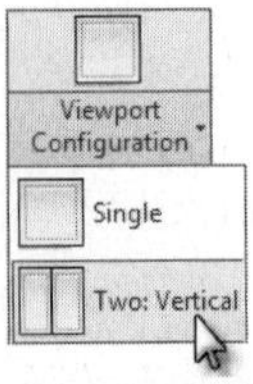

2. Split your model space into two views for easier working: Go to the View tab ➢ Model Viewports panel and select Viewport Configuration ➢ Two: Vertical.

   Pan in the drawing to show the profile view for CAB-Syrah EOP W. Your screen should look similar to Figure 11.56.

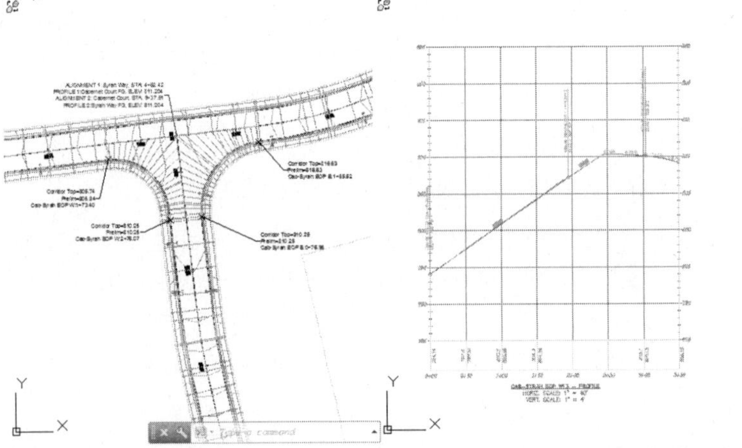

FIGURE 11.56
Use a split screen to see the plan and profile simultaneously.

3. Pick the Cab-Syrah EOP W – FG profile (the blue line in the profile view), and from the contextual Ribbon, click Geometry Editor.

4. On the Profile Layout Tools toolbar, click Insert PVIs-Tabular.

5. In the Insert PVIs dialog, set the vertical curve type to None.

6. Type **173.4′** (**52.85** m) for the station. (Station notation is not needed.) Type **804.84′** (**245.62** m) for the elevation. Click OK.

7. Select the corridor and choose Rebuild Corridor from the contextual tab.

   Your first curb return label will now report a perfect match between the Corridor Surface elevation and the FG Profile elevation (Figure 11.57).

FIGURE 11.57
Add PVI stations and elevations to match the desired FG elevations.

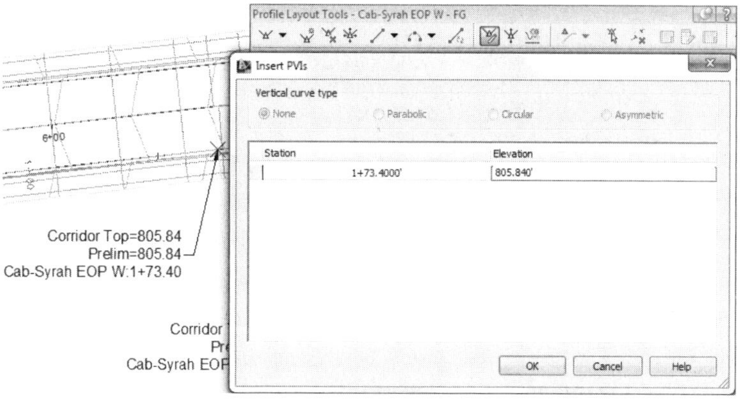

8. Pick the Corridor Top surface, and use the Object Viewer to study the TIN in the intersection area.

Study the contours in the intersection area. You may wish to add feature lines as breaklines to the surface model, as you learned in Chapter 10.

Real World Scenario

WHEN AUTOMATIC SURFACE BOUNDARIES ARE NOT AVAILABLE

You will come to a point in the corridor modeling where you need to add a corridor surface boundary, but the best options (Add Automatically ➤ Daylight or Corridor Extents As Outer Boundary) are not available. There will also be situations where Corridor Extents As Outer Boundary does not go where you want it to go. In those situations, you will need to use the Add Interactively tool.

Add Interactively allows you to direct Civil 3D to use the correct feature line as the boundary. You will trace your desired outer bounds with a temporary graphic called a *jig*.

1. Open the `Concord Commons Corridor.dwg` (`Concord Commons Corridor_METRIC.dwg`) file, which you can download from the book's web page.

You will see a surface in this file that needs some serious reining in.

2. Select the corridor, and click Corridor Properties on the contextual Ribbon. Select the Boundaries tab.
3. Right-click on the Corridor Topsurface and select Corridor Extents as the outer boundary. Click OK and rebuild the corridor.
4. Pan around the drawing.

   The contours are tidier, but the surface inside the loop formed by Frontenac Drive has not been cleaned up. In the next steps of the exercise, you will clean up the surface using the Add Interactively tool.

5. Set your AutoCAD object snaps to use only Endpoint and Nearest.

   These will be the most helpful osnaps to steer Civil 3D in the right direction.

6. In the Corridor Properties dialog ➢ Boundaries tab, right-click Concord Commons Corridor Top and select Add Interactively.
7. Start near station 0+00 (0+000 for metric users) on the inside of the loop and click the Sidewalk_Out feature line. Trace it with your cursor.
8. When the jig is going to the correct location, click to commit the boundary.

   The following image shows what the jig will look like:

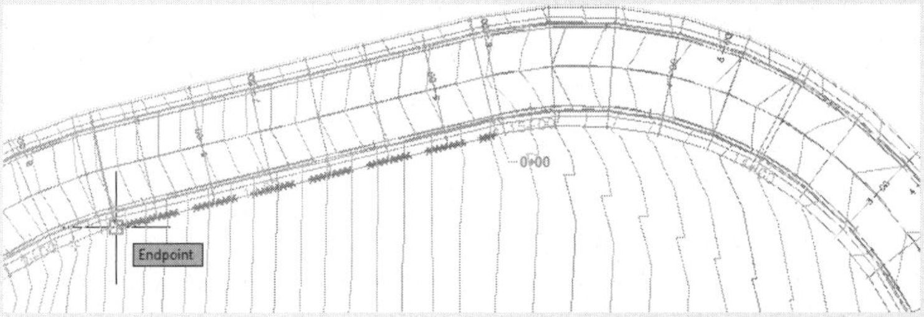

As you move around, you will see the jig tends to move unexpectedly at region boundaries. Steer the jig where you want it to go by clicking and tracing. If you made a mistake, type **U** for undo at the command line.

If you click a location where there is more than one feature line, or you are zoomed out to where your pickbox encompasses two lines, you will see a dialog like this one asking you to pick the line you are after:

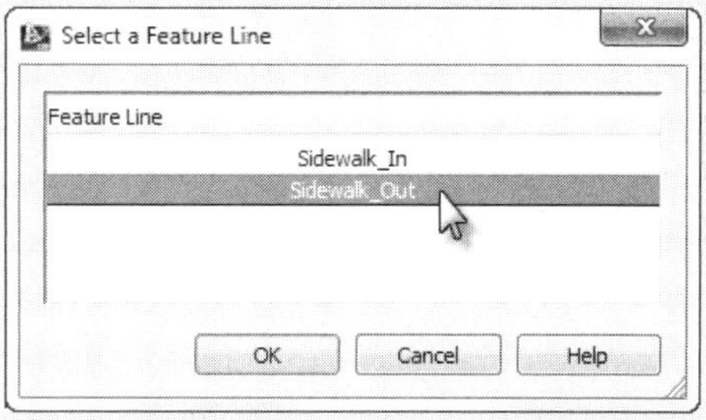

9. After you have finally come back around to where you started the boundary, type **C** (for close) at the command line, and press ↵.
10. When you complete the inner loop and return to the Corridor Properties dialog, set the Corridor Boundary (2) boundary type to Hide Boundary, as shown here. Click OK and rebuild the corridor.

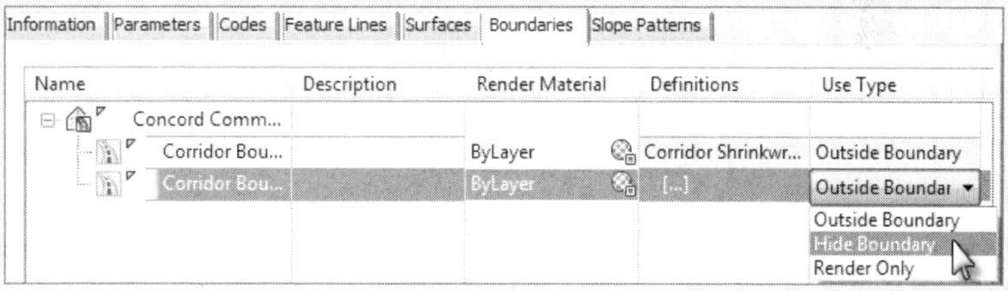

Yes, this process can be tedious. However, your work will pay off when changes are made to your design. These boundaries are dynamic to the design, unlike a polyline boundary.

Using an Assembly Offset

In Chapter 10, "Basic Corridors," you completed a road-widening example with a simple lane transition. Earlier in this chapter, you worked with a roadside-ditch transition, intersections, and cul-de-sacs. These are just a few of the techniques for adjusting your corridor to accommodate a widening, narrowing, interchange, or similar circumstances. There is no single method for building a corridor model; every method discussed so far can be combined in a variety of ways to build a model that reflects your design intent.

Another tool in your corridor-building arsenal is the assembly offset. In Chapter 10, you had your first glimpse of an offset assembly, but in the example that follows you will have a chance to use one for a bike path design.

Notice in Figure 11.58 how the frequency lines in the corridor are running perpendicular to the main alignment. The bike path is an alignment that is not a constant offset through the length of the corridor. In this scenario, the cross section of the bike path itself is skewed. This could prove problematic when computing end area volumes for the bike path pavement. This is the result of using an assembly where all of the design is based off one main baseline assembly.

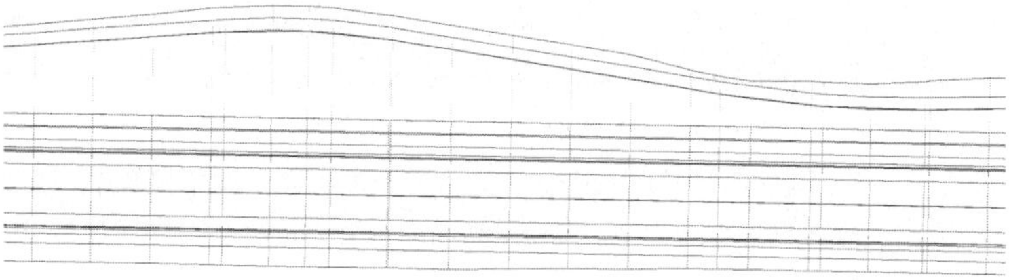

FIGURE 11.58
A bike path modeled with a traditional assembly

There are several advantages to using an offset assembly instead. The offset assembly requires its own alignment and profile for design. In the corridor that results, a secondary set of frequency lines is generated perpendicular to the offset alignment, as shown in Figure 11.59. Additionally, you can use a marked point assembly to model the ditch between the bike path and the main road.

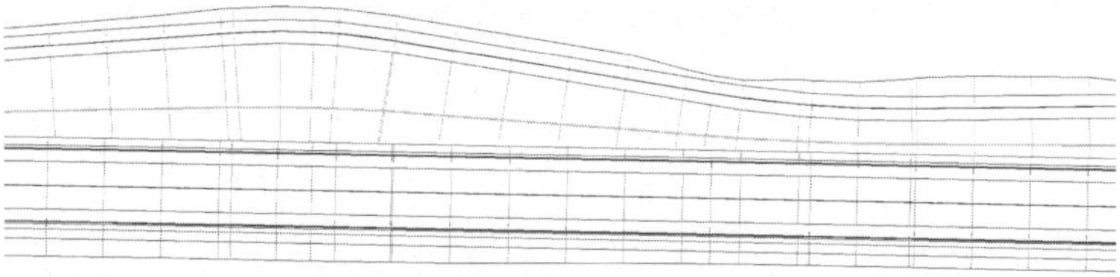

FIGURE 11.59
Modeling a bike path with an assembly offset

There are many uses of offset assemblies besides bike paths. Typical examples of when you'll use an assembly offset include transitioning ditches, divided highways, and interchanges. The assembly in Figure 11.60, for example, includes two assembly offsets. The assembly could be used for transitioning roadside swales, similar to the first exercise in this chapter.

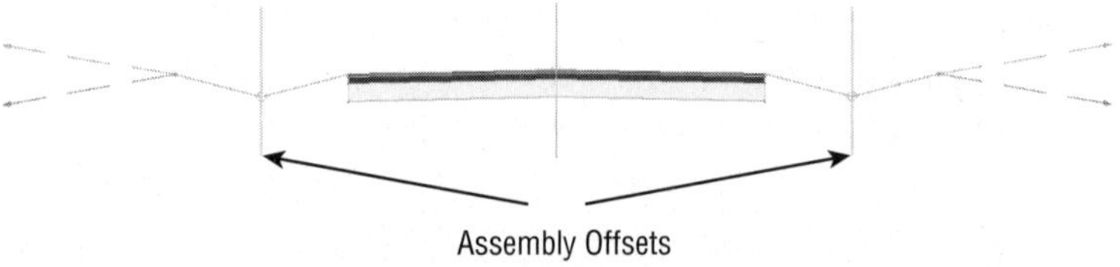

FIGURE 11.60
An assembly with two offsets representing roadside swale centerlines

When you use an assembly using an offset in your corridor, you must assign an alignment and profile to it. The only restriction to the offset assembly is that it can't use the same alignment or profile as the main part of the assembly. In the case where you want your offsets to follow the same elevation, you will need to use the Superimposed Profile tool to effectively make a copy of the desired profile.

In this exercise, you will model a bike path with an assembly offset:

1. Open the `HWY10-BikePath.dwg` (`HWY10-BikePath_METRIC.dwg`) file, which you can download from this book's web page.
2. Zoom to the area of the drawing where the assemblies are located.

You'll see an incomplete assembly called Road With GR And Bikepath.

3. Click on the main assembly marker, and then from the contextual Ribbon tab, click Add Offset in the Modify Assembly panel.
4. At the `Specify offset location:` prompt, click to the left of the Road With GR And Bikepath assembly, leaving enough room for the bike lane and ditch.

   Your result should look like Figure 11.61. The warning symbols indicate that you can no longer use the assembly for roads where superelevation occurs at points other than the centerline. Superelevation at the crown will still work, which is the situation used here.

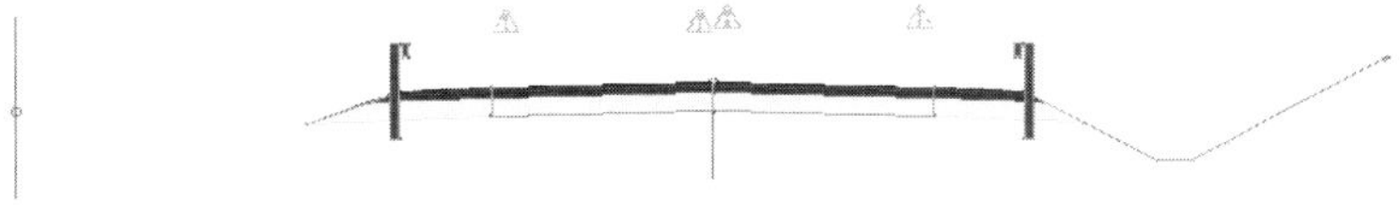

FIGURE 11.61
Road With GR And Bikepath assembly, so far

5. Open the Civil Imperial Subassembly tool palette.
6. Switch to the Basic tab, and select the BasicLane subassembly.
7. In the Advanced Parameters in the Properties dialog, set Side to Right and Width to **5′** (**1** m).
8. Click to place the subassembly on the offset assembly, and click the offset assembly again to form the left side.

   Your assembly should now look like Figure 11.62.

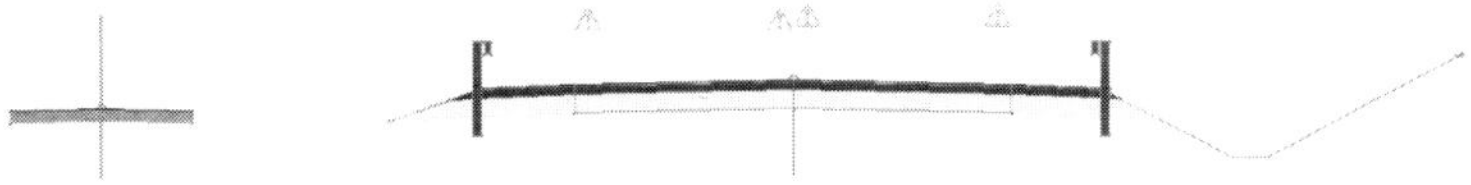

FIGURE 11.62
Road With GR And Bikepath assembly with the BasicLane as a bike path

Next, you will use a MarkPoint assembly to set the stage for building a ditch between the bike path and the main road.

9. Switch to the Generic tab in the subassemblies tool palette, and click the MarkPoint subassembly.
10. Change the Point Name to **BIKE** (use all capital letters).
11. Click on the outermost point of the left shoulder subassembly.

    Your marker will look like Figure 11.63.

FIGURE 11.63
A close-up of the MarkPoint subassembly

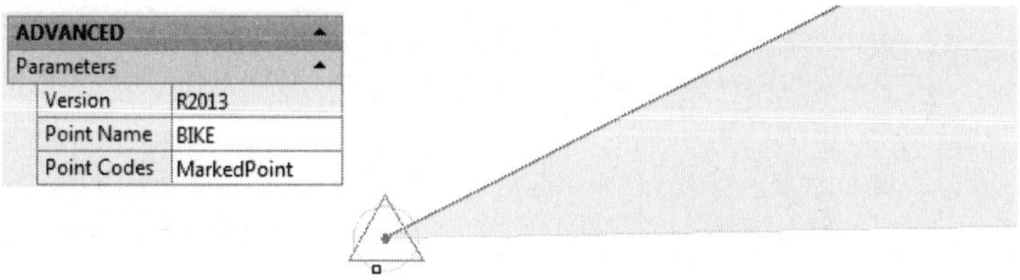

12. From the Generic tab of the Subassemblies tool palette, click the LinkSlopesBetweenPoints subassembly, and do the following:

    A. Set Marked Point Name to **BIKE** (again, use all capital letters).

    B. Set Ditch Width to **0.5′** (**0.15** m).

    C. Click on the right side of the bike path.

    The offset assembly will now look like Figure 11.64.

FIGURE 11.64
The LinkSlopesBetweenPoints subassembly in layout mode

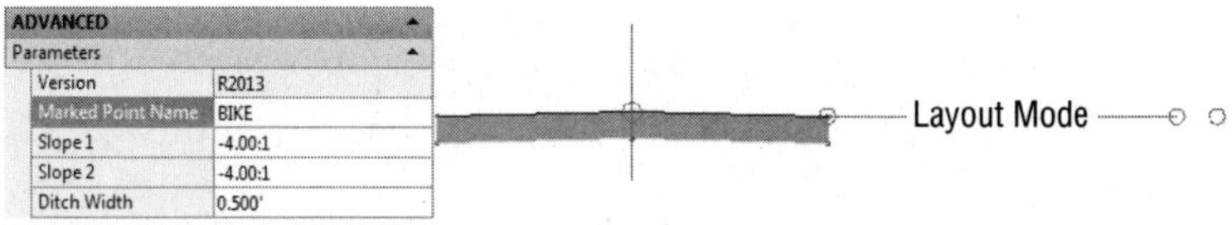

13. Add a LinkSlopeToSurface generic link subassembly, and do the following:

    A. Set the Side to Left and Slope to 25%.

    B. Place the subassembly on the left side of the bike path. Press Esc to complete adding subassemblies and save the drawing.

    The completed assembly will look like Figure 11.65.

FIGURE 11.65
The completed assembly with offset

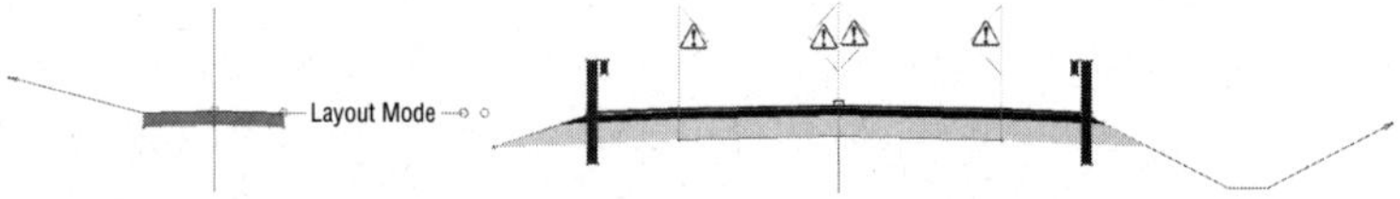

Next, you will create a corridor using this new assembly. You need to have completed the previous exercise before proceeding.

1. Continue working in your drawing from the previous exercise.

2. From the Home tab ➤ Create Design Panel, click Corridor and do the following:

    A. Name the corridor **Bike Path**.

    B. Set Alignment to USH 10.

C. Set Profile to USH 10 Roadway CL Prof.

D. Set Assembly to Road W GR And Bikepath.

E. Set Target Surface to Existing Intersection.

F. Verify that there is a check next to Set Baseline And Region Parameters.

G. Click OK.

In the Baseline And Region Parameters dialog, notice the Offset - (1) (your numbers may vary) is not associated with an alignment.

3. Click the alignment field for Offset - (1), select Bike Path from the drop-down list, and click OK.

4. Click the profile field for Offset - (1).

A. From the Select An Alignment drop-down list, select Bike Path.

B. From the Select A Profile drop-down list select Bike Path FG, and click OK.

The Bike Path alignment is slightly shorter than the main USH 10 alignment, which would cause the "waterfall" effect explained in Chapter 10.

5. To prevent corridor errors, do the following:

A. Set the start station for the Offset - (1) region to **0+25.00** (**0+010.0** for metric users).

B. Set the end station to **40+00** (**1+220** for metric users).

C. Click OK and rebuild the corridor.

Your completed corridor will resemble the example shown earlier in Figure 11.59.

6. Select the corridor by clicking on one of the frequency lines anywhere near the middle of the alignment, and click Section Editor.

You may want to change your annotation scale to 1″ = 1′ (1:1 for metric users) to get an unobstructed view of your masterpiece.

7. Explore the finished design by clicking through the Corridor Editor.

At each station, the offset assembly ties back to the main assembly because of the use of the LinkSlopeBetweenPoints subassembly. Your design in the Section Editor should resemble Figure 11.66.

FIGURE 11.66
Inside the corridor Section Editor

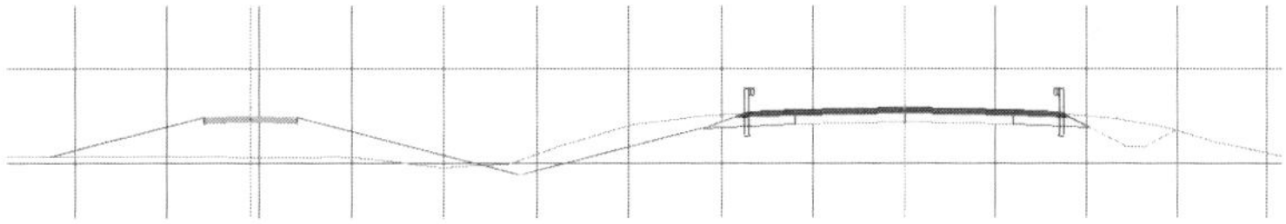

Don't forget to close the Section Editor before closing the drawing!

The Trouble with Bowties

In your adventures with corridors, chances are pretty good that you'll create an overlapping link or two. These overlapping links are known not-so-affectionately as *bowties*. Here's an example:

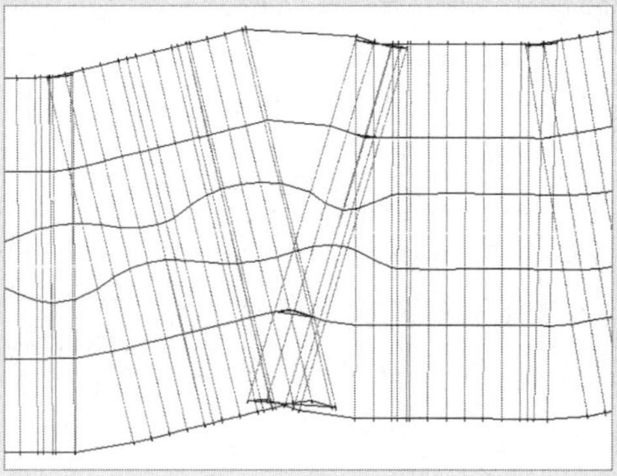

Bowties are problematic for several reasons. In essence, the corridor model has created two or more points at the same *x* and *y* locations with a different *z*, making it difficult to build surfaces, extract feature lines, create a boundary, and apply codeset styles that render or hatch.

When your corridor surface is created, the TIN has to make some assumptions about crossing breaklines that can lead to strange triangulation and incoherent contours, such as in the following graphic:

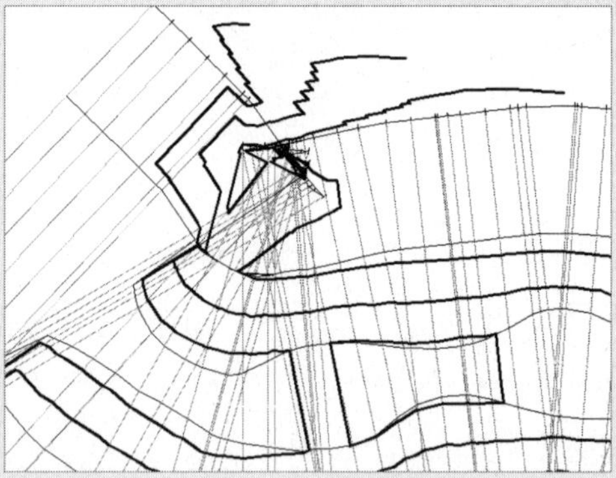

When you create a corridor that produces bowties, the corridor won't behave as expected. Choosing Corridors ➤ Utilities to extract polylines or feature lines from overlapping corridor areas yields an entity that is difficult to use for additional grading or manipulation because of extraneous, overlapping, and invalid vertices. If the corridor contains many overlaps, you may have trouble even executing the extraction tools. The same concept applies to extracted alignments, profiles, and COGO points.

If you try to add an automatic or interactive boundary to your corridor surface, either you'll get an error or the boundary jig will stop following the feature line altogether, making it impossible to create an interactive boundary.

To prevent these problems, the best plan is to try to avoid link overlap. Be sure your baseline, offset, and target alignments don't have redundant or PI locations that are spaced excessively close.

If you initially build a corridor with simple transitions that produce a lot of overlap, try using an assembly offset and an alignment besides your centerline as a baseline. Another technique is to split your assembly into several smaller assemblies and to use your target assemblies as baselines, similar to using an assembly offset. This method was used to improve the river corridor shown in the previous graphics. The following graphic shows the two assemblies that were created to attach at the top of bank alignments instead of the river centerline:

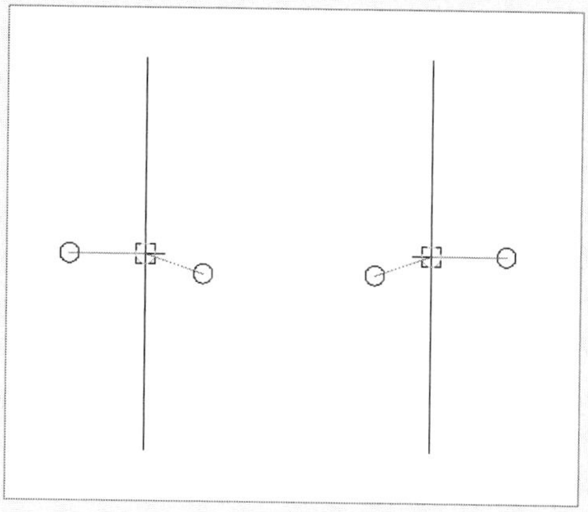

The resulting corridor is shown here:

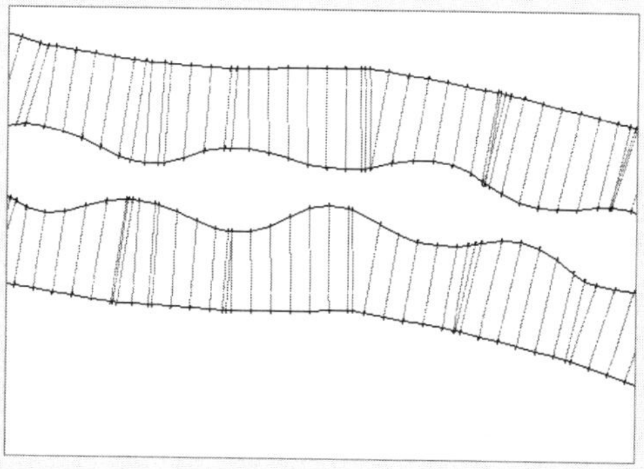

The TIN connected the points across the flat bottom and modeled the corridor perfectly, as you can see in the following image:

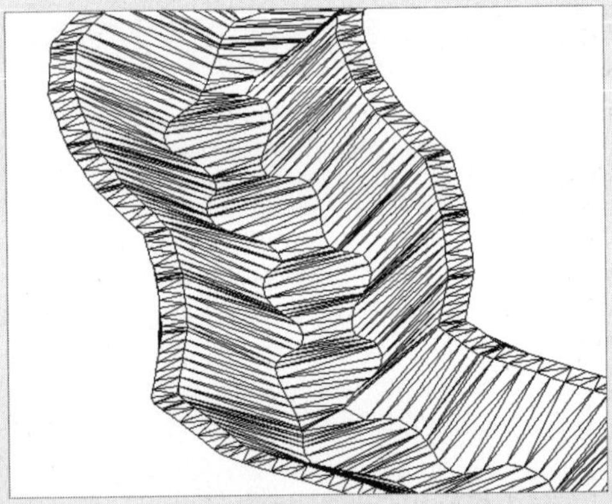

Another method for eliminating bowties is to notice the area where they seem to occur and then adjust the regions. If your daylight links are overlapping, perhaps you can create an assembly that doesn't include daylighting and create a region to apply that new assembly.

If overlap can't be avoided in your corridor, don't panic. If your overlaps are minimal, you should still be able to extract a polyline or feature line — just be sure to weed out vertices and clean up the extracted entity before using it for projection grading. You can create a boundary for your corridor surface by drawing a regular polyline around your corridor and adding it as a boundary to the corridor surface under the Surfaces branch in Prospector. The surface-editing tools, such as Swap Edge, Delete Line, and Delete Point, can also prove useful for the final cleanup and contour improvement of your final corridor surface.

As you gain more experience building corridors, you'll be able to prevent or fix most overlap situations, and you'll also gain an understanding of when they aren't having a detrimental effect on the quality of your corridor model and resulting surface.

Using a Feature Line as a Width and Elevation Target

You've gained some hands-on experience using alignments and profiles as targets for swale, intersection, and cul-de-sac design. Civil 3D adds options for corridor targets beyond alignments and profiles. You can use grading feature lines, survey figures, or polylines to drive horizontal and/or vertical aspects of your corridor model.

Dynamic Feature Lines Cannot Be Used as Targets

It's important to note that dynamic feature lines extracted using the Feature Lines From Corridor tool can't be used as targets. The possibility of circular references would be too difficult for the program to anticipate and resolve.

Imagine using an existing polyline that represents a curb for your lane-widening projects without duplicating it as an alignment, or grabbing a survey figure to assist with modeling an existing road for a rehabilitation project. The next exercise will lead you through an example where a lot-grading feature line is integrated with a corridor model:

1. Open the `Feature Line Target.dwg` (`Feature Line Target_METRIC.dwg`) file, which you can download from this book's web page.

   This drawing includes a completed assembly and a partially completed corridor. You task will be to use the yellow feature lines that run through the project as targets in the corridor.

2. Zoom to the corridor and select it. From the Corridor contextual tab ➢ Modify Corridor panel, choose Corridor Properties.
3. Switch to the Parameters tab in the Corridor Properties dialog.
4. Click the ellipsis button in the Targets field in the RG-Subdivision region. Note that the number following the region will vary. The higher the number, the more previous attempts at building the corridor the authors have made before handing the drawing over to you.

   The Target Mapping dialog appears.

5. Click the <None> field next to Target Alignment for the Slope - Left subassembly.

   The Set Width Or Offset Target dialog appears.

6. Choose Feature Lines, Survey Figures And Polylines from the Select Object Type To Target drop-down list (see Figure 11.67).

FIGURE 11.67 The Select Object Type To Target drop-down list as seen in both the Set Width Or Offset Target and Set Slope Or Elevation Target dialogs

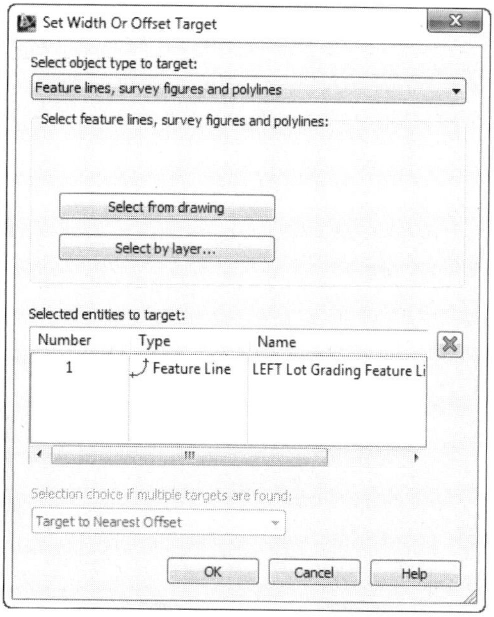

7. Click the Select From Drawing button, and when you see the command-line prompt `Select feature lines, survey figures or polylines to target:`, select the yellow feature line to the north of the alignmentand then press ↵.

   The Set Width Or Offset Target dialog reappears, with an entry in the Selected Entries To Target area.

8. Click OK to return to the Target Mapping dialog.

   If you stopped at this point, the horizontal location of the feature line would guide the Slope-Left subassembly, and the vertical information would be driven by the slope set in the subassembly properties. Although this has its applications, most of the time you'll want the feature line elevations to direct the vertical information. The next few steps will teach you how to dynamically apply the vertical information from the feature line to the corridor model.

9. In the Slope or Elevation Targets section, click the <None> field next to Target Profile for the Slope-Left subassembly.

   The Set Slope Or Elevation Target dialog appears.

10. Make sure Feature Lines, Survey Figures And Polylines is selected in the Select Object Type To Target drop-down.

11. Click the Select From Drawing button, and when you see the command-line prompt `Select feature lines, survey figures or 3D polylines to target:`, select the yellow feature line to the north again and then press ↵.

    The Set Slope Or Elevation Target dialog reappears, with an entry in the Selected Entries To Target area.

12. Click OK to return to the Target Mapping dialog.

13. Click OK to return to the Corridor Properties dialog.

14. Repeat steps 3–10 for Slope - Right on the right side of the corridor using the feature line to the south of the alignment.

15. Click OK to exit the Corridor Properties dialog and choose rebuild corridor.

    The corridor will rebuild to reflect the new target information and should look similar to Figure 11.68.

FIGURE 11.68
The corridor now uses the grading feature lines as width and elevation targets.

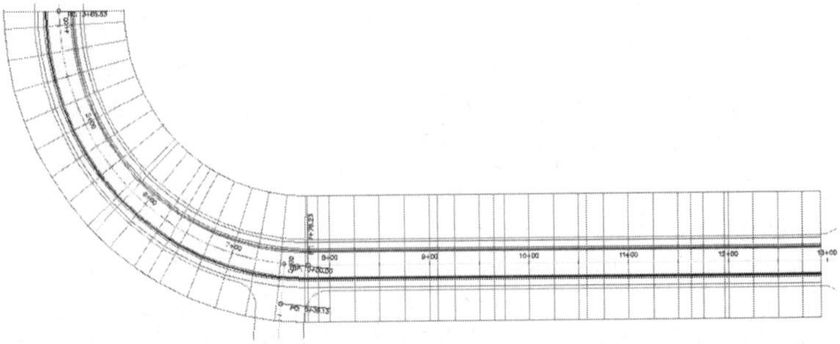

Once you've linked the corridor to this feature line, any edits to this feature line will be incorporated into the corridor model. You can establish this feature line at the beginning of the project and then make horizontal edits and elevation changes to perfect your design. The next few steps will lead you through making some changes to this feature line and then rebuilding the corridor to see the adjustments.

16. Select the corridor, right-click, and choose Display Order ➢ Send To Back.

    Doing so sends the corridor model behind the target feature line.

17. Select one of the feature lines so that you can see its grips. Experiment with the feature lines by moving several grips.

18. Select the corridor, and then right-click and choose Rebuild Corridor.

    The corridor will rebuild to reflect the changes to the target feature lines.

Edits to targets — whether they're feature lines, alignments, profiles, or other Civil 3D objects — drive changes to the corridor model, which in turn drives changes to any corridor surfaces, sections, section views, associated labels, and other objects that are dependent on the corridor model.

Roundabouts: The Mount Everest of Corridors

If you really understand what went on earlier in this chapter, you are almost ready for roundabout design. You may want to wait to tackle your first roundabout until after reading Chapter 16, "Grading."

The same concepts apply to a roundabout as for a standard road junction, but you will have several more regions, baselines, and corresponding profiles.

This section will help you prepare files for roundabout design. This section will not take you through every detail of corridor creation, but once you master the topics of intersection design, a roundabout is an extension of the same concepts.

A roundabout is best done in several corridors:

Preliminary Corridor for Circulatory Road This is a corridor used to determine the elevations of the approach road profiles. The circular portion of the roadway will set the elevations for all of the alignments leading into it. Using similar techniques as earlier in the chapter, you create a corridor surface from this corridor and use it as a tie-in for approach road profiles.

Main Corridor with Approaches and Circulatory Road This corridor is the main part of your design. You will spend lots of time in the corridor properties tweaking stations, adding baselines and regions and targeting the appropriate locations.

Curb Island Corridors (Optional) There are many different philosophies about the best way to show curb islands. If showing them in a section is not important, you can omit this altogether. Some people prefer to use grading feature lines (as discussed in Chapter 16). In this section, you will go "all out" and use the dynamic capabilities of the corridor model to make curb islands.

Drainage First

Based on your existing ground surface, determine the general direction that you want water to flow away from the center of the roundabout. Use grading tools and a feature line to create the general drainage direction of the roundabout.

Chapter 16 will go into much more depth on creating grading. You will certainly want to have an understanding of grading basics before you tackle a roundabout.

Create a feature line that represents the highest elevations. In the example shown in Figure 11.69, the feature line slopes downward and acts as a ridge to separate water flow. The grading tools are then used to create grading objects and a corresponding surface model called "Roundabout Grading."

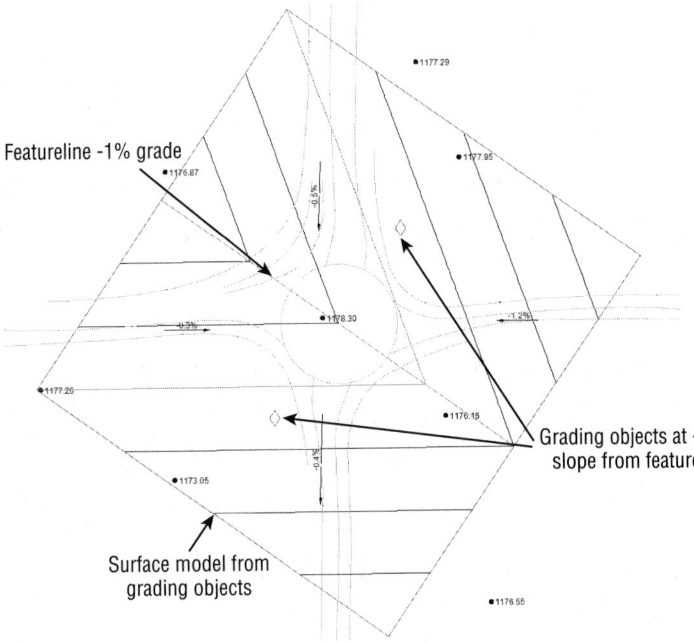

FIGURE 11.69 Feature lines and grading create a preliminary surface to ensure proper drainage through the roundabout.

The Roundabout Grading surface will be the basis for our profile elevations through the rest of the design process.

Roundabout Alignments

Roundabouts need alignments to guide the design for the same reasons that an intersection needs them. Alignments will be baselines and targets for the approaches and rotary. Create alignments manually with the tools you learned earlier in this chapter, or start with the handy roundabout layout tool.

The roundabout layout tools create horizontal data based on the location of the center of the roundabout and the approach alignments.

In the exercise that follows, you will create the Civil 3D alignments needed to create a roundabout:

1. Open the `Roundabout Layout.dwg` (`Roundabout Layout_METRIC.dwg`) file, which you can download from this book's web page.
2. From the Home tab ➤ Create Design panel, select Intersections ➤ Create Roundabout, as shown in Figure 11.70.

FIGURE 11.70
Access the Create Roundabout tool from the Home tab.

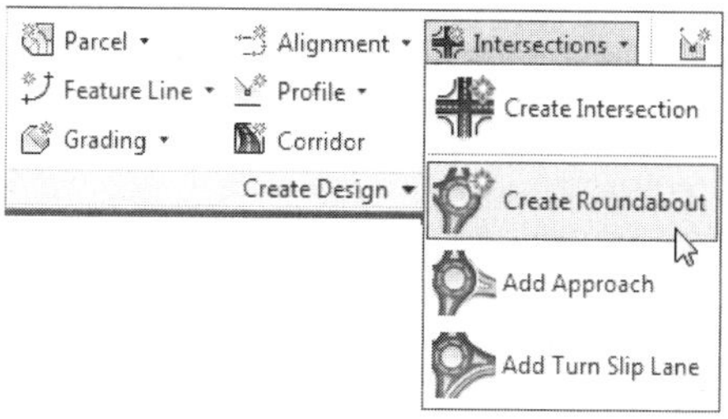

3. At the `Specify roundabout centerpoint:` prompt, use your Intersection osnap to select the point where the approaches meet.
4. At the `Select approach road:` prompt, select all four alignments leading into the roundabout, and press ↵ when you are done.

   You now see the Create Roundabout – Circulatory Road screen, shown in Figure 11.71.

FIGURE 11.71
The first roundabout layout screen for designing the main circulatory road

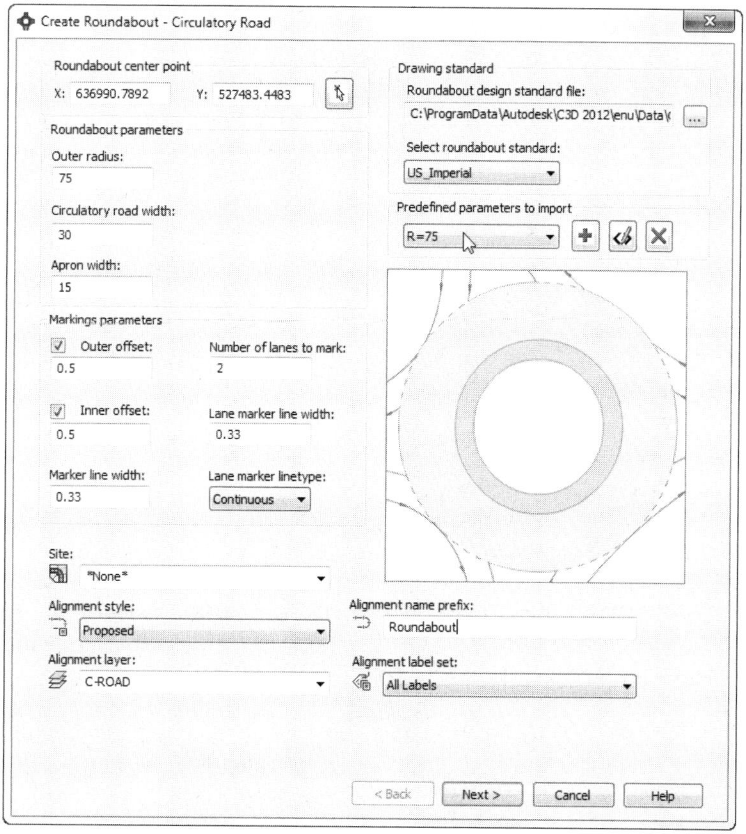

5. Verify that you are using the correct standards for your units. Click on the ellipsis next to the Roundabout Design Standard File field, and verify that you are using the correct XML standards file.

   A. If it is set correctly, click Cancel and continue to step 6.

   B. If it is not set correctly, browse to the `C:\ProgramData\Autodesk\C3D 2013\enu\Data\Corridor Design Standards\` folder. Browse to either the Imperial or metric units folder and select the correct roundabout presets XML file. Click Open to return to the Create Roundabout dialog.

6. From the Predefined Parameters To Import drop-down, choose R=75 (Rg=25m).

   This will be the radius from the center of the roundabout to the outermost circular edge of pavement.

7. Set Alignment Layer to C-ROAD and Alignment Label Set to Major And Minor Only. Click Next.

   Now, you'll design the approach road exit and entry geometry. The options in the Create Roundabout – Approach Roads screen (see Figure 11.72) can be set independently for each approach, or you can click Apply To All, which will set the geometry for all four approaches.

FIGURE 11.72
Approach road widths at entry and exit

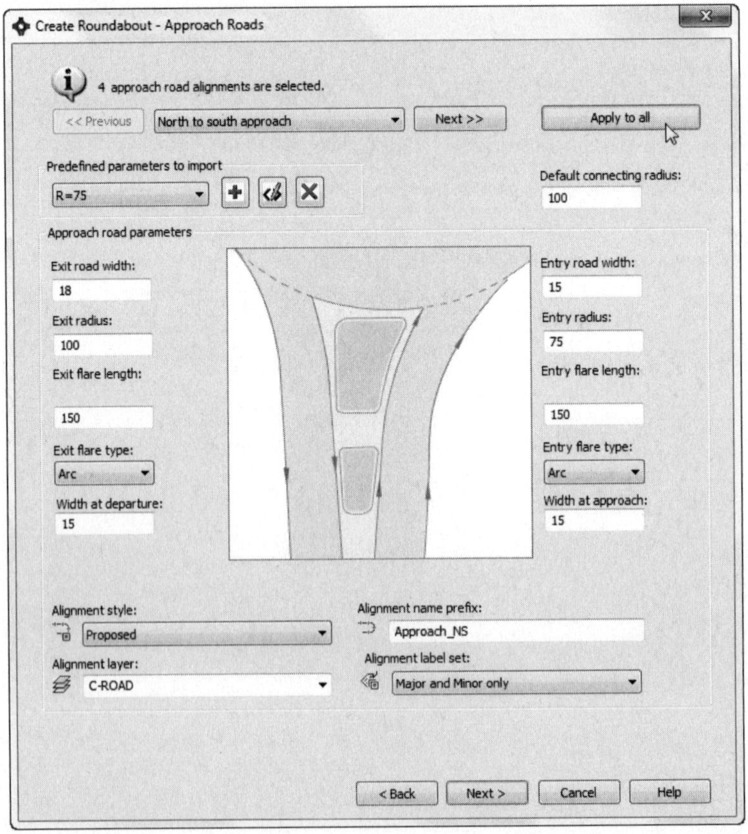

8. Set Predefined Parameters To Import to R=75 (Rg=25m). For Alignment Label Set, choose Major And Minor Only. Leave all other settings at their defaults.

9. Click the Apply To All button, and click Next.

10. In the Create Roundabout – Islands screen (see Figure 11.73), again set Predefined Parameters To Import to R=75 (Rg=25m).

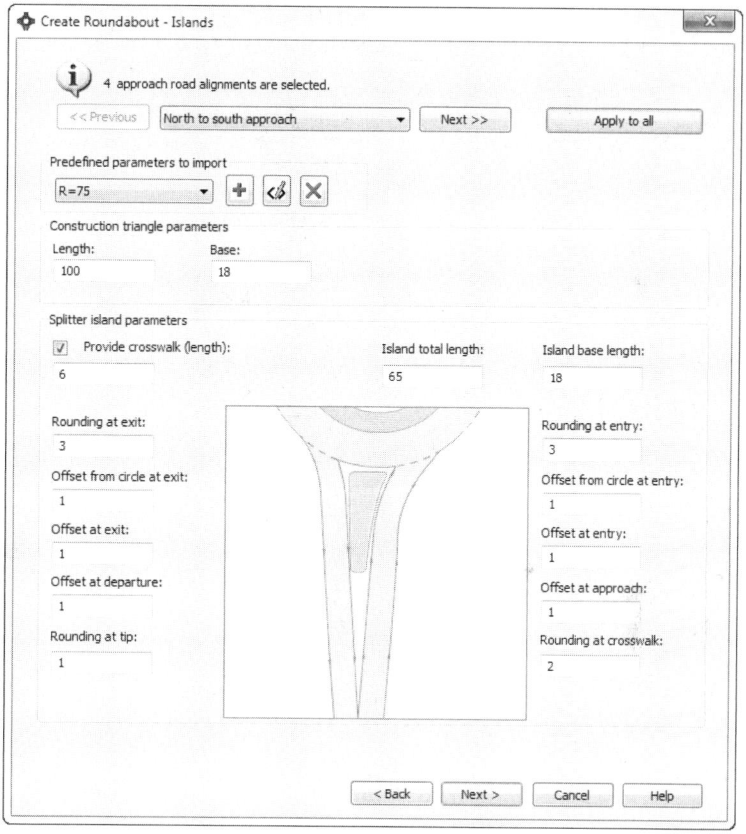

FIGURE 11.73
Roundabout Islands parameters

11. Click Apply To All, and then click Next.

The final screen of the Create Roundabout wizard deals with pavement markings and signage. Notice that you can specify your own blocks for the signs that will be placed in this process.

Everything created in this last step is an AutoCAD polyline or block. The polylines have a global width set to indicate pavement marking thicknesses. These thicknesses are set in the Markings And Signs screen (Figure 11.74).

12. Leave all defaults in the Create Roundabout – Markings And Signs screen (Figure 11.74), and click Finish.

Your roundabout should resemble Figure 11.75. Since standards vary by region, the metric drawing will have slightly different default pavement markings.

FIGURE 11.74
Pavement markings galore!

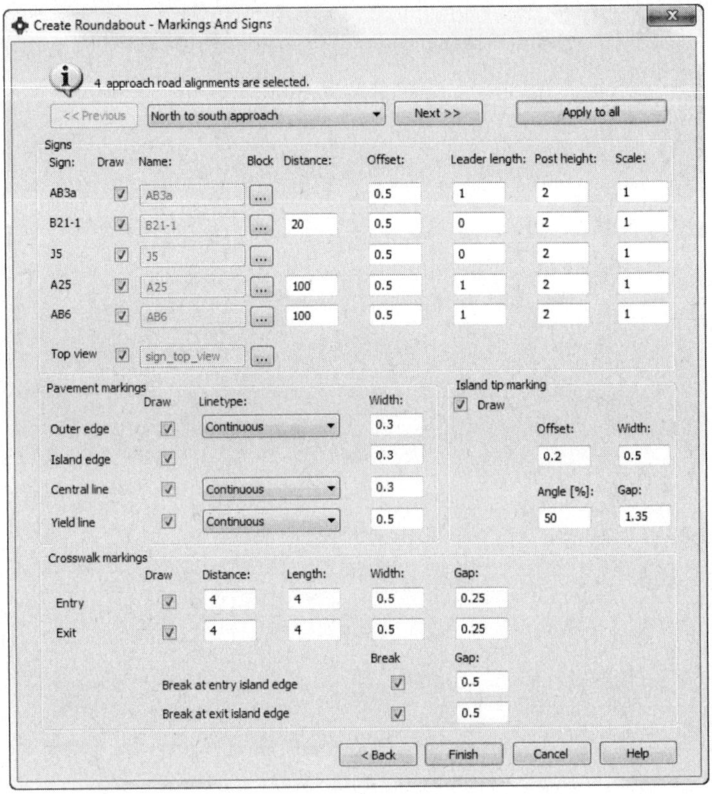

FIGURE 11.75
Completed roundabout alignment layout

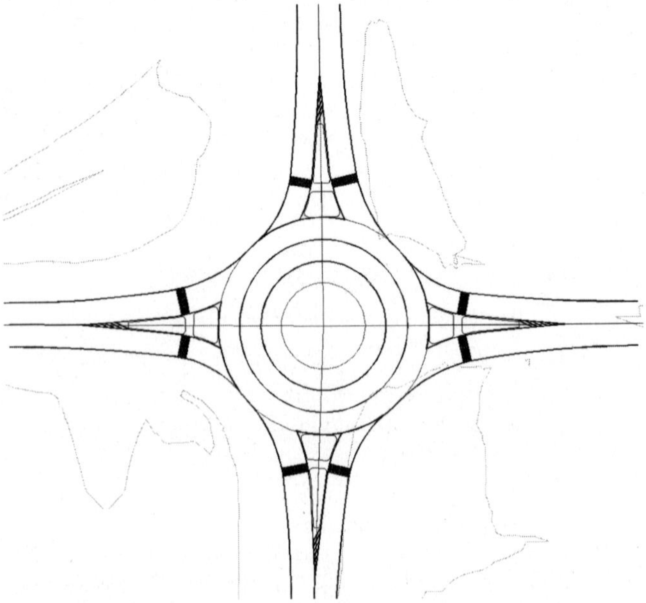

Labels omitted for clarity

Finally, you will add a turn lane in the NW quadrant of the roundabout. When you're creating slip turn lanes, remember that the turn radius must be large enough to fillet the exit and entry roads without overlapping the other alignments.

When selecting the approach entry and exit alignments, you need to click the shorter approach alignments created by Civil 3D rather than the original approach road. For this reason, the exercise has you select inside the islands, just to be sure.

13. From the Home tab ➢ Create Design panel, select Intersections ➢ Add Turn Slip Lane.
14. When prompted to select the entry approach, select the north approach alignment inside the curb island.
15. When prompted for the exit approach, select the west approach alignment inside the curb island, as shown in Figure 11.76.

FIGURE 11.76
Entry and exit approach alignments for the slip lane

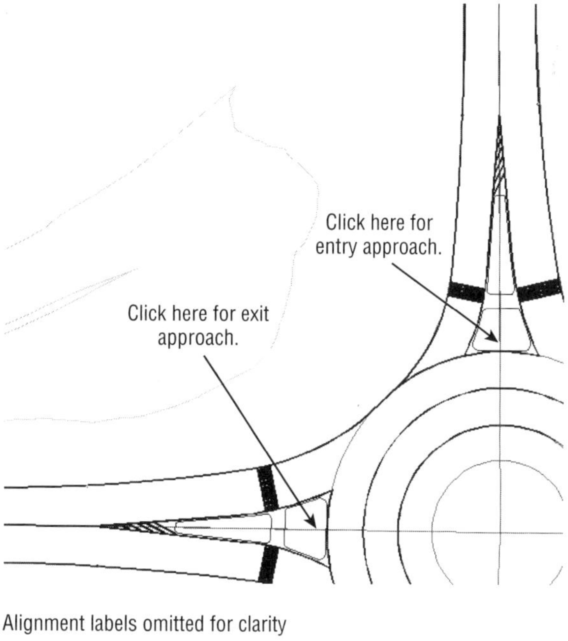

16. In the Draw Slip Lane screen shown in Figure 11.77, set the lane width to **14′** (**4** m) and the radius to **150′** (**45** m).
17. For the alignment layer, choose C-ROAD and for Alignment Label Set, choose Major And Minor Only. Click OK.

    Your roundabout will now look like Figure 11.78.

18. Save and close the drawing.

You now have all the alignments you need to start your roundabout design. At this point, you can add geometry to the alignments and modify what Civil 3D has created for you.

The horizontal layout is complete, but the roundabout design is far from done. No vertical data has been created; that is up to you.

This is as far as we will take you in this book in regard to building a roundabout step-by-tedious-step. Rest assured, however, that if you have truly mastered corridors, the techniques for completing a roundabout are similar to that of any intersection.

The remainder of this chapter gives you an overview of how to accomplish the rest on your own.

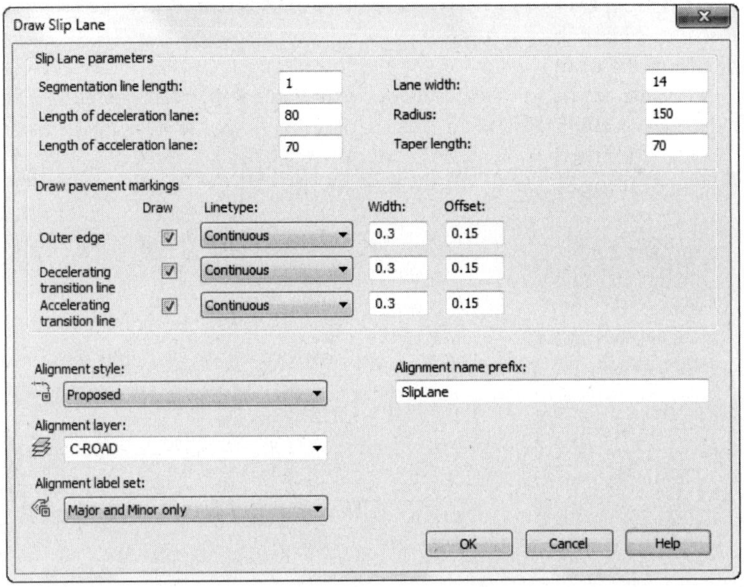

FIGURE 11.77
Adding a slip lane

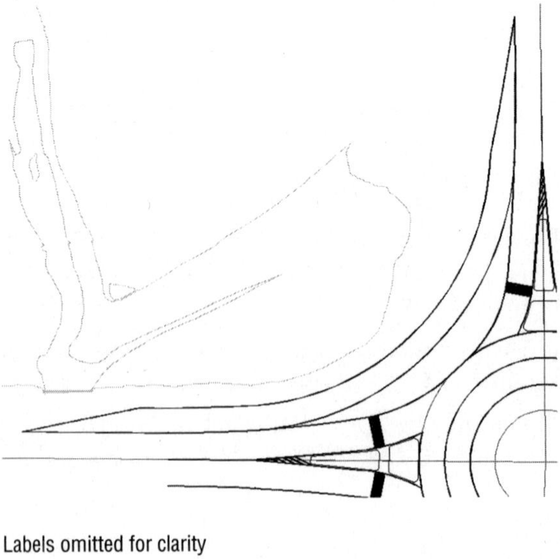

FIGURE 11.78
The completed slip lane alignments

Center Design

All profiles need to meet at the elevations inside the traveled way in the circular pavement area of the roundabout. Therefore, the main circle design comes first.

To see an example of a completed roundabout corridor, open the drawing `RoundaboutExample_FINISHED.dwg` or `RoundaboutExample_METRIC_FINISHED.dwg`. Use these examples as a guide to "reverse engineer" your own roundabouts when you've mastered other forms of intersections.

You can use any of the circular alignments created by Civil 3D as the basis for this step, as long as your assembly works with the design. Remember to make note of which direction the alignment goes, to ensure the assembly you create is not backwards.

Extract profiles for the main circle design from the Existing Ground and Roundabout Grading surfaces, as shown in Figure 11.79.

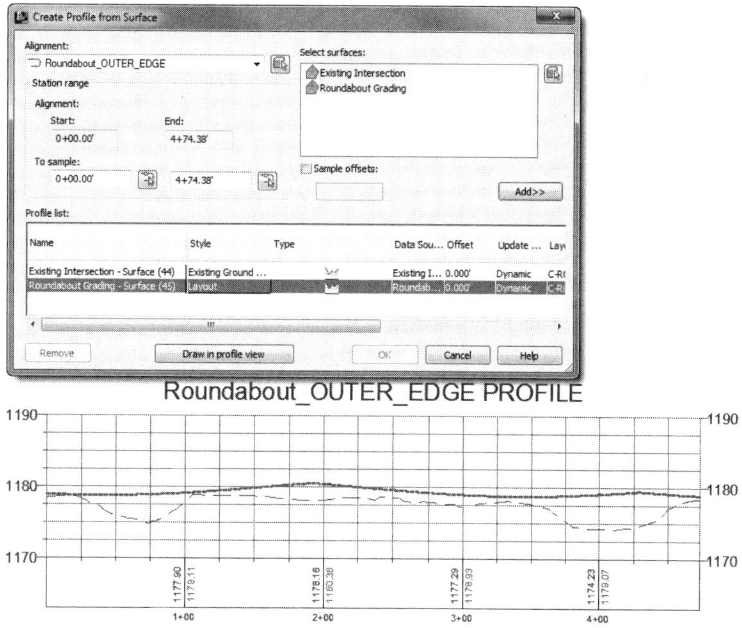

FIGURE 11.79 Extract surface profiles around the main circular alignment.

The assembly you create for this preliminary design will also be used in the main design. Decide which alignment will be used as the circular design basis, and create an assembly based on your alignment location and desired geometry, as shown in Figure 11.80.

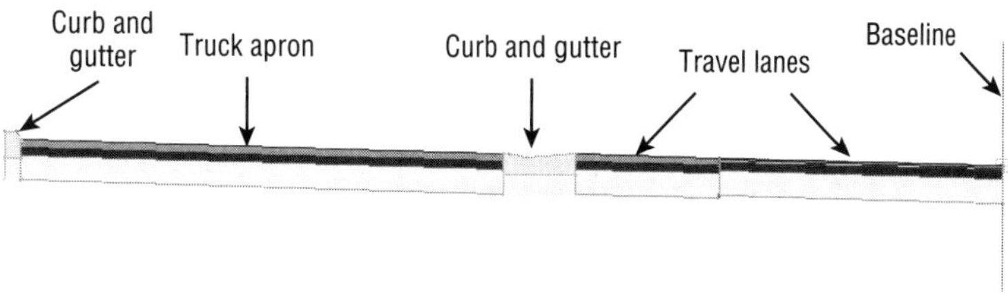

FIGURE 11.80 Center assembly for roundabout

This assembly can be used in several steps of the process. First it is used in a preliminary corridor called **RAB Center**. It can also be recycled to be the centerpiece of your main corridor. In the example, the main circle alignment created by Civil 3D, Roundabout_OUTER_EDGE, was used as the baseline for this initial corridor. There are no targets or frequencies set in this corridor. Like the cul-de-sac example earlier in this chapter, this is a preliminary corridor, used to ensure that profiles from the approach roads tie in at the correct elevations.

In the example drawings, a Top link surface was created from the RAB Center corridor. At this point, the roundabout will resemble Figure 11.81.

FIGURE 11.81
Preliminary center corridor and surface

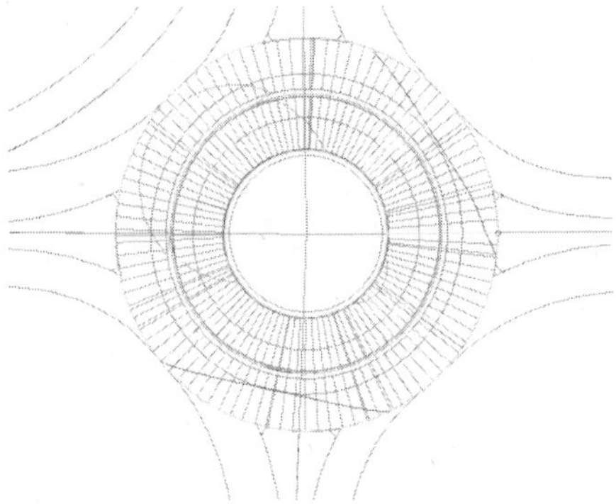

Profiles for All

You have all the preliminary surfaces in place, and you have all the alignments you need, so it is time to extract profiles from your various surfaces.

In the `RoundaboutExample_FINISHED.dwg` and `RoundaboutExample_METRIC_FINISHED.dwg` files, the surface profiles for all the approach alignments were created by sampling the existing ground, drainage surface, and preliminary RAB Center corridor surface. These profiles will look something like Figure 11.82.

FIGURE 11.82
Surface profiles needed for design

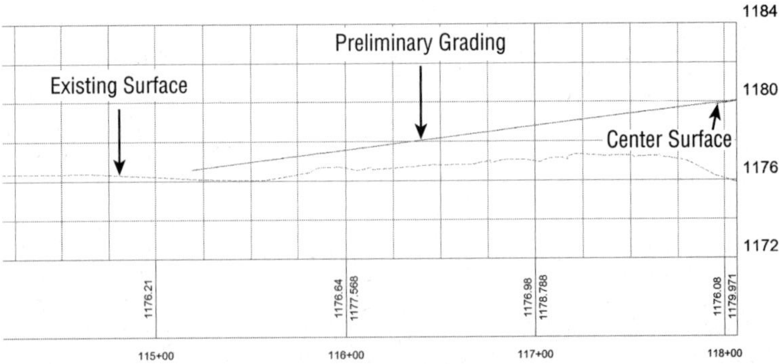

When you create your design, you will see all your design considerations in the profile views. No matter how you decide to tie into existing ground and slope upward toward the surface, your design must tie into the preliminary center surface, as shown in Figure 11.83.

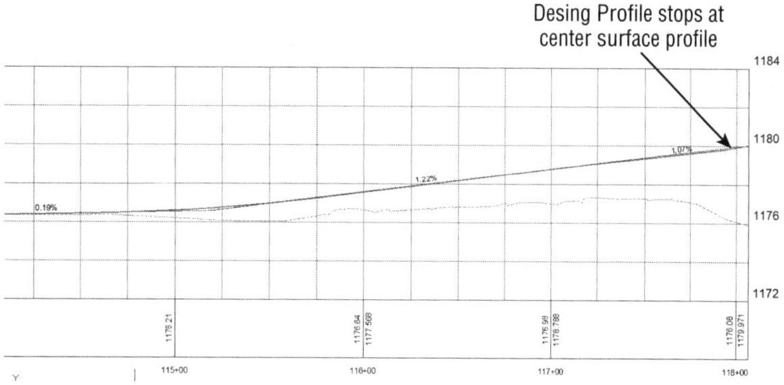

FIGURE 11.83
Design profiles must tie into the center.

Use techniques you learned earlier in this chapter to assist you. Labels are an especially valuable tool for roundabouts. Keep your profile views organized, as you will have at least three for each approach. If you have a slip turn lane, you will have a profile for that as well.

Tie It All Together

Stretch your legs and go for more coffee. It is time to put this thing together into a completed corridor. When you model the corridor initially, ignore curb islands — you will add them as individual corridors in a later step.

A simple roundabout can be completed using as few as three assemblies. In our example, however, the slip turn lane necessitates a total of four assemblies. In addition to the RAB Center assembly you saw in Figure 11.80, you will need three more assemblies, as shown in Figure 11.84.

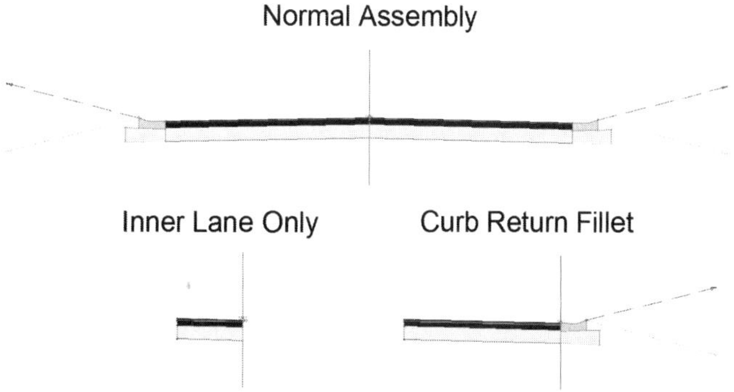

FIGURE 11.84
Assemblies needed for the main roundabout corridor

These assemblies will be tied to the EOP alignments and profiles as baselines, similar to a traditional intersection. Each quadrant of the roundabout will target at least two alignments and profiles, as shown in Figure 11.85.

FIGURE 11.85
Roundabout corridor regions and targets

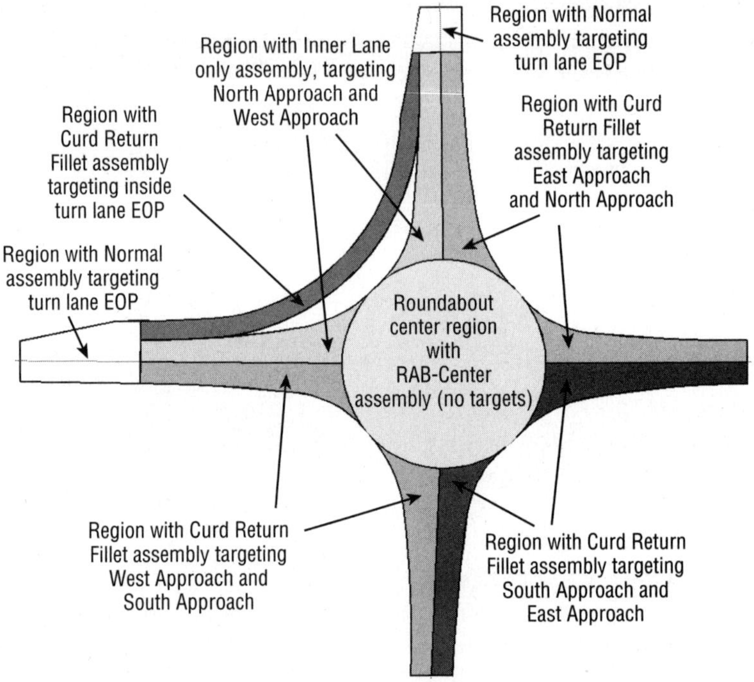

Keep in mind the direction of your alignments. If you build the corridor in stages, check the corridor periodically to make sure it is building correctly.

Create a corridor surface from your completed roundabout lanes. You will likely have to use the Add Interactively tool to add the boundary correctly.

Finishing Touches

The median islands are the last parts to go on the corridor. You can create them using simple grading objects, but since this is a book about mastering skills, and this is a chapter about corridors, you should examine the dynamic way.

Create a simple assembly containing the curb and gutter for the curb islands. This will be the assembly that you use with the median corridors (Figure 11.86).

FIGURE 11.86
Simple assembly containing just the curb and gutter on the median islands

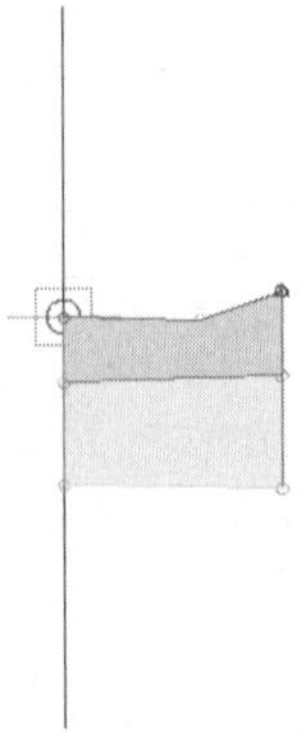

Each median island will need its own alignment. Take note of the direction of the alignments to make sure they are compatible with the curb island assembly. If necessary, change the direction of the alignments using the Reverse Direction tool in the Modify panel of the contextual Ribbon tab. Figure 11.87 shows the bypass island and north island with directions.

FIGURE 11.87
Several median island alignments with direction shown

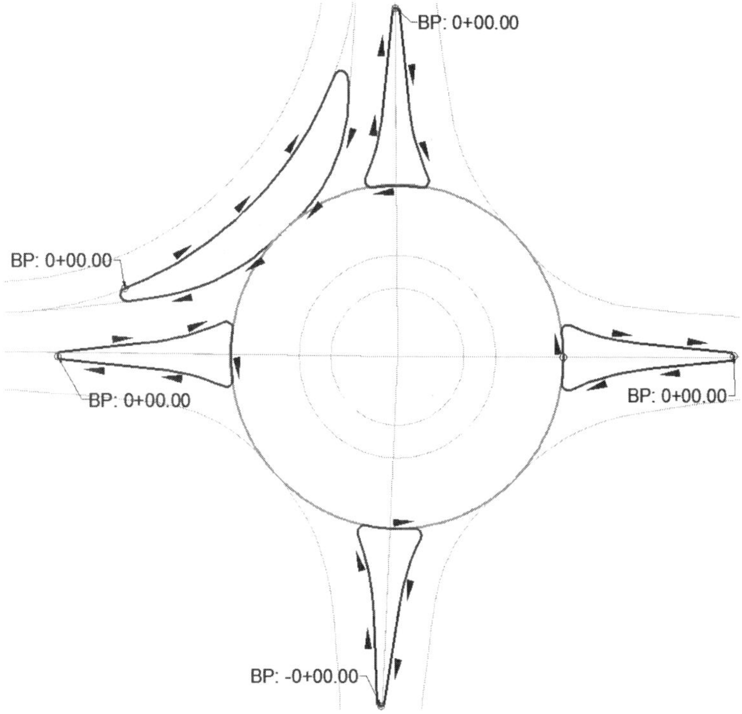

The good news is that the elevation data for the medians is already complete. Your main corridor's surface will act as the profile for each individual median. This also means that once the curb return corridors are created, they will be dynamic to the main corridor. After these little corridors are created and surface model information has been obtained, you can set them to Rebuild – Automatic and forget all about them.

Extract a profile for all the curb return alignments from the Top link surface model from the main roundabout corridor, as shown in Figure 11.88. You do not need to see this profile in a view, so you can click OK to extract.

When the design roundabout corridors are complete and surfaces are made, the next step is to merge the surfaces together. Create a final surface model and paste the main roundabout design in first. After the main corridor is pasted in, paste the smaller median corridor surfaces (see Figure 11.89).

Your next step is to use the corridor fine-tuning techniques you learned earlier in this chapter to ensure your grades are correct and the design is correct. To see an example of a completed corridor using these steps, take a look at `RoundaboutExample_FINISHED.dwg` (`RoundaboutExample_FINISHED_METRIC.dwg`), which you can download from `www.sybex.com/go/masteringcivil3d2013`.

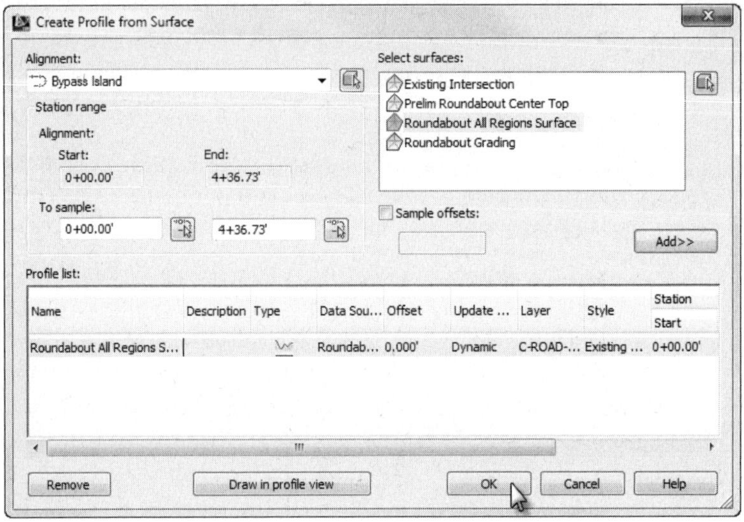

FIGURE 11.88
Extract the profile for medians from your main roundabout surface.

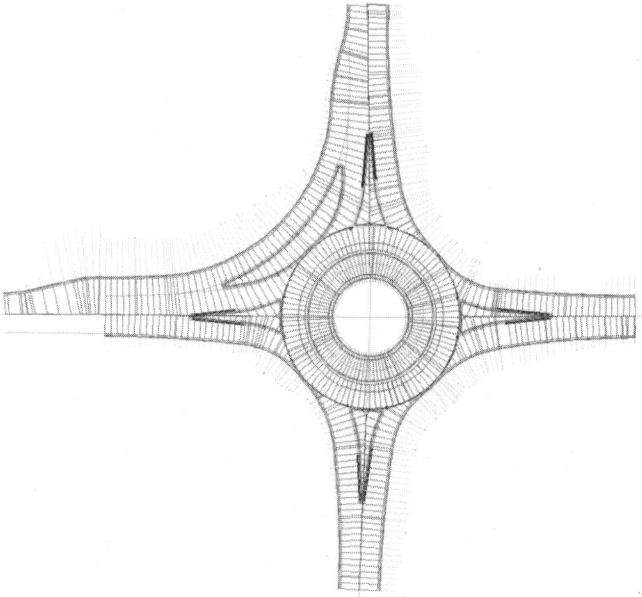

FIGURE 11.89
The completed roundabout

The Bottom Line

Create corridors with noncenterline baselines. Although for simple corridors you may think of a baseline as a road centerline, other elements of a road design can be used as a baseline. In the case of a cul-de-sac, the EOP, the top of curb, or any other appropriate feature can be converted to an alignment and profile and used as a baseline.

Master It Open the `MasteringAdvancedCorridors.dwg` (`MasteringAdvancedCorridors_METRIC.dwg`) file, which you can download from `www.sybex.com/go/masteringcivil3d2013`. Add the cul-de-sac alignment and profile to the corridor as a baseline. Create a region under this baseline that applies the Intersection Typical assembly.

Add alignment and profile targets to a region for a cul-de-sac. Adding a baseline isn't always enough. Some corridor models require the use of targets. In the case of a cul-de-sac, the lane elevations are often driven by the cul-de-sac centerline alignment and profile.

Master It Continue working in the `Mastering Advanced Corridors.dwg` (`Mastering Advanced Corridors_METRIC.dwg`) file. Add the Second Road alignment and Second Road FG profile as targets to the cul-de-sac region. Adjust Assembly Application Frequency to 5′ (1 m), and make sure the corridor samples are profile PVIs.

Create a surface from a corridor and add a boundary. Every good surface needs a boundary to prevent bad triangulation. Bad triangulation creates inaccurate contours, and can throw off volume calculations later in the process. Civil 3D provides several tools for creating corridor surface boundaries, including an Interactive Boundary tool.

Master It Continue working in the `Mastering Advanced Corridors.dwg` (`Mastering Advanced Corridors_METRIC.dwg`) file. Create an interactive corridor surface boundary for the entire corridor model.

Chapter 12

Superelevation

Superelevation and cant tools have matured greatly since the early releases, and theAutoCAD® Civil 3D® 2013 software has added even more functionality.

Before you can understand the superelevation tools, you should have a good grasp of alignments, assemblies, and corridors. You may want to refer to the corresponding chapters for a refresher on those topics before attempting to put superelevation or cant into practice.

In this chapter, you will learn to:

- Add superelevation to an alignment
- Create a superelevation assembly
- Create a rail corridor with cant
- Create a superelevation view

Getting Ready for Super

Several pieces need to be in place before superelevation can be applied to the design. You will need a *design criteria file* appropriate for your locale, design speeds applied to an alignment, and an assembly that is capable of superelevating.

There are a lot of abbreviations and terminology thrown around when it comes to superelevation, so let's take a look at that first.

When an assembly is applied without superelevation, the geometry comes directly from the original design, as shown in Figure 12.1.

As the assembly begins its entrance into a curve with superelevation applied to it, it will first start to lose its normal crown. Figure 12.2 shows the same assembly at the End Normal Crown (ENC) station.

When the assembly has one side flattened out, as shown in Figure 12.3, it is called Level Crown (LC).

When the assembly straightens out into a plane that matches the inside lane, it is called Reverse Crown (RC), as shown in Figure 12.4.

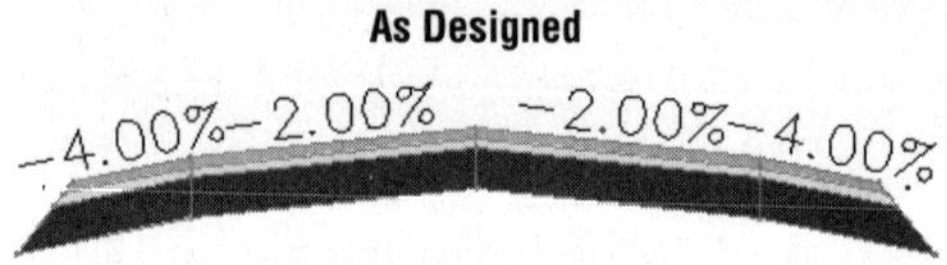

FIGURE 12.1
An example assembly without superelevation

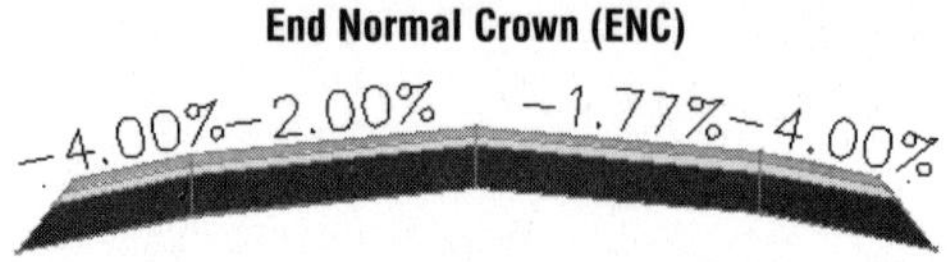

FIGURE 12.2
At the End Normal Crown (ENC) station, the default lane slope starts to change.

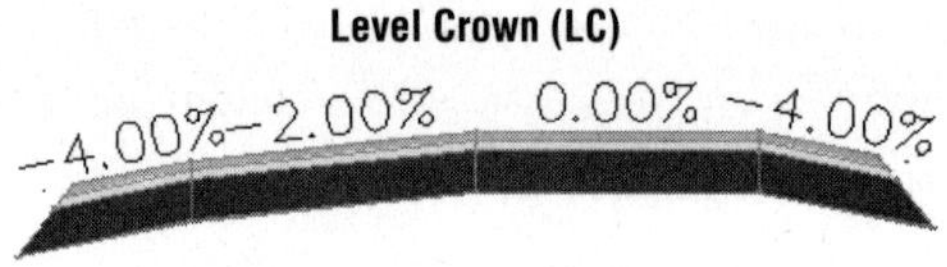

FIGURE 12.3
Level Crown entering a left-hand turn

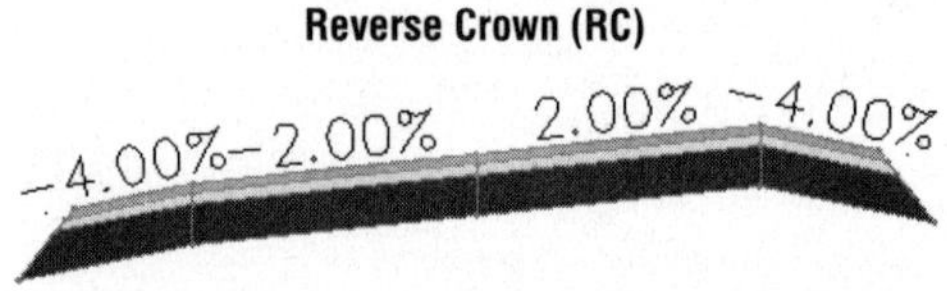

FIGURE 12.4
Reverse Crown (RC)

If accommodations have been made for shoulder slope rollover and breakover removal, the shoulder will shift as well. The left image of Figure 12.5 shows where the superelevated lane becomes steep enough to cause an outside rollover problem with the outside shoulder. The shoulder begins to adjust to increase the safety of the road. The right image of Figure 12.5 shows the change in the lower shoulder.

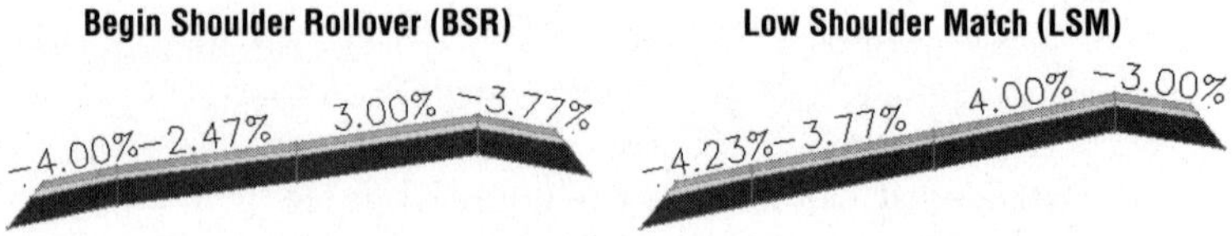

FIGURE 12.5
Begin Shoulder Rollover (BSR) (left) and Low Shoulder Match (LSM) (right)

Finally, the lane gets to its maximum slope at the Begin Full Super (BFS) station. As you can see in Figure 12.6, all the geometry adjustment has taken place.

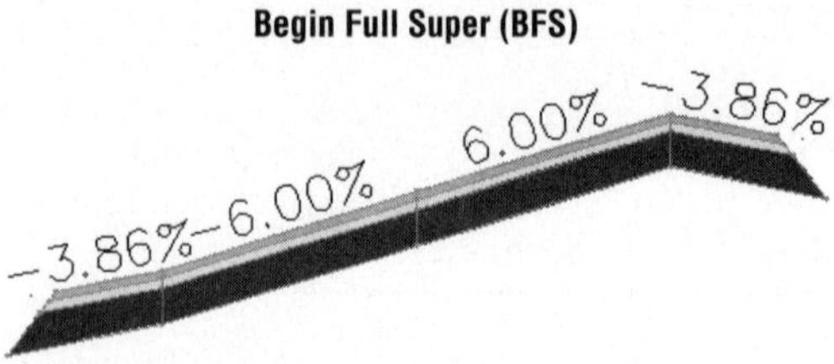

FIGURE 12.6
Begin Full Super (BFS)

On the way out of the curve, you will see the geometry transitioning back to its original design. It will pass through End Full Super (EFS), Low Shoulder Match (LSM), Reverse Crown (RC), Level Crown (LC), Begin Normal Crown (BNC), and finally back to Begin Normal Shoulder (BNS).

Now that you are familiar with the terminology and abbreviations Civil 3D uses, let's get started on some design.

Design Criteria Files

Having the correct design criteria file in place is the first step to applying superelevation to your corridor. These XML-based files contain instructions to the software on when to flag your design for geometry problems both horizontally and vertically. Design criteria files are the brains behind how your road behaves when superelevation is applied to the design.

Several design criteria files are supplied with Civil 3D upon installation. The out-of-the-box standards include AASHTO 2001 and AASHTO 2004 for both metric and Imperial units. Several of the country kits include design criteria files for your locality if you are outside of the United States. If country or state kits do not exist for your situation, you can create your own, user-defined files.

To create your own design criteria:

1. Select any alignment.
2. Click the Design Criteria Editor icon on the context tab.

   It is easiest to modify an existing table in your desired units rather than starting from an empty file.
3. Be sure to click the Save As icon before making any changes.

USER-DEFINED CRITERIA FILES

If you create design criteria files for your region or design scenario, you will need to be able to share the file with anyone who will be working with your design.

Inside your organization, the best way to handle the design criteria file is to move it to a shared network location. If you are worried that a less experienced person may accidentally modify the file, you can set the XML file to read-only.

The default location for road design standards is:

```
C:\ProgramData\Autodesk\C3D 2013\enu\Data\Corridor Design Standards\
```

The default location for rail design standards is:

```
C:\ProgramData\Autodesk\C3D 2013\enu\Data\Railway Design Standards\
```

Once you re-path to the design criteria file for an alignment, the location will be saved with the alignment.

If you are collaborating with someone without direct access to your design criteria files, you will need to send that person the XML file with your DWG. The XML file will not automatically go along for the ride if you use the eTransmit command.

Inside the Design Criteria Editor (Figure 12.7), you will see three headings: Units, Alignments, and Profiles. The Units page tells Civil 3D what type of values it will be using in the file. The Alignments page is used for checking design, creating superelevation, and widening outside curves. The Profiles page provides tabular data for minimum K value used to check vertical design.

FIGURE 12.7
Inside the Design Criteria Editor

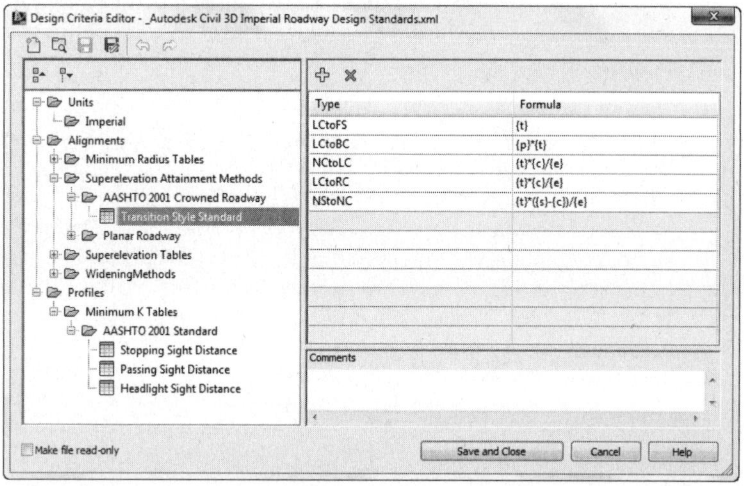

Civil 3D will graphically flag alignments when the design speed specified in the alignment properties has a radius less than the value specified in the Minimum Radius Table. The Minimum Radius Tables from AASHTO use superelevation rates in the table names, but this does not lock you into that rate for applying superelevation to the corridor. In other words, just because you use a more conservative value in your radius check, that doesn't mean you can't superelevate at a steeper rate. The tables are independent of one another.

Also in the Alignments page you will find the superelevation attainment equations. These equations determine the distance between superelevation critical stations. Familiarize yourself with the terminology and locations represented by these stations, as shown in Figure 12.8.

FIGURE 12.8
Superelevation critical stations and regions calculated by Civil 3D

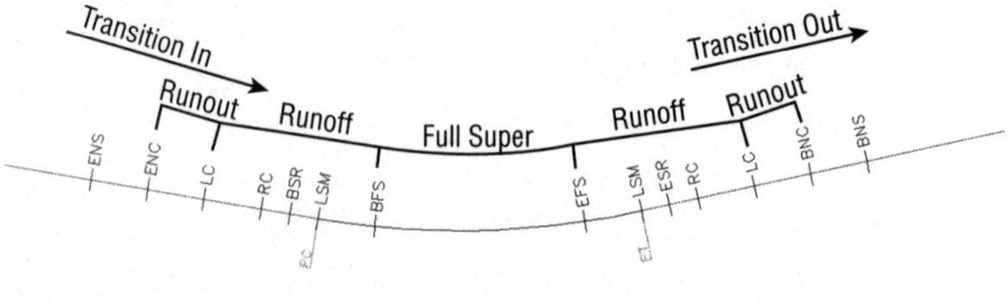

In the following exercise, you will modify an example design criteria file and save it:

1. Open the `Highway 10 Criteria.dwg` (`Highway 10 Criteria_METRIC.dwg`) file, which you can download from this book's web page at `www.sybex.com/go/masteringcivil3d2013`.

2. Select the Highway 10 alignment that already exists in the drawing.

Design Criteria Editor

3. From the Modify panel on the context tab, select Design Criteria Editor.
4. Click the Open button at the top of the dialog, and browse to the `_MasteringExample.xml` (`_MasteringExample_METRIC.xml`) file, which is part of the dataset you downloaded for this chapter.
5. Expand the Alignments category, and expand the Superelevation Tables area.

   There is only one superelevation table in this example for 4% maximum slope.
6. Right-click on Superelevation Tables, and select New Superelevation Table.
7. Double-click the new table and rename it **Example 6% Super**.
8. Right-click on Example 6% Super, and select New SuperelevationTypeByTable.
9. Expand the new section, right-click SuperelevationTypeByTable, and select New Superelevation Design Speed, as shown in Figure 12.9.

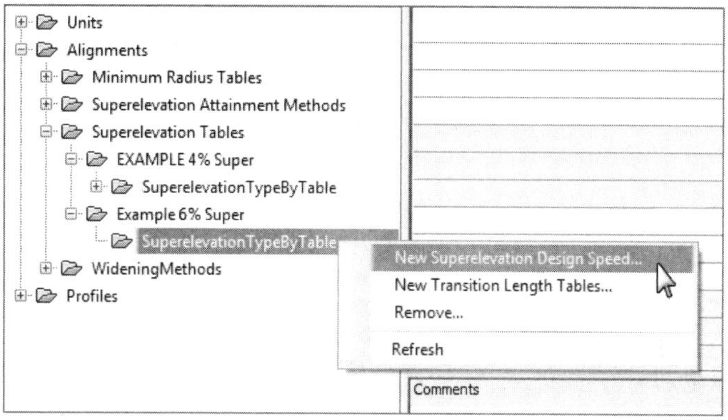

FIGURE 12.9
Adding a design speed to the design criteria file

   When you expand the SuperelevationTypeByTable section, you will see that Civil 3D has placed a new design speed with a default of 10mph in the listing.
10. Double-click the Design Speed 10 and change the design speed to **30** mi/hr (**50** km/hr).

    Notice that you only need the numeric value; Civil 3D fills in the rest of the table name for you.
11. Highlight your example design speed.

    The right side of the dialog will have an empty table containing columns for Radius and Superelevation Rate.
12. Select the first row of the table. Click the first field in the Radius column to start entering data. Add several radius and superelevation values, as follows:

CURVE RADIUS (FEET)	CURVE RADIUS (METERS)	SUPERELEVATION %
300	90	6
1000	300	4.5
1500	450	3.2
2000	600	2.6
4000	1250	NC

When you have completed your data entry from the table, your Design Criteria Editor will resemble Figure 12.10.

FIGURE 12.10
Adding example data using the Design Criteria Editor

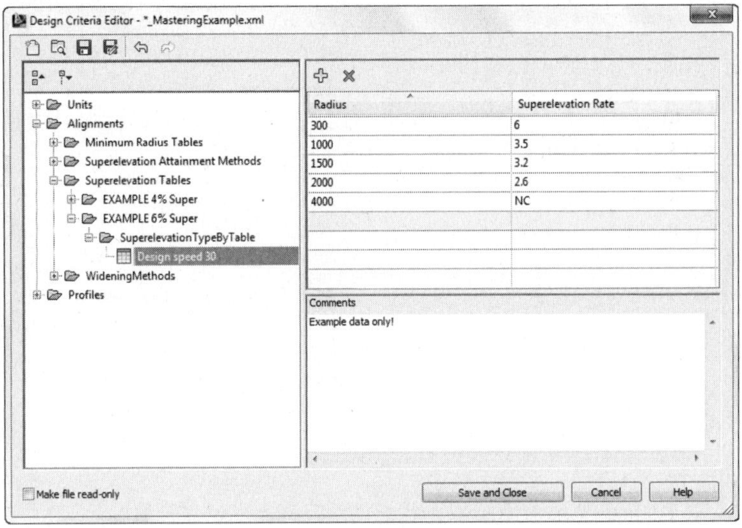

13. Add a note in the Comment field that reads **Example Data Only!**
14. Click the Save And Close button at the bottom of the editor.
15. Click Save Changes and exit when prompted.

Ready Your Alignment

Superelevation stations are connected to alignment curves. The design speed from the alignment properties is needed at each curve to specify which superelevation rate tables to use from the design criteria. The design speed has an effect on the distance between superelevation critical stations and the cross-slope used when the road is at full-super.

It is a good idea to get your alignment geometry and design speed locations finalized before attaching superelevation. If a change is made to your alignment, the superelevation stationing will be marked as out of date.

Super Assemblies

As a general rule, if the lane subassembly has the word "super" somewhere in its name, it will respond to superelevation. If you want to verify that the lane you are choosing will behave the way you want it to in a superelevation situation, you can right-click it from the tool palette and access the subassembly help.

As long as you stay away from the Basic tab, all of the shoulder and curb subassemblies have parameters you can set to dictate how the assembly is to behave when an adjacent lane superelevates.

Most subassemblies that are capable of superelevating are intended for use where the pivot point for the cross section is at the center crown of the road. When the pivot point is at the center of the road, the baseline profile dictates the final elevation of the crown of the road. Figure 12.11 shows an example of a two-lane highway (top image) and a four-lane divided highway (bottom image) that are designed to be used with the superelevation tools.

FIGURE 12.11 Two-lane road ready for super; four-lane road used in super

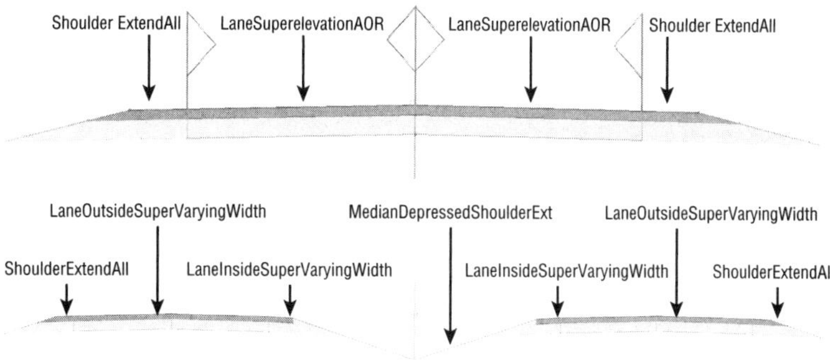

AXIS OF ROTATION SUPPORT

The axis of rotation (AOR) subassembly can be used when the centerline of the road is not the pivot point for superelevation. The "flag" symbols (as shown in Figure 12.12) indicate potential pivot points on the assembly.

FIGURE 12.12 AOR subassemblies used on an undivided, crowned roadway

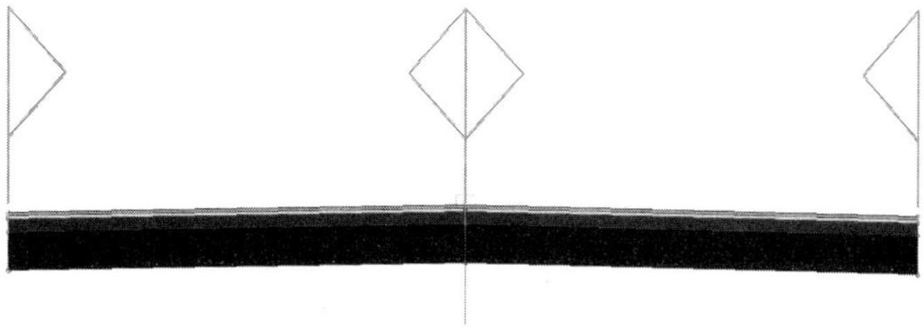

The flag symbols on LaneSuperelevationAOR indicate where the lane can be pinned down and used as a pivot point. When the axis of rotation is not the centerline of the road, the lane geometry is used to determine the change in elevation that will occur as a result.

AXIS OF ROTATION (AOR)

The following is a list of the limitations and other factors to be aware of if you decide to take these new assemblies for a spin:

- Don't use an offset assembly with an axis of rotation superelevation assembly. Doing so can throw off the superelevation calculation.
- When working with curbs, medians, and shoulders, keep an eye out for the parameter Superelevation Axis Of Rotation. If this parameter reads Unsupported, it means that the subassembly will not adjust for superelevations other than at the center. None of the curb and gutter subassemblies will adjust for breakover or rollover, but most of the shoulder assemblies do. This is a "hard-coded" parameter that cannot be changed by the end user.

When building assemblies with LaneSuperelevationAOR you may see warnings appear, as shown in Figure 12.13.

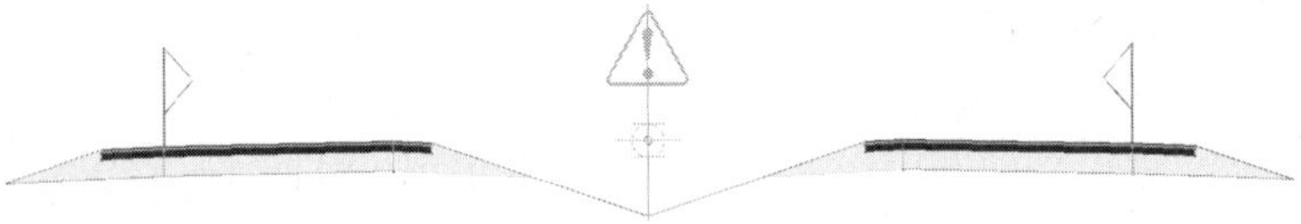

FIGURE 12.13
A warning symbol on an assembly using LaneSuperelevationAOR

Here are some of the warnings you may encounter:

Center Pivots Not Applied When Only One Group This usually occurs when you start your assembly with a median at the center. The construction of the assembly then does not have a distinct left and right group. The fix is to build your assembly with left and right sides, and add the median last.

Unsupported Subassemblies This warning will appear when you're attempting to use an assembly that has the Superelevation Axis Of Rotation parameter absent from its parameters.

Unsupported in Assemblies Containing Offsets Using an offset assembly will interfere with the software's ability to calculate the correct slope and curve widening on a superelevated road. Therefore, offset assemblies are not recommended for use with the AOR subassembly. You can still use offsets in traditional center-pivot based superelevation.

No Center Pivots Found Make sure your assembly properties list the assembly as the correct type. For example, if you accidentally set Assembly Type to Divided Crowed Road when it is actually an Undivided Crowned Road, you will receive this message. Check the Construction tab of the assembly Properties dialog to verify.

You can still add assemblies with warnings to a corridor; however, the superelevation may not behave as expected.

Real World Scenario

THE TIPPING POINT

In some design situations that use superelevation, the crown of the road is not the ideal pivot point. In the following example, you will create a four-lane divided highway that pivots inside the curve rather than at the crowns during superelevation.

1. Open the file `AOR Assembly.dwg` (`AOR Assembly_METRIC.dwg`).
2. Zoom into the assembly that was started for you in this drawing.

   Note that a generic link has been placed on each side as a spacer for the subassemblies that you will be adding in the next steps.
3. Open your subassembly tool palette (Ctrl+3 will work) and find the Lanes tab.
4. Select the LaneSuperelevationAOR subassembly.
5. In the Advanced Parameters of the properties, set Side to Right and Use Superelevation to Right Lane Outside.
6. Place the subassembly on the right side, using the generic link marker as its "hanger." Press Esc to complete the placement of the generic link.

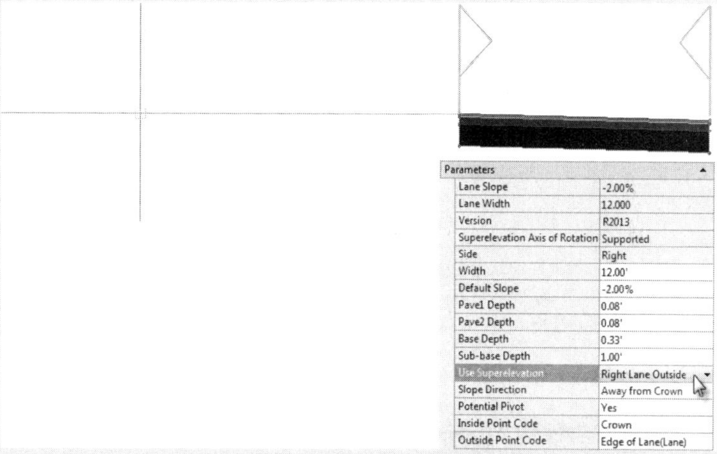

7. Select the newly placed subassembly, choose the Subassembly: LaneSuperelevationAOR contextual tab ➢ Modify Subassembly panel, and click Mirror.
8. Click to attach the mirrored subassembly to the left side of the subassembly you just placed.

   A dialog will appear confirming that you want to place the mirrored subassembly on the same side. Click OK.
9. Select the newly placed subassembly and view its AutoCAD properties.
10. Change the Use Superelevation parameter to Right Lane Inside, as shown here.

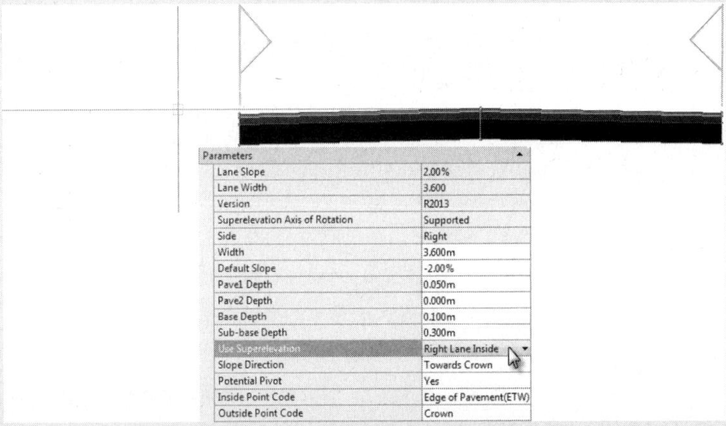

11. Repeat steps 4–10 for the left side of the assembly.

    Be sure to use the subassembly parameters to the correct superelevation property. Note that the inside lane must have Use Superelevation set to Left Lane Inside and the outside lane must be set to Left Lane Outside.

12. On the tool palette, switch to the Generic tab. Select the MarkPoint subassembly, change Point Name to **MEDIAN**, and place it by clicking the red circle on the inside right edge of the pavement.

13. On the Medians tab, select Median Depressed, and change Marked Point Name to MEDIAN.

14. Leave all other parameters at their defaults and place the median on the inside left of the subassembly as shown here. Press Esc when complete.

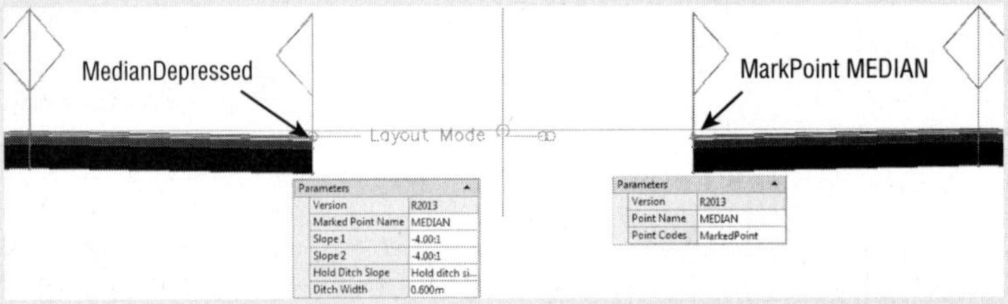

15. As the final step in building this assembly, select both of the original generic link spacers and set the Omit Link property to Yes.

    If time permits, add the ShoulderExtendAll subassembly to the outermost edges using the default settings. Press Esc when complete.

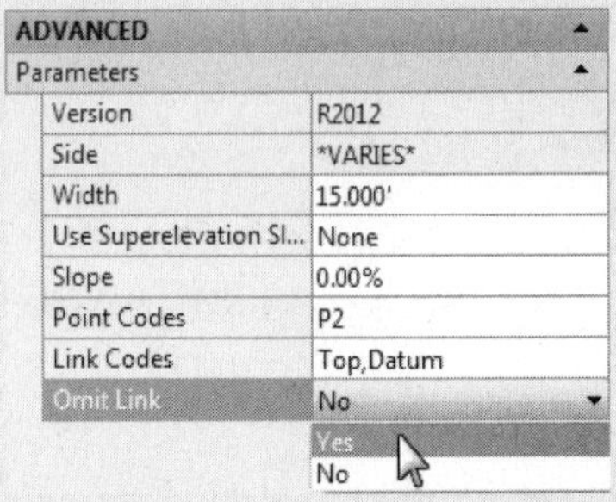

16. Select the alignment that runs through the project.
17. On the Alignment contextual tab, click Superelevation ➤ Calculate/Edit Superelevation, and click the Calculate Superelevation Now option when notified that no data exists.
18. Set Roadway Type to Divided Crown With Median.
19. Set the pivot method to Inside Of Curve.
20. Set the median treatment to Distorted Median, and click Next.

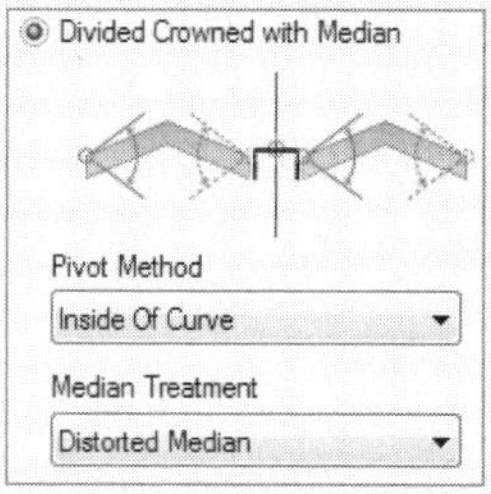

21 On the Lanes page, set the normal lane width to **12′** (**4** m) and the normal lane slope to -2.00%.
22. Place a checkmark next to Symmetric Roadway, and click Next.
23. On the Shoulder Control page, leave the settings at their defaults, and click Next.
24. On the Attainment page, set the active criteria file to `Autodesk Civil 3D Imperial (2004) Roadway Design Standards.xml` (`Autodesk Civil 3D Metric(2004) Roadway Design Standards.xml`).
25. Set the superelevation rate table to `AASHTO 2004 Customary eMax 6%` (`AASHTO 2004 Metric eMax 6%`) using the 4 Lane Transition Length table.
26. Toggle on Automatically Resolve Overlap, and click Finish.

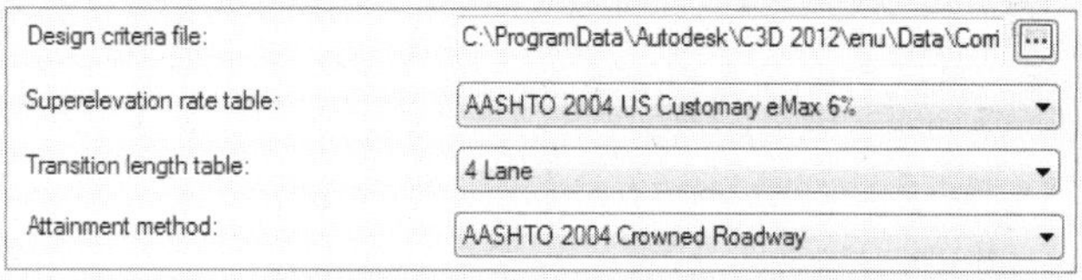

27. In Prospector, select the corridor Route 66, right-click, and select Rebuild.
28. Select the corridor in plan and enter the Section Editor to examine your corridor.

You should observe that superelevation is occurring based on our specified pivot point of inside the curves. Close the Section Editor when you can't take anymore awesomeness.

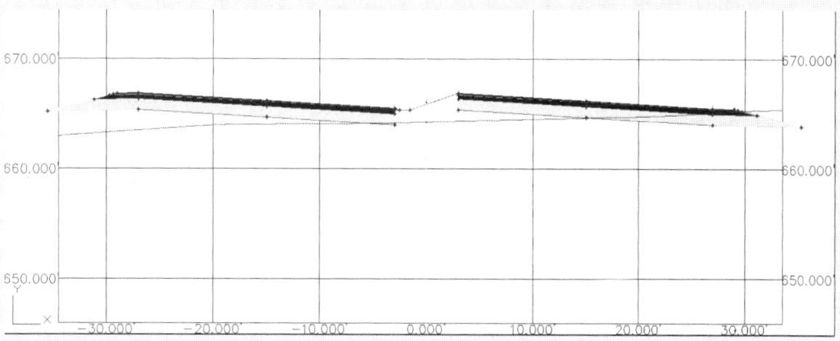

Applying Superelevation to the Design

Civil 3D takes into account other factors such as curve station locations and assembly geometry. Superelevation information is associated with the alignment, but is handed in a separate calculation area. In this section you will put all the pieces in place that are needed for the software to dynamically apply superelevation or cant to your design.

Start with the Alignment

To begin applying superelevation to the design, select your alignment:

1. Open the `Highway 10 Super.dwg` (`Highway 10 Super_METRIC.dwg`) file, which you can download from this book's web page.
2. Select the Highway 10 alignment that already exists in the drawing.
3. From the Alignment contextual tab, select Superelevation ➤ Calculate/Edit Superelevation.
4. When prompted by the dialog, click Calculate Superelevation Now.
5. On the Calculate Superelevation – Roadway Type page, select the Undivided Crowned radio button.
6. From the Pivot Method drop-down, choose Center Baseline, as shown in Figure 12.14, and click Next.

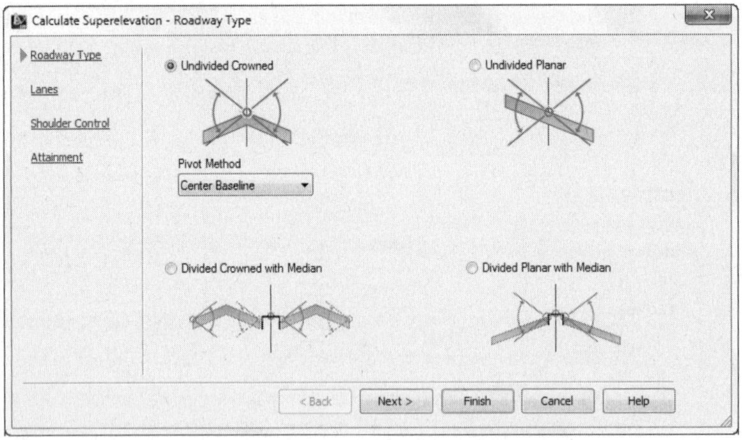

FIGURE 12.14 Roadway type specification for superelevation

7. On the Lanes screen, verify that the Symmetric Roadway check box is selected.
8. Set Normal Lane Width to **12′** (**4** m), and set Normal Lane Slope to **-2.00%**, as shown in Figure 12.15, and click Next.
9. On the Shoulder Control screen, make sure the Calculate check box is selected on the right side of the page.

Only the Outside Edge Shoulders should be active. Since this is an undivided road, the options for the Inside Median shoulders are grayed out.

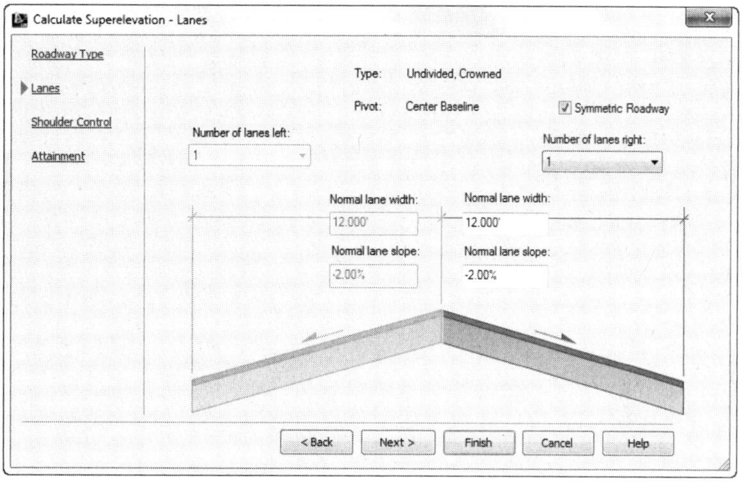

FIGURE 12.15
Lane information

10. Set Normal Shoulder Width to **6′** (**2** m), and set Normal Shoulder Slope to **-5.00%**.
11. For Shoulder Slope Treatment, set the following:
    - Set the Low Side option to Breakover Removal.
    - Set the High Side option to Default Slopes.
    - Place a check mark next to Maximum Shoulder Rollover and set the value to **8.00%**.

Your Shoulder Control screen should look like Figure 12.16.

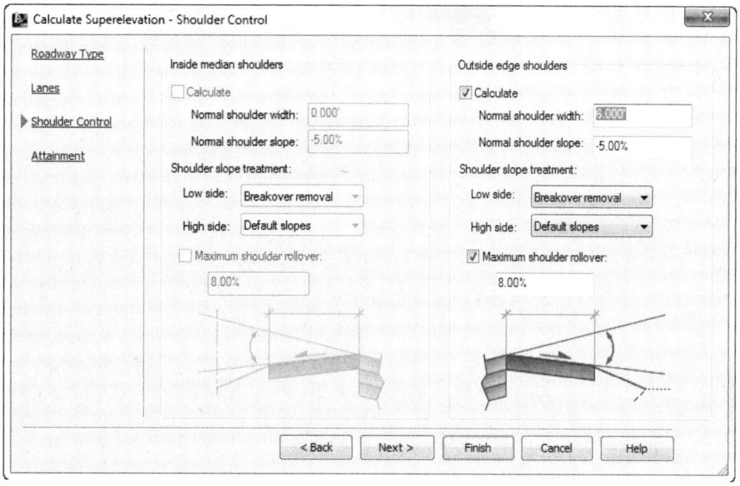

FIGURE 12.16
Shoulder Control and Breakover Removal parameters

12. Click Next.
13. On the Attainment screen, click the ellipsis button to verify that the design criteria file is set to `Autodesk Civil 3D Imperial (2004) Roadway Design Standards .xml` (`Autodesk Civil 3D Metric (2004) Roadway Design Standards.xml`).
14. Set the superelevation rate table to AASHTO 2004 US Customary eMax 4% (AASHTO 2004 Metric eMax 4%).
15. Set the transition length table to 2 Lane.
16. Set the Attainment method to AASHTO 2004 Crowned Roadway.
17. Place a checkmark next to Curve Smoothing and set the Curve Length to **50′** (**20** m).

    Leave all other settings at their defaults, as shown in Figure 12.17.
18. Click Finish.

FIGURE 12.17
Finalizing the superelevation on the Attainment screen

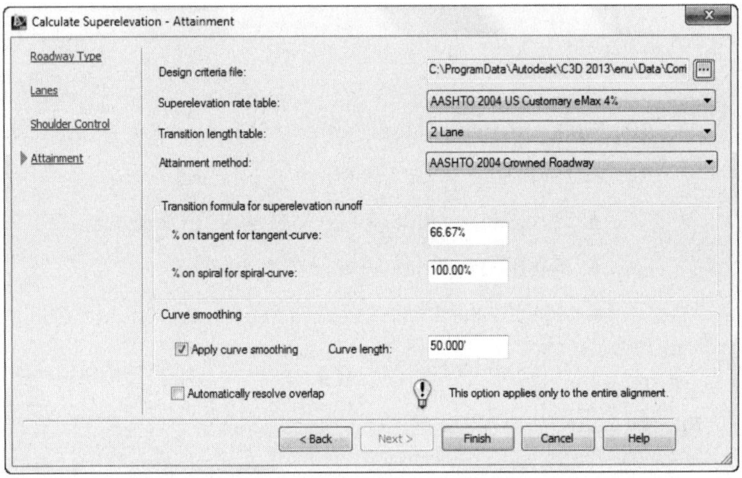

You should now seethe superelevation table appear inside Panorama with the data resulting from the wizard. Examine your alignment; you should now have labels showing the superelevation critical stations created by the wizard.

As you click in the table, you will see helpful glyphs showing you which superelevation station and corresponding curve you are editing, as shown in Figure 12.18.

Transition Station Overlap

It is not uncommon to have overlap warnings in your superelevation table. You should resolve the transition station overlap before you continue your design.

Overlap occurs when there is not enough room between curves to fully transition out of one curve and back into the next. Transition station overlap will always occur when a reverse curve or compound curve exists in your alignment. As you can see in Figure 12.19, Curve 1 does not complete its transition out until station 16+82.56, but according to the attainment calculations, Curve 2 will begin affecting the shoulder starting at station 15+35.24.

FIGURE 12.18 Superelevation table with glyphs in the graphic

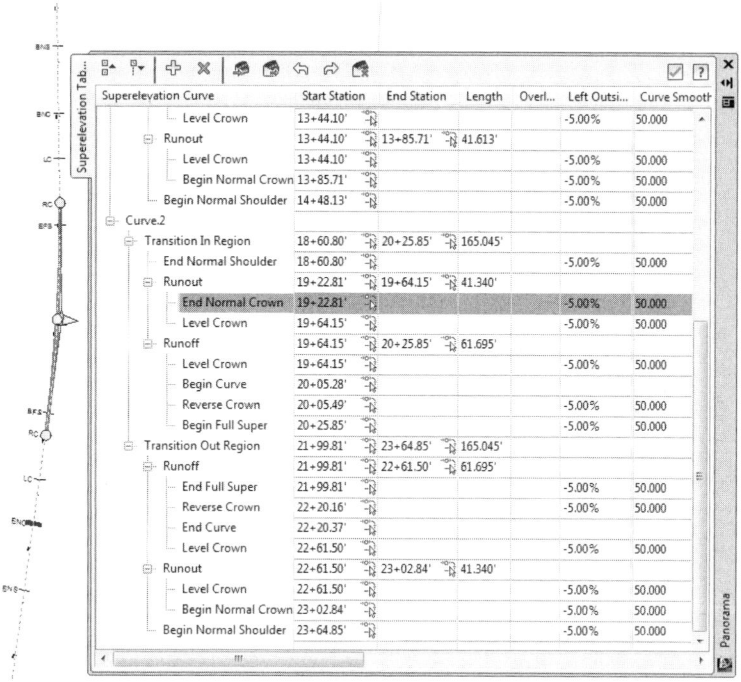

Superelevation Curve	Start Station	End Station	Length	Overl...	Left Outsi...	Curve Smooth
Level Crown	13+44.10'				-5.00%	50.000
Runout	13+44.10'	13+85.71'	41.613'			
Level Crown	13+44.10'				-5.00%	50.000
Begin Normal Crown	13+85.71'				-5.00%	50.000
Begin Normal Shoulder	14+48.13'				-5.00%	50.000
Curve.2						
Transition In Region	18+60.80'	20+25.85'	165.045'			
End Normal Shoulder	18+60.80'				-5.00%	50.000
Runout	19+22.81'	19+64.15'	41.340'			
End Normal Crown	19+22.81'				-5.00%	50.000
Level Crown	19+64.15'				-5.00%	50.000
Runoff	19+64.15'	20+25.85'	61.695'			
Level Crown	19+64.15'				-5.00%	50.000
Begin Curve	20+05.28'					
Reverse Crown	20+05.49'				-5.00%	50.000
Begin Full Super	20+25.85'				-5.00%	50.000
Transition Out Region	21+99.81'	23+64.85'	165.045'			
Runoff	21+99.81'	22+61.50'	61.695'			
End Full Super	21+99.81'				-5.00%	50.000
Reverse Crown	22+20.16'				-5.00%	50.000
End Curve	22+20.37'					
Level Crown	22+61.50'				-5.00%	50.000
Runout	22+61.50'	23+02.84'	41.340'			
Level Crown	22+61.50'				-5.00%	50.000
Begin Normal Crown	23+02.84'				-5.00%	50.000
Begin Normal Shoulder	23+64.85'				-5.00%	50.000

FIGURE 12.19 Superelevation table showing overlap between two curves

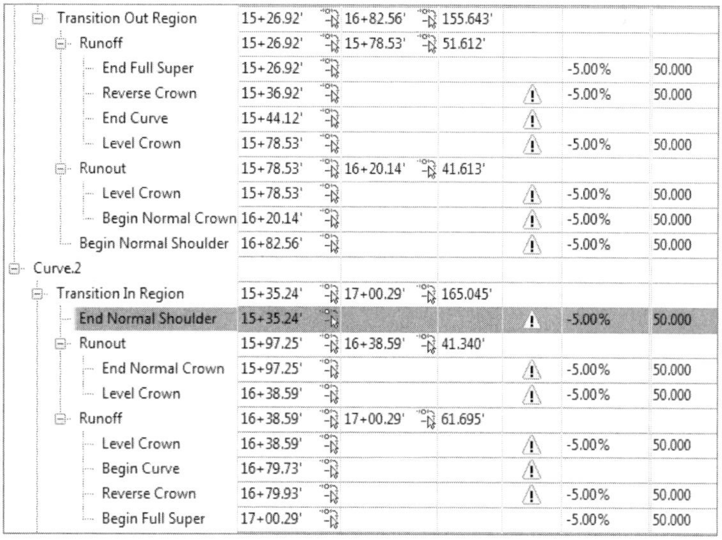

Transition Out Region	15+26.92'	16+82.56'	155.643'			
Runoff	15+26.92'	15+78.53'	51.612'			
End Full Super	15+26.92'				-5.00%	50.000
Reverse Crown	15+36.92'			⚠	-5.00%	50.000
End Curve	15+44.12'			⚠		
Level Crown	15+78.53'			⚠	-5.00%	50.000
Runout	15+78.53'	16+20.14'	41.613'			
Level Crown	15+78.53'			⚠	-5.00%	50.000
Begin Normal Crown	16+20.14'			⚠	-5.00%	50.000
Begin Normal Shoulder	16+82.56'			⚠	-5.00%	50.000
Curve.2						
Transition In Region	15+35.24'	17+00.29'	165.045'			
End Normal Shoulder	15+35.24'			⚠	-5.00%	50.000
Runout	15+97.25'	16+38.59'	41.340'			
End Normal Crown	15+97.25'			⚠	-5.00%	50.000
Level Crown	16+38.59'			⚠	-5.00%	50.000
Runoff	16+38.59'	17+00.29'	61.695'			
Level Crown	16+38.59'			⚠	-5.00%	50.000
Begin Curve	16+79.73'			⚠		
Reverse Crown	16+79.93'			⚠	-5.00%	50.000
Begin Full Super	17+00.29'				-5.00%	50.000

You have several options for fixing superelevation overlap:

- You can choose to have Civil 3D rectify the overlap for you.
- You can manually modify the stations in the table.
- You can change the stationing for superelevation by modifying the superelevation view, which we will discuss later in this chapter.

To have Civil 3D clear the overlap for you, click the warning symbol that appears in the Superelevation Tabular Editor. Civil 3D resolves overlap by omitting noncritical stations and/or by compressing the transition length between certain stations. In the case of a reverse curve, Civil 3D will pivot the road from full-super to full-super, without transitioning back to normal crown. Be sure to verify that the software has made the update that meets the requirements of your locale.

PRESERVING YOUR MANUAL EDITS

You may spend hours getting your superelevation stations to work out correctly. However, it just takes one wrong button click to blow away all your time-consuming edits to the tabular input. To save you from yourself (or the click-happy intern), back up your superelevation tables by exporting them to a file. You'll find the Export Superelevation Data button at the top of the tabular input.

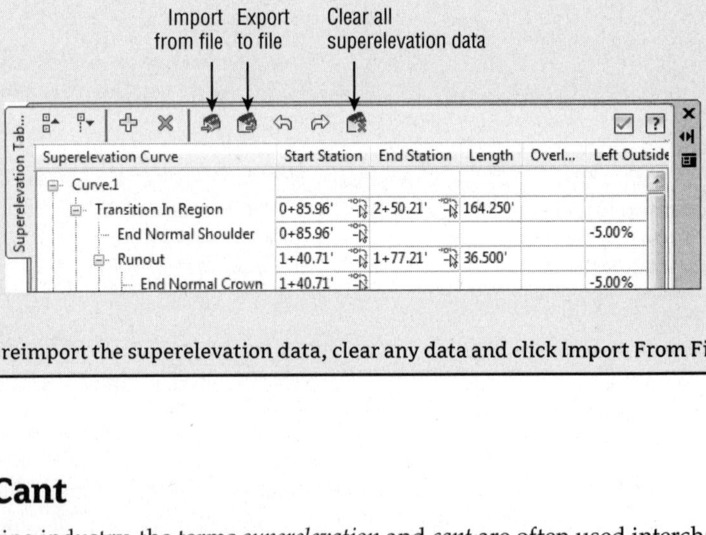

When you need to reimport the superelevation data, clear any data and click Import From File.

Oh Yes, You Cant

In the civil engineering industry, the terms *superelevation* and *cant* are often used interchangeably. Inside Civil 3D, the terms have distinct meanings (Figure 12.20).

Superelevation tools in Civil 3D are used for roads where the cross-slope changes within a curve are expressed by a percentage. Unlike superelevation, cant is expressed as a difference in height between the outer and inner rails.

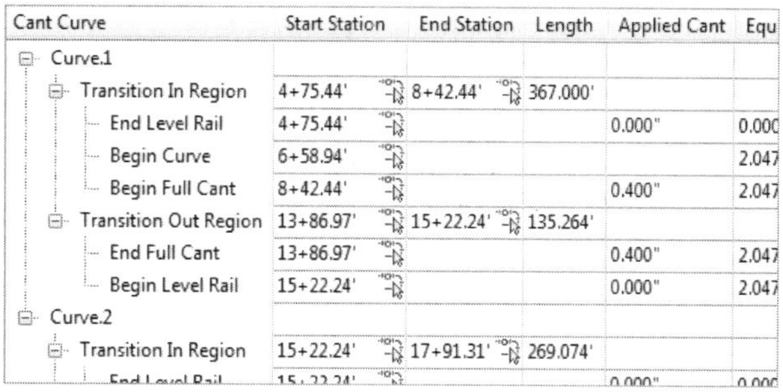

Cant Curve	Start Station	End Station	Length	Applied Cant	Equ
Curve.1					
Transition In Region	4+75.44'	8+42.44'	367.000'		
End Level Rail	4+75.44'			0.000"	0.000
Begin Curve	6+58.94'				2.047
Begin Full Cant	8+42.44'			0.400"	2.047
Transition Out Region	13+86.97'	15+22.24'	135.264'		
End Full Cant	13+86.97'			0.400"	2.047
Begin Level Rail	15+22.24'			0.000"	2.047
Curve.2					
Transition In Region	15+22.24'	17+91.31'	269.074'		
End Level Rail	15+22.24'			0.000"	0.000

FIGURE 12.20
Example cant data table

Workin' on the Railroad

Cant tools are new to Civil 3D 2013 and are specifically designed for rail. In order to work with cant, the following must be in place:

Alignment Type Set to Rail Normally, you would set the alignment type when you first define the alignment. If you forget to set this initially, you can change it at any time in the alignment properties on the Information tab.

Alignment Design Speed Like motorways and superelevation, rail requires a design speed in order to apply cant. Cant design standards are provided with the software, and can be edited in the same manner as other design criteria files.

Cant Calculated Like superelevation, cant is attached to an alignment and its curves. As seen in Figure 12.21, the icon's location in the Alignment contextual tab should be reminiscent of the superelevation button.

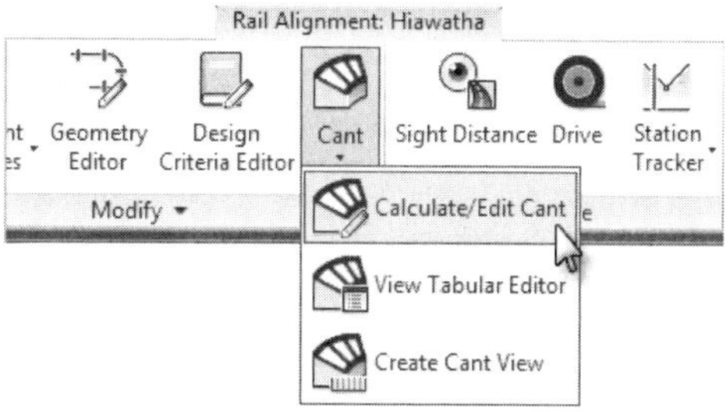

FIGURE 12.21
Accessing the Cant Calculation tools from the Rail Alignment contextual tab

Rail Assembly When working with an assembly for rail, the type must be set to Railway. You can set this on assembly creation, or in the assembly properties after the fact. There is one subassembly that is purpose-built for cant, the Rail subassembly, as shown in Figure 12.22. If you are working with a drawing created in a previous release, make sure you replace the old assemblies with this new rail subassembly to ensure cant takes place.

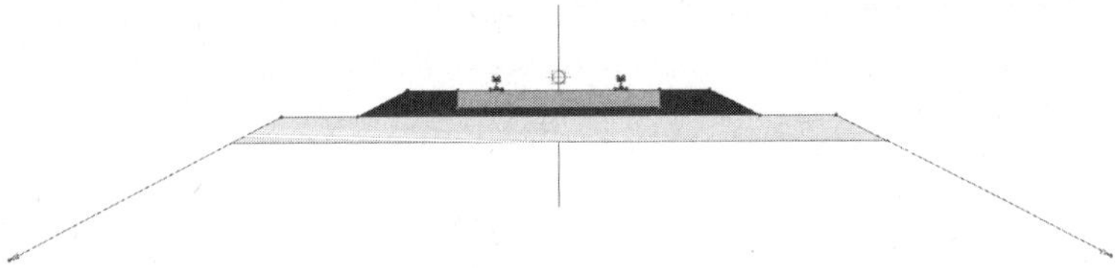

FIGURE 12.22
The new rail subassembly

Creating a Rail Assembly

You will find the Rail assembly in the Bridge And Rail tab of the tool palettes. The following exercise will walk you through creating a typical rail bed design:

1. Open the drawing `Rail.dwg` (`Rail_METRIC.dwg`), which you can download from this book's website.

   This drawing contains an alignment and design profile.

2. On the Home tab ➢ Create Design panel ➢ Assembly, click Create Assembly.
3. Name the assembly **Rail w Service Road**.
4. Set Assembly Type to Railway, and click OK.
5. Click to place the assembly anywhere in the graphic.
6. Open the tool palettes if they are not already open. In the subassembly tool palettes, locate the Bridge And Rail palette, and click Rail Single. In the Properties palette:

   A. Set the Subballast width to **20′** (**6** m).

   B. Set the Subballast Side slope to **0.001:1**.

   Leave all other parameters at their defaults.

7. Click the assembly to place the RailSingle subassembly.
8. Switch to the Basic palette, and click GenericPavementStructure.

   This will be our service road, constructed out of the same material as the subballast.

9. Enter the following data into the Properties palette, remembering that the codes are case-sensitive:

   A. Set Side to Left.

   B. Set Width to **8′** (**2.5** m).

   C. Set Depth to **1′** (**0.3** m).

   D. Set DeflectOuterVerticalFace to Yes.

   E. Set Outer Edge Slope to **2:1**.

   F. Set TopLink Codes to Top, and set BottomLink Codes to Datum.

   G. Set Shape Codes to Subballast.

10. When your Parameters resemble Figure 12.23, click to place the GenericPavementStructure to the left of the rail subassembly.

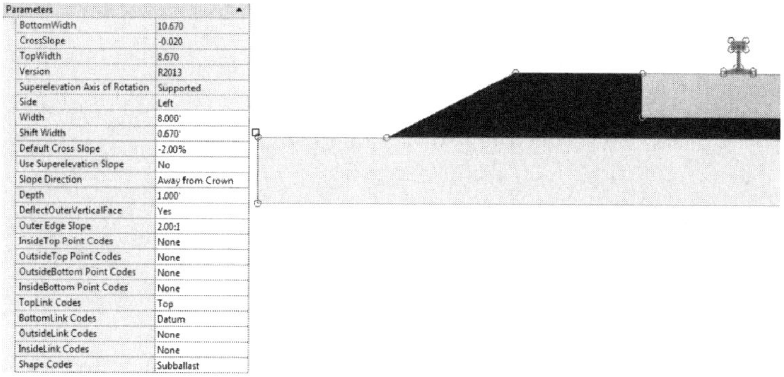

FIGURE 12.23
Placing an assembly to represent additional subballast for a service road next to the rail bed

11. Remain in the GenericPavementStructure tool, but change the width to **0.1'** (**0.03** m).
12. Click to place the structure on the right side of the rail subassembly.
13. Switch to the Generic palette, and click LinkSlopeToSurface.
14. Set Slope to **-50** and click the "tip" of the generic pavement structure on both sides of the assembly.

    Your completed assembly will look like Figure 12.24.

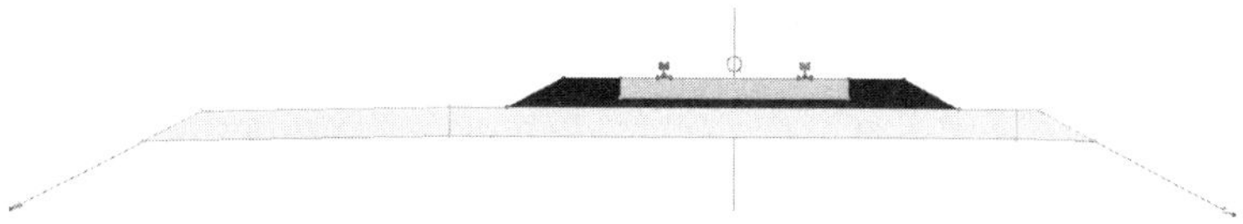

FIGURE 12.24
Your completed rail assembly

15. Save and close the drawing.

Applying Cant to the Alignment

Like superelevation in a roadway alignment, *cant* is related to a rail alignment. The following exercise will walk you through applying cant to the alignment. You should experience a distinct *déjà-vu* feeling if you completed earlier exercises involving applying superelevation to the alignment.

1. Open the drawing `RailAlignment.dwg` (`RailAlignment_METRIC.dwg`), which you can download from this book's website.

   This drawing contains a corridor with rail; your task is to update the alignment to apply cant. Observe the section views in the drawing. Each is created at a location on the curve that should have cant applied. However, these views show flat rails.

2. Select the alignment. From the Rail Alignment contextual tab ➢ Modify panel, click Cant ➢ Calculate/Edit Cant.

3. Click Calculate Cant Now.
4. Keep the pivot method as Low Side Rail, and click Next.
5. Place a check mark next to Automatically Resolve Overlap.
6. Set Applied Cant Table to Freight Train Applied Cant Table (metric users, use Mixed Passenger And Freight Cant).
7. Leave all other settings at their defaults, and click Finish.
8. In Prospector, rebuild the Hiawatha corridor.
9. Save the drawing.

You should observe that your section views show the cant applied to the design.

Superelevation and Cant Views

Superelevation and cant views are a graphic representation of the roadway or rail superelevation. Grip edits to the graphical view will also edit the superelevation stations. The view itself is not intended for plotting. The superelevation view plots station against lane slope to form a sort of profile of the left and right edges of the pavement.

At first glance, the superelevation view may seem difficult to read, but with a little explanation it can shed a lot of light on what is going on with your lane slopes. Figure 12.25 shows the superelevation diagram for a two-lane road that contains a reverse curve. The superelevation graphic plots the station value against the percent cross-slope of each edge of pavement. The upper line shows the behavior of the right edge of the pavement, and the lower line shows the left edge of the pavement.

FIGURE 12.25
The superelevation view

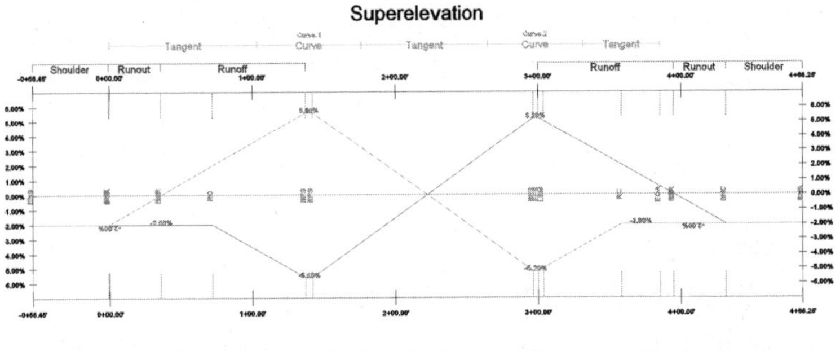

Where no superelevation is applied, the graph data is flat. As the assembly twists into position during superelevation, the distances between the lines become greater as the right edge slopes up to a maximum superelevation of 5.6%. You can also see in Figure 12.25 how Civil 3D has resolved overlap between the two curves. Neither line goes back to the default position at -2.00% until the end of the alignment. The crisscross in the center indicates that the roadway transitions directly from one max super to the other.

Place the view into Civil 3D by selecting the alignment and clicking Superelevation ➢ Create Superelevation View. To simplify the view and make it easier to edit, you can choose which lanes to show or hide at the bottom of the view, as shown in Figure 12.26. You should also change the colors for left and right edge of pavement by clicking the ByBlock option and changing it to the color of your choice.

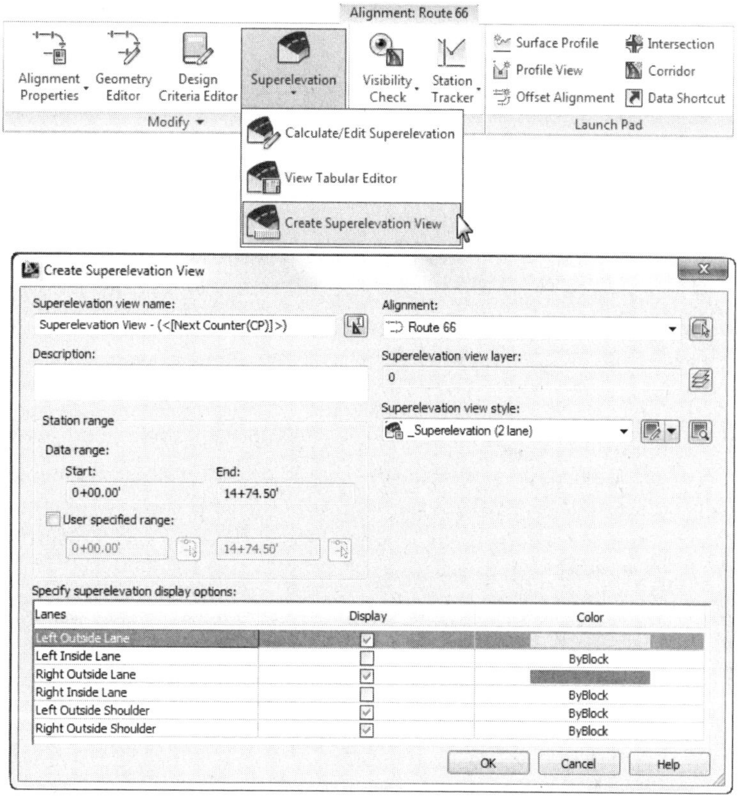

FIGURE 12.26
Creating a superelevation view

To modify superelevation data using the graphic, click on it to activate the numerous grips. Figure 12.27 shows a close-up of some of the grips you will see.

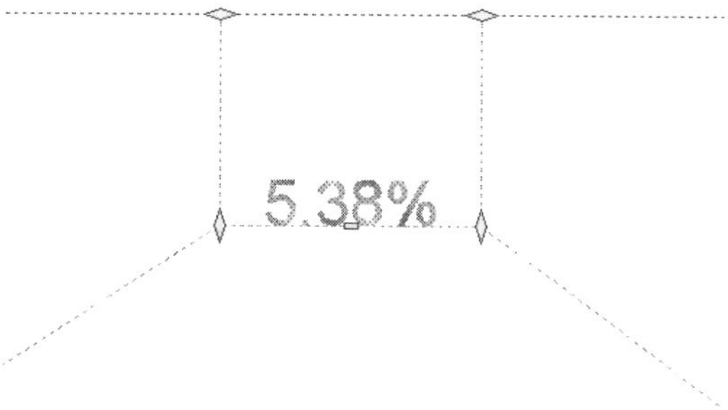

FIGURE 12.27
Grip-editing is another method for fine-tuning your superelevation stationing.

The diamond-shaped grips can be slid in one axis to modify stationing (the horizontally oriented grips) or slope (the vertically oriented grips). The rectangular grip can be moved to reduce the maximum lane slope when it is in a full-super state.

The Bottom Line

Add superelevation to an alignment. Civil 3D has convenient and flexible tools that will apply safe, correct superelevation to an alignment curve.

Master It Open the`Master Super.dwg` (`Master Super_METRIC.dwg`)file, which you can download from `www.sybex.com/go/masteringcivil3d2013`. Set the design speed of the road to 20 miles per hour (35 km per hour) and apply superelevation to the entire length of the alignment. Use AASHTO 2004 Design Criteria with an eMax of 6% 2-Lane.

Create a superelevation assembly. In order for superelevation to happen, you need to have an assembly that is capable of superelevating.

Master It Continue working in `Master Super.dwg` (`Master Super_METRIC.dwg`). Create an assembly similar to the one in the top image of Figure 12.11. Set each lane to be 14′ (4.5 m) wide, and each shoulder to be 6′ (2 m) wide. Leave all other options at their defaults. If time permits, build a corridor based on the alignment and assembly.

Create a rail corridor with cant. Cant tools are new to Civil 3D 2013 and allow users to create corridors that meet design criteria specific to rail needs.

Master It In the drawing `MasterCant.dwg` (`MasterCant_METRIC.dwg`), create a Railway assembly with the RailSingle subassembly using the default parameters for width and depth. Add a LinkSlopetoSurface generic link with 50% slope to each side. Add cant to the alignment in the drawing using the default settings for attainment. Create a corridor from these pieces.

Create a superelevation view. Superelevation views are a great place to get a handle on what is going on in your roadway design. You can visually check the geometry as well as make changes to the design.

Master It Open the drawing `MasterView.dwg` (`MasterView_METRIC.dwg`). Create a superelevation view for the alignment. Show only the Left and Right Outside Lanes as blue and red, respectively.

Chapter 13

Cross Sections and Mass Haul

Cross sections are used in the AutoCAD® Civil 3D® program to allow the user to have a graphic confirmation of design intent, as well as to calculate the quantities of materials used in a design. Sections must have at least two types of Civil 3D objects: an alignment and a surface. Other objects, such as pipes, structures, and corridor components, can be sampled in a sample line group, which is used to create the graphical section displayed in a section view. These section views and sections remain dynamic throughout the design process, reflecting any changes made to the sampled information. This process reduces potential errors in materials reports, keeping often costly mistakes from happening during the construction process.

In this chapter, you will learn to:

- Create sample lines
- Create section views
- Define and compute materials
- Generate volume reports

Section Workflow

When the time comes that you wish to see how the information along your alignment will appear plotted, it is time to create *sample lines*. If your goal is to show your completed design, at this point you should have a completed corridor, corridor Top surface, and corridor Datum surface.

Sample Lines vs. Frequency Lines

Sample lines are created at any stations where you wish to create a *section view*. Sample lines are also used to compute end area volumes.

A common point of confusion with new users is the difference between frequency lines and sample lines. Table 13.1 explains the differences.

TABLE 13.1: Sample lines vs. frequency lines

SAMPLE LINES	FREQUENCY LINES
Can be created without a corridor present (e.g., you wish to see existing surface sections along the alignment).	Are part of a corridor.
Occur at any station where a section view is needed.	Occur anywhere the design needs to be modified (e.g., a driveway).
Used for end area volume computation.	Used to apply assembly calculations to the design (e.g., locating slope-intercept).
Can be skewed at an angle other than 90° from the baseline.	Are always perpendicular to the baseline.
Swath width is usually uniform and dependent on user plotting needs.	Length from baseline depends on assembly and will vary from station to station.
Section views are read-only reflections of the design.	Design can be modified in the Section Editor at each frequency line.
Section views are readily adapted to plotting.	Plotting should never occur from the section editor.

When you create your sample line group, you will have the option to sample any surface in your drawing, including corridor surfaces, the corridor assembly itself, and any pipes in your drawing. The sections are then sampled along the alignment with the left and right widths specified and at the intervals specified. Once the sample lines are created, you can then choose to create section views or to define materials.

Creating Sample Lines

Sample lines are the engine underneath both sections and materials and are held in a collection called sample line groups. A sample line group is always associated with an alignment and can be found under the associated alignment in Prospector, as shown in Figure 13.1.

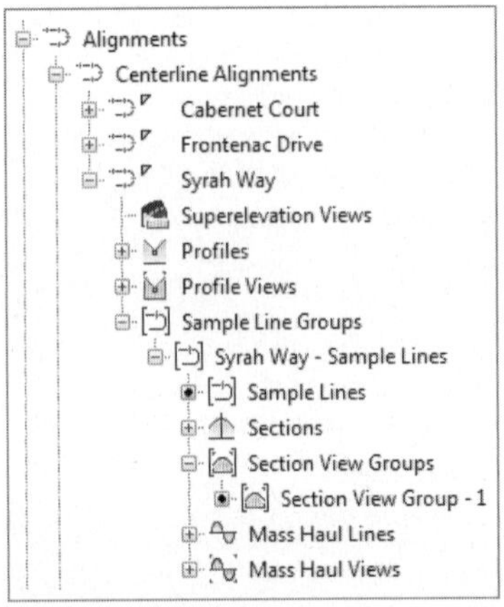

FIGURE 13.1
A view of Prospector; sample lines are held in a collection called a sample line group and are dependent on an alignment.

You can also see in Figure 13.1 that quite a few additional items are dependent on sample lines, such as sections, section views, mass haul lines, and mass haul views. If you click a sample line, you will see it has three types of grips, as shown in Figure 13.2.

FIGURE 13.2
The three grips on a sample line

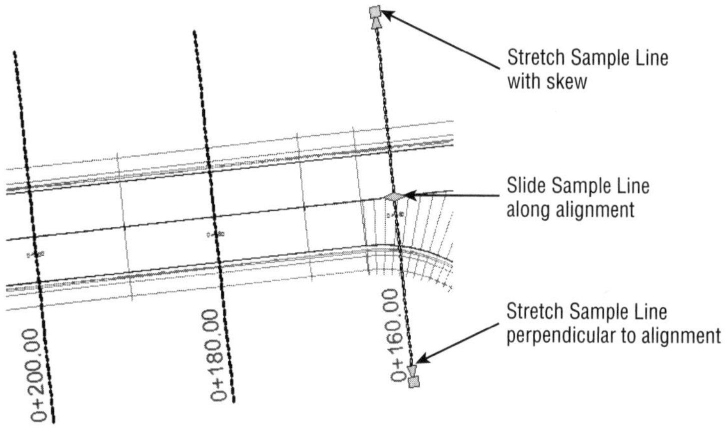

To create a sample line group, change to the Home tab and choose Sample Lines from the Profile & Section Views panel. After selecting the appropriate alignment, the Create Sample Line Group dialog shown in Figure 13.3 will display. You should name the sample line group and verify the sample line and label style.

FIGURE 13.3
The Create Sample Line Group dialog

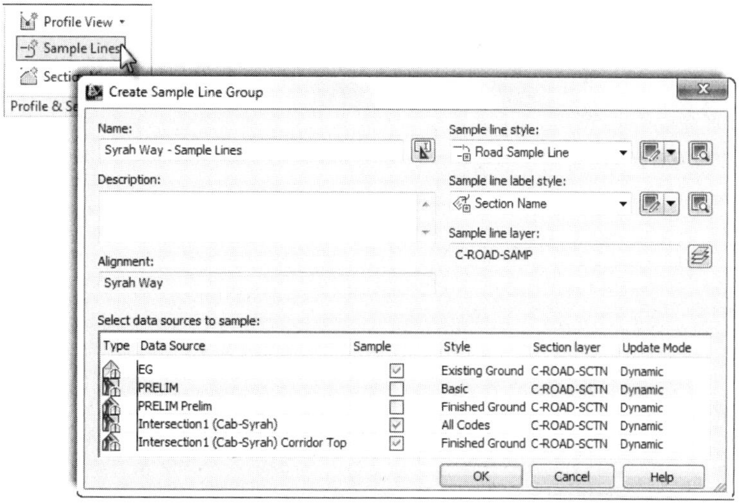

Every source object that is available will be displayed at the bottom of this box. If you wish to omit specific data from the section view, you can clear the check box. Set the applicable style for each item by clicking in the column to the right of the object. The section layer should be preset as specified in your template settings.

Once you've selected the sample data, the Sample Line Tools toolbar will appear, as shown in Figure 13.4.

FIGURE 13.4
From the Sample Line Tools toolbar, choose By Range Of Stations.

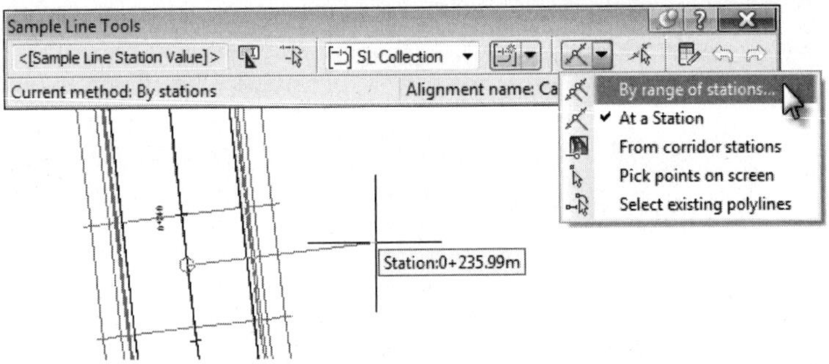

The By Range Of Stations option is used most often. You can use At A Station to create one sample line at a specific station. From Corridor Stations allows you to insert a sample line at each corridor assembly insertion. Pick Points On Screen allows you to pick any two points to define a sample line. This option can be useful in special situations, such as sampling a pipe on a skew. The last option, Select Existing Polylines, lets you define sample lines from existing polylines.

To define sample lines, you need to specify a few settings. Figure 13.5 shows these settings in the Create Sample Lines – By Station Range dialog. Right Swath Width is the width from the alignment that you sample. Most of the time this distance is greater than the ROW distance. You can also select Sampling Increments and choose whether to include special stations, such as horizontal geometry (PC, PT, and so on), vertical geometry (PVC, high point, low point, and so on), and superelevation critical stations.

FIGURE 13.5
Create Sample Lines – By Station Range dialog

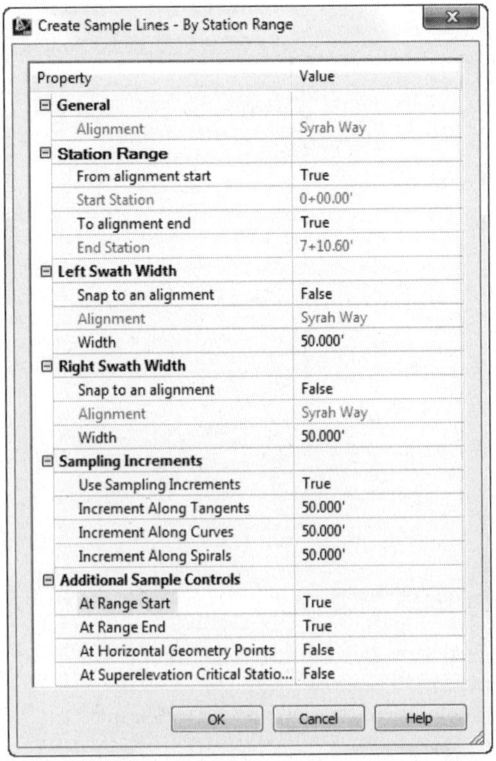

CERT OBJECTIVE

In the following exercise, you'll create sample lines for Cabernet Court:

1. Open the `SampleLines.dwg` (`SampleLines_METRIC.dwg`) file, which you can download from this book's web page at `www.sybex.com/go/masteringcivil3d2013`.
2. Select the Home tab ➢ Profile & Section Views panel and click Sample Lines.
3. At the `Select an alignment, or press enter key to select from list>:` prompt, press ↵ to display the Select Alignment dialog.
4. Select the Syrah Way alignment and click OK.

   The Create Sample Line Group dialog opens.
5. In the Create Sample Line Group dialog:
   a. Name the group **Syrah Way-Sample Lines**.
   b. Clear the check boxes to the right of the PRELIM corridor and its surface named PRELIM Prelim.
   c. Set surface EG to style Existing Ground by clicking in the Style column to the right.
   d. Set the style for surface Intersection1 (Cab-Syrah) Corridor Top to Finished Ground.

      The column is too narrow to view the full names of the items, so to see the name of the item pause your cursor over the name or expand the column.
   e. Set the corridor Intersection1 (Cab-Syrah) style to All Codes.
6. When your dialog looks similar to Figure 13.3 (your sampled items may be listed in a different order), click OK.
7. On the Sample Line Tools toolbar, click the Sample Line Creation Methods drop-down arrow and then select By Range Of Stations.
8. In the Create Sample Lines – By Station Range dialog, leave the swath widths and sampling increments at their defaults (this will be 50′ for Imperial units and 20 m for metric units).
9. Change both the At Range Start and At Range End options to True, as shown earlier in Figure 13.5.
10. Click OK, and press ↵ to end the command.
11. If you receive a Panorama view telling you that your corridor is out of date and may require rebuilding, dismiss it by clicking the green check box.

    You should now have dashed lines at even station intervals; these are your new sample lines.
12. Save the drawing for use in the next exercise.

Editing the Swath Width of a Sample Line Group

There may come a time when you need to show information outside the limits of your section views or not show as much information. To edit the width of a section view, you will have to change the swath width of a sample line group. These sample lines can be edited manually on an individual basis, or you can edit the entire group at once.

Sample Line and Section View Words of Wisdom

It is extremely rare to encounter situations where you need to touch each sample line or section view individually. Select a sample line or section view to access many time- (and sanity-) saving tools from the contextual tab.

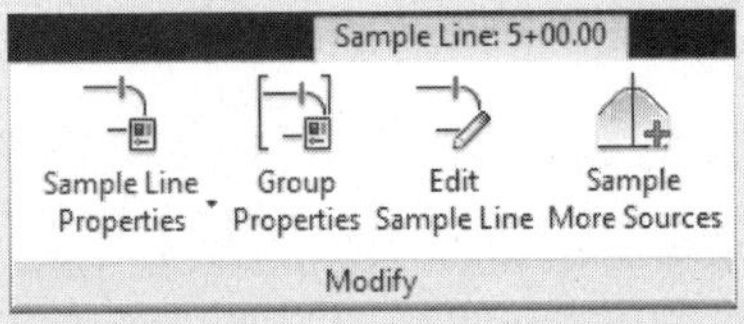

You want to Change All the Sample Line Lengths To change many swath widths at once:

1. Open the Sample Line Group Properties ➤ Sample Lines tab, as shown here:

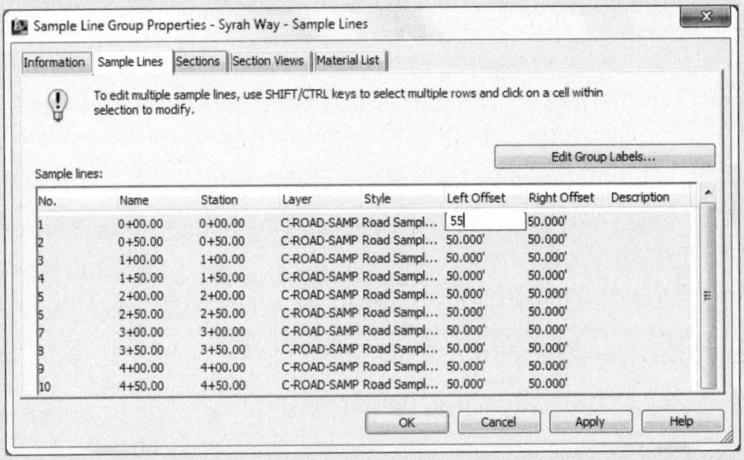

2. Select the first station you wish to change and hold down Shift as you click the last row.
3. Change the swath widths as desired.

   You will need to change both the left and the right, because they are independent from each other.

   You Want to Add or Remove Section Data in Many Views A common situation with sections is that surface or corridor data you've created after you created sample lines is not appearing in your cross sections. This is easily rectified:

1. Click Sample More Sources.

   Data that exists but that is not recognized by the sample lines will appear to the left, as shown here:

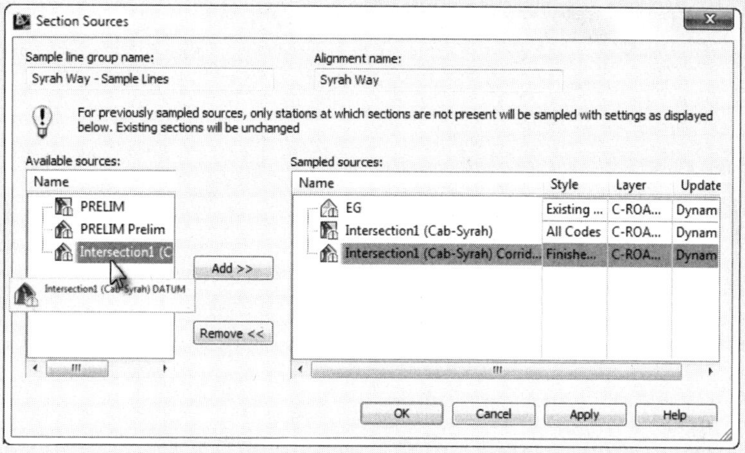

2. Highlight the data you wish to add, and click Add.

In this exercise, you'll edit the widths of an entire sample line group. You must complete the previous exercise before proceeding.

1. Continue working on the `SampleLines.dwg` (`SampleLines_METRIC.dwg`) file.
2. Select a sample line.

3. From the Sample Line contextual tab ➤ Modify panel, click Group Properties.
4. Switch to the Sample Lines tab.
5. Click to select the first sample line in the listing at station 0+00.00 (0+000.00).
   a. Scroll down to the bottom of the list.
   b. While holding the Shift key, select the last station in the listing. This will highlight all rows.
   c. Click your cursor in the Left Offset column and change it to 100′ (30 m).
   d. Press ↵ to complete the edit.

      After a moment, the left offsets should change.
   e. With all of the rows still highlighted, make the same change to the Right Offset column.
6. Click OK.

   After a moment the sample lines will resize to reflect your change.
7. Save the drawing.

Creating Section Views

Once the sample line group is created, it is time to create views. You can create a single view or many views arranged together (Figure 13.6).

FIGURE 13.6
Section views arranged to plot by page

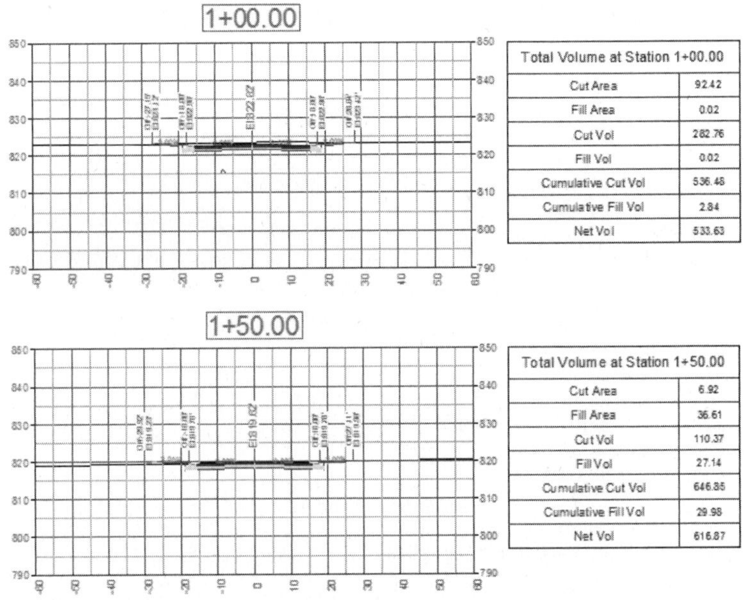

A section view is nothing more than a window showing the section. The view contains horizontal and vertical grids, tick marks for axis annotation, the axis annotation itself, and a title. Views can also be configured to show horizontal geometry, such as the centerline of the section, edges of pavement, and right of way. Tables displaying quantities or volumes can also be shown for individual sections.

Creating a Single-Section View

CERT OBJECTIVE

There are occasions when all section views are not needed. In these situations, a single-section view can be created. In this exercise, you'll create a single-section view of station 0 + 00.00 (0+000.00) from sample lines:

1. Open the `SectionViews.dwg` (`SectionViews _METRIC.dwg`) file, which you can download from this book's web page.

   You will want to switch to this drawing, since there are a few steps completed for you that you will learn later in the chapter.

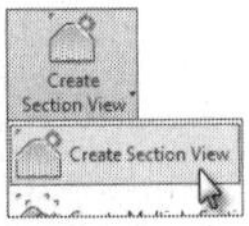

2. Select the sample line at 0+50.00 (0+020.00).
3. From the Sample Line contextual tab ➤ Launch Pad ➤ Create Section view, select Create Section View.

4. Verify that your alignment, sample line group name, and station are correct on the General page of the wizard, as shown in Figure 13.7.

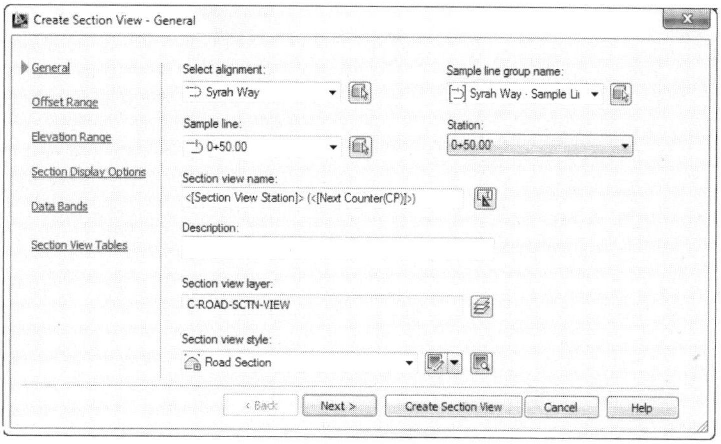

FIGURE 13.7
The General page of the Create Section View wizard

You can navigate from one page to another by either clicking Next at the bottom of the screen or clicking the links on the left side of the screen.

5. Click Next to view the Offset Range page.

The top of Figure 13.8 shows the Offset Range page, which should match your sample line swath width. In this case, as in the vast majority of cases, you will leave this set to Automatic. No action is needed on the Offset Range page.

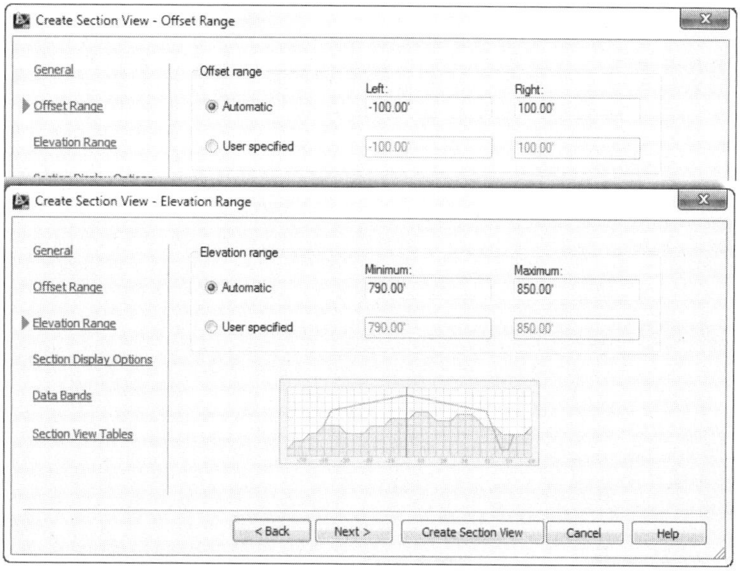

FIGURE 13.8
The Offset Range page (top) and Elevation Range (bottom) of the Create Section View wizard

6. Click Next to view the Elevation Range page.

   The bottom of Figure 13.8 shows the Elevation Range page. The values shown here are taken from your design max and min elevations. In this case, as in the vast majority of cases, you will leave this set to Automatic. No action is needed on the Elevation Range page.

7. Click Next to view the Section Display Option page.

   The fourth page contains the section display options, as shown in Figure 13.9. This page reflects the styles and data you selected when creating your sample lines. If you forgot to set a style or wish to omit additional data, you can change your options here. No action is needed on the Section Display Options page.

FIGURE 13.9
The Section Display Options page of the Create Section View wizard

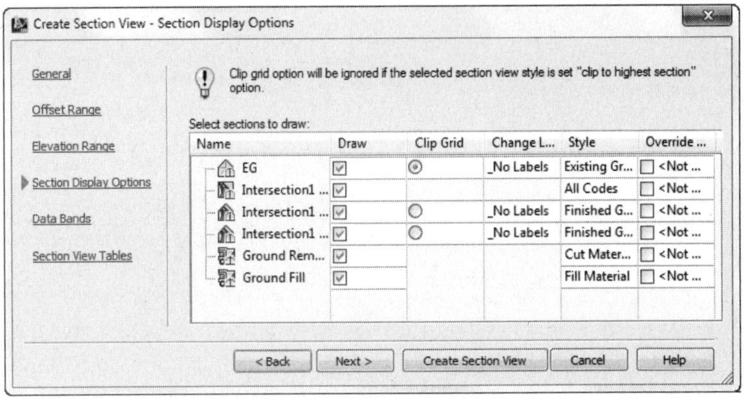

8. Click Next to view the Data Bands page.

   The fifth page, shown in Figure 13.10, lets you specify the data band options. Here, you can select band sets to add to the section view, pick the location of the band, and choose the surfaces to be referenced in the bands. No action is needed on the Data Bands page.

FIGURE 13.10
The Data Bands page of the Create Section View wizard

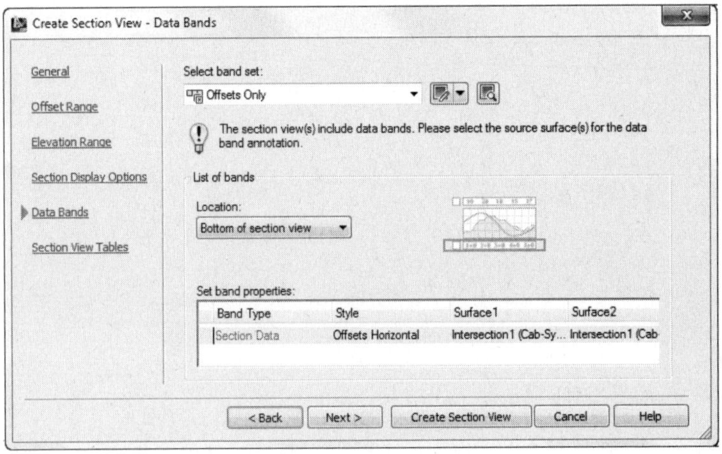

9. Click Next to view the Section View Tables page.

   The sixth and last page, shown in Figure 13.11, is where you set up the section view tables. Note that this screen will be available only if you have already computed materials for the sample line group. On this page, you can select the type of table and the table style, and select the position of the table relative to the section view. Notice the graphic on the right side of the window that illustrates the current settings.

FIGURE 13.11
The Section View Tables page of the Create Section View wizard

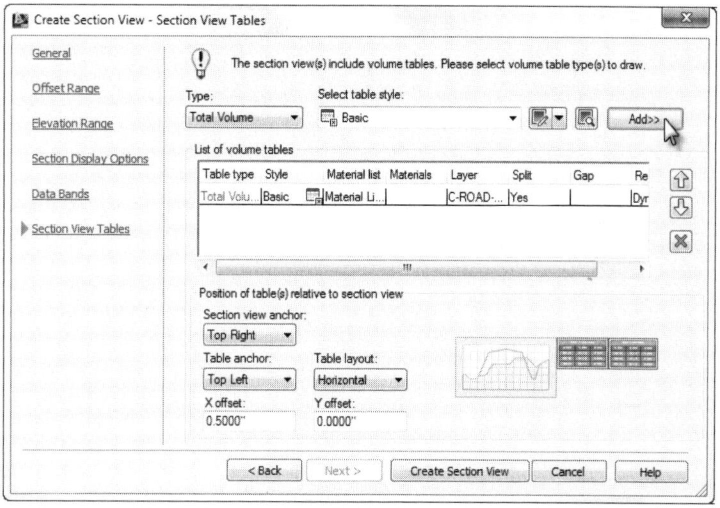

10. With the table type set to Total Volume and the Select Table Style set to Basic, click Add.

    As shown in Figure 13.11, you will have a line indicating that a material table will come in to the right of your view.

11. Click Create Section View.

12. Pick any point in the drawing area to place your section view.

13. Examine your section view.

    The display should resemble Figure 13.12.

FIGURE 13.12
The finished section view

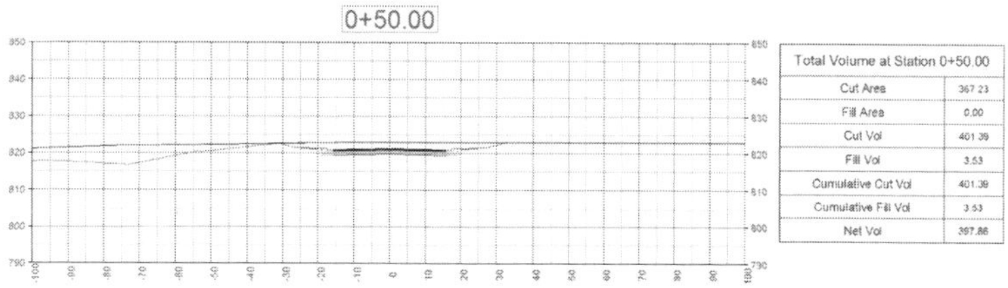

14. Save the drawing for use in the next exercise.

SECTION VIEW OBJECT PROJECTION

Civil 3D has the ability to project AutoCAD® points, blocks, 3D solids, 3D polylines, Civil 3D COGO points, feature lines, and survey figures into section and profile views. Each of the objects listed can be projected to a section view and labeled appropriately. See Chapter 7, "Profiles and Profile Views," to learn more. You access the command by changing to the Home tab's Profile And Section Views panel and selecting Section Views ➢ Project Objects To Section View or Project Objects To Multiple Section Views.

Creating Multiple Section Views

Section views belong in packs. In the exercise that follows, you will create section views intended to plot together on a sheet. (It is not necessary to have completed the previous exercise to continue.)

1. Continue working in `SectionViews.dwg` (`SectionViews_METRIC.dwg`).
2. Select any sample line.
3. From the Sample Line contextual tab ➢ Modify Panel ➢ Create Section view, select Create Multiple Section Views.

   Alternately, you can access this tool from the Home tab ➢ Profile & Section Views ➢ Section Views ➢ Create Multiple Views.
4. On the General page, set the Section View Style option to Road Section–No Grids, and click Next.
5. On the Section Placement page, click the ellipsis next to the path for Template For Cross Section Sheet.
6. From the default cross section sheet template, select ARCH D Section 20 Scale (ISO A1 Section 1 to 500), as shown in Figure 13.13, and click OK.
7. Set Group Plot Style to Grid By Page, and click Next.

   Figure 13.13 is a page you did not see when placing a single view. Setting the Production radio button allows you to use the Create Section Sheets tool from the Output tab. The Draft option forces Civil 3D to behave like version 2010 and prior.

   Group Plot Style controls how the views are arranged on a page. In this example, the grid will come from the Group Plot style rather than the Section View style.

   You can skip the rest of the wizard because you will keep the default input for the remainder of the settings.
8. Click Create Section Views.

9. Click anywhere off to the right of the graphic to place the views.

   You should see views arranged on the screen (Figure 13.14). Metric users will see one page of section views, and Imperial unit users will see three pages of section views.

10. Save the drawing for use in the next exercise.

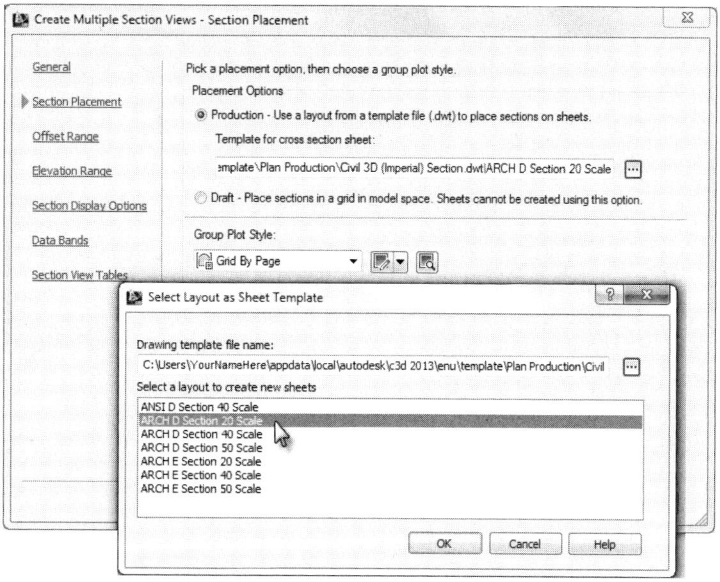

FIGURE 13.13
When you're creating multiple views, options allow for neat-looking groups.

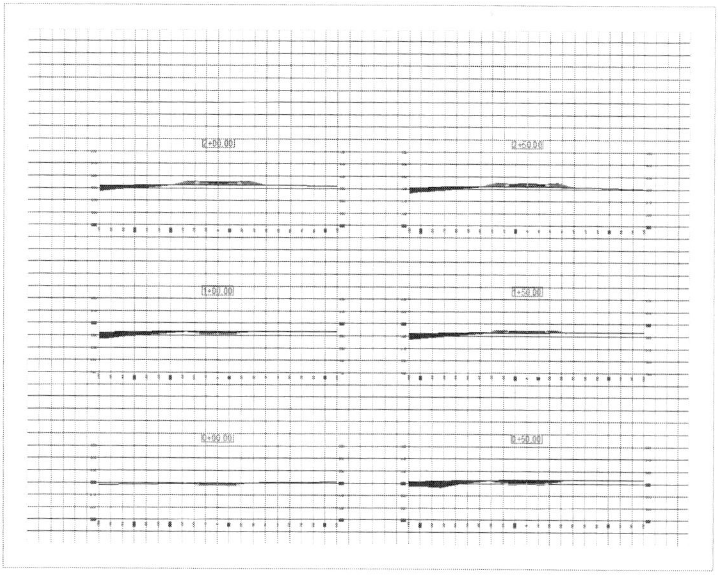

FIGURE 13.14
One of the pages of cross-section views

DELETE SECTION VIEWS THE PAIN-FREE WAY

The best way to remove an unwanted batch of section views from your graphic is to delete them through Prospector. In the following graphic you see the intrepid user tried (and apparently failed) seven times before keeping the view group.

If you delete views graphically, you'll create more work for yourself. First, the section view group stays behind in Prospector, as you see in the following image. As indicated by the dot in front of the group, only group 8 and the individual section view contain information.

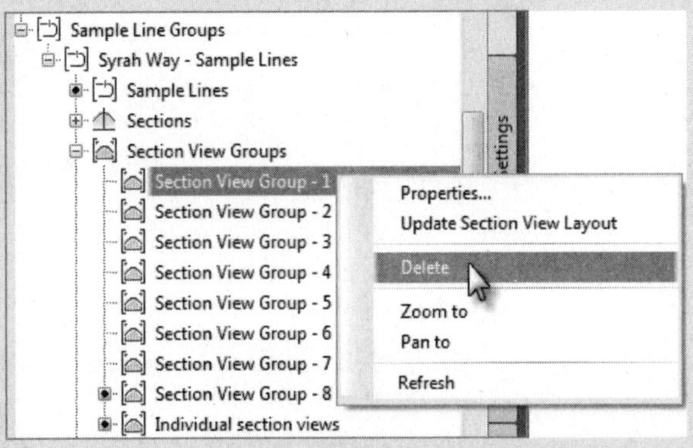

The second problem with graphically deleting your section views is that Civil 3D interprets this maneuver as removing data from the sample lines. The next time you create sections, not all of the data you expect will be there. In this case, you would need to use the Sample More Sources tool to bring back the missing data.

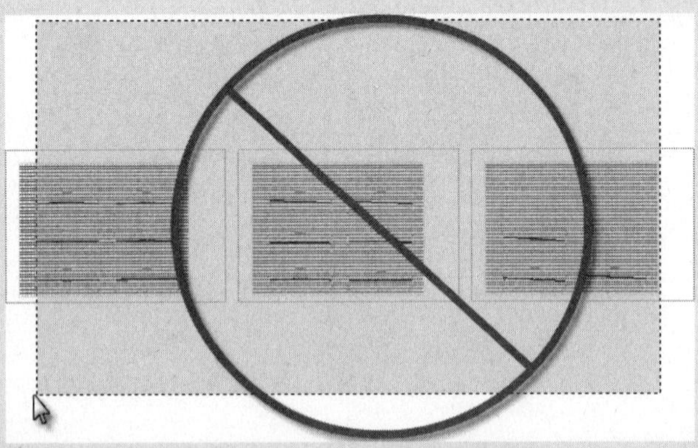

The best way to delete section view groups is by right-clicking the group from Prospector and selecting Delete.

Section Views and Annotation Scale

In the previous exercise, you created section views using a cross-section sheet template. In step 6, you chose a scale. Coincidentally, the scale you used in the exercise was already set as the annotation scale. Since both scales agreed with each other, everything came out nicely.

Depending on what portion of the design you are working on, the scale of the sections and the scale you are comfortable working with may not agree. Additionally, you may just forget to set the scale ahead of placing section views.

Ideally, your section views and their sheets should be in a drawing separate from the corridor and the rest of the design. In Chapter 18, "Advanced Workflows," you will learn how to do this using data shortcuts.

In the meantime, it is helpful to learn how to work with annotation scale and section views. In the following exercise, you will go through a brief lesson in reorganizing sections. You need to have completed the previous exercise before continuing.

1. Continue working in `SectionViews.dwg` (`SectionViews_METRIC.dwg`).
2. Select the Annotation scale in the lower-right portion of the screen.
3. Change the scale to **1"=10'** (**1:250** mm).

   The views probably look pretty funky — and not in a good way. The page has gotten smaller, but the views have not updated.
4. Select any section view by clicking on the station value.
5. From the Section View contextual tab ➢ Modify View panel, click Update Group Layout.

   Update Group Layout

   The views should now be reorganized to a better-looking state.

Real World Scenario

SHOWING ROW LINES IN CROSS SECTION VIEWS

The trick you are about to learn will come in handy for more than just showing right-of-way locations in cross section. Any item that is alignment-based can work the same way.

1. Open the drawing `SectionROW.dwg` (`SectionROW_METRIC.dwg`).
2. Select any sample line, and then from the Sample Line contextual tab ➢ Modify panel, click Group Properties.
3. Switch to the Section Views tab.
4. Locate the Profile Grade column for the section view group. You may need to expand the columns to read what they are.
5. Click the ellipsis, as shown here:

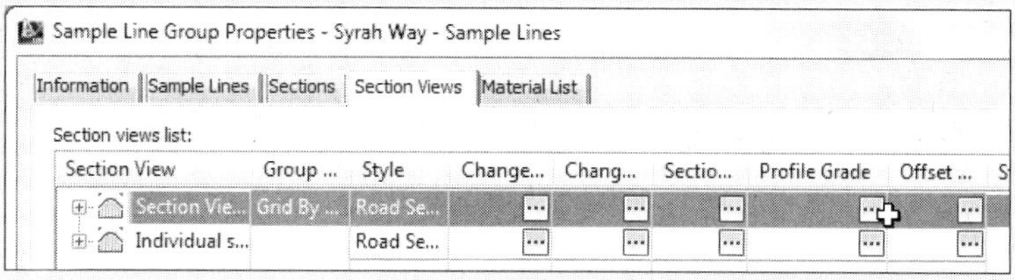

The offset alignments represent the ROW location. An existing ground profile has been created for you as well. You will use this information to place the ROW symbol in the cross-section graphic.

6. Set Syrah Way Right 33.000 (Syrah Way Right 11.000) as the active alignment in the listing, and click Add.
7. Repeat step 6 for Syrah Way Left 33.000 (Syrah Way Left 11.000).
8. Set the Marker Style option for both left and right alignments to ROW Marker.

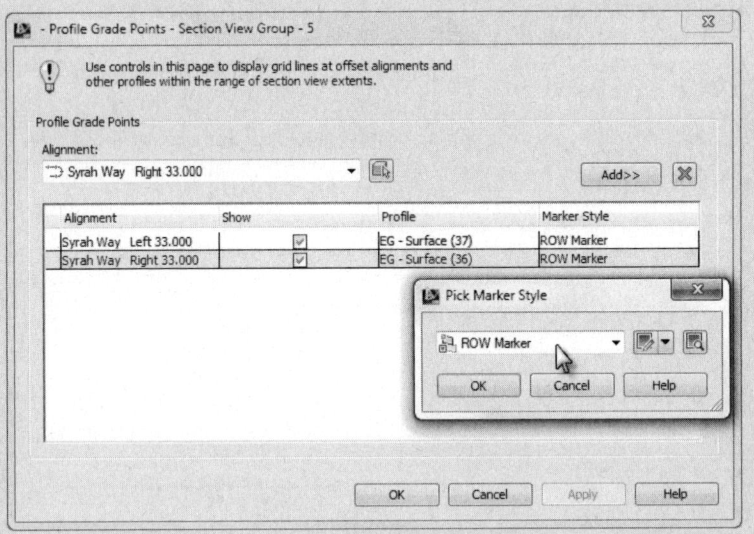

9. Click OK several times until you are out of all dialog boxes.

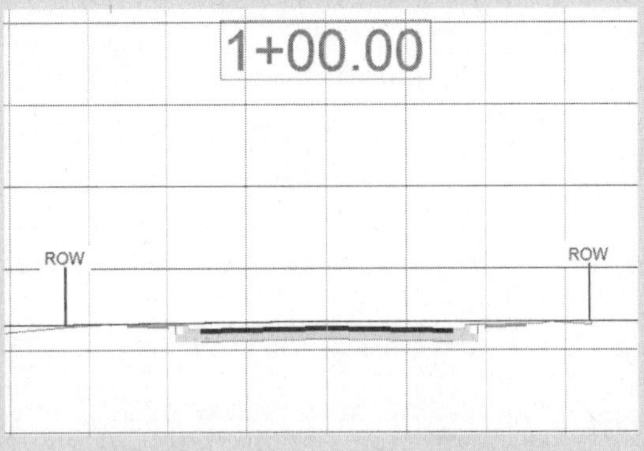

You should now have a ROW symbol whose insertion offset was determined by the alignment, and whose elevation was from the surface profile.

It's a Material World

Once alignments are sampled, volumes can be calculated from the sampled surface or from the corridor section shape. These volumes are calculated in a materials list and can be displayed as a label on each section view or in an overall volume table, as shown in Figure 13.15.

The volumes can also be displayed in an XML report, as shown in Figure 13.16.

FIGURE 13.15
A total volume table inserted into the drawing

Total Volume Table

Station	Fill Area	Cut Area	Fill Volume	Cut Volume	Cumulative Fill Vol	Cumulative Cut Vol
0+00.00	3.81	66.27	0.00	0.00	0.00	0.00
0+50.00	0.00	162.07	3.53	211.43	3.53	211.43
1+00.00	0.02	92.42	0.02	235.64	3.55	447.06
1+50.00	36.61	6.92	33.92	91.98	37.47	539.04
2+00.00	126.37	0.00	150.91	6.40	188.38	545.44
2+50.00	124.23	0.00	232.04	0.00	420.42	545.44
3+00.00	73.59	0.00	182.55	0.00	602.97	545.44
3+50.00	26.71	30.59	92.06	28.45	695.03	573.89
4+00.00	4.17	74.57	28.26	97.76	723.30	671.65
5+00.00	17.59	88.75	40.29	302.45	763.58	974.11
5+50.00	23.72	44.92	38.25	123.77	801.83	1097.88
6+00.00	30.73	37.47	50.42	76.29	852.25	1174.16
6+50.00	30.41	41.93	56.61	73.52	908.86	1247.68
7+00.00	21.98	67.39	48.51	101.23	957.37	1348.91
7+10.60	19.38	77.52	8.12	28.46	965.49	1377.36

FIGURE 13.16
A total volume XML report shown in Microsoft Internet Explorer

Volume Report

Project: C:\Users\-Name-\appdata\local\temp\Materials_FINISHED_1_1_2418.sv$

Alignment: Syrah Way
Sample Line Group: Syrah Way - Sample Lines
Start Sta: 0+00.000
End Sta: 7+10.604

Station	Cut Area (Sq.ft.)	Cut Volume (Cu.yd.)	Reusable Volume (Cu.yd.)	Fill Area (Sq.ft.)	Fill Volume (Cu.yd.)	Cum. Cut Vol. (Cu.yd.)	Cum. Reusable Vol. (Cu.yd.)	Cum. Fill Vol. (Cu.yd.)	Cum. Net Vol. (Cu.yd.)
0+00.000	66.27	0.00	0.00	3.81	0.00	0.00	0.00	0.00	0.00
0+50.000	162.07	253.71	253.71	0.00	2.82	253.71	253.71	2.82	250.89
1+00.000	92.42	282.76	282.76	0.02	0.02	536.48	536.48	2.84	533.63
1+50.000	6.92	110.37	110.37	36.61	27.14	646.85	646.85	29.98	616.87
2+00.000	0.00	7.68	7.68	126.37	120.73	654.53	654.53	150.70	503.83
2+50.000	0.00	0.00	0.00	124.23	185.63	654.53	654.53	336.33	318.20
3+00.000	0.00	0.00	0.00	73.59	146.04	654.53	654.53	482.38	172.15
3+50.000	30.59	34.14	34.14	26.71	73.65	688.67	688.67	556.03	132.64
4+00.000	74.57	117.32	117.32	4.17	22.61	805.98	805.98	578.64	227.35
5+00.000	88.75	362.94	362.94	17.59	32.23	1168.93	1168.93	610.87	558.06
5+50.000	44.92	148.52	148.52	23.72	30.60	1317.45	1317.45	641.47	675.98
6+00.000	37.47	91.55	91.55	30.73	40.33	1409.00	1409.00	681.80	727.20
6+50.000	41.93	88.22	88.22	30.41	45.29	1497.22	1497.22	727.09	770.13
7+00.000	67.39	121.47	121.47	21.98	38.81	1618.69	1618.69	765.89	852.80
7+10.604	77.52	34.15	34.15	19.38	6.50	1652.84	1652.84	772.39	880.45

Once a materials list is created, it can be edited to include more materials or to make modifications to the existing materials. For example, soil expansion (fluff or swell) and shrinkage factors can be entered to make the volumes more accurately match the true field conditions. This can make cost estimates more accurate, which can result in fewer surprises during the construction phase of any given project.

Creating a Materials List

CERT OBJECTIVE

Materials can be created from surfaces or from corridor shapes. Surfaces are great for earthwork because you can add cut or fill factors to the materials, whereas corridor shapes are great for determining quantities of asphalt or concrete. In this exercise, you practice calculating earthwork quantities for the Syrah Way corridor:

1. Open the `Materials.dwg` (`Materials _METRIC.dwg`) file, which you can download from this book's web page.
2. Select the Analyze tab ➢ Volumes And Materials panel and click Compute Materials.

   The Select A Sample Line Group dialog appears.
3. In the Select Alignment field, verify that the alignment is set to Syrah Way, and then click OK.

   The Compute Materials dialog appears.
4. In the Quantity Takeoff Criteria drop-down box, select Earthworks from the drop-down menu.
5. Click the Object Name cell for the Existing Ground surface, and select EG from the drop-down menu.
6. Click the Object Name cell for the Datum surface, and select Concord Commons Corridor Concord Commons Datum (this is the name of the corridor followed by the name of the surface) from the drop-down menu.
7. Verify that your settings match those shown in Figure 13.17, and then click OK.

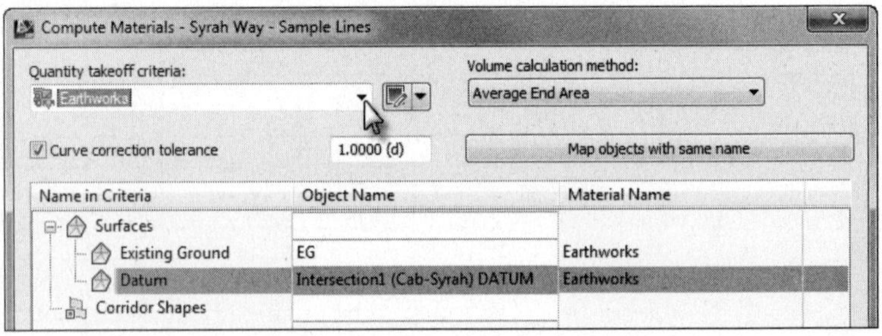

FIGURE 13.17 The settings for the Compute Materials dialog

Graphically, nothing will happen. However, in the background Civil 3D has computed material data.

8. Save the drawing for use in the next exercise.

Creating a Volume Table in the Drawing

In the preceding exercise, materials were created that represent the total dirt to be moved or used in the sample line group. In the next exercise, you insert a table into the drawing so you can inspect the volumes. You need to have successfully completed the previous exercise before proceeding.

1. Continue working in `Materials.dwg` (`Materials _METRIC.dwg`).
2. Select the Analyze tab ➢ Volumes And Materials and click Total Volume Table.

   The Create Total Volume Table appears.

3. Verify that your settings match those shown in Figure 13.18.

FIGURE 13.18
The Create Total Volume Table dialog settings

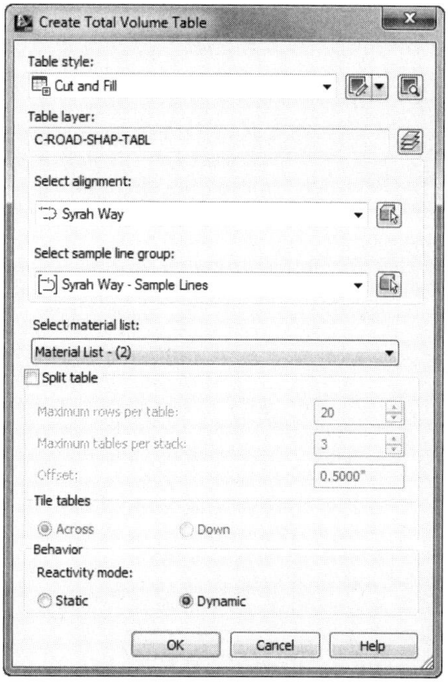

4. Verify that Reactivity Mode at the bottom of the dialog is set to Dynamic.

   This will cause the table to update if any changes are made to the corridor or related objects. Split Table should be enabled by default, so you can leave that as is.

5. Click OK.
6. Pick a point in the drawing to place the volume table.

   The table indicates a Cumulative Fill Volume of 965.49 cubic yards (715.20 cubic meters) and a Cumulative Cut Volume of 1377.36 cubic yards (901.56 cubic meters), as shown previously in Figure 13.15.
7. Save the drawing for use in the next exercise.

Adding Soil Factors to a Materials List

CERT OBJECTIVE

The material calculation indicates that this design has an excessive amount of fill. The materials need to be modified to bring them closer in line with true field numbers. For this exercise, the shrinkage factor will be assumed to be 0.80 and 1.20 for the expansion factor (20 percent shrink and swell). In addition to these numbers (which Civil 3D represents as Cut Factor for swell and Fill Factor for shrinkage), you can specify a Refill Factor value. This value specifies how much cut can be reused for fill. For this exercise, assume a Refill Factor value of 1.00. You need to have successfully completed the previous exercises before proceeding.

1. Continue working in `Materials.dwg` (`Materials _METRIC.dwg`).
2. Select the Analyze tab ➤ Volumes And Materials panel and click Compute Materials.

   The Select A Sample Line Group dialog appears.
3. Select the Cabernet Court alignment and the SL Collection–3 sample line group, if not already selected, and click OK.

   The Edit Material List dialog appears.
4. Enter a Cut Factor of **1.20** and a Fill Factor of **0.80**.
5. Verify that all other settings are the same as in Figure 13.19, and click OK.
6. Examine the Total Volume table again.

   Notice that the new Cumulative Fill Volume is 772.39 cubic yards (572.16 cubic meters) and the new Cumulative Cut Volume is 1652.84 cubic yards (1081.87 cubic meters).
7. Save the drawing.

Generating a Volume Report

Civil 3D provides you with a way to create a report that is suitable for printing or for transferring to a word processing or spreadsheet program. In this exercise, you'll create a volume report for the Concord Commons corridor. You need to have successfully completed the previous exercises before proceeding.

FIGURE 13.19
The Edit Material List dialog

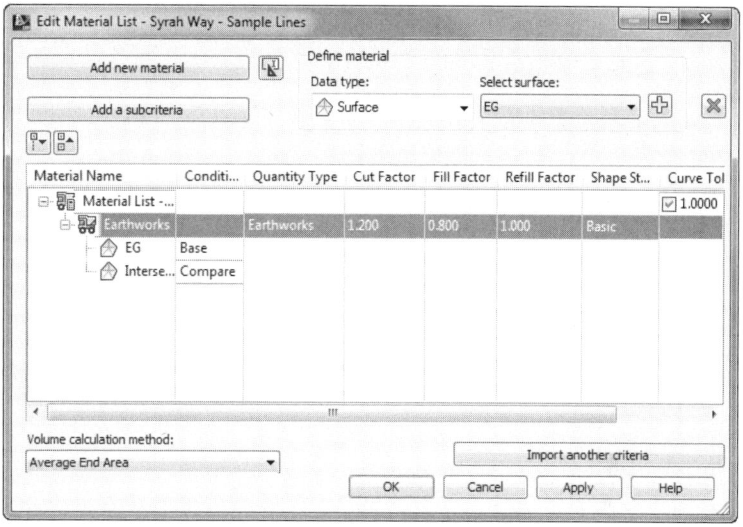

1. Continue working in `Materials.dwg` (`Materials _METRIC.dwg`).

2. Select the Analyze tab ➢ Volumes And Materials panel and click Volume Report.

   The Report Quantities dialog appears.

3. Verify that Material List–(2) is selected in the dialog, and click OK.

4. You may get a warning message that says "Scripts are usually safe. Do you want to allow scripts to run?" Click Yes.

   Civil 3D temporarily takes over your web browser to display the volume report. This is the same information that you placed in your drawing, but since it's in this form you can more readily copy and paste it into Excel, Word, or another program.

5. Note the cut-and-fill volumes and compare them to your volume table in the drawing.

6. Close the report when you are done viewing it.

7. Save and close the drawing.

Section View Final Touches

Before you move away from section views, a few last touches are needed in your sections. First you will add last-minute data to the sections. You will also add labels to the sections.

Sample More Sources

It is a fairly common occurrence that data is created from the design after sample lines have been generated. For example, you may need to add surface data or pipe network data to existing section views.

To accomplish this, you need to add that data to the sample line group. In this exercise, you'll add a pipe network to a sample line group and inspect the existing section views to ensure that the pipe network was added correctly:

1. Open the `FinalTouches.dwg` (`FinalTouches_METRIC.dwg`) file, which you can download from this book's web page.
2. Select one of the section views by clicking on the station label.
3. In the Section View contextual tab ➢ Modify Section panel, click Sample More Sources.

Sample More Sources

4. Click Sanitary Network on the left side of the dialog to highlight it, as shown in Figure 13.20, and click Add.

FIGURE 13.20 Adding sewer data to the cross-section view via Sample More Sources

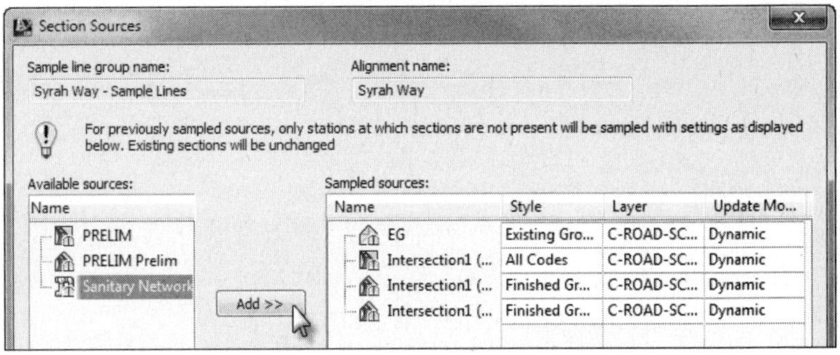

The Sanitary network will now appear on the right side of the dialog.

5. Click OK to dismiss the Section Sources dialog.
6. Save the drawing for use in the next exercise.

Cross-Section Labels

The best way to label cross sections that contain corridor data is using the code set style. Using the code set style you can control which parts of the corridor are labeled. You can learn more about creating code set styles in Chapter 20, "Label Styles." In the meantime, you will look at how to change the code set style active on a cross-section view group.

In the following exercise, you will use the view group properties to add labels. It is *not* necessary to have completed the previous exercise before continuing.

1. Continue working in the `FinalTouches.dwg` (`FinalTouches_METRIC.dwg`) file.
2. Select one of the section views by clicking on its station label.
3. In the Section View contextual tab ➢ Modify View panel, click View Group Properties.

View Group Properties

4. On the Sections tab, locate the Style column for the corridor and click the field that currently reads All Codes.

   This is the code set style that is current in the views.

5. Switch the style to All Codes w Labels, as shown in Figure 13.21, and click OK.

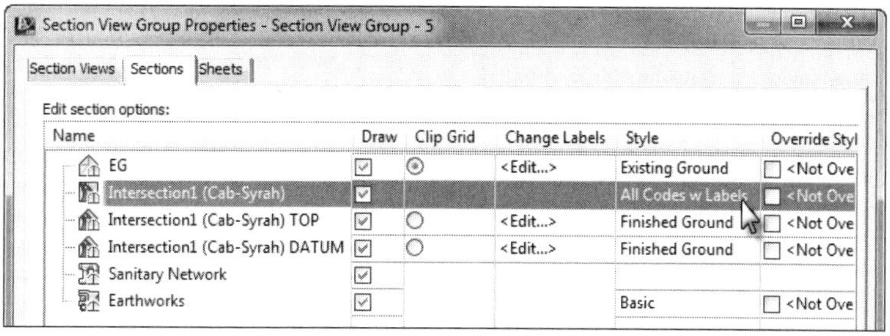

FIGURE 13.21
Changing the active code set style for all views in the group

You should now have lovely little labels on all of the views, as shown in Figure 13.22.

6. Save and close the drawing.

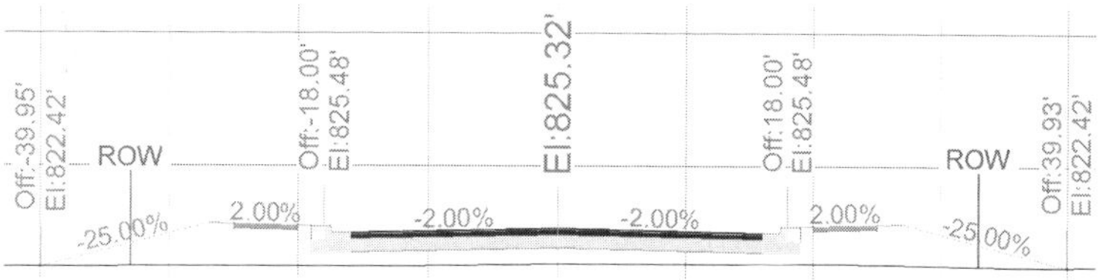

FIGURE 13.22
Slope, elevation, and offset labels from the code set style

Mass Haul

Mass haul diagrams help designers and contractors gauge how far and how much soil needs to be moved around a site. Figure 13.23 shows the mass haul diagram for Syrah Way. The Free Haul Area is material the contractor has agreed to move at no extra charge. The fact that the mass haul line is always above 0 indicates that the project is in a net cut situation through the length of the Syrah Way alignment.

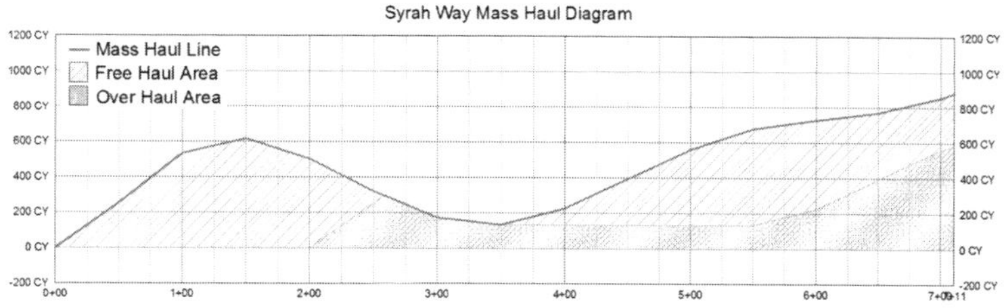

FIGURE 13.23
Syrah Way Mass Haul diagram

Taking a Closer Look at the Mass Haul Diagram

In an ideal design situation, there is no leftover cut material and no extra material needs to be brought in. This would mean that net volume = 0. When the line appears above the zero volume point, it is showing net cut values.

As the mass haul diagram continues, it shows the cumulative effect of net cut and fill for the alignment. When the net cut and net fill converge at the zero volume, the earthwork along the alignment is balanced. When the line appears below the zero volume point, it is showing net fill values, as you can see in Figure 13.24.

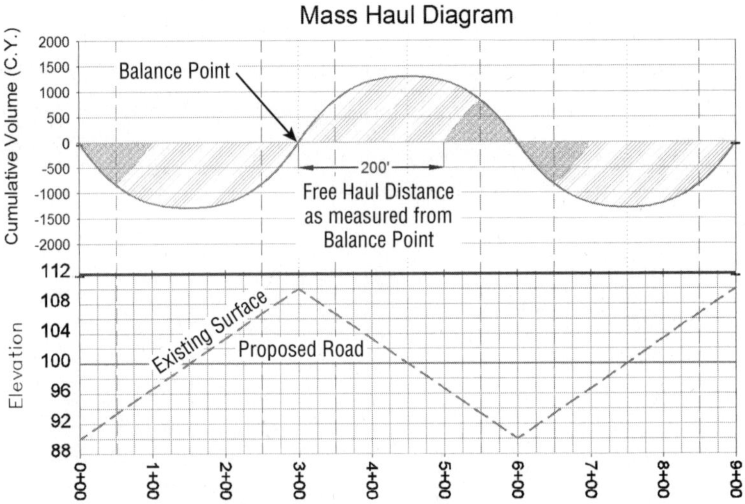

FIGURE 13.24
The volume, net cut, and net fill on an idealized mass haul diagram shown with profile

Here is some of the terminology you will encounter:

Balanced The state where the cumulative cut and fill volumes are equal.

Origin Point The beginning of the mass haul diagram, typically at station 0 + 00, but it can vary depending on your stationing.

Borrow A negative value, typically at the end of the mass haul diagram, that indicates fill material that will need to be brought into the site.

Waste A positive value, typically at the end of the mass haul diagram, that indicates cut material that will need to be hauled out of the site.

Free Haul Earthwork that a contractor has contractually agreed to move. This typically specifies a contracted distance.

Over Haul Earthwork that the contractor has not contractually agreed to move. This excess can be used for borrow pits or waste piles.

Create a Mass Haul Diagram

Now, let's put it all together and build a mass haul diagram in Civil 3D for Syrah Way and you'll see how easy it is:

1. Open the `MassHaul.dwg` (`MassHaul_METRIC.dwg`) file.

   Remember, you can download all the data files from this book's web page.

2. From the Analyze tab ➢ Volumes And Materials panel, select Mass Haul.

   The Create Mass Haul Diagram dialog opens.

3. On the General page:

   a. Verify that you are creating a mass haul diagram for Syrah Way alignment.

   b. Verify that you are using the Syrah Way–Sample Lines group.

   c. Leave all other options at their defaults, as shown in Figure 13.25, and click Next.

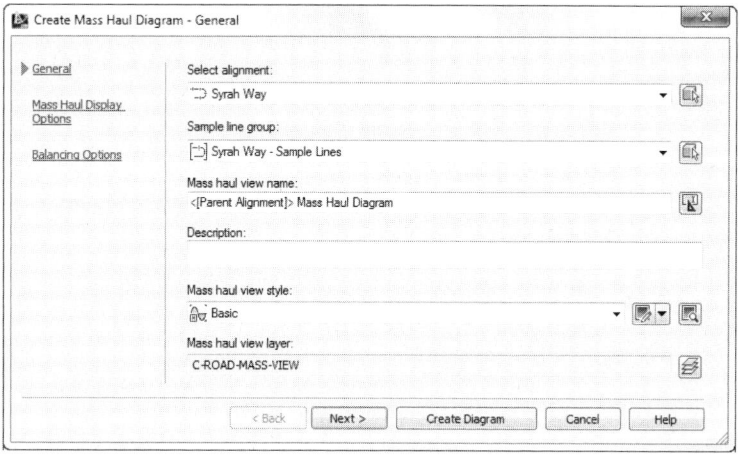

FIGURE 13.25
The Create Mass Haul Diagram dialog, General options

   On the Mass Haul Display Options page, no changes are needed.

4. Click Next.

5. In the Balancing Options area, place a check box next to Free Haul Distance.

6. Set the distance to **200′** (**60** m), as shown in Figure 13.26, and click Create Diagram.

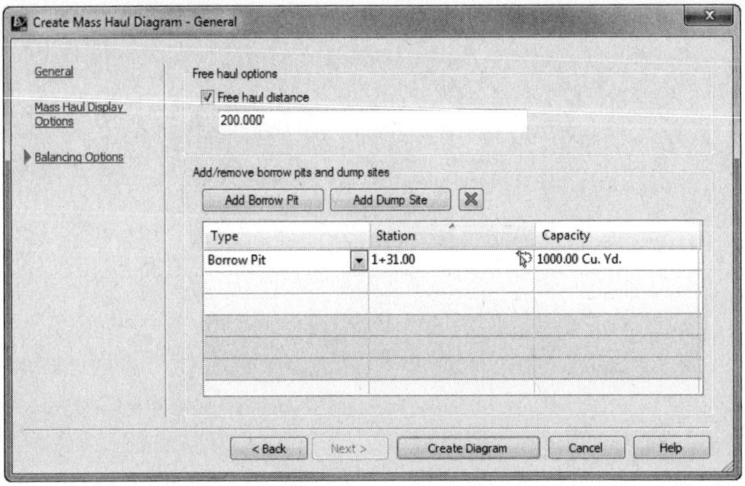

FIGURE 13.26
The Create Mass Haul Diagram dialog, Balancing Options page

7. Find a clear spot on your drawing to place the mass haul diagram.

   The diagram should look similar to Figure 13.23 (shown previously).

Editing a Mass Haul Diagram

When you create a mass haul diagram, you can easily modify parameters and get instant feedback on your diagram. Follow these steps to see how (you must have completed the previous exercise before continuing):

1. Continue working in the `MassHaul.dwg` (`MassHaul_METRIC.dwg`) file.
2. Select anywhere on the mass haul object or grid.

   If you look at your mass haul diagram, you can see that nearly all the earthwork for Syrah Way involves net cut, which means hauling away dirt.

   The Balancing options give you a chance to change or add waste and borrow pits. You will now make a few changes and observe what happens to the mass haul diagram.

Balancing Options

3. From the Mass Haul View contextual tab ➤ Modify panel, select Balancing Options.

   The Mass Haul Line Properties dialog opens. It looks similar to part of the Balancing Options page shown earlier.
4. The Free Haul Distance is presently set for 200′ (60 m). Change it to **500′** (**150** m).

5. Drag the dialog away from the screen to see the changes and then click Apply.

   You can further tweak this amount to cut down on the net cut values by adding a dump site.

6. Click the Add Dump Site button.

   a. Set the station to **1+31.00** (**0+040.00**).

   b. Set the capacity to **850** cubic yards. (**500** cubic meters).

7. When your dialog resembles the top of Figure 13.27, click OK.

   The mass haul diagram will immediately update to reflect the changes, as shown at the bottom of Figure 13.27.

8. Save the drawing.

For this stretch of the project the mass haul diagram ends near zero. The changes you made also decrease the amount of overhaul in the project, decreasing potential cost.

FIGURE 13.27
The Mass Haul Line Properties dialog: adding a dump site (top). The mass haul diagram adjusted for the dump site (bottom)

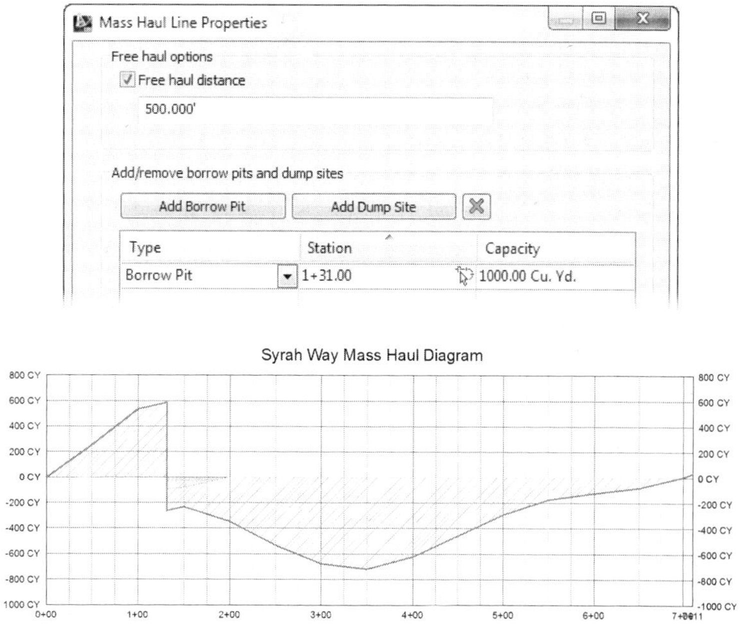

The Bottom Line

Create sample lines. Before any section views can be displayed, sections must be created from sample lines.

Master It Open `MasterSections.dwg` (`MasterSections_METRIC.dwg`) and create sample lines along the USH 10 alignment every 50' (20 m). Set the left and right swath widths to **50'** (**20** m).

Create section views. Just as profiles can only be shown in profile views, sections require section views to display. Section views can be plotted individually or all at once. You can even set them up to be broken up into sheets.

Master It In the previous exercise, you created sample lines. In that same drawing, create section views for all the sample lines. Use all the default settings and styles.

Define and compute materials. Materials are required to be defined before any quantities can be displayed. You learned that materials can be defined from surfaces or from corridor shapes. Corridors must exist for shape selection, and surfaces must already be created for comparison in materials lists.

Master It Using `MasterSections.dwg` (`MasterSections_METRIC.dwg`), create a materials list that compares Existing Intersection with HWY 10 DATUM Surface. Use the Earthworks Quantity takeoff criteria.

Generate volume reports. Volume reports give you numbers that can be used for cost estimating on any given project. Typically, construction companies calculate their own quantities, but developers often want to know approximate volumes for budgeting purposes.

Master It Continue using `MasterSections.dwg` (`MasterSections_METRIC.dwg`. Use the materials list created earlier to generate a volume report. Create a web browser–based report and a Total Volume table that can be displayed on the drawing.

Chapter 14

Pipe Networks

In this chapter, we will look at pipe networks (gravity-based systems) and pressure networks. First, we will look at the setup needed to use these networks successfully. Planning is a key part of the process.

In this chapter, you will learn to:

- Create a pipe network by layout
- Create an alignment from network parts and draw parts in profile view
- Label a pipe network in plan and profile
- Create a dynamic pipe table

Parts Lists

Before you can draw pipes in your project, some setup is needed. Ideally the setup will be done in your AutoCAD® Civil 3D® template so that you don't have to repeat the process more than necessary.

A *parts list* contains the pieces needed to complete a pipe design. Both pipe networks and pressure networks use parts lists to help you organize design elements and determine how they will appear in a drawing. For example, you'll want different parts available when working with sanitary sewers than you'll want when working with storm sewers.

Parts lists for gravity pipe networks vary significantly from pressure pipe networks. Pipe network parts lists contain pipes, structures, pipe rules, structure rules, styles, and the ability to associate a pay item to each item.

Examples of pipe network parts lists include:

- Storm sewer
  - Catch basin/inlet structures
  - Manhole structures
  - Concrete pipe
  - HDPE pipe
- Sanitary sewer (gravity) as shown in Figure 14.1:
  - Manhole structures
  - HDPE pipe
  - Ductile iron pipe

FIGURE 14.1
Example sanitary sewer parts list

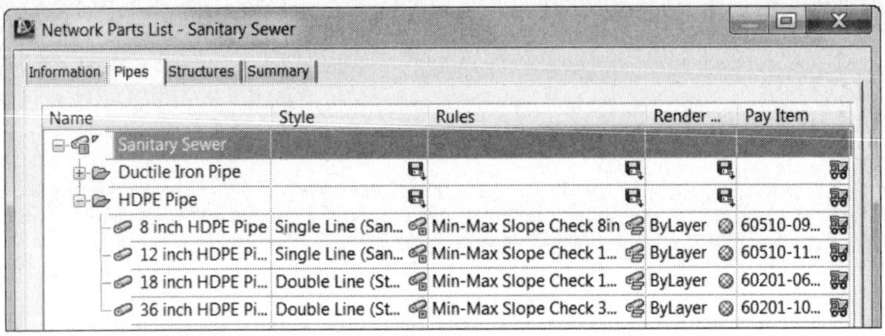

Pressure parts lists contain pipes, fittings, and appurtenances. You'll find the style for each object in the parts list, but instead of rules, pressure pipe design checks are tucked into the command settings.

Additionally, the ability to assign a pay item with the parts list is not available in this release.

Examples of pressure pipe network parts lists include:

- Water main and service connections
  - Ductile iron pipe
  - Tees, elbows, and crosses
  - Valves
- Gas main
  - PVC pipe
  - Valves

In the upcoming section, you will explore planning and creating a parts list. You'll start by examining your local requirements, and then you will compare those needs with what is available in the software.

Planning a Typical Pipe Network

Let's look at a typical sanitary sewer design. You typically start by going through the sewer specifications for the jurisdiction in which you're working. There is usually a published list of allowable pipe materials, manhole details, slope parameters, and cover guidelines. Perhaps you have concrete and PVC pipe manufacturer catalogs that contain pages of details for different manholes, pipes, and junction boxes. There is usually a recommended symbology for your submitted drawings—and, of course, you have your own in-house CAD standards. Assemble this information, and make sure you address these issues:

- Recommended structures, including materials and dimensions (be sure to attach detail sheets)
- Structure behavior, such as required sump, drops across structures, and surface adjustment
- Structure symbology

- Recommended pipes, including materials and dimensions (again, be sure to attach detail sheets)
- Pipe behavior, such as cover requirements; minimum, maximum, and recommended slopes; velocity restrictions; and so on
- Pipe symbology

The following is an example of a completed checklist for our example jurisdiction, Sample County:

- ❑ Sanitary Sewers in Sample County
- ❑ Recommended Structures: Standard concentric manhole, small-diameter cleanout.
- ❑ Structure Behavior: All structures have 1.5′ (0.46 m) sump, rims, a 0.10′ (0.03 m) invert drop across all structures. All structures designed at finished road grade.
- ❑ Structure Symbology: Manholes are shown to scale in plan view as a circle with an S inside. Cleanouts are shown to scale as a hatched circle. (See Figure 14.2.)

FIGURE 14.2 Sanitary sewer manhole in plan view (left) and a cleanout in plan view (right)

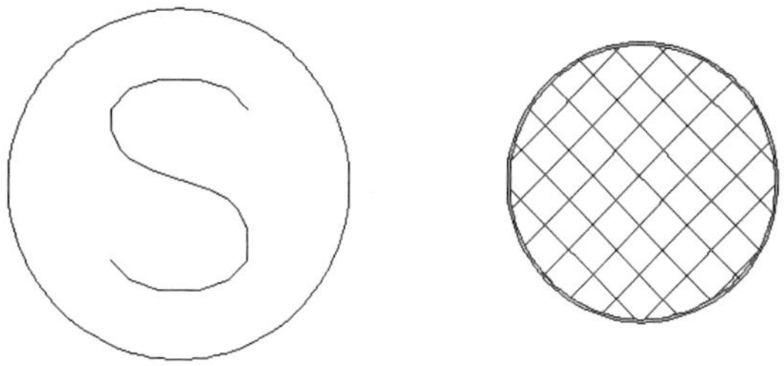

Manholes are shown in profile view with a coned top and rectangular bottom. Cleanouts are shown as a rectangle (see Figure 14.3).

FIGURE 14.3 Profile view of a sanitary sewer manhole (left) and a cleanout (right)

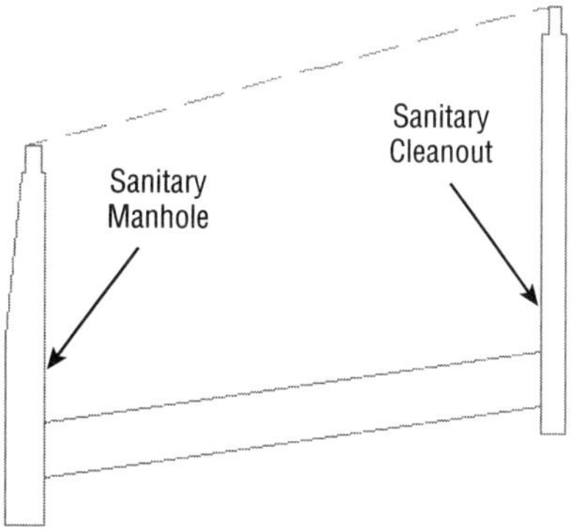

- ❑ Recommended Pipes: 8″ (200 mm), 10″ (250 mm), and 12″ (300 mm) PVC pipe, per manufacturer specifications.

- ❑ Pipe Behavior: Pipes must have cover of 4′ (1.22 m) to the top of the pipe; the maximum slope for all pipes is 10 percent, although minimum slopes may be adjusted to optimize velocity as follows:

SEWER SIZE	MINIMUM SLOPE
8" (200 mm)	0.40%
10" (250 mm)	0.28%
12" (300 mm)	0.22%

- ❑ Pipe Symbology: In plan view, pipes are shown with a CENTER2 linetype line that has a thickness corresponding to the inner diameter of the pipe. In profile view, pipes show both inner and outer walls, with a hatch between the walls to highlight the wall thickness (see Figure 14.4).

FIGURE 14.4
Sanitary pipe in plan view (a) and in profile view (b)

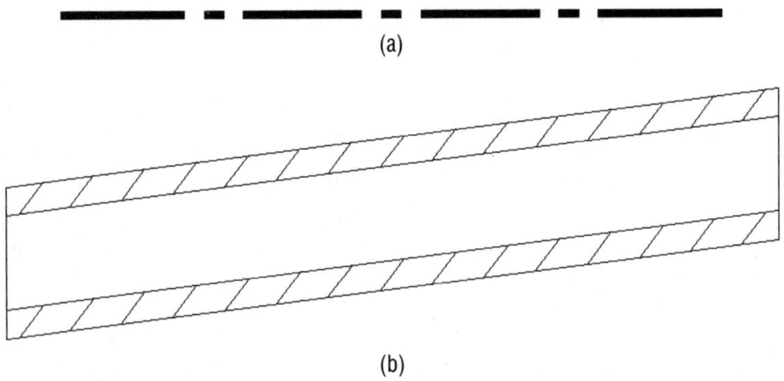

Now that you know your requirements for Sample County, the next step is to begin entering this information into your Civil 3D template file.

Part Rules

At the beginning of this chapter, you made notes about your hypothetical municipality having certain requirements for how structures and pipes behave—things like minimum slope, sump depths, and pipe-invert drops across structure. Depending on the type of network and the complexity of your design, there may be many different constraints on your design. Civil 3D allows you to establish structure and pipe rules that will assist in respecting these constraints during initial layout and edits. Some rules don't change the pipes or structures during layout but provide a "violation only" check that you can view in Prospector.

Structures and pipes have separate rule sets. When creating rules, don't be thrown off by the fact that the category always reads Storm Sewer, as you see in Figure 14.5. You can use these rules regardless of the type of parts list you are creating. You can then add these rule sets to specific parts in your parts list, which you'll build later in this chapter.

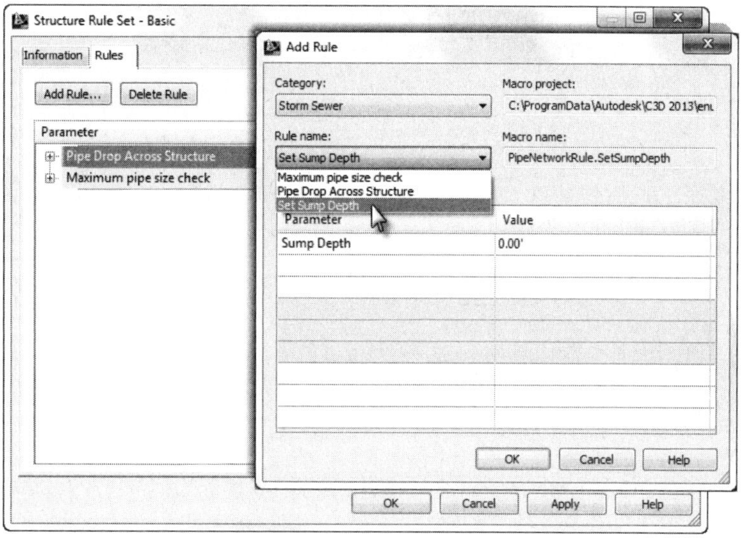

FIGURE 14.5
In the Add Rule dialog, the category always shows Storm Sewer but you can use rules for whatever type you wish.

STRUCTURE RULES

Structure rule sets are located on the Settings tab of Toolspace, under the Structure tree. To locate these rules in any drawing file:

1. Right-click Basic and click Edit.
2. Click the Add Rule button on the Rules tab in the Structure Rule Set dialog.

   The Add Rule dialog appears, which allows you to access all the various structure rules (see Figure 14.5). Although it looks like you can, you won't be able to change the values until you finalize adding the rule.

You will have a chance to work with this firsthand in the upcoming exercise.

Maximum Pipe Size Check

The Maximum Pipe Size Check rule (see Figure 14.6) examines all pipes connected to a structure and flags a violation in Prospector if any pipe is larger than your rule. This is a violation-only rule—it won't change your pipe size automatically.

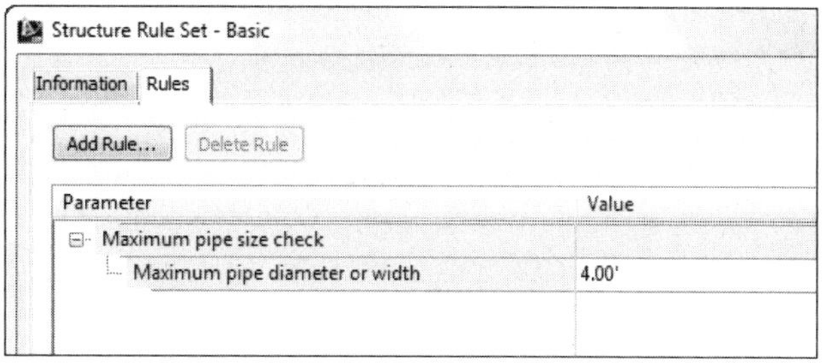

FIGURE 14.6
The Maximum Pipe Size Check rule option

Pipe Drop Across Structure

The Pipe Drop Across Structure rule (see Figure 14.7) tells any connected pipes how their inverts (or, alternatively, their crowns or centerlines) must relate to one another.

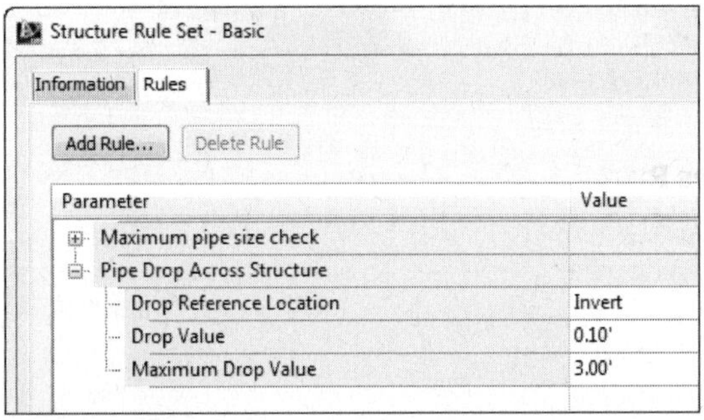

FIGURE 14.7 The Pipe Drop Across Structure rule options

When a new pipe is connected to a structure that has the Pipe Drop Across Structure rule applied, the following checks take place:

- A pipe drawn to be exiting a structure has an invert equal to or lower than the lowest pipe entering the structure.
- A pipe drawn to be entering a structure has an invert equal to or higher than the highest pipe exiting the structure.
- Any minimum specified drop distance is respected between the lowest entering pipe and the highest exiting pipe.

In the hypothetical sanitary sewer example, you're required to maintain a 0.10′ (3 cm) invert drop across all structures. You'll use this rule in your structure rule set in the next exercise.

Set Sump Depth

Sump depth is additional structure depth below the lowest pipe invert (known in this author's area as "Mosquito Breeders"). The Set Sump Depth rule (Figure 14.8) establishes sump depth for structures.

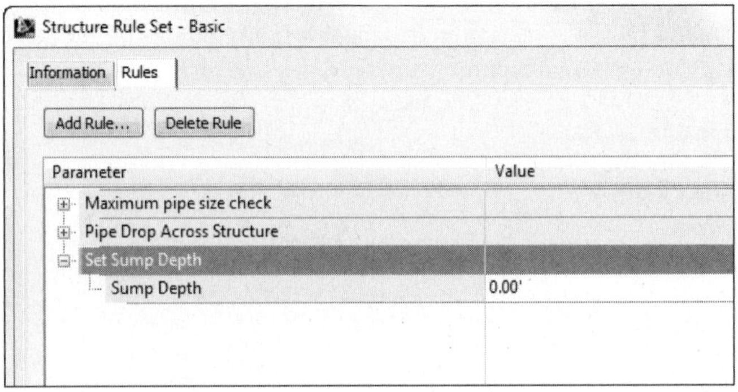

FIGURE 14.8 The Set Sump Depth rule options

It's important to add a sump depth rule to all of your structure rule sets. If no sump rule is used, Civil 3D will assume a 2′ sump for English units and 2 meters in metric units! If you forget to set this before placing structures, there is no way globally to change the sump value. After the structures are placed, you would need to edit the individual structure properties or make the rule and retroactively apply it to each structure.

In the hypothetical sanitary sewer example, all the structures have a 1.5′ (0.5 m) sump depth. You'll use this rule in your structure rule set in the next exercise.

Pipe Rules

Pipe rule sets are located on the Settings tab of Toolspace, under the Pipe tree. For a detailed breakdown of pipe rules and how they're applied, including images and illustrations, please see the Civil 3D Users Guide.

After you right-click on a Pipe Rule Set and click Edit, you can access all the pipe rules by clicking the Add Rule button on the Rules tab of the Pipe Rule Set dialog.

Cover and Slope Rule

The Cover And Slope rule (Figure 14.9) allows you to specify your desired slope range and cover range. As you place your pipe network, Civil 3D tries to use the minimum and maximum depth and minimum and maximum slopes to set the initial pipe depths and slope.

Depending on your site conditions, applying this rule to every pipe may not be feasible. In this situation, the rule becomes what is referred to as violation only. You are still able to place pipes, and the rule will cause a violation message to show in the Pipe Network panel.

If part of your design changes and you'd like Civil 3D to make another attempt to enforce the cover and slope rule, you can use the Apply Rules feature.

You'll create one Cover And Slope rule for each size pipe in the hypothetical sanitary sewer example.

FIGURE 14.9
The Cover And Slope rule options

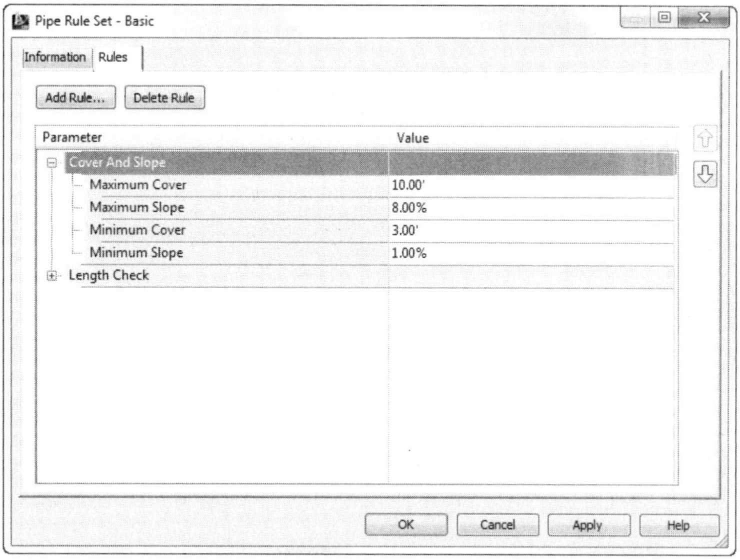

Cover Only Rule

The Cover Only rule (Figure 14.10) is designed for use with pipe systems where slope can vary or isn't a critical factor. Like Cover And Slope, this rule is used on first placement. Manual edits can cause rules to be violated. The rule will show as a violation in the Pipe Network panel, but not changes to your design take place until you use the Apply rules command.

FIGURE 14.10
The Cover Only rule options

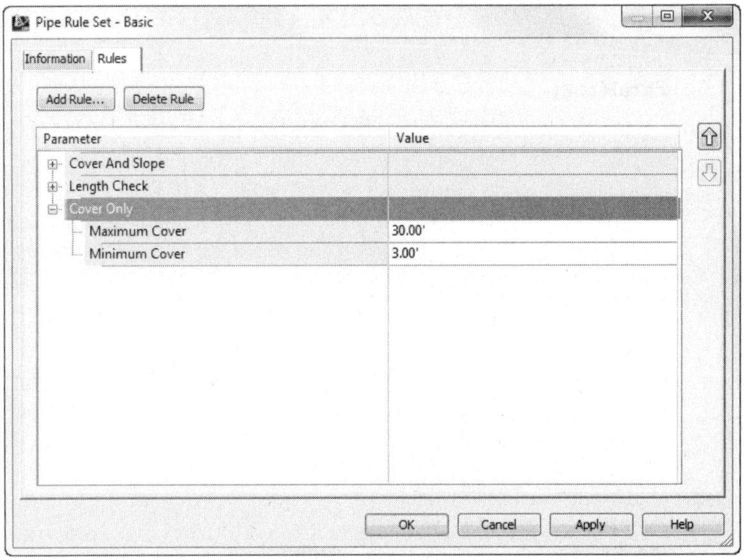

Length Check

Length Check is a violation-only rule; it won't change your pipe length size automatically. The Length Check options (see Figure 14.11) allow you to specify a minimum and maximum pipe length.

FIGURE 14.11
The Length Check rule options

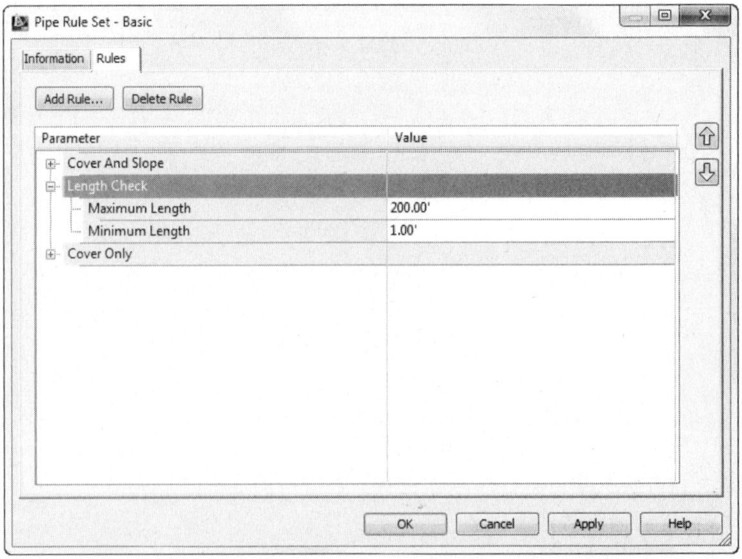

Pipe To Pipe Match Rule

The Pipe To Pipe Match rule (Figure 14.12) is also designed for use where there are no true structures (only null structures), including situations where pipe is placed to break into an existing pipe. This rule determines how pipe inverts are assigned when two pipes come together, similar to the Pipe Drop Across Structure rule.

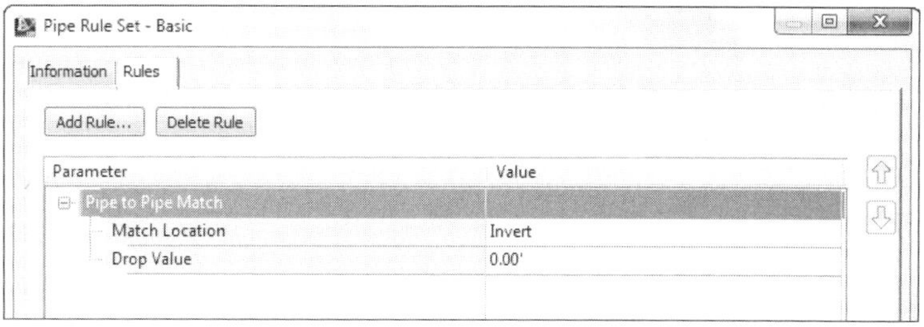

FIGURE 14.12
The Pipe To Pipe Match rule options

Set Pipe End Location Rule

Without the Set Pipe End Location rule (Figure 14.13), Civil 3D assumes you are measuring pipes from center of structure to center of structure. With the rule in place, you have the capability to set where the pipe end is located on the structure. The options are Structure Center (the default without the rule), Structure Inner Wall, or Structure Outer Wall.

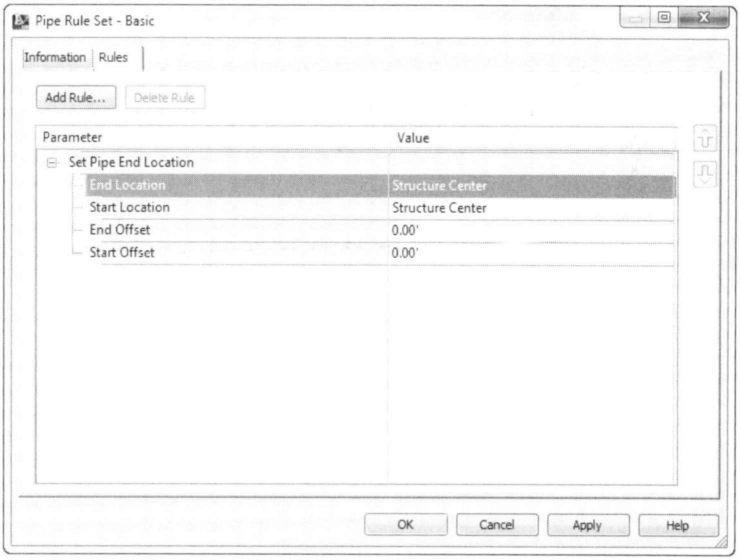

FIGURE 14.13
The Set Pipe End Location rule options

End Offset and Start Offset are used if you would like to have your pipes protrude past the inner or outer wall on the respective ends. The offset is ignored if the location for the end is set to the center of the structure. The value must be positive and will be ignored if the additional length causes the pipe end to be located past the center of the structure.

Creating Structure and Pipe Rule Sets

In this exercise, you'll create one structure rule set and three pipe rule sets for a hypothetical sanitary sewer project:

1. Open the drawing `Civil 3D Template-in-progress_Rules.dwg` (`Civil 3D Template-in-progress_Rules_METRIC.dwg`), which you can download from this book's web page at `www.sybex.com/go/masteringcivil3d2013`.
2. Locate the Structure Rule Set on the Settings tab of Toolspace under the Structure branch.
3. Right-click the Structure rule set, and choose New.
4. Switch to the Information Tab and enter **Sanitary Structure Rules** in the Name text box.
5. Switch to the Rules tab, and click the Add Rule button.
6. In the Add Rule dialog, choose Pipe Drop Across Structure in the Rule Name drop-down list. Click OK.

   You can't change the parameters until the next step.
7. Expand the new rule and confirm that the parameters in the Structure Rule Set dialog are the following:

Drop Reference Location	Invert
Drop Value	0.1' (0.03 m)
Maximum Drop Value	3' (1 m)

   These parameters establish a rule that will match your hypothetical municipality's standard for the drop across sanitary sewer structures.
8. Click the Add Rule button again.
9. In the Add Rule dialog, choose Set Sump Depth in the Rule Name drop-down list. Click OK.

   You can't change the parameters until the next step.
10. Expand the rule. Change the Sump Depth parameter to **1.5'** (**0.5** m) in the Structure Rule Set dialog to meet the hypothetical municipality's standard for sump in sanitary sewer structures, and click OK.
11. On the Settings tab of Toolspace ➤ Pipe, locate Pipe Rule Set.
12. Right-click the Pipe Rule Set, and choose New.
13. Switch to the Information tab and enter **8 Inch Sanitary Pipe Rule** (for metric, **200 mm Sanitary Pipe Rule**) for the name.
14. Switch to the Rules tab. Click Add Rule.
15. In the Add Rule dialog, choose Cover And Slope in the Rule Name drop-down list. Click OK.

    You can't change the parameters until the next step.

16. Expand the Cover And Slope rule and then modify the parameters to match the constraints established by your hypothetical municipality for 8″ (200 mm) pipe, as follows:

Maximum Cover	10′ (3m)
Maximum Slope	10%
Minimum Cover	4′ (1.5m)
Minimum Slope	0.40%

17. Click OK.
18. Select the rule set you just created. Right-click, and choose Copy.
19. Switch to the Information tab and enter **10 Inch Sanitary Pipe Rule** (for metric, **250 mm Sanitary Pipe Rule**) in the Name text box.
20. Expand the Cover And Slope rule and then modify the parameters to match the constraints established by your hypothetical municipality for a 10″ (250 mm) pipe, as follows:

Maximum Cover	10′ (3 m)
Maximum Slope	10%
Minimum Cover	4′ (1.5 m)
Minimum Slope	0.28%

21. Click OK when you have completed modifying the rule set. Repeat the process to create a rule set for the 12″ (300 mm) pipe using the following parameters:

Maximum Cover	10′ (3 m)
Maximum Slope	10%
Minimum Cover	4′ (1.5 m)
Minimum Slope	0.22%

22. You should now have one structure rule set and three pipe rule sets.
23. Save your drawing—you'll use it in the next exercise.

Putting Your Parts List Together

The next step is to put your efforts from earlier in the chapter in a consolidated place. Everything you've done in this chapter up to this point is leading up to the creation of the parts list.

Your parts lists should reside in your Civil 3D template file. There is quite a bit of setup involved, and having this information in your template will ensure you need to do it only once. You will have multiple parts lists for each type of system you are creating and possibly for each jurisdiction you work in.

You should complete the previous exercise before continuing, as we will reference the rules you created there.

1. Open the drawing `Civil 3D Template-in-progress_Rules.dwg` (`Civil 3D Template-in-progress_Rules_METRIC.dwg`).
2. From the Settings tab, expand the Pipe Network branch, and expand Parts Lists.

   In the drawing there are currently two parts lists: Standard and Storm Sewer.
3. Right-click on Parts Lists and select Create Parts List.
4. Switch to the Information tab and name the parts list **Example County Sanitary**.
5. Switch to the Pipes tab.
6. Right-click on the New Parts List and select Add Part Family, as shown in Figure 14.14.

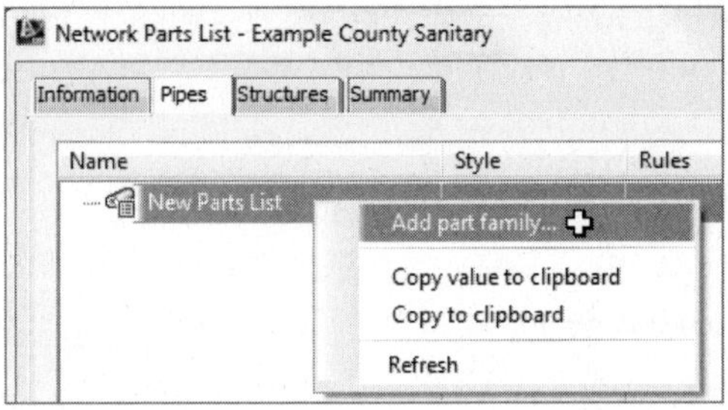

FIGURE 14.14 Add a part family for your new parts list.

   The Part Catalog dialog will appear.
7. In the list of available pipes, place a check mark next to PVC Pipe (PVC Pipe SI in the metric catalog) and click OK.

   At this step the Parts List name should appear at the top of the pipe list.
8. Expand the pipe category to see the new PVC Pipe family you just added.
9. Right-click on PVC Pipe and select Add Part Size, as shown in Figure 14.15.

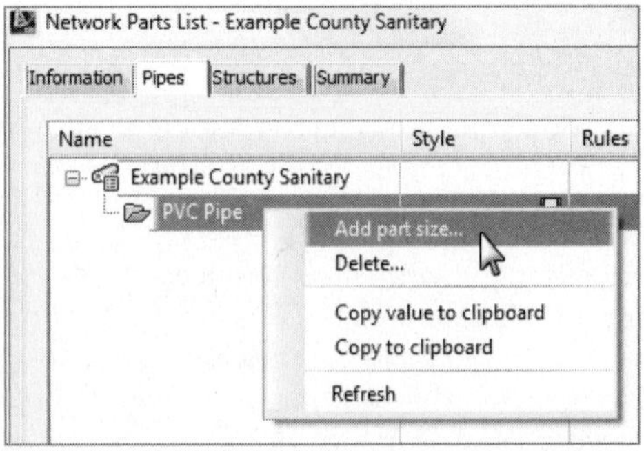

FIGURE 14.15 Add a new part size to the part family PVC Pipe.

10. In the Part Size Creator, click the pull-down in the Inner Pipe Diameter value field, as highlighted in Figure 14.16.

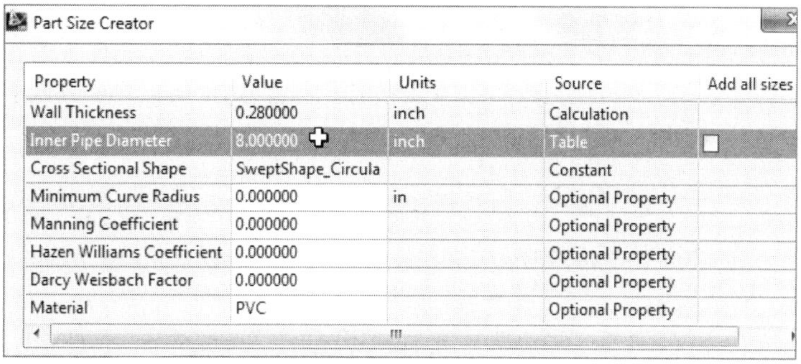

FIGURE 14.16 Set diameter and material for all the needed pipes.

11. Select 8 Inch (metric: 200 mm).
12. In the Material field, set the material to **PVC**, and click OK.

    For Material, you can use the drop-down to select from several preexisting material types, or you can type in your own material name. The material name is used frequently in labels.

13. Expand the new PVC category to examine the result. Repeat steps 10–12 for 10″ (250 mm) and 12″ (300 mm) pipes.

    In this example, the pipes will share the same style and rule. Using the disk icon in the PVC Pipe part family row of the table, you will apply your style choice to the entire PVC family.

    For the columns Render Material and Pay Item, leave the defaults. Render materials and pay items do not affect the design portion of the pipe network. Render materials are used if you wish to give a realistic material to the object for visualization purposes. You will take an in-depth look at assigning pay items to a parts list in Chapter 19, "Quantity Takeoff."

14. Click the disk icon in the Style column.
15. Set Pipe Style to Single Line Sanitary and click OK.

16. For each pipe size, set the respective rule by clicking the Pipe Rule Set icon.

    At the end of the process your Pipes tab will look like Figure 14.17.

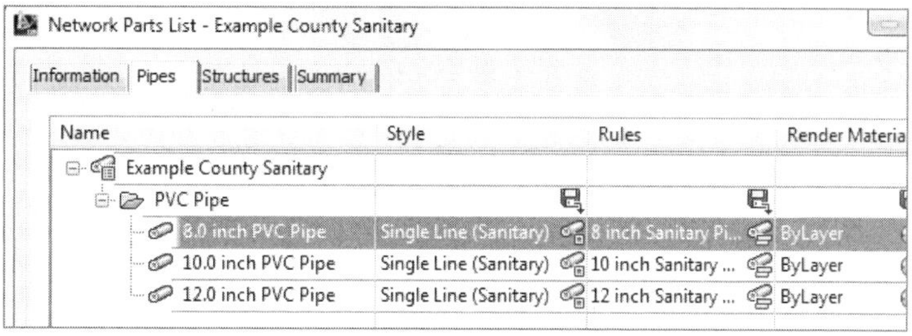

FIGURE 14.17 The completed Pipes tab

17. Switch to the Structures tab and expand New Parts List.

    Notice there is already a null structure in the listing.

18. Expand the null structure part family and click the Structure Style icon for the Null structure.
19. Change the null structure style to Null and click OK.
20. Right-click on the main heading New Parts List and select Add Part Family.
21. In the Part Catalog listing, locate the Junction Structures With Frames grouping and put a check mark next to Concentric Cylindrical Structure and Cylindrical Structure Slab Top Circular Frame. Click OK.

    Note that in the metric drawings, the structure descriptions end with SI.

22. Right-click on Concentric Cylindrical Structure and select Add Part Size.
23. In the Part Size Creator, select an Inner Structure diameter of 48″ (1200 mm).
24. Leave all other size options at their defaults, and click OK.
25. Repeat steps 23–24 to add the 60″ (1500 mm) structure.
26. Use the disk icon to set the style for both concentric cylindrical structures to Sanitary Sewer Manhole.
27. Right-click on Cylindrical Structure Slab Top Circular Frame, and select Add Part Size.
28. Set the Inner Structure Diameter to 15″ (450 mm), and click OK.
29. Set Structure Style to Cleanout using the same process you used in step 26.
30. Use a similar procedure to set the rules for the three new structures to Sanitary Structure rules.

    If you expand all the structure part families, your network parts list will look like Figure 14.18.

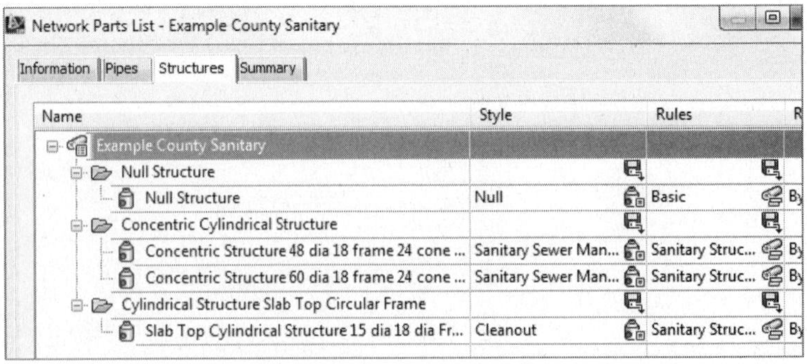

FIGURE 14.18
Completed Structures tab in your new parts list

31. Click OK to close the network parts list.
32. Save the drawing.

ALL IN THE FAMILY

To get a better idea of what parts are available to you "out of the box," you may want to take a peek at the pipes catalog folder. This folder can be found on your hard drive at:

```
C:\ProgramData\Autodesk\C3D 2013\enu\Pipes Catalog\
```

Outside of Civil 3D, use Windows Explorer to browse to this path. By default, the `ProgramData` folder is hidden, so you'll need to change your Windows folder view options to see this path. If you are not sure how to change this setting, you can also type in the path listed here to access it. To learn more about how a catalog is organized, let's explore the `US Imperial Structures` (or `metric Structures`) folder.

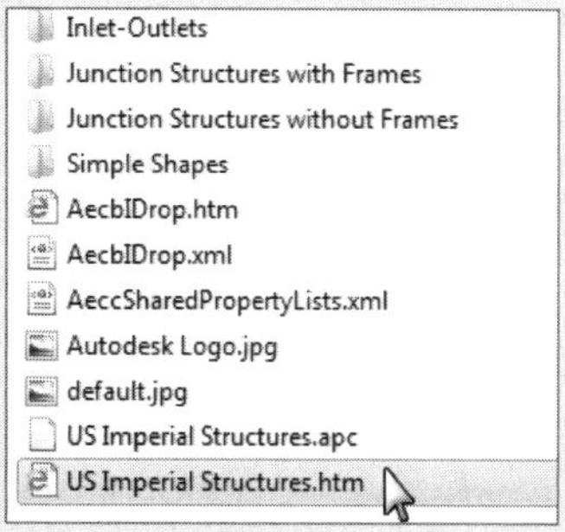

In the `US Imperial Structures` (or `Metric Structures`) folder is a document called `US Imperial Structures.htm` (`Metric Structures.htm`). Double-click this file to open it in Internet Explorer.

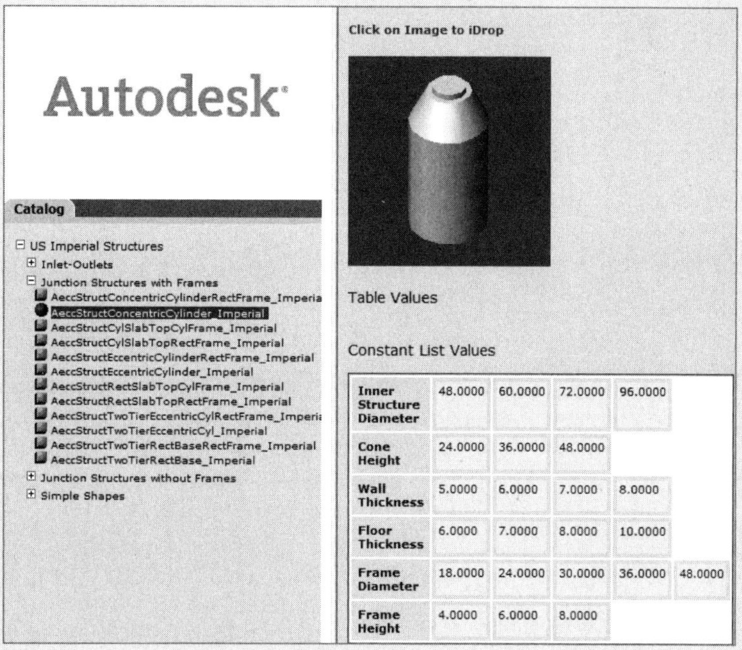

For both Imperial and metric units, there is a `pipes` folder and a `structures` folder. You may see these different categories referred to as a *domain* as you browse the catalogs. Each domain has an HTML document like the one shown here that you can look at before committing parts to your template parts list.

On the left side of the screen, you see the different parts belonging to the domain. On the right, you see a picture and a table of the various allowable values. Go ahead and explore! There's no way to accidentally edit something from here.

Whenever possible, use a part that is already in the catalog. Keep in mind that these are building blocks and are not meant to be photorealistic representations of the 3D objects. If you must create a part or add a size that is not already part of the default catalog, Part Builder is available.

For those of you brave enough to tackle Part Builder, look at the special section at the end of this chapter.

Exploring Pipe Networks

Parts in a pipe network have relationships that follow a network paradigm. A pipe network, such as the one in Figure 14.19, can have many branches. In most cases, the pipes and structures in your network will be connected to each other; however, they don't necessarily have to be physically touching to be included in the same pipe network.

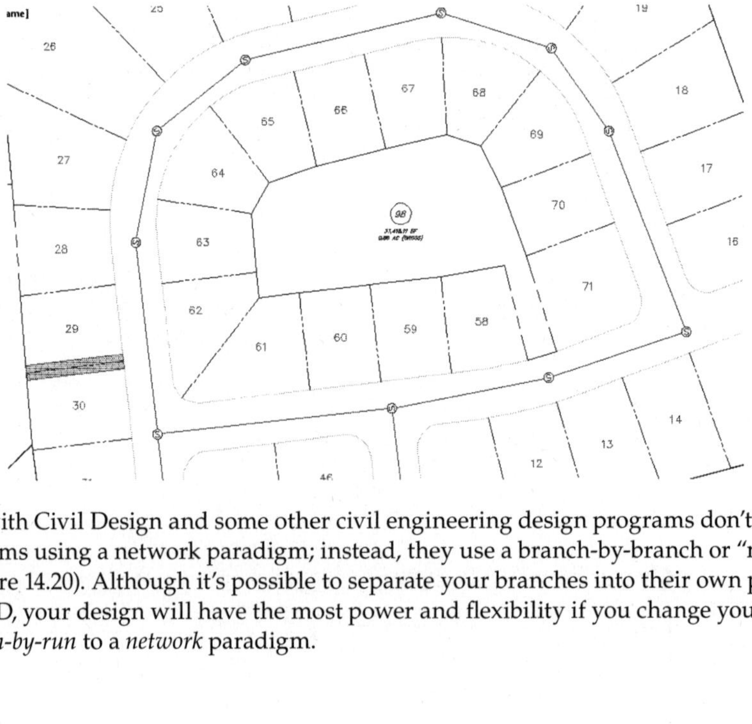

FIGURE 14.19 A typical Civil 3D pipe network

Land Desktop with Civil Design and some other civil engineering design programs don't design piping systems using a network paradigm; instead, they use a branch-by-branch or "run" paradigm (see Figure 14.20). Although it's possible to separate your branches into their own pipe networks in Civil 3D, your design will have the most power and flexibility if you change your thinking from a *run-by-run* to a *network* paradigm.

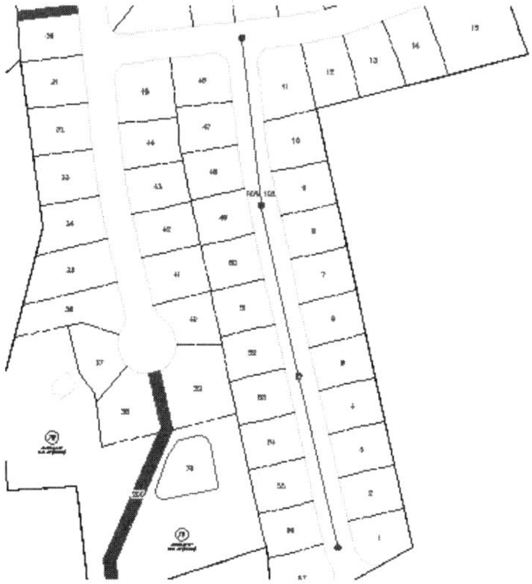

FIGURE 14.20
A pipe network with a single-pipe run

Pipe networks have the following object types:

Pipes *Pipes* are components of a pipe network that primarily represent pipes, but can be used to represent any type of conduit such as culverts, gas lines, or utility cables. They can be straight or curved, and although primarily used to represent gravity systems, they can be adapted and customized to represent pressure and other types of systems such as water networks and force mains.

The standard catalog has pipe shapes that are circular, elliptical, egg-shaped, and rectangular and are made of materials that include PVC, RCP, DI, and HDPE. You can use Part Builder (discussed later in this chapter) to create your own shapes and materials if the default shapes and dimensions can't be adapted for your design.

Structures *Structures* are the components of a pipe network that represent manholes, catch basins, inlets, joints, and any other type of junction between two pipes. The standard catalog includes inlets, outlets, junction structures with frames (such as manholes with lids or catch basins with grates), and junction structures without frames (such as simple cylinders and rectangles). Again, you can use Part Builder to create your own shapes and materials if needed.

Null Structures *Null structures* are needed when two pipes are joined together without a structure; they act as a placeholder for a pipe endpoint. They have special properties, such as allowing pipe cleanup at pipe intersections. Most of the time, you'll create a style for them that doesn't plot or is invisible for plotting purposes.

Creating a Sanitary Sewer Network

Earlier, you prepared a parts list for a typical sanitary sewer network. This chapter will lead you through several methods for using that parts list to design, edit, and annotate a pipe network.

There are several ways to create pipe networks. You can do so using the Civil 3D pipe layout tools. You can also create pipe networks from certain AutoCAD and Civil 3D objects, such as lines, polylines, alignments, and feature lines.

Creating a Pipe Network with Layout Tools

Creating a pipe network with layout tools is much like creating other Civil 3D objects. After naming and establishing the parameters for your pipe network, you're presented with a special toolbar that you can use to lay out pipes and structures in plan, which will also drive a vertical design.

Establishing Pipe Network Parameters

This section will give you an overview of establishing pipe network parameters. Use this section as a reference for the exercises in this chapter. When you're ready to create a pipe network, select the Home tab ➢ Create Design panel and choose Pipe Network ➢ Pipe Network Creation Tools. The Create Pipe Network dialog appears (see Figure 14.21), and you can establish your settings.

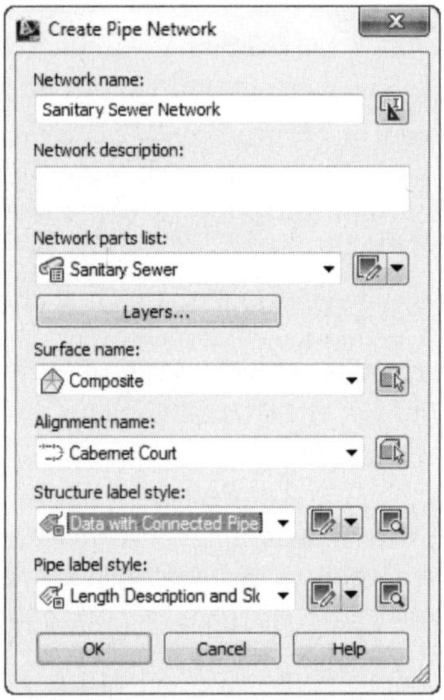

FIGURE 14.21 The Create Pipe Network dialog

Before you can create a pipe network, you must give your network a name, but more important, you need to assign a parts list for your network. As you saw earlier, the parts list provides a toolkit of pipes, structures, rules, and styles to automate the pipe network design process. It's also important to select a reference surface in this interface. This surface will be used for rim elevations and rule application.

When creating a pipe network, you're prompted for the following options:

Network Name Choose a name for your network that is meaningful and will help you identify it in Prospector and other locations.

Network Description The description of your pipe network is optional. You might make a note of the date, the type of network, and any special characteristics.

Network Parts List Choose the parts list that contains the parts, rules, and styles you want to use for this design.

Surface Name Choose the surface that will provide a basis for applying cover rules as well as provide an insertion elevation for your structures (in other words, rim elevations). You can change this surface later or for individual structures. For proposed pipe networks, this surface is usually a finished ground surface.

Alignment Name Choose an alignment that will provide station and offset information for your structures in Prospector as well as any labels that call for alignment stations and/or offset information. Because most pipe networks have several branches, it may not be meaningful for every structure in your network to reference the same alignment. Therefore, you may find it better to leave your Alignment option set to None in this dialog and set it for individual structures later using the layout tools or structure list in Prospector.

Using the Network Layout Creation Tools

CERT OBJECTIVE

After establishing your pipe network parameters in the Create Pipe Network dialog (shown earlier in Figure 14.21), click OK; the Network Layout Tools toolbar appears (see Figure 14.22).

FIGURE 14.22
The Network Layout Tools toolbar

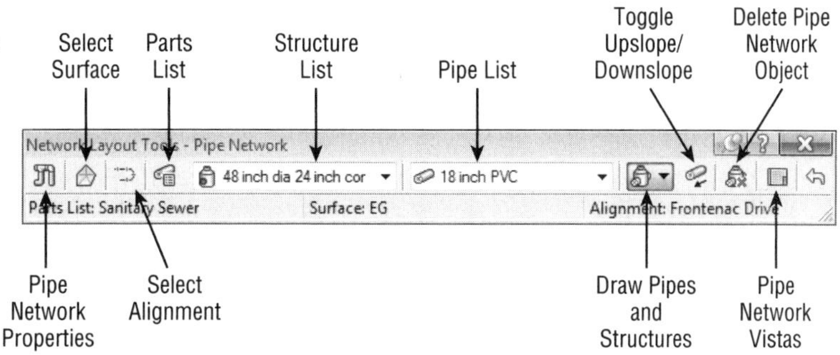

Clicking the Pipe Network Properties tool displays the Pipe Network Properties dialog, which contains the settings for the entire network. If you mistyped any of the parameters in the original Create Pipe Network dialog, you can change them here. In addition, you can set the default label styles for the pipes and structures in this pipe network.

The Pipe Network Properties dialog contains the following tabs:

Information On this tab, you can rename your network, provide a description, and choose whether you'd like to see network-specific tooltips.

Layout Settings Here you can change the default label styles, parts list, reference surface and alignment, master object layers for plan pipes and structures, as well as name templates for your pipes and structures (see Figure 14.23).

FIGURE 14.23
The Layout Settings tab of the Pipe Network Properties dialog

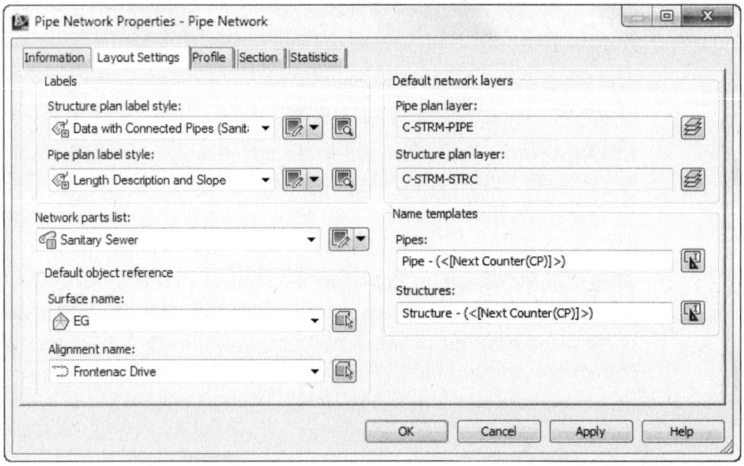

Profile On this tab, you can change the default label styles and master object layers for profile pipes and structures (see Figure 14.24).

FIGURE 14.24
The Profile tab of the Pipe Network Properties dialog

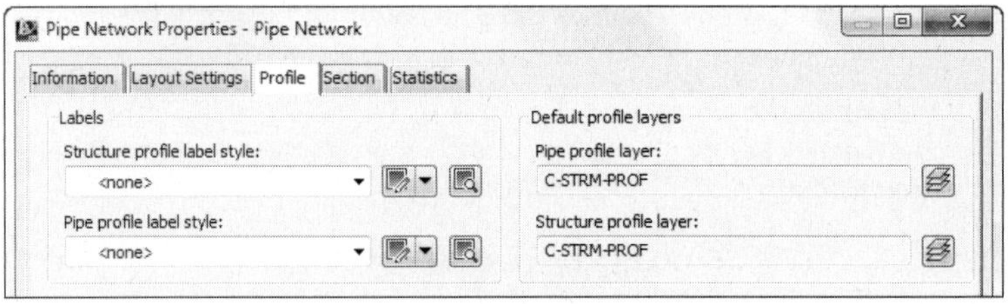

Section Here you can change the master object layers for network parts in a section (see Figure 14.25).

FIGURE 14.25
The Section tab of the Pipe Network Properties dialog

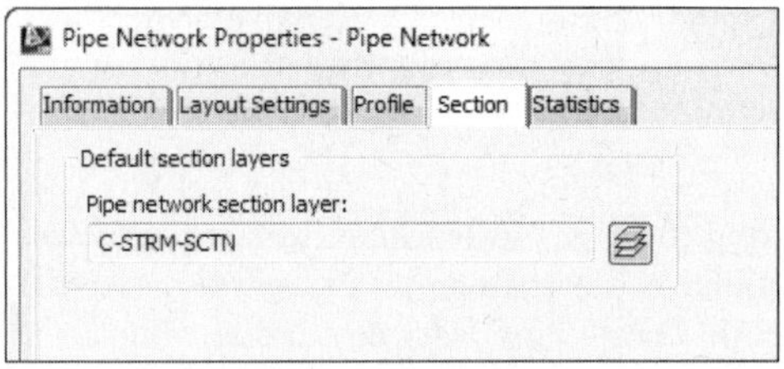

Statistics This tab gives you a snapshot of your pipe network information, such minimum and maximum as elevation information, pipe and structure quantities, and references in use such as alignments and surfaces (see Figure 14.26).

FIGURE 14.26
The Statistics tab of the Pipe Network Properties dialog

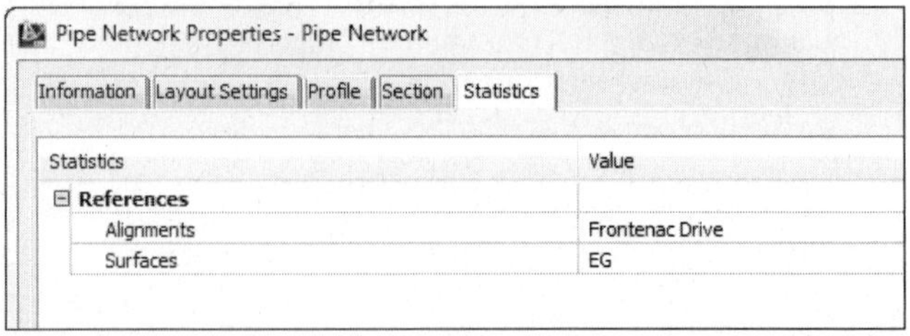

The Select Surface tool on the Network Layout Tools toolbar allows you to switch between reference surfaces while you're placing network parts. For example, if you're about to place a structure that needs to reference the existing ground surface but your network surface was set to a proposed ground surface, you can click this tool to switch to the existing ground surface.

> **USING A COMPOSITE FINISHED GRADE SURFACE FOR YOUR PIPE NETWORK**
>
> It's cumbersome to constantly switch between a patchwork of different reference surfaces while designing your pipe network. You should create a finished grade-composite surface that includes components of your road design, finished grade, and even existing ground. You can create this finished grade-composite surface by pasting surfaces together, so it's dynamic and changes as your design evolves. The surfaces being dynamic means that the rim elevations and reporting of min/max cover will update automatically. The pipes, however, will not raise and lower automatically in response to changes in the composite surface.

The Select Alignment tool on the Network Layout Tools toolbar lets you switch between reference alignments while you're placing network parts, similar to the Select Surface tool.

The Parts List tool allows you to switch parts lists for the pipe network.

The Structure drop-down list (the image on the left in Figure 14.27) lets you choose which structure you'd like to place next, and the Pipes drop-down list (the image on the right in Figure 14.27) allows you to choose which pipe you'd like to place next. Your choices come from the network parts list.

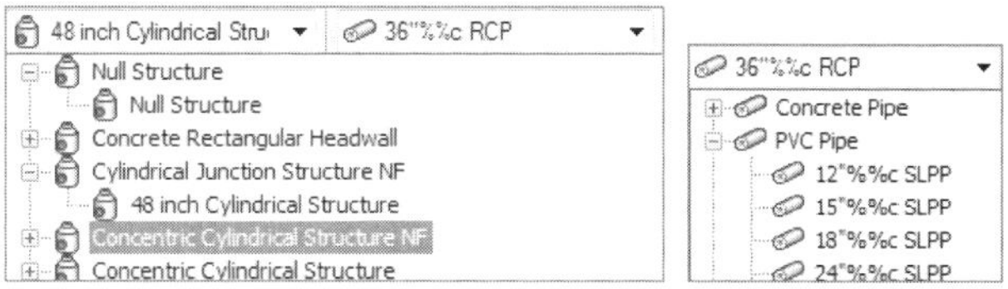

(a) (b)

FIGURE 14.27
(a) The Structure drop-down list, and (b) the Pipes drop-down list

The options for the Draw Pipes and Structures category let you choose what type of parts you'd like to lay out next. You can choose Pipes And Structures, Pipes Only, or Structures Only.

PLACING PARTS IN A NETWORK

You place parts much as you do other Civil 3D objects or AutoCAD objects such as polylines. You can use your mouse, transparent commands, dynamic input, object snaps, and other drawing methods when laying out your pipe network.

If you choose Pipes And Structures, a structure is placed wherever you click, and the structures are joined by pipes. If you choose Structures Only, a structure is placed wherever you click, but the structures aren't joined. If you choose Pipes Only, you can connect previously placed structures. If you have Pipes Only selected and there is no structure where you click, a null structure is placed to connect your pipes.

While you're actively placing pipes and structures, you may want to connect to a previously placed part. For example, there may be a service or branch that connects into a structure along the main trunk. Begin placing the new branch. When you're ready to tie into a structure, you get a circular connection marker (shown at the top of Figure 14.28) as your cursor comes within connecting distance of that structure. If you click to place your pipe when this marker is visible, a structure-to-pipe connection is formed. If you're placing parts and you'd like to connect to a pipe, hover over the pipe you'd like to connect to until you see a connection marker that has two square shapes.

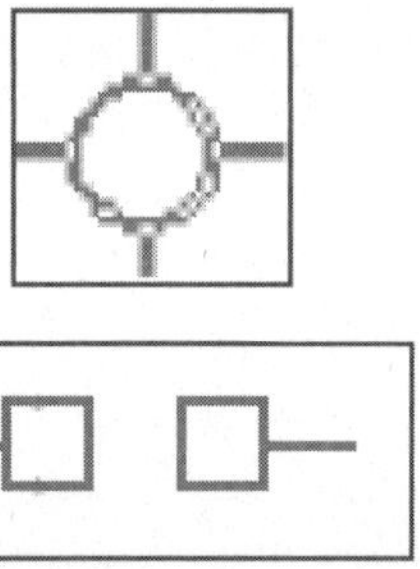

FIGURE 14.28
The Structure Connection marker and the Pipe Connection marker

Clicking to connect to the pipe breaks the pipe in two pieces and places a structure (or null structure) at the break point.

The Toggle Upslope/Downslope tool changes the flow direction of your pipes as they're placed. In Figure 14.29, Structure 9 was placed before Structure 10.

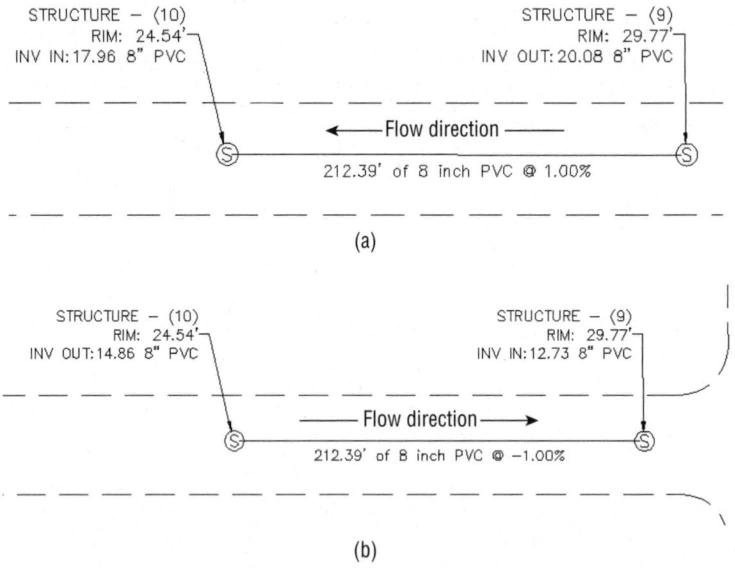

FIGURE 14.29
Using (a) the Downslope toggle and (b) the Upslope toggle to create a pipe network leg

OPTIMIZING THE COVER BY STARTING UPHILL

If you're using the Cover And Slope rule for your pipe network, you'll achieve better cover optimization if you begin your design at an upstream location and work your way down to the connection point.

The Cover And Slope rule prefers to hold minimum slope over optimal cover. In practice, this means that as long as minimum cover is satisfied, the pipe will remain at minimum slope. If you start your design from the upstream location, the pipe is forced to use a higher slope to achieve minimum cover. The following graphic shows a pipe run that was created starting from the upstream location (right to left):

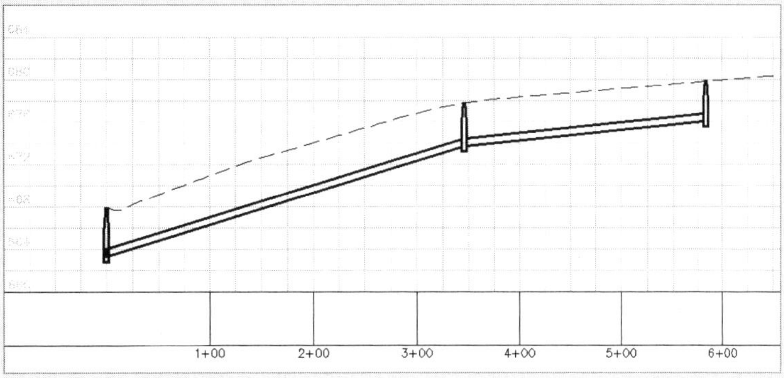

When you start from the downhill side of your project, the Minimum Slope Rule is applied as long as minimum cover is achieved. The following graphic shows a pipe run that was created starting from the downstream location (left to right):

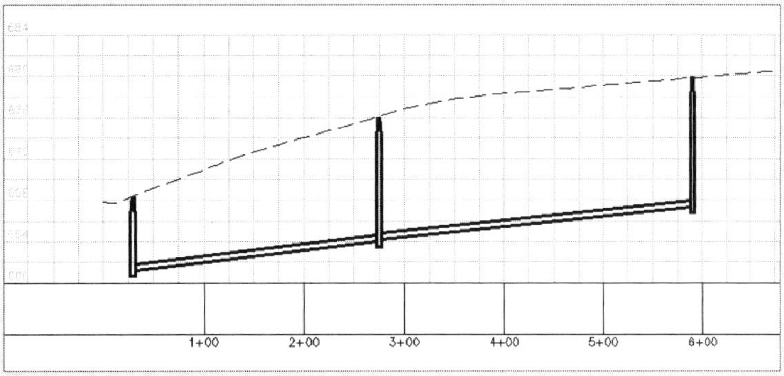

Notice how the slope remains constant even as the pipe cover increases. Maximum cover is a *violation-only* rule, which means it never forces a pipe to increase slope to remain within tolerance; it only provides a warning that maximum cover has been violated.

Click Delete Pipe Network Object to delete pipes or structures of your choice. AutoCAD Erase can also delete network objects, but using it requires you to leave the Network Layout Tools toolbar.

Clicking Pipe Network Vistas brings up Panorama (see Figure 14.30), where you can make tabular edits to your pipe network while the Network Layout Tools toolbar is active.

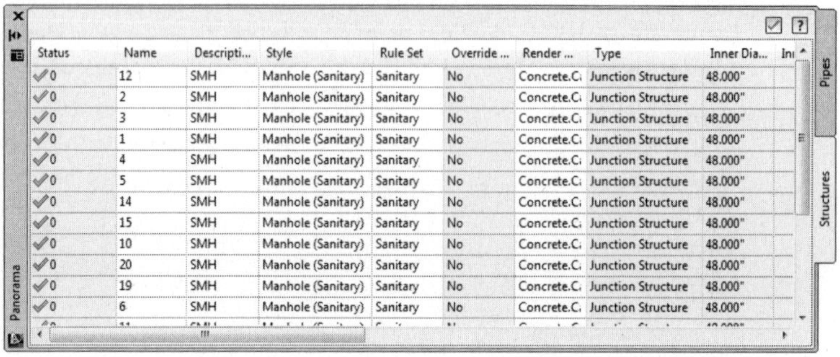

Status	Name	Descripti...	Style	Rule Set	Override ...	Render ...	Type	Inner Dia...	Ini
0	12	SMH	Manhole (Sanitary)	Sanitary	No	Concrete.C	Junction Structure	48.000"	
0	2	SMH	Manhole (Sanitary)	Sanitary	No	Concrete.C	Junction Structure	48.000"	
0	3	SMH	Manhole (Sanitary)	Sanitary	No	Concrete.C	Junction Structure	48.000"	
0	1	SMH	Manhole (Sanitary)	Sanitary	No	Concrete.C	Junction Structure	48.000"	
0	4	SMH	Manhole (Sanitary)	Sanitary	No	Concrete.C	Junction Structure	48.000"	
0	5	SMH	Manhole (Sanitary)	Sanitary	No	Concrete.C	Junction Structure	48.000"	
0	14	SMH	Manhole (Sanitary)	Sanitary	No	Concrete.C	Junction Structure	48.000"	
0	15	SMH	Manhole (Sanitary)	Sanitary	No	Concrete.C	Junction Structure	48.000"	
0	10	SMH	Manhole (Sanitary)	Sanitary	No	Concrete.C	Junction Structure	48.000"	
0	20	SMH	Manhole (Sanitary)	Sanitary	No	Concrete.C	Junction Structure	48.000"	
0	19	SMH	Manhole (Sanitary)	Sanitary	No	Concrete.C	Junction Structure	48.000"	
0	6	SMH	Manhole (Sanitary)	Sanitary	No	Concrete.C	Junction Structure	48.000"	

FIGURE 14.30
Pipe Network Vistas via Panorama

The Pipe Network Vistas interface is similar to what you encounter in the Pipe Networks branch of Prospector. The advantage of using Pipe Network Vistas is that you can make tabular edits without leaving the Network Layout Tools toolbar. You can edit pipe properties, such as Invert and Slope, on the Pipes tab, and you can edit structure properties, such as Rim and Sump, on the Structure tab.

CREATING A SANITARY SEWER NETWORK

This exercise will apply the concepts taught in this section, and give you hands-on experience using the Network Layout Tools toolbar:

1. Open `Pipes-Exercise1.dwg` (`Pipes-Exercise1_METRIC.dwg`), which you can download from this book's web page.
2. Expand the Surfaces branch in Prospector.

   This drawing has several surfaces which have a _No Display style applied to simplify the drawing. The surface you will be working off of is a composite of the existing conditions, corridor surfaces, and grading surfaces.
3. Expand the Alignments and Centerline Alignments branches, and notice that there are several road alignments. (No action is required).

4. On the Home tab ➢ Create Design panel ➢ Pipe Network, select Pipe Network Creation Tools.
5. In the Create Pipe Network dialog (shown previously in Figure 14.21), give your network the following information:
   - Network Name: **Sanitary Sewer Network**
   - Network Parts List: **Sanitary Sewer**
   - Surface Name: **Composite**

- Alignment Name: **Cabernet Court**
- Structure Label Style: **Data with Connected Pipes (Sanitary)**
- Pipe Label Style: **Length Description and Slope**

6. Click OK.

   The Network Layout Tools toolbar appears.

7. From the Structure list, Concentric Cylindrical Structure collection, choose Concentric Structure 48 Dia 18 Frame 24 Cone (metric: Concentric Structure 1,200 dia 450 frame 600 cone) from the drop-down list in the Structure menu and 8 Inch PVC (200 mm PVC) from the Pipe list.

8. Click the Draw Pipes And Structures tool. Working right-to-left, click the X labeled 1 in the drawing to place the first structure. Click the X labeled 2 to place the second structure.

9. Without exiting the command, go back to the Network Layout Tools toolbar and change the Pipe drop-down from 8 Inch PVC to 10 Inch PVC (200 mm PVC to 250 mm PVC) and then place structures at the Xs labeled 3, 4, and 5.

   The labels show that the diameter of the pipe between these structures is 10" (250 mm).

10. Press ↵ to exit the command.

    Next, you'll add a branch of the network from Frontenac Drive. You may wish to use the label grip to drag the label off to the side. This will form the leaders as shown in Figure 14.31, making the next step easier.

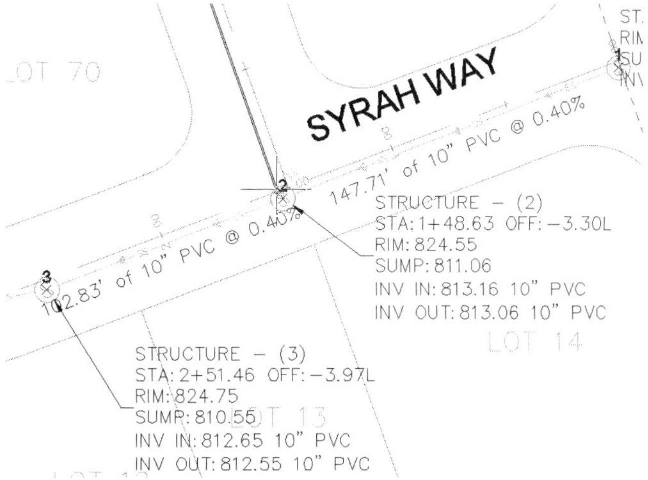

FIGURE 14.31
The connection marker appears when your cursor is near the existing structure.

11. Go back to the Network Layout Tools toolbar, and select 8 Inch PVC from the Pipe list.

12. Click the Draw Pipes And Structures tool button again.

13. Working north to south, click the next structure at the X labeled 6.
14. Tie into the Syrah Way branch by moving your cursor near Structure 2.

    You'll know you're about to connect when you see the connection marker (a round, golden-colored glyph shaped like the one in Figure 14.31) appear next to your previously inserted structure.
15. Press ↵ to exit the command.

    Observe your pipe network, including the labeling that automatically appeared as you drew the network.
16. Expand the Pipe Networks branch in Prospector in Toolspace, and locate your sanitary sewer network.
17. Click the Pipes branch.

    The list of pipes appears in the preview pane.
18. Click the Structures branch; the list of structures appears in the preview pane.
19. Experiment with tabular and graphical edits, drawing parts in profile view, and other tasks described throughout this chapter.

Creating a Storm Drainage Pipe Network from a Feature Line

CERT OBJECTIVE

If you already have an object in your drawing that represents a pipe network (such as a polyline, an alignment, or a feature line), you may be able to take advantage of the Create Pipe Network From Object command in the Pipe Network drop-down menu.

This option can be used for applications such as converting surveyed pipe runs into pipe networks and bringing forward legacy drawings that used AutoCAD linework to represent pipes. Because of some limitations described later in this section, it isn't a good idea to use this in lieu of Create Pipe Network By Layout for new designs.

It's often tempting in Civil 3D to rely on your former drawing habits and try to "convert" your AutoCAD objects into Civil 3D objects. You'll find that the effort you spend learning Create Pipe Network By Layout pays off quickly with a better-quality model and easier revisions.

The Create Pipe Network From Object option creates a pipe for every linear segment of your object and places a structure at every vertex of your object. For example, the polyline with three line segments and three arcs, shown in Figure 14.32, is converted into a pipe network containing three straight pipes, three curved pipes, and seven structures—one at the start, one at the end, and one at each vertex.

This option is most useful for creating pipe networks from long, single runs. It can't build branching networks or append objects to a pipe network. For example, if you create a pipe network from one feature line and then, a few days later, receive a second feature line to add to that pipe network, you'll have to use the pipe-network editing tools to trace your second feature line; no tool lets you add AutoCAD objects to an already-created pipe network. However, once a pipe network exists, you can merge it into another using the Merge Networks command.

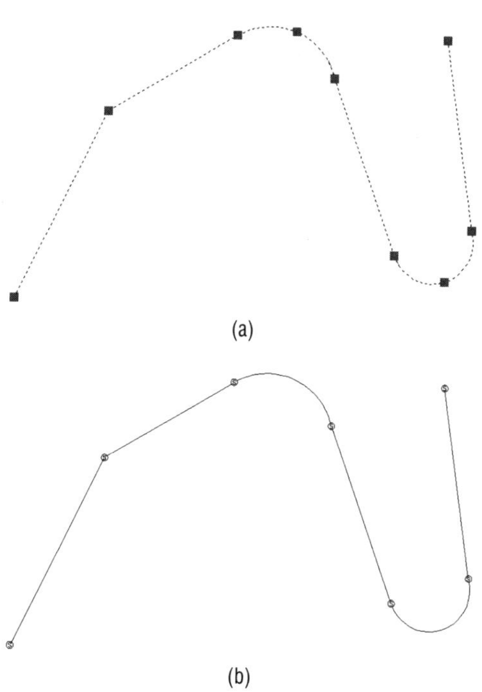

FIGURE 14.32
(a) A polyline showing vertices and (b) a pipe network created from the polyline

This exercise will give you hands-on experience building a pipe network from a feature line with elevations:

1. Open the `Pipes-Exercise2.dwg` (`Pipes-Exercise2_METRIC.dwg`) file.

   This drawing contains the feature line you will use to generate a storm network.

2. Expand the Surfaces branch in Prospector.

   This drawing has several surfaces which have a _No Display style applied to simplify the drawing. (No action is required).

3. Expand the Alignments and Centerline Alignments branches, and notice that there are several road alignments.

   In the drawing, a cyan feature line runs through Syrah Way (the horizontal road) and then goes onto Frontenac Drive. This feature line represents utility information for a storm-drainage line. The elevations of this feature line correspond with centerline elevations that you'll apply to your pipe network.

4. Choose Create Pipe Network From Object from the Pipe Network drop-down.

5. At the `Select Object or [Xref]:` prompt, select the cyan feature line near the eastern side of the line.

You're given a preview (see Figure 14.33) of the pipe-flow direction that is based on where you selected the line. In this case, the east end of the feature line is considered the upstream end.

FIGURE 14.33
Flow-direction preview

6. At the Flow Direction [OK/Reverse] <Ok>: prompt, press ↵ to choose OK.

   The Create Pipe Network From Object dialog appears.

7. In the dialog, give your pipe network the following information:

   - Network Name: **Storm Network**
   - Network Parts List: **Storm Sewer**
   - Pipe To Create: **12 Inch Concrete Pipe** (**300 mm Concrete Pipe**)
   - Structure To Create: From the Rectangular Junction Structure NF collection **2 × 2 Rectangular Structure** (**750 × 750 mm Rectangular Structure**)
   - Surface Name: **Composite**
   - Alignment Name: **<none>**
   - Erase Existing Entity check box: Selected
   - Use Vertex Elevations check box: Selected
   - Set Vertex Elevation Reference to **Invert**

   If you select Use Vertex Elevations, the pipe rules for your chosen parts list will be ignored. Note the many options you have for how the vertex elevation is used.

8. Click OK. A pipe network is created.

   Next you will merge this network with another network that exists in the drawing.

9. Select a pipe or structure from the newly created network. From the Pipe Networks contextual tab ➢ Modify panel, select Merge Networks.

10. In the Select Pipe Network dialog, highlight Existing Storm Network and click OK.
11. In the Select Destination Pipe Network dialog, highlight Storm Network and click OK.

This has combined all your object-created networks into a more manageable single network.

Changing Flow Direction

To change flow direction, select any part from a pipe network to open the Pipe Networks Contextual tab. Select Change Flow Direction from the Modify panel. Change Flow Direction allows you to reverse the pipe's understanding of which direction it flows, which comes into play when you're using the Apply Rules command and when you're annotating flow direction with a pipe label-slope arrow.

Changing the flow direction of a pipe doesn't make any changes to the pipe's invert. By default, a pipe's flow direction depends on how the pipe was drawn and how the Toggle Upslope/Downslope tool was set when the pipe was drawn:

- If the toggle was set to Downslope, the pipe flow direction is set to Start To End, which means the first endpoint you placed is considered the start of flow and the second endpoint is established as the end of flow.
- If the toggle was set to Upslope when the pipe was drawn, the pipe flow direction is set to End To Start, which means the first endpoint placed is considered the end for flow purposes and the second endpoint the start.

After pipes are drawn, you can set four additional flow options—Start To End, End To Start, Bi-directional, and By Slope—in Pipe Properties:

Start To End A pipe label-flow arrow shows the pipe direction from the first pipe endpoint drawn to the second endpoint drawn, regardless of invert or slope.

End To Start A pipe label-flow arrow shows the pipe direction from the second pipe endpoint drawn to the first pipe endpoint drawn, regardless of invert or slope.

Bi-directional Typically, this is a pipe with zero slope that is used to connect two bodies that can drain into each other, such as two stormwater basins, septic tanks, or overflow vessels. The direction arrow is irrelevant in this case.

By Slope A pipe label-flow arrow shows the pipe direction as a function of pipe slope. For example, if End A has a higher invert than End B, the pipe flows from A to B. If B is edited to have a higher invert than A, the flow direction flips to be from B to A.

Editing a Pipe Network

You can edit pipe networks in several ways:

- Using drawing layout edits such as grip, move, and rotate
- Grip-editing the pipe size
- Using vertical movement edits using grips in profile (see the "Vertical Movement Edits Using Grips in Profile" section later in this chapter)

- Using tabular edits in the Pipe Networks branch in Prospector, or from Panorama in the pipe toolbar
- Right-clicking a network part to access tools such as Swap Part or Pipe/Structure Properties
- Returning to the Network Layout Tools toolbar by right-clicking the object and choosing Edit Network
- Selecting the pipe network in Toolspace-Prospector, right-clicking, and choosing Edit Network
- Selecting a network part to access the Pipe Networks contextual tab on the Ribbon

You will have the chance to explore most of these methods in the following sections.

Editing Your Network in Plan View

When selected, a structure has two types of grips, shown in Figure 14.34. The first is a square grip located at the structure insertion point. You can use this grip to grab the structure and stretch/move it to a new location using the insertion point as a base point. Stretching a structure results in the movement of the structure as well as any connected pipes. You can also scroll through Stretch, Move, and Rotate by using your spacebar once you've grabbed the structure by this grip.

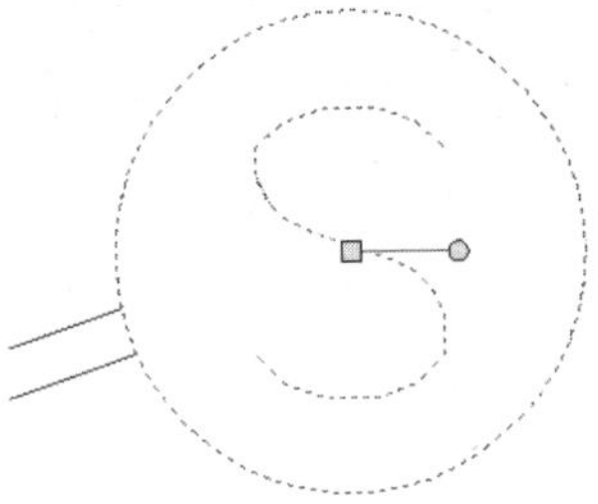

FIGURE 14.34
Two types of structure grips

The second structure grip is a rotational grip that you can use to spin the structure about its insertion point. This is most useful for aligning eccentric structures, such as rectangular junction structures.

In plan view, many common AutoCAD Modify commands work with structures. You can execute the following commands normally: Move, Copy, Rotate, Align, and Mirror. (Scale doesn't have an effect on structures.) Keep in mind that the Modify commands are applied to the structure model itself; depending on how you have your style established, it may not be clear that you've made a change.

You can use the AutoCAD Erase command to erase network parts. Note that erasing a network part in plan completely removes that part from the network. Once erased, the part disappears from plan, profile view, Prospector, and so on.

When selected, a pipe end has two types of grips (see Figure 14.35). The first is a square endpoint-location grip. Using this grip, you can change the location of the pipe end without constraint. You can move it in any direction; make it longer or shorter; and take advantage of Stretch, Move, Rotate, and Scale by using your spacebar.

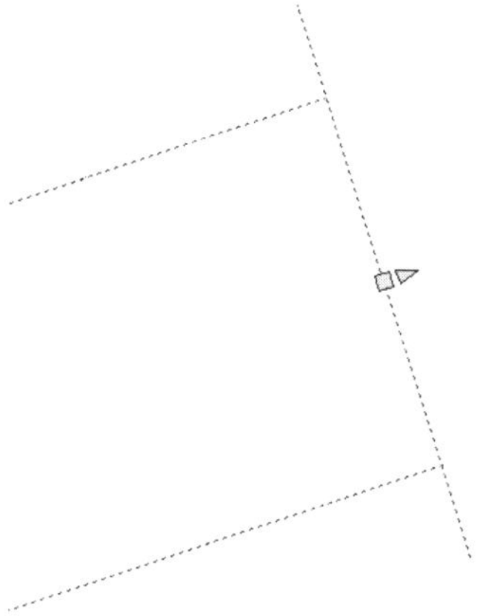

FIGURE 14.35
Two types of pipe-end grips

The second grip is a pipe-length grip. This grip lets you extend a pipe along its current bearing.

A pipe midpoint also has two types of grips (see Figure 14.36). The first is a square location grip that lets you move the pipe using its midpoint as a base point. As before, you can take advantage of Stretch, Move, Rotate, and Scale by using your spacebar.

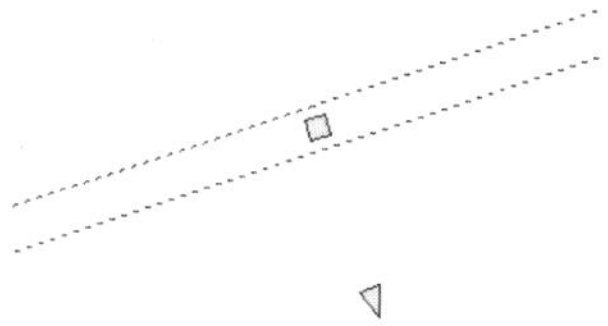

FIGURE 14.36
Two types of pipe midpoint grips

The second grip is a triangular-shaped pipe-diameter grip. Stretching this grip gives you a tooltip showing allowable diameters for that pipe, which are based on your parts list. Use this grip to make quick, visual changes to the pipe diameter.

A WORD ABOUT PAY ITEMS

Note that if you have a pay item specified in your parts list, changing the pipe diameter graphically does not change the associated pay item. For more information on associating pipe parts to pay items, see Chapter 19.

Dynamic Input and Pipe Network Editing

Dynamic input has been in AutoCAD-based products for many years, but many Civil 3D users aren't familiar with it. Dynamic input shows command-line information at your cursor in the form of a tooltip. In many cases, it is a quick way to obtain information and change objects. Especially for pipe network edits, it provides a visual way to interactively edit your pipes and structures.

To turn the Dynamic Input feature on or off at any time, press F12 or click the icon at the bottom of your Civil 3D window.

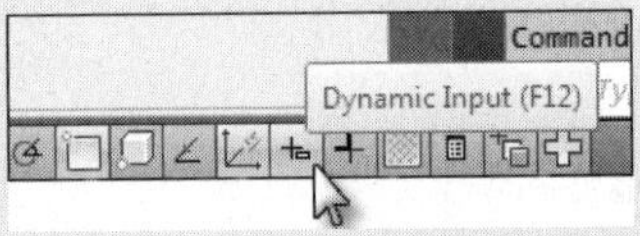

Structure Rotation While dynamic input is active, you get a tooltip that tracks rotation angle.

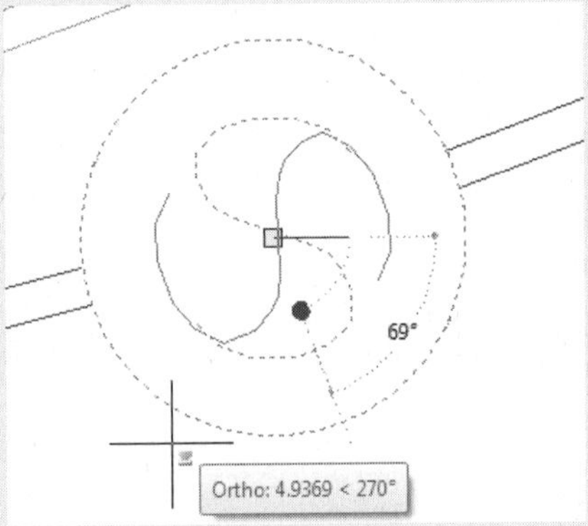

Press your down arrow key to get a pop-up menu that allows you to specify a base point followed by a rotation angle, as well as options for Copy and Undo. Dynamic input combined with the Rotate command is beneficial when you're rotating eccentric structures for proper alignment.

Pipe Diameter When you're using the pipe-diameter grip edit, dynamic input gives you a tooltip to assist you in choosing your desired diameter. Note that the tooltip depends on your drawing units.

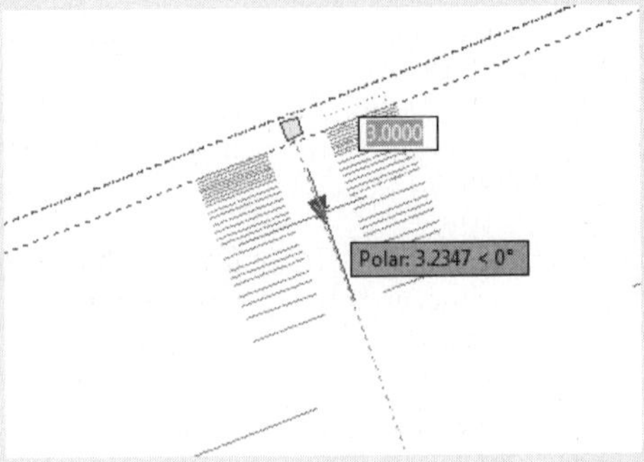

Pipe Length This is probably the most common reason to use dynamic input for pipe edits. Choosing the pipe-length grip when dynamic input is active shows tooltips for the pipe's current length and preview length, as well as fields for entering the desired pipe total length and pipe delta length. Use your Tab key to toggle between the Total Length and Delta Length fields. One of the benefits of using dynamic input in this interface is that even though you can't visually grip-edit a pipe to be shorter than its original length, you can enter a total length that is shorter than the original length. Note that the length shown and edited in the dynamic input interface is the 3D center-to-center length.

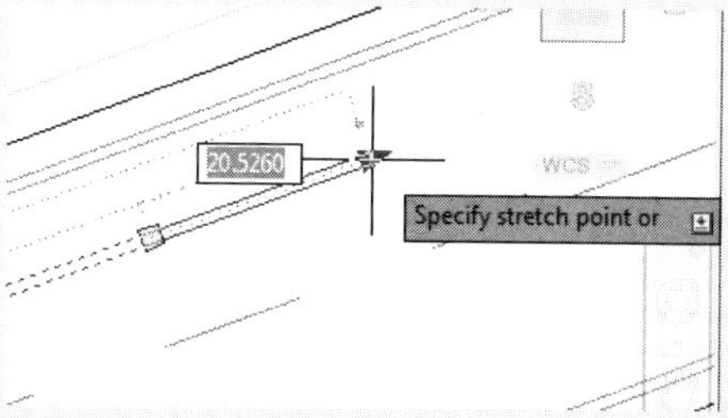

Pipe Endpoint Edits Similar to pipe length, pipe endpoint location edits can benefit from using dynamic input. The active fields give you an opportunity to input x- and y-coordinates.

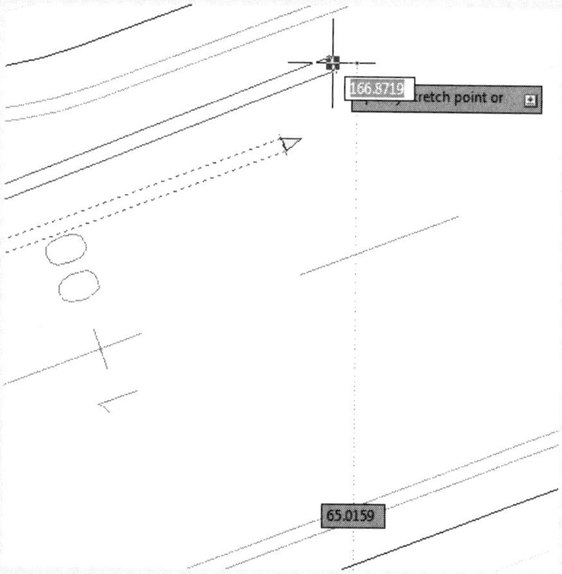

Pipe Vertical Grip Edits in Profile Using dynamic input in profile view lets you set exact invert, centerline, or top elevations without having to enter the Pipe Properties dialog or Prospector. Choose the appropriate grip, and note that the active field is Tracking Profile Elevation. You may need to zoom out to see the elevation field. Enter your desired elevation, and your pipe will move as you specify.

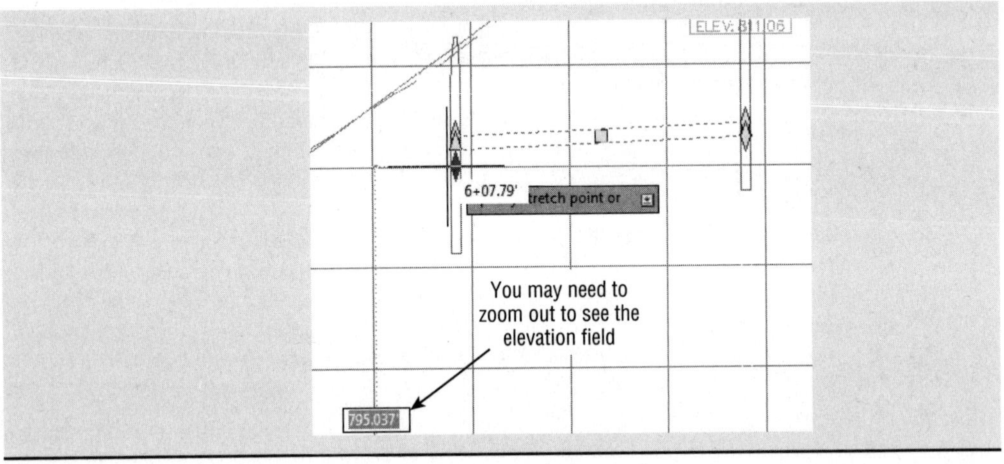

Making Tabular Edits to Your Pipe Network

Another method for editing pipe networks is in a tabular form using the Pipe Networks branch in Prospector (see Figure 14.37).

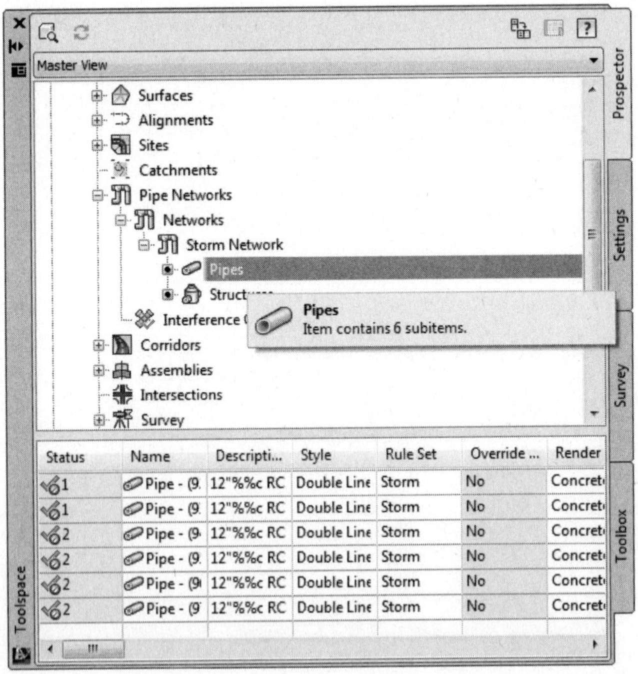

FIGURE 14.37
The Pipe Networks branch in Prospector

To edit pipes in Prospector, highlight the Pipes entry under the appropriate pipe network. For example, if you want to edit your sanitary sewer pipes, expand the Sanitary Sewers branch and select the Pipes entry. You should get a preview pane that lists the names of your pipes and some additional information in a tabular form. The same procedure can be used to list the structures in the network.

White columns can be edited in this interface. Gray columns are considered calculated values and therefore can't be edited.

You can adjust many things in this interface, but you'll find it cumbersome for some tasks. The interface is best used for the following:

Batch Changes to Styles, Render Materials, Reference Surfaces, Reference Alignments, Rule Sets, and So On Use your Shift or Ctrl key to select the desired rows, and then right-click the column header of the property you'd like to change. Choose Edit, and then select the new value from the drop-down menu. If you find yourself doing this on every project for most network parts, confirm that you have the correct values set in your parts list and in the Pipe Network Properties dialog.

Batch Changes to Pipe Description Use your Shift key to select the desired rows, and then right-click the Description column header. Choose Edit, and then type in your new description. If you find yourself doing this on every project for most network parts, check your parts list. If a certain part will always have the same description, you can add it to your parts list and prevent the extra step of changing it here.

Changing Pipe or Structure Names You can change the name of a network part by typing in the Name field. If you find yourself doing this on every project for every part, check that you're taking advantage of the Name templates in your Pipe Network Properties dialog (which can be further enforced in your Pipe Network command settings).

You can Shift-click and copy the table to your Clipboard and insert it into Microsoft Excel for sorting and further study. (This is a static capture of information; your Excel sheet won't update along with changes to the pipe network.)

This interface can be useful for changing pipe inverts, crowns, and centerline information. It's not always useful for changing the part rotation, insertion point, start point, or endpoint. It isn't as useful as many people expect because the pipe inverts don't react to each other. If Pipe A and Pipe B are connected to the same structure, and Pipe A flows into Pipe B, changing the end invert of Pipe A does *not* affect the start invert of Pipe B automatically. If you're used to creating pipe design spreadsheets in Excel using formulas that automatically drop connected pipes to ensure flow, this behavior can be frustrating.

Pipe Network Contextual Tab Edits

You can perform many edits at the individual part level by selecting it. The Pipe Network contextual tab will appear for the object you are working with.

If you realize you placed the wrong part at a certain location—for example, if you placed a catch basin where you need a drainage manhole—use the Swap Part option on the contextual tab (see Figure 14.38). You're given a list of all the parts from all the parts lists in your drawing.

The same properties listed in Prospector (or Pipe Network Vista) can be accessed on an individual part level by clicking and choosing Pipe Properties or Structure Properties from the contextual tab. A dialog like the Structure Properties dialog in Figure 14.39 opens, with several tabs that you can use to edit that particular part.

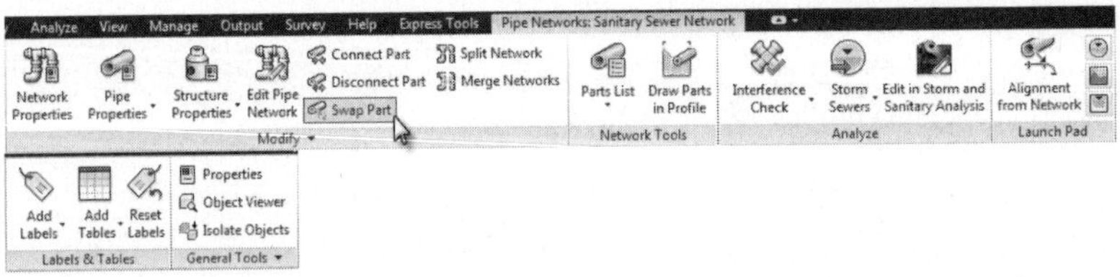

FIGURE 14.38 Selecting a network part brings up a contextual tab with many options, including Swap Part.

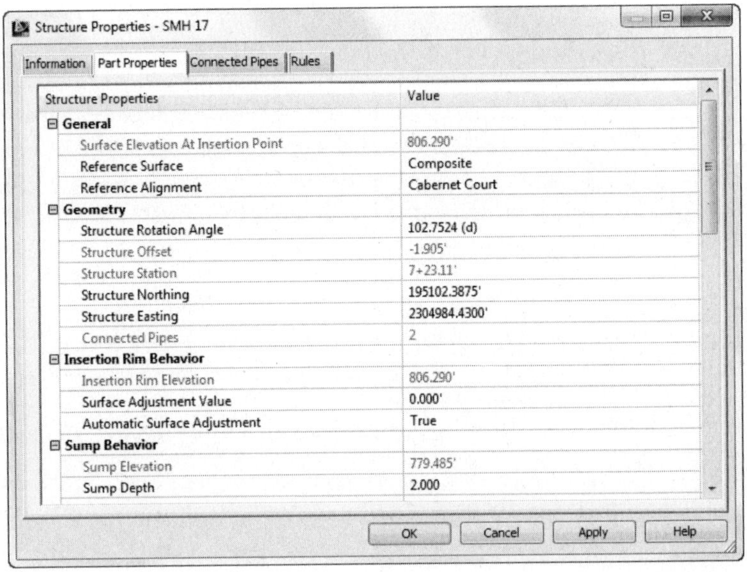

FIGURE 14.39 The Part Properties tab in the Structure Properties dialog gives you the opportunity to perform many edits and adjustments.

THE LABELS & TABLES PANEL

The Labels & Tables panel is where you do the most annotation labeling for pipes. You can access Labels & Table from the Annotate tab, or if you click on a pipe network, you will find it on the contextual menu.

Add Labels For the best control over what styles are used in your label, use the Add Pipe Network Labels option. From there, you can add labels for the selected pipe network using the suboptions such as Entire Network Plan, Entire Network Profile, Entire Network Section, Single Part Plan, Single Part Profile, Single Part Section, Spanning Pipes Plan, and Spanning Pipes Profile.

Add Tables You can add a structure or pipe table.

Reset Labels This Reset Labels is different from the Reset Label option you see when clicking directly on a label. In this version of the command, pipe networks that are in the drawing via data reference will be labeled using styles from the source drawing. For more information on source drawings and data references, see Chapter 18, "Advanced Workflows."

The General Tools Panel

While not specifically for pipe networks or even Civil 3D for that matter, these tools are placed here for convenience.

Properties Toggles the Properties palette on and off.

Object Viewer Allows you to view the selected object or objects in 3D via a separate viewer.

Isolate Objects Selected objects will be the only objects visible on the screen. This is useful if you are working in a tight area and do not wish to see extraneous objects, and it is much quicker than clicking to turn off or freeze layers.

These tools are available via the General Tools panel drop-down:

Select Similar When this tool is invoked, it will select all similar type objects.

Quick Select or QSelect Opens the Quick Select dialog, which allows custom filtering.

Draw Order Icons Allow you to move objects either to the front or back of other objects, or above or behind a specific object.

The Modify Panel

The Modify Panel is where you do editing of an existing pipe network. It can also be selected by clicking on a pipe network and selecting it from the right-click contextual menu.

Network Properties Tool Opens the Pipe Network Properties dialog.

Pipe Properties Tool Opens the Pipe Properties dialog. If you are seeing the Modify panel because you have selected a structure, you will be prompted to select a pipe.

Edit Pipe Style Tool Opens the Pipe Style dialog. If you are seeing the Modify panel because you have selected a structure, you will be prompted to select a pipe. For more on pipe styles, see Chapter 19.

Structure Properties Tool Opens the Structure Properties dialog. If you are seeing the Modify panel because you have selected a pipe, you will be prompted to select a structure.

Edit Structure Style Tool Opens the Structure Style dialog. For more on structure styles, see Chapter 19.

Edit Pipe Network Tool Opens the Network Layout Tools toolbar.

Connect And Disconnect Part Tools Allow you to connect pipes to structures that may have been disconnected and to disconnect pipes from structures.

Swap Part Tool Allows you to replace a structure or pipe type with another one from a Swap Part Size dialog.

Split and Merge Network Tools Allow you to take an existing network and split it into two networks, or take an existing pipe network and merge it into another pipe network.

The Modify panel drop-down list contains the following tools:

Rename Parts Tool Opens the Rename Pipe Network Parts dialog. Here, you can rename pipes and structures, modify pipe numbering, and decide how you want to handle conflicting names or numbers.

Apply Rules Tool Forces the set rule on the selected objects.

Change Flow Direction Tool Changes the path of the selected objects. It is very important to have this option set correctly if you are going to use any of the analysis programs.

The Network Tools Panel

The Parts List drop-down on the Network Tools panel contains the following tools:

Create Parts List Allows you to create new parts list via the Network Parts List – New Parts List dialog.

Create Full Parts List Takes all the parts available in the parts catalog and creates a list called Full Catalog.

Edit Parts List Opens the Parts List dialog, where you can select the network to add to an existing or new pipe network.

Set Network Catalog Opens the Pipe Network Settings dialog, where you can select whether you wish to use Imperial or metric parts, and also sets the location of the Network Catalog.

Part Builder Opens the Part Builder tool.

Draw Parts In Profile View Adds the selected pipe network objects into an existing profile.

The Analyze Panel

The Analyze Panel contains the following tools for doing various checks on a pipe network.

Interference Check Properties Allows you to create an interference check between parts, whether or not they are on the same network. The following tools are available via the submenu:

Create Interference Check Tool This is the same as the Interference Check Properties tool.

When you select the Interference Check Properties tool, you are prompted to select Interference and then the Interference Check Properties dialog opens:

- The Information tab displays basic information such as the name of the interference set, style, rendering material, and layer.
- The Criteria tab contains specific editable items, such as whether to use 3D proximity check, distance, and scale factors.
- The Statistics tab is a combined listing of the other tabs but also allows you to see the networks used for the interference check.

Interference Properties Tool Interference Properties will give you information about a single interference location in your drawing. This differs from the Interference Check Properties in that you can't change the criteria from here. You only see the following information about a specific instance of interference:

- Information tab (where you can name the individual interference objects)
- Statistics tab (which contains the X-Y location of the selected interference object)

Edit Interference Style Tool Opens the Interference Style dialog, where you can change the visual parameters for the interference object. For more on styles, see Chapter 19.

Storm Sewers Tool These commands interact with the Hydraflow Storm Sewers Extension. Hydraflow is a separate program that is included with Civil 3D. An in-depth discussion of Hydraflow is beyond the scope of this book; see the Hydraflow Storm Sewers Extension help files for more information.

Edit In Storm And Sanitary Analysis Tool Opens the Storm and Sanitary Analysis (SSA) program. For more on SSA, see Chapter 15, "Storm and Sanitary Analysis."

The Launch Pad Panel

The Launch Pad panel contains the following tools:

Alignment From Network Allows you to create an alignment from pipe network parts.

Storm Sewers Tool Opens the Hydroflow Storm Sewers program.

Hydrographs Tool Opens the Hydraflow Hydrographs program. Hydraflow Hydrographs is a separate program that is included with Civil 3D. An in-depth discussion of Hydraflow Hydrographs is beyond the scope of this book; see the Hydraflow Hydrographs Extension help files for more information.

Express Tool Opens the Hydraflow Express program. Hydraflow Express is a separate program that is included with Civil 3D. An in-depth discussion of Hydraflow Express is beyond the scope of this book; see the Hydraflow Express help files for more information.

Editing with the Network Layout Tools Toolbar

You can also edit your pipe network by retrieving the Network Layout Tools toolbar. This is accomplished by selecting a pipe network object and choosing Edit Network from the context tab. You can also go to the Modify tab and click Pipe Network on the Design panel.

Once the toolbar is up, you can continue working exactly the way you did when you originally laid out your pipe network.

This exercise will give you hands-on experience in making a variety of edits to a sanitary and storm-drainage pipe network:

1. Open the `EditingPipesPlan.dwg` (`EditingPipesPlan_Metric.dwg`) file.

   This drawing includes a sanitary sewer network and a storm drainage network as well as some surfaces and alignments. For metric users, the structure family names end with SI.

2. Select the structure STM STR 3 in the drawing. It is labeled with the structure name showing.
3. From the context tab, choose Swap Part.
4. Select the 2 × 4 (1000 × 750 mm) structure from the Rectangular Junction Structure NF, and click OK.

Toggling Numbers

When you grip the triangular Endpoint grip, dynamic input shows the amount that the pipe is being lengthened. You want to lengthen the entire pipe run. Pressing the Tab key toggles between the additional pipe lengthened versus the entire pipe length.

5. Select the newly placed catch basin so that you see the two structure grips.
6. Use the rotational grip and your nearest Osnap to align the catch basin, as shown in Figure 14.40.

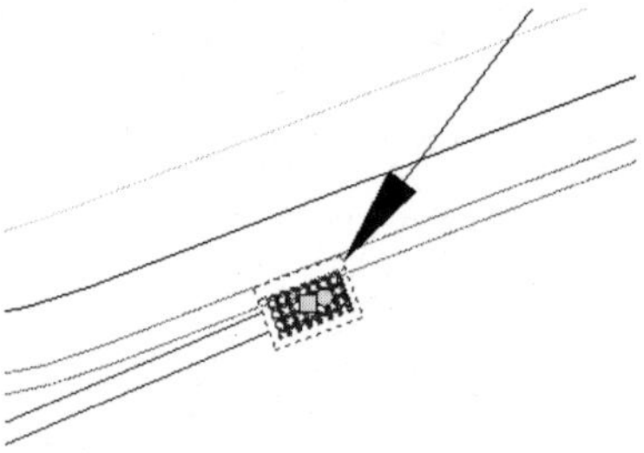

Figure 14.40 Rotate and move the catch basin into place along the curb.

7. Use the AutoCAD Erase command to erase SAN STR 7.
8. If dynamic input is not already on, press F12 to enable it. Select the sanitary sewer pipe that is labeled in the north of the drawing.
9. Use the triangular Endpoint grip and the Dynamic Input tooltip to lengthen it to a total length of 200′ (60.96 m), as shown in Figure 14.41.

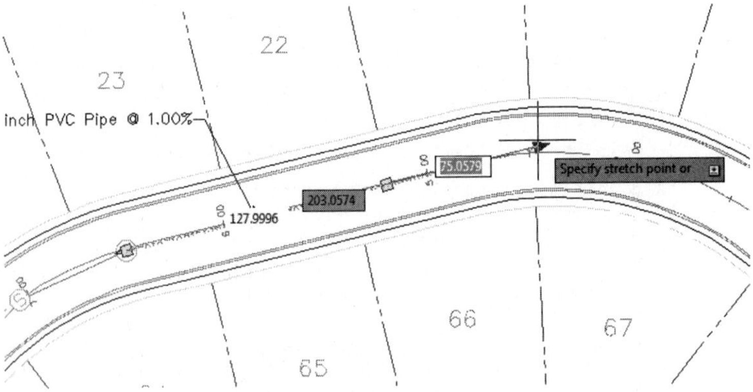

Figure 14.41 Lengthen the pipe to 200′ (60.96 m).

Note that this is the 3D Center To Center Pipe Length. Next you will add a structure to the end of the pipe you just modified.

10. Select any pipe in the sanitary network. In the Pipe Networks contextual tab, choose Edit Pipe Network.

11. From the Structures List, set the structure to SMH from the Concentric Cylindrical Structures NF family.
12. From the Draw Pipes and Structures pull-down, set the option to Structures Only. Place the structure in the drawing at the end of the pipe you lengthened in step 9.
13. Select STM STR 3 in the drawing, and from the context tab, choose Structure Properties.
14. Switch to the Part Properties tab, scroll down to the Sump Depth field, and change the value to **0′** (**0** m).
15. Click OK to exit the Structure Properties dialog.
16. On the Prospector tab of Toolspace, expand the Pipe Networks ➢ Networks ➢ Sanitary Network branch, and select the Structures entry.
17. Use the tabular interface in the preview pane area of Prospector to change the names of SAN STR 1 through SAN STR 4 to **MH1** through **MH4**.

Real World Scenario

I Thought I Took Care of That Sump?

In steps 13–15, you were instructed to make the sump depth 0. So why isn't the bottom of the structure even with the pipe in the Profile Graphic? Structures are created using the outermost edges, and in the case of structures, this includes a 6″ (150 mm) sump.

After you learn about Part Builder at the end of the chapter, come back and try this to add a zero value to the Floor Thickness list:

1. Start Part Builder.
2. Select the Concentric Cylinder part and click Edit.
3. Right-click Size Parameters and select Edit Values.
4. Click your cursor in the FTh field. This will enable the Edit Values button.
5. Click the Edit Value button. In the Edit Values dialog, click Add.
6. Type **0** for the new value and click OK.
7. Exit Part Builder, saving the part.
8. Back in Prospector, choose Settings ➢ Pipe Network ➢ Parts Lists ➢ Storm Sewer.
9. Expand the part you changed earlier, right-click, and choose Edit.
10. Notice that there is now a Floor Thickness value of 0.

Creating an Alignment from Network Parts

On some occasions, certain legs of a pipe network require their own stationing. Perhaps most of your pipes are shown on a road profile but the legs that run offsite or through open space require their own profiles. Whatever the reason, it's often necessary to create an alignment from network parts. Follow these steps:

1. Open the `AlignmentFromNetworkParts.dwg` (`AlignmentFromNetworkParts_METRIC.dwg`) file.
2. Select the CB1 structure, which will be the first structure and will be station 0+00 (0+000) on the alignment.
3. On the Pipe Networks contextual tab ➢ Launch Pad panel, select Alignment From Network.

   The command line reads `Select next Network Part or [Undo]`.
4. Select the STM STR 7 structure, which will be the last structure on the alignment.
5. Press ↵, and a dialog appears that is almost identical to the one you see when you create an alignment from the Alignments menu.
   - Name your alignment **Storm CL**.
   - Notice the Create Profile And Profile View check box on the last line of the dialog. Leave the box selected and click OK.

   The Create Profile From Surface dialog appears (see Figure 14.42).

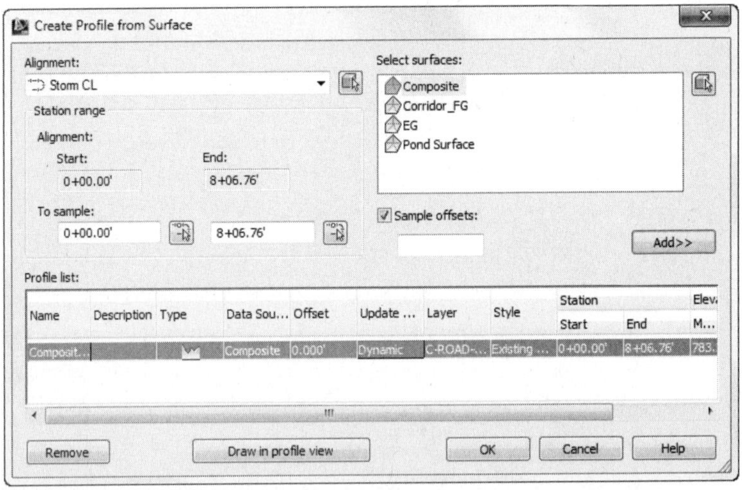

FIGURE 14.42
The Create Profile From Surface dialog

6. Click to highlight the Composite surface, and click Add.
7. Click Draw In Profile View.

   You see the Create Profile View Wizard (see Figure 14.43).

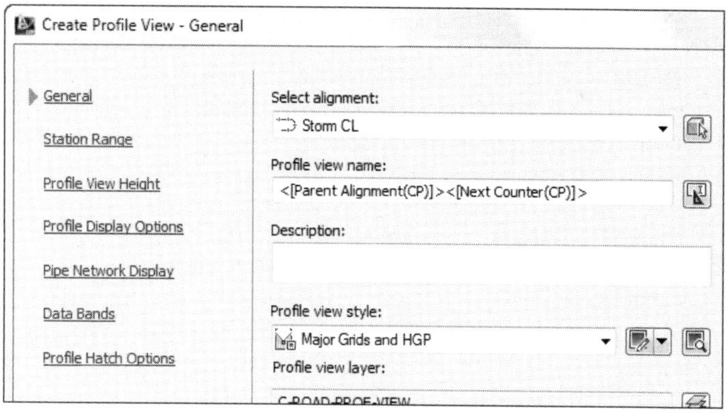

FIGURE 14.43
The Create Profile View Wizard

8. Click the link next to Pipe Network Display to jump to that page.

   You should see a list of pipes and structures in your drawing.

9. Verify that Yes is selected for each pipe and structure in the Storm Network only.

10. Click Create Profile View, and place the profile view to the right of the site plan.

You see three structures and two pipes drawn in a profile view, which is based on the newly created alignment (see Figure 14.44).

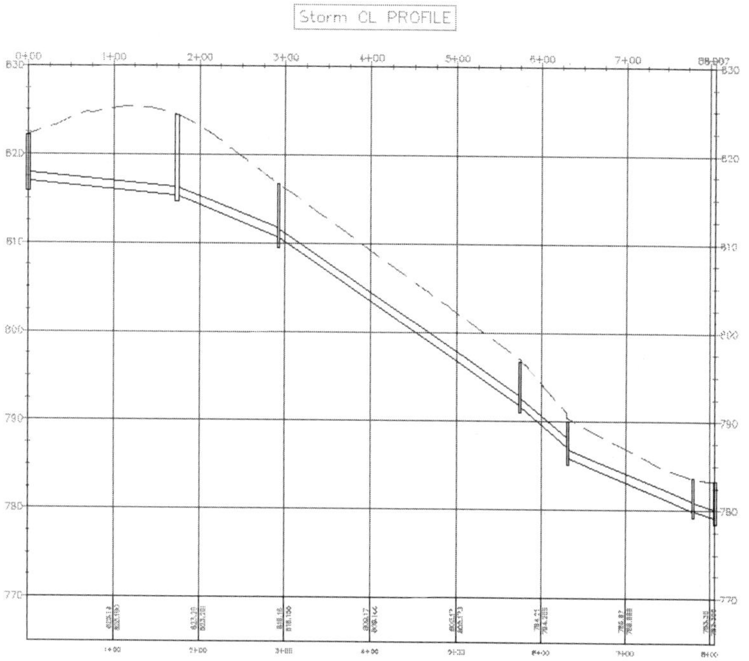

FIGURE 14.44
Creating a profile view

Drawing Parts in Profile View

CERT OBJECTIVE

To add pipe network parts to an existing profile view, select a network part, and choose Draw Parts In Profile from the Network Tools panel on the contextual tab. When you're using this command, it's important to note that only selected parts are drawn in your chosen profile view.

Profiles and profile views are always cut with respect to an alignment. Therefore, pipes are shown in the profile view on the basis of how they appear along that alignment, or how they cross that alignment. Unless your alignment *exactly follows the centerline of your network parts,* your pipes will likely show some drafting distortion.

Let's look at Figure 14.45 as an example. This particular jurisdiction requires that all utilities be profiled along the road centerline. There's a road centerline, a storm network that jogs across the road to connect with another catch basin.

FIGURE 14.45
These pipe lengths will be distorted in profile view.

At least two potentially confusing elements show up in your profile view. First, the distance between structures (2D Length - Center To Center) isn't the same between the plan and the profile (see Figure 14.46) because the storm pipe doesn't run parallel to the alignment. Because the labeling reflects the network model, all labeling is true to the 2D Length - Center To Center or any other length you specify in your label style.

The second potential issue is that the invert of your crossing storm pipe is shown at the point where the storm pipe *crosses the alignment,* and not at the point where it crosses the sanitary pipe (see Figure 14.47).

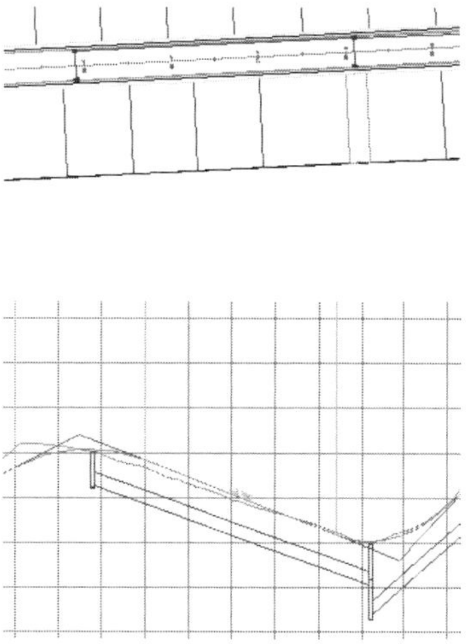

FIGURE 14.46
Pipe labels in plan view (top) and profile view (bottom)

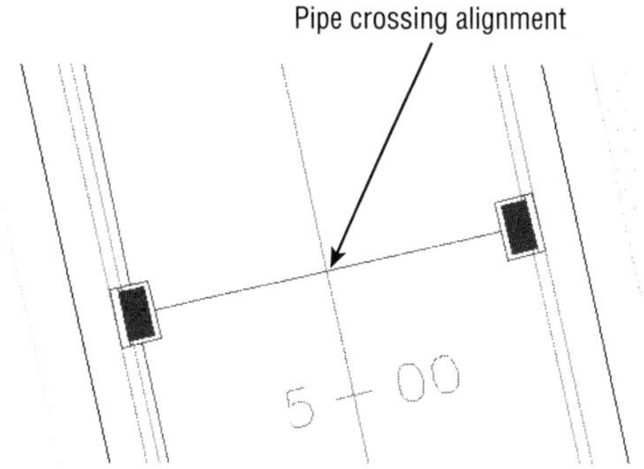

FIGURE 14.47
The invert of a crossing pipe is drawn at the location where it crosses the alignment.

Vertical Movement Edits Using Grips in Profile

Although you can't make changes to certain part properties (such as pipe length) in profile view, pipes and structures both have special grips for changing their vertical properties in profile view.

When selected, a structure has two grips in profile view (see Figure 14.48). The first is a triangular-shaped grip representing a rim insertion point. This grip can be dragged up or down and affects the model structure-insertion point.

FIGURE 14.48
A structure has two grips in profile view.

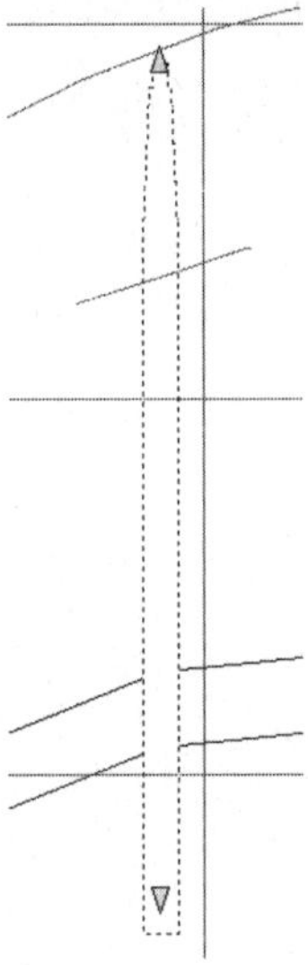

Moving this grip can affect your structure insertion point two ways, depending on how your structure properties were established:

- If your structure has Automatic Surface Adjustment set to True, grip-editing this Rim Insertion Point grip changes the surface adjustment value. If your reference surface changes, then your rim will change along with it, plus or minus the surface adjustment value.
- If your structure has the Automatic Surface Adjustment set to False, grip-editing this Rim grip modifies the insertion point of the rim. No matter what happens to your reference surface, the rim will stay locked in place.

Typically, you'll use the Rim Insertion Point grip only in cases where you don't have a surface for your rims to target to, or if you know there is a desired surface adjustment value. It's tempting to make a quick change instead of making the improvements to your surface that are fundamentally necessary to get the desired rim elevation. One quick change often grows in scope. Making the necessary design changes to your target surface will keep your model dynamic and, in the long run, will make editing your rim elevations easier.

The second grip is a triangular grip located at the sump depth. This grip doesn't represent structure invert. In Civil 3D, only pipes truly have invert elevation. The structure uses the connected pipe information to determine how deep it should be. When the sump has been set at a depth of 0, the sump elevation equals the invert of the deepest connected pipe.

This grip can be dragged up or down. It affects the modeled sump depth in one of two ways, depending on how your structure properties are established:

Controlling Sump By Depth If your structure is set to control sump by depth, editing with the Sump grip changes the sump depth.

The depth is measured from the structure insertion point. For example, if the original sump depth was 0, grip-editing the sump 0.5′ (15.24 cm) lower would be the equivalent of creating a new sump rule for a 0.5′ (15.24 cm) depth and applying the rule to this structure. This sump will react to hold the established depth if your reference surface changes, your connected pipe inverts change, or something else is modified that would affect the invert of the lowest connected pipe. This triangular grip is most useful in cases where most of your pipe network will follow the sump rule applied in your parts lists, but selected structures need special treatment.

Controlling Sump By Elevation If your structure is set to control sump by elevation, adjusting the Sump grip changes that elevation.

When sump is controlled by elevation, sump is treated as an absolute value that will hold regardless of the structure insertion point. For example, if you grip-edit your structure so its depth is 8.219′ (2.51 m), the structure will remain at that depth regardless of what happens to the inverts of your connected pipes. The Control Sump By Elevation parameter is best used for existing structures that have surveyed information of absolute sump elevations that won't change with the addition of new connected pipes.

When selected, a pipe end has three grips in profile view (see Figure 14.49). You can grip-edit the invert, crown, and centerline elevations at the structure connection using these grips, resulting in the pipe slope changing to accommodate the new endpoint elevation.

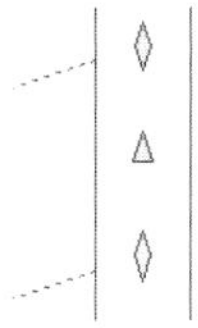

FIGURE 14.49
Three grips for a pipe end in profile view

When selected, a pipe in profile view has one grip at its midpoint (see Figure 14.50). You can use this grip to move the pipe vertically while holding the slope of the pipe constant.

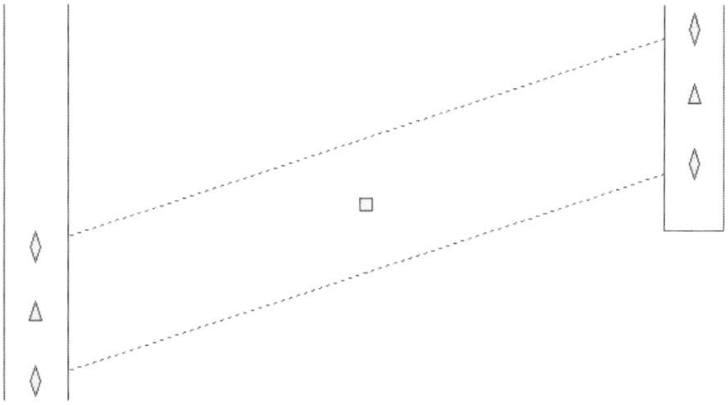

FIGURE 14.50
Use the Midpoint grip to move a pipe vertically.

You can access pipe or structure properties by choosing a part, right-clicking, and choosing Pipe or Structure Properties.

Removing a Part from Profile View

If you have a part in profile view that you'd like to remove from the view but not delete from the pipe network entirely, you have a few options.

AutoCAD Erase can remove a part from profile view; however, that part is then removed from every profile view in which it appears. If you have only one profile view, or if you're trying to delete the pipe from every profile view, this is a good method to use.

Be careful when using the Del key or the Erase command on objects in plan view. Keep in mind that deleting any object from plan view removes the object outright, which includes pipe network parts. Use your Esc key liberally before selecting items to remove; this will help you avoid unintended deletion of items. Of course, if you accidentally blow away something, Undo will bring it back.

A better way to remove parts from a particular profile view is through the Profile View Properties. You can access these properties by selecting the profile view, right-clicking, and choosing Profile View Properties.

The Pipe Networks tab of the Profile View Properties dialog (see Figure 14.51) provides a list of all pipes and structures that are shown in that profile view. You can deselect the check boxes next to parts you'd like to omit from this view.

At the bottom of the Profile View Properties dialog is a checkbox for Show Only Parts Drawn In Profile View. This checkbox is off by default so that you can see every possible pipe and structure. When this is checked on, the Pipe Network tab will hide any pipes that are not visible in the profile view you are examining.

FIGURE 14.51
Deselect parts to omit them from a view.

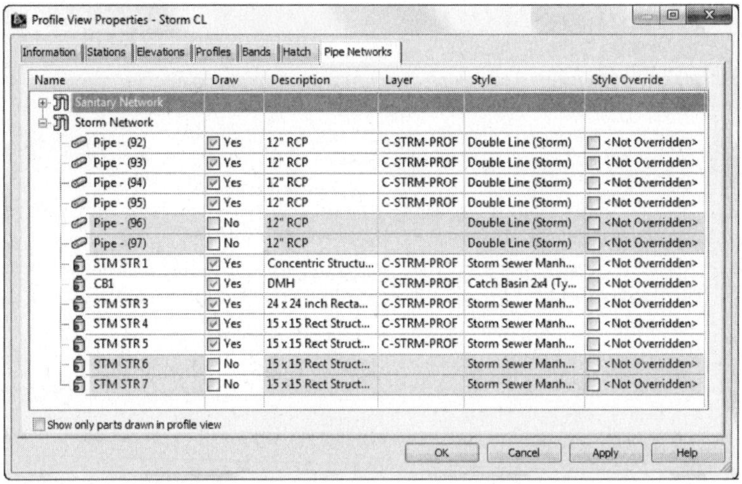

Showing Pipes That Cross the Profile View

If you have pipes that cross the alignment related to your profile view, you can show them with a crossing style. A pipe must cross the parent alignment to be shown as a crossing in the profile view. The location of a crossing pipe is always shown *at the elevation where it crosses the alignment* (see Figure 14.52).

When pipes enter directly into profiled structures, they can be shown as ellipses through the Display tab of the Structure Style dialog (see Figure 14.53). See Chapter 21, "Object Styles," for more information about creating structure styles.

FIGURE 14.52
A pipe crossing a profile

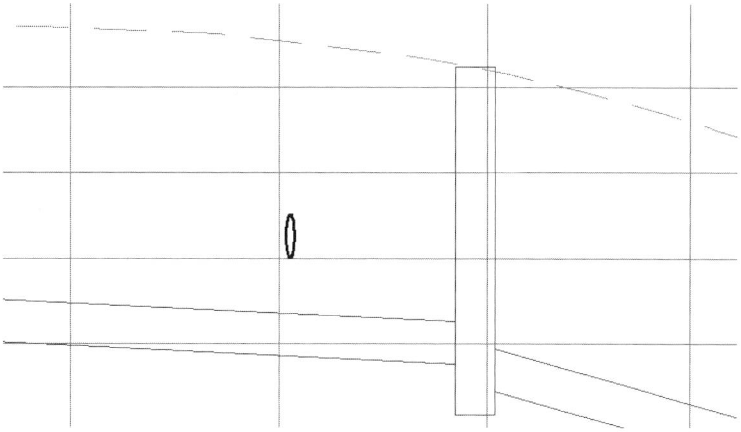

FIGURE 14.53
Pipes that cross directly into a structure can be shown as part of the structure style.

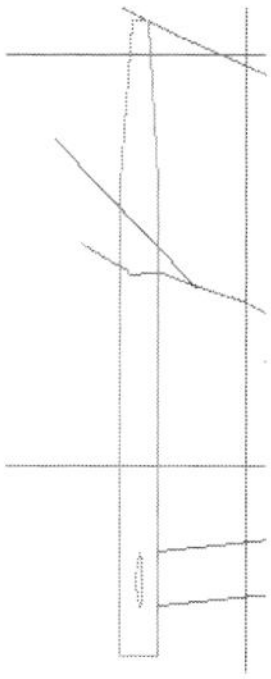

The first step to display a pipe crossing in profile is to add the pipe that crosses your alignment to your profile view by either selecting the pipe, right-clicking, and selecting Draw Parts In Profile from the Network Tools panel and selecting the profile view, or by checking the appropriate boxes on the Pipe Network tab of the Profile View Properties dialog. When the pipe is added, it's distorted when it's projected onto your profile view—in other words, it's shown as if you wanted to see the entire length of pipe in profile (see Figure 14.54).

FIGURE 14.54
The pipe crossing is distorted.

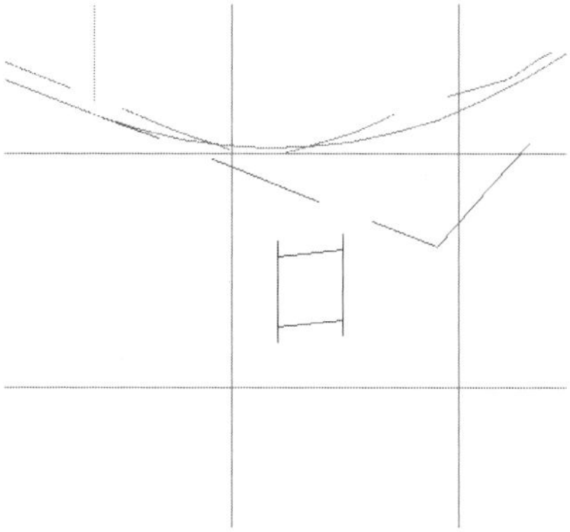

The next step is to override the pipe style *in this profile view only*. Changing the pipe style through pipe properties won't give you the desired result, because that will affect the visibility of every single instance of the pipe. You must override the style on the Pipe Networks tab of the Profile View Properties dialog (see Figure 14.55).

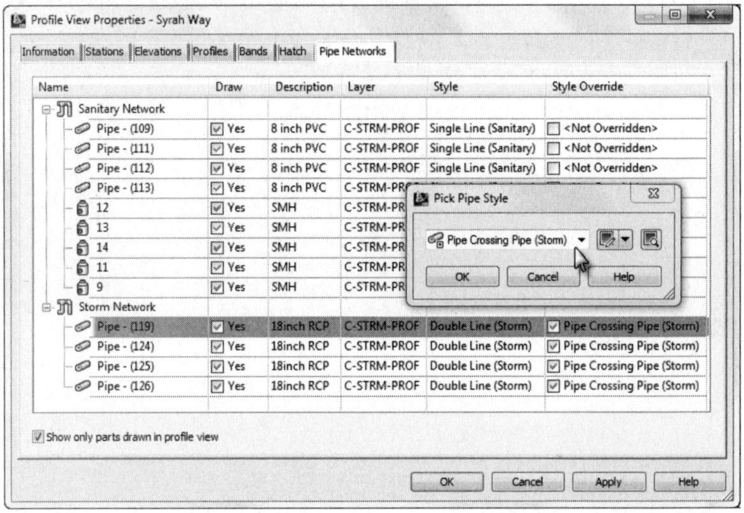

FIGURE 14.55
Correct the representation in the Profile View Properties dialog.

Locate the pipe you just added to your profile view, and scroll to the last column on the right (Style Override). Select the Style Override check box, and choose your pipe crossing style. Click OK. Your pipe should now appear as an ellipse.

If your pipe appears as an ellipse but suddenly seems to have disappeared in the plan and other profiles, chances are good that you didn't use the Style Override but accidentally changed the pipe style. Go back to the Profile View Properties dialog and make the necessary adjustments; your pipes will appear as you expect.

Adding Pipe Network Labels

Once you've designed your network, it's important to annotate the design. This section focuses on pipe network–specific label components in plan and profile views (see Figure 14.56).

Like all Civil 3D objects, the Pipe and Structure label styles can be found in the Pipe and Structure branches of the Settings tab in Toolspace and are covered in Chapter 20, "Label Styles."

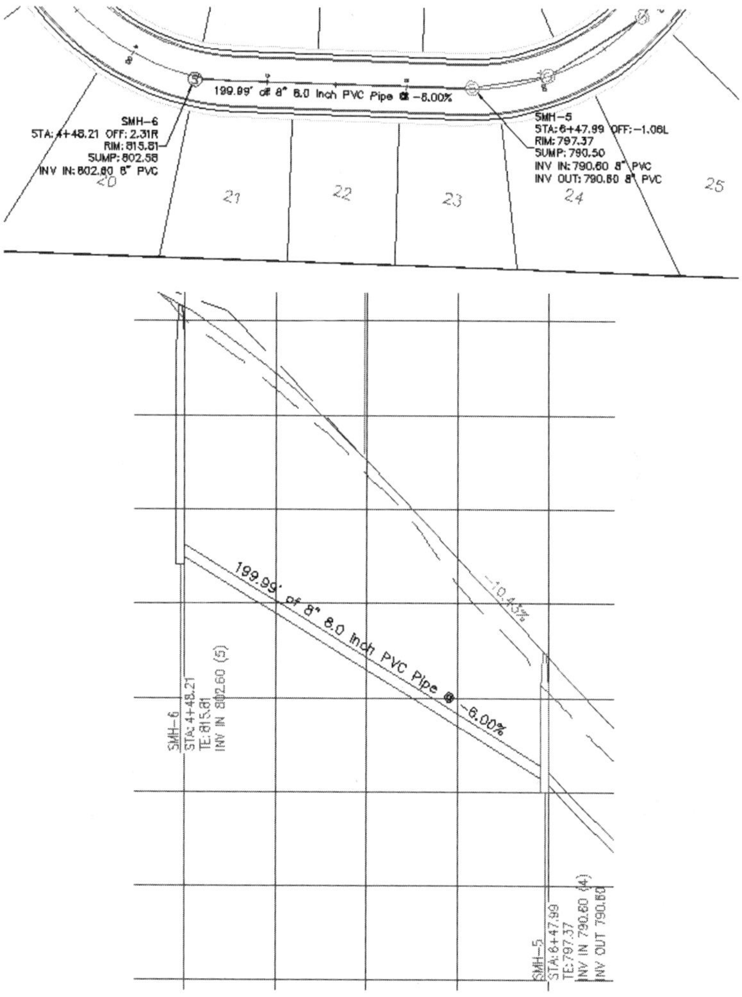

FIGURE 14.56
Typical pipe network labels (top) in plan view and (bottom) in profile view

Creating a Labeled Pipe Network Profile Including Crossings

This exercise will apply several of the concepts in this chapter to give you hands-on experience producing a pipe network profile that includes pipes that cross the alignment. You will have another chance to practice creating an alignment from a pipe network and associating the new alignment to the parts. You will also practice overriding pipe display in the Profile Properties. There are storm pipes that cross the centerline nearly perpendicular to the alignment. You will show these as ellipses in the profile view.

1. Open the `Pipes-Exercise3.dwg` (`Pipes-Exercise3_METRIC.dwg`) file. (It's important to start with this drawing rather than use the drawing from an earlier exercise.)
2. Locate sanitary structure 16 (near station 5+15 or 0+160 of Cabernet Court), select it to open the Pipe Networks contextual tab, and then select Alignment From Network on the Launch Pad panel.
3. When prompted to select next network part, select Structure 18 and press ↵ to create an alignment from the sanitary sewer network.
4. Choose the following options in the Create Alignment From Network Parts dialog:
   - Site: <none>
   - Name: **SMH16 to SMH18 Alignment**
   - Alignment Type: Miscellaneous
   - Alignment Style: _None
   - Alignment Label Set: _No Labels
   - Create Profile And Profile View check box: Selected
5. Click OK.
6. Sample the EG and Corridor FG surfaces in the Create Profile From Surface dialog.
7. Change the Style for Corridor FG - Surface (2) to Design Profile.
8. Click Draw In Profile View to open the Create Profile View dialog.
9. Click the Next button in the Create Profile View Wizard until you reach the Pipe Network Display page.

   You should see a list of pipes and structures in your drawing. Make sure Yes is selected for each pipe and structure under Sanitary Sewer Network.
10. Click Create Profile View, and place the profile view to the right of the site plan.
11. Select either a pipe or a structure to open the Pipe Network contextual tab. On the Labels & Tables panel, select Add Labels ➢ Entire Network Profile.

    The alignment information is missing from your structure labels, because the alignment was created after the pipe network.
12. Expand the Sanitary Network branch in the Prospector tab of Toolspace to set your new alignment as the reference alignment for these structures.
13. Click the Structures entry, and find the Structures list in the preview pane.
14. Select the structures 16–18 in the preview pane using Shift+click. You may need to left-click on the Name column header to sort by name.

15. Scroll to the right until you see the Reference Alignment column.
16. Right-click the column header and select Edit. Choose SMH16 To SMH18 Alignment from the dialog and click OK.
17. You may need to type **REGEN ALL** at the command line to see your updated structure labels.
18. Pan your drawing until you see the Storm layout located on Syrah Way.

    The sanitary pipe network has already been laid out on the Syrah Way profile. You want to show the storm where it crosses the sanitary so you can adjust the pipe elevations if necessary.

19. Select the four pipes that cross perpendicular to the Syrah Way alignment.
20. In the Pipe Networks contextual tab ➤ Network Tools panel, select Draw Parts In Profile.
21. Select the Syrah Way profile view.
22. Press Esc to ensure no extra objects are selected, and then select the Profile view and pick Profile View Properties from the contextual tab.
23. On the Pipe Network tab of the Profile View Properties, place a check mark next to Show Only Parts Drawn In Profile View.
24. For all four Storm Network pipes shown, override the pipe style in this profile view by placing a check mark next to <Not Overridden> in the Style Override column. You will be prompted to pick the pipe style as shown previously in Figure 14.55.
25. When all four pipes have had overrides set, click OK.

    Your crossing pipes will resemble Figure 14.57.

26. Adjust the elevations of the crossing pipes as needed. Save and close the drawing.

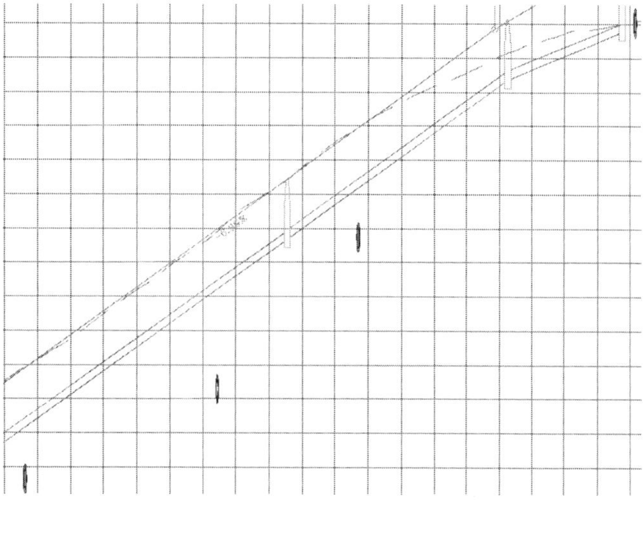

FIGURE 14.57
The completed exercise with crossing pipes

Pipe and Structure Labels

In earlier exercises, you have already had a sneak peek at adding labels for structures. Civil 3D makes no distinction between a plan label and a profile label. The same label style can be used both places. In this chapter, we will make use of label styles that are already part of the drawing.

To create your own pipe labels from scratch, see Chapter 20.

Spanning Pipe Labels

In addition to single-part labels, pipes shown in either plan or profile view can be labeled using the Spanning Pipes option. This feature allows you to choose more than one pipe; the length that is reported in the label is the cumulative length of all pipes you choose.

Unlike Parcel spanning labels, no special label-style setting is required to use this tool. The Spanning Pipes option is on the Annotate tab. Select Add Labels ➢ Pipe Network ➢ Spanning Pipes Profile to access the command.

Creating an Interference Check

In design, you must make sure pipes and structures have appropriate separation. You can perform some visual checks by rotating your model in 3D and plotting pipes in profile and section views (see Figure 14.58). Civil 3D also provides a tool called Interference Check that makes a 3D sweep of your pipe networks and lets you know if anything is too close for comfort.

FIGURE 14.58 Two pipe networks may interfere vertically where crossings occur (top). Viewing your pipes in profile view can also help identify conflicts (bottom).

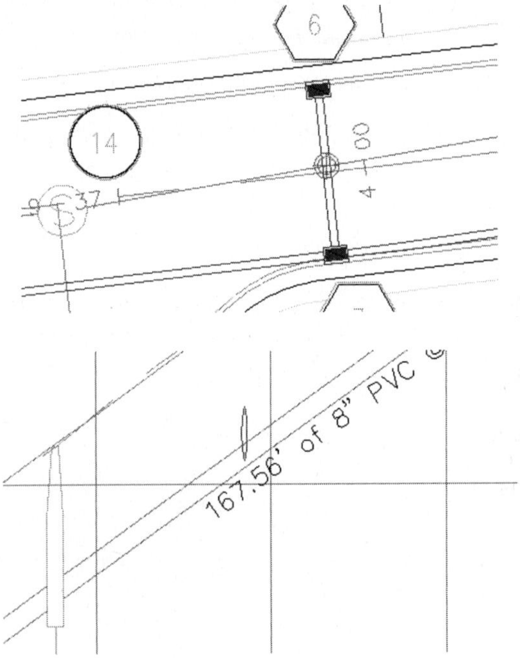

The following exercise will lead you through creating a pipe network and using Interference Check to scan your design for potential pipe network conflicts:

1. Open the `Interference.dwg` (`Interference_METRIC.dwg`) file. The drawing includes a sanitary sewer pipe network and a storm drainage network.
2. Select a part from either network and choose Interference Check from the Analyze panel.

   You'll see the prompt `Select a part from the same network or different network:`.
3. Select a part from the network not chosen.

   The Create Interference Check dialog appears (Figure 14.59).

FIGURE 14.59
The Create Interference Check dialog

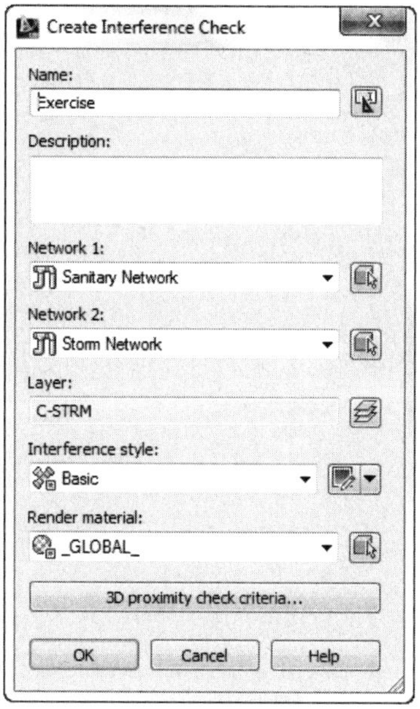

4. Name the Interference Check **Exercise**, and confirm that Sanitary Network and Storm Network appear in the Network 1 and Network 2 boxes.
5. Click 3D Proximity Check Criteria, and the Criteria dialog appears (see Figure 14.60).

FIGURE 14.60
Criteria for the 3D proximity check

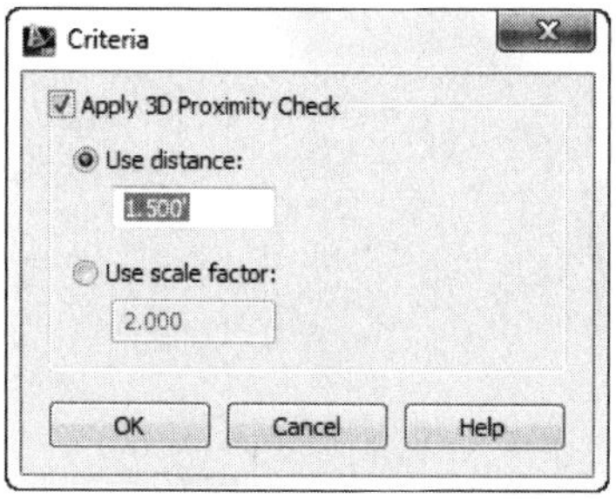

You're interested in finding all network parts that are within a certain tolerance of one another, so enter **1.5′** (**0.5** m) in the Use Distance box.

This setting creates a buffer to help find parts in all directions that might interfere. If you forget to check Apply 3D Proximity Check, you would only get direct, physical collisions listed as collisions.

6. Click OK to exit the Criteria dialog, and click OK to run the Interference Check.

   You see a dialog that alerts you to three interferences.

7. Click OK to dismiss this dialog.

8. On the Prospector tab of Toolspace ➤ Pipe Networks, expand Interference Checks. Right-click on Exercise and select Zoom To.

   A small marker has appeared at each location where interference occurs, as shown in Figure 14.61.

FIGURE 14.61
The interference marker in plan view

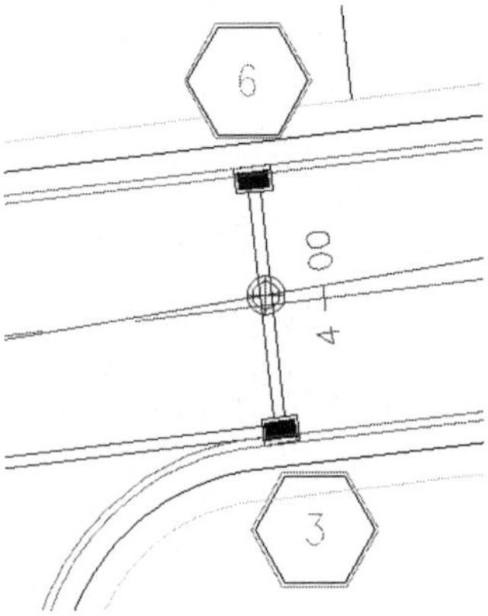

9. Select any one of the interference markers, the pipes that intersect, and the nearby inlets.

Object Viewer

10. From the Multiple contextual tab, click Object Viewer.

    The interference marker appears in 3D, as shown in Figure 14.62.

FIGURE 14.62
The interference marker in 3D

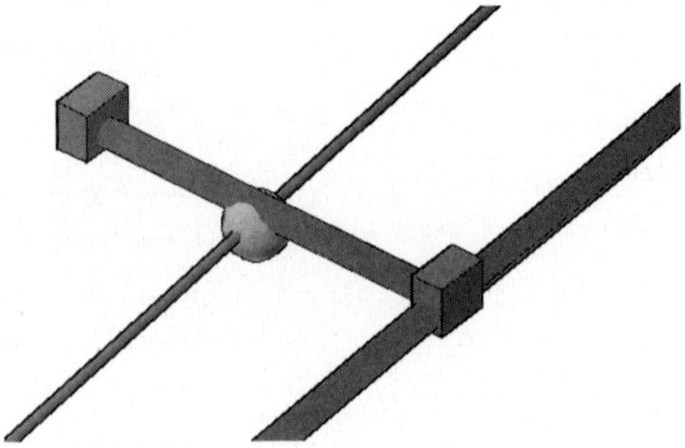

11. Use the ViewCube in Object Viewer to navigate in 3D.

12. Once you have examined the 3D objects, close Object Viewer. Save the drawing.

    Note that each instance of interference is listed in the preview pane for further study.

Making edits to your pipe network flags the interference check as "out of date." You can rerun Interference Check by right-clicking Interference Check in Prospector. You can also edit your criteria in this right-click menu.

Creating Pipe Tables

Just like with parcels and labels, labeling pipes can turn into a mess with all the labels set on the plan, such as pipes and structures (see Figure 14.63). In this section, you will explore table creation for pipes and structures.

FIGURE 14.63
Pipes and structure labels crowded on a plan

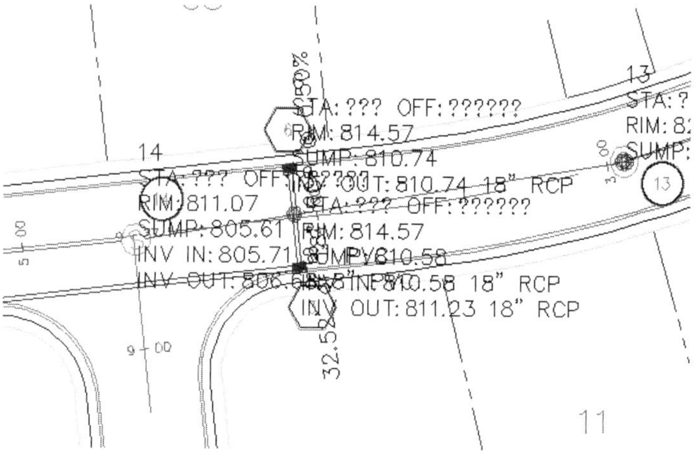

Exploring the Table Creation Dialog

Since the structure and pipe table creation dialogs are similar, we will cover both of them in this section:

- The Table Style option allows you to select a table look or style. You can select the available styles from the arrow to the right of the table style name. You can also create new, copy, edit, or pick a table style. For more on table styles, see Chapter 20.
- The Table Layer option will set the overall layer for the table.
- With the By Network radio button selected, you can select the network to create a table from the drop-down list, or use the pick icon to select the network on the drawing.
- With the Multiple Selection radio button selected and by using the pick icon, you can select structures or pipes (depending on which table type is selected) from the drawing. You can pick pipes or structures regardless of the network.
- The Split Table check box will allow you to split the table if it gets too large. You can specify the maximum number of rows per table and the maximum number of tables per stack. Additionally, you can set the offset distance between the stacked tables.

- You can choose whether you want the split tables tiled across or down.
- The last option to choose defines the behavior you want for the table: Static or Dynamic. A static table will not update if any changes are made to the pipe network, such as swapping a part. Dynamic will update the table to those changes.

In the following exercise, you will create a pipe network table for the sanitary sewer structures:

1. Open the `Pipes-Exercise3.dwg` (`Pipes-Exercise3_METRIC.dwg`) file.

   It is not necessary to have completed the previous exercise that uses the same file.

2. Click on the Sanitary Sewer Manhole 1.
3. From the Pipe Networks contextual tab ➢ Labels & Tables panel, select Add Tables ➢ Add Structure.

   The Structure Table Creation dialog opens.

4. Click OK to accept the default settings (shown previously in Figure 14.64).

FIGURE 14.64
Structure Table Creation dialog box

5. Place the table to the right of your plan.

   The table should look similar to Figure 14.65.

FIGURE 14.65
The finished structure table

STRUCTURE TABLE			
STRUCTURE NAME:	DETAILS:	PIPES IN:	PIPES OUT
12	48" RIM = 825.60 SUMP = 821.8 INV OUT = 821.91		Pipe – (111), 8" PVC INV OUT =821.91
2	48" RIM = 824.52 SUMP = 817.7 INV IN = 817.78 INV OUT = 817.78	Pipe – (101), 8" PVC INV IN =817.78	Pipe – (102), 8" PVC INV OUT =817.78
3	48" RIM = 822.81 SUMP = 815.3 INV IN = 815.37 INV OUT = 815.37	Pipe – (102), 8" PVC INV IN =815.37	Pipe – (103), 8" PVC INV OUT =815.37
1	48" RIM = 822.17 SUMP = 818.4 INV OUT = 818.48		Pipe – (101), 8" PVC INV OUT =818.48
4	48" RIM = 819.88 SUMP = 814.2 INV IN = 814.31 INV OUT = 814.31	Pipe – (103), 8" PVC INV IN =814.31	Pipe – (104), 8" PVC INV OUT =814.31
5	48" RIM = 816.20 SUMP = 809.3 INV IN = 809.44 INV OUT = 809.44	Pipe – (104), 8" PVC INV IN =809.44	Pipe – (105), 8" PVC INV OUT =809.44
	48"		

The table creation for pipes is similar to the structure table creation process. Continue working in `Pipes-Exercise3.dwg` (`Pipes-Exercise3_METRIC.dwg`):

1. Click on a Sanitary Sewer pipe.
2. From the Labels & Tables panel, select Add Tables ➢ Add Pipe.

   The Pipe Table Creation dialog opens (Figure 14.66).

FIGURE 14.66
The Pipe Table Creation dialog

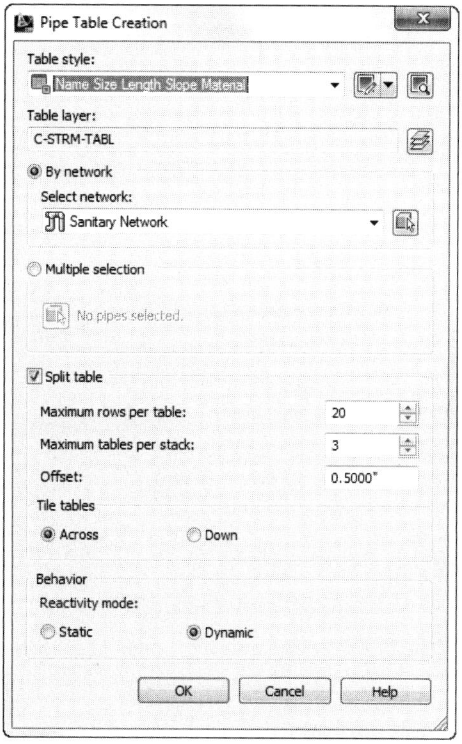

3. Click OK to accept the default settings.
4. Place the table to the right of the structure table.

   The pipe table should look similar to Figure 14.67.

FIGURE 14.67
The finished pipe table

Pipe Table				
NAME	SIZE	LENGTH	SLOPE	MATERIAL
Pipe - (101)	8"	70.05'	1.00%	PVC
Pipe - (102)	8"	240.56'	1.00%	PVC
Pipe - (103)	8"	106.00'	1.00%	PVC
Pipe - (104)	8"	93.53'	5.21%	PVC
Pipe - (105)	8"	225.24'	8.76%	PVC
Pipe - (106)	8"	100.34'	8.00%	PVC
Pipe - (107)	8"	99.66'	3.57%	PVC
Pipe - (108)	8"	122.61'	0.99%	PVC
Pipe - (109)	8"	213.96'	3.93%	PVC
Pipe - (110)	8"	225.12'	5.01%	PVC
Pipe - (111)	8"	87.48'	4.00%	PVC
Pipe - (112)	8"	167.56'	8.00%	PVC
Pipe - (113)	8"	249.06'	6.97%	PVC
Pipe - (114)	8"	340.81'	2.83%	PVC
Pipe - (115)	8"	247.51'	0.74%	PVC
Pipe - (116)	8"	194.22'	3.34%	PVC
Pipe - (117)	8"	160.24'	7.89%	PVC
Pipe - (118)	8"	250.78'	6.70%	PVC

The Table Panel Tools

When you click on a table, the Table contextual tab opens and has several tools available (Figure 14.68). We'll look at each in this section.

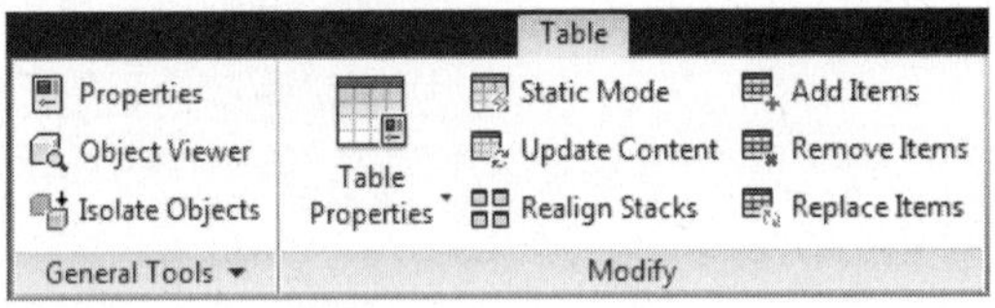

FIGURE 14.68
The Table contextual tab

THE GENERAL TOOLS PANEL

The tools here are the same as mentioned earlier when we discussed the pipe network tools.

THE MODIFY PANEL

The Modify Panel contains the following tools:

Table Properties Tool Opens the Table Properties dialog. You can set the table style and choose whether to split the table with all the options mentioned earlier. In addition, you can force realignment of stacks and force content updating.

Edit Table Style Tool Opens the Table Style dialog. For more on editing table styles, see Chapter 20.

Static Mode Tool Turns a dynamic table into a static table.

Update Content Tool Forces an update on a table.

Realign Stacks Tool Readjusts the table columns back to the default setting.

Add Items Tool Adds pipe data to the table that was added after the table was created.

Remove Items Tool Removes pipe or structure objects from a table.

Replace Items Tool Allows you to swap pipe or structures in the table.

Under Pressure

New to Civil 3D 2013 are pressure networks. Pressure pipes work differently than gravity flow pipe systems within Civil 3D. Much of the need for custom parts like valves or hydrants is eliminated with these systems.

Pressure Network Parts List

Like gravity-based networks, a pressure network starts with a parts list. All of the parts available in Civil 3D are based on standards established by the American Water Works Association (AWWA), and are listed in both inches and millimeters.

Under the Hood of the Pressure Network

Before you can create a pressure parts list, you must decide which catalog you will be working from. Set the pressure network catalog by going to the Home tab of the ribbon ➢ Create Design panel flyout, and select Set Pressure Network Catalog, as shown in Figure 14.69.

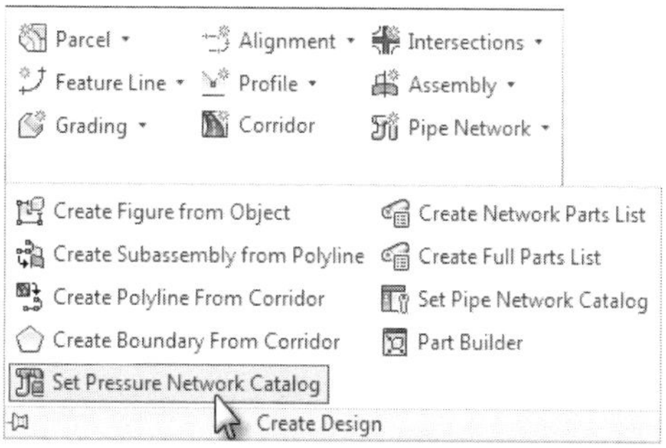

Figure 14.69
Setting the pressure network catalog

Set the path to the `Pressure Pipes Catalog` folder to `C:\ProgramData\Autodesk\C3D 2013\enu\Pressure Pipes Catalog`. Click the folder icon to choose either the Metric or the Imperial database depending on your needs.

The catalog database file determines the join type between pressure network parts. The most commonly used join type in modern water main construction is the push-on type, which is the default pressure database. As shown in Figure 14.70, Imperial units have a choice of three options:

- Imperial_AWWA_Flanged
- Imperial_AWWA_Mechanical
- Imperial_AWWA_PushOn

FIGURE 14.70
Setting your catalog database file

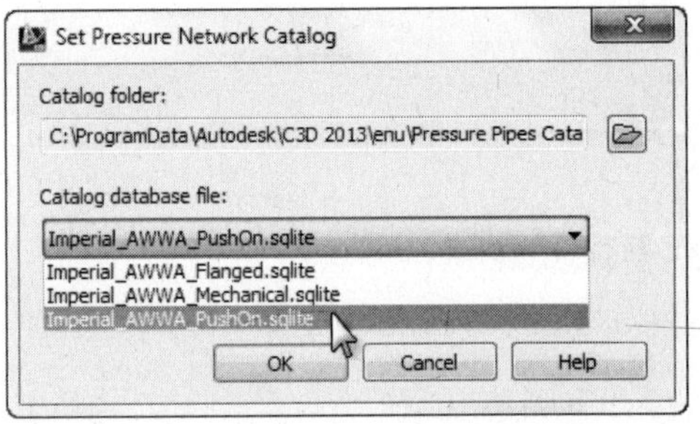

Metric_AWWA_PushOn is currently the only available option for metric users.

Each of these differs slightly in their options for pipes, fittings, and appurtenances. Only one type of pressure network catalog can be active at a time. A parts list can have parts from only one catalog in it; for example, you cannot mix and match push-on with mechanical parts. You can, however, have multiple pressure parts lists in your template; each can pull parts from the various catalog database files.

Creating a Pressure Parts List

In the Settings tab of Toolspace you will find the listing for pressure networks. The Pressure Networks branch is where you will create a parts list. In the case of pressure networks, a parts list contains three components:

Pressure Pipes Ductile iron of various sizes can be added. Like the gravity networks, each part can have a style. Furthermore, different sizes within the part families can have style, which can help you identify them in the graphic.

Fittings Fittings such as T's, crosses, and elbows are specified in this tab.

Appurtenances Valves are specified in the Appurtenances tab.

Bursting with Parts

To create a full list of all pressure network parts available in your active part catalog, type **CreatePressurePartListFull** at the command line and press ↵.

Keep in mind that the three different catalogs available to US Imperial units varies with regard to which parts are available. For example, the following graphic shows the variation of the fittings available to Push-on, Flanged, and Mechanical catalogs.

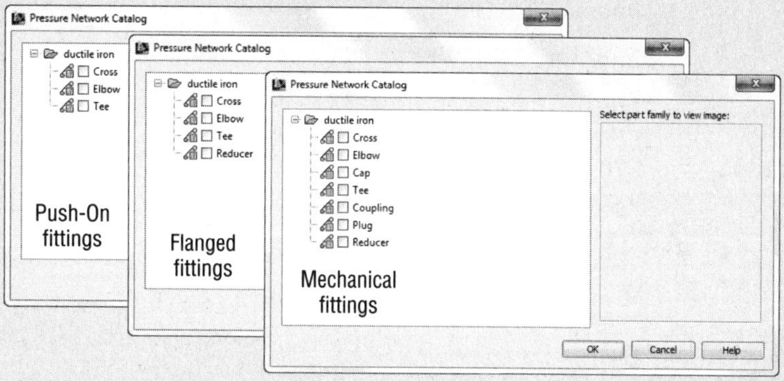

The next question on your lips is probably, "Okay, so can I make my own parts?" The answer is, "Well, sort of, but it is currently unsupported."

Pressure pipe network parts are 3D solids with a few extra bells and whistles that tell the program how they connect to other parts. If you are comfortable drafting 3D solids using the tools from the 3D modeling workspace, the procedure is fairly painless.

The Part Publishing Wizard tools built into AutoCAD will allow you to create a `*.CONTENT` package, for example, `Hydrant.CONTENT`. Once the content package is created, you can import it into the pressure pipe catalog of your choice by launching the Content Catalog editor.

The Content Catalog editor is automatically installed with Civil 3D 2013 and can be found off your Windows Start menu along with the other AutoCAD Civil 3D 2013 utilities.

For more information on creating parts through the Part Publishing Wizard, see `PartPublishingWizardUsersGuide.docx` located in `C:\Program Files\Autodesk\AutoCAD Civil 3D 2013\Sample\Civil 3D API\Part Publishing Wizard`.

In the following exercise, you will create a pressure network parts list:

1. Open the drawing `Pressure.dwg` (`Pressure_METRIC.dwg`), which you can download from this book's web page.

This file is set up with a layer state that makes other objects gray. This will help you focus on the placement of pressure pipe network objects.

2. On the Home tab ➢ Create Design panel flyout, select Set Pressure Network Catalog.
3. Verify that the Catalog Database File is set to `Imperial_AWWA_PushOn.sqlite` (`Metric_AWWA_PushOn.sqlite`). Click OK.
4. On the Settings tab of Toolspace ➢ Pressure Network, expand Pressure Network, right-click Parts Lists, and select New.
5. On the Information tab, rename the Pressure Network Parts List to **Watermain**.
6. Switch to the Pressure Pipes tab. Right-click New Parts List and select Add Material.
7. Place a check mark next to Ductile Iron and click OK.
8. The name of the pipe parts list will update to Watermain. Expand the Watermain branch.
9. Right-click on Ductile Iron and select Add Size.
10. Set the Nominal Diameter value to **10″** (**250** mm).
11. Set the Cut Length value to **20′** (**6** m). Leave all other defaults values, as shown in Figure 14.71, and click OK.

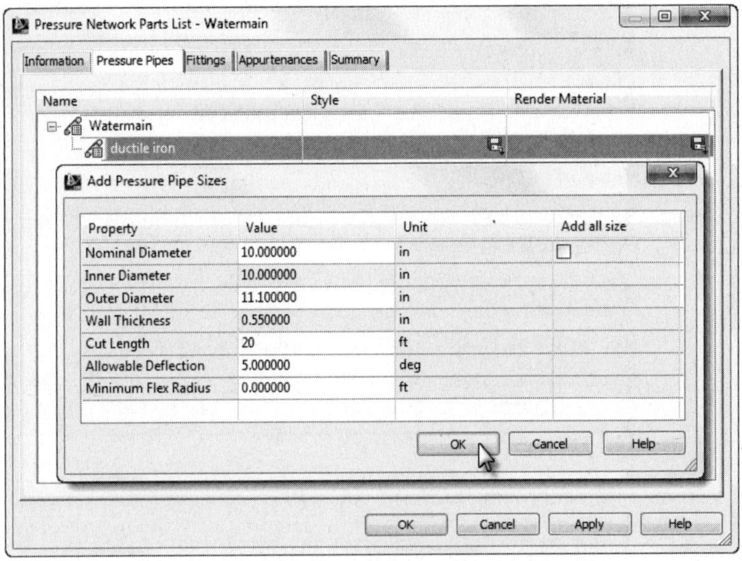

FIGURE 14.71
Adding ductile iron pipe to the pressure network parts list

12. Switch to the Fittings tab, right-click New Parts List, and select Add Type.
13. Place a check mark next to all three fitting types—Cross, Elbow, and Tee—as shown in Figure 14.72, and then click OK.

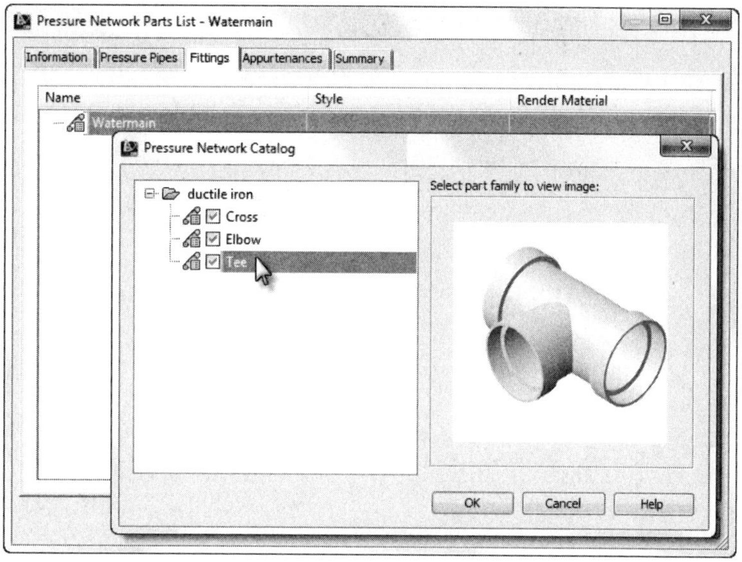

FIGURE 14.72
Adding fittings to the pressure network parts list

Metric users will just have the options for Elbow and Tee and can skip to step 16.

14. Expand the new Watermain Fittings listing if it is not already expanded.
15. Right-click on Ductile Iron Cross and select Add Size.
16. Change Nominal Diameter to 14 × 14 × 10 × 10. Leave the allowable deflection as 5 degrees, and click OK.
17. Right-click Ductile Iron Elbow, and click Add Size.
18. Set the bend angle to 11.25 and set the nominal diameter to **10″** (**250** mm). Leave the Allowable Deflection value at 5°, which is the default for the size you picked. Click OK.
19. Repeat steps 17–19 for 22.5° and 45° elbows, with a nominal diameter of **10″** (**250** mm).
20. Right-click Ductile Iron Tee and select Add Size.
21. Set the nominal diameter to **10″** (**250** mm). Leave the allowable deflection as 5°. Click OK.
22. Switch to the Appurtenances tab.
23. Right-click New Parts List and select Add Type.
24. Place a check mark next to Gate Valve-Push On-Ductile Iron-200psi (Gate Valve-Push On-Ductile Iron-16 Bar) and click OK.
25. Expand the Appurtenance listing.
26. Right-click the new gate valve and select Add Size.
27. Change the Nominal Diameter value to **10″** (**250** mm) and click OK.

28. Click OK again to complete the creation of the Watermain pressure network parts list, and then save the drawing for use in the next exercise.

Creating a Pressure Network

After you have set your pressure network catalog, created your pressure network parts list, and set your design parameters, it is time to draw your first network.

Pressure Networks in Plan View

As you work with pressure pipes, you will see some useful glyphs appear as you draw.

As shown in Figure 14.73, selecting a pressure pipe will give you tools to modify and continue your design.

FIGURE 14.73
Glyphs on a pressure pipe end

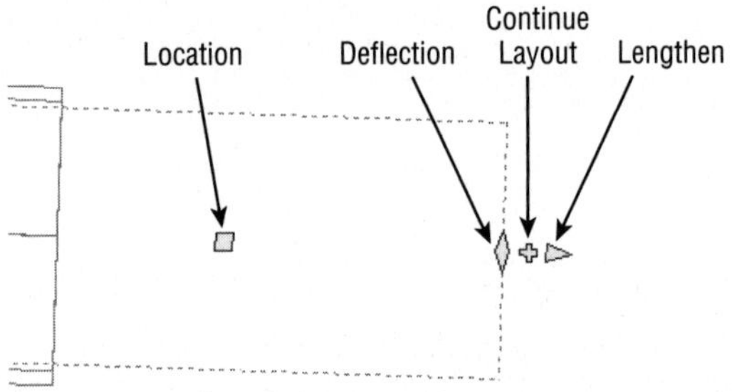

Location The location glyph moves the pipe both horizontally and vertically and will disconnect it from the adjoining fittings or appurtenances. Changing the location of a fitting or appurtenance will move the pipe and maintain the connection.

Deflection The deflection glyph will change the angle at which the pipe sits in the adjoining fitting. When this glyph is active, you will see a fan-shaped guide indicating the allowable deflection from the fitting properties in the pressure network parts list. You are able to move the pipe beyond the guide, but will receive design check errors when analyzing the network. You will take a closer look at design checks later in this chapter.

Continue Layout Continue Layout will help you pick up where you left off when working with pressure pipes. When you use this glyph from the end of a pipe, it will create a bend using the elbows from your pressure parts list. When you use this grip from a fitting or appurtenance, you are restricted to creating your pipe within the object's deflection tolerance.

Lengthen Lengthen will allow you to stretch or shorten a pipe. When used with dynamic input, you can set pipes to a specific desired 3D length. See the section on working with pipes and dynamic input earlier in the section "Editing with the Network Layout Tools Toolbar."

As shown in Figure 14.74, you will encounter more glyphs when working with fittings and appurtenances.

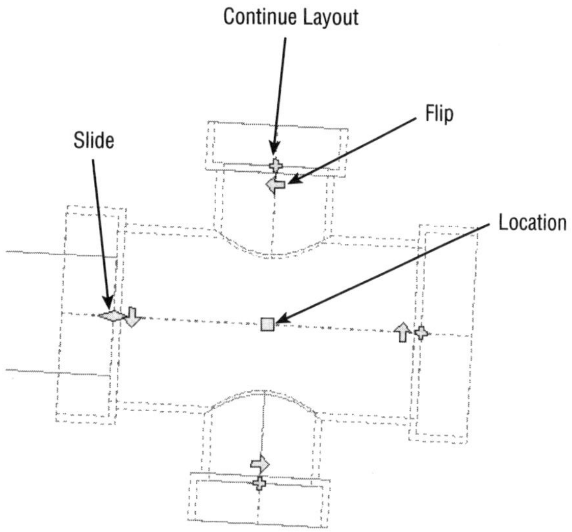

FIGURE 14.74
Glyphs on a Tee fitting

Slide The slide glyph is similar to lengthen, but can modify a pipe's length when a fitting or appurtenance is already attached.

Flip The flip glyph will mirror the part at its center. Flip glyphs are especially handy when working with Tees, as it is often necessary to change the outlet direction of a Tee after it is inserted. Be cautious when using the flip glyph with elbows, since it may disconnect the adjoining pipe.

In this exercise, you will create a pressure network. Use the X's as a guide for placement, but don't worry if your pipe network is slightly off from the guides. Due to the 3D nature of the pipes, the restrictions on placement angles within the pressure network parts, and object snap behavior, duplicating an example network exactly would be quite tedious. Get a feel for the pressure network creation tools and have fun! You need to have completed the previous exercise before continuing.

1. Continue working in the drawing `Pressure.dwg` (`Pressure_METRIC.dwg`).

   As you work through this exercise, you will have best results if you turn off object snaps, object snap tracking, polar tracking, and/or ortho. Because the pressure pipe tools already have restrictions on how they can be drafted, sometimes these tools conflict with where you want to place the pipe.

2. From the Home tab ➢ Create Design panel, click Pipe Network ➢ Pressure Network Creation Tools, as shown in Figure 14.75.

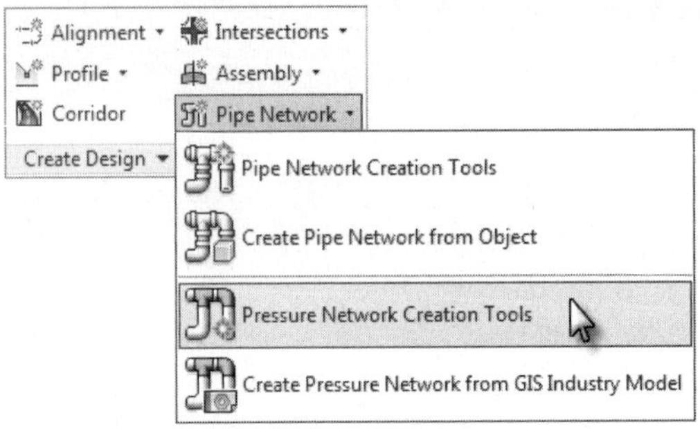

FIGURE 14.75
Selecting Pressure Network Creation Tools

3. In the Create Pressure Pipe Network dialog, name the network **Watermain North**, and then do the following:

   A. Set the parts list to Watermain.

   B. Set the surface name to Composite.

   C. Set Alignment Name to Syrah Way.

   D. Set the pipe, fitting, and appurtenance labels as shown in Figure 14.76.

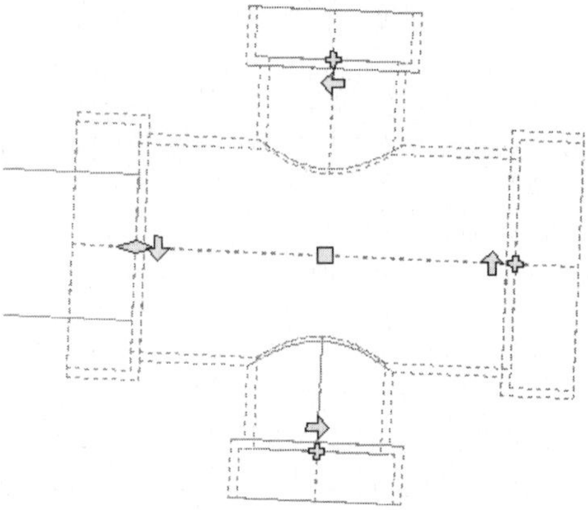

FIGURE 14.76
Creating your new Watermain system using pressure pipe network tools

4. Click OK.

   The ribbon now changes to show you the Pressure Network Plan Layout ribbon tab, as shown in Figure 14.77.

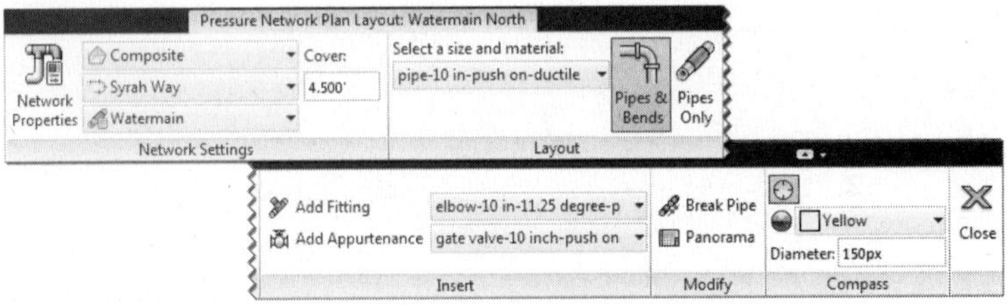

FIGURE 14.77
The Pressure Network Plan Layout toolbar

5. On the Network settings tab, set the default cover to **4.5′** (**1.5** m).
6. On the Layout panel, set your Size and Material pull-down to Pipe-10 In-Push On-Ductile Iron350 psi-AWWA C251. (Pipe-250 mm-Push On-Ductile Iron-25 Bar-AWWA C251).
7. Start to place a waterline by clicking the X labeled 1 toward the east end of Syrah Way.
8. Place the first bend by clicking near the X labeled 2, to the left.

   At the bend, you are restricted to the bend angles listed in your part network. The compass glyph (Figure 14.78) that appears represents the elbow angles to the left and right of your pipe. If you had not included multiple elbow angles in you pressure pipe network parts list, only the default elbow angle of 11.25 degrees would be available.

FIGURE 14.78
The pressure pipe fitting glyph reflects your elbow angles.

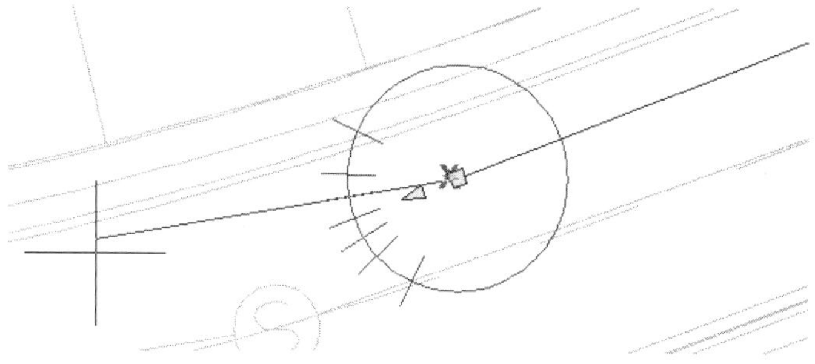

9. Click the X labeled 3 to place the next pipe end.
10. Continue working east to west until you click near the X labeled 8. Press Esc.

    To keep the pressure pipe on track, you will use the allowable deflection of the elbow to move the pipe closer to the edge of the road.

    Make sure your object snaps off for the next steps, as they will interfere with the pipe modification glyphs.

11. Select the pipe section between the X labeled 7 and the X labeled 8, and then use the deflection glyph to move the end of the pipe up to the approximate center of the green circle, as shown in Figure 14.79.

FIGURE 14.79
Use the deflection glyph to move the pipe.

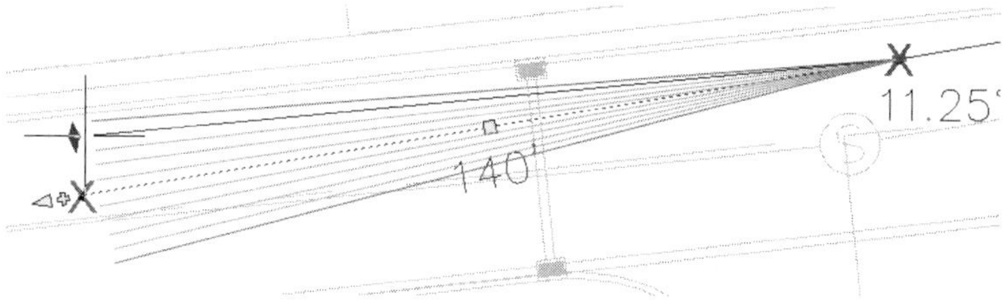

12. Click the + glyph to continue the layout.
13. Place the next pipe end at the X labeled 9.

14. Continue working west until you reach near the X labeled 10.

    Don't worry if you are off the desired location, as you can always use the glyphs to edit the pipe location after the fact.

15. Press Esc after placing this pipe.

16. On the Pressure Network Plan Layout contextual tab ➤ Insert panel, change the fitting to **Tee-10 in × 10 in** (**Tee-250 mm × 250 mm**).

17. Click Add Fitting.

    Add Fitting

    As you hover your cursor near the end of the pipe, you will see the Add Fitting glyph, as shown in Figure 14.80.

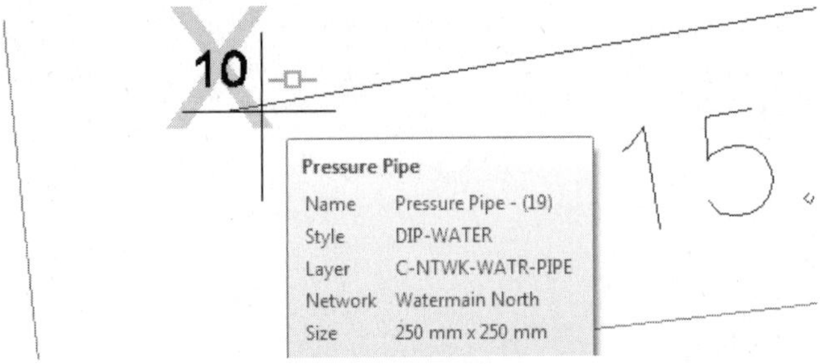

FIGURE 14.80
Add fitting to the end of the pipe

18. Click to add the tee.

19. Press Esc to complete the task.

    At this point, the tee is in the graphic but it is not positioned in such a way that would make it useful in continuing the design north and south along the intersecting road (Frontenac Drive). To fix this situation, you will disconnect it from the pipe, rotate it, and then reconnect the pipe.

20. Click the tee in the graphic to select it, and then right-click the part and select Disconnect From Pressure Part, as shown in Figure 14.81.

21. At the prompt `Select connected pressure part:`, select the pipe connected to the tee.

    Now that the part is disconnected, you are free to rotate it into place. Select the part to reveal its glyphs. There is a rotation glyph visible on the object, which you will use to rotate the part 90° counterclockwise.

22. Press F12 to turn on dynamic input if it is not already on.

23. As shown in Figure 14.82, click the rotation glyph and enter **-90**↵.

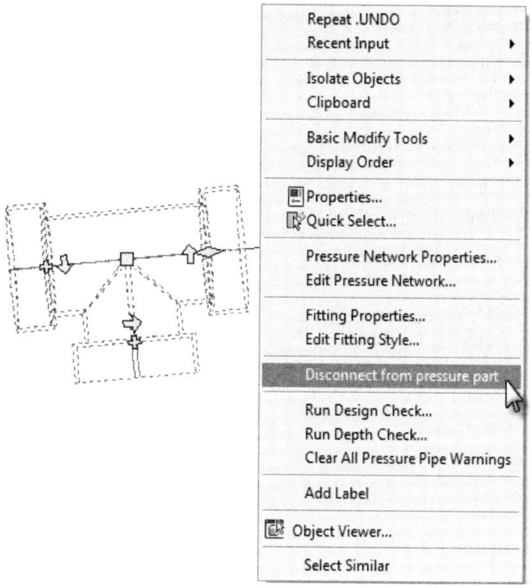

FIGURE 14.81
Disconnecting the part in preparation for rotating it

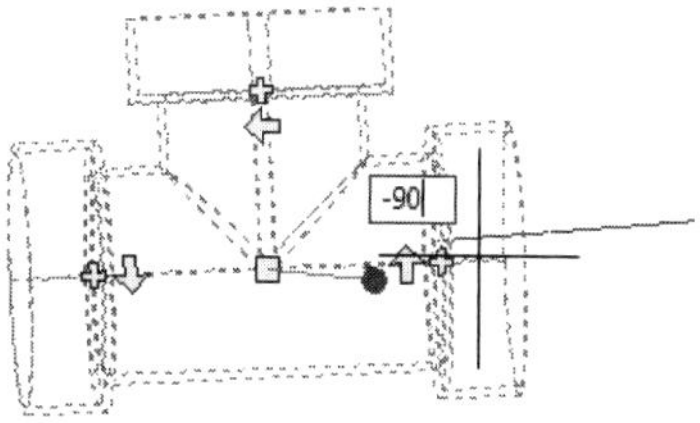

FIGURE 14.82
Rotate the tee to correct its position before reconnecting.

The tee is now in the correct position, but it must be reconnected to the pipe.

24. Select the location glyph, and move the tee to the east until you see the connection glyph similar to the one you saw in Figure 14.80.

25. When you see the correct glyph, click.

    You will know the tee is connected properly when the rotation glyph no longer appears on the selected fitting. Another graphic indication that the tee is connected will be the reappearance of the slide glyph on the connected end.

26. Select the tee to reveal the grips. Working north from the tee, click the + glyph to continue the layout.

27. Click to place an elbow at the X labeled 11 and finally the X labeled 12. Press Esc.
28. Working south from the tee, click the + glyph to continue the layout.
29. Click to place a pipe ending at the X labeled 13, and press Esc.

30. From the Home tab ➤ Layers panel, click Layer Freeze.
31. Click one of the X's to freeze the _Placement Labels layer, and press ↵ to complete.
32. On the Pressure Network Plan Layout contextual tab ➤ Insert panel, verify that the Appurtenance is set to Gate Valve 10" (Gate Valve 250 mm).

33. Click Add Appurtenance.
34. Place the appurtenance in the drawing by clicking near the end of the north pipe.

    Be sure to look for the attachment glyph before clicking, as shown in Figure 14.82. Don't be shy about zooming in close to get a good look at the object you are working with.
35. Place another valve at the south end.
36. Save the drawing.

Importing an Industry Model

One of the options you have with Civil 3D 2013 is to create a pressure network from an industry model.

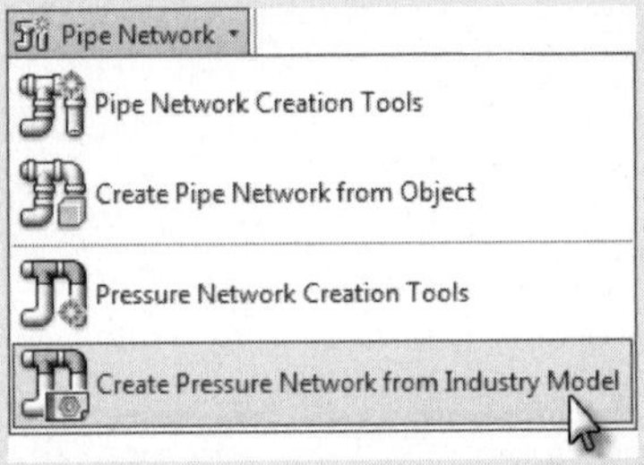

Industry models are a type of AutoCAD drawing file that contain data from one of the handful of Autodesk-created templates, most commonly from Autodesk Map 3D. The Map 3D standalone product comes with premade infrastructure data classes such as wastewater, water, gas, and electric, just to name a few. Not just any Map 3D file will do. A drawing containing pipe and valve information will only be recognized as an industry model if it was created based off one of the industry-specific templates. The ability to create an industry model is not available on the Map 3D workspace that is part of Civil 3D.

Pressure Pipe Networks in Profile View

Pressure pipe networks can do things in profile view that gravity pipes cannot. With pressure pipes, the profile view can be used to add on to the pipe layout, change straight pipes to curves, and delete parts from the project altogether. To these tools, select any pressure part and from the

Pressure Networks contextual tab ➢ Modify panel, pick Edit Network ➢ Profile Layout Tools, as shown in Figure 14.83.

FIGURE 14.83
Locating the Profile Layout Tools

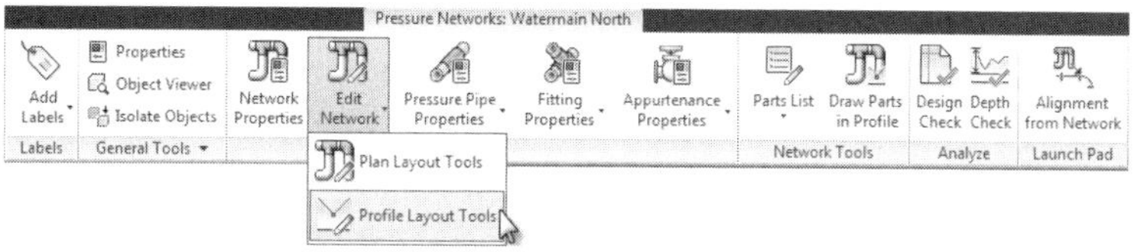

In the following exercise, you will draw the pressure pipe network in profile view and modify the layout using the Follow Surface command.

1. Open the drawing `PressureProfile.dwg` (`PressureProfile_METRIC.dwg`), which you can download from this book's web page.
2. Select any pressure network part in the drawing.
3. From the Pressure Networks contextual tab ➢ Launch Pad panel, select Alignment From Network.
4. At the prompt `Select first Pressure Network Part (Pipe or Fitting or Appurtenance):`, click the Gate Valve at the far right of the drawing.
5. At the prompt `Select next Pressure Network Part or [Undo]:`, select the Gate Valve at the southern part of the project.
6. Press ↵ to continue.

   The next few steps are exactly the same as when you created an alignment and profile from a gravity pipe network. You will be prompted to create an alignment, sample the surface, and create a profile view.

7. In the Create Alignment dialog, change the name to **Syrah Water**.
8. Verify that Create Profile And Profile View is checked. Leave all other styles and settings at their defaults and click OK.
9. In the Create Profile From Surface dialog, highlight the composite surface and click Add.
10. Click Draw In Profile View.
11. In the Create Profile View wizard, leave all settings at their defaults and click Create Profile View.
12. Place the view by entering the coordinate **230000, 200000** (**700000, 60000** for metric users).

    You now see the profile view with your pressure pipe network present in all its glory. As you can see in Figure 14.84, the pipe looks good, except it appears that the pipe cover is inadequate toward the end of the alignment. You will fix this in the steps that follow.

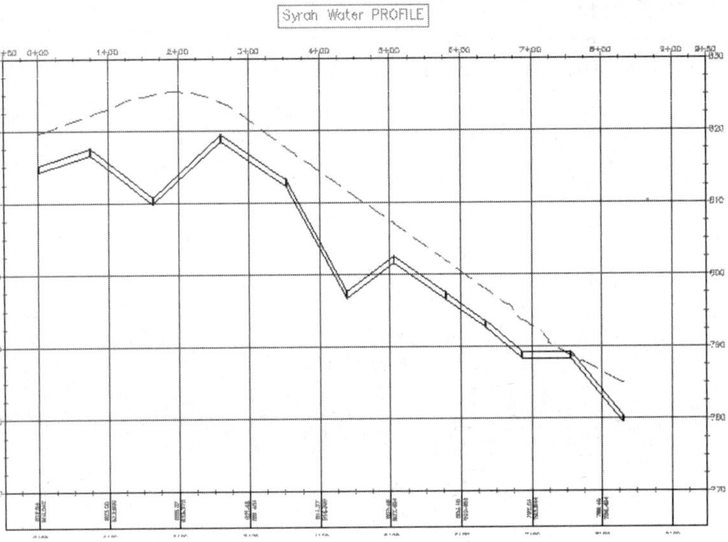

FIGURE 14.84
Pressure network in profile

13. Select a part from the network if you do not already have the Pressure Networks contextual tab visible.

14. From the Pressure Networks contextual tab ➤ Modify panel, Edit Network ➤ Profile Layout Tools.

    Now the Pressure Network Profile Layout contextual tab appears, as shown in Figure 14.85.

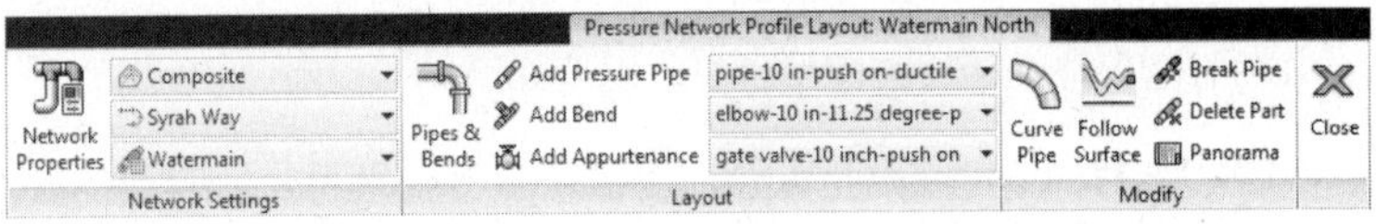

FIGURE 14.85
Pressure Network Profile Layout contextual tab

Follow Surface

15. From the Pressure Network Profile Layout contextual tab ➤ Modify panel, select Follow Surface.

16. At the prompt `Select first pressure part in profile:`, select the leftmost valve in the profile view.

17. At the prompt `Select next pressure part in profile [Enter to finish]:`, click the rightmost valve in the profile view and then press ↵ to finish selecting parts. All connected parts in between will become selected.

18. At the prompt `Enter depth below surface <0.0000>:`, enter **4.5′** (**1.5 m**)↵.

    Your profile view will change to resemble Figure 14.86.

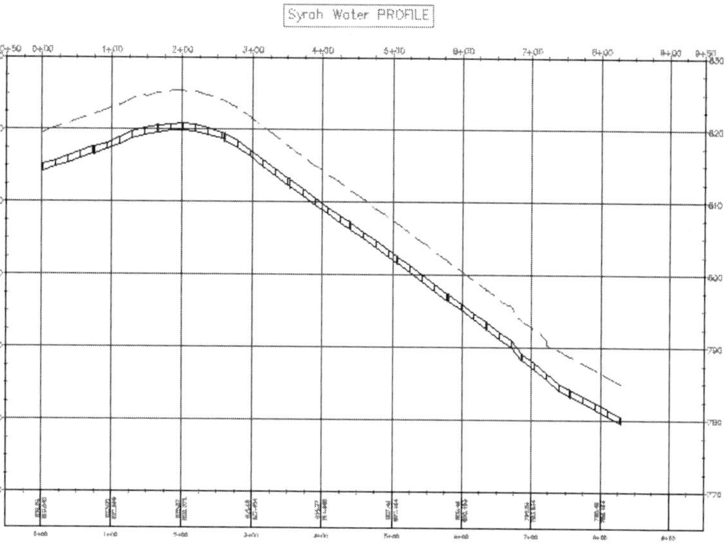

FIGURE 14.86
Pressure pipe following the surface

19. Save and close the drawing.

Design Checks

Pressure networks differ from the networks you created earlier in this chapter. Since the fluid in a pressure network can go uphill, the rules you saw in gravity systems no longer apply. The main concerns for a pressure network are pressure loss and depth of cover.

You can locate the depth check values on the Settings tab of Toolspace. Locate the Pressure Network branch and expand the Commands listing. Double-click RunDepthCheck to edit the command settings. A dialog like that in Figure 14.87 will open.

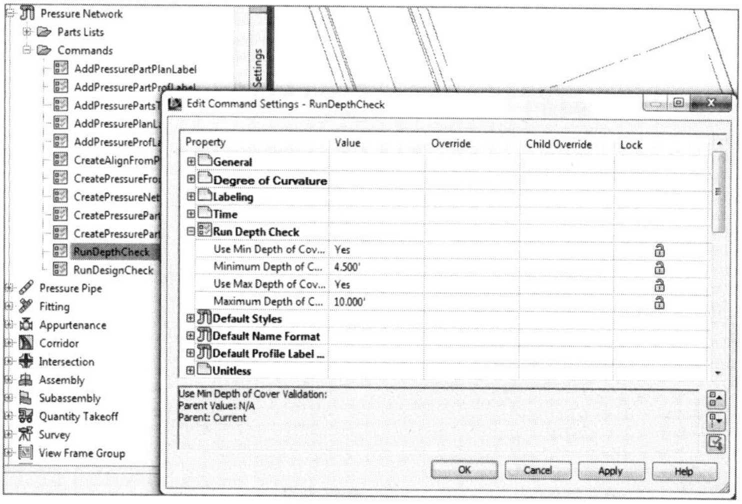

FIGURE 14.87
Edit the RunDepthCheck command settings to validate your design.

Civil 3D will not compute the pressure loss at each junction, but you can set an acceptable range of values for pipe bends and radius of curvature for curved pipes. Deflection Validation settings are found under RunDesignCheck, as shown in Figure 14.88.

FIGURE 14.88
Turning on deflection validation settings

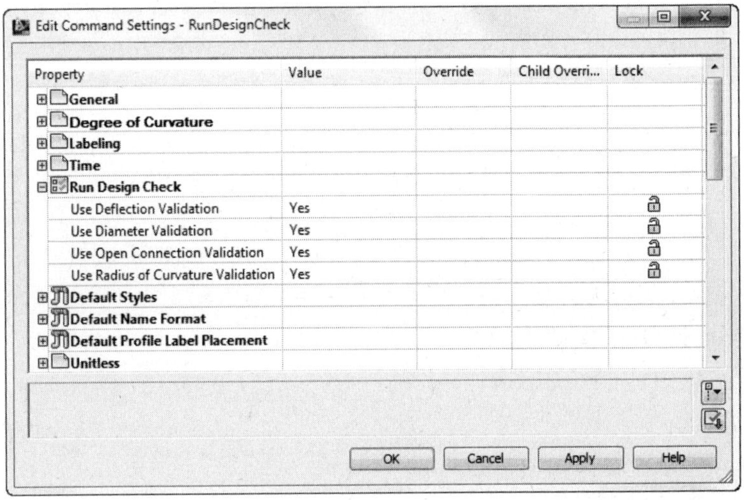

Once you have created your pressure network, you should check your initial design for flaws. From the Pressure Networks contextual tab, you can check your design to see if it meets the requirements you set up in the command settings.

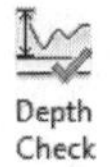

The depth check verifies that all pipes and fittings are within the acceptable range of values for depth.

Design Check will check for improperly terminating pipes, mismatched pipe and fitting diameters, any curved pipe whose radius has exceeded acceptable values, and pipes that have exceeded the maximum deflection you set up in the parts list.

In the following exercise, you will modify command settings and run a depth check on the pipe network:

1. Open the drawing `PressureDesignCheck.dwg` (`PressureDesignCheck_METRIC.dwg`), which you can download from this book's web page.
2. In the Settings tab of Toolspace expand Pressure Network ➤ Commands, right-click RunDepthCheck, and select Edit Command Settings.
3. Expand Run Depth Check Property, and verify that Minimum Depth Of Cover is set to 4.5′ (1.0 m).
4. Click in the Value column and change Use Max Depth Of Cover Validation to Yes.
5. Click OK.

6. Select a pressure network part if you do not already see the Pressure Networks contextual tab.

7. From the Pressure Networks contextual tab ➤ Analyze panel, select Depth Check.

   The Depth Check command will allow you to perform the analysis either in plan or profile view. Either way you choose to select your pressure pipe network, the result will be the same.

8. At the prompt `Select a path along a Pressure Network in plan or profile view:`, click the Gate Valve to the far left in the profile view. You will need to zoom in close to the end to see the bowtie-shaped valve.

9. At the prompt `Select next point on path [Enter to finish]:`, select the Gate Valve to the far right in the profile view and press ↵.

   You will now see the Depth Check dialog. The settings should be the same values as those in steps 3 and 4.

10. Click OK.

    In both plan view (Figure 14.89, top) and profile view (Figure 14.89, bottom), warnings will appear if any Depth Check violations are found.

FIGURE 14.89 Depth Check result in plan (top) and profile (bottom)

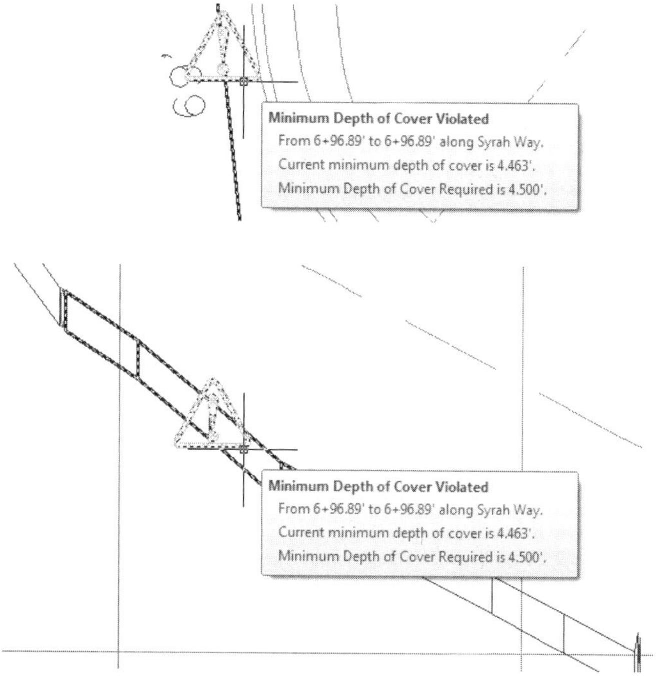

11. Save the drawing.

Part Builder

Part Builder is an interface that allows you to build and modify pipe network parts. You access Part Builder by selecting the drop-down list under Create Design from the Home tab. At first, you may use Part Builder to add a few missing pipes or structure sizes. As you become more familiar with the environment, you can build your own custom parts from scratch.

There is currently no tool available for creating custom pressure network parts, so this section only applies to gravity networks.

Real World Scenario

A Word of Caution about Part Builder

Every time one of our clients asks about Part Builder, we cringe. Our advice is, "If all you need to do is add a size to an existing part, no problem. If you want to create new geometry from scratch? Run. Run, screaming."

This is not a tool that you can master by dabbling in it for a few hours. Even after you understand Part Builder, it can take days to build a complex part that functions correctly. Think of Part Builder as a half-step away from programming, and you will have more realistic expectations.

Before you buckle down to learn about Part Builder, ask yourself a few questions:

- Do I really understand pipes and structures well enough to tackle this? Is it worth taking extra time from a billable project to learn this feature?
- Is the part I need unique and not available from another source (such as `http://www.cad-1.com/Solutions.aspx?id=1&SoftwareId=42`)?

If you can answer yes to all these questions, you have our blessing. In many cases, the new pressure network tools eliminate the need for custom parts.

This section is intended to be an introduction to Part Builder and a primer in some basic skills required to navigate the interface. It isn't intended to be a robust "how-to" for creating custom parts. Civil 3D includes three detailed tutorials for creating three types of custom structures. The tutorials lead you through creating a cylindrical manhole structure, a drop inlet manhole structure, and a vault structure. You can find these tutorials by going to Help ➢ Tutorials and then navigating to Part Builder Tutorials.

Back Up the Part Catalogs

Here's a warning: Before exploring Part Builder in any way, it's critical that you make a backup copy of the part catalogs. Doing so will protect you from accidentally removing or corrupting default parts as you're learning, and will provide a means of restoring the original catalog.

The catalog (as discussed in the previous section) can be found by default at:

```
C:\ProgramData\Autodesk\C3D 2013\enu\Pipes Catalog\
```

To make a backup, copy this entire directory and then save that copy to a safe location, such as another folder on your hard drive or network, or to a CD.

We recommend that you do this and use the backup file for the exercises here. To do this, you will have to point to the new location by clicking the drop-down list in the Create Design panel and selecting Set Pipe Catalog Location. Select the icon next to the catalog folder and point it to your saved location.

The parts in the Civil 3D pipe network catalogs are *parametric*. Parametric parts are dynamically sized according to a set of variables, or parameters. In practice, this means you can create one part and use it in multiple situations.

For example, in the case of circular pipes, if you didn't have the option of using a parametric model, you'd have to create a separate part for each diameter of pipe you wanted, even if all other aspects of the pipe remained the same. If you had 10 pipe sizes to create, that would mean 10 sets of `partname.dwg`, `partname.xml`, and `partname.bmp` files, as well as an opportunity for mistakes and a great deal of redundant editing if one aspect of the pipe needed to change.

Fortunately, you can create one parametric model that understands how the different dimensions of the pipe are related to each other and what sizes are allowable. When a pipe is placed in a drawing, you can change its size. The pipe will understand how that change in size affects all the other pipe dimensions such as wall thickness, outer diameter, and more; you don't have to sort through a long list of individual pipe definitions.

Part Builder Orientation

Each drawing "remembers" which part catalog it is associated with. If you're in a metric drawing, you need to make sure the catalog is mapped to metric pipes and structures, whereas if you're in an Imperial drawing, you'll want the Imperial catalog. By default, the Civil 3D templates should be appropriately mapped, but it's worth the time to check. Set the catalog by changing to the Home tab and selecting the drop-down list on the Create Design panel. Verify the appropriate folder and catalog for your drawing units in the Pipe Network Catalog Settings dialog (see Figure 14.90), and you're ready to go.

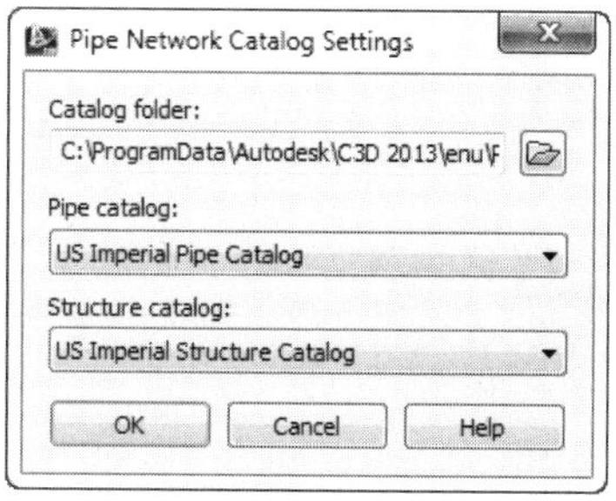

FIGURE 14.90 Choose the appropriate folder and catalog for your drawing units.

Understanding the Organization of Part Builder

The vocabulary used in the Part Builder interface is related to the vocabulary in the HTML catalog interface that you examined in the previous section, but there are several differences that are sometimes confusing.

Open Part Builder by going to the Home tab ➤ Create Design panel and selecting the Part Builder icon from the drop-down list.

The first screen that appears when you start the Part Builder is Getting Started – Catalog Screen (see Figure 14.91).

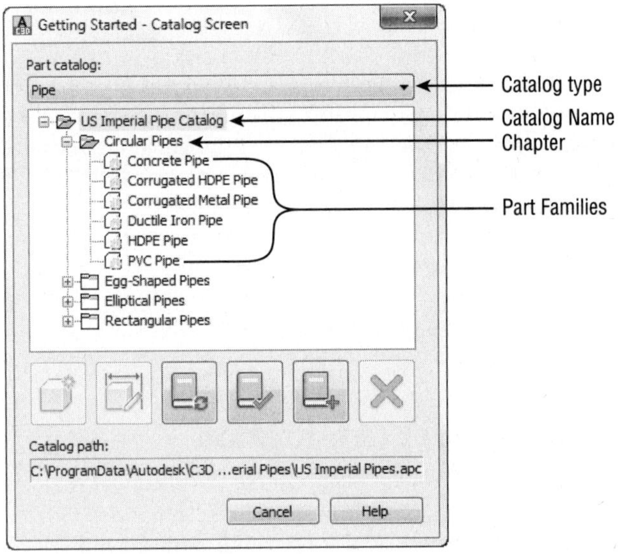

FIGURE 14.91
The Getting Started – Catalog Screen

At the top of this window is a drop-down menu for selecting the part catalog. Depending on what you need to modify, you can set this to ether Pipe or Structure. Once you pick the catalog type, you will see the main catalog name. In Figure 14.91, you see US Imperial Pipe Catalog.

Below the part catalog is a listing of chapters. In Part Builder vocabulary, a chapter is a grouping based on the shape. Inside the chapters you will see the part families listed. The US Imperial Pipe Catalog has four default chapters:

- Circular Pipes
- Egg-Shaped Pipes
- Elliptical Pipes
- Rectangular Pipes

You can create new chapters for different-shaped pipes, such as Arch Pipe.

The US Imperial Structure Catalog also has four default chapters: Inlets-Outlets, Junction Structures With Frames, Junction Structures Without Frames, and Simple Shapes. You can create new chapters for custom structures. You can expand each chapter folder to reveal one or more part families. For example, the US Imperial Circular Pipe Chapter has six default families:

- Concrete Pipe
- Corrugated HDPE Pipe
- Corrugated Metal Pipe
- Ductile Iron Pipe
- HDPE Pipe
- PVC Pipe

Pipes that reside in the same family typically have the same parametric behavior, with only differences in size.

As Table 14.1 shows, a series of buttons on the Getting Started – Catalog Screen lets you perform various edits to chapters, families, and the catalog as a whole.

TABLE 14.1: The Part Builder catalog tools

ICON	FUNCTION
	The New Parametric Part button creates a new part family.
	The Modify Part Sizes button allows you to edit the parameters for a specific part family.
	The Catalog Regen button refreshes all the supporting files in the catalog when you've finished making edits to the catalog.
	The Catalog Test button validates the parts in the catalog when you've finished making edits to the catalog.
	The New Chapter button creates a new chapter.
	The Delete button deletes a part family. Use this button with caution, and remember that if you accidentally delete a part family, you can restore your backup catalog as mentioned in the beginning of this section.

Exploring Part Families

The best way to get oriented to the Part Builder interface is to explore one of the standard part families. In this case,

1. Click the Part Catalog drop-down list and select Pipe.
2. Drill down through the US Imperial Pipe Catalog ➤ Circular Pipe ➤ Concrete Pipe family by highlighting Concrete Pipe and clicking the Modify Part Sizes button.

   It is normal to receive a message that the file contains previous version AEC objects.

3. Click Close to continue.

   A Part Builder task pane appears with `AeccCircularConcretePipe_Imperial.dwg` on the screen along with the content builder toolspace, as shown in Figure 14.92.

FIGURE 14.92
Content builder toolspace and related DWG file

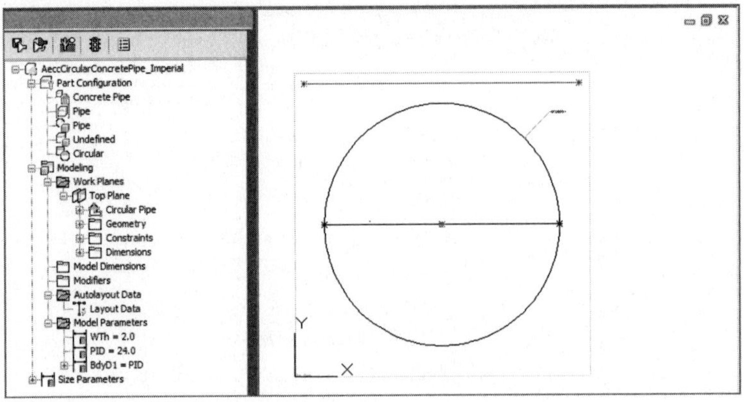

4. Close Part Builder by clicking on the X on the Part Builder task pane.
5. Click No when asked to save the changes to the Concrete Pipe.
6. Click No when asked to save the drawing.

The Part Builder task pane, or Content Builder (Figure 14.93), is well documented in the Civil 3D User's Guide. Please refer to the User's Guide for detailed information about each entry in Content Builder.

FIGURE 14.93
Content Builder

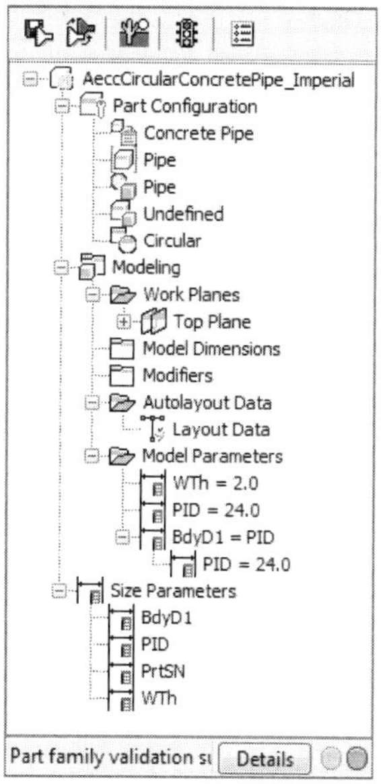

IS THAT ALL THERE IS TO PART BUILDER?

There is much more to Part Builder than this book can cover. As a matter of fact, a separate book could be written just on that subject. So if you are hungering for more information on Part Builder, we'd like to point you the blog of Cyndy Davenport. Cyndy has been a repeat speaker at Autodesk University on Part Builder and Civil 3D. You can find her blog at `http://c3dcougar.typepad.com/`. You can access Part Builder and other resources at `au.autodesk.com`.

Adding a Part Size Using Part Builder

The hypothetical municipality requires a 12″ (300 mm) sanitary sewer cleanout. After studying the catalog, you decide that Concentric Cylindrical Structure NF is the appropriate shape for your model, but the smallest inner diameter size in the catalog is 48″ (1200 mm). The following exercise gives you some practice in adding a structure size to the catalog—in this case, adding a 12″ (300 mm) structure to the US Imperial Structures catalog.

You can make changes to the US Imperial Structures catalog from any drawing that is mapped to that catalog, which is probably any imperial drawing you have open, as follows:

1. For this exercise, start a new drawing from `_AutoCAD Civil 3D (Imperial) NCS.dwt` (`_AutoCAD Civil 3D (Metric) NCS.dwt`).
2. On the Home tab, select the drop-down list on the Create Design panel and select Part Builder.
3. Choose Structure from the drop-down list in the Part Catalog selection box.
4. Expand the Junction Structure Without Frames chapter.
5. Highlight the Concentric Cylindrical Structure NF (no frames) part family.
6. Click the Modify Part Sizes button.

   The Part Builder interface opens `AeccStructConcentricCylinderNF_Imperial.dwg` along with the Content Builder task pane.
7. Close the message that the part was created with a previous version. Zoom extents if necessary.
8. Expand the Size Parameters tree.
9. Right-click the SID (Structure Inner Diameter) parameter, and choose Edit.

   The Edit Part Sizes dialog appears.
10. Locate the SID column (see Figure 14.94), and double-click inside the box.

FIGURE 14.94
Examining available part sizes

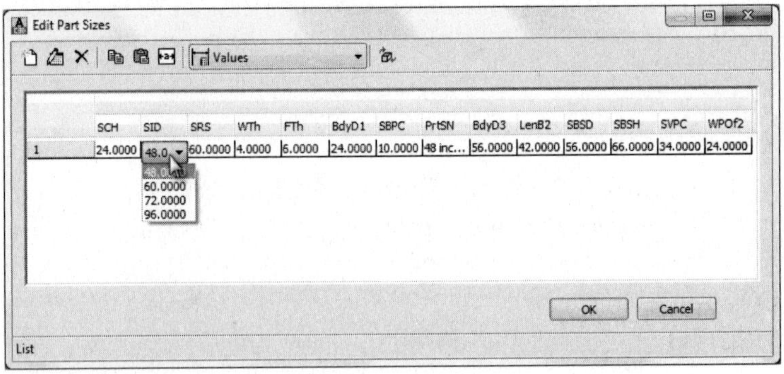

Note that a drop-down menu shows the available inner diameter sizes 48″, 60″, 72″, and 96″ (1200 mm, 1400 mm, 1600 mm, and 1800 mm).

11. Locate the Edit button. Make sure you're still active in the SID column cell, and then click Edit.

    The Edit Values dialog appears.

12. Click Add, and type **12″** (**300** mm), as shown in Figure 14.95.

FIGURE 14.95
Add the 12″ (300 mm) value to the Edit Values dialog

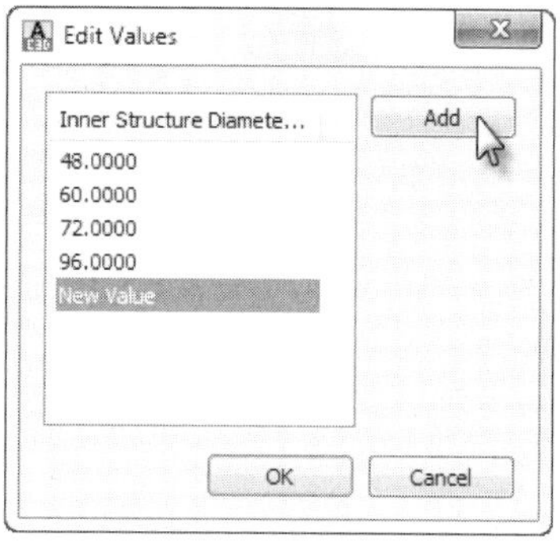

13. Click OK to close the Edit Values dialog, and click OK again to close the Edit Part Sizes dialog.
14. Click the small X in the upper-right corner of the Content Builder task pane to exit Part Builder.
15. When you see the message "Save Changes To Concentric Cylindrical Structure NF?", click Yes.

    You could also click Save in Content Builder to save the part and remain active in the Part Builder interface.

You're back in your original drawing. If you created a new parts list in any drawing that references the US Imperial Structures catalog, the 12″ (300 mm) structure would now be available for selection.

Sharing a Custom Part

You may find that you need to go beyond adding pipe and structure sizes to your catalog and build custom part families or even whole custom chapters. Perhaps instead of building them yourself, you're able to acquire them from an outside source.

The following section can be used as a reference for adding a custom part to your catalog from an outside source, as well as sharing custom parts that you've created. The key to sharing a part is to locate the three files mentioned earlier.

Adding a custom part size to your catalog requires these steps:

1. Locate the `partname.dwg`, `partname.xml`, and (optionally) `partname.bmp` files of the part you'd like to obtain.
2. Make a copy of the `partname.dwg`, `partname.xml`, and (optionally) `partname.bmp` files.
3. Insert the `partname.dwg`, `partname.xml`, and (optionally) `partname.bmp` files in the correct folder of your catalog.
4. Run the **`PARTCATALOGREGEN`** command in Civil 3D.

Adding an Arch Pipe to Your Part Catalog

This exercise will teach you how to add a premade custom part to your catalog. You can make changes to the US Imperial Pipes catalog from any drawing that is mapped to that catalog, which is probably any imperial drawing you have open.

1. For this exercise, start a new drawing from the `_AutoCAD Civil 3D (Imperial) NCS.dwt` file (`_AutoCAD Civil 3D (Metric) NCS.dwt`).
2. In Windows Explorer, create a new folder called **`Arch Pipes`** in your backup directory.

   Out of the box, it is located here:

   `C:\ProgramData\Autodesk\C3D 2013\enu\Pipes Catalog\US Imperial Pipes\`

   `(C:\ProgramData\Autodesk\C3D 2013\enu\Pipes Catalog\Metric Pipes\)`

   This directory should now include five folders: `Arch Pipes`, `Circular Pipes`, `Egg-Shaped Pipes`, `Elliptical Pipes`, and `Rectangular Pipes`.
3. Copy the `Concrete Arch Pipe.dwg` and `Concrete Arch Pipe.xml` files into the `Arch Pipe` folder.

   Note that there is no optional bitmap for this custom pipe.
4. Return to your drawing, and enter **`PARTCATALOGREGEN`** in the command line.
5. Press P to regenerate the Pipe catalog, and then press ↵ to exit the command.

   If you created a new parts list at this point in any drawing that references the US Imperial Pipes catalog, the arch pipe would be available for selection.
6. To confirm the addition of the new pipe shape to the catalog, locate the catalog HTML file at:

   `C:\ProgramData\Autodesk\C3D 2013\enu\Pipes Catalog\US Imperial Pipes\Imperial Pipes.htm`

   Explore the catalog as you did in the "All in the Family" section earlier.

The Bottom Line

Create a pipe network by layout. After you've created a parts list for your pipe network, the first step toward finalizing the design is to use Pipe Network By Layout.

Master It Open the `MasteringPipes.dwg` or `Mastering pipes_METRIC.dwg` file. From the Home tab ➤ Create Design panel ➤ Pipe Network, select Pipe Network Creation Tools to create a sanitary sewer pipe network. Use the Composite surface, and name only structure and pipe label styles. Don't choose an alignment at this time. Create 8" (200 mm) PVC pipes and concentric manholes. There are blocks in the drawing to assist you in placing manholes. Begin at the START HERE marker, and place a manhole at each marker location. You can erase the markers when you've finished.

Create an alignment from network parts and draw parts in profile view. Once your pipe network has been created in plan view, you'll typically add the parts to a profile view based on either the road centerline or the pipe centerline.

> **Master It** Continue working in the `MasteringPipes.dwg` (`Mastering pipes_METRIC.dwg`) file. Create an alignment from your pipes so that station zero is located at the START HERE structure. Create a profile view from this alignment, and draw the pipes on the profile view.

Label a pipe network in plan and profile. Designing your pipe network is only half of the process. Engineering plans must be properly annotated.

> **Master It** Continue working in the `MasteringPipes.dwg` (`Mastering pipes_METRIC.dwg`) file. Add the Length Description And Slope style label to profile pipes and the Data With Connected Pipes (Sanitary) style to profile structures. Add the alignment created in the previous Master It to all pipes and structures.

Create a dynamic pipe table. It's common for municipalities and contractors to request a pipe or structure table for cost estimates or to make it easier to understand a busy plan.

> **Master It** Continue working in the `MasteringPipes.dwg` (`Mastering pipes_METRIC.dwg`) file. Create a pipe table for all pipes in your network. Use the default table style.

Chapter 15

Storm and Sanitary Analysis

One of the trickiest and most controversial parts of a civil designer's job is determining where water and wastewater will go in a site. Civil 3D offers a variety of tools for designing and analyzing the hydraulic and hydrology portions of your project. Hydraulic calculations for pipe design take into account many factors that influence the end product. Hydrology is part science and part art, with a dash of voodoo. Luckily, Storm and Sanitary Analysis (SSA) is here to help you sort it all out.

SSA is a separate program that launches from the Analysis tab in AutoCAD® Civil 3D® or from the Autodesk folder in the Windows Start menu. This is the portion of the product that can compute site runoff, pipe sizing calculations, and design pressure pipe systems.

In this chapter, you will learn to:

- Create a catchment object
- Export pipe data from Civil 3D into SSA
- Export a stage-storage table from Civil 3D
- Model drainage systems in SSA

Getting Started on the CAD Side

When you are getting ready to perform hydraulic or hydrologic computations using SSA, there are several steps that can be completed on the CAD side of Civil 3D.

You will use Civil 3D tools to find drainage points, catchment areas, and flow paths.

Water Drop

In any hydrologic system, it is important to know the direction water will flow on a surface. The Water Drop command is an excellent tool for determining where your outfall location should be. You will want to locate preliminary flow paths before defining the drainage basin.

Water Drop

To use the Water Drop command, select the surface you wish to analyze. Then select the Water Drop command from the Analyze panel of the Surface contextual tab.

In the following exercise, you will use the Water Drop command on the existing site to determine the current flow pattern:

1. Open the `WaterDrop.dwg` (or `WaterDrop_METRIC.dwg`) file, which you can download from `www.sybex.com/go/masteringcivil3d2013`.
2. For this exercise, you will find it helpful to turn off object snaps and polar tracking.
3. Select the surface EG by clicking on any contour or portion of its boundary.
4. From the Surface contextual tab ➢ Analyze panel click Water Drop.

   The Water Drop settings will appear.
5. Verify that the Place Marker At Start Point option is set to Yes, as shown in Figure 15.1, and click OK.

   Note that you can choose a 3D or 2D polyline as the path type. You will keep this as 2D Polyline.

FIGURE 15.1
Water Drop command options

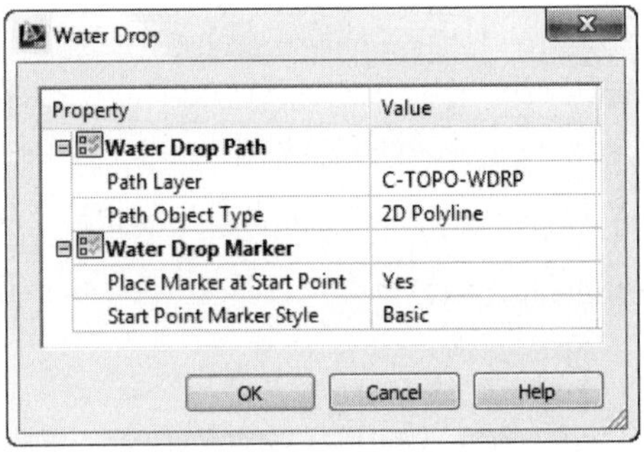

6. Get a feel for the site drainage by clicking on various areas throughout the site.

   There is no wrong place to click, as long as you are inside the surface boundary. An X represents the place your cursor clicked. The place you click is a location on the surface where water hits the surface. You will see lines forming in the direction water flows from the area. If you see an X but no path, it could mean you are in a flat area or inside a small depression. After you click in several locations, your drawing will resemble Figure 15.2.
7. Click around as many places as you would like to examine.
8. Press Esc to complete the Water Drop command.
9. Press Esc a second time to clear the surface selection.

   The water drop paths that are created are polylines without any special intelligence. The following optional steps demonstrate how to remove them if your surface model changes or you need to clear them from the screen.

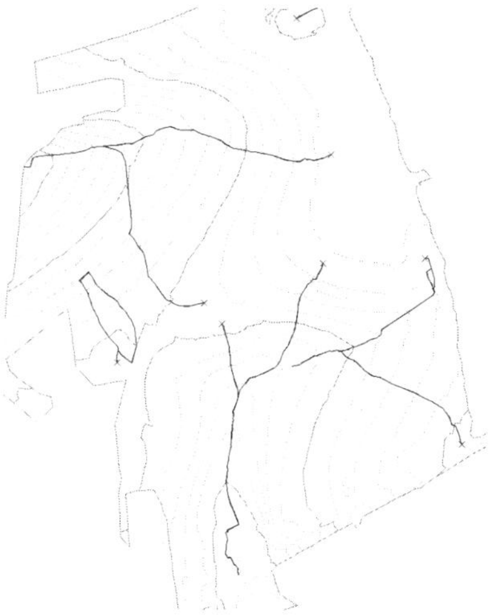

FIGURE 15.2
Surface with water drop paths

10. Select one of the X symbols and its corresponding water drop path. Be sure your surface is not selected.
11. Right-click and choose Select Similar.

    All of the flow paths and corresponding Xs will be highlighted.
12. Remove them by pressing Delete on your keyboard.

Catchments

The first step in any runoff computation is finding the area of the watersheds (also known as drainage basins or *catchments*) that contribute to the flow.

Catchments work with your surface models to determine the area draining to a point you specify. They are also used to find the drainage path within the catchment to compute time of concentration (Tc).

Catchment Groups

You must create a catchment group before creating catchments. Catchment groups are a way to separate pre- and post-development watershed areas. Similar to sites, a catchment in one catchment group will not interfere with objects in another catchment group. Catchment groups do not interact with each other, allowing catchments from differing groups to overlay each other.

Catchment Creation

In Figure 15.3 you see a catchment with all its related features. You will need a grasp of the terminology used in the software before proceeding. It is useful to think about a hypothetical raindrop hitting the catchment to grasp what the various components represent.

FIGURE 15.3
A catchment with several flow path segments

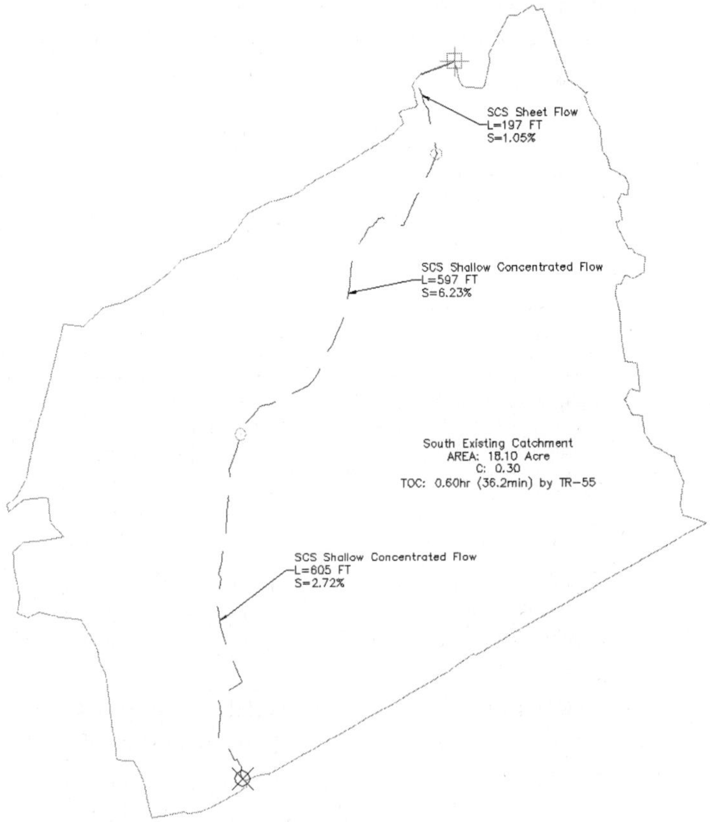

Catchment Boundary The outline represents the catchment boundary, often called the basin divide or watershed boundary. This represents the outermost points of the catchment object. Any raindrops that hit inside this boundary will flow to the same point.

Discharge Point Toward the bottom of the boundary in Figure 15.3 is the discharge point. This is the main point of analysis used for the catchment. All raindrops that hit the basin ultimately end up at this location.

Hydraulically Most Distant Point Near the top of the catchment boundary in Figure 15.3 is a marker indicating the hydraulically most distant point. A raindrop that lands at this point will take the longest to arrive at the discharge point.

Catchment Flow Path In Figure 15.3, the catchment flow path is shown as a dashed line running through the catchment. This linear path represents the course the raindrop takes on its journey, from the hydraulically most distant point to the discharge point. The catchment flow path is used to determine the slope and length for computing the Tc for hydrology calculations.

Flow Path Segments By default, a flow path is created with an average slope through its overall length. Using flow-path segments, you can break up the flow path into smaller pieces with the average slope computed per segment. Each segment can have different computation methods associated with them for Tc computations. The computation methods are as follows:

- SCS Shallow Concentrated Flow
- SCS Sheet Flow
- SCS Channel Flow

Time of Concentration Time of concentration is the time it takes for a drop of rain to travel from the hydraulically most distant point to the discharge point. The default method for determining this value is TR-55. Civil 3D uses the flow type, slope, and user-specified Manning's roughness to determine the time. Alternately, you can input a user-defined time of concentration.

SCS, WHAT? TR, WHO?

TR-55 is short for a technical document put out in 1986 by a branch of the US Department of Agriculture. The agency prior to 1996 was called SCS (Soil Conservation Service), but is now known as NRCS (National Resources Conservation Service). Technical Release 55 "Urban Hydrology for Small Watersheds" is the seminal work behind hydrology computations in the United States.

This document, which has been added to the chapter dataset (which you can download from this book's web page) for your reference, describes the procedures needed for determining the curve number (a.k.a. runoff coefficient), when to apply the different flow types, and how to compute Tc for each flow type.

Civil 3D incorporates the Tc computation directly in the catchment object. However, TR-55 does have some limitations that you need to keep an eye out for in Civil 3D:

- Minimum time of concentration = 0.1 hour (6 minutes).
- If you're using sheet flow, it should be your first flow segment and should not exceed 300′ (91 m). Some newer literature says sheet flow should not exceed 200′ (60 m). Check with your local regulating authority for specific requirements. Typically, sheet flow morphs into shallow concentrated flow before 200′ (60 m).
- Each Civil 3D catchment supports one entry for curve number. However, the curve number is not used for determining the time of concentration. If you have nonhomogeneous land and/or soil types, leave the default, but be sure to compute the composite curve number in the SSA portion of the software.
- If you need to use a method other than TR-55 for Tc, you can specify User-Defined and leave the entry blank until you export to the SSA portion of the product.

Finally, keep in mind that Civil 3D catchments are not dynamic to the surface model. If the surface model changes, you will need to re-create your catchment.

CATCHMENT OPTIONS

You can choose to create a catchment from a surface model or by converting a closed polyline. You will get the best results when defining your catchment by converting a polyline to a catchment area. Manually delineate your watershed areas using the Water Drop command you used in the previous section as a guide.

Create Catchment From Object In most cases, this will be the method you should use for watershed creation. Use Create Catchment From Object if you have watershed data created ahead of time (i.e., from an imported GIS file or existing DWG). You will also need to use the Create Catchment From Object option for watersheds that are adjacent to your project's limits of disturbance. Even if you choose to use an existing polyline as the catchment boundary, you can still use the surface model to determine the flow path location and slope.

Create Catchment From Surface Use the Create Catchment From Surface option if you have a well-formed surface model with adequate data for Civil 3D to compute the watershed area. Surface models with many flat areas, sparse data, or multiple hide boundaries will not work well for this tool. If you do not have good luck with this tool, don't be surprised. Make sure the surface model used for catchment creation is complete before defining a catchment object. The catchment is not dynamically linked to the surface model; therefore, it will not update if changes are made to the surface model.

If you are using the Hydrology tool on an existing ground surface (predevelopment), make sure all surveyed shots and breaklines are added before proceeding. If you are using the tools on a proposed site (postdevelopment), make sure corridor surface models, grading surface models, and any proposed elevation data are pasted together in the surface model used for analysis.

In the following exercise, you will use multiple tools to delineate several watershed areas for a predeveloped site. You will also use the site characteristics to specify Manning's roughness for SCS Sheet flow and SCS Channel flow. The site is covered in a dense grass.

1. Open the `Catchment.dwg` (or `Catchment _METRIC.dwg`) file, which you can download from this book's web page.

   For this exercise you will find it helpful to turn on the Insert Osnap.

2. On the Analyze tab ➤ Ground Data panel, choose Catchments ➤ Create Catchment Group, as shown in Figure 15.4.
3. Name the group **Pre-developed**. Click OK.

FIGURE 15.4
Accessing catchment tools from the Analyze tab

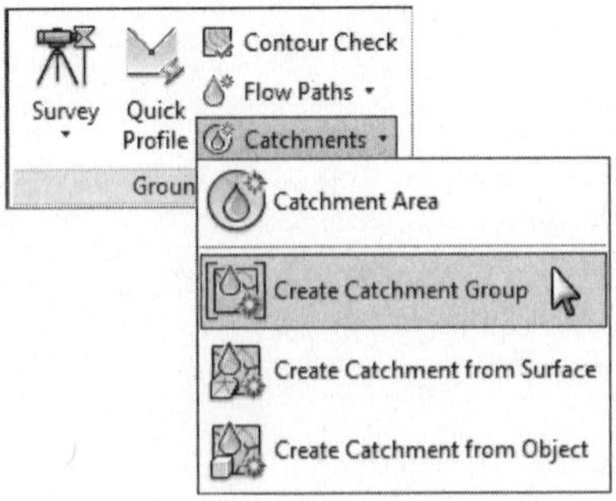

4. On the Analyze tab ➤ Ground Data panel, choose Catchments ➤ Create Catchment From Surface.

5. When you see the prompt `Specify the Discharge Point:`, select the insertion point of the storm structure labeled Existing Inlet (S).
6. When the Create Catchment From Surface dialog appears:
   a. Name the catchment **South**.
   b. Set the surface to Drainage GE-Survey.
   c. Click the Select Structure icon to set the Reference Pipe network structure to Existing Inlet (S). (Alternately, you can right-click to pick structures from a list.)
   e. Set Catchment Label Style to Name Area And Properties.
   f. Set Runoff Coefficient to **0.4**. (This is the unitless runoff coefficient used in the Rational method.)

   If your dialog looks like Figure 15.5, click OK.

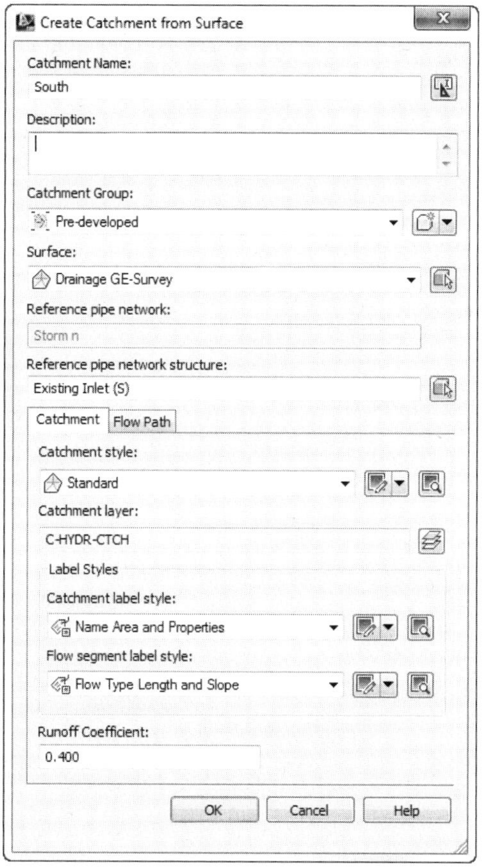

FIGURE 15.5
Create Catchment From Surface settings

After a moment, a red line representing the catchment boundary will form.

7. Press Esc on your keyboard to dismiss the command.
8. On the Analyze tab ➢ Ground Data panel, choose Catchments ➢ Create Catchment From Object.
9. Select the closed polyline north of the first catchment.
10. At the prompt `Select a polyline on the uphill end to use as a flow path (or Press Esc to skip):`, click the eastern end of the blue, dashed polyline that runs across the catchment.
11. In the Create Catchment From Object dialog that appears:
    a. Name the catchment **North**.
    b. Click the Select Structure icon to set the Reference Pipe network structure to Existing Inlet (N).
    c. Set Catchment Label Style to Name Area And Properties.
    d. Set Runoff Coefficient to **0.4**. (This is the same for both metric and Imperial units.)
    e. Uncheck Erase Existing Entities. Your dialog will now look like the image on the left in Figure 15.6.

FIGURE 15.6
Create Catchment From Object settings (left) and the Flow Path tab (right)

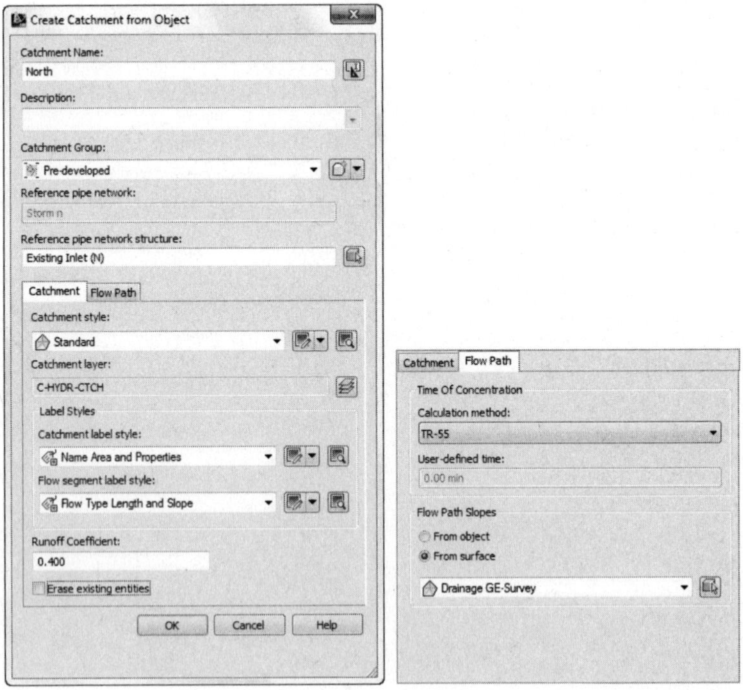

f. Switch to the Flow Path tab.

g. Set Flow Path Slopes to From Surface.

h. Set Surface to Drainage GE-Survey.

Your Flow Path tab will now look like the image on the right in Figure 15.6.

i. Click OK, and press Esc on your keyboard to exit the command.

Your new catchment should resemble Figure 15.7.

FIGURE 15.7
The North catchment and flow path (catchment area label has been moved)

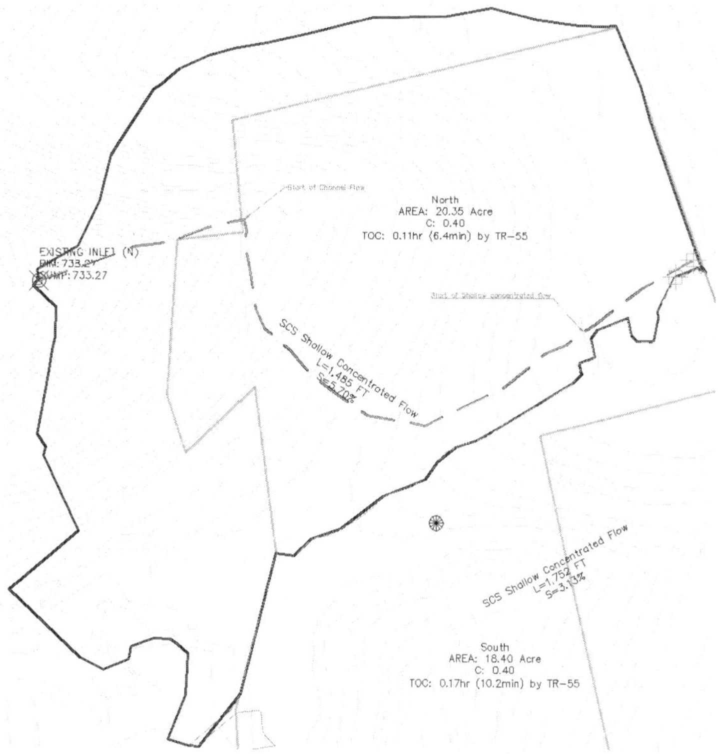

12. Select the North catchment by clicking anywhere on the red boundary, or on the dashed flow path line.

13. Right-click and select Edit Flow Segments, as shown in Figure 15.8.

Panorama will appear showing the continuous path as SCS Shallow Concentrated Flow.

14. Change Surface Type to Short Grass Pasture.

15. Click the plus sign in the upper-left corner to create a flow segment.

FIGURE 15.8
Right-click to access Edit Flow Segments

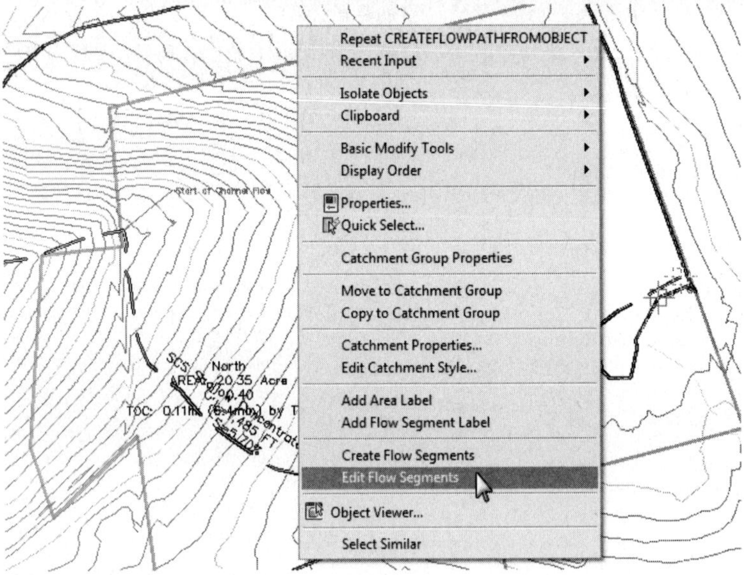

16. Create the new flow segment by clicking at the location on the flow line labeled "Start of Shallow Concentrated Flow."

    The new segment is placed in the Panorama listing from uphill to downhill. Segment 1 is the segment you just created and Segment 2 represents everything else downhill. This new segment represents sheet flow.

17. Change the flow type for Segment 1 to SCS Sheet Flow. Note the input fields change to accommodate this type of flow.

    a. Set the 2Yr-24Hr Rainfall value to **2.9**″ (**74** mm).

    b. Set the Manning's Roughness value to **0.24** (Manning's roughness is a unitless value so this is the same for both metric and Imperial drawings).

       The rainfall information used in this step was found in the PennDOT drainage design manual, which you can check for yourself at `ftp://ftp.dot.state.pa.us/public/bureaus/design/PUB584/PDMChapter07A.pdf`.

       The Manning's roughness for this area is based on coefficients for long grass found in the TR-55 document.

18. Click the plus sign and add the next segment at the abrupt bend in the flow path near the site boundary.

    This is where our flow enters a swale.

19. For Segment 3, change Flow Type to SCS Channel Flow.

  a. Set the Manning's Roughness value to **0.05**.

  b. Set the Cross-Sectional Area value to **27** ft^2 (**2.5** m^2).

  c. Set the Wetted Perimeter value to **28.2′** (**8.6** m).

  Manning's roughness was determined based on an earthen swale from TR-55. The cross-sectional area and wetted perimeter are site-specific measurements of the swale.

  Your Panorama should now look like Figure 15.9. You have used Civil 3D to compute the Tc for the north catchment.

FIGURE 15.9 Three flow segments with different characteristics

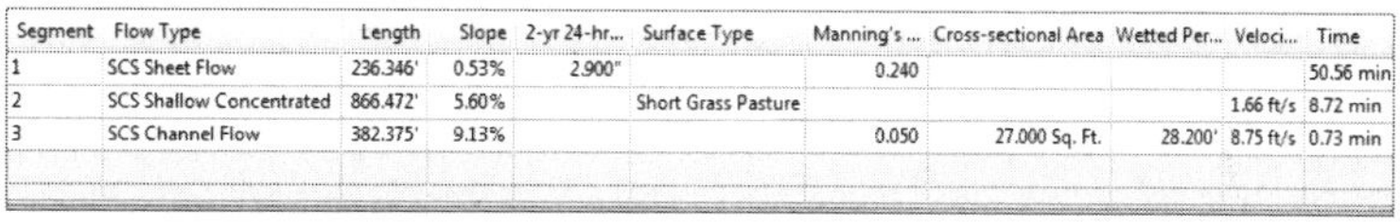

Segment	Flow Type	Length	Slope	2-yr 24-hr...	Surface Type	Manning's ...	Cross-sectional Area	Wetted Per...	Veloci...	Time
1	SCS Sheet Flow	236.346'	0.53%	2.900"		0.240				50.56 min
2	SCS Shallow Concentrated	866.472'	5.60%		Short Grass Pasture				1.66 ft/s	8.72 min
3	SCS Channel Flow	382.375'	9.13%			0.050	27.000 Sq. Ft.	28.200'	8.75 ft/s	0.73 min

20. Dismiss the Panorama by clicking the green check box in the upper-right corner. Save the drawing for use in the next exercise.

Exporting Pond Designs

One of the capabilities of SSA is to evaluate the performance of a detention pond during a storm event. Information about the pond's surface area at various depths must be collected. Civil 3D can easily compute stage-storage information from surface contours.

In Chapter 16, "Grading" you will learn all about creating proposed pond surfaces. In the following exercise, you will prepare a pond surface for export and generate the file needed to pass the information to SSA:

1. Open the file `Proposed Pond.dwg` or `Proposed Pond_METRIC.dwg`.

2. In the Prospector tab of Toolspace, expand Surfaces. Right-click Pond Surface and select Zoom To.

3. Click on one of the contours to bring up the Surface contextual tab.

4. From the Surface Tools panel, click Extract Objects:

  a. Uncheck Major Contour.

  b. Change the Minor Contour Value drop-down to Select From Drawing, as shown in Figure 15.10.

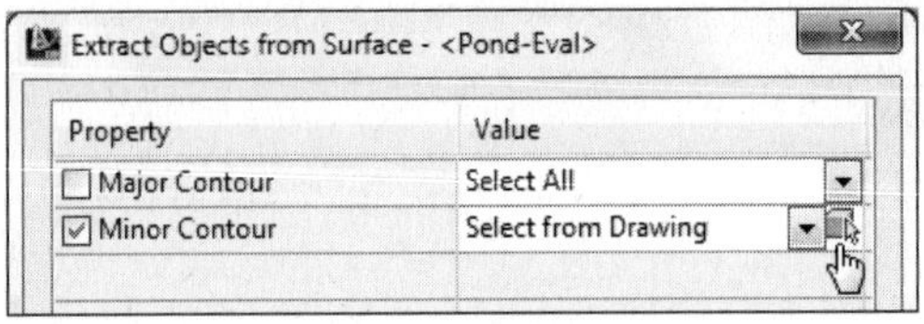

FIGURE 15.10
Use the Extract Objects tool to create a polyline from a contour. This will be used as a surface boundary.

   c. Click the Select From Drawing icon.

   d. Pick the 784.0′ (239.0 m) elevation contour labeled with a red leader at the edge of the basin and press ↵.

5. Click OK.

   A polyline is now generated from this contour.

6. Press Esc to clear the surface selection.

7. Expand Prospector ➢ Surfaces ➢ Pond-Eval Surface ➢ Definition, and right-click Boundaries. Select Add.

8. Name the boundary **Pond Top**, clear the Non-destructive Breakline check box, and click OK.

9. Click the newly extracted polyline.

   The surface is now narrowed down to the portion that is capable of holding water. The contours outside of this area have been omitted from the boundary, as shown in Figure 15.11.

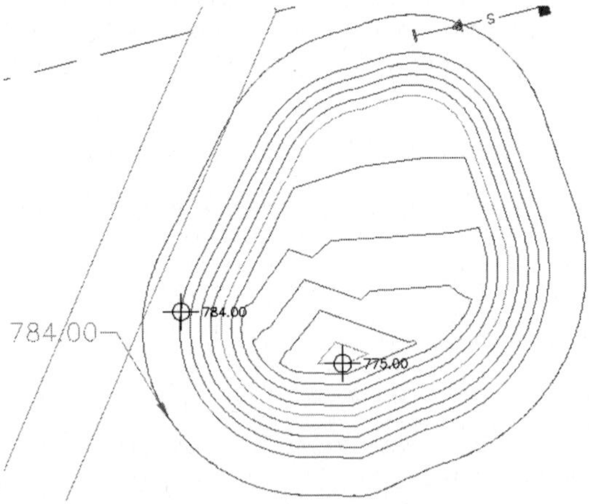

FIGURE 15.11
The surface restricted to inside pond contours

10. Again, select the surface to bring up the contextual tab.

11. From the Analyze panel flyout, select Stage Storage, as shown in Figure 15.12.

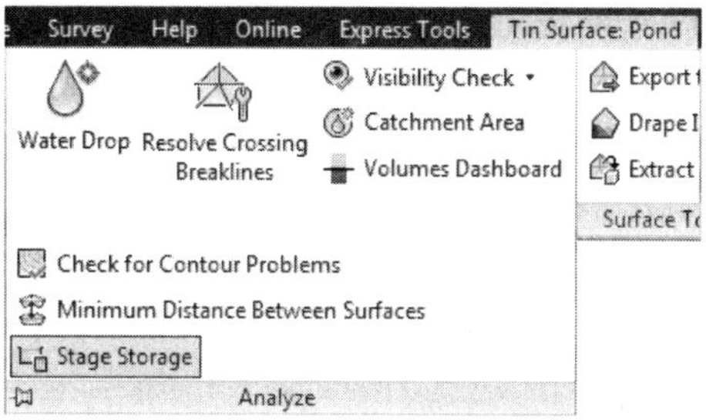

FIGURE 15.12 Access the Stage Storage tool from the Analyze panel of the Surface contextual tab

12. In the Report Title field, type **Pond-Eval**. Leave all other options as default.
13. Click Define Basin.
14. In the Basin Name field, type **Pond1**.
15. Verify that Define Basin From Surface Contours is toggled on, as shown in Figure 15.13.

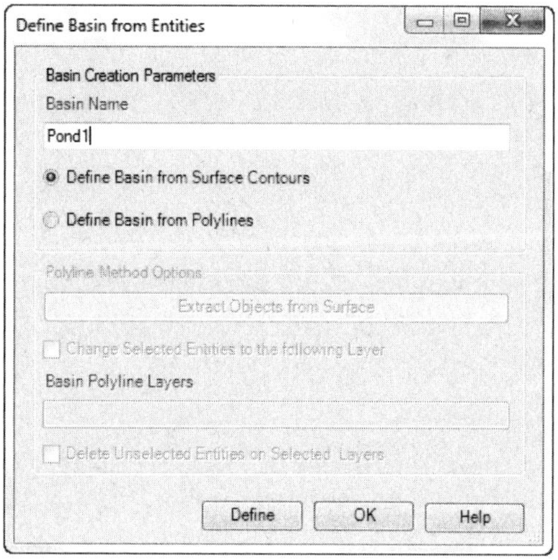

FIGURE 15.13 Define Basin From Entities dialog

16. Click Define.
17. Select the surface by clicking any of the contours.

    Your table result will look like Figure 15.14.
18. Click Save Table, and save the file to your Mastering Civil 3D folder as **Pond-Eval**.

    The file type will be AECCSST, which is a special format used by SSA.
19. Close the Stage Storage dialog by clicking Cancel.

    In the SSA-focused portion of this chapter, you will use the stage-storage table to see if your pond floods during a storm event.

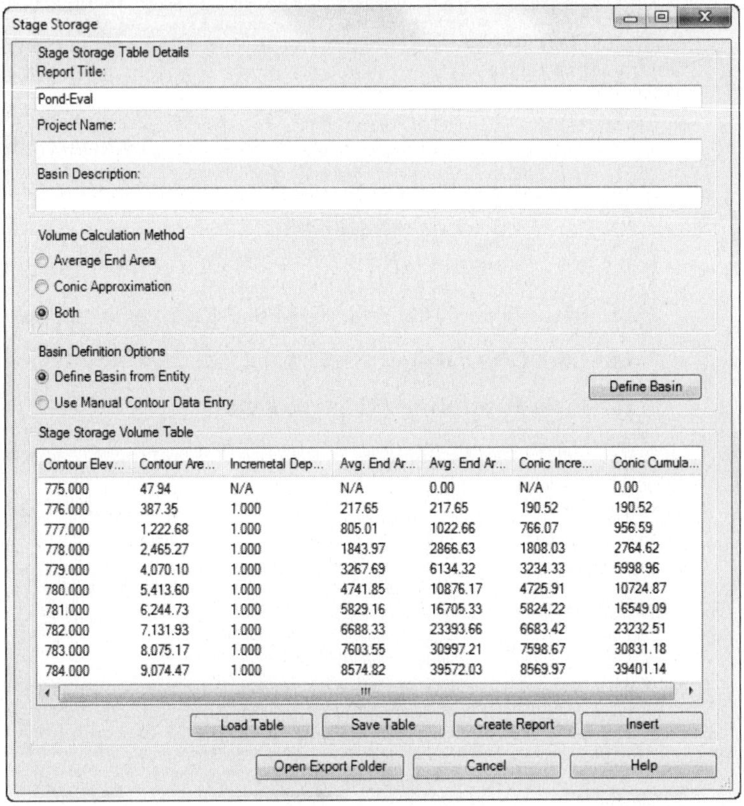

FIGURE 15.14
The Stage Storage tabular results

Exporting Pipes to SSA

When your catchments and your pipe networks are ready to be analyzed, switch to the Analyze tab and select Edit In Storm And Sanitary Analysis. You must have at least one pipe to export.

When you export, the pipe network information, inlets, pipes, and elevations convert over to SSA. Catchment data is shown with area, Tc, and runoff coefficient as data only. The graphic area that you define on the Civil 3D end is not needed.

When you first export a project over, SSA will ask if you would like to create a new project or open an existing project. If you are adding pipes to an existing network that has been analyzed in SSA, click the Open Existing Project radio button and browse to the location of your stored SPF file.

When you create a new project, SSA does a few grunt-work things for you. First, it converts all your junctions and pipes to the SSA format (via Hydraflow's STM file, oddly enough) and asks if you would like to save the log file. Generally, saving the log file is not necessary unless you are having difficulty importing certain objects. In most cases, you can click No to continue.

Once you are in the SSA interface, you can access your active project by clicking the Plan View tab at the top of the window. You will see your DWG file as an underlay to the project you are working on.

How Does SSA Decide What's an Inlet and What's a Manhole?

Buried deep in the bowels of your command settings are the Part Matchup settings.

1. Start a new drawing based on the template of your choice. On the Settings tab, choose Pipe Networks ➤ Commands and double-click the EditInSSA option.
2. Once you are there, expand the Storm Sewers Migration Defaults area.
3. Click the Part Matching Defaults field.
4. Click the ellipsis to enter the Part Matchup Settings dialog.

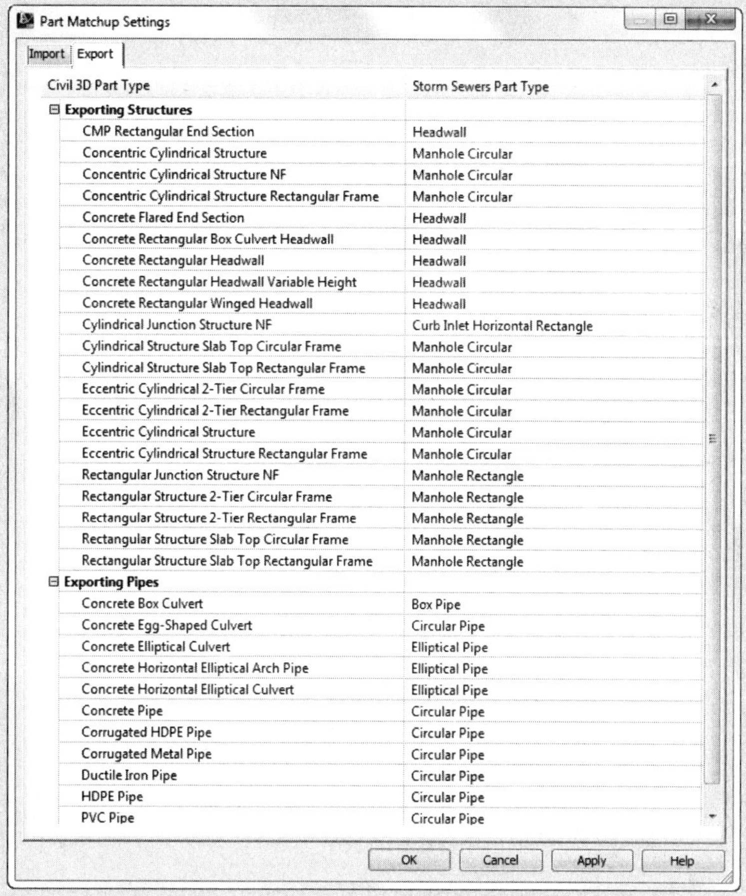

Civil 3D Part Type	Storm Sewers Part Type
⊟ **Exporting Structures**	
CMP Rectangular End Section	Headwall
Concentric Cylindrical Structure	Manhole Circular
Concentric Cylindrical Structure NF	Manhole Circular
Concentric Cylindrical Structure Rectangular Frame	Manhole Circular
Concrete Flared End Section	Headwall
Concrete Rectangular Box Culvert Headwall	Headwall
Concrete Rectangular Headwall	Headwall
Concrete Rectangular Headwall Variable Height	Headwall
Concrete Rectangular Winged Headwall	Headwall
Cylindrical Junction Structure NF	Curb Inlet Horizontal Rectangle
Cylindrical Structure Slab Top Circular Frame	Manhole Circular
Cylindrical Structure Slab Top Rectangular Frame	Manhole Circular
Eccentric Cylindrical 2-Tier Circular Frame	Manhole Circular
Eccentric Cylindrical 2-Tier Rectangular Frame	Manhole Circular
Eccentric Cylindrical Structure	Manhole Circular
Eccentric Cylindrical Structure Rectangular Frame	Manhole Circular
Rectangular Junction Structure NF	Manhole Rectangle
Rectangular Structure 2-Tier Circular Frame	Manhole Rectangle
Rectangular Structure 2-Tier Rectangular Frame	Manhole Rectangle
Rectangular Structure Slab Top Circular Frame	Manhole Rectangle
Rectangular Structure Slab Top Rectangular Frame	Manhole Rectangle
⊟ **Exporting Pipes**	
Concrete Box Culvert	Box Pipe
Concrete Egg-Shaped Culvert	Circular Pipe
Concrete Elliptical Culvert	Elliptical Pipe
Concrete Horizontal Elliptical Arch Pipe	Elliptical Pipe
Concrete Horizontal Elliptical Culvert	Elliptical Pipe
Concrete Pipe	Circular Pipe
Corrugated HDPE Pipe	Circular Pipe
Corrugated Metal Pipe	Circular Pipe
Ductile Iron Pipe	Circular Pipe
HDPE Pipe	Circular Pipe
PVC Pipe	Circular Pipe

The Import tab handles how parts from SSA or Hydraflow behave on the way into Civil 3D. The Export tab handles how SSA or Hydraflow handles parts from Civil 3D. As you can tell from poking around here, there isn't always a perfect match for every part you use.

There are a few quirks and limitations with the intended SSA workflow:

- Pipe networks that contain parts from multiple-part families (for example, a pipe network may contain a part from the HDPE family and another part from the Concrete family) will reimport back into Civil 3D with only one part family.
- To prevent SSA and Civil 3D from changing your pipe sizes, make a single part family for all circular pipes. Be sure that the "master" part family contains all possible sizes. Once it is reimported back to Civil 3D, use Swap Part to change back to the original part family type.
- For advanced users, there is a super-secret XML file located in `C:\Program Files(x86)\Autodesk\SSA 2013\Samples\Part Matching`. For more information about using and editing this file, see the Part Matching help file.
- Culverts are treated the same as a single pipe in SSA. Once you export a culvert into SSA, you'll need to set entrance and exit conditions.

A little part tune-up will be needed after the round-trip; however, the time savings in using the export commands are still very much worth it.

Storm and Sanitary Analysis

SSA can perform a number of water- and wastewater-related tasks. In the remainder of this chapter, you will be going through several examples of what SSA can do, but you are just seeing the tip of the proverbial iceberg.

In this section, you will focus on the workflow between Civil 3D and SSA. First, you will learn how to create subbasins (catchments) directly in SSA. You'll also see examples of the Rational method and TR-55, and learn about SSA reports.

Guided Tour of SSA

SSA is a separate program that is optionally installed with Civil 3D. To get into SSA you can launch it from the desktop icon, from the Start menu, or from the Analyze tab in Civil 3D.

SSA is not at all like AutoCAD®. It is a true modeling software program that shows a schematic of your design in a plan view. On the left side of the screen you will see the data tree (Figure 15.15). Like Prospector in Civil 3D, the SSA data tree gives you access to tools and information about objects.

The easiest way to add new structures, pipes, or catchments to the project is to do so from the Elements toolbar along the top of the screen. Here, you will find everything you need to build a project from scratch if you did not start out in Civil 3D.

The next stop on our tour is the Project Options dialog (Figure 15.16). Access this dialog by double-clicking on the Project Options listing in the data tree. The General tab of the Project Options dialog is where you set your units, hydrology method, routing method, and force main equation. Depending on the scope of your project, you may not need to worry about every setting.

FIGURE 15.15
Overview of the SSA interface

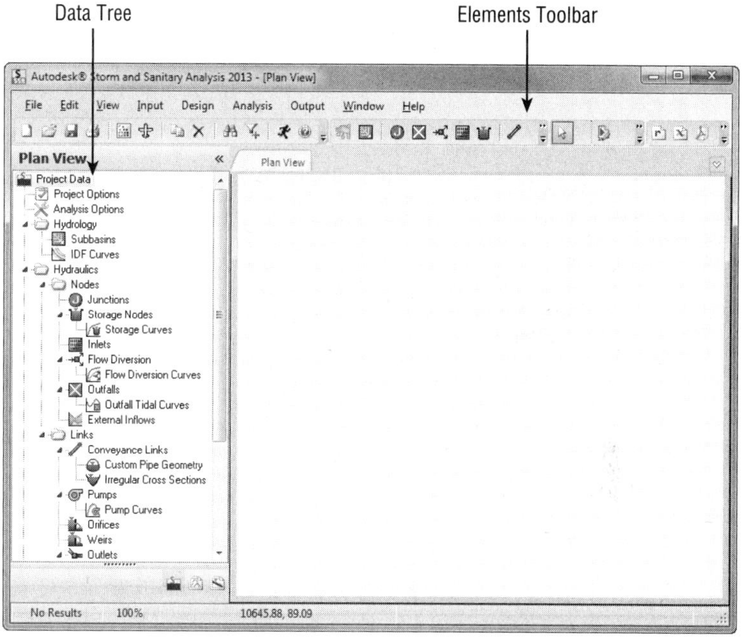

A Few Hints for Working in SSA

Since this is not traditional CAD, the program behaves differently than you might expect. This list contains a few hints to help you along the way:

- Make sure you download and install the Civil 3D 32-bit object enabler. SSA needs the object enabler to see Civil 3D objects such as pipes, structures, and nodes. Since SSA is a 32-bit application under the hood, you should download the 32-bit version of the object enabler even if your computer is on a 64-bit operating system.
- Unlike AutoCAD, SSA keeps you in a command until you press Esc or click the Select Element tool from the toolbar.
- Be mindful of the tabs along the top of the screen. To dismiss them, use the red X that appears on the tab itself. It is a common mistake to dismiss the program accidentally by clicking the incorrect red X.

- Don't forget to save! Unlike CAD, there is no auto-save.
- Work upstream to downstream.
- There is no Undo command in SSA. Luckily, adding and removing elements is a simple process. If something gets goofed up beyond repair, you could always close without saving.

It may take a while to get used to, but SSA is a powerhouse analysis tool.

In one of the upcoming examples, you will use the Rational method with TR-55 Tc calculations. Since you are calculating runoff only, you do not need to set the other project options. Once you have finished entering the options shown in Figure 15.16, click OK.

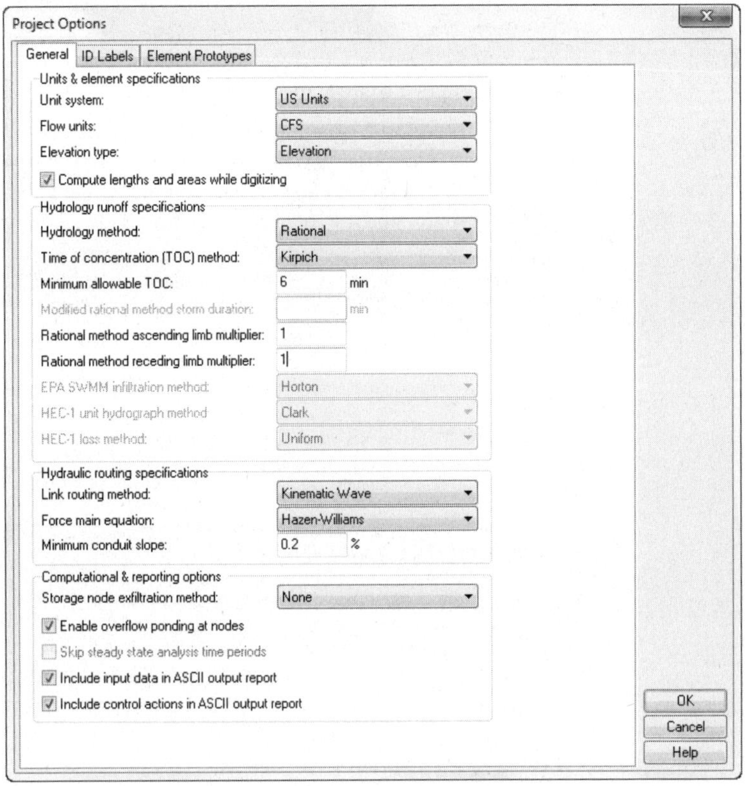

FIGURE 15.16
SSA Project Options dialog

Whenever you are designing in SSA, always work upstream to downstream. If your system includes a pump, you should work from highest elevation to lower elevations.

An SSA project consists of three main object types: subbasins, nodes, and links:

Subbasins Subbasins are important for runoff computations, but are not mandatory if you are inputting flows manually. Subbasins can be defined directly in SSA or brought over as data from Civil 3D. SSA can also turn an SHP file directly into a subbasin.

Nodes Nodes are where all calculations are done in SSA. If you want information along a pipe, for example, you will need to place a junction node at the location of interest.

Nodes are the workhorses of SSA, and can take several forms:

Junction Node When SSA converts structures from Civil 3D, it assumes a junction node by default. These can be used as manholes, null structures, water, or pollutant infiltration or treatment points. Junctions are frequently used to consolidate links that would flow to the same outfall. Junctions can connect many links, but outfalls cannot.

Outfall Node Outfall node is the end of your system. An outfall can be a culvert discharge location, where a proposed sewer ties into an existing one, or where sanitary sewer enters a treatment facility. An outfall can be attached to only one link.

Flow Diversion Node If a node has two outgoing links, you will need to use a flow diversion. Flow diversions most commonly represent multiple outflow points on a detention basin, such as a weir or orifice. They can also represent flow diversion in a combined sewer system or anywhere an overflow check is in use.

Inlets Inlets are used in storm systems and can collect flow from a subbasin. You can also manually enter flows.

Inlets are either on grade or in sag. For inlets on grade, a downstream link must be specified. The downstream link (i.e., curb and gutter between storm inlets) tells the software where water goes if the current inlet cannot handle the rainfall.

Inlets in sag are assumed to pond in a storm event and do not need downstream links associated with them.

Storage Nodes Anywhere water is detained on your site, a storage node can model the scenario. Storage nodes can represent ponds, wet wells, underground storage, or a residential rain barrel.

Later in this chapter you will use a storage node to model headwater effects on a culvert.

Links Any time water is moved from place to place, it does so via links. Links connect nodes to each other. Links represent pipes, channels, curbs and gutters, or culverts.

Pumps Pumps can be used to push water uphill. They can be built to start functioning when the water at a storage node reaches a certain depth using operational controls. When using a pump link, you can work in design mode when specific pump data is not available. As pump data becomes available, you should enter pump curve data to tell SSA how much "oomph" the pump has.

Weirs and Orifices Weirs and orifices both allow water to flow out of a storage structure at a controlled rate.

Outlet An outlet is a node type that can be used where weirs or orifices are too simple to model discharge from a storage node. An outlet can have a rating curve or discharge function associated with it.

Hydraflow, We Haven't Forgotten You

When it comes to flexibility in storm design and routing, SSA beats the pants off Hydraflow any day. In general, there is a lot of overlap between the two products. For basic storm sewer calculations, either product will work great. When your designs get more complex and require hydrograph and storage routing with infiltration and exfiltration, SSA has a much better solution.

You can still access all the Hydraflow functions from the Analyze tab ➤ Design panel flyout.

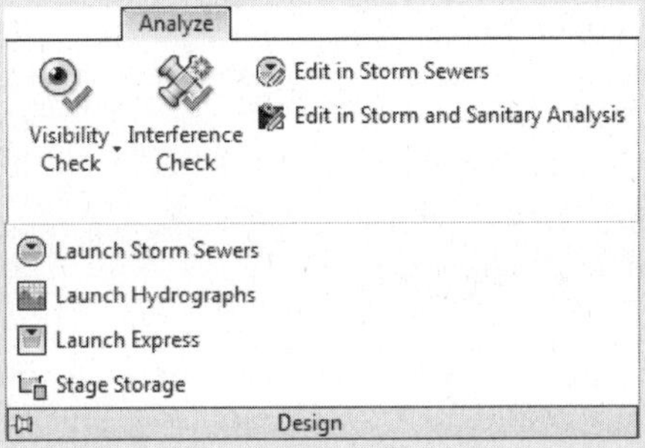

Hydraflow is still part of the picture when it comes to exporting and importing data to and from Civil 3D. Hydraflow's file format does a much better job than LandXML at keeping the data integrity of your pipes and structures.

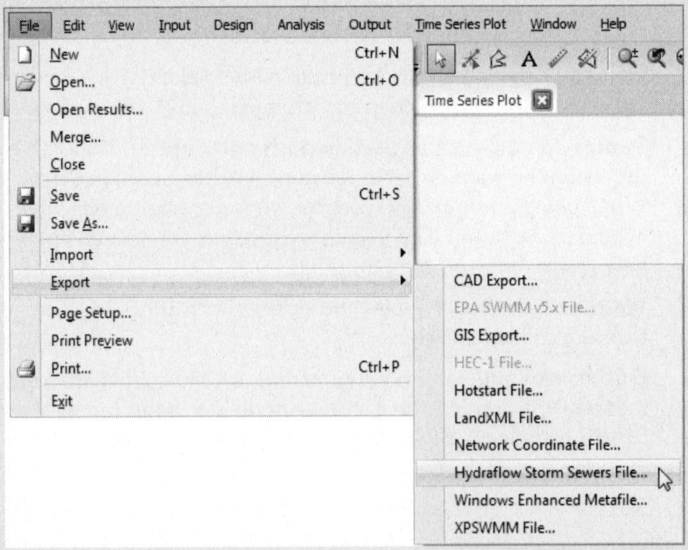

When exporting from SSA, choose File ➤ Export ➤ Hydraflow Storm Sewers File for the best path back to Civil 3D.

Hydrology Methods

SSA is capable of computing rainfall runoff using any of the following methods:

Rational Method The Rational method is used to approximate the peak discharge (or flow, Q) from an impervious area (A) using the tried and true Q=ciA. This is the most common method used for sizing sewer pipes.

Imagine yourself outside in a rainstorm, staring at a single point on your driveway (assuming your driveway is impervious and on a nice, even slope). At first, the water on your driveway is coming right from the sky. After a while, water flows in from rain falling elsewhere in your neighborhood. Now imagine that the storm stops as soon as raindrops from the farthest point in your neighborhood reach the point on your driveway.

The Rational method is based on the assumption that it takes the same amount of time for those farthest raindrops to get to you as they do to run off your driveway completely.

The first step to using the Rational method is to delineate the drainage basin, as described earlier in this chapter, and assign composite runoff coefficients. SSA needs an intensity-duration-frequency (IDF) curve for the study area to compute the rainfall.

Modified Rational and DeKalb Rational These hydrology methods are similar to the Rational method, except that they are intended to approximate the storage volume needed on small, simple detention basins. Instead of the storm duration equaling two times the time of concentration, the storm duration is longer. The longer storm duration results in a larger volume than the Rational method but results in a lower peak discharge.

The same information is needed for the Modified and DeKalb Rational methods as for the Rational method, and you must have a value in the Project Options dialog for Modified Rational Storm Duration.

Both the Rational method and the Modified Rational method are intended for small watersheds, usually less than 20 acres (8.1 hectares).

SCS TR-55 Method and TR-20 Method More gifts from the USDA, the SCS TR-55 runoff method is a more complex method that is appropriate for study areas up to 2,000 acres (809 hectares).

For the SCS TR-55 and TR-20 methods, SSA contains tools for looking up computation variables. You will find a curve number (CN) lookup table inside the subbasin. SSA will even help you determine the composite curve number. The TR-55 hydrology method is also where you will use SSA rain gauges.

HEC-1 The US Army Corps of Engineers' Hydraulic Engineering Center has a computer program called HEC-1 for hydrology calculations. The HEC-1 method in SSA duplicates the results from the Army Corps' program. Their method is used in situations where there is an existing body of water, such as a stream. For an explanation of this method directly from the source, go to `http://www.hec.usace.army.mil/publications/TrainingDocuments/TD-15.pdf`.

EPA-SWMM The US Environmental Protection Agency (EPA) Storm Water Management Model (SWMM) method is used for developed watershed areas and can take into account many more factors than the other analysis methods. SWMM allows for localized ponding effects, which can be entered into the subbasin physical properties area. This method also allows for the effects of existing soil moisture, evaporation, and snowmelt in the system. When using EPA-SWMM, be sure to set all the methods you intend to use for infiltration and exfiltration in the Project Options dialog.

In the following example, you will work through a simplified example to get the feel for the SSA interface:

1. Launch SSA from the stand-alone icon in your computer's Start menu or from the desktop. Open the file `SSA Intro.spf` or `SSA Intro_METRIC.spf`, which you can download from this book's web page.
2. In the SSA interface, choose File ➢ Import ➢ Layer Manager (DWG/DXF/TIF/More), as shown in Figure 15.17.

   This will bring up the Layer Manager, where you can import a DWG as a background.

FIGURE 15.17
To show the AutoCAD information as an underlay, use the Layer Manager tool

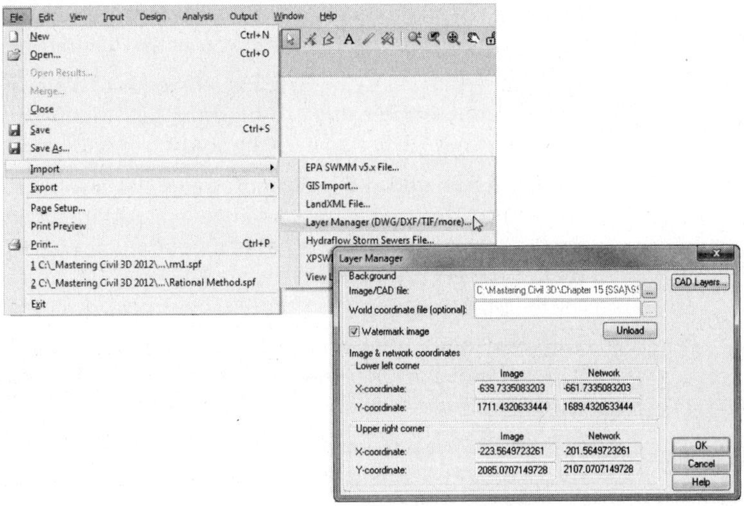

3. Click the ellipsis next to the Image/CAD File field.
4. Browse to the file `SSA Intro.dwg` (or `SSA Intro_METRIC.dwg`) file, which you can download from this book's web page.
5. Place a check mark next to Watermark Image and click OK.

6. Click the green Add Subbasin button from the elements toolbar along the top of the window.
7. Using this tool, click around the boundary lines that represent the example watersheds.
8. When you are close to completing a shape, right-click and select Done to close the shape.

   There are no snap or curve drawing tools in SSA. Digitize as closely as possible using multiple small segments. If you make a mistake, you can right-click and select Delete Last Segment.

   When you have completed both shapes, your graphic will resemble Figure 15.18.

FIGURE 15.18
New subbasins in SSA

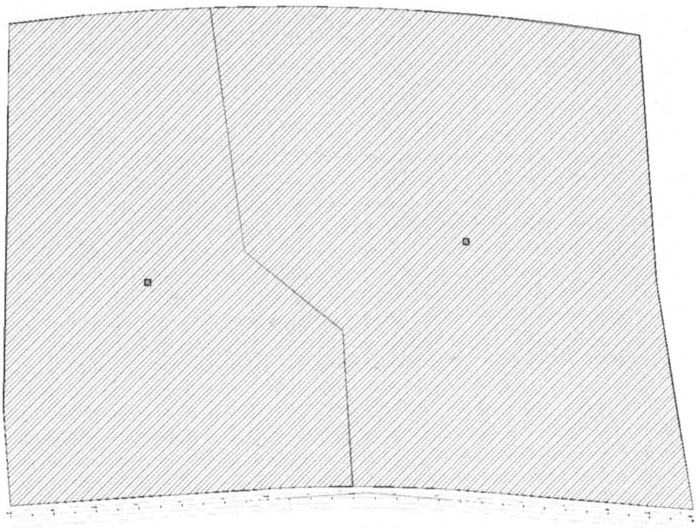

9. Switch to the Inlet tool by clicking the Inlet button at the top of the SSA window.
10. Working upstream to downstream, (left to right in this case), place two inlets on the south side of the site.

    An X has been placed at each inlet location to help you locate the recommended positions.
11. When you have finished placing inlets, press Esc on your keyboard or click the arrow icon in SSA to get back to Selection mode.

    If you accidentally place more inlets than you wanted, you can right-click the wayward inlet while in Selection mode and select Delete from the menu.
12. Double-click one of the inlets to enter the Inlets dialog:
    - **a.** Rename the first inlet you placed to **West Inlet**.
    - **b.** Set Number Of Inlets to **2**.
    - **c.** Change Invert Elevation to **500′** (**152.4** m).
    - **d.** Change Rim Elevation to **510′** (**155.4** m).
    - **e.** Use the table at the bottom of the dialog to switch to the second inlet, and rename the second inlet to **East Inlet**.
    - **f.** Set Number Of Inlets to **2**.
    - **g.** Change Invert Elevation to **485′** (**147.8** m).
    - **h.** Change Rim Elevation to **495′** (**150.9** m).

    Leave all other settings at their defaults, as shown in Figure 15.19, and close the Inlets dialog.

FIGURE 15.19
Setting inlet properties

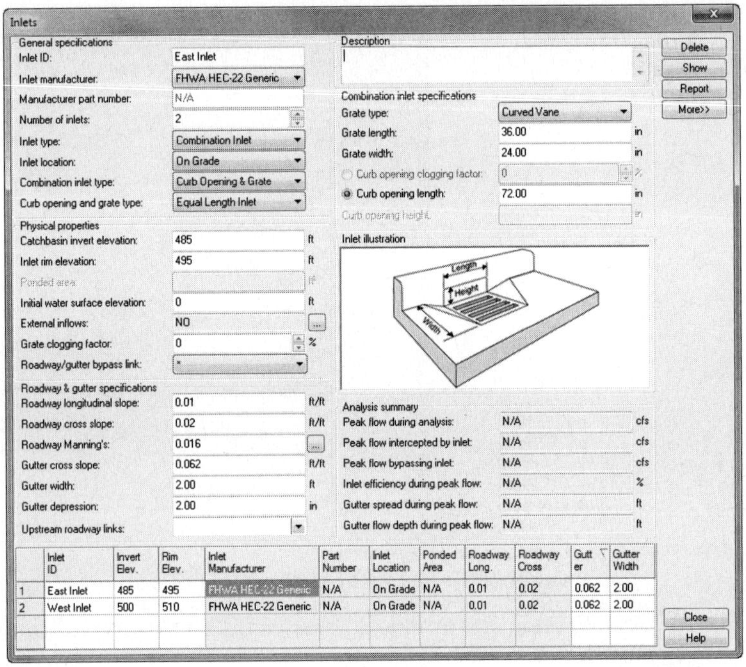

13. Click the Outfall icon. Click to place the outfall structure in the location marked with a circle at the southeastern corner of the site.
14. Press Esc to return to the selection tool.
15. Double-click the new outfall to enter the Outfalls dialog:
    a. Set the Invert Elevation value of the outfall to **480′** (**146.3** m).
    b. Set the Boundary Condition type to Free, as shown in Figure 15.20.
    c. Leave all other options at their defaults, and click Close.

16. Switch to the Conveyance Link tool.
17. Working from left to right across the project, click the upstream inlet. (Resist the temptation to click and drag.)
18. Connect the first inlet with a straight segment to the second inlet.
19. Connect the second inlet to the outfall.

    At this point, your SSA file should resemble Figure 15.21.
20. Press Esc to return to the selection tool.

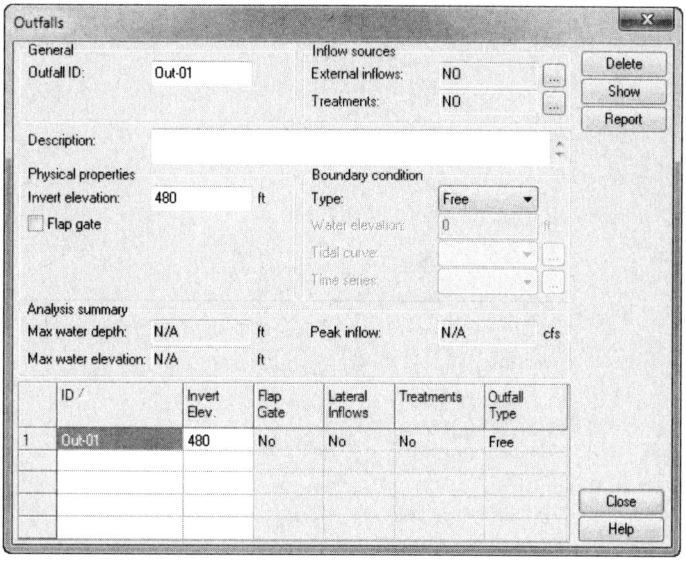

FIGURE 15.20
Setting the outfall elevation

FIGURE 15.21
Pipes connecting inlets to each other and the outfall

21. Double-click the first conveyance link you created.

You will see the Conveyance Links dialog. The easiest place to rename the pipes is in the small grid at the bottom of the box, as shown in Figure 15.22. When your cursor is in the field to rename a pipe, a dark black square will appear on the pipe in the main graphic, helping to indicate which pipe you are editing. Your initial numbering may vary, so you will rename the pipes in the steps that follow.

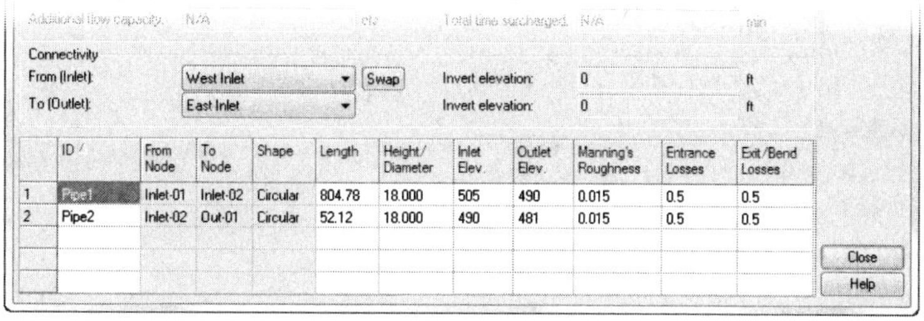

FIGURE 15.22
The Conveyance Links dialog is the main pipe and channel control tool

22. Rename the first link to **Pipe1**. Rename the second conveyance link to **Pipe2**, and then click Close.

    So far, you've told SSA what your drainage areas are and you've established a pipe network. Next, you need to tell SSA which subbasin drains to which inlet.

23. Switch to the selection tool by pressing Esc if it is not already active.

24. Right-click the icon that represents the western subbasin and select Connect To, as shown in Figure 15.23. If you receive a dialog box message with the "Connect To" help message, clear the check box and click OK.

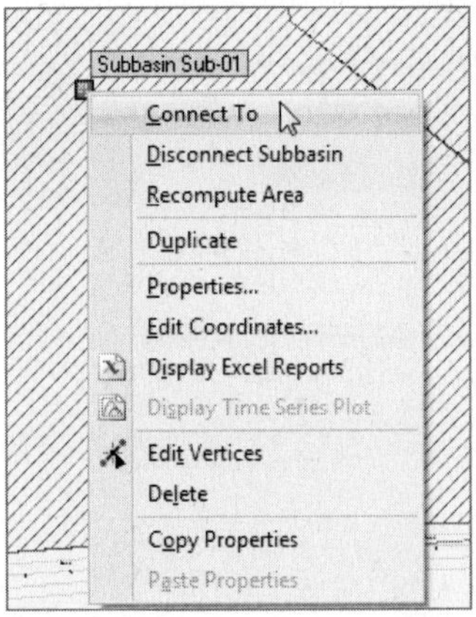

FIGURE 15.23
Connecting a subbasin to an inlet

25. Connect the subbasin to the western inlet by clicking the inlet symbol in plan view.

    A line will form to indicate the connection was successfully created.

26. Repeat the process for the eastern subbasin and inlet.

    At the end of the process your SSA screen will resemble Figure 15.24.

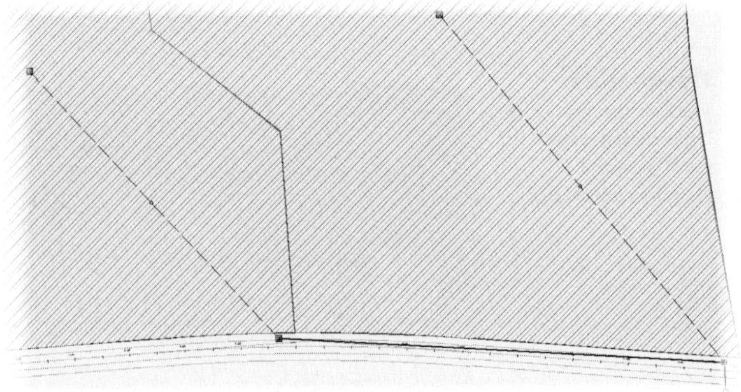

FIGURE 15.24
Subbasins assigned to inlets

Before SSA can run any analysis on the project, you must make one last crucial connection. Both of these inlets are on grade, which means water will flow toward them, but in a big storm, some water may flow past into the next downstream inlet. The flow along the gutter of the road between inlets is called a *bypass link*.

27. Click the Add Conveyance Link button.

28. Connect the west to the east again, but this time follow the curve of the road.

This conveyance link represents the gutter flow of any water that doesn't make it into the west inlet and continues on to the eastern, downstream inlet.

29. Add a conveyance link between the east inlet and the outfall.

To visually differentiate the culvert from the bypass link, you can put a small kink in the line, as shown in Figure 15.25. When you modify link properties, you will have the opportunity to compensate for the change in length this technique causes.

FIGURE 15.25
The bypass link between the east inlet and the outfall

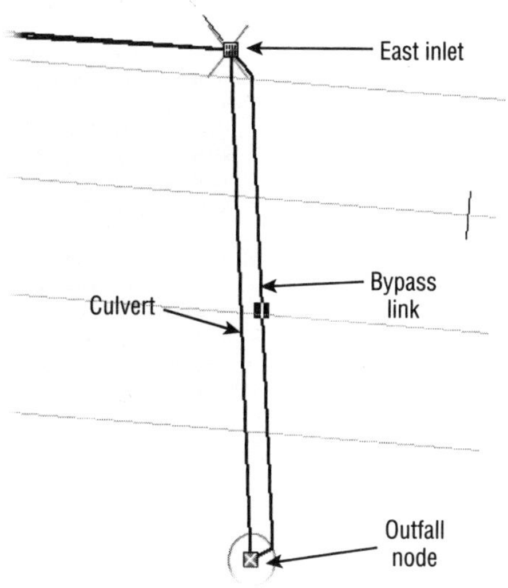

30. Rename the new links to **Overland1** and **Overland2**, respectively.

Overland1 and Overland2 represent surface flow along the gutter and over the road. For this example, you will approximate the flow using a rectangular open channel.

31. Set the shapes of both Overland1 and Overland2 from Pipe to Open Channel, as shown in Figure 15.26.

a. Set Height to **0.5**′ (**0.15** m).

b. Set Width to **5**′ (**1.5** m).

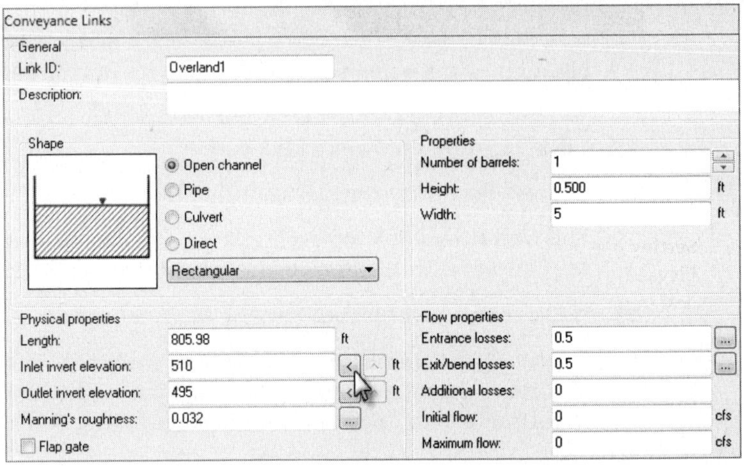

FIGURE 15.26 Set the channel type and width as shown. Use the arrows next to the elevation fields to force the correct elevation.

Now that these conveyance links are set to Open Channel, SSA recognizes that they must be surface-flowing channels. In the next step you will update the curb and gutter elevation to reflect the top of the inlets.

32. For both Overland1 and Overland2:

    a. Click the arrow button next to Inlet Invert Elevation.

    b. Click the arrow button next to Outlet Invert Elevation.

    The elevations should update to reflect the rim elevations of the inlets. (You set these values back in step 12.)

33. Using the tabular input at the bottom screen, enter the values for Diameter, Inlet Elevation, Outlet Elevation, and Manning's Roughness, as shown in Figure 15.27. Leave all other options at their defaults, and click Close when you are done.

English

	ID	From Node	To Node	Shape	Length	Height/ Diameter	Inlet Elev.	Outlet Elev.	Manning's Roughness
1	Overland1	Inlet-01	Inlet-02	Rectangu	805.02	0.500	510	495	0.025
2	Overland2	Inlet-02	Out-01	Rectangu	50	0.500	495	480	0.025
3	Pipe1	Inlet-01	Inlet-02	Circular	804.78	24.000	505	490	0.015
4	Pipe2	Inlet-02	Out-01	Circular	52.12	24.000	490	481	0.015

Metric

	ID	From Node	To Node	Shape	Length	Height/ Diameter	Inlet Elev.	Outlet Elev.	Manning's Roughness
1	Overland1	Inlet-01	Inlet-02	Rectangu	245.37	1.000	155.45	150.9	0.025
2	Overland2	Inlet-02	Out-01	Rectangu	15.24	1.000	150.9	146.3	0.025
3	Pipe1	Inlet-01	Inlet-02	Circular	245.2	600	153.92	149.35	0.015
4	Pipe2	Inlet-02	Out-01	Circular	15.9	600	149.35	146.6	0.015

FIGURE 15.27 Set the pipe and gutter elevation data as shown

All the pieces are here; next you will verify the connections to the pipes, inlets, outfalls, and curb flows.

34. Double-click one of the inlets to return to the Inlets dialog.

- **a.** For West Inlet, set Roadway/Gutter Bypass Link to Overland.
- **b.** For East Inlet, set Roadway/Gutter Bypass Link to Overland2, as shown in Figure 15.28.
- **c.** Click Close when complete.

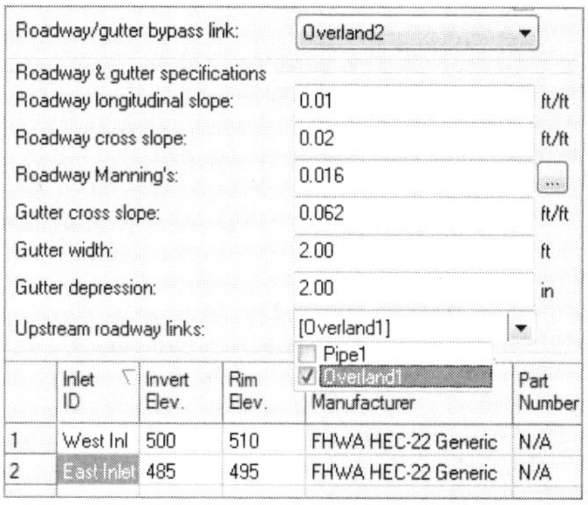

FIGURE 15.28 Setting Invert Elevation and design settings

Now that all the elevations are set, you must tell SSA how much rain you expect to see in your system.

35. In the data tree on the left side of the SSA window, in the Hydrology branch, double-click IDF Curves, as shown in Figure 15.29.

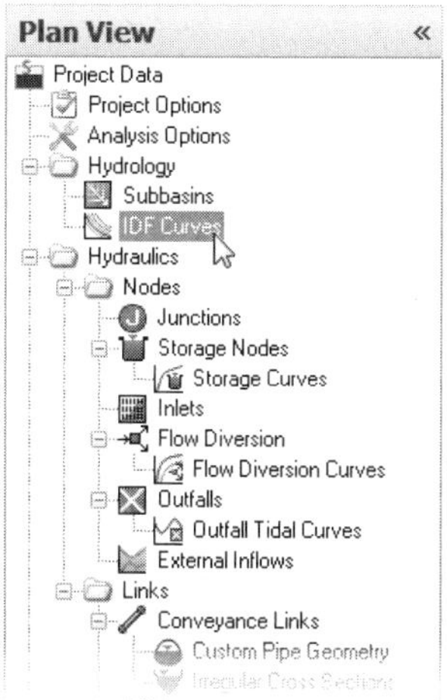

FIGURE 15.29 IDF Curves are found in the data tree

36. Click Load, and browse to the file `York Co PA.idfdb` (or `York Co PA_METRIC.idfdb`) and choose Open.

This is the rainfall data for our example file, which you can download from the book's web page.

37. After loading the file, set the ID to **York, PA, USA**, as shown in Figure 15.30. Click Close.

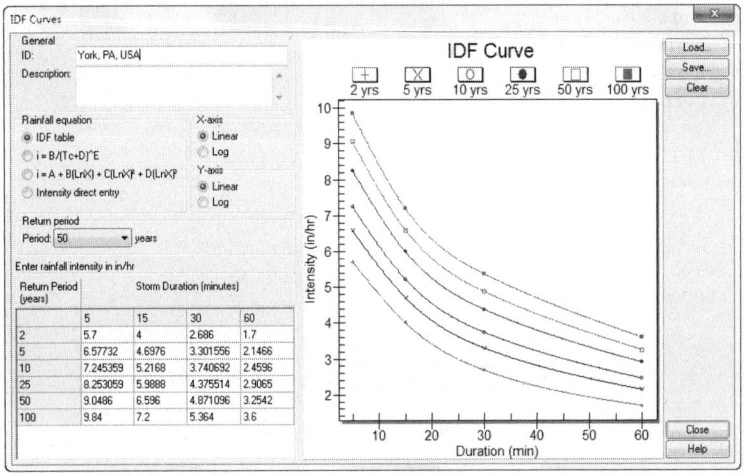

FIGURE 15.30
IDF curves entered in SSA

38. In the data tree on the far left, double-click Project Options.
    - **a.** Set Hydrology Method to Rational.
    - **b.** Set Time Of Concentration (TOC) Method to Kirpich.
    - **c.** Set your minimum allowable TOC to **6** minutes.
    - **d.** When your Project Options dialog resembles Figure 15.31, click OK.

39. Double-click Analysis Options from the data tree.
    - **a.** On the General tab, set the End Analysis On value to **02:00:00**.

      Your dates will vary from what is shown in Figure 15.32, but the important thing is that the duration of the analysis is long enough to accommodate the Rational method.
    - **b.** On the Storm Selection tab, set Return Period to 25 years for a single storm analysis.
    - **c.** Click OK when complete.

    The last step before you are ready to run the design is setting the subbasin information for use with the Kirpich method of time of concentration. The Kirpich method calculates Tc based on the average slope and area characteristics of the site.

40. Double-click the green subbasin icon located in the centroid of the western subbasin.
    - **a.** For the western subbasin, set the flow length to **750′** (**228.6** m), and set the average slope to **3**%.
    - **b.** Switch to the Runoff Coefficient tab and set the runoff coefficient to **0.37**.
    - **c.** Switch to the eastern subbasin by selecting its row at the bottom of the dialog.
    - **d.** Set the eastern subbasin flow length to **700′** (**213.4** m), and set its average slope to **2.5**%.
    - **e.** Switch to the Runoff Coefficient tab and set the runoff coefficient to **0.37**.
    - **f.** When your dialog resembles Figure 15.33, click Close.

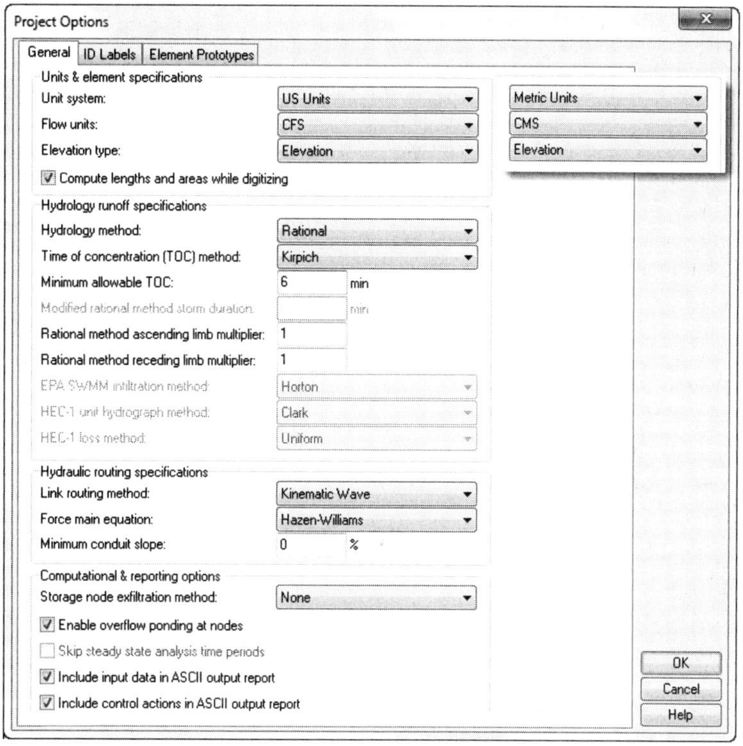

FIGURE 15.31 Verify your settings and click OK (metric units are shown for illustration purposes).

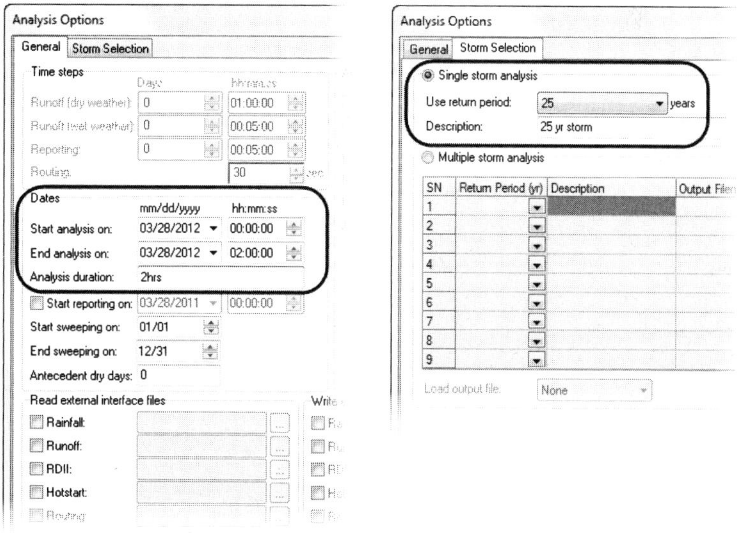

FIGURE 15.32 Setting the design storm in Analysis Options

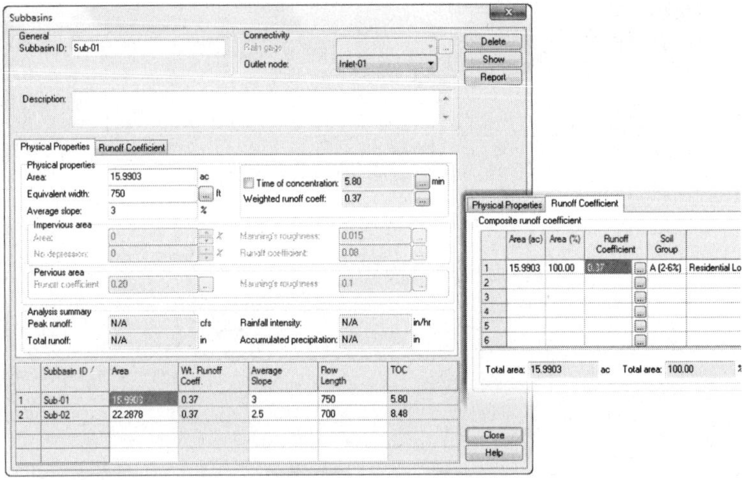

FIGURE 15.33
Final preparation of the subbasin data

41. Click Perform Analysis.

    After a moment, the analysis will complete.

42. Click OK.

    In the SSA graphic, you will see a heavy, blue line indicating that conveyance link Overland1 is flooded. Later on in the chapter you will look at reporting capabilities to see the details of this predicament.

43. Save the file as **`SSA1.spf`** and close. Close SSA in preparation for the next exercise.

From Civil 3D, with Love

Earlier in this chapter, you saw some of the preparations that must take place before you click the Edit In Storm And Sanitary Analysis button. Once inside SSA, a few more pieces of information need to be added before SSA can do its analysis.

In the following example, you will see what becomes of catchment objects and pipes once they are imported into SSA:

1. From Civil 3D, open the file `C3DtoSSA.dwg` (or `C3dtoSSA_METRIC.dwg`). You can download this file from the book's web page.

2. Select the Analyze tab ➤ Design panel and click Edit In Storm And Sanitary Analysis.
3. Verify that there is a check mark next to the STORM pipe network.
4. Click OK in the Export To Storm Sewers dialog, as shown in Figure 15.34.
5. SSA launches. Click OK to create a new project.

    After a moment, SSA will let you know that you have successfully imported the Hydraflow Storm Sewers file.

6. Click No—you do not need to save the log file.

FIGURE 15.34
The pipe network on its way to SSA

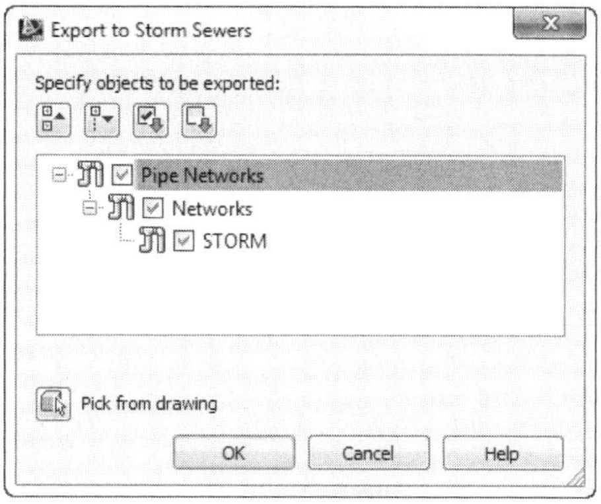

7. At the top of the SSA window, click Plan View.
8. Examine the inlets and catch basins that have been imported.

   Notice that an offsite outfall has been added to each inlet. This is because you don't model the overland links in Civil 3D. When inlets are first imported into SSA, the program assumes they are on grade. Water that bypasses on grade inlets needs to go somewhere, so SSA automatically throws in an outfall to take care of the excess water. In the next steps you will remove the outfalls and create the overland links.
9. Right-click on the outfall that has been added to Inlet Structure - (1).
10. Click Delete, as shown in Figure 15.35.

FIGURE 15.35
Right-click and click Delete to remove the extraneous outfall.

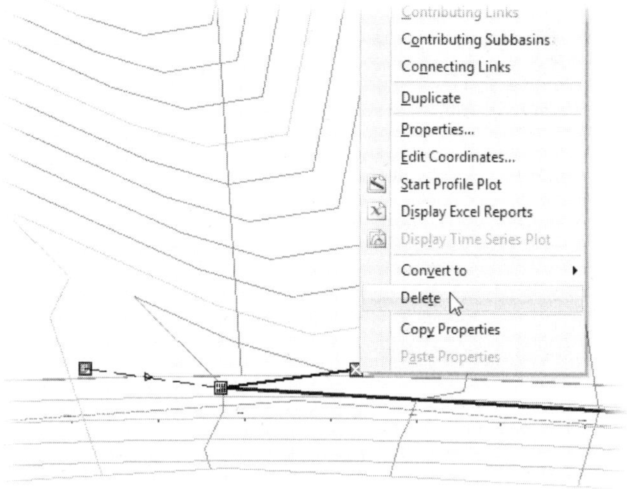

11. Click Yes to confirm the deletion.
12. Right-click and delete the outfall to the far right. Click Yes to confirm the deletion.

13. Add overland conveyance links similar to the previous exercise:
    a. Add a link between Inlet Structure - (1) and Inlet Structure - (2).
    b. Add a link between Inlet Structure - (2) and the outfall.
14. Double-click one of the conveyance links to edit the links in tabular form.
15. Set the diameters of Pipe - (1) and Pipe - (2) to **24"** (**600** mm), as shown in Figure 15.36.

FIGURE 15.36
Design values for links and pipes

English Units

	ID	From Node	To Node	Shape	Length	Height/ Diameter	Inlet Elev.	Outlet Elev.	Manning's Roughness	Entr Los
1	Link-01	Structur	Structur	User-Defi	813.85	0.359	468.1	453.5	.013	0.5
2	Link-02	Structur	Out-1Pip	User-Defi	61	0.359	453.5	434.6	.013	0.5
3	Pipe - (1)	Structur	Structur	Circular	811.2469	24.000	459.745	443.520	.013	0.5
4	Pipe - (2)	Structur	Out-1Pip	Circular	60.71659	24.000	439.520	434.662	.013	0.5

Metric Units

	ID	From Node	To Node	Shape	Length	Height/ Diameter	Inlet Elev.	Outlet Elev.	Manning's Roughness	Entr Los
1	Link-01	Structur	Structur	User-Defi	428.06	0.359	142.683	138.226	.013	0.5
2	Link-02	Structur	Out-1Pip	User-Defi	18.59	0.359	138.249	132.466	.013	0.5
3	Pipe - (1)	Structur	Structur	Circular	247.26	600	140.130	135.184	.013	0.5
4	Pipe - (2)	Structur	Out-1Pip	Circular	18.5	600	133.966	132.48	.013	0.5

16. Select the row for either one of the overland links.
    a. Change the channel type to Open Channel.
    b. Set the shape type to User-Defined, as shown in Figure 15.37.

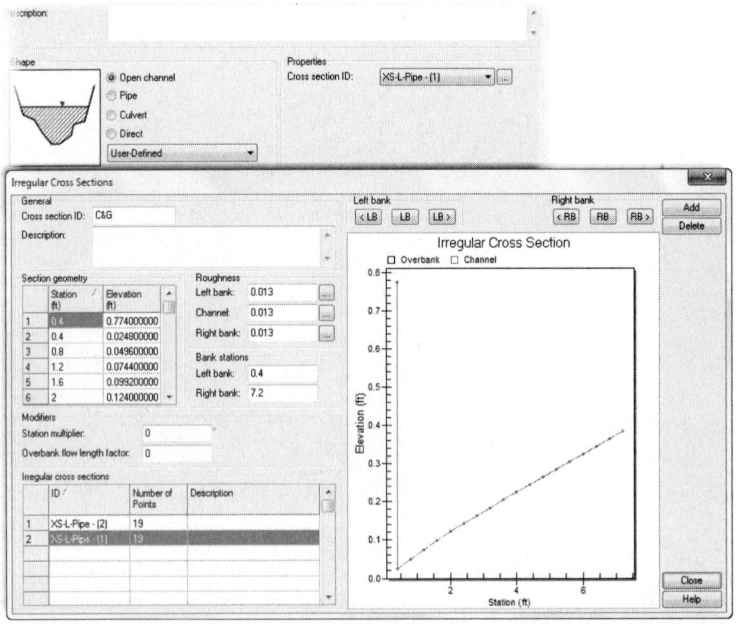

FIGURE 15.37
Gutter slope in cross-section; this will be the overland channel shape

c. Set the Cross Section ID to XS-L-Pipe - (1).

d. Click the ellipsis next to XS-L-Pipe - (1) to open the Irregular Cross Sections dialog.

The shape you see in Figure 15.37 is an exaggerated view of a curb and gutter. In the previous example, you approximated the curb and gutter channel using a rectangular open channel. In this exercise, you will use one of the cross sections created for you by the export process.

17. Rename the Cross Section ID to **C&G**. Close the Irregular Cross Section dialog.

18. Make sure both Link-01 and Link-02 are using the C&G cross section for open channel flow by setting the Cross Section ID for each.

19. Click Close when your conveyance links are complete.

20. Double-click Project Options in the data tree.

a. Set the Hydrology method to SCS TR-55.

b. Set the Time Of Concentration method to SCS TR-55.

c. Set Minimum Allowable TOC to **6** minutes.

At the end of this step, your Project Options dialog should look like Figure 15.38.

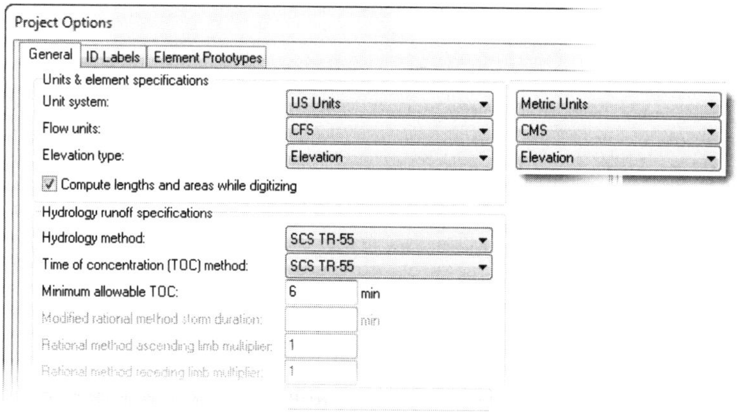

FIGURE 15.38
Project Options for this example (metric is shown for illustration purposes)

21. Click OK.

22. Back in plan view, double-click the west subbasin icon.

23. Switch to the SCS TR-55 TOC tab. You will enter the following information to compute TOC for the western subbasin:

a. On the Sheet Flow subtab, set the Manning's roughness value for Subarea A to **0.17**.

This value is based off the TR-55 listing for different land conditions. The listing of values can be found by selecting the ellipsis next to the entry field.

b. Set Flow Length to **150′** (**45.7** m).

c. Set Slope to **5**%.

This is another site-specific value and represents the average overall slope of the subbasin.

d. Click the ellipsis next to the 2Yr-24Hr Rainfall, and in the resulting dialog, choose Pennsylvania from the State drop-down, and York from the County drop-down, as shown in Figure 15.39. (Metric users, enter a value of **79** mm for rainfall value, because this specification is only reported in inches.)

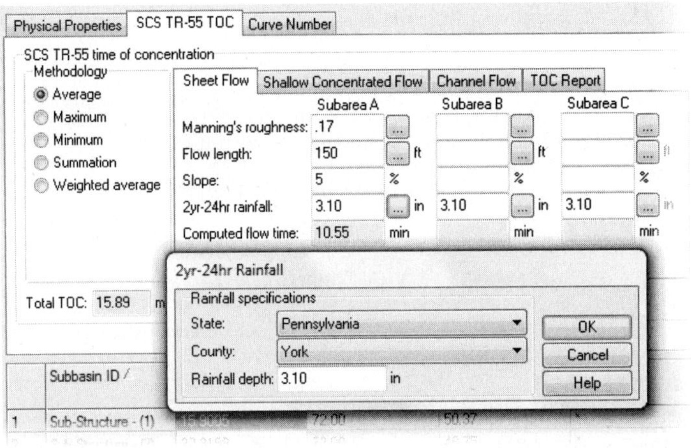

FIGURE 15.39 Sheet Flow input and rainfall location

e. Click OK.

Selecting the rainfall location at this point is helping you determine the Tc. Your actual design storm duration may vary. Later in this chapter you'll learn more about rainfall.

24. Switch to the Shallow Concentrated Flow subtab for Subarea A.

a. Set Flow Length to **250′** (**76.2** m).

b. Set Slope to **5**%.

c. Set the surface type to Short Grass Pasture.

The velocity for sheet flow will calculate based off the nomograph, which you can examine by clicking the ellipsis button next to Velocity. Your results should resemble the top of Figure 15.40.

25. Switch to the Channel Flow subtab and set the Manning's roughness value for Subarea A to **0.05**.

a. Set the flow length to **433′** (**131.9** m).

b. Set the channel slope to **3**%.

c. Set Cross Section Area to **14.5** ft^2 (**1.35** m^2).

d. Set Wetted Perimeter to **12.2′** (**3.7** m).

Wetted Perimeter and Cross Section Area are site-specific values and are needed to calculate the hydraulic radius used in the Manning's roughness equation. Figure 15.41 shows a schematic of what these values mean in channel flow.

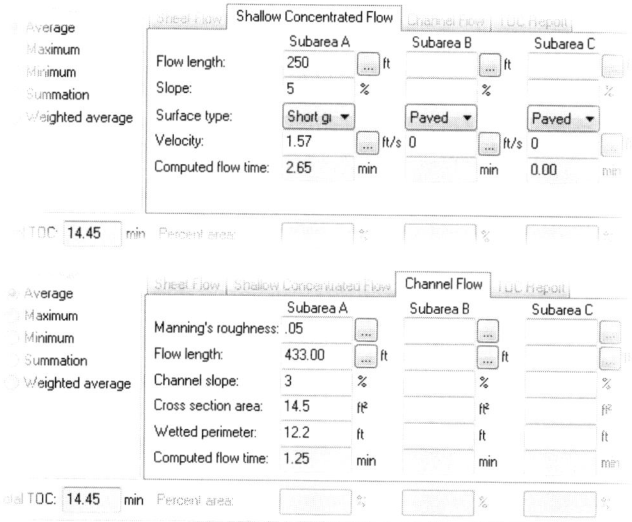

FIGURE 15.40 Shallow concentrated flow and channel flow for TR-55

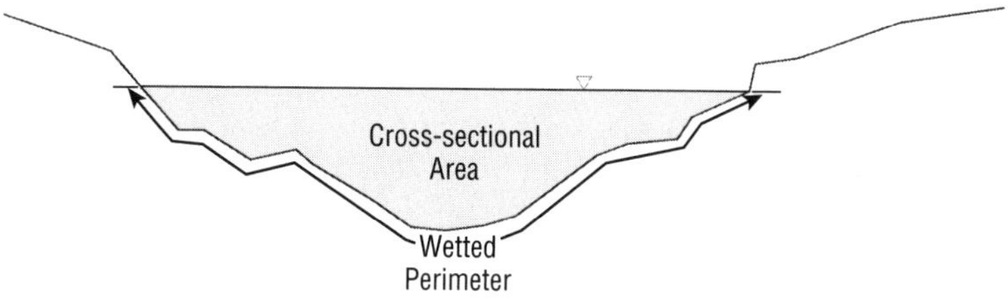

FIGURE 15.41 The Cross Section Area and Wetted Perimeter values are used to find the Tc for channel flow.

After entering the values for channel flow, your dialog will now resemble the bottom of Figure 15.40.

26. Still working with Sub-Structure - (1), switch to the Curve Number tab.
27. Set the curve number values for the subbasin as shown in Figure 15.42.

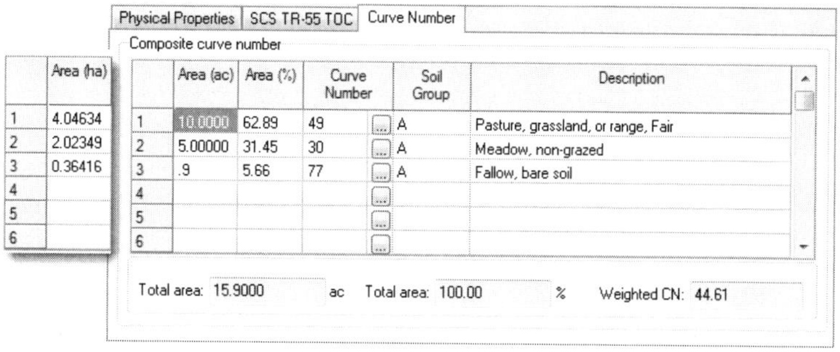

FIGURE 15.42 Use these values on the Curve Number tab to create a weighted curve number (metric areas are shown for illustration).

28. Switch to the eastern subbasin by highlighting Sub-structure - (2) in the tabular listing near the bottom of the dialog. The Tc parameters for this subbasin are:

   - Sheet Flow: Manning's Roughness of **0.17**, Flow Length of **200′** (**61.0** m), and Slope of **4.5%**
   - Shallow Concentrated Flow: Flow Length of **412′** (**125.6** m), Slope of **4%**, and Surface Type set to Short Grass Pasture
   - Channel Flow: Manning's Roughness of **0.05**, Flow Length of **305′** (**93.0** m), Channel Slope of **2%**, Cross Section Area of **12.2** ft2 (**1.13** m^2), Wetted Perimeter of **8.4′** (**2.5** m).

29. Still working with Sub-Structure - (2), switch to the Curve Number tab and set the curve number to **45** for the entire area.

30. Click Close. Save the file as **`C3DtoSSA.spf`** for use in the next exercise.

Make It Rain

There are several methods for telling SSA the rainfall information for the model. A typical rainfall event in Herten, the Netherlands, is going to be vastly different from Phoenix, Arizona.

In the previous exercise, you set up many of the physical characteristics of the hydrologic system. You did not, however, tell SSA what type of rain to expect on the site.

Different hydrology methods require different assumptions about rainfall. In the Rational method, you used an intensity-based rainfall model (IDF Curve) because you were primarily interested in peak runoff. Because this is a TR-55 example, you will need to use rainfall information appropriate for this method.

1. Continue working in the file from the previous exercise or open `SSA-Rain.spf` (or `SSA-Rain_METRIC.spf`) in the Storm And Sanitary Analysis program.

   You can download this file from the book's web page.

2. From the top of the SSA screen, select Add Rain Gauge.
3. Click anywhere in plan view to place the rain gauge in the project.
4. Press Esc on your keyboard to return to Selection mode.
5. Double-click the New Rain gauge, and rename the Rain Gauge to **Design Storm**.
   a. Set Rain Data Format Type to Intensity.
   b. Set Incremental Interval to **0:20**.

   At this step your Rain Gauge dialog will look like Figure 15.43.
6. Click the ellipsis next to Time Series.
7. In the Time Series dialog, click Add.
8. Rename the Time Series to **York, PA 50**.

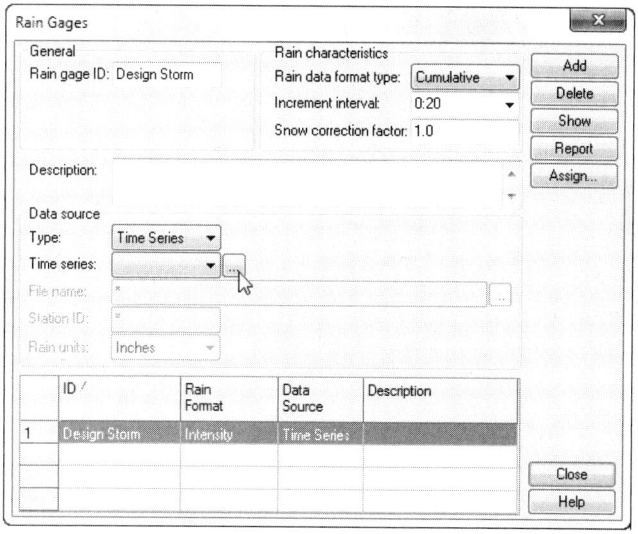

FIGURE 15.43 The Rain Gauge dialog

9. Set the data type to Standard Rainfall and click the Rainfall Designer button.
   a. Set Rainfall Type to Intensity.
   b. Set State to Pennsylvania and County to York.
   c. Set Return Period to 50 years.
10. Make sure that the Unit Intensity check box next to SCS Type II 24-hour is selected, as shown in Figure 15.44. (Metric users, override the Rainfall Depth by entering **157** mm.) Click OK.

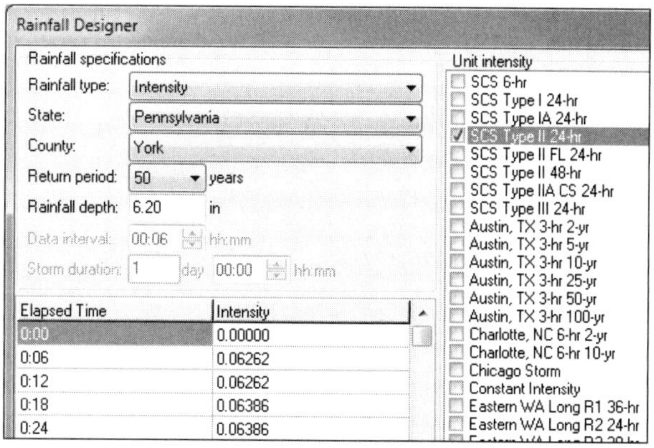

FIGURE 15.44 Rainfall Designer dialog

11. Once your Time Series dialog resembles Figure 15.45, click Close.

FIGURE 15.45
The completed time series

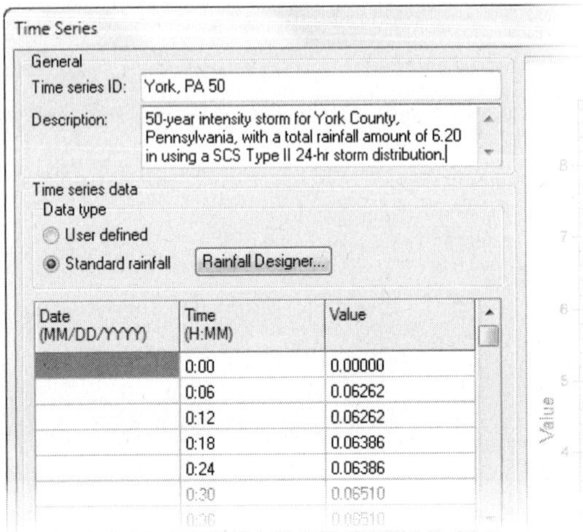

12. In the Rain Gauge dialog, click Assign.
    a. Click Yes when asked if you'd like to assign the rain gauge to all subbasins.
    b. Close the Rain Gauge dialog.
13. Double-click Inlet Structure - (1). Expand the Inlet ID column to get a better view of the table.
14. Set Roadway Gutter Bypass Link to Link-01.
15. Switch to Structure - (2) by selecting it from the table at the bottom of the Inlets dialog.
16. Set Roadway Gutter Bypass Link to Link-02, and click Close.

17. Double-click Analysis Options on the data tree.
18. On the General tab, set the Analysis Duration to **1d** by changing the date in the End Analysis On field to one day after the value in the Start Analysis On field On (the exact date will vary depending on when you open the file).
19. On the Storm Selection tab, set Single Storm to Use Assigned Rain Gauge, and click OK.

20. Click Perform Analysis.
21. After the analysis runs successfully, click OK. Save and close the project.

Culvert Design

In SSA, culverts can be modeled with great accuracy by treating them like a dam. If you think about it, that is exactly what is happening; the road acts like a dam between flow locations and the culvert is a hole connecting the two sides. During a storm event, water backs up and is stored until the culvert can release it or the road floods. If the water elevation on the high side goes over the road, this is similar to a weir releasing water over a dam.

In this exercise, you will work through the steps of designing a culvert. Several important steps have already been completed for you:

- The Project options have been set to use the Modified Rational hydrology method.
- Watershed areas have been delineated and associated with respective inlets.
- The local IDF curve has already been imported and saved with the drawing.

1. From SSA, open the file `Culvert.spf` (or `Culvert_METRIC.spf`), which you can download from this book's web page.

   This file contains four subbasins that will drain through a culvert to the final outfall. The culvert portion of the project is incomplete.

2. Right-click on the node Junction1, and select Convert To ➤ Storage Node, as shown in Figure 15.46.

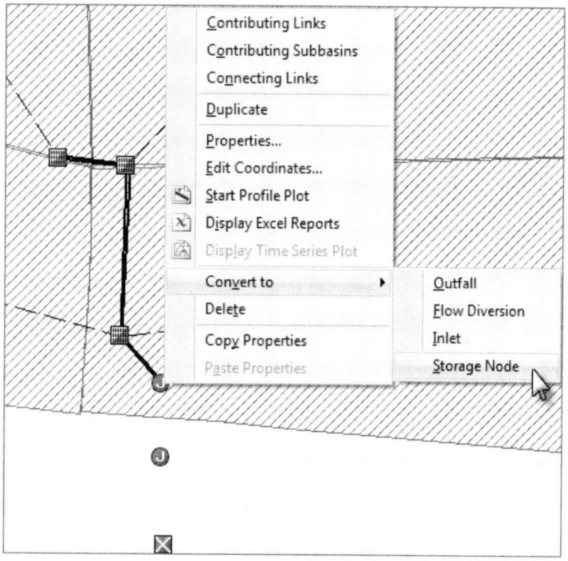

FIGURE 15.46 Converting the existing junction to a storage node

3. Once the storage node has been converted, double-click the node.

   a. Rename the storage node to **Culvert Headwater**.

   b. Verify that the Storage shape type is Functional.

      This means that the area of the storage node can be described as a function of the depth.

   c. Set the Constant Area to **4** ft^2 (**0.37** m^2).

      We are assuming that the last junction before the culvert has storage capability. A constant area is indicating that the junction is a vertical walled structure with a floor area of 4 ft^2 (0.37 m^2).

   d. Set the Coefficient to **0**.

   e. Set the Exponent to **0**.

4. When the dialog looks like Figure 15.47 click Close.

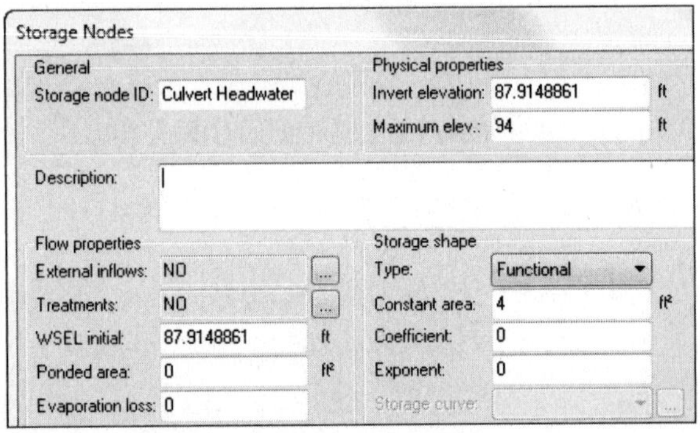

FIGURE 15.47
Rename the storage node for easy identification

5. Click the Add Conveyance Link tool.
6. Connect the storage node to the remaining junction.
7. Connect the remaining junction to the outfall.
8. Press Esc to exit the conveyance link tool.
9. Double-click the link that connects the storage node to the junction.
    a. Rename the Link ID to **Culvert**.
    b. Set the Shape toggle to Culvert.
    c. Leave the diameter to the default **18″** (**450** mm), as shown in Figure 15.48.
10. Close the Conveyance Link dialog.

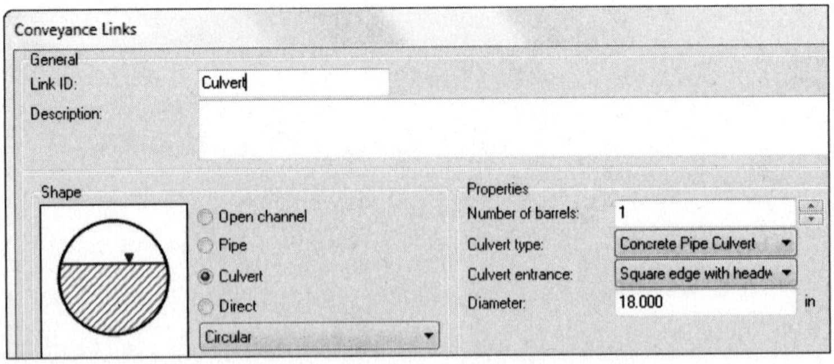

FIGURE 15.48
Your new culvert

Next, you will add a weir to the storage structure. A weir is used to model what happens when water rises above the maximum storage elevation and runs over the street.

11. From the Elements toolbar, click Add Weir.
12. Draw the weir in the same manner as you drew the overland links in the previous exercise.

13. Add an extra segment to differentiate it visually from the main culvert link. Press Esc.
14. Double-click the new weir:
    a. Rename the weir to **Road Overflow**.
    b. Set the crest invert elevation to **94.3′** (**28.7** m).

       This elevation represents the roadway crown elevation at the culvert crossing.
    c. Set the crest length to **5′** (**1.5** m).
    d. Set the weir total height to **5′** (**1.5** m).
    e. Set the type to Trapezoidal.
    f. Set the side slope to **1:50**.
    g. Set the discharge coefficient to **2.9**.
    h. Set the contraction type to No Ends.
    i. Set the trapezoidal end coefficient to **0**.
15. When your dialog looks like Figure 15.49, click Close.

FIGURE 15.49 Entering weir data to model road overflow

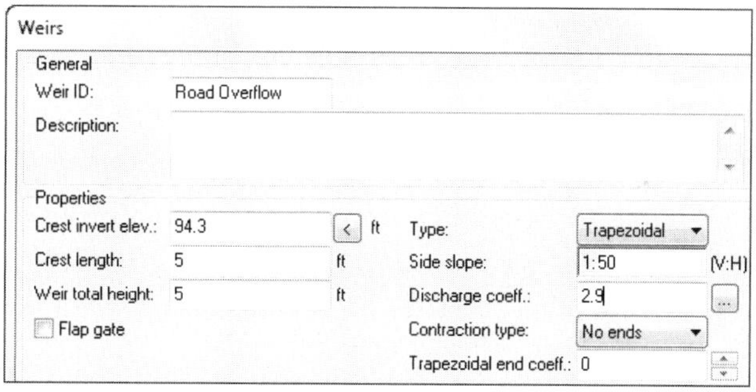

Because this not a true weir or spillway, you are entering the above values to mimic the long, flat characteristics of a roadway.

16. Double-click the southernmost link that connects the junction to the outfall.

    The extra link to the outfall is needed because an outfall can only be connected to one link in SSA. You will use the Direct option to indicate that no head loss or time is contributed to the system in this link.
17. Change Shape to Direct. Leave all other options at their defaults, and click Close.

    Now that all the physical characteristics of the model are in place, it is time to check the environmental aspects. You already have the IDF curve in place, but you still need to enter time of concentration and runoff coefficients to the subbasins.
18. Double-click any of the green subbasin icons.

19. Set the Runoff Coefficient to **0.88** for all four subbasins to reflect a commercial (highly impermeable) site.

20. Set the time of concentration for each subbasin as shown in Figure 15.50. Times will be the same for both metric and Imperial, but the area value will differ.

FIGURE 15.50
Subbasin data

	Subbasin ID	Area	Wt. Runoff Coeff.	TOC
1	Sub-01	0.8097	0.88	15.00
2	Sub-02	1.2421	0.88	17.00
3	Sub-03	0.4534	0.88	6.00
4	Sub-04	0.5490	0.88	6.00

21. Close the Subbasin dialog when complete.

22. Double-click Analysis Options.

  a. On the General tab, change the End Analysis time to **01:00:00**, as shown in the left image in Figure 15.51.

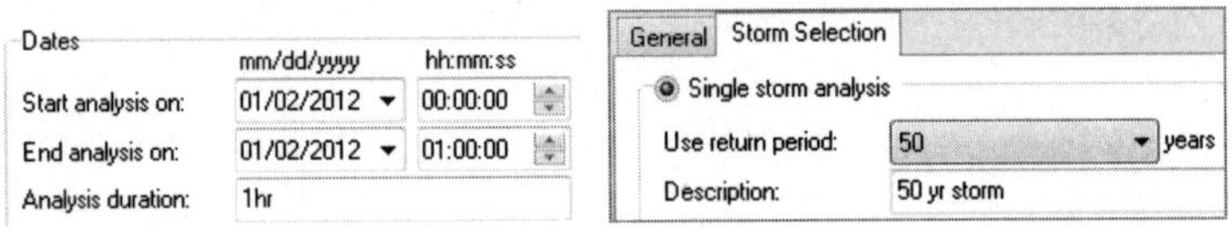

FIGURE 15.51
Set the analysis duration on the General tab (left). Storm selection of a 50-year return period (right)

  Note that you cannot change the Analysis duration field directly—you need to change the time.

  b. On the Storm Selection tab, set the storm return period to 50 years, as shown in the right image in Figure 15.51.

23. Click OK to close the Analysis Options dialog.

24. Click the Perform Analysis icon.

25. When the analysis is complete, click OK to dismiss the Perform Analysis dialog.

  If you receive an error; check to make sure you have followed all the steps correctly and try again.

  You will see a dark blue line indicating that the storage node north of the culvert is flooded. In the next steps you will fix this and rerun the analysis.

26. Double-click the Culvert link.

27. Change the culvert diameter to **24"** (**600** mm) then click Close.

28. Click Perform Analysis again to recompute the result. Click OK.

The dark blue line should disappear, indicating that the flooding no longer occurs Save and close the culvert file.

Pond Analysis

SSA can model a number of pond and detention scenarios. Using the same humble storage node tool from the previous example, with SSA you can model exfiltration ponds, underground storage, and sedimentation basins.

In an earlier section of this chapter, you created stage storage information for a pond. That type of information tells you the geometry of the pond, but it does not tell you whether the size is adequate for all the rain that may go in it. In this section, you will use SSA to verify that your pond is sized correctly for the area.

In the following exercise, you will go through the steps to model a detention basin that has water loss due to exfiltration.

Already completed for you in this exercise are the following steps:

- The Project options have been set to use the Modified Rational hydrology method with the Kirpich method for developing time of concentration.
- The IDF curve for the local area has been imported.
- Runoff coefficient, average slope, and flow length have been entered for all of the subbasin. Slope and flow length are needed to compute the time of concentration via the Kirpich method.
- Every inlet on the grade has been assigned a Roadway/Gutter bypass link to account for overflow, and upstream roadway links to ensure overland flows accumulate.
- All the pipes that drain to the pond have been tested for a 50-year storm event.

All you need to focus on is the detention pond. You will also learn a few additional SSA tools along the way.

1. In SSA, open the file `Pond.spf` (or `Pond_METRIC.spf`), which you can download from this book's web page. If you are prompted to locate the background file, browse to the same folder and select `Pond_Underlay.dwg`.

The area of focus is the southern area that drains to the pond. There are a few stray inlets that can be removed from the drawing.

2. With the selection tool active, locate the Inlet named Orphan1.

As you pause your cursor over the various nodes and junctions, the tooltip will tell you which item you are examining.

3. Right-click on Inlet Orphan1 and click Delete, as shown in Figure 15.52.

4. Click Yes to verify the deletion.

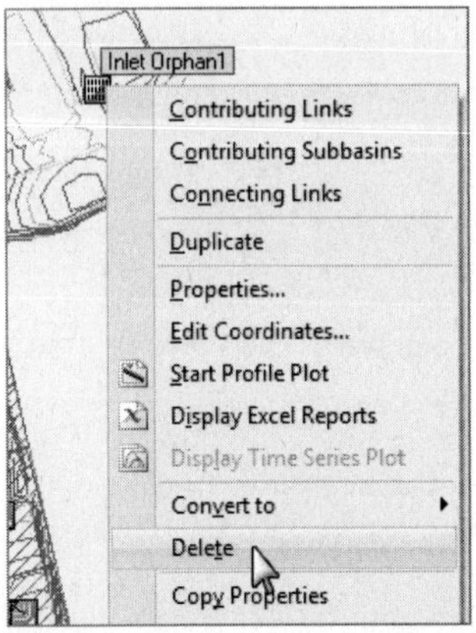

FIGURE 15.52
Deleting a single entity using the right-click menu

Deleting a single entity will come in handy, but you may find that after import from Civil 3D, you still have many disconnected and unneeded nodes. For this situation, you will use Delete Orphan nodes.

5. To delete the remaining disconnected inlets, right-click anywhere there is not an entity in the drawing and click Delete Orphan Nodes, as shown in Figure 15.53.

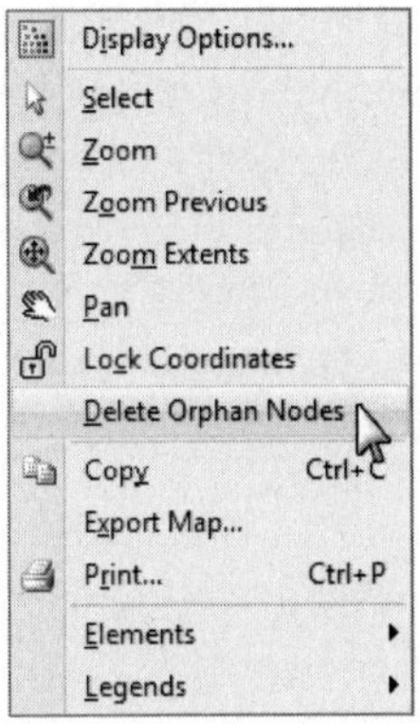

FIGURE 15.53
Deleting orphan nodes

6. Click Yes when prompted to verify.

   Next, you will allow for exfiltration from your pond. As water sits, it will leave the pond as it seeps into the ground water below.

7. Double-click Project Options.

   a. Verify that the Link Routing method is Hydrodynamic.

   b. Change the storage node exfiltration method to Constant Rate, Wetted Area.

      This option means that the loss rate changes as a function of the wet surface of the earth that forms the basin.

8. When the project options look like Figure 15.54, click OK.

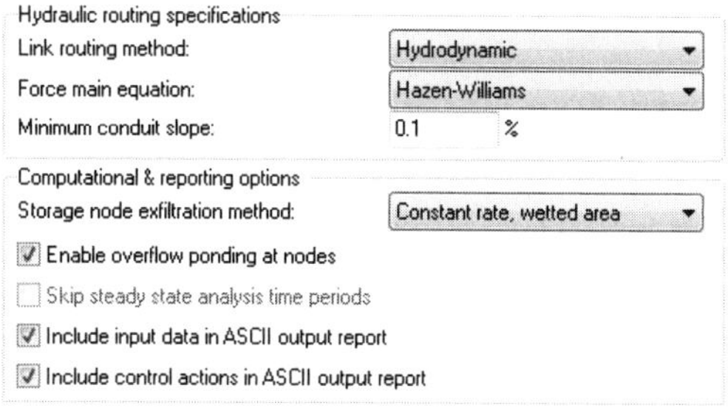

FIGURE 15.54
Project Options dialog

9. Click the storage node tool and place a storage node in the graphic in the pond area, south of the outfalls.
10. Press Esc to return to the selection tool.
11. Double-click the new storage node.
    a. Change the Storage Node ID to **Pond**.
    b. Set the Invert elevation to **775′** (**236.2** m).
    c. Set the Maximum elevation to **784** (**239.0** m).
    d. Change Storage Shape Type to Storage Curve.
    e. Set Exfiltration to At All Elevations.
    f. Set Exfiltration Rate to **0.04** in/hr (**1** mm/hr). This value is the typical exfiltration rate for a silty clay soil.
12. Click the ellipsis to the right of the Storage Curve selection. This will open the Storage Curves dialog.
    a. Click Add.
    b. Rename the Curve ID to **Detention**.
    c. Click Load. Make sure the file type is set to AECCSST.

       This is the file you created in Civil 3D when you created a stage-storage table. If you were able to export this file earlier in the chapter, browse to the file you created. If not, use the file `Pond-Eval_FINISHED.aeccsst` (`Pond-Eval_METRIC_FINISHED.aeccsst`) provided at the web page.
13. When your storage curve looks like Figure 15.55, click Close.

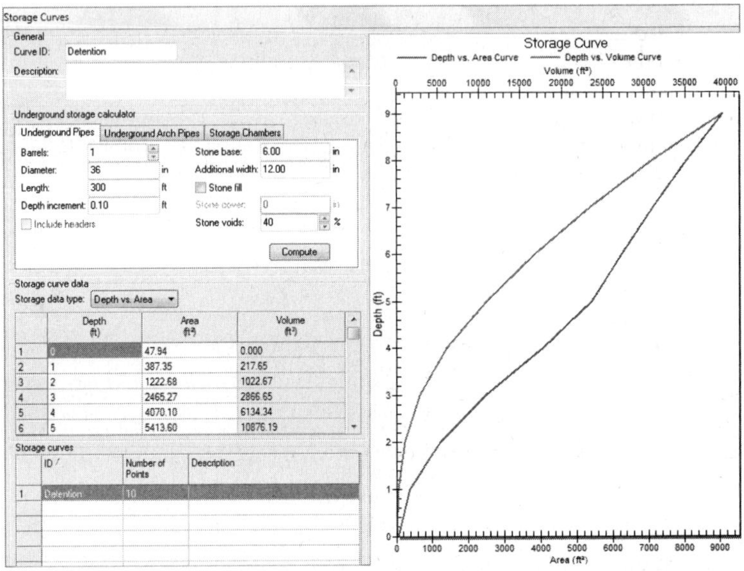

FIGURE 15.55
The imported pond data

14. Verify that the storage curve listed in the Storage Node dialog is set to Detention and click Close.

    You will complete the model by connecting the ends of both pipe runs to the storage node. You will need to convert both outfalls to junctions and specify a new outfall.

15. Right-click one of the outfalls and select Convert To ➤ Junction.

16. Repeat this for the second outfall.

    Both outfalls are now junctions. (If time permits, rename the node to reflect that they are now junctions).

17. Before you place the links, be sure to set the invert elevation for the outfall.

    Pipes always inherit the invert elevations from the junctions they are connected to.

18. Click the outfall tool and place a new outfall to the south of the storage node.

19. Double-click the new outfall and set its invert elevation to **773′** (**235.6** m), then click Close.

20. Start the Add Conveyance Link tool.

21. Working upstream to downstream, connect each junction to the storage node.

22. Double-click one of the conveyance links that connects the junction to the storage node.

23. Set Shape to Direct. Repeat this for the other conveyance link as shown in Figure 15.56.

    Next, you will model an outflow riser using the Orifice tool.

24. Click Add Orifice.

25. Connect the storage node to the final outfall and press Esc.

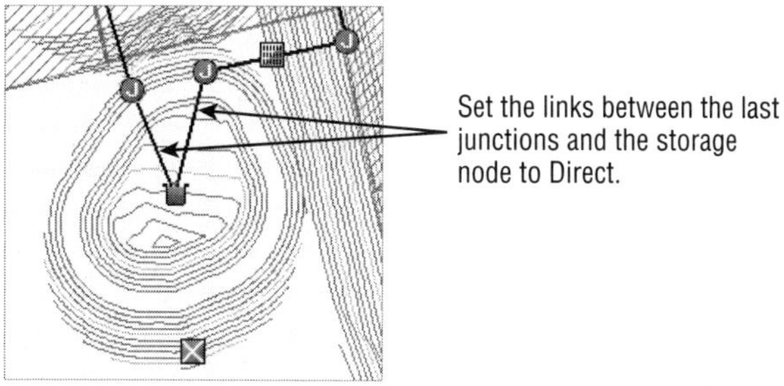

FIGURE 15.56 Set the links to Direct to indicate that water flows unimpeded to the pond.

26. Double-click the new Orifice link.
    a. Change the Orifice ID to **Riser**.
    b. Set Diameter to **8**″ (**200** mm).
    c. Set the crest elevation to **776**′ (**236.5** m). This will allow for 1′ (0.3 m) of standing water in the pond.
27. Leave all other options as shown in Figure 15.57, and click Close.

FIGURE 15.57 Orifice is added to release water from the pond slowly.

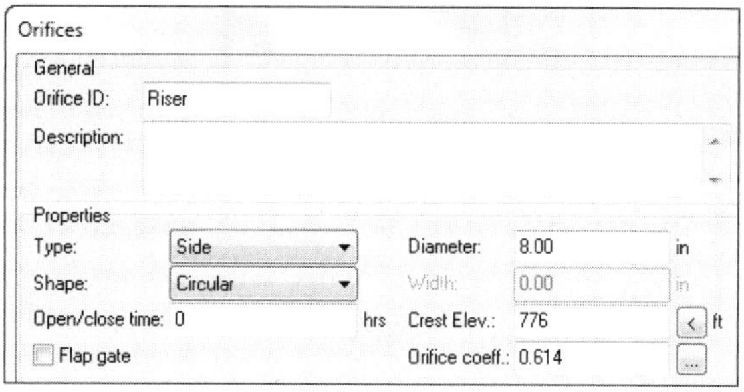

28. Click Perform Analysis.

    If the analysis does not run correctly, review the steps and try again.
29. When the analysis is complete, click OK.
30. Right-click on the orifice and select Display Time Series Plot.

    The result will resemble Figure 15.58.
31. Save and close the project.

FIGURE 15.58 Time series plot shows the max outflow from the pond

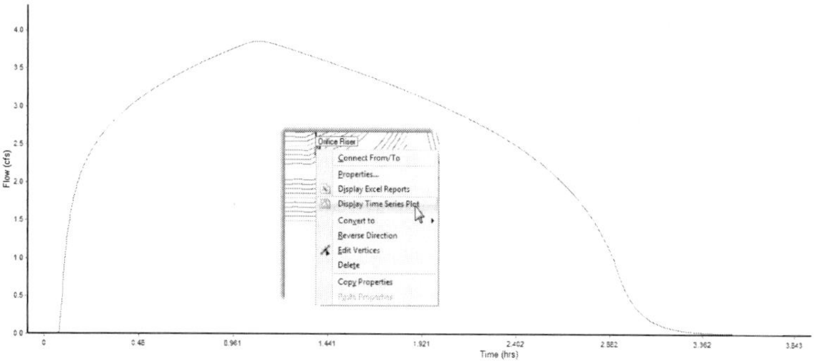

Real World Scenario

Stormwater Best Management Practices and EPA SWMM

Stormwater best management practices are intended to remove sediment and other pollutants from drainage systems. Many municipalities require using the EPA SWMM hydrology method to determine the efficacy of these methods.

In the following example, you will walk through how to account for sediment removal at a junction. Again, some important (but time-consuming) tasks have been performed for you already:

- The Hydrology method in the Project Options dialog has been set to EPA SWMM.
- The curve number and physical properties of the subbasins have been set.
- A rain gauge indicating an SCS Type II storm has been created and assigned to all subbasins.

1. Open the file `Sediment Removal.spf` (or `Sediment Removal_METRIC.spf`), which you can download from this book's web page.
2. From the data tree on the left side of the SSA window, double-click Pollutants.

3. In the Pollutants dialog, click Add.
4. Change Pollutant ID to **TSS**.
5. In the Description area, type **Total Suspended Solids,** and click Close.
6. Double-click Pollutant Land Types, and click Add.
7. Rename Land Type ID to **Residential**.

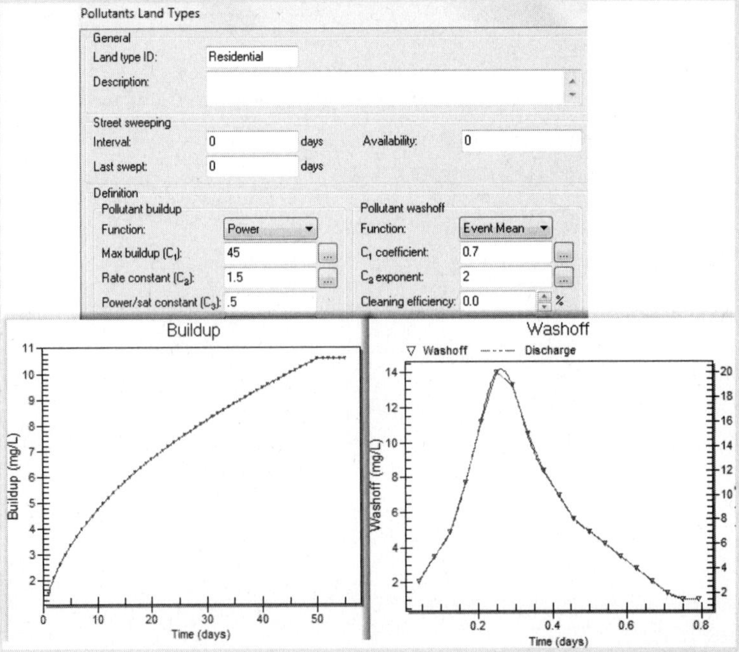

8. Set Pollutant Buildup Function to Power.
9. Set Max Buildup (C1) to **45**.
10. Set Rate Constant (C2) to **1.5**.
11. Set Power/Sat Constant to **0.5**.
12. Set Pollutant Washoff Function to Event Mean.
13. Set C1 Coefficient to **0.7**. Set C2 Coefficient to **2**.
14. Close the Pollutant Land Types dialog.
15. Double-click the junction named Sediment Removal.
16. Click the ellipsis next to Treatments.
17. In the Treatment Expression column for TSS, type **R=0.8**.

    This expression indicates that 80 percent of TSS is removed by this treatment.
18. Click OK when complete.
19. Close the Junctions dialog.
20. Double-click one of the subbasins to open the Subbasins dialog.
21. Switch to the Flow Properties tab.
22. Click the ellipsis next to the Land Type field.
23. Double-click next to Residential and type **100**, and then click OK.

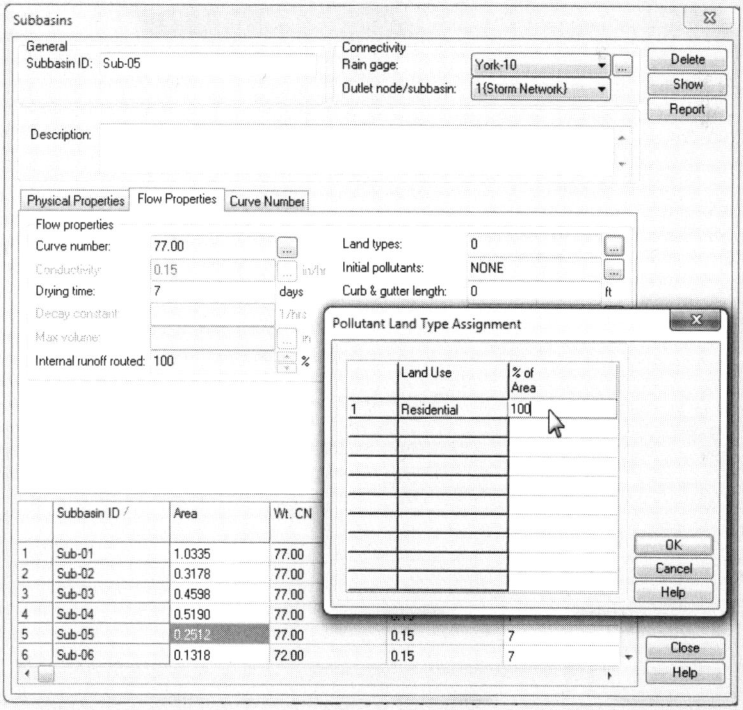

24. Repeat steps 22 and 23 for all of the subbasins.

25. For each subbasin, set 100 percent of the land area to Residential. Click OK after each entry. Close the Subbasins dialog.

    This is the pollutant land type you defined in the previous steps.

26. Click Perform Analysis; then click OK. Save and close the project.

You can now show the efficacy of your treatment as it relates to the rest of the storm design.

Running Reports from SSA

There are many places in SSA to view the results of your work. After you've run the analysis, the link and node listing will tell you if your pipe is underdesigned or if there is water backing up in the manholes.

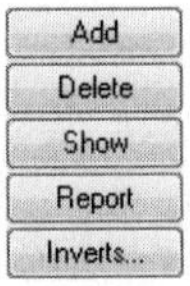

From virtually every dialog in SSA, clicking the Report button will create an Excel spreadsheet from the data you are observing. You will see the Report button for links, nodes, junctions, outfalls, and subbasins.

After running the analysis, you will want to see performance information and graphical representations of the relationship between the parts of your design. The Reports toolbar is at the top of the SSA screen. Figure 15.59 shows the different types of reports that you can choose.

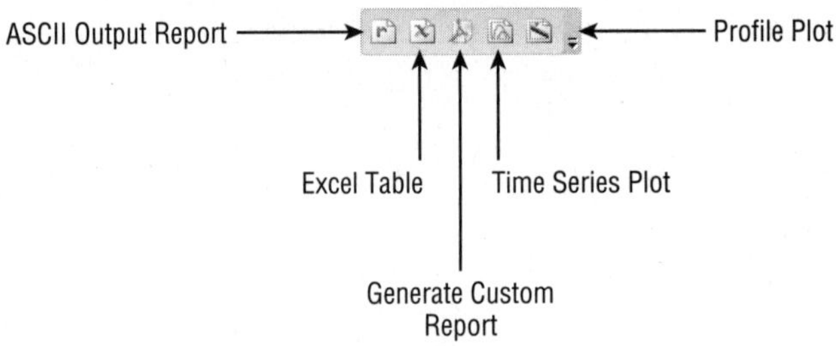

FIGURE 15.59
The Reports toolbar

First, let's take a look at the hydrograph that has been created for the site:

1. From SSA, open the file `SSA-Reports.spf`, which you can download from this book's web page.
2. Click the Perform Analysis button. Click OK.
3. Click the Time Series Plot button.
4. On the left side of the screen, expand the `Subbasins` folder.
5. Click the word Runoff.

   A plus sign will appear.

6. Click the plus sign to expand the listing.

7. Put a check mark next to both subbasins, as shown in Figure 15.60.

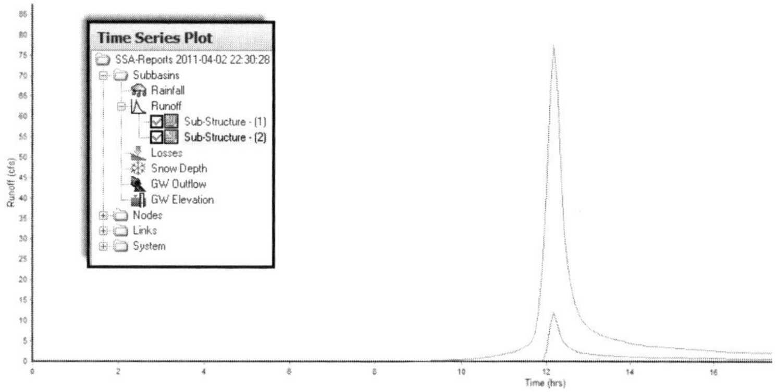

FIGURE 15.60 The resulting hydrograph from the TR-55 Hydrology analysis

8. Right-click anywhere in the graph area.

9. Select Export Time Series Plot ➤ CAD Export, as shown in Figure 15.61.

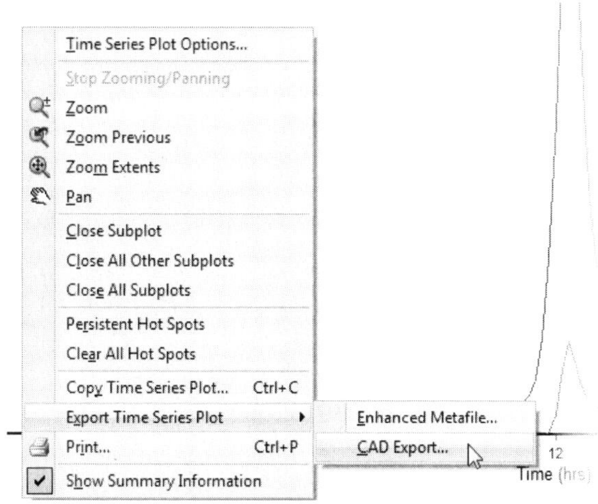

FIGURE 15.61 Exporting to CAD

10. Save the resulting DWG to your desktop as **SSA-OUT.dwg**.

11. Close the Time Series Plot tab.

Remain in the SSA project for the next exercise.

A common analysis you will want to examine is the profile plot. On the left side of the SSA window you will see the Profile Plot option. Once the profile plot settings appear, you can graphically select the nodes you wish to include in your output.

In the next exercise, you will examine a profile plot for the pipes in the system:

1. From SSA, continue working in `SSA-Reports.spf`.

   You do not need to have completed the previous exercise to continue—you can download this file from the book's web page.

2. Click the Perform Analysis button, and then click OK.

3. From the Reports toolbar, select Profile Plot.

   On the left side you can specify Starting Node and Ending Node. You can also specify the link you'd like to plot by clicking on it in the graphic.

4. Click the link Pipe - (1).

   The pipe will highlight in the plan view.

5. Click Plot Options.

6. In the Plot Options dialog, place a check mark next to Maximum Flow, Maximum Velocity, and Maximum Depth.

7. In the Other Display Specifications, place a check mark next to Maximum EGL, Critical Depth, HGL Markers, and Show Flooded Node.

   Your Profile Plot Options dialog should resemble Figure 15.62.

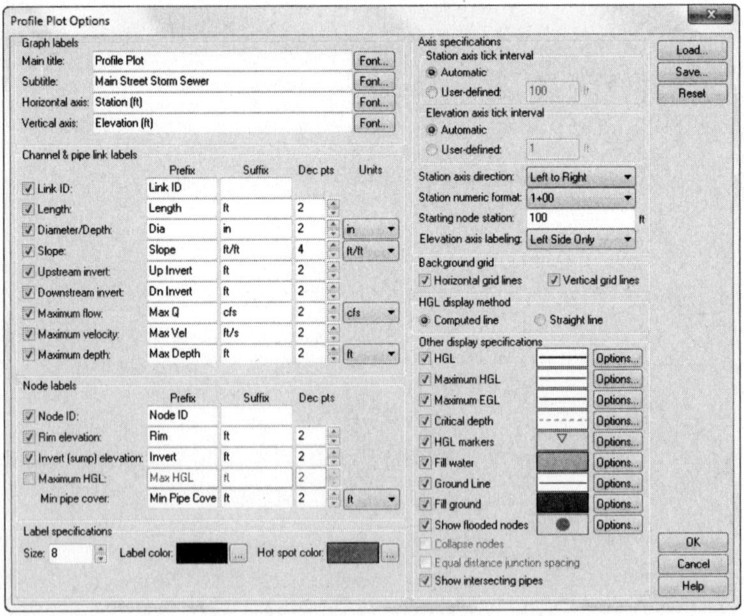

FIGURE 15.62 Use the Profile Plot Options dialog to control the display of elements in the graphic.

8. Click OK.

9. Click Show Plot.

Your graphic should resemble Figure 15.63.

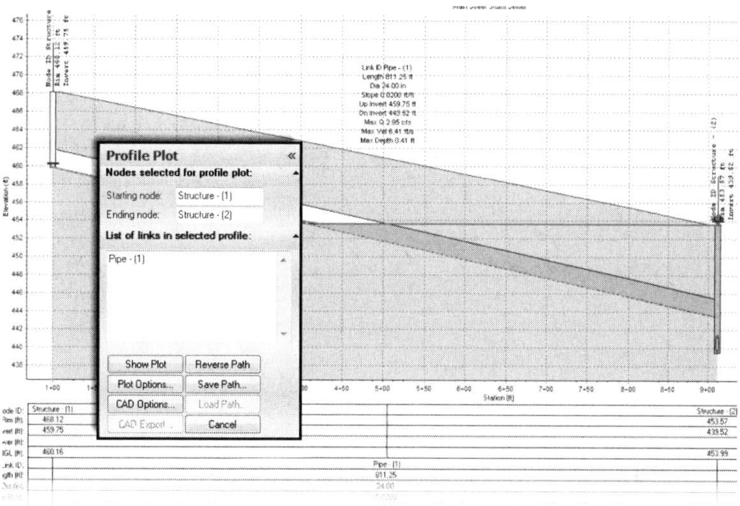

FIGURE 15.63
A sample profile plot

After your break from the world of Civil 3D, it is time to export SSA modified data back to CAD. At first glance, you may be thinking that CAD Export or LandXML may be the way to go. However, those options are not the best way to maintain the integrity of your pipe and structure objects. The best option for getting your edits back into Civil 3D is to use File ➢ Export ➢ Hydraflow Storm Sewers File.

The following exercise will walk you through the steps needed to get back to Civil 3D:

1. From SSA, open the file `SSAtoC3D.spf`.

   You can download this file from this book's web page.

2. Choose File ➢ Export ➢ Hydraflow Storm Sewers File, as shown in Figure 15.64.

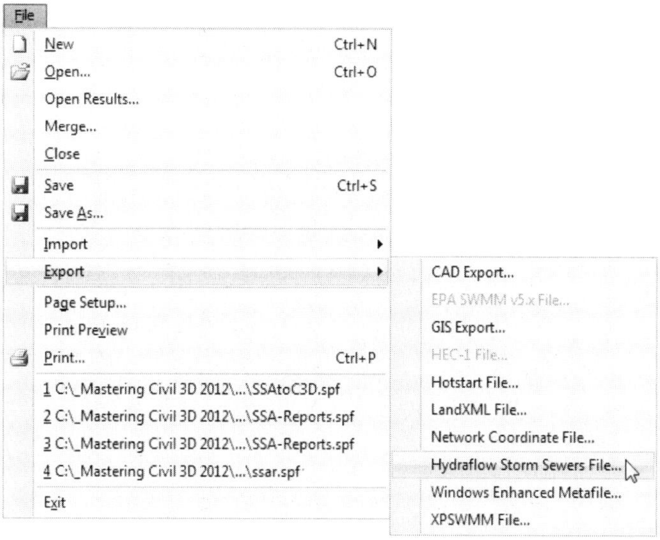

FIGURE 15.64
Export back to Civil 3D via Hydraflow

3. Save the file as **`SSAtoC3D.stm`** on your computer's desktop.
4. Close SSA.
5. In Civil 3D, open the file `C3DtoSSA.dwg`.
6. From the Import panel on the Insert tab, select Storm Sewers as shown in Figure 15.65.

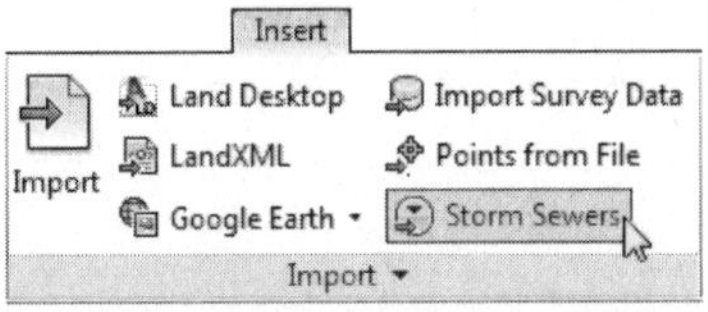

FIGURE 15.65
Import the file as if it were a storm sewers file.

7. Browse to the STM file you exported and click Open.

   You will get a warning that there is already a pipe network with the same name in the file.
8. Click Update The Existing Pipe Network.

After a moment, your pipes and structures should reflect any size or elevation changes that may have occurred in SSA.

The Bottom Line

Create a catchment object. Catchment objects are the newest object type that Civil 3D can use to determine the area and time of concentration of a site. You can create a catchment from a surface, but in most cases you will use the option to create from a polyline.

Master It Create a catchment for predeveloped conditions on the site.

Export pipe data from Civil 3D into SSA. Civil 3D by itself cannot do pipe flow or runoff calculations. For this reason, it is important to export Civil 3D pipe networks into the Storm And Sanitary Analysis portion of the product.

Master It Verify your pipe export settings and then export to SSA.

Export a stage-storage table from Civil 3D. When designing detention basins in Civil 3D, you will often need to export surface data for analysis in SSA.

Master It Generate a stage storage table for a detention basin surface. Determine how much water the pond can hold.

Model drainage systems in SSA. SSA is capable of modeling everything from complex drainage systems to simple culverts.

Master It Determine which links flood during a 2-hour, 50-year storm event. Model the example drainage design using SSA.

Chapter 16

Grading

Beyond creating streets and sewers, cul-de-sacs, and inlets, much of what happens to the ground as a site is being designed must still be determined. Describing the final plan for the earthwork of a site is a crucial part of bringing the project together. This chapter examines feature lines and grading groups, which are the two primary tools of site design. These two functions work in tandem to provide the site designer with tools for completely modeling the land.

In this chapter, you will learn to:

- Convert existing linework into feature lines
- Model a simple linear grading with a feature line
- Model planar site features with grading groups

Working with Grading Feature Lines

There are two types of feature lines: corridor feature lines and grading feature lines. Corridor feature lines are discussed in Chapter 10, "Basic Corridors," and grading feature lines are the focus of this chapter. It's important to note that grading feature lines can be extracted from corridor feature lines, and you can choose whether or not to dynamically link them to the host object.

As discussed in Chapter 4, "Surfaces," terrain modeling can be defined as the manipulation of triangles created by connecting points and vertices to achieve Delaunay triangulation. In the AutoCAD® Civil 3D® software, the creation of the feature line object adds a level of control and complexity not available to 3D polylines. In this section, you look at the feature line, various methods of creating feature lines, some simple elevation edits, planar editing functionality, and labeling of the newly created feature lines.

Accessing Grading Feature Line Tools

The Feature Line creation tools can be accessed from the Home tab's Create Design panel, as shown in Figure 16.1.

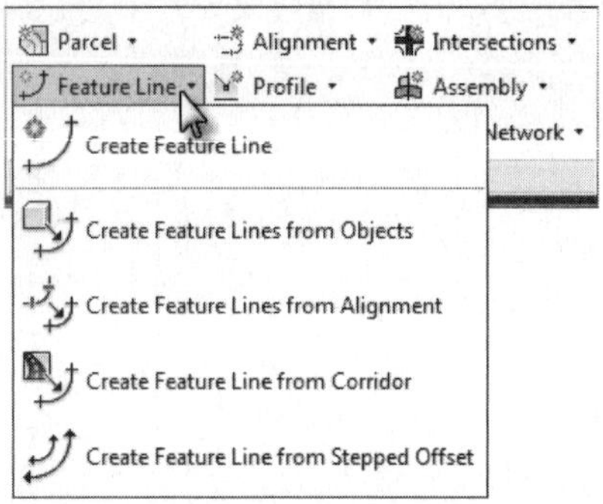

FIGURE 16.1
The Feature Line drop-down menu on the Create Design panel

The Feature Line editing tools can be accessed via the Feature Line contextual tab (see Figure 16.2). To access the Feature Line contextual tab you can select a feature line and the Feature Line contextual tab is activated, or from the Modify tab ➤ Design panel choose Feature Line.

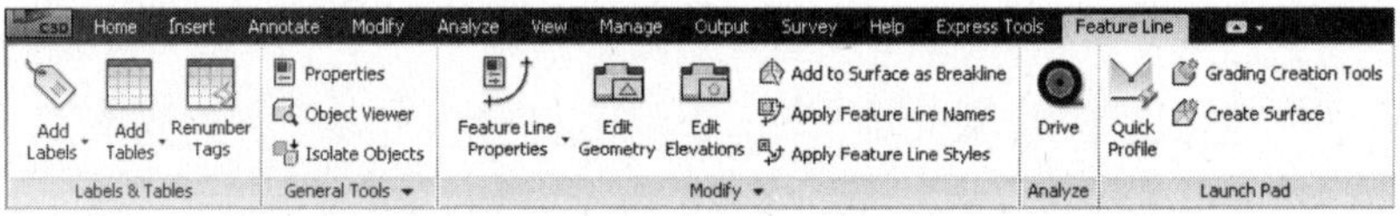

FIGURE 16.2
The Feature Line contextual tab accessed by selecting an existing feature line.

One thing to remember when working with feature lines is that they do belong in a site, as shown in Figure 16.3. Feature lines within the same site snap to each other in the vertical direction and can cause some confusion when you're trying to build surfaces.

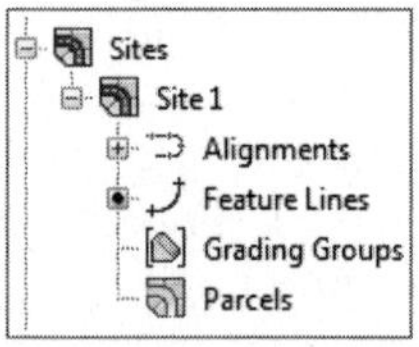

FIGURE 16.3
Feature lines are located in the Sites branch in Prospector.

UNDERSTANDING PARENT SITES

If you're experiencing some weird elevation data along your feature line, be sure to check out the parent site. If the concept of a parent site or sites in general doesn't make much sense to you, be sure and look at Chapter 5, "Parcels," before going too much further. Sites are a major part of the way feature lines interact with each other, and many users who have problems with grading are the same users who are ignoring sites.

The next few sections break down the various tools in detail. You'll use almost all of them in this chapter, so in each section you'll spend some time getting familiar with the available tools and the basic concepts behind them.

Creating Grading Feature Lines

CERT OBJECTIVE

There are five primary methods for creating feature lines, as shown previously in Figure 16.1. They generate similar results but have some key differences:

Create Feature Line The Create Feature Line tool allows you to create a feature line from scratch, assigning elevations as you go. These elevations can be based on direct data input at the command line, slope information, or surface elevations.

Create Feature Lines From Objects The Create Feature Lines From Objects tool converts lines, arcs, polylines, and 3D polylines into feature lines. This process also allows elevations to be a constant elevation, assigned from a grading group, or assigned from a surface. You are given the option as to whether you want the original objects to be deleted or whether to weed points.

Create Feature Lines From Alignment The Create Feature Lines From Alignment tool allows you to build a new feature line from an alignment, using a profile to assign elevations. This feature line can be dynamically tied to the alignment and the profile, which limits your ability to edit it directly, but makes it easy to generate 3D design features based on horizontal and vertical controls of other objects. If the feature line created from an alignment is not dynamically linked, then any of the feature line editing commands can be used.

Create Feature Line From Corridor The Create Feature Line From Corridor tool is used to export a grading feature line from a corridor feature line. The feature line created from a corridor feature line keeps a dynamic link to the corridor.

Create Feature Line From Stepped Offset The Create Feature Line From Stepped Offset tool is used to create a feature line from an offset and the difference in elevation from a feature line, survey figure, polyline, or 3D polyline. The feature line created from a stepped offset does not keep a dynamic link to the original feature line.

You explore each of these methods in the next few exercises. In this exercise, you will be creating a swale from feature lines.

1. Open the `CreatingFeatureLines.dwg` or `CreatingFeatureLines_METRIC.dwg` file. (Remember, all data can be downloaded from `www.sybex.com/go/masteringcivil3d2013`.)

   This drawing has a polyline inset from the outer perimeter of the subdivision, and a polyline at the proposed centerline of an overland swale.

2. From the Home ➢ Create Design panel, choose Feature Line ➢ Create Feature Line to display the Create Feature Lines dialog.

   You need to create a new site in which to put the swale. Just like parcels, grading objects will react with like objects in a site. So by isolating this swale, you can ensure that everything will be drawn properly before committing it to a site.

3. Click the Create New button to the right of the site name. The Site Properties dialog appears.
4. Enter **Swale** for the name of this site, and click OK to dismiss the Site Properties dialog.

   The Create Feature Lines dialog should now look like Figure 16.4.

FIGURE 16.4
The Create Feature Lines dialog

5. Click OK to accept the Create Feature Lines dialog.
6. At the `Specify start point:` prompt, use an Endpoint Osnap to pick the right end of the line (closest to the red line).
7. At the `Specify elevation or [Surface]:` prompt, enter **S** ↵ to use a surface to set elevation information.

   If there were multiple surfaces in the drawing, then the Select Surface dialog would appear and you would use the drop-down to choose EG, and click OK. However, since you only have one surface in this drawing currently, there is no need to state what surface to use.
8. At the `Surface elevation:` prompt, press ↵ to accept the default surface elevation offered on the command line.
9. At the `Specify the next point or [Arc]:` prompt, use an Endpoint Osnap to pick the left end of the line.

10. At the `Specify grade or [SLope Elevation Difference SUrface Transition]:` prompt, enter **-5.2** ↵ to set the grade between points.

11. Press ↵ to end the command.

    Your screen should look like Figure 16.5.

FIGURE 16.5
Setting the grade between points

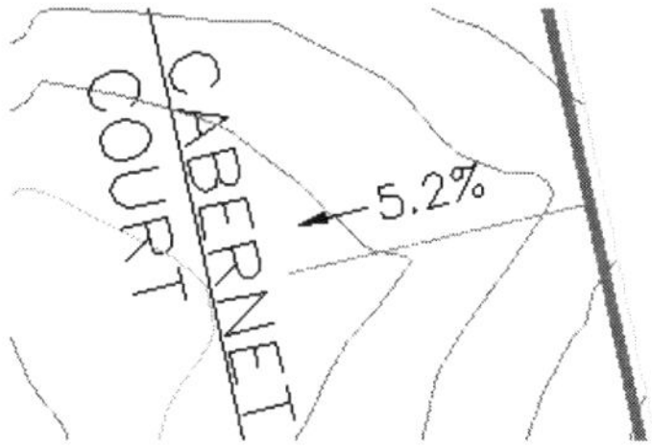

> **ASSIGNING NAMES**
>
> **Looking back at Figure 16.4, note that there is an unused option for assigning a Name value to each feature line. Using names does make it easier to pick feature line objects if you decide to use them for building a corridor object. The Name option is available for each method of creating feature line objects, but will generally be ignored for this chapter, except for one exercise covering the renaming tools available.**

This method of creating a feature line connecting a few points seems pretty tedious to most users. In the next portion of the exercise, you convert an existing polyline to a feature line and set its elevations on the fly.

12. From the Home tab ➢ Create Design panel, choose Feature Line ➢ Create Feature Lines From Objects.

13. At the `Select lines, arcs, polylines or 3d polylines to convert to feature lines or [Xref]:` prompt, select the red closed polyline representing the limits of grading.

14. Right-click and select Enter, or press ↵.

    The Create Feature Lines dialog will appear.

15. Click the Create New button to the right of the site name to display the Site Properties dialog.

16. Enter **Rough Grading** for the name of this site, and click OK to dismiss the Site Properties dialog.

There are some differences in this Create Feature Lines dialog from that shown in Figure 16.4. Notably, the Conversion Options near the bottom of the dialog are now active, so take a look at the options presented:

Erase Existing Entities This option deletes the object and replaces it with a feature line object. This avoids the creation of duplicate linework, but could be harmful if you wanted your linework for planimetric purposes.

Assign Elevations This option lets you set the feature line elevations from a surface or grading group, essentially draping the feature line on the selected object.

Weed Points Weed Points decreases the number of nodes along the object. This option is handy when you're converting digitized information into feature lines.

17. Check the Assign Elevations box.

    The Erase Existing Entities option is checked by default, and you will not select the Weed Points option.

18. Click OK to dismiss the Create Feature Lines dialog.

    The Assign Elevations dialog appears. Here you can assign a single elevation for the feature line, assign the elevations from a grading if one is present in the drawing, or select a surface to pull elevation data from.

19. Verify that the EG surface is selected.
20. Verify that the Insert Intermediate Grade Break Points option is checked.

    This inserts a vertical point of intersection (VPI) at every point along the feature line where it crosses an underlying TIN line.

21. Click OK to dismiss the Assign Elevations dialog.
22. Select the feature line that was just created, and the grips will look like Figure 16.6.

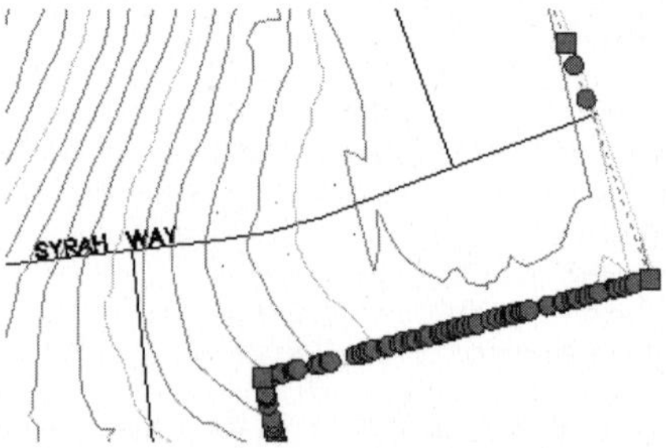

FIGURE 16.6
Conversion to a feature line object

When this exercise is complete, you may close the drawing. A saved copy of this drawing is available from the book's web page with the filename CreatingFeatureLines_FINISHED.dwg or CreatingFeatureLines_METRIC_FINISHED.dwg.

SQUARE GRIPS VS. CIRCULAR GRIPS

Note that in Figure 16.6 there are two types of grips: square and circular. Feature lines offer feedback via the grip shape:

Square Grips square feature line grip indicates a full PI. This node can be moved in the *x*, *y*, and *z* directions, manipulating both the horizontal and vertical design.

Circular Grips Circular grips are elevation points only. In this case, the elevation points are located at the intersection of the original polyline, and the TIN lines existing in the underlying surface. Elevation points can be slid along a given feature line, adjusting the vertical design, but cannot be moved in a horizontal plane.

This combination of PIs and elevation points makes it easy to set up a long element with numerous changes in design grade that will maintain its linear design intent if the endpoints are moved.

Both of the methods used so far assume static elevation assignments for the feature line. They're editable but are not physically related to other objects in the drawing. This is generally acceptable, but sometimes it's necessary to have a feature line that is dynamically related to an object. For grading purposes, it is often ideal to create a horizontal representation of a vertical profile along an alignment. Rather than build a corridor model as discussed in Chapter 10, "Basic Corridors," and Chapter 11, "Advanced Corridors, Intersections, and Roundabouts," a dynamic feature line can be extracted from a profile along an alignment, offset both horizontally and vertically, and used for grading. In the following example, a dynamic feature line is extracted from an alignment. Elevations for the vertices of the feature line are extracted from a profile, and finally, offset using the Create Feature Line From Stepped Offset tool to represent a swale.

1. Open the SteppedOffset.dwg or SteppedOffset_METRIC.dwg file.

2. From the Home tab ➢ Create Design panel, choose Feature Line ➢ Create Feature Lines From Alignment.
3. At the Select an alignment <or press enter key to select from list>: prompt, select the Cabernet Court alignment at the lower portion of the site (the north-south alignment) to display the Create Feature Line From Alignment dialog.
4. From the Create Feature Line From Alignment dialog, create a new site named **Grading ROW**.
5. In the Create Feature Line From Alignment dialog, deselect Weed Points.
6. Leave the other default settings as shown in Figure 16.7.

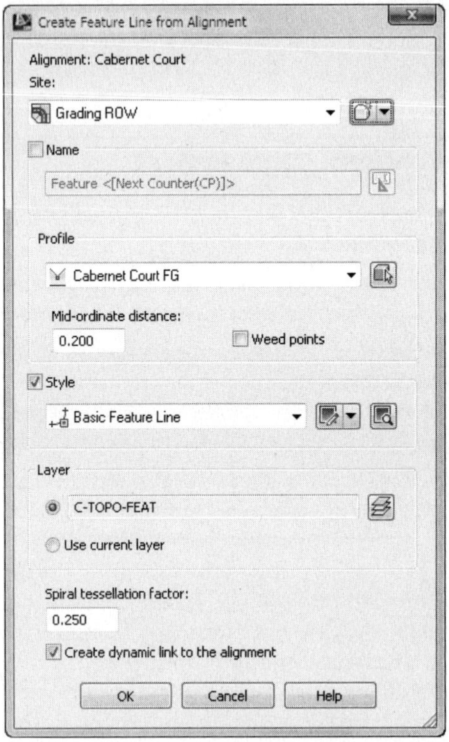

FIGURE 16.7
The Create Feature Line From Alignment dialog

Note that the Create Dynamic Link To The Alignment check box is selected near the bottom.

7. Click OK to dismiss the Create Feature Line From Alignment dialog. Note that if you select the feature line you just created, the Properties list it as an Auto Feature Line, whereas the feature line created in the first exercise is just a Feature Line.

8. From the Home tab ➢ Create Design panel, choose Feature Line ➢ Create Feature Line From Stepped Offset.

9. At the `Specify offset distance or [Through Layer]:` prompt, enter **25** ↵ (or **7.5** ↵ for metric users).

10. At the `Select an object to offset:` prompt, select the auto feature line along the alignment.

11. At the `Specify side to offset or [Multiple]:` prompt, select a point to the right of the alignment.

12. At the `Specify elevation difference or [Grade Slope Elevation Variable]:` prompt, enter **0** ↵ as the elevation difference at the command line.

    This simply offsets the line from the road centerline to the right-of-way line while holding the same elevation.

13. Repeat steps 10 through 12, but this time, pick a point to the left of the alignment.
14. Once complete, press ↵ to end the command.

    Your results should appear as shown in Figure 16.8.

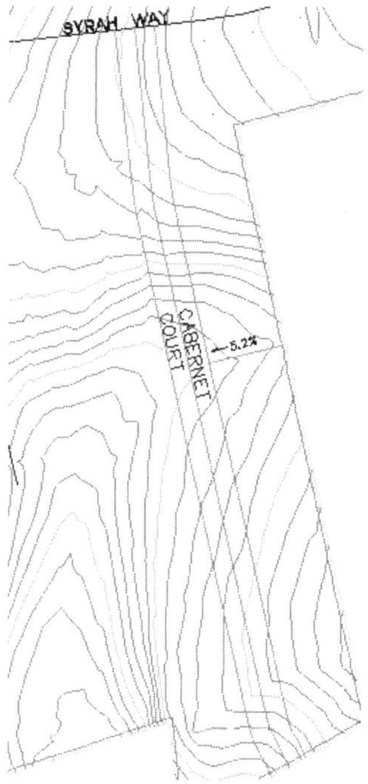

FIGURE 16.8
The completed alignment with offsets in place

Keep this drawing open to use in the next portion of the exercise.

If you click the alignment now, you will select the feature line that is stacked on top of the alignment. The feature line created by the alignment will not have any grips available. This is because the feature line is dynamically linked to the design profile along the alignment and can't be modified. If either the alignment or the profile changes, the dynamic feature line will automatically update. Simply repeat the preceding procedure to create new offsets if needed. These three feature lines can be included in a new surface definition as breaklines, as discussed in Chapter 4.

Because dynamically linked auto feature line objects are slightly different, you'll look at them in this next portion of the exercise.

15. Pan or zoom to view the feature line along the Cabernet Court alignment.
16. Select the auto feature line created from alignment to activate the Feature Line contextual tab.

17. From the Feature Line contextual tab ➤ Modify panel, choose Feature Line Properties to display the Feature Line Properties dialog, as shown in Figure 16.9.

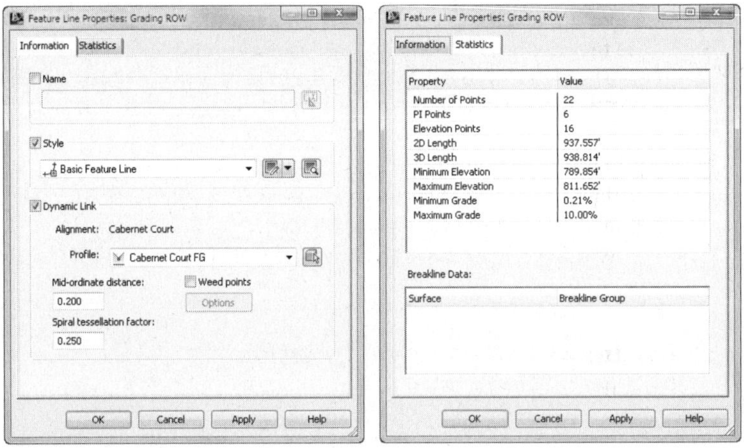

FIGURE 16.9
The Information tab (left) and Statistics tab (right) of the Feature Line Properties dialog for an alignment-based feature line

The information displayed on the Information tab is unique to the dynamic feature line, and you still have some level of control over the linking options.

18. On the Information tab, uncheck the Dynamic Link option and click OK to dismiss the Feature Line Properties dialog.

Notice the grips appear; they weren't previously visible when the feature line was dynamically linked. The dynamic relationship between the feature line and the alignment has been severed.

19. With the feature line still selected, from the Feature Line contextual tab ➤ Modify panel, choose Feature Line Properties and notice the Dynamic Link options have disappeared.

Once the dynamic link is severed, it cannot be reinstated.

20. Click OK to dismiss the Feature Line Properties dialog and press Esc to deselect the feature line.

21. Select the two feature lines offset from the alignment feature line.

Notice that with two feature lines selected the Feature Line contextual tab still appears but the panels are limited.

22. With the feature lines still selected, from the Feature Line contextual tab ➤ Modify panel, choose Apply Feature Line Styles to display the Apply Feature Line Style dialog.

23. Select the Grading Ditch style from the Style drop-down list and click OK to dismiss the dialog

24. Press Esc to deselect the feature lines.

USING STYLES WITH FEATURE LINE OBJECTS

Many people don't see much advantage in using styles with feature line objects, but there is one major benefit: linetypes. Zoom in on the feature lines on either side of the alignment, and you'll see the Grading Ditch style has a dashed linetype. 3D polylines do not display linetypes, but feature lines do. If you need to show the linetypes in your grading, feature line styles are your friend.

When this exercise is complete, you may close the drawing. A saved copy of this drawing is available from the book's web page with the filename `SteppedOffset_FINISHED.dwg` or `SteppedOffset_METRIC_FINISHED.dwg`.

Now that you've created a couple of feature lines, you'll edit and manipulate them some more.

Editing Feature Line Information

From the Feature Line contextual tab, several more commands can be found on the Modify panel, as shown in Figure 16.10, that are worth examining before you get into editing objects.

FIGURE 16.10
The Modify panel on the Feature Line contextual tab

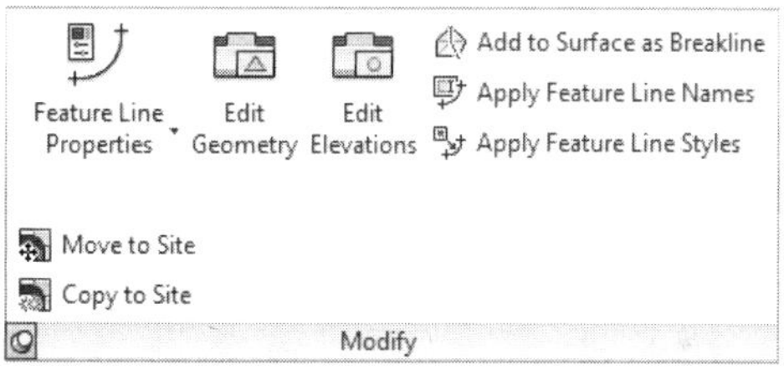

The Modify panel of the Feature Line contextual tab provides commands for editing various properties of the feature line, the feature line style, and the feature line geometry as follows:

Feature Line Properties The top half of the Feature Line Properties tool is a button that will access the Feature Line Properties dialog. The Feature Line Properties dialog has two tabs: Information and Statistics, as shown previously in Figure 16.9. The Information tab of the Feature Line Properties dialog allows you to edit the name or feature line style. The Statistics tab of the Feature Line Properties dialog allows you to access various physical properties such as minimum and maximum grade.

The bottom half of the Feature Line Properties tool is a drop-down menu containing two commands:

Feature Line Properties This command accesses the same Feature Line Properties dialog as discussed earlier when you click the button.

Edit Feature Line Style This command is used to access various display characteristics of the feature line such as color and linetype. Feature line styles will be discussed further in Chapter 21, "Object Styles."

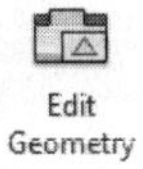

Edit Geometry The Edit Geometry toggle opens the Edit Geometry panel on the Feature Line contextual tab (see Figure 16.11). This panel will remain open until the Edit Geometry button is toggled off (it's highlighted when toggled on). The Edit Geometry panel will be discussed later in this section.

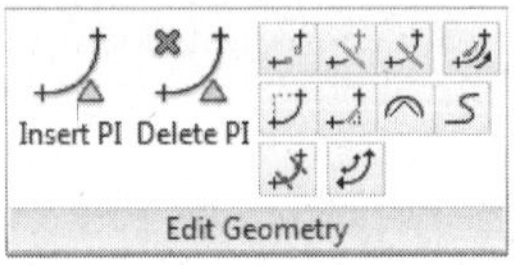

FIGURE 16.11
The Edit Geometry panel on the Feature Line contextual tab

Edit Elevations The Edit Elevations toggle opens the Edit Elevations panel on the Feature Line contextual tab (see Figure 16.12). This panel will remain open until the Edit Elevations button is toggled off (it's highlighted when toggled on). The Edit Elevations panel will be discussed later in this chapter in the "Editing Feature Line Elevations" section.

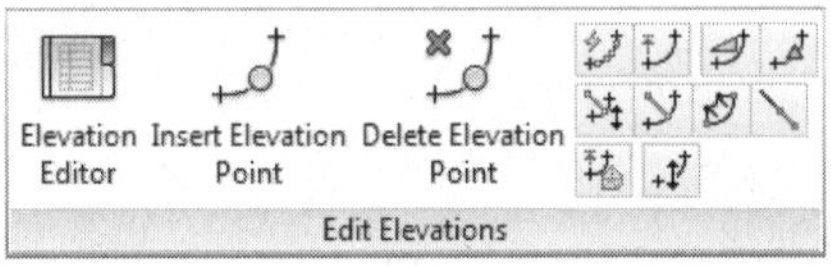

FIGURE 16.12
The Edit Elevations panel on the Feature Line contextual tab

Add To Surface As Breakline The Add To Surface As Breakline tool allows you to select a feature line or feature lines to add to a surface as breaklines. Once feature lines are added to the surface, they will be listed as a Breakline Set in the Definition branch of the surface in Prospector.

Apply Feature Line Names The Apply Feature Line Names tool allows you to change a series of feature lines en masse based on a new naming template. This tool can be helpful when you want to rename a group or just assign names to feature line objects. This tool cannot be used on an auto feature line that is dynamically linked to an alignment and profile.

Apply Feature Line Styles The Apply Feature Line Styles tool allows you to change feature line objects and their respective styles en masse. Many users don't apply styles to their feature line objects because the feature lines are found in grading drawings, and not meant to be seen in construction documents. But if you need to make a global change, you can.

Move To Site If you expand the Modify panel, you will notice two additional commands. The first command is Move To Site. This command allows you to associate the selected feature line with a new site.

Copy To Site The other command in the extended Modify panel is Copy To Site. This command allows you to duplicate the selected feature with a new site while leaving the original feature line in its current site. The two feature lines do not remain dynamically linked.

Once the Feature Line contextual tab has been activated, the Quick Profile tool is available on the Launch Pad panel. The Quick Profile tool generates a temporary profile of the feature line based on user parameters found in the Create Quick Profiles dialog (Figure 16.13).

FIGURE 16.13
The Create Quick Profiles dialog

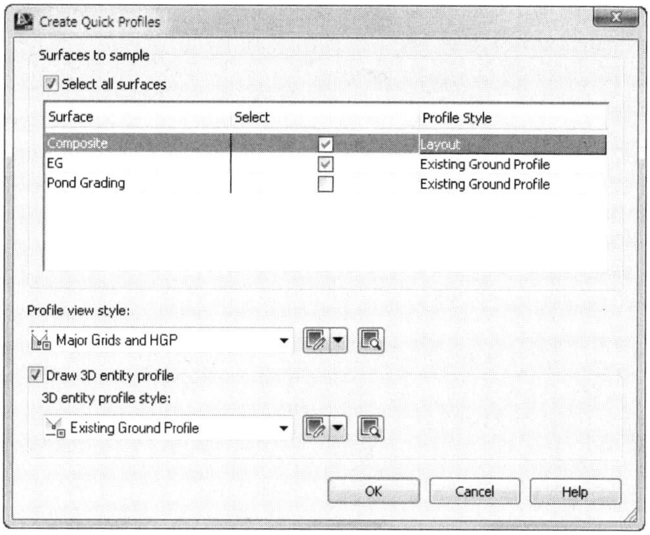

A few notes on this operation:

- Civil 3D creates a temporary phantom alignment that will not display in Prospector as the basis for a quick profile. A unique alignment number is assigned to this alignment.
- Panorama will display a message to tell you that a quick profile has been generated and reminds you that "this is a temporary object and will be deleted on save command or on exit from drawing." You can close Panorama or move the Panorama palette out of the way if necessary.

Feature lines aren't the only things you can create a quick profile from. You can also create a quick profile for 2D or 3D lines or polylines, lot lines, survey figures, or even a series of points. When a quick profile is created from 3D objects, there is an additional option to draw the 3D entity profile.

The Edit Geometry and Edit Elevations toggles provide even more commands, which make them considerably more powerful than a standard 3D polyline. The Edit Geometry functions and the Edit Elevations functions are described in the next sections. While these sections reference the Edit Geometry and Edit Elevations panels of the Feature Line contextual tab, these panels are also available on the Modify tab. When used from the Modify tab, many of these commands can also be used to edit parcel lines, survey figures, 3D polylines, and 2D polylines, in addition to feature lines as discussed in the sections that follow.

Editing feature-line geometry grading revisions often requires adding PIs, breaking apart feature lines, trimming, and performing other planar operations without destroying the vertical information. To access the commands for editing feature line horizontal information, select the feature line to access the Feature Line contextual tab and toggle on the Edit Geometry panel, as shown in Figure 16.14.

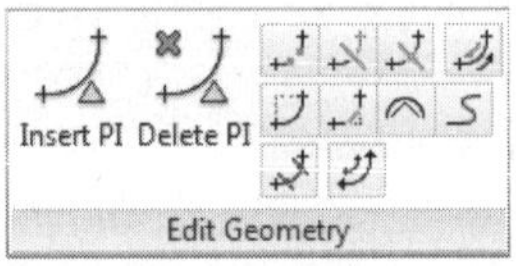

FIGURE 16.14
The Edit Geometry panel on the Feature Line contextual tab

The first two tools are designed to manipulate the PI points that make up a feature line:

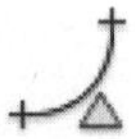

Insert PI The Insert PI tool allows you to insert a new PI, controlling both the horizontal and vertical design.

Delete PI The Delete PI tool removes a PI. The feature line will mend the adjoining segments if possible, attempting to maintain similar geometry.

The next few tools act like their AutoCAD® counterparts, but are specifically for use with feature lines since they understand that elevations are involved and will add PIs accordingly:

Break The Break tool operates much like the AutoCAD Break command, allowing two feature lines to be created from one. Additionally, if a feature line is part of a surface definition, both new feature lines are added to the surface definition to maintain integrity. Elevations at the new PIs are assigned on the basis of an interpolated elevation.

Trim The Trim tool operates much like the AutoCAD Trim command, trimming a feature line and adding a new end PI on the basis of an interpolated elevation.

Join The Join tool creates one feature line from two, making editing and control easier. You can set the tolerance distance from the settings associated with the Join tool on the Settings tab of Toolspace by doing the following:

1. Expand the Grading ➤ Commands branch.
2. Right-click JoinFeatures and select Edit Command Settings.
3. In the Edit Command Settings – JoinFeatures dialog, expand the Feature Line Join property to change the tolerance.

Reverse The Reverse tool changes the direction of a feature line. This will change the labeling.

Edit Curve The Edit Curve tool allows you to modify the radius that has been applied to a feature line object. Once the feature line is selected, the Edit Feature Line Curve dialog will display. This dialog will allow you to step through each of the curves along the feature line. For each curve you can modify the radius while viewing information on the curve length, chord length, and tangent length. There is also an option to maintain tangency.

Fillet The Fillet tool inserts a curve at PIs along a feature line and will join feature lines sharing a common PI that are not actually connected.

The last few tools refine feature lines, making them easier to manipulate and use in surface building:

Fit Curve The Fit Curve tool analyzes a number of elevation points and attempts to define a working arc through them all. This tool is often used when the corridor utilities are used to generate feature lines. These derived feature lines can have a large number of unnecessary PIs in curved areas. You can modify the tolerance and the minimum number of segments by entering **O** to select Options on the command line during the prompt, and display the Fit Curve Options dialog.

Alternatively you can set the default values for these options from the settings associated with the Fit Curve tool on the Settings tab of Toolspace by doing the following:

1. Expand the Grading ➢ Commands branch.
2. Right-click FitCurveFeature and select Edit Command Settings.
3. In the Edit Command Settings – FitCurveFeature dialog, expand the Feature Line Fit Curve property to change the tolerance and specify the minimum number of segments.

Smooth The Smooth tool takes a series of disjointed feature line segments and creates a best-fit curve. This tool is great for creating streamlines or other natural terrain features that are known to curve, but there's often not enough data to fully draw them that way. You can also straighten previously-smoothed feature lines on the Modify tab ➢ Edit Geometry panel by choosing the Smooth command. Notice that the Smooth command accessed through the Feature Line contextual tab does not give you the straighten option.

Weed The Weed tool allows the user to remove elevation points and PIs on the basis of various criteria. Once you select the feature line or multiple feature lines, or a partial feature line, the Weed Vertices dialog will display. This dialog will allow you to weed based on any combination of angle, grade, or length; in addition you can remove points based on their 3D distance between one another. At the bottom of the Weed Vertices dialog it states how many of the total number of vertices will be weeded. This is great for cleaning up corridor-generated feature lines as well.

Similar to the Join and Fit Curve tools, you can set the default settings associated with the Weed command on the Settings tab of Toolspace by doing the following:

1. Expand the Grading ➢ Commands branch.
2. Right-click WeedFeatures and select Edit Command Settings.
3. In the Edit Command Settings – WeedFeatures dialog, expand the Feature Line Weed property to change the various values.

Stepped Offset As discussed in detail earlier, the Stepped Offset tool allows offsetting in a horizontal and vertical manner, making it easy to create stepped features such as stairs or curbs.

By using these controls, it's easier to manipulate the design elements of a typical site while still using feature lines for surface design. In this exercise, you manipulate a number of feature lines that were created by corridor operations:

1. Open the `FeatureLineGeometry.dwg` or `FeatureLineGeometry_METRIC.dwg` file.

   This drawing is a continuation of the `SteppedOffset.dwg` file but has been populated with some more feature lines.

2. Select the southern east-west feature line (which is the offset from the Syrah Way alignment feature line), as shown in Figure 16.15.

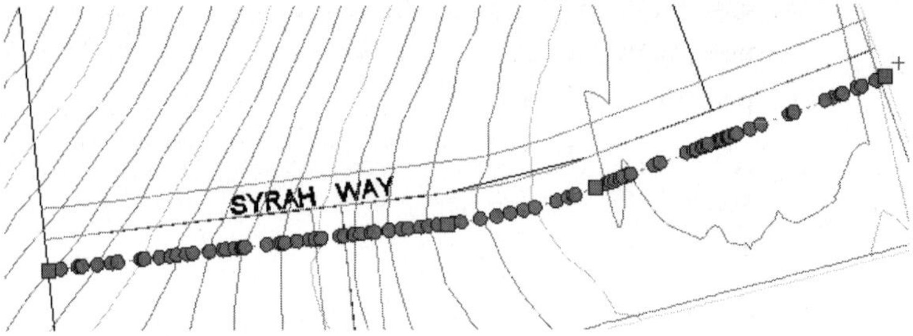

FIGURE 16.15
Picking the southern feature line on Syrah Way

3. From the Feature Line contextual tab ➢ Modify panel, choose Edit Geometry to toggle on the Edit Geometry panel.

4. Click the Break tool.
5. At the `Select an object to break:` prompt, select the lower feature line again.
6. At the `Specify second break point or [First point]:` prompt, enter **F** ↵ to pick the first point of the break.
7. At the `Specify first break point:` prompt, using an Intersection Osnap select the intersection of the east-west feature line to the south of Syrah Way, and the north-south feature line to the west of Cabernet Court, as shown in Figure 16.16.

FIGURE 16.16
Using the Intersection object snap to select a point

8. At the `Specify second break point:` prompt, using a Nearest Osnap select a point on the east-west feature line to the east of the eastern north-south feature line, leaving a gap, as shown in Figure 16.17.

FIGURE 16.17
The feature line after executing the Break command

9. Select the right east-west feature line that was previously broken and notice the large number of grips.

10. From the Feature Line contextual tab ➤ Edit Geometry panel, click the Weed tool.
11. At the `Select a feature line, 3d polyline or [Multiple Partial]:` prompt, select the feature line again.

    The Weed Vertices dialog will be displayed.

12. Change the Angle to **0.2** and the Grade to **0.25%**.

    Watch the glyphs on the feature line and notice the number of vertices that will be removed from the feature line. Additionally, the glyphs on the feature line itself will change from green to red to reflect nodes that will be removed under the current setting, as shown in Figure 16.18.

FIGURE 16.18
Feature lines to be weeded

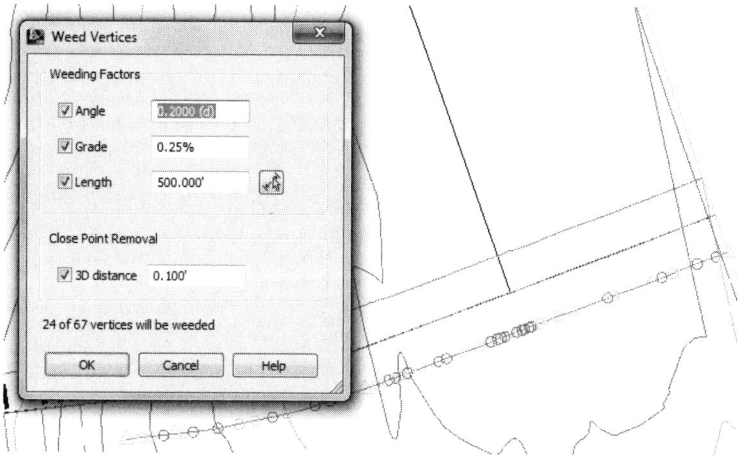

13. Click OK to complete the command and dismiss the Weed Vertices dialog.
14. Using a standard AutoCAD Extend command, extend the right east-west feature line to the eastern Cabernet Court offset feature line.

    There is no unique feature line extending tool.

15. Select the left east-west feature line that was previously broken to activate the Feature Line contextual tab.

16. From the Feature Line contextual tab ➤ Edit Geometry panel, click the Trim tool from the Edit Geometry panel.

    A standard AutoCAD trim will not work in this case.

17. At the `Select cutting edges:` prompt, select the left east-west feature line as the cutting edge and press ↵.
18. At the `Select objects to trim:` prompt, select the western Cabernet Court offset feature line above the cutting edge and press ↵ to end the command and review the results (see Figure 16.19).

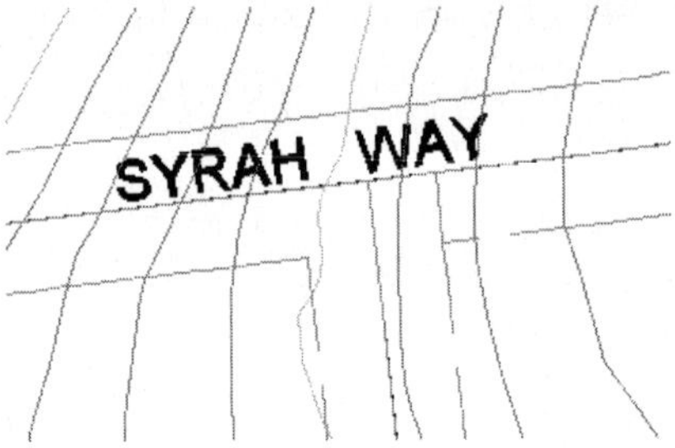

FIGURE 16.19
The left east-west feature line trimmed

19. From the Feature Line contextual tab ➢ Edit Geometry panel, click the Join tool.
20. At the `Select the connecting feature line, polyline or 3d polyline or [Multiple]:` prompt, select the left north-south feature line and press ↵.

    Notice the grips and that the two feature lines have been joined.

21. From the Feature Line contextual tab ➢ Edit Geometry panel, click the Fillet tool.
22. At the `Specify corner or [All Join Radius]:` prompt, enter **R** ↵ to adjust the radius value.
23. At the `Specify radius:` prompt, enter **15** ↵ (or **4.5** ↵ for metric users).
24. Move your cursor toward the join corner just created until the glyph appears and select near the glyph to fillet the two feature lines.
25. Press ↵ to end the command.

26. From the Feature Line contextual tab ➢ Edit Geometry panel, click the Edit Curve tool
27. At the `Select feature line curve to edit or [Delete]:` prompt, select the curve you just create.

    The Edit Feature Line Curve dialog opens, as shown in Figure 16.20.

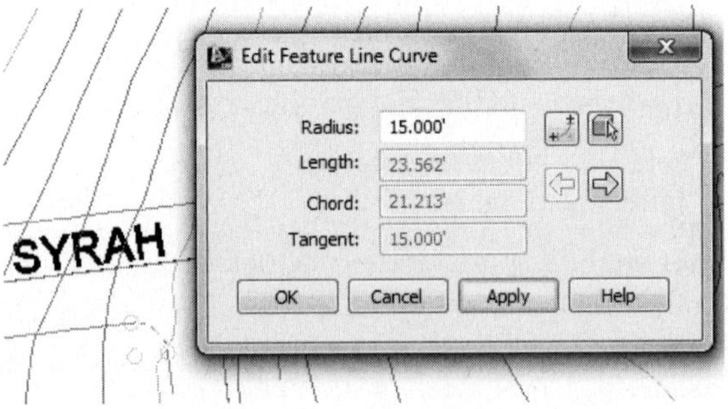

FIGURE 16.20
The Edit Feature Line Curve dialog

28. Enter a Radius value of **34** ↵ (or **10** ↵ for metric users) , and click OK to close the dialog.
29. Press ↵ to end the command. The drawing should now look similar to Figure 16.21.

FIGURE 16.21
Filleted feature lines

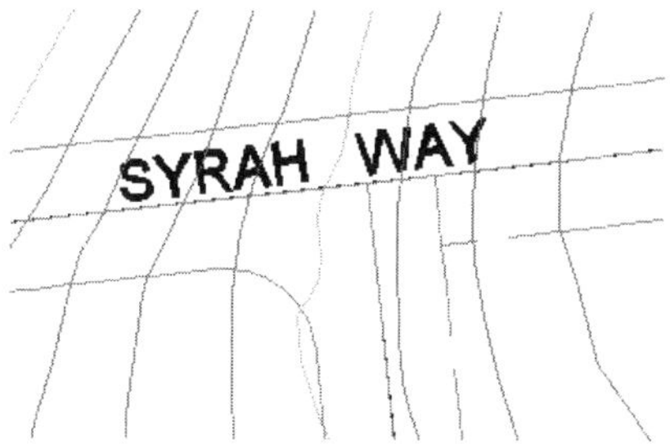

When this exercise is complete, you may close the drawing. A saved copy of this drawing is available from the book's web page with the filename `FeatureLineGeometry_FINISHED.dwg` or `FeatureLineGeometry_METRIC_FINISHED.dwg`.

> **MODIFYING FEATURE LINES**
>
> When modifying the radius of a feature line curve, it's important to remember that you must have enough tangent feature line on either side of the curve segment to make a curve fit, or the program will not make the change. In that case, tweak the feature line on either side of the arc until there is a mathematical solution.
>
> Sometimes it is necessary to use the Weed Vertices tool to remove vertices, and create enough room to fillet feature lines. In some cases, it may be necessary to plan ahead when creating feature lines to ensure that vertices will not be placed too closely together.

EDITING FEATURE LINE ELEVATIONS

To access the commands for editing feature line elevation information, from the Feature Line contextual tab ➢ Modify panel, choose the Edit Elevations toggle to open the Edit Elevations panel.

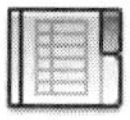

The first tool in this panel is the Elevations Editor, which will activate a palette in Panorama to display and edit station, elevation, length, and grade information about the feature line selected in a tabular grid format. When you are in a row in the Elevation Editor, the corresponding point will be shown with a temporary marker on the plan.

Quite a few tools are available to modify and manipulate feature lines. You won't use all of the tools, but at least you'll have some concept of what is available. The next exercises give you a look at a few of them. In this first exercise, you'll take a brief look at the Grading Elevation Editor tools:

1. Open the `EditingFeatureLineElevations.dwg` or `EditingFeatureLineElevations_METRIC.dwg` file.

   This drawing contains a sample layout with some curb and gutter work.

2. Select the feature line representing the left flowline of the curb and gutter area to activate the Feature Line contextual tab.

3. On the Feature Line contextual tab ➤ Edit Elevations panel, choose the Elevation Editor tool

   The Grading Elevation Editor in Panorama will open, as shown in Figure 16.22.

FIGURE 16.22 The Grading Elevation Editor

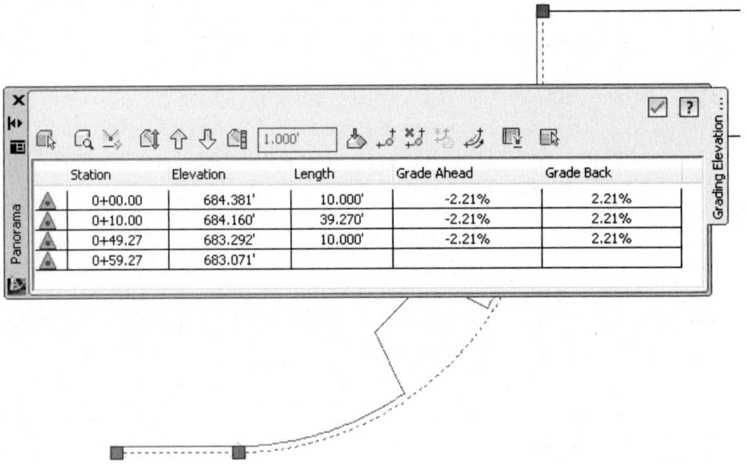

In the far left column of the Grading Elevation Editor is a series of symbols; these are the same glyph symbols used on the feature line grips. A triangular symbol denotes a PI, and a circular symbol denotes an elevation point. In this example you don't have any elevation points.

4. Click in the Grade Ahead column for the first PI at Station 0+00.00 (or 0+000.00).

   It's hard to see in the images, but as a row is selected in the Grading Elevation Editor, the PI or the elevation point that was selected will be highlighted on the screen with a small glyph.

   You can use the Grading Elevation Editor to make changes to the Station, Elevation, Length, Grade Ahead, and Grade Back settings. The exception is that you cannot edit stationing for primary geometry points, as indicated by the triangle glyph in the Grading Elevation Editor.

   Across the top of the Grading Elevation Editor are multiple tools that will be used as you edit the data in the table:

   Select Clicking the Select tool allows you to select the feature line on the screen for editing in the Grading Elevation Editor.

Zoom To The Zoom To tool will do exactly as it says. If you have a station highlighted in the Panorama and select the Zoom To tool, your plan view will be zoomed to that station on the feature line on the screen.

Quick Profile The Quick Profile tool will generate a temporary profile based on the feature line selected. This is the same tool discussed earlier that was available on the Feature Line contextual tab ➤ Launch Pad panel. The Create Quick Profiles dialog will open, allowing you to select what surface(s) you wish to display, as well as what profile view style and 3D entity profile you want.

Raise/Lower Clicking the Raise/Lower tool will activate the Set Increment text box. This allows you to raise or lower selected elevation points, or if no elevation points are selected, it will raise or lower the entire feature line elevation points by the amount displayed in the text box.

Raise Incrementally and Lower Incrementally The Raise Incrementally/Lower Incrementally tools will raise or lower the elevation point or points by the amount listed in the Set Increment text box.

Set Increment The Set Increment tool works in tandem with the text box. You can enter a value that will be used by other tools for raising or lowering elevation point or points.

Flatten Grade Or Elevations The Flatten Grade Or Elevations tool will make all of the selected elevation points the value of the first selected point, or if no points are selected, the value of the first entry in the cells. When this tool is selected, the Flatten dialog will open asking if you want to flatten by constant elevation or by constant grade.

Insert Elevation Point The Insert Elevation Point will let you select a spot on the feature line, and will create an intermediate elevation point. The Insert PVI dialog will open, allowing you to fine-tune the station and enter an elevation.

Delete Elevation Point The Delete Elevation Point will delete a point or points that are highlighted in the Grading Elevations Editor. Note that this tool will allow you to delete only intermediate points.

Elevations From Surface Clicking the Elevations From Surface tool will open the Select Surface dialog if there are multiple surfaces to choose from. If you have an elevation point or points selected, it will affect only those points, or if nothing is selected it will use all elevation points and drape them onto the selected surface.

Reverse The Direction The Reverse The Direction tool will do exactly as it says; it will reverse the direction of the feature line, thus changing the stationing and the grade ahead/grade back directions.

Show Grade Breaks Only The Show Grade Breaks Only tool is a toggle (click, it's on and click, it's off) that will display only the rows where the grade varies or breaks on the feature line.

Unselect All Rows The Unselect All Rows tool does exactly as it says; it will deselect any rows that have been highlighted for editing.

5. Click the green check mark in the upper-right corner to dismiss the Panorama.

Using the Grading Elevation Editor is the most basic way to manipulate elevation information. Keep this drawing open for use in the next exercise.

More Feature Line Elevation Editing Tools

To access the commands for editing feature line elevation information, select the feature line to access the Feature Line contextual tab and toggle on the Edit Elevations panel as shown in Figure 16.23. Many of the tools in the Elevation Editor may seem redundant from the Grading Elevation Editor tools just discussed, but they are placed here for ease of use.

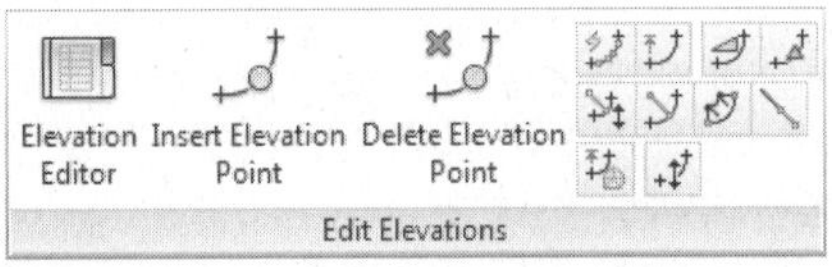

Figure 16.23
The Edit Elevations panel on the Feature Line contextual tab

Moving across the panel beyond the Elevation Editor tool, which was just discussed, you find the following tools for modifying or assigning elevations to feature lines:

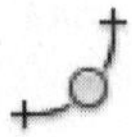

Insert Elevation Point The Insert Elevation Point tool inserts an elevation point at the point selected or multiple elevation points at a specified increment. Note that elevation points can control only elevation information; it does not act as a horizontal control point.

Delete Elevation Point The Delete Elevation Point tool deletes the selected elevation point; the points on either side then become connected linearly on the basis of their current elevations. You can also delete all elevation points on a feature line with this command.

Quick Elevation Edit The Quick Elevation Edit tool allows you to use onscreen cues to set elevations and slopes quickly between PIs on any feature lines.

Edit Elevations The Edit Elevations tool steps through the selected feature line, much like working through a polyline edit at the command line, allowing you to change elevations and grades or insert, move, and delete elevation points.

Set Grade/Slope Between Points The Set Grade/Slope Between Points tool sets a continuous slope along the feature line between selected points.

Insert High/Low Elevation Point The Insert High/Low Elevation Point tool places a new elevation point on the basis of two picked points and the forward and backward slopes. This calculated point is simply placed at the intersection of two vertical slopes.

Raise/Lower By Reference The Raise/Lower By Reference tool allows you to adjust a feature line elevation based on a given slope from another location. This relationship isn't dynamic!

Set Elevation By Reference The Set Elevation By Reference tool sets the elevation of a selected point along the feature line by picking a reference point, and then establishing a relationship to the selected feature line point. This relationship isn't dynamic!

Adjacent Elevations By Reference The Adjacent Elevations By Reference tool allows you to adjust the elevation on a feature line by coming at a given slope or delta from another point or feature line point. This relationship isn't dynamic!

Grade Extension By Reference The Grade Extension By Reference tool allows you to apply the same grades to different feature lines across a gap. For example, you might use this tool along the back of curbs at locations such as driveways or intersections. This relationship isn't dynamic!

Elevations From Surface The Elevations From Surface tool sets the elevation at each PI and elevation point on the basis of the selected surface. It will optionally add elevation points at any point where the feature line crosses a surface TIN line.

Raise/Lower The Raise/Lower tool simply moves the entire feature line in the *z* direction by an amount entered at the command line.

Some of the relative elevation tools are a bit harder to understand, so you'll look at them in our next exercise and see how they function in some basic scenarios:

1. If not opened for the previous exercise, open the `EditingFeatureLineElevations.dwg` or `EditingFeatureLineElevations_METRIC.dwg` file.

   This drawing contains a sample layout with some curb and gutter work.

2. Zoom to the ramp shown on the left-hand side of the intersection.
3. Select the feature line describing the ramp at the left-hand side to activate the Feature Line contextual tab.
4. If the Edit Elevations panel isn't already displayed, from the Feature Line contextual tab ➤ Modify panel choose Edit Elevations.
5. From the Feature Line contextual tab ➤ Edit Elevations panel, choose the Elevation Editor tool to display the Grading Elevation Editor tab in Panorama.

   Notice that the entire feature line is at elevation 0.000′ (0.000 m).

6. Click the green check mark in the upper right to close the Panorama.
7. Press Esc to deselect the left ramp feature line.
8. Select the feature line representing the flowline of the curb and gutter area at the left-hand side to activate the Feature Line contextual tab.
9. From the Feature Line contextual tab ➤ Edit Elevations panel, choose the Adjacent Elevations By Reference tool.
10. At the `Select object to edit or [Name]:` prompt, select the left ramp feature line.

    Civil 3D will display a number of glyphs and lines to represent what points along the flowline it is using to establish elevations from.

11. At the `Specify elevation difference or [Grade Slope]:` prompt, enter **0.5** ↵ (or **0.15** ↵ for metric users) at the command line to update the ramp elevations.
12. Press ↵ again to end the command.
13. Press Esc to deselect the gutter feature line, and then select the ramp feature line.

14. From the Feature Line contextual tab ➤ Edit Elevations panel, choose the Elevation Editor tool to display the Grading Elevation Editor tab in Panorama.

    Notice that the PIs now each have an elevation, as shown in Figure 16.24.

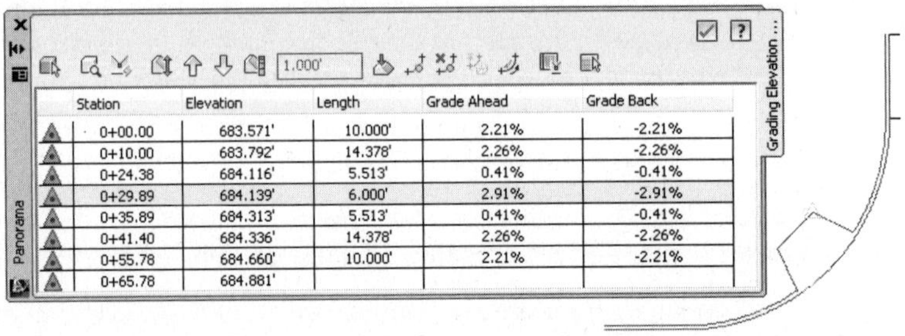

Station	Elevation	Length	Grade Ahead	Grade Back
0+00.00	683.571'	10.000'	2.21%	-2.21%
0+10.00	683.792'	14.378'	2.26%	-2.26%
0+24.38	684.116'	5.513'	0.41%	-0.41%
0+29.89	684.139'	6.000'	2.91%	-2.91%
0+35.89	684.313'	5.513'	0.41%	-0.41%
0+41.40	684.336'	14.378'	2.26%	-2.26%
0+55.78	684.660'	10.000'	2.21%	-2.21%
0+65.78	684.881'			

FIGURE 16.24 Completed editing of the curb ramp feature line

15. Click the green check mark in the upper right of the Panorama to dismiss it.

16. Save and keep this drawing open for the next portion of the exercise.

    Next, you'll need to extend the grade along the line representing the flowline of the curb and gutter area on the left of the screen to determine the elevation to the south of your intersection on the right of the screen. You'll use the Grade Extension By Reference tool in the following portion of the exercise to accomplish this.

17. Select the feature line representing the flowline of the curb and gutter area on the left side of the drawing.

18. From the Feature Line contextual tab ➤ Edit Elevations panel, choose the Grade Extension By Reference tool.

19. At the `Select reference segment:` prompt, select the left flowline feature line again (see Figure 16.22).

    This tool will evaluate the feature line as if it were three separate components (two lines and an arc). Because you are extending the grade of the flowline to the east and across the intersection, it is important to select the tangent segment, as shown in Figure 16.25. If you select the wrong segment, press Esc and then select the correct segment.

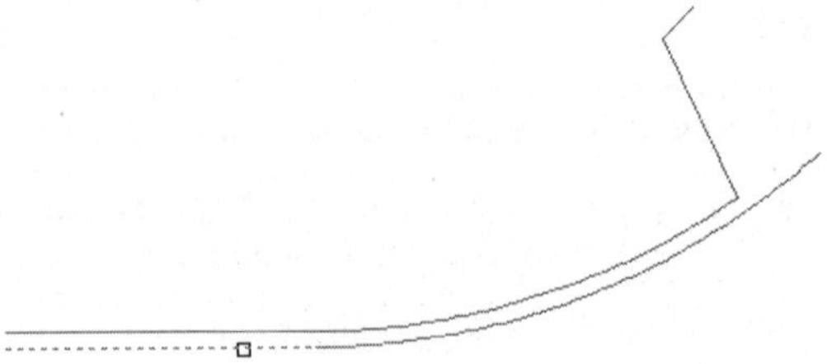

FIGURE 16.25 Selecting the flowline of the curb and gutter section at the left

20. At the `Select object to edit or [Name]:` prompt, select the line representing the flowline of the curb and gutter area to the right of your screen, as shown in Figure 16.26.

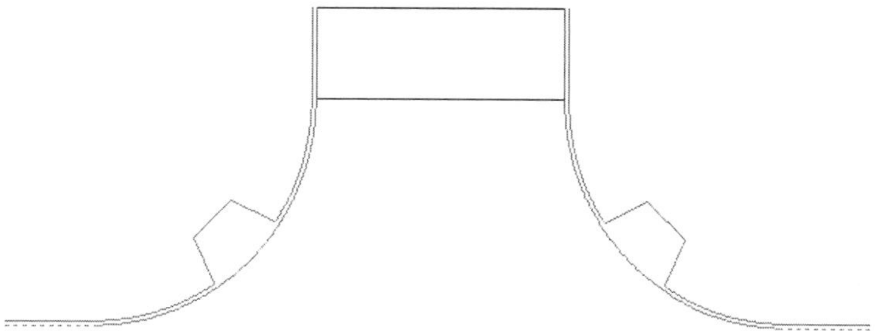

FIGURE 16.26
Selecting the flowline at the right side of the tangent segment

21. At the `Specify point:` prompt, pick the PI at the left side of the tangent, as shown in Figure 16.27.

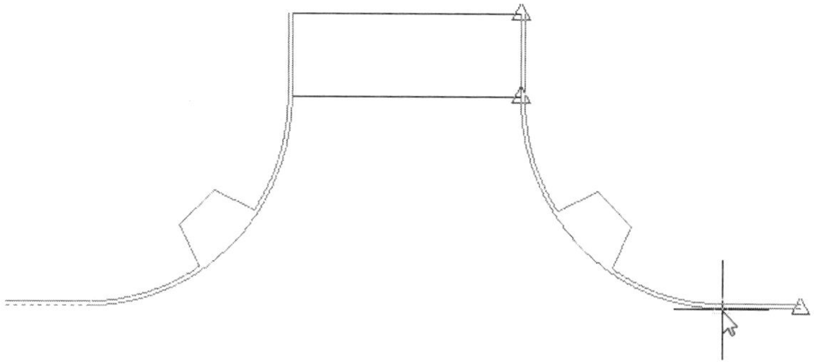

FIGURE 16.27
Selecting the PI at the left side of the tangent segment representing the flowline of the curb and gutter section

22. At the `Specify grade or [Slope Elevation Difference] <-2.21>:` prompt, press ↵ to accept the default value of -2.21 (this is the grade of the flowline of the curb and gutter section to the left).
23. Press ↵ to end the command, and press Esc to cancel grips.
24. Select the feature line representing the flowline of the curb and gutter section to the right of your screen to enable grips.
25. Move your cursor over the top of the PI, as shown in Figure 16.28, but do not click.

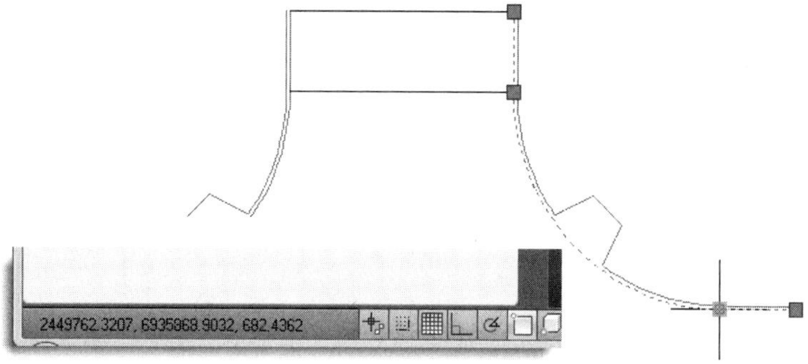

FIGURE 16.28
The x-, y-, and z-coordinates of the PI displayed on the status bar

Your cursor will automatically snap to the grip, and the grip will change color. Notice that the elevation of the PI is displayed on the status bar, as shown in Figure 16.28. This is a quick way to check elevations of vertices when modeling terrain.

If your coordinates are not displayed,

a. Enter the AutoCAD command **COORDS**.

b. Set the value to **1** to display them.

Leave this drawing open for the last portion of the exercise.

With the elevation of a single point determined, the grade of the feature line representing the curb and gutter section can be modified to ensure positive water flow. In the last portion of the exercise, you'll use the Set Grade/Slope Between Points tool to modify the grade of the feature line.

26. Select the feature line representing the flowline of the curb and gutter section to the right of your screen.

27. From the Feature Line contextual tab ➤ Edit Elevations panel, choose the Set Grade/Slope Between Points.

28. At the `Specify the start point:` prompt, select the PI as shown previously in Figure 16.27.

The elevation of this point has been established and will be used as the basis for grading the entire feature line.

29. At the `Specify elevation:` prompt, press ↵ to accept the default value.

This is the current elevation of the PI as established earlier.

30. At the `Specify the end point:` prompt, select the PI as shown in Figure 16.29.

FIGURE 16.29
Specifying the PI to establish the elevation at the flowline of the curb and gutter section

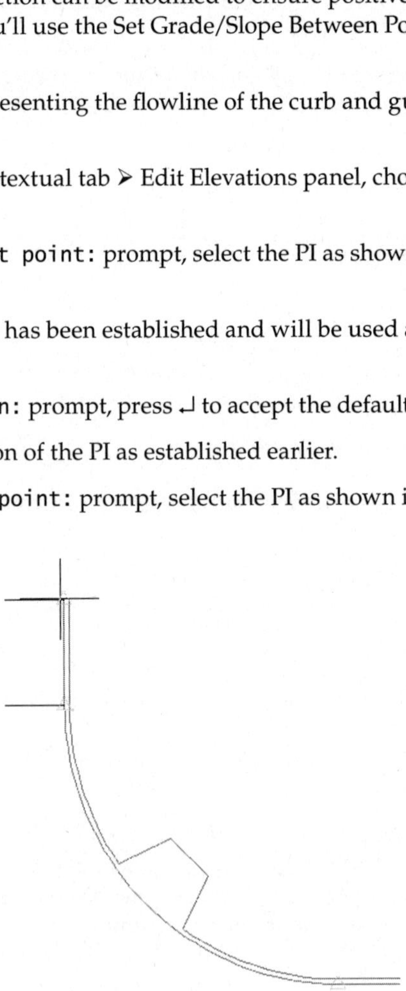

31. At the `Specify grade or [Slope Elevation Difference]:` prompt, enter **2** ↵ to set the grade between the points at 2 percent. Do not end the command.

32. At the `Select object:` prompt, select the feature line representing the flowline of the curb and gutter section to the right again.

33. At the `Specify the start point:` prompt, pick the PI (shown earlier in Figure 16.27) again.

34. At the `Specify elevation:` prompt, press ↵ to accept the default value.

    This is the current elevation of the PI as established earlier.

35. When you see the `Specify the end point:` prompt, pick the PI at the far bottom right along the feature line currently being edited.

36. At the `Specify grade or [Slope Elevation Difference]:` prompt, enter **-2** ↵ to set the grade between the points at negative 2 percent.

37. Press ↵ to end the command but do not cancel grips.

38. Select the Elevation Editor tool from the Edit Elevations panel to display Panorama.

    Notice the values in both the Grade Ahead and Grade Back columns, as shown in Figure 16.30.

FIGURE 16.30
The grade of the feature line set to 2 percent

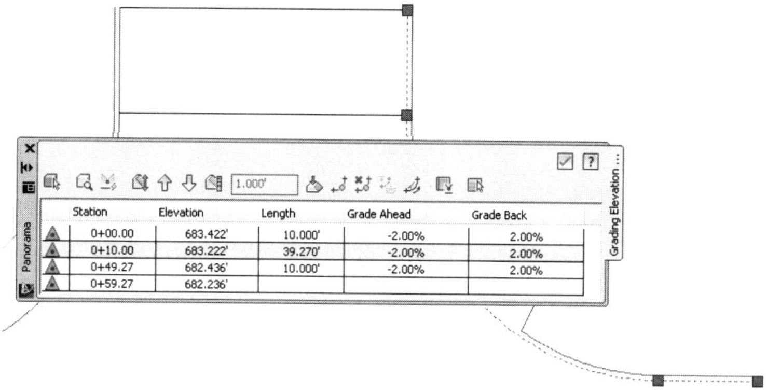

To finish the intersection, you could use the Adjacent Elevations By Reference tool again to define the right ramp.

When this exercise is complete, you may close the drawing. A saved copy of this drawing is available from the book's web page with the filename `EditingFeatureLineElevations_FINISHED.dwg` or `EditingFeatureLineElevations_METRIC_FINISHED.dwg`.

Using these feature line elevation editing tools, the possibilities are endless. In using feature lines to model proposed features, you are limited only by your creative approach. You've seen many of the tools in action, so you can now put a few more of them together and grade a pond.

DRAINING THE POND

You need to use a combination of feature line tools and options to pull your pond together, and get the most flexibility should you need to update the bottom area or manipulate the pond's general shape. In this exercise, you will use feature lines to design your pond:

1. Open the `PondDrainageDesign.dwg` or `PondDrainageDesign_METRIC.dwg` file.

   The engineer gave us a bit more information about the pond design, as shown in Figure 16.31.

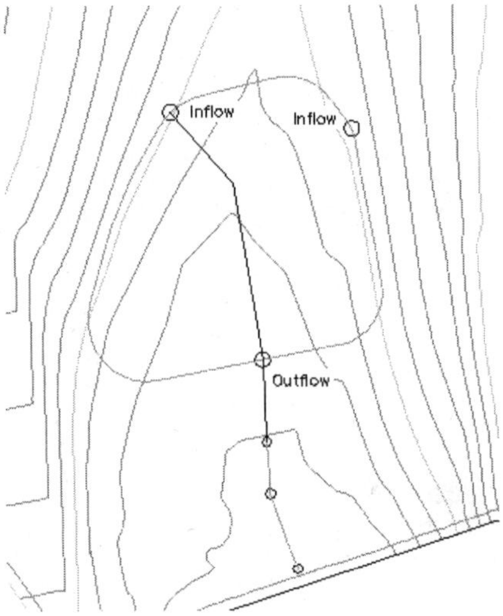

FIGURE 16.31
A polyline at the bottom of pond and a polyline for the outlet channel

   There is a feature line depicting the pond bottom, which has been assigned a constant elevation of 0, and a feature line depicting the outlet channel, which has been assigned elevations from the existing surface. In addition, the engineer has provided us with the Inflow and Outflow locations.

2. Select the pond basin feature line to activate the Feature Line contextual tab.
3. From the Feature Line contextual tab ➢ Edit Elevation panel (toggle it on if needed), choose the Insert Elevation Point tool.
4. Use the Center Osnap to insert the elevation points in the center of each circle.
5. Enter **779.5′** (**237.6** m) as the elevation of each inflow, and accept the default elevation at the outflow.

   You will change the elevation at the outflow in a moment.

6. Press ↵ to end the command, and press Esc to cancel grips.
7. From the Home tab ➢ Create Design panel, choose Feature Line ➢ Create Feature Lines From Objects.

8. Pick the polyline that represents the pilot channel from the northwestern inflow to the outflow channel centerline and press ↵.

   The Create Feature Lines dialog appears.

9. Verify that Site is set to Pond and that the Assign Elevations check box is unchecked, and click OK.

   Because this feature line was created in the same site as the bottom of the pond, the elevation of the PI at the southernmost inflow will reset to match the elevation of the endpoint of the pilot channel.

10. Select the feature line you just created representing the pilot channel to activate the Feature Line contextual tab.

11. From the Feature Line contextual tab ➤ Edit Geometry panel (toggle it on if needed), choose the Fillet tool.

12. At the `Specify corner or [All Join Radius]:` prompt, enter **R** ↵, and enter **25** ↵ (**7.5** ↵ for metric users) for the radius.

13. Select the corner of the feature line in the center of the pond to fillet all PIs.

14. Press ↵ to exit the Fillet command.

15. From the Feature Line contextual tab ➤ Edit Elevations panel, choose the Set Grade/Slope Between Points tool.

16. At the `Specify the start point:` prompt, select the PI at the inflow.

17. At the `Specify elevation:` prompt, enter **779.5** ↵ (**237.6** ↵ for metric users) to set this elevation.

18. At the `Specify the end point:` prompt, select the other end of the feature line, as shown in Figure 16.32.

FIGURE 16.32
Setting grade between points

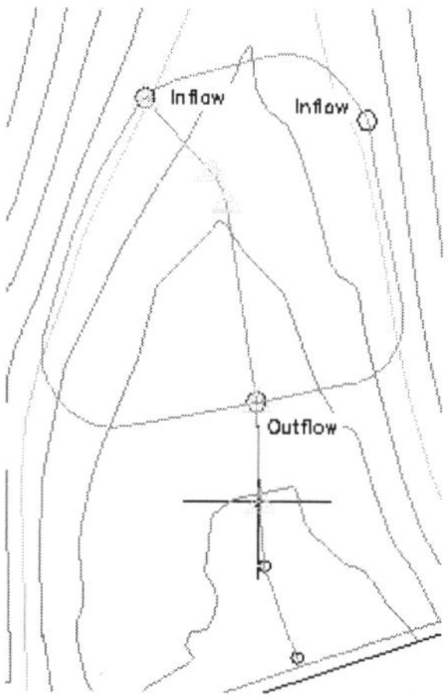

All the PIs will highlight, and Civil 3D will display the total length, elevation difference, and average slope at the command line.

19. At the `Specify grade or [Slope Elevation Difference]:` prompt, press ↵ again to accept the grade as shown.

    This completes a linear slope from one end of the feature line to the other, ensuring drainage through the pond and outfall structure.

20. Press ↵ to end the command.

21. Press Esc to cancel grips.

22. Select the feature line representing the bottom of the pond to activate the Feature Line contextual tab.

23. From the Feature Line contextual tab ➤ Edit Elevations panel, choose the Elevation Editor tool display the Grading Elevation Editor in Panorama.

24. Click in the Station cells in the Grading Elevation Editor to highlight and ascertain the elevation at the outfall, as shown in Figure 16.33.

FIGURE 16.33
The Grading Elevation Editor

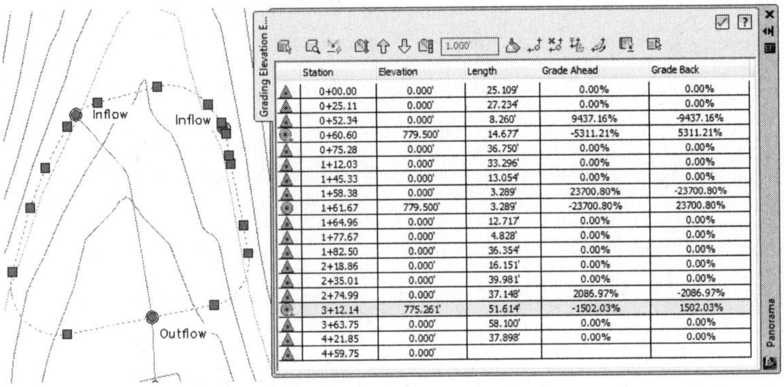

In this case, the elevation is 775.261′ (236.302 m). When done observing the elevations, you may close the Panorama.

25. From the Feature Line contextual tab for the feature line representing the bottom edge of the pond, choose the Set Grade/Slope Between Points tool from the Edit Elevations panel.

26. At the `Specify the start point:` prompt, move your cursor over the PI at the northwestern inflow and select it.

27. Press ↵ to accept the default elevation.

28. At the `Specify the end point:` prompt, move your cursor counterclockwise and select the PI at the outflow.

29. Press ↵ to accept the grade.

This sets the elevations between the one inflow and the outflow PIs to all fall at the same grade.

30. At the `Select object:` prompt, select the feature line representing the pond bottom again.

31. Repeat steps 26–29 for the PI on the other inflow (located on the northeast side of the pond) and the outflow located clockwise from the start point.

They will set the elevations between the other inflow and the outflow to a constant slope.

32. When done, press ↵ to end the command.

The entire outline of the pond bottom is graded except the area between the two inflows, as shown in Figure 16.34.

FIGURE 16.34
Zero elevation remains between the two inflows

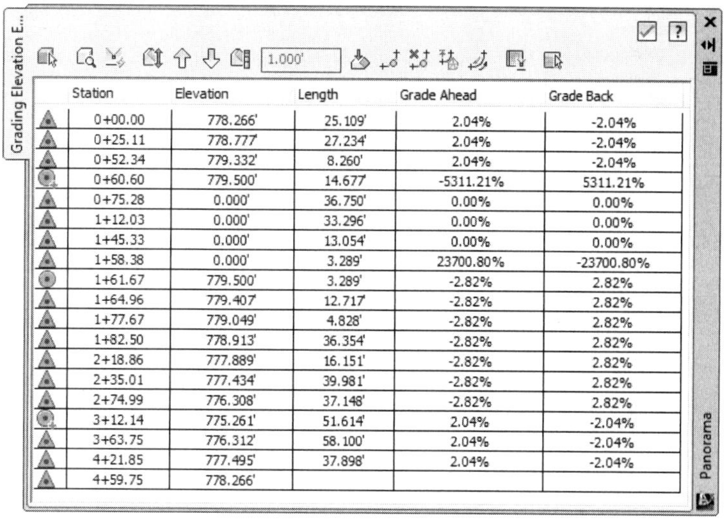

Station	Elevation	Length	Grade Ahead	Grade Back
0+00.00	778.266'	25.109'	2.04%	-2.04%
0+25.11	778.777'	27.234'	2.04%	-2.04%
0+52.34	779.332'	8.260'	2.04%	-2.04%
0+60.60	779.500'	14.677'	-5311.21%	5311.21%
0+75.28	0.000'	36.750'	0.00%	0.00%
1+12.03	0.000'	33.296'	0.00%	0.00%
1+45.33	0.000'	13.054'	0.00%	0.00%
1+58.38	0.000'	3.289'	23700.80%	-23700.80%
1+61.67	779.500'	3.289'	-2.82%	2.82%
1+64.96	779.407'	12.717'	-2.82%	2.82%
1+77.67	779.049'	4.828'	-2.82%	2.82%
1+82.50	778.913'	36.354'	-2.82%	2.82%
2+18.86	777.889'	16.151'	-2.82%	2.82%
2+35.01	777.434'	39.981'	-2.82%	2.82%
2+74.99	776.308'	37.148'	-2.82%	2.82%
3+12.14	775.261'	51.614'	2.04%	-2.04%
3+63.75	776.312'	58.100'	2.04%	-2.04%
4+21.85	777.495'	37.898'	2.04%	-2.04%
4+59.75	778.266'			

Because you want to avoid a low spot, you'll now force a high point:

33. From the Feature Line contextual tab for the pond feature line, choose the Insert High/Low Elevation Point tool from the Edit Elevations panel.

34. At the `Specify the start point:` prompt, select the PI at the northwestern inflow.

35. At the `Specify the end point:` prompt, select the PI at the northeastern inflow as the endpoint.

36. At the `Specify grade ahead or [Slope]:` prompt, enter **1.0** ↵.

37. At the `Specify grade back or [Slope]:` prompt, enter **1.0** ↵ as the grade behind. A new elevation point will be created, as shown in Figure 16.35.

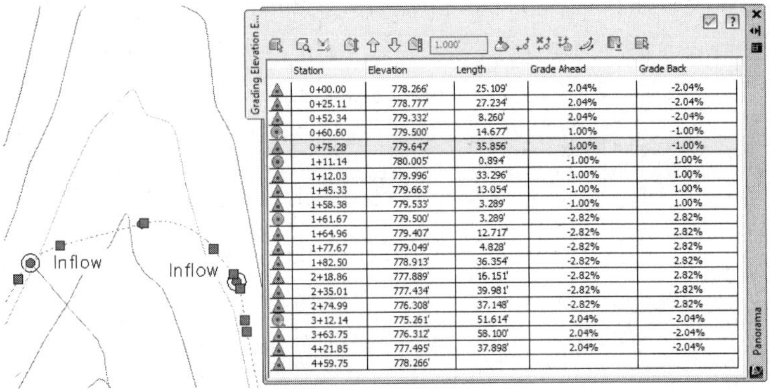

FIGURE 16.35
Using the Grade Back column in the Grading Elevation Editor

38. When done, press ↵ to end the command and press Esc to deselect the feature line.

When this exercise is complete, you may close the drawing. A saved copy of this drawing is available from the book's web page with the filename `PondDrainageDesign_FINISHED.dwg` or `PondDrainageDesign_METRIC_FINISHED.dwg`.

By using all the tools in the Feature Lines toolbar, you can quickly grade elements of your design and pull them together. If you have difficulty getting all the elevations in this exercise to set as they should, slow down, and make sure you are moving your mouse in the right direction when setting the grades by slope. It's easy to get the calculation performed around the other direction—that is, clockwise versus counterclockwise. It helps to move your mouse along the feature line in the direction you want the slopes to be maintained. This procedure seems to involve a lot of steps, but it takes less than a minute in practice.

There are roughly 25 ways to modify feature lines using both the Edit Geometry and Edit Elevations panels of the Feature Line contextual tab. Take a few minutes and experiment with them to understand the options and tools available for these essential grading elements. By manipulating the various pieces of the feature line collection, it's easier than ever to create dynamic modeling tools that match the designer's intent.

Labeling Feature Lines

Though it's not common, feature lines can be stylized to reflect particular uses, and labels can be applied to help a reviewer understand the nature of the object being shown. In the next couple of exercises, you'll label a few critical points on your pond design to help you better understand the drainage patterns.

Feature lines do not have their own unique label styles, but instead share with general lines and arcs. You can learn more about label styles in Chapter 20, "Label Styles." The templates that ship with Civil 3D contain styles for labeling segment slopes, so you'll label the grades of feature line segments in the following exercise:

1. Open the `LabelingFeatureLines.dwg` or `LabelingFeatureLines_METRIC.dwg` file.
2. From the Annotate tab ➢ Labels & Tables panel, click the Add Labels button to display the Add Labels dialog.

3. In the Add Labels dialog, do the following:
   a. Set the Feature drop-down to Line And Curve.
   b. Set Label Type to Single Segment
   c. Set Line Label Style to Grade Only.
   d. Set Curve Label Style to Grade Only.

   When complete, the dialog should look like Figure 16.36.

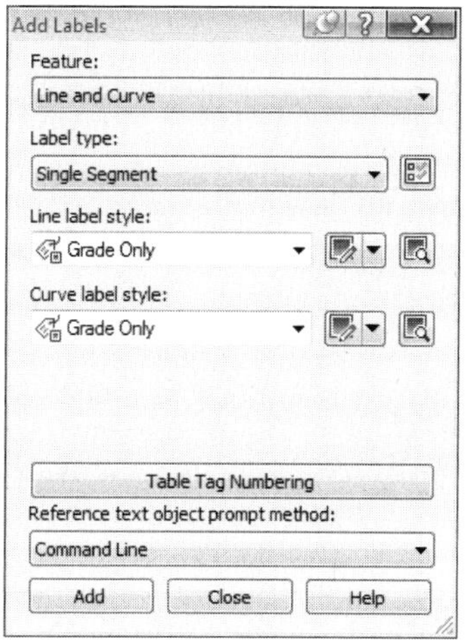

FIGURE 16.36
Adding feature line grade labels

4. Click the Add button.
5. At the `Select point on entity:` prompt, pick a few points along the pilot channel feature line (the feature line that goes across the bottom of the pond) tangents to create labels, as shown in Figure 16.37.

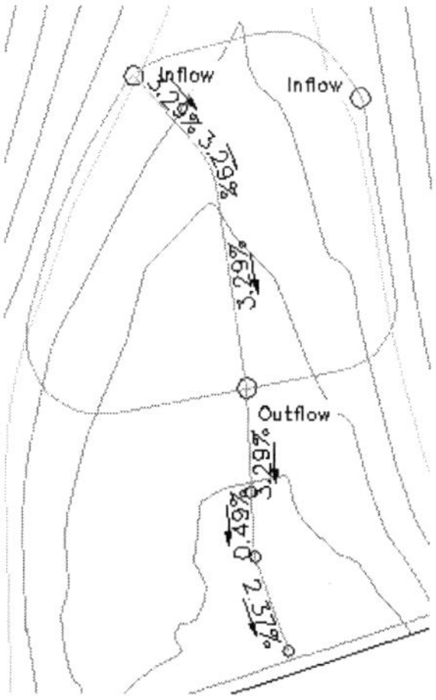

FIGURE 16.37
Feature line grade labels in the Imperial drawing

When this exercise is complete, you may close the drawing. A saved copy of this drawing is available from the book's web page with the filename `LabelingFeatureLines_FINISHED.dwg` or `LabelingFeatureLines_METRIC_FINISHED.dwg`.

Although it would be convenient to label the feature line elevations as well, there's no simple method for doing so. In practice, you would want to label the surface that contains the feature line as a component.

Grading Objects

Once a feature line is created, there are two main uses. One is to incorporate the feature line itself directly into a surface object as a breakline; the other is to create a grading object (referred to hereafter as simply a grading or gradings) using the feature line as a baseline. A grading consists of a baseline with elevation information, and a criteria set for projecting outward from that baseline based on distance, slope, or other criteria. These criteria sets can be defined and stored in grading criteria sets for ease of management. Finally, gradings can be stylized to reflect plan production practices or convey information such as cut or fill.

In this section, you'll use a number of methods to create gradings, edit those gradings, and finally convert the grading group into a surface.

Creating Gradings

CERT OBJECTIVE

Let's look at grading the pond as designed. In this section, you'll look at grading groups and then create the individual gradings within the group. Grading groups act as a collection mechanism for individual gradings, and let Civil 3D understand the daisy chain of individual gradings that are related and act in sync with each other.

One thing to be careful of when working with gradings is that they are part of a site. Any feature line within that same site will react with the feature lines created by the grading. For that reason, the exercise drawing has a second site called Pond Grading to be used for just the pond grading.

1. Open the `GradingThePond.dwg` or `GradingThePond_METRIC.dwg` file.
2. Select the feature line representing the bottom of the pond to activate the Feature Line contextual tab.
3. From the Feature line contextual tab ➢ expanded Modify panel, choose Move To Site to display the Move To Site dialog.
4. On the Information tab of the Site Properties dialog, set Name to **Pond Grading** and click OK.
5. Set the Destination Site drop-down list to the new site named Pond Grading and click OK.

   This will avoid interaction between the pond banks and the pilot channel you laid out earlier.

6. With the feature line still selected, on the Feature Line contextual tab ➢ Launch Pad panel, choose Grading Creation Tools.

   The Grading Creation Tools toolbar, shown in Figure 16.38, appears. The left section is focused on settings, the middle on creation, and the right on editing.

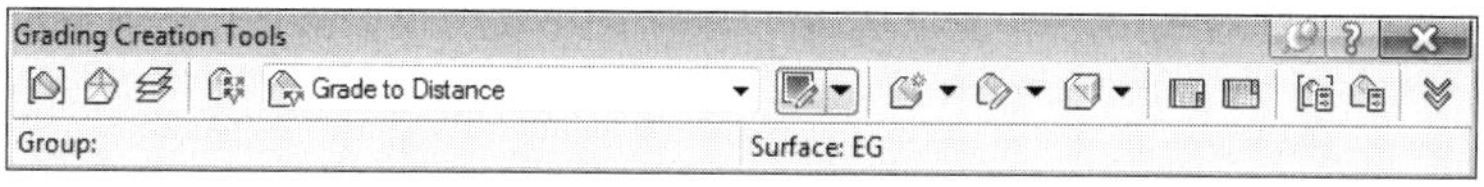

FIGURE 16.38
The Grading Creation Tools toolbar

7. On the Grading Creation Tools toolbar, click the Set The Grading Group tool to the far left of the toolbar to display the Site dialog.
8. Choose the Pond Grading site and click OK.

   The Create Grading Group dialog is displayed.
9. Enter **Pond Grading** in the Name text box, as shown in Figure 16.39, and click OK.

FIGURE 16.39
Assign the name Pond Grading in the Create Grading Group dialog.

You'll revisit the surface creation options in a bit.

10. On the Grading Creation Tools toolbar, click the Select Target Surface tool located next to the Select Grading Group tool to display the Select Surface dialog.
11. Select the EG surface and click OK.

12. On the Grading Creation Tools toolbar, verify that the Grading Criteria is set to Grade To Elevation and click the Create Grading tool, or click the down arrow next to the Create Grading tool and select Create Grading, as shown in Figure 16.40.

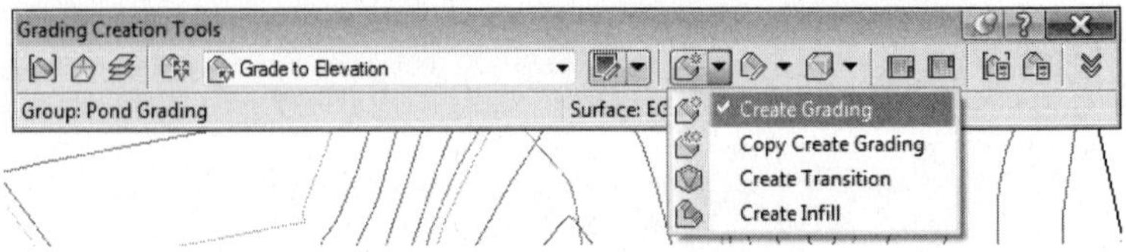

FIGURE 16.40
Creating a grading using the 3:1 To Elevation criteria

13. At the `Select the feature:` prompt, select the pond outline.

    If you get a dialog asking you to weed the feature line,

    a. Select the Weed The Feature Line option.

    b. Click OK in the Weed Vertices dialog.

14. At the `Select the grading side:` prompt, pick a point on the outside of the pond to indicate the direction of the grading projections.

15. At the `Apply to entire length? [Yes No]:` prompt, enter **Y** ↵ to apply the grading to the entire length of the pond outline.

16. Enter **784** ↵ (or **239** ↵ for metric users) at the command line as the target elevation.

17. At the `Cut Format [Grade Slope]:` prompt, enter **S** ↵.

18. At the `Cut Slope:` prompt, enter **3** ↵ for a 3 horizontal to 1 vertical slope.

19. At the `Fill Format [Grade Slope]:` prompt, enter **S** ↵.

20. At the `Fill Slope:` prompt, enter **3** ↵ for a 3 horizontal to 1 vertical slope.

    The first grading is complete. The lines onscreen are part of the Grading style.

21. Press Esc to end the command.

22. On the Grading Creation Tools toolbar, verify that the Grading Criteria is set to Grade To Distance, and click the Create Grading tool.

23. At the `Select the feature:` prompt, select the upper boundary of the grading made in step 20, as shown in Figure 16.41.

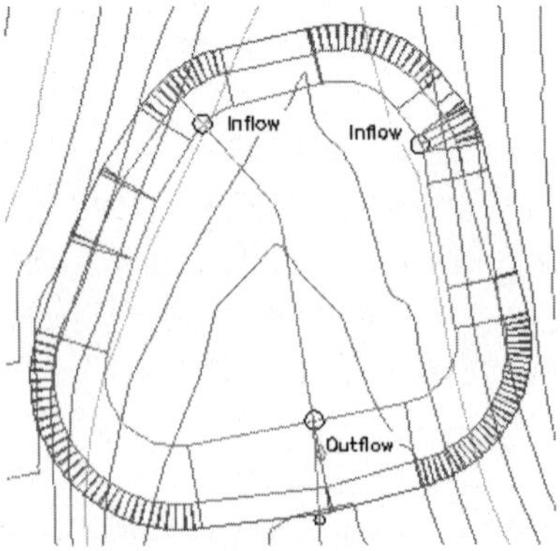

FIGURE 16.41
Creating a daisy chain of gradings

24. At the `Apply to entire length? [Yes/No]:` prompt, enter **Y** ↵ to apply to the whole length. Notice that you did not have to select a side, because there is already a grading object on one side of the selected feature line.

25. At the `Specify distance:` prompt, enter **10** ↵ (**3** ↵ for metric users) for the target distance to build the safety ledge.

26. At the `Format [Grade Slope]:` prompt, enter **G** ↵.

27. At the `Grade:` prompt, enter **0** ↵.

28. Press Esc to end the command.

29. On the Grading Creation Tools toolbar, verify that Grading Criteria is set to Grade To Surface, and click the Create Grading tool.

30. At the `Select the feature:` prompt, select the outer edge of the safety ledge just created.

31. At the `Apply to entire length? [Yes/No]:` prompt, enter **Y** ↵ to apply to the whole length.

32. At the `Cut Format [Grade Slope]:` prompt, enter **S** ↵.

33. At the `Cut Slope:` prompt, enter **3** ↵ for a 3 horizontal to 1 vertical slope.

34. At the `Fill Format [Grade Slope]:` prompt, enter **S** ↵.

35. At the `Fill Slope:` prompt, enter **3** ↵ for a 3 horizontal to 1 vertical slope.

36. Press Esc to end the command.

    Your drawing should look similar to Figure 16.42.

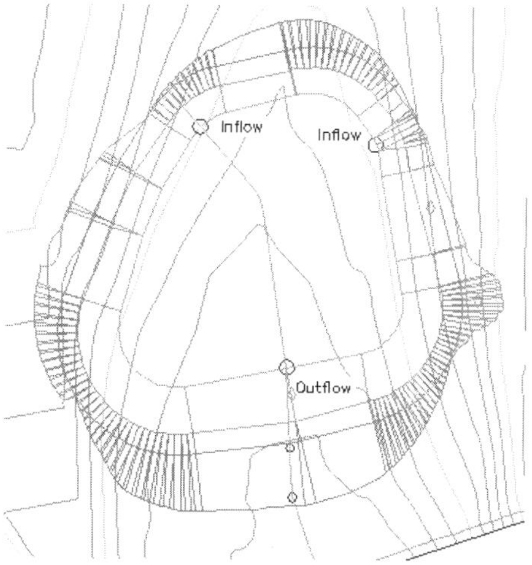

FIGURE 16.42
Complete pond feature line grading

When this exercise is complete, you may close the drawing. A saved copy of this drawing is available from the book's web page with the filename `GradingThePond_FINISHED.dwg` or `GradingThePond_METRIC_FINISHED.dwg`.

Each piece of this pond is tied to the next, creating a dynamic model of your pond design on the basis of the designer's intent. What if that intent changes? The next section describes editing the various gradings.

Editing Gradings

Once you've created a grading, you often need to make changes. A change can be as simple as changing the slope or changing the geometric layout. In this exercise, you'll make a simple change, but the concept applies to all the gradings you've created in your pond.

1. Open the `EditingGrading.dwg` or `EditingGrading_METRIC.dwg` file.
2. Pick one of the interior projection lines or the small diamond on the interior slope of the pond in order to select the grading object.
3. From the Grading contextual tab ➤ Modify panel, choose the Grading Editor tool to display the Grading Editor in Panorama. If the correct grading object was selected in step 2, then at the top of Panorama it should state Criteria: Grade To Elevation.
4. Change the Fill Slope Projection and Cut Slope Projection both to **4**. Note that after you click out of the cell, the software will change this to read as 4.00:1, as shown in Figure 16.43.

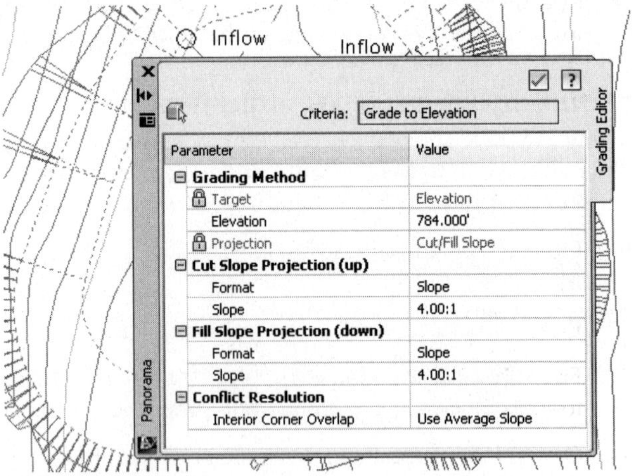

FIGURE 16.43
Editing the Cut and Fill Slope values

The grading may take a few seconds to update between edits.

5. Close Panorama.

   Your display should look like Figure 16.44. (Compare this to Figure 16.42 if you'd like to see the difference.)

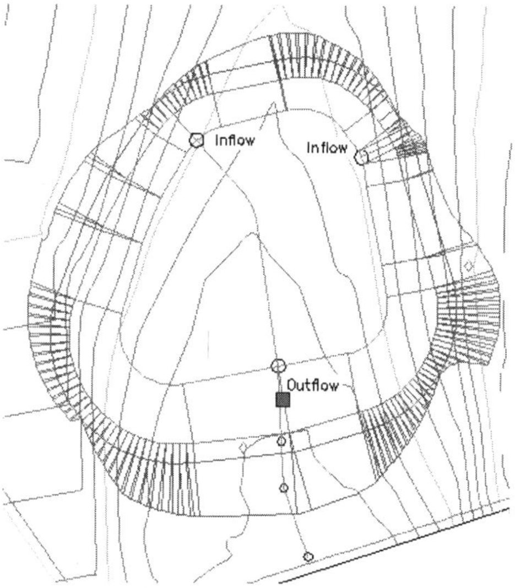

FIGURE 16.44
Completed grading edit

When this exercise is complete, you may close the drawing. A saved copy of this drawing is available from the book's web page with the filename `EditingGrading_FINISHED.dwg` or `EditingGrading_METRIC_FINISHED.dwg`.

Editing any aspect of the grading will reflect instantly, and if other gradings within the group are dependent on the results of the modified grading, they will recalculate as well.

Creating Surfaces from Grading Groups

Grading groups work well for creating the model, but you have to use a TIN surface to go much further with them. In this section, you'll look at the conversion process, and then use the built-in tools to understand the impact of your grading group on site volumes.

1. Open the `CreatingGradingSurfaces.dwg` or `CreatingGradingSurfaces_METRIC.dwg` file.
2. Pick one of the diamonds in the grading group.
3. From the Grading contextual tab ➢ Modify panel, choose the Grading Group Properties tool to display the Grading Group Properties - Pond Grading dialog.

4. On the Information tab, check the box for Automatic Surface Creation. The Create Surface dialog appears.
5. Click in the Style field, and then click the ellipsis button.

   The Select Surface Style dialog appears.
6. Select Contours 1′ And 5′ (Design) (or Contours 1 m And 5 m (Design) for metric users) from the drop-down list in the selection box and click OK to return to the Create Surface dialog.

7. Click OK to accept the settings in the Create Surface dialog.
8. In the Grading Group Properties - Pond Grading dialog, check the Volume Base Surface option and select the EG surface to perform a volume calculation, as shown in Figure 16.45.

FIGURE 16.45
Automatic surface creation through the Grading Group Properties

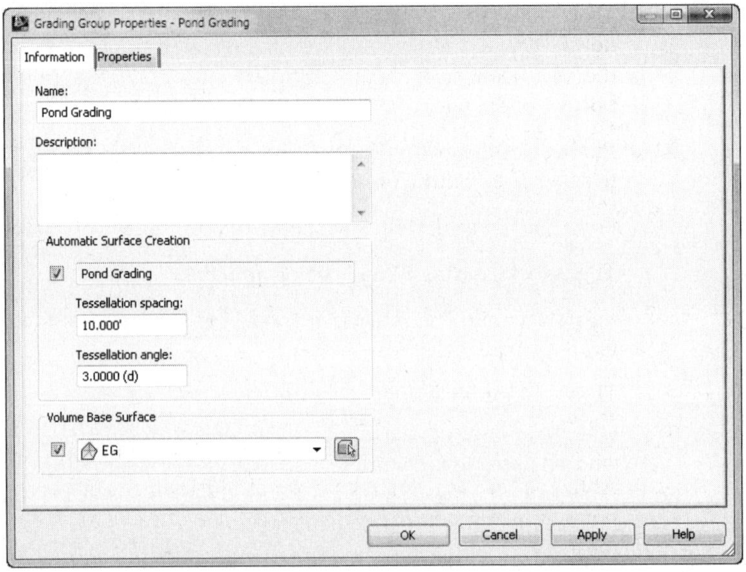

Note that this is not generating a volume surface; you will look at what this check box does in a few steps.

9. Click OK to dismiss the Grading Group Properties - Pond Grading dialog. If Panorama appears, you may close it.
10. Click OK to dismiss the Select Surface Style dialog and return to the Grading Group Properties.

You're going through this process now because you didn't turn on the Automatic Surface Creation option when you created the grading group. If you're performing straightforward gradings, that option can be a bit faster and simpler. There are two options available when creating a surface from a grading group. They both control the creation of projection lines in a curved area:

- The Tessellation Spacing value controls how frequently along an arced feature line TIN points are created and projection lines are calculated. A TIN surface cannot contain any true curves the way a feature line can, because it is built from triangles. The default values typically work for site mass grading, but might not be low enough to work with things such as parking lot islands where the 10′ (10 m) value would result in too little detail.
- The Tessellation Angle value is the degree measured between outside corners in a feature line. Corners with no curve segment have to have a number of projections swung

in a radial pattern to calculate the TIN lines in the surface. The tessellation angle is the angular distance between these radial projections. The typical values work most of the time, but in large grading surfaces a larger value might be acceptable, lowering the amount of data to calculate without significantly altering the final surface created.

There is one small problem with this surface. If you examine the bottom of the pond, you'll notice there are no contours running through this area. If you move your mouse to the middle, you also won't see any Tooltip elevation because there is no surface data in the bottom portion of the pond. To fix that (and make the volumes accurate), you need a grading infill.

11. Pick one of the projection lines, or the small diamond on the inside of the pond in order to select the grading object.

12. From the Grading contextual tab ➤ Modify panel, choose the Create Grading Infill tool.

    The Select Grading Group dialog appears.

13. Verify that Site Name is set to Pond Grading, and Group Name is also set to Pond Grading, and click OK.

    The Grading Style dialog appears.

14. Verify that Grading Style is set to Cut Slope Display, and click OK.

15. At the `Select an area to infill:` prompt, hover your cursor over the middle of the pond and the pond feature line created earlier will be highlighted, indicating a valid area for infill.

16. Click once to create the infill, and press ↵ to apply.

    Civil 3D will calculate. If Panorama appears, you may close it. You should now have some contours running through the pond base area, as shown in Figure 16.46.

FIGURE 16.46
The pond after applying an infill grading

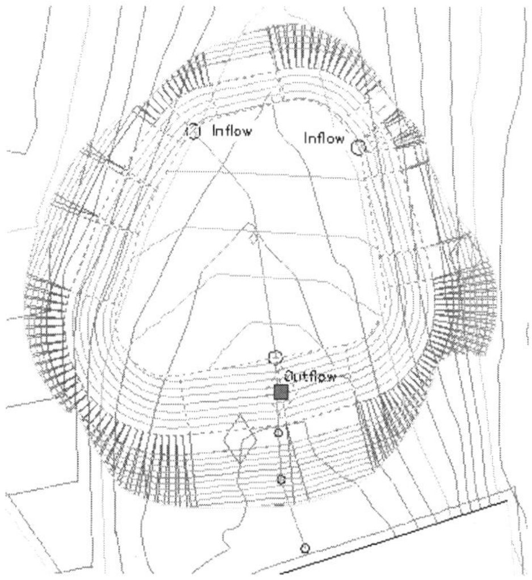

17. Zoom in if needed and pick one of the grading diamonds again to select one of the gradings.

    Make sure you grab one of the gradings, and not the surface contours that are being drawn on top of them.

18. From the Grading contextual tab ➢ Modify panel, choose Grading Group Properties to display the Grading Group Properties - Pond Grading dialog.

19. Switch to the Properties tab to display the volume information for the pond, as shown in Figure 16.47.

FIGURE 16.47
Reviewing the grading group volumes

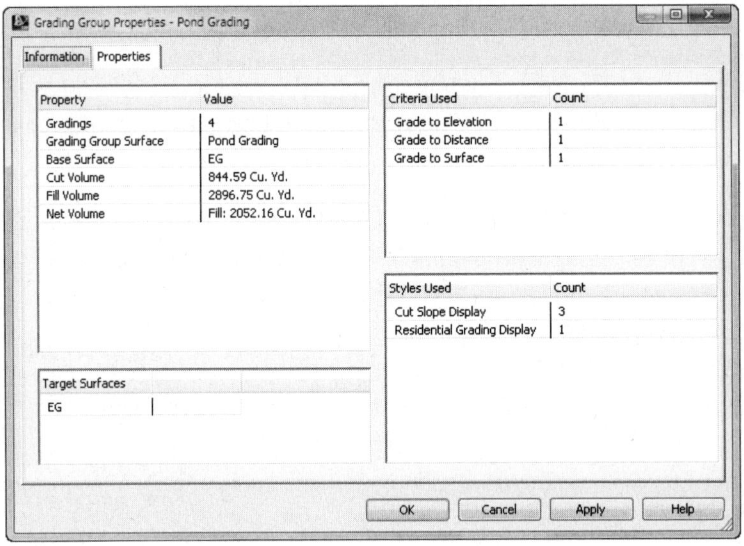

    This tab also allows you to review the criteria and styles being used in the grading group.

20. Click OK to dismiss the Grading Group Properties - Pond Grading dialog.

This new surface is listed in Prospector and is based on the gradings created. A change to the gradings would affect the grading group, which would, in turn, affect the surface and these volumes.

When this exercise is complete, you may close the drawing. A saved copy of this drawing is available from the book's web page with the filename `CreatingGradingSurfaces_FINISHED.dwg` or `CreatingGradingSurfaces_METRIC_FINISHED.dwg`.

In the last exercise, you'll pull it all together and generate a composite surface from your grading surface and existing surface:

1. Open the `CreatingCompositeSurfaces.dwg` or `CreatingCompositeSurfaces_METRIC.dwg` file.

2. Right-click Surfaces in Prospector and select Create Surface.

3. In the Create Surface dialog, enter **Composite** in the Name text box.
4. Click in the Style field, and then click the ellipsis to display the Select Surface Style dialog.
5. Select the Elevation Banding (2D) option from the drop-down list box, and click OK to dismiss the Select Surface Style dialog.
6. Click OK to dismiss the Create Surface dialog and create the surface in Prospector.
7. In Prospector, expand the Surfaces ➢ Composite ➢ Definition branches.
8. Right-click Edits and select Paste Surface.

   The Select Surface To Paste dialog appears.
9. Select EG from the list and click OK. Dismiss Panorama if it appears.
10. Right-click Edits again and select Paste Surface one more time.
11. Select Pond Grading and click OK.
12. Right-click the new Composite surface, and select Display Order ➢ Send To Back.

    The drawing should look like Figure 16.48.

FIGURE 16.48
Completed composite surface

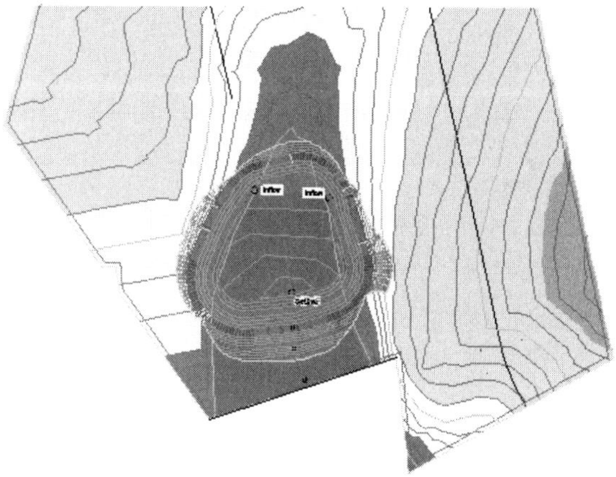

When this exercise is complete, you may close the drawing. A saved copy of this drawing is available from the book's web page with the filename `CreatingGradingSurfaces_FINISHED.dwg` or `CreatingGradingSurfaces_METRIC_FINISHED.dwg`.

By creating a composite surface consisting of pasted-together surfaces, the TIN triangulation cleans up any gaps in the data, making contours that are continuous from the original grade, through the pond, and out the other side. With the grading group still being dynamic and editable, this composite surface reflects a dynamic grading solution that will update with any changes. Using a composite surface such as this as a reference for pipe networks is very useful, since the network may occur under both existing and proposed surfaces.

The Bottom Line

Convert existing linework into feature lines. Many site features are drawn initially as simple linework for the 2D plan. By converting this linework to feature line information, you avoid a large amount of rework. Additionally, the conversion process offers the ability to drape feature lines along a surface, making further grading use easier.

Master It Open the `MasteringGrading.dwg` or `MasteringGrading_METRIC.dwg` file from the book's web page. Convert the magenta polyline, describing a proposed temporary swale, into a feature line and drape it across the EG surface to set elevations, and set intermediate grade break points.

Model a simple linear grading with a feature line. Feature lines define linear slope connections. This can be the flow of a drainage channel, the outline of a building pad, or the back of a street curb. These linear relationships can help define grading in a model, or simply allow for better understanding of design intent.

Master It Edit the curve on the feature line you just created to be 100′ (30 m). Set the grade from the west end of the feature line to the next PI to 4 percent, and the remainder to a constant slope to be determined in the drawing. Draw a temporary profile view to verify the channel is below grade for most of its length.

Model planar site features with grading groups. Once a feature line defines a linear feature, gradings collected in grading groups model the lateral projections from that line to other points in space. These projections combine to model a site much like a TIN surface, resulting in a dynamic design tool that works in the Civil 3D environment.

Master It Use the two grading criteria to define the pilot channel, with grading on both sides of the sketched centerline. Define the channel using a Grading to Distance of 5′ (1.5 m) with a slope of 3:1 and connect the channel to the EG surface using a grading with slopes that are 4:1. Generate a surface from the grading group. If prompted, do not weed the feature line.

Chapter 17

Plan Production

So you've toiled for days, weeks, or maybe months creating your design in the AutoCAD® Civil 3D® program, and now it's time to share it with the world — or at least your corner of it. Even in this digital age, paper plan sets still play an important role. You generate these plans in Civil 3D using the Plan Production feature. This chapter takes you through the steps necessary to create a set of sheets, from initial setup, to framing and generating sheets, to data management and plotting.

In this chapter, you will learn to:

- Create view frames
- Edit view frames
- Generate sheets and review Sheet Set Manager
- Create section views

Preparing for Plan Sets

Before you start generating all sorts of wonderful plan sets, you must address a few concepts and prerequisites. Civil 3D takes advantage of several features and components to build a plan set. Some of these components have existed in AutoCAD and Civil 3D software for years (for example, layout tabs, drawing templates, alignments, and profiles). Others are newer properties of existing features (such as Plan and Profile viewport types), view frames, match lines, and view frame groups. Let's look at what you need to have in place before you can create your plotted masterpieces.

Prerequisite Components

The Plan Production feature draws on several components to create a plan set. Here is a list of these components and a brief explanation of each. Later, this chapter will explore these elements in greater detail:

Drawing Template Plan Production creates new layouts for each sheet in a plan set. To do this, the feature uses drawing templates with predefined viewports. These viewports have their Viewport Type property set to either Plan or Profile.

For the exercises in this chapter, the default location for the final sheets will refer to the `C:\Mastering\CH 17\Final Sheets` location. It is recommended that you download all of the files (including the `Final Sheets` folder) for this chapter from the book's web page (`www.sybex.com/go/masteringcivil3d2013`) and place them in the `C:\Mastering\CH 17` folder.

Object and Display Styles Like every other feature in Civil 3D, Plan Production uses objects. Specifically, these objects are view frames, view frame groups, and match lines. Before creating plan sheets, you'll want to make sure you have styles set up for each of these objects.

Alignments and Profiles In Civil 3D, the Plan Production feature is designed primarily for use in creating plan and profile views. Toward that end, your drawing must contain (or data reference) at least one alignment. If you're creating sheets with both plan and profile views or just profile views, a profile must also be present.

Sample Lines and Sections Creating section sheets requires an alignment, a sample line group, cross sections, and a sheet template with associated section viewports.

With these elements in place, you're ready to dive in and create some sheets. The general steps in creating a set of plans are as follows:

1. Meet the prerequisites listed previously.
2. Create view frames.
3. Create plan or plan-profile sheets.
4. (Optional) Create section view groups.
5. (Optional) Create section sheets.
6. Manage the sheets using the Sheet Set Manager.
7. Plot or publish (hardcopy or digitally).

The next section describes this process in detail and the tools used in Plan Production. The Sheet Set Manager, which is found in basic AutoCAD, is an integral part of this process.

Using View Frames and Match Lines

When you create sheets using the Plan Production tools, Civil 3D first automatically helps you divide your alignment into areas that will fit on your plotted sheet and display at the desired scale. To do this, Civil 3D creates a series of rectangular frames placed end to end (or slightly overlapping) along the length of alignment, like those in Figure 17.1. These rectangles are referred to as *view frames* and are automatically sized and positioned to meet your plan sheet requirements. This collection of view frames is referred to as a *view frame group*. Where the view frames overlap one another, Civil 3D creates *match lines* that establish continuity from frame to frame by referring to the previous or next sheet in the completed plan set. View frames and match lines are created in modelspace, using the prerequisite elements described in the previous section.

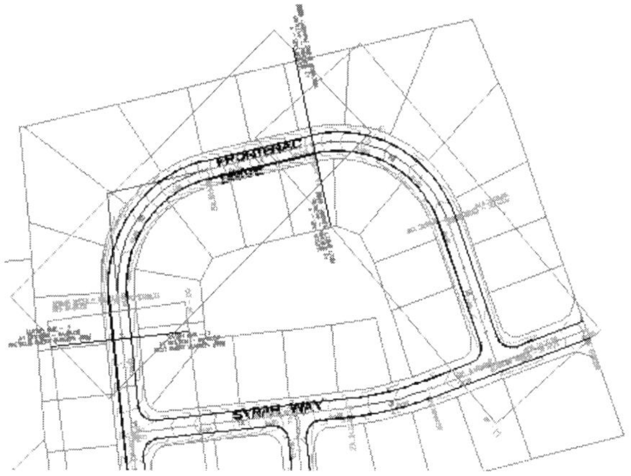

FIGURE 17.1
View frames and match lines

The Create View Frames Wizard

CERT OBJECTIVE

The first step in the process of creating plan sets is to generate view frames. Civil 3D provides an intuitive wizard that walks you through each step of the view frame creation process. Let's look at the Create View Frames wizard and the various page options. After you've seen each page, you'll have a chance to put what you've learned into practice in an example.

From the Output tab ➢ Plan Production panel, choose Create View Frames to launch the Create View Frames wizard (Figure 17.2). The wizard consists of several pages. A list of these pages is shown along the left sidebar of the wizard, and an arrow indicates which page you're currently viewing. You move among the pages using the Next and Back navigation buttons along the bottom of each page. Alternatively, as with all wizards, you can jump directly to any page by clicking its name in the list on the left. The following sections walk you through the pages of the wizard and explain their features.

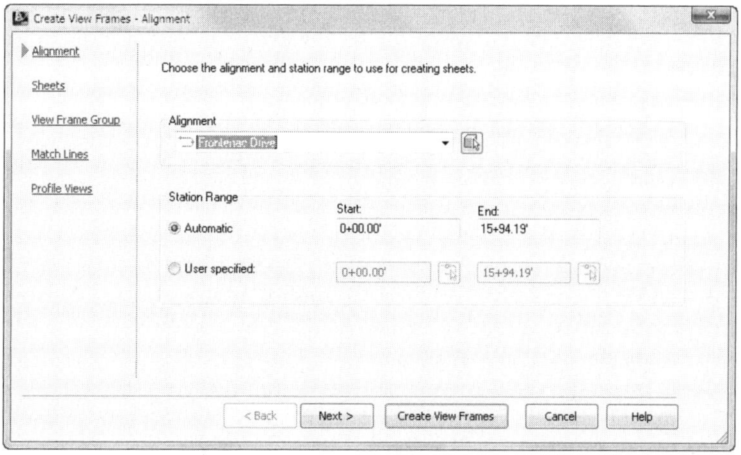

FIGURE 17.2
The Create View Frames – Alignment wizard page

Create View Frames – Alignment Page

You use the first page of the Create View Frames wizard (shown previously in Figure 17.2) to select the alignment and station range, along which the view frames will be created.

Alignment In the top area of this page, you select the alignment along which you want to create view frames. You can either select it from the drop-down list or click the Select From The Drawing button to select the alignment on screen.

Station Range In the Station Range area of the page, you define the station range over which the frames will be created. Selecting Automatic creates frames from the Alignment Start to the Alignment End. Selecting User Specified lets you define a custom range, by either keying in start and end station values in the appropriate box or by clicking the button to the right of the station value fields and graphically selecting the station from the drawing.

An example of when you would want to select specific stations is if you have a subdivision that will be constructed in phases. You have designed an entire roadway but only need to create specific sheets for a specific phase.

Create View Frames – Sheets Page

You use the second page of the Create View Frames wizard (Figure 17.3) to establish the sheet type and the orientation of the view frames along the alignment. A plan production *sheet* is a layout tab in a drawing file. To create the sheets, Civil 3D references a predefined drawing template (with the file extension `.dwt`). As mentioned earlier, the template must contain layout tabs, and in each layout tab the viewport's Viewport Properties must be set to either Plan or Profile. Later in this chapter, you'll learn about editing and modifying templates for use in plan production.

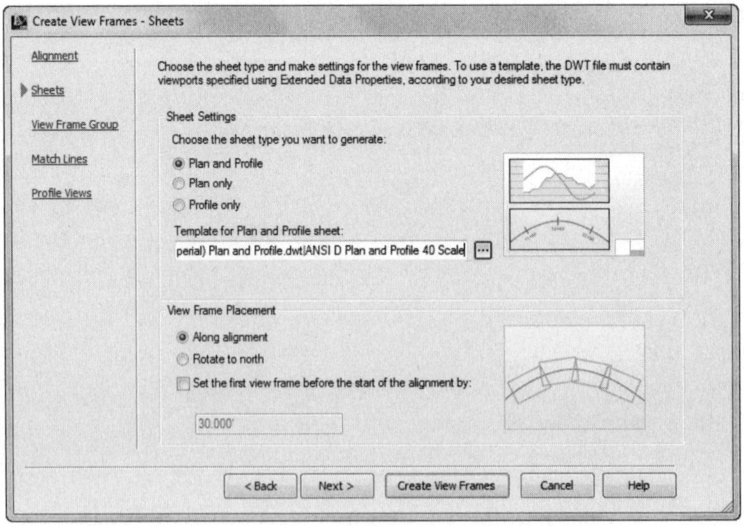

FIGURE 17.3
Create View Frames – Sheets wizard page

Sheet Settings

The Plan Production feature provides options for creating three types of sheets:

Plan And Profile This option generates a sheet with two viewports; one viewport shows a plan view and the other shows a profile view of the section of the selected alignment segment.

Plan Only As the name implies, this option creates a sheet with a single viewport showing only the plan view of the selected alignment segment.

Profile Only Similar to Plan Only, this option creates a sheet with a single viewport, showing only the profile view of the selected alignment segment.

INFORMATIONAL GRAPHICS

Did you notice the nifty graphic to the right of the sheet-type options in Figure 17.3? This image changes depending on the type of sheet you've selected. It provides a schematic representation of the sheet layout to further assist you in selecting the appropriate sheet type. You'll see this type of graphic image throughout the Create View Frame wizard and in other wizards used in Civil 3D.

After choosing the sheet type, you must define the template file and the layout tab within the selected template that Civil 3D will use to generate your sheets. Several predefined templates ship with Civil 3D and are part of the default installation. Be sure to choose the sheet type before selecting the template, because if the sheet type doesn't match the viewport types in the template then you won't be able to select the necessary layout.

1. Click the ellipsis button on this page to display the Select Layout As Sheet Template dialog, as shown in Figure 17.4.

FIGURE 17.4
Use the Select Layout As Sheet Template dialog to choose which layout you would like to apply to your newly created sheets

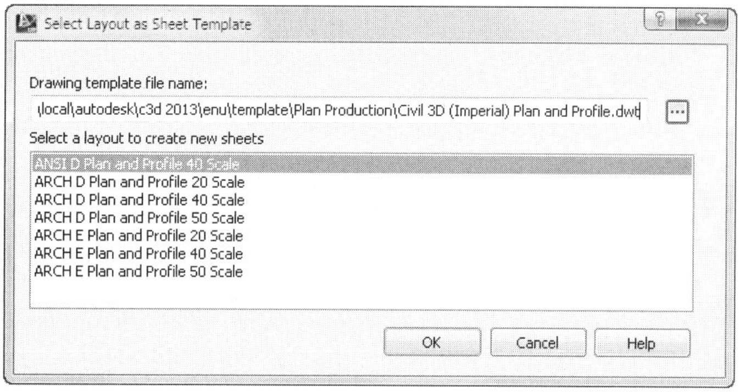

This dialog provides the option to select the DWT file and the layout tab within the template.

2. Click the ellipsis button in the Select Layout As Sheet Template dialog to browse to the desired template location.

Typically the default template location is:

```
C:\Users\<username>\AppData\Local\Autodesk\C3D2013\
enu\Template\Plan Production\
```

Alternatively, if you are working in a network environment, your templates can be kept in a common folder on the network.

After you select the template you want to use, a list of the layouts contained in the DWT file appears in the Select Layout As Sheet Template dialog.

3. Choose the appropriate layout.

Notice in the template selected (see Figure 17.4) there are layouts for various sheet sizes as well as various scales that are included in the Plan Production templates that ship with Civil 3D.

View Frame Placement

Your view frames can be placed in one of two ways: either along the alignment or rotated to north. Use the bottom area of the Sheets page of the wizard to establish the placement.

Along Alignment This option aligns the long axes of the view frames parallel to the alignment. Refer to the graphic to the right of the radio buttons in the dialog for a visual representation. This graphic is shown at the left in Figure 17.5.

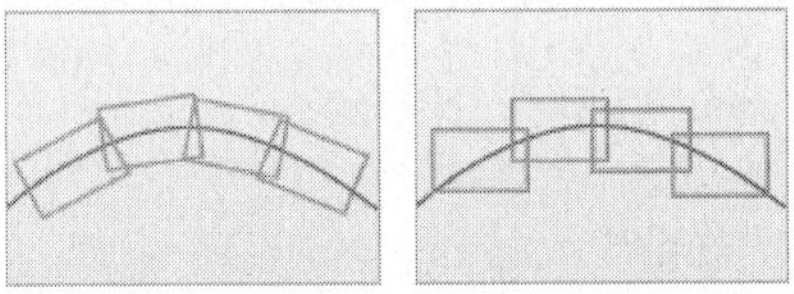

FIGURE 17.5 View Frame Placement shown using the Along Alignment option (left) and the Rotate To North option (right)

Rotate To North As the name implies, this option aligns the view frames so they're all rotated to the north direction, regardless of the changing rotation of the alignment centerline. *North* is defined by the orientation of the drawing. This graphic is shown at the right in Figure 17.5.

TWISTED NORTH

If you want the north arrow to rotate according to the view twist of the viewport, the block that is being used for the north arrow must be included in the template, and must be located in the layout.

Set The First View Frame Before The Start Of The Alignment By

Regardless of the view frame placement you choose, you have the option to place the first view frame some distance before the start of the alignment. This option is useful if you want to show a portion of the site, such as an existing offsite road, in the plan view. When this option is selected, the text box becomes active, letting you enter the desired distance.

CREATE VIEW FRAMES – VIEW FRAME GROUP PAGE

You use the third page of the Create View Frames wizard (Figure 17.6) to define creation parameters for your view frames and the view frame group to which they'll belong. The page is divided into two areas: the top for the View Frame Group and the bottom for the View Frames themselves.

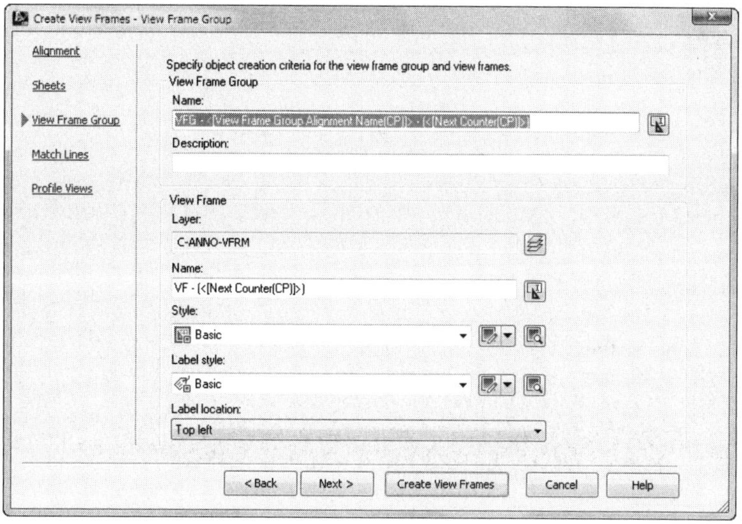

FIGURE 17.6
Create View Frames – View Frame Group wizard page

View Frame Group Use this area of the View Frame Group page to set the name and an optional description for the view frame group. The name can consist of manually entered text, text automatically generated based on the Name template settings, or a combination of both.

Name Template

To adjust the Name template settings, click the Edit View Frame Group Name button to open the Name Template dialog, shown here:

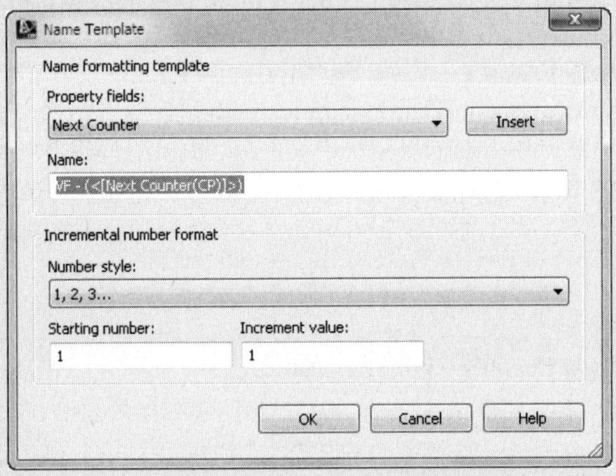

With the settings shown previously in Figure 17.6, the name will include manually defined prefix text (VFG -) followed by automatically generated text, which inserts the view frame group alignment name and a sequential counter number. This will result in a view frame group name of VFG - Frontenac Drive - 1.

The Name Template dialog isn't unique to the Plan Production feature. However, the property fields available vary depending on the features to be named. If you need to reset the incremental number counter, use the text box in the lower area of the Name Template dialog. You can change the increment value in the Name Template dialog as well.

View Frame Use this area of the View Frame Group page to set various parameters for the view frames, including the layer for the frames, view frame names, view frame object and label styles, and the label location. Each view frame can have a unique name (using an incremental counter), but the other parameters are the same for all view frames.

Layer This option defines the layer on which the view frames are created. This layer is defined in the Drawing Settings, but you can override it by clicking the Layer button and selecting a different layer. Setting the view frame layer to No-Plot will ensure your drawing does not end up plotting with unwanted rectangles.

Name The Name setting is nearly identical in function to that for the View Frame Group Name discussed earlier. With the settings shown previously in Figure 17.6, the default name results in VF - 1, VF - 2, and so on.

Style Like nearly all objects in Civil 3D, view frames have styles associated with them. The View Frame Style is simple, with only one component: the View Frame Border. You use the drop-down list to select a predefined style.

Label Style Also like most other Civil 3D objects, view frames have label styles associated with them. And like other label styles, the View Frame Labels are created using the Label Style Composer and can contain a variety of components. The label style used in Figure 17.6 includes the View Frame Name placed at the top of the frame.

Label Location The last option on this page lets you set the label location. The default feature setting places the label at the top left of the view frame.

For view frame labels are placed at the top of the frame, the term *top* is relative to the frame's orientation. For alignments that run left to right across the page, the top of the frame points toward the top of the screen. For alignments that run right to left, the top of the frame points toward the bottom of the screen.

CREATE VIEW FRAMES – MATCH LINES PAGE

You use the fourth page of the Create View Frames wizard (Figure 17.7) to establish settings for match lines. Match lines are used to maintain continuity from one sheet to the next. They're typically placed at or near the edge of a sheet, with instructions to "See Sheet XX" for continuation.

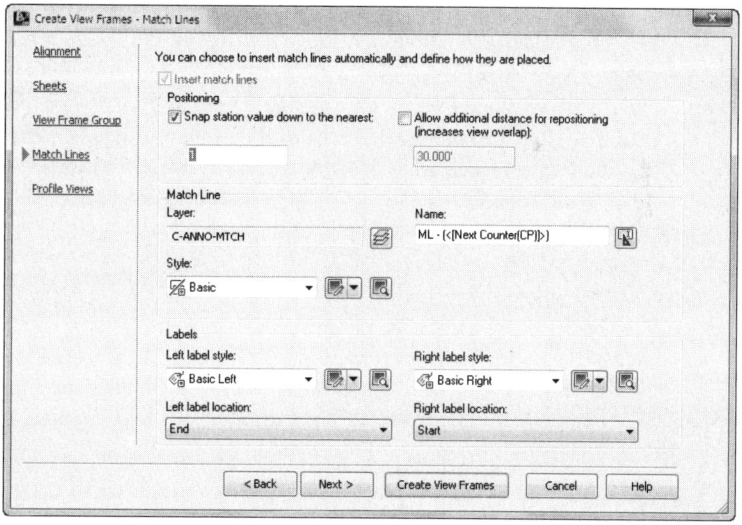

FIGURE 17.7
Create View Frames – Match Lines wizard page

Insert Match Lines You have the option whether to automatically insert match lines. Match lines are used only for plan views, so if you're creating Plan And Profile or Plan Only sheets, the option is automatically selected and can't be deselected.

Positioning Use this area of the Match Lines page to define the initial location of the match lines and provide the ability to later move or reposition the match lines.

Snap Station Value Down To The Nearest By selecting this option, you override the drawing station settings and define a rounding value specific to match line placement. With the settings shown previously in Figure 17.7, a value of 1 is entered, resulting in the match lines being placed at the nearest whole station. For example, if the station was 8 + 14.83 and the value was 1, it would round down to 8 + 14, but for the same station if the value was set to either 50 or 100 it would round down to 8 + 00.

This feature always rounds down (snap station down as opposed to snap station up). The exception to this is that if the rounding would put the match line at an undesirable location (such as before the previous match line or before the beginning of the alignment), then no rounding would be performed and the calculated station would be used.

Allow Additional Distance For Repositioning (Increases View Overlap) Selecting this option activates the text box, allowing you to enter a distance by which the views on adjacent sheets will overlap, and the maximum distance that you can move a match line from its original position within the overlap area. While the match line locations are originally created automatically, there are going to be instances when you want to move the match line if it bisects a critical location.

Match Line Use this area of the Match Lines page to provide the settings for the match line. This area is similar to those for view frames described on the previous page of the wizard. You can define the Layer, the Name, and the Match Line Style.

With the settings shown previously in Figure 17.7, the match lines will be named using a predefined text (ML -) and a next counter: ML - 1, ML - 2, ML - 3, etc.

Labels The options in this area of the Match Line page are also similar to those for View Frames. Different label styles are used to annotate match lines located at the left and right side of a frame. This lets you define match-line label styles that reference either the previous or next station adjacent to the current frame. You can also set the location of each label independently using the Left and Right Label Location drop-down lists. You have options for placing the labels at the start, end, or middle of the match line, or at the point where the match line intersects the alignment.

With the settings shown previously in Figure 17.7, the label style for use at the left match line is shown in Figure 17.8. This Basic Left label style uses the Match Line Number, Match Line Station Value, and Previous Sheet Number.

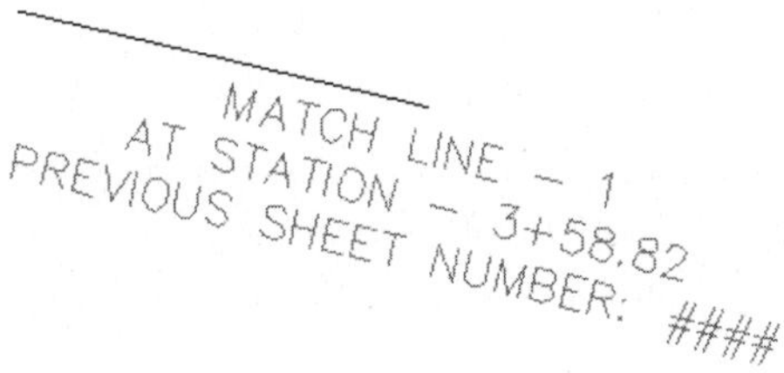

FIGURE 17.8
An example match line label style

Create View Frames – Profile Views Page

The final page of the Create View Frames wizard (Figure 17.9) is optional and will be disabled and skipped if you chose to create Plan Only sheets on the Sheets page of the wizard. Use the drop-down lists to select both the Profile View Style and the Band Set Style. These styles were discussed previously in Chapter 7, "Profiles and Profile Views."

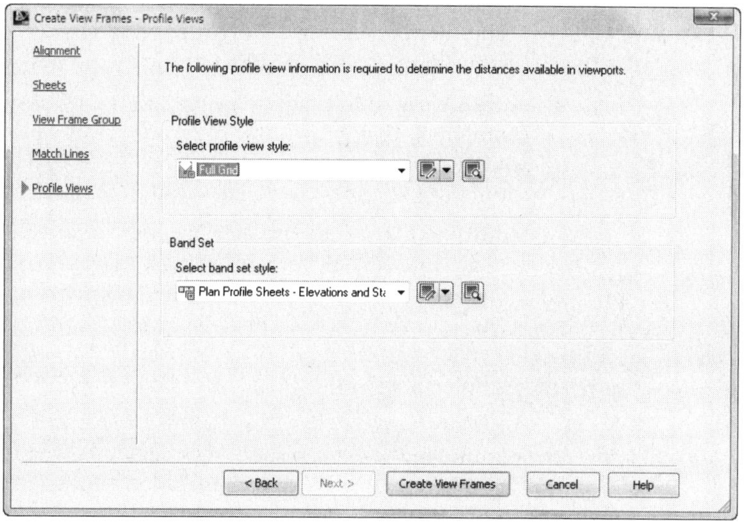

Figure 17.9
Create View Frames – Profile Views wizard page

Civil 3D has difficulty determining the proper extents of profile views. If you find that your profile view isn't positioned correctly in the viewport (for example, the annotation along the sides or bottom is clipped), you may need to create *buffer* areas in the profile-view band set style by modifying the text box width. The `_Autodesk Civil 3D NCS.dwt` (both Imperial and metric) file contains styles with these buffers created.

This last page of the wizard has no Next button. To complete the wizard, click the Create View Frames button.

Creating View Frames

Now that you understand the wizard pages and available options, you'll try them out in this exercise:

1. Open the `ViewFrameWizard.dwg` file or `ViewFrameWizard_METRIC.dwg` file. (Remember, all data can be downloaded from `www.sybex.com/go/masteringcivil3d2013`.)

   This drawing contains several alignments and profiles as well as styles for view frames, view frame groups, and match lines.

2. To launch the Create View Frames wizard, from the Output tab ➢ Plan Production panel choose Create View Frames.

3. On the Alignment page, do the following:
   - A. Select Frontenac Drive from the Alignment drop-down list.
   - B. For Station Range, verify that Automatic is selected.
   - C. Click Next to advance to the next page.
4. On the Sheets page, do the following:
   - A. Select the Plan And Profile option.
   - B. Click the ellipsis button to display the Select Layout As Sheet Template dialog.
   - C. In the Select Layout As Sheet Template dialog, click the ellipsis button and browse to the Plan Production subfolder in the default template file location. Typically, this is:

     ```
     C:\Users\<username>\AppData\Local\Autodesk\C3D2013\
     enu\Template\Plan Production\
     ```

   - D. Select the template named `Civil 3D (Imperial) Plan and Profile.dwt` (or `Civil 3D (Metric) Plan and Profile.dwt`), and click Open.

     A list of the layouts in the DWT file appears in the Select Layout As Sheet Template dialog.
   - E. Select the layout named ARCH D Plan And Profile 20 Scale (or ISO A1 Plan and Profile 1 to 500 for metric users), and click OK to dismiss the Select Layout As Sheet Template dialog.
   - F. In the View Frame Placement area, select the Along Alignment option.
   - G. Select the Set The First View Frame Before The Start Of The Alignment By option.

     Note that the default value for this particular drawing is 30′ (or 10 m for metric users).
   - H. Click Next to advance to the next page.
5. On the View Frame Group page, confirm that all settings are as follows (these are the same settings shown previously in Figure 17.6), and then click Next to advance to the next page:

SETTING	VALUE
View Frame Group Name	VFG - <[View Frame Group Alignment Name(CP)]> - (<[Next Counter(CP)]>)
View Frame Name	VF - (<[Next Counter(CP)]>)
Style	Basic
Label Style	Basic
Label Location	Top Left

6. On the Match Lines page, confirm that all settings are as follows (these are the same settings shown previously in Figure 17.7), and then click Next to advance to the next page.

Setting	Value
Snap Station Value Down To The Nearest	1
Layer	C-ANNO-MTCH
Name	ML - (<[Next Counter(CP)]>)
Style	Basic
Left Label Style	Basic Left
Left Label Location	End
Right Label Style	Basic Right
Right Label Location	Start

7. On the Profile Views page, confirm that the settings are as follows (these are the same settings shown previously in Figure 17.9), and then click Create View Frames:

Setting	Value
Select Profile View Style	Full Grid
Select Band Set Style	Plan Profile Sheets - Elevations And Stations

The view frames and match lines are created as shown in Figure 17.10.

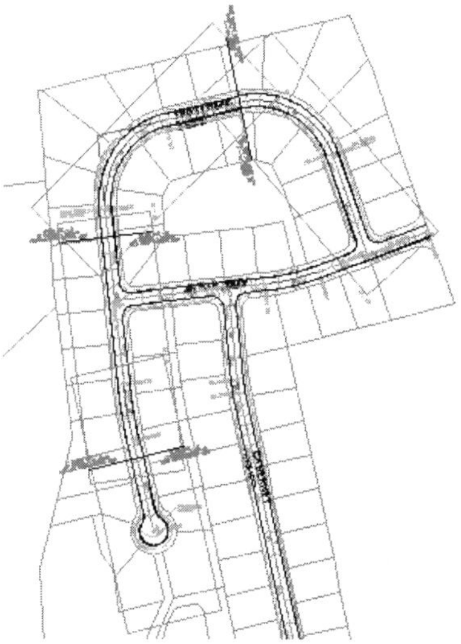

Figure 17.10
Finished view frames and match lines in the drawing

Due to the sheet sizes and scales, the Imperial drawing in this example generates four view frames while the metric drawing generates only two view frames.

THE EFFECTS OF INCREMENTAL COUNTING

The numbering for your view frames, view frame groups, and match lines may not identically match that shown in the images. This is due to the incremental counting Civil 3D performs in the background. As previously mentioned, each time you create one of these objects, the counter increments. You can reset the counter by modifying the Name template.

When this exercise is complete, you may close the drawing. A saved copy of this drawing is available from the book's web page with the filename `ViewFrameWizard_FINISHED.dwg` or `ViewFrameWizard_METRIC_FINISHED.dwg`.

Editing View Frames and Match Lines

After you've created view frames and match lines, you may need to edit them. Edits to some view frame and match line properties can be made via the Prospector tab in the Toolspace palette by expanding the View Frame Groups branch, as shown in Figure 17.11.

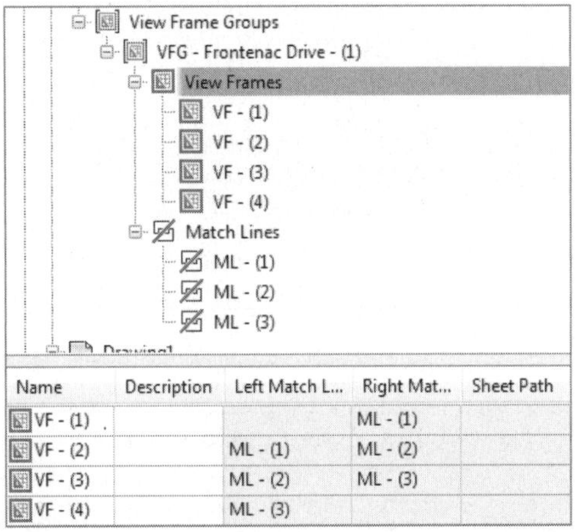

FIGURE 17.11
View Frame Groups in Prospector

You can change some information in the Preview area of Prospector when you highlight either the View Frames branch or the Match Lines branch. Alternatively, you can make further edits from the View Frame Properties dialog or the Match Line Properties dialog. One way of accessing these dialogs is through the View Frame contextual tab or the Match Line contextual tab. Another method is by right-clicking on the desired object in Prospector and selecting Properties. For both view frames and match lines, you can only change the object's name and/or style via the Information tab in their Properties dialog. All other information displayed on the other tabs is read-only.

You make changes to geometry and location graphically using special grip edits (Figure 17.12). Like many other Civil 3D objects with special editing grips (such as profiles and Pipe Network objects), view frames and match lines have editing grips you use to modify the objects' location, rotation, and geometry. Let's look at each separately.

FIGURE 17.12
View frame and match line grips

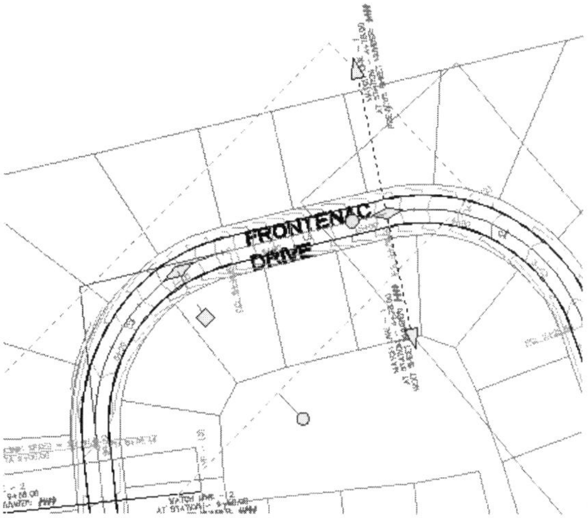

View frames can be graphically edited in three ways. Once you select a view frame object (the rectangular object selected at the left in Figure 17.12), you can move them, slide them along the alignment, and rotate them as follows:

To Move a View Frame The first grip is the standard square grip that is used for most typical edits, including moving the object.

To Slide a View Frame The diamond-shaped grip at the center of the frame lets you move the view frame in either direction along the alignment while maintaining the orientation (Along Alignment or Rotated North) you originally established for the view frame when it was created.

To Rotate a View Frame The circular handle grip works like the one on pipe-network structures. Using this grip, you can rotate the frame about its center.

DON'T FORGET YOUR AUTOCAD FUNCTIONS!

While you're getting wrapped up in learning all about Civil 3D and its great design tools, it can be easy to forget you're sitting on an incredibly powerful AutoCAD application.

AutoCAD features add functionality beyond what you can do with Civil 3D commands alone. First, make sure the Dynamic Input option is enabled by either clicking on the button at the bottom left of your screen as shown here or by pressing F12:

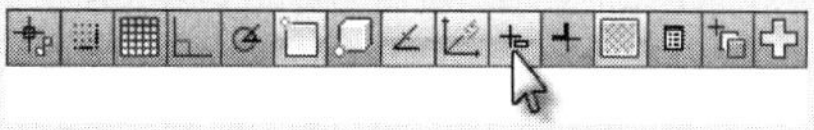

This gives you additional functionality when you're moving a view frame. With Dynamic Input enabled, you can enter an exact station value to precisely locate the frame where you want it. Similar to moving a view frame, with Dynamic Input active you can enter an exact rotation angle. Note that this rotation angle is relative to your drawing settings (for example, 0 degrees is to the left, 90 degrees is straight up, and so on).

Also, selecting multiple objects and then selecting their grips while holding Shift makes each grip "hot" (usually a red color). This allows you to grip-edit one object and all of the "hot" objects will also experience the same grip edit, like sliding a group of view frames along the alignment. You can edit a match line's location and length using special grips. As with view frames, you can slide them along the alignment and rotate them. They can also be lengthened or shortened. Unlike view frames, they can't be moved to an arbitrary location. Once you select a match line object (the object selected at the right in Figure 17.12), you can edit them as follows:

To Slide a Match Line The diamond-shaped grip at the center of the match line lets you move the match line in either direction along the alignment while maintaining the orientation (Along Alignment or Rotated North) that you originally established for the view frame.

Note that the match line can only be moved in either direction a distance equal to or less than that entered on the Match Line page of the wizard at the time the view frames were created. For example, if you entered a value of 50′ (15 m) for the Allow Additional Distance For Repositioning option, your view frames are overlapped 50′ (15 m) to each side of the match line, and you can slide the match line only 50′ (15 m) in either direction from its original location.

To Rotate a Match Line The circular handle grip works like the one on a view frame.

To Change a Match Line's Length When you select a match line, a triangular grip is displayed at each end. You can use these grips to increase or decrease the length of each half of the match line. For example, moving the grip on the top end of the match line changes the length of only the top half of the match line; the other half of the match line remains unchanged. See the sidebar "Don't Forget Your AutoCAD Functions!" for tips on using AutoCAD features. If you select a match line and click one of the triangular grips and then hold Shift and select the other triangular grip, then as you lengthen the match line on one end it will shorten on the other, and vice versa.

The following exercise lets you put what you've learned into practice as you change the location and rotation of a view frame, and change the location and length of a match line:

1. Open the `EditViewFramesAndMatchLines.dwg` or `EditViewFramesAndMatchLines_METRIC.dwg` file from this book's web page. This drawing contains view frames and match lines.
2. Confirm that Dynamic Input is enabled; if it is not, press F12.
3. Select the view frame, which consists of stations 9+56 to 14+34 (or 0+267 to the end for metric users), and select its diamond-shaped sliding grip.
4. Slide this diamond grip so the overlap with the lower view frame isn't so large. Either graphically slide it to station 11+00 (or 0+360 for metric users) or enter **1100↵** (**360↵** for metric users) in the Dynamic Input text box.
5. Press Esc to clear your selection.
6. Select the view frame, which consists of stations 14+34 to the end (or 0+267 to the end for metric users), and select the circular rotation grip.
7. Rotate the view frame slightly to better encompass the road. In the Dynamic Input text box, enter **8↵**.

8. Press Esc to clear your selection.

   Next you will adjust the match line's location.

9. Select the match line, which is presently at Station 14+34 (or 0+267 for metric users), and then select its diamond sliding grip.

10. Either graphically slide it to station 13+25 (or 0+250 for metric users) or enter **1325**↵ (**250**↵ for metric users) in the Dynamic Input text box.

    Notice that the match line label is updated with the revised station.

    Next, you'll adjust the length of the match line to fit within the view frame extents.

11. Select the triangular lengthen grip at the west end of the Match Line, and either graphically shorten it to 75′ (or 50 m for metric users) or enter **75**↵ (or **50**↵ for metric users) in the Dynamic Input text box.

12. Select the triangular lengthen grip at the east end of the same match line, and either graphically shorten it to 75′ (or 50 m for metric users) or enter **75**↵ (or **50**↵ for metric users) in the Dynamic Input text box.

13. Press Esc to clear your selection.

When this exercise is complete, you may close the drawing. A saved copy of this drawing is available from the book's web page with the filename `EditViewFramesAndMatchLines_FINISHED.dwg` or `EditViewFramesAndMatchLines_METRIC_FINISHED.dwg`.

Now that you have generated the view frames and match lines in the drawing and you have placed them where you want them, let's look at using these objects to generate sheets.

Creating Plan and Profile Sheets

The Plan Production feature uses the concept of *sheets* to generate the pages that make up a set of plans. Simply put, *sheets* are layout tabs with viewports showing a given portion of your design model, based on the view frames previously created. The viewports have special Viewport Properties set that define them as either Plan or Profile viewports. These viewports must be predefined in a drawing template (DWT) file to be used with the Plan Production feature. You manage the sheets using the standard AutoCAD Sheet Set Manager feature.

The Create Sheets Wizard

CERT OBJECTIVE

After you've created view frames and match lines, you can proceed to the next step of creating sheets. Like view frames, sheets are created using a wizard. Let's look at the Create Sheets wizard and the various page options. After you've seen each page, you'll have a chance to put what you've learned into practice in an example.

From the Output tab ➢ Plan Production panel, you launch the Create Sheets wizard by choosing Create Sheets. Like the Create View Frames wizard, a list of the Create Sheets wizard's pages is shown along the left sidebar, and an arrow indicates which page you're currently viewing. You move among the pages using the Next and Back navigation buttons along the bottom of each page. Alternatively, you can jump directly to any page by clicking its name in the list on the left. The following sections walk you through the pages of the wizard and explain their features.

Create Sheets – View Frame Group And Layouts Page

You use the first page of the Create Sheets wizard (Figure 17.13) to select the view frame group for which the sheets will be created. It's also used to define how the layouts for these sheets will be generated and which sheets will be created (all or a range).

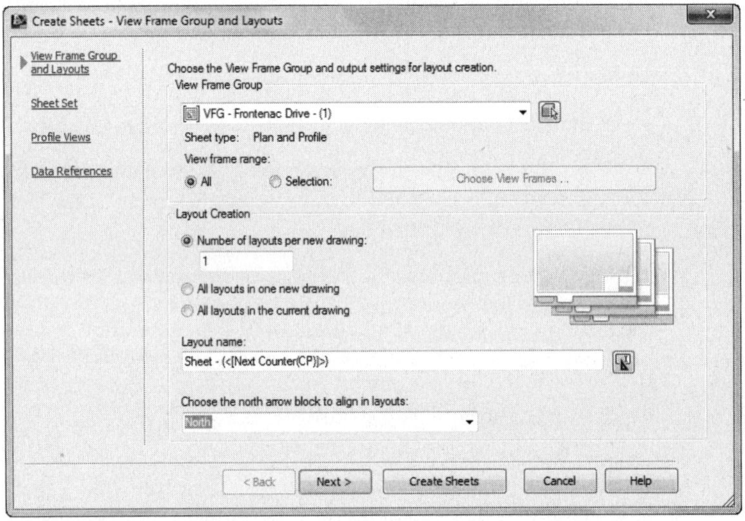

Figure 17.13
Create Sheets – View Frame Group And Layouts wizard page

View Frame Group In the top area of this page, you select the view frame group. You can either select it from the drop-down list or click the Select From The Drawing button to select the view frame group on screen. After you've selected the group, you use the View Frame Range option to create sheets for all frames in the group or only for specific frames of your choosing.

All Select this option when you want sheets to be created for all view frames in the view frame group.

Selection Selecting this option activates the Choose View Frames button. Click this button to display the Select View Frames dialog, where you can select specific view frames from a list. You can select a range of view frames by using the standard Windows selection technique of clicking the first view frame in the range and then holding Shift while you select the last view frame in the range. You can also select individual view frames in nonsequential order by holding Ctrl while you make your view frame selections. Figure 17.14 shows two of the four view frames selected in the Select View Frames dialog.

Layout Creation In this section, you define where and how the new layouts for each sheet are created, as well as the name format for these sheets, and information about the alignment of the north arrow block.

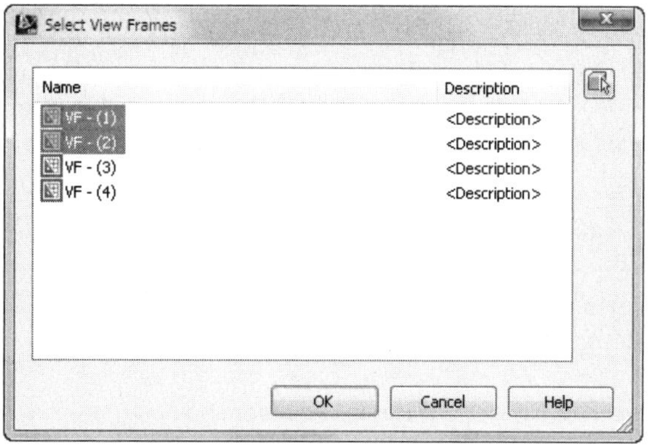

FIGURE 17.14
Select view frames by using standard Windows techniques

There are three options for creating layout sheets: the layouts are created in multiple new drawing files (with a limit to the maximum number of layout sheets created in each file), all the new layouts are created in a new drawing file, or all the layout tabs are created in the current drawing (the drawing you're in while executing the Create Sheets wizard).

Number Of Layouts Per New Drawing This option creates layouts in new drawing files and limits the maximum number of layouts per drawing file to the value you enter in the text box. For best performance, Autodesk recommends that a drawing file contain no more than 10 layouts. On the last page of this wizard you're given the option to select the objects for which data references will be made. These data references are then created in the new drawings.

All Layouts In One New Drawing As the name implies, this option creates all layouts for each view frame in a single new drawing. Use this option if you have fewer than 10 view frames, to ensure best performance. If you have more than 10 view frames, the previous option is recommended. On the last page of this wizard you're given the option to select the objects for which data references will be made. These data references are then created in the new drawing.

All Layouts In The Current Drawing When you choose this option, all layouts are created in the current drawing. You need to be aware of two scenarios when working with this option. (As explained later, you can share a view frame group via data shortcuts and reference it into other drawings as a data reference.)

- When creating sheets, it's possible that your drawing references the view frame group from another drawing (rather than having the original view frame group in your current drawing). In this case, you're given the option to select the additional objects for which data references will be made (such as alignments, profiles, pipes, and so on). These data references are then created in the current drawing. You select these objects on the last page of the wizard.

- If you're working in a drawing in which the view frames were created (therefore, you're in the drawing in which the view frame group exists), the last page of this wizard is disabled. This is because in order for you to create view frames (and view frame groups), the alignment (and possibly the profile) must either exist in the current drawing or be referenced as a data reference (recall the prerequisites for creating view frames, mentioned earlier).

Layout Name Use this text box to enter a name for each new layout. As with other named objects in Civil 3D, you can use the Name template to create a name format that includes information about the object being named. With the settings shown previously in Figure 17.14, the layouts will be named using a predefined text (Sheet -) and a next counter: Sheet - 1, Sheet - 2, Sheet - 3, etc. Using the Name template you could alternatively use the Parent Drawing Name, View Frame Start/End Raw Station, View Frame Start/End Station Value, View Frame Group Alignment Name, and View Frame Group Name options.

Real World Scenario

WHERE AM I?

We strongly recommend that you set up the Name template for the layouts so that it includes the View Frame Group Alignment Name option and the station range (View Frame Start Station Value and View Frame End Station Value). This conforms to the way many organizations create sheets, helps automate the creation of a sheet index, and generally makes it easier to navigate a DWG file with several layout tabs.

This would result in a name that is listed as:

```
<[View Frame Group Alignment Name]> <[View Frame Start Station Value]> to <[View
Frame End Station Value]>
```

This will generate a layout named similar to: Frontenac Drive 0 + 00.00 to 4 + 78.00.

Choose The North Arrow Block To Align In Layouts If the template file you've selected contains a north arrow block, it can be aligned so that it points north on each layout sheet. The block must exist in the template and be located in the layout. If there are multiple blocks, select the one you want to use from the drop-down list.

CREATE SHEETS – SHEET SET PAGE

You use the second page of the Create Sheets wizard (Figure 17.15) to determine whether a new or existing sheet set (with the file extension `.dst`) is used and the location of the DST file. The sheet storage location and sheet name are also defined here. Additionally, on this page you decide whether to add the sheet set file (with the file extension `.dst`) and the sheet files (with the file extension `.dwg`) to the project vault.

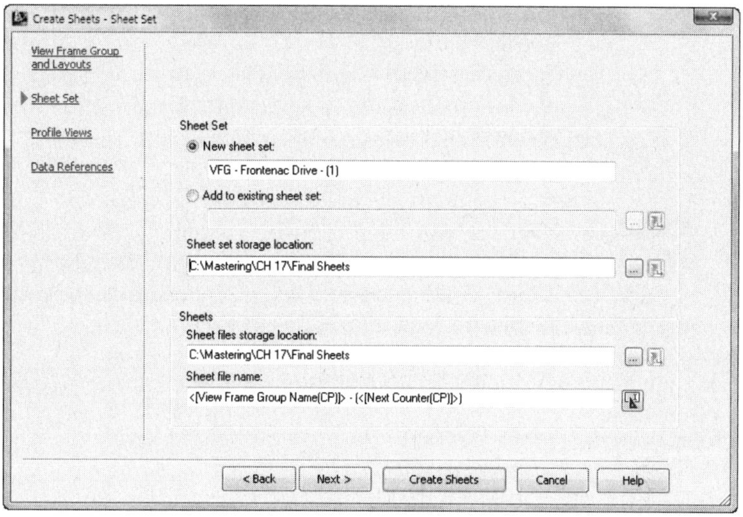

FIGURE 17.15
Create Sheets – Sheet Set wizard page

Sheet Set In the Sheet Set area of the page, you select whether to create a new DST file or add the sheets created by this wizard to an existing DST file.

New Sheet Set By selecting this option, you create a new sheet set. You must enter a name for the DST file and specify a sheet set storage location. By default, the sheet set is created in the same folder as the current drawing. You can change this by clicking the ellipsis and selecting a new location.

Add To Existing Sheet Set Selecting this option lets you select an existing sheet-set file to which the new sheets created by this wizard will be added. Click the ellipsis to browse to the existing DST file location. Selecting this option will disable the Sheet Set Storage Location selection on this page.

Sheets You use the bottom area of the page to set the name and storage location for any new DWG files created by this wizard. On the previous page of the wizard, you had the choice of creating new files or creating the sheet layout in the current drawing. If you chose the latter option, the Sheets area on this page of the wizard is inactive. If you chose the former, here you specify the sheet-files storage location and enter the sheet-file name (with the `.dwg` extension).

Like many of the previous names, you can use a Name template to specify how the sheet-file name is generated. With the settings shown previously in Figure 17.15, the sheet files will be named using the View Frame Group Name entry and a next counter. Using the Name template you could alternatively use the Parent Drawing Name, View Frame Start/End Raw Station, View Frame Start/End Station Value, and View Frame Group Alignment Name options.

> **WHAT IS A SHEET?**
>
> This page can be a little confusing due to the way the word *sheets* is used. In some places, *sheets* refers to layout tabs in a given drawing (DWG) file. On this page, though, the word *sheets* is used in the context of Sheet Set Manager and refers to the DWG file itself.

CREATE SHEETS – PROFILE VIEWS PAGE

The third page of the Create Sheets wizard (Figure 17.16) lists the profile view style and the band set selected in the Create View Frames wizard. You can't change these selections. You can, however, make adjustments to other profile settings.

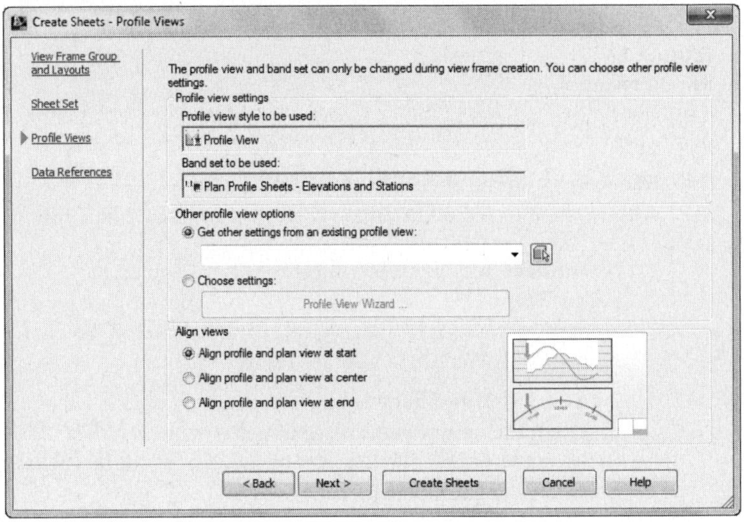

FIGURE 17.16
Create Sheets – Profile Views wizard page

Other Profile View Options The Other Profile View Options area lets you modify certain profile view options either by using an existing profile view in your drawing as an example or by running the Profile View wizard. Regardless of what option you choose, the "other options" you can change are limited to the following in the Profile View wizard:

- Profile View Datum By Minimum Elevation or Mean Elevation on the Profile View Height page
- Split Profile View options from the Profile View Height page
- All options on the Profile Display Options page
- If available, all options of the Pipe Network Display page
- Most of the settings on the Data Bands page
- All options on the Profile Hatch Options page
- All settings on the Multiple Plot Options page

See Chapter 7 for details on each of these settings. The inactive settings are ones that were previously set when you were going through the settings of the Create View Frames wizard. If you need to change these settings, you will have to delete your current view frame group and regenerate the view frames with the desired settings.

Align Views In the Align Views area of the page, you can choose Align Profile And Plan View At Start, Align Profile And Plan View At Center, or Align Profile And Plan View At End.

CREATE SHEETS – DATA REFERENCES PAGE

The final page of the Create Sheets wizard (Figure 17.17) is used to create data references in the drawing files that contain your layout sheets.

FIGURE 17.17
Select Create Sheets – Data References wizard page

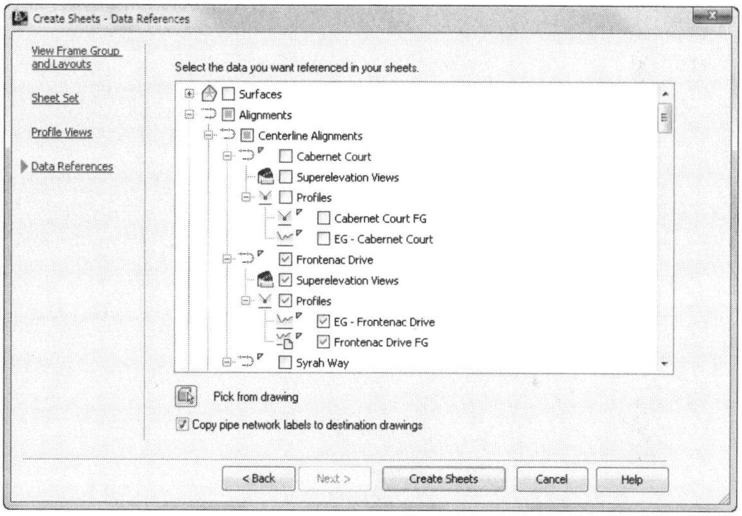

Based on the view frame group used to create the sheets and the type of sheets (plan, profile, plan and profile), certain objects are selected by default. You have the option to select additional objects for which references will be made. You can either pick them from the list or click the Pick From The Drawing button and select the objects in the drawing.

It's common to create references to pipe networks that are to be shown in plan and/or profile views. If you choose to create references for pipe network objects, you can also copy the labels for those network objects into the sheet's drawing file. This is convenient in that you won't need to relabel your networks.

Managing Sheets

After you've completed all pages of the Create Sheets wizard, you create the sheets by clicking the Create Sheets button. Doing so completes the wizard and starts the creation process. If you're creating sheets with profile views, you're prompted to select a profile view origin. Civil 3D then displays several dialogs, indicating the process status for the various tasks, such as creating the new sheet drawings and creating the DST file. Once complete the Panorama Event Viewer vista will list two new events, in this example one stating "Sheets created were added to the sheet set file `C:\Mastering\CH 17\Final Sheets\VFG - Frontenac Drive - (1).dst`." and one stating "4 layout(s) created in path `C:\Mastering\CH 17\Final Sheets`."

If the Sheet Set Manager isn't currently open, it opens with the newly created DST file loaded. The sheets are listed, and the details of the drawing files for each sheet appear (Figure 17.18).

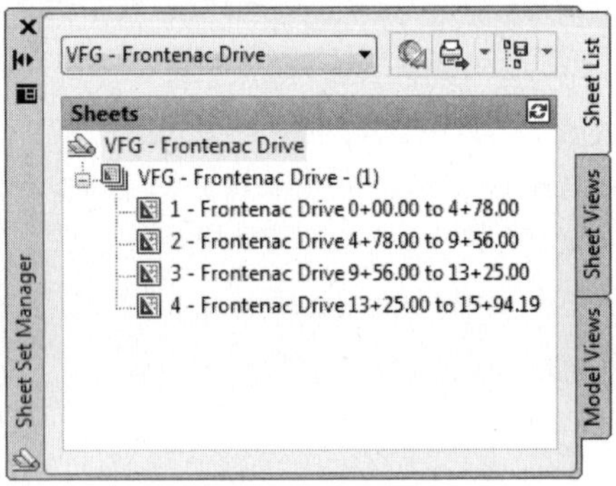

FIGURE 17.18
New sheets in the Sheet Set Manager

If you double-click to open the new drawing file that contains the newly created sheets, you'll see layout sheet tabs created for each of the view frames as selected in the Create Sheets wizard. The sheets are named using the Name template as defined in the Create Sheets wizard. Figure 17.19 shows the names that result from the following template:

```
<[View Frame Group Alignment Name]> <[View Frame Start Station Value]>
to <[View Frame End Station Value]>
```

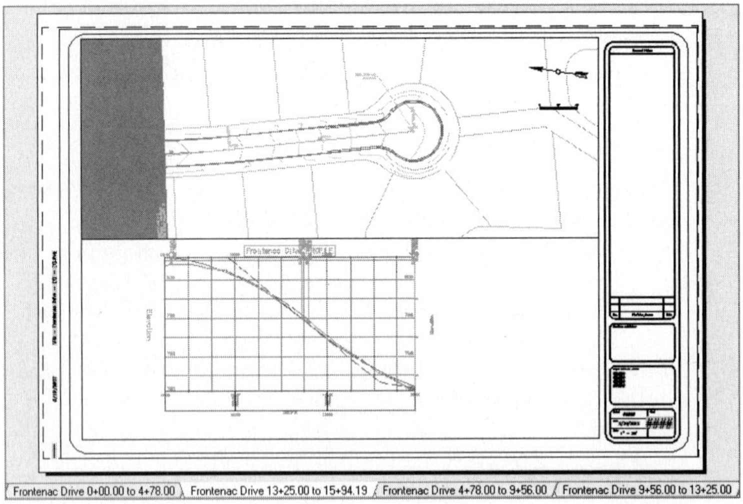

FIGURE 17.19
The template produces the Frontenac Drive tab names shown here.

To create the final sheets in this new drawing, Civil 3D externally references (XRefs) the drawing containing the view frames; creates data references (DRefs) for the alignments, profiles, and any additional objects you selected in the Create Sheets wizard; and, if profile sheet types were selected in the wizard, creates profile views in the final sheet drawing.

The following exercise pulls all these concepts together:

1. Open the `SheetsWizard.dwg` or `SheetsWizard_METRIC.dwg` file.

   This drawing contains the view frame group, alignment, and profile for Frontenac Drive. Note that the drawing doesn't have profile views.

2. To launch the Create Sheets wizard, from the Output tab ➢ Plan Production panel, choose Create Sheets.

3. On the View Frame Group And Layouts page, do the following:

   A. Verify that View Frame Range is set to All.

   B. Verify that Number Of Layouts Per New Drawing is set to **10**.

      Since this view frame group has less than 10 view frames, only one drawing will be generated, but it is good practice to set this value to 10 nonetheless.

   C. Set Layout Name to **`<[View Frame Group Alignment Name]> <[View Frame Start Station Value]> to <[View Frame End Station Value]>`**.

USE A NAME TEMPLATE

To set the Layout Name in step 3 or any of the other steps that require the use of the Name template, don't type all that text into the text box. Instead, follow these simple steps:

1. Click the Edit Layout Name button to the right of the Layout Name text box to display the Name Template dialog.
2. Delete all of the text in the Name text box that you do not want.
3. Using the Property Fields drop-down list, select the first piece of text that you want. In this example it will be View Frame Group Alignment Name.
4. Click Insert.
5. Repeat steps 3 and 4 for each of the pieces of text that you want.

   At any time you can also insert manual text into the Name text box such as a space or hyphen between property fields or the word "to," as shown in this example.

   Make certain to choose Start/End Station Value and not Start/End Raw Station as required.

6. Once complete, click OK to dismiss the Name Template dialog.

The Layout Name option is now filled in based on the Name template.

   D. From the Choose The North Arrow Block To Align In Layouts drop-down list, select North.

   E. Click Next to advance to the next page.

4. On the Sheet Set page, do the following:

   A. Select the New Sheet Set option and set Name to **VFG - Frontenac Drive** (or **VFG - Frontenac Drive_METRIC** for metric users).

   B. For the Sheet Set File Storage Location, use the ellipsis to browse to `C:\Mastering\CH 17\Final Sheets`, and click Open to dismiss the Browse For Sheet Set Folder dialog.

   Notice that by changing this location, the Sheet Files Storage Location entry automatically changes to match.

   C. Verify that Sheet File Name is set to: `<[View Frame Group Name(CP)]> - (<[Next Counter(CP)]>)` for Imperial users, or `<[View Frame Group Name(CP)]> - (<[Next Counter(CP)]>)_METRIC` for metric users.

   D. Click Next to advance to the next page.

5. On the Profile Views page, do the following:

   A. For Other Profile View Options, select Choose Settings and then click the Profile View Wizard button.

   The Create Multiple Profile Views dialog opens.

   B. On the left side of the Create Multiple Profile Views dialog, click Profile Display Options to jump to that page.

   C. On the Create Multiple Profile Views – Profile Display Options page, verify that the Draw option is only selected for the EG – Surface and the Frontenac Drive – FG surface.

   D. Scroll to the right and verify that the Labels setting for the EG – Surface is set to _No Labels and that the Frontenac Drive – FG surface is set to Complete Label Set.

   E. On the left side of the Create Multiple Profile Views dialog, click Data Bands to jump to that page.

   F. On the Create Multiple Profile Views – Data Bands page, change Profile2 to Frontenac Drive FG for both bands.

   G. Click Finish to dismiss the Create Multiple Profile Views wizard and return to the Create Sheets wizard.

   H. Verify that Align Views is set to Align Profile And Plan View At Start.

   I. Click Next to advance to the next page.

6. On the Data References page, confirm that at minimum, the Frontenac Drive and both the EG and FG profiles are selected, and then click Create Sheets to complete the wizard.

   Before creating the sheets, Civil 3D must save your current drawing.

7. Click OK when prompted to save.

   The drawing is saved, and you're prompted for an insertion point for the profile view. The location you pick represents the lower-left corner of the profile view grid.

8. Select an open area in the drawing, above the right side of the site plan.

   Civil 3D displays a progress dialog, and then the Panorama palette is displayed with information about the results of the sheet creation process.

9. Close the Panorama palette.

INVISIBLE PROFILE VIEWS

Note that the profile views are created in the current drawing only if you selected the option to create all layouts in the current drawing. Because you didn't do that in this exercise, the profile views aren't created in the current drawing. Rather, they're created in the sheet drawing files in modelspace in a location relative to the point you selected in this step.

After the sheet creation process is complete, the Sheet Set Manager window opens.

10. Click the first sheet, as shown in Figure 17.20, named Frontenac Drive 0+00.00 to 4+78.00 (or Frontenac Drive 0+000.00 to 0+250.00_METRIC for metric users).

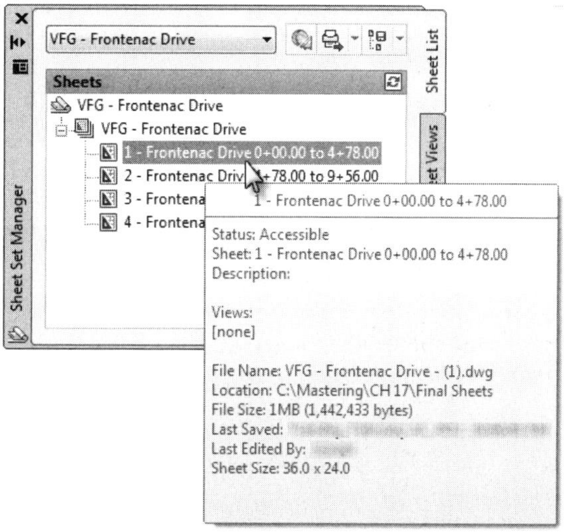

FIGURE 17.20
The Sheet Set Manager once the sheet creation process is complete

Notice that the name conforms to the Name template and includes the alignment name and the station range for the sheet.

11. Review the details listed for the sheet. In particular, note the filename and storage location.
12. Double-click this sheet to open the new sheets drawing and display the layout tab for Frontenac Drive 0 + 00.00 to 4 + 78.00 (or Frontenac Drive 0+000.00 to 0+250.00_METRIC for metric users).
13. Review the multiple tabs created in this drawing file as previously shown in Figure 17.19.

    The template used also takes advantage of AutoCAD fields, some of which don't currently have values assigned.

When this exercise is complete, you may close the drawing. A saved copy of this drawing is available from the book's web page with the filename `SheetsWizard_FINISHED.dwg` or `SheetsWizard_METRIC_FINISHED.dwg`. In addition, the downloadable dataset includes the `Final Sheets` folder, which includes additional files created in this exercise.

Now that you have the Plan and Profile sheets generated, let's look at generating some section sheets using the Plan Production feature.

Creating Section Sheets

Similar to the process of creating plan and profile sheets, creating section sheets is a two-step process:

1. You have to create a section view group that determines the layout, labeling, and styles of the profile views.
2. You have to generate the actual sheets and add them to the AutoCAD sheet set.

Creating Section View Groups

The process of creating the section view group is where you will determine how your section views will be laid out on the page, what labels will be used, and what styles will be used to represent the various components of the model. If you have any questions about section styles or section view styles, refer to Chapter 21, "Object Styles."

In this exercise, you'll walk through setting up a basic section view group for the main road of our sample set:

1. Open `MultipleSectionViews.dwg` or the `MultipleSectionViews_METRIC.dwg` file from the provided dataset.

    In this drawing, sample lines have been added along the Frontenac Drive alignment. These lines are sampling the existing and proposed surfaces.

2. From the Home tab ➢ Profile & Section Views panel, choose Section Views ➢ Create Multiple Views to display the Create Multiple Section Views wizard, shown in Figure 17.21.

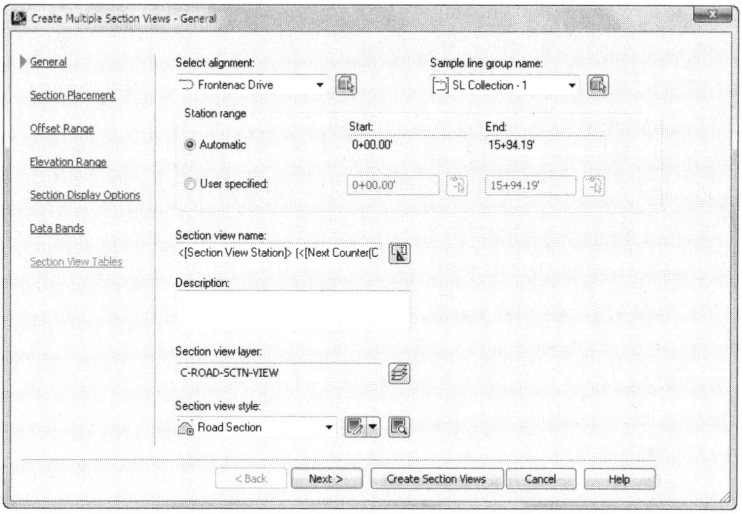

FIGURE 17.21
Create Multiple Section Views – General wizard page

3. On the General page, do the following:
   - A. Verify that Section View Style is set to Road Section.
   - B. Click Next to advance to the next page.

The Section Placement page (Figure 17.22) allows you to specify the section placement options using either a layout from a template file to place production sections on sheet or place draft sections in a grid in modelspace. The Production template and button allow you to navigate to your own sheet template that suits your needs. Notice that if you generate draft sections, no sheets can be created.

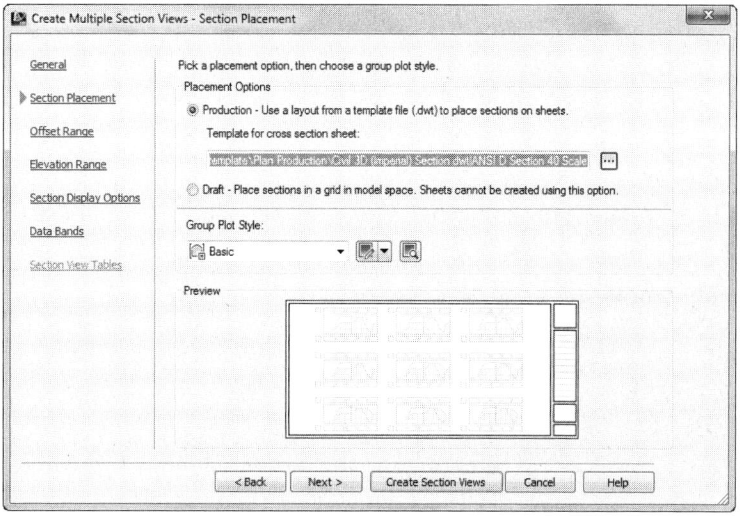

FIGURE 17.22
Create Multiple Section Views – Section Placement wizard page

4. On the Section Placement page, do the following:
   - **A.** Verify that you are generating Production sections.
   - **B.** Change Template For Cross Section Sheet to **ARCH D Section 50 scale** (or **ISO A1 Section 1 to 500** for metric users).
   - **C.** Verify that Group Plot Style is set to Basic.
   - **D.** Click Next to advance to the next page.

   The Offset Range page (Figure 17.23) determines the width of the section views.

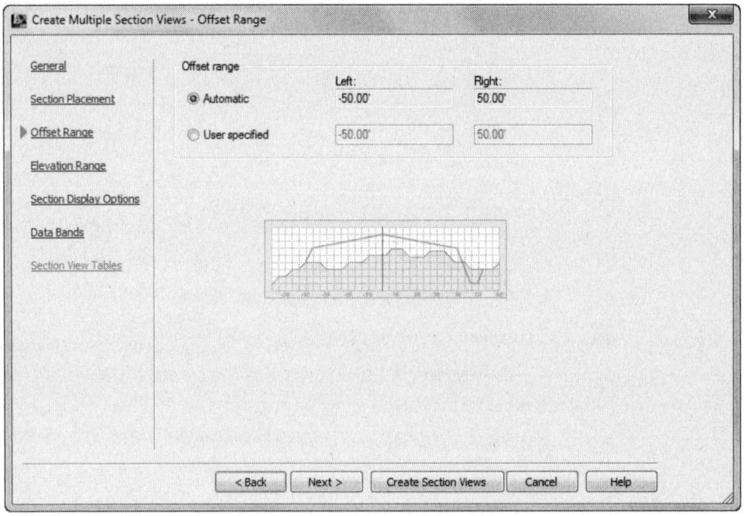

FIGURE 17.23 Create Multiple Section Views – Offset Range wizard page

5. On the Offset Range page, do the following:
   - **A.** Using the User Specified Offset option, change Left Offset Range to **-50′** (or **-15** m for metric users).
   - **B.** Change the Right Offset Range to **50′** (or **15** m for metric users).
   - **C.** Click Next to advance to the next page.

   The Elevation Range page (Figure 17.24) determines the height of the section views. The options on the Elevation Range page help if you have extra tall sections, allowing you to set some limits manually.

6. On the Elevation Range page, click Next to advance to the next page.

FIGURE 17.24 Create Multiple Section Views – Elevation Range wizard page

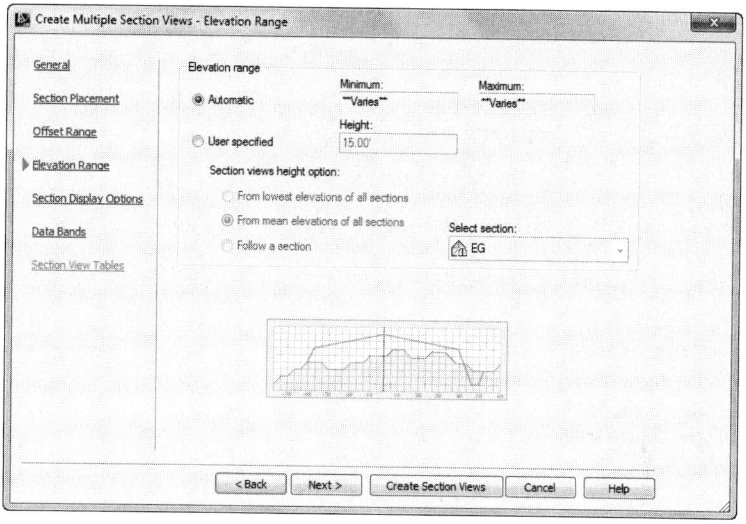

7. On the Section Display Options page, do the following:
   - **A.** Verify that in the Change Labels column the EG labels are set to _No Labels and the FG surface labels are set to _No Labels.
   - **B.** Verify that in the Style column the EG style is set to Existing Ground, the FG style is set to Finished Ground, and each of the corridor surface styles are set to Basic, as shown in Figure 17.25.
   - **C.** Click Next to advance to the next page.

FIGURE 17.25 Changing styles for the Create Multiple Section View – Section Display Options wizard page

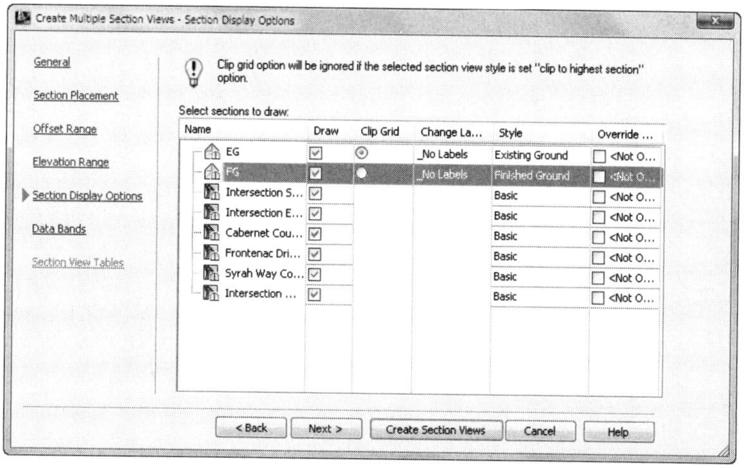

8. On the Data Bands page, do the following:
   A. Verify that the Select Band Set drop-down list is set to Offsets Only.
   B. Verify that Surface1 is the EG surface and Surface2 is the FG surface.
   C. Click Create Section Views to dismiss the wizard and place your section views in the drawing.
9. Click a point to the east of the plan view to draw the section views and sheet outlines.

   Your drawing should look something like Figure 17.26.

FIGURE 17.26
The finished multiple section views operation

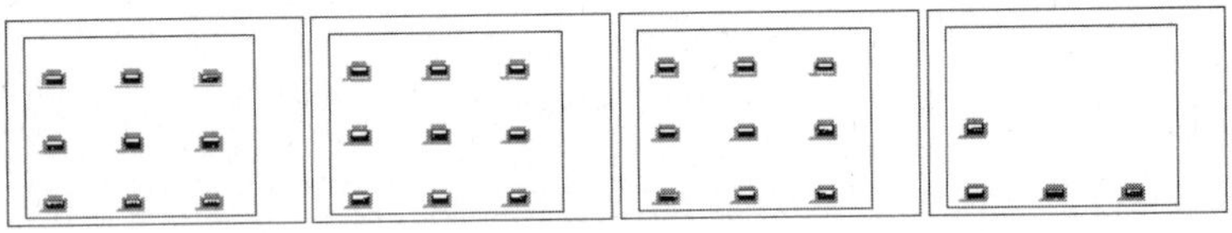

When this exercise is complete, you may close the drawing. A saved copy of this drawing is available from the book's web page with the filename `MultipleSectionViews_FINISHED.dwg` or `MultipleSectionViews_METRIC_FINISHED.dwg`.

Now that you have a section view group, you can begin the process of creating section sheets for plotting.

Creating Section Sheets

Many long transportation projects such as highways, light-rail, or canals require the production of many section sheets. While Civil 3D could produce the views prior to Autodesk Civil 3D 2012, the sheet creation process improved greatly in that release. In this exercise, you'll convert a section view group into a collection of sheets and place them in a new sheet set.

1. Open the `CreatingSectionSheets.dwg` or `CreatingSectionSheets_METRIC.dwg` file.

   This file is the result of the previous exercise and contains the section view group for the Frontenac Drive alignment.

2. From the Output tab Plan Production Panel, choose Create Section Sheets to display the Create Section Sheets dialog.
3. In the Create Section Sheets dialog, do the following:
   A. Verify that New Sheet Set name is set to **CreatingSectionSheets** (or **CreatingSectionSheets_METRIC**).
   B. Click the ellipsis to set the Sheet Set Storage Location to `C:\Mastering\CH 17\Final Sheets`.

   The Create Section Sheets dialog should now look similar to Figure 17.27.

FIGURE 17.27
The Create Section Sheets dialog

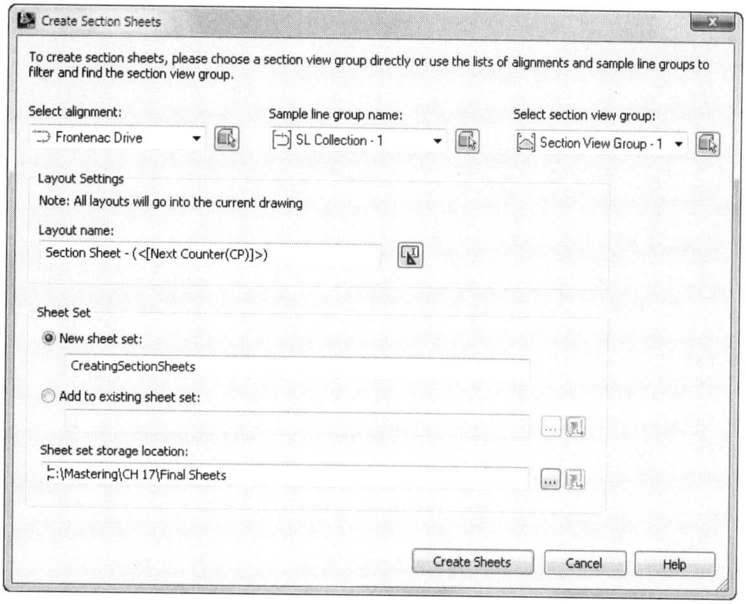

Note that in a drawing with more section view groups you would be able to select each one and create sheets quickly and easily.

4. Click Create Sheets to dismiss the dialog and generate sheets.

   Before creating the sheets, Civil 3D must save your current drawing.

5. Click OK when prompted to save.

   The drawing is saved and Civil 3D will generate new layouts and sheets in a sheet set. The Sheet Set Manager will appear.

6. Switch to the Section Sheet - (2) layout tab.

   Your layout should look something like Figure 17.28.

7. Close the Sheet Set Manager palette.

When this exercise is complete, you may close the drawing. A saved copy of this drawing is available from the book's web page with the filename `CreatingSectionSheets_FINISHED.dwg` or `CreatingSectionSheets_METRIC_FINISHED.dwg`.

While there are still some tweaks to be made to any sheet, large portions of the mundane details are handled by the wizards and tools. There are some elements that you can modify to customize these details for your organization, and you'll look at those in the next section.

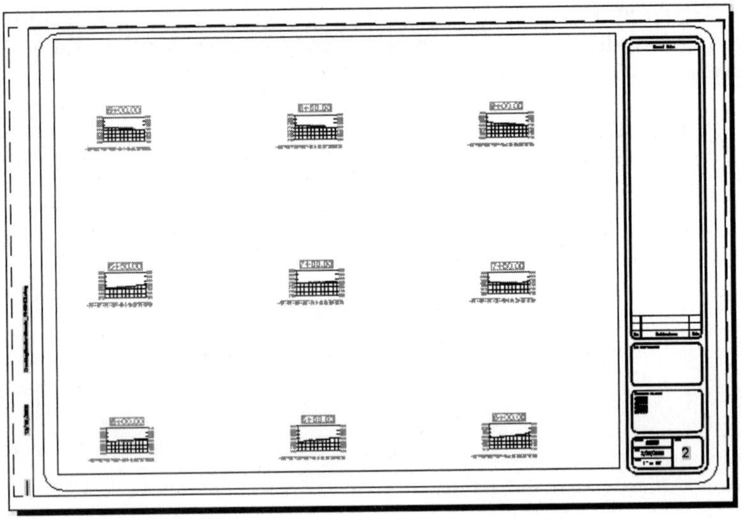

FIGURE 17.28
A completed section sheet

Drawing Templates

The beginning of this chapter mentioned that there are several prerequisites to using the Plan Production tools in Civil 3D. The list includes drawing templates (DWT) set up to work with the Plan Production feature, and styles for the objects generated by this feature. In this section of the chapter, you'll learn how to prepare these items for use in creating your finished sheets.

Civil 3D ships with several predefined template files for various types of sheets that Plan Production can create. By default, these templates are installed in a subfolder called `Plan Production`, which is located in the standard `Template` folder. You can see the `Template` folder location by opening the Files tab of the Options dialog, as shown in Figure 17.29.

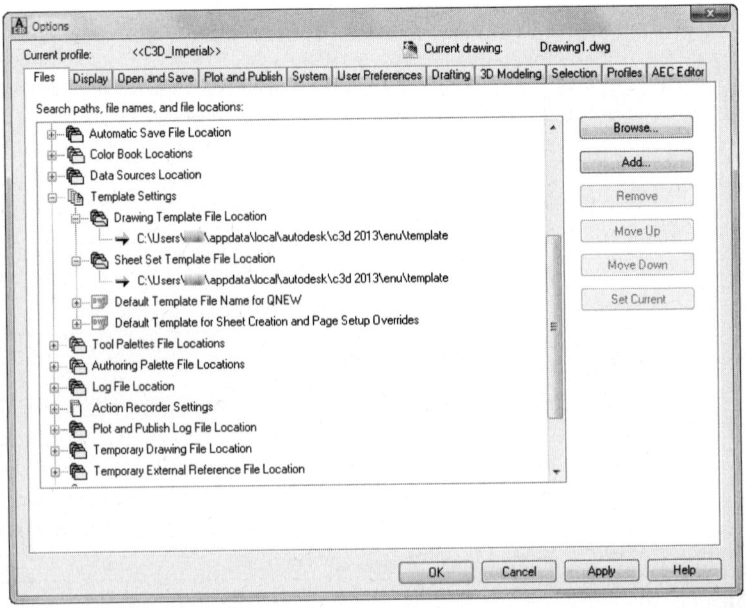

FIGURE 17.29
Template files location

Figure 17.30 shows the default contents of the `Plan Production` subfolder. Notice the templates for Plan, Profile, and Plan And Profile sheet types. There are Imperial and metric versions of each.

FIGURE 17.30
Plan Production DWT files

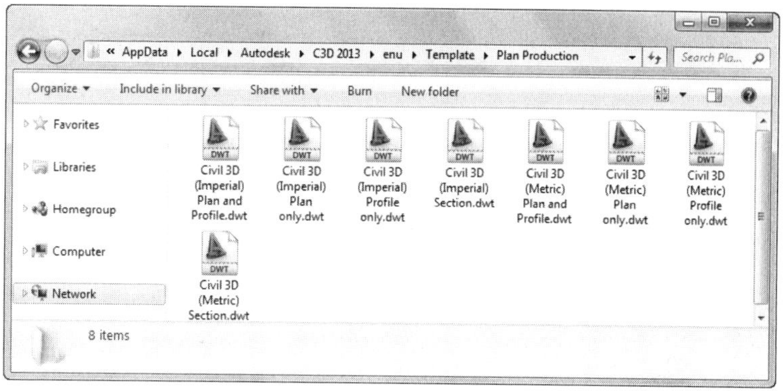

As previously discussed, each template contains layout tabs with pages set to various sheet sizes and plan scales. For example, the `Civil 3D (Imperial) Plan and Profile.dwt` template has layouts created at various ANSI and ARCH sheets sizes and scales, as shown in Figure 17.31.

FIGURE 17.31
Various predefined layouts in standard DWT

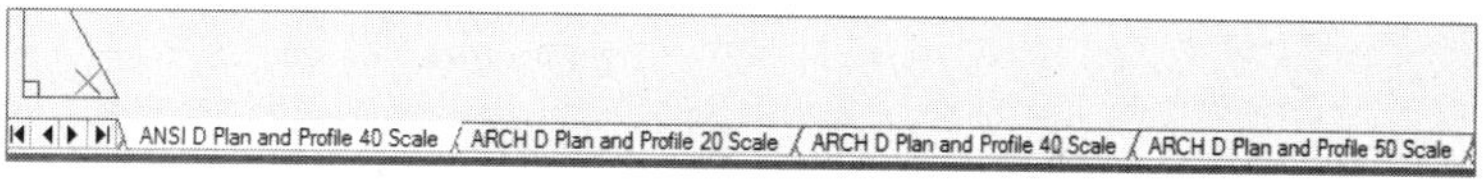

If you decide to make your own Plan Production templates, it is good practice to provide multiple drawing sizes and scales so that you have them available when you go to make your sheets. But beyond just having them available, make sure that the layout names that you provide in your Plan Production template are descriptive enough that you know which one to select.

The viewports in these templates must be rectangular in shape and must have Viewport Type set to Plan, Profile, or Section, depending on the intended use. You set Viewport Type on the Design tab of the Properties dialog, as shown in Figure 17.32.

IRREGULAR VIEWPORT SHAPES

Just because the viewports must start out rectangular doesn't mean they have to stay that way. Experiment with creating viewports from rectangular polylines that have vertices at the midpoint of each side of the viewport (not just at the corners). After you've created your sheets using the Plan Production tool, you can stretch your viewport into irregular shapes. You can also convert a rectangular viewport into an irregular viewport by selecting the viewport, right-clicking, selecting Viewport Clip, and following the prompts.

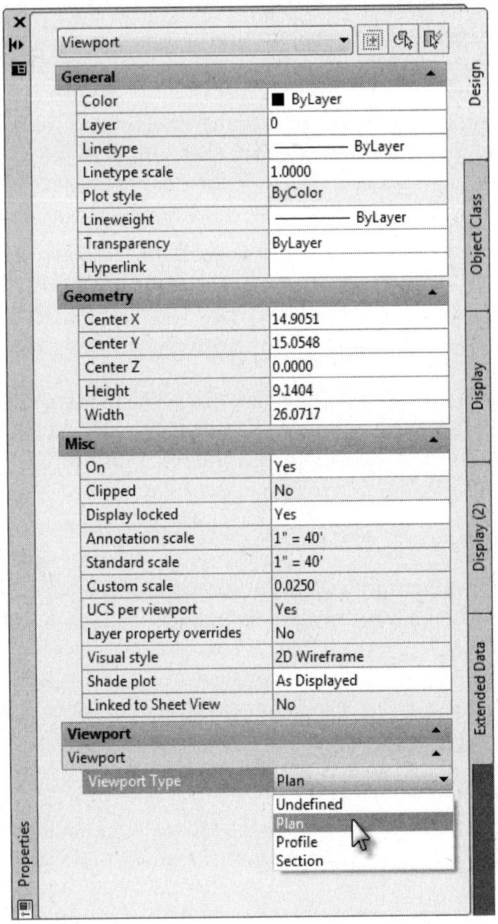

FIGURE 17.32 Viewport Properties – Viewport Type

The Bottom Line

Create view frames. When you create view frames, you must select the template file that contains the layout tabs that will be used as the basis for your sheets. This template must contain predefined viewports. You can define these viewports with extra vertices so you can change their shape after the sheets have been created.

Master It Open the `MasteringPlanProduction.dwg` or `MasteringPlanProduction_METRIC.dwg` file. Run the Create View Frames wizard to create view frames for Alignment A in the current drawing. (Accept the defaults for all other values.) These view frames will be used to generate Plan and Profile sheets on ARCH D (ISO A1) sheets at 20 scale (1:200 scale) using the plan and profile template `MasteringPandPTemplate.dwt` or `MasteringPandPTemplate_METRIC.dwt`. All files should be saved in `C:\Mastering\CH 17\`.

Edit view frames. The grips available to edit view frames allow the user some freedom on how the frames will appear.

Master It Open the `MasteringEditViewFrames.dwg` or `MasteringEditViewFrames_METRIC.dwg` file, and move the VF- (1) view frame to Sta. 2+20 (or Sta. 0+050 for metric users) to lessen the overlap. Then adjust Match Line 1 (or Match Line 2 for metric users) so that it is now at Sta. 4+25 (or Sta. 0+200 for metric users) and shorten it so that the labels are completely within the view frames.

Generate sheets and review Sheet Set Manager. You can create sheets in new drawing files or in the current drawing. The resulting sheets are based on the template you chose when you created the view frames. If the template contains customized viewports, you can modify the shape of the viewport to better fit your sheet needs.

Master It Open the `MasteringCreateSheets.dwg` or `MasteringCreateSheets_METRIC.dwg` file. Run the Create Sheets wizard to create plan and profile sheets in the current drawing for Alignment A using the using the plan and profile template `MasteringPandPTemplate.dwt` or `MasteringPandPTemplate_METRIC.dwt`. Make sure to choose a north arrow. (Accept the defaults for all other values.) All files should be saved in `C:\Mastering\CH 17\`.

Create section views. More and more municipalities are requiring section views. Whether this is a mile-long road or a meandering stream, Civil 3D can handle it nicely via Plan Production.

Master It Open the `MasteringSectionSheets.dwg` or `MasteringSectionSheets_METRIC.dwg` file. Create section views and Plan Production section sheets in a new sheet set for Alignment A using the using the Road Section section style and the section template `MasteringSectionTemplate.dwt` or `MasteringSectionTemplate_METRIC.dwt`. Make sure the sections are set to be generated on ARCH D (ISO A1) sheets at 20-scale (1:200 scale). (Accept the defaults for all other values.) All files should be saved in `C:\Mastering\CH 17\`.

Chapter 18

Advanced Workflows

The AutoCAD® Civil 3D® program is unique in that with a few exceptions, the data you create is stored in the DWG. Using data shortcuts will allow you to collaborate with coworkers and prevent the DWG size from getting unwieldy. Even if you work on small projects, data shortcuts can make creating cross-section views a more straightforward process.

As survey technology advances and sources for accurate data become more varied, the ability to transform between coordinate systems and collaborate with other versions of software is imperative. For the solutions to many interoperability questions, you will look to LandXML and the Map functionality to help you along.

In this chapter, you will learn to:

- Create a data shortcut folder
- Create data shortcuts
- Export to earlier releases of AutoCAD
- Export to LandXML
- Transform coordinates between drawings

Data Shortcuts

CERT OBJECTIVE

A *data shortcut* is a link between drawings, allowing specific types of Civil 3D data to be shared. The shortcut itself does not contain data, but it is a pointer, directing Civil 3D to read data from a common pool of data. A *data shortcut* is created in the source drawing and a *data reference* is the manifestation of the data in a host drawing.

There are many situations in which you need data or information to link between drawings. Connections between drawings can be in the form of external references (XREFs), data shortcuts, or a combination of the two. These two options are similar but not the same. Let's compare (Table 18.1).

TABLE 18.1: XREF vs. data shortcut

XREF	DATA SHORTCUT
For most objects, XREF is a graphic-only representation of objects created in another drawing.	Information-only link to Civil 3D data created in another DWG.
Any objects (base AutoCAD or Civil 3D) are displayed in XREFs.	Only specific types of Civil 3D data can be used with data shortcuts.
Visibility of objects controlled by original drawing layers.	Visibility of data controlled by host drawing styles and layers.
A drawing containing a Civil 3D surface or parcels can be XREF'ed into a host drawing and labeled dynamically.	Surface data must be referenced using a data shortcut for use in design (i.e., creating an existing ground profile)
A drawing containing a Civil 3D corridor can be XREF'ed into a host drawing. Sample lines can pull corridor data from the XREF.	A data shortcut to an alignment in conjunction with an XREF to the corridor enables sample lines to be created in a separate drawing.

As noted, only Civil 3D objects can be used with shortcuts, and even then some objects are not available through shortcuts. The following objects are available for use through data shortcuts:

- Alignments
- Surfaces
- Profile data
- Pipe networks (but not pressure networks)
- View frame groups

A NOTE ABOUT THE EXERCISES IN THIS CHAPTER

This chapter is about workflow—therefore, it is difficult to jump in partway through the chapter. In order to get the most out of the exercises you should work from beginning to end, as all of the exercises build on each other. If you skip any exercises (including the Real World Scenario, "Sending Cross-Section Sheets to Their Own Drawing"), the subsequent steps will not work.

Use the recommended names for objects and drawings you create on your own. Doing so will make things go much smoother for you.

Remember that you can always get the original files from the chapter folder that you can download from `www.sybex.com/go/masteringcivil3d2013`.

Getting Started

Before making your first project, you should make a *project template*. A project template is simply a set of folders, subfolders, and files (if you wish). When you start a new project using data shortcuts, we recommend that you use a project template to keep files organized.

In the following exercise, you will create folders for use when creating a data shortcut project:

1. Open Windows Explorer and navigate to `C:\Civil 3D Project Templates`.
2. Create a new folder titled **`MasteringC3D2013`**.
3. Inside `MasteringC32013`, create subfolders called **`Design, Documents, Sheets,`** and **`Survey`** as shown in Figure 18.1.
4. Locate the file `Project Checklist.doc` from this chapter's files that you downloaded from `www.sybex.com/go/masteringcivil3d2013.`

FIGURE 18.1 Folder structure and file to be included in the project template

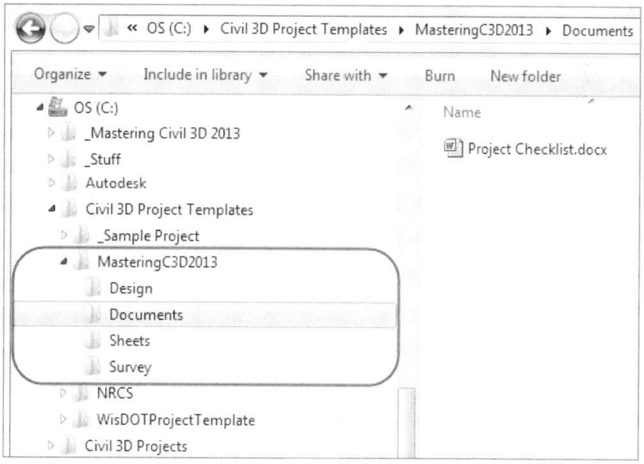

5. Place the file in the `Documents` folder you created in step 3.

This structure will appear inside Civil 3D and in the working folder when a project is created. A `Documents` folder is included as an example of other, non-Civil 3D–related folders you might have in your project. The `Project Checklist.doc` file will also be automatically copied to each project you create from this template.

Setting a Working Folder and Data Shortcuts Folder

You can think of the working folder as a project directory. The working folder can contain a number of projects, each with a data shortcut folder where the shortcut files reside.

In this exercise, you'll set the working folder and create a new project:

1. Create a new blank drawing using the template of your choice.
2. From the Manage tab ➢ Data Shortcuts panel, click Set Working Folder, as shown in Figure 18.2.

FIGURE 18.2
Creating a new working folder

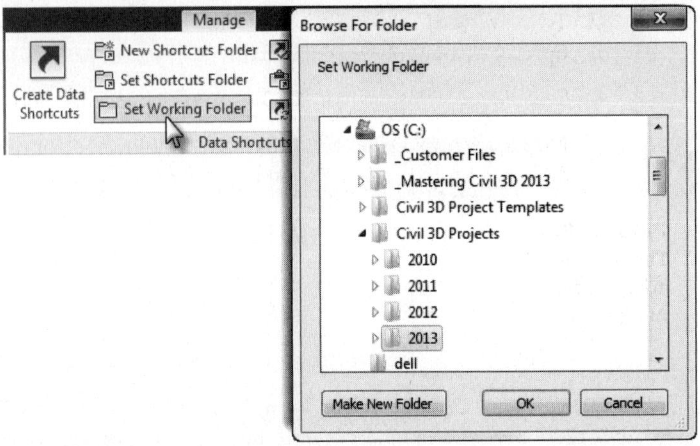

3. Browse to your local C drive ➢ `Civil 3D Projects` folder.

   This folder is created automatically on your local drive when Civil 3D is initially installed.

4. With the `Civil 3D Projects` folder highlighted, click Make New Folder, and make a new folder called 2013.

5. Highlight the new folder, and click OK.

6. From the Manage tab ➢ Data Shortcuts Panel, click New Shortcuts Folder to display the New Data Shortcut Folder dialog shown in Figure 18.3.

FIGURE 18.3
Creating a new shortcut folder

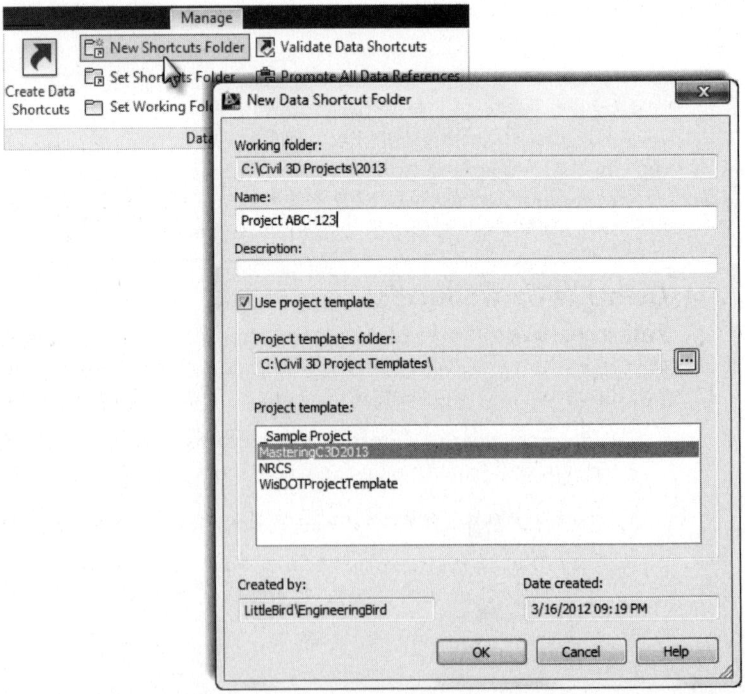

7. Type **Project ABC-123** for the folder name, and toggle on the Use Project Template option, as shown in Figure 18.3.

8. Select the MasteringC3D2013 folder from the list, and click OK to dismiss the dialog.

   Notice that the Data Shortcuts branch in Prospector now reflects the path of the Project ABC-123 project, as shown in Figure 18.4.

FIGURE 18.4
The Data Shortcut area listed in Prospector

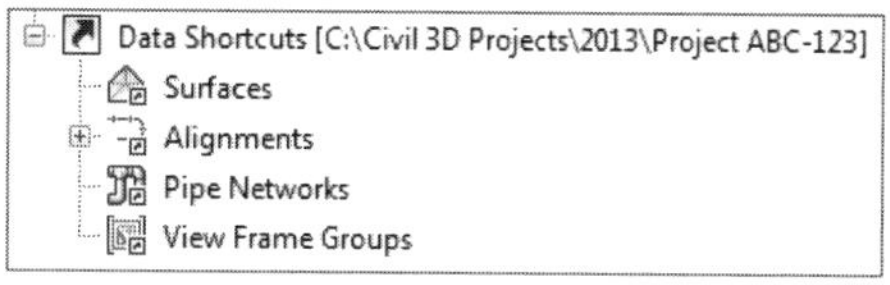

If you open Windows Explorer and navigate to C:\Civil 3D Projects\2013\Project ABC-123, you'll see the folder from the Mastering project template plus a special folder named _Shortcuts, as shown in Figure 18.5. Civil 3D creates this folder for you and is where the data shortcuts will live.

FIGURE 18.5
Your new project shown in Windows Explorer

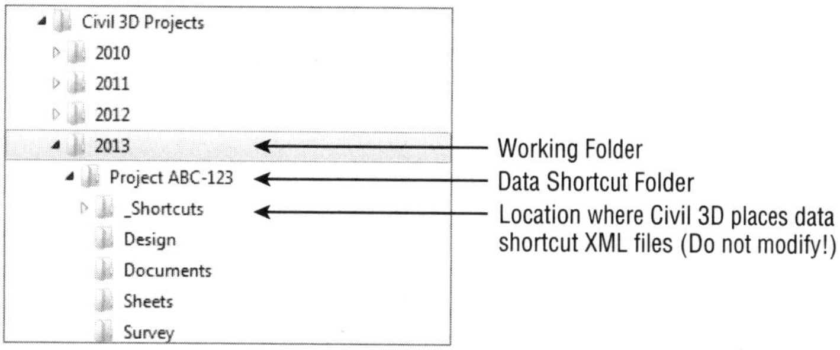

In real life, your working folder and data shortcut folder will probably be on a local network server. If you have an established workflow for creating project folders and do not wish to use the Civil 3D project template, you can manually create a folder named _Shortcuts. This _Shortcuts folder needs to be inside the folder set as your Data Shortcut folder in Civil 3D (i.e., one folder level down, as shown in Figure 18.5).

Creating Data Shortcuts

With a shortcut folder in place, it's time to use it. It's a best practice to keep the drawing files in the same location as your shortcut files, just to make things easier to manage. In this exercise, you'il publish data shortcuts for the alignments and layout profiles in your project:

1. Open the Points-Surface.dwg (Points-Surface_METRIC.dwg) file, which you can download from this book's web page.

   This drawing contains points and an existing ground surface.

2. From the Application menu, click Save As, and save a copy of this drawing to C:\Civil 3D Projects\2013\Project ABC-123\Survey.

Location, Location, Location

When working with data shortcuts, location matters! You save the project in its permanent home. Making a data shortcut to the surface and moving the file afterward would cause a broken reference.

Create Data Shortcuts

3. From the Manage tab ➢ Data Shortcuts Panel, click Create Data Shortcuts.
4. Place a check mark next to Surfaces and EXISTING SURFACE, as shown in Figure 18.6, and click OK.

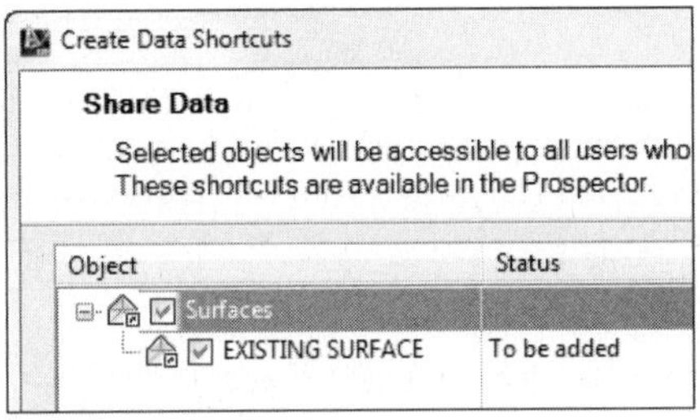

Figure 18.6 Adding surface data to data shortcuts

5. Save and close the current drawing file.
6. Open the `Alignments-Profiles.dwg` (`Alignments-Profiles_METRIC.dwg`) file from your class data.

   This file contains three alignments that you will add to the list of data shortcuts.
7. From the Application menu, click Save As, and save a copy of this file to `C:\Civil 3D Projects\2013\Project ABC-123\Design`.
8. From the Manage tab ➢ Data Shortcuts panel, click Create Data Shortcuts.
9. Place a check mark next to all the alignments (as shown in Figure 18.7), and click OK.

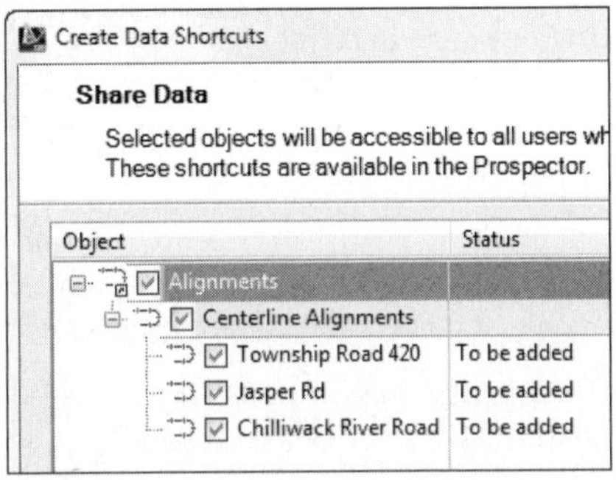

Figure 18.7 Adding alignment data to the pool of data shortcuts

You should now have surfaces and alignments available for use in the data shortcuts listing in Prospector, as shown in Figure 18.8. Notice in Figure 18.8 how highlighting the data shortcut in the listing reveals the Source Location at the bottom of Toolspace.

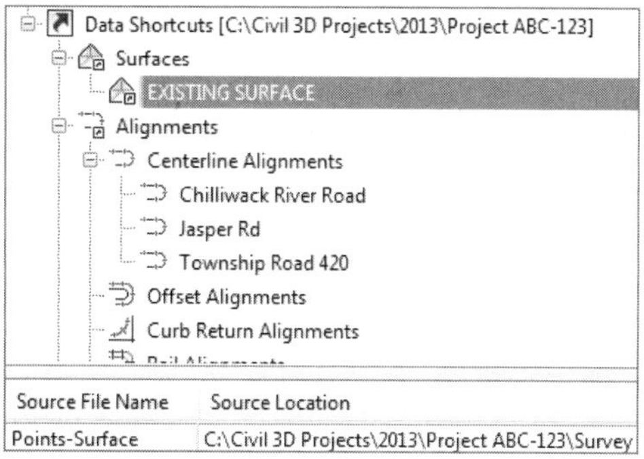

FIGURE 18.8 List of Civil 3D data shortcuts available to the current project

Keep `Alignments-Profiles.dwg` (`Alignments-Profiles_METRIC.dwg`) open for use in the next exercise.

Creating a Data Reference

Now that you've created the shortcut files to act as pointers back to the original drawing, you'll use them in other drawings. In this section you will use and create data references. You will also use data references in conjunction with a traditional XREF to create cross-section sheets.

CERT OBJECTIVE

Shortcut references are made using the Data Shortcuts branch within Prospector. In this exercise, you'll create references to the objects you previously published to the Concord project.

You need to have completed the previous exercises to continue.

1. Keep working in `Alignments-Profiles.dwg` (`Alignments-Profiles_METRIC.dwg`).
2. In Prospector ➤ Data Shortcuts collection, right-click EXISTING SURFACE and select Create Reference, as shown in Figure 18.9.
3. Leave all surface options at their defaults, and click OK. Zoom extents to view the surface.

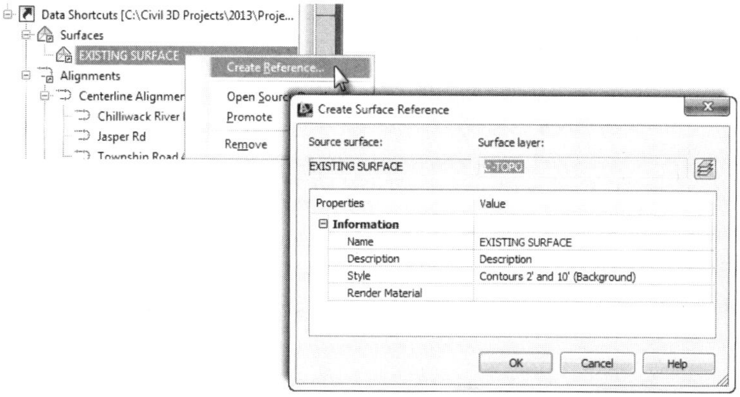

FIGURE 18.9 The Create Surface Reference dialog

4. Save the current drawing and keep it open for the next exercise.

You should now see the surface in the drawing. At first glance, it does not look different from other surfaces you've worked with within Civil 3D. If you examine the Surfaces branch of Prospector, however, you'll see that the Definition category under EXISTING SURFACE is missing (as shown in Figure 18.10). This is because you cannot edit the surface data in the host drawing. If a change needs to be made to the surface, you would need to open the file where EXISTING SURFACE was originally created.

FIGURE 18.10
You can see and use the surface, but you cannot edit the surface definition.

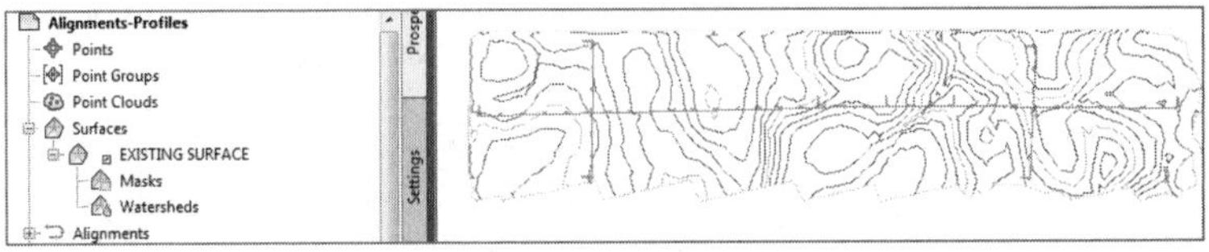

Any changes to the surface will be communicated through the data shortcut to the drawings where it is referenced.

> **RENAMING OBJECTS ON REFERENCE CREATION**
>
> Avoid it. Not only is it confusing to people who open the file after you, it is simply bad practice. You may experience broken references if the name of an object changes during the Create Reference part of the process.

In the exercise that follows, you will use the skills you've learned in Chapter 7, "Profiles and Profile Views," to create an existing ground profile and a design profile. After you have all the information, you will use skills from Chapter 10, "Basic Corridors," to put the information together. You need to have completed the previous exercises to continue.

1. Keep working in `Alignments-Profiles.dwg` (`Alignments-Profiles_METRIC.dwg`).
2. Select the longest alignment in the drawing, Township Road 420.
3. In the Alignment contextual tab ➢ Launch Pad panel, click Surface Profile, and then do the following:
   - **A.** Click Add.
   - **B.** Click Draw In Profile View.
4. Click Create Profile View, and click to place the profile view anywhere in the drawing.
5. Select the Profile view. From the Profile View contextual tab ➢ Launch Pad, click Profile Creation Tools.

Profile Creation Tools

6. Name the new profile **TR 420 Design**.
7. Leave all other options at their defaults, and click OK.
8. From the Profile Layout Tools, click Insert PVIs Tabular.

9. Enter the data for your unit system, as shown in Figure 18.11.

   You can close the Profile Layout Tools toolbar when complete.

FIGURE 18.11
Profile data Imperial (left) and metric (right)

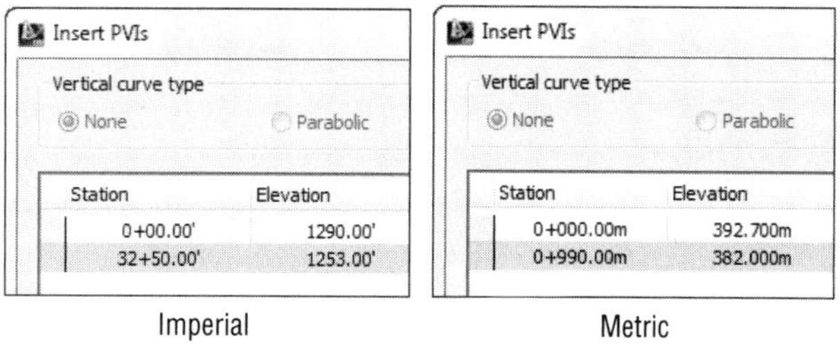

10. Save the drawing.
11. From the Manage tab ➤ Data Shortcuts panel, click Create Data Shortcuts.
12. Place a check mark next to Profiles, as shown in Figure 18.12, and click OK.

FIGURE 18.12
Adding additional drawing data to the data shortcuts

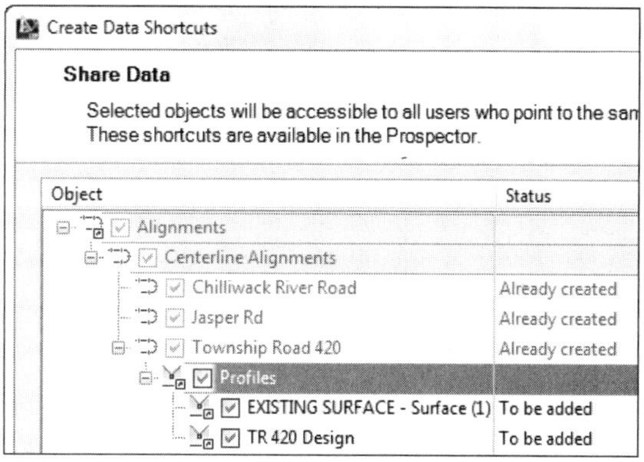

The alignments that you added earlier are listed but are grayed out. Once you add an item to the data shortcut list, you do not need to re-add them. Only new data will be added to the list of data shortcuts.

13. Save and close `Alignments-Profiles.dwg` (`Alignments-Profiles_METRIC.dwg`).
14. From your class files, open `Corridor.dwg` (`Corridor_Metric.dwg`).
15. From the Application menu, click Save As, and save the drawing to `C:\Civil 3D Projects\2013\Project ABC-123\Design`.
16. From the Prospector tab of Toolspace, expand the Data Shortcuts branch, and then expand Surfaces.
17. Right-click EXISTING SURFACE, and click Create Reference.

18. Leave all default settings, and click OK.

19. From the Prospector tab of the toolbar, expand Alignments ➤ Centerline Alignments ➤ Township Road 420 ➤ Profiles.

20. Right-click TR 420 Design and select Create Reference, as shown in Figure 18.13.

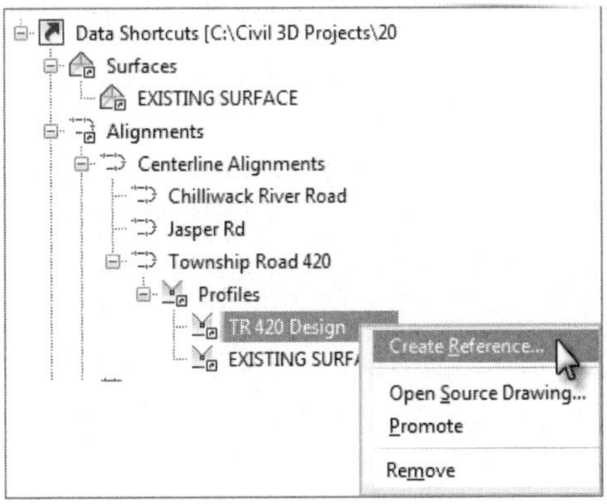

FIGURE 18.13
Creating a reference to a profile will automatically bring its alignment along.

21. Leave all settings at their default, and click OK to the Create Profile Reference dialog.

    If you zoom extents in the drawing, you should now see the alignment and existing surface. Because all profiles must be associated with an alignment, creating a reference to a profile will automatically bring the alignment along for the ride.

    The profile data is referenced but is not currently visible. If you wanted to, you could create a profile view. However, we do not need the profile to be visible to create a corridor.

22. From the Home tab of the Ribbon ➤ Create Design panel, click Corridor and do the following:

    A. Name the corridor **Township Road 420**.

    B. Verify the alignment is set to Township Road 420.

    C. Verify that the profile is set to TR 420 Design.

    D. Set the assembly to Typical Section.

    E. Set Target Surface to EXISTING SURFACE.

    F. Clear the check box for Set Baseline And Region Parameters.

    G. Click OK.

23. After the corridor builds successfully, save and close the drawing as you will be coming back to it later on.

Real World Scenario

SENDING CROSS-SECTION SHEETS TO THEIR OWN DRAWING

Unlike plan and profile sheets, cross-section sheets do not give you any options for automatically creating layouts in new sheets.

It is a great idea to send cross sections to a drawing separate from the corridor. If you don't use data shortcuts for anything else, we implore you to use them for cross sections.

There are two main reasons for separating sections from the herd:

- You don't need to worry that the annotation scale that looks good for cross sections looks strange for other modelspace items.
- Corridor drawings already contain lots of data; send section sheets to their own drawing to keep file size down.

The following exercise builds from what you have already created in this chapter. You will step through the procedure of creating cross-section sheets using a combination of XREF and data shortcut functionality. You may wish to review Chapter 13, "Cross Sections and Mass Haul," if you have trouble completing the steps in this exercise.

1. Start a new drawing with the Civil 3D template of your choice.
2. Save the drawing to the folder `C:\Civil 3D Projects\2013\Project ABC-123\Sheets` as **`Sections.dwg`**.
3. From the Prospector tab of Toolspace ➢ Data Shortcuts ➢ Alignments ➢ Centerline Alignments, right-click Township Road 420 and click Create Reference.
4. From the Insert tab of the Ribbon ➢ Reference panel, click Attach.
5. At the bottom of the Select Reference File dialog, set the Files Of Type to Drawing (`*.dwg`).
6. Browse to `C:\Civil 3D Projects\2013\Project ABC-123\Design` and select `Corridor.dwg` (`Corridor_METRIC.dwg`) and click Open.
   - **A.** Set the reference type to Overlay.
   - **B.** Set the path type to Relative Path.
   - **C.** Clear the check box for Insertion Point Specify On Screen.
   - **D.** Click OK.

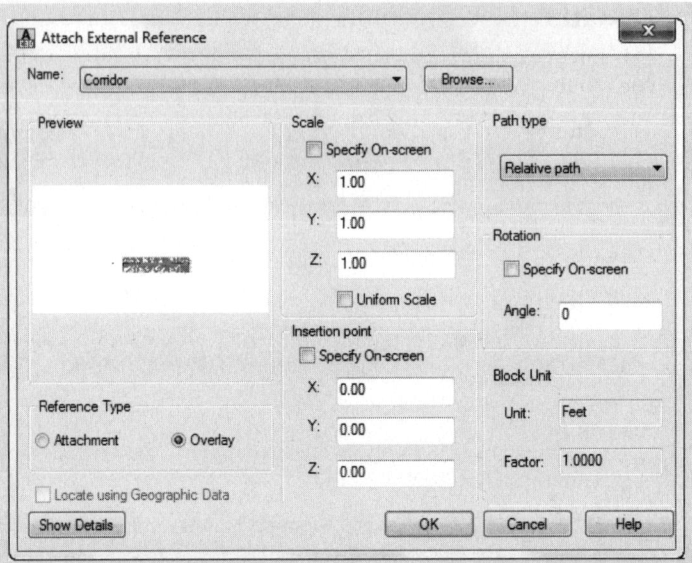

You will receive an Unreconciled New Layers message, which you can dismiss by clicking the X.

Zoom extents to get a look at your drawing so far. You now have everything you need to create sample lines in this new drawing.

7. From the Home tab ➤ Profile & Section Views panel, click Sample Lines, press Enter to pick Township Road 420 from a list, and click OK.
8. In the Create Sample Line Group dialog, keep all default options and styles.

   Pause here and take notice of a few peculiar things. The EXISTING SURFACE is appearing even though there is no data reference to it. Additionally, you should see the corridor listed. Data references to corridors do not exist in Civil 3D, but you can still link to the data without it. Sample lines are "strong enough" to pull this data out of an external drawing reference. If you had created corridor surfaces, those too would appear in the sample lines.
9. Click OK when you are done having your mind blown.
10. From the Sample Line Tools, click By Range Of Stations.
11. Set the left and right swath widths to **100′** (**30** m).
12. Keep the default sampling increments, and click OK.
13. Press Esc to complete the command.
14. Set your annotation scale to 1″=20′ (1:500).
15. From the Home tab ➤ Profile And Section Views panel ➤ Section Views, click Create Multiple Views.
16. Leave all the defaults, and click Create Section Views.
17. Click to place the sheets in the drawing.
18. Save the drawing and keep it open for the next exercise.

After you have successfully created sample lines, the procedure for creating section sheets is exactly the same as you learned in previous chapters.

Updating References

When it's necessary to make a change, you can use the tools in the Data Shortcut menu to jump back to that file, make the changes, and refresh the reference:

1. If it is not open from the exercise in the Real-World Scenario "Sending Cross-Section Sheets to Their Own Drawing," open the file `Sections.dwg`.

   Hopefully, you saved it to `C:\Civil 3D Projects\2013\Project ABC-123\Sheets`.

2. Keep this drawing open and also open the file `Corrior.dwg`.

3. In Prospector, expand Data Shortcuts ➢ Surfaces.

4. Right-click EXISTING SURFACE and select Open Source Drawing, as shown in Figure 18.14.

FIGURE 18.14
Open Source Drawing is a fast way to jump to the drawing you are after.

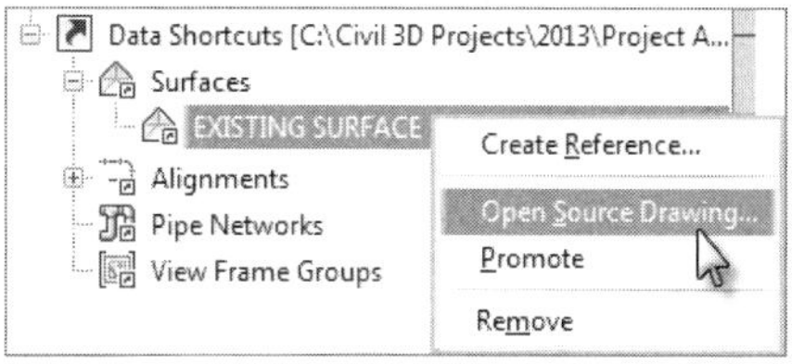

At this point you should have three drawings open (Corridors, Points-Surface, and Sections). You can use Ctrl+Tab on your keyboard to cycle through open drawings. You can also use the Quick View Drawings button at the bottom of the AutoCAD window to switch between open drawings.

`Points-Surface.dwg` (`Points-Surface_METRIC.dwg`) should be the active drawing and ready to make updates. Remember that the surface is read-only in all other project files; this is the only drawing where changes can be made to the surface.

5. On the Insert tab ➢ Import panel, click Points From File.

6. In the Import Points dialog, click the plus sign and browse for `NewSurfacePoints.txt` (`NewSurfacePoints_METRIC.txt`). Select this file and click Open.

   This file is part of the chapter data set, and can be downloaded from this book's web page.

7. Set Point File Format to PNEZD (Comma Delimited), leave all other settings at their defaults, and click OK.

8. In Prospector, right-click Point Groups and select Update.

9. In Prospector ➢ Surfaces, right-click EXISTING SURFACE and select Rebuild.

10. Save and close `Points-Surface.dwg` (or `Points-Surface_METRIC.dwg`).

11. Switch drawings so that `Corridor.dwg` (`Corridor_METRIC.dwg`) is the active drawing.

    Shortly after bringing this drawing to the forefront, you should see a bubble message appear indicating that Data Shortcut Definitions may have changed, as shown in Figure 18.15.

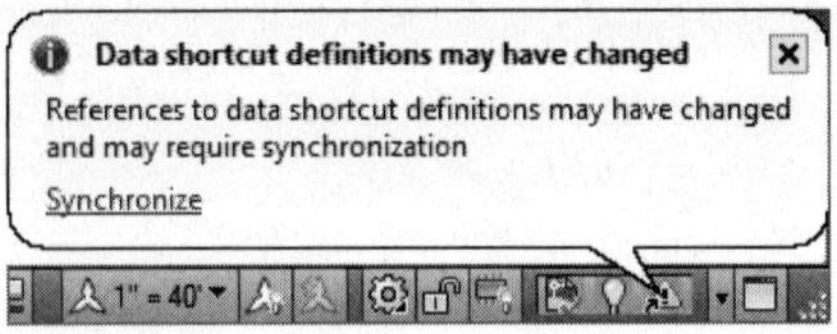

FIGURE 18.15
Thanks for the head's up! Civil 3D will send the user a message when a data referenced object has changed.

12. Click the Synchronize link in the message.

In some situations you may not see the large bubble message. You may just see a warning symbol in the reference indicator at the bottom of the screen. In that case, you can always synchronize directly from the right-click menu of the referenced object, as shown in Figure 18.16.

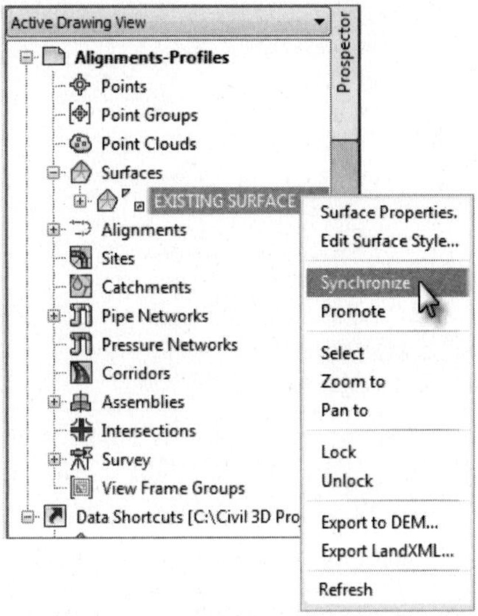

FIGURE 18.16
Synchronizing from the object's right-click menu

You will receive a message in Panorama indicating that the item was synchronized.

13. Dismiss this message by clicking the green check box.

14. In Prospector ➢ Corridors, right-click the Township Road 420 corridor, and select Rebuild.

15. Save and close `Corridor.dwg` (`Alignments-Corridor.dwg`).

16. Switch to `Sections.dwg`.

This time, you will receive a message that your XREF is out of date, as shown in Figure 18.17. If you accidentally dismiss this message without updating, a warning symbol will remain in the tray area at the bottom of your screen. You can always right-click to reload the reference from this icon.

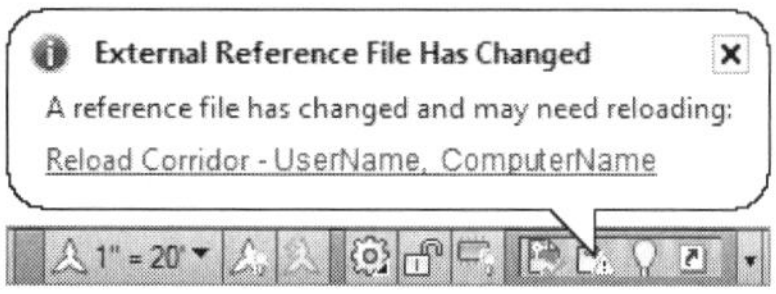

FIGURE 18.17
Reload the XREF to ensure your cross sections reflect the design update.

17. Click Reload Corridor.

    Your sections should now reflect all the design changes. If you receive another message regarding unreconciled layers, click the X to dismiss it.

18. Save and close all drawings.

Real World Scenario

BEST PRACTICES FOR EMPLOYING DATA SHORTCUTS

When people first add data shortcuts into their workflow, they often have the following questions:

I only work on small sites by myself. Do I need to use data shortcuts? Even if you work on very small projects, you should use data shortcuts for cross-section sheets.

How should I break up my project? Every project is different, of course, but use this as a guideline:

Existing Survey and Surface The first drawing you create is usually the existing conditions. Keep survey points, figures, and the existing surface together in the first drawing. Create a data shortcut to the existing surface. Points can't be shared via data shortcut, but you can always use a point query to pull specific points between drawings.

Alignments and Profiles While the existing surface is being fine-tuned, you can get to work on developing the alignments and profiles right away. Alignments and profiles are fairly lightweight in terms of how much data they consume. You can usually store all of your project's alignment and profile data in the same DWG.

Corridors When in doubt, use multiple corridor drawings. This way several designers can be working on different segments of the project at once. Be mindful not to put too much data in one file—you may want to keep interchanges and roundabouts in their own drawings. At the end of the process, data shortcuts will allow you to bring proposed surface data back together.

Pipe Networks and Pressure Networks Pipe networks can be used with data shortcuts, but currently pressure networks cannot. Keep this in mind when splitting up your projects.

Site Grading Parking lots, building pads, and other site items are good candidates for data shortcuts.

Where is the data stored? The data for a data shortcut is stored in the source DWG file. The data shortcut file itself is simply a pointer that keeps track of filenames and paths. No object data is stored with the data shortcut.

Can I store my working folder and data shortcut folder on a network? Yes! This is an excellent way to work. You will want to keep your working folder, data shortcut folder, and project drawings on a local network drive, however (i.e., within your building).

What if I need to share the project data with another office? This question is where data shortcuts get a little tricky. How to handle multiple offices depends on the scenario:

Example 1 Your Houston office is working on the road design while the Chicago office is working on grading. You want a live, remote workflow with frequent syncing between offices. For this type of scenario, you would be better served by Autodesk® Vault Collaboration:

`http://usa.autodesk.com/adsk/servlet/pc/index?siteID=123112&id=4502718`

Example 2 Your Houston office has completed this stage of the road design and the next phase will be done in the Chicago office. In other words, no live data exchange needs to take place. You can use eTransmit (discussed later in this chapter), or you can move the entirety of the project's `Data Shortcut` folder to the new location.

Fixing Broken References

A broken reference is fairly easy to fix, but you will not be able to continue working in your design until broken references are rectified.

Actions that will cause broken references include:

- Renaming a source file
- Moving a source file
- Missing a source file
- Renaming project folders

In the following exercise you will connect to an existing project and fix the broken data references that you find. You *do not* need to have completed the previous exercises to continue.

1. Using Windows Explorer, locate the folder called `Project XYZ-789`.

   This is part of the download for this chapter at the book's web page.

2. Copy this folder and its contents to `C:\Civil 3D Projects\2013`.

3. Open the file `XYZ-789 Corridor.dwg`, located in `C:\Civil 3D Projects\2013\Project XYZ-789\Design`.

   This file contains a number of references pointing to a file that was renamed. Panorama will appear with the messages regarding the problems that it found.

4. Close the Panorama window by clicking the green check mark.

   In Prospector ➢ Alignments ➢ Centerline Alignments you will see that all three alignments have a warning chevron next to them.

5. Right-click Donner Pass and select Repair Broken References, as shown in Figure 18.18.

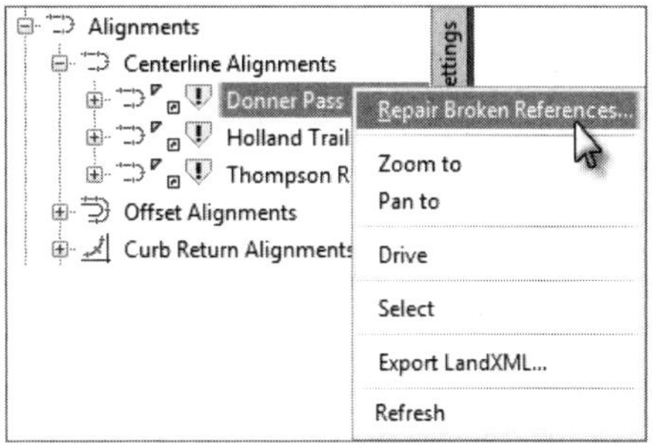

FIGURE 18.18
Choosing Repair Broken References

6. Navigate to the `XYZ-789 Alignment.dwg` file in the ...`\Project XYZ-789\Design` folder, and click Open.

   After fixing the first broken reference, Civil 3D determines whether this drawing can be used to fix other reference issues. If it finds more fixes, you will see the message shown in Figure 18.19.

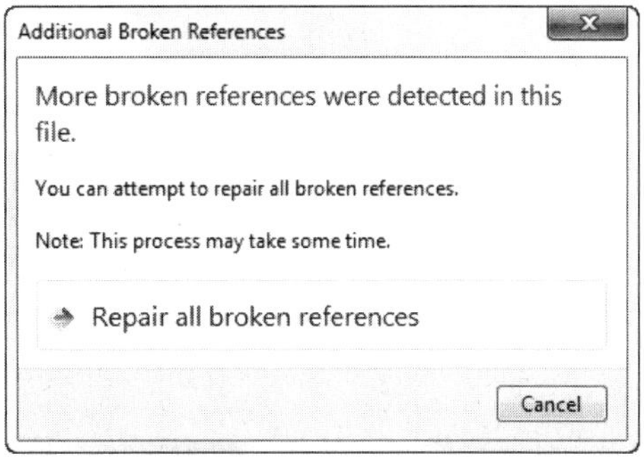

FIGURE 18.19
The Additional Broken References message

7. Click Repair All Broken References.

   Civil 3D will match the Civil 3D objects with broken references to objects in the selected drawing.

The ability to repair broken links helps make file management a bit easier, but there will be times when you need to completely change the path of a shortcut to point to a new file. To do so, you must use the Data Shortcuts Editor.

THE DATA SHORTCUTS EDITOR

Civil 3D tenaciously tries to hang on to links to the information it uses. The Data Shortcuts Editor is used to update or change the file to which a shortcut points. You may want to do this when an alternative design file is approved, or when you move from preliminary to final design.

In the following exercise, you'll move the example project and ensure that the data shortcuts are updated (the procedure is the same for both metric and Imperial units):

1. Save and close any drawings you have open and close Civil 3D.
2. Using Windows Explorer, browse to `C:\Civil 3D Projects\2013`.

3. Inside the 2013 folder, create a new folder called **`Submitted`**.
4. In the download for this chapter is a folder called `Project XYZ-789`. Copy `Project XYZ-789` to the `Submitted` folder. Rename the folder `Project XYZ-789` to **`Project XYZ-789 SUB`**.

   Your resulting folder structure should look like Figure 18.20.

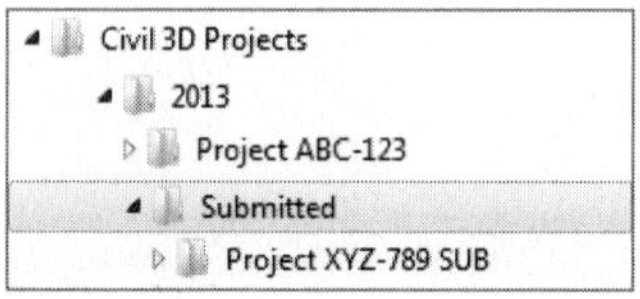

FIGURE 18.20
Simulating a new project phase by copying the example project to a new folder

You are keeping the original intact for archive purposes. If you attempted to open drawings that reference data in the new location, you would see that the data references are looking at the original path.

Next, you will ensure that Civil 3D will use the new path in the `Submitted` version.

5. In Windows, choose Start ➢ All Programs ➢ Autodesk ➢ Autodesk Civil 3D 2013, and click the Data Shortcuts Editor to load it.
6. Select File ➢ Open Data Shortcuts Folder to display the Browse For Folder dialog.
7. Navigate to `C:\Civil 3D Projects\2013\Submitted\Project XYZ-789 SUB`, and with the `Project XYZ-789 SUB` folder highlighted, click OK.

   Your Data Shortcuts Editor should resemble Figure 18.21.

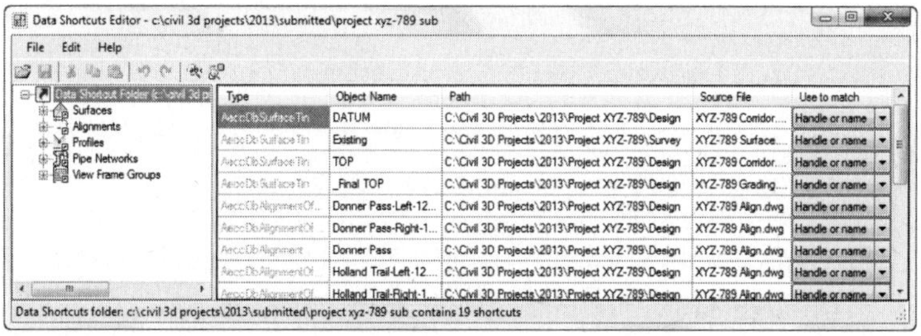

FIGURE 18.21
Inside the Data Shortcuts Editor

There are several things that need to be changed before the new phase of the project can begin. Some of the data references are looking for the incorrect file.

Notice that the Path column of the table still refers to the old path. In order to make a clean break from the old project to move forward into the `Submitted` version, you need to change all these.

8. Choose Edit ➢ Find And Replace.

   Unfortunately, there is no browse option here but you can use the basic Windows Copy and Paste tools to make this a little easier. You can also click your cursor in one of the fields you wish to change for a starting point.

9. Type **C:\Civil 3D Project\2013\Project XYZ-789** in the Find field, and **C:\Civil 3D Projects\2013\Submitted\Project XYZ-789 SUB** in the Replace With field, as shown in Figure 18.22.

FIGURE 18.22
Updating the paths to the new project

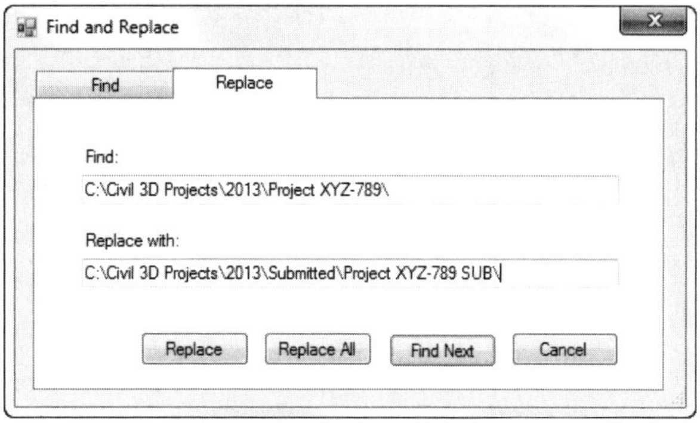

10. Click Replace All.

    Your data shortcut paths and source files are now correct (as shown in Figure 18.23) and ready for more action.

FIGURE 18.23
Updated paths

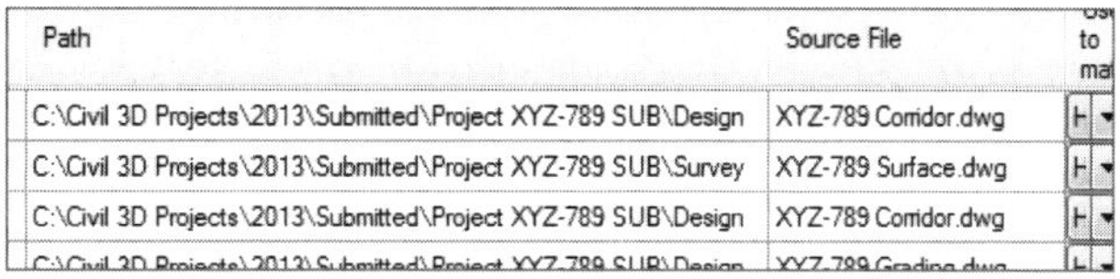

11. Highlight the Alignments branch on the left side of the Data Shortcuts Editor.

    This will narrow the focus to alignment information.

12. Choose Edit ➤ Find And Replace.

13. In the Find field, enter **XYZ-789 Align.dwg**.

14. In the Replace field, enter **XYZ-789 Alignment.dwg**, as shown in Figure 18.24.

FIGURE 18.24
Ensuring the new project does not contain references to an old filename by using Find And Replace

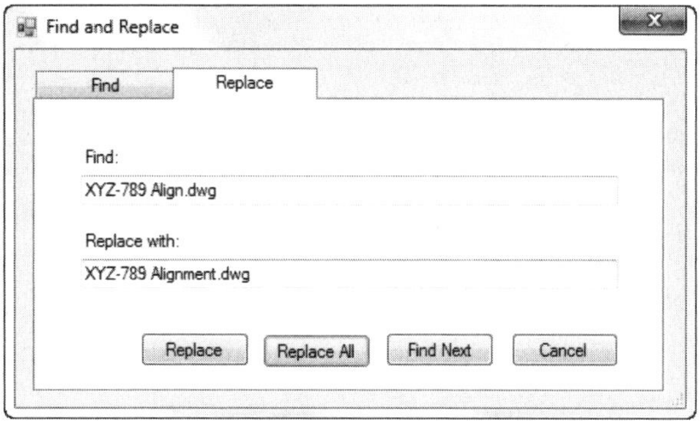

15. Click Replace All.

    All of the source file paths for alignments should be updated.

16. You can now close the Data Shortcuts Editor and return to Civil 3D.

ETRANSMIT + DATA SHORTCUTS = AWESOMESAUCE!

If you need to pass a project on to another designer with all the information they need, use the eTransmit command. If a file contains a data shortcut, the eTransmit command will recognize the link and include the necessary drawings. You can even include non-AutoCAD files such as Word or Excel documents by clicking the Add File button.

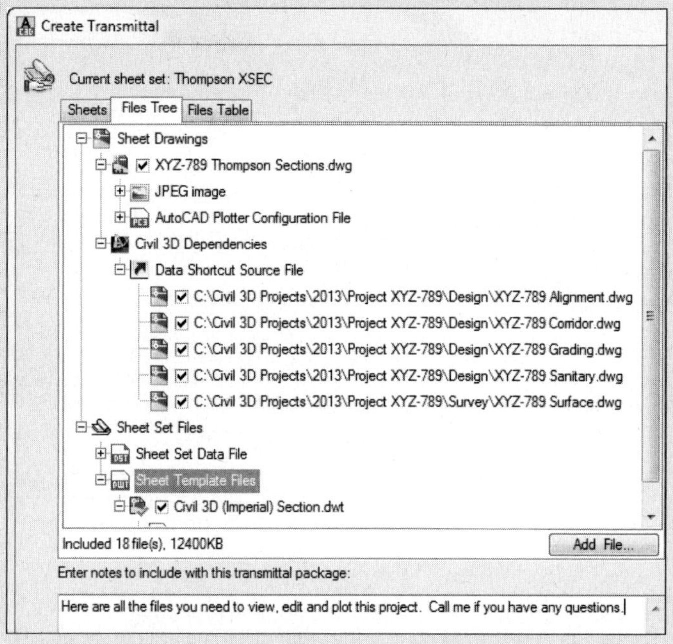

AutoCAD will generate a zip file that contains any XREF files (including DWF, DGN, PDF, or TIF references), plot configurations, images, and templates used in sheet creation. When your recipient unzips the file, the connections between drawings are maintained.

eTransmit is the best way to send Civil 3D files with the assurance that you are not sending a pile of broken references.

Sharing with Earlier Versions of Civil 3D

As crazy as it sounds, not everybody is using Civil 3D 2013, so you need to know how to work with the "world outside."

Civil 3D is not backward compatible. While the Save As command works fine for base AutoCAD objects (such as lines, arcs, and circles), the intelligence behind Civil 3D objects is inaccessible to previous versions of both Civil 3D and AutoCAD.

Contrary to what many people claim, this is not an evil conspiracy. Civil 3D objects evolve year to year. As functionality changes, objects get leaner and smarter. Under the proverbial hood, the underlying components of things like surfaces, alignments, and corridors are too different from those of previous years to work in older versions.

If possible, avoid the need to "go backward." Be mindful that the moment you save a previous version drawing in a newer version of Civil 3D, it will no longer work properly in the old version of Civil 3D.

Alas, the need to send data to older versions of Civil 3D is not always avoidable. In the following exercise, you will step through an example of what needs to happen to force Civil 3D to previous versions:

1. Open `LandXML-OUT.dwg` (`LandXML-OUT_METRIC.dwg`), which can be found at this book's web page.

Export to LandXML

2. From the Output tab ➢ Export panel, select Export To LandXML.

   The Export To LandXML dialog shown in Figure 18.25 opens.

FIGURE 18.25
The Export To LandXML dialog

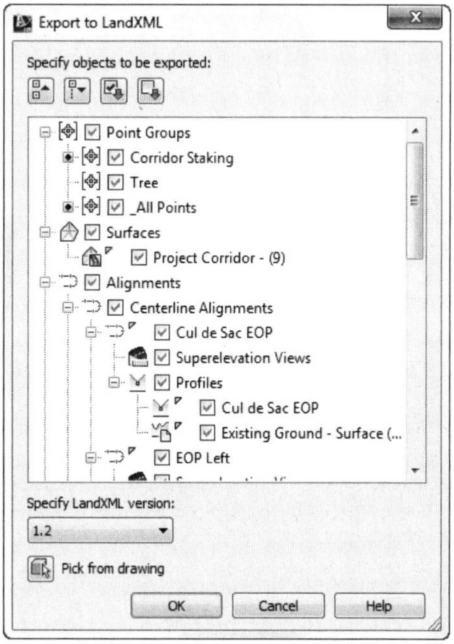

3. Click OK.
4. In the Export LandXML dialog, browse to the same folder as the source drawing.
5. Keep the default name (the dialog will pick up the name of the drawing you are exporting from), and click Save.

   You now have as much Civil 3D as possible packed up into the LandXML file. This file contains:

   - Alignments
   - Point groups and points
   - Surfaces
   - Profiles
   - Pipe data

   The next few steps will break the Civil 3D objects down into base AutoCAD components and save them to a previous version format.

6. From the Application menu ➢ Export ➢ DWG, select 2010.
7. In the Export Drawing Name dialog, browse to the same folder as the source drawing.
8. Keep the default name (the dialog will pick up the name of the drawing you are exporting and add an `ACAD-` prefix), and click Save.

   The resulting file is a "dumbed-down" version of your original Civil 3D file.

   Next you will go through the process of reassembling the data as if you were working on a previous version of Civil 3D.

9. Start a new drawing based on the template of your choice.
10. Save the file as **`LandXML-IN.dwg`** in the location of your choice.

    You need to start with a Civil 3D template, because exporting to earlier versions will destroy Civil 3D–specific styles and settings.

Insert

11. On the Insert tab ➢ Block panel, click Insert.
12. In the Insert dialog, clear the Insertion Point, Scale, and Rotation check boxes.
13. Place a check mark next to Explode.
14. Click Browse and locate the file `ACAD-LandXML-OUT.dwg` or `ACAD-LandXML-OUT_METRIC.dwg` that you created in steps 7 and 8 and click Save.
15. Click OK.
16. On the Insert tab ➢ Import panel, click LandXML.
17. In the Import LandXML dialog, browse for the XML file you created in steps 4 and 5, and click Open.

Figure 18.26 shows the Import LandXML dialog.

FIGURE 18.26
Importing LandXML data

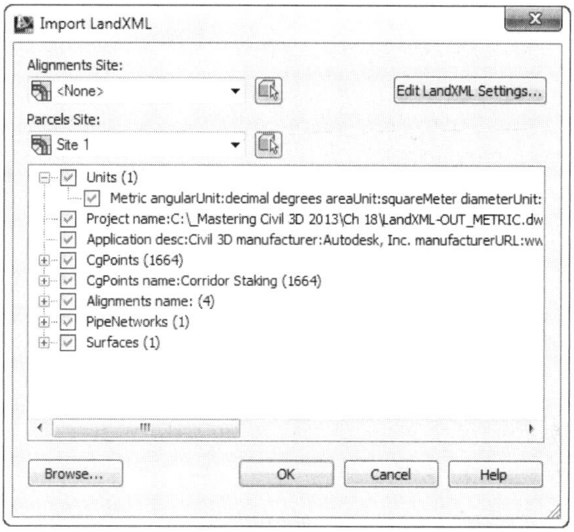

18. Click OK.

You may receive several messages in Panorama because of the Parts List differences between the source file and this new file.

19. Dismiss Panorama, and save and close the drawing.

WHAT IS LANDXML?

LandXML is not specific to Civil 3D. It is a consortium of fellow Civil 3D users wanting to standardize a way of sharing files in a nonproprietary format.

In the Import LandXML dialog was a drop-down list that let you select the version of LandXML. This is a result of that consortium. The latest as of this writing is the 1.2 schema. Most CAD programs have methods of importing and exporting LandXML files, and this is one way to tackle that barrier.

You can read up on LandXML at `www.landxml.org`.

Map Tools: A Transformative Experience

A common occurrence in Civil 3D is needing to show an aerial image as it relates to your design. The coordinate system of the photo is not always in the same units or coordinate system as your drawing.

Standard image attachment tools will not automatically transform the image. Even if both your drawing and the image are assigned coordinate systems, Civil 3D will not presume to move one into the other. You need to use the Map portion of Civil 3D to force the coordinate transformation.

Here are the general steps you will need to follow in order to transform an image from its original coordinate system to the coordinate system of your Civil 3D drawing:

1. Start a new drawing and set it to the same coordinate system as the image.

   This will be referred to as the *source coordinate system*.

2. Get to the Map workspace.
3. Insert the image to the new, empty drawing using the MAPIINSERT command.
4. Save and close the initial drawing.
5. Open your project drawing and ensure that it is set to the correct coordinate system.

   This will be referred to as the *destination coordinate system* in this section.

6. Use the Map 3D Attach command to attach the source drawing.
7. Use the ADEQUERY command to pull the image across to the destination coordinate system.

In the following sections, you will work through these steps in greater detail.

Finding Your Map Tools

Autodesk® Map 3D is a product that bridges the gap between GIS and CAD. Because infrastructure designers often dip their toes in the GIS waters, Civil 3D contains all the tools that come with Map 3D. To access the tools you will work with, you will need to change your active workspace.

In the following exercise, you will change workspaces.

1. Start a new drawing based on the `_Autocad Civil 3D (Metric) NCS.dwt`.

   Even if your preferred working units are feet, the first step in the process will be working with the TIFF image in its native coordinate system. The TIFF you will be working with was obtained from the United States Geological Survey (USGS) website (`http://viewer.nationalmap.gov/viewer/`). Like most of the data from USGS, the file is in UTM NAD83 and is in metric units.

2. Save the file in your example file folder as **`SOURCE.dwg`**.
3. From the Settings tab of Toolspace, right-click on the name of the drawing, and select Edit Drawing Settings.
4. On the Units And Zone tab, enter **`UTM83-16`** in the Selected Coordinate System Code field.

   This will set the drawing's coordinate system to UTM83 Zone 16, Meter.

5. Click OK to dismiss the Edit Drawing Settings dialog.
6. Click the Workspaces icon in the lower-right corner of the screen and select Planning And Analysis, as shown in Figure 18.27.

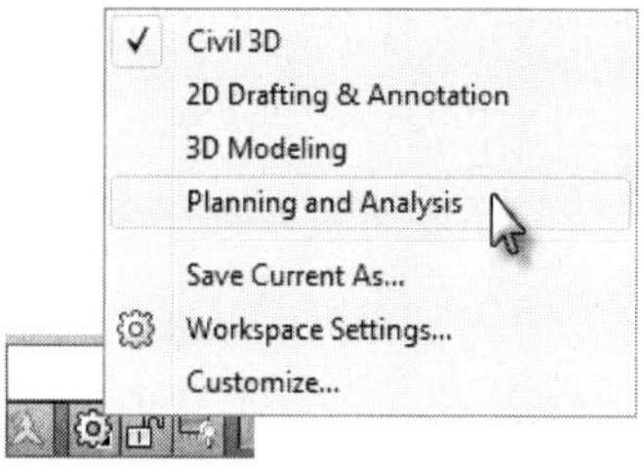

FIGURE 18.27 Switching to the Planning And Analysis workspace to access Map 3D tools

7. Save the drawing and keep it open for the next exercise.

Inserting the Image

Now that you have your source drawing set, it is time to insert the file. In this example you'll use a TIFF, but the process is the same regardless of the type of image file you are working with.

1. Continue working in the file `SOURCE.dwg` that you created in the previous exercise. Remain in the Planning And Analysis workspace.

2. From the Home tab ➤ Data panel, click the MAPIINSERT button.
3. In the Insert Image dialog, set Files Of Type to Tagged Image File Format (TIFF).
4. Browse to the files you downloaded for this chapter, select `MKE94-43-41.tif`, and click Open.
5. In the Image Correlation dialog, click OK.

   The insertion point and units are coming from the TFW file that accompanies the image. The two files have the same name but different file extensions.
6. Zoom extents to see the image in your file.
7. Remain in the Planning And Analysis workspace, and save and close `SOURCE.dwg`.

If you happen to receive a TIFF that is in the same coordinate system as the rest of your project, you could use the MAPIINSERT command directly in the project. A coordinate system match may happen periodically, but more often than not you will need to continue the process as outlined in the next section.

Transforming to Local Coordinates

The first thing you should do in any project drawing is set your coordinate system, as seen in Figure 18.28. Be sure you have a coordinate system set before you import your survey and create Civil 3D data. Changing coordinate systems or units partway through a project will not adjust any existing information and is generally a bad idea.

The technique outlined in this section works on images and any base AutoCAD object. If you have AutoCAD objects (lines, arc, circles, etc.) in a file that has a coordinate system assigned to it, you can attach the file and query-in data to do an exact transformation.

FIGURE 18.28
Set Units and Zone before you start importing data.

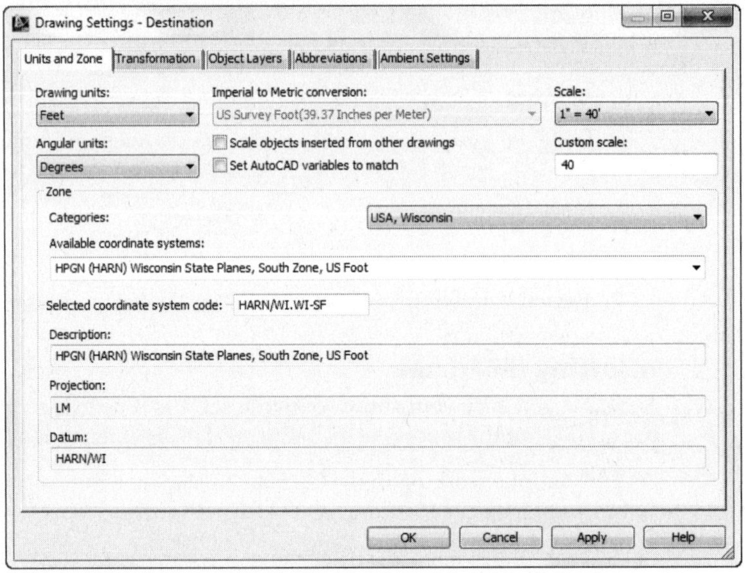

LOCATE USING GEOGRAPHIC DATA

For XREFs and block insertion, you will see an option called Locate Using Geographic Data.

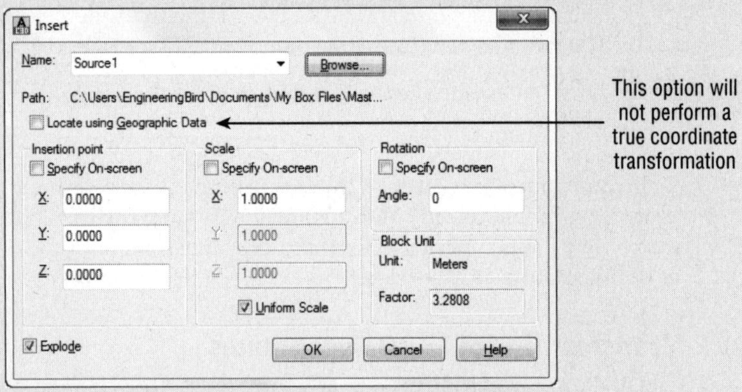

Be wary of this option for Civil 3D and Map applications. It is intended for base AutoCAD users who simply want to line up drawings; it does not perform a true coordinate transformation.

1. Open the drawing `Destination.dwg` (`Destination_METRIC.dwg`), which you can download from this book's web page.

   These files have been set to a Wisconsin State Plane coordinate system.

2. In the Planning And Analysis workspace, select the Home tab ➤ Data panel, and click Attach.
3. Use the Select Drawings To Attach dialog to browse to your file.
4. Highlight `SOURCE.dwg`, and click Add. Click OK.

   The file will appear at the bottom of the dialog in the Selected Drawings section, as shown in Figure 18.29.

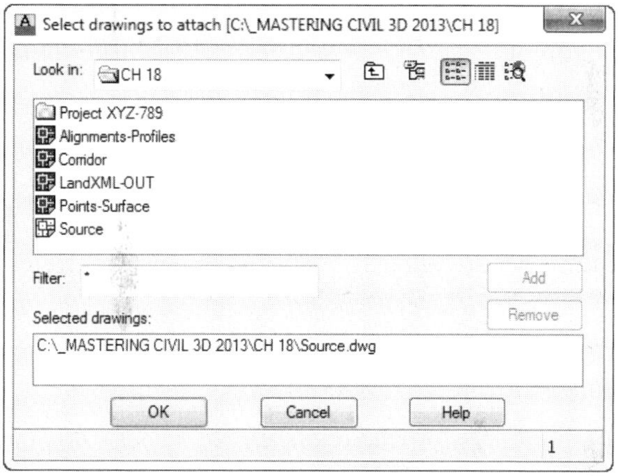

FIGURE 18.29 Locate the source and click Add.

If you are using a drive other than the C drive of your computer to store the files for this process, you will need to click the Create/Edit Aliases button to access additional drives.

5. From the Home tab ➤ Data panel, click Define Query.
6. Click the Location button.
7. Set the boundary type to All, and click OK.
8. Set the query mode to Draw. Leave all other options at their defaults.
9. When your dialog resembles Figure 18.30, click Execute Query.

FIGURE 18.30
Setting query properties

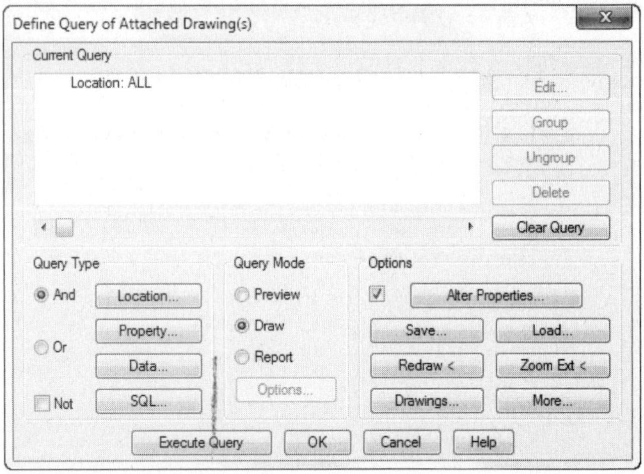

You should now see the image translated to the correct coordinate system.

10. Select the image, right-click, and select Display Order ➤ Send To Back, as shown in Figure 18.31.

FIGURE 18.31
Change the display order to get a better look at the alignments.

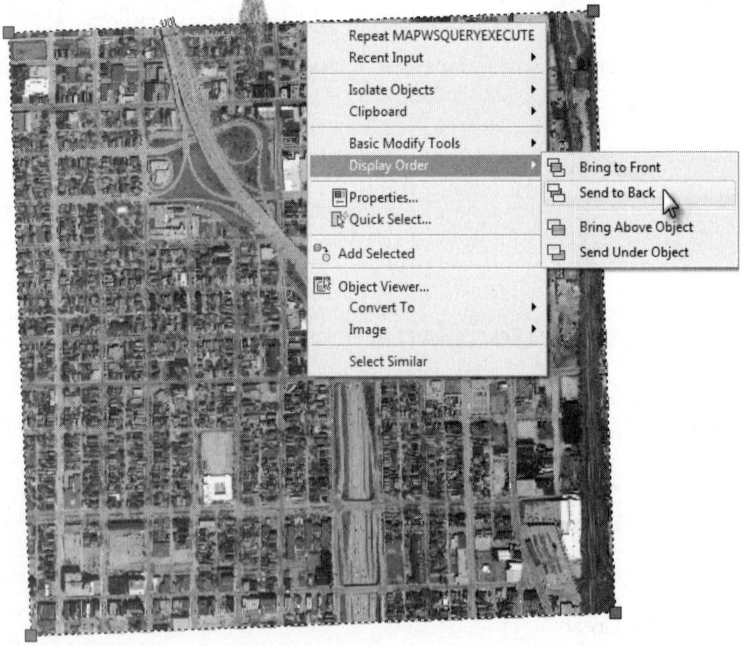

11. Save the drawing.

12. Because you have used Map 3D functionality to attach drawings, you will see a message about maintaining the association between drawings. Click OK to dismiss this warning.

AN ALTERNATE METHOD

To insert and transform a georeferenced image into Civil 3D, you could also use the Map Data Connect tool from the Planning And Analysis workspace in the Home ➢ Data panel (or type **MAPCONNECT**). This method works for various data types, including raster images as shown here. The advantage of this method is that it involves fewer steps.

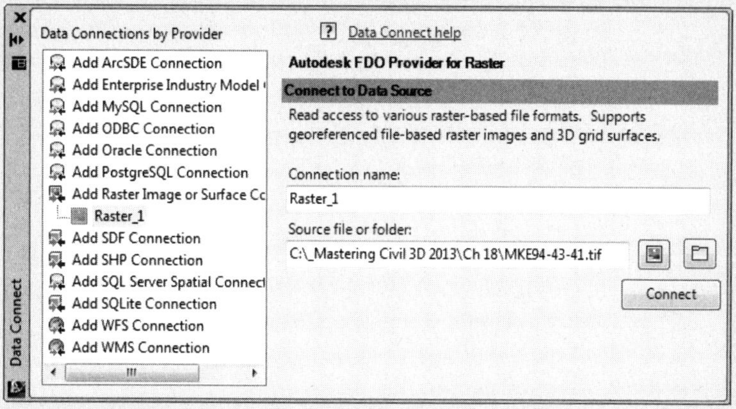

There are several reasons we made you do it the "hard way":

- If an image is inserted via the attach and query method, as you did in the previous exercise, the image can be manipulated in all the same ways as an image attached via base AutoCAD—that is, you can crop it, make it transparent, and adjust it.
- Images brought in through the MAPCONNECT command can be color adjusted, but that's it.
- If you use the attach and query method, you can transform anything—images and entire CAD files, if necessary.

The Bottom Line

Create a data shortcut folder. The ability to load design information into a project environment is an important part of creating an efficient team. The main design elements of the project are available to the data shortcut mechanism via the working folder and data shortcut folder.

> **Master It** Using the drawing `MasterWorkflow.dwg` (`MasterWorkflow_METRIC.dwg`), create a new data shortcut folder called `Master Data Shortcuts`. Use the `_Sample Project` project template.

Create data shortcuts. To allow sharing of the data, shortcuts must be made before the information can be used in other drawings.

> **Master It** Save the drawing to the `Source Drawings` folder in the project you created in the previous exercise. Create data shortcuts to all the available data in the file `MasterWorkflow.dwg` (`MasterWorkflow_METRIC.dwg`).

Export to earlier releases of AutoCAD. Being able to export to earlier base AutoCAD versions is sometimes necessary.

Master It Using `MasterWorkflow.dwg` (`MasterWorkflow_METRIC.dwg`), export the Civil 3D file so it can be used by a user working in base AutoCAD 2010.

Export to LandXML. Being able to work with outside clients or even other departments within your firm who do not have Civil3D is an important part of collaboration.

Master It Using `MasterWorkflow.dwg` (`MasterWorkflow_METRIC.dwg`), create a LandXML file will all of the exportable information.

Transform coordinates between drawings. Most project locations have the ability to use several coordinate systems or units. Being able to correctly manipulate a drawing's coordinate system is extremely important.

Master It The file `TransformThis.dwg` is in Universal Transverse Mercator (UTM) North American Datum (NAD) 83, Zone 10, in meters. Start a new drawing based on `_Autocad Civil 3D (Imperial) NCS.dwt` or `_Autocad Civil 3D (Metric) NCS.dwt`. Set the new drawing's coordinates to AND 83 California State Plane Zone 3 (US Foot or Meters, depending on which template you used).

Chapter 19

Quantity Takeoff

The goal of every project is eventually construction. Before the first bulldozer can be fired up and the first pile of dirt moved, the owner, city, or developer has to know how much all of this paving, pipe, and dirt are going to cost. Although contractors and construction managers are typically responsible for creating their own estimates for contracts, the engineers often perform an estimate of cost to help judge and award the eventual contract. To that end, many firms have entire departments that spend their days counting manholes, running planimeters around paving areas, and measuring street lengths to figure out how much striping will be required.

The AutoCAD® Civil 3D® software includes a Quantity Takeoff (QTO) feature to help relieve that tedious burden. You can use the model you've built as part of your design to measure and quantify the pieces needed to turn your project from paper to reality. You can export this data to a number of formats and even to other applications for further analysis.

In this chapter, you will learn to:

- Open and review a list of pay items along with their categorization
- Assign pay items to AutoCAD objects, pipe networks, and corridors
- Use QTO tools to review what items have been tagged for analysis
- Generate QTO output to a variety of formats for review or analysis

Employing Pay Item Files

Before you can begin running any sort of analysis or quantity, you have to know what items you are trying to quantify. Various municipalities, states, and review agencies have their own lists of items and methods of breaking down the quantities involved in a typical development project.

There are three main files associated with quantity takeoff in Civil 3D; the *pay item list*, the *pay item categorization file*, and the *formula file*. Each file serves a different purpose and is stored externally to the project. Once a file has been associated to a DWG, the path to the external file is stored with the DWG.

Pay Item List When preparing quantities, different types of measurements are used based on the items being counted. Some are simple individual objects such as light posts, fire hydrants, or manholes. Only slightly more complicated are linear objects such as road striping, or area items such as grass cover. These measurements are also part of the pay item list.

At minimum the pay item list will provide you with three main pieces of information for each item:

- The pay item number
- The pay item description
- The pay item unit of measure

There are three file types allowable for the pay item list file:

- CSV (Comma Delimited)
- AASHTO TransXML
- Florida DOT

Pay Item Categorization File In any project, there can be thousands of items to tabulate. To make this process easier, most organizations have built pay categories, and Civil 3D makes use of this system in a categorization file. Civil 3D includes an option to create favorite items that are used most frequently. This file, which is always formatted as an XML file, is optional.

Formula File The formula file is used when the unit of measure for the pay item needs to be converted or calculated based on another value. For example, bituminous concrete for a parking lot is usually paid for by the ton, but you will generally be obtaining an area from CAD. You can set up a formula that converts square feet into tons, taking into account an assumed depth and density. Later in this chapter, you will create a formula that relates length to light poles along the road.

A command is available if you need to disconnect your pay item and formula files from the drawing. At the command line you can type **DETACHQTOFILES**. The Undo command will not reconnect them if you type this in accidentally.

In the next section, you will learn how to connect the needed pay item files and create favorites.

Pay Item Favorites

Your Civil 3D file will "remember" which files it needs to correctly assess quantities. The Favorites list is a separate category that can be populated for quick access to commonly used pay items. This list is also stored as part of the DWG.

In this exercise, you'll look at how to open a pay item and categorization files and add a few items to the Favorites list for later use:

1. Create a new drawing using the `_AutoCAD Civil 3D (Imperial) NCS` template. For metric users, use the `_AutoCAD Civil 3D (Metric) NCS` template.
2. From the Analyze tab ➤ QTO panel, choose QTO Manager (see Figure 19.1) to display the QTO Manager.

FIGURE 19.1
The QTO tools on the Analyze tab

3. Click the Open button at the top left of the QTO Manager to display the Open Pay Item File dialog.
4. In the Open Pay Item File dialog, verify that the Pay Item File Format drop-down list is set to CSV (Comma Delimited).

5. Click the Open button next to the Pay Item File text box to browse for the file.
6. Navigate to the `Getting Started` folder and select the `Getting Started.csv` file.

   In Windows Vista and Windows 7, this file is found in `C:\ProgramData\Autodesk\C3D 2013\enu\Data\Pay Item Data\Getting Started\`.

C:\PROGRAMDATA

If you do not see the `ProgramData` folder listed in `C:\`, follow these simple steps:

1. **Open Windows Explorer.**
2. **Click the Organize button at the upper left.**
3. **Select Folder And Search Options to display the Folder Options dialog.**
4. **On the View tab, in Advanced Settings select the radio button Show Hidden Files, Folders, And Drives.**
5. **Click OK to accept the settings in the Folder Options dialog.**

   For metric users, use the `Getting Started_METRIC.csv` downloadable from the book's web page (`www.sybex.com/go/masteringcivil3d2013`).

7. Click Open to select this CSV pay item file.
8. Click the Open button next to the Pay Item Categorization File text box to browse for the file.
9. Navigate to the `Getting Started` folder and select the `Getting Started Categories.xml` file.

   This file can be used by both Imperial and metric users.

10. Click Open to select the file.

Your display should look similar to Figure 19.2.

FIGURE 19.2
Open Pay Item File dialog

11. Click OK to accept the settings in the Open Pay Item File dialog.

    The QTO Manager will now be populated with a collection of divisions, as shown in Figure 19.3. These divisions came from the `Getting Started Categories.xml` file.

FIGURE 19.3
The QTO Manager in Panorama populated within categories (Divisions and Groups)

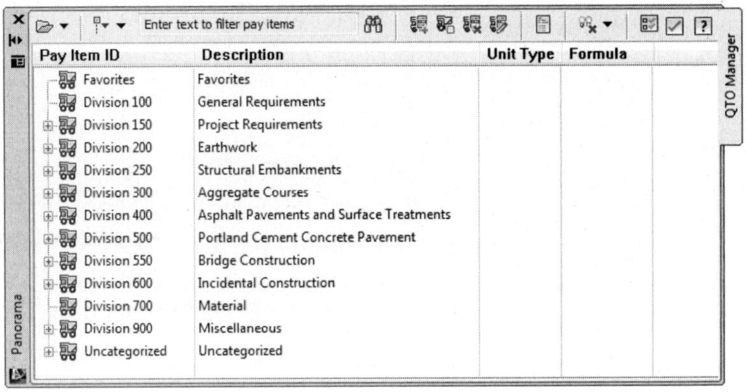

12. Expand the Division 200 ➤ Group 201 ➤ Section 20101 branches, and select the item 20101-0000 Clearing And Grubbing.
13. Right-click and select Add To Favorites List, as shown in Figure 19.4.
14. Expand a few other branches to familiarize yourself with this parts list.
15. Add the following to your Favorites list in preparation for the next exercise:
    - 60404-2000 Catch Basin, Type 2
    - 61106-0000 Fire Hydrant
    - 61203-0000 Manhole, Sanitary Sewer
    - 63401-0600 Pavement Markings, Type C, Broken

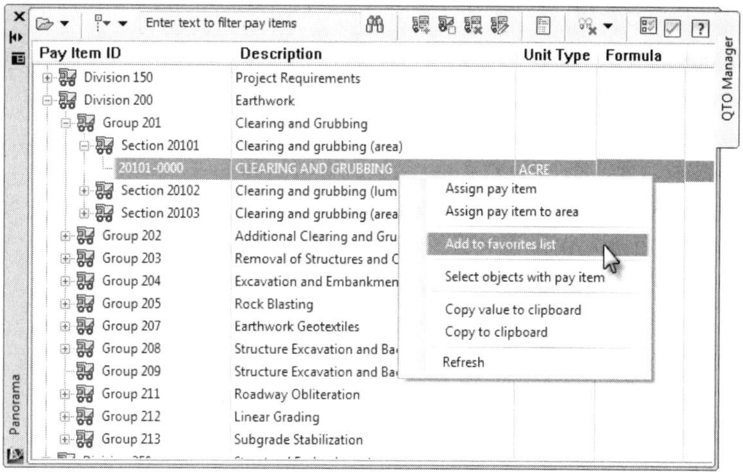

FIGURE 19.4 Selecting a pay item to add as a favorite

16. Save the drawing as **`QTOPractice.dwg`** or **`QTOPractice_METRIC.dwg`** and keep it open for the next exercise.

When complete, scroll to the top of the Pay Item ID list to expand the Favorites category. Your QTO Manager should look similar to Figure 19.5.

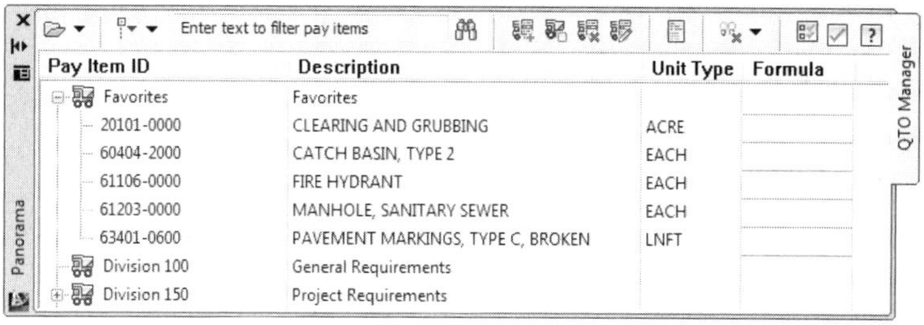

FIGURE 19.5 A list of favorites within the QTO Manager

Civil 3D ships with a number of pay item list categorization files but only one actual pay item list. This avoids any issue with out-of-date data. One commonly used categorization type found in the `C:\ProgramData\Autodesk\C3D2013\enu\Data\Pay Item Data\CSI\` folder is MasterFormat. In the `United States` folder, you can also find categorization files for AASHTO, Federal Highway Administration, and several states' DOT formats. The pay item files that install with the software are intended to be examples. Contact your reviewing agency for access to their pay item list and categorization files if they're not already part of the Civil 3D product.

Once you have pay items to choose from, it's time to assign them to your model for analysis.

Searching for Pay Items

In the top of the QTO Manager is an extremely handy filter tool. If you don't know what category an item is listed under, you can type it in the field to the left of the binoculars icon.

When you first click the binoculars icon to execute the filter, it will appear as if nothing happened. That's because the category headings are getting in the way of the listing. To see an uncategorized look at the list your filter produced, select Turn Off Categorization from the drop-down to the left of the filter field, as shown in Figure 19.6.

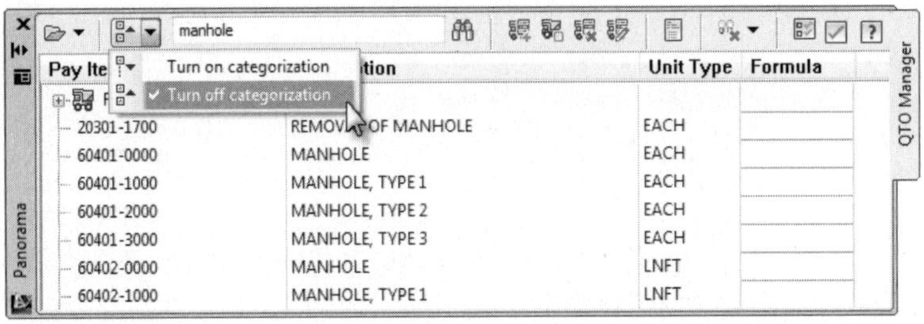

FIGURE 19.6
Choose Turn Off Categorization to see the results of your filter

In the following exercise, you will use the filter functionality to add items to your Favorites list:

1. If not still open from the previous exercise, open the `QTOPractice.dwg` or the `QTOPractice_METRIC.dwg` file.
2. If the QTO Manager is not already open, from the Analyze tab ➢ QTO panel choose QTO Manager to open it.
3. To the left of the filter field, click the drop-down to select Turn Off Categorization.

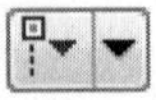

4. In the filter field of the QTO Manager, type **grandi** and then click the binoculars icon.

   Your filter should result in three types of trees.
5. Right-click on the first item, Fagus Grandiflora, and select Add To Favorites List.
6. Using the same filter technique, add the following item to the Favorites list: 63620-0500 Pole, Type Washington Globe No. 16 Light Standard.

You will find that by adding your most frequently used items to the Favorites list, they will be at your fingertips instead of you having to dig through branch after branch of pay items.

When this exercise is complete, you may close the drawing. A saved copy of this drawing is available from the book's web page with the filename `QTOPractice_FINISHED.dwg` or `QTOPractice_METRIC_FINISHED.dwg`.

Real World Scenario

CREATING YOUR OWN CATEGORIZATION FILE

One of the challenges in automating any quantity takeoff analysis is getting the pay item list and categories to match up with your local requirements. Getting a pay item list is pretty straightforward — many reviewing agencies provide their own list for public use to keep all bidding on an equal footing.

Creating the category file can be slightly more difficult, and it will probably require experimentation to get just right. In this example, you'll walk through creating a couple of categories to be used with a provided pay item list.

Note that modifying the QTO Manager affects all open drawings. You might want to finish the other exercises and come back to this when you need to make your own file in the real world.

1. Create a new drawing using the `_AutoCAD Civil 3D (Imperial) NCS` or `_AutoCAD Civil 3D (Metric) NCS` template.
2. From the Analyze tab ➢ QTO panel, choose QTO Manager to display the QTO Manager.
3. Click the Open button at the top left of the QTO Manager to display the Open Pay Item File dialog.
4. Verify that Pay Item File Format is set to CSV (Comma Delimited).
5. Click the Open button next to the Pay Item File text box and navigate to open the `Mastering.csv` or `Mastering_METRIC.csv` file.

   Remember, all files can be downloaded from `www.sybex.com/go/masteringcivil3d2013`.
6. Click the Open button next to the Pay Item Categorization File text box to display the Open Pay Item Categorization File dialog.
7. Browse to open the `MasteringCategories.xml` file (the same file is acceptable for both Imperial and metric users), and then click OK to accept the settings in the dialog.

   Your QTO Manager should look like this:

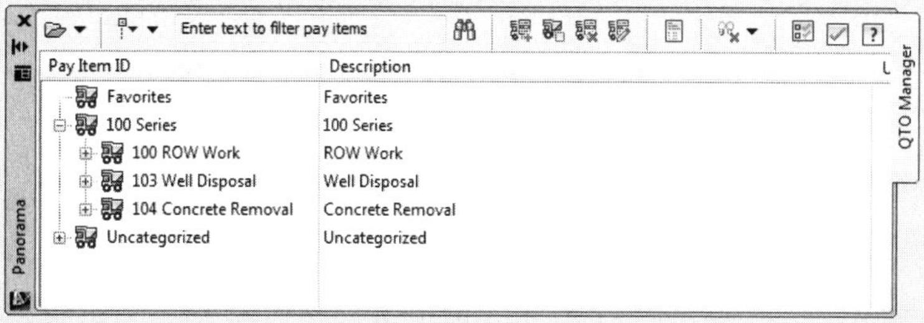

8. Launch XML Notepad. (This is a simple XML editor that you can download for free from Microsoft.)
9. Browse and open the `MasteringCategories.xml` file from within XML Notepad.
10. Expand the PayItemCategorizationRules ➢ Categories branch, right-click the Category branch, and select Duplicate, as shown here:

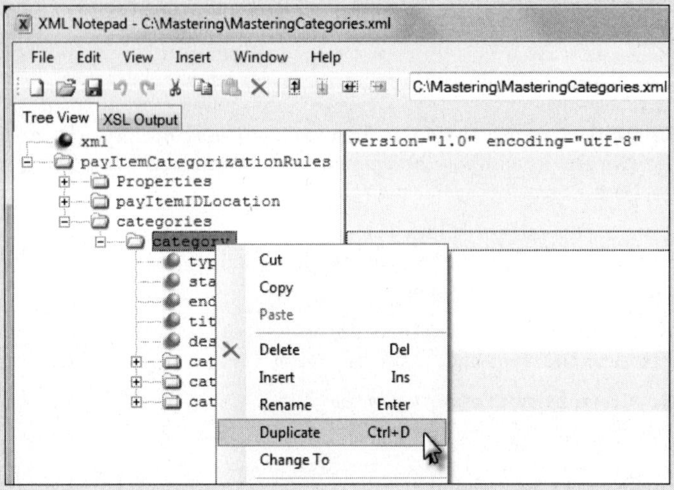

11. Modify the values to match the following:

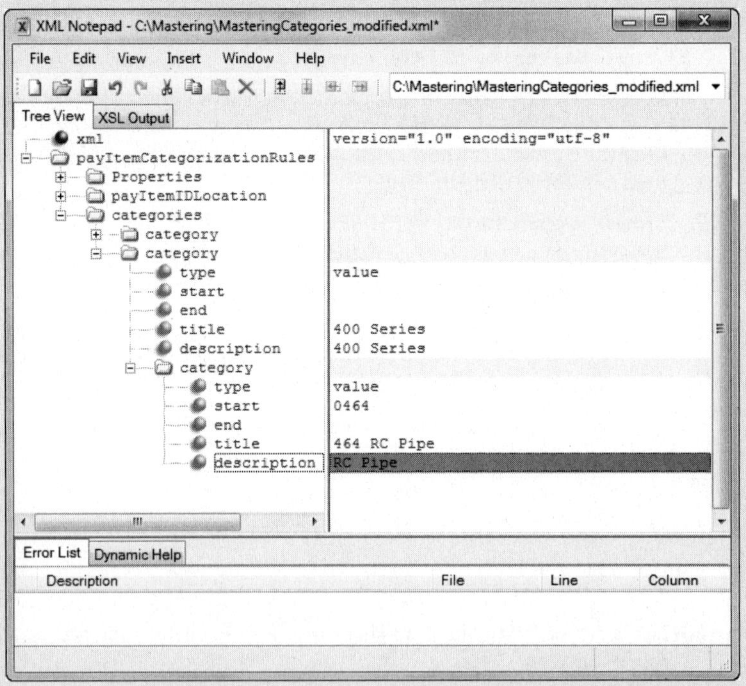

12. Delete the extra category branches under the new branch by right-clicking on the two extra category listings and clicking Delete (the previous image is shown with these extra branches already deleted).

13. Save and close the modified `MasteringCategories.xml` file.
14. Switch back to the Civil 3D QTO Manager.

    Next, you will need to reload the categorization file to view the changes.
15. Click the Open drop-down menu in the top left of the QTO Manager, and select Open ➢ Categorization File.
16. Browse to the `MasteringCategories.xml` file again and open it.

    Once Civil 3D processes your XML file, your QTO Manager should look similar to this:

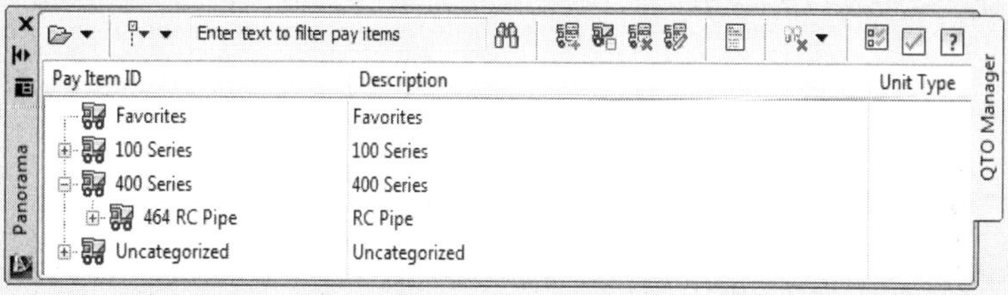

When this exercise is complete, you may close the drawing. A saved copy of this drawing is available from the book's web page with the filename `QTOCategories_FINISHED.dwg` or `QTOCategories_METRIC_FINISHED.dwg`. In addition, a saved copy of the modified XML file is available with the filename `MasteringCategories_modified.xml`.

Creating a fully-developed category list is a bit time-consuming, but once it's done, the list can be shared with your entire office so everyone has the same data to use.

Keeping Tabs on the Model

Once you have a list of items that should be accounted for in your project, you have to assign them to items in your drawing file. You can do so in any of the following ways:

- Assign pay items to simple AutoCAD® objects such as blocks and lines
- Assign pay items to corridor components
- Assign pay items to pipe network pipes and structures

In the next few sections, you'll explore each of these methods, along with some formula tools that can be used to convert things such as linear items to individual quantity counts.

AutoCAD Objects as Pay Items

The most basic use of the QTO tools is to assign pay items to things like blocks and linework within your drawing file. The QTO tools can be used to quantify any number of things, including tree plantings, signposts, or area items such as clearing and grubbing. In the following exercise, you'll assign pay items to blocks as well as to some closed polylines. Be sure the

Getting Started.csv (or Getting Started_METRIC.csv) and Getting Started Categories .xml files are loaded as described in the first exercise.

1. Open the AcadObjectsInQTO.dwg or AcadObjectsInQTO_METRIC.dwg file.
2. From the Analyze tab ➤ QTO panel, choose QTO Manager to display the QTO Manager.
3. Expand the Favorites branch, right-click on Clearing And Grubbing, and select Assign Pay Item To Area, as shown in Figure 19.7.

FIGURE 19.7
Assigning an area-based pay item

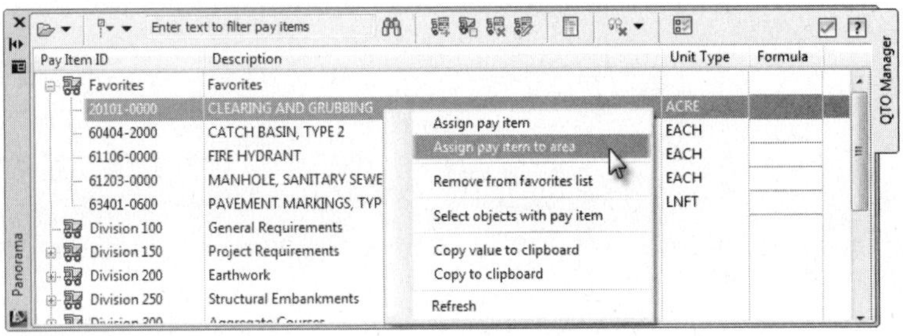

4. At the Select Point or [select Object]: prompt, type **O**↵ to activate the Object option for assignment.
5. Click on the polyline around the outer edge of the site, as shown in Figure 19.8.

FIGURE 19.8
Selecting a closed polyline for an area-based quantity

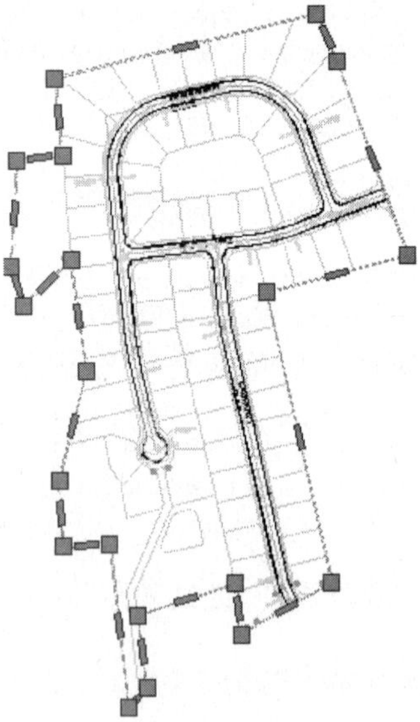

Notice the entire polyline highlights to indicate what object is being picked. The command line should also echo Pay item 20101-0000 assigned to object when you pick the polyline. Note that this will create a hatch object reflecting the area being assigned to the pay item.

6. Press ↵ again to end the command.
7. Select the hatch to activate the Hatch Editor contextual tab.
8. From the Hatch Editor contextual tab ➢ Properties panel, change the Hatch Transparency to **80**.
9. Right-click on the hatch and select Display Order ➢ Send To Back.
10. Move the QTO Manager to the side and zoom in on an area of the road where you can see one or more of the blocks representing trees.
11. Select one of the tree blocks and then select all of the tree blocks in the drawing, right-click, and pick Select Similar.
12. Back on the QTO Manager, under the Favorites branch select 62606-0150 Fagus Grandiflora, American Beech.
13.  Near the top of the panel, click the Assign The Selected Pay Items To Object(s) In The Drawing button.

    Notice the great tooltips on these buttons.
14. Press ↵ to end the assigning command.
15. Repeat steps 11 through 14 to assign the pay item 61106-0000 Fire Hydrant to all of the fire hydrant blocks on the site.

When this exercise is complete, you may close the drawing. A saved copy of this drawing is available from the book's web page with the filename `AcadObjectsInQTO_FINISHED.dwg` or `AcadObjectsInQTO_METRIC_FINISHED.dwg`.

ASSIGNING PAY ITEMS

The two assignment methods described in this section, by object or by area, are essentially interchangeable. Pay items can be assigned to any number of AutoCAD objects, meaning you don't have to redraw the planners' or landscape architects' work in Civil 3D to use the QTO tools.

Keep in mind that the pay item tag is saved with the block. This is a good thing if you assign a pay item to a block and copy the block because the copies will be automatically tagged. If you assign a pay item to an object and use the **WBLOCK** command to copy that object out of your current drawing and into another, the pay item assignment goes along as well.

You'll find out how to unassign pay items after we discuss all the ways to assign them.

Pricing Your Corridor

The corridor functionality of Civil 3D is invaluable. You can use it to model everything from roads to streams to parking lots. With the QTO tools, you can also use the corridor object to quantify much of the project construction costs.

Assemblies and QTO Are Related

Be mindful of what parts of an assembly are available when creating quantity takeoff assignments in the code set style.

If you plan to price an item based on the centerline of your road, make sure you have set a Crown Point Code in your assembly properties. If you used the LaneSuperelevationAOR subassembly, this is defined by the Inside Point Code code if your Slope Direction is set to Away From Crown.

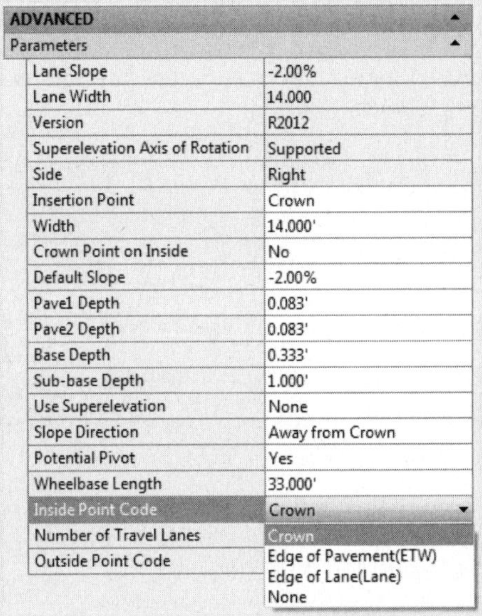

Changes to QTO information will not cause your corridor to flag itself as out of date. Before you run a takeoff report, it is a good idea to rebuild your corridors to ensure the most up-to-date information is accounted for.

In this example, you'll use the pay item list along with a formula to convert the linear curb measurement to an incremental count of light poles required for the project:

1. Open the `QTOCorridors.dwg` or `QTOCorridors_METRIC.dwg` file.
2. From the Analyze tab ➢ QTO panel, choose QTO Manager.
3. Expand the Favorites branch to display your 63620-0500 light standard pay item.
4. Scroll to the right to the Formula column and click in the empty cell on the light standard row.

   When you do, Civil 3D will display the alert box shown in Figure 19.9, which warns you that formulas must be stored to an external file.

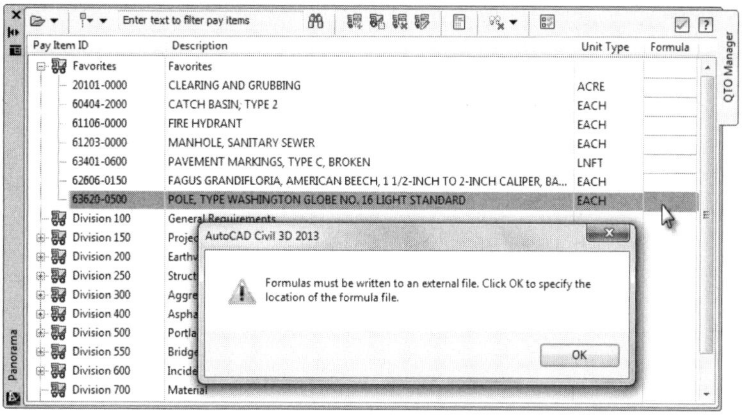

FIGURE 19.9
Click in the Formula cell to display this warning dialog.

5. Click OK to dismiss the warning, and Civil 3D will present the Select A Quantity Takeoff Formula File dialog.
6. Navigate to `C:\Mastering\`, change the filename to **`Mastering`** (or **`Mastering_METRIC`**), and click Save.

   Civil 3D will display the Pay Item Formula: 63620-0500 dialog, as shown in Figure 19.10. (The expression will be empty when you first open it, but you'll take care of that in the next steps.)

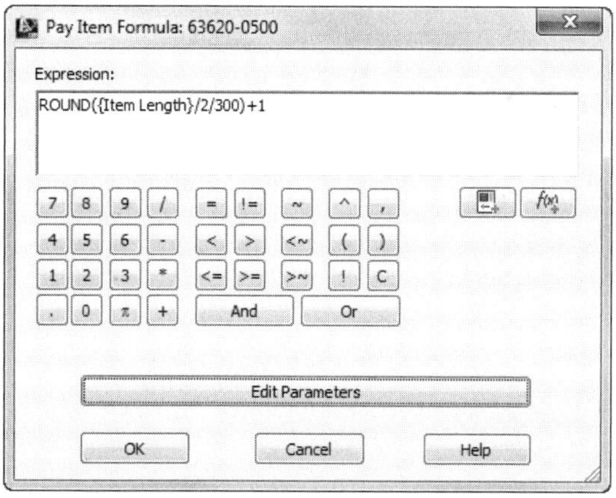

FIGURE 19.10
The completed pay item expression

Assume that you need a street light every 300 feet, but only on one side of the street. To do this, you add up all the lengths of curb, and then divide by 2 because you want only one half of the street to have lights. You then divide by 300 (or 100 for metric) because you are running lights in an interval. Finally, you round to the nearest integer and add 1 to make the number conservative.

7. Enter the formula using the following steps:

a. Click the Property button and select Item Length from the drop-down list.

b. Use the buttons in the dialog or your keyboard to divide this value by 2 to get half of the curb length.

c. Use the buttons in the dialog or your keyboard to divide this value by 300 (or 100), which is the spacing of the light standards.

d. To round this value, move your cursor to the beginning of the expression, click the Function button, and select ROUND.

Note that there are also ROUNDUP and ROUNDDOWN expressions available if these were better suited for your calculations.

e. Move your cursor to the end of the expression and add the close parentheses for the ROUND command.

f. Use the buttons in the dialog or your keyboard to add 1 to this count to be conservative.

The formula should now match the one shown previously in Figure 19.10.

8. Click OK.

Note that the pay item list now shows a small calculator icon on that row to indicate a formula is in use.

Now that you've modified the way the light poles will be quantified from your model, you can assign the pay items for light poles and road striping to your corridor object. This is done by modifying the code set, as you'll see in the next steps.

9. In the Toolspace Settings tab, expand General ➢ Multipurpose Styles ➢ Code Set Styles, right-click on All Codes, and select Edit to display the Code Set Style – All Codes dialog.

10. On the Codes tab, scroll to the Point branch and find the row for Crown, as shown in Figure 19.11.

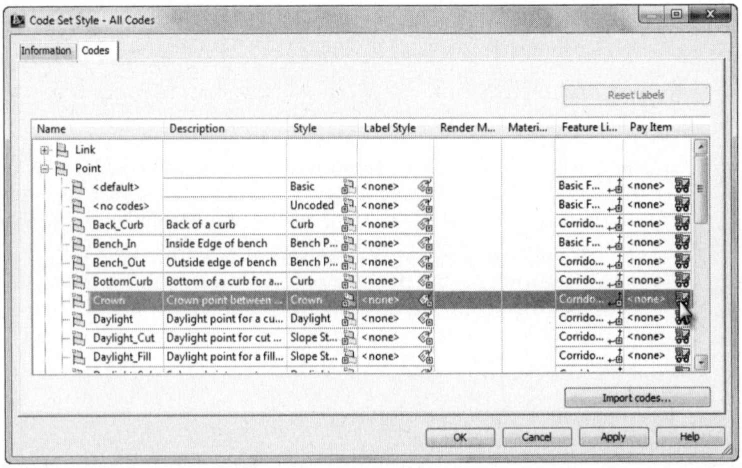

FIGURE 19.11
Select the Crown row in the Point section in the Code Set Style – All Codes dialog

As always, you may need to widen some of the columns to view the necessary text.

11. Click the truck icon in the Pay Item column, as shown in Figure 19.11, to open the Pay Item List dialog.
12. Expand the Favorites branch, and select Pavement Markings, Type C, Broken.
13. Click OK to return to the Code Set Style – All Codes dialog.

    Your dialog should now reflect the pay item number of 63401-0600.

14. Again in the Point branch, find the row for Back_Curb, and repeat steps 10 and 12 to assign 63620-0500 to the Back_Curb code.

    Remember, this is your light-standard pay item with the formula from the previous steps. Your Code Set Style – All Codes dialog should look like Figure 19.12, which lists a pay item for both of these point codes.

FIGURE 19.12 Completed code set editing for pay items

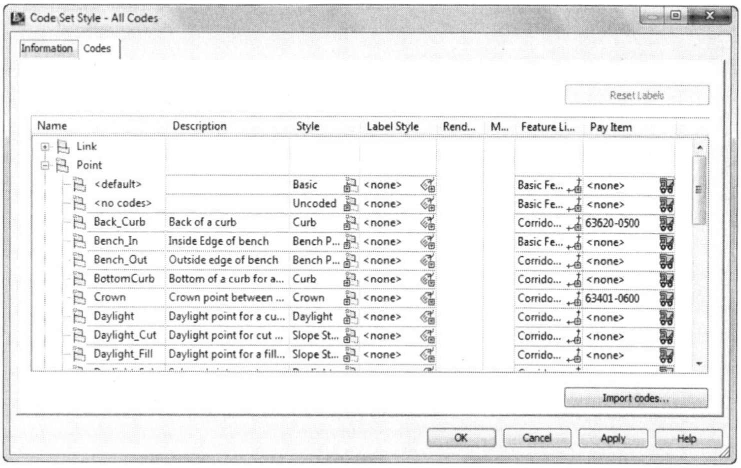

15. Click OK to accept the settings in this dialog.
16. Switch to Prospector, and expand the Corridor branch.
17. Right-click Frontenac Drive Corridor, and select Properties to open the Corridor Properties dialog.
18. On the Codes tab, scroll down to the Point branch.

    Notice that Back_Curb and Crown codes reflect pay items in the far-right column.

19. Click OK to accept the settings in the dialog.

When this exercise is complete, you may close the drawing. A saved copy of this drawing is available from the book's web page with the filename `QTOCorridors_FINISHED.dwg` or `QTOCorridors_METRIC_FINISHED.dwg`.

BEST PRACTICE: STORING A FORMULA FILE WITH THE PROJECT FILES

Every drawing in which you utilize QTO tools is intended to have its own, unique formula file. We recommend that when you create a new formula file, you store it in the same folder as the rest of the project.

If you need to send a drawing to another firm and you want them to have your quantity takeoff formulas, use eTransmit to export everything they need to work with the drawing. You can access the eTransmit command by clicking the Application menu in the upper-left corner and selecting Publish ➢ eTransmit. The eTransmit command will attach the pay item file, the categorization file, and the formula file as part of the transmission.

Corridors can be used to measure a large number of items. You've always been able to manage pure quantities of material, but now you can add to that the ability to measure linear and incremental items as well. Although we didn't explore every option, you can also use shape and link codes to assign pay items to your corridor models.

Now that you've looked at AutoCAD objects and corridors, it's time to examine the pipe network objects in Civil 3D as they relate to pay items.

Pipes and Structures as Pay Items

One of the easiest items to quantify in Civil 3D is the pipe network. There are numerous reports that will generate pipe and structure quantities. This part of the model has always been fairly easy to account for; however, with the ability to include it in the overall QTO reports, it's important to understand how parts get pay items assigned. There are two methods: via the parts lists and via the part properties.

ASSIGNING PAY ITEMS IN THE PARTS LIST

Ideally, you'll build your model using standard Civil 3D parts lists that you've set up as part of your template. These parts lists contain information about pipe sizes, structure thicknesses, and so on. They can also contain pay item assignments. This means that the pay item property will be assigned as each part is created in the model, skipping the assignment step later.

In this exercise, you'll see how easy it is to modify parts lists to include pay items:

1. Open the `QTOPipeNetworks.dwg` or `QTOPipeNetworks_METRIC.dwg` file.
2. On the Settings tab of Toolspace, expand the Pipe Network ➢ Parts Lists branch.
3. Highlight and right-click on Sanitary Sewer and select Edit to display the Network Parts List – Sanitary Sewer dialog.
4. On the Pipes tab, expand the Sanitary Sewer ➢ PVC Pipe part family.

   Notice that the far-right column is the Pay Item assignment column.

5. Click the truck icon in the 8-inch PVC (or 200 mm PVC Pipe) row to display the pay item list.
6. Enter **PVC** in the text box, as shown in Figure 19.13, and press ↵ to filter the dialog.

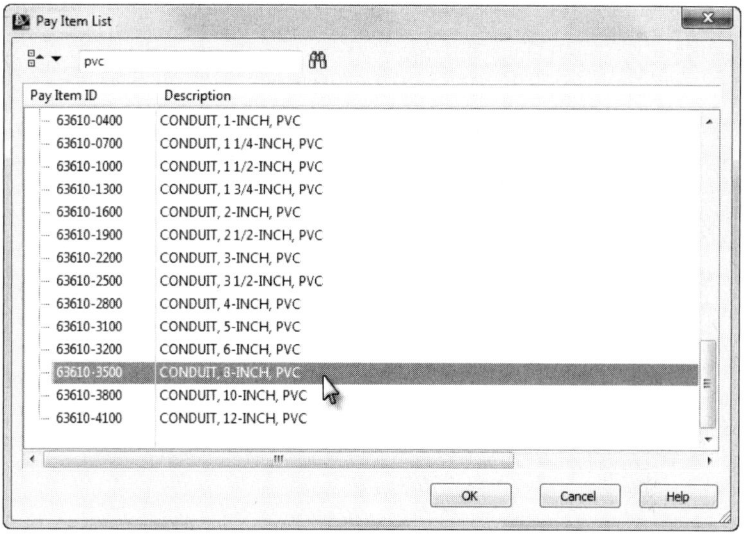

FIGURE 19.13 Filtering and selecting the 8-inch PVC conduit as a pay item

Remember that you can turn off categories with the button to the left of the text box.

7. Select the CONDUIT, 8-INCH, PVC item (CONDUIT, 200 MM PVC) as shown, and click OK to assign this pay item to the applicable pipe part.
8. Repeat steps 4 through 7 to assign pay items to 10- and 12-inch (250 mm and 300 mm) PVC conduits.

Your dialog should look like Figure 19.14.

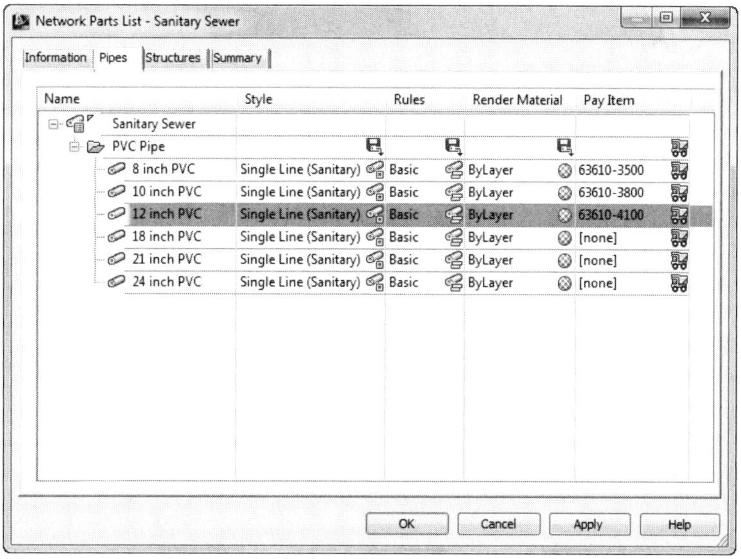

FIGURE 19.14 Completed pipe parts pay item assignment

9. On the Structure tab of the Parts List dialog, expand the Sanitary Sewer ➤ Concentric Cylindrical Structure branch.
10. Click the truck icon on the Concentric Structure row to display the Pay Item List dialog.
11. In the Pay Item List dialog, search for **Manhole**, select MANHOLE, SANITARY SEWER, and click OK.

    The Network parts list should now look similar to Figure 19.15.

FIGURE 19.15
Completed structure parts pay item assignment

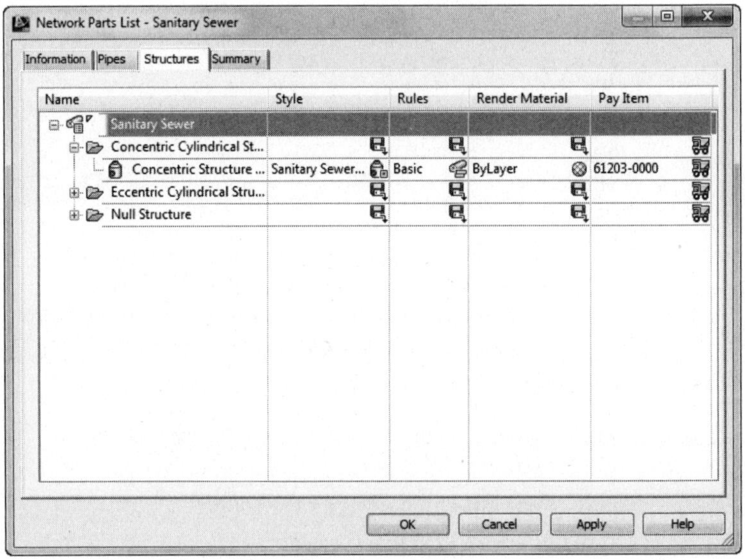

12. Click OK to accept the settings in the Network parts list.

When this exercise is complete, you may save and keep the drawing open to continue on to the next exercise. Or you may use the saved copy of this drawing available from the book's web page (`QTOPipeNetworksPart1_FINISHED.dwg` or `QTOPipeNetworksPart1_METRIC_FINISHED.dwg`).

PAY ITEMS AS PART PROPERTIES

If you have existing Civil 3D pipe networks that were built before your parts list had pay items assigned, or if you change out a part during your design, you need a way to review and modify the pay items associated with your network. Unfortunately, doing so isn't as simple as just telling Civil 3D to reprocess some data, but it's not too complicated either. You simply remove the pay item association in place and then add new ones.

In this exercise, you'll add pay item assignments to a number of parts already in place in the drawing:

1. Open or continue working with the `PipeNetworksPart1_FINISHED.dwg` or `QTOPipeNetworksPart1_METRIC_FINISHED.dwg` file.
2. If the QTO Manager is not already open, from the Analyze tab ➢ QTO panel choose QTO Manager to open it.
3. Slide the QTO Manager to one side, and then select one of the manholes in the Sanitary Sewer Network (they have an S symbol on them).
4. Right-click and choose Select Similar, as shown in Figure 19.16. Twenty manholes should highlight.

FIGURE 19.16
Use Select Similar to find all sanitary manhole structures

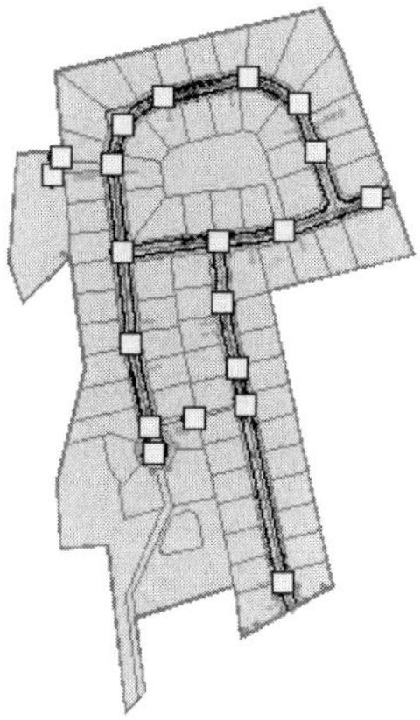

5. In the QTO Manager, expand Favorites and select MANHOLE, SANITARY SEWER.
6. Click the Assign The Selected Pay Items To Object(s) In The Drawing button.
7. At the command line, press ↵ to complete the assignment.

   You can pause over one of the manholes, and the tooltip will reflect a pay item now, in addition to the typical information found on a manhole.

When this exercise is complete, you may save and keep the drawing open to continue on to the next exercise. Or you may use the saved copy of this drawing available from the book's web page (`QTOPipeNetworksPart2_FINISHED.dwg` or `QTOPipeNetworksPart2_METRIC_FINISHED.dwg`).

Heads Up on Pipe and Structure QTO Assignments

Pay items change when pipes and structures change if the new part has a pay item assigned in the parts list. However, if you swap to a part that does not have a pay item assigned, Civil 3D drops the QTO tag.

The best way to keep a proper count of your pipes and structures is to make sure every part in your network parts list has an appropriate pay item.

If you graphically change a pipe property (such as its diameter), this will not cause a change in the pay item assignment. You should always use Swap Part to change sizes, as you learned in Chapter 14, "Pipe Networks."

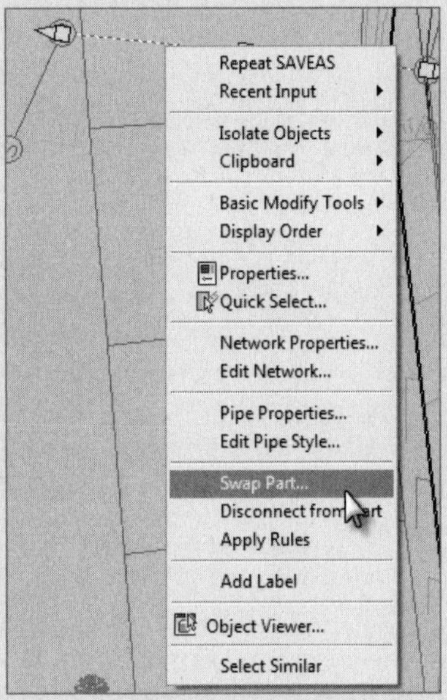

This is something you definitely want to keep an eye on. So, if you do need to change a pay item assignment to a part that's already in the network, how do you do it? You'll find that out in the next section.

Assigning pay items to existing structures and pipes is similar to adding data to standard AutoCAD objects. As mentioned before, the pay item assignments sometimes get confused in the process of changing parts and pipe properties, and they should be manually updated. To do so, you'll need to remove pay item data and then add it back in, as demonstrated in this exercise:

1. Open or continue working with the `QTOPipeNetworksPart2_FINISHED.dwg` or `QTOPipeNetworksPart2_METRIC_FINISHED.dwg` file.
2. Pan to the northwest area of the site where the sanitary sewer network terminates.
3. Pause your cursor over the pipe.

The tooltip will appear indicating the pipe information but no pay item, as shown in Figure 19.17.

FIGURE 19.17
Tooltip for a pipe without an assigned pay item

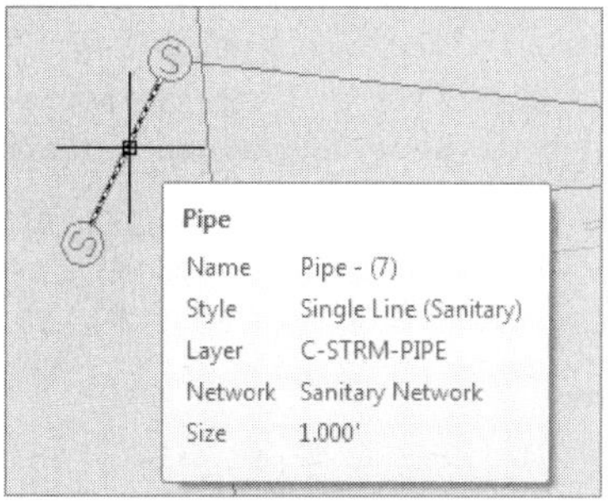

4. Select the pipe, right-click, and select Swap Part.

   This will display the Swap Part Size dialog.

5. Select 12 Inch PVC (or 300 mm PVC Pipe) from the list of sizes (the same size of pipe that it was originally — you are not changing the diameter), and then click OK.
6. Pause near the newly-sized pipe and notice that the tooltip now shows the pay item, as shown in Figure 19.18.

FIGURE 19.18
Tooltip after the pipe part has been swapped to a part with an assigned pay item

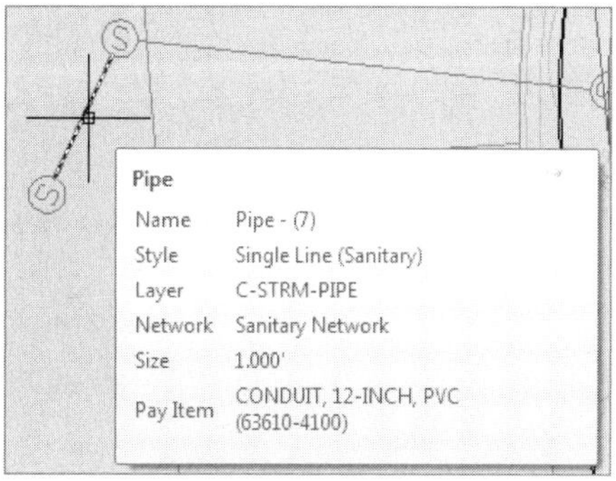

When this exercise is complete, you may close the drawing. A saved copy of this drawing is available from the book's web page with the filename `QTOPipeNetworksPart3_FINISHED.dwg` or `QTOPipeNetworksPart3_METRIC_FINISHED.dwg`.

You might wonder why Civil 3D allows you to have multiple pay items on a single object. For example, linear feet of striping and tree counts can both be derived from street lengths; bedding and pipe material can both be calculated from pipe objects. You can also add related tasks to an item. For instance, a tree is usually a pay item by itself, but the labor to install the tree may be treated as a separate pay item.

You've now built up a list of pay items, tagged your drawing a number of ways, updated and modified pay item data, and looked at formulas in pay items. In the next section, you'll make a final check of your assignments before running reports.

Highlighting Pay Items

Before you run any reports, it's a good idea to make a cursory pass through your drawing and look at what items have had pay items assigned and what items have not. This review will allow you to hopefully catch missing items (such as hydrants added after the pay item assignment was done), as well as see any items that perhaps were blocked in with unnecessary pay items already assigned. In this exercise, you'll look at tools for highlighting objects with and without pay item assignments:

1. Open the `QTOHighlighting.dwg` or the `QTOHighlighting_METRIC.dwg` file.

2. From the Analyze tab ➤ QTO panel, choose QTO Manager to display the QTO Manager.
3. In the QTO Manager, select Highlight Objects With Pay Items, as shown in Figure 19.19.

FIGURE 19.19 Turning on highlighting for items with pay items assigned

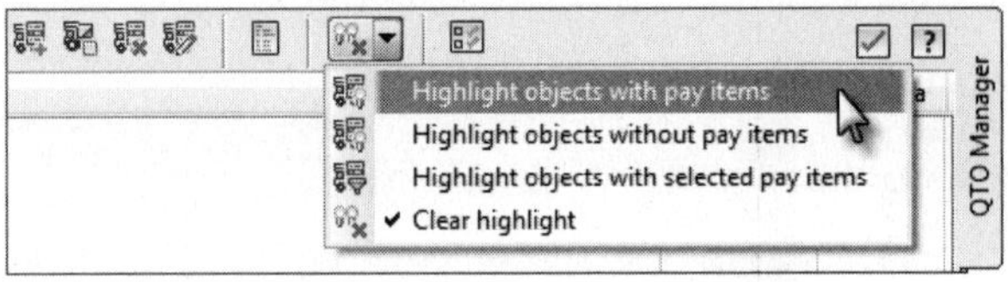

   Notice that from this same drop-down menu you can also Highlight Objects Without Pay Items, Highlight Objects With Selected Pay Items, or Clear Highlight.

4. Pan around the drawing and zoom in on a tree.
5. In the QTO Manager, select Highlight Objects Without Pay Items.

   Notice that the trees turn from green to muted. This means they have a pay item assigned.

6. In the QTO Manager, switch back to Highlight Objects With Pay Items.

   Next, you will change the pay item assignment for the trees at the cul-de-sac since the planner has decided to use a different tree type.

7. In the QTO Manager, click the Remove Pay Item(s) From Specified Objects button.
8. At the `Select object(s):` prompt, select the tree to the southwest of the cul-de-sac at the end of Frontenac Drive and press ↵ to end the command.

   Notice that the tree goes from green to a muted gray, indicating that the tree no longer has a pay item associated with it since the highlighting is enabled.

9. In the QTO Manager, enter **Maple** in the text box to filter.
10. Turn the categorization option off to more easily see the filtered results.
11. Select item 62601-0100.

12. Click the Assign The Selected Pay Items To Objects In The Drawing button.
13. At the `Select pay item(s) from master pay item list or [Enter]:` prompt, press ↵.
14. At the `Select object(s):` prompt, select the same tree to the southwest of the cul-de-sac at the end of Frontenac Drive that you just unassigned a pay item from, and press ↵ to end the command.

Notice that the tree goes from a muted gray to green, indicating that the tree once again has a pay item associated with it since the highlighting is enabled.

Unassigning and then reassigning a pay item may seem cumbersome. Instead, you may find it simpler to edit the pay item.

15. In the QTO Manager, click the Edit Pay Items On Specified Object button.
16. At the `Select object(s):` prompt, select the other tree to the southeast of the cul-de-sac at the end of Frontenac Drive to display the Edit Pay Items dialog.

    While we are only selecting one tree for this exercise, you could select multiple objects.
17. Select the Fagus Grandiflora row, and then click the red X in the upper right to remove this pay item from the tree, as shown in Figure 19.20.

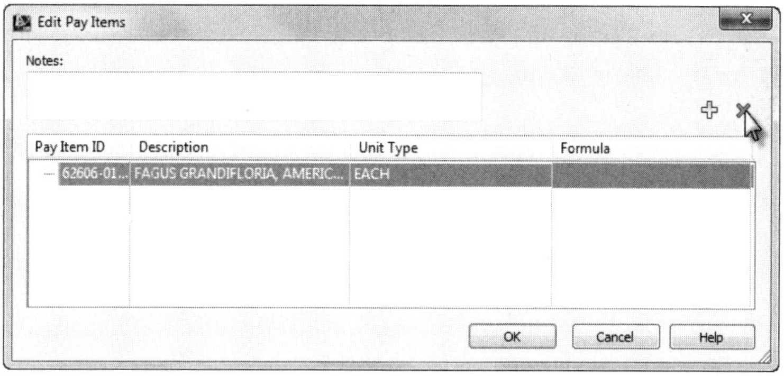

FIGURE 19.20
Editing pay item assignments: deleting the tree pay item

18. Click the plus in the upper right to display the Pay Item List dialog in order to add a new pay item to the tree.
19. Enter **Maple** in the text box to filter.
20. Select item 62601-0100 and click OK.
21. Click OK to accept the new pay item designation shown in the Edit Pay Items dialog.

22. In the QTO Manager, select Highlight Objects Without Pay Items from the drop-down menu shown previously in Figure 19.19.

    This switches the highlighting from objects that do have pay items to objects that do not have pay items.

23. In the QTO Manager, select Clear Highlight to return the drawing view to normal.

When this exercise is complete, you may close the drawing. A saved copy of this drawing is available from the book's web page with the filename `QTOHighlighting_FINISHED.dwg` or `QTOHighlighting_METRIC_FINISHED.dwg`.

While objects are highlighted, you can add, remove, or edit pay items using the tools at the top of the QTO Manager as well as any other normal AutoCAD work you might perform. This makes it easier to correct any mistakes made during the assignment phase of the process. Finally, always be sure to clear highlighting before exiting the drawing, or your peers might wind up awfully confused when they open the file!

Inventorying Your Pay Items

At the end of the process, you need to generate some sort of report that shows the pay items in the model, the quantities of each item, and the units of measurement. This data can be used as part of the plan set in some cases, but it's often requested in other formats to make further analysis possible. In this exercise, you'll look at the Quantity Takeoff tool that works in conjunction with the QTO Manager to create reports:

1. Open the `QTOReporting.dwg` or `QTOReporting_METRIC.dwg` file.
2. From the Analyze tab ➢ QTO panel, choose Takeoff to display the Compute Quantity Takeoff dialog, shown in Figure 19.21.

FIGURE 19.21
The Compute Quantity Takeoff dialog with default settings

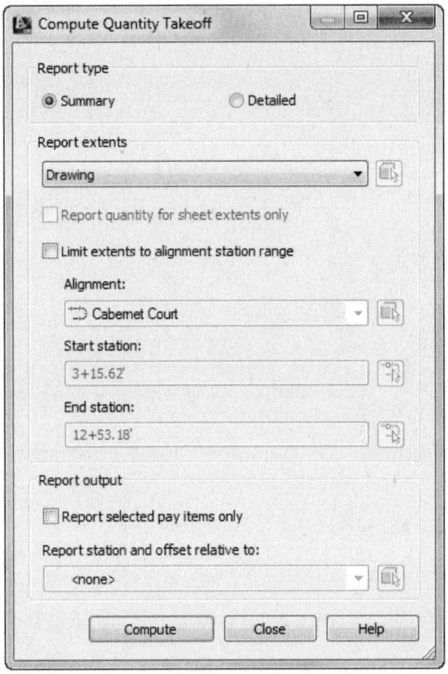

> **LIMITING THE REPORT EXTENTS**
>
> Note that you can limit the Report Extents by drawing; by sheets, if done from paperspace; by selection set; or by alignment station ranges. Most of the time, you'll want to run the full drawing. You can set the Report Output for only selected pay items if, for instance, you just want a table of pipe and structures.

3. Click Compute to open the Quantity Takeoff Report dialog, as shown in Figure 19.22.

   The report is shown in the default Extensible Stylesheet Language (XSL) format.

4. From the drop-down menu on the lower left of the dialog, select Summary (TXT).xsl to change the format to something more understandable, as shown in Figure 19.23.

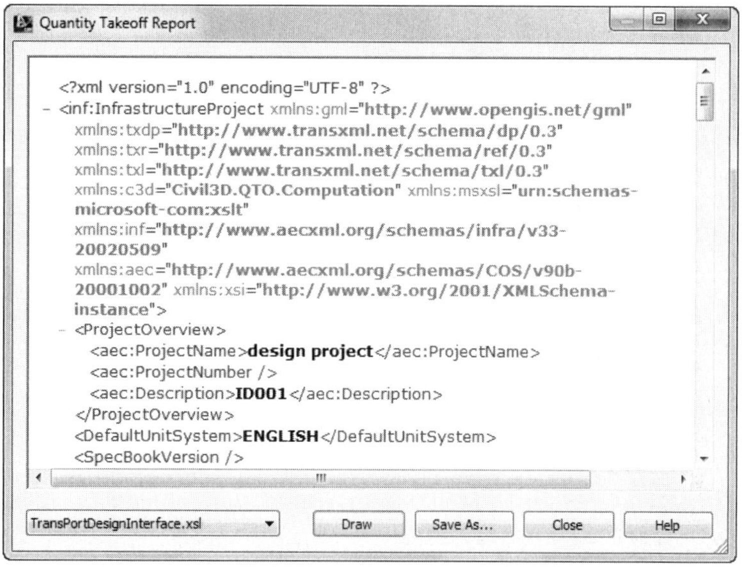

FIGURE 19.22
Quantity Takeoff Report in the default XSL format

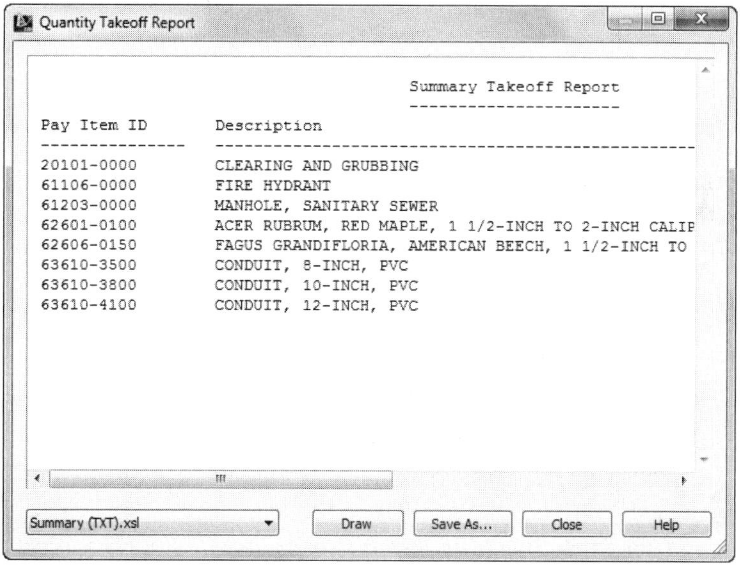

FIGURE 19.23
Quantity Takeoff Report in the Summary TXT format

At this point, you can export this data out as a text file, but for the purposes of this exercise, you'll simply insert it into the drawing.

5. Click the Draw button at bottom of the dialog, and Civil 3D will prompt you to pick a point in the drawing for the report table.
6. Click near some clear space, and you'll be returned to the Quantity Takeoff Report dialog.

7. Click Close to dismiss this dialog, and then click Close again to dismiss the Compute Quantity Takeoff dialog.
8. Zoom in where you picked in step 6, and you should see something like Figure 19.24.

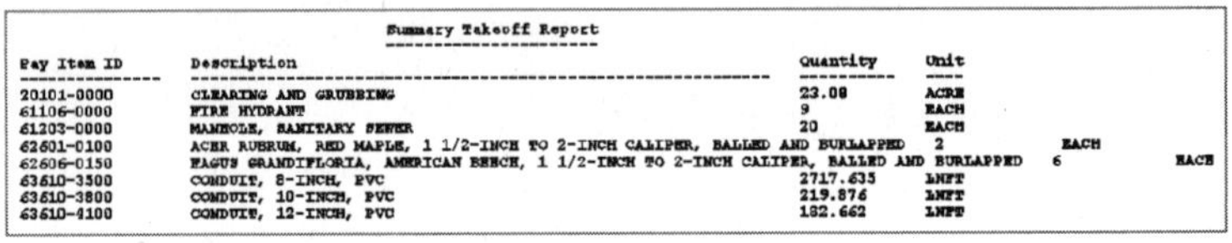

Summary Takeoff Report

Pay Item ID	Description	Quantity	Unit
20101-0000	CLEARING AND GRUBBING	23.08	ACRE
61106-0000	FIRE HYDRANT	9	EACH
61203-0000	MANHOLE, SANITARY SEWER	20	EACH
62601-0100	ACER RUBRUM, RED MAPLE, 1 1/2-INCH TO 2-INCH CALIPER, BALLED AND BURLAPPED	2	EACH
62606-0150	FAGUS GRANDIFLORIA, AMERICAN BEECH, 1 1/2-INCH TO 2-INCH CALIPER, BALLED AND BURLAPPED	6	EACH
63610-3500	CONDUIT, 8-INCH, PVC	2717.635	LNFT
63610-3800	CONDUIT, 10-INCH, PVC	219.876	LNFT
63610-4100	CONDUIT, 12-INCH, PVC	182.662	LNFT

FIGURE 19.24
Summary Takeoff data inserted into the drawing

When this exercise is complete, you may close the drawing. A saved copy of this drawing is available from the book's web page with the filename QTOReporting_FINISHED.dwg or QTOReporting_METRIC_FINISHED.dwg.

That's it! The hard work in preparing QTO data is in assigning the pay items. The reports can be saved to HTML, TXT, or XLS format for use in almost any analysis program.

The Bottom Line

Open and review a list of pay items along with their categorization. The pay item list is the cornerstone of quantity takeoffs. You should download and review your pay item list and compare it against the current reviewing agency list regularly to avoid any missed items.

Master It Using the template of your choice, open the `Getting Started.csv` (or `Getting Started_Metric.csv`) pay item file and add the 12-, 18-, and 24-Inch Pipe Culvert (or 300 mm, 450 mm, and 600 mm Pipe Culvert) pay items to your Favorites list in the QTO Manager.

Assign pay items to AutoCAD objects, pipe networks, and corridors. The majority of the work in preparing quantity takeoffs is in assigning pay items accurately. By using the linework, blocks, and Civil 3D objects in your drawing as part of the process, you reduce the effort involved in generating accurate quantities.

Master It Open the `MasteringQTO.dwg` or `MasteringQTO_Metric.dwg` file and assign the CLEARING AND GRUBBING pay item to the polyline that was originally extracted from the border of the corridor. Change the hatch to have a transparency of 80.

Use QTO tools to review what items have been tagged for analysis. By using the built-in highlighting tools to verify pay item assignments, you can avoid costly errors when running your QTO reports.

Master It Verify that the area in the previous exercise has been assigned a pay item.

Generate QTO output to a variety of formats for review or analysis. The Quantity Takeoff Reports give you a quick understanding of what items have been tagged in the drawing, and they can generate text in the drawing or external reports for uses in other applications.

Master It Display the amount of Type C Broken markings in a Quantity Takeoff Report Summary (TXT) using the `MasteringQTOReporting.dwg` or `MasteringQTOReporting_Metric.dwg` file.

Chapter 20

Label Styles

Styles control the display properties of AutoCAD® Civil 3D® objects and labels. The creation of proper styles and settings can make or break your experience with Civil 3D. Understanding and applying styles correctly can mean the difference between getting a job out in several hours, and fighting with your project drawing for days.

This chapter is organized by style type. First, read the chapter to be introduced to the style concepts in a logical manner. Later, when you use this chapter as a reference to build styles, you will likely jump around to the examples that meet your needs.

In this chapter, you will learn to:

- Override individual labels with other styles
- Create a new label set for alignments
- Create and use expressions
- Apply a standard label set to profiles

Label Styles

The best design in the universe is not worth anything unless it is labeled properly. Civil 3D labels are smart objects that are dynamically linked to the object they are labeling. Civil 3D labeling is customizable to fit the needs of your design and local requirements.

When talking about labels in Civil 3D, keep in mind that you are not just talking about text. Labels can contain lines and blocks if desired. Label styles control the size, contents, and precision of text. Label styles also control how leaders are applied when the label is dragged away from its initial position. You will even work with some labels that contain no text at all!

General Labels

CERT OBJECTIVE

On the Settings tab, you see a complete list of objects that Civil 3D uses to build its design model. Each of these has special features unique to the object being described, but there are some common features as well. The General collection contains settings and styles that are applied to various objects across the entire product.

The General collection serves as the catchall for styles that apply to multiple objects, and for settings that apply to *no* objects. For instance, the Civil 3D General Note object doesn't really belong with the Surface or Pipe collections. It can be used to relate information about those objects, but because it can also relate to something like "Don't Dig Here!" or a Northing-Easting of an arbitrary location, it falls into the General category.

The Label Styles collection allows Civil 3D users to place general text notes or label single entities while still taking advantage of the flexibility and scaling properties. With the various Label Styles shown in Figure 20.1, you can get some idea of their use.

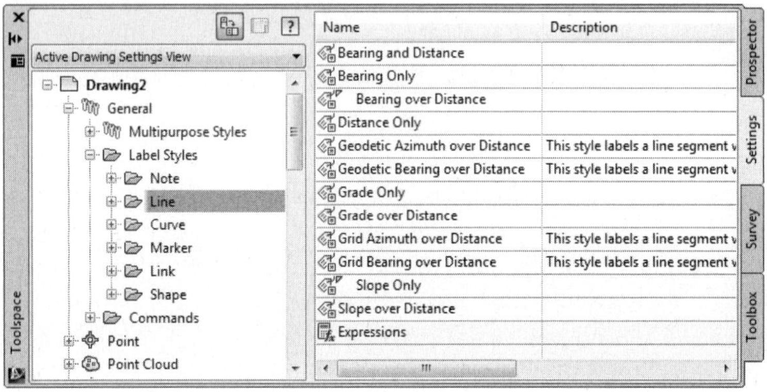

FIGURE 20.1
Line Label Styles

Label styles are a critical part of producing plans with Civil 3D. In this chapter, you'll learn how to build a new basic label and explore some of the common components that appear in every label style throughout the product.

Frequently Seen Tabs

To get into a label style, find the appropriate label type in the Settings tab of Toolspace. Right-click the style you wish to modify (or copy) and select Edit.

All styles, regardless of type, have a few things in common:

Information Tab The Information tab, shown in Figure 20.2, controls the name of the style.

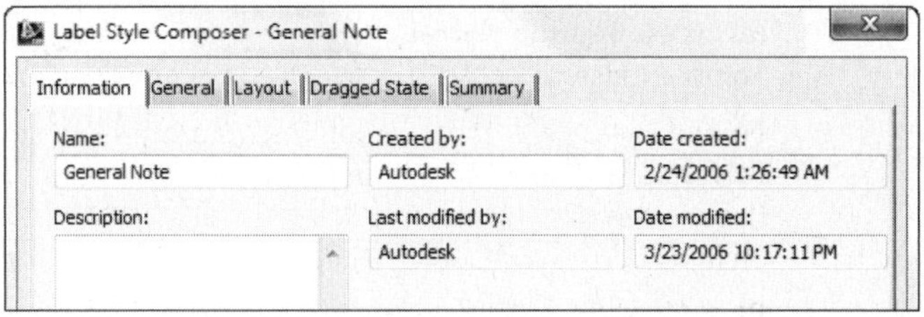

FIGURE 20.2
The Information tab exists for all object and label styles.

If desired, you can create a description. This information will appear as a tooltip as you browse through the Settings tab.

On the right side of the dialog, you will see the name of the style, the date when it was created, and the last person who modified the style. These names are initially pulled from the Windows login information and only the Created By field can be edited.

General Tab The General tab (Figure 20.3) is where the Text Style, Visibility, and Layer options are set for the text item. The Behavior and Plan Readability options control how the text is positioned as it is placed and whether rotation is applied to the object or view it is in. Other options include:

FIGURE 20.3
General tab

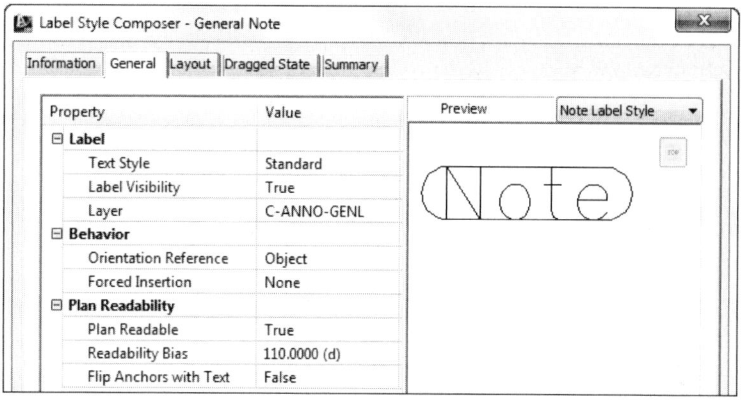

Text Style This option refers to the base AutoCAD text style used in the label. The available text styles come from the base AutoCAD styles.

Layer This option defines the layer on which the components of a label are inserted. The overall label may appear on a completely different layer depending on the layers set up in the Drawing Settings.

Orientation Reference This option controls how text rotation is controlled. Figure 20.4a shows the most common option where the label is aligned with the object. Figure 20.4b shows the text rotated to the view. Even though the view has been rotated, the text still appears parallel with the bottom of the screen. The last, and least used, option for text is the World Coordinate System option, shown in Figure 20.4c. The view is rotated, and the direction of the text rotates with the coordinate system.

Forced Insertion This option makes more sense in other objects and will be explored further. The Forced Insertion feature allows you to dictate the insertion point of a label on the basis of the object being labeled. Figure 20.5 shows the effects of the various options on a bearing and distance label.

Plan Readable When this option is enabled, text maintains the up direction in spite of view rotation. Rotating 100 labels is a tedious, thankless task, and this option handles it with one click.

Readability Bias This option specifies the angle at which readability kicks in. This angle is measured from the 0 degree of the x-axis that is common to AutoCAD angle measurements. When a piece of text goes past the readable bias angle, the text flips direction to maintain vertical orientation, as shown in Figure 20.6. The default Readability angle is 110. A common desired default is 90.1, which would flip any text just after passing a vertical direction.

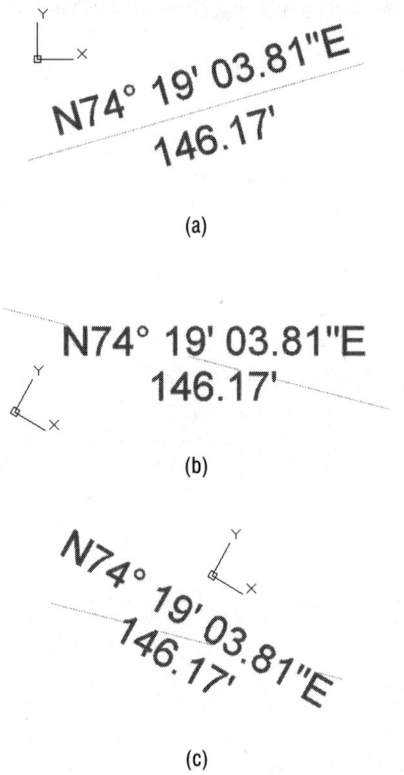

FIGURE 20.4
Orientation reference options set to Object (a), View (b), and World Coordinate System (c)

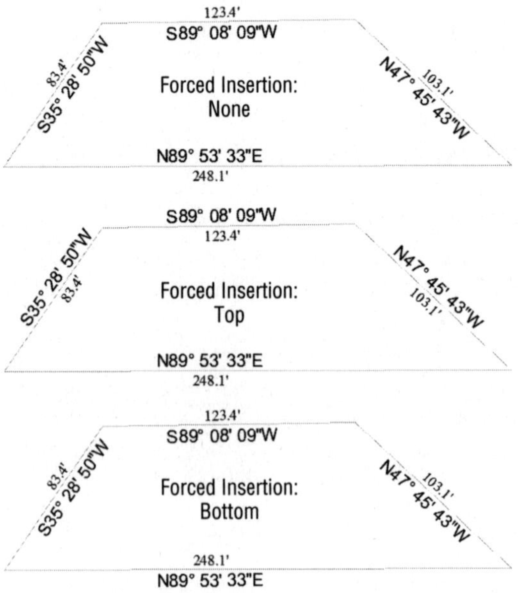

FIGURE 20.5
Forced Insertion options for parcel segments

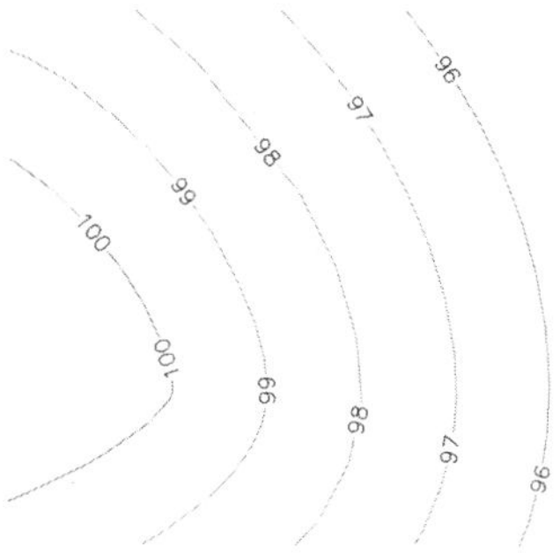

FIGURE 20.6
Plan-readable text shown on contours; note the direction difference between the top grouping and the bottom grouping.

Flip Anchors With Text Most users leave this obscure setting at its default, False, and never give it another thought. Figure 20.7a shows what happens to the text insertion points when this option is set to False and readability kicks in. Notice how the text flips to the bottom of the line it is labeling when this setting is set to True, as shown in Figure 20.7b.

Layout Tab Each label can be made up of several components. A label component can be text, a block, or a line, and the top row of buttons controls the selection, creation, and deletion of these components, as seen in Figure 20.8.

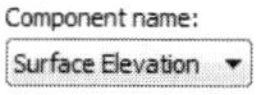

Component Name Choose the component you want to modify from the Component Name drop-down menu. The components are listed in the order in which they were created. Pay attention to which component is active when you make changes to properties on the Layout tab. The changes you make to properties will only apply to the active component.

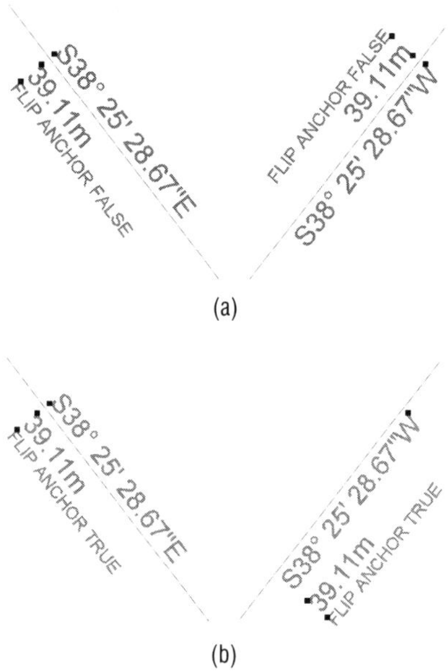

FIGURE 20.7
Flip anchors with text: for the truly persnickety user

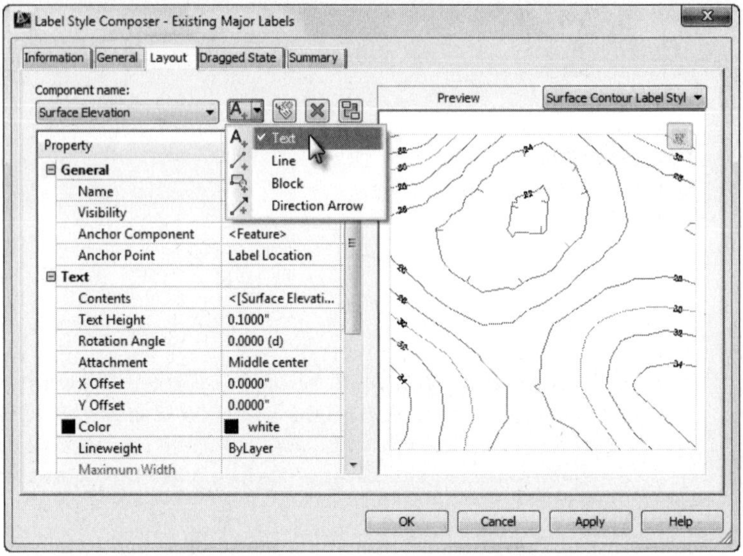

FIGURE 20.8
The Layout tab

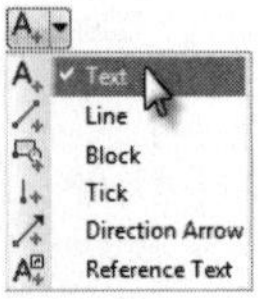

Create Component The Create Component button lets you add new elements to enhance your labels. A component can be either Text, Line, Block, Tick, Direction Arrow, or Reference Text. The options will vary depending on the object you are labeling. Not every option is available in every label.

The ability to label one object while referencing another (reference text) is one of the most powerful labeling features of Civil 3D. This is what allows you to label a spot elevation for both an existing and a proposed surface at the same time, using the same label. Alignments, COGO points, parcels, profiles, surfaces, and survey figures can all be used as reference text.

Copy Component The Copy Component button copies the component currently selected in the Component Name drop-down menu. This will be helpful when you're creating label styles that contain multiple pieces of similar information.

Delete Component The Delete Component button deletes components. Elements that act as the basis for other components can be deleted, but you will receive a warning. Table tag components, found in curve, line, and parcel segment labels, cannot be deleted.

Component Draw Order The Component Draw Order button lets you shuffle components up and down within the label. This feature is especially important when you're using masks or borders as part of the label.

You can work your way down the component properties and adjust them as needed for a label:

Name This option defines the name used in the Component Name drop-down menu and when selecting other components. When you're building complicated labels, a descriptive name goes a long way.

Visibility When set to True, this option means the component shows on screen. Why create a component that can't be seen? There are cool tricks you can do with styles, as you'll see in the section "Pipe Labels" later in this chapter.

Anchor Component and Anchor Point Use these options to control text placement. Anchor Component lets you tell Civil 3D how text (or other label components) relate to the main object you are labeling. See Figure 20.9 and Figure 20.11 for a graphical explanation of how these options relate to each other. The circles represent anchor points and the squares represent attachment.

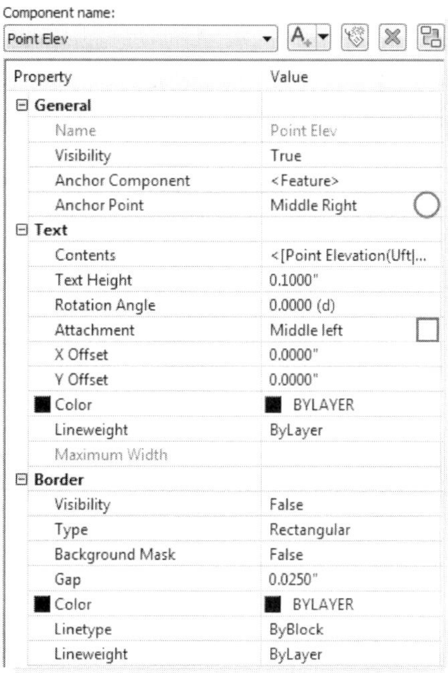

FIGURE 20.9
Closer look at the Layout tab options. The circle and square is shown for illustration to indicate Anchor Point and Attachment, respectively. See Figure 20.11 to see how these relate to each other.

Text Height This option determines the plotted height of the label. Text placed by Civil 3D is always annotative. Figure 20.10 shows two viewports showing some COGO points along a road. Even when the viewport scales differ, the text is the same size.

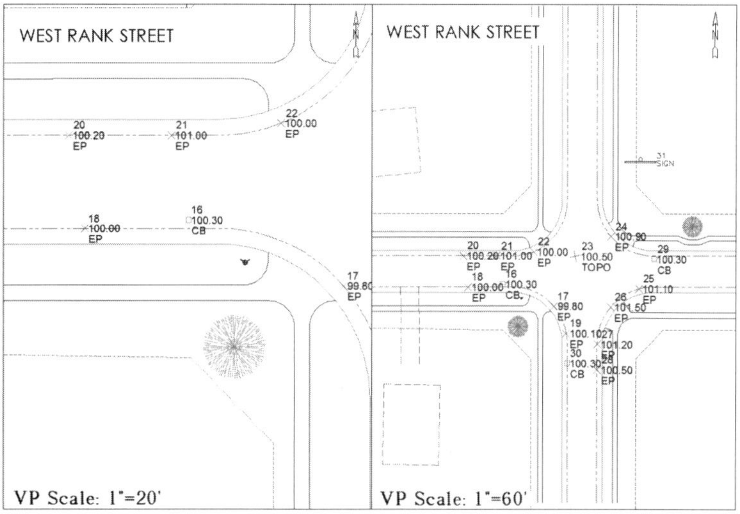

FIGURE 20.10
Annotative text shown at multiple scales. Civil 3D text is always annotative.

Rotation Angle, X Offset, and Y Offset These options give you the ability to refine the placement of the component by rotating or displacing the text in an *x* or *y* direction. Set your text as close as possible using the anchors and attachments, and use the offsets as additional spacing.

Attachment This option determines which of the nine points on the label components bounding box are attached to the anchor point. See Figure 20.11 for an illustration.

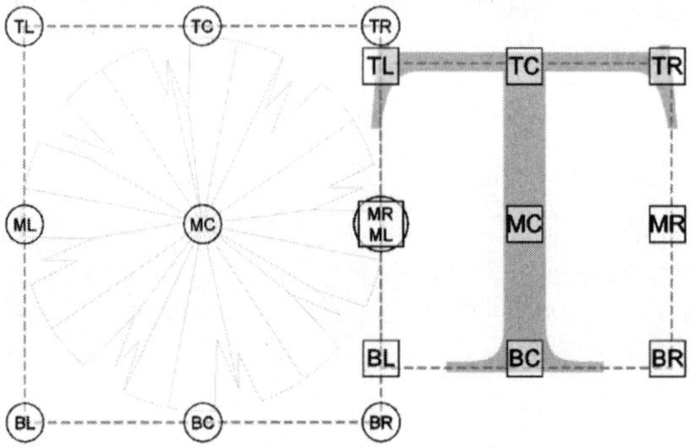

FIGURE 20.11 Schematic showing the relationship between anchor points (circles) and attachments (squares)

Color and Lineweight These options allow you to force the color and lineweight if desired.

Maximum Width Some labels can get rather lengthy. Instead of letting the label component continue indefinitely across the page, you can use Maximum Width to force word wrap after a specified plotted length. The default setting of 0.00 will not force word wrap.

The final piece of the component puzzle is the Border option. These options are as follows:

Visibility Use this option to turn the border on and off for the component. Remember that component borders shrink to the individual component; if you're using multiple components in a label, they all have their own borders.

Type This option allows you to select a rectangle, a rounded rectangle (slot), or a circle border. Figure 20.12 shows examples of the three types of borders.

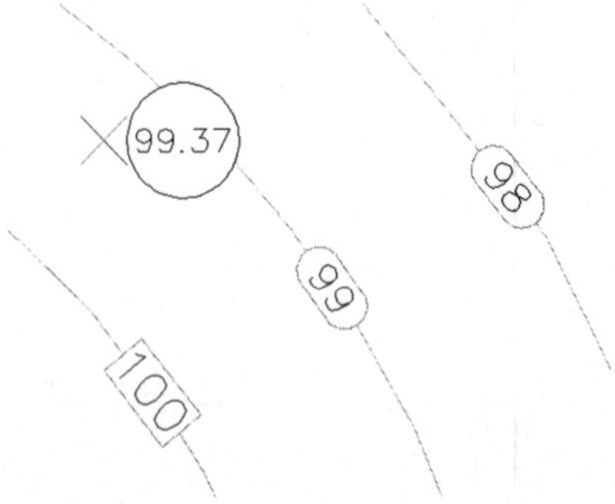

FIGURE 20.12 Border types shown on various surface label styles

Background Mask This option lets you determine whether linework and text behind this component are masked. This option can be handy for construction notes in place of the usual wipeout tools. The surface labels in Figure 20.12 show the background mask in action.

Gap This option determines the offset from the component bounding box to the outer points on the border. Setting this to half of the text size usually creates a visually pleasing border.

Linetype and Lineweight These options give you control of the border lines.

Dragged State When a label is dragged in Civil 3D, it typically creates a leader, and text is rearranged. The settings that control these two actions appear on this tab (see Figure 20.13).

FIGURE 20.13
The Dragged State tab

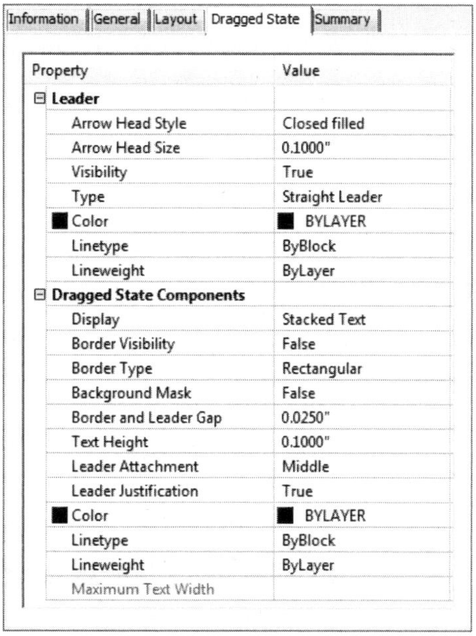

Here are some of the unique options:

Arrow Head Style and Size These options control the tip of the leader. Note that Arrow Head Size also controls the tail size leading to the text object.

Type This option specifies the leader type. Options are Straight Leader and Spline Leader. Civil 3D can only have one leader coming from a label.

Display In this option you can choose Stacked Text or As Composed. The Stacked Text option will remove any blocks, ticks, lines, or borders that are components of the label and will realign the text based on the direction you drag. The As Composed option leaves blocks, ticks, lines, or borders intact and adds a leader.

Figure 20.14a shows an alignment label as it was originally placed. Figure 20.14b shows the same label in a dragged state with the Stacked Text option set. Figure 20.14c shows the label in a dragged state with the As Composed option set.

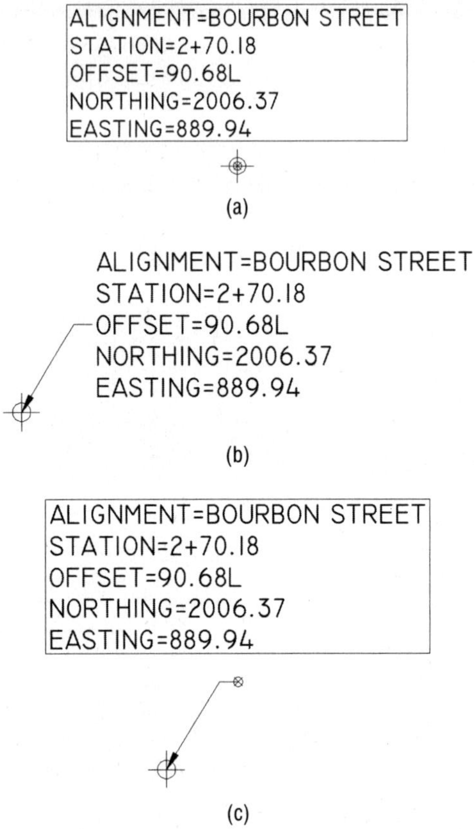

FIGURE 20.14
An alignment label as originally placed (a); dragged state, Stacked Text (b); and dragged state, As Composed (c)

Summary Tab The Summary tab is exactly what it sounds like—that is, a summary of all other settings that exist in the style. The information from other tabs in list form as well as their override status is shown in Figure 20.15.

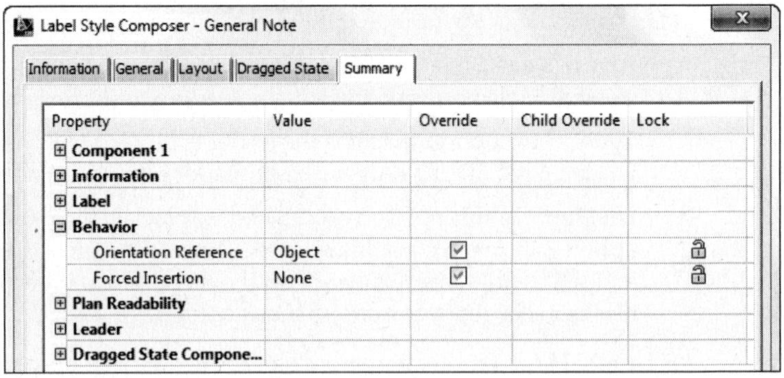

FIGURE 20.15
Summary tab of a label style

CERT OBJECTIVE

Like other settings in Civil 3D, there is a hierarchy that helps the user and the software decide which styles take precedence over other styles. There are also items that can be changed at a drawing-wide level and pushed out to object-level labels. Use this ability with caution, as most labels will use their own layers and behave differently depending on the object it is labeling.

In the first exercise, you will set all labels to use the same initial text style:

1. Open the drawing `Label Basics.dwg` (`Label Basics_METRIC.dwg`), which you can download from this book's web page at `www.sybex.com/go/masteringcivil3d2013`.
2. From the Settings tab of Toolspace, right-click on the name of the drawing and select Edit Label Style Defaults, as shown in Figure 20.16.

FIGURE 20.16
Accessing the global text settings

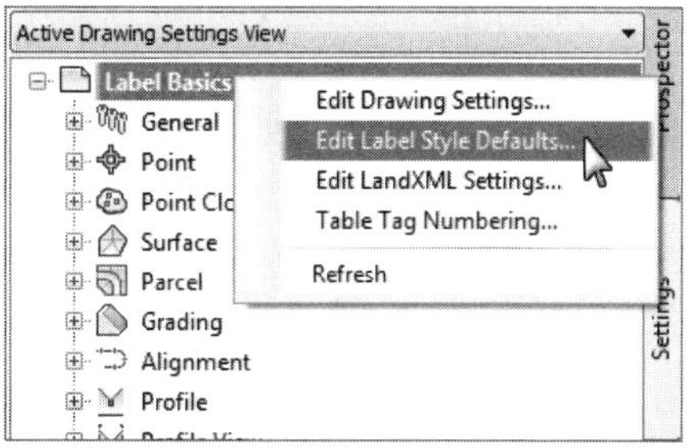

3. Expand the Label category and change Text Style to Arial.
4. Click the arrow in the Child Override column to force all label styles in this drawing to use the same text, as shown in Figure 20.17.

FIGURE 20.17
Label placement options as shown at the drawing level

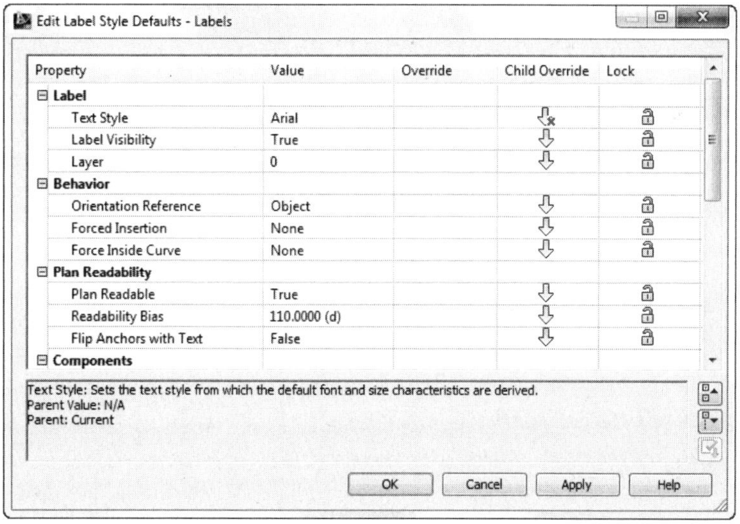

5. Do not make any more changes to this dialog, but examine the various options and settings, and click OK.
6. Save the drawing for use in the next exercise.

Getting to Know the Text Component Editor

Few dialogs in Civil 3D inspire expletives the way the Text Component Editor can. The interface is very logical and will do exactly what you tell it to do—not necessarily what you *want* it to do. Hopefully, with some explanation you will see the reasoning behind its behavior and master it!

To enter the Text Component Editor, from the Layout tab of any text style, click the ellipsis that pops up when you click in the Text Contents area.

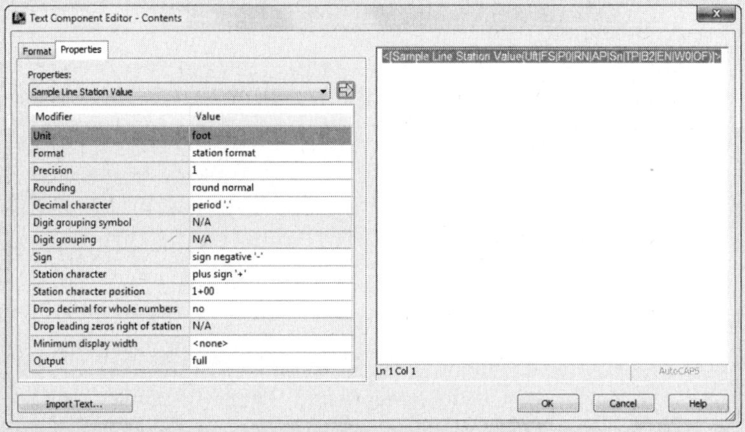

Within the Text Component Editor are two main areas:

- ◆ The left side is where you select what aspect of the Civil 3D object you want to pull into your label.
- ◆ The right side shows what is already in the label.

To modify an existing label, highlight the chunk of text on the right side of the dialog. In Civil 3D, "smart" text will always highlight as a unit. Each of the cryptic codes in the label represents a setting for units, precision, or other ways the text can display. What appears on the left is a decoded list of what is currently part of your label. After making changes, don't forget to click the arrow to update the information on the right.

Before adding new text to a label, make sure you don't have any existing text highlighted on the right, to avoid overwriting it.

The Properties list shows everything that can be labeled about the item you are working with. The length of the list varies depending on the object. Here you see the available properties for a contour (left) and a pipe-network structure (right):

Once you pick what you'd like to label, the units, precision, and any other special rounding or formatting you would like to see.

When everything is set, click the arrow next to the Properties list.

To further fine-tune the text, a Format tab sits inconspicuously behind the Properties list. Use this to add special symbols, or to override the color and text style.

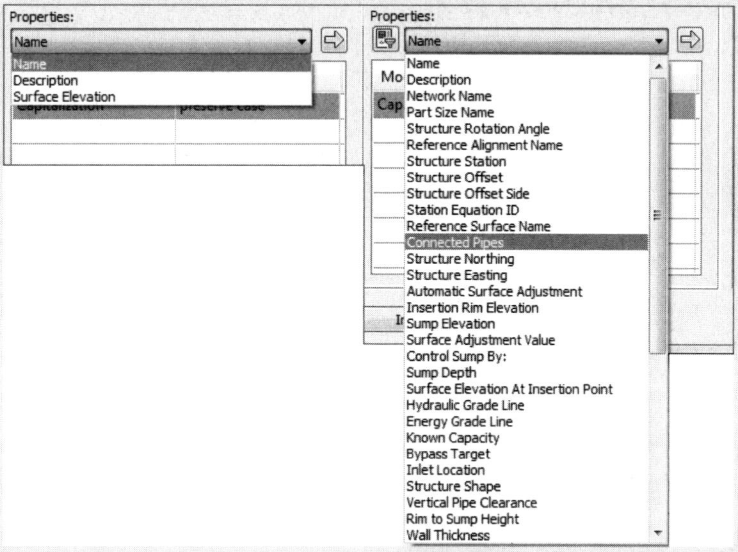

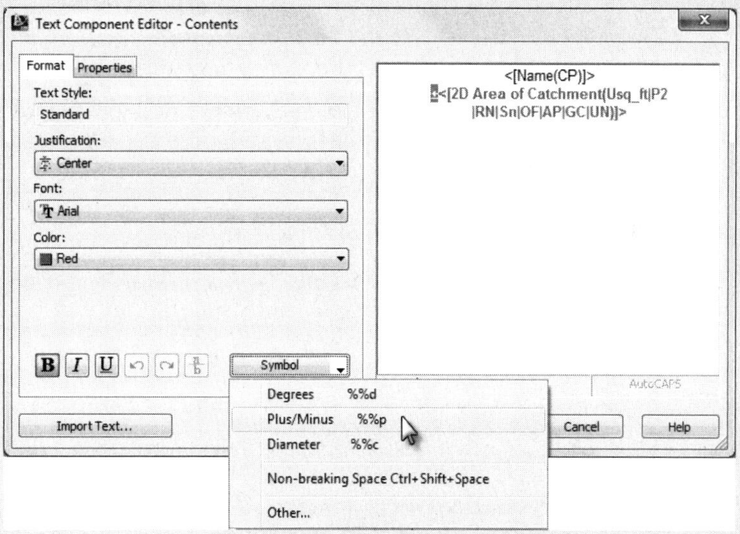

Inevitably, you will forget the arrow that adds or updates the text. You may even click it twice and end up with duplicates! Now that you've been given the heads-up you'll be able to laugh it off, knowing that it happens to even the most seasoned users.

General Note Labels

General Note labels are versatile, non-object-specific labels that can be placed anywhere in the drawing. There are several advantages to using these instead of base-AutoCAD MTEXT. Notes will leader off and scale the same as the rest of your Civil 3D text, and best of all, they can contain reference text.

In the following example, you will create an alternate parcel label that contains reference text. You do not need to have completed the previous exercise to proceed.

1. Continue working in `Label Basics.dwg` (`Label Basics_METRIC.dwg`), which you can download from this book's web page.
2. From the Settings tab of Toolspace, choose General ➢ Label Styles ➢ Note.
3. Right-click on Note and select New (as shown in Figure 20.18).

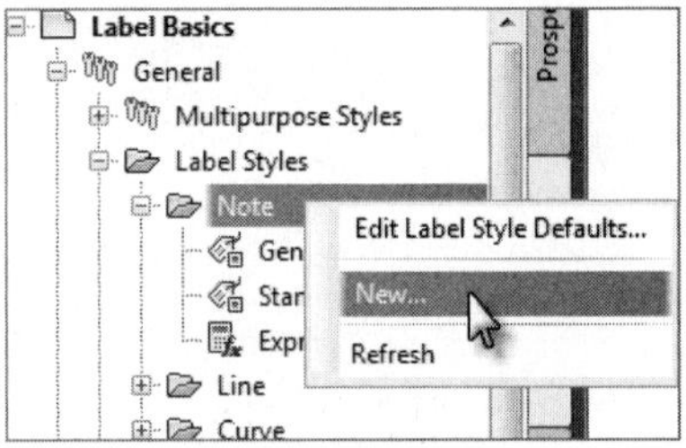

FIGURE 20.18 Creating your first new style from the Settings tab

4. On the Information tab, name the style **Easement Parcel Text**.
5. On the General tab, set Layer to C-PROP-TEXT, and set Orientation Reference to View.
6. On the Layout tab, click the field next to Contents under the Text category.

   Clicking this will cause an ellipsis button to appear.
7. Click the ellipsis button to enter the Text Component Editor.
8. On the right side of the Text Component Editor, delete the existing text and replace it with **Drainage Easement**, as shown in Figure 20.19. Click OK.
9. Back in the Layout tab, set Border Visibility to False.
10. Click the flyout next to Create Text Component and select Reference Text.

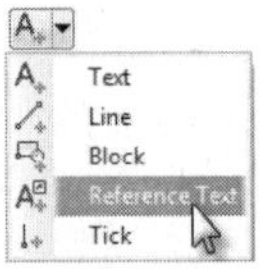

    You will be prompted to select the type of reference text, as shown in Figure 20.20.
11. Select Parcel and click OK.
12. Rename the Reference Text.1 to **Parcel Area** and make the following changes:
    - **A.** Set Anchor Component to Text.
    - **B.** Set Anchor Point to Bottom Center.
    - **C.** Set Attachment to Top Center.
    - **D.** Click the Contents Label Text field ellipsis and enter the Text Component Editor.

FIGURE 20.19
Entering the Text Component Editor for basic text

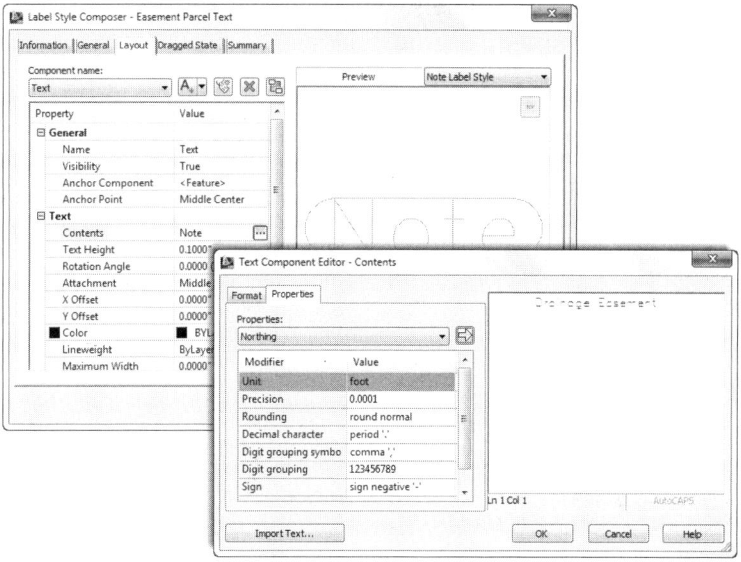

FIGURE 20.20
Picking the reference type

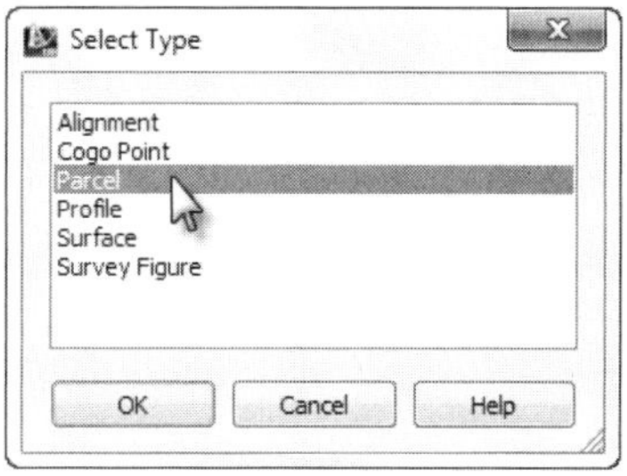

13. Remove the existing label text.
14. On the left side of the Text Component Editor, do the following:
    - **A.** Set the Properties drop-down to Parcel Area.
    - **B.** Set Unit to Acre (Hectares).
    - **C.** Set Precision to **0.01**.

15. When you have set the properties, click the arrow icon to place the text in the right side of the editor.
16. Click to put your cursor after the previously inserted text. Add **ACRES** (or **HECTARES**) after the coding.

    The Text Component Editor will resemble Figure 20.21.
17. Click OK.

    You will still have question marks in the preview, but this is completely expected.

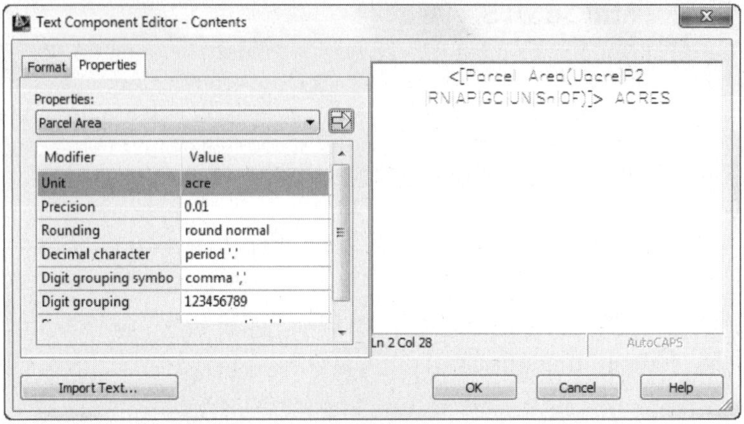

FIGURE 20.21
Adding "smart" text to the Text Component Editor

18. Click OK to complete the command.
19. On the Annotate ribbon tab ➢ Labels & Tables panel, click Add Labels.
20. With Feature and Label Type set to Note, change Note Label Style to Easement Parcel Text, and click Add.
21. Click anywhere in the example parcel.
22. At the prompt `Select parcel for label style component Parcel Area:`, click the label on Property : 1.
23. Press Esc on your keyboard to finish placing labels.

    Your completed and placed label should resemble Figure 20.22.

24. Save and close the drawing.

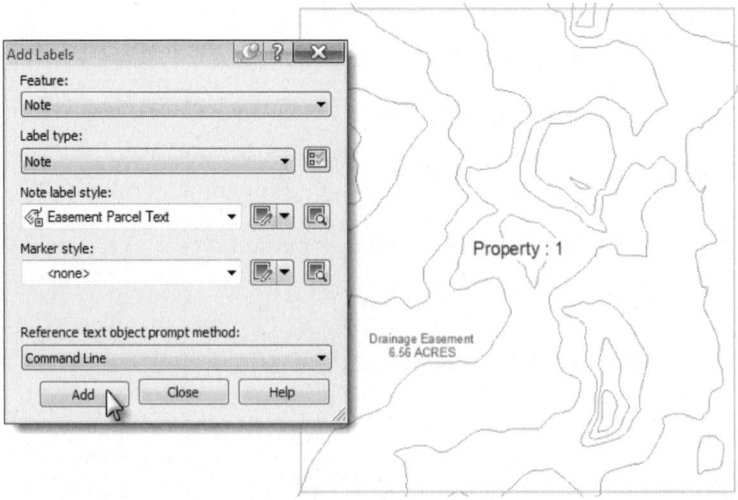

FIGURE 20.22
Your first label! Referencing a parcel area

Point Label Styles

CERT OBJECTIVE

If you have used other software packages for surveying work, odds are that you controlled point label text with layers, but Civil 3D is different. In Civil 3D, you must control what information is showing next to a point by swapping the label style applied to a group of points.

In the following exercise, you will create a new point label style. Your first point label style will show only Point Number and Description, so you will need to delete the default Elevation component.

1. Open the drawing `Point Labels.dwg` (`Point Labels_METRIC.dwg`), which you can download from this book's web page.
2. From the Settings tab of Toolspace, choose Point ➢ Label Styles.
3. Right-click Label Styles and select New.
4. On the Information tab, name the style **Point Number & Description**.
5. On the General tab, set the layer to V-NODE-TEXT, and leave all other General tab options at their defaults.
6. On the Layout tab, do the following:
    - A. Set the active component to Point Number.
    - B. Change the anchor component to <Feature>.
    - C. Set Anchor Point to Middle Right.
    - D. Set Attachment to Middle Left.
    - E. Change the Active component to Point Elevation.
    - F. Click the red X to delete this component.

      You will receive a warning that reads "This label component is used an as anchor in this style or in a child style. Do you want to delete it?"
    - G. Click Yes.
    - H. Change the active component to Point Description.
    - I. Change the anchor component to Point Number.
    - J. Change the anchor point to Middle Right.
    - K. Change the attachment to Middle Left.
    - L. Change the X Offset to **0.05**" (**1** mm).
7. Click OK to complete the label style.
8. On the Prospector tab of Toolspace, locate the point group named TOPO; right-click on TOPO, and select Properties.
9. Set the Point Label style to **Point Number & Description**.

   All the points in the group will change to resemble Figure 20.23.

FIGURE 20.23
Completed point label style

In the previous exercise, you created a simple new label style and made some modifications to the default components. In the following exercise, you will remove all the default components and add Northing and Easting values to the label using the Text Component Editor.

1. Continue working in the drawing `Point Labels.dwg` (`Point Labels_METRIC.dwg`).
2. From the Settings tab of Toolspace, choose Point ➢ Label Styles.
3. Right-click Label Styles, and select New.
4. On the Information tab, name the style **Northing & Easting**.
5. On the General tab, set the layer to V-NODE-TEXT, and leave all other General tab options at their defaults.
6. On the Layout tab, do the following:
    - **A.** Click the red X until all three default components are gone.
    - **B.** When you are asked about deleting the anchor, click Yes.
    - **C.** Click the button to create a new text component, and rename the component to **N-E**.
    - **D.** Set the anchor point to Bottom Center.
    - **E.** Set the attachment to Top Center.
    - **F.** Click the Label Text field and enter the Text Component Editor by clicking the ellipsis.
    - **G.** Highlight the default label text and delete it by pressing the Delete key on your keyboard.
    - **H.** From the Properties list, select Northing.
    - **I.** Set the Precision to **0.01** (two decimal places).
    - **J.** Click the arrow to place the text to the right.
    - **K.** After the label text, place an **N** as a static label after the Northing value. Press ↵ to move to the next line.
    - **L.** From the Properties list, select Easting, and set the Precision to **0.01** (two decimal places).
    - **M.** Click the arrow to place the text to the right.
    - **N.** After the label text, place an **E** as a static label after the Easting value.
    - **O.** Click OK to close the text editor and click OK again to close the Label Style Composer.

      The Text Component Editor will now resemble Figure 20.24.

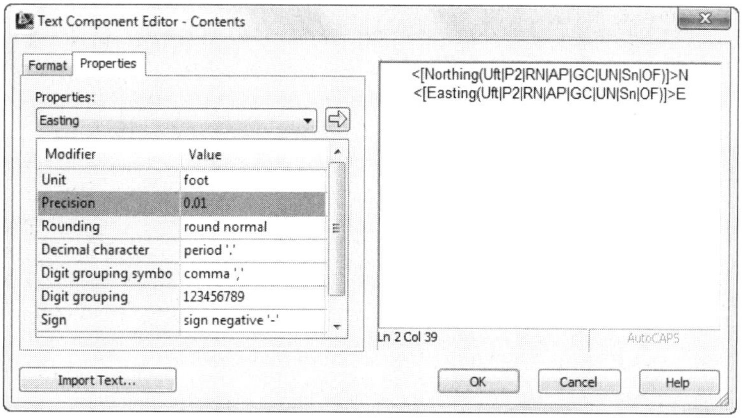

FIGURE 20.24
The Northing & Easting label in progress

7. On the Prospector tab of Toolspace, locate the point group named Group2; right-click it and select Properties.
8. Set the Point Label style to **Northing & Easting**, and then click OK.

The labels will now resemble Figure 20.25.

FIGURE 20.25
Northing & Easting in the completed exercise

SANITY-SAVING SETTINGS

In your Civil 3D template, you will have your AutoCAD styles, linetypes, layers, Civil 3D styles, and a plethora of helpful goodies that make doing your job easier. There are a few drawing-specific AutoCAD variables you may not have thought of that will improve your relationship with Civil 3D:

MSLTSCALE This variable stands for modelspace linetype scale. We strongly recommend that you have this set to 1 in your template. This setting makes linetypes react to your annotation scale. All your other Civil 3D objects are doing it, so having your linetypes follow suit will help! A general rule of thumb is that MSLTSCALE, PSLTSCALE, and LTSCALE should be set to 1.For more information on what these control, check out the Help system.

LAYEREVALCTL Set this to 0 to avoid the annoying pop-up that flags users when there are new layers in a drawing. Civil drafters are constantly using XRefs and inserting blocks, both of which cause the pop-up to occur.

GEOMARKERVISIBILITY This is the control for that weird thing in your drawing that appears any time a coordinate system is set up on your drawing. The geomarker looks like a block, but you can't select it and it keeps rescaling as you zoom. Set the GEOMARKERVISIBILITY variable to 0 in your template and the geomarker won't appear. Be aware that if you set it to 0, the option for Locate Using Geographic Data will not be available in XRef and block insert commands.

AUNITS This variable defines the angular units for the base AutoCAD part of the world. Keep this set at 0 (decimal degrees) to help differentiate base AutoCAD angular entry from Civil 3D angular entry.

Line and Curve Labels

Many Civil 3D objects can have a bearing and distance label added to them. Anything from plain lines and polylines, to parcels and alignment tangent segments, can use nearly identical label types.

The examples in the following exercises will use parcels for labeling, but the tools can be applied to all other types of line labels.

Single Segment Labels

In the following exercise, you will create a new line label style that uses default components. You will remove the direction arrow and change the display precision of the direction component.

1. Open the drawing file `Line and Curve Labels.dwg` (`Line and Curve Labels_METRIC.dwg`), which you can download from this book's web page.
2. From the Settings tab of Toolspace, choose Parcel ➢ Label Styles ➢ Line; right-click Line and select New.
3. On the Information tab, name the style **Parcel Segment**.
4. On the General tab, set the layer to C-PROP-LINE-TEXT.

   Leave all other General tab options at their defaults.
5. On the Layout tab, do the following:
   - A. Change the active component to Direction Arrow, and click the red X to delete this component.
   - B. Change the active component to Distance. Enter the Text Component Editor by clicking in the Text Contents area and clicking the ellipsis.
6. Highlight the Segment Length contents on the right by clicking on the text.

   All of the text should highlight as a unit.
7. On the left side of the Text Component Editor, change the Precision value to **0.01**.
8. Click the arrow to update the text.

9. Click OK to dismiss the Text Component Editor, and click OK to complete the style.
10. Add labels to the parcels by selecting the Annotate tab ➢ Labels & Tables panel and doing the following:
    - **A.** Click Add Labels and set Feature to Parcel.
    - **B.** Set the label type to Single Segment.
    - **C.** Set Line Label Style to Parcel Segment.
    - **D.** Click Add and select several straight segments for labeling.

The completed label will resemble Figure 20.26.

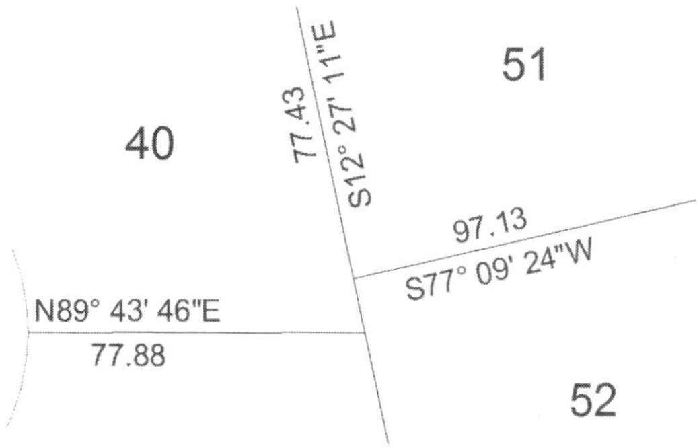

FIGURE 20.26
Your new bearing and distance label style in action on parcel segments

SPANNING SEGMENT LABELS

Looking at the labels you created in the previous exercise, you may notice that they stop at each parcel vertex. If the back property line is the same bearing, most plats show these as a single label with an overall length rather than many shorter lengths at the same bearing. To label these in Civil 3D, a separate label style is needed.

Spanning labels can be used in both line and curve parcel labels. In the following exercise, you will create line labels that span across multiple parcel segments:

1. Continue working in the drawing `Line and Curve Labels.dwg` (`Line and Curve Labels_METRIC.dwg`).
2. From the Settings tab of Toolspace, expand Parcel ➢ Label Styles ➢ Line, right-click Parcel Segments, and select Copy.
3. On the Information tab, name the style **Spanning Segment**.
4. On the Layout tab:
    - **A.** Change the active component name to Table Tag.
    - **B.** Change the Span Outside Segments setting to True.
5. Change to the Bearing component, and do the following:
    - **A.** Change the Span Outside Segments setting to True.

  B. Change the active component to Distance.

  C. Change the Span Outside Segments setting to True.

6. Click OK to complete the style.

7. Add labels to the parcels using a similar technique you used in the previous exercise. This time use Spanning Segment as the line label style.

Flip Label

When you initially place a spanning label in the drawing, the overall length will not be reported unless the distance portion of the label is on the outside of a row of parcels. You may need to use the Flip Label command to see the spanning option take effect. Your completed label will resemble Figure 20.27.

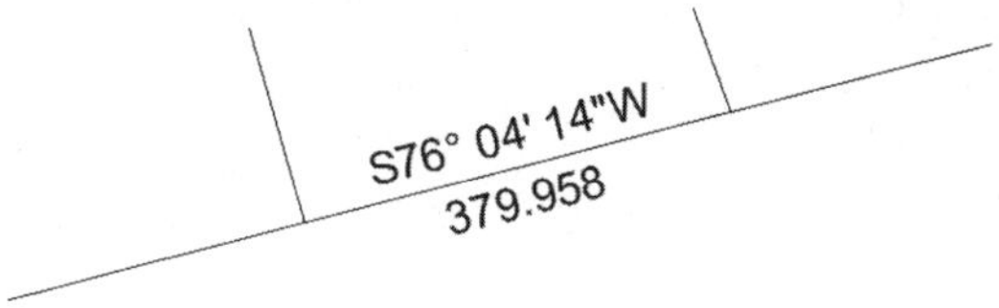

FIGURE 20.27
A spanning label shown on the outside of parcel segments

CURVE LABELS

In base AutoCAD, creating curve labels is a chore. If you want text to align to curved objects, it is no longer usable as traditional MTEXT. Luckily, Civil 3D gives you the ability to add curved text without compromising the usability.

In the following exercise, you will create a curve label style with a delta symbol and text that curves with the parcel segment:

1. Continue working in the drawing file `Line and Curve Labels.dwg` (`Line and Curve Labels_METRIC.dwg`).

2. From the Settings tab of Toolspace, choose Parcel ➤ Label Styles ➤ Curve, right-click on Curve, and select New.

3. On the Information tab, rename the style to **Delta Length & Radius**.

4. On the General tab, set Layer to C-PROP-LINE-TEXT.

5. On the Layout tab, do the following:

  A. Change the active component to Distance And Radius.

  B. Enter the Text Component Editor.

  C. Highlight the Segment Length property on the right and change the precision to **0.01**.

  D. Click the arrow to update the style.

6. Highlight the Segment Radius property and change the precision to **0.01**.

7. Click the arrow to update the text.

8. Delete the comma that appears as static text in the Text Component Editor, and click OK.

9. Click the Create Text Component button and in the resulting window:
   - **A.** Rename the new text component to **Delta**.
   - **B.** Change Attachment to Bottom Center.
   - **C.** Change the Y offset to **0.025"** (**1** mm).
   - **D.** Set Allow Curved Text to True.
10. Enter the Text Component Editor, and remove the default label text.
11. Switch to the Format tab, click the Symbol button, and select Other.

    You should now see the Character Map dialog (Figure 20.28).

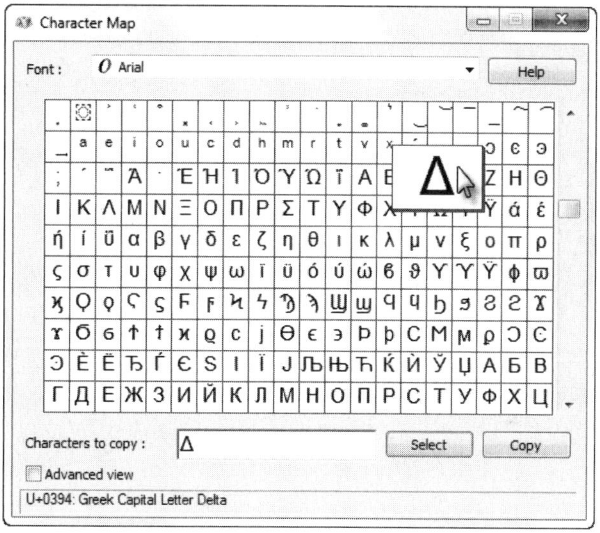

FIGURE 20.28 Browse for special symbols using the Windows Character Map.

12. Browse through the dialog to find the Delta symbol; when you locate it, click on the symbol, click the Select button, and then click Copy. Click the red X to close the Character Map dialog.
13. Back in the Text Component Editor, click on the right side of the dialog, right-click, and select Paste.
14. Press Backspace on your keyboard if the cursor jumps to the next line of text.

    You should now see the delta symbol appear in the Text Component Editor.
15. Type in = as static text after the delta.
16. Switch back to the Properties tab.
17. From the Properties drop-down, select Segment Delta Angle, and do the following:
    - **A.** Set the Format to DD°MM'SS.SS".

B. Set the Precision value to **1 Second.**

C. Click the arrow to place the text in the right side.

D. Click OK to exit the Text Component Editor.

18. Click OK to complete the style.
19. Add labels to the parcels using the same technique you used in the previous exercise.
20. Set the active curve label style to Delta Length & Radius.

Your completed label when applied to the design should resemble Figure 20.29.

FIGURE 20.29
Completed curve labels with delta symbol and curved text

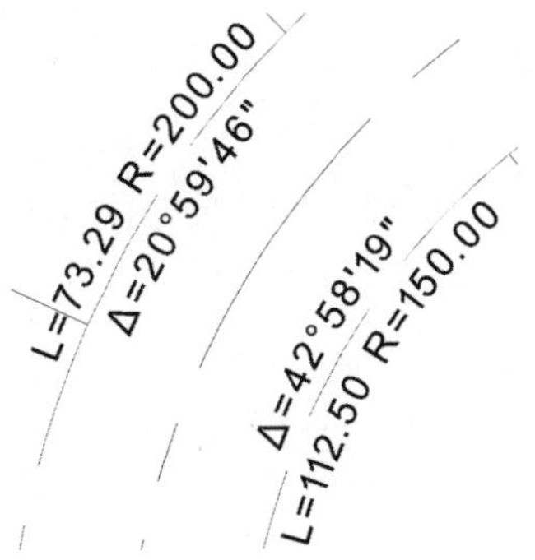

Pipe and Structure Labels

It seems like no two municipalities label their pipes and sewer structures exactly the same way. Luckily, Civil 3D offers lots of flexibility in how you label these items.

Pipe Labels

Pipe labels have two separate label types: Plan Profile and Crossing Section. Both label types have many of the same options, but are used in different view directions.

In the following exercise, you will use a common "trick" in Civil 3D labels where a nonvisible component acts as an anchor to visible objects. In this case, you will use the flow direction arrow to force text to be placed at the ends and middle of the pipe regardless of the pipe length.

1. Open the drawing file `Pipe and Structure Labels.dwg` (`Pipe and Structure Labels_METRIC.dwg`), which you can download from this book's web page.
2. From the Settings tab of Toolspace, choose Pipe ➤ Label Styles ➤ Plan Profile, and then right-click Plan Profile and select New.
3. On the Information tab, name the style **Length Diameter Slope**.
4. On the General tab, set the layer to C-STRM-TEXT.
5. On the Layout tab, delete the existing Pipe Text component.

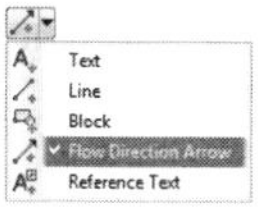

6. Click the Add New Component button and select Flow Direction Arrow; then adjust these settings:
   - A. Set Visibility to False.
   - B. Set Anchor Point to Top Outer Diameter.
   - C. Set a Y offset of **0.1**" (**0.3** mm).
7. Click the Add New Component button and select Text. Change these settings:
   - A. Rename the Text Component to **Length**.
   - B. Set Anchor Component to Flow Direction Arrow.1.
   - C. Set Anchor Point to Start.
   - D. Set Attachment to Bottom Left.
   - E. In the Text Component Editor, remove the default label text.
   - F. From the Properties list, set 2D Length - Center to Center current.
   - G. Set Precision to **1** and Rounding to round up. This causes the pipe length value to round to the next highest whole unit.
   - H. Click the arrow to place the text in the editor.
   - I. Add a foot symbol (or **m** for meters) after the text component.
8. Click OK.

9. With Length as the current component, click Copy Component. Then change these settings:
   - A. Rename the component to **Diameter**.
   - B. Change Anchor Point to Middle.
   - C. Change Attachment to Bottom Center.
   - D. Enter the Text Component Editor and remove all of the text.
   - E. From the Properties list, select Inner Pipe Diameter.
   - F. Set the Precision to **1** and click the arrow to add the text.
   - G. Add the inch symbol (or **mm** for millimeters) after the text component.
10. Click OK.
11. With Diameter as the current component, click Copy Component. Then change these settings:
    - A. Rename the component to **Slope**.
    - B. Set Anchor Point to End.
    - C. Set Attachment to Bottom Right.

D. Enter the Text Component Editor and remove all of the text that is already there. Set Pipe Slope as the current property.

E. Click the arrow to update the text.

12. Click OK.

13. Click OK to complete the style.

14. Add the label to the pipe by doing the following:

A. On the Annotate tab ➢ Labels & Tables panel, click Add Labels.

B. Set the Feature option to Pipe Network and Label Type to Single Part Plan.

C. Set the pipe label style to Length Diameter Slope and the label style to Current.

D. Click Add. Select any pipe in plan view.

Your labeled pipe will resemble Figure 20.30.

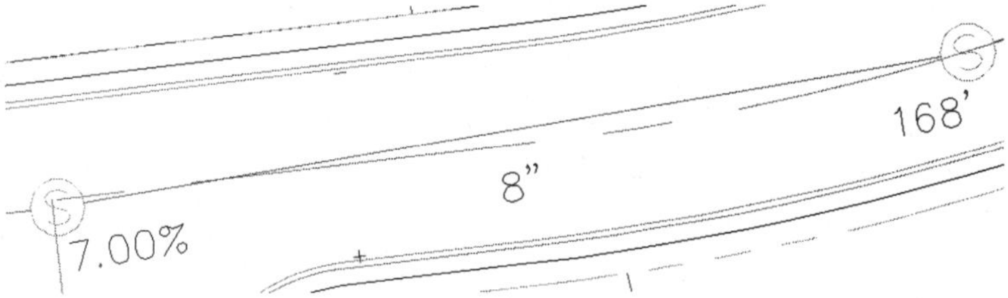

FIGURE 20.30 Labeled pipe using the invisible arrow trick

Structure Labels

Structure labels are nifty because they have a component no other label style has: the Text For Each component. The number of pipes entering and exiting a structure will vary depending on where the structure is in a network, and the Text For Each option accommodates that unique feature.

Real World Scenario

ADDING EXISTING GROUND ELEVATION TO STRUCTURE LABELS

In design situations, it's often desirable to track not only the structure rim elevation at finished grade, but also the elevation at existing ground. This gives the designer an additional tool for optimizing the earthwork balance.

This exercise will lead you through creating a structure label that includes surface-reference text. It assumes you're familiar with Civil 3D label composition in general:

1. Continue working in the file `Pipe and Structure Labels.dwg` (`Pipe and Structure Labels_METRIC.dwg`).

2. Locate the Structure Label Styles branch on the Settings tab of Toolspace.

3. Right-click Label Styles and select New, and then do the following:
    - **A.** On the Information tab, name the label **Structure w Surface**.
    - **B.** On the General tab, set the layer to C-STRM-TEXT.
    - **C.** On the Layout tab is a default text component called Structure Text. Set the Y offset to **-0.25**" (**-1** mm).
    - **D.** Click in the Contents box to bring up the Text Component Editor.
    - **E.** Delete the `<[Description(CP)]>` text string.
    - **F.** Set the Properties drop-down menu to <Name> and click the arrow button.
    - **G.** Press ↵ to move to the next line and type **RIM** with a space after it.
    - **H.** Set the Properties drop-down to `<Insertion Rim Elevation>`, set the Precision to two decimal places and click the arrow button.
4. Click OK to leave the Text Component Editor.
5. In the Label Style Composer, choose Reference Text from the Add Component drop-down menu.
6. In the Select Type dialog, choose Surface, and then click OK.
7. Rename the component from Reference Text.1 to **Existing Ground**.
8. Click in the Contents box, and then click the ellipsis to bring up the Text Component Editor.
9. Delete the Label Text text string, and then use the Properties drop-down menu to add the EG and a space <Surface Elevation> (two decimal places). Click the arrow to add the surface information to the label.
10. Click OK to dismiss the Text Component Editor.
11. In the Label Style Composer, do the following:
    - **A.** Change Anchor Component For Existing Ground to Structure Text.
    - **B.** Change Anchor Point to Bottom Center.
    - **C.** Change Attachment to Top Center.
12. Choose the Text For Each option from the Add Component drop-down menu.
13. In the Select Type dialog, choose Structure All Pipes, and then click OK.
14. Click in the Contents box, and then click the ellipsis to bring up the Text Component Editor.
15. Delete the Label Text text string, and then use the Properties drop-down menu to add the INV and a space <Connected Pipe Invert Elevation> (two decimal places). Click the arrow to add the surface information to the label.
16. Click OK to dismiss the Text Component Editor.
17. In the Label Style Composer, do the following:
    - **A.** Change Anchor Component For Text for Each.1 to Existing Ground.
    - **B.** Change Anchor Point to Bottom Center.
    - **C.** Change Attachment to Top Center.

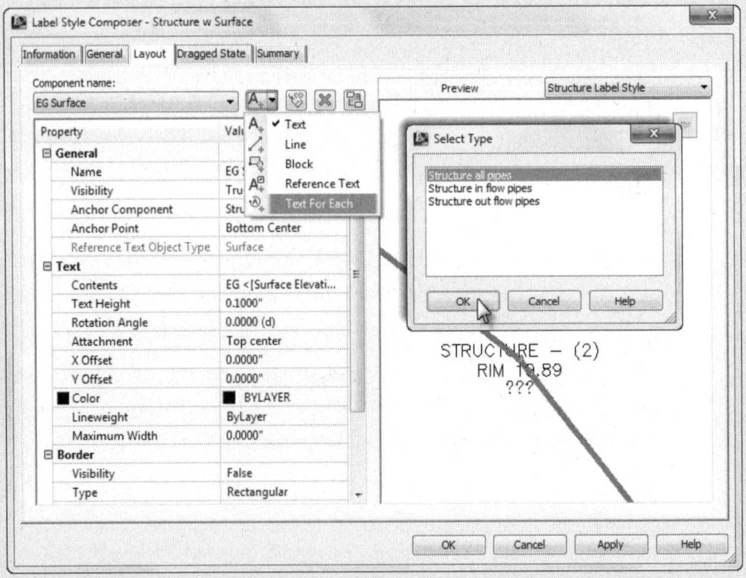

18. Click OK to dismiss the Label Style Composer.

19. Select the Annotate tab ➤ Labels & Tables panel and click Add Labels.

A. Set the Feature option to Pipe Network.

B. Set the label type to Single Part Plan.

C. Set the structure label style to Structure w Surface.

D. Click Add.

20. Click the structure you want to label.

You will immediately see a prompt at the command line that reads `Select surface for label style component Existing Ground Surface:`.

21. Press ↵ to select EG from the surface listing and then click OK.

Your label is now complete.

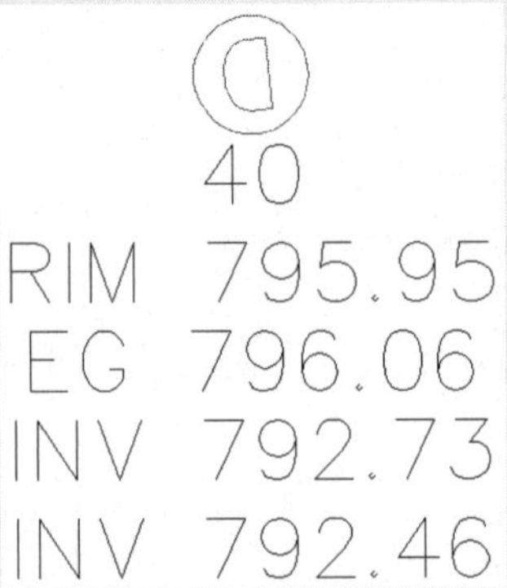

Profile and Alignment Labels

Profile and alignment labels can take on many forms. On an alignment you may want to show labels every 100' (25 m) in addition to PC, PT, and PI information. On a profile, you will want tangent grades, curve information, and grade breaks.

For every element you wish to label, a separate style controls the look and behavior of that label.

Label Sets

A *label set* is a grouping of labels that apply to the same object. In lieu of having one big style that accounts for multiple aspects of an object, the labels are broken out into specific pieces to allow you more control.

Label sets come into play with alignments and design profiles. When you look at an alignment or profile and see labels, you are usually seeing multiple label styles in action.

Consider the alignment shown in Figure 20.31. How many labels are there on this alignment? The geometry points, the major ticks, the minor ticks, superelevation stations, and design speed are all separate labels.

FIGURE 20.31
One alignment, five label styles in play

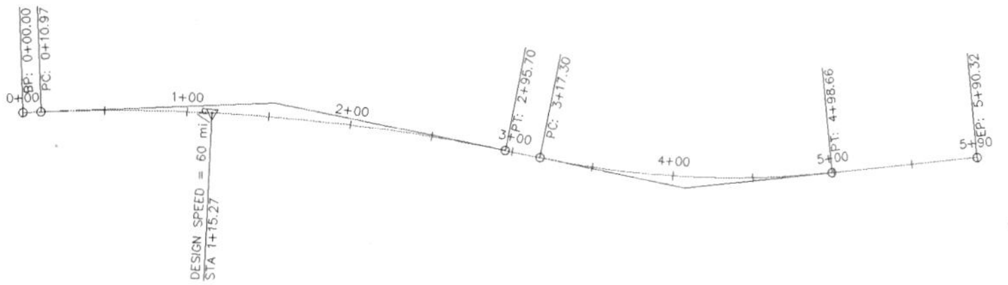

How did those labels get here? When you first created the alignment or profile, one of the options was to specify a label set (as shown in Figure 20.32a). The labels from the set are applied as a batch to the object.

FIGURE 20.32
Specifying an alignment label set upon creation (a) and accessing the label list after creation (b)

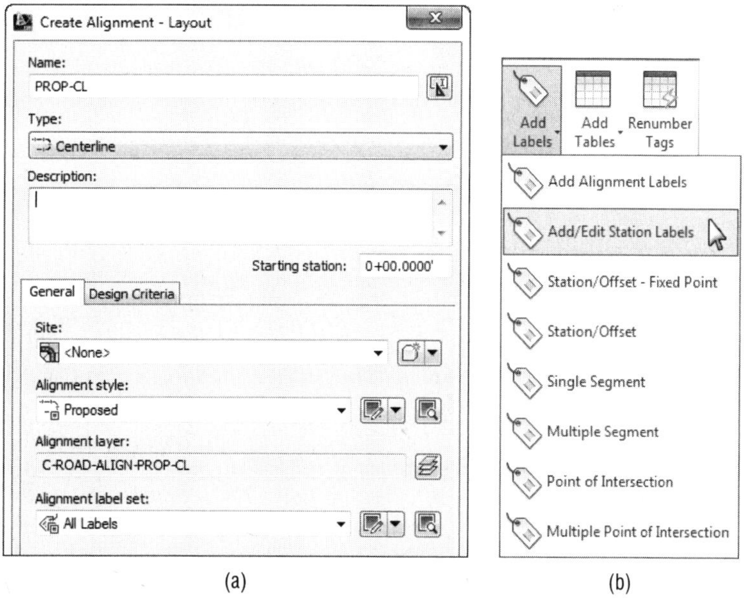

To edit which labels are applied to an alignment or profile, select Add/Edit Station Labels from the context tab (as shown in Figure 20.32b).

Label sets also dictate the placement of the annotation. An alignment label set controls the major and minor station labeling increment. A profile label set has a big job to do in helping you place labels in the correct location with respect to both the design and the profile view. In Figure 20.33, Dim Anchor Opt and Dim Anchor Val control where the label is placed on the object, or in relation to the graph.

FIGURE 20.33 Profile labels and placement

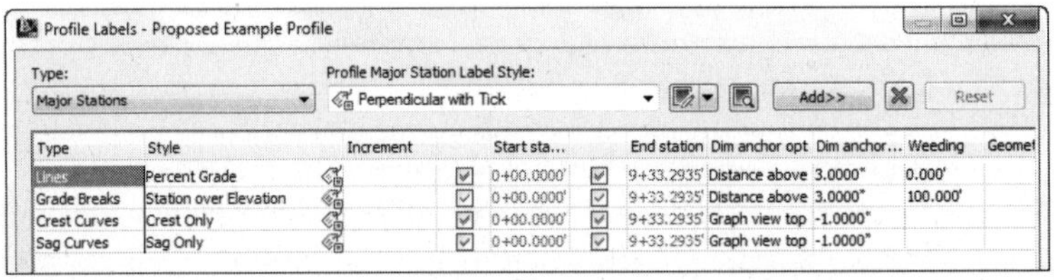

Alignment Labels

You'll create individual label styles over the next couple of exercises and then pull them together with a label set. At the end of this section, you'll apply your new label set to the alignments.

MAJOR STATION

Major station labels typically include a tick mark and a station callout. In this exercise, you'll build a style to show only the station increment and run it parallel to the alignment:

1. Open the `Alignment&ProfileLabels.dwg` (`Alignment&ProfileLabels_METRIC.dwg`) file, which you can download from this book's web page.
2. Switch to the Settings tab, and expand the Alignment ➤ Label Styles ➤ Station ➤ Major Station branch.
3. Right-click the Parallel With Tick style and select Copy.

   The Label Style Composer dialog appears.
4. On the Information tab, type **Index Station** in the Name field.
5. Switch to the Layout tab.
6. Click in the Contents Value field, under the Text property, and then click the ellipsis button to open the Text Component Editor dialog.
7. Click in the preview area, and delete the text that's already there.
8. With station Value as the active property, click in the Output Value field, and click the down arrow to open the drop-down list.

   You will need to scroll down to see the Output option.
9. Select the Left Of Station Character option, as shown in Figure 20.34, and click the insert arrow circled in the figure. (Metric users, the resulting label will be more interesting if you use the Right Of Station Character option instead. Because this alignment is less than 300 m, the left option will give you all 0s!)

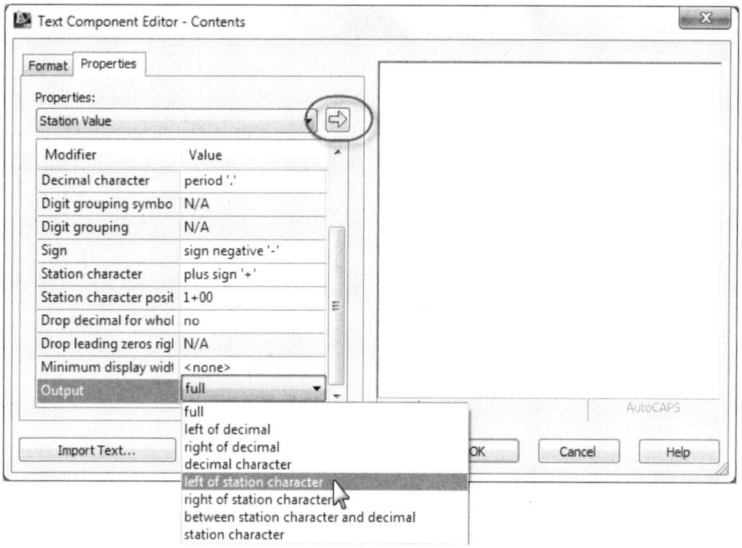

FIGURE 20.34
Modifying the Station Value Output value in the Text Component Editor dialog

10. Click OK to close the Text Component Editor dialog.
11. Click OK to close the Label Style Composer dialog.

The label style now shows in your label styles, but it's not applied to any alignments yet.

GEOMETRY POINTS

Geometry points reflect the PC, PT, and other points along the alignment that define the geometric properties. The existing label style doesn't reflect a plan-readable format, so you'll copy it and make a minor change in this exercise:

1. Continue working in the `Alignment&ProfileLabels.dwg` (`Alignment&ProfileLabels_METRIC.dwg`) file.
2. Expand the Alignment ➢ Label Styles ➢ Station ➢ Geometry Point branch.
3. Right-click the Perpendicular With Tick style, and select Copy to open the Label Style Composer dialog.
4. On the Information tab, change the name to **Perpendicular with Line**.
5. Switch to the General tab.
6. Change the Readability Bias setting to **90**.

   This value will force the labels to flip at a much earlier point.
7. Switch to the Layout tab and make these changes:
   - **A.** Set the Component Name field to the Tick option.
   - **B.** Click the Delete Component button (the red X button).

   - **C.** Click the Create New Line Component button.

D. Change the Line Angle to **90**.

E. Change the component to Geometry Point & Station.

F. Change the Anchor Component to Line.1.

G. Change the Anchor Point to End.

H. Change Rotation Angle to **0**.

8. Click OK.

9. Click OK to close the Label Style Composer dialog.

10. Save the drawing.

This new style flips the plan-readable labels sooner and includes a line with the label. Next, you will put the styles together in a set.

Alignment Label Set

Once you have several labels you wish to use on an alignment, it is time to save them as an alignment label set.

1. Continue working in `Alignment&ProfileLabels.dwg` (`Alignment&ProfileLabels_METRIC.dwg`). On the Settings tab of Toolspace, expand the Alignment ➤ Label Styles ➤ Label Sets branch.
2. Right-click Label Sets, and select New to open the Alignment Label Set dialog.
3. On the Information tab, change the name to **Paving**.
4. Switch to the Labels tab.
5. Set the Type field to the Major Stations option and the Major Station Label Style field to the Index Station Only style that you just created; then click the Add button.
6. Set the Type field to the Minor Stations option and the Minor Station Label Style field to the Tick option; then click the Add button.
7. Set the Type field to Geometry Points and the Geometry Point Label Style field to Perpendicular With Line. Then click the Add button to open the Geometry Points dialog, as shown Figure 20.35.
8. Deselect the Alignment Beginning and Alignment End check boxes as shown, and then click OK to dismiss the dialog.

   Three label types will now be shown in the Alignment Label Set dialog.
9. Click OK to dismiss this dialog.

In the next exercise, you'll apply your label set to the example alignment and then see how an individual label can be changed from the set.

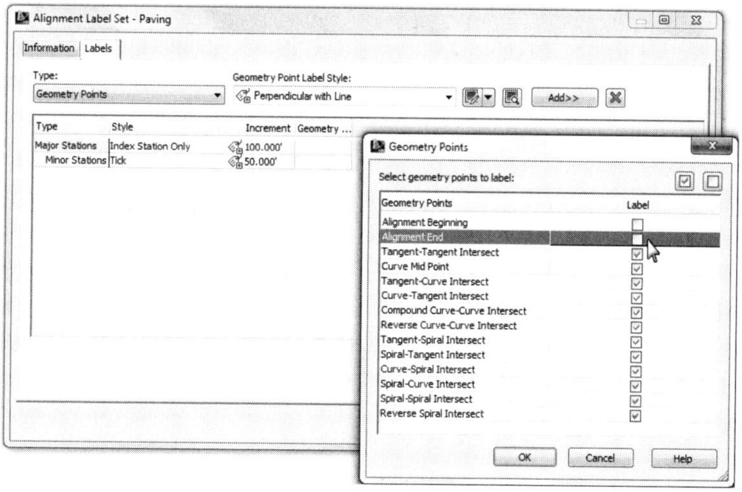

FIGURE 20.35 Deselecting the Alignment Beginning and Alignment End geometry point check boxes

1. Continue working in the `Alignment&ProfileLabels.dwg` (`Alignment&ProfileLabels_METRIC.dwg`) file. Select the Example alignment on screen.
2. Right-click, and select Edit Alignment Labels to display the Alignment Labels dialog.

   This dialog shows which labels are currently applied to the alignment. Initially, it will be empty.
3. Click the Import Label Set button near the bottom of this dialog.

   Any labels appearing in this listing will be replaced by the labels in the set that is imported.
4. In the Select Style Set drop-down list, select the Paving Label Set and click OK.

   The alignment labels list populates with the option you selected.
5. Click OK to dismiss the dialog.
6. When you've finished, press Esc twice to be sure you have exited the command and zoom in on any of the major station labels.
7. Hold down the Ctrl key, and select one of the major station labels.

   Notice that a single label is selected, not the label set group.
8. Now that the single label is selected, drag the grip to leader it off.
9. Right-click and select Label Properties.

   The Label Properties dialog appears, allowing you to pick another label style from the Major Station Label Style drop-down list.
10. Change the Label Style value to Parallel With Tick, and change the Flip Label value to True, as shown in Figure 20.36.

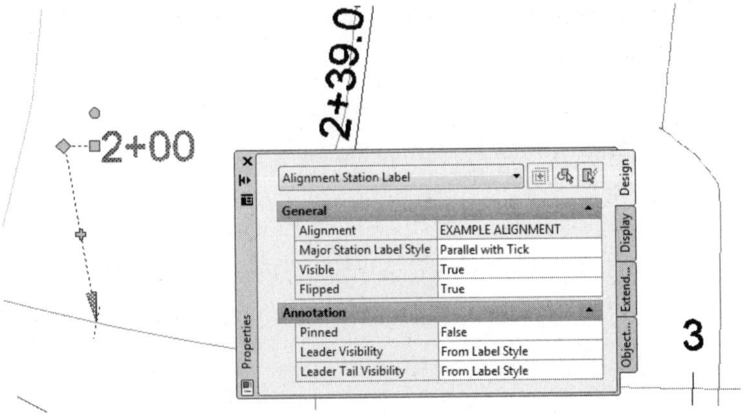

FIGURE 20.36 Modifying a single label's properties through base AutoCAD properties

11. Press Esc to deselect the label item.

If you add labels to an alignment and like the look of the set, use the Save Label Set option. By using alignment label sets, you'll find it easy to standardize the appearance of labeling and stationing across alignments. Building label sets can take some time, but it's an easy, effective way to enforce standards.

Station Offset Labeling

Beyond labeling an alignment's basic stationing and geometry points, you may want to label points of interest in reference to the alignment. Station offset labeling is designed to do just that. In addition to labeling the alignment's properties, you can include references to other object types in your station-offset labels. The objects available for referencing are as follows:

- Other alignments
- COGO points
- Parcels
- Profiles
- Surfaces

In Chapter 11, "Advanced Corridors, Intersections, and Roundabouts," you used special alignment labels that referenced other alignments to make adjusting your design easier. In this exercise, you will make a similar type of label. The label you create in the following exercise finds the intersection of two alignments.

1. Open the `Alignment&ProfileLabels.dwg` (`Alignment&ProfileLabels_METRIC.dwg`) drawing file.
2. On the Settings tab, expand Alignment ➢ Label Styles ➢ Station Offset.
3. Right-click Station And Offset, and select Copy to open the Label Style Composer dialog.
4. On the Information tab, change the name of your new style to **Alignment Intersection**.

5. Switch to the Layout tab. In the Component Name field, delete the Marker component.
6. In the Component Name field, select the Station Offset component.
7. Change the name to **Main Alignment**.
8. In the Contents Value field, click the ellipsis button to bring up the Text Component Editor.
9. Select the text in the preview area and delete it all.
10. Type **Sta.** in the preview area; be sure to leave a space after the period.
11. In the Properties drop-down field, select Station Value, and set the Precision to **0.01**.

12. Click the insert arrow in the Text Component Editor dialog; press the right arrow on your keyboard to move your cursor to the end of the line, and type one space.
13. In the Properties drop-down field, select Alignment Name.
14. Click the insert arrow to add this bit of code to the preview.
15. Click your mouse in the preview area, or press the right arrow or End key. Move to the end of the line, and type a space and an equal sign (=).

    Your Text Component Editor should look like Figure 20.37.

FIGURE 20.37
The start of the alignment label style

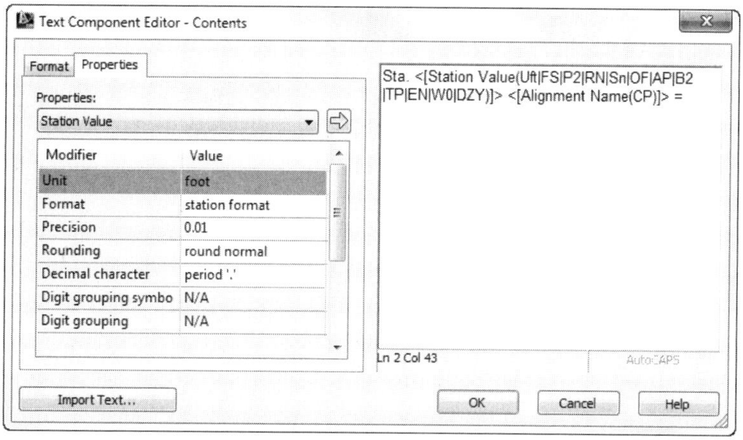

16. Click OK to return to the Label Style Composer dialog.
17. Under the Border Property, set the Visibility field to False.

18. Select Reference Text from the drop-down list next to the Add Component tool.
19. In the Select Type dialog that appears, select Alignment and click OK.
20. Change the name to **Intersecting Alignment**.
21. In the Anchor Component field, select Main Alignment.

22. In the Anchor Point field, select Bottom Left.
23. In the Attachment field, select Top Left.

    When you choose the anchor point and attachment point in this fashion, the bottom left of the Main Alignment text is linked to the top left of the Intersection Alignment text.
24. Click in the Contents field, and click the ellipsis button to open the Text Component Editor.
25. Delete the generic label text that currently appears in the preview area.
26. Type **Sta.** in the preview area; be sure to leave a space after the period.
27. In the Properties drop-down list, select Station Value.
28. Click the insert arrow in the Text Component Editor dialog, and then move your cursor using the right arrow on your keyboard or End key, and add a space.
29. In the Properties drop-down list, select Alignment Name.
30. Click the insert arrow to add this bit of code to the preview.
31. Click OK to exit the Text Component Editor, and click OK again to exit the Label Style Composer dialog.
32. Add the label to the drawing by selecting the Annotate tab and clicking Add Labels.
33. Change the label settings to match those shown in Figure 20.38 and click Add.

FIGURE 20.38
Adding the new alignment label

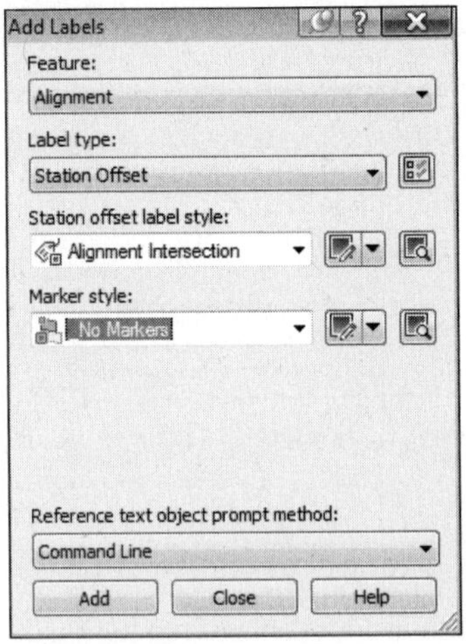

    Watch the command line for placement instructions. You will be prompted to select the main alignment, the station along the alignment, and the offset. You will then be prompted to click the intersecting alignment.
34. Press Esc to complete the labeling command.

35. Click the label to select it and reveal the grips. Select the square grip and drag it away from the current location to form a leader.

Your completed label should look like Figure 20.39.

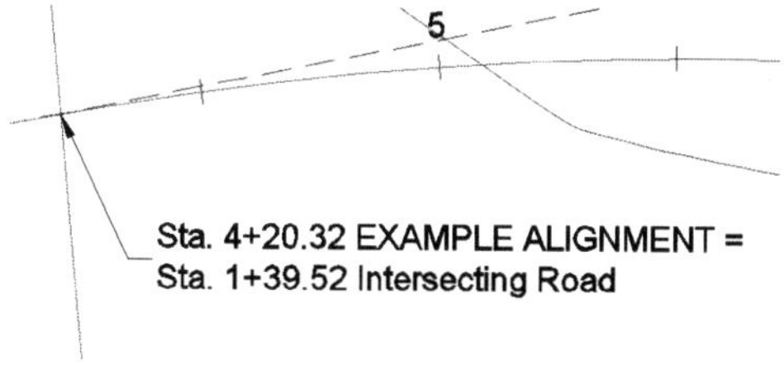

FIGURE 20.39 The completed alignment label with reference text

PROFILE LABELS

It's important to remember that the profile and the profile view aren't the same thing. The labels discussed in this section are those that relate directly to the profile. This usually means station-based labels, individual tangent and curve labels, or grade breaks. You'll look at individual label styles for these components and then at the concept of the label set.

PROFILE LABEL SETS

As with alignments, you apply labels as a group of objects separate from the profile in the form of profile label sets. In this exercise, you'll learn how to add labels along a profile object:

1. Open the `Alignment&ProfileLabels.dwg` (`Alignment&ProfileLabels_METRIC.dwg`) file.
2. Pick the blue layout profile (the profile with two vertical curves) to activate the profile object.
3. From the Profile contextual tab ➢ Labels panel, select Edit Profile Labels to display the Profile Labels dialog (see Figure 20.40).

Edit Profile Labels

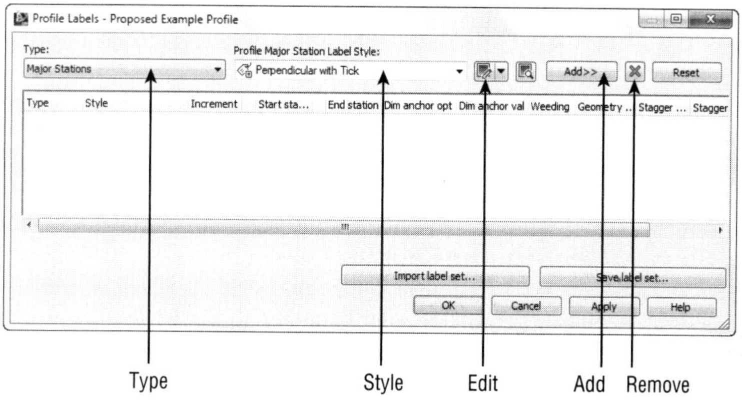

FIGURE 20.40 An empty Profile Labels dialog

Selecting the type of label from the Type drop-down menu changes the Style drop-down menu to include styles that are available for that label type. Next to the Style drop-down menu are the usual Style Edit/Copy button and a preview button. Once you've selected a style from the Style drop-down menu, clicking the Add button places it on the profile. The middle portion of this dialog displays information about the labels that are being applied to the profile selected; you'll look at that in a moment.

4. Choose the Major Stations option from the Type drop-down menu.

   The name of the second drop-down menu changes to Profile Major Station Label Style to reflect this option.

5. Set the style to Perpendicular With Tick in this menu.
6. Click Add to apply this label to the profile.
7. Choose Horizontal Geometry Points from the Type drop-down menu.

   The name of the Style drop-down menu changes to Profile Horizontal Geometry Point.

8. Select the Horizontal Geometry Station style, and click Add again to display the Geometry Points dialog shown in Figure 20.41.

FIGURE 20.41
The Geometry Points dialog appears when you apply labels to horizontal geometry points.

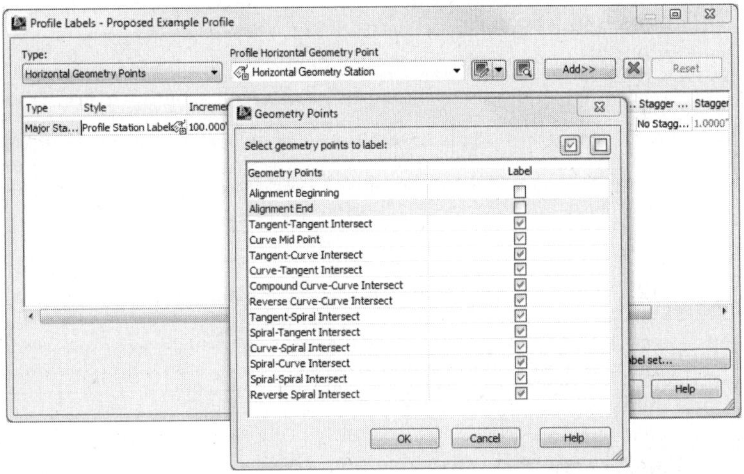

This dialog lets you apply different label styles to different geometry points if necessary.

9. Deselect the Alignment Beginning and Alignment End rows, as shown in Figure 20.41, and click OK to close the dialog.
10. Click the Apply button, and then drag the dialog out of the way to view the changes to the profile (see Figure 20.42).

12. Change the Type field to the Lines option. The name of the Style drop-down menu changes to Profile Tangent Label Style. Select the Length And Percent Grade option.

13. Click the Add button, and then click OK to exit the dialog. The profile view should look like Figure 20.45.

FIGURE 20.45
A new line label applied to the lay-out profile

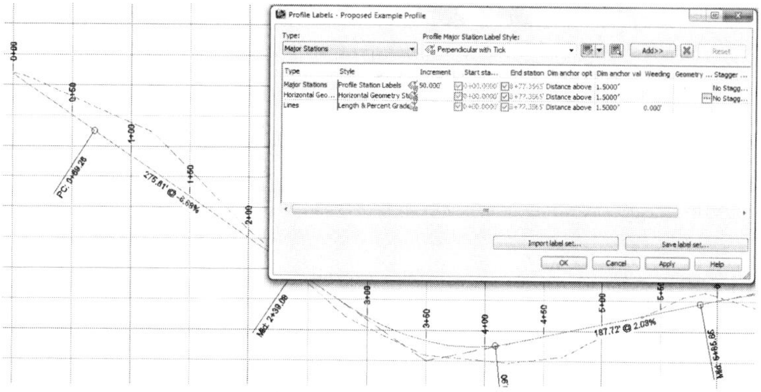

WHERE IS THAT DISTANCE BEING MEASURED?

The *tangent slope length* is the distance along the horizontal geometry between vertical curves. This value doesn't include the tangent extensions. There are a number of ways to label this length; be sure to look in the Text Component Editor if you want a different measurement.

CURVE LABELS

Vertical curve labels are one of the most confusing aspects of profile labeling. Many people become overwhelmed rapidly, because there's so much that can be labeled and there are so many ways to get all the right information in the right place. In this quick exercise, you'll look at some of the special label anchor points that are unique to curve labels and how they can be helpful:

1. Open the `Alignment&ProfileLabels.dwg` (`Alignment&ProfileLabels_METRIC.dwg`) file.

2. Pick the layout profile. From the Profile contextual tab ➢ Labels panel, select Edit Profile Labels to display the Profile Labels dialog.

3. Choose the Crest Curves option from the Type drop-down menu.

   The name of the Style drop-down menu changes.

4. Select the Crest Only option.
5. Click the Add button to apply the label.
6. Choose the Sag Curves option from the Type drop-down menu.

   The name of the Style drop-down menu changes to Profile Sag Curve Label Style.
7. Select and add the Sag Only label style. Click the Add button to apply the label.
8. Click OK to close the dialog.

Your profile should look like Figure 20.46.

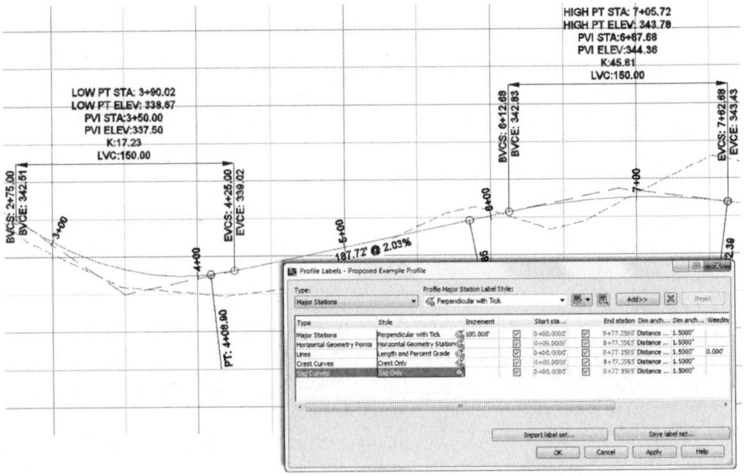

FIGURE 20.46 Curve labels applied with default Dim Anchor values

Most labels are applied directly on top of the object being referenced. Because typical curve labels contain a large amount of information, putting the label right on the object can yield undesired results. In the following exercise, you'll modify the label settings to review the options available for curve labels:

1. Continue working in the `Alignment&ProfileLabels.dwg` (`Alignment&ProfileLabels_METRIC.dwg`) file.
2. Pick the layout profile and select Edit Profile Labels from the Labels panel to display the Profile Labels dialog.
3. Scroll to the right, and change both Dim Anchor Opt values for the Crest and Sag Curves to Graph View Top.
4. Change the Dim Anchor Val for both curves to **-2.25**" (-**10** mm), and click OK to close the dialog.

Your drawing should look like Figure 20.47.

The labels can also be grip-modified to move higher or lower as needed. By using the top or bottom of the graph as the anchor point, you can apply consistent and easy labeling to the curve, regardless of the curve location or size.

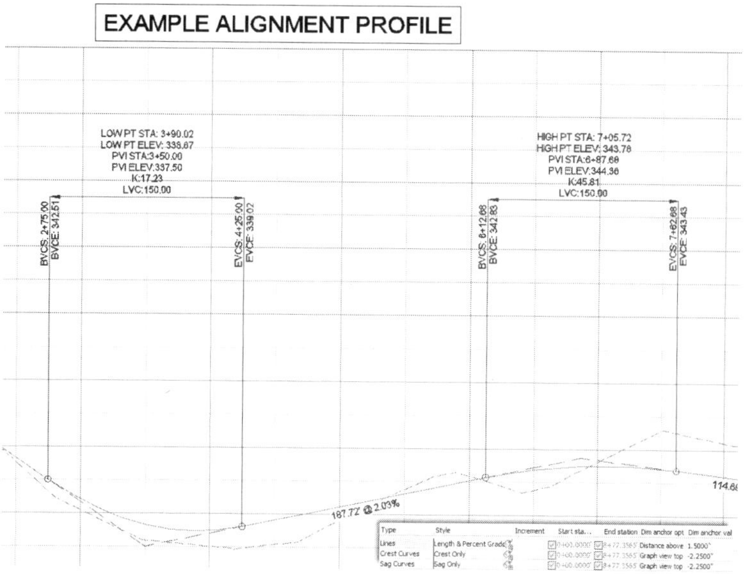

FIGURE 20.47
Curve labels anchored to the top of the graph

THOSE CRAZY CURVE LABELS!

A profile curve label can be as intricate or simple as you desire. Civil 3D gives you many options for where along the curve feature you want your label component to appear. The following illustration shows where some of the commonly used curve locations are in a label:

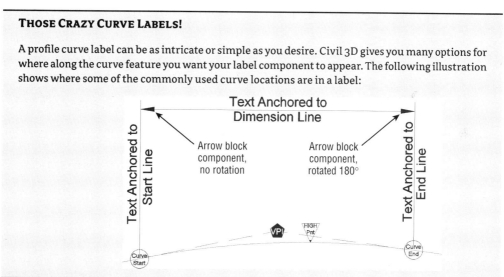

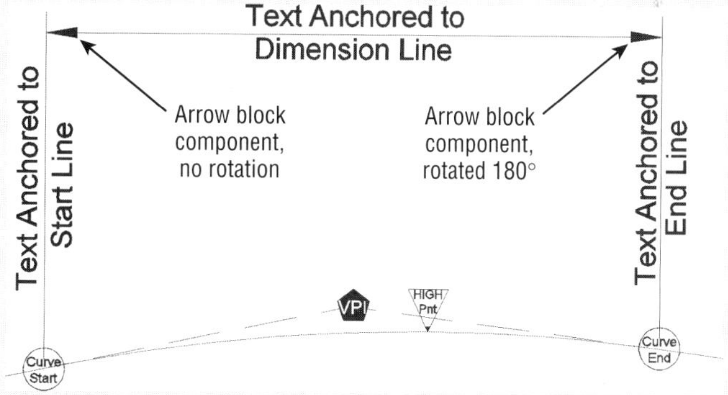

GRADE BREAKS

The last label style typically involved in a profile is a grade-break label at PVI points that don't fall inside a vertical curve, such as the beginning or end of the layout profile. Additional uses include things like water-level profiling, where vertical curves aren't part of the profile information or existing surface labeling. In this exercise, you'll add a grade-break label and look at another option for controlling how often labels are applied to profile data:

1. Open the `Alignment&ProfileLabels.dwg` (`Alignment&ProfileLabels_METRIC.dwg`) file.
2. Pick the green surface profile (the irregular profile). From the Profile contextual tab ➤ Labels panel, select Edit Profile Labels to display the Profile Labels dialog.
3. Choose Grade Breaks from the Type drop-down menu.

   The name of the Style drop-down menu changes.
4. Select the Station Over Elevation style and click the Add button.
5. Click Apply, and drag the dialog out of the way to review the change.

   A sampled surface profile has grade breaks every time the alignment crosses a surface TIN line. Why wasn't your view coated with labels?
6. Scroll to the right, and change the Weeding value to **150'** (**45 m**). Click Apply.
7. Select one of the new grade break labels. Use the square grip at the location where the label touches the profile to form a leader and clean up any labels that overlap.

   The profile labels should appear as shown in Figure 20.48.

FIGURE 20.48
It's starting to get a little crowded in here: grade-break labels on a sampled surface

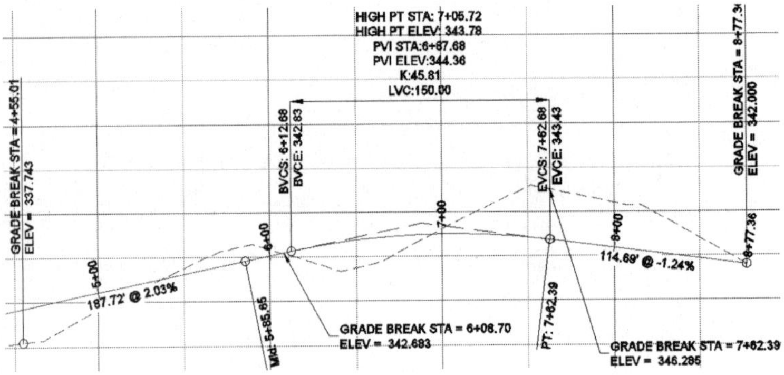

8. Click OK to dismiss the dialog.

Weeding lets you control how frequently grade-break labels are applied. This makes it possible to label dense profiles, such as a surface sampling, without being overwhelmed or cluttering the view beyond usefulness.

As you've seen, there are many ways to apply labeling to profiles, and applying these labels to each profile individually could be tedious. In the next section, you'll build a label set to make this process more efficient.

PROFILE LABEL SETS

Applying labels to both crest and sag curves, tangents, grade breaks, and geometry with the label style selection and various options can be monotonous. Thankfully, Civil 3D gives you the ability to use label sets, as in alignments, to make the process quick and easy. In this exercise, you'll apply a label set, make a few changes, and export a new label set that can be shared with team members or imported to the Civil 3D template. Follow these steps:

1. Open the `Alignment&ProfileLabels.dwg` (`Alignment&ProfileLabels_METRIC.dwg`) file.
2. To tidy things up, select one of the grade-break labels from the previous exercise. Press Delete on your keyboard to remove the labels.
3. Pick the layout profile. From the Profile contextual tab ➢ Labels panel, select Edit Profile Labels to display the Profile Labels dialog.
4. Click the Import Label Set button near the bottom of the dialog to display the Select Style Set dialog.
5. Select the Complete Label Set option from the drop-down menu, and click OK.
6. Click OK again to close the Profile Labels dialog and see the profile view.

   The label set you chose contains curve labels, grade-break labels, and line labels.
7. Pick the layout profile. From the Profile contextual tab ➢ Labels panel, select Edit Profile Labels to display the Profile Labels dialog.
8. Click Import Label Set to display the Select Style Set dialog.
9. Select the _No Labels option from the drop-down menu, and click OK.

   All the labels from the listing will be removed.

   In the next steps, you will add labels to the listing and save the listing as its own label set for future use.
10. Set the active type to Lines. Set Profile Tangent Label Style to Length And Percent Grade, and click Add.
11. Set the active type to Grade Breaks. Set Profile Grade Break Label Style to Station Over Elevation, and click Add.
12. Set the type to Crest Curves. Set Profile Crest Curve Label Style to Crest Only, and click Add.
13. Set the type to Sag Curves. Set Profile Sag Curve Label Style to Sag Only, and click Add.
14. Set the Crest Curve and Sag Curve label types to use the Graph View Top as the Dim Anchor Opt.
15. Set both Dim Anchor Val fields to **-1.5**" (**-40** mm), as shown in Figure 20.49.

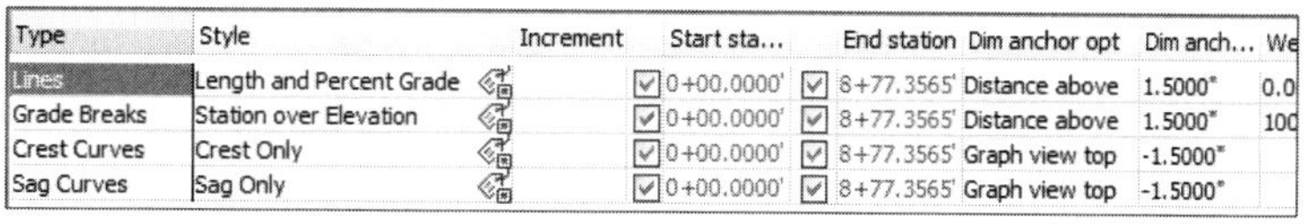

Type	Style	Increment	Start sta...	End station	Dim anchor opt	Dim anch...	We
Lines	Length and Percent Grade		0+00.0000'	8+77.3565'	Distance above	1.5000"	0.0
Grade Breaks	Station over Elevation		0+00.0000'	8+77.3565'	Distance above	1.5000"	100
Crest Curves	Crest Only		0+00.0000'	8+77.3565'	Graph view top	-1.5000"	
Sag Curves	Sag Only		0+00.0000'	8+77.3565'	Graph view top	-1.5000"	

FIGURE 20.49
Four label types and dimension anchor settings in the label set to be saved

16. Click the Save Label Set button to open the Profile Label Set dialog and create a new profile label set.
17. On the Information tab, change the name to **Road Profile Labels**.

18. Click OK to close the Profile Label Set dialog.

19. Click OK to close the Profile Labels dialog.

20. On the Settings tab of Toolspace, select Profile ➤ Label Styles ➤ Label Sets.

    Note that the Road Profile Labels set is now available for sharing or importing to other profile label dialogs.

Label sets are the best way to apply profile labeling uniformly. When you're working with a well-developed set of styles and label sets, it's quick and easy to go from sketched profile layout to plan-ready output.

Advanced Style Types

Now that you are familiar with the basics of object styles and label styles, you are ready to take your skills to the next level. The styles in the following section combine aspects of object styles and label styles.

You have a great deal of control over every detail, even ones that may seem trivial. Instead of being bogged down trying to understand every option, don't be afraid to try a "trial and error" approach. If you make a change you don't like, you can always edit the style until you get it right.

Table Styles

Civil 3D does a beautiful job of placing dynamically-linked data tables that relate to your objects. The tables use the Text Component Editor to grab dynamic information from your objects. You also have control over fill colors, table headings, and how the data is sorted.

For the table style, the Data Properties tab contains all the column information. You can add columns by clicking the plus sign. You can remove columns by highlighting the column you want to remove and clicking the Delete button. Change column order by dragging them around and dropping them where you want them to go, as shown in Figure 20.50.

FIGURE 20.50 Modifying table columns

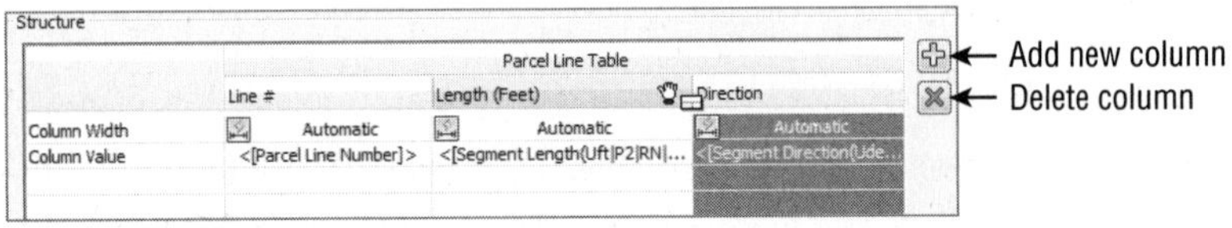

In the following exercise you will see the basic steps of modifying a table style. Now that you understand the ins and outs of the Label Style Composer, this procedure should be a breeze:

1. Open the `Tables.dwg` (`Tables_METRIC.dwg`) file, which you can download from this book's web page.

   This file contains a parcel line table whose style you will modify.

2. Zoom into the parcel line table.

   Notice there are several things you will want to change:

   - The line numbers are out of order.

- The directions have far too much precision.
- The length does not display units.

All that is about to change.

Edit Table Style

3. Click the table to select it. From the context tab, select the Table Properties drop-down and choose Edit Table Style.
4. Switch to the Data Properties tab.

The Data Properties tab (Figure 20.51) is the main control area for all table styles. This is where you set behavior, text styles, and sizes for fields.

FIGURE 20.51
Data Properties tab for table styles

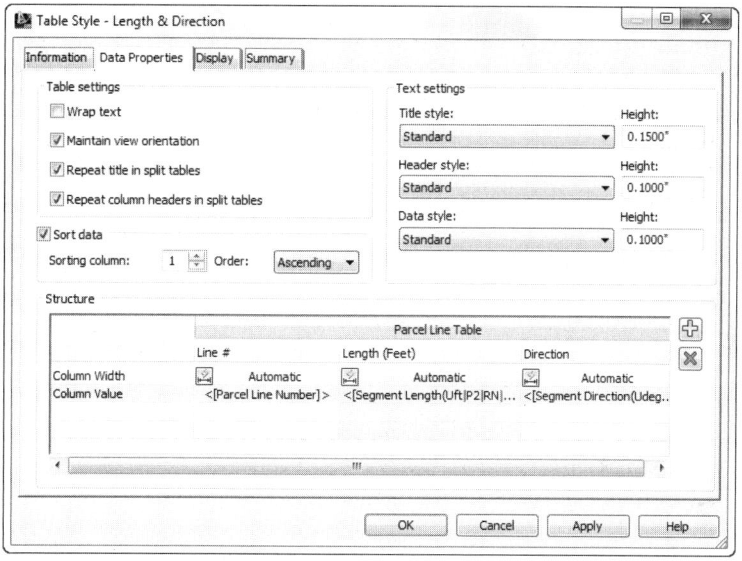

5. Place a check mark next to Sort Data, and set the Sorting column to **1**.

   Column 1 corresponds to the Column Value containing the Parcel Line Number. Set the order option to Ascending. The Ascending option will ensure that the parcel numbers are listed in the table from the lowest to highest value.

6. Double-click the column heading field for Length.

   Doing so opens a stripped-down version of the Text Component Editor. Headings and table titles are static text only; therefore only the text formatting tools are shown.

7. Add **(Feet)** (**(meters)**) after the column heading to add a proper units heading to the column, as shown in Figure 20.52, and click OK.
8. Double-click the Direction Column value field below the column heading.

   Doing so opens the Text Component Editor, similar to what you've used in earlier exercises.

FIGURE 20.52
Adding static text to a table column heading

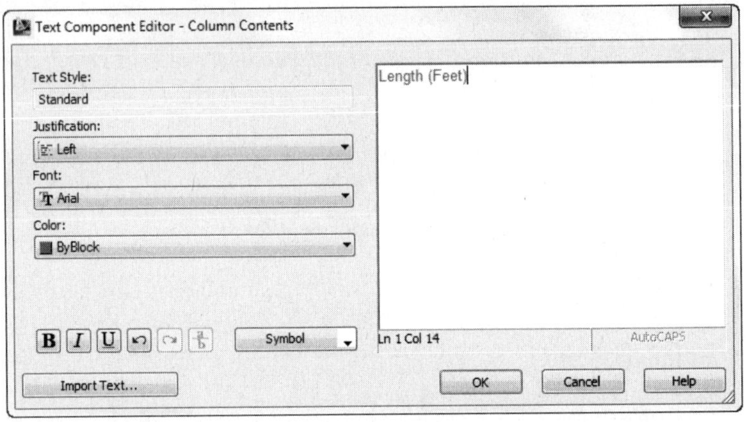

9. Click on the text in the preview area so that it becomes highlighted.
10. Change Precision to **1 Second**.
11. Click the arrow to update the text. Click OK.
12. Click OK to complete the table style modifications.

Code Set Styles

Code set styles determine how your assembly design will appear. Code set styles are used in many places. A code set style is in play when you first create your assembly. One is used in corridor creation and in the Section Editor. The most apparent use of a code set style is in section views like those that you created in the previous exercise.

Code set styles are a collection of many other styles. In a code set style you will find:

- Links and link label styles
- Points, point label styles, and feature line styles
- Shapes and shape label styles
- Quantity takeoff pay items
- Render materials for visualization tasks

It is helpful in naming your code set style to have the name of the set reflect its use. Multiple code set styles are needed because of different applications of its use. When you are designing an assembly, you may want to see more labels than when you are getting ready to plot the assembly in a cross-section sheet. Labeling that is useful in a cross-section sheet may obstruct your view of the design when working with it in the corridor cross-section editor.

The hardest part of working with code set styles is figuring out the name of the link or point you wish to label. Luckily, it is unusual for users to label shapes, so you won't need to worry about those. The names of each point or link can be found in the subassembly properties. Most of the links and points are logically named, but there's no harm in a little trial and error if you are not sure.

Shapes

Shapes are the areas that define materials. Because it is not common for people to label these materials in a section view, you will not be experiencing these in an exercise.

One heads-up, however: Resist the temptation to use a hatch pattern on shapes where multiple cross-section views will be created. Solid fills and no patterns are your best bet to avoid performance issues and the annoying "Hatch pattern is too dense" warning.

Links and Link Labels

You learned in Chapter 8, "Assemblies and Subassemblies," that a link is the linear part of a subassembly. The object style for the link itself is very simple—just a single linear component. The label for a link is usually expressed as a percent grade or as a slope ratio.

In the following exercise, you will modify a code set style to apply link labels to an assembly.

1. Open the `Code Set Styles.dwg` (`Code Set Styles_METRIC.dwg`) file, which you can download from this book's web page.

   This file contains a corridor and cross-section views. Zoom into one of the cross-section views so you can observe the changes as you apply them to the code set style.

2. From the Settings tab of Toolspace, expand General ➤ Multipurpose Styles ➤ Code Set Styles; right-click on All Codes, and select Edit.

3. In the Codes tab Link list, locate Pave, and click the Label Tag icon in the Label Style column, as shown in Figure 20.53.

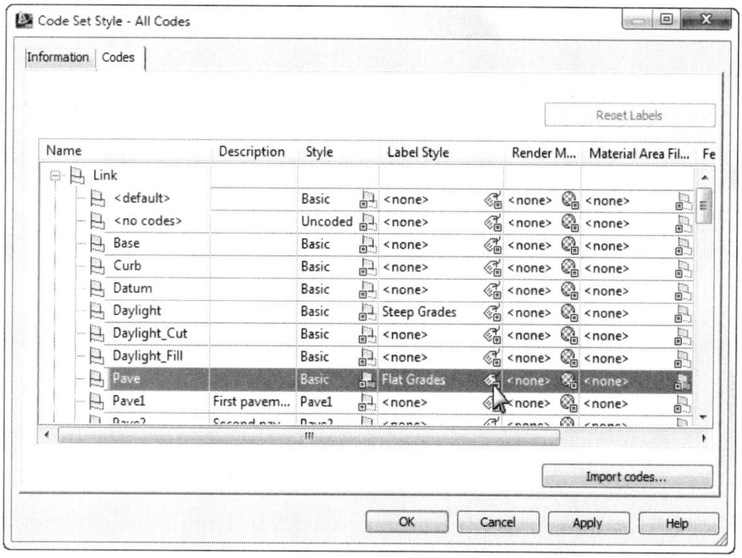

FIGURE 20.53 Adding labels to the link codes in the code set style

4. Select the Flat Grades label style and click OK; then click Apply to examine the change on the cross sections.

You should see that the lanes now have slope information labeled.

5. In the Link list, locate Daylight; click the Label Tag icon in the Label Style column, and select Steep Grades as the label style.
6. Click OK, and then click OK to dismiss the All Codes label style and see what is happening with the cross section.

The cross sections should resemble Figure 20.54.

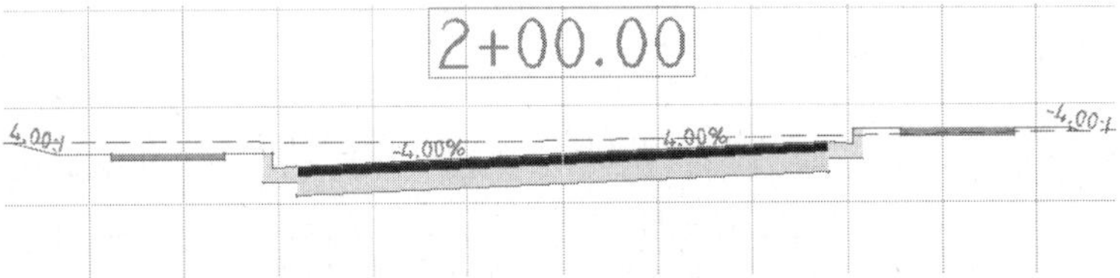

FIGURE 20.54
Cross section with link labels applied to pave and daylight links

IT'S JUST AN EXPRESSION

Expressions allow you to use properties available in a label and perform calculations. The resulting expression can then be used as part of a displayed label, or as any of the numerical text settings such as text height, rotation angle, or width.

A frequently seen use of expressions is with surface spot elevations. In the following image, an expression called `Slant Curb` contains `{Surface Elevation}-0.125`. The label style then uses the Slant Curb as one of the properties displayed in the spot elevation label as shown here:

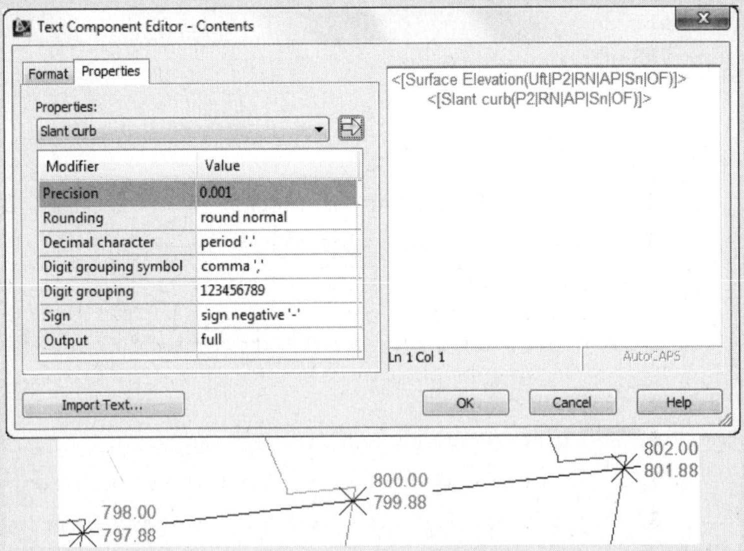

Another common use of expressions is in code set styles. You want to label all of the Top codes, but on short segments the labels would appear too crowded.

In the following example, an expression is used as the text height in a code set style. When the length of the link is large enough, the text appears normally. If the link is shorter than 0.5' (0.1 m), the text is set to a value that makes it too small to see.

1. Open the drawing `Expression.dwg` (`Expression_METRIC.dwg`).
2. Zoom into one of the cross-sections views.

   You will see that the labels are crowded on the assembly. An expression can help clean this up by omitting labels for segments that are too short.

3. In the Settings tab of Toolspace ➤ General ➤ Label Styles ➤ Link, right-click Expressions and select New.

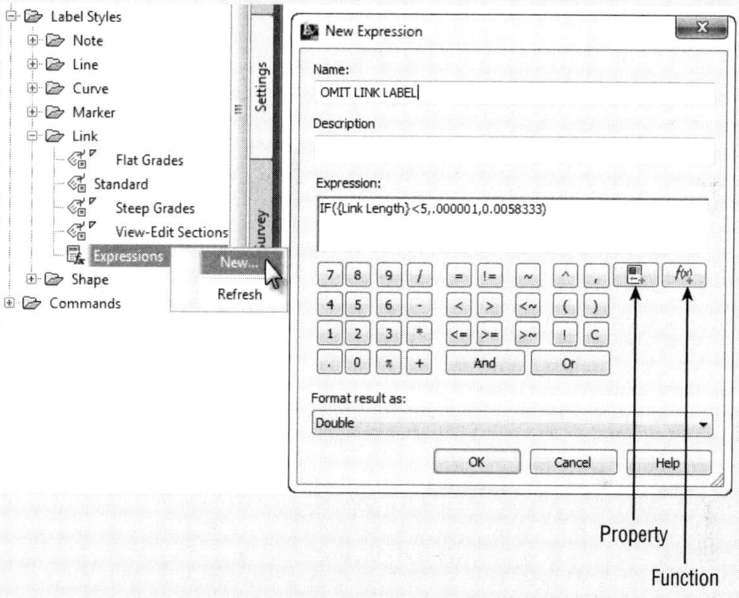

4. Name the expression **OMIT LINK LABEL**.
5. Click the function button to the right and select `IF`.

   A peek into the Civil 3D Help will tell you what it is expecting after the `IF`. According to help, the syntax is `IF(test,true_val,false_val)`. The "test" in this example is if the link is less than 5' (1.5 m). If the link is less than the desired length, the text size will be very small so that it is effectively hidden. If the link is the desired length or longer, then a text height of 0.005833' (0.002 m) will be used.

6. Click the Property button and select Link Length.
7. Enter a **<** sign or click the < operator button in the New Expression dialog.
8. Enter **5**' (**1.5** m) and the test portion of the expression is complete.
9. Add a comma and add **0.000001**' (**0.0000001** m).

   This is the extremely small text height that will hide the text. The units are only shown in these examples to distinguish unit systems; do not add them to the expression.

10. Add another comma and add **0.0058333**' (**0.0015** m).

    This is the text height that will display. Note that the value is in feet (or meters).

11. Add a close parenthesis and verify that Format Result As is set to Double.
12. When your new expression resembles the following image, click OK.
13. In the same branch of Prospector, right-click Flat Grades and select Edit.
14. On the Layout tab of the Label Style Composer, click in the text height field and set it to OMIT LINK LABEL. Click OK.

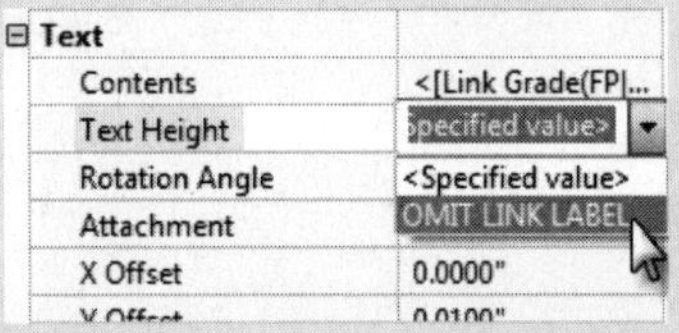

15. Run REGEN on your drawing if you don't see the change right away.

The labels on the section should now look much neater.

POINTS AND POINT LABELS

A common frustration with new users of Civil 3D are the marker styles and their labels. For display in cross-section views, you may not want points to display at all. In the following exercise, you will create a new code set style, modify point codes, and add more labels to the sections:

1. Continue working in `Code Set Styles.dwg` (`Code Set Styles_METRIC.dwg`).
2. From the Settings tab of Toolspace, expand General ➤ Multipurpose Styles Code Set Styles. Right-click on All Codes and select Copy.
3. On the Information tab, rename the style to **All Codes-Plotting**.
4. On the Codes tab, locate the Point category. Locate the Back_Curb point and click the Tag icon in the Label Style column. Set the style to Offset Elevation. Click OK.
5. Repeat step 4 to set the label style for Sidewalk_Out to Offset Elevation.
6. Click the first point name, <default>. While holding down the Shift key on your keyboard, scroll down to the last point listing, Top_Curb. With all of the points selected, click the Tag icon in the style column and change the style to _No Markers. Click OK.

   The Code Set Style dialog should resemble Figure 20.55.

7. Click OK.

   You do not see any changes to your cross sections yet because the style is not active.

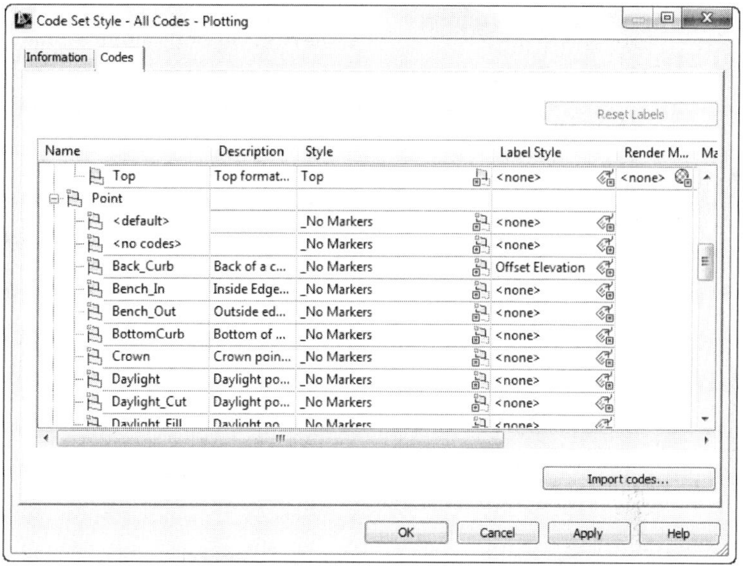

FIGURE 20.55 Points set to _No Markers and labels set to Offset Elevation

8. Select a section view. From the Section Views contextual tab ➢ Modify View panel, select View Group Properties.
9. On the Sections tab, change the style of the EXAMPLE corridor to the All Codes-Plotting style you created, as shown in Figure 20.56, and click OK.
10. Click OK again to complete the changes to the Section View Group Properties dialog.

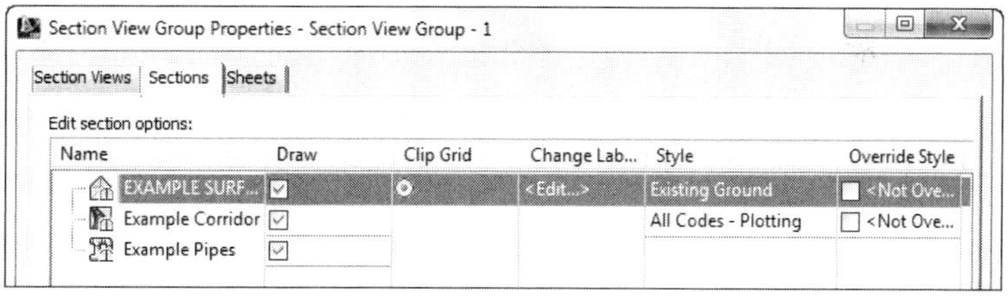

FIGURE 20.56 Setting the code set style current on the section views

Your section view should now resemble Figure 20.57.

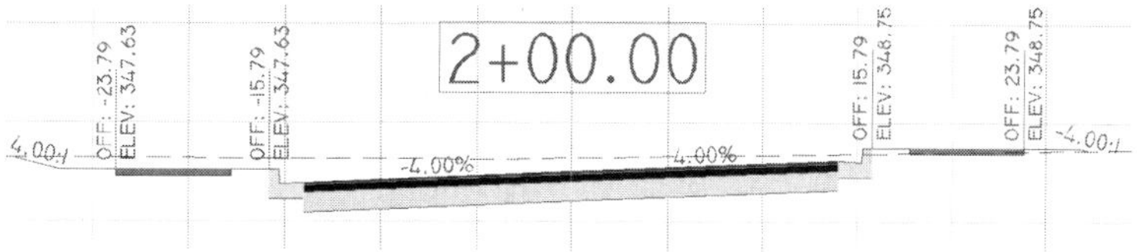

FIGURE 20.57 New code set style applied to the section view

The Bottom Line

Override individual labels with other styles. In spite of the desire to have uniform labeling styles and appearances between alignments within a single drawing, project, or firm, there are always exceptions. Using the Ctrl+click method for element selection, you can access commands that let you modify your labels and even change their styles.

Master It Open the drawing `MasteringLabelStyles.dwg` (`MasteringLabelStyles_METRIC.dwg`). Create a copy of the Perpendicular With Tick Major Station style called **Major With Marker**. Change Tick Block Name to **Marker Pnt**. Replace some (but not all) of your major station labels with this new style.

Create a new label set for alignments. Label sets let you determine the appearance of an alignment's labels and quickly standardize that appearance across all objects of the same nature. By creating sets that reflect their intended use, you can make it easy for a designer to quickly label alignments according to specifications with little understanding of the requirement.

Master It Within the `Mastering LabelStyles.dwg` (`Mastering Label Styles_METRIC.dwg`) file, create a new label set containing only major station labels, and apply it to all the alignments in that drawing.

Create and use expressions. Expressions give you the ability to add calculated information to labels or add logic to label creation.

Master It In the `Mastering LabelStyles.dwg` (`Mastering Label Styles_METRIC.dwg` file, create an expression that adds 0.5¢ (0.15 m) to a surface elevation. Use the expression in a spot elevation label that shows both the surface elevation and the expression-based elevation.

Apply a standard label set to profiles. Standardization of appearance is one of the major benefits of using Civil 3D styles in labeling. By applying label sets, you can quickly create plot-ready profile views that have the required information for review.

Master It In the `Mastering LabelStyles.dwg` (`Mastering Label Styles_METRIC.dwg`) file, apply the Road Profiles label set to all layout profiles.

Chapter 21

Object Styles

As you learned in the previous chapter, styles control the display properties of labels but they also control the display properties of objects. Object styles control the starting display of AutoCAD® Civil 3D® objects such as points, surfaces, alignments, pipes, sections, and so on.

Consider a Civil 3D surface. Sometimes you want to see the surface as contours, other times you want to see the surface as triangles, and sometimes you don't want to see it at all. With styles changing what you see is not usually just a matter of freezing and thawing layers, but it is a matter of changing the active style. Think of the active style on an object as a wrapper. A surface model will have many potential wrappers depending on what you want to see. The underlying data does not change, but the representation of the data will change.

Understanding and applying object styles correctly can mean the difference between getting a job out in several hours, and fighting with your CAD drawing for days.

In this chapter, you will learn to:

- Override object styles with other styles
- Create a new surface style
- Create a new profile view style

Getting Started with Object Styles

Before you get your hands on specific object styles, it is nice to understand some general things all styles have in common.

CERT OBJECTIVE

There are several ways to enter the various dialogs used for editing styles. The easiest, most direct way into any style is from the Settings tab. Right-click on any style you see listed and select Edit, as shown in Figure 21.1.

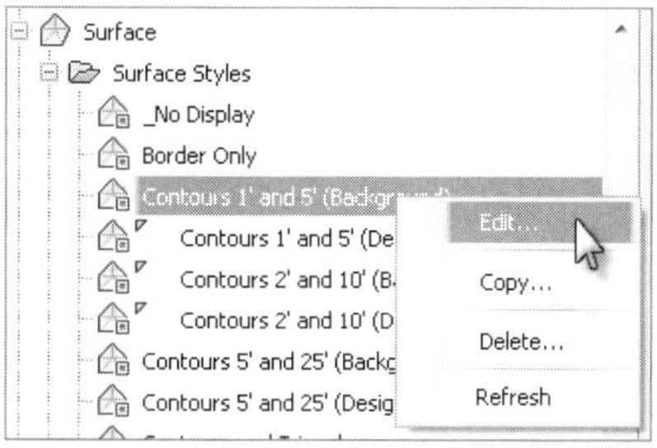

FIGURE 21.1 Every style can be edited by right-clicking on the style name from the Settings tab of Toolspace.

You can also enter the styles' dialogs from the properties of any object by clicking the Edit button. Figure 21.2 shows the active styles on a surface, with the Edit button to the right. Editing at this level affects all objects that use the style, the same as it would if you had entered the style from the Settings tab. The downside to editing a style in this manner is that you will not immediately be able to click Apply to see your change. You'll need to exit the style's dialog and click Apply at the object level before seeing your style update.

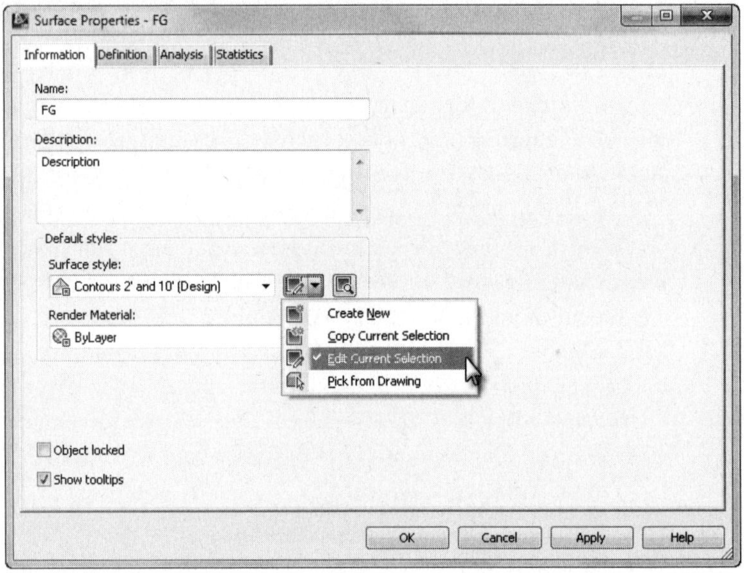

FIGURE 21.2 An object's properties reveal the current style, which can be edited, as in this Surface Properties dialog.

Real World Scenario

MAKING SENSE OF CHILD STYLES AND OVERRIDES

Civil 3D styles are customizable at several levels. In the Drawing Settings area discussed in Chapter 1, "The Basics," you saw overall settings that initially affect the entire drawing. The Drawing Settings are the highest level of the settings hierarchy (the main parent settings). You will find the same settings at the object level that can diverge from the overall drawing settings. Farther down the hierarchy are the command settings. The term "child" in Civil 3D refers to any style that can also be controlled at a higher level.

For example, for most objects a precision of two decimal places (0.00) is adequate for your station work. However, for corridor creation it is advantageous to see more decimal places of precision. At the drawing level, keep the station precision to 0.00. At the Corridor level, right-click on the object and select Edit Feature Settings.

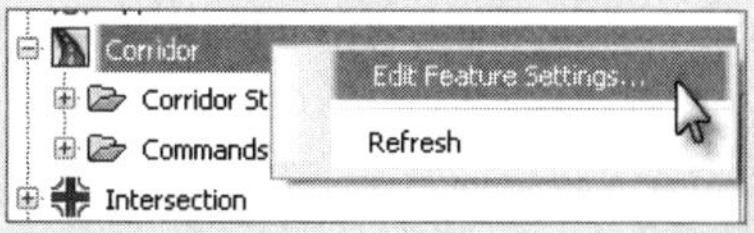

Inside the Edit Feature Settings – Corridor dialog, you will have a feeling of déjà vu from when you edited the Ambient Settings tab in the Drawing Settings dialog. All the same settings (and a few more object-specific ones) are here. The difference is that changing the setting here will only affect corridors.

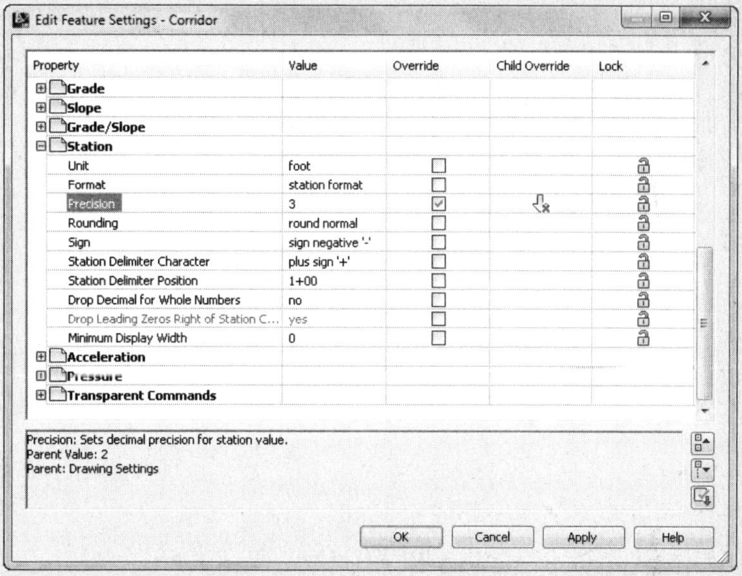

The check mark in the Override column indicates that this setting differs from settings higher up in the hierarchy of settings. At the bottom of the dialog further information is provided regarding the parent and its value.

An arrow in the Child Override column indicates that further down the chain of command a style or setting differs. To force these subordinate styles to match the style of the parent, click the arrow so that a red X appears. The red X indicates that the change you make, in the dialog you are looking at, will get pushed to its child styles or settings.

The Lock column allows you to lock a setting, which prevents it from changing any of its child styles.

As you are designing the styles for use in your office, be sure to keep an eye on the overrides that are set.

Frequently Seen Tabs

Object styles control the display of Civil 3D objects such as points, surfaces, alignments, pipes, sections, and so on.

Consider a Civil 3D surface. Sometimes you want to see the surface as contours, other times you want to see the surface as triangles, and sometimes you don't want to see it at all. Changing what you see is not a matter of freezing and thawing layers, but it is a matter of changing the active style. Think of the active style on an object as a wrapper. A surface model will have many potential wrappers depending on what you want to see. The underlying data does not change, but the representation of the data will change.

On the Settings tab of Toolspace, if you click the Expand button next to the drawing name, you see the full array of objects that Civil 3D uses to build its design model. Each of these has special features unique to the object being described, but there are some common features as well.

Information Tab

The Information tab (Figure 21.3) controls the name of the style. You will see this tab in every object style dialog.

FIGURE 21.3
The Information tab exists for all object styles.

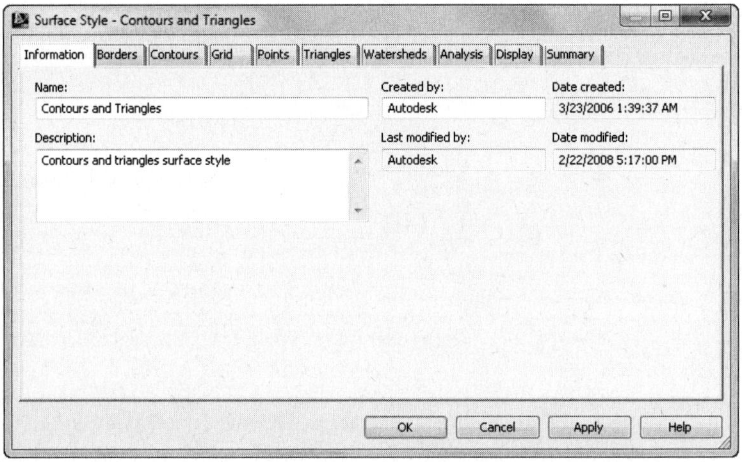

The description is always optional, but we recommend that you add some information to the style. The description can be seen in tooltip form as you search through the Settings tab, as shown in Figure 21.4.

FIGURE 21.4
Tooltip showing style information, including the description

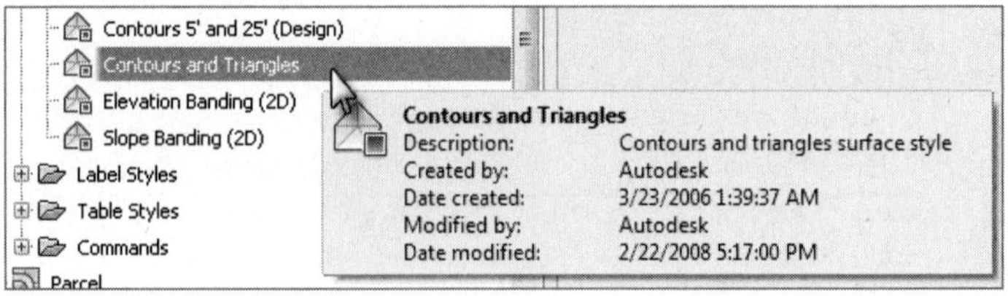

On the right side of the dialog you will see the name of the style, the date when it was created, and the last person who modified the style. These names are initially pulled from the Windows login information and only the Created By field can be edited.

Display Tab

On the Display tab, you will see a list of the various components that can be displayed for the object you are working with. You will see this tab in every object style dialog. The Display tab controls the look of the object in Plan, Model, Profile, and Section views. Not every object type will have all of these views available. In Figure 21.5 you can see that a general-purpose marker contains only one component (the marker itself), and a surface model has components for contours, triangles, points, and more.

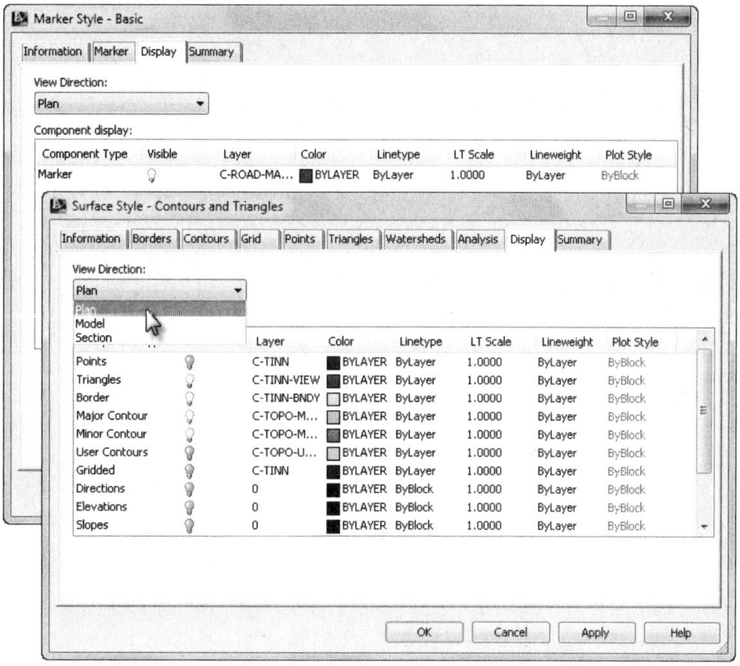

FIGURE 21.5
The display of a simple marker style (above) and a more complex surface style (below)

While other tabs in an object style control the specifics of *how* certain components look, the Display tab controls *if* the component displays at all. The visibility lightbulb indicates whether the component will be displayed when the style is applied to the object. In the Surface Style dialog shown in Figure 21.5, you can see that triangles, border, major contours, and minor contours will all display for the Surface object style but the other components will not be displayed in plan view.

Each component will have a layer designation. You may be wondering, "Didn't I set a surface layer back in the Drawing Settings on the Object Layers tab back in Chapter 1?" Yes, you did. Those were overall object layers. The layers you see here are component layers. Using the analogy of a base AutoCAD® block, the Object layer can be thought of in the same way as a block insertion layer. The object component layers can be thought of in the same way as layers inside the block definition.

This is Civil 3D, of course, so the objects exist in three dimensions. View Direction controls the display of an object depending on how you are looking at it. Certain items, such as a profile view, are intended to only be seen in plan, so they do not have multiple view directions listed. Surface models make an appearance in plan, model, and section, as shown previously in Figure 21.5. A marker, on the other hand, can be shown in plan, model, profile, or section, as seen in Figure 21.6.

FIGURE 21.6
View Directions for the marker style

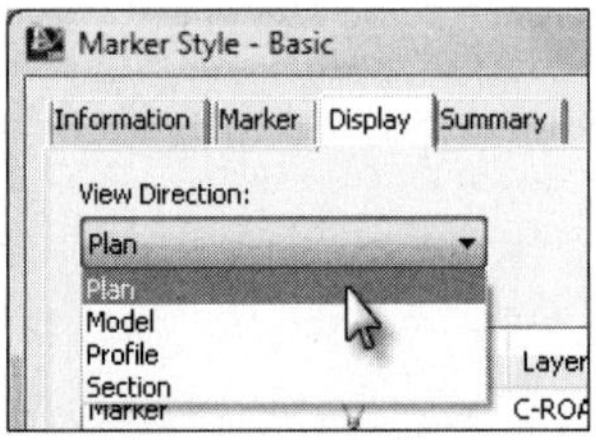

SUMMARY TAB

The Summary tab is exactly what it sounds like — that is, a summary of all other settings that exist in the style. You will see this tab in every object style dialog. The information from other tabs in list form as well as their override status is shown in Figure 21.7.

FIGURE 21.7
Summary tab of a marker style

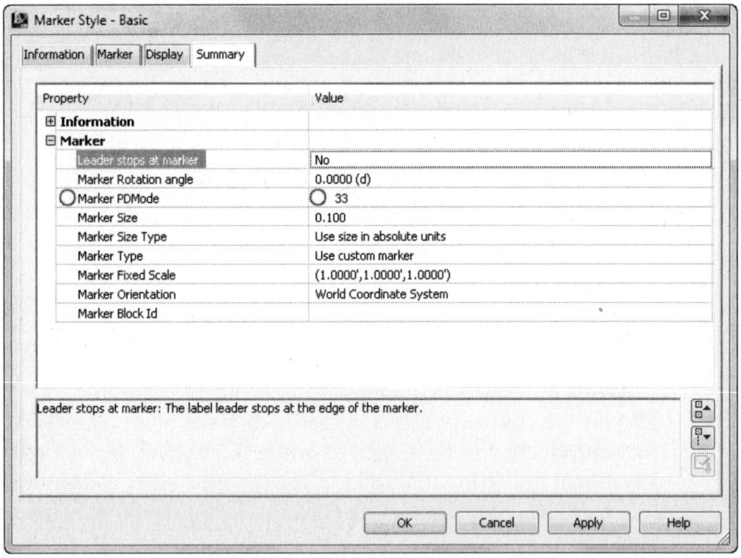

You can click the + or – button next to each category branch to expand to see further settings. At the bottom-right corner of the dialog are three buttons. The top button collapses all of the category branches, the middle button expands all of the category branches, and the bottom button (which is sometimes not selectable) overrides all child dependencies. In addition, at the bottom of this dialog, additional information will be shown for the property selected.

General Settings

The General collection contains settings and styles that are applied to various objects across the entire product. The General collection serves as the catchall for styles that apply to multiple objects and for settings that apply to no objects. The General collection has three collections (or branches):

- Multipurpose Styles
- Label Styles
- Commands

You learned about the Label Styles collection in Chapter 20, "Label Styles," but let's look a little closer at the Multipurpose Styles and Commands collections.

MULTIPURPOSE STYLES

If you expand the Multipurpose Styles collection, you will see seven folders, as shown in Figure 21.8.

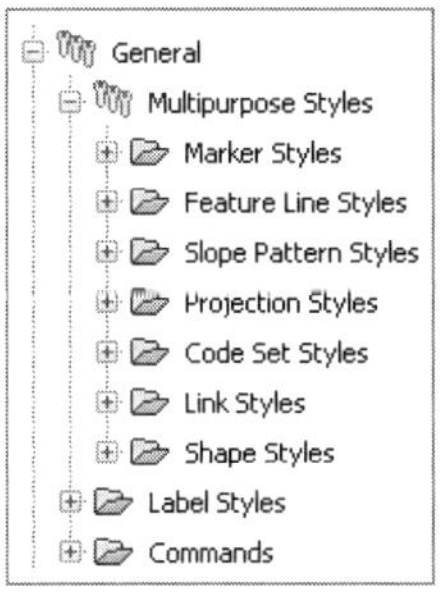

FIGURE 21.8
General ➢ Multipurpose Styles

These styles are used in many objects to control the display of components in objects. For example, the Marker Styles and Link Styles collections are typically used in cross-section views, whereas the Feature Line Styles collection, shown in Figure 21.9, is used in grading and other commands.

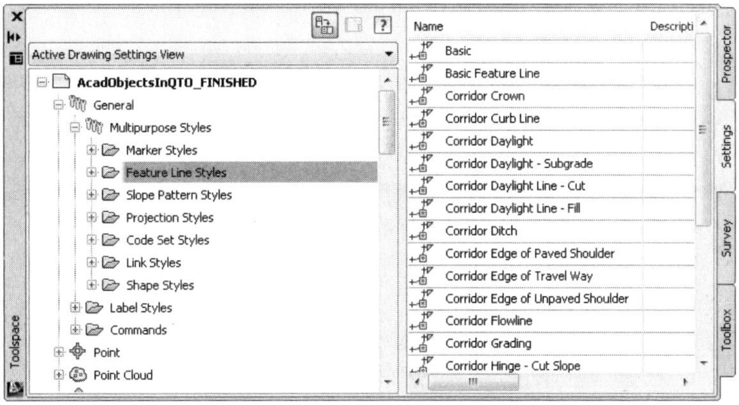

FIGURE 21.9
Feature Line Styles collection

TOGGLE TOOLSPACE ORIENTATION

Toolspace is shown horizontally in Figure 21.9 and in other figures throughout this chapter for illustration purposes. If you like the way this looks, you can also set your Toolspace like this by clicking the orientation toggle at the top of Toolspace when it is floating (not docked). This will bring the item view to the right side of Toolspace instead of the default location at the bottom.

COMMANDS

Almost every branch in the Settings tree contains a `Commands` folder. Expanding this folder, as shown in Figure 21.10, shows the typical long, unspaced command names that refer to the parent object. Various commands have been discussed throughout this book.

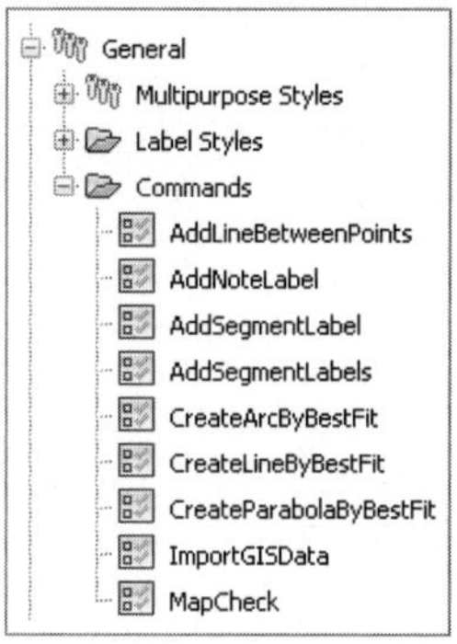

FIGURE 21.10
`Commands` folder

Point and Marker Object Styles

Markers are used in many places throughout Civil 3D. They are called from other styles to show vertices on Civil 3D objects, as label location marks, or even as a marker to indicate the start of a flow line.

MARKER TAB

Marker styles and point styles both contain a Marker tab (Figure 21.11). The Marker tab controls what symbol or block is used, its rotation, and how it should be sized when it is placed in the drawing.

Three symbol types can be used. Use Custom Marker and Use AutoCAD BLOCK Symbol For Marker are the best options. Use AutoCAD POINT For Marker is a poor choice because it uses the symbol specified by the DDPTYPE setting and is difficult to control. When you choose the Use Custom Marker option, you can select a combination of symbols to mix and match, one symbol from the left of the vertical and either or none of the symbols from the right of the vertical line.

When you choose the Use AutoCAD BLOCK Symbol For Marker option, you will be able to access a listing of blocks in your drawing. If the block you want to use does not yet exist in the drawing, you can right-click in the block listing and choose Browse, as shown in Figure 21.12.

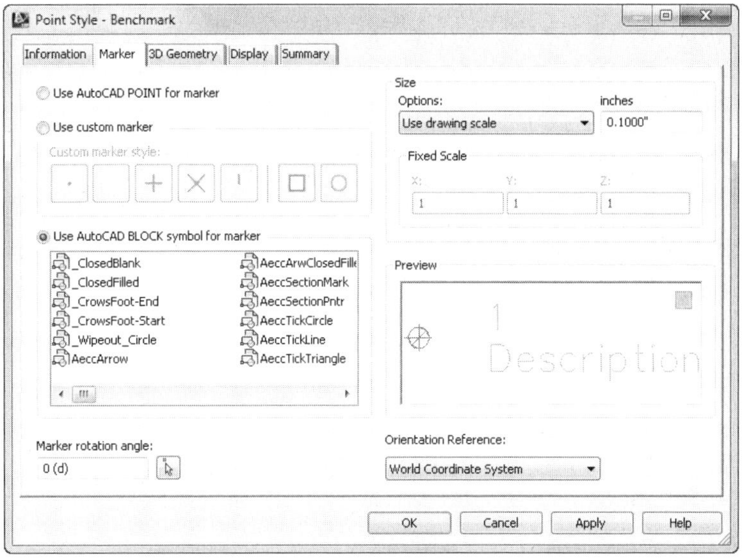

FIGURE 21.11 Marker tab for the Benchmark point style.

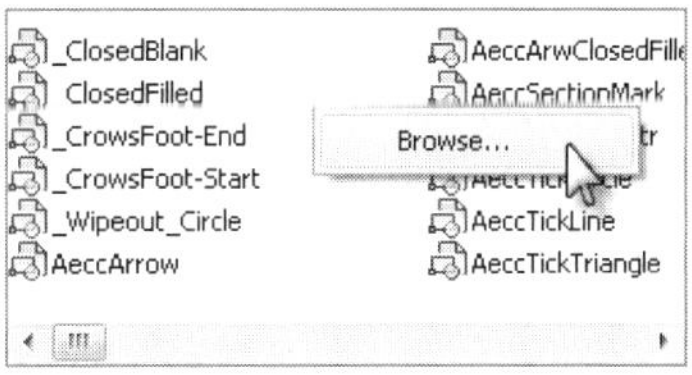

FIGURE 21.12 Right-click to browse for a block if it is not already defined in your drawing.

The Size options control how the symbol is scaled when inserted in the drawing (Figure 21.13):

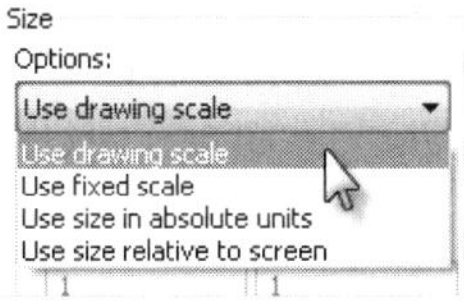

FIGURE 21.13 Size options for marker display

Use Drawing Scale Use Drawing Scale allows you to specify the plotted size of the symbol. The modelspace size of the symbol will be the size specified in the style multiplied by your annotation scale.

Use Fixed Scale Use Fixed Scale will scale the symbol based on the X, Y, and Z scale set in the style. This option will also use the Fixed scale factor in the description key set when used with a survey point.

Use Size In Absolute Units Use Size In Absolute Units is the option you will use in point styles that vary in size based on a survey description. For example, if this is a TREE symbol and you want the size of the symbol to reflect the trunk diameter in feet, set this to 1.

Size Relative To Screen Size Relative To Screen allows you to specify the size of the symbol as a percentage of your screen. This setting can quickly become annoying, because the marker will constantly change size as you zoom in or out. We do not recommend that you enable it.

Orientation Reference (as seen in the bottom-right corner of Figure 21.11) controls whether the symbol stays rotated with the view, world coordinate system, or, in the case of survey points, the object. In most cases, the Orientation Reference setting should be set to World Coordinate System or View.

Create a Marker Style

Now to get your hands on some object styles. You will start with simple styles and work your way up in complexity as this chapter progresses. In this first exercise you will create a marker style.

1. Start a new blank drawing from the `_AutoCAD Civil 3D (Imperial) NCS` template that ships with Civil 3D. For metric users, use the `_AutoCAD Civil 3D (Metric) NCS` template.
2. From the Settings tab of Toolspace, expand General ➢ Multipurpose Styles ➢ Marker Styles.
3. Right-click Marker Styles and select New.
4. On the Information tab, do the following:
   a. Set Name to **VPI Marker**.
   b. Add the description **Use to indicate vertical PI in profiles**.

   Your login name will be listed in the Created By and Last Modified By fields.
5. On the Marker tab, do the following:
   a. Click the Use AutoCAD BLOCK Symbol For Marker radio button.
   b. In the block listing, highlight STA by clicking on it.
   c. Verify that Size Options is set to Use Drawing Scale.
   d. Set the size to **0.2**″ (or **5** mm).

   Leave all other Marker settings at their defaults.
6. On the Display tab with the View Direction set to Plan, do the following:
   a. Click in the Layer column for the Marker component to display the Layer Selection dialog.
   b. In the Layer Selection dialog, set the layer to C-ROAD-PROF-STAN-GEOM and click OK.

Note that you may need to widen the column to view the full layer name.

c. Click in the Color column for the Marker component to display the Select Color dialog.

d. In the Select Color dialog, click the ByLayer button and click OK.

7. Repeat step 6 for View Direction set to Model, Profile, and Section.
8. Click OK to complete the creation of a new marker style.
9. From the application menu select Save As ➢ Drawing Template.
10. Set File Name to **`PointObjectTemplate.dwt`** (or **`PointObjectTemplate_METRIC.dwt`**) and click Save.
11. Set the description in the Template Options dialog to **Mastering Template with Point Object Styles**, verify that Measurement is set to either English or metric as applicable, and click OK.

Your VPI marker will now be listed in the General ➢ Multipurpose Styles ➢ Marker Styles branch. By creating this marker style in a drawing template, you have made it accessible for any future projects that use this template to start from. Keep this drawing template open for the next portion of the exercise.

Create a Survey Point Style

Survey point styles contain many of the same options as marker styles. As you work through the following example, you will perform many of the same steps you did in the previous exercise:

1. Continue working in the drawing template from the previous exercise.
2. From the Settings tab of Toolspace, expand Point ➢ Point Styles.
3. Right-click Point Styles and select New.
4. On the Information tab, do the following:

   a. Set Name to **TRUNK**.

   b. Add the description **Simple circle representing trunk diameter in inches** (or **Simple circle representing trunk diameter in mm**).

5. On the Marker tab, do the following:

   a. Click the Use Custom Marker radio button.

   b. Click the blank marker option from the group of symbols on the left.

   c. Click to add the circle option on the right.

   d. Verify that Size Options is set to Use Size In Absolute Units and set the size to **0.0833** (or **0.001** for metric).

   This value will scale down the symbol so the trunk diameter represents inches (or millimeters). Leave all other Marker settings at their defaults. Instead of entering these values as a decimal, you could alternatively enter it as a fraction such as 1/12 or 1/1000.

6. On the Display tab with View Direction set to Plan, do the following:

   a. Click in the Layer column for the Marker component to display the Layer Selection dialog.

   b. In the Layer Selection dialog, set the Marker layer to V-NODE-TREE and click OK.

   c. Use steps a and b to set the Layer for the Label component to V-NODE-TREE.

   Leave the other settings at their defaults.

   Hint! To save time in this step you could alternatively use the Shift key to multiselect the Marker and Label components, as shown in Figure 21.14. By selecting both first, when you select the layer for one component, it will apply to both components.

FIGURE 21.14 Use the Shift key on your keyboard as you click the components to multiselect.

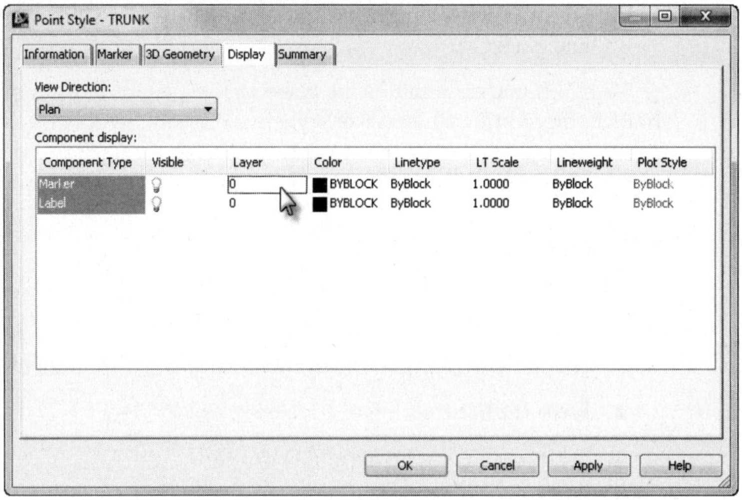

7. Repeat step 6 for View Direction set to Model, Profile, and Section.

8. Click OK to complete the creation of a new point style.

You can save and keep this drawing template open to continue on to the next exercise, or use the saved copy of this drawing template (`PointObjectTemplate_FINISHED.dwt` or `PointObjectTemplate_METRIC_FINISHED.dwt`) available from the book's web page at `www.sybex.com/go/masteringcivil3d2013`.

Linear Object Styles

In this section, you will see some linear styles such as alignments, profiles, and parcels. Hopefully, you are already seeing that concepts from one type of style often apply to other types of object styles. For alignment styles and profile styles, this is especially true.

Both alignment styles and profile styles have a Design tab, as shown in Figure 21.15. In the case of the alignment styles, the Enable Radius Snap option restricts the grip-edit behavior of alignment curves. If you enable this option and set a value of 0.5′, the resulting radius value of curves will be rounded to the nearest 0.5′. In the case of profiles, the curve tessellation distance is a little more abstract. Curve tessellation refers to the smoothing factor applied to the profile when viewing it in 3D. Most users leave these settings at their default values.

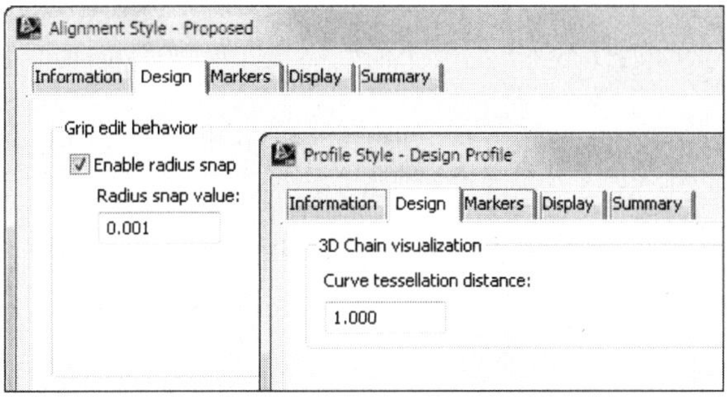

FIGURE 21.15
Design tabs exist in both alignment and profile object styles.

Alignment styles and profile styles have very similar Markers tabs. This tab is where you can place markers (like the one you created in the first exercise of the chapter) at specific geometry points. Figure 21.16 shows the markers assigned to various locations along an alignment.

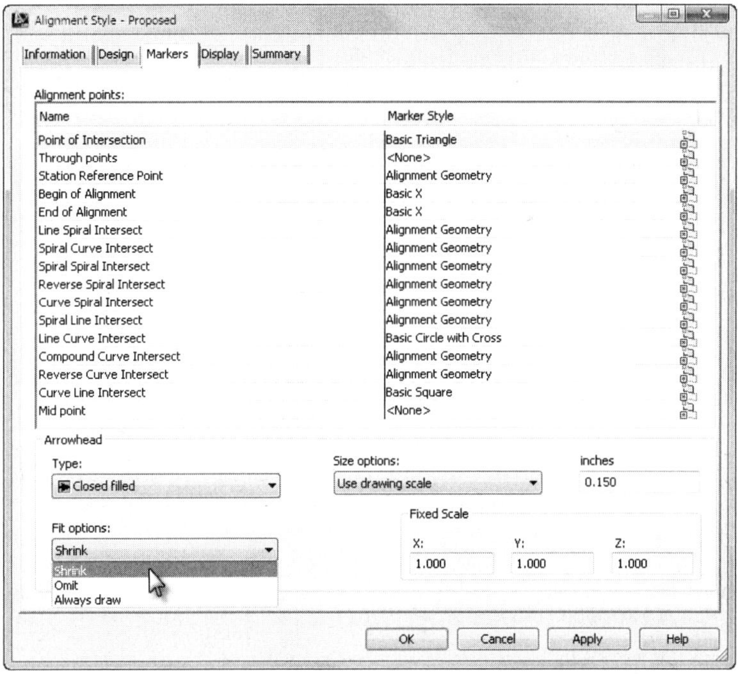

FIGURE 21.16
Markers for an alignment style.

At the bottom of Figure 21.16 you see Arrowhead information. Both alignment styles and profile styles have the option of showing a direction arrow on each segment. Most people choose to omit this by turning the Arrow component off in the Display tab or setting the component to a layer that is set to No Plot. You should consider the latter, as knowing the direction of an alignment comes in handy when designing roundabouts, as you saw in Chapter 11, "Advanced Corridors, Intersections, and Roundabouts."

Figure 21.17 shows some commonly highlighted geometry points and components in an alignment. The markers in Figure 21.17 correspond to the markers defined in Figure 21.16.

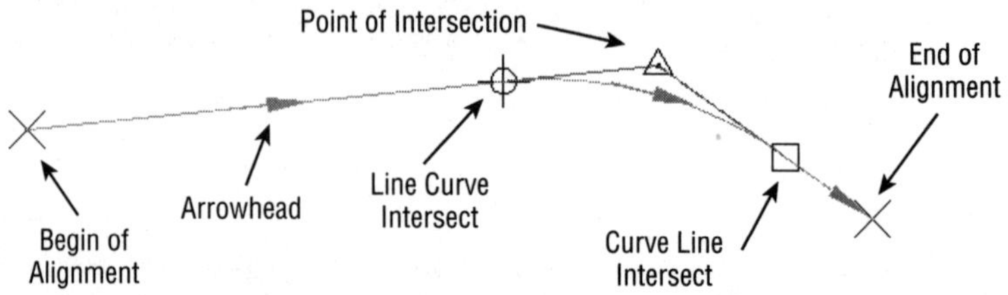

FIGURE 21.17 Example alignment with alignment points labeled

Figure 21.18 shows the Markers tab for the profile style. It looks pretty similar to the alignment Markers tab, don't you think?

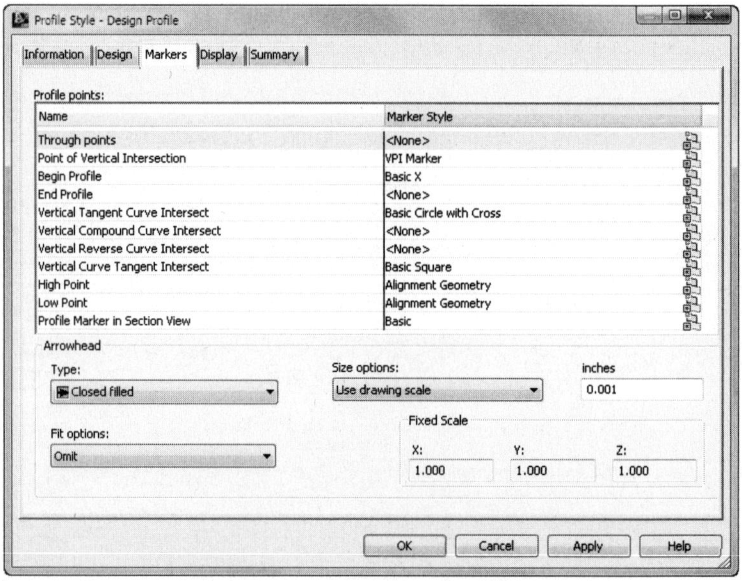

FIGURE 21.18 Markers for a profile style

The markers in Figure 21.19 correspond to the markers defined in Figure 21.18.

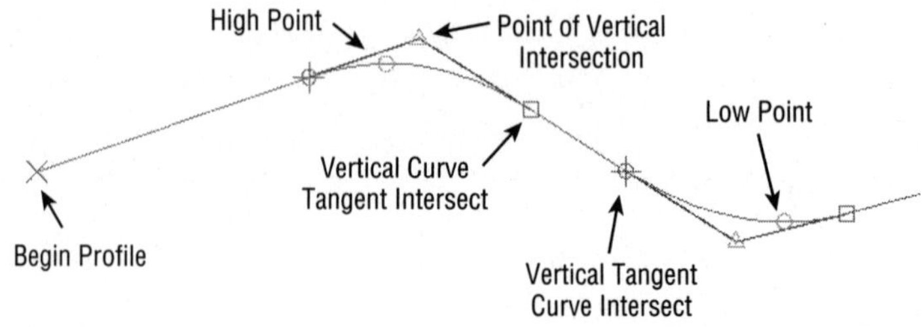

FIGURE 21.19 Example profile with profile points labeled

Alignment Style

Alignment styles can be helpful in identifying key design components, as well as showing the stationing direction. Use multiple alignment styles to visually differentiate centerline alignments from supplemental alignments such as offset alignments and curb return alignments.

In the following exercise, you will create a style that restricts radius grip edits to five-foot increments and displays basic alignment components:

1. If not still open from the previous exercise, open `PointObjectTemplate_FINISHED.dwt` or `PointObjectTemplate_METRIC_FINISHED.dwt`.

   You can download either file from this book's web page.

2. From the Settings tab of Toolspace, expand Alignment ➤ Alignment Styles.
3. Right-click Alignment Styles and select New.
4. On the Information tab, set Name to **Centerline**.
5. On the Design tab, do the following:
   a. Verify that Enable Radius Snap is selected.
   b. Verify that Radius Snap is set to **5** (or **1** for metric).

      To see the Help document about any dialog, press F1 on your keyboard to be taken directly to the help section for that topic.

6. On the Markers tab, do the following:
   a. Set the Point Of Intersection marker by double-clicking the current value of <None>.
   b. In the Pick Marker Style dialog, select VPI Point from the marker listing, and click OK.
   c. Using the same procedure, set the Through Points marker and the Station Reference Point marker both to <None>.

      Note that <None> is located near the top of the list.

7. On the Display tab with the View Direction set to Plan, do the following:
   a. Using the Shift key to select for Arrow, Line Extensions, and Curve Extensions together, use one of their lightbulb icons to turn off the display of these three components.
   b. Using the Shift key to select Line, Curve, and Spiral together, click in the Layer column to display the Layer Selection dialog.
   c. In the Layer Selection dialog, set the layer to C-ROAD-CNTR and click OK. This will set the layer for all three of the selected components.

      Leave the Model and Section view directions at their defaults.

8. Click OK to complete the creation of a new alignment style.

   This alignment will show the line, curve, and spiral on layer C-ROAR-CNTR and a marker will be shown at the Point Of Vertical Intersection, as shown in Figure 21.20.

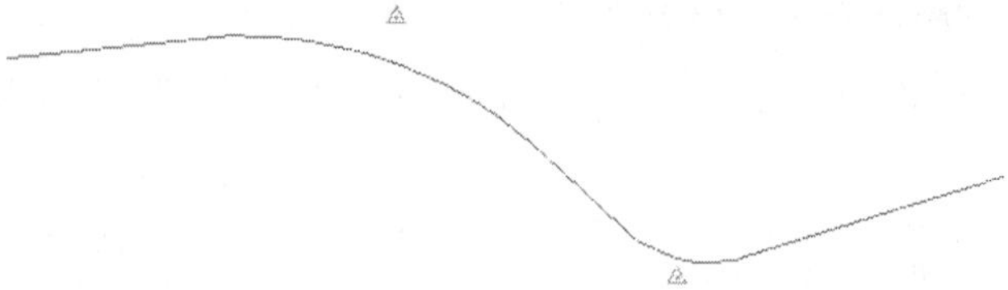

FIGURE 21.20
Centerline alignment style.

When this exercise is complete, you can close the drawing template. A saved copy of this drawing template is available from the book's web page with the filename `AlignmentObjectTemplate_FINISHED.dwt` or `AlignmentObjectTemplate_METRIC_FINISHED.dwt`.

Parcel Styles

The parcel styles have several unique features that make them different from other styles.

In the Design tab of a parcel (shown at the top in Figure 21.21), you see parcel-specific options. A fill distance can be specified to place a hatch pattern along the perimeter of the parcel. This setting is useful to differentiate "special" parcels such as parks, limits of disturbance, or environmentally sensitive areas.

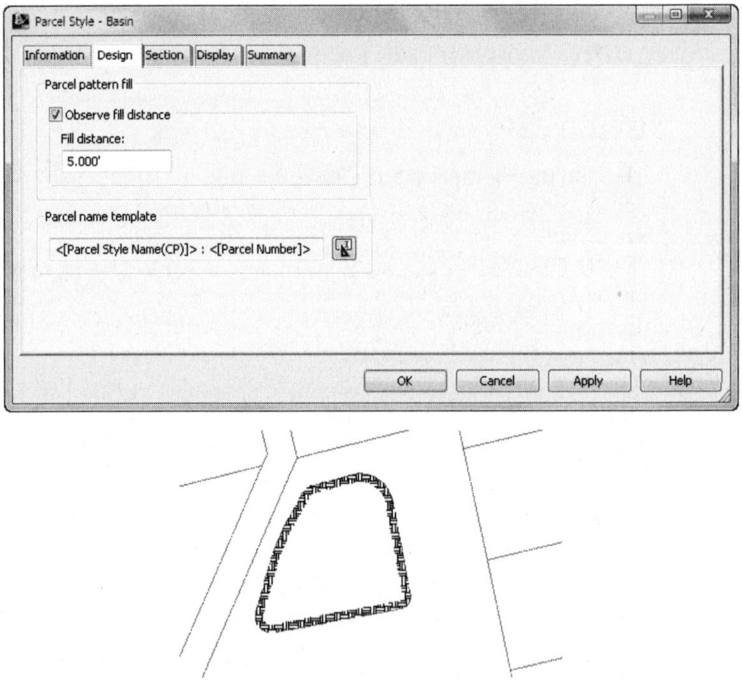

FIGURE 21.21
Parcel style options (above) and the resulting parcel graphic (below)

The fill distance indicates the width of the hatch pattern. The Component Hatch Display characteristics, including the pattern, angle, and scale, are specified at the bottom of the Display tab. Be sure to put the hatch component on its own layer so that it can be turned off independently from the parcel itself, as hatches tend to slow down the graphics. The bottom of Figure 21.21 shows the parcel graphic resulting from the design settings shown.

Feature Line Styles

Feature lines are found in quite a few different places. They are created automatically as part of corridors, created as a result of grading, or can be created independently by the user. Because their scope crosses functionality, you will find feature line styles in the General ➤ Multipurpose Styles collection in Settings.

By definition, a feature line is a 3D object; therefore, its style can be controlled in plan, profile, and section. As shown in Figure 21.22, a feature line can use markers at geometry points.

FIGURE 21.22
Feature line profile marker options (left) and section marker option (right)

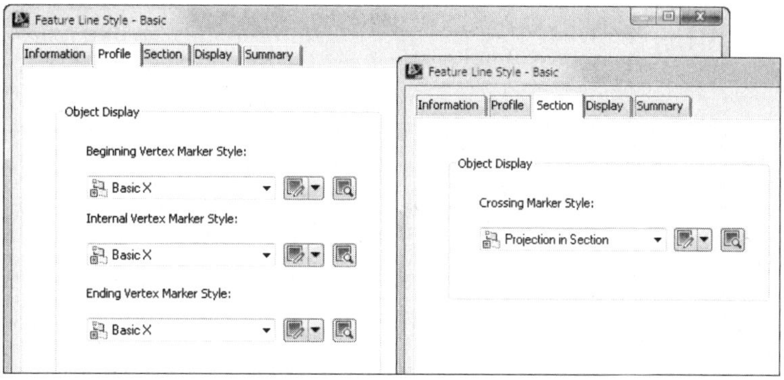

Surface Styles

CERT OBJECTIVE

Surface styles are the most widely used styles in any Civil 3D project. Depending on which style is active, certain editing options may be restricted. For example, you need to see points before Civil 3D will let you use the Delete Point command.

Each tab in the Surface Style dialog corresponds to components listed in the Display tab. Remember, the Display tab determines the component's layer, color, linetype, and so forth, but it does not control the specifics of the component's properties (such as contour interval or point symbol).

Once a surface is created, you can display information in many ways. The most common so far has been contours and triangles, but these are the basics. By using varying styles, you can show a large amount of data with one single surface. Not only can you do simple things such as adjust the contour interval, but also Civil 3D can apply a number of analysis tools to any surface:

Contours Allows the user to specify a more specific color scheme or linetype as opposed to the typical minor-major scheme. Commonly used in cut-fill maps to color negative colors one way, positive contours another, and the balance or zero contours yet another color.

Directions Draws arrows showing the normal direction of the surface face. This tool is typically used for aspect analysis, helping site planners review the way a site slopes with regard to cardinal directions and the sun.

Elevations Creates bands of color to differentiate various ranges of elevations. You can use this tool to create a simple weighted distribution to help in creation of marketing materials, hard-coded elevations to differentiate floodplain, and other elevation-driven site concerns, or ranges to help a designer understand the earthwork involved in creating a finished surface.

Slopes Colors the face of each triangle on the basis of the assigned slope values. While a distributed method is the normal setup, a common use is to check site slopes for compliance with Americans with Disabilities Act (ADA) requirements or other site slope limitations, including vertical faces (where slopes are abnormally high).

Slope Arrows Displays the same information as a slope analysis, but instead of coloring the entire face of the TIN, this option places an arrow pointing in the downhill direction and colors that arrow on the basis of the specified slope ranges. This is useful in confirming surface flow direction for site drainage.

User-Defined Contours Refers to contours that typically fall outside the normal intervals. These user-defined contours are useful to draw lines on a surface that are especially relevant but don't fall on one of the standard levels. A typical use is to show the normal water surface elevation on a site containing a pond or lake.

Watersheds Used for watershed analysis, this style allows you to examine how water flows along and off of a surface. Using the surface TIN, the drain targets and watersheds are defined. An example of creating a watersheds surface style is provided later in this section.

In the following exercises, you will walk through the steps of creating or modifying surface styles.

Contour Style

Contouring is the standard surface representation on which land development plans are built. In this example, you'll copy an existing surface contouring style and modify the interval to a setting more suitable for commercial site design review:

1. Open the `SurfaceStylesContours.dwg` or `SurfaceStylesContours_METRIC.dwg` file, which you can download from this book's web page.
2. From the Settings tab of Toolspace, expand Surface ➢ Surface Styles.
3. Right-click Surface Styles and select New.
4. On the Information tab, do the following:
   a. Set Name to **Exaggerated Existing Contours**.
   b. Add the description **2′ minor contours with a 5× exaggeration when viewed in 3D** (or **1m minor contours with a 5× exaggeration when viewed in 3D** for metric).
5. On the Contours tab, do the following:
   a. Expand the Contour Intervals category.
   b. Set the Minor Interval to **2′** (or **1** m).

      Notice that the Major Interval automatically adjusts to **10′** (or **5** m).
   c. Expand the Contour Smoothing category (you may have to scroll down).
   d. Set Smooth Contours to True, which activates the Contour Smoothing slider bar near the bottom.

      Don't change this Smoothing value, but keep in mind that this gives you a level of control over how much Civil 3D modifies the contours it draws.

SURFACE VS. CONTOUR SMOOTHING

Remember, contour smoothing is not surface smoothing. Contour smoothing applies smoothing at the individual contour level but not at the surface level. If you want to make your surface contouring look fluid, you should be smoothing the surface.

The Contours tab will now look similar to Figure 21.23.

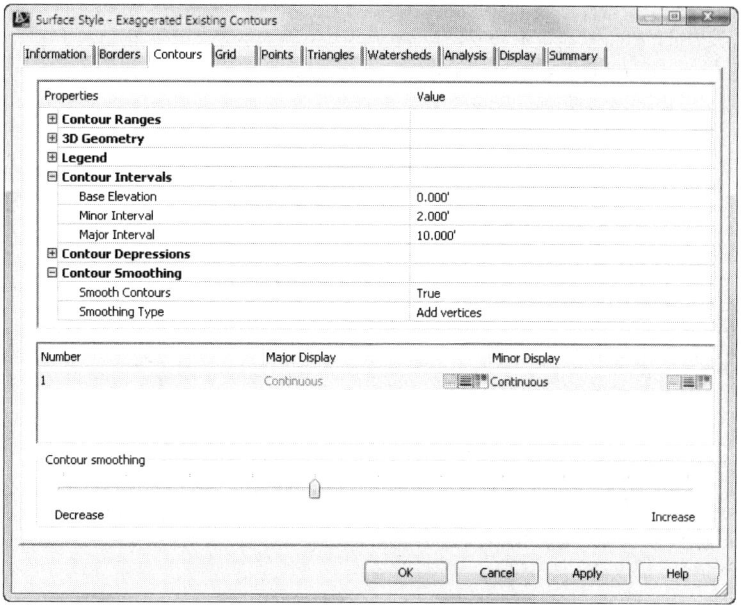

FIGURE 21.23 The Contours tab in the Surface Style dialog

6. On the Triangles tab, do the following:
   a. Change Triangle Display Mode to Exaggerate Elevation.
   b. Set Exaggerate Triangles By Scale Factor to **5** (see Figure 21.24).

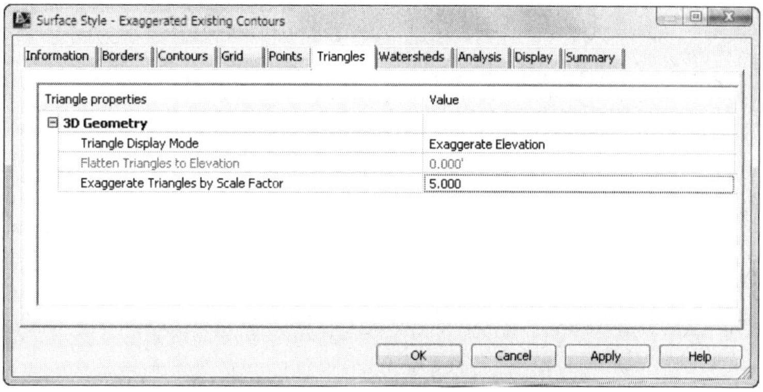

FIGURE 21.24 Exaggerate the elevations seen in the Object Viewer.

7. On the Display tab with the View Direction set to Plan, do the following:
   a. Shift+click to highlight all the components.
   b. Click in the Color column for one of the components to display the Color Selection dialog.
   c. In the Select Color dialog, click the ByLayer button and then click OK.
   d. While all the components are still highlighted, set Linetype to ByLayer using a similar procedure to steps b and c.

BYLAYER OR BYSTYLE

If you are a true CAD stickler, you will try to make as many items as possible set to ByLayer. This approach greatly simplifies things if you need to change color or linetype by use of the Layer Manager. Alternatively, you can define the color and linetype independently through the style.

You will only be using the Minor Contour in this style, but later on you will copy this style and your efforts will be carried forward.

   e. Use on the lightbulb icons to turn on the visibility for Minor Contour and turn it off for all other components.
   f. Set the Minor Contours layer to C-TOPO-MINR.

Your Display tab should now resemble Figure 21.25.

FIGURE 21.25
Minor Contour flying solo in plan

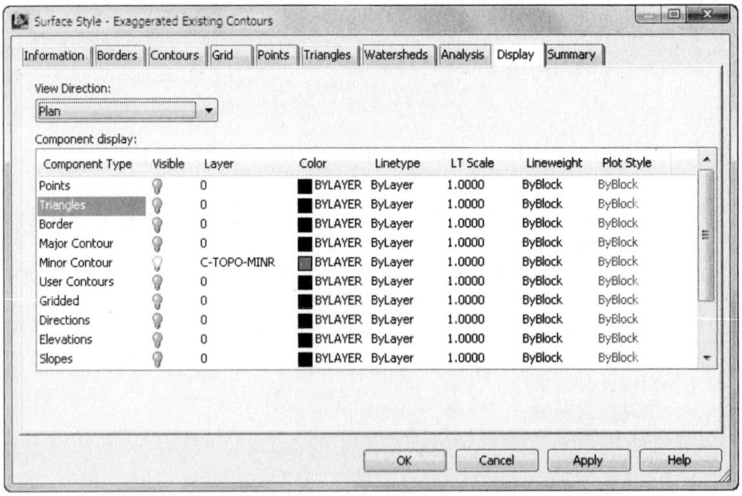

8. On the Display tab, change View Direction to Model and do the following:
   a. Shift+click to highlight all the components.
   b. Using the same procedure previously used, set all the Colors and Linetypes to ByLayer.

c. Verify that Triangles is the only component turned on.

d. Set the Triangles layer to C-TINN-VIEW.

To see the surface in the Object Viewer or in any other 3D view, you must have triangles set to display in Model.

9. Click OK to complete the style. Save the drawing.
10. From the Prospector tab of Toolspace, expand Surfaces, right-click the EG surface, and select Surface Properties.
11. Set Surface Style to Exaggerated Existing Contours, and click OK.

The surface should be rendered faster than you can read this sentence, even with the contour interval you've selected with only the minor contours displayed. After the style is applied to the surface model, you should see simple contours in plan view (the left image in Figure 21.26) and an exaggerated surface model in the Object Viewer (the right image in Figure 21.26).

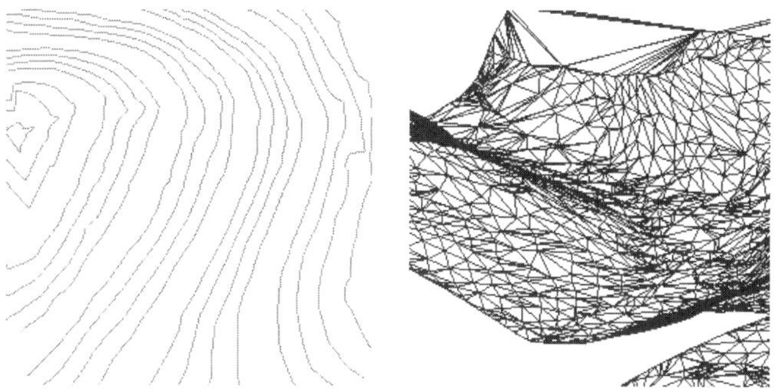

FIGURE 21.26
A portion of the surface showing your new style shown in plan (left) and in model (right) as shown in the Object Viewer

You skipped over one portion of the surface contours that many people consider a great benefit of using Civil 3D: depression contours. If this option is turned on via the Contours tab, ticks will be added to the downhill side of any closed contours leading to a low point. This is a stylistic option, and usage varies widely.

You can save and keep this drawing open to continue on to the next exercise or use the saved copy of this drawing available from the book's web page (`SurfaceStyleContours_FINISHED.dwg` or `SurfaceStyleContours_METRIC_FINISHED.dwg`).

Triangle and Points Surface Style

The next style you create will help facilitate surface editing. To work with the Swap Edge or Delete Line Surface edits, you must be able to see triangles. To work with the Delete Point, Modify Point, and Move Point commands, you must be able to see triangle vertices.

It is important to note that the points you see in the surface style do not refer to survey points. The points you are working with in this exercise are triangle vertices. The triangle vertices and survey points will initially be in the same locations for a surface built from points. However, as breaklines are added or edits are made to the triangle vertices, the surface model will differ from the original survey.

1. If it's not still open from the previous exercise, open `SurfaceStyleContours_FINISHED.dwg` or `SurfaceStyleContours_METRIC_FINISHED.dwg`

   You can download either file from this book's web page.

2. From the Settings tab of Toolspace, expand Surface ➤ Surface Styles.
3. Right-click the Exaggerated Existing Contours style you created in the previous exercise, and select Copy, as shown in Figure 21.27.

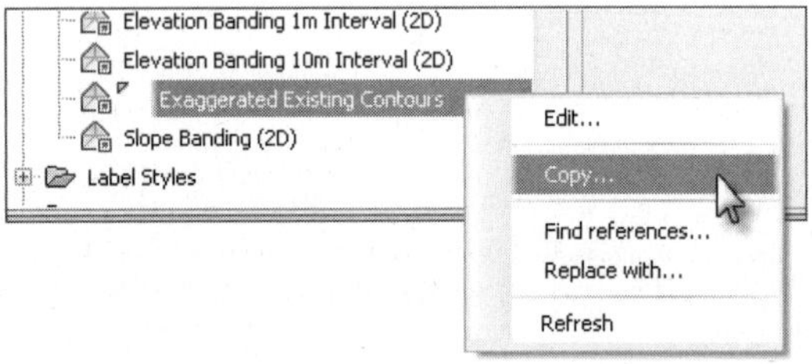

FIGURE 21.27 Access Copy by right-clicking directly on the style name.

4. On the Information tab, do the following:
   a. Rename the surface style to **Surface Editing**.
   b. Add the description **Points and triangles**.
5. On the Points tab, do the following:
   a. Expand Point Display and click the ellipsis for Data Point Symbol.
   b. Set the point type to the X symbol, as shown in Figure 21.28.

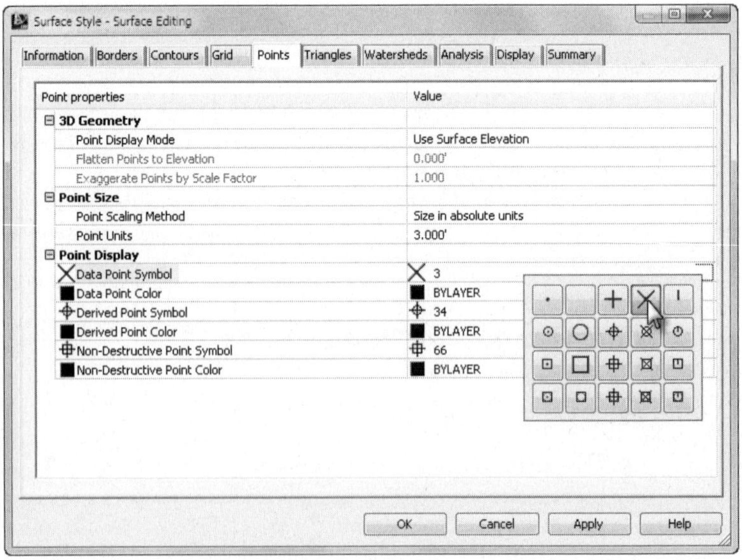

FIGURE 21.28 Change the point display so triangle vertices stand out.

6. On the Triangles tab, remove the exaggeration by setting the Triangle display mode back to Use Surface Elevation.

7. On the Display tab with the View Direction set to Plan, do the following:

   a. Use the lightbulb icons to turn off visibility for Minor Contour, and turn it on for Points and Triangles.

   b. Click in the Layer column for the Points components to display the Layer Selection dialog.

   c. In the Layer Selection dialog, click the New button to create a new layer for the Points.

   d. In the Create Layer dialog, set Layer Name to **C-TINN-PNTS**, set Color to Red, and click OK.

   e. In the Layer Selection dialog, set the layer to C-TINN-PNTS and click OK.

   f. Click in the Layer column for the Triangles components to display the Layer Selection dialog.

   g. In the Layer Selection dialog, set the Triangles layer to C-TINN-VIEW.

   Your Display tab will resemble Figure 21.29.

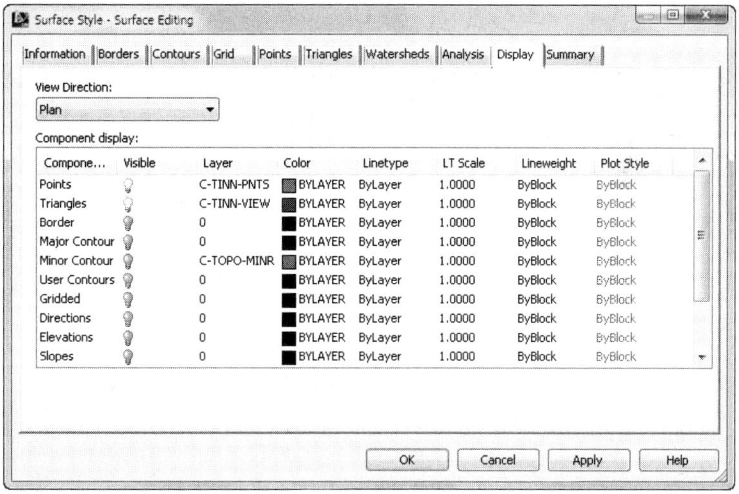

FIGURE 21.29
Change the Points and Triangles layers in the Display tab.

8. Click OK to complete the style.

9. Use the same procedure from the previous exercise to apply the Surface Editing style to the surface.

   Your surface will now resemble Figure 21.30.

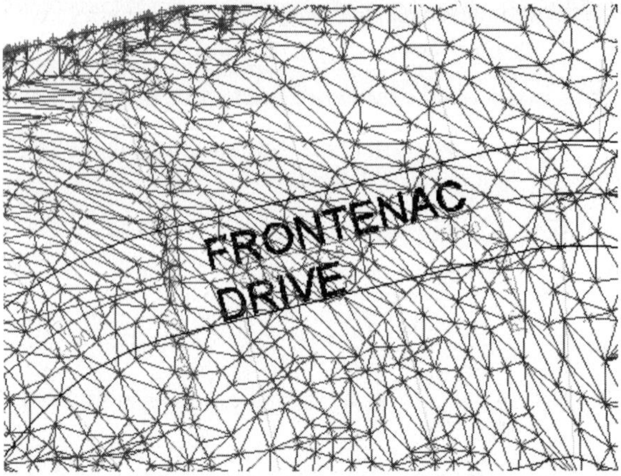

FIGURE 21.30
Triangles and points shown using the new surface style

With the points and triangles displayed, you can now edit the surface, such as deleting points or swapping edges.

You can save and keep this drawing open to continue on to the next exercise or use the saved copy of this drawing available from the book's web page (`SurfaceStyleTriangles_FINISHED.dwg` or `SurfaceStyleTriangles_METRIC_FINISHED.dwg`).

Analysis Styles

Analysis styles are unique in several ways. To see the style applied to your design, you must run the analysis in the surface properties in addition to applying the style to the surface. The elevation and slope analysis styles do not let you set a layer for the components because the colors and behavior are set on the Analysis tab.

These distribution methods show up in nearly all the surface analysis methods. Here's what they mean:

Equal Interval This method uses a stepped scale, created by taking the minimum and maximum values and then dividing the delta into the number of selected ranges. For example, if the surface has elevations from 0 to 10, with four ranges they will be 0 to 2.5, 2.5 to 5.0, 5.0 to 7.5, and 7.5 to 10. This method can create real anomalies when extremely large or small values skew the total range so that much of the data falls into one or two intervals, with almost no sampled data in the other ranges.

Quantile This method is often referred to as an equal count distribution and will create ranges that are equal in sample size. These ranges will not be equal in linear size but in distribution across a surface. For example, if the surface has elevations from 0 to 10 with most of the surface at the higher elevations, with four ranges they may be 0 to 4.78 (25 percent of the surface), 4.78 to 6.25 (25 percent of the surface), 6.25 to 7.95 (25 percent of the surface), and 7.95 to 10 (25 percent of the surface). This method is best used when the values are relatively equally spaced throughout the total range, with no extremes to throw off the group sizing.

Standard Deviation Standard Deviation is the bell curve that most engineers are familiar with, suited for when the data follows the bell distribution pattern. It generally works well for slope analysis, where very flat and very steep slopes are common and would make another distribution setting unwieldy.

You looked at an elevations, slopes, and slope arrows earlier in Chapter 4, "Surfaces." In the following exercise, you will create a surface style for watershed analysis. To apply the new style to the surface, you must also run the analysis.

1. If not still open from the previous exercise, open `SurfaceStyleTriangles_FINISHED.dwg` or `SurfaceStyleTriangles_METRIC_FINISHED.dwg`.

   You can download either file from this book's web page.

2. From the Settings tab of Toolspace, expand Surface ➢ Surface Styles.
3. Right-click the Exaggerated Existing Contours style you created in the earlier exercise and select Copy.
4. On the Information tab, do the following:
   a. Rename the surface style to **Watershed Analysis**.
   b. To the end of the current description add **Display watersheds and slope arrows**.
5. On the Triangles tab, change Triangle Display Mode to Use Surface Elevation.
6. On the Watersheds tab, do the following:
   a. Expand the Boundary Segment Watershed category.
   b. Set Use Hatching to False.
   c. Expand the Multi-Drain Watershed category.
   d. Set Use Hatching to False.

   Your Watersheds tab will now look like Figure 21.31.

7. On the Analysis tab, do the following:
   a. Expand the Slope Arrows category.
   b. Change Scheme to Hydro.
   c. Change Arrow Length to **2′** (or **1 m**).

   The Analysis tab will now resemble Figure 21.32.

8. On the Display tab with the View Direction set to Plan, do the following:
   a. Turn off visibility for the Minor Contour component, and turn it on for Slope Arrows and Watersheds — you may have to scroll.
   b. Set the Watershed layer to C-TINN-VIEW.
9. Click OK to complete the surface style.
10. Select the surface and open Surface Properties.

FIGURE 21.31
Changing the hatch options for watershed areas

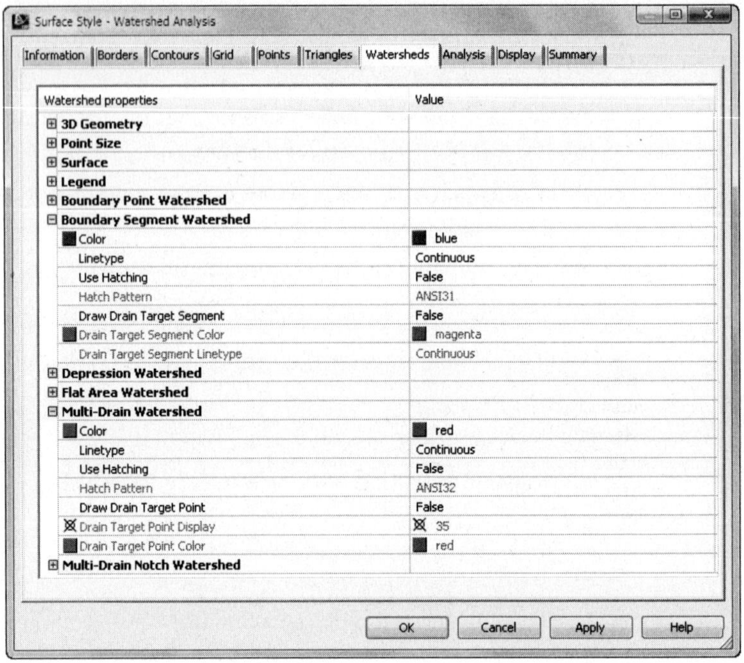

FIGURE 21.32
Set the color scheme and arrow length on the Analysis tab.

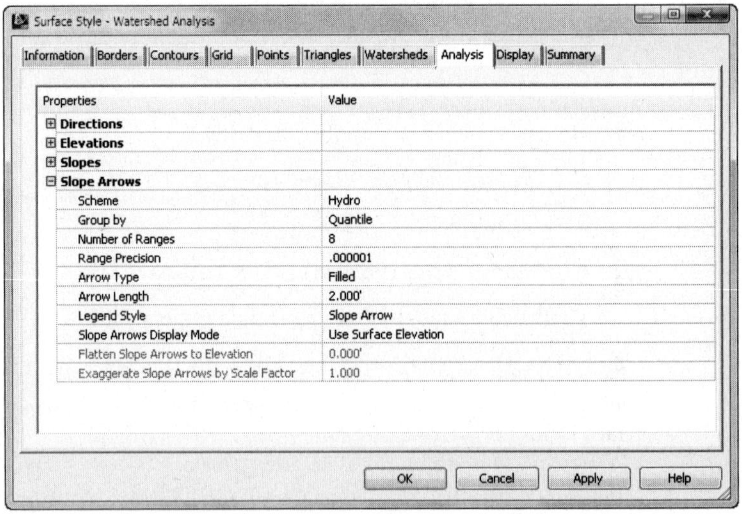

11. On the Information tab, set Surface Style to Watershed Analysis.
12. On the Analysis tab, do the following:
    a. Set Analysis type to Watersheds.
    b. Set Merge Depressions to **0.4′** (or **0.1 m**).
    c. Place a check mark next to Merge Adjacent Boundary Watersheds.

    d. Click the Run Analysis arrow in the middle of the dialog to populate the Range Details area.

    Your Surface Properties Analysis tab should now resemble Figure 21.33.

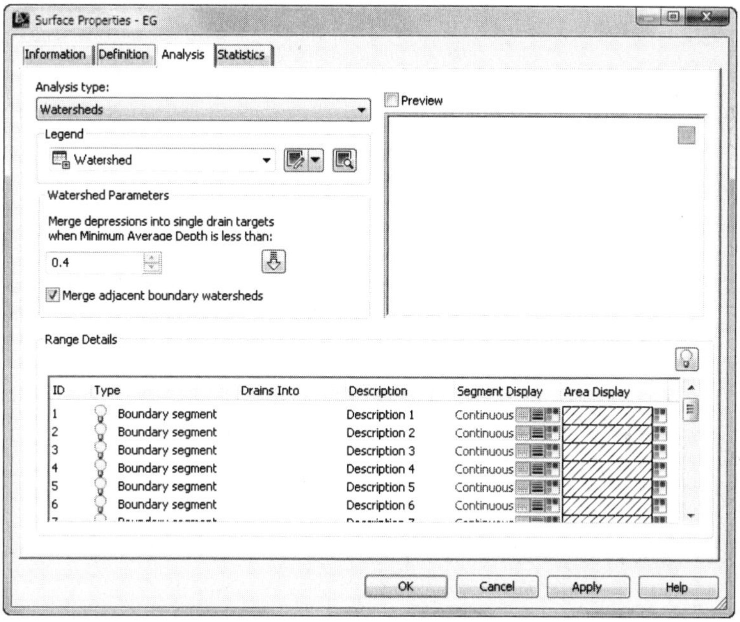

FIGURE 21.33
You must run the analysis in surface properties before the watershed style kicks in.

13. Click OK to close the Surface Properties dialog.

    Your surface model should now resemble Figure 21.34.

Now that the watershed analysis has been run, you could provide further information by generating a dynamic Watershed table similar to how you generated a table showing other surface information in Chapter 4.

When this exercise is complete, you can close the drawing. A saved copy of this drawing is available from the book's web page with the filename `SurfaceStyleWatershed_FINISHED.dwg` or `SurfaceStyleWatershed_METRIC_FINISHED.dwg`.

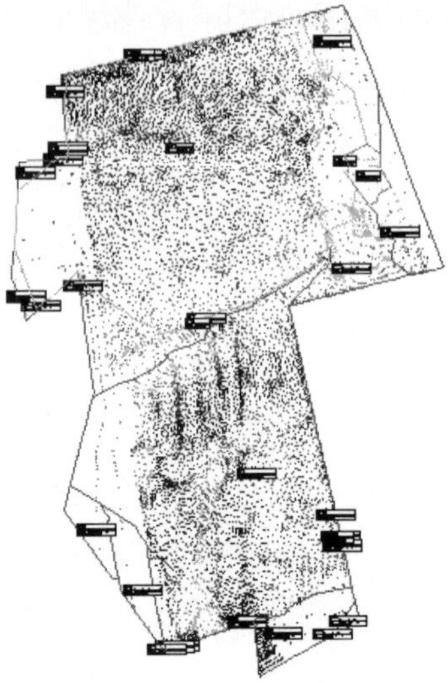

FIGURE 21.34
The isolated surface using the Watershed Analysis style

Pipe and Structure Styles

In Chapter 14, "Pipe Networks," you learned that the best way to manage pipes and structures was in a parts list. One of the functions of a parts list is to associate styles to pipes and structures. In this section, you will learn how to create the pipe and structure styles that are used by a parts list.

In your template, you will have multiple styles for the various parts lists. You will want to have separate styles for water systems, storm sewers, and sanitary sewers. Additionally, you may want to have separate styles for existing and proposed systems. The main difference between the styles for the different systems will be the layers you set in the Display tab.

Pipe Styles

It seems like no two municipalities want pipes displayed the same way on construction documents. Luckily, Civil 3D is very flexible in how pipes are shown. With a single pipe style, you can control how a pipe is displayed in plan, profile, and section views. You can use multiple pipe styles to graphically differentiate larger pipes from smaller ones. This section explores all the options.

Plan Tab The tab you see in Figure 21.35 controls what represents your pipe when you're working in plan view.

Options on the Plan tab include the following:

Pipe Wall Sizes You have a choice of having the program apply the part size directly from the catalog part (that is, the literal pipe dimensions as defined in the catalog), or specifying your own constant or scaled dimensions.

Pipe Hatch Options If you choose to show pipe hatching, this part of the tab gives you options to control that hatch. You can hatch the entire pipe to the inner or outer walls, or you can hatch the space between the inner and outer walls only, as shown in

Figure 21.36. There are also options to align the hatch to the pipe, and to clean up pipe-to-pipe connections.

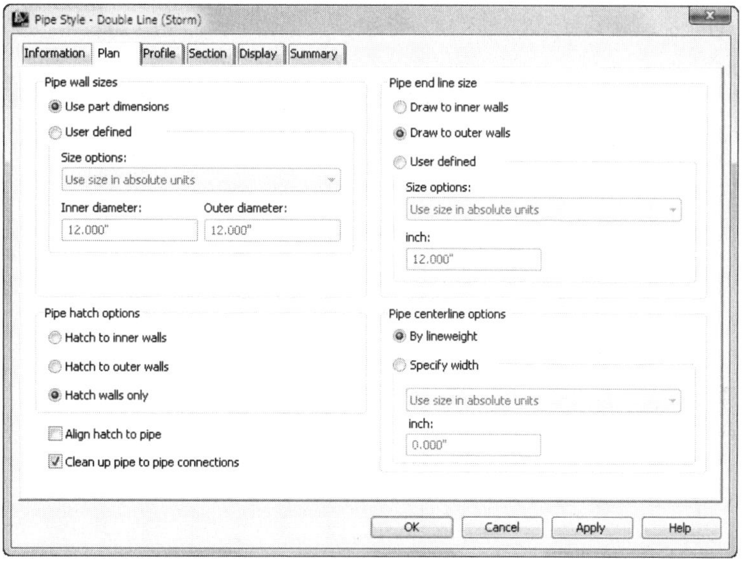

FIGURE 21.35
The Plan tab in the Pipe Style dialog

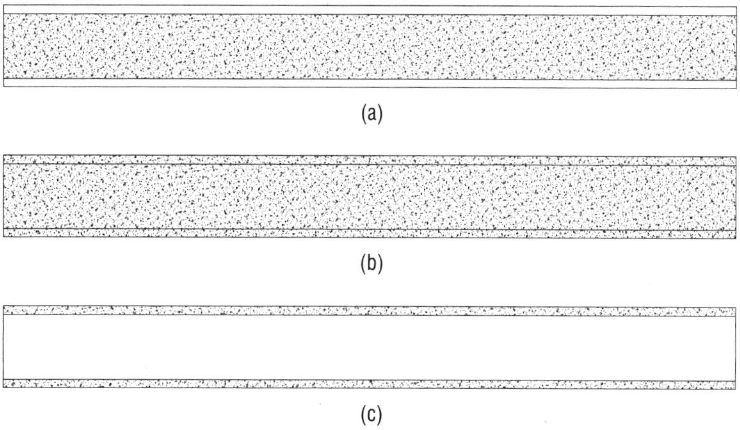

FIGURE 21.36
Pipe hatch to inner walls (a), outer walls (b), and hatch walls only (c)

Pipe End Line Size If you choose to show an end line, you can control its length with these options. An end line can be drawn connecting the outer walls or the inner walls, or you can specify your own constant or scaled dimensions. The pipes from Figure 21.36 are all shown with pipe end lines drawn to outer walls.

Pipe Centerline Options If you choose to show a centerline, you can display it by the lineweight established in the Display tab, or you can specify your own part-driven, constant, or scaled dimensions. You could use this option for your sanitary pipes in places where the width of the centerline widens or narrows on the basis of the pipe diameter.

Profile Tab The Profile tab (see Figure 21.37) is almost identical to the Plan tab, except the controls here determine what your pipe looks like in profile view. The only additional settings on this tab are the crossing-pipe hatch options. If you choose to display crossing pipe with a hatch, these settings control the location of that hatch.

FIGURE 21.37
The Profile tab in the Pipe Style dialog

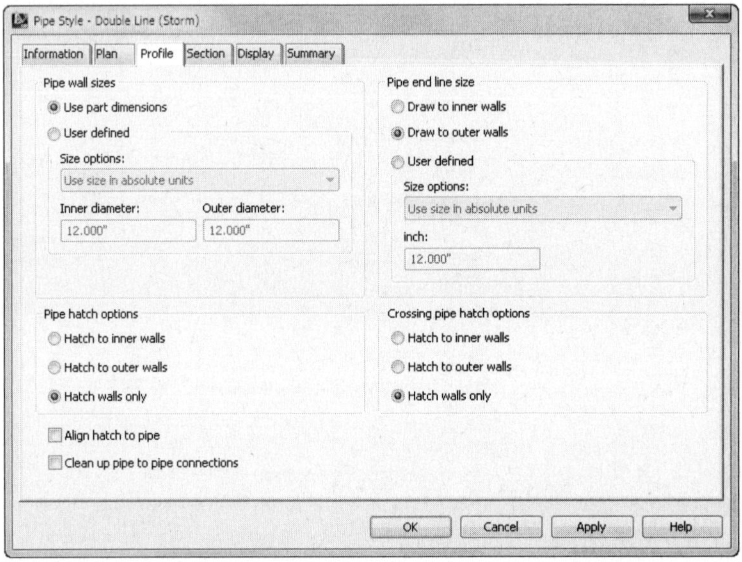

Section Tab If you choose to show a hatch on your pipes in section, you control the hatch location on this tab (see Figure 21.38).

FIGURE 21.38
The Section tab in the Pipe Style dialog

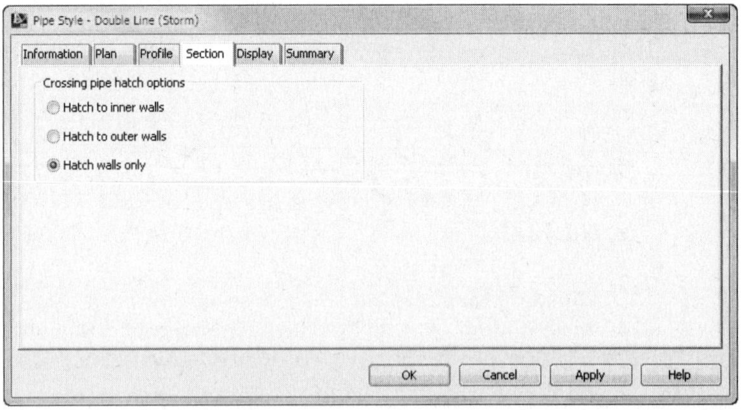

Pressure Pipe Styles

Pressure pipes are a new object for Civil 3D 2013, and with a new object comes new styles. You will find that the Pressure Pipe Styles dialog looks very similar to the Pipe Style dialog, with a Plan and Profile tab that have similar behavior to that discussed for regular pipes.

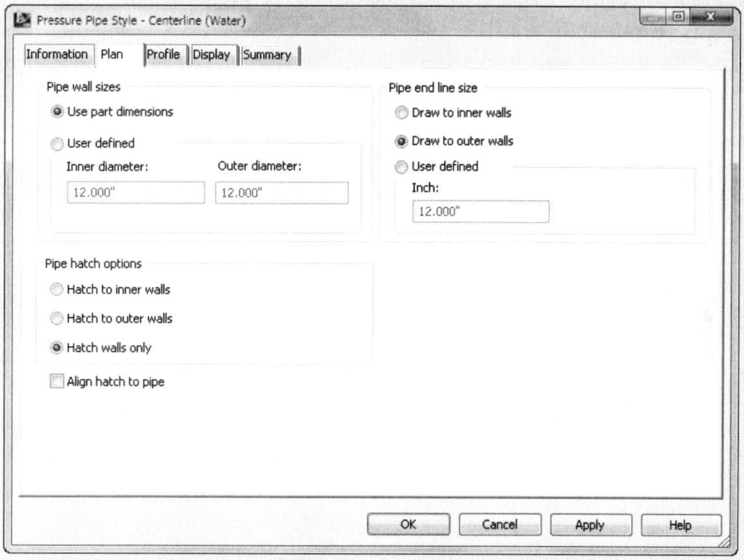

The biggest two differences are the lack of Section tab and Crossing Pipe Hatch options on the Plan tab. There are a few other minor changes to how the user can define sizes.

In the examples that follow, you will create various types of pipe styles.

The first style is for a situation where the pipe must be shown in plan view with a single line, the thickness of which matches the pipe inner diameter. In profile, the pipe will show the inner diameter lines, and in section it will show as a hatch-filled ellipse.

1. Open the `PipeStyle.dwg` or `PipeStyle_METRIC.dwg` file, which you can download from this book's web page.
2. From the Settings tab of Toolspace, expand Pipe ➢ Pipe Styles.
3. Right-click Pipe Styles and select New.
4. On the Information tab, set Name to **Proposed Sanitary CL**.
5. On the Plan tab, do the following:
   a. Verify that Pipe Centerline Options is set to Specify Width.
   b. Set Specify Width to Draw To Inner Walls.
6. On the Profile tab, no changes are needed.

7. On the Section tab, verify that Crossing Pipe Hatch Options is set to Hatch To Inner Walls.
8. On the Display tab with the View Direction set to Plan, do the following:
   a. Turn off the display for all components except Pipe Centerline.
   b. Set the Pipe Centerline layer to C-SSWR-PIPE.
9. On the Display tab, change View Direction to Profile and do the following:
   a. Turn off the display for all components except Inside Pipe Walls.
   b. Set the layer to C-SSWR-PROF.
10. On the Display tab change View Direction to Section and do the following:
    a. Set Crossing Pipe Inside Wall to the C-SSWR-PIPE layer and make it visible.
    b. Set Crossing Pipe Hatch to the C-SSWR-PIPE-PATT layer and make it visible.
    c. Turn off Crossing Pipe Outside Wall.
    d. At the bottom of the dialog, click in the Pattern column for the Crossing Pipe Hatch component type to display the Hatch Pattern dialog.
    e. Set the hatch pattern to Solid Fill, as shown in Figure 21.39, and click OK.

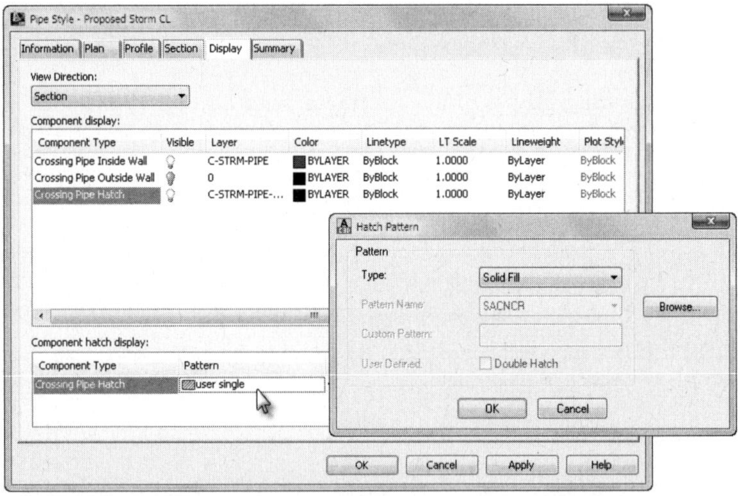

FIGURE 21.39 Setting the hatch pattern display for the section view direction

11. Click OK to complete creation of a new pipe style.
12. From the Prospector tab of Toolspace, expand Pipe Networks ➢ Networks ➢ Sanitary Network ➢ Pipes.

13. Select all of the pipes in the Item List using the Shift key.
14. Right-click the heading of the Style column and select Edit.
15. In the Select Pipe Style dialog, select Proposed Sanitary CL and click OK.
16. Examine the pipe in plan, profile, and cross section, as shown in Figure 21.40.

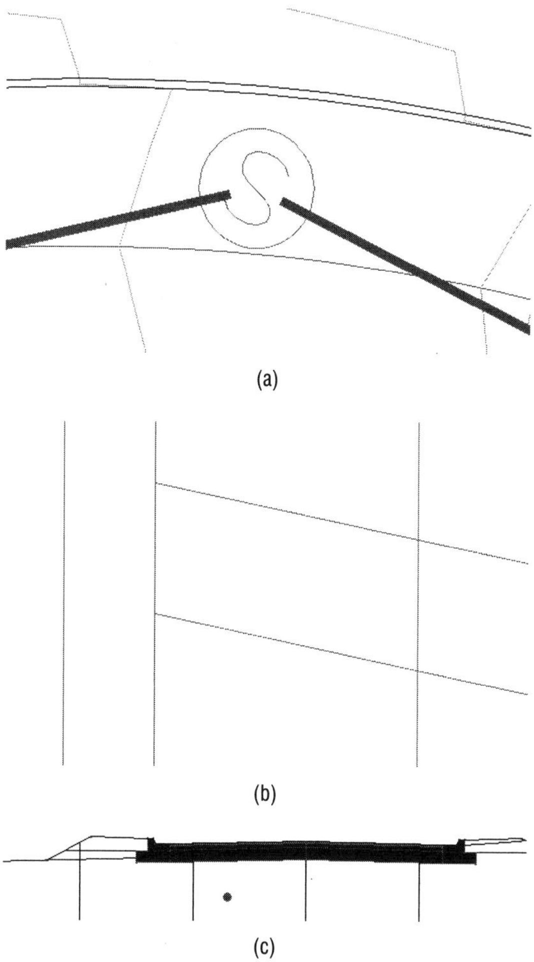

FIGURE 21.40
Proposed Sanitary CL pipe style shown in plan (a), profile (b), and section (c)

You can save and keep this drawing open to continue to the next exercise, or use the saved copy of this drawing available from the book's web page (`PipeStyle_FINISHED.dwg` or `PipeStyle_METRIC_FINISHED.dwg`).

HEY! WHY DOES MY PIPE LOOK LIKE AN OCTAGON?

Civil 3D helps itself perform better on large drawings by knocking down the resolution of 3D curved objects, such as the pipe in the cross section view in the previous steps.

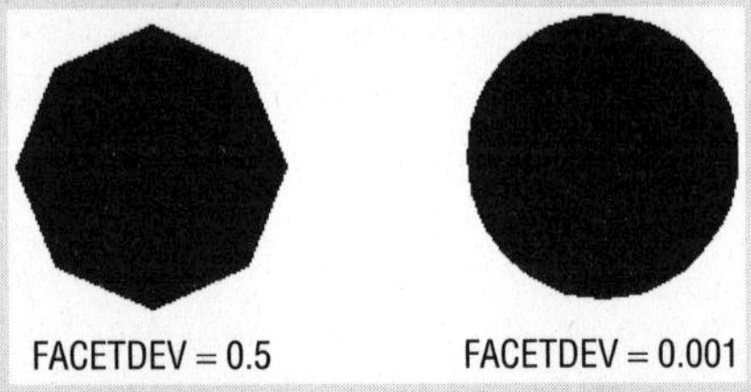

The system variable you can use to make these pipes look nicer is Facet Deviation, or FACETDEV. The default FACETDEV value for any Imperial unit drawing is 0.5 inches. In metric drawings, the default is 10 millimeters. The lower the FACETDEV value, the smoother the 3D curve.

Note that another variable, FACETMAX, controls the maximum number of facets on any curved object. The Civil 3D default of 500 facets is usually more than enough to display a Civil 3D pipe smoothly.

In the next pipe style example, you will create a style that uses several options for hatching pipe walls for a pipe:

1. If not still open from the previous exercise, open `PipeStyle_FINISHED.dwg` or `PipeStyle_METRIC_FINISHED.dwg`.
2. From the Settings tab of Toolspace, expand Pipe ➢ Pipe Styles.
3. Right-click Pipe Styles and select New.
4. On the Information tab, set Name to **Proposed Hatch Wall**.
5. On the Plan tab, verify that the Pipe Hatch Options is set to Hatch Walls Only.
6. On the Profile tab, do the following:
    a. Verify that the Pipe Hatch Options is set to Hatch Walls Only.
    b. Verify that Align Hatch To Pipe is selected.
7. On the Display tab with View Direction set to Plan, do the following:
    a. Verify that the only components turned on are Inside Pipe Walls, Outside Pipe Walls, Pipe End Line, and Pipe Hatch components.
    b. Set all four layers to C-SSWR-PIPE.
8. On the Display tab, change View Direction to Profile and do the following:
    a. Verify that the only components turned on are Inside Pipe Walls, Outside Pipe Walls, Pipe End Line, and Pipe Hatch.
    b. Set all four of these layers to C-SSWR- PROF.

9. Click OK to complete creation of a new pipe style.
10. From the Prospector tab of Toolspace, expand Pipe Networks ➤ Networks ➤ Sanitary Network ➤ Pipes.
11. Select all of the pipes in the Item List using the Shift key.
12. Right-click the heading of the Style column and select Edit.
13. In the Select Pipe Style dialog, select Proposed Hatch Wall, and click OK.
14. Examine the pipe in plan and profile as shown in Figure 21.41.

    As you can see depending on your drawing's scale, it may or may not be worth it to you to add a pipe wall hatch since with thin walls it may be hard to see, as shown in Figure 21.41a.

FIGURE 21.41 Proposed Hatch Wall pipe style shown in plan (a), profile (b), and section (c)

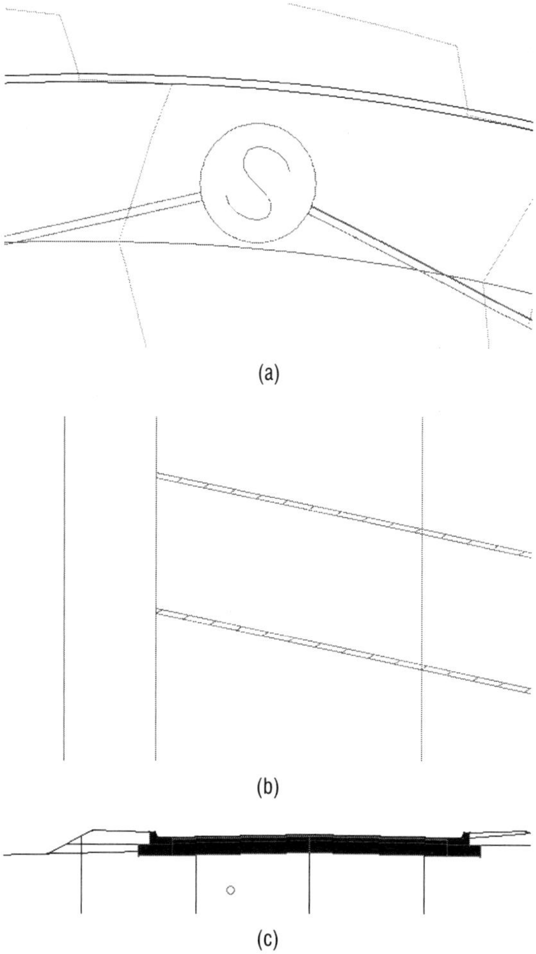

When this exercise is complete, you can close the drawing. A saved copy of this drawing is available from the book's web page with the filename `PipeStyleHatch_FINISHED.dwg` or `PipeStyleHatch_METRIC_FINISHED.dwg`.

You could also do this same exercise for a pressure pipe style. The only difference would be setting the layers to C-WATR-PIPE instead of C-SSWR-PIPE, or C-WATR-PROF instead of C-SSWR-PROF.

Structure Styles

The following tour through the structure-style interface can be used for reference as you create company-standard styles:

Model Tab The Model tab (Figure 21.42) controls what represents your structure when you're working in 3D. Typically, you want to leave the Use Catalog Defined 3D Part radio button selected so that when you look at your structure, it looks like your flared end section or whatever you've chosen in the parts list.

FIGURE 21.42
The Model tab in the Structure Style dialog

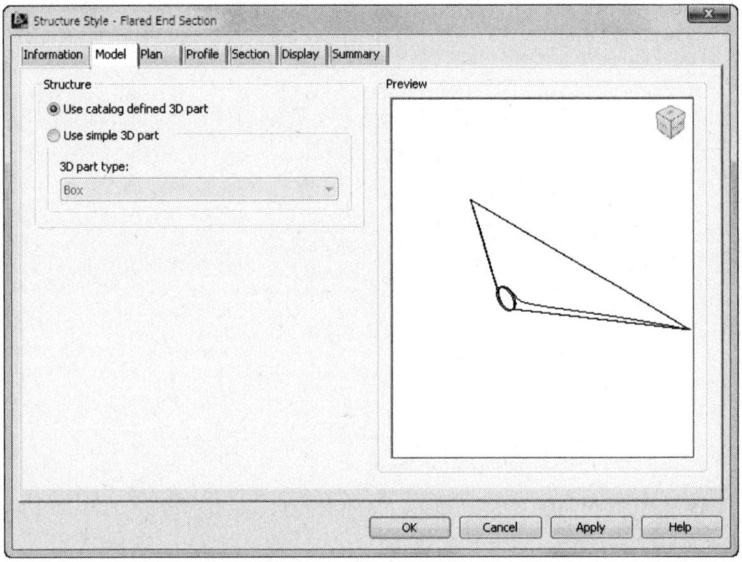

Plan Tab The Plan tab (Figure 21.43) enables you to compose your object style to match that specification.

Options on the Plan tab include the following:

Use Outer Part Boundary This option uses the limits of your structure from the parts list and shows you an outline of the structure as it would appear in the plan.

User Defined Part This option uses any block you specify. In the case of your flared end section, you chose a symbol to match the CAD standard. When using a User Defined Part you also must provide Size Options. The options in this drop-down are similar to what you see in other places in Civil 3D. Use Size In Absolute Units is a common way to represent a manhole at actual size. Use Drawing Scale will treat the object like an annotative block.

Enable Part Masking This option creates a wipeout or mask inside the limits of the structure. Any pipes that connect to the center of the pipe appear trimmed at the limits of the structure.

Profile Tab Once you know what your structure must look like in the profile, you use the Profile tab (Figure 21.44) to create the style.

FIGURE 21.43
The Plan tab in the Structure Style dialog

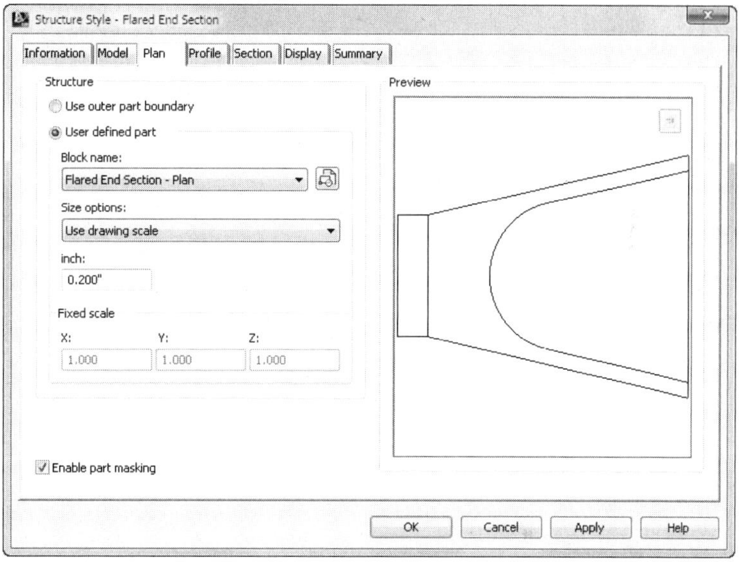

FIGURE 21.44
The Profile tab in the Structure Style dialog

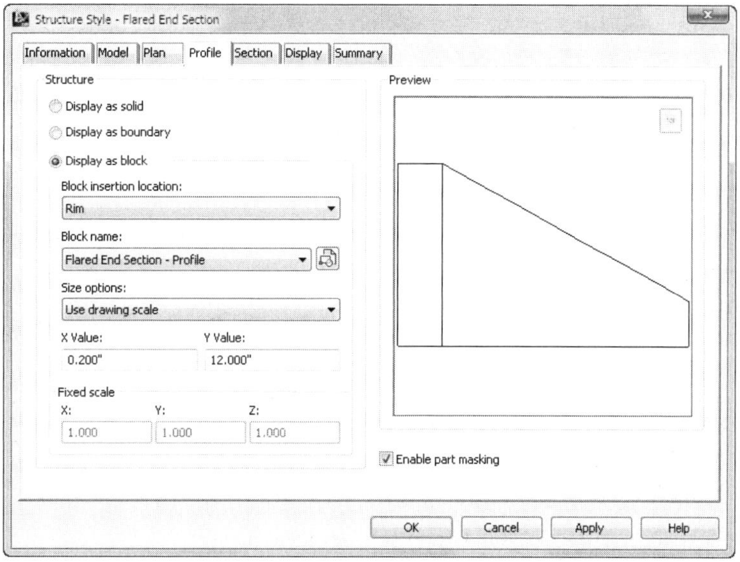

Options on the Profile tab include the following:

Display As Solid This option uses the limits of your structure from the parts list, and shows you the mesh of the structure as it would appear in profile view.

Display As Boundary This option uses the limits of your structure from the parts list, and shows you an outline of the structure as it would appear in profile view. You'll use this option for the sanitary manhole.

Display As Block This option uses any block you specify. When using a Structure displayed as a block you also must provide Size Options.

Enable Part Masking This option creates a wipeout or mask inside the limits of the structure. Any pipes that connect to the center of the pipe appear trimmed at the limits of the structure.

Section Tab Once you know what your structure must look like in section view, you use the Section tab (Figure 21.45) to create the style.

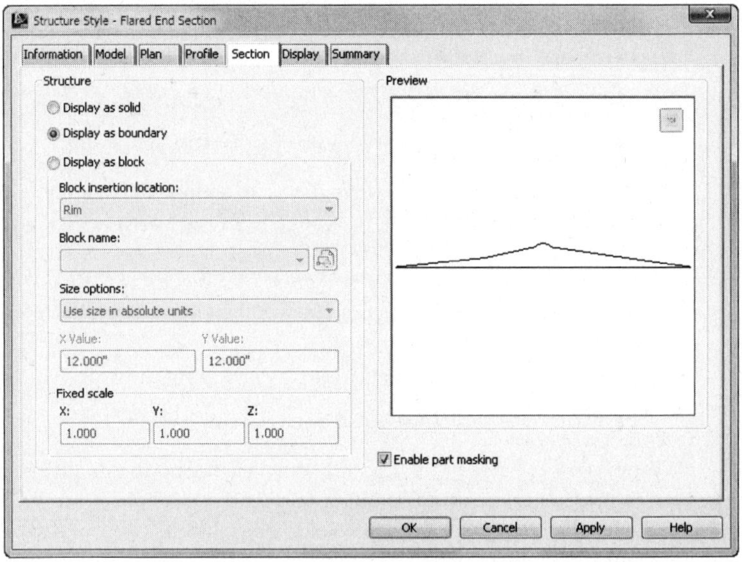

FIGURE 21.45
The Section tab in the Structure Style dialog

You will notice that these options look (and behave) very similar to the Profile tab options listed earlier.

In the following exercise, you'll create a new structure style that uses a block in plan view to represent a sanitary manhole. Because the block is drawn at actual size, you will use the size option Use Fixed Scale.

1. Open the `StructureStyle.dwg` or `StructureStyle_METRIC.dwg` file, which you can download from this book's web page.
2. From the Settings tab of Toolspace, expand Structure ➤ Structure Styles.

3. Right-click Structure Styles and select New.
4. On the Information tab, rename the style to **Simple Sanitary Manhole**.
5. On the Plan tab, do the following:
    a. Verify that User Defined Part is selected.
    b. Set the Block Name to **_Wipeout_Circle** using the list box.
    c. Set Size to Use Fixed Scale.
    d. Set the X and Y scale factors set to **3** (or **1** for metric).
6. On the Display tab with View Direction to Plan, set the Structure layer to C-SSWR-STRC.
7. Repeat step 6 with View Direction set to Profile and then set to Section.
8. Click OK to complete the style.
9. From the Prospector tab of Toolspace, expand Pipe Networks ➢ Networks ➢ Sanitary Network ➢ Structures.
10. Select all of the structures in the Item List using the Shift key.
11. Right-click the heading of the Style column and select Edit.
12. In the Select Pipe Style dialog, select Simple Sanitary Manhole, and click OK.

When this exercise is complete, you can close the drawing. A saved copy of this drawing is available from the book's web page with the filename `StructureStyle_FINISHED.dwg` or `StructureStyle_METRIC_FINISHED.dwg`.

While this example discussed creating object styles for a structure, you will find that the same procedure is applicable to the Appurtenances and Fittings styles used in pressure pipe networks.

Profile View Styles

CERT OBJECTIVE

When you are looking at a profile view that contains data, you are seeing many styles in play. The profiles themselves (existing and proposed) have a profile object style applied to them. The labeling has several styles applied to it, as you learned in Chapter 20. Additionally, you will see profile view styles and band styles in action.

This section focuses on the profile view. A profile view controls many aspects of the display. The *profile view style* has a large role in determining:

- Vertical exaggeration
- Grid spacing
- Elevation annotation
- Title annotation

Figure 21.46 shows the basic anatomy of the profile view you have been working with throughout this book.

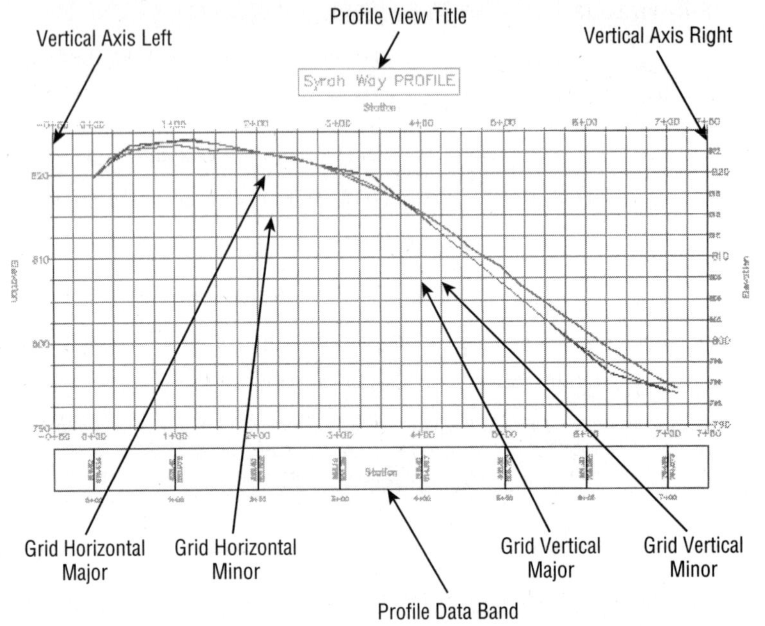

FIGURE 21.46
Profile view style and basic anatomy

In the example that follows, you will be making major modifications to a profile view style. The profile view that you will be practicing with does not contain any bands. Later on in this section, you will learn the ins and outs of band creation and modification.

1. Open the `ProfileViewStyles.dwg` or `ProfileViewStyles_METRIC.dwg` file.

   This file contains a profile view of Cabernet Court with a very ugly style applied to it. You will be performing a complete makeover on this style.

2. Select the profile view by clicking anywhere on a grid line or axis.
3. From the contextual tab ➢ Modify View panel, select Profile View Properties ➢ Edit Profile View Style, as shown in Figure 21.47.

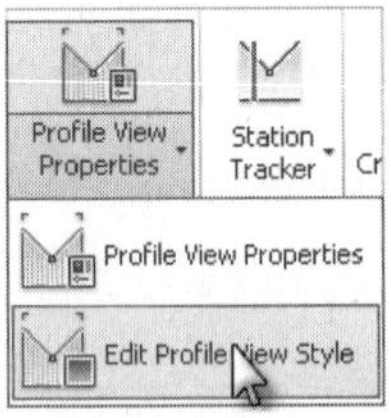

FIGURE 21.47
Accessing the profile view style

4. Position the dialog on your screen so you can make changes to the style and observe the changes in the profile view behind it.
5. On the Graph tab, change the Vertical Exaggeration value to **10**, as shown in Figure 21.48, and click Apply.

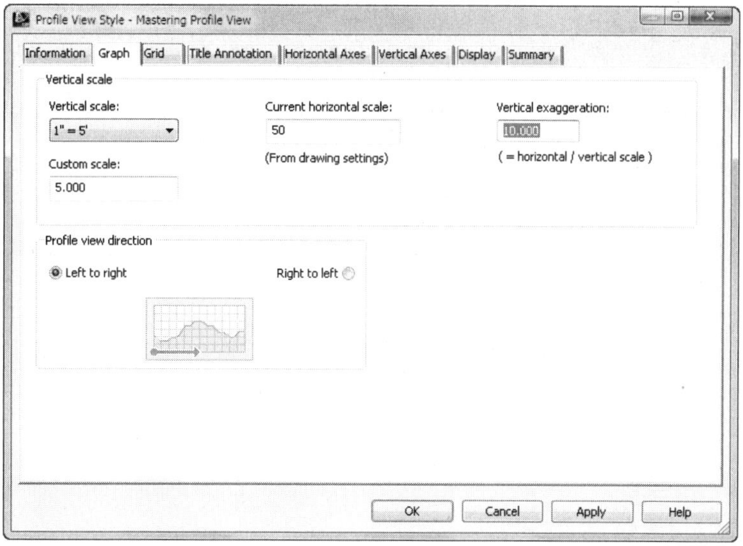

FIGURE 21.48 Change Vertical Exaggeration on the Graph tab of the Profile View Style dialog.

When you do so, you will notice that the vertical scale listed in the dialog automatically changes from 1″ = 50′ to 1″ = 5′ (or from 1:500 to 1:50 for metric users). In addition, if you can see your profile view in the background, you will notice that it now has expanded vertically by a factor of 10.

6. On the Grid tab, do the following:

a. Verify that both Clip Vertical Grid and Clip Horizontal Grid options are unchecked.

b. Set Grid Padding (Major Grids) to **0.5** for Above Maximum Elevation and **0.5** for Below Datum.

This setting will create additional space above and below the design data at least 0.5 times the vertical major tick interval (you will set the major tick interval later on).

c. Set all values for Axis Offset (Plotted Units) to **0**.

This will ensure that the axes and the grid lines coincide around the edges of the view. The settings on the Grid tab should now match what is shown in Figure 21.49.

7. On the Title Annotation tab, do the following:

a. In the Graph View Title area, change Text Height to **0.4**″ (or **10** mm).

b. In the Graph View Title area, click the Edit Mtext button.

c. In the Text Component Editor, remove all the text in the Preview area.

You will be starting over with a blank Text Component Editor.

d. From the Properties drop-down list, select Parent Alignment and make these changes:

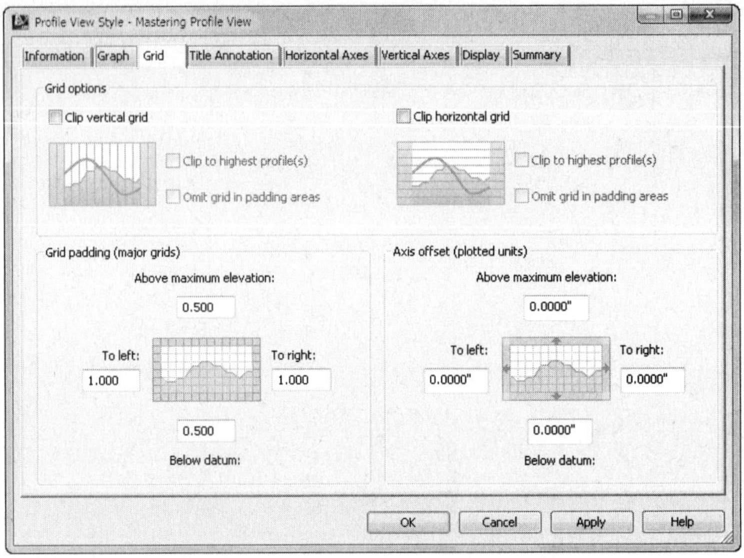

FIGURE 21.49
The Grid tab of the Profile View Style dialog

- *a.* Set Capitalization to Upper Case.
- *b.* Click the arrow button to add the text to the Preview area.
- *c.* After the dynamic text, add a space and type **PROFILE VIEW**.

- **e.** Click OK to accept the entry in the Text Component Editor.
- **f.** Change the Y Offset for the Title Position to **2**″ (or **10** mm).

8. Click Apply and examine your changes in the background.

The Title Annotation tab should now match the settings shown in Figure 21.50.

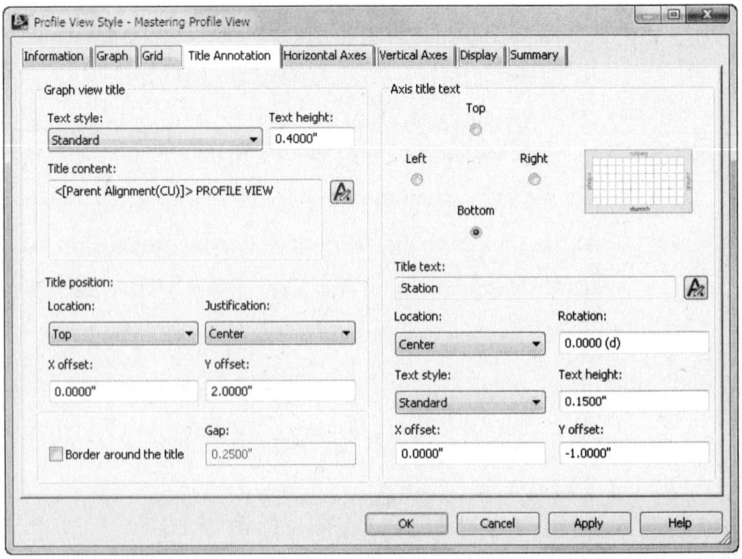

FIGURE 21.50
Working with the Graph View Title size and placement

You will not bother to adjust any settings for the Axis title text on the right side of Figure 21.50 because the display will be turned off for all four of these possible elements.

9. On the Horizontal Axes tab, do the following:

a. Verify that the Axis To Control radio button is set to Bottom.

b. In the Major Tick Details area, set Interval to **100′** (or **20** m).

c. In the Major Tick Details area, click the Edit Mtext button.

d. In the Text Component Editor, do the following:

a. Remove all the text in the Preview area.

b. From the Properties drop-down list, select Station Value.

c. Change the precision to **1**.

d. Click the arrow button to add the text to the Preview area.

e. Click OK to accept the entry in the Text Component Editor.

f. In the Major Tick Details area, change Rotation to **90**.

g. In the Major Tick Details area, change the X offset to **0″** (**0** mm) and the Y offset to **−0.25″** (**−10** mm).

h. In the Minor Tick Details area, set Interval to **50′** (**10** m).

No other changes are needed in the Minor Tick Details area.

The Horizontal Axes tab should match the settings shown in Figure 21.51.

FIGURE 21.51
The bottom axis controls grid spacing.

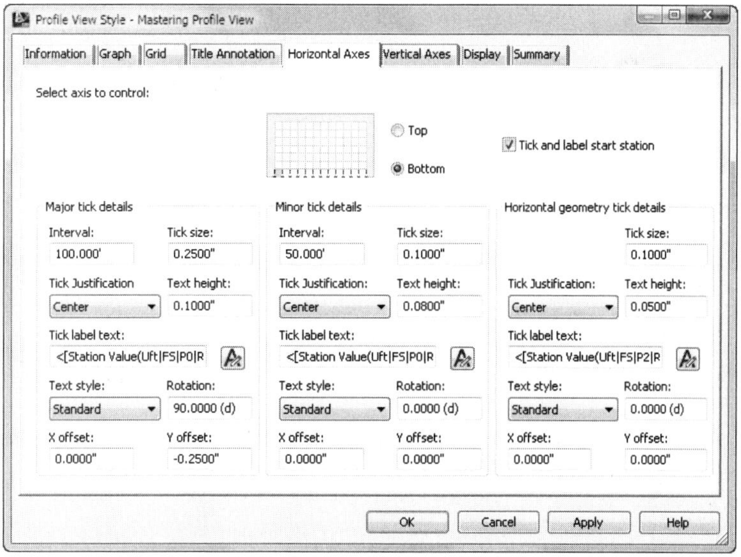

10. On the Vertical Axes tab, do the following:

a. Verify that the Select Axis To Control radio button is set to Left.

b. In the Major Tick Details area, click the Edit Mtext button.

c. In the Text Component Editor, do the following:

a. Remove all the text in the Preview area.

b. From the Properties drop-down list, select Profile View Point Elevation.

c. Change Precision to **1**.

d. Click the arrow button to add the text to the Preview area.

d. Click OK to exit the Text Component Editor.

e. In the Major Tick Details area, change the X offset to **−0.25″** (**−5** mm) and the Y offset to **0″** (**0** mm).

All the changes made up to this point on the Vertical Axes tab apply to the Left axis. You will now do similar modifications to the Right axis.

f. Change the Select Axis To Control radio button to Right.

g. In the Major Tick Details area, click the Edit Mtext button.

h. In the Text Component Editor, do the following:

a. Remove all the text in the Preview area.

b. From the Properties drop-down list, select Profile View Point Elevation.

c. Change Precision to **1**.

d. Click the arrow button to add the text to the Preview area.

i. Click OK to exit the Text Component Editor.

j. In the Major Tick Details area, change the X offset to **0.25″** (**5** mm) and the Y offset to **0″** (**0** mm).

11. Click Apply and examine your changes.

Figure 21.52 shows the Vertical Axes tab as yours should look at this point in the exercise.

12. On the Display tab, do the following:

a. Select all of the components and use the lightbulb icon to turn off visibility for all of the components.

b. Turn back on the visibility for the following components:

- Graph Title
- Left Axis

- Left Axis Annotation Major
- Right Axis
- Right Axis Annotation Major
- Top Axis
- Bottom Axis
- Bottom Axis Annotation Major
- Grid Horizontal Major
- Grid Horizontal Minor
- Grid Vertical Major
- Grid Vertical Minor

13. Click OK to complete the profile view style.

Your profile view should resemble Figure 21.53.

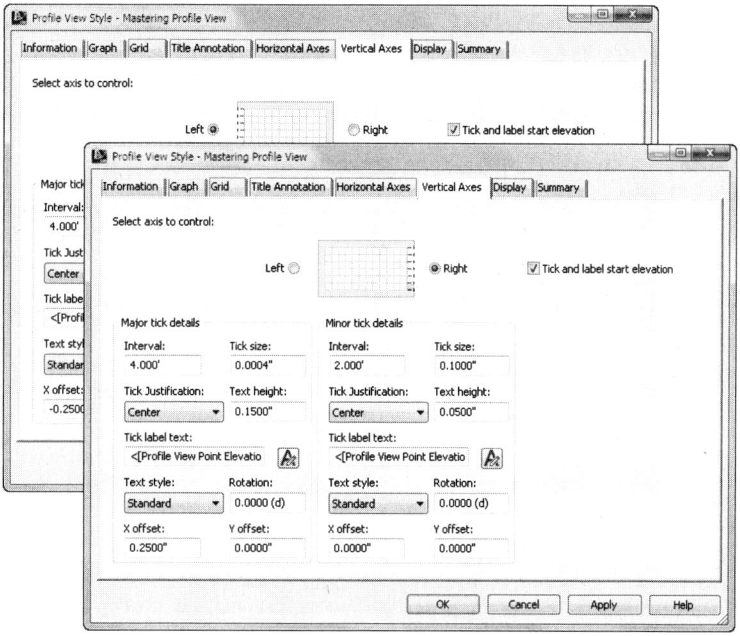

FIGURE 21.52
Don't forget to change the settings for both Left and Right axes in this tab.

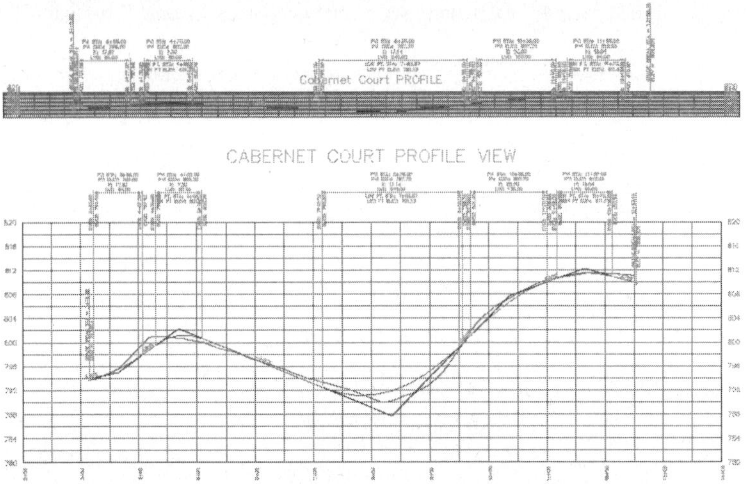

FIGURE 21.53
Profile view that you started with (above) and after the style is completed (below)

WHAT DRIVES PROFILE VIEW GRID SPACING?

On the Horizontal Axes tab and Vertical Axes tab of the Profile View Style dialog, you may notice that you can control opposing axes separately. Each tab has a toggle for Select Axis To Control; Top or Bottom and Left or Right.

It is the Bottom and Left options in the respective tabs that control grid spacing. For both axes, you will find options for Major Tick Details. The Interval values for the major ticks are the key to getting the grid spacing to look the way you want. Changes to either horizontal or vertical major tick intervals will affect the height and length of the Profile view, as well as grid spacing. Changing the Interval on Minor Tick Details will affect the grid spacing, but will not affect the aspect ratio of the profile view.

A good rule of thumb is to set your bottom horizontal major tick Interval value equal to the view's Vertical Exaggeration multiplied by the left vertical axis Interval value. For example, if you set Vertical Exaggeration for the view to 10 on the Graph tab and your left vertical axis major tick Interval is 10', the Horizontal Major Tick Interval would be 100'.

Even if you don't turn on the ticks or grid lines on these axes, the spacing increment will be reflected in the profile view.

Profile View Bands

Data bands are horizontal elements that display additional information about the profile or alignment that is referenced in a profile view. The most common band type is the profile data band. The other band types are not frequently used, but can be helpful to designers by showing schematic representations of design as it relates to the profile stationing.

Bands can be applied to both the top and bottom of a profile view, and there are six band types: Profile Data Bands, Vertical Geometry Bands, Horizontal Geometry

Bands, Superelevation Bands, Sectional Bands, and Pipe Data Bands. These band types were discussed in Chapter 7, "Profiles and Profile Views," but a graphic reminder of the various band types is shown in Figure 21.54 through Figure 21.59.

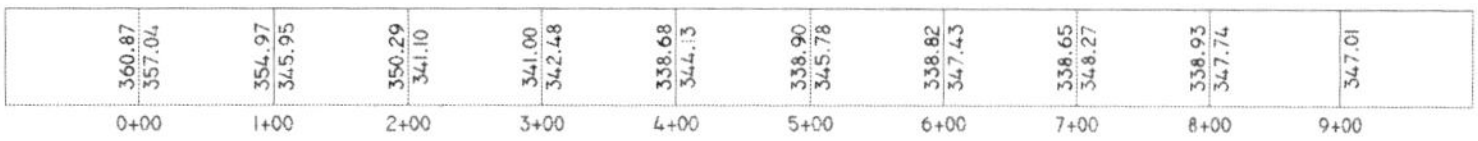

FIGURE 21.54 Profile Data Band showing existing and proposed elevation in addition to major stations

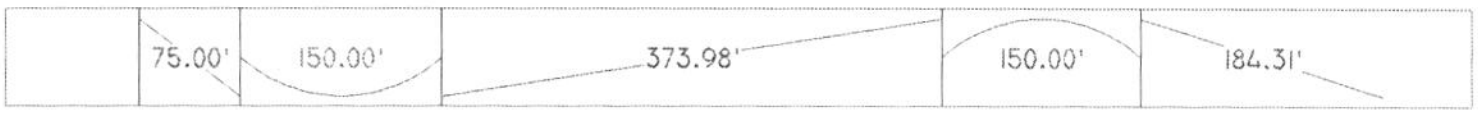

FIGURE 21.55 Vertical Geometry Band

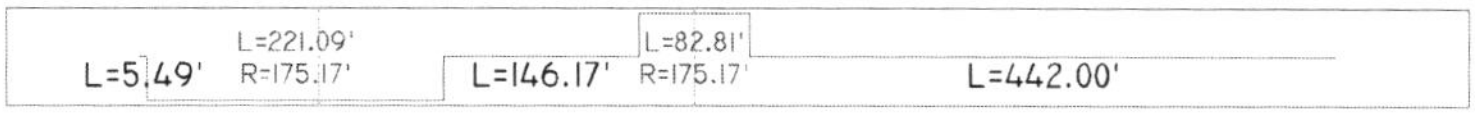

FIGURE 21.56 Horizontal Geometry Band

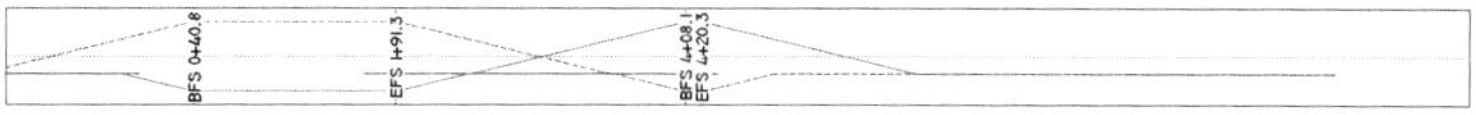

FIGURE 21.57 Superelevation Data Band

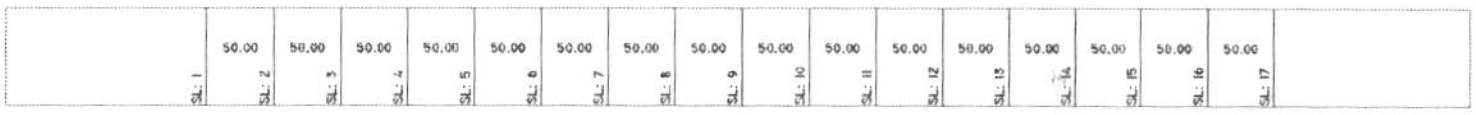

FIGURE 21.58 Section Data Band

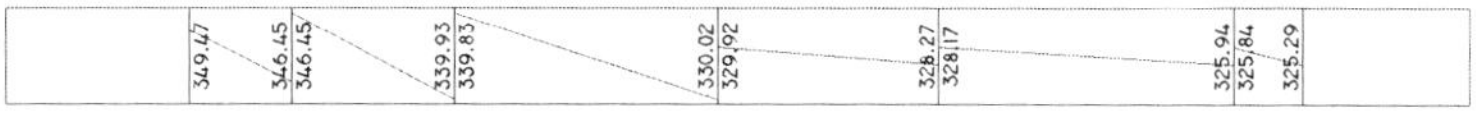

FIGURE 21.59 Pipe Data Band showing invert elevations and slope schematic

Bands can be saved in *band sets*. Like alignment label sets and profile label sets, a band set determines which bands are applied to a profile view and will dictate the placement. The most common use for a band set is to create a single, viewable band and an additional nonvisible band for spacing purposes in plan and profile sheet generation.

In the following exercise, you will create a band that contains existing and proposed profile elevations:

1. Open the `ProfileBands.dwg` or `ProfileBands_METRIC.dwg` file.

   This file contains the profile view from the previous exercise and an empty band on the bottom of the view. You will be adding information to the profile data band.

2. From the Settings tab of Toolspace, expand Profile View ➤ Band Styles ➤ Profile Data.
3. Right-click the style called Mastering Band and select Edit.
4. On the Band Details tab, do the following:
    a. Change Band Height to **1.0″** (or **25** mm).
    b. On the right side of the Band Details tab, highlight Major Station, and click the Compose Label button, as shown in Figure 21.60.

FIGURE 21.60
Band Details tab

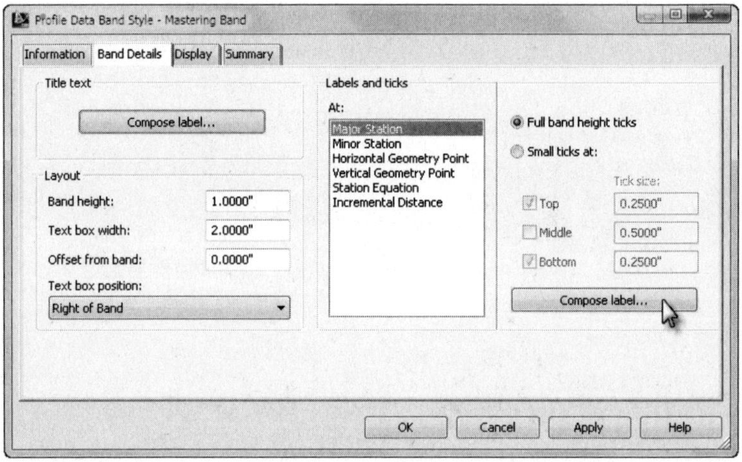

c. On the Layout tab of the Label Style Composer – Major Station dialog, click the Create Text Component button and make these changes:
    a. Set Name to **Station**.
    b. Use the drop-down list to change Anchor Point to Band Bottom.
    c. Use the drop-down list to change Attachment to Top Center.
    d. Change the Y offset to **−0.02″** (or **−0.5** mm).
    e. Click the Contents value which currently reads Label Text, and click the ellipsis to display the Text Component Editor dialog.
    f. In the Text Component Editor, remove all text in the Preview area, and select Station Value from the Properties list.
    g. Set Precision to **1**.
    h. Click the arrow button to add the text to the Preview area.
d. Click OK to accept the Text Component Editor.
e. Click the Create Text Component button again, and make these changes:
    a. Set Name to **Existing El**.
    b. Use the drop-down list to change Anchor Point to Band Middle.

   c. Set Rotation Angle to **90**.

   d. Use the drop-down list to change Attachment to Bottom Center.

   e. Change the X offset to **–0.02″** (or **–0.5** mm).

   f. Click the Contents value, which currently reads Label Text, and click the ellipsis to display the Text Component Editor dialog.

   g. In the Text Component Editor, remove all text in the Preview area, and select Profile1 Elevation from the Properties list.

   h. Set Precision to **0.01** (or **0.001** for metric).

   i. Click the arrow button to add the text to the Preview area.

f. Click OK to accept the Text Component Editor.

g. Click the Create Text Component button a third time, and make these changes:

   a. Set Name to **Proposed El**.

   b. Use the drop-down list to change Anchor Point to Band Middle.

   c. Set Rotation Angle to **90**.

   d. Use the drop-down list to change Attachment to Top Center.

   e. Change the X offset to **0.02″** (or **0.5** mm).

   f. Click the Contents value, which currently reads Label Text, and click the ellipsis to display the Text Component Editor dialog.

   g. In the Text Component Editor, remove all text in the Preview area and select Profile2 Elevation from the Properties list.

   h. Verify the Precision is **0.01** (or **0.001** for metric).

   i. Click the arrow button to add the text to the Preview area.

h. Click OK to exit the Text Component Editor.

i. Click OK to finish working with the Major Stations Label Composer and return to the Band Details tab.

5. On the Display tab, use the lightbulb icon to turn off visibility for Minor Tick, and click OK.

The completed band should resemble Figure 21.61.

FIGURE 21.61
Text along the bottom of your profile view in the form of a band

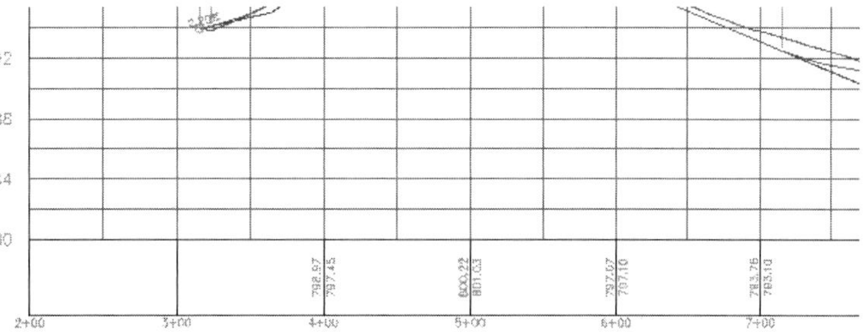

As you can see, profile bands can provide a lot of information in a compact manner. In this example you only provided information at the major stations, but you could also provide information at minor stations, horizontal geometry points, vertical geometry points, station equations, and incremental distances.

When this exercise is complete, you can close the drawing. A saved copy of this drawing is available from the book's web page with the filename `ProfileBands_FINISHED.dwg` or `ProfileBands_METRIC_FINISHED.dwg`.

Section View Styles

Section view styles share many of the same concepts as creating profile view styles. In fact, the Section View Style dialog has all the same tabs and looks nearly identical to the Profile View style.

In this section, you will walk through the creation of a section view style suitable for creating a section sheet:

1. Open the `SectionStyles.dwg` or `SectionStyles_METRIC.dwg` file.

   This file contains section views created with the default settings for section views.

2. From the Settings tab of Toolspace, expand Section View ➤ Section View Style ➤ Road Section.
3. Right-click Road Section and select Edit.
4. On the Grid tab, set Grid Padding (Major Grids) to **0** for the Above Maximum Elevation and Below Datum options.
5. On the Display tab, turn off the visibility for all components except:
   - Graph Title
   - Left Axis Annotation Major
   - Right Axis Annotation Major
   - Bottom Axis Annotation Major
6. Click OK.

   Your section views should resemble Figure 21.62.

FIGURE 21.62
Yes, this is correct! A very stripped-down section view.

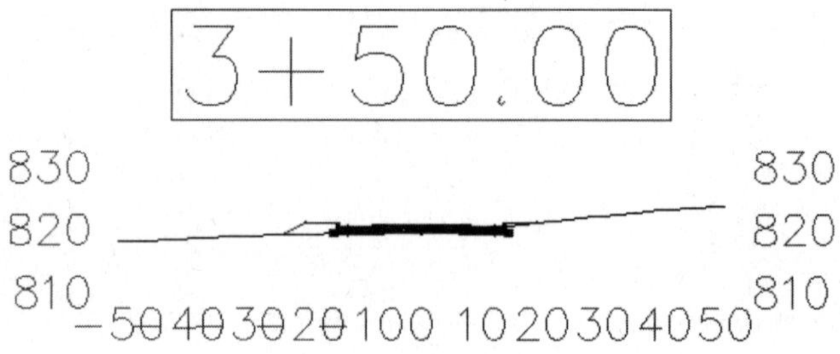

7. Select one of the views by clicking on the station label.

8. From the Section View contextual tab ➤ Modify View panel, select Update Group Layout.

   The section views will rearrange to fit more sections per page.

Why the bare-bones section view? The section view grid will come from the group plot style, so the only information you really need is in this simple style.

You can save and keep this drawing open to continue on to the next exercise or use the saved copy of this drawing available from the book's web page (`SectionStyles_FINISHED.dwg` or `SectionStyles_METRIC_FINISHED.dwg`).

Group Plot Styles

The driver behind how section sheets get laid out on a page is the *group plot style*. Upon section creation, the group plot style takes the Section template file and fits sections inside the paper-space viewport:

1. If not still open from the previous exercise, open `SectionStyles_FINISHED.dwg` or `SectionStyles_METRIC_FINISHED.dwg`

   You can download either file from this book's web page. This file contains section views created with the default settings.

2. From the Settings tab of Toolspace, expand Section View ➤ Group Plot Styles.

3. Right-click Basic and select Edit.

4. On the Array tab, change the column spacing to **4″** (or **100** mm). Change the row spacing to **2″** (or **50** mm), as shown in Figure 21.63.

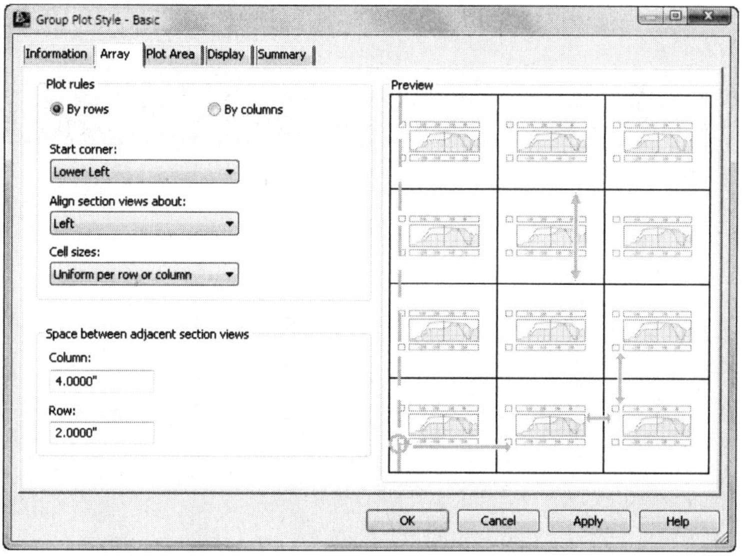

FIGURE 21.63 The Array tab controls section view spacing.

These spacing changes will allow more cross sections per page and additional room for scooting the views up and right to fit on the page better in the upcoming steps.

5. On the Plot Area tab, leave all the default settings as shown in Figure 21.64.

   This is where the grid spacing you will be using on your sheets comes from.

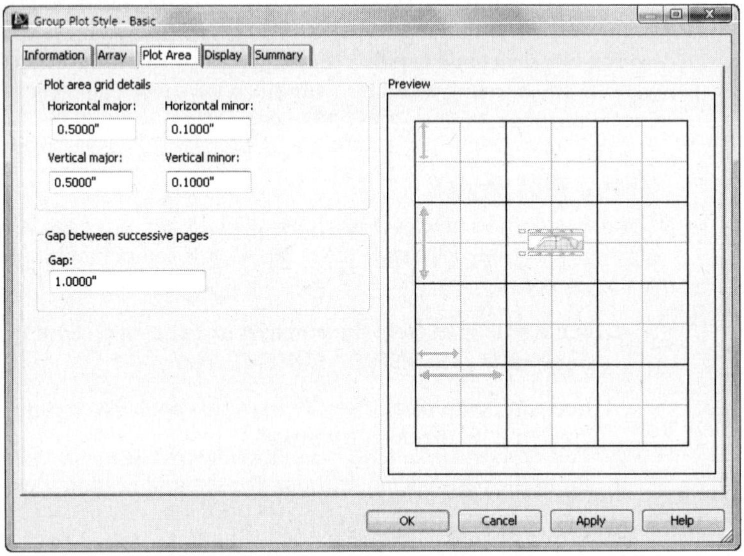

FIGURE 21.64
The Plot Area tab is where grid spacing on sheets comes from.

6. On the Display tab, do the following:
   - **a.** Verify that the visibility is turned on for Major Horizontal Grid and Major Vertical Grid.
   - **b.** Verify that the visibility is turned off for Minor Horizontal Grid and Minor Vertical Grid.
   - **c.** Verify that the layers for all of the grid components are set to C-ROAD-SCTN-GRID.
   - **d.** Verify that the layers for Print Area and Sheet Border are set to C-ROAD-SCTN-TTLB.
   - **e.** Set Color for Major Horizontal Grid and Major Vertical Grid to **9**.
   - **f.** Set Color for Print Area to Cyan and Sheet Border to Green. Your Display tab will look like Figure 21.65.
7. Click OK to complete the group plot style edits.

   Your section view sheets should be shaping up to the point where you could almost generate sheets. There may be instances when some text oozes out of the cyan line that represents the viewport border. In the next steps you will use a nonvisible section band to prevent this from happening and push the views onto the page.
8. From the Settings tab of Toolspace, expand Section View ➢ Band Styles ➢ Section Data.
9. Right-click Section Data and select New.

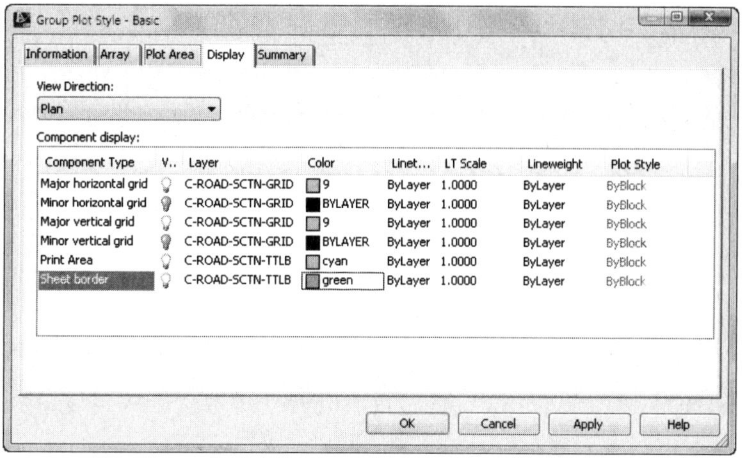

FIGURE 21.65 Group plot style display components

10. On the Information tab, set Name to **_NO DISPLAY**.

Prefixing the style name with the underscore ensures it will be alphabetized to the top of the style list.

11. On the Band Details tab, do the following:

a. Set Band Height to **0.2″** (or **5** mm).

b. Set Text Box Width to **0.2″** (or **10** mm).

c. Set Offset From Band to **0″** (or **0** mm).

Even though the band is not visible, Civil 3D still accounts for this spacing when placing the views on the sheet. In this step, you are using this to your advantage.

12. On the Display tab, verify that the visibility is turned off for all components.

13. Click OK to complete the creation of a new section data band style.

14. Select any section view by clicking on the station label or the elevation labels.

15. From the Section View contextual tab ➢ Modify View panel, choose View Group Properties to display the Section View Group Properties dialog.

16. On the Section Views tab, do the following:

a. Click the ellipsis in the Change Band Set column, as shown in Figure 21.66.

You may need to widen the columns to view the full titles.

The Section View Group Bands dialog will appear.

b. Verify Band Type is set to Section Data.

c. Set the band style as _NO DISPLAY and click Add.

d. Set the Gap distance to **0**, as shown in Figure 21.67, and click OK to dismiss the Section View Group Bands dialog.

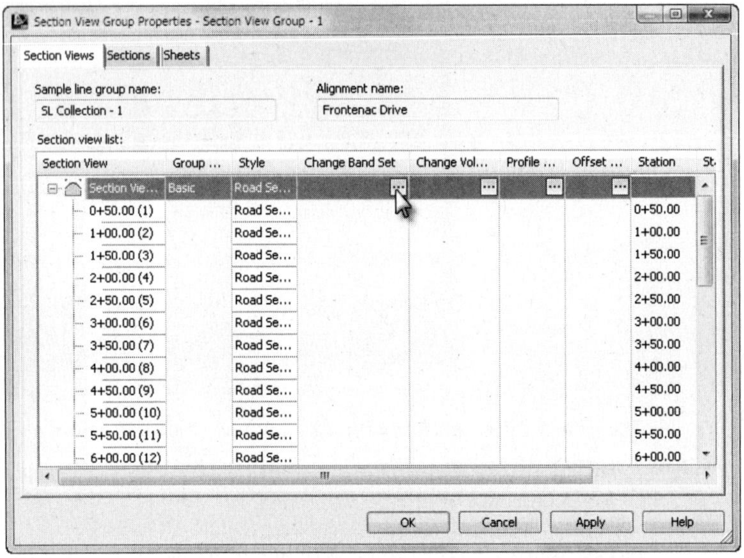

FIGURE 21.66
Changing the band set in use for all section views

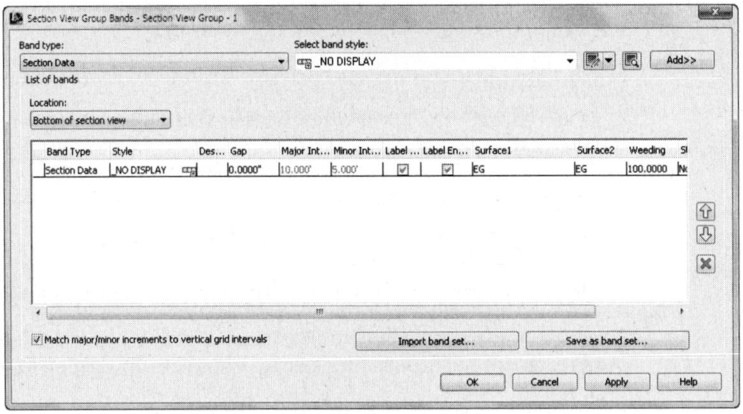

FIGURE 21.67
Add the data band and set the gap to 0.

e. Click OK to exit the View Group Properties dialog.

17. From the Section View contextual tab ➤ Modify View panel, select Update Group Layout.

The cross section sheets should look like Figure 21.68.

When this exercise is complete, you can close the drawing. A saved copy of this drawing is available from the book's web page with the filename `GroupPlotStyles_FINISHED.dwg` or `GroupPlotStyles_METRIC_FINISHED.dwg`.

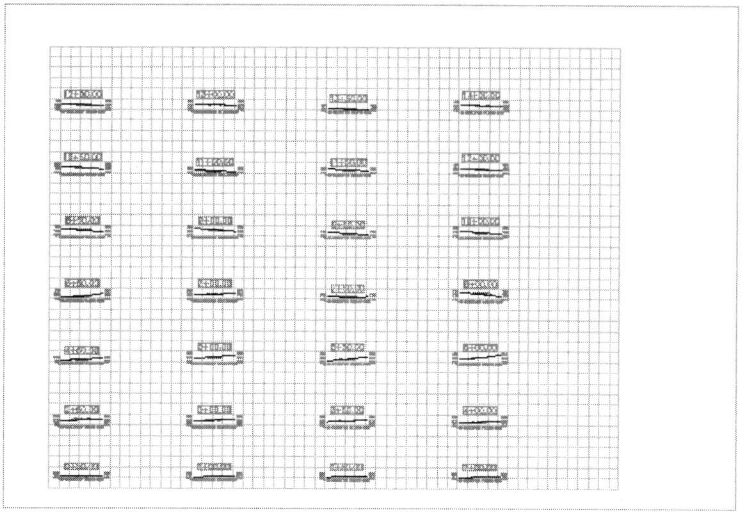

FIGURE 21.68
The completed exercise

With all of the object styles, you have a great deal of control over every detail, even ones that may seem trivial. Instead of being bogged down trying to understand every option, don't be afraid to try a "trial and error" approach. If you make a change you don't like, you can always edit the style until you get it right.

The Bottom Line

Override object styles with other styles. In spite of the desire to have uniform styles and appearances between objects within a single drawing, project, or firm, there are always going to be changes that need to be made.

Master It Open the `MasteringStyles.dwg` or `MasteringStyles_METRIC.dwg` file and change the alignment style associated with Alignment B to Layout. In addition, change the surface style used for the EG surface to Contours And Triangles, but change the contour interval to be 1′ and 5′ (or 0.5 m and 2.5 m) and the color of the triangles to be yellow.

Create a new surface style. Almost every set of plans that you send out of the office is going to include a surface, so it is important to be able to generate multiple surface styles that match your company standards. In addition to surface styles for production, you may find it helpful to have styles to use when you are designing that show a tighter contour spacing as well as the points and triangles needed to make some edits.

Master It Open the `MasteringSurfaceStyle.dwg` or `MasteringSurfaceStyle_METRIC.dwg` file and create a new surface style named **Micro Editing**. Set this style to display contours at 0.5′ and 1.0′ (or 0.1 m and 0.2 m), as well as triangles and points. Set the EG surface to use this new surface style.

Create a new profile view style. Everyone has their preferred look for a profile view. These styles can provide a lot of information in a small space, so it is important to be able to create a profile view that will meet your needs.

Master It Open the `MasteringProfileViewStyle.dwg` or `MasteringProfileViewStyle_METRIC.dwg` file and create a new profile view style named **Mastering Profile View**. Set this style to not clip the vertical or horizontal grid. Set the bottom horizontal ticks at 50′ and 10′ intervals (25 m and 5 m). Set the left and right vertical ticks at 10′ and 2′ intervals (5 m and 1 m). In addition, turn off the visibility of the Graph Title, Bottom Axis Annotation Major, and Bottom Axis Annotation Horizontal Geometry Point. Set the profile view in the drawing to use this new profile view style.

Appendix A

The Bottom Line

Each of The Bottom Line sections in the chapters suggest exercises to deepen skills and understanding. Sometimes there is only one possible solution, but often you are encouraged to use your skills and creativity to create something that builds on what you know and lets you explore one of many possible solutions.

Chapter 1: The Basics

Find any Civil 3D object with just a few clicks. By using Prospector to view object data collections, you can minimize the panning and zooming that are part of working in a CAD program. When common subdivisions can have hundreds of parcels or a complex corridor can have dozens of alignments, jumping to the desired one nearly instantly shaves time off everyday tasks.

Master It Open `BasicSite.dwg` from `www.sybex.com/go/masteringcivil3d2013`, and find parcel number 18 without using any AutoCAD commands or scrolling around on the drawing screen. (Hint: Take a look at Figure 1.5.)

Solution

1. In Prospector, expand Sites ➢ Proposed ➢ Parcels.
2. Right-click on LOT:18 and select Zoom To.

Modify the drawing scale and default object layers. Civil 3D understands that the end goal of most drawings is to create hard-copy construction documents. By setting a drawing scale and then setting many sizes in terms of plotted inches or millimeters, Civil 3D removes much of the mental gymnastics that other programs require when you're sizing text and symbols. By setting object layers at a drawing scale, Civil 3D makes uniformity of drawing files easier than ever to accomplish.

Master It Change the Annotation scale in the model tab of `BasicSite.dwg` from the 100-scale drawing to a 40-scale drawing. (For metric users: Use `BasicSite_METRIC.dwg` and change the scale from 1:1000 to 1:500.)

Solution

1. In the lower-right corner of the application window, select 1″ = 40 (1:500) from the Annotation Scale list.
2. Type **REA** and press ↵ to regenerate the screen and show the labels at the new scale.

Navigate the Ribbon's contextual tabs. As with AutoCAD, the Ribbon is the primary interface for accessing Civil 3D commands and features. When you select an AutoCAD Civil 3D object, the Ribbon displays commands and features related to that object. If several object types are selected, the Multiple contextual tab is displayed.

Master It Continue working in the file `BasicSite.dwg` (`Basic Site_METRIC.dwg`). It is not necessary to have completed the previous exercise to continue. Using the Ribbon interface, access the Alignment properties for Road A.

Solution

1. Select the Road A alignment to display the contextual Alignment tab Ribbon.
2. In the Alignment Properties menu, select the Information tab.
3. Click the drop-down list next to the Proposed object style and select Edit Current Selection.

Create a curve tangent to the end of a line. It's rare that a property stands alone. Often, you must create adjacent properties, easements, or alignments from their legal descriptions.

Master It Open the drawing `MasterIt.dwg` (`MasterIt_METRIC.dwg`). Create a curve tangent to the east end of the line labeled in the drawing. The curve should meet the following specifications:

- Radius: 200.00′ (60 m)
- Arc Length: 66.580′ (20 m)

Solution

1. Select Home ➢ Draw ➢ Curves ➢ Create Curve From End Of Object.
2. Select the east side of the line that is labeled "Create a curve tangent to this line."
3. On the command line, press ↵ to confirm that you will enter a radius value.
4. On the command line, type **200.00** (**60**), and then press ↵
5. Type **L** to specify the length, and then press ↵
6. Type **66.580** (**20**), and then press ↵

Label lines and curves. Although converting linework to parcels or alignments offers you the most robust labeling and analysis options, basic line- and curve-labeling tools are available when conversion isn't appropriate.

Master It Add line and curve labels to each entity created in `MasterIt.dwg` or `MasterIt_Metric.dwg`. It is recommended you complete the previous exercise so you will have a curve to work with. Choose a label that specifies the bearing and distance for your lines and length, radius, and delta of your curve.

Solution

1. Change to the Annotate tab in the Ribbon.
2. Click the Add Labels button from the Labels & Tables panel.

- Set Feature to Line And Curve.
- Set Label Type to Single Segment.
- Set Line Label Style to Bearing Over Distance.
- Set Curve Label Style to Distance-Radius And Delta.

3. Choose each line and curve.

The default label should be acceptable. If not, perform the following steps:

1. Select a label.
2. From the General Segment Label contextual tab on the Modify panel, click Label Properties.
3. In the resulting AutoCAD Properties dialog, select an alternative label in the General section.

Chapter 2: Survey

Properly collect field data and import it into Civil 3D. Once survey data has been collected, you will want to pull it into Civil 3D via the Survey Database. This will enable you to create lines and points that correctly reflect your field measurements.

Master It Create a new drawing based on the template of your choice and a new survey database and import the `MASTER_IT_C2.txt` (or `master_IT_C2_METRIC.txt`) file into the drawing. The format of this file is PNEZD (Comma Delimited).

Solution

1. Create a new drawing using a template of your choice.
2. On the Survey tab, create a new local survey database.
3. Create a new network in the newly created survey database.
4. Import the `Master_It_C2.txt` (or `Master_It_C2_METRIC.txt`) file and edit the options to insert both the figures and the points.

Set up description key and figure databases. Proper setup is key to working successfully with the Civil 3D survey functionality.

Master It Create a new description key set and the following description keys using the default styles. Make sure all description keys are going to layer V-Node:

- CL*
- EOP*
- TREE*
- BM*

Change the description key search order so that the new description key set takes precedence over the default.

Create a figure prefix database called MasterIt containing the following codes:

- CL
- EOP
- BC

Solution

1. Open a new drawing based on a Civil 3D template or continue working in the drawing from the previous exercise.
2. On the Settings tab, locate the description key area. Right-click Description Key Sets and select New. Give the description key set the name of your choice and click OK.
3. Right-click the new description key set and select Edit Keys. Add the description keys including the asterisk, as shown in the list.
4. Set the layer for each item in the table to V-NODE.
5. Close the description key table and save the drawing.
6. Right-click Description Key Sets and select Properties.
7. Move your new description key set to the top of the listing using the arrows. Click OK.
8. In the Survey tab, right-click Figure Prefix Databases and select New.
9. Create a new figure prefix database called **MasterIt**.
10. Add the required codes to the list. Leave all options as default.
11. Import the file `Codetest.txt`. When importing, verify that the current figure prefix database is set to MasterIt.

    Your file should now contain linework that reflects your efforts.

Translate surveys from assumed coordinates to known coordinates. Understanding how to manipulate data once it is brought into Civil 3D is important to making your field measurements match your project's coordinate system.

Master It Create a new drawing and survey database. Start a new drawing based on the template of your choice. Import `traverse.fbk` (or `traverse_METRIC.fbk`). Translate the database based on:

- Base Point 1
- Rotation Angle of 10.3053°

Solution

1. Create a new drawing and survey database and import the `traverse.fbk` (or `traverse_METRIC.fbk`) file into a network.

2. Using the Translate Survey Network command, rotate the network based on the point number and rotation you were given.

3. Save the drawing and leave it open.

   Your drawing should look like `mastering2.dwg` (`mastering2_METRIC.dwg`), which you can download from this book's web page, `www.sybex.com/go/masteringcivil3d2013`.

Perform traverse analysis. Traverse analysis is needed for boundary surveys to check for angular accuracy and closure. Civil 3D will generate the reports that you need to capture these results.

Master It Use the survey database and network from the previous Master It. Analyze and adjust the traverse using the following criteria:

- Use an Initial Station of value 2 and an Initial Backsight value of 1.
- Use the Compass Rule for Horizontal Adjustment.
- Use the Length Weighted Distribution Method for Vertical Adjustment.
- Use a Horizontal Closure Limit value of 1:25,000.
- Use a Vertical Closure Limit value of 1:25,000.

Solution

1. Continue working in the drawing from the previous Master It. Create a new survey database and network, and import the `traverse.fbk` (`traverse_METRIC.fbk`) field book.

2. Create a new traverse from the four points.

3. Perform a traverse analysis on the newly created traverse, and apply the changes to the survey database.

   The following files will be generated as a result of the analysis:

   - `Trav1 Balanced Angles.trv.txt` (`METRIC_Trav1 Balanced Angles.trv.txt`)
   - `Trav1 Raw Closure.trv.txt` (`METRIC_Trav1 Raw Closure.trv.txt`)
   - `Trav1 Vertical Adjustment.trv.txt` (`METRIC_Trav1 Vertical Adjustment .trv.txt`)
   - `Trav1.lso.txt` (`METRIC_Trav1.lso.txt`)

Chapter 3: Points

Import points from a text file using description key matching. Most engineering offices receive text files containing point data at some time during a project. Description keys provide a way to automatically assign the appropriate styles, layers, and labels to newly imported points.

Master It Create a new drawing from `_AutoCAD Civil 3D (Imperial) NCS.dwt` (or `_AutoCAD Civil 3D (Metric) NCS.dwt`). Revise the Civil 3D description key set to contain only the parameters listed here:

CODE	POINT STYLE	POINT LABEL STYLE	FORMAT	LAYER
GS*	Basic	Elevation Only	Ground Shot	V-NODE
GUY*	Guy Pole	Elevation and Description	Guy Pole	V-NODE
HYD*	Hydrant (existing)	Elevation and Description	Existing Hydrant	V-NODE-WATR
TOP*	Basic	Point#-Elevation-Description	Top of Curb	V-NODE
TREE*	Tree	Elevation and Description	Existing Tree	V-NODE-TREE

Import the PNEZD (space delimited) file `Concord.txt` (`Concord_METRIC.txt`). Confirm that the description keys made the appropriate matches by looking at a handful of points of each type. Do the trees look like trees? Do the hydrants look like hydrants?

Save the resulting file.

Solution

1. Select File ➤ New, and create a drawing from `_AutoCAD Civil 3D (Imperial) NCS Extended.dwt` (or `_AutoCAD Civil 3D (Metric) NCS.dwt`).
2. Switch to the Settings tab of Toolspace, and locate the description key set called Civil 3D.
3. Right-click this set and choose Edit Keys.
4. Delete the first two keys in this set by right-clicking each one and choosing Delete.
5. Revise the remaining key to match the GS specifications listed under the "Master It" instructions.
6. Right-click the GS key, and choose Copy.
7. Create the four additional keys listed in the instructions, and exit Panorama.
8. On the Create Ground Data panel, select Points ➤ Point Creation Tools and then click the Import Points button on the toolbar.
9. Navigate out to the `Concord.txt` (`Concord_METRIC.txt`) file and click Open.
10. Select PNEZD Space Delimited from the listing create a point group with the name of your choosing and click OK.

11. Zoom in to see the points.

12. Save the drawing for use in the next exercise.

Note that each description key parameter (style, label, format, and layer) has been respected. Your hydrants should appear as hydrants on the correct layer, your trees should appear as trees on the correct layer, and so on. Compare your work to `EndofChapter_FINISHED.dwg` (`EndofChapter_METRIC_FINISHED.dwg`).

Create a point group. Building a surface using a point group is a common task. Among other criteria, you may want to filter out any points with zero or negative elevations from your Topo point group.

Master It Create a new point group called Topo that includes all points *except* those with elevations of zero or less. Use the DWG created in the previous Master It or start with `Master_It.dwg` (`Master_It_METRIC.dwg`).

Solution

1. In Prospector, right-click Point Groups and choose New.
2. On the Information tab, enter **Topo** as the name of the new point group.
3. Switch to the Exclude tab.
4. Click the Elevation check box to turn it on, and enter **<0** in the field.
5. Click OK to close the dialog.

Export points to LandXML and ASCII format. It's often necessary to export a LandXML or ASCII file of points for stakeout or data-sharing purposes. Unless you want to export every point from your drawing, it's best to create a point group that isolates the desired point collection.

Master It Create a new point group that includes all the points with a raw description of TOP. Export this point group via LandXML to a PNEZD comma-delimited text file.

Use the DWG created in the previous Master It or start with `Master_It.dwg` (`Master_It_METRIC.dwg`).

Solution

1. In Prospector, right-click Point Groups and choose New.
2. On the Information tab, enter **Top of Curb** as the name of the new point group.
3. Switch to the Include tab.
4. Select the With Raw Descriptions Matching check box, and type **TOP** in the field.
5. Click OK, and confirm in Prospector that all the points have the description TOP and click OK.
6. Right-click the Top Of Curb point group, and choose Export To LandXML.
7. Click OK in the Export LandXML dialog.

8. Choose a location to save your LandXML file, and then click Save.
9. Navigate out to the LandXML file to confirm it was created.
10. Right-click the Top Of Curb point group, and choose Export Points.
11. Choose the PNEZD comma-delimited format and a destination file, and confirm that the Limit To Points In Point Group check box is selected for the Top of Curb point group. Click OK.
12. Navigate out to the ASCII file to confirm it was created.
13. Save the file `Master_It.dwg` (`Master_It_METRIC.dwg`) for use in the next Master It.

Create a point table. Point tables provide an opportunity to list and study point properties. In addition to basic point tables that list number, elevation, description, and similar options, you can customize point table formats to include user-defined property fields.

Master It Continue working in `Master_It.dwg` (`Master_It_METRIC.dwg`). Create a point table for the Topo point group using the PNEZD format table style.

Solution

1. Change to the Annotate tab, and select Add Tables ➢ Add Point table.
2. Choose the PNEZD format for the table style.
3. Click Point Groups, and choose the Topo point group.
4. Click OK.

   The command line prompts you to choose a location for the upper-left corner of the point table.
5. Choose a location on your screen somewhere to the right of the project.
6. Zoom in, and confirm your point table.

Chapter 4: Surfaces

Create a preliminary surface using freely available data. Most land development projects involve a surface at some point. During the planning stages, freely available data can give you a good feel for the lay of the land, allowing design exploration before money is spent on fieldwork or aerial topography. Imprecise at best, this free data should never be used as a replacement for final design topography, but it's a great starting point.

Master It Create a new drawing from the Civil 3D template and set the Coordinate System to NAD83 Connecticut State Plane Zone, US Foot (CT83F) or NAD83 Connecticut State Plane Zone, Meter (CT83). Create a surface named MarlboroughCT_DEM. Add the `Marlborough_CT.DEM` file (UTM Zone 18, NAD27 datum, meters) downloadable from the book's web page.

Solution

1. In Civil 3D, create a new drawing from the `_AutoCAD Civil 3D (Imperial) NCS` or `_AutoCAD Civil 3D (Metric) NCS` template.
2. Change to the Settings tab and right-click the drawing name to open the Drawing Settings dialog.
3. Select an appropriate coordinate system.
4. From the Home tab ➢ Create Ground Data panel, choose Surfaces ➢ Create Surface.
5. Set Name to **MarlboroughCT_DEM** and accept the defaults in the Create Surface dialog, and click OK.
6. Expand the Surfaces ➢ MarlboroughCT_DEM ➢ Definition branch.
7. Right-click DEM Files and select the Add option.
8. Use the button to the right of the DEM File Name area to navigate to the `Marlborough_CT.dem` file and click Open.
9. In the Add DEM File dialog, click the Value column next to CS Code to display the ellipsis button; click that button to display the Select Coordinate Zone dialog.
10. Set the Coordinate System Code to match the DEM file by selecting UTM with NAD27 datum, Zone 18, Meter; Central Meridian 75d W (UTM27-18) and click OK.
11. Click OK in the Add DEM File dialog.
12. Zoom extents to see the surface.

Modify and update a TIN surface. TIN surface creation is mathematically precise, but sometimes the assumptions behind the equations leave something to be desired. By using the editing tools built into Civil 3D, you can create a more realistic surface model.

Master It Open the `MasteringBoundary.dwg` or the `MasteringBoundary_METRIC.dwg` file. Use the irregular-shaped polyline and apply it to the surface as an outer boundary of the surface.

Solution

1. Draw a polyline that includes the desired area.
2. Expand the Surfaces ➢ MarlboroughCT_DEM ➢ Definition branch.
3. Right-click Boundaries and select the Add option.
4. Select the newly created polyline and click Add to complete the boundary addition.

Prepare a slope analysis. Surface analysis tools allow users to view more than contours and triangles in Civil 3D. Engineers working with nontechnical team members can create strong meaningful analysis displays to convey important site information using the built-in analysis methods in Civil 3D.

Master It Open the `MasteringSlopeAnalysis.dwg` or the `MasteringSlopeAnalysis_METRIC.dwg` file. Create a Slope Banding analysis showing slopes under and over 10 percent and insert a legend to help clarify the image.

Solution

1. Right-click the surface and bring up the Surface Properties dialog.
2. Set the Surface Style field to Slope Banding (2D).
3. Switch to the Analysis tab for the Elevation analysis type.
4. Verify that Create Ranges By is set to Number Of Ranges and that the value is set to **2**, and then click the Run Analysis arrow.
5. Change the Maximum Slope for ID 1 and the Minimum Slope for ID 2 both to **10%**.
6. Click OK to close the Surface Properties dialog.
7. Select the surface to display the TIN Surface contextual tab.
8. From the TIN Surface contextual tab ➤ Labels & Tables panel, choose Add Legend Table.
9. Enter **S**↵ and then **D**↵ at the command line and pick a placement point on the screen to create a dynamic elevations legend.

Label surface contours and spot elevations. Showing a stack of contours is useless without context. Using the automated labeling tools in Civil 3D, you can create dynamic labels that update and reflect changes to your surface as your design evolves.

Master It Open the `MasteringLabelSurface.dwg` or the `MasteringLabelSurface_METRIC.dwg` file. Label the major contours on the surface at 2′ and 10′ (Background) or 1 m and 5 m (Background).

Solution

1. Change the Surface Style to Contours 2′ and 10′ (Background) or Contours 1m and 5m (Background).
2. From the Annotate tab ➤ Labels & Tables panel, click the Add Labels button
3. Set the Feature to Surface and the Label Type to Contour – Multiple.
4. Set the Major Count Label Style to Existing Major Labels and the Minor Contour Label Style to <none>.
5. Click Add.
6. Pick a point on one side of the site, and draw a contour label line across the entire site.

Import a point cloud into a drawing and create a surface model. As point cloud data becomes more common and replaces other large-scale data-collection methods, the ability to use this data in Civil 3D is critical. Intensity helps postprocessing software determine the ground cover type. While Civil 3D can't do postprocessing, you can see the intensity as part of the point cloud style.

Master It Import an LAS format point cloud `Denver.las` into the Civil 3D template (with a coordinate system) of your choice. As you create the point cloud file, set the style to Elevation Ranges. Use a portion of the file to create a Civil 3D surface model.

Solution

1. Start a new file by using the default Civil 3D template of your choice. Save the file before proceeding as **`DenverUSA.dwg`**.
2. In Prospector, right-click the Point Clouds and select the Create Point Cloud option to display the Create Point Cloud wizard.
3. Set the name of the Point Cloud to **Denver**.
4. Set the Point Cloud Style to Elevation Ranges, and click the Next button.
5. Using the white plus sign to browse to the LAS file.
6. Select `Denver.las`. Click Finish.

   This file contains 4.7 million data points, so be patient while the file imports.
7. Select the bounding box representing the point cloud to display the Point Cloud contextual tab.
8. Select the Add Points To Surface command.
9. Name the surface, set a surface style, and click the Next button.
10. Choose the Window radio button, and click Define Region In Drawing.
11. Define the region by creating a window around the western half of the point cloud.
12. Click Next to see the Summary page and click the Finish button.

Chapter 5: Parcels

Create a boundary parcel from objects. The first step to any parceling project is to create an outer boundary for the site.

Master It Open the `MasteringParcels.dwg` (`MasteringParcels_METRIC.dwg`) file, which you can download from `www.sybex.com/go/masteringcivil3d2013`. Convert the line segments in the drawing to a parcel.

Solution

1. From the Home tab's Create Design panel, select Parcel ➢ Create Parcel From Objects.
2. At the `Select lines, arcs, or polylines to convert into parcels or [Xref]:` prompt, pick the lines that represent the site boundary, and press ↵

   The Create Parcels – From Objects dialog appears.
3. From the drop-down menus, select Subdivision; Property; and Name, Square Foot & Acres (Name Square Meter & HA) in the Site, Parcel Style, and Area Label Style selection boxes, respectively.

   Keep the default values for the remaining options.

4. Click OK to dismiss the dialog.

   The boundary polyline forms parcel segments that react with the alignment. The label is placed at the newly created parcel centroid.

Create a right-of-way parcel using the right-of-way tool. For many projects, the ROW parcel serves as frontage for subdivision parcels. For straightforward sites, the automatic Create ROW tool provides a quick way to create this parcel. A cul-de-sac serves as a terminal point for a cluster of parcels.

Master It Continue working in the `Mastering Parcels.dwg` (`MasteringParcels_METRIC.dwg`) file. Create a ROW parcel that is offset by 25′ (10 m) on either side of the road centerline with 25′ (10 m) fillets at the parcel boundary and alignment ends. Then add the circles representing the cul-de-sac as a parcel.

Solution

1. From the Home tab's Create Design panel, select Parcel ➤ Create Right Of Way.
2. At the `Select parcels:` prompt, pick your newly created parcel on screen.
3. Press ↵ to stop picking parcels.

   The Create Right Of Way dialog appears.

4. Expand the Create Parcel Right Of Way parameter, and enter **25′** (**10** m) in the Offset From Alignment text box.
5. Expand the Cleanup At Parcel Boundaries parameter. Enter **25′** (**10** m) in the Fillet Radius At Parcel Boundary Intersections text box.
6. Select Fillet from the drop-down menu in the Cleanup Method selection box.
7. Set the Fillet Radius at Alignment Intersections to **25′** (**10** m).
8. Click OK to dismiss the dialog and create the ROW parcels.
9. Trim the two circles at the ROW line to create arcs.
10. From the Home tab's Create Design panel, select Parcel ➤ Create Parcel From Objects.
11. Pick the two arcs and accept the default settings.

    Two new parcels are created.

12. Pick a label at one of the newly created parcels. From the Parcel contextual tab ➤ Modify panel, select Parcel Layout Tools.
13. Select the Delete Sub-Entity tool and pick the two ROW lines and ROW arc, leaving the outer arc alone.

    The cul-de-sac is created and is part of the ROW parcel.

14. Repeat steps 11 and 12 for the other cul-de-sac arc. Press Esc when complete.

Create subdivision lots automatically by layout. The biggest challenge when creating a subdivision plan is optimizing the number of lots. The precise sizing parcel tools provide a means to automate this process.

Master It Continue working in the `Mastering Parcels.dwg` (`MasteringParcels_METRIC.dwg`) file. Create a series of lots with a minimum of 8,000 sq. ft. (700 m^2) and 75′ (20 m) frontage. Set the Use Minimum Offset option to No. Leave all other options at their defaults.

Solution

1. From the Home Tab's Create Design panel, select Parcel ➤ Parcel Creation Tools.
2. Expand the Parcel Layout Tools toolbar.
3. Change the value of the following parameters by clicking in the Value column and typing in the new values:
   - Minimum Area: **8000** Sq. Ft. (**700** square meters)
   - Minimum Frontage: **75**′ (**20** m)
4. Change the following parameters by clicking in the Value column and selecting the appropriate option from the drop-down menu:
   - Automatic Mode: On
   - Remainder Distribution: Redistribute Remainder
5. Click the Slide Line – Create tool.

   The Create Parcels – Layout dialog appears.
6. Select Subdivision; Single Family; and Name Square Foot & Acres (Name Square Meter & HA) from the drop-down menus in the Site, Parcel Style, and Area Label Style selection boxes, respectively.

   Keep the default values for the rest of the options.
7. Click OK to dismiss the dialog.
8. At the `Select Parcel to be subdivided:` prompt, pick the Property: 1 label for your property parcel.
9. At the `Select start point on frontage:` prompt, use your Endpoint osnap to pick the point of curvature on the north end of the project along the ROW parcel segment. The side of the road you start with is up to you.

   The parcel jig appears.
10. Move your cursor slowly along the ROW parcel segment, and notice that the parcel jig follows the segment.
11. At the `Select end point on frontage:` prompt, loop back to the opposite side of the street from where you started. Use your Endpoint osnap to pick the point of curvature along the ROW parcel segment.
12. At the `Specify angle at frontage:` prompt, type **90**, and press ↵. If you receive the message `No Solution Found` on your command line, try again. This may mean you snapped to the wrong spot.
13. At the `Accept Result:` prompt, press ↵ to accept the lot layout.

Note that the parcels are not going to line up properly. For extra credit, fix them to your liking.

Add multiple-parcel segment labels. Every subdivision plat must be appropriately labeled. You can quickly label parcels with their bearings, distances, direction, and more using the segment labeling tools.

Master It Continue working in the `MasteringParcels.dwg` (`MasteringParcels_METRIC.dwg`) file. Place Bearing Over Distance labels on every parcel line segment and Delta Over Length And Radius labels on every parcel curve segment using the Multiple Segment Labeling tool.

Solution

1. From the Annotate tab, select Add Labels ➢ Parcels ➢ Add Parcel Labels.
2. From the drop-down menus, in the Add Labels dialog, select Multiple Segment, Bearing Over Distance, and Delta Over Length And Radius in the Label Type, Line Label Style, and Curve Label Style selection boxes, respectively.
3. Click Add.
4. At the `Select parcel to be labeled by clicking on area label:` prompt, pick the area label for each of your single-family parcels. Press ↵ to accept Clockwise as the default.
5. Press Esc to exit the command.

Chapter 6: Alignments

Create an alignment from an object. Creating alignments based on polylines is a traditional method of building engineering models. With built-in tools for conversion, correction, and alignment reversal, it's easy to use the linework prepared by others to start your design model. These alignments lack the intelligence of crafted alignments, however, and you should use them sparingly.

Master It Open the `MasteringAlignments-Objects.dwg` or `MasteringAlignments-Objects_METRIC.dwg` file, and create alignments from the linework found there with the All Labels label set.

Solution From the Home tab ➢ Create Design panel, choose Alignment ➢ Create Alignment From Objects. Select the lines and arc. Make sure you reverse the alignment if necessary to match the start point indicated in the drawing.

Create a reverse curve that never loses tangency. Using the alignment layout tools, you can build intelligence into the objects you design. One of the most common errors introduced to engineering designs is curves and lines that aren't tangent, requiring expensive revisions and resubmittals. The free, floating, and fixed components can make smart alignments in a large number of combinations available to solve almost any design problem.

Master It Open the `MasteringAlignments-Reverse.dwg` or the `MasteringAlignments-Reverse_METRIC.dwg` file, and create an alignment from the linework on the right. Create a reverse curve with both radii equal to 200 (or 60 for metric users) and with a pass-through point at the intersection of the two arcs.

Solution

1. From the Home tab ➤ Create Design panel, choose Alignment ➤ Alignment Creation Tools.
2. In the Create Alignment – Layout dialog, accept the defaults and click OK to display the Alignment Layout Tools toolbar.
3. Use the Fixed Line (Two Points) tool to trace both lines and press ↵ when complete to end the command.
4. Use the Floating Curve (From Entity, Radius, Through Point) tool to draw an arc from the endpoint of the line with a radius of 200 (or 60 for metric users) to a pass-through point at the intersection of the two sketched arcs.
5. Press ↵ when complete to end the command.
6. Use the Free Curve Fillet (Between Two Entities, Radius) tool to fillet the floating curve created in the previous step and the last fixed segment with a reverse curve with a radius of 200 (or 60 for metric users).
7. Close the Alignment Layout Tools toolbar.

Replace a component of an alignment with another component type. One of the goals in using a dynamic modeling solution is to find better solutions, not just the first solution. In the layout of alignments, this can mean changing components out along the design path, or changing the way they're defined. The ability of Civil 3D to modify alignments' geometric construction without destroying the object or forcing a new definition lets you experiment without destroying the data already based on an alignment.

Master It Convert the reverse curve indicated in the `MasteringAlignments-Rcurve.dwg` or the `MasteringAlignments-Rcurve_METRIC.dwg` file to a floating arc that is constrained by the next segment. Then change the radius of the curves to 150 (or 45 for metric users).

Solution

1. Select the alignment to activate the contextual tab.
2. From the Alignment contextual tab ➤ Modify panel, choose Geometry Editor to display the Alignment Layout Tools toolbar.
3. Select the Alignment Grid View tool.
4. Starting with the first segment, click in the Tangency Constraint field and change it to Constrained By Next (Floating).

5. Repeat for the other segments except the last one, which cannot be modified because it is dependent on the previous constraint.
6. Change the radii of the two curves to **150′** (or **45** m for metric users).

Create alignment tables. Sometimes there is just too much information that is displayed on a drawing, and to make it clearer, tables are used to show bearings and distances for lines, curves, and segments. With their dynamic nature, these tables are kept up to date with any changes.

Master It Open the `MasteringAlignments-Table.dwg` or `MasteringAlignments-Table_METRIC.dwg` file, and generate a line table, a curve table, and a segment table. Use whichever style you want to accomplish this.

Solution

For lines:

1. Select the alignment to activate the contextual tab.
2. From the Alignment contextual tab ➢ Labels & Tables panel, choose Add Labels ➢ Alignment ➢ Multiple Segments and select the alignment.
3. On the Alignment contextual tab ➢ Labels & Tables panel choose Add Tables ➢ Add Line.
4. Using the Pick On-Screen button at the bottom of the dialog, select the line segments of the alignment.

   If a warning comes up regarding child styles, select the Convert All Selected Label Styles To Tag mode.
5. Click OK to accept the settings in the dialog.
6. Place the table anywhere on your drawing. The bearings and distances are now replaced by tag labels.

For curves:

1. If not done during the lines portion of the exercise, select the alignment and on the Alignment contextual tab ➢ Labels & Tables panel, choose Add Labels ➢ Alignment ➢ Multiple Segments and select the alignment.
2. From the Alignment contextual tab Labels & Tables panel select Add Tables ➢ Add Curve.
3. Using the Pick On-Screen button at the bottom of the dialog, select the curve segments of the alignment.

   If a warning comes up regarding child styles, select the Convert All Selected Label Styles To Tag mode.
4. Click OK to accept the settings in the dialog.
5. Place the table anywhere on your drawing. The bearings and distances are now replaced by tag labels.

For segments:

1. If not done during the lines portion of the exercise, select the alignment and on the Alignment contextual tab ➤ Labels & Tables panel, choose Add Labels ➤ Alignment ➤ Multiple Segments and select the alignment.
2. From the Alignment contextual tab ➤ Labels & Tables panel choose Add Tables ➤ Add Segment.
3. In the By Alignment section, select the alignment you want to label and click OK.
4. Place the table anywhere on your drawing. The bearings and distances are now replaced by tag labels.

Chapter 7: Profiles and Profile Views

Sample a surface profile with offset samples. Using surface data to create dynamic sampled profiles is an important advantage of working with a three-dimensional model. Quick viewing of various surface centerlines and grip-editing alignments makes for an effective preliminary planning tool. Combined with offset data to meet review agency requirements, profiles are robust design tools in Civil 3D.

Master It Open the `MasteringProfiles.dwg` file (or `MasteringProfiles_METRIC.dwg` file) and sample the ground surface along Alignment A, along with offset values at 15′ left and 15′ right (or 4.5 m left and 4.5 m right) of the alignment. Generate a profile view showing this information using the Major Grids profile view style with no data band sets.

Solution

1. From the Home tab ➤ Create Design panel, choose Profile ➤ Create Surface Profile.
2. Verify that Alignment A and the EG surface are selected and then click the Add button to add the EG surface.
3. Check the Sample Offsets check box and enter **15, –15** (or **4.5, –4.5** for metric users) in the box below the sample offsets and then click the Add button.
4. Click the Draw In Profile button to open the Create Profile View dialog.
5. On the General wizard page, verify that the profile view style is set to Major Grids.
6. On the Data Bands wizard page, verify that Select Band Set is set to _No Bands
7. Click the Create Profile View button.
8. Place the profile anywhere on the drawing.

Lay out a design profile on the basis of a table of data. Many programs and designers work by creating pairs of station and elevation data. The tools built into Civil 3D let you input this data precisely and quickly.

Master It In the `MasteringProfiles.dwg` file (or the `MasteringProfiles_METRIC.dwg` file), create a layout profile on Alignment A using the Layout profile style and a Complete Label Set with the following information for Imperial users:

STATION	PVI ELEVATION	CURVE LENGTH
0+00	822.00	
1+80	825.60	300′
6+50	800.80	

Or the following information for metric users:

STATION	PVI ELEVATION	CURVE LENGTH
0+000	250.400	
0+062	251.640	100 m
0+250	244.840	

Solution

1. Create a surface profile for Alignment A and generate a profile view (if not done in the previous example) or use the `MasteringProfiles_SolutionA.dwg` or `MasteringProfiles_SolutionA_METRIC.dwg` file.
2. From the Home tab ➢ Create Design panel, choose Profile ➢ Profile Creation Tools.
3. Select the profile view that shows the surface profile.
4. Verify that Profile Style is set to Layout and Profile Label Set is set to Complete Label Set
5. Click OK to dismiss the Create Profile – Draw New dialog.
6. In the Profile Layout Tools toolbar, set the L-value of the Curve settings to the specified curve length.
7. Use the Draw Tangents With Curves button and the Transparent Commands toolbar to enter station elevation data.
8. If needed, you may move the labels to be legible.

Alternatively, you can import a text file.

Add and modify individual entities in a design profile. The ability to delete, modify, and edit the individual components of a design profile while maintaining the relationships is an important concept in the 3D modeling world. Tweaking the design allows you to pursue a better solution, not just a working solution.

Master It In the `MasteringProfiles.dwg` file (or the `MasteringProfiles_METRIC.dwg` file) used in the previous exercise, on profile A modify the original curve so that it is 200′

(or 60 m for metric users). Then insert a PVI at Station 4+90, Elevation 794.60 (or at Station 0+150, Elevation 242.840 for metric users) and add a 300′ (or 96 m for metric users) parabolic vertical curve at the newly created PVI.

Solution

1. Using the file from the previous example or `MasteringProfiles_SolutionB.dwg` or `MasteringProfiles_SolutionB_METRIC.dwg` file, pick the Design profile, and from the Profile contextual tab ➢ Modify Profile panel, select the Geometry Editor button.
2. In the Profile Layout Tools toolbar, select the Profile Grid View button.
3. In the Profile Grid View, change the Profile Curve Length field to **200** (or **60** for metric users).
4. In the Profile Layout Tools toolbar, select the Insert PVI button.
5. Using the Profile Station Elevation transparent command, select the profile grid, enter **490** for the station and **794.60** for the elevation (or **150** for the station and **242.840** for the elevation for metric users). Press Esc twice.
6. Back in the Profile Layout Tools toolbar, click the drop-down arrow next to the Vertical Curve Creation button and select More Free Vertical Curves ➢ Free Vertical Curve (PVI Based).
7. Pick the newly created PVI and enter **300** (or **96** for metric users) for Curve Length. Press ↵ twice.

Apply a standard band set. Standardization of appearance is one of the major benefits of using styles in labeling. By applying band sets, you can quickly create plot-ready profile views that have the required information for review.

Master It In the `MasteringProfiles.dwg` (or the `MasteringProfiles_METRIC.dwg`) file, apply the Cut and Fill band set to the layout profile created in the previous exercise with the appropriate profiles referenced in each of the bands.

Solution

1. Using the file from the previous example or the `MasteringProfiles_SolutionC.dwg` or `MasteringProfiles_SolutionC_METRIC.dwg` file, pick the profile view, and from the Profile View contextual tab ➢ Modify View panel, choose Profile View Properties to display the Profile View Properties dialog.
2. On the Bands tab, click the Import Band Set, and select the Cut and Fill band set.
3. Select Bottom Of Profile View from the Location drop-down list.
4. Scroll over and change the Profile2 to Layout (1) for both rows.
5. Select Top Of Profile View from the Location drop-down list.
6. Scroll over and change the Profile2 to Layout (1).

Chapter 8: Assemblies and Subassemblies

Create a typical road assembly with lanes, curbs, gutters, and sidewalks. Most corridors are built to model roads. The most common assembly used in these road corridors is some variation of a typical road section consisting of lanes, curb, gutter, and sidewalk.

Master It Create a new drawing from the DWT of your choice. Build a symmetric assembly using LaneSuperelevationAOR, UrbanCurbGutterValley2, and LinkWidthAndSlope for terrace and buffer strips adjacent to the UrbanSidewalk. Use widths and slopes of your choosing.

Solution

1. Create a new drawing from the DWT of your choice.
2. From the Home tab ➢ Create Design panel choose Assembly ➢ Create Assembly.
3. Name your assembly and set styles as appropriate.
4. Pick a location in your drawing for the assembly.
5. Locate the Lanes tab on the Tool Palettes window.
6. Click the LaneSuperelevationAOR button on the Lanes tab.
7. Use the AutoCAD Properties palette to edit the subassembly parameters, and follow the command-line prompts to set the LaneSuperelevationAOR on the left and right sides of your assembly.
8. Repeat the process with UrbanCurbGutterValley2, LinkWidthAndSlope, and UrbanSidewalk.
9. Complete this portion of the exercise by placing a final LinkWidthAndSlope on the outside of the UrbanSidewalk. (Refer to the "Subassemblies" section of Chapter 8 for additional information.)
10. Save the drawing for use in the next Master It exercise.

Edit an assembly. Once an assembly has been created, it can be easily edited to reflect a design change. Often, at the beginning of a project you won't know the final lane width. You can build your assembly and corridor model with one lane width, and then change the width and rebuild the model immediately.

Master It Working in the same drawing, edit the width of each LaneSuperelevationAOR to 14′ (4.3 m), and change the cross slope of each LaneSuperelevationAOR to –3.00%.

Solution

1. Select both lane subassemblies.

   Be sure these are the only two elements selected.
2. From the contextual tab ➢ Modify Subassembly panel choose Subassembly Properties.
3. In the Advanced Parameters, change the width to **14′** (**4.3** m).

Note that width will be listed twice. The topmost width reports the default value. You will change the second occurrence.

4. Change the slope to **–3.00%**.
5. Save the drawing for use in the next Master It exercise.

Add daylighting to a typical road assembly. Often, the most difficult part of a designer's job is figuring out how to grade the area between the last engineered structure point in the cross section (such as the back of a sidewalk) and existing ground. An extensive catalog of daylighting subassemblies can assist you with this task.

Master It Working in the same drawing, add the DaylightMinWidth subassembly to both sides of your typical road assembly. Establish a minimum width between the outermost subassembly and the daylight offset of 10′ (3 m).

Solution

1. Locate the Daylight tab on the tool palette.
2. Click the DaylightMinWidth button on the tool palette.
3. Use the AutoCAD Properties palette to verify that Min Width is set to 10′.
4. Follow the command-line prompts to set the DaylightMinWidth on the right side of your assembly.
5. Press Esc on your keyboard to complete the command.
6. Select the right daylight subassembly.
7. From the contextual tab ➢ Modify Subassembly panel choose Mirror.
8. Click the outermost left point on the LinkWidthAndSlope link.

You should now have daylighting subassemblies visible on both sides of your assembly.

Chapter 9: Custom Subassemblies

Define input and output parameters with default values. By providing detailed input parameters, you let the Civil 3D user edit your custom subassembly. In addition, by providing detailed output parameters you let them use the subassembly's characteristics to edit the subsequent subassemblies in the assembly.

Master It Generate a new subassembly PKT file for a subassembly named MasteringLane (or MasteringLaneMetric). For input parameters, define the following:

- Side with a default value of Right
- LaneWidth provided as a decimal precise number with a default value of 12 feet or 3.6 meters
- LaneSlope provided as a percentage with a default value of –3%
- Depth provided as a decimal precise number with a default value of 0.67 feet or 0.2 meters

For output parameters, define the following:

- CalcLaneWidth provided as a decimal precise number
- CalcLaneSlope provided as a percentage

Solution

1. On the Packet Settings tab, set the Subassembly Name to **MasteringLane** (or **MasteringLaneMetric**).
2. On the Input/Output Parameters tab, change Default Value for the Side input parameter to Right.
3. Click Create Parameter to add a new parameter for each of the input and output parameters with the following settings (note that the metric values are shown in parentheses if different from the Imperial values):

NAME	TYPE	DIRECTION	DEFAULT VALUE
LaneWidth	Double	Input	12 (3.6)
LaneSlope	Grade	Input	-3%
Depth	Double	Input	0.67 (0.2)
CalcLaneWidth	Double	Output	
CalcLaneSlope	Grade	Output	

Define target offsets, target elevations, and/or target surfaces. Targets allow the subassembly to reference unique items within the drawing. They allow for varying widths, tapered elevations, and daylighting to a surface.

Master It For the same subassembly from the previous Master It, define an elevation target named TargetLaneElevation with a preview value of –0.33 feet (or –0.1 meters) and an offset target named TargetLaneOffset with a preview value of 14 feet (or 4 meters).

Solution

1. On the Target Parameters tab, click Create Parameter to add a new parameter.
2. Change Name to **TargetLaneElevation**, verify that Type is set to Elevation, and set Preview Value to **–0.33** (or **–0.1** for metric).
3. Create another parameter, change Name to **TargetLaneOffset**, verify that Type is set to Offset, and set Preview Value to **14** (or **4** for metric).

Generate a flowchart of subassembly logic using the elements in the Tool Box. The Tool Box in Subassembly Composer is full of all sorts of tools. But like a tool box in your garage, the tools are only useful if you know their capabilities and practice using them. The simplest tools are points, links, and shapes, but there are also tools for more complex flow charts.

Master It Define the subassembly using the width, slope, and depth input parameters defined earlier. Make certain that if the slope of the lane changes based on the targets, then the top and the datum links have matching slopes.

The subassembly should have four points: P1 ("Crown") located at the origin, P2 ("ETW") located at the upper-right corner of the lane cross section (with both the TargetLaneElevation and TargetLaneOffset), P3 ("Crown_Subbase") at the lower left of the lane cross section, and P4 ("ETW_Subbase") at the lower right of the lane cross section.

The subassembly should have four links: L1 ("Top", "Pave") connecting P1 and P2, L2 connecting P1 and P3, L3 ("Datum", "Subbase") connecting P3 and P4, and L4 connecting P2 and P4.

The subassembly should have a shape: S1 ("Pave1").

Solution

1. From the Tool Box ➤ Geometry branch, drag a Point element (P1) below the Start element.
2. Define point P1 as follows:
   - **A.** Set Point Codes to **"Crown"**.
   - **B.** Verify that From Point is set to Origin.
3. From the Tool Box ➤ Geometry branch, drag a Point element (P2) to below P1.
4. Define point P2 as follows:
   - **A.** Set Point Codes to **"ETW"**.
   - **B.** Verify that Type is set to Slope And Delta X.
   - **C.** Verify that From Point is set to P1.
   - **D.** Set Slope to **LaneSlope**.
   - **E.** Set Delta X to **LaneWidth**.
   - **F.** Set Offset Target (Overrides Delta X) to **TargetLaneOffset**.
   - **G.** Set the Elevation Target (Overrides Slope And Superelevation) to **TargetLaneElevation**.
   - **H.** Verify that the Add Link To From Point check box is selected. This link will automatically be numbered L1.
   - **I.** Set the Link ➤ Codes to **"Top", "Pave"**.
5. From the Tool Box ➤ Geometry branch, drag a Point element (P3) to below P2&L1.
6. Define point P3 as follows:
   - **A.** Set Point Codes to **"Crown_Subbase"**.
   - **B.** Verify that Type is set to Delta X and Delta Y.

C. Verify that From Point is set to P1.

D. Set the Delta X to **0**.

E. Set the Delta Y to **-Depth**.

F. Verify that the Add Link To From Point checkbox is selected. This link will automatically be numbered L2.

7. From the Tool Box ➤ Geometry branch, drag a Point element (P4) to below P3 and L2.
8. Define point P4 as follows:

A. Set Point Codes to **"ETW_Subbase"**.

B. Verify that Type is set to Delta X and Delta Y.

C. Verify that From Point is set to P2.

D. Set the Delta X to **0**.

E. Set the Delta Y to **-Depth**.

G. Verify that the Add Link To From Point check box is selected. This link will automatically be numbered L3.

H. Change the name for this link to **L4**.

9. From the Tool Box ➤ Geometry branch, drag a Link element (automatically numbered L5) to below P4 and L4.
10. Define link currently numbered L5 as follows:

A. Change the Link Number to **L3**.

B. Set Link Codes to **"Datum", "Subbase"**.

C. Verify that Start Point is set to P3.

D. Verify that End Point is set to P4.

11. From the Tool Box ➤ Geometry branch, drag a Shape element (S1) to below L3.
12. Define shape S1 as follows:

A. Set Shape Codes to **"Pave1"**.

B. Click the Select Shape In Preview button.

C. Select inside of the closed shape representing the lane.

13. From the Tool Box ➤ Miscellaneous branch, drag Set Output Parameter to below S1.
14. Define Set Output Parameter as follows:

A. Verify that Output Parameter is set to CalcLaneWidth.

B. Set Value to **L1.xlength**.

15. From the Tool Box ➤ Miscellaneous branch, drag Set Output Parameter to below CalcLaneWidth.

16. Define Set Output Parameter as follows:

    A. Verify that Output Parameter is set to CalcLaneSlope.

    B. Set Value to **L1.slope**.

Note that there are other ways to define the points and links that will have the same end results.

Import a custom subassembly made with Subassembly Composer into Civil 3D. A custom subassembly is no good if it is just made and not used. By knowing how to import your PKT file into Civil 3D, you open up the world of building and sharing tool palettes full of subassemblies to help your office's workflow.

Master It Import your MasteringLane or MasteringLaneMetric subassembly into Civil 3D.

Solution

1. In Subassembly Composer, save your PKT file.
2. Open Civil 3D and create a new drawing using the `_AutoCAD Civil 3D (Imperial) NCS` or the `_AutoCAD Civil 3D (Metric) NCS` template.
3. From the Insert tab ➤ expanded Import panel, choose Import Subassemblies.
4. In the Import Subassemblies dialog, click the folder button to display the Open dialog.
5. Navigate to your PKT file and click Open.
6. Verify that the Import To: Tool Palette check box is selected.
7. Using the drop-down list, either select an existing tool palette or select Create New Palette from the bottom of the list to display the New Tool Palette dialog to define a new palette using the Import Subassemblies dialog
8. Click OK to accept the settings in the Import Subassemblies dialog.

Chapter 10: Basic Corridors

Build a single baseline corridor from an alignment, profile, and assembly. Corridors are created from the combination of alignments, profiles, and assemblies. Although corridors can be used to model many things, most corridors are used for road design.

Master It Open the `MasteringCorridors.dwg` or `MasteringCorridors_METRIC.dwg` file. Build a corridor named Corridor A on the basis of the Alignment A alignment, the Project Road Finished Ground profile, and the Basic Assembly. Set all frequencies to **10'** (or **3** m for metric users).

Solution

1. From the Home tab ➤ Create Design panel, choose Corridors.
2. In the Name text box, name your corridor **Corridor A**.

   Keep the default values for Corridor Style and Corridor Layer.

3. Verify that Alignment is set to Alignment A and Profile is set to FG.
4. Verify that Assembly is set to Basic Assembly.
5. Verify that Target Surface is set to the EG surface.
6. Verify that Set Baseline And Region Parameters is checked.
7. Click OK to accept the settings in the Create Corridor dialog, and to display the Baseline And Region Parameters dialog.
8. Click the Set All Frequencies button to display the Frequency To Apply Assemblies dialog.
9. Change the value for all of the frequencies to **10′** (or **3** m for metric users).
10. Click OK to accept the settings in the Frequency To Apply Assemblies dialog.
11. Click OK to accept the settings in the Baseline And Region Parameters dialog.
12. You may receive a dialog warning that the corridor definition has been modified. If you do, select the Rebuild The Corridor option.

Use targets to add lane widening. Targets are an essential design tool used to manipulate the geometry of the road.

Master It Open the `MasteringCorridorTargets.dwg` or `MasteringCorridorTargets_METRIC.dwg` file. Set Right Lane to target the Alignment A-Left alignment.

Solution

1. From the Corridor contextual tab Modify Corridor panel, choose Corridor Properties.
2. On the Parameters tab, in the Targets column in the baseline row click the ellipsis button to display the Target Mapping dialog.
3. In the Target Mapping dialog, click <None> in the Width Alignment for Right Assembly to display the Set Width Or Offset Target dialog.
4. Select Alignment A-Right and click Add.
5. Click OK to dismiss the Set Width Or Offset Target dialog.
6. Click OK to accept the settings in the Target Mapping dialog box.
7. Click OK to accept the settings in the Corridor Properties and allow the corridor to rebuild.

Create a corridor surface. The corridor model can be used to build a surface. This corridor surface can then be analyzed and annotated to produce finished road plans.

Master It Open the `MasteringCorridorSurface.dwg` or `MasteringCorridorSurface_METRIC.dwg` file. Create a corridor surface for the Alignment A corridor from Top links. Name the surface **Corridor A-Top**.

Solution

1. From the Corridor contextual tab ➤ Modify Corridor panel, choose Corridor Properties.

2. On the Surfaces tab, click the Create A Corridor Surface button in the upper-left corner of the dialog.
3. Click the surface item under the Name column and change the default name of your surface to **Corridor A-Top**.
4. Verify that Links has been selected from the drop-down list in the Data Type selection box.
5. Verify that Top has been selected from the drop-down list in the Specify Code selection box.
6. Click the Add Surface Item button to add Top Links to the surface definition.
7. Click OK to accept the settings in the dialog; then choose Rebuild The Corridor when prompted.

   The corridor and surface will build.

Add an automatic boundary to a corridor surface. Surfaces can be improved with the addition of a boundary. Single-baseline corridors can take advantage of automatic boundary creation.

Master It Open the `MasteringCorridorBoundary.dwg` or `MasteringCorridorBoundary_METRIC.dwg` file. Use the Automatic Boundary Creation tool to add a boundary using the Daylight code.

Solution

1. Open the Corridor Properties dialog and switch to the Boundaries tab.
2. Right-click on the surface entry and click Add Automatically ➢ Daylight.
3. Click OK to accept the settings in the dialog; then choose Rebuild The Corridor when prompted.

   The corridor and surface will build.

Chapter 11: Advanced Corridors, Intersections, and Roundabouts

Create corridors with noncenterline baselines. Although for simple corridors you may think of a baseline as a road centerline, other elements of a road design can be used as a baseline. In the case of a cul-de-sac, the EOP, the top of curb, or any other appropriate feature can be converted to an alignment and profile and used as a baseline.

Master It Open the `MasteringAdvancedCorridors.dwg` (`MasteringAdvancedCorridors_METRIC.dwg`) file, which you can download from `www.sybex.com/go/masteringcivil3d2013`. Add the cul-de-sac alignment and profile to the corridor as a baseline. Create a region under this baseline that applies the Intersection Typical assembly.

Solution

1. Select the corridor, right-click, and choose Corridor Properties.
2. Switch to the Parameters tab.

3. Click Add Baseline, choose Cul de Sac EOP in the Create Corridor Baseline dialog, and click OK.
4. In the Profile column, click inside the <Click here...> box, choose Cul de Sac EOP FG in the Select A Profile dialog, and click OK.
5. Right-click the new baseline, and choose Add Region.
6. Select Intersection Typical in the Create Corridor Region dialog.
7. Click OK to leave the Corridor Properties dialog and rebuild the corridor.

Add alignment and profile targets to a region for a cul-de-sac. Adding a baseline isn't always enough. Some corridor models require the use of targets. In the case of a cul-de-sac, the lane elevations are often driven by the cul-de-sac centerline alignment and profile.

Master It Continue working in the `Mastering Advanced Corridors.dwg` (`Mastering Advanced Corridors_METRIC.dwg`) file. Add the Second Road alignment and Second Road FG profile as targets to the cul-de-sac region. Adjust Assembly Application Frequency to 5′ (1 m), and make sure the corridor samples are profile PVIs.

Solution

1. Select the corridor. From the Corridor contextual tab Modify Corridor panel, click Corridor Properties.
2. Switch to the Parameters tab.
3. Click the Target Mapping button in the Cul de Sac EOP region.
4. In the Target Mapping dialog, assign Second Road as Width Alignment for Lane - L and Second Road FG profile as Outside Elevation Profile; then click OK.
5. Click OK to leave the Target Mapping dialog.
6. Click the Frequency button in the appropriate region.
7. Change the Along Curves value to **5′** (**1** m) and the At Profile High/Low Point value to Yes.
8. Click OK to exit the Frequency To Apply To Assemblies dialog.
9. Click OK to leave the Corridor Properties dialog and rebuild the corridor.

Create a surface from a corridor and add a boundary. Every good surface needs a boundary to prevent bad triangulation. Bad triangulation creates inaccurate contours, and can throw off volume calculations later in the process. Civil 3D provides several tools for creating corridor surface boundaries, including an Interactive Boundary tool.

Master It Continue working in the `Mastering Advanced Corridors.dwg` (`Mastering Advanced Corridors_METRIC.dwg`) file. Create an interactive corridor surface boundary for the entire corridor model.

Solution

1. Select the corridor. From the Corridor contextual tab ➢ Modify Corridor panel, click Corridor Properties.

2. Switch to the Boundaries tab.
3. Select the corridor surface, right-click, and choose Add Interactively.
4. Follow the command-line prompts to add a feature line–based boundary all the way around the entire corridor.
5. Press C to close the boundary, and then press ↵ to end the command.
6. Click OK to leave the Corridor Properties dialog and build the corridor.

An example of the finished exercise can be found in `Mastering Advanced Corridors Finished.dwg` (`Mastering Advanced Corridors_METRIC.dwg`) at the book's web page.

Chapter 12: Superelevation

Add superelevation to an alignment. Civil 3D has convenient and flexible tools that will apply safe, correct superelevation to an alignment curve.

Master It Open the `Master Super.dwg` (`Master Super_METRIC.dwg`) file, which you can download from `www.sybex.com/go/masteringcivil3d2013`. Set the design speed of the road to 20 miles per hour (35 km per hour) and apply superelevation to the entire length of the alignment. Use AASHTO 2004 Design Criteria with an eMax of 6% 2-Lane.

Solution

1. Select the alignment. From the Alignment contextual tab ➢ Modify panel, choose Alignment Properties.
2. On the Design Criteria tab, place a check mark next to Use Criteria Based Design.
3. Set the design criteria file and superelevation eMax from the right side of the dialog.
4. Set the design speed to **20** mph (**35** kmph) on the left side, and click OK.
5. From the contextual tab, click Superelevation ➢ Calculate/Edit Superelevation.
6. Click Calculate Superelevation Now.
7. Step through the superelevation wizard, taking all the defaults for pivot and shoulder control.
8. On the Attainment page, place a check mark next to Automatically Resolve Overlap. Click Finish.

   You should now have superelevation applied to the design with no overlap.

Create a superelevation assembly. In order for superelevation to happen, you need to have an assembly that is capable of superelevating.

Master It Continue working in `Master Super.dwg` (`Master Super_METRIC.dwg`). Create an assembly similar to the one in the top image of Figure 12.11. Set each lane to be 14′ (4.5 m) wide, and each shoulder to be 6′ (2 m) wide. Leave all other options at their defaults. If time permits, build a corridor based on the alignment and assembly.

Solution

1. From the Home tab ➤ Create Design panel ➤ Assembly, choose Create Assembly.
2. Name the assembly **AOR** and set the assembly type to Undivided Crowned Road. Click OK.
3. Click to place the assembly in the graphic.
4. From the Lanes palette, click the LaneSuperelevationAOR subassembly.
5. Click the assembly to place one lane on the right; click again to place the assembly to the left.
6. Select the right subassembly and set its Use Superelevation parameter to Right Lane Outside.
7. Select the left subassembly and set its Use Superelevation parameter to Left Lane Outside.
8. Place the shoulders on each side.
9. If time permits, create a corridor based on the alignment and assembly you just created.
10. Save the drawing.

Create a rail corridor with cant. Cant tools are new to Civil 3D 2013 and allow users to create corridors that meet design criteria specific to rail needs.

Master It In the drawing `MasterCant.dwg` (`MasterCant_METRIC.dwg`), create a Railway assembly with the RailSingle subassembly using the default parameters for width and depth. Add a LinkSlopetoSurface generic link with 50% slope to each side. Add cant to the alignment in the drawing using the default settings for attainment. Create a corridor from these pieces.

Solution

1. From the Home tab ➤ Create Design panel ➤ Assembly, choose Create Assembly.
2. Name the assembly **Rail** and set the type to Railway.
3. From the Bridge And Rail palette, click the rail single subassembly.
4. Click the assembly in the drawing to place the rail design.
5. From the Generic palette, click LinkSlopetoSurface.
6. Set the slope to –50% and click once on each side of the assembly to place the link.
7. Press Esc to complete the process.
8. Select the alignment. From the alignment contextual tab ➤ Modify panel, click Cant ➤ Calculate/Edit Cant.
9. Click Calculate Cant Now, and click Finish.

10. Build a corridor from the alignment, assembly, and the proposed profile (which has been designed for you ahead of time).

    If you need assistance building your corridor, review Chapters 10 and 11. But hopefully you've figured it out by this point!

Create a superelevation view. Superelevation views are a great place to get a handle on what is going on in your roadway design. You can visually check the geometry as well as make changes to the design.

Master It Open the drawing `MasterView.dwg` (`MasterView_METRIC.dwg`). Create a superelevation view for the alignment. Show only the Left and Right Outside Lanes as blue and red, respectively.

Solution

1. Select the alignment and choose Superelevation ➢ Create Superelevation View.
2. In the Create Superelevation View dialog, toggle off Left Outside Shoulder and Right Outside Shoulder. The remaining check boxes will be for the lane views.
3. Set Left Outside Lane Color to Blue, set Right Outside Shoulder Color to Red, and click OK.
4. Place the view in the drawing.

Chapter 13: Cross Sections and Mass Haul

Create sample lines. Before any section views can be displayed, sections must be created from sample lines.

Master It Open `MasterSections.dwg` (`MasterSections_METRIC.dwg`) and create sample lines along the USH 10 alignment every 50′ (20 m). Set the left and right swath widths to **50′** (**20** m).

Solution

1. From the Home tab ➢ Profile & Section Views, click Sample Lines.
2 Select the USH_10 alignment and sample all data.
3. In the Sample Line tools, select the By Range Of Station option.
4. Create sample lines by station range and set your sample line distance to **50′** (**20** m); then click OK and ↵ to complete the command.

Create section views. Just as profiles can only be shown in profile views, sections require section views to display. Section views can be plotted individually or all at once. You can even set them up to be broken up into sheets.

Master It In the previous exercise, you created sample lines. In that same drawing, create section views for all the sample lines. Use all the default settings and styles.

Solution

1. Continue working in `MasterSections.dwg` (`MasterSections_METRIC.dwg`).
2. Select one of the sample lines.
3. From the Sample Line contextual tab, click Create Section View ➢ Create Multiple Section Views.
4. Leave all options at their defaults, and click Create Section Views.
5. Click in the graphic to place the views.

Define and compute materials. Materials are required to be defined before any quantities can be displayed. You learned that materials can be defined from surfaces or from corridor shapes. Corridors must exist for shape selection, and surfaces must already be created for comparison in materials lists.

Master It Using `MasterSections.dwg` (`MasterSections_METRIC.dwg`), create a materials list that compares Existing Intersection with HWY 10 DATUM Surface. Use the Earthworks Quantity takeoff criteria.

Solution

1. Continue working in `MasterSections.dwg` (`MasterSections_METRIC.dwg`).
2. Select one of the sample lines.
3. From the Sample Lines contextual tab ➢ Launch Pad panel, click Compute Materials. Select the alignment and sample line group, and then click OK.
4. Set the quantity takeoff criteria to Earthworks.
5. Set the existing ground surface to Existing Intersection.
6. Set the datum to HWY 10 DATUM, and click OK.

   Graphically nothing will appear. Continue to the Master It to see the results of your work.

Generate volume reports. Volume reports give you numbers that can be used for cost estimating on any given project. Typically, construction companies calculate their own quantities, but developers often want to know approximate volumes for budgeting purposes.

Master It Continue using `MasterSections.dwg` (`MasterSections_METRIC.dwg`. Use the materials list created earlier to generate a volume report. Create a web browser–based report and a Total Volume table that can be displayed on the drawing.

Solution

1 Continue working in `MasterSections.dwg` (`MasterSections_METRIC.dwg`).
2. Without any object selected, select Analyze ➢ Volumes And Materials panel, and click Volume Report.
3. Leave all options at their defaults, and click OK.

4. If asked "Do you want to allow scripts to run?" click Yes.

   Your report will display.

5. Close the browser window.

6. In Civil 3D, select Analyze ➢ Volumes And Materials panel, and click Total Volume Table.

7. Leave all options at their defaults and click OK.

Chapter 14: Pipe Networks

Create a pipe network by layout. After you've created a parts list for your pipe network, the first step toward finalizing the design is to use Pipe Network By Layout.

Master It Open the `MasteringPipes.dwg` or `Mastering pipes_METRIC.dwg` file. From the Home tab ➢ Create Design panel ➢ Pipe Network, select Pipe Network Creation Tools to create a sanitary sewer pipe network. Use the Composite surface, and name only structure and pipe label styles. Don't choose an alignment at this time. Create 8″ (200 mm) PVC pipes and concentric manholes. There are blocks in the drawing to assist you in placing manholes. Begin at the START HERE marker, and place a manhole at each marker location. You can erase the markers when you've finished.

Solution

1. From the Home tab's Create Design panel, select Pipe Network ➢ Pipe Network Creation Tools.

2. In the Create Pipe Network dialog, set the following parameters:

   - Network Name: **Mastering**
   - Network Parts List: Sanitary Sewer
   - Surface Name: Corridor FG
   - Alignment Name: <none>
   - Structure Label Style: Name Only (Sanitary)
   - Pipe Label Style: Name Only

3. Click OK. The Pipe Layout Tools toolbar appears.

4. Set the structure to SMH and the pipe to 8 Inch PVC (200 mm PVC).

5. Click Draw Pipes And Structures, and use your Insertion Osnap to place a structure at each marker location.

6. Press ↵ to exit the command.

7. Move the structure labels.

8. Select a marker, right-click, choose Select Similar, and click Delete.

Create an alignment from network parts and draw parts in profile view. Once your pipe network has been created in plan view, you'll typically add the parts to a profile view based on either the road centerline or the pipe centerline.

Master It Continue working in the `MasteringPipes.dwg` (`Mastering pipes_METRIC.dwg`) file. Create an alignment from your pipes so that station zero is located at the START HERE structure. Create a profile view from this alignment, and draw the pipes on the profile view.

Solution

1. Select the structure labeled START HERE to display the Pipe Networks contextual tab and select Alignment From Network on the Launch Pad panel.
2. Select the last structure in the pipe run, and press ↵ to accept the selection.
3. In the Create Alignment dialog, name the alignment **Mastering** and make sure the Create Profile And Profile View check box is selected.
4. Accept the other defaults, and click OK.
5. In the Create Profile dialog, sample both the EG and Corridor FG surfaces for the profile.
6. Click Draw In Profile View.
7. In the Create Profile View dialog, click Create Profile View and choose a location in the drawing for the profile view.

   A profile view showing your pipes appears.

Label a pipe network in plan and profile. Designing your pipe network is only half of the process. Engineering plans must be properly annotated.

Master It Continue working in the `MasteringPipes.dwg` (`Mastering pipes_METRIC.dwg`) file. Add the Length Description And Slope style label to profile pipes and the Data With Connected Pipes (Sanitary) style to profile structures. Add the alignment created in the previous Master It to all pipes and structures.

Solution

1. Select one of the pipe or structure objects. From the Pipe Networks contextual tab, select Add Pipe Network Labels.
2. In the Add Labels dialog, change Feature to Pipe Network, and then change Label Type to Entire Network Profile.
   - For pipe labels, choose Length Description And Slope.
   - For structure labels, choose Data With Connected Pipes (Sanitary).
3. Click Add, and choose any pipe or structure in your profile view.
4. Drag or adjust any profile labels as desired.
5. In the Prospector tab of Toolspace, expand Pipe Networks ➤ Networks ➤ Mastering and select Pipes.

6. Select all pipes in Prospector, right-click on Reference Alignment column header, and select Edit.
7. Choose the Mastering alignment.
8. Repeat steps 6 and 7 but choose Structures.

Create a dynamic pipe table. It's common for municipalities and contractors to request a pipe or structure table for cost estimates or to make it easier to understand a busy plan.

Master It Continue working in the `MasteringPipes.dwg` (`Mastering pipes_METRIC.dwg`) file. Create a pipe table for all pipes in your network. Use the default table style.

Solution

1. Select one of the pipe or structure objects. From the Pipe Networks contextual tab ➤ Add Tables ➤ Add Pipe.
2. In the Pipe Table Creation dialog, make sure your pipe network is selected.
3. Accept the other defaults, and click OK.
4. Place the table in your drawing.

Chapter 15: Storm and Sanitary Analysis

Create a catchment object. Catchment objects are the newest object type that Civil 3D can use to determine the area and time of concentration of a site. You can create a catchment from a surface, but in most cases you will use the option to create from polyline.

Master It Create a catchment for predeveloped conditions on the site.

Solution

1. Open the file `Mastering Catchment Creation.dwg` (or `Mastering Catchment Creation_METRIC.dwg`).
2. On the Analyze tab, choose Catchments ➤ Create Catchment Group.
3. Name the new catchment group **Predeveloped** and click OK.
4. On the Analyze tab, select Catchments ➤ Create Catchment From Object.
5. Select the red closed polyline that represents the catchment.
6. When prompted for the flow path, select the dashed polyline that runs through the site. Be sure to select it on the north part of the site (the uphill side).
7. In the Create Catchment dialog, name the catchment **Basin A**.
8. Leave all other styles and settings at their defaults and click OK.

   You now have a completed catchment.
9. Press Esc when complete.

Export pipe data from Civil 3D into SSA. Civil 3D by itself cannot do pipe flow or runoff calculations. For this reason, it is important to export Civil 3D pipe networks into the Storm And Sanitary Analysis portion of the product.

Master It Verify your pipe export settings and then export to SSA.

Solution

1. Open `Export to SSA.dwg` (or `Export to SSA_METRIC.dwg`).
2. Select the Settings tab of Toolspace; then expand Pipe Networks and expand Commands.
3. Double-click the EditInSSA option.
4. Expand the Storm Sewers Migration Defaults area.
5. Click the field for part matching defaults.
6. Click the ellipsis to open the Part Matchup Settings dialog.
7. Examine the Import tab and verify that the settings make sense.
8. Switch to the Export tab and examine your options for bringing files back into Civil 3D.
9. Click OK — no changes are needed.
10. Click OK to exit the command settings.
11. Switch to the Analyze tab in Civil 3D.
12. Click Edit In Storm And Sanitary Analysis.
13. Click OK to export the Highway Drainage storm network.
14. When Storm Sewers launches, click OK to create a new project.
15. Click No to saving the log file.
16. Switch to the Plan View tab.
17. Double-click one of the structures. Make sure it contains elevation data.
18. Save the SSA file and exit.

Export a stage-storage table from Civil 3D. When designing detention basins in Civil 3D, you will often need to export surface data for analysis in SSA.

Master It Generate a stage storage table for a detention basin surface. Determine how much water the pond can hold.

Solution

1. In Civil 3D, open `Master_It_Pond.dwg` (or `Master_It_Pond_METRIC.dwg`).
2. Select the surface in the drawing.
3. From the Analyze flyout panel of the context tab, select Stage Storage.

4. From the Stage Storage dialog, click Define Basin.
5. On the Define Basin From Entities dialog, name the basin **New Pond**. Click Define.
6. When prompted, select the surface from the graphic.

   You should be returned to the Stage Storage dialog.
7. Make note of the maximum storage of the pond.

   Your result should be 346,489.16 cubic feet (9,670.99 cubic meters).
8. Click Save Table.
9. Save the table as **New Pond.AeccSST**.
10. Close the Stage Storage dialog by clicking Cancel.
11. Save and close the drawing.

Model drainage systems in SSA. SSA is capable of modeling everything from complex drainage systems to simple culverts.

Master It Determine which links flood during a 2-hour, 50-year storm event. Model the example drainage design using SSA.

Solution

1. In SSA, open `Master_It_Storm.spf` (or `Master_It_Storm_METRIC.spf`).
2. Right-click Subbasin 1 and select Connect To. Connect Subbasin 1 to Inlet 9-W.
3. Repeat step 2 for the three remaining subbasins.

   The subbasins drain to the following inlets:

   - Subbasin 2 drains to Inlet 2-E.
   - Subbasin 3 drains to Inlet 7-W.
   - Subbasin 4 drains to Inlet 3-E.
4. Double-click Analysis Options.
5. On the General tab, set the storm duration to 2 hours.
6. On the Storm Selection tab, choose a single 50-year storm. Click OK when complete.
7. Click Run Analysis.

At the end of the process, the analysis will show C&G 2 and C&G 4 as flooded.

Chapter 16: Grading

Convert existing linework into feature lines. Many site features are drawn initially as simple linework for the 2D plan. By converting this linework to feature line information, you avoid a large amount of rework. Additionally, the conversion process offers the ability to drape feature lines along a surface, making further grading use easier.

Master It Open the `MasteringGrading.dwg` or `MasteringGrading_METRIC.dwg` file from the book's web page. Convert the magenta polyline, describing a proposed temporary swale, into a feature line and drape it across the EG surface to set elevations, and set intermediate grade break points.

Solution

1. From the Home tab ➤ Create Design panel, choose Feature Lines ➤ Create Feature Lines From Objects.
2. Select the polyline.
3. Toggle the Assign Elevations check box on and then click OK.
4. Select the EG surface in the Assign Elevations dialog.
5. Verify that Insert Intermediate Grade Break Points is selected.
6. Click OK to close the dialog and return to your model.

Model a simple linear grading with a feature line. Feature lines define linear slope connections. This can be the flow of a drainage channel, the outline of a building pad, or the back of a street curb. These linear relationships can help define grading in a model, or simply allow for better understanding of design intent.

Master It Edit the curve on the feature line you just created to be 100′ (30 m). Set the grade from the west end of the feature line to the next PI to 4 percent, and the remainder to a constant slope to be determined in the drawing. Draw a temporary profile view to verify the channel is below grade for most of its length.

Solution

1. Select the feature line to activate the Feature Line contextual tab.
2. From the Feature Line contextual tab ➤ Modify panel, toggle on the Edit Geometry panel if not already visible.
3. From the Feature Line contextual tab ➤ Edit Geometry panel, choose the Edit Curve tool.
4. Select the feature line curve.
5. In the Edit Feature Line Curve dialog, change the radius to **100′** (**30** m) and click OK.
6. Press ↵ to end the command.
7. From the Feature Line contextual tab ➤ Modify panel, toggle on the Edit Elevations panel if not already visible.
8. From the Feature Line contextual tab ➤ Edit Elevations panel, choose the Set Grade/Slope Between Two Points tool.
9. Use the End Osnap to pick the western end of the feature line, and press ↵ to accept the elevation.
10. Select the next PI at the start of the curve.

11. At the `Specify grade or [SLope Elevation Difference SUrface Transition]:` prompt, enter **4**↵ to set the grade.
12. Select the feature line again.
13. Pick the PI at the start of the curve, and press ↵ to accept the elevation.
14. Pick the PI at the upstream (eastern) end of the channel, and press ↵ to accept the grade.
15. Press ↵ to end the command.
16. Select the feature line, and from the Feature Line contextual tab ➤ Launch Pad panel, choose Quick Profile.
17. Click OK to accept the defaults and pick a point on the screen to draw the quick profile view.
18. Dismiss Panorama to view the Quick Profile.

Model planar site features with grading groups. Once a feature line defines a linear feature, gradings collected in grading groups model the lateral projections from that line to other points in space. These projections combine to model a site much like a TIN surface, resulting in a dynamic design tool that works in the Civil 3D environment.

Master It Use the two grading criteria to define the pilot channel, with grading on both sides of the sketched centerline. Define the channel using a Grading to Distance of 5′ (1.5 m) with a slope of 3:1 and connect the channel to the EG surface using a grading with slopes that are 4:1. Generate a surface from the grading group. If prompted, do not weed the feature line.

Solution

1. From the Home tab ➤ Create Design panel, choose Grading ➤ Grading Creation Tools to activate the Grading Creation Tools toolbar.
2. Click the Create A Grading Group tool to create a grading group.
3. Verify that the Automatic Surface Creation option is checked, and click OK.
4. Click OK to accept the surface creation options.
5. Click the Set The Target Surface tool to set the target surface to EG, and click OK.
6. Change Grading Criteria to Grade To Distance.
7. Click the Create Grading tool, and pick the feature line.
8. If the Weed Feature Line dialog appears, select Continue Grading Without Feature Line Weeding
9. Pick the left or right side, and press ↵ to model the full length.
10. Enter **5′** (**1.5** m) for the distance.
11. Press ↵ to accept entering the slope.

12. Enter **3** for the slope value.
13. Pick the main feature line again, and grade the other side using the same steps.
14. Change Grading Criteria to Grade To Surface, and then create a grading object on both the left and right sides with slopes of 4:1.
15. Press Esc to complete the gradings.

Chapter 17: Plan Production

Create view frames. When you create view frames, you must select the template file that contains the layout tabs that will be used as the basis for your sheets. This template must contain predefined viewports. You can define these viewports with extra vertices so you can change their shape after the sheets have been created.

Master It Open the `MasteringPlanProduction.dwg` or `MasteringPlanProduction_METRIC.dwg` file. Run the Create View Frames wizard to create view frames for Alignment A in the current drawing. (Accept the defaults for all other values.) These view frames will be used to generate Plan and Profile sheets on ARCH D (ISO A1) sheets at 20 scale (1:200 scale) using the plan and profile template `MasteringPandPTemplate.dwt` or `MasteringPandPTemplate_METRIC.dwt`. All files should be saved in `C:\Mastering\CH 17\`.

Solution

1. From the Output tab ➤ Plan Production panel, choose Create View Frames.
2. On the Alignment page, select Alignment A from the Alignment drop-down list and click Next.
3. On the Sheets page, select the Plan And Profile option.
4. Click the ellipsis button to display the Select Layout As Sheet Template dialog.
5. In this dialog, click the ellipsis button, browse to `C:\Mastering\CH 17\`, select the template named `MasteringPandPTemplate.dwt` (or `MasteringPandPTemplate_METRIC.dwt`), and click Open.
6. Select the layout named ARCH D Plan And Profile 20 Scale (or ISO A1 Plan and Profile 1 to 200 for metric users), and click OK.
7. Click Create View Frames.

Edit view frames. The grips available to edit view frames allow the user some freedom on how the frames will appear.

Master It Open the `MasteringEditViewFrames.dwg` or `MasteringEditViewFrames_METRIC.dwg` file, and move the VF- (1) view frame to Sta. 2+20 (or Sta. 0+050 for metric users) to lessen the overlap. Then adjust Match Line 1 (or Match Line 2 for metric users) so that it is now at Sta. 4+25 (or Sta. 0+200 for metric users) and shorten it so that the labels are completely within the view frames.

Solution

1. Click on the VF- (1) view frame.
2. Make sure you have Dynamic Input on.
3. Click on the diamond grip, and type **220**↵ (or **50**↵ for metric users).
4. Press Esc to clear the selection.
5. Click on the Match Line 1 (or Match Line 2 for metric users) to show its grips.
6. Click on the diamond grip, and type **425**↵ (or **200**↵ for metric users).
7. Click on the triangular grip on one end of the match line and shorten it so that the labels are completely within the view frames.
8. Repeat step 7 for the triangular grip on the opposite end of the match line.

The match line is now centered better between the two view frames.

Generate sheets and review Sheet Set Manager. You can create sheets in new drawing files or in the current drawing. The resulting sheets are based on the template you chose when you created the view frames. If the template contains customized viewports, you can modify the shape of the viewport to better fit your sheet needs.

Master It Open the `MasteringCreateSheets.dwg` or `MasteringCreateSheets_METRIC.dwg` file. Run the Create Sheets wizard to create plan and profile sheets in the current drawing for Alignment A using the using the plan and profile template `MasteringPandPTemplate.dwt` or `MasteringPandPTemplate_METRIC.dwt`. Make sure to choose a north arrow. (Accept the defaults for all other values.) All files should be saved in `C:\Mastering\CH 17\`.

Solution

1. From the Output tab ➢ Plan Production panel, choose Create Sheets.
2. On the View Frame Group And Layouts page, under the Layout Creation section select All Layouts In The Current Drawing
3. Verify that the North arrow is selected from the drop-down list.
4. Click Create Sheets.
5. Click OK to save the drawing.
6. Click a location as the profile origin.
7. Dismiss the events Panorama.

Create section views. More and more municipalities are requiring section views. Whether this is a mile-long road or a meandering stream, Civil 3D can handle it nicely via Plan Production.

Master It Open the `MasteringSectionSheets.dwg` or `MasteringSectionSheets_METRIC.dwg` file. Create section views and Plan Production section sheets in a new sheet set for Alignment A using the using the Road Section section style and the section template MasteringSectionTemplate.dwt or MasteringSectionTemplate_METRIC.dwt. Make sure the sections are set to be generated on ARCH D (ISO A1) sheets at 20-scale (1:200 scale). (Accept the defaults for all other values.) All files should be saved in C:\Mastering\CH 17\.

Solution

1. From the Home tab ➤ Profile & Section Views panel, choose Section Views ➤ Create Multiple Views.
2. On the General page, verify that Section View Style is set to Road Section and click Next.
3. On the Section Placement page, select the Production option.
4. Click the ellipsis button to display the Select Layout As Sheet Template dialog.
5. In this dialog, click the ellipsis button, browse to C:\Mastering\CH 17\, select the template named MasteringSectionTemplate.dwt (or MasteringSectionTemplate_METRIC.dwt), and click Open.
6. Select the layout named ARCH D Section 20 Scale (or ISO A1 Section 1 to 200 for metric users), and click OK.
7. Click Create Section Views.
8. Click a location as the section origin.

   The multiple section views are created.
9. From the Output tab ➤ Plan Production panel, choose Create Section Sheets.
10. In the Create Section Sheets dialog, verify that New Sheet Set is selected and set Sheet Set Storage Location to C:\Mastering\CH 17\Final Sheets.
11. Click Create Sheets.
12. Click OK to save the drawing.

Chapter 18: Advanced Workflows

Create a data shortcut folder. The ability to load design information into a project environment is an important part of creating an efficient team. The main design elements of the project are available to the data shortcut mechanism via the working folder and data shortcut folder.

Master It Using the drawing MasterWorkflow.dwg (MasterWorkflow_METRIC.dwg), create a new data shortcut folder called Master Data Shortcuts. Use the _Sample Project project template.

Solution

1. Open the drawing MasterWorkflow.dwg (MasterWorkflow_METRIC.dwg).
2. On the Manage tab, click New Shortcuts Folder.

3. Name the project and place a check mark next to Use Project Template.
4. With `_Sample Project` highlighted, click OK.

The data shortcut folder is now complete.

Create data shortcuts. To allow sharing of the data, shortcuts must be made before the information can be used in other drawings.

Master It Save the drawing to the `Source Drawings` folder in the project you created in the previous exercise. Create data shortcuts to all the available data in the file `MasterWorkflow.dwg` (`MasterWorkflow_METRIC.dwg`).

Solution

1. Continue working in the drawing from the previous Master It.
2. From the Application menu, use Save As to save the drawing to `C:\Civil 3d projects\2013\Master Data Shortcuts\Source Drawings`.
3. On the Manage tab, click Create Data Shortcuts.
4. Place a check mark next to All Available Data, and click OK.

Export to earlier releases of AutoCAD. Being able to export to earlier base AutoCAD versions is sometimes necessary.

Master It Using `MasterWorkflow.dwg` (`MasterWorkflow_METRIC.dwg`), export the Civil 3D file so it can be used by a user working in base AutoCAD 2010.

Solution

1. Continue working in the drawing from the previous Master It.
2. From the Application menu, select Export ➤ DWG ➤ 2010.
3. Save the file with the default name in the same directory as `MasterWorkflow.dwg` (`MasterWorkflow_METRIC.dwg`).

Export to LandXML. Being able to work with outside clients or even other departments within your firm who do not have Civil3D is an important part of collaboration.

Master It Using `MasterWorkflow.dwg` (`MasterWorkflow_METRIC.dwg`), create a LandXML file will all of the exportable information.

Solution

1. Continue working in the drawing from the previous Master It.
2. From the Output tab ➤ Export panel, click Export To LandXML. Use all the default settings and click OK.
3. Save the file with the default name in the same directory as `MasterWorkflow.dwg` (`MasterWorkflow_METRIC.dwg`).

Transform coordinates between drawings. Most project locations have the ability to use several coordinate systems or units. Being able to correctly manipulate a drawing's coordinate system is extremely important.

Master It The file `TransformThis.dwg` is in Universal Transverse Mercator (UTM) North American Datum (NAD) 83, Zone 10, in meters. Start a new drawing based on `_Autocad Civil 3D (Imperial) NCS.dwt` or `_Autocad Civil 3D (Metric) NCS.dwt`. Set the new drawing's coordinates to AND 83 California State Plane Zone 3 (US Foot or Meters, depending on which template you used).

Solution

1. Start a new drawing based on `_Autocad Civil 3D (Imperial) NCS.dwt` or `_Autocad Civil 3D (Metric) NCS.dwt`.
2. From the Settings tab of Toolspace, right-click on the name of the drawing and select Edit Drawing Settings.
3. On the Units And Zone tab, set the category to USA, California.
4. Set the available coordinate system to NAD83 California State Planes, Zone III US Foot (NAD83 California State Planes, Zone III Meter), and click OK.
5. Save the drawing to the same location as the rest of your chapter files using the filename **`CAStatePlane.dwg`**.
6. Switch to the Planning And Analysis workspace, and click Attach.
7. Attach the file `TransformThis.dwg`.
8. Click Define Query, and set the location to All and Query Mode to Draw.
9. Click Execute Query.

Chapter 19: Quantity Takeoff

Open and review a list of pay items along with their categorization. The pay item list is the cornerstone of quantity takeoffs. You should download and review your pay item list and compare it against the current reviewing agency list regularly to avoid any missed items.

Master It Using the template of your choice, open the `Getting Started.csv` (or `Getting Started_Metric.csv`) pay item file and add the 12-, 18-, and 24-Inch Pipe Culvert (or 300 mm, 450 mm, and 600 mm Pipe Culvert) pay items to your Favorites list in the QTO Manager.

Solution

1. Start a new file by using the default Civil 3D template of your choice.
2. Open the QTO Manager.
3. Click the Open button at the top left of the QTO Manager.
4. Verify that the Pay Item File Format drop-down list is set to CSV (Comma Delimited).
5. Click the Open button next to the Pay Item File text box.

6. Navigate to the `Getting Started` folder and select the `Getting Started.csv` file. For metric users, use the `Getting Started_METRIC.csv` file, which is downloadable from the book's web page.
7. Click Open to select this CSV pay item file.
8. Click OK.
9. Enter **12-Inch Pipe** (or **300 mm Pipe**) in the text box to filter.
10. Right-click on the 12-Inch Pipe Culvert item (or the 300 mm Pipe Culvert item), and select Add To Favorites.
11. Repeat for the other sizes.

Assign pay items to AutoCAD objects, pipe networks, and corridors. The majority of the work in preparing quantity takeoffs is in assigning pay items accurately. By using the linework, blocks, and Civil 3D objects in your drawing as part of the process, you reduce the effort involved in generating accurate quantities.

Master It Open the `MasteringQTO.dwg` or `MasteringQTO_Metric.dwg` file and assign the CLEARING AND GRUBBING pay item to the polyline that was originally extracted from the border of the corridor. Change the hatch to have a transparency of 80.

Solution

1. Open the QTO Manager.
2. Expand the Favorites branch and select the CLEARING AND GRUBBING item.
3. Right-click and select Assign Pay Item To Area.
4. Switch to the Object option by entering **O**↵ at the command line.
5. Select the polyline representing the limits of corridor surface.
6. Press ↵ again to end the command.
7. Select the hatch to activate the Hatch Editor contextual tab.
8. From the Hatch Editor contextual tab ➤ Properties panel, change the Hatch Transparency to **80**.

Use QTO tools to review what items have been tagged for analysis. By using the built-in highlighting tools to verify pay item assignments, you can avoid costly errors when running your QTO reports.

Master It Verify that the area in the previous exercise has been assigned a pay item.

Solution

1. Turn on Highlight Objects With Pay Items in the QTO Manager.
2. Pan and hover over the hatch to confirm that the tooltip indicates a pay item assignment.

Generate QTO output to a variety of formats for review or analysis. The Quantity Takeoff Reports give you a quick understanding of what items have been tagged in the drawing, and they can generate text in the drawing or external reports for uses in other applications.

Master It Display the amount of Type C Broken markings in a Quantity Takeoff Report Summary (TXT) using the `MasteringQTOReporting.dwg` or `MasteringQTOReporting_Metric.dwg` file.

Solution

1. From the Analyze tab ➢ QTO panel, choose Takeoff Command, and click Compute to run the report with default settings.
2. In the lower-left corner of the Quantity Takeoff Report dialog, change the report style to `Summary(TXT).xsl`.
3. Click the Draw button at the bottom of the dialog.
4. Click near some clean space and you'll be returned to the Quantity Takeoff Report dialog.
5. Click Close to dismiss this dialog, and then click Close again to dismiss the Compute Quantity Takeoff dialog
6. The calculated amount for Type C Broken Pavement Markings should be 3163.30′ (or 1000.528 m).

Chapter 20: Label Styles

Override individual labels with other styles. In spite of the desire to have uniform labeling styles and appearances between alignments within a single drawing, project, or firm, there are always exceptions. Using the Ctrl+click method for element selection, you can access commands that let you modify your labels and even change their styles.

Master It Open the drawing `MasteringLabelStyles.dwg` (`MasteringLabelStyles_METRIC.dwg`). Create a copy of the Perpendicular With Tick Major Station style called **Major With Marker**. Change Tick Block Name to **Marker Pnt**. Replace some (but not all) of your major station labels with this new style.

Solution

1. On the Settings tab, expand the Alignment ➢ Label Styles ➢ Station ➢ Major Station branch.
2. Right-click Perpendicular With Tick, and select Copy.
3. Change the name to **Major with Marker**.
4. Change to the Layout tab.
5. Change to the Tick component.
6. Change AeccTickLine Block Selection to Marker Pnt.

7. Click OK to close the dialog.
8. Open the AutoCAD Properties palette.
9. Ctrl+click a major station label.
10. Change the style to **Major With Marker**.

Create a new label set for alignments. Label sets let you determine the appearance of an alignment's labels and quickly standardize that appearance across all objects of the same nature. By creating sets that reflect their intended use, you can make it easy for a designer to quickly label alignments according to specifications with little understanding of the requirement.

Master It Within the `Mastering LabelStyles.dwg` (`Mastering Label Styles_METRIC.dwg`) file, create a new label set containing only major station labels, and apply it to all the alignments in that drawing.

Solution

1. On the Settings tab, expand the Alignments ➢ Label Styles ➢ Label Sets branch.
2. Right-click Major And Minor Only, and select Copy.
3. Change the name to **Major Only**.
4. Delete the Minor label on the Labels tab.
5. Right-click each alignment, and select Edit Alignment Labels.
6. Import the Major Only label set. Click OK until you are out of the dialog.
7. Repeat for each alignment.

Solutions may vary!

Create and use expressions. Expressions give you the ability to add calculated information to labels or add logic to label creation.

Master It In the `Mastering LabelStyles.dwg` (`Mastering Label Styles_METRIC.dwg`) file, create an expression that adds 0.5′ (0.15 m) to a surface elevation. Use the expression in a spot elevation label that shows both the surface elevation and the expression-based elevation.

Solution

1. In the Settings tab of Toolspace, expand Surface Label Styles ➢ Spot Elevation, right-click Expressions, and click New.
2. Name the expression anything that makes sense.

   The Expression will read `{Surface Elevation}+0.5` (for metric `{Surface Elevation}+0.15`). Format as Double.
3. Click OK.
4. In the same branch of Settings, right-click Spot Elevation, and select New.

5. Give the label a name.
6. On the Layout tab of the Label Styles Composer, click the Contents field and open the Text Component Editor.
7. Without removing the existing text, add the new expression into the label with the surface elevation. Press ↵ to ensure the text appears on two lines with the expression-based label on the bottom.
8. Place the new label in the drawing to check your work.

Apply a standard label set to profiles. Standardization of appearance is one of the major benefits of using Civil 3D styles in labeling. By applying label sets, you can quickly create plot-ready profile views that have the required information for review.

Master It In the `Mastering LabelStyles.dwg` (`Mastering Label Styles_METRIC.dwg`) file, apply the Road Profiles label set to all layout profiles.

Solution

1. Pick one of the layout profiles, right-click, and select the Edit Labels option.
2. Click Import Label Set, and select the Road Profile Labels set.
3. Repeat this procedure for all layout profiles.

Chapter 21: Object Styles

Override object styles with other styles. In spite of the desire to have uniform styles and appearances between objects within a single drawing, project, or firm, there are always going to be changes that need to be made.

Master It

Open the `MasteringStyles.dwg` or `MasteringStyles_METRIC.dwg` file and change the alignment style associated with Alignment B to Layout. In addition, change the surface style used for the EG surface to Contours And Triangles, but change the contour interval to be 1′ and 5′ (or 0.5 m and 2.5 m) and the color of the triangles to be yellow.

Solution

1. Select then right-click Alignment B and select Alignment Properties.
2. Set Alignment Object Style to Layout and click OK.
3. Select and then right-click the EG surface and select Surface Properties.
4. Set Surface Style to Contours And Triangles and click OK.
5. From the Settings tab of Toolspace, expand Surface ➤ Surface Styles.
6. Right-click Contours And Triangles and select Edit.
7. On the Contours tab, do the following:

A. Expand the Contour Intervals category.

B. Set the Minor Interval to **1′** (or **0.5** m).

Notice that the Major Interval automatically adjusts to **5′** (or **2.5** m).

8. On the Display tab with the View Direction set to Plan, set the color of the Triangles component to yellow.
9. Click OK to complete the revisions to the style.

Create a new surface style. Almost every set of plans that you send out of the office is going to include a surface, so it is important to be able to generate multiple surface styles that match your company standards. In addition to surface styles for production, you may find it helpful to have styles to use when you are designing that show a tighter contour spacing as well as the points and triangles needed to make some edits.

Master It Open the `MasteringSurfaceStyle.dwg` or `MasteringSurfaceStyle_METRIC.dwg` file and create a new surface style named **Micro Editing**. Set this style to display contours at 0.5′ and 1.0′ (or 0.1 m and 0.2 m), as well as triangles and points. Set the EG surface to use this new surface style.

Solution

1. From the Settings tab of Toolspace, expand Surface ➤ Surface Styles.
2. Right-click Surface Styles and select New.
3. On the Information tab, set Name to **Micro Editing**.
4. On the Contours tab, do the following:

   A. Expand the Contour Intervals category.

   B. Set the Minor Interval to **0.5**′ (or **0.1** m).

   Notice that the Major Interval automatically adjusts to **1.0**′ (or **0.2** m).

5. On the Display tab with the View Direction set to Plan, verify that the only components turned on are Points, Triangles, Minor Contours, and Major Contours.
6. Click OK to complete creation of a new surface style.
7. Select and then right-click the EG surface and select Surface Properties.
8. Set Surface Style to Micro Editing, and click OK.

Create a new profile view style. Everyone has their preferred look for a profile view. These styles can provide a lot of information in a small space, so it is important to be able to create a profile view that will meet your needs.

Master It Open the `MasteringProfileViewStyle.dwg` or `MasteringProfileViewStyle_METRIC.dwg` file and create a new profile view style named **Mastering Profile View**. Set this style to not clip the vertical or horizontal grid. Set the bottom horizontal ticks at 50′ and 10′ intervals (25 m and 5 m). Set the left and right vertical ticks at 10′ and 2′ intervals (5 m and 1 m). In addition, turn off the visibility of the Graph Title, Bottom Axis Annotation Major, and Bottom Axis Annotation Horizontal Geometry Point. Set the profile view in the drawing to use this new profile view style.

Solution

1. From the Settings tab of Toolspace, expand Profile View ➢ Profile View Styles.
2. Right-click Profile View Styles and select New.
3. On the Information tab, set Name to **Mastering Profile View**.
4. On the Grid tab, verify that Clip Vertical Grid and Clip Horizontal Grid are not selected.
5. On the Horizontal Axes tab, do the following:
   A. Verify that the Axis To Control radio button is set to Bottom.
   B. In the Major Tick Details area, set Interval to **50′** (or **25** m).
   C. In the Minor Tick Details area, set Interval to **10′** (or **5** m).
6. On the Vertical Axes tab, do the following:
   A. Verify that the Axis To Control radio button is set to Left.
   B. In the Major Tick Details area, set Interval to **10′** (or **5** m).
   C. In the Minor Tick Details area, set Interval to **2′** (or **1** m).
   D. Change the Axis To Control radio button to Right.
   E. In the Major Tick Details area, set Interval to **10′** (or **5** m).
   F. In the Minor Tick Details area, set Interval to **2′** (or **1** m).
7. On the Display tab with View Direction set to Plan, turn off the visibility of Graph Title, Bottom Axis Annotation Major, and Bottom Axis Annotation Horizontal Geometry Point.
8. Click OK to complete creation of a new profile view style.
9. Select and then right-click the profile view in the drawing, and select Profile View Properties.
10. Set Profile View Style to Mastering Profile View, and click OK.

Appendix B

Autodesk Civil 3D 2013 Certification

Autodesk® certifications are industry-recognized credentials that can help you succeed in your design career, providing benefits to both you and your employer. Getting certified is a reliable validation of skills and knowledge, and it can lead to accelerated professional development, improved productivity, and enhanced credibility.

This Autodesk Official Training Guide can be an effective component of your exam preparation. Autodesk highly recommends (and we agree!) that you schedule regular time to prepare, review the most current exam preparation roadmap and objectives available at http:// `www.autodesk.com/certification`, use Autodesk Official Training Guides, take a class at an Authorized Training Center (find ATCs near you here: http://`www.autodesk.com/atc`), and use a variety of resources to prepare for your certification — including plenty of actual hands-on experience.

CERT OBJECTIVE

To help you focus your studies on the skills you'll need for these exams, Table B.1 shows the objectives that could potentially appear on an exam and in what chapter you can find information on that topic — and when you go to that chapter, you'll find certification icons like the one in the margin here. The sections and exam objectives listed in the table are from the Autodesk Certification Exam Guide.

Good luck preparing for your certification!

TABLE B.1: CERTIFIED PROFESSIONAL EXAM SECTIONS AND OBJECTIVES

TOPIC	LEARNING OBJECTIVE	CHAPTER
User Interface	Navigate the user interface	1
	Use the functions on the Prospector tab	1
	Use functions on the Settings tab	1
Styles	Create and use object styles	21
	Create and use label styles	20
Lines & Curves	Use the line and curve commands	1
	Use the Transparent command	1
Points	Create points using the Point Creation command	3

TABLE B.1: CERTIFIED PROFESSIONAL EXAM SECTIONS AND OBJECTIVES *(CONTINUED)*

Topic	Learning Objective	Chapter
Points (continued)	Create points by importing point data	3
	Use point groups to control the display of points	3
Surfaces	Create and edit surfaces	4
	Use styles and settings to display surface information	4 and 21
	Create a surface by assembling fundamental data	4
	Use styles to analyze surface display results	4 and 21
Parcels	Create parcels using parcel layout tools	5
	Design a parcel layout	5
	Select parcel styles to change the display of parcels	5
	Select styles to annotate parcels	5
	Create alignments	5
Alignments	Design a geometric layout	6
Profiles and Profile Views	Create a surface profile	7
	Design a profile	7
	Create a layout profile	7
	Create a profile view style	21
	Create a profile view	7
Corridors	Design and create a corridor	10
	Derive information and data from a corridor	10
	Design and create an intersection	11

TABLE B.1: CERTIFIED PROFESSIONAL EXAM SECTIONS AND OBJECTIVES *(CONTINUED)*

Topic	Learning Objective	Chapter
Sections and Section Views	Create and analyze sections and section views	13
Pipe Networks	Design and create a pipe network	14
Grading	Design and create a grading model	16
	Create a grading model feature line	16
Managing and Sharing Data	Use data shortcuts to share/manage data	18
	Create a data sharing setup	18
Plan Production	Generate a sheet set using plan production	17
	Create a sheet set	17
Survey	Use description keys to control the display of points created from survey data	2
	Use figure prefixes to control the display of linework generated from survey data	2
	Create a topographic/boundary drawing from field data	2

Index

Note to the reader: Throughout this index **boldfaced** page numbers indicate primary discussions of a topic. *Italicized* page numbers indicate illustrations.

D

F

G

H

I

J

M

N

Q

R

S

U

V

W